Second Edition

Topics in Contemporary
Mathematical Physics

Second Edition

Topics in Contemporary Mathematical Physics

Kai S Lam

California State Polytechnic University, Pomona, USA

World Scientific

NEW JERSEY · LONDON · SINGAPORE · BEIJING · SHANGHAI · HONG KONG · TAIPEI · CHENNAI · TOKYO

Published by

World Scientific Publishing Co. Pte. Ltd.

5 Toh Tuck Link, Singapore 596224

USA office: 27 Warren Street, Suite 401-402, Hackensack, NJ 07601

UK office: 57 Shelton Street, Covent Garden, London WC2H 9HE

Library of Congress Cataloging-in-Publication Data

Lam, Kai S. (Kai Shue), 1949–

  Topics in contemporary mathematical physics / by Kai S. Lam (California State Polytechnic University, Pomona, USA). -- Second edition.

    pages cm

  Includes bibliographical references and index.

  ISBN 978-9814667791 (hardcover : alk. paper) -- ISBN 978-9814667807 (pbk. : alk. paper)

  1. Mathematical physics. I. Title.

  QC20.L14 2015

  530.15--dc23

                              2015026506

British Library Cataloguing-in-Publication Data

A catalogue record for this book is available from the British Library.

Printed in Singapore

to my family,
and
Shiing-Shen Chern

Preface to the Second Edition

In this new addition I have taken the opportunity to add fourteen new chapters, but have also decided to leave the contents of the first edition intact, as the first forty four chapters.

The new material, which appears to be quite disparate from that contained in the first edition, both in contents and style, is entirely focused on an introduction to quantum field theory in a simple scalar field setting, with emphasis on the Feynman path (functional) integral approach and the renormalization group. With respect to the latter, the focus is on an introduction of its application to critical phenomena in statistical physics, following the trail-blazing contributions of Kenneth Wilson. One of the subtexts, of course, is also to draw students' attention to the deep connections between high energy physics and statistical mechanics. In pursuing material that has the aura of a bygone period, which a casual glance at the contents page for the new chapters might convey, I run the risk of belying the title of the text, especially with regard to the adjective "contemporary". My justification for giving it prominence of place in the present venue, among many other interesting and worthy topics that could have been chosen, is that many more recent developments at the intersections of physics and mathematics, such as non-perturbative field theories, duality and supersymmetric models of various sorts, not to speak of several incarnations of string theories, have all pointed to possible deeper meeting grounds between these disciplines yet to emerge from a better appreciation of the "potency" of quantum field theory. This optimism may appear to be unwarranted, especially since many instances of the kind of mathematics used in quantum field theory have superficially the flavor of clever tricks and gimmicks which still, by and large, lack a firm mathematical foundation. In fact, it is a truism among some of the mathematical cognoscenti that quantum field theories may not exist in a mathematical sense! On the other hand, one is also reminded of the old quip by legendary physicist Pauli expressing his unease towards ad hoc renormalization schemes in the early days of quantum field theory, that "just because something is infinite doesn't mean it has to be zero"! Yet, perhaps precisely because of its unsettled but rich mathematical nature, quantum field theory may hold out a special promise for revolutionary developments, in *both* physics and mathematics. It is with these considerations in mind that I wish to introduce some basics of this rich subject to students of mathematical physics, who may not necessarily aim at pursuing it as their specialty, but who are interested in acquiring enough background to appreciate exciting future developments. This purpose is perhaps the strongest tie between this edition and the first one. I would be remiss if I did not also mention that, in writing this second edition, I did so as someone with no special expertise in the subject matter presented therein, but only as a theoretical physicist guided by an urgent sense of its (somewhat neglected) fundamental importance in the general physics curriculum. For this I would like to beg the reader's indulgence. In approaching the various topics, I have relied heavily on the many authoritative sources cited in the references, and have attempted to indicate more locally within the text from which particular source (or sources) the text material had been drawn from. I would again like to reiterate that material here is entirely introductory in character and by no means complete in any sense. Readers who wish to pursue the subject matter further should consult the vast literature available, some of which is cited in the Bibliography section.

The contents of this new edition are quite independent of the first one, and readers do not need to be well grounded in the material there, in particular the introduction to differential forms, differential geometry and topology, or even group theory (despite the prominence of the term "renormalization group" in the new chapters), in order to pursue the new topics. In fact, the approach here is decidedly more "physical", in the sense that hardly any theorems are proved or even presented carefully. (The reader who is interested in a much more mathematical review of quantum field theory should consult, for example, Deligne et. al. 1999.) Again, this may perhaps be due to the above-mentioned lack of firm mathematical foundations for the subject. Nor is any prior knowledge in quantum field theory, as usually covered in advanced graduate level physics courses, presumed. The necessary background is quite modest, and conventional: On the physics side, a good exposure to non-relativistic quantum mechanics and Lagrangian/Hamiltonian dynamics; on the mathematics side, a working knowledge of the usual topics of mathematical physics covered in the upper-division physics curriculum, such as asymptotic expansions, special functions, and complex variable theory. Above all, a willingness to engage in "stretching the limits of validity" of heuristic arguments, precisely the kind demonstrated by Feynman's formulation of the functional integral approach to quantum mechanics, would be very helpful. Although this approach is viewed with some distrust by the mathematical purist, its role in the spectacular development in theoretical physics in the latter half of the twentieth century, at least on the practical level, is undeniable. With historical hindsight, one can even advance the claim that it has stimulated prodigious developments in pure mathematics also. As in the first edition, I have attempted to provide at particular junctures, as much as possible, explicit references to items discussed in earlier parts of the texts (by means of specific equation numbers and text locations). Derivational details and motivational discussions are also given special attention. These measures have been taken in order to ensure continuity of reading, and to avoid trying the reader's patience.

Since the publication of the first edition, I have received many kind words of encouragement and suggestions from various sources, both students and trained professionals. I would like to take this opportunity to express my heartfelt gratitude to them, and also to apologize for any shortcomings in my attempts to address the various constructive criticisms directed to my attention. In particular, I would like to thank Mr. Steven Huntsman, who took the trouble to go through the entire first edition with a piercing eye and forwarded his valuable comments to me, chapter by chapter, soon after its appearance. A more attentive and appreciative reader I could not have hoped for. I must also convey a special note of appreciation to the many students whom I have had the fortune to meet over these past decades as a physics instructor, who, in their special ways, gave me the courage and confidence, and even a sense of mission, to put in book form material that is not deemed to be particularly useful or fashionable in a text aimed, at least in part, to an introductory audience. Very belatedly, I like to thank Professor Kerson Huang, who made many significant contributions to the topics discussed in the pages of this new edition, and who, as one of my physics professors at M.I.T. in the early nineteen-seventies when I was a beginning graduate student there, opened my eyes to the wonders of quantum mechanics, quantum field theory, and statistical mechanics in the many memorable classes that he taught, where he imparted to his students, besides the physics itself, a sense of how to do physics calculations in an elegant manner. In addition, Professor Huang's exemplary textbooks have set inimitable but truly worthwhile standards for emulation. Much material in this book have been adapted from Professor Huang's writings, and I would like to pay tribute to them, and also to the many authoritative sources cited in the references that I have drawn from in the preparation of this book. Here as elsewhere, I have to pay special homage to the late eminent geometer, Professor S. S. Chern, as someone who exerted great influence in the second half of my teaching career. Professor Chern has been a constant source of knowledge and inspiration through the years that I have striven to include a modicum of differential geometry in the physics courses that I have taught, and to work on textbooks reflecting that desire, of which the first edition of this book was a humble product. Ms. L. F. Kwong and Mr. R. Babu of World Scientific have been most helpful and understanding, the former in directing me to the completion of the editorial process for the second edition and making it a most pleasant experience, and the latter in his technical assistance. To them, therefore, I would like to say thank you also. Almost twelve years have passed since I expressed my gratitude to my family – my wife Dr. Bonnie Buratti and our three sons Nathan, Reuben and Aaron, now all young adults – in the first edition of this book for their unflinching support. The intervening years have only deepened that gratitude, in every respect.

Kai S. Lam,
Department of Physics and Astronomy,
California State Polytechnic University, Pomona   *January, 2015*

Preface to the First Edition

Physics and mathematics have undergone an intensely symbiotic period over the last two decades, and are poised for an even more exciting and productive relationship into the twenty-first century. The present text is the result of this physicist's conviction (which I believe is shared by a growing number of physicists) that it is beneficial to prepare physics students, especially those with a theoretical inclination, with the necessary background to appreciate and take advantage of this development, starting as early as practically feasible in their physics education. Various parts of the first two-thirds or so of the text, dealing mainly with things algebraic, such as the theory of linear transformations, group theory, and Lie algebra theory, have been used as instructional material in an advanced undergraduate level Mathematical Physics three-quarter sequence that I have taught several times at the California State Polytechnic University, Pomona. The last third, mainly on differential geometry, is probably more suitable at the beginning graduate level; but the ambitious undergraduate should by no means be deterred. Mathematics students who are interested in seeing how some fundamental mathematical ideas and techniques can be applied to broad swaths of physics may also find the book useful. Because of its wide coverage, the book may, in addition, serve as a reference volume, useful hopefully to student and researcher alike.

The choice of material is dictated by the desire to communicate to a mainly advanced undergraduate and beginning graduate audience those topics that 1) play a significant role in contemporary applications of mathematics to physics, and 2) are not usually given prominence in conventional texts at the same level. Thus a major part of the text is focused on group representation theory, Lie groups and Lie algebras, exterior algebra, and finally, differential geometry, at the expense of more traditional (but no less important) topics such as differential equations, complex function theory, special functions, and functional analysis. This choice is limited, necessarily, by the author's knowledge or lack thereof, and also, needless to say, by space. Still, it is hoped that the reader will find between the two covers a more or less coherent body of information that is also reasonably complete, and that, above all, possesses a certain degree of thematic unity. Many excellent texts already exist which deal with the applications of *either* group theory *or* differential geometry to physics, but rarely simultaneously. In the present book we bring these two vital strands of contemporary mathematical physics together, not only for convenience, but also to demonstrate some of the deep connections between them.

The organization of the book may be described as functionally (but not logically) modular – with each chapter serving as a distinct module whose contents can be clearly discerned from the title, and which is topically (but not logically) independent of the others. Yet if the book is read from beginning to end in the order presented (although this is by no means obligatory), an unbroken thread may be seen to run through all the chapters, in a loosely thematic sense. This thread weaves together linear spaces and linear operators, representations of groups and algebras, algebraic structures built on differentiable manifolds, vector and principal bundles, and finally, the algebraic objects (characteristic classes) constructed from analytical data (curvatures) that unify the local and global properties of fiber bundles. At various points, detours are taken to show carefully how these notions have relevance in physics.

In principle, the book is self-contained, the only prerequisites being sophomore-level calculus, differential equations, (three-dimensional) vector analysis, and some linear algebra. In its entirety, there is probably enough material for a four-quarter or three-semester sequence of courses. However, students engaged in self-study and instructors may select different subsets of the book as meaningful units, according to their individual predilections and needs, although they should be forewarned that, for the most part, later chapters depend logically on earlier ones. In order to make the book maximally useful, a copious amount of cross-references (backward and forward) have been incorporated. This feature, together with a rather detailed index, will hopefully eliminate most sources of ambiguity. Whenever calculations are presented, they tend to be quite explicit with step-by-step justifications and relatively few steps left out. This practice has no doubt increased the physical size of the book, but hopefully will substantially decrease the frustration level of the reader. Numerous exercises are inserted at strategic locations throughout the text. In principle, they should be completely workable once the material in the text is comprehended. They mainly serve to amplify, concretize, and reinforce things learned, but never to intimidate.

I have decided to adopt a somewhat mathematical style of presentation, at odds with the usual practice in the physics literature. This by and large means frequent definition-theorem-proof sequences, with a level of rigor somewhere between the mathematical and the physical. There are primarily three reasons for doing this. The first is that the mathematical style affords a certain compactness, precision, generality, and economy of presentation that is quite indispensable for a text of this size. The second is that this style, when used with moderation, will often facilitate comprehension of deep and general concepts significantly, especially those that find very diversified applications in physics. Most physicists usually learn the mathematics that they need through specific, multiple, and contextual applications. While this approach has the definite advantage of making abstract ideas concrete, and thus initially less intimidating, the many different physical guises under which a single mathematical notion may appear frequently tend to obscure the essential unity of the latter. We need only mention two related examples, one elementary (and assumed familiar to the reader), the other relatively more advanced (but dealt with exhaustively in this book): the derivative and the covariant derivative (connection on a fiber bundle). The last, perhaps most controversial, reason is my belief that even physicists should learn to "speak", with a reasonable degree of fluency, the language of mathematics. This belief in turn stems from the observation that the period of "acrimonious divorce" (in Freeman Dyson's words) between physicists and mathematicians seems to be drawing to an end, and the two groups will find it increasingly rewarding to communicate with each other, not just on the level of trading applications, but also on the deeper one of informing each other of their different but complementary modes of thinking. In this book, however, rigor is never pursued for rigor's sake. Proofs of theorems are only presented when they help clarify abstract concepts or illustrate special calculational techniques. On the other hand, when they are omitted (usually without apologies), it can be assumed that they are either too lengthy, too technically difficult, or simply too distracting.

A good many complete chapters deal exclusively with physics applications. These tend to be in close proximity to the exposition of the requisite mathematics, and one may notice a somewhat abrupt change in style from the mathematical to the physical, and vice versa. This is again done with some deliberation, in order to prepare the reader for the "culture shock" that she/he may experience on going from the standard literature in one discipline to the other. In some cases, the physics applications are presented even before the necessary mathematics has been completely explained. This may disrupt the logical flow of the presentation, but I suspect that the physicist reader's anxiety level (and blood pressure!) may be considerably lowered on being reassured frequently that there are serious physics applications to rather esoteric pieces of mathematics. Indeed, if I have succeeded in this volume to convince some physics students (or even practicing physicists) that the mathematical style and contents therein are not just fancy garb and window dressing, that they are there not to obfuscate, but rather to clarify, unify, and even to lend depth and hence generality to a host of seemingly disconnected physics ideas, my purpose would have been more than well-served. Unfortunately, my lack of training and knowledge does not permit me to relate the story of the other direction of flow in this fascinating two-way traffic: that physical reasoning and techniques (for example, in quantum field theory) have recently provided significant insights and tools for the solution of long-standing problems in pure mathematics.

The writing of much of this text would not have been possible without a recent unique collaborative experience which was the author's great fortune to enjoy. Over the course of about two years, Professor S. S. Chern generously and patiently guided me through the translation and expansion of his introductory text "Lectures on Differential Geometry" (Chern, Chen and Lam, 1999). This immensely valuable learning experience deeply enhanced not only my technical knowledge, but perhaps more importantly, my appreciation of the mysteriously fruitful but sometimes tortuous relationship between mathematics and physics. It also provided a degree of much-needed confidence for a physicist with relatively little formal training in mathematics. The last third of the book, which focuses on differential geometry, bears strongly the imprint of what I learned from Professor Chern, particularly of the material in the above-mentioned "Lectures". It is to S. S. Chern, therefore, that I owe my first and foremost note of deep gratitude. On the algebraic side, the author benefited greatly from S. Okubo's lectures on group theory, which he attended some two decades ago while he was a research associate at the University of Rochester. These lectures systematically distilled for the novice the essentials of the application of group theory to physics, and despite the proliferation of texts on this topic, still constitute a most valuable source (S. Okubo, 1980). During what seemed to be an interminable period in which this book gradually took shape, numerous colleagues in both the Physics and the Mathematics Departments at Cal Poly Pomona provided indispensable support, encouragement, and inspiration. To one and all of these kind individuals I extend my heartfelt gratitude, but would especially like to thank John Fang, Soumya Chakravarti, and Antonio Aurilia in Physics, and Bernard Banks, Martin Nakashima, and Weiqing Xie in Mathematics. Many of my former students who were made captive audience to the less than definitive form of my lecture notes provided frank and constructive feedback, which became the driving force urging me to produce a useful yet user-friendly volume. To them I would also like to express my gratitude. All the resolutions and planning for this book would have meant little if a sustained period of focused time had not been made available. For this I owe my debt to the Faculty Sabbatical Program and the Research, Scholarship and Creative Activity Summer Fellowship Program, both of the California State University.

I am greatly indebted to Dr. Sen Hu of World Scientific for initially taking interest in my project, for making all the necessary arrangements to get it started, and once started, for putting up graciously with my many requests and changes of plans, all in the most expedient manner. Towards the final stages, the kind assistance of Drs. Daniel Cartin and Ji-tan Lu of World Scientific in guiding the project to completion is also greatly appreciated. Andres Cardenas, my former student and good friend, expertly prepared all the figures, and patiently acted as my computer guru as I clumsily latexed my way through the manuscript. Without his help, this manuscript would probably still be languishing in some nether-zone of near completion. To him I owe a special note of thanks. Last but not least, I am very grateful to my wife, Dr. Bonnie Buratti, and our three boys, Nathan, Reuben, and Aaron, for their always being there for me, as part of a wonderful and supportive family.

Kai S. Lam
California State Polytechnic University, Pomona

Contents

Contents   xvii

Chapter 1

Vectors and Linear Transformations

The theory of linear transformations on vector spaces forms the cornerstone of large areas of mathematical physics. The basic notions and facts of this theory will be used repeatedly in this book. We begin our development by using an elementary example to provide a concrete and familiar context for the introduction of some of these notions. A more abstract and formal discussion will be presented in Chapter 5. In this chapter we will also establish a notational scheme for vectorial (tensorial) quantities which will be adhered to as much as possible throughout the entire text.

Consider a vector x in the plane $\mathbb{R}^2$ written in terms of its components x^1 and x^2:

$$x = x^1 e_1 + x^2 e_2 = x^i e_i \quad . \tag{1.1}$$

The vectors e_1 and e_2 in (1.1) form what is called a **basis** of the **linear vector space** $\mathbb{R}^2$, and the components of a vector x are determined by the choice of the basis. In Fig. 1.1, we have chosen (e_1, e_2) to be an **orthonormal basis**. This term simply means that e_1 and e_2 are each of unit length and are orthogonal (perpendicular) to each other. We will see later that, in a general vector space, the notions of length and orthogonality only make sense after the imposition of a scalar product on the vector space (Chapter 8).

Notice that in (1.1) we have used a superscript (upper index) for components and a subscript (lower index) for basis vectors, and repeated indices (one upper and one lower) are understood to be summed over (from 1 to the **dimension** of the vector space under consideration, which is 2 in the present case). This is called the **Einstein summation convention**. The orthonormal basis (e_1, e_2) is called a **reference frame** in physics. Note also that we have not used an arrow (or a boldface type) to represent a vector in order to avoid excessive notation.

Now consider the same vector x with components x'^i with respect to a

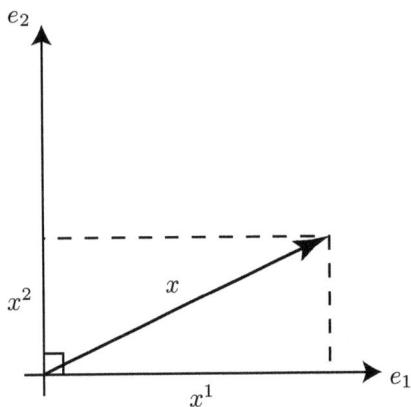

FIGURE 1.1

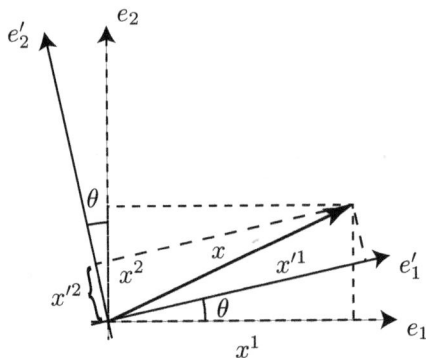

FIGURE 1.2

rotated orthonormal frame (e'_1, e'_2):

$$x = x'^i e'_i \quad . \tag{1.2}$$

It is simple to show that

$$(x')^1 = (\cos\theta)x^1 + (\sin\theta)x^2 , \quad (x')^2 = (-\sin\theta)x^1 + (\cos\theta)x^2 \quad . \tag{1.3}$$

Exercise 1.1 Verify (1.3) by considering the geometry of Fig. 1.2.

Eq. (1.3) can also be obtained by working with the basis vectors, instead of the components, directly. It is evident that

$$e_1 = (\cos\theta)e'_1 - (\sin\theta)e'_2 , \quad e_2 = (\sin\theta)e'_1 + (\cos\theta)e'_2 \quad . \tag{1.4}$$

Thus, (1.1) and (1.2) together, that is,

$$x = x^i e_i = x'^i e'_i$$

imply that

$$x^1\cos\theta e'_1 - x^1\sin\theta e'_2 + x^2\sin\theta e'_1 + x^2\cos\theta e'_2 = x'^1 e'_1 + x'^2 e'_2 \quad . \tag{1.5}$$

Comparison of the coefficients of e'_1 and e'_2 on both sides of (1.5) immediately yields (1.3).

Let us now write (1.3) and (1.4) in matrix notation. Equations (1.3) can be written as the single matrix equation

$$(x'^1, x'^2) = (x^1, x^2)\begin{pmatrix} \cos\theta & -\sin\theta \\ \sin\theta & \cos\theta \end{pmatrix} , \tag{1.6}$$

while Eqs. (1.4) can be written as

$$\left(\begin{array}{c} e_1 \\ e_2 \end{array} \right) = \left(\begin{array}{cc} \cos\theta & -\sin\theta \\ \sin\theta & \cos\theta \end{array} \right) \left(\begin{array}{c} e'_1 \\ e'_2 \end{array} \right) \quad . \tag{1.7}$$

Denote the 2×2 matrix in both (1.6) and (1.7) by

$$\left(a_i^j \right) = \left(\begin{array}{cc} a_1^1 & a_1^2 \\ a_2^1 & a_2^2 \end{array} \right) = \left(\begin{array}{cc} \cos\theta & -\sin\theta \\ \sin\theta & \cos\theta \end{array} \right) \quad , \tag{1.8}$$

where the lower index is the row index and the upper index is the column index. (This convention for denoting matrix elements will be used for the entire text.) Eqs. (1.6) and (1.7) can then be compactly written using the Einstein summation convention as

$$x'^i = a_j^i \, x^j \, , \tag{1.9}$$

$$e_i = a_i^j \, e'_j \, . \tag{1.10}$$

Note again that repeated pairs of indices, one upper and one lower, are summed over.

Eqs. (1.9) and (1.10) are equivalent, in the sense that either one is a consequence of the other. By way of illustrating the usefulness of the index notation, we again derive (1.9) and (1.10) as follows:

$$x = x^j e_j = x^j a_j^i e'_i = x'^i e'_i \quad . \tag{1.11}$$

The last equality implies (1.9).

We have presented all the above using what physicists call the passive viewpoint, in which the same vector ends up with different components when the reference frame is changed. Obviously (1.9) can also describe the situation where a vector is mapped to a different vector, with both old and new vectors described under the same reference frame or basis set. This is called the active viewpoint. In physics applications, the passive viewpoint is more natural, although we will have occasion to use both frequently.

Under the active viewpoint, (1.9) describes a so-called **linear transformation** of a vector x to another vector x'. We can express this linear transformation by

$$A : V \longrightarrow V \quad , \quad x' = A(x) \quad , \tag{1.12}$$

where x and x' are vectors in the same vector space V (written $x \in V, x' \in V$) and A is represented by the matrix (a_i^j). It is quite obvious that the linear transformation A satisfies the following property:

$$A(ax + by) = aA(x) + bA(y) \tag{1.13}$$

where a, b are scalars and $x, y \in V$.

$\boxed{\text{Exercise 1.2}}$ Verify (1.13) for the matrix (a_i^j) given by (1.8).

In fact (1.13) defines the notion of **linearity**, and we can introduce more generally the concept of a **linear map** between two different vector spaces.

Definition 1.1. *Given two vector spaces V and W over the same field $\mathbb{F}$ (of scalars), a map $f : V \longrightarrow W$ is called a **linear map** if it is a **homomorphism** that preserves vector addition and scalar multiplication, that is, if f satisfies*

$$f(ax + by) = af(x) + bf(y) \tag{1.14}$$

for all $x, y \in V$ and $a, b \in \mathbb{F}$.

How does one obtain a concrete **matrix representation** of an abstract linear transformation $A : V \longrightarrow V$? The answer is that a particular matrix representation arises from a particular choice of basis $\{e_i\}$ for V. (Now we can be more general and consider an n-dimensional vector space V.) For any $x \in V$ given by $x = x^i e_i$, the linearity condition (1.13) implies that

$$A(x) = x^i A(e_i) \quad . \tag{1.15}$$

Thus the action of A on any $x \in V$ is completely specified by $A(e_i), i = 1, \ldots, n$. Since $A(e_i)$ is a vector in V, it can be expressed as a **linear combination** of the e_i, that is

$$A(e_i) = a_i^j e_j \quad , \tag{1.16}$$

where the a_i^j are scalars in the field $\mathbb{F}$. Similar to (1.8) the quantities $a_i^j, i = 1, \ldots, n, j = 1, \ldots, n$, can be displayed as an $n \times n$ matrix

$$(a_i^j) = \begin{pmatrix} a_1^1 & a_1^2 & \cdots & a_1^n \\ a_2^1 & a_2^2 & \cdots & a_2^n \\ \cdots\cdots\cdots\cdots\cdots \\ a_n^1 & a_n^2 & \cdots & a_n^n \end{pmatrix} \quad . \tag{1.17}$$

Now suppose that, under the action of A, $x \in V$ is transformed into $x' \in V$. So

$$x' = A(x) = A(x^i e_i) = x^i A(e_i) = x^i a_i^j e_j = x^j a_j^i e_i \quad , \tag{1.18}$$

where in the last equality we have performed the interchange $(i \leftrightarrow j)$ since both i and j are dummy indices that are summed over. Eq. (1.18) then implies

$$x'^i = a_j^i x^j , \tag{1.19}$$

while Eqs. (1.4) can be written as

$$
\begin{pmatrix} e_1 \\ e_2 \end{pmatrix} = \begin{pmatrix} \cos\theta & -\sin\theta \\ \sin\theta & \cos\theta \end{pmatrix} \begin{pmatrix} e_1' \\ e_2' \end{pmatrix} . \tag{1.7}
$$

Denote the 2×2 matrix in both (1.6) and (1.7) by

$$
\left(a_i^j \right) = \begin{pmatrix} a_1^1 & a_1^2 \\ a_2^1 & a_2^2 \end{pmatrix} = \begin{pmatrix} \cos\theta & -\sin\theta \\ \sin\theta & \cos\theta \end{pmatrix} , \tag{1.8}
$$

where the lower index is the row index and the upper index is the column index. (This convention for denoting matrix elements will be used for the entire text.) Eqs. (1.6) and (1.7) can then be compactly written using the Einstein summation convention as

$$
x'^i = a_j^i \, x^j \,, \tag{1.9}
$$

$$
e_i = a_i^j \, e_j' \,. \tag{1.10}
$$

Note again that repeated pairs of indices, one upper and one lower, are summed over.

Eqs. (1.9) and (1.10) are equivalent, in the sense that either one is a consequence of the other. By way of illustrating the usefulness of the index notation, we again derive (1.9) and (1.10) as follows:

$$
x = x^j e_j = x^j a_j^i e_i' = x'^i e_i' \,. \tag{1.11}
$$

The last equality implies (1.9).

We have presented all the above using what physicists call the passive viewpoint, in which the same vector ends up with different components when the reference frame is changed. Obviously (1.9) can also describe the situation where a vector is mapped to a different vector, with both old and new vectors described under the same reference frame or basis set. This is called the active viewpoint. In physics applications, the passive viewpoint is more natural, although we will have occasion to use both frequently.

Under the active viewpoint, (1.9) describes a so-called **linear transformation** of a vector x to another vector x'. We can express this linear transformation by

$$
A : V \longrightarrow V \,, \quad x' = A(x) \,, \tag{1.12}
$$

where x and x' are vectors in the same vector space V (written $x \in V, x' \in V$) and A is represented by the matrix (a_i^j). It is quite obvious that the linear transformation A satisfies the following property:

$$
A(ax + by) = aA(x) + bA(y) \tag{1.13}
$$

where a, b are scalars and $x, y \in V$.

$\boxed{\text{Exercise 1.2}}$ Verify (1.13) for the matrix (a_i^j) given by (1.8).

In fact (1.13) defines the notion of **linearity**, and we can introduce more generally the concept of a **linear map** between two different vector spaces.

Definition 1.1. *Given two vector spaces V and W over the same field $\mathbb{F}$ (of scalars), a map $f : V \longrightarrow W$ is called a **linear map** if it is a **homomorphism** that preserves vector addition and scalar multiplication, that is, if f satisfies*

$$f(ax + by) = af(x) + bf(y) \tag{1.14}$$

for all $x, y \in V$ and $a, b \in \mathbb{F}$.

How does one obtain a concrete **matrix representation** of an abstract linear transformation $A : V \longrightarrow V$? The answer is that a particular matrix representation arises from a particular choice of basis $\{e_i\}$ for V. (Now we can be more general and consider an n-dimensional vector space V.) For any $x \in V$ given by $x = x^i e_i$, the linearity condition (1.13) implies that

$$A(x) = x^i A(e_i) \quad . \tag{1.15}$$

Thus the action of A on any $x \in V$ is completely specified by $A(e_i), i = 1, \ldots, n$. Since $A(e_i)$ is a vector in V, it can be expressed as a **linear combination** of the e_i, that is

$$A(e_i) = a_i^j e_j \quad , \tag{1.16}$$

where the a_i^j are scalars in the field $\mathbb{F}$. Similar to (1.8) the quantities $a_i^j, i = 1, \ldots, n, j = 1, \ldots, n$, can be displayed as an $n \times n$ matrix

$$(a_i^j) = \begin{pmatrix} a_1^1 & a_1^2 & \cdots & a_1^n \\ a_2^1 & a_2^2 & \cdots & a_2^n \\ \cdots\cdots\cdots\cdots\cdots \\ a_n^1 & a_n^2 & \cdots & a_n^n \end{pmatrix} \quad . \tag{1.17}$$

Now suppose that, under the action of A, $x \in V$ is transformed into $x' \in V$. So

$$x' = A(x) = A(x^i e_i) = x^i A(e_i) = x^i a_i^j e_j = x^j a_j^i e_i \quad , \tag{1.18}$$

where in the last equality we have performed the interchange $(i \leftrightarrow j)$ since both i and j are dummy indices that are summed over. Eq. (1.18) then implies

$$x'^i = a_j^i x^j , \tag{1.19}$$

which is formally the same as (1.9). Thus we have shown explicitly how *the matrix representation of a linear transformation depends on the choice of a basis set.*

Let us now note some important properties of the **rotation matrix**

$$a = (a_i^j) = \begin{pmatrix} \cos\theta & -\sin\theta \\ \sin\theta & \cos\theta \end{pmatrix} .$$

One can easily check that

$$a^{-1} = a^T = \begin{pmatrix} \cos\theta & \sin\theta \\ -\sin\theta & \cos\theta \end{pmatrix} , \tag{1.20}$$

where a^{-1} denotes the **inverse** of a and a^T the **transpose** of a. Recall that if a matrix a is **invertible**, then there exists a matrix a^{-1}, called the inverse of a, such that

$$aa^{-1} = a^{-1}a = 1 . \tag{1.21}$$

In the above equation, 1 represents the **identity matrix**. The matrix a in (1.8) also satisfies

$$\det(a) = 1 . \tag{1.22}$$

The properties (1.20) and (1.22) are in fact satisfied by all rotation matrices, not just those representing rotations in $\mathbb{R}^2$ but also those representing rotations in $\mathbb{R}^n$ (n-dimensional Euclidean space) for $n \geq 3$. We can understand this geometrically as follows. The rotation of a vector certainly does not change its length. In the **Euclidean metric** (see Chapter 8) the square of the length of a vector $x = x^i e_i$ is given by $\sum_i x^i x^i$. (We write the summation sign explicitly because the Einstein summation convention is not used here.) Suppose under a rotation A, x is rotated to $x' = A(x)$, where

$$x'^i = a_j^i x^j . \tag{1.23}$$

Thus, **invariance** of the length of a vector under rotation implies

$$\sum_j x^j x^j = \sum_i x'^i x'^i = \sum_i a_j^i x^j a_k^i x^k = \sum_k a_j^i (a^T)_i^k x^j x^k = \sum_k (aa^T)_j^k x^j x^k . \tag{1.24}$$

Comparing the first and the last expressions in the string of equalities above, we see that

$$\left(aa^T\right)_j^k = \delta_j^k , \tag{1.25}$$

where δ_j^k is the **Kronecker delta**:

$$\delta_j^k = \begin{cases} 1 & , \quad j = k \\ 0 & , \quad j \neq k. \end{cases} \tag{1.26}$$

Remark on Notation: $a = (a_i^j)$ denotes a matrix whose elements are a_i^j ($i =$ row index and $j =$ column index), while A stands for the linear transformation represented by a.

Eq. (1.25) is equivalent to the matrix equation

$$aa^T = 1 \quad , \tag{1.27}$$

or

$$a^T = a^{-1} \quad . \tag{1.28}$$

Hence we have

$$a^T a = aa^T = 1 \quad . \tag{1.29}$$

Matrices satisfying (1.29) are called **orthogonal matrices**. All $n \times n$ matrices satisfying (1.29) form a **group**, of great importance in physics, called the **orthogonal group** of dimension n, denoted by $O(n)$. The formal definition of a group will be given in Chapter 4 (Def. 4.1).

Exercise 1.3   Write down explicitly the 3×3 orthogonal matrices representing rotations in 3-dimensional Euclidean space by

    1. $45°$ about the z-axis,

    2. $45°$ about the x-axis,

    3. $45°$ about the y-axis.

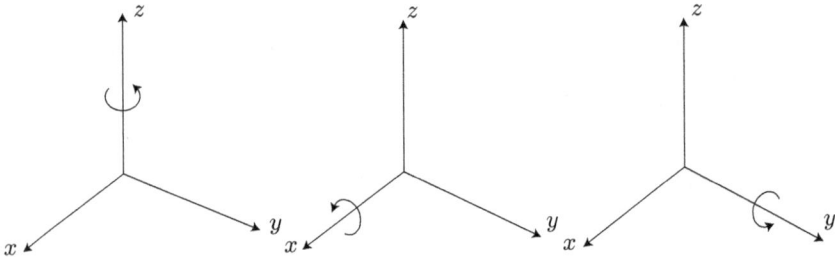

Exercise 1.4   Show that $SO(2)$, the group of rotations in 2 dimensions, is a **commutative group**, that is, any two 2×2 matrices in $SO(2)$ commute with each other.

Commutative groups are also called **abelian groups**.

Exercise 1.5   Show by an explicit example that $O(3)$ is not an abelian group.

Making use of the facts (which we will prove later in Chapter 16) that, if a and b are square matrices of the same size,

$$\det(ab) = \det(a)\det(b) \quad , \tag{1.30}$$

and

$$\det(a^T) = \det(a) \quad , \tag{1.31}$$

it follows immediately from (1.29) that, if a is an orthogonal matrix, then

$$\det(a) = \pm 1 \quad . \tag{1.32}$$

We have shown above that if a is a rotation matrix, then a is orthogonal. But rotations do not change the **orientation** of a space (we will explain this concept more clearly in Chapter 29), and thus rotation matrices must satisfy

$$\det(a) = +1 \quad . \tag{1.33}$$

[The relationship between determinants and orientations will be treated in detail later (in Chapter 29).] Thus we see that orthogonal matrices form a larger group than the group of rotation matrices. The group of $n \times n$ matrices which satisfy (1.29) and (1.33) is a very important **subgroup** of $O(n)$, called the **special orthogonal group** of dimension n, denoted by $SO(n)$. What, then, are the effects of matrices with determinant -1 in $O(n)$ on vectors in n-dimensional spaces? The answer is that they are either **pure inversions**, where some or all of the components of a vector change sign, or products of pure inversions and rotations.

In summary, an orthogonal matrix is one that satisfies (1.29). If its determinant is $+1$, then it represents a rotation; if its determinant is -1, then it represents either a pure inversion, or a combination of a pure inversion and a rotation.

Exercise 1.6   Construct explicitly an orthogonal matrix in $O(3)$ whose determinant is -1 and describe its action.

It is important to note that the concept of orthogonality introduced above applies not just to a matrix representing a linear transformation, but also to the linear transformation itself. The same is true of the concept of linearity. Thus we can speak of orthogonal linear transformations and invertible linear transformations.

It is perhaps most natural to think of the vector $x = x^i e_i$ as a **position vector** in space, so that the **coordinates** (components) of this vector transform as (1.9) under a change of reference (coordinate) frame, or equivalently, a change of basis. But there are many other kinds of vectors (other than position vectors) that transform in this way also under a change of reference frame. We can think of, in the physics context for example, the linear momentum p, the force F, etc. Thus we arrive at the physics way (mathematicians do it differently, see Def. 5.5) of defining vectors in general. In fact, a vector is a mathematical object whose components transform in a certain way under a change of reference frame (the passive viewpoint). More specifically:

Definition 1.2. *An element*

$$v = v^i e_i \tag{1.34}$$

*of an n-dimensional linear vector space V (expressed with respect to a choice
of basis vectors $e_i, i = 1, \ldots, n$) is said to be a **contravariant vector** if its
components v^i transform according to*

$$v'^i = a^i_j v^j \tag{1.35}$$

under a basis change

$$e_i = a^j_i e'_j \quad . \tag{1.36}$$

The term "contravariant" is used to emphasize the fact that the components
of a vector transform differently than the basis vectors. [In the transformation
equation for the components, it is the row index of the transformation matrix
(a^j_i) that is summed over, while in the transformation equation for the basis
vectors, it is the column index that is summed over.]

The above definition is a "physics" definition which is aimed at being more
physically intuitive than mathematically rigorous. It suggests the picture of
"vectors living in an underlying coordinate space", such as a force field $\boldsymbol{F}(\boldsymbol{r})$
depending on the position $\boldsymbol{r}$. This last object is an example of a **vector field**,
in which a vector is assigned at each point in space. The general mathematical
set-up describing this situation is called a **vector bundle**, in which a vector
space is attached to each point in an underlying **manifold** (space). A vector
field is then known as a **section of a vector bundle**. In the general case, the
underlying manifold may not have a **linear** (vector space) **structure**, as in the
case of a tangent vector field on a two-dimensional spherical surface. A detailed
consideration of these topics will entail the study of differential geometry, which
will be introduced later, starting with Chapter 30.

For invertible linear transformations, (1.35) and (1.36) imply

$$e'_i = (a^{-1})^j_i e_j \quad , \tag{1.37}$$

and

$$v^i = (a^{-1})^i_j v'^j \quad , \tag{1.38}$$

where a^{-1} is the inverse of the matrix a.

We summarize the transformation properties of a contravariant vector in
Table 1.1.

$v = v^i e_i = v'^i e'_i$	
$e_i = a^j_i e'_j$	$e'_i = (a^{-1})^j_i e_j$
$v'^i = a^i_j v^j$	$v^i = (a^{-1})^i_j v'^j$

TABLE 1.1

Let us now investigate the transformation properties of the matrix representation of a linear transformation A under a change of basis (1.36):

$$e_i = s^j_i e'_j \quad .$$

The required matrix a' is given by [(c.f. (1.16)]

$$A(e'_i) = a'^j_i e'_j \quad . \tag{1.39}$$

Using (1.37), we have

$$A(e'_i) = A((s^{-1})^l_i e_l) = (s^{-1})^l_i A(e_l) = (s^{-1})^l_i a^k_l e_k = (s^{-1})^l_i a^k_l s^j_k e'_j , \tag{1.40}$$

where in the second equality we have used the linearity property of A, in the third equality, Eq. (1.16), and in the fourth equality, Eq. (1.36). Comparison with (1.39) gives the desired result:

$$\boxed{a'^j_i = (s^{-1})^l_i a^k_l s^j_k} \quad . \tag{1.41}$$

In matrix notation (1.41) can be written

$$\boxed{a' = s^{-1}as} \quad . \tag{1.42}$$

The transformation $a \to a'$ given by (1.42) is called a **similarity transformation**. These transformations are of great importance in physics. Two matrices related by an invertible matrix s as in (1.42) are said to be **similar**.

Chapter 2

Tensors

According to (1.41) the upper index of a matrix (a_i^j) transforms like the upper index of a contravariant vector v^i (c.f. Table 1.1), and the lower index of (a_i^j) transforms like the lower index of a basis vector e_i (c.f. Table 1.1 also). In general a multi-indexed object

$$T^{i_1 \ldots i_r}_{j_1 \ldots j_s}$$

(with r upper indices and s lower indices) which transforms under a change of basis

$$e_i = s_i^j e_j'$$

[(1.36)] according to the following rule

$$\boxed{(T')^{i_1 \ldots i_r}_{j_1 \ldots j_s} = s^{i_1}_{k_1} \ldots s^{i_r}_{k_r} (s^{-1})^{l_1}_{j_1} \ldots (s^{-1})^{l_s}_{j_s} T^{k_1 \ldots k_r}_{l_1 \ldots l_s}} \qquad (2.1)$$

is called an (r, s)-**type tensor**, where r is called the **contravariant order** and s the **covariant order** of the tensor. Thus a matrix (a_i^j) which transforms as (1.41) is a (1,1)-type tensor. (r, s)-type tensors with $r \neq 0$ and $s \neq 0$ are called **tensors of mixed type**. A (0,0)-type tensor is a scalar. The term "covariant" means that the transformation is the same as that of the basis vectors:

$$(e')_i = (s^{-1})^j_i e_j' \quad ,$$

while "contravariant" means that the indexed quantity transforms according to the inverse of the transformation of the basis vectors.

As an example let us consider the **moment of inertia** tensor of a rigid body in classical mechanics I_i^j, which satisfies

$$L^i = I_j^i \omega^j \quad , \quad (i, j = 1, 2, 3) \quad , \qquad (2.2)$$

where L^i and ω^i are the components of the angular momentum and the angular velocity, respectively.

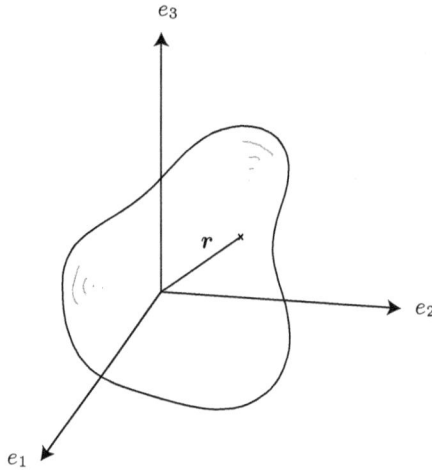

FIGURE 2.1

With respect to a given frame (e_1, e_2, e_3), the components of the inertia tensor are given by

$$I^i_j = \int \mathrm{d}^3 r\, \rho(\mathbf{r}) \left\{ r^2 \delta^i_j - \delta_{jk} x^k x^i \right\} \quad , \tag{2.3}$$

where $\mathbf{r}$ is the position vector of a point in the rigid body given by (see Fig. 2.1)

$$\mathbf{r} = x^i e_i \quad , \tag{2.4}$$

$\rho(\mathbf{r})$ is the mass density (position dependent), $\mathrm{d}^3 r$ is the volume element

$$\mathrm{d}^3 r = \mathrm{d}x^1 \mathrm{d}x^2 \mathrm{d}x^3 = \mathrm{d}x \mathrm{d}y \mathrm{d}z \quad , \tag{2.5}$$

and the integral is over the volume occupied by the rigid body. It is quite obvious that I^j_i is a **symmetric tensor**, i.e., $I^j_i = I^i_j$.

Exercise 2.1   Show that I^j_i transforms as a (1,1)-type tensor according to (2.1).

We will show later (in Chapter 17) that *a real, symmetric matrix is always diagonalizable*, that is, there exists an orthogonal transformation s such that $I' = s^{-1}Is$ is a diagonal matrix. In other words, one can always choose an orthonormal set of axes such that the inertia tensor is diagonal. These are called the **principal axes** of the rigid body and the corresponding diagonal elements are called the **principal moments of inertia**.

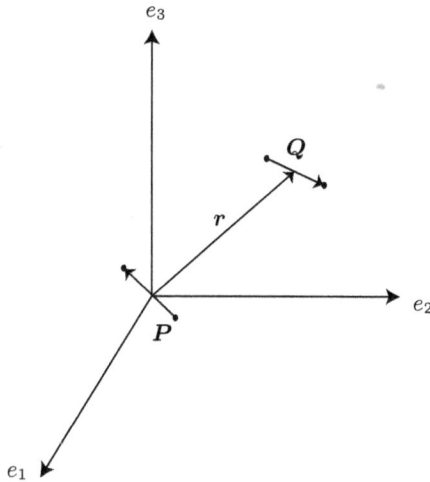

$$e_3$$

$$Q$$

$$r$$

$$e_2$$

$$P$$

$$e_1$$

FIGURE 2.2

Exercise 2.2 Verify that the inertia tensor can be displayed explicitly in matrix form as

$$I_i^j = \int d x d y d z \, \rho(x,y,z) \begin{pmatrix} y^2 + z^2 & -xy & -xz \\ -yx & x^2 + z^2 & -yz \\ -zx & -zy & x^2 + y^2 \end{pmatrix}, \qquad (2.6)$$

where $x^1 = x$, $x^2 = y$, and $x^3 = z$ are the Cartesian spatial coordinates.

Exercise 2.3 From electrostatics we know that the electrostatic potential energy of two interacting dipoles with dipole moments $\boldsymbol{P}$ and $\boldsymbol{Q}$ is given by

$$U = \frac{1}{r^3} \left\{ \boldsymbol{P} \cdot \boldsymbol{Q} - \frac{3(\boldsymbol{P} \cdot \boldsymbol{r})(\boldsymbol{Q} \cdot \boldsymbol{r})}{r^2} \right\} \equiv \frac{1}{r^3} P_i T_j^i Q^j, \qquad (2.7)$$

where r is the position vector from $\boldsymbol{P}$ to $\boldsymbol{Q}$, $\boldsymbol{P} = P^i e_i$, $\boldsymbol{Q} = Q^i e_i$, and $P_i = \delta_{ij} P^j$ (Fig. 2.2). Give an expression for the **dipole-coupling tensor** T_i^j in terms of x^1, x^2 and x^3.

Exercise 2.4 Show that under orthogonal transformations
1) δ^{ij} transforms as a (2,0)-type tensor;
2) δ_{ij} transforms as a (0,2)-type tensor;
3) δ_i^j transforms as a (1,1)-type tensor.

Exercise 2.5 Define the **Levi-Civita tensor** (also known as the **completely**

antisymmetric tensor) by

$$
\varepsilon^i{}_{jk} = \begin{cases} 0 & \text{;} \quad i,j,k \quad \text{not all distinct} \\ +1 & , \quad (ijk) \text{ is an even permutation of } (123) \\ -1 & , \quad (ijk) \text{ is an odd permutation of } (123) \end{cases} \tag{2.8}
$$

Thus

$$
\varepsilon^1{}_{23} = \varepsilon^2{}_{31} = \varepsilon^3{}_{12} = 1,
$$
$$
\varepsilon^1{}_{32} = \varepsilon^3{}_{21} = \varepsilon^2{}_{13} = -1,
$$
$$
\varepsilon^1{}_{12} = \varepsilon^1{}_{11} = \varepsilon^2{}_{12} = \cdots = 0.
$$

Show that $\varepsilon^i{}_{jk}$ transforms as a (1,2)-type tensor under a rotation.

The **vector (cross) product** (which is only well-defined for Euclidean 3-vectors) between two vectors $\boldsymbol{A}$ and $\boldsymbol{B}$ is given by

$$
\boxed{(\boldsymbol{A} \times \boldsymbol{B})^i = \varepsilon^i{}_{jk} A^j B^k} \tag{2.9}
$$

$\boxed{\text{Exercise 2.6}}$ For the unit vectors along the x, y and z axes in $\mathbb{R}^3$, we have

$$
e_1 = (1,0,0) , \quad e_2 = (0,1,0) , \quad e_3 = (0,0,1) .
$$

Use (2.9) to show that

$$
e_i \times e_j = -e_j \times e_i \quad , \tag{2.10}
$$
$$
e_1 \times e_2 = e_3 , \quad e_2 \times e_3 = e_1 , \quad e_3 \times e_1 = e_2 , \tag{2.11}
$$

and thus that the formula given by (2.9) concurs with the elementary definition of the vector product in 3-dimensional Euclidean space.

$\boxed{\text{Exercise 2.7}}$ Show that

$$
\varepsilon^k{}_{ij} \varepsilon_{klm} = \delta_{il}\delta_{jm} - \delta_{im}\delta_{jl} \quad , \tag{2.12}
$$

where

$$
\varepsilon_{klm} = \delta_{kn}\varepsilon^n{}_{lm} = \varepsilon^k{}_{lm} \quad . \tag{2.13}
$$

$\boxed{\text{Exercise 2.8}}$ Use (2.9) and (2.12) to prove the "triple cross-product" identity

$$
\boldsymbol{A} \times (\boldsymbol{B} \times \boldsymbol{C}) = \boldsymbol{B}(\boldsymbol{A} \cdot \boldsymbol{C}) - \boldsymbol{C}(\boldsymbol{A} \cdot \boldsymbol{B}) \quad , \tag{2.14}
$$

where the **scalar product** $\boldsymbol{A} \cdot \boldsymbol{B}$ is defined by

$$
\boldsymbol{A} \cdot \boldsymbol{B} = A^i B_i \quad , \tag{2.15}
$$

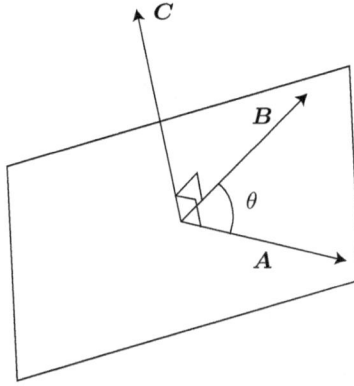

FIGURE 2.3

$$B_i = \delta_{ij} B^j = B^i \quad . \tag{2.16}$$

Exercise 2.9 Use (2.15) to show that

$$\boldsymbol{A} \cdot \boldsymbol{B} = AB \cos \theta \quad .$$

Exercise 2.10 Use (2.9) [or equivalently, (2.10) and (2.11)] to verify the "right hand rule" for cross products, and that if $\boldsymbol{C} = \boldsymbol{A} \times \boldsymbol{B}$, then $C = AB \sin \theta$ (Fig. 2.3).

Exercise 2.11 Use (2.15) and (2.9) to show that, for any three (3-dimensional Euclidean) vectors $\boldsymbol{A}, \boldsymbol{B}$ and $\boldsymbol{C}$,

$$\boldsymbol{A} \cdot (\boldsymbol{B} \times \boldsymbol{C}) = \boldsymbol{B} \cdot (\boldsymbol{C} \times \boldsymbol{A}) = \boldsymbol{C} \cdot (\boldsymbol{A} \times \boldsymbol{B}) , \tag{2.17}$$

and that each quantity above is equal to the **oriented volume** spanned by the vectors $\boldsymbol{A}, \boldsymbol{B}$ and $\boldsymbol{C}$ (Fig. 2.4).

As another important example of tensors let us consider the **electromagnetic field tensor** $F_{\mu\nu}$ in **Maxwell's electrodynamics**, which is a (0,2)-type tensor defined by

$$F_{\mu\nu} = \frac{\partial A_\nu}{\partial x^\mu} - \frac{\partial A_\mu}{\partial x^\nu} \equiv \partial_\mu A_\nu - \partial_\nu A_\mu \quad , \tag{2.18}$$

where $\mu, \nu = 0, 1, 2, 3$ are indices for the 4-dimensional **Minskowski space**,

$$A_\mu = \eta_{\mu\nu} A^\nu \quad , \tag{2.19}$$

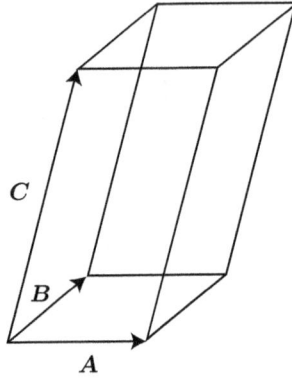

FIGURE 2.4

$\eta_{\mu\nu}$ is the so-called **Lorentz metric tensor**, a (0,2)-type tensor given by

$$\eta_{\mu\nu} = \begin{pmatrix} -1 & 0 & 0 & 0 \\ 0 & 1 & 0 & 0 \\ 0 & 0 & 1 & 0 \\ 0 & 0 & 0 & 1 \end{pmatrix} , \tag{2.20}$$

and $A^{\mu} = (A^0, A^1, A^2, A^3) = (\phi, \boldsymbol{A})$ is the **4-vector potential** of electrodynamics satisfying

$$\boldsymbol{B} = \nabla \times \boldsymbol{A} \ , \tag{2.21}$$

$$\boldsymbol{E} = -\frac{\partial \boldsymbol{A}}{\partial t} - \nabla \phi \ . \tag{2.22}$$

In the above, $\boldsymbol{B}$ and $\boldsymbol{E}$ are the magnetic and electric vector fields, respectively. According to (2.18), $F_{\mu\nu}$ is obviously an antisymmetric tensor. F_{01}, for example, is given by

$$F_{01} = \frac{\partial A_1}{\partial x^0} - \frac{\partial A_0}{\partial x^1} \ . \tag{2.23}$$

Now

$$A_1 = \eta_{11}A^1 = A^1 \ , \tag{2.24}$$

$$A_0 = \eta_{00}A^0 = -A^0 \ . \tag{2.25}$$

Hence

$$F_{01} = \frac{\partial A^1}{\partial t} + \frac{\partial A^0}{\partial x^1} = \frac{\partial A^1}{\partial t} + \frac{\partial \phi}{\partial x^1} = -E^1 \ . \tag{2.26}$$

Similarly,

$$F_{02} = -E^2 \ , \quad F_{03} = -E^3 \ . \tag{2.27}$$

Exercise 2.12   Eqs. (2.21) and (2.9) imply that

$$B^i = \varepsilon^{ij}{}_k \frac{\partial A^k}{\partial x^j} = \varepsilon^{ij}{}_k \partial_j A^k \quad , \tag{2.28}$$

where

$$\varepsilon^{ij}{}_k = \delta^{jl}\varepsilon^i{}_{lk} = \varepsilon^i{}_{jk} \quad . \tag{2.29}$$

Use (2.28) and (2.18) to show that

$$F_{12} = B^3 , \quad F_{13} = -B^2 , \quad F_{23} = B^1 . \tag{2.30}$$

Finally, we have in matrix form, with $\mu =$ the row index and ν the column index,

$$F_{\mu\nu} = \begin{pmatrix} 0 & -E^1 & -E^2 & -E^3 \\ E^1 & 0 & B^3 & -B^2 \\ E^2 & -B^3 & 0 & B^1 \\ E^3 & B^2 & -B^1 & 0 \end{pmatrix} . \tag{2.31}$$

In the examples introduced above, the dipole-coupling tensor T^j_i and the electromagnetic tensor $F_{\mu\nu}$ are **tensor fields**, since these tensors "live" on an underlying space and depend on the position in that space.

Chapter 3

Symmetry and Conservation: the Angular Momentum

Conservation laws are the most fundamental as well as the most useful in physics. Examples are the conservation of linear and angular momentum and the conservation of energy in classical mechanics and classical field theories, and the conservation of various types of quantum numbers in quantum mechanics. A principle of the most fundamental importance and far-reaching consequences is that *every conserved quantity arises from a certain symmetry of the physical system.* In this chapter we will demonstrate this principle as applied to the case of the angular momentum in classical mechanics. We will see that the vector and tensor formalism developed in the previous chapters will be most suitable for this purpose.

One begins with the construction of a quantity called the **classical action,** usually denoted by S. For a point particle of mass m moving under the influence of a potential $V(x^i)(i = 1, 2, 3)$, S is given by

$$S[x^i(t)] = \int_{t_1}^{t_2} dt \left\{ \frac{1}{2} m \frac{dx^i}{dt} \frac{dx_i}{dt} - V(x^i) \right\} \quad , \tag{3.1}$$

where t_1, t_2 are two fixed times, the Einstein summation convention has been used, and indices are raised and lowered by the Kronecker delta:

$$x_i = \delta_{ij} x^j = x^i \quad . \tag{3.2}$$

The square brackets on the LHS of (3.1) is the standard notation to signify that the classical action S is a **functional** of the classical trajectories $x^i(t)$. The quantity inside { } is called the **Lagrangian** L:

$$L = \frac{1}{2} m \frac{dx^i}{dt} \frac{dx_i}{dt} - V(x^i) \quad , \tag{3.3}$$

19

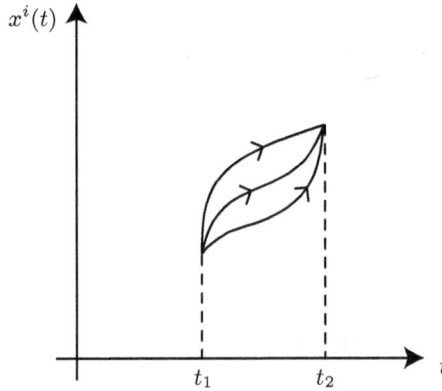

FIGURE 3.1

which is seen to be the difference between the kinetic and potential energies of the mass m.

What happens when $x^i(t)$ is varied slightly so that the endpoints remain fixed? (See Fig. 3.1.) In other words we consider the small **variations** $\delta x^i(t)$:

$$x^i(t) \longrightarrow x^i(t) + \delta x^i(t) \tag{3.4}$$

such that

$$\delta x^i(t_1) = \delta x^i(t_2) = 0 \quad . \tag{3.5}$$

The response of $S[x^i]$ to a small variation δx^i can be easily calculated. We have, to first order in δx^i,

$$
\begin{aligned}
S[x^i + \delta x^i] &= \int_{t_1}^{t_2} dt \left\{ \frac{1}{2} m \frac{d}{dt}(x^i + \delta x^i) \frac{d}{dt}(x_i + \delta x_i) - V(x^i + \delta x^i) \right\} \\
&= \int_{t_1}^{t_2} dt \left\{ \frac{1}{2} m \left(\frac{dx^i}{dt} \frac{dx_i}{dt} + 2 \frac{dx^i}{dt} \frac{d(\delta x_i)}{dt} + \frac{d}{dt}(\delta x^i) \frac{d}{dt}(\delta x_i) \right) \right. \\
&\qquad \left. - \left(V(x^i) + \delta x^i \frac{\partial V}{\partial x^i} \right) \right\} \\
&= \int_{t_1}^{t_2} dt \left\{ \frac{1}{2} m \left(\frac{dx^i}{dt} \frac{dx_i}{dt} + 2 \frac{d}{dt} \left(\delta x_i \frac{dx^i}{dt} \right) - 2 \delta x_i \frac{d^2 x^i}{dt^2} \right) \right. \\
&\qquad \left. - V(x^i) - \delta x^i \partial_i V \right\} \quad , \tag{3.6}
\end{aligned}
$$

where

$$\partial_i V \equiv \frac{\partial V}{\partial x^i} \quad . \tag{3.7}$$

Thus

$$
\begin{aligned}
\delta S = S[x^i + \delta x^i] - S[x^i] &= \int_{t_1}^{t_2} dt \left\{ \delta x_i \left(-m \frac{d^2 x^i}{dt^2} \right) - \delta x^j \partial_j V \right\} \\
&= \int_{t_1}^{t_2} dt \left\{ \delta x_i \left(-m \frac{d^2 x^i}{dt^2} - \delta^{ij} \partial_j V \right) \right\} . \tag{3.8}
\end{aligned}
$$

Defining the so-called **functional derivative** $\dfrac{\delta S}{\delta x^i}$ by

$$S[x^i + \delta x^i] = S[x^i] + \int dt\, \delta x^i\, \frac{\delta S}{\delta x^i} + \dots \quad , \tag{3.9}$$

we have, from (3.8),

$$\frac{\delta S}{\delta x^i} = -\left(m\frac{d^2 x^i}{dt^2} + \delta^{ij}\partial_j V\right) \quad . \tag{3.10}$$

Setting $\dfrac{\delta S}{\delta x^i} = 0$, that is, **extremizing the classical action** S, we arrive at the **equations of motion** for classical mechanics (Newton's second law of motion):

$$m\frac{d^2 x^i}{dt^2} = -\frac{\partial V}{\partial x^i} \quad, i = 1, 2, 3. \tag{3.11}$$

Note that the term on the RHS of (3.6) proportional to the total derivative, $\dfrac{d}{dt}(\delta x_i \dfrac{dx^i}{dt})$, called the **surface term**, does not contribute to the integral due to the assumption (3.5).

Now let us assume the validity of the equations of motion (3.11) and consider the variation of S under an infinitesimal rotation of the position vector $x = x^i e_i$. We also assume the case of a **central potential** so that

$$V(x^i) = V\left(\sqrt{x^i x_i}\right) \quad , \tag{3.12}$$

that is, V is a function of the length of the position vector only. Since both

$$v^2 = \frac{dx^i}{dt}\frac{dx_i}{dt} \tag{3.13}$$

and $V(x^i)$ remain invariant under a rotation of $x^i e_i$, $\delta S = 0$ under a rotation also. Let us now calculate δS explicitly under a rotation.

First recall from (1.9) and (1.29) that, under a rotation,

$$x'^i = a^i_j x^j \quad , \tag{3.14}$$

with

$$aa^T = a^T a = 1 \quad , \tag{3.15}$$

where a^T is the transpose of the rotation matrix a. Consider an infinitesimal rotation

$$a = 1 + \epsilon \quad . \tag{3.16}$$

Eq. (3.15) implies

$$(1 + \epsilon)(1 + \epsilon^T) = 1 \quad . \tag{3.17}$$

Thus, to first order in the infinitesimal matrix ϵ, we have

$$\epsilon^T = -\epsilon \quad . \tag{3.18}$$

For an infinitesimal rotation

$$x'^i = (\delta^i_j + \epsilon^i_j)x^j = x^i + \epsilon^i_j x^j = x^i + \delta x^i \ . \tag{3.19}$$

Equivalently,

$$\delta x^i = \epsilon^i_j x^j \quad , \tag{3.20}$$

where

$$\epsilon^i_j = -\epsilon^j_i \quad . \tag{3.21}$$

We now return to (3.6):

$$\delta S = \int_{t_1}^{t_2} dt\, \delta x_i \left(-m\frac{d^2 x^i}{dt^2} - \delta^{ij}\partial_j V \right) + \int_{t_1}^{t_2} dt\, \frac{d}{dt}\left(\delta x_i \frac{dx^i}{dt} \right)m \quad . \tag{3.22}$$

The equations of motion (3.11) imply that the first term on the RHS of the above equation vanishes. Since, as discussed above, $\delta S = 0$ under a rotation, we then have

$$\int_{t_1}^{t_2} dt\, \frac{d}{dt}\left(m\frac{dx^i}{dt}\delta x_i \right) = 0 \quad , \tag{3.23}$$

or

$$\left. m\frac{dx^i}{dt}\delta x_i \right|_{t_1}^{t_2} = 0 \quad , \tag{3.24}$$

which in turn implies

$$m\frac{dx^i}{dt}\delta x_i = \text{constant (in time)} \quad . \tag{3.25}$$

Using (3.20),

$$m\frac{dx^i}{dt}\delta x_i = m\delta_{ij}\epsilon^j_k x^k \frac{dx^i}{dt} = m\epsilon^j_k t^k_j \quad , \tag{3.26}$$

where

$$t^k_j \equiv \delta_{ij} x^k \frac{dx^i}{dt} \quad . \tag{3.27}$$

Eq. (3.21) implies that the matrix (ϵ^i_j) can be displayed explicitly as the antisymmetric matrix

$$(\epsilon^i_j) = \begin{pmatrix} 0 & \epsilon^2_1 & \epsilon^3_1 \\ -\epsilon^2_1 & 0 & \epsilon^3_2 \\ -\epsilon^3_1 & -\epsilon^3_2 & 0 \end{pmatrix} \quad . \tag{3.28}$$

Thus the sums over j and k in (3.26) can be written explicitly as

$$m\epsilon^j_k t^k_j = m\epsilon^2_1(t^1_2 - t^2_1) + m\epsilon^3_1(t^1_3 - t^3_1) + m\epsilon^3_2(t^2_3 - t^3_2) \quad .$$

Now define

$$l^j_i \equiv -m(t^j_i - t^i_j) \quad . \tag{3.29}$$

Since the ϵ_i^j's are arbitrary [as long as they satisfy (3.21)], the individual l_i^j's must be conserved (in time). To put this fact in more familiar form, we will raise all indices. We thus have the conserved quantities

$$l^{ij} = m(t^{ij} - t^{ji}) \quad .$$

According to (3.27)

$$t^{ij} = \delta^{ik} t_k^j = \delta^{ik} \delta_{lk} x^j \frac{\mathrm{d}x^l}{\mathrm{d}t} = \delta_l^i x^j \frac{\mathrm{d}x^l}{\mathrm{d}t} = x^j \frac{\mathrm{d}x^i}{\mathrm{d}t} \quad . \tag{3.30}$$

The conserved quantities

$$l^{ij} \equiv m \left(x^i \frac{\mathrm{d}x^j}{\mathrm{d}t} - x^j \frac{\mathrm{d}x^i}{\mathrm{d}t} \right) \tag{3.31}$$

are in fact the components of the angular momentum $\boldsymbol{L} = \boldsymbol{r} \times \boldsymbol{p}$. This is seen from (2.9) for the vector product. Thus

$$L^1 = l^{23}, \quad L^2 = l^{31}, \quad L^3 = l^{12} . \tag{3.32}$$

Note that l^{ij} is an antisymmetric (2,0)-type tensor, which has three independent components when the underlying manifold is 3-dimensional. The fact that l^{ij} can be considered as the three components of a 3-dimensional vector is an accident of the dimensionality of the manifold in which the tensor l^{ij} lives.

This is an important point. We will see later that the vector product is not a meaningful mathematical construct in spaces of dimension other than three. The generalization of the vector product to spaces of arbitrary dimension is called an **exterior product**, which will be developed systematically in Chapter 28.

| Exercise 3.1 | Show by the action principle that the linear momentum $\boldsymbol{p} = m\boldsymbol{v}$ is conserved if the external force is zero. The relevant action S is given by (3.1) with $V(x^i) = 0$. You need to show that $\boldsymbol{p}$ is conserved as a result of the fact that $\delta S = 0$ under a uniform translation of coordinates, that is,

$$x^i \longrightarrow x^i + \epsilon^i \quad ,$$

where ϵ^i is an infinitesimal quantity.

The conservation laws of linear and angular momentum are special cases of the celebrated **Noether's Theorem**, which states, roughly, that *to every symmetry leading to the invariance of the action functional, there corresponds a conserved physical quantity* (see Y. Choquet-Bruhat and C. DeWitt-Morette 1989).

Chapter 4

The Angular Momentum as Generators of Rotations: Lie Groups and Lie Algebras

We saw in the last chapter that in classical mechanics, *conservation of linear momentum is a consequence of translational invariance* and *conservation of angular momentum is a consequence of rotational invariance.* The same applies to quantum mechanics. This fact underlies the fundamental importance of symmetry considerations in the formulation of the laws of physics and the solution of physics problems.

In quantum mechanics the **observable** of linear momentum is replaced by the **generator of translations**, and the angular momentum by the **generator of rotations**. Consider the infinitesimal translation

$$x^i \longrightarrow x^i + \delta x^i \quad , i = 1, 2, 3, \tag{4.1}$$

where

$$\delta x^i = \epsilon^i \quad . \tag{4.2}$$

We can rewrite

$$\delta x^k = i\epsilon^j p_j x^k \quad , \tag{4.3}$$

where p_j is the **linear momentum operator**

$$\boxed{p_j = -i\frac{\partial}{\partial x^j}} \quad , \tag{4.4}$$

or

$$\boldsymbol{p} = -i\nabla \quad ,$$

where ∇ is the gradient in vector calculus. (More precisely, $p_j = -i\hbar\frac{\partial}{\partial x^j}$; but we have set Planck's constant $\hbar = 1$.) The factor $i = \sqrt{-1}$ in (4.3) and (4.4) is

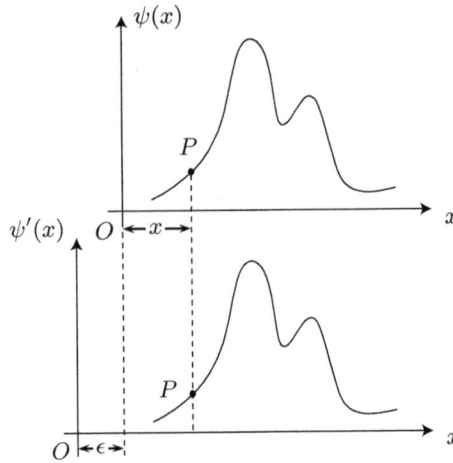

FIGURE 4.1

present due to the fact that p_j has to be a **hermitian operator** (representing physical observables) in quantum theory. The significance of the **hermiticity** of operators will be discussed in Chapter 10.

How does a scalar **wave function** transform under translations? The general rule (2.1) for transformations of (r, s)-type tensors applied to scalars [$(0, 0)$-type tensor] is clearly

$$\psi'(x'^{\,i}) = \psi(x^i) \quad . \tag{4.5}$$

This is also evident from Fig. 4.1: the height of a certain fixed spot P on the hillside is invariant with respect to the translation of a coordinate frame. Using

$$x'^{\,i} = x^i + \epsilon^i \quad , \tag{4.6}$$

(4.5) implies

$$\psi'(x^i) \equiv U(\epsilon^i)\psi(x^i) = \psi(x^i - \epsilon^i) = \psi(x^i) - \epsilon^i \partial_i \psi + O(\epsilon^2) \, , \tag{4.7}$$

or

$$U(\epsilon^i)\psi(x^i) = \psi(x^i) - i\frac{\epsilon^i}{\hbar}p_i\psi + O(\epsilon^2) \quad . \tag{4.8}$$

This equation suggests the following result (which is in fact true, although we will not prove it) for a **finite translation** $x^i \longrightarrow x^i + a^i$:

$$\psi'(x^i) = U(a^i)\psi(x^i) \quad , \tag{4.9}$$

where

$$\boxed{U(a^i) = \exp\left(-\frac{i}{\hbar}a^j p_j\right)} \quad , \tag{4.10}$$

and

$$p_j = -i\hbar\partial_j \quad . \tag{4.11}$$

The symbol U has been chosen to remind us of the fact that it represents a **unitary operator** (transformation), which plays the same role as an orthogonal transformation when one deals with **complex vector spaces**— the kind required in quantum mechanics. Eq. (4.10) is the reason behind the fact that the linear momentum p_j is called the **generator** of translations.

The fact that in quantum mechanics, p_j is a hermitian operator rather than a number has the most profound physical consequences. It is easy to establish mathematically that x^i and p^j do not commute. In fact

$$\boxed{[x^j, p_k] = i\hbar\delta^j_k}\quad . \tag{4.12}$$

This **commutation relation** is the mathematical basis of **Heisenberg's uncertainty principle**.

Exercise 4.1   Verify (4.12) by using (4.11).

We also have, trivially,

$$[x^i, x^j] = 0\quad, \tag{4.13}$$

$$[p_i, p_j] = 0\quad. \tag{4.14}$$

Let us now consider infinitesimal rotations, described by (3.20) and (3.21):

$$\delta x^i = \epsilon^i_j x^j\quad, \tag{4.15}$$

$$\epsilon^i_j = -\epsilon^j_i\quad. \tag{4.16}$$

Again, using (4.5) and following (4.7), we have, on recalling (3.14) to (3.16),

$$\psi'(x^i) = U(\epsilon^i_j)\psi(x^i) = \psi(a^{-1}x) = \psi(a^T x) = \psi((1+\epsilon^T)x) = \psi((1-\epsilon)x)$$
$$= \psi(x - \epsilon x) = \psi(x^i) - (\epsilon x)^i \partial_i \psi + O(\epsilon^2) = \psi(x^i) - \epsilon^i_j x^j \partial_i \psi + O(\epsilon^2)\,. \tag{4.17}$$

The second term can be readily written in terms of the components of the angular momentum, if we recall that ϵ^i_j is an antisymmetric matrix [(3.28)] and has only three independent components. Thus

$$\epsilon^i_j x^j \partial_i \psi = i\epsilon^i_j x^j p_i \psi$$
$$= i\epsilon^2_1(x^1 p_2 - x^2 p_1)\psi + i\epsilon^1_3(x^3 p_1 - x^1 p_3)\psi + i\epsilon^3_2(x^2 p_3 - x^3 p_2)\psi\,. \tag{4.18}$$

The quantities in parentheses on the RHS are recognized to be the components of the angular momentum $\boldsymbol{L} = \boldsymbol{r} \times \boldsymbol{p}$ [see (2.9)]:

$$L^3 = x^1 p_2 - x^2 p_1\,, \quad L^2 = x^3 p_1 - x^1 p_3\,, \quad L^1 = x^2 p_3 - x^3 p_2\,. \tag{4.19}$$

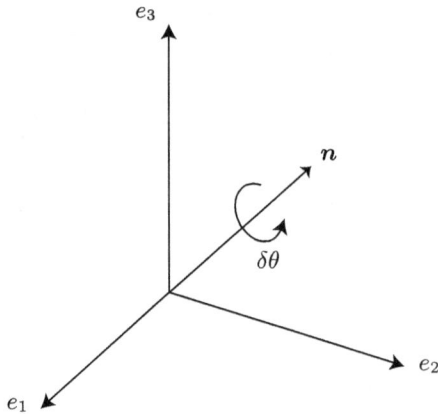

FIGURE 4.2

The three small parameters $(\epsilon_2^3, \epsilon_3^1, \epsilon_1^2)$ can be regarded as the components of the axis of rotation $\delta\theta n$ (n being a unit vector and $\delta\theta$ the angle of rotation in the positive sense) (see Fig. 4.2).

Analogous to (4.9) and (4.10), a wave function transforms under a finite rotation in 3-dimensional Euclidean space as

$$\psi'(x^i) = U(n, \theta)\psi(x^i) \quad , \tag{4.20}$$

$$\boxed{U(n, \theta) = \exp\left(-\frac{i}{\hbar}\theta\, n \cdot L\right)} \quad , \tag{4.21}$$

where θ is a finite angle of rotation. The angular momentum L is thus the generator of rotations.

$\boxed{\text{Exercise 4.2}}$ Work out explicit expressions for $\epsilon_1^2, \epsilon_3^1, \epsilon_2^3$ for the case of an infinitesimal rotation about the z-axis by an angle $\delta\theta$; and thus express $\psi'(x^i)$ in terms of $\psi(x^i)$ for this rotation.

As in the classical mechanical case, the components of the angular momentum operator

$$\begin{cases} L^3 = l_1^2 = -i\hbar(x^1\partial_2 - x^2\partial_1) \quad , \\ L^2 = l_3^1 = -i\hbar(x^3\partial_1 - x^1\partial_3) \quad , \\ L^1 = l_2^3 = -i\hbar(x^2\partial_3 - x^3\partial_2) \quad , \end{cases} \tag{4.22}$$

are actually the three independent components of an antisymmetric (1,1)-type tensor operator. In general, the number of independent components of a rank 2 antisymmetric tensor [where the rank of an (r, s)-type tensor is defined to be $r+s$] defined on an underlying space (manifold) of dimension n is equal to $n(n-$

1)/2. This is easily seen from Fig. 4.3, which displays an antisymmetric matrix with its zero diagonal elements, and upper right and lower left triangular blocks with elements that are negatives of each other. The number of independent elements in such a matrix is just equal to the number of elements in either the upper right or the lower left blocks. This number is clearly

$$\frac{n^2 - n}{2} = \frac{n(n-1)}{2} \ .$$

Thus for $n = 3$, $n(n-1)/2 = 3$ also.

$$\begin{pmatrix} 0 & l_1^2 & \cdots & \cdots & l_1^n \\ -l_1^2 & 0 & \cdots & \cdots & l_2^n \\ \cdots\cdots\cdots\cdots\cdots\cdots\cdots\cdots \\ \cdots & \cdots & \cdots & 0 & l_{n-1}^n \\ -l_1^n & -l_2^n & \cdots & -l_{n-1}^n & 0 \end{pmatrix}$$

FIGURE 4.3

As discussed in Chapter 1, the set of all pure rotations in 3 dimensions forms the group $SO(3)$. This is a special case of the more general $SO(n)$, which are very important examples of so-called **Lie groups**. A Lie group, roughly speaking, is a composite mathematical object which is both a group and a **differentiable manifold** (see Chapter 34). $SO(3)$ is a 3-dimensional **compact manifold** which can be characterized geometrically as a solid sphere of radius π with **antipodal points** on the spherical surface identified (see Fig. 4.4). Let us explain this.

Each $g \in SO(3)$ can be specified by 3 parameters: $0 \le \theta \le \pi, 0 \le \phi \le 2\pi$ (which are the polar and azimuth angles giving the orientation of the rotation axis $\boldsymbol{n}$), and $0 \le \alpha \le \pi$ (the angle of rotation). Thus each point in the interior of the aforementioned sphere represents a distinct rotation $g \in SO(3)$. The identity (no rotation at all) is represented by the center of the sphere. Antipodal (diametrically opposite) points on the surface of the sphere have to be identified since a rotation by π about $\boldsymbol{n}$ is exactly the same as a rotation by π about $-\boldsymbol{n}$. Also, a rotation by α about $\boldsymbol{n}$ is the same as a rotation by $2\pi - \alpha$ about $-\boldsymbol{n}$.

| Exercise 4.3 | Convince yourself of the validity of the preceding two sentences.

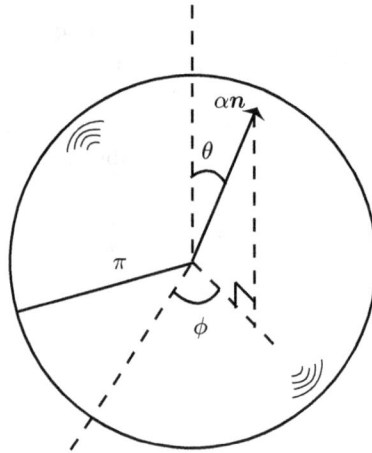

FIGURE 4.4

For further discussion of $O(3)$ and $SO(3)$ see Chapters 21 and 22. Before proceeding further it will be useful to give a general definition of a group and then introduce the idea of a **group representation**.

Definition 4.1. *A set G with a multiplication rule that satisfies the following properties is called a **group**:*
(1) if $g_1 \in G$ and $g_2 \in G$, then $g_1 g_2 \in G$ (closure under multiplication);
(2) $g_1(g_2 g_3) = (g_1 g_2)g_3$, for all $g_1, g_2, g_3 \in G$ (associativity);
*(3) there exists an element $e \in G$, called the **identity**, such that*

$$eg = ge = g \quad \text{for all} \quad g \in G \quad ;$$

and
*(4) for every $g \in G$, there exists an element $g^{-1} \in G$, called the **inverse** of g, such that*

$$gg^{-1} = g^{-1}g = e \quad .$$

Note that as mentioned earlier (in Chapter 1), group multiplication is in general noncommutative. Up to now we have dealt only with Lie groups such as $O(3)$ and $SO(3)$, whose elements are parametrized by continuous variables. There are also **discrete groups**, whose elements can be counted. A discrete group of fundamental importance in quantum statistics is the **symmetry group** (or **permutation group**) $\mathcal{S}(n)$, which is the group of all permutations of n objects. We will encounter this group later in Chapter 16. Among the Lie groups some of the most important ones in physics are examples of the so-called **classical groups**, such as $O(n), SO(n), U(n)$ and $SU(n)$, the latter two being the unitary group and the special unitary group of dimension n, respectively.

$U(n)$ can be regarded as the group of unitary $n \times n$ matrices, and $SU(n)$ the group of unitary $n \times n$ matrices with determinant $+1$. The designation "unitary" means that $gg^\dagger = g^\dagger g = e$, where $g^\dagger$ is the transpose, complex-conjugated matrix (or **hermitian conjugate**) of the matrix g. Thus unitary matrices have complex entries in general (see Chapter 10).

Definition 4.2. *A **group homomorphism** from a group G to another group G' is a mapping (not necessarily one-to-one) $f : G \longrightarrow G'$ which preserves group multiplication, that is, if $g_1, g_2 \in G$ and $g_1 g_2 = g_3$, then $f(g_1)f(g_2) = f(g_3)$.*

In physics, we often consider the action of group elements on scalars, vectors, tensors, etc. We have already seen examples of this in our discussion of the action of translations and rotations on wave functions, which can be considered as scalar fields (ψ is not indexed). The relevant notion here is that of a group representation.

Definition 4.3. *A **group representation** of a group G is a group homomorphism $f : G \longrightarrow \mathcal{T}(V)$, where $\mathcal{T}(V)$ is a group of linear transformations on a vector space V. A representation is said to be **faithful** if the homomorphism is also an **isomorphism** [one-to-one (injective) and onto (surjective)].*

Our previous examples (4.10) and (4.21), for the abelian (additive) group of translations and the nonabelian group of rotations $SO(3)$, respectively, are important examples of faithful representations.

A most important problem in mathematical physics is to find all the so-called **inequivalent irreducible representations** of a certain group. We will elaborate on these notions later (Chapter 18). Suffice it to say at this point that the solution of this problem has led to great progress in diverse fields of physics, from atomic, molecular and nuclear spectroscopy, to the classification of elementary particles.

We now return to the relationship between elements of a Lie group $g \in G$ and their generators. According to (4.10) and (4.21), the group elements are obtained by exponentiation of the generators. Without elaboration, we point out that the generators of a Lie group G are the vectors of a certain vector space $\mathcal{G}$, called the **Lie algebra of the Lie group** G. The dimension of $\mathcal{G}$ is the same as the dimension of G, when the latter is considered as a differentiable manifold. In fact, *the Lie algebra $\mathcal{G}$ is defined to be the space of left- (or right-) invariant vector fields on G, and is isomorphic to the tangent space of G at the identity.* A more detailed study of all these concepts will be deferred until Chapter 34. For an intuitive discussion of the notion of the tangent space to a differentiable manifold, we refer the reader to Chapter 30.

In general, for a compact Lie group G, one can define an **exponential map**, $\exp : \mathcal{G} \longrightarrow G$, such that for any $g \in G$, there exists an $A \in \mathcal{G}$ satisfying

$$g = \exp(A) \quad . \tag{4.23}$$

For matrix Lie groups, such as $O(n)$, elements in G and $\mathcal{G}$ are matrices of the same size and

$$\exp(A) = 1 + A + \frac{A^2}{2!} + \cdots \quad . \tag{4.24}$$

In (4.24) the sum always converges. For $SO(3)$, (4.16) implies that $\mathcal{SO}(3)$ (the Lie algebra of a certain Lie group is usually written with the designation of the group in script letters) is the set of all 3×3 antisymmetric matrices. The multiplication rule for $\mathcal{G}$ is given by the **Lie bracket** (or commutator):

$$[A, B] = AB - BA \quad , \tag{4.25}$$

for all $A, B \in \mathcal{G}$. It is not hard to prove the so-called **Jacobi identity**:

$$[A, [B, C]] + [B, [C, A]] + [C, [A, B]] = 0 \quad . \tag{4.26}$$

$\boxed{\text{Exercise 4.4}}$ Prove (4.26) using (4.25).

$\boxed{\text{Exercise 4.5}}$ Show that if A and B are both antisymmetric matrices, $[A, B]$ is also an antisymmetric matrix. Thus $\mathcal{SO}(n)$ is indeed closed under the Lie bracket multiplication rule.

The Lie bracket multiplication rule obviously also satisfies the distributive law:

$$[\alpha A + \beta B, C] = \alpha[A, C] + \beta[B, C] \quad , \tag{4.27}$$

for all $\alpha, \beta \in \mathbb{R}$; and the antisymmetric law:

$$[A, B] = -[B, A] \quad . \tag{4.28}$$

Indeed, any n-dimensional real vector space having a multiplication rule that satisfies the distributive law, the antisymmetric law and the Jacobi identity is called an n-dimensional Lie algebra. As an example, it is interesting to note that the 3-dimensional Euclidean space equipped with a vector (cross) product [defined by (2.9)] is a 3-dimensional Lie algebra.

$\boxed{\text{Exercise 4.6}}$ Verify the last statement.

Since the elements of the Lie algebra $\mathcal{G}$ are infinitesimal generators of the corresponding Lie group G near the identity, the structure of G near the identity is specified by the structure of $\mathcal{G}$. The structure of a Lie algebra is completely determined by the so-called structure constants, defined below.

Definition 4.4. *Let $(e_1, \ldots, e_n)$ be a basis of the Lie algebra $\mathcal{G}$ of a Lie group G. The **structure constants** $c^k{}_{ij} \in \mathbb{R}$ are defined by*

$$[e_i, e_j] = c^k{}_{ij}\, e_k \quad . \tag{4.29}$$

Exercise 4.7 Find the structure constants for $\mathbb{R}^3$ equipped with the ordinary cross product, i.e., find all $c^k{}_{ij}$ defined by

$$e_i \times e_j = c^k{}_{ij}\, e_k \quad . \tag{4.30}$$

Exercise 4.8 Use the Jacobi identity in the form

$$[e_i, [e_j, e_k]] + [e_j, [e_k, e_i]] + [e_k, [e_i, e_j]] = 0 \tag{4.31}$$

and (4.29) to show that the structure constants $c^k{}_{ij}$ satisfy the identity

$$c^h{}_{im}c^m{}_{jk} + c^h{}_{km}c^m{}_{ij} + c^h{}_{jm}c^m{}_{ki} = 0 \quad , \tag{4.32}$$

for all i, j, h, k.

We will now calculate the structure constants of $SO(n)$ and hence determine the Lie algebra $\mathcal{SO}(n)$. For $g \in SO(n)$, let $U(g)$ be an N-dimensional representation of g. [$U(g)$ can thus be considered as an $N \times N$ matrix, where N is in general different from n]. Recall that an element g near the identity $e \in SO(n)$ is specified by an antisymmetric matrix ϵ^j_i of small parameters [with $n(n-1)/2$ independent elements]. (See discussion immediately preceding Fig. 4.3.) Analogous to (4.21) we write

$$U(g) = 1 - \frac{1}{2}J_i^j\,\epsilon^i_j + O(\epsilon^2) \quad , \tag{4.33}$$

where the J_i^j are $N \times N$ matrices, ϵ^i_j are numbers, and

$$J_i^j = -J_j^i \quad . \tag{4.34}$$

The factor $1/2$ is introduced to take care of the antisymmetric nature of ϵ^j_i and J_i^j. The J_i^j are then elements of the Lie algebra $\mathcal{SO}(n)$, and actually constitute a basis of $\mathcal{SO}(n)$. To calculate the structure constants $c^k{}_{ij}$ we need to determine the commutation relations for the J_i^j. We will present the details of this rather long but instructive calculation in due course. First we give the result and illustrate its application to $SO(3)$. The Lie algebra structure of $\mathcal{SO}(n)$ is given in its entirety by the following commutation relations:

$$[J_i^j, J_k^l] = \delta_k^j J_i^l - \delta_i^l J_k^j + \delta_i^k J_l^j - \delta_l^j J_i^k \quad , \quad i, j, k, l = 1, \ldots, n \quad . \tag{4.35}$$

For $SO(3)$ there are only 3 independent components for J_i^j, namely, J_1^2, J_3^1 and J_2^3. According to (4.35)

$$[J_2^3, J_3^1] = J_2^1 \quad , \tag{4.36}$$

$$[J_3^1, J_1^2] = J_3^2 \quad , \tag{4.37}$$

$$[J_1^2, J_2^3] = J_1^3 \quad . \tag{4.38}$$

The above three equations give the "multiplication table" for $SO(3)$. The angular momentum operators L^1, L^2, L^3 [(4.22)] are actually related to the J_i^j by

$$\hbar J_1^2 = iL^3 \ , \quad \hbar J_3^1 = iL^2 \ , \quad \hbar J_2^3 = iL^1 \ . \tag{4.39}$$

Thus the angular momentum operators in quantum theory obey the familiar commutation relations:

$$[L^1, L^2] = i\hbar L^3 \ , \quad [L^2, L^3] = i\hbar L^1 \ , \quad [L^3, L^1] = i\hbar L^2 \ . \tag{4.40}$$

Let us now establish (4.35). Let $a, b \in SO(n)$ be elements that are close to the identity e. Suppose

$$a = 1 + \epsilon + O(\epsilon^2) \quad , \tag{4.41}$$

and

$$b = 1 + \epsilon' + O(\epsilon^2) \quad , \tag{4.42}$$

where both ϵ and ϵ' are antisymmetric $n \times n$ matrices of small parameters. Note that in the above equations 1 represents the $n \times n$ identity matrix. Since $U(g)$ is a group homomorphism (see Def. 4.2), we have

$$U(b^{-1}ab) = U(b^{-1})U(a)U(b) \quad . \tag{4.43}$$

On the other hand, by (4.41),

$$b^{-1}ab = b^{-1}(1 + \epsilon + O(\epsilon^2))b \approx 1 + b^{-1}\epsilon b + O(\epsilon^2) = 1 + \bar{\epsilon} + O(\epsilon^2) \ , \tag{4.44}$$

where we have defined

$$\bar{\epsilon} \equiv b^{-1}\epsilon b = b^T \epsilon b \quad , \tag{4.45}$$

with the last equality holding on account of the fact that b is an orthogonal matrix. Thus, by (4.33),

$$U(b^{-1}ab) = 1 - \frac{1}{2} J_i^j \bar{\epsilon}_j^i + O(\epsilon^2) = U^{-1}(b)\left(1 - \frac{1}{2} J_i^j \epsilon_j^i + O(\epsilon^2)\right)U(b)$$

$$= 1 - \frac{1}{2} U^{-1}(b) J_i^j \epsilon_j^i U(b) + O(\epsilon^2) \ . \tag{4.46}$$

Note again that, in the above equation, 1 represents the $N \times N$ identity matrix and all the J_i^j are $N \times N$ matrices, if $U(g)$ is an N-dimensional representation of $SO(n)$. All ϵ_i^j's are numbers. By (4.45)

$$\bar{\epsilon}_\beta{}^\alpha = (b^T \epsilon b)_\beta^\alpha = (b^T)_\beta^i \, \epsilon_i^j \, b_j^\alpha \quad . \tag{4.47}$$

Eq. (4.46) then implies

$$J_\alpha^\beta \epsilon_i^j (b^T)_\beta^i b_j^\alpha = \epsilon_j^i U^{-1}(b) J_i^j U(b) \quad . \tag{4.48}$$

Interchanging i and j ($i \leftrightarrow j$) on the LHS of the above equation (we can do this since they are both dummy indices that are summed over), (4.48) reads

$$\epsilon_j^i \left\{ b_i^\alpha (b^T)_\beta^j J_\alpha^\beta - U^{-1}(b) J_i^j U(b) \right\} = 0 \quad . \tag{4.49}$$

The quantity in braces in the above equation,

$$F_i^j \equiv b_i^\alpha (b^T)_\beta^j J_\alpha^\beta - U^{-1}(b) J_i^j U(b) \quad , \tag{4.50}$$

is an antisymmetric tensor:

$$F_i^j = -F_j^i \quad . \tag{4.51}$$

Indeed, on interchanging i and j ($i \leftrightarrow j$),

$$
\begin{aligned}
F_i^j &= b_j^\alpha (b^T)_\beta^i J_\alpha^\beta - U^{-1}(b) J_j^i U(b) \\
&= b_j^\beta (b^T)_\alpha^i J_\beta^\alpha - U^{-1}(b) J_j^i U(b) \quad (\alpha \leftrightarrow \beta) \\
&= -b_i^\alpha (b^T)_\beta^j J_\alpha^\beta + U^{-1}(b) J_i^j U(b) \quad (\text{since } J_\beta^\alpha = -J_\alpha^\beta) \\
&= -F_i^j \quad .
\end{aligned}
\tag{4.52}
$$

Eq. (4.49) can then be written

$$\epsilon_j^i F_i^j = 0 \quad , \tag{4.53}$$

or, since ϵ_j^i is antisymmetric ($\epsilon_j^i = -\epsilon_i^j$),

$$\sum_{j<i} \epsilon_j^i (F_i^j - F_j^i) = 0 \quad . \tag{4.54}$$

It follows that, since all the ϵ_j^i ($j < i$) are independent,

$$F_i^j = F_j^i \quad . \tag{4.55}$$

Eqs. (4.55) and (4.51) then necessarily imply that

$$F_i^j = 0 \quad . \tag{4.56}$$

Thus, from (4.50),

$$U^{-1}(b) J_i^j U(b) = b_i^\alpha (b^T)_\beta^j J_\alpha^\beta \quad . \tag{4.57}$$

Now choose

$$b = 1 + \epsilon' + O(\epsilon^2) \quad ,$$

with

$$(\epsilon')^T = -\epsilon' \quad . \tag{4.58}$$

Then,

$$U(b) = 1 - \frac{1}{2} J_\alpha^\beta \epsilon_\beta'^\alpha + O(\epsilon'^2) \quad , \tag{4.59}$$

and

$$U^{-1}(b) = 1 + \frac{1}{2} J_\alpha^\beta \epsilon_\beta'^\alpha + O(\epsilon'^2) \quad . \tag{4.60}$$

Also,

$$b_i^\alpha = \delta_i^\alpha + \epsilon_i'^\alpha + O(\epsilon'^2) \quad , \tag{4.61}$$

$$(b^T)_\beta^j = \delta_\beta^j - \epsilon_\beta'^j + O(\epsilon'^2) \quad . \tag{4.62}$$

Putting the above four equations in (4.57), we have

$$\left(1 + \frac{1}{2} J_\alpha^\beta \epsilon_\beta'^\alpha + O(\epsilon'^2) \right) J_i^j \left(1 - \frac{1}{2} J_\alpha^\beta \epsilon_\beta'^\alpha + O(\epsilon'^2) \right)$$

$$= \left(\delta_i^\alpha + \epsilon_i'^\alpha + O(\epsilon'^2) \right) \left(\delta_\beta^j - \epsilon_\beta'^j + O(\epsilon'^2) \right) J_\alpha^\beta , \tag{4.63}$$

or,

$$J_i^j + \frac{1}{2} \epsilon_\beta'^\alpha [J_\alpha^\beta, J_i^j] + O(\epsilon'^2) = J_i^j + \epsilon_i'^\alpha \delta_\beta^j J_\alpha^\beta - \delta_i^\alpha \epsilon_\beta'^j J_\alpha^\beta + O(\epsilon'^2) \quad , \tag{4.64}$$

or,

$$\frac{1}{2} \epsilon_\beta'^\alpha [J_\alpha^\beta, J_i^j] = \epsilon_i'^\alpha J_\alpha^j - \epsilon_\beta'^j J_i^\beta \quad . \tag{4.65}$$

We manipulate the right hand side further as follows:

$$
\begin{aligned}
\epsilon_i'^\alpha J_\alpha^j - \epsilon_\beta'^j J_i^\beta &= \epsilon_\beta'^\alpha \delta_i^\beta J_\alpha^j - \epsilon_\beta'^\alpha \delta_\alpha^j J_i^\beta \\
&= \frac{1}{2} \left(\epsilon_\beta'^\alpha \delta_i^\beta J_\alpha^j - \epsilon_\beta'^\alpha \delta_\alpha^j J_i^\beta \right) + \frac{1}{2} \left(\epsilon_\alpha'^\beta \delta_i^\alpha J_\beta^j - \epsilon_\alpha'^\beta \delta_\beta^j J_i^\alpha \right) \\
&= \frac{1}{2} \sum_{\alpha,\beta} \epsilon_\beta'^\alpha \left\{ \delta_i^\beta J_\alpha^j - \delta_\alpha^j J_i^\beta - \delta_\alpha^j J_i^\beta + \delta_\beta^j J_i^\alpha \right\} \quad , \tag{4.66}
\end{aligned}
$$

where, in the third and fourth terms on the RHS of the second equality, we have performed the interchange $(\alpha \leftrightarrow \beta)$ compared to the first and second terms. Eqs. (4.65) and (4.66) then imply

$$[J_\alpha^\beta, J_i^j] = \delta_i^\beta J_\alpha^j - \delta_\alpha^j J_i^\beta - \delta_i^\alpha J_\beta^j + \delta_\beta^j J_i^\alpha = \delta_i^\beta J_\alpha^j - \delta_\alpha^j J_i^\beta + \delta_i^\alpha J_j^\beta - \delta_\beta^j J_\alpha^i , \tag{4.67}$$

which is exactly the same equation as (4.35). This equation specifies the Lie algebra of $SO(n)$.

Chapter 5

Algebraic Structures

Up to this point we have introduced and worked with some important examples in the **algebraic categories** of linear vector spaces, groups, and algebras. These are the most important ones in physics applications, and will be the main ones considered in this book. In general, each algebraic category entails a special **algebraic structure** characterized by a specific set of relational rules (operations) on elements of any object in the category. For example, Def. 4.1 formalizes the structure for the category of groups. In this chapter we will do the same for the categories of linear vector spaces and algebras. We will also introduce two others, rings and modules, in order to establish a certain hierarchy among the different algebraic categories. In the next two chapters we will return to a more detailed look at groups and Lie algebras, to set the stage for further applications.

The group structure is the simplest, involving only one internal operation — group multiplication. In the case of abelian groups, this multiplication is sometimes written as an addition. Thus ab is written as $a + b$, ab^{-1} as $a - b$, and the identity e is written as 0 (zero).

$\boxed{\text{Exercise 5.1}}$ Check that all the group properties spelled out in Def. 4.1 are valid when multiplication is written as addition for abelian groups.

To differentiate formally the operations of multiplication and addition, we have to introduce the next level of complication in algebraic structures — the ring structure.

Definition 5.1. *A **ring** is a set R together with two **internal operations** $(x, y) \mapsto xy$ and $(x, y) \mapsto x + y$, called respectively multiplication and addition, such that*

   a) R is an abelian group under addition;

b) *multiplication is associative, and distributive with respect to addition, that is, for all $x, y, z \in R$,*

(1)
$$(xy)z = x(yz), \quad (associativity)$$

(2)
$$\begin{cases} x(y+z) = xy + xz, & (distributivity) \\ (y+z)x = yx + zx. \end{cases}$$

As in group multiplication, ring multiplication need not be commutative. If a ring R is commutative under multiplication, then R is said to be an **abelian (commutative) ring**.

Definition 5.2. *A ring R with an element $e \in R$, called the **identity** in R, such that $ex = xe = x$ for all $x \in R$ is called a **ring with identity**.*

Definition 5.3. *A ring F with identity is called a **field** if every element $x \in F$ except zero has an inverse, that is, to every non-zero $x \in F$, there corresponds a unique $x^{-1} \in F$, called the inverse of x, such that $xx^{-1} = x^{-1}x = e$.*

The most important examples of fields are the field of real numbers $\mathbb{R}$ and the field of complex numbers $\mathbb{C}$.

Exercise 5.2   Is the set of all integers $\mathbb{Z}$
(a) a ring?
(b) a field?
Why?

Exercise 5.3   Verify that the set of smooth real functions on $\mathbb{R}$ is a ring with identity.

Exercise 5.4   Justify that $\mathbb{R}$ and $\mathbb{C}$ are both fields.

The next level of complication in algebraic structures distinguishes between internal and external operations.

Definition 5.4. *A **module** M over a ring R is an abelian group M under addition together with an **external operation**, called **scalar multiplication**, $R \times M \longrightarrow M$, $(\alpha, x) \mapsto \alpha x$, such that*

$$\alpha(x + y) = \alpha x + \alpha y \quad,$$
$$(\alpha + \beta)x = \alpha x + \beta y \quad,$$
$$(\alpha\beta)x = \alpha(\beta x) \quad,$$

for all $\alpha, \beta \in R$ and $x, y \in M$. If R is a ring with identity e, then $ex = x$ for all $x \in M$.

A linear vector space is a module with a special requirement on the ring associated with the module.

Definition 5.5. *A **linear vector space** V over a field $\mathbb{F}$ is a module over a ring $\mathbb{F}$ which is also a field.*

Elements of a vector space V over $\mathbb{F}$ are called **vectors** and the elements of $\mathbb{F}$ are called **scalars**. If $\mathbb{F} = \mathbb{R}$, then V is called a **real vector space**; if $\mathbb{F} = \mathbb{C}$, then V is called a **complex vector space**. $\mathbb{R}, \mathbb{R}^2$ and $\mathbb{R}^3$ over $\mathbb{R}$ (with scalar multiplication by $x \in \mathbb{R}$) are the most familiar examples of real vector spaces, in which the notions of vector addition and scalar multiplication (being those of elementary vector arithmetic) have very intuitive, geometric interpretations. These notions can be easily generalized to $\mathbb{R}^n$ (the n-dimensional real coordinate space) and $\mathbb{C}^n$ (the n-dimensional complex coordinate space), where n is any positive integer. The vector space structure of $\mathbb{C}^n$ over $\mathbb{C}$ is specified as follows. $\mathbb{C}^n$, $n = 1, 2, \ldots$, is defined to be the set of all n-tuples of complex numbers. For any $z = (z_1, \ldots, z_n) \in \mathbb{C}^n$, $w = (w_1, \ldots, w_n) \in \mathbb{C}^n$, and $\alpha \in \mathbb{C}$, we define

$$z + w \equiv (z_1 + w_1, \ldots, z_n + w_n) \qquad \text{(vector addition)},$$
$$\alpha z \equiv (\alpha z_1, \ldots, \alpha z_n) \qquad \text{(scalar multiplication)}.$$

Note that the zero-vector is given by

$$0 \equiv \underbrace{(0, \ldots, 0)}_{n \text{ times}} \quad .$$

The same definitions apply to $\mathbb{R}^n$, where $\mathbb{C}$ is replaced by $\mathbb{R}$ in the above definition.

| Exercise 5.5 | Consider the equation of motion for the classical damped oscillator

$$\frac{d^2 x}{dt^2} + 2\gamma \frac{dx}{dt} + \omega_0^2 x = 0 \quad ,$$

where x is the displacement of the oscillator from equilibrium, ω_0 is the natural frequency of the oscillator, and γ is the damping constant. Show that the set of solutions to this equation constitutes a complex 2-dimensional vector space.

Finally, we come to the structure of an algebra, which is also a special kind of module in which special restrictions are placed on both the module and the ring associated with the module (instead of just the ring, as for vector spaces).

Definition 5.6. *An **algebra** A is a module over a ring R with identity such that*

1) A itself is a ring,

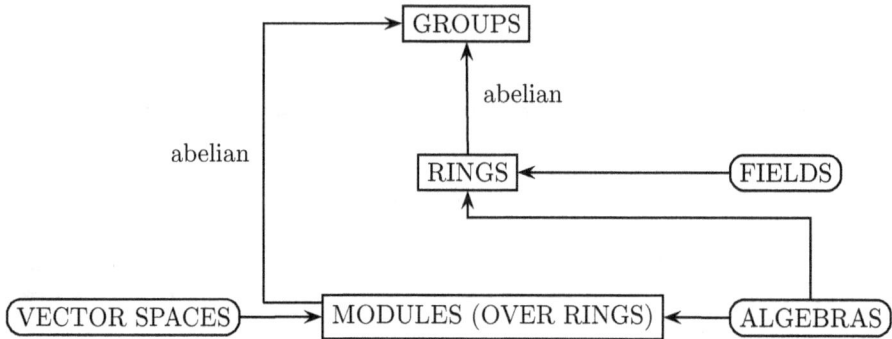

FIGURE 5.1

2) the scalar multiplication $(\alpha, x) \mapsto \alpha x$ satisfies

$$\alpha(xy) = (\alpha x)y = x(\alpha y) \quad ,$$

for all $\alpha \in R$, and $x, y \in A$.

Note that the above equation involves two kinds of multiplication — internal multiplication between elements of A (from the ring structure of A), and external (scalar) multiplication between elements of R (the ring associated with the module A) and those of A. In the Lie algebras studied in the last chapter, the internal multiplications are given by the Lie brackets (commutators) [c.f. (4.25)]. Hence for matrix groups the Lie algebra (internal) multiplication is not the same as ordinary matrix multiplication. [c.f. Exercise 4.5.]

Figure 5.1 summarizes the relationship between the algebraic categories introduced in this chapter. The rectangular boxes indicate the most general categorical entities with given structures whereas the oval boxes indicate special cases with restrictions. The structures increase in complexity as one moves vertically downwards. The arrows reaching upwards from a category A to a category B means that an object in A is a set that contains the structure of objects in B, but at the same time carries additional structure. Thus a ring is not only an abelian group (under addition), but also has a multiplicative structure, etc.

Chapter 6

Basic Group Concepts

Recall Def. 4.1, which gives the formal definition of a group.

Definition 6.1. *A subset H of a group G ($H \subset G$) which is itself a group with the same multiplication rule as that of G is called a **subgroup** of G.*

Thus $SO(n)$ is a subgroup of $O(n)$ and is isomorphic to a subgroup of $SO(n+1)$. In particular, $SO(2)$ is isomorphic to an abelian subgroup of $O(3)$. Note that a subgroup H of a group G must share the identity element with G.

The elements of any group can be partitioned into disjoint **conjugacy (equivalence) classes.**

Definition 6.2. *Two group elements $g_1, g_2 \in G$ are said to be **conjugate** to each other if there exists a $p \in G$ such that $g_2 = p\,g_1 p^{-1}$. Elements of a group which are conjugate to each other are said to form a **conjugacy class**.*

The conjugacy relation is an equivalence relation and thus the conjugacy classes of a group are necessarily disjoint. Recall that an **equivalence relation** R in a set S is defined as a subset R of $S \times S$ such that

   i) $(x, x) \in R$, for all $x \in S$ (reflexivity);

   ii) $(x, y) \in R \implies (y, x) \in R$, for all $x, y \in S$ (symmetry);

   iii) $(x, y) \in R$ and $(y, z) \in R \implies (x, z) \in R$, for all $x, y, z \in S$ (transitivity).

We usually denote $(x, y) \in R$ by writing $x \sim y$. An equivalence relation necessarily partitions a set into disjoint subsets.

| Exercise 6.1 | Verify the assertion immediately following Def. 6.2.

Two facts follow directly from the above definition:

1) the identity of a group is always a one-element conjugacy class;

2) for an abelian group, every element is in a conjugacy class by itself.

Exercise 6.2   Verify the above two facts.

Exercise 6.3   Use (1.41) to verify that all rotations by the same angle (about any axis) belong to the same conjugacy class in $SO(3)$.

Definition 6.3. *Let $H \subset G$ be a subgroup of G. Introduce an equivalence relation in G by the following criterion. Two elements $g_1, g_2 \in G$ are said to be equivalent mod H, written $g_1 \sim g_2 (mod\ H)$ if $g_1^{-1} g_2 \in H$. The disjoint equivalence classes of G induced by this equivalence relation are called **left cosets** of H. Each such equivalence class (left coset) is denoted by gH, where g is some element in G. Similarly we can define **right cosets** of H by the equivalence relation: $g_1 \sim g_2 (mod\ H)$ if $g_1 g_2^{-1} \in H$. Each right coset is denoted by Hg, where g is some element in G.*

Several things need to be clarified. Given a subgroup $H \subset G$, the group G can always be partitioned into either disjoint left cosets of H or disjoint right cosets of H. In general, however, these two partitions are not the same (see Fig. 6.1). The designations 'left' and 'right', as indicated by the notations gH and Hg (g written to the left and right of H), respectively, come from the following consideration. The notation gH (g being a fixed element in G) for a particular left coset implies that all elements in this coset can be exhausted by the set $\{gh|\ h \in H\}$; similarly, $Hg = \{hg|\ h \in H\}$. Indeed, suppose $h_1, h_2 \in H$ (elements in the subgroup H of G). Then

$$(gh_1)^{-1} gh_2 = h_1^{-1} g^{-1} gh_2 = h_1^{-1} h_2 \in H \ ,$$

which implies that gh_1 and gh_2 are in the same left coset gH. Similarly,

$$h_1 g(h_2 g)^{-1} = h_1 gg^{-1} h_2^{-1} = h_1 h_2^{-1} \in H,$$

implying that $h_1 g$ and $h_2 g$ are in the same right coset Hg. It is also clear that two left(right) cosets of a subgroup $H \subset G$ either coincide completely, or else have no elements in common (disjoint). Indeed, let $g_1 H$ and $g_2 H$ be two left cosets of H. Suppose $g_1 h_1 = g_2 h_2$ for some $h_1, h_2 \in H$. Then $g_2^{-1} g_1 = h_2 h_1^{-1} \in H$. Thus $g_2^{-1} g_1 H = H$, which implies $g_1 H = g_2 H$ (the two left cosets $g_1 H$ and $g_2 h$ coincide). Identical considerations apply to the case of right cosets.

We represent the partitions (by left and right cosets) of a group G schematically by Fig. 6.1.

It is quite obvious that the subgroup H is both a left and a right coset of itself. (It can, in fact, be written as $H = eH = He$.) However, aside from H, neither the left nor right cosets are subgroups of G (simply because none of

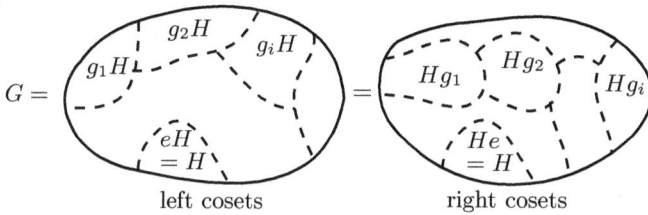

FIGURE 6.1

them contain the identity). We also note that each coset of H (left and right) has exactly the same number of elements as H.

A very special and important situation arises when a subgroup $H \subset G$ is such that

$$gH = Hg, \quad \text{for all } g \in G. \tag{6.1}$$

Equivalently, $gHg^{-1} = H$. Such a subgroup is called an **invariant subgroup** (or **normal subgroup**) of G. Any group has at least two trivial invariant subgroups: $H = \{e\}$ and $H = G$. It is also clear that *any subgroup of an abelian group is an invariant subgroup*. Furthermore, if $H \subset G$ is an invariant subgroup, G is partitioned in exactly the same way by left and right cosets of H. One then simply refers to the cosets generated by an invariant subgroup. In fact, we have the following theorem.

Theorem 6.1. *Let $H \subset G$ be an invariant subgroup of G. Then the distinct cosets of H, $\{gH = Hg \mid g \in G\}$, form a group, called the **quotient group** G/H.*

Proof. We have to define a multiplication rule to multiply different cosets by each other and show that it satisfies the group properties given in Def. 4.1.

Since different cosets are disjoint, a coset can be identified by any one of its elements g_1, say, also called a **representative**, and be labelled by $[g_1]$. We introduce the multiplication rule:

$$[g_1][g_2] \equiv [g_1 g_2]. \tag{6.2}$$

We have to check that this definition makes sense by verifying that it is independent of the representative. Suppose $[g_1'] = [g_1]$ and $[g_2'] = [g_2]$. Then, according to Def. 6.3, $g_1' = g_1 h_1$ and $g_2' = g_2 h_2$, with $h_1, h_2 \in H$. Thus

$$\begin{aligned}
[g_1'][g_2'] &= [g_1' g_2'] \quad \text{[by (6.2)]} \\
&= [g_1 h_1 g_2 h_2] \\
&= [g_1 g_2 h h_2] \quad \text{for some } h \in H \quad \text{[by (6.1)]} \tag{6.3} \\
&= [g_1 g_2] \\
&= [g_1][g_2],
\end{aligned}$$

showing that the multiplication (6.2) is indeed well-defined. The identity in G/H is the subgroup (also coset) H itself. Since, for $h \in H, g \in G$, there exists $h' \in H$ such that

$$[h][g] = [hg] = [gh'] = [g][h'] = [g][h] = [g] . \tag{6.4}$$

Associativity of the multiplication (6.2) is obvious.

| Exercise 6.4 | Prove that the multiplication rule for the quotient group G/H defined by (6.2) is associative.

Finally the inverse of $[g]$ is given by

$$[g]^{-1} = [g^{-1}] . \tag{6.5}$$

This is easily checked by noting that

$$[g][g]^{-1} = [g][g^{-1}] = [gg^{-1}] = [g^{-1}][g] = [g]^{-1}[g] = [e] = [h] , h \in H . \tag{6.6}$$

$\square$

Note that for a subgroup H of an abelian group G, the fact that $g_1 \sim g_2 \,(mod\, H)$, or $[g_1] = [g_2]$, is usually expressed by $g_1 - g_2 = h$, for some $h \in H$. [c.f. Exercise 5.1.]

At this point, let us recall the notion of a group homomorphism (Def. 4.2). We have the following important theorem.

Theorem 6.2 (The Group Homomorphism Theorem). *Let $f : G \longrightarrow G'$ be a group homomorphism, K be the **kernel** of f:*

$$K = ker(f) = \{g \in G|\, f(g) = e' \text{ (identity in } G')\} ,$$

*and I be the **image** of f:*

$$I = im(f) = \{g' \in G'|\, \text{there exists } g \in G \text{ such that } f(g) = g'\} .$$

Then

1) K is a normal subgroup of G,

2) G/K is isomorphic to I.

Proof. 1) Let $a, b \in K$. Then $f(ab) = f(a)f(b) = e'e' = e'$. Hence $ab \in K$. For any $g \in G$, $f(g) = f(ge) = f(g)f(e)$. Hence $f(e) = e'$, implying $e \in K$. Furthermore, if $a \in K$, $f(a^{-1}) = (f(a))^{-1} = (e')^{-1} = e'$; thus $a^{-1} \in K$. This proves that K is a subgroup of G. To show that it is a normal subgroup, we consider the set gKg^{-1} for some $g \in G$. Observe that for $k \in K$,

$$f(gkg^{-1}) = f(g)e'f(g^{-1}) = f(g)(f(g))^{-1} = e' ,$$

which implies $gkg^{-1} \in K$, or $gKg^{-1} \subset K$. Conversely, any $k \in K$ can be expressed as $g(g^{-1}kg)g^{-1} = gk'g^{-1}$, $k' \in K$, which implies $K \subset gKg^{-1}$. Thus

$$gKg^{-1} = K \qquad \text{for all } g \in G,$$

which is the criterion for K to be a normal subgroup of G.

2) Define the mapping $\varphi : G/K \longrightarrow I$ by

$$\varphi([g]) \equiv f(g), \quad g \in G, \tag{6.7}$$

where $[g] \in G/K$ is the coset $\{gk \mid k \in K\}$. We have

$$\varphi([g_1][g_2]) = \varphi([g_1 g_2]) = f(g_1 g_2) = f(g_1)f(g_2) = \varphi([g_1])\varphi([g_2]) . \tag{6.8}$$

Since $I \subset G'$ is a subgroup of G', φ gives a group homomorphism. Now if $\varphi([g_1]) = \varphi([g_2])$, then by (6.7), $f(g_1) = f(g_2)$, and thus $f(g_1)f(g_2^{-1}) = f(g_1 g_2^{-1}) = e'$. This implies that $g_1 g_2^{-1} \in K$, or $[g_1] = [g_2]$. Hence φ is injective (one-to-one). φ is also obviously surjective (onto) since, for any $g' \in I$, there exists a $g \in G$ such that $g' = f(g) = \varphi([g])$. Being bijective, the group homomorphism φ is thus also an isomorphism. □

As an example of the above theorem, consider the quotient group $O(3)/SO(3)$.

Exercise 6.5   Show that $SO(3)$ is a normal subgroup of $O(3)$.

There is an obvious homomorphism between $O(3)$ and $\mathbb{Z}_2$, the additive group of integers modulo 2, which can be realized as the set $\{1, -1\}$ with the usual multiplication. The homomorphism $f : O(3) \longrightarrow \mathbb{Z}_2$ is simply given by [c.f. (1.32)]:

$$f(a) = \det(a), \quad a \in O(3) . \tag{6.9}$$

Exercise 6.6   Show that the above equation indeed defines a group homomorphism.

Since in the group $\mathbb{Z}_2$, the identity is the integer 1, the kernel of the homomorphism f is given by

$$ker(f) = SO(3) . \tag{6.10}$$

Now it is easy to see that

$$O(3)/SO(3) = \{SO(3), sSO(3)\} , \tag{6.11}$$

where $s \in O(3)$ is the spatial inversion

$$s = \begin{pmatrix} -1 & 0 & 0 \\ 0 & -1 & 0 \\ 0 & 0 & -1 \end{pmatrix} \, , \tag{6.12}$$

and $im(f) = \mathbb{Z}_2$ (f is surjective). The isomorphism $\varphi : O(3)/SO(3) \longrightarrow \mathbb{Z}_2$ of Theorem 6.2 is thus given by (6.7) as

$$\varphi(SO(3)) = 1 \, , \tag{6.13a}$$

$$\varphi(sSO(3)) = -1 \, . \tag{6.13b}$$

We introduce two more important notions which are of fundamental importance in the classification of groups, and which have direct counterparts in Lie algebra theory.

Definition 6.4. *i) A group that does not have any invariant (normal) subgroups other than $\{e\}$ and itself is called a **simple group**. [e.g. $SU(2)$ is simple.] ii) A group that does not have any abelian invariant (normal) subgroups other than $\{e\}$ is called a **semisimple group**. [e.g. $SO(3)$, $SO(4)$, and the Lorentz group (considered in Chapter 11) are all semisimple.]*

We have already introduced the all-important notion of group representations in Chapter 4 (Def. 4.3), the main idea of which is to consider the action of group elements on vector spaces. This notion can be generalized to that of the action of a group on an arbitrary set. Suppose G is a group and M is a set, a **group action** by G on M is a mapping $\sigma : G \times M \longrightarrow M, (g, x) \mapsto gx$, which satisfies the associative rule

$$g_1(g_2 x) = (g_1 g_2)x \, , \tag{6.14}$$

and the condition

$$ex = x \, , \qquad \text{for all } x \in M \, , \tag{6.15}$$

where e is the identity in G.

In physics applications, the most common situation is when M is a vector space (Euclidean space, Minkowski spacetime, various subspaces of Hilbert spaces in quantum mechanics, etc.). When G is a Lie group, it often happens that M is a differentiable manifold, in which case σ is also required to be a differentiable map. (This is a most important situation in differential geometry.) Obviously M can also be a group.

Exercise 6.7   Show that the **left action** of a group on itself defined by

$$\sigma_L(g, g') = L_g g' = gg' \tag{6.16}$$

is a group action. How about the **right action** defined by

$$\sigma_R(g, g') = R_g g' = g'g \quad ? \tag{6.17}$$

The following are important notions regarding group actions.

Definition 6.5. *A group action* $\sigma : G \times M \longrightarrow M$ *is said to be **effective** if the identity* $e \in G$ *is the only element that produces the trivial action on* M*, that is, if* $\sigma(g, x) = x$ *for all* $x \in M$*, then* $g = e$*.*

Definition 6.6. *A group action* $\sigma : G \times M \longrightarrow M$ *is said to be **free** if every element that is not the identity of* G *(*$g \neq e$*) has no **fixed points** in* M*, that is, if there exists an element* $x \in M$ *such that* $\sigma(g, x) = x$*, then* $g = e$*.*

Definition 6.7. *A group action* $\sigma : G \times M \longrightarrow M$ *is said to be **transitive** if, for any* $x_1, x_2 \in M$*, there exists a* $g \in G$ *such that* $\sigma(g, x_1) = x_2$*.*

| Exercise 6.8 | Is the action of $O(n)$ on the Euclidean space $\mathbb{R}^n$ effective? free? transitive? [Consider the geometrically intuitive case of $O(3)$ acting on $\mathbb{R}^3$.] Answer the same question for $O(n)$ acting on S^{n-1} [the $(n-1)$-dimensional sphere].

Definition 6.8. *Given a group action* $\sigma : G \times M \longrightarrow M$*, the **isotropy group** (**stabilizer**, or **little group**) of an element* $x \in M$ *is the subgroup of* G *defined by*

$$H(x) = \{g \in G \,|\, \sigma(g, x) = x\} \,. \tag{6.18}$$

| Exercise 6.9 | Show that $H(x)$ defined above is indeed a subgroup of G.

It follows from Def. 6.6 that if a group action is free, then the isotropy group $H(x) = \{e\}$ for any $x \in M$.

| Exercise 6.10 | Prove the above statement.

Consider again $G = SO(3)$ acting on $\mathbb{R}^3$ and a fixed point $x \in \mathbb{R}^3$. The isotropy subgroup $H(x)$ is obviously the group of rotations about the axis x, the vector pointing from the origin to x. It is thus isomorphic to $SO(2)$. Denote the space of left cosets (the following considerations apply to the space of right

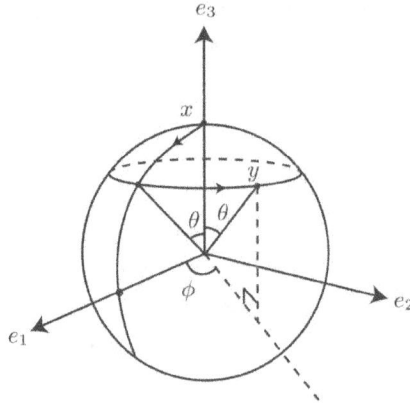

FIGURE 6.2

cosets also) by $SO(3)/SO(2)$. Since $SO(2)$ is a subgroup, but not an invariant (normal) subgroup of $SO(3)$, $SO(3)/SO(2)$ does not admit a group structure. This coset space, however, does have a manifold structure (see Chapter 30), and is in fact homeomorphic (topologically equivalent) to S^2 (the 2-sphere). To see this, we recall from Def. 6.3 that two elements g and $g' \in SO(3)$ belong to the same left coset (an element in $SO(3)/SO(2)$) provided $g' = gh$, where $h \in H(x)$. It is obvious that g and g' have the same effect on x, since

$$g'x = ghx = gx \ . \tag{6.19}$$

Let x be the north pole $(0,0,1)$ on the unit circle in $\mathbb{R}^3$, and suppose $gx = y$, another point on the unit circle (see Fig. 6.2).

The coset $[g]$ then consists of all the elements in $SO(3)$ which take the north pole to the point y, which, of course, is specified by the polar coordinates (θ, ϕ). Thus each point on the unit circle corresponds to a coset and we have

$$SO(3)/SO(2) \sim S^2 \ , \tag{6.20}$$

where the symbol '$\sim$' in the above equation is used in the sense of 'homeomorphic to'. The following generalizations are also true for all positive integers n (but we will omit the proofs):

$$O(n+1)/O(n) \sim SO(n+1)/SO(n) \sim S^n \ , \tag{6.21a}$$
$$U(n+1)/U(n) \sim SU(n+1)/SU(n) \sim S^{2n+1} \ , \tag{6.21b}$$

where S^n is the n-sphere; and $U(n)$ and $SU(n)$ are the groups of $n \times n$ unitary matrices and $n \times n$ unitary matrices with determinant $+1$, respectively.

In general the coset space G/H, where G is a Lie group and H is a subgroup of G, admits a differentiable structure (see Chapter 30) and becomes a differentiable manifold, called a **homogeneous space**.

We will now state and prove a very useful fact on isotropy subgroups.

Lemma 6.1. *Given a group action* $\sigma : G \times M \longrightarrow M, g \in G$ *and* $x \in M$, *the isotropy subgroups at* x *and* gx *are related by*

$$\boxed{H(gx) = gH(x)g^{-1}} \ . \tag{6.22}$$

Proof. Let $g' \in H(gx)$. Then $g'gx = gx$, and so

$$g^{-1}g'gx = g^{-1}gx = x \ .$$

Hence

$$g'' \equiv g^{-1}g'g \in H(x) \ ,$$

whence $g' = gg''g^{-1}$, which implies

$$g' \in gH(x)g^{-1} \ .$$

Thus

$$H(gx) \subset gH(x)g^{-1} \ . \tag{6.23}$$

Conversely, let $g'' \in H(x)$. Then $g''x = x$, and

$$(gg''g^{-1})gx = gg''x = gx \ .$$

Consequently $gg''g^{-1} \in H(gx)$, and thus,

$$gH(x)g^{-1} \subset H(gx) \ . \tag{6.24}$$

Eqs. (6.23) and (6.24) together imply the lemma. □

We end this chapter with an interesting application of the above lemma in the proof of the so-called Euler's theorem, which is of basic importance in the kinematics of rigid bodies.

Theorem 6.3 (Euler's Theorem). *Every rotation* $R \in SO(3)$ *can be written as a product*

$$\boxed{R = R_3(\phi)R_2(\theta)R_3(\psi)} \ , \tag{6.25}$$

where R_2 *and* R_3 *are rotations about the* e_2 *and* e_3 *axes, respectively, and the angles* ϕ, θ, ψ *(called the **Euler angles**) fall in the ranges* $0 \le \phi, \psi \le 2\pi$, $0 \le \theta \le \pi$.

Proof. Consider an arbitrary rotation $R \in SO(3)$ on $\mathbb{R}^3$ and its action on the point $x = (0, 0, 1)$ (north pole of the unit sphere). Suppose $Rx = y$, where y is the point (θ, ϕ) on the unit sphere (see Fig. 6.2). We see that

$$Rx = R_3(\phi)R_2(\theta)x \equiv gx \ , \tag{6.26}$$

where

$$g = R_3(\phi)R_2(\theta) \in SO(3) \ . \tag{6.27}$$

Write
$$R = g'g, \quad \text{where } g' = Rg^{-1} \in SO(3) . \tag{6.28}$$

Then $g'(gx) = gx$, and hence
$$g' \in H(gx) = H(Rx) ,$$

where $H(gx)$ is the isotropy subgroup of gx, or the group of rotations about an axis through the point gx on the unit sphere. By Lemma 6.1, there exists a rotation about the z-axis e_3, $R_3(\psi)$, such that
$$g' = gR_3(\psi)g^{-1} .$$

By (6.28),
$$R = gR_3(\psi)g^{-1}g = gR_3(\psi) ,$$

and by (6.27),
$$R = R_3(\phi)R_2(\theta)R_3(\psi) . \tag{6.29}$$

$\square$

The rotations in Euler's theorem are about axes fixed in space. These are called **space-fixed axes** in the physics literature. Very often in physics, it is convenient to express an arbitrary rotation R of a rigid body in terms of axes moving with the rigid body, the so-called **body-fixed axes** (or **moving frames**) (see Chapter 31). We will establish the following result:

$$\boxed{R = R_{3'}(\psi)R_{2'}(\theta)R_3(\phi) , \quad \text{body-fixed axes} ,} \tag{6.30}$$

where the sequence of rotations involves the following sets of body-fixed axes [Figs. 6.3(a) through 6.3(d)]:

$$(e_1, e_2, e_3) \xrightarrow{R_3(\phi)} (e_1', e_2', e_3) \xrightarrow{R_{2'}(\theta)} (e_1'', e_2', e_3') \xrightarrow{R_{3'}(\psi)} (e_1''', e_2'', e_3') .$$

In Fig. 6.3, the body-fixed axes are drawn as solid lines.
Indeed, using Lemma 6.1,
$$R_{3'}(\psi) = R_{R_{2'}(\theta)e_3}(\psi) = R_{2'}(\theta)R_3(\psi)R_{2'}^{-1}(\theta) , \tag{6.31}$$

and
$$R_{2'}(\theta) = R_{R_3(\phi)e_2}(\theta) = R_3(\phi)R_2(\theta)R_3^{-1}(\phi) . \tag{6.32}$$

It follows from (6.31) that
$$R_{3'}(\psi)R_{2'}(\theta)R_3(\phi) = R_{2'}(\theta)R_3(\psi)R_3(\phi) . \tag{6.33}$$

Eq. (6.32) then finally yields
$$\begin{aligned} R_{3'}(\psi)R_{2'}(\theta)R_3(\phi) &= R_3(\phi)R_2(\theta)R_3^{-1}(\phi)R_3(\psi)R_3(\phi) \\ &= R_3(\phi)R_2(\theta)R_3(\psi) , \end{aligned} \tag{6.34}$$

where in the second equality we have used the fact that $R_3^{-1}(\phi)$ and $R_3(\psi)$ commute since these are rotations about the same axis. Note that the orders of the angles of rotation in (6.25) and (6.30) are the reverse of each other.

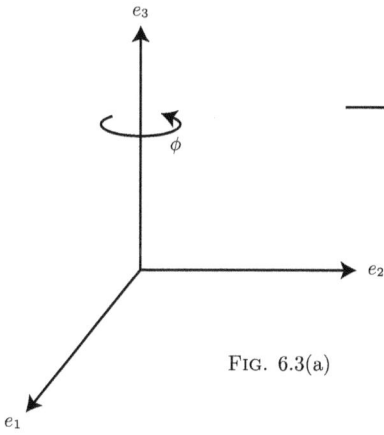

FIG. 6.3(a)

$R_3(\phi)$

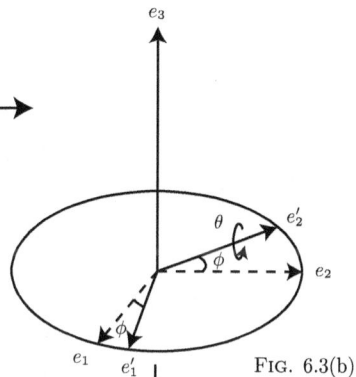

FIG. 6.3(b)

$R_{2'}(\theta)$

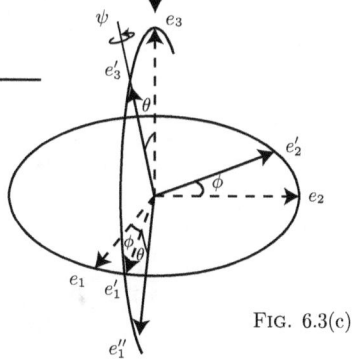

FIG. 6.3(c)

$R_{3'}(\psi)$

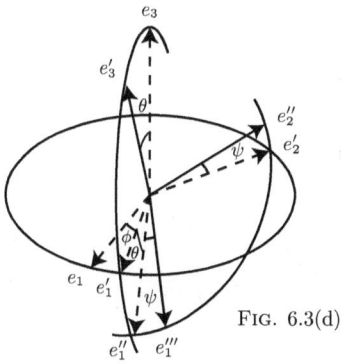

FIG. 6.3(d)

Chapter 7

Basic Lie Algebra Concepts

In Chapter 4 we introduced the Lie algebra of a Lie group as the linear space of infinitesimal generators of the Lie group, equipped with the Lie bracket (commutator) as the internal multiplication. We further introduced the abstract notion of a Lie algebra [see discussion following (4.28)]. In physics applications, the most important Lie algebras are the classical Lie algebras, which are the Lie algebras of the corresponding classical (Lie) groups. These are always real Lie algebras, which means that they are also vector spaces over the field of real numbers. One can also speak of complex Lie algebras (over the field $\mathbb{C}$).

Since Lie groups are also differentiable manifolds, there is a differential geometric side of the theory, which we will defer until Chapter 34. Here we will only focus on the algebraic aspects.

Corresponding to the group concepts of subgroups, invariant(normal) subgroups, and quotient groups, we have the Lie algebra concepts of subalgebras, ideals, and quotient algebras. The notions of simplicity and semi-simplicity of groups also have direct counterparts in Lie algebras.

Definition 7.1. *A subset S of a Lie algebra $\mathcal{G}$ is a **Lie subalgebra** of $\mathcal{G}$ if S is closed under the Lie bracket (commutator), that is, if $[s_1, s_2] \in S$ whenever $s_1, s_2 \in S$. We abbreviate this by writing $[S, S] \subset S$.*

Definition 7.2. *A Lie algebra $\mathcal{G}$ is the **direct sum** of two Lie algebras $\mathcal{A}$ and $\mathcal{B}$ if $\mathcal{G}$ is the direct sum of $\mathcal{A}$ and $\mathcal{B}$ as a vector space (direct sum of vector spaces) and if $[\mathcal{A}, \mathcal{B}] = 0$. We write $\mathcal{G} = \mathcal{A} \oplus \mathcal{B}$.*

$\boxed{\text{Exercise 7.1}}$ Show that if $\mathcal{G} = \mathcal{A} \oplus \mathcal{B}$, then $\mathcal{A}$ and $\mathcal{B}$ are both subalgebras of $\mathcal{G}$.

Definition 7.3. *A Lie algebra is the **semi-direct sum** of two subalgebras $\mathcal{A}$ and $\mathcal{B}$ if $\mathcal{G}$ is the direct sum of $\mathcal{A}$ and $\mathcal{B}$ as a vector space, and if $[\mathcal{A}, \mathcal{B}] \subset \mathcal{A}$. We write $\mathcal{G} = \mathcal{A} \oplus_S \mathcal{B}$.*

Definition 7.4. *A Lie subalgebra S is an **ideal** (or an **invariant subalgebra**) of a Lie algebra $\mathcal{G}$ if $[S, \mathcal{G}] \subset S$, that is, $[s, g] \in S$ whenever $s \in S$ and $g \in \mathcal{G}$. Obviously, if $\mathcal{G} = \mathcal{A} \oplus_S \mathcal{B}$, then $\mathcal{A}$ is an ideal of $\mathcal{G}$.*

| Exercise 7.2 | In a direct sum,

$$\mathcal{G} = \mathcal{A}_1 \oplus \mathcal{A}_2 \oplus \cdots \oplus \mathcal{A}_n \ ,$$

show that each $\mathcal{A}_i$, $i = 1, \ldots, n$, is an ideal of $\mathcal{G}$.

Definition 7.5. *The **center** of a Lie algebra $\mathcal{G}$ is the unique largest ideal $\mathcal{C}$ such that $[\mathcal{G}, \mathcal{C}] = 0$.*

In group theory, we have subgroups generating left and right coset spaces, and normal subgroups inducing quotient groups (see Chapter 6). There is a similar situation in Lie algebras. Suppose $S \subset \mathcal{G}$ is a Lie subalgebra of a Lie algebra $\mathcal{G}$. For $A \in \mathcal{G}$ define the equivalence class $[A]$ to be the set $A + s$ for all $s \in S$. Thus two elements A and B are in the same equivalence class if $A - B = s \in S$, written $A = B \,(mod \ s)$. Any $A \in \mathcal{G}$ in this class is called a representative of $[A]$ and specifies $[A]$. These equivalence classes correspond to the cosets in group theory; but there is no distinction between left and right since addition in an algebra is vector addition and hence commutative.

| Exercise 7.3 | Show that the above definition for $[A]$, $A \in \mathcal{G}$, is indeed an equivalence relation.

Suppose we try to define a Lie bracket among the different equivalence classes by

$$[[A], [B]] \equiv [[A, B]] \ . \tag{7.1}$$

In order for this definition to make sense, $[A, B]$ must indeed be a representative of the right hand side of (7.1), which means that it must differ from any other member in the same class by an element $s \in S$. Now, for $s_1, s_2 \in S$,

$$[A + s_1, B + s_2] = [A, B] + [s_1, B] + [A, s_2] + [s_1, s_2] \ , \tag{7.2}$$

which is equal to $[A, B] + s$, $s \in S$, only if S is an ideal. Thus we have

Definition 7.6. *If $S \subset \mathcal{G}$ is an ideal of the Lie algebra $\mathcal{G}$, then the equivalence classes in $\mathcal{G}$ defined by*

$$[A] = [A + s] \ , \quad A \in \mathcal{G}, \ s \in S, \tag{7.3}$$

*forms a Lie algebra called the **quotient algebra**, denoted by $\mathcal{G}/S$. The Lie bracket of the quotient algebra is given by (7.1).*

Exercise 7.4 | Show that for the quotient algebra the vector addition

$$[A] + [B] = [A + B] \tag{7.4}$$

is well-defined.

Exercise 7.5 | Let G be a Lie group and $H \subset G$ be a normal subgroup of G. Let $\mathcal{G}$ and $\mathcal{H}$ be the corresponding Lie algebras of G and H, respectively. Show that $\mathcal{H}$ is an ideal of $\mathcal{G}$.

The set of commutators of a Lie algebra $\mathcal{G}$, denoted by $[\mathcal{G}, \mathcal{G}]$, is obviously a subalgebra of $\mathcal{G}$. It is in fact also an ideal of $\mathcal{G}$, since, for any $A, A_1, A_2 \in \mathcal{G}$,

$$[[A_1, A_2], A] = [A_3, A] \in [\mathcal{G}, \mathcal{G}], \tag{7.5}$$

where $A_3 = [A_1, A_2]$. We write

$$\mathcal{G}^{(1)} \equiv [\mathcal{G}, \mathcal{G}]. \tag{7.6}$$

Similarly,

$$\mathcal{G}^{(2)} \equiv [\mathcal{G}^{(1)}, \mathcal{G}^{(1)}] \tag{7.7}$$

is an ideal of $\mathcal{G}^{(1)}$, and by induction,

$$\mathcal{G}^{(n+1)} \equiv [\mathcal{G}^{(n)}, \mathcal{G}^{(n)}] \tag{7.8}$$

is an ideal of $\mathcal{G}^{(n)}$, $n = 0, 1, 2, \dots$, $\mathcal{G}^{(0)} \equiv \mathcal{G}$.

Exercise 7.6 | Justify the above statement by induction.

Definition 7.7. *If the series of ideals of $\mathcal{G}$ given by (7.8) terminates in 0 (the zero vector in $\mathcal{G}$), $\mathcal{G}$ is said to be **solvable**.*

Exercise 7.7 | Is the Lie algebra $\mathcal{SO}(3)$ solvable? Consult equations (4.36) to (4.38). Is $\mathcal{SO}(n)$ solvable?

Exercise 7.8 | Show that the set of all $n \times n$ **upper triangular matrices**

$$\begin{pmatrix} a_1^1 & a_1^2 & \cdots & a_1^n \\ 0 & a_2^2 & \cdots & a_2^n \\ \cdots \cdots \cdots \cdots \\ 0 & 0 & \cdots & a_n^n \end{pmatrix}$$

is a solvable Lie algebra.

An important theorem proved by Lie states that *every complex solvable matrix algebra is isomorphic to a subalgebra of triangular matrices* (see D. H. Sattinger and O. L. Weaver 1986).

The Heisenberg algebra (introduced in Chapter 13), which is of basic importance in quantum mechanics, is an example of a solvable Lie algebra.

A stronger condition than solvability is that of nilpotency.

Definition 7.8. *A Lie algebra $\mathcal{G}$ is said to be **nilpotent** if the nested sequence*

$$\mathcal{G}_{(n+1)} \equiv [\mathcal{G}, \mathcal{G}_{(n)}] , \ n = 0, 1, 2, \ldots, \tag{7.9}$$

$(\mathcal{G}_{(0)} \equiv \mathcal{G})$, *in which*

$$\mathcal{G}_{(n+1)} \subset \mathcal{G}_{(n)} \subset \cdots \subset \mathcal{G}_{(1)} = \mathcal{G}^{(1)} \subset \mathcal{G}_{(0)} = \mathcal{G}$$

terminates in 0 (the zero vector in $\mathcal{G}$).

Eqs. (7.8) and (7.9) imply $\mathcal{G}^{(n)} \subset \mathcal{G}_{(n)}$. Thus *nilpotency implies solvability.* The Heisenberg algebra mentioned above is also nilpotent in addition to being solvable.

Definition 7.9. *The **radical** $\mathcal{R}$ of a Lie algebra $\mathcal{G}$ is the unique maximum solvable ideal of $\mathcal{G}$.*

The radical of a Lie algebra always exists (proof omitted here).

Definition 7.10. *A Lie algebra is said to be **simple** if it contains no ideals other than $\{0\}$ and itself.*

Definition 7.11. *A Lie algebra is said to be **semisimple** if it contains no abelian ideals other than $\{0\}$.*

As in the case of groups, simplicity of algebras implies semi-simplicity. An example of a semisimple Lie algebra is $\mathcal{SO}(3)$.

Semisimple Lie algebras play a fundamental role in geometry and physics. Cartan has given a complete classification of the complex semisimple Lie algebras (see Chapter 24). All of the Lie algebras of the classical groups are semisimple with the exception of $\mathcal{GL}(n, \mathbb{C})$.

The notions of solvability and semi-simplicity are linked by the following central fact in the theory of the classification of Lie algebras, which will be stated without proof (see N. Jacobson 1962).

Theorem 7.1 (Levi's Decomposition Theorem). *Every Lie algebra is the semi-direct sum of its radical and a semisimple Lie algebra.*

As an example of Levi's decomposition consider the set of six operators p_1, p_2, p_3 (linear momentum) and L^1, L^2, L^3 (angular momentum) defined by (4.4) and (4.22), respectively. Recall that the p_j are generators of translations in $\mathbb{R}^3$ and the L^j are generators of rotations in $\mathbb{R}^3$. Separately, they satisfy the commutation relations (4.14) and (4.40), and constitute the bases for the Lie algebras $\mathcal{T}(3)$ and $\mathcal{SO}(3)$, which are the Lie algebras corresponding to $T(3)$ (the abelian group of translations in $\mathbb{R}^3$) and $SO(3)$, respectively. It is not hard to show that, using (4.12), (4.14), (4.22) and (4.40),

$$[p_j, L^k] = [L^j, p_k] = i\hbar\, \varepsilon_{jk}{}^l p_l , \tag{7.10}$$

where $\varepsilon_{jk}{}^l = \varepsilon_{jkl}$ is the Levi-Civita tensor given by (2.8).

Exercise 7.9 Prove (7.10).

Eq. (7.10), together with the commutation relations [(4.14) and (4.40)]:

$$[p_i, p_j] = 0 , \tag{7.11}$$
$$[L^j, L^k] = i\hbar\, \varepsilon^{jk}{}_l L^l , \tag{7.12}$$

determine a six-dimensional Lie algebra $\mathcal{E}(3)$, the Lie algebra of Euclidean motions in $\mathbb{R}^3$. Due to (7.10) and (7.11), $\mathcal{T}(3)$ is the radical of $\mathcal{E}(3)$, and $\mathcal{SO}(3)$ is semisimple. Thus

$$\mathcal{E}(3) = \mathcal{T}(3) \oplus_S \mathcal{SO}(3) \tag{7.13}$$

is an example of Levi's decomposition.

As in the case of groups, we can represent the elements of a Lie algebra by linear operators on vector spaces.

Definition 7.12. *A representation of a Lie algebra $\mathcal{G}$ on a vector space V is a homomorphism $f : \mathcal{G} \longrightarrow \mathcal{T}(V)$, where $\mathcal{T}(V)$ is the algebra of linear transformations on V. f is a homomorphism means that it is structure-preserving:*

$$f(\alpha A + \beta B) = \alpha f(A) + \beta f(B) , \tag{7.14a}$$
$$f[A, B] = [f(A), f(B)] , \tag{7.14b}$$

where $\alpha, \beta \in \mathbb{F}$ (the field associated with $\mathcal{G}$), and $A, B \in \mathcal{G}$.

Since $\mathcal{G}$ itself is a vector space, we can represent $\mathcal{G}$ by operators on itself. A particularly important representation of $\mathcal{G}$ on itself is the so-called **adjoint representation**, $A \in \mathcal{G} \mapsto Ad_A$, defined by

$$\boxed{Ad_A(B) \equiv [A, B]} . \tag{7.15}$$

For this definition to make sense, we must have

$$Ad_{\alpha A + \beta B} = \alpha Ad_A + \beta Ad_B , \tag{7.16a}$$
$$Ad_{[A,B]} = [Ad_A, Ad_B] . \tag{7.16b}$$

The first condition is obvious; the second is guaranteed by Jacobi's identity (4.26).

Exercise 7.10   Verify Eqs. (7.16).

Using (1.16) and (4.29) we can derive the matrix representation of the adjoint representation of a Lie algebra in terms of the structure constants. Let $(e_1, \ldots, e_n)$ be a basis of $\mathcal{G}$. Then

$$Ad_{e_k}(e_i) = (Ad_{e_k})_i^j e_j = [e_k, e_i] = c^j{}_{ki} e_j . \tag{7.17}$$

Thus

$$(Ad_{e_k})_i^j = c^j{}_{ki} . \tag{7.18}$$

As mentioned before, there is an important differential geometric interpretation of the adjoint representation of a Lie algebra, which we will explain in Chapter 34.

As an example, consider the adjoint representation of $\mathcal{SO}(3)$. Recalling the commutation relations (4.36) to (4.38), choose the basis of $\mathcal{SO}(3)$ given by

$$e_1 = J_3^2 , \quad e_2 = J_1^3 , \quad e_3 = J_2^1 . \tag{7.19}$$

So we have

$$[e_i, e_j] = \varepsilon_{ij}{}^k e_k . \tag{7.20}$$

Thus the structure constants for $\mathcal{SO}(3)$ with respect to the basis (7.19) are given by

$$c^k{}_{ij} = \varepsilon^k{}_{ij} . \tag{7.21}$$

This is precisely the algebra of the ordinary vector (cross) product in $\mathbb{R}^3$ [c.f. Exercise 4.7]. Furthermore, from (7.18),

$$Ad_{e_i}(e_j) = \varepsilon^k{}_{ij} e_k . \tag{7.22}$$

Thus $\boldsymbol{A} \times \boldsymbol{B}$ can be interpreted as the adjoint representation of $\boldsymbol{A}$ operating on $\boldsymbol{B}$ when $\boldsymbol{A}$ and $\boldsymbol{B}$ are considered as vectors in the Lie algebra $\mathbb{R}^3$ (equipped with the cross product).

For a given representation $f : \mathcal{G} \longrightarrow \mathcal{T}(V)$ of a Lie algebra $\mathcal{G}$, the image of f in $\mathcal{T}(V)$ are linear operators on the representation space V, for which internal addition and multiplication are well-defined. The sums and products of elements in $im(f) \subset \mathcal{T}(V)$, with the customary distributive and associative laws, form an algebra, called the **enveloping algebra** of $\mathcal{G}$ with respect to the representation f, denoted by $\mathcal{A}_f(\mathcal{G})$.

Consider $\mathcal{SO}(3)$ again. With a slight abuse of notation, let us write, for a given faithful representation f,

$$L_x = f(L^1) , \quad L_y = f(L^2) , \quad L_z = f(L^3) .$$

The square of the total angular momentum

$$L^2 = L_x^2 + L_y^2 + L_z^2 \tag{7.23}$$

is an element of the enveloping algebra, and in fact lies in the center of this algebra (commutes with all the elements of the algebra). It is equal to $C/2$, where C, called the **Casimir operator** of the representation $f : \mathcal{G} \longrightarrow \mathcal{T}(V)$, is of great importance in the theory of angular momentum in quantum mechanics. Casimir operators will be considered in greater detail in Chapter 8, after we have introduced the notion of a particular metric in $\mathcal{G}$, called the Killing form.

Chapter 8

Inner Products, Metrics, and Dual Spaces

In the elementary vector algebra of three-dimensional Euclidean vectors we learned that the scalar product (or inner product) of two vectors $\boldsymbol{A}$ and $\boldsymbol{B}$ with components A^i and B^i is defined to be

$$\boldsymbol{A} \cdot \boldsymbol{B} = \sum_i A^i B^i \quad . \tag{8.1}$$

Using the Einstein summation convention we can write

$$\boldsymbol{A} \cdot \boldsymbol{B} = \delta_{ij} A^i B^j = A^i B_i = A_j B^j \ . \tag{8.2}$$

The above equation may appear as just a notational gimmick; but (8.1) actually hides several important mathematical notions which are explicated by (8.2).

First is the notion of the dual space to a given vector space. Then there is the notion of a metric, which specifies rules for measuring distances between points in a vector space and for measuring angles between vectors in the same space. Finally, the metric gives a one-to-one correspondence between objects in a vector space and objects in the dual space. This correspondence is obtained by the procedure of lowering and raising indices using the so-called metric tensor and its inverse.

The Kronecker delta in (8.2) is a very special case of a metric tensor, which, in a more general setting (e.g. of a **Riemannian manifold**), is a (0,2)-type tensor field $g_{ij}(x)$, depending on the position x in the underlying manifold. The case when $g_{ij}(x) = \delta_{ij}$ of a constant tensor field is known as the **Euclidean metric**. When g_{ij} is constant, the metric is called **flat**. A very important example of a flat but non-Euclidean metric in physics is the **Minkowski metric** $\eta_{\mu\nu}$ in

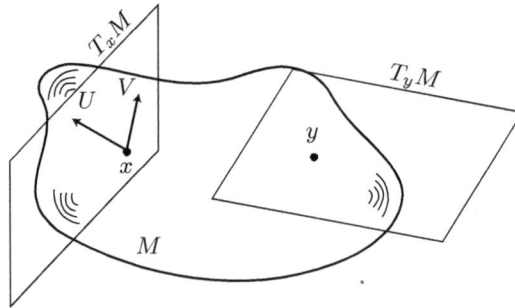

FIGURE 8.1

special relativity [already encountered in (2.20)]:

$$\eta_{\mu\nu} = \begin{pmatrix} -1 & 0 & 0 & 0 \\ 0 & 1 & 0 & 0 \\ 0 & 0 & 1 & 0 \\ 0 & 0 & 0 & 1 \end{pmatrix} , \tag{8.3}$$

where $\mu, \nu = 0, 1, 2, 3$; the "0" component is the time component; and the "1", "2", "3" components are the spatial components. The length squared of a 4-dimensional spacetime vector with components (x^0, x^1, x^2, x^3) is then

$$\eta_{\mu\nu} x^\mu x^\nu = -(x^0)^2 + (x^1)^2 + (x^2)^2 + (x^3)^2 \quad , \tag{8.4}$$

where c (speed of light) has been set equal to one. When $g_{ij}(x)$ depends on the spacetime point x, curvature is present and we are in the realm of general relativity. Note that the Minkowski metric $\eta_{\mu\nu}$ is not **positive definite**, since the RHS of (8.4) can be either positive, zero, or negative. In physics terminology, a 4-vector is said to be **space-like**, **light-like**, or **time-like** depending on whether $\eta_{\mu\nu} x^\mu x^\nu$ is positive, zero, or negative, respectively.

In the general setting, we have a **differentiable manifold** M at each point $x \in M$ of which there exists a vector space called a **tangent space**. (We will not be able to give the precise meanings to these terms here; the reader will find more detailed discussions in Chapter 30.) The collection of all tangent spaces (one for each point in M) is called the **tangent bundle** TM on M. (Fig. 8.1.)

In Fig. 8.1, U, V are vectors in $T_x M$, the tangent space at $x \in M$. A **vector field** on M is a continuous assignment of a vector at every point of M. As mentioned before, mathematically, a vector field on M is known as a **section** on the tangent bundle TM.

Let us focus our attention on a particular tangent space $T_x M$. The length squared of a vector $U \in T_x M$ is given by [generalizing (8.2)]

$$\|U\|^2 = g_{ij}(x) U^i U^j \quad , \tag{8.5}$$

and the **inner product** of two vectors $U, V \in T_x M$ is given by

$$G(U,V) = g_{ij}(x)U^i V^j \qquad .\tag{8.6}$$

A metric tensor G is said to be **non-degenerate** at the point x if, whenever $U \in T_x M$ and

$$G(U,V) = 0 \tag{8.7}$$

for all $V \in T_x M$, it must be true that $U = 0$. We will only work with non-degenerate and **symmetric** ($g_{ij} = g_{ji}$) metric tensors. The above condition for non-degeneracy immediately implies that the system of n algebraic equations

$$g_{ij}U^i = 0 \quad , \quad j = 1, \ldots, n, \tag{8.8}$$

where n is the dimension of the manifold M and thus of any tangent space on M, only has the zero-solution ($U^i = 0$ for all i). This in turn implies (as will be seen in Chapter 16) that

$$\det(g_{ij}) \neq 0 \qquad . \tag{8.9}$$

In fact $\det(g_{ij}) \neq 0$ is the necessary and sufficient condition that G is non-degenerate. Define a **covariant vector** [a (0,1)-type tensor] with components U_i by

$$U_i = g_{ij}U^j \qquad . \tag{8.10}$$

U_i is a vector living in a different space, called the **dual space** to the tangent space $T_x M$, or the **cotangent space** at $x \in M$, denoted by $T_x^* M$. From the above discussion, the non-degeneracy of G implies $\det(g_{ij}) \neq 0$, which means that the linear transformation from $T_x M$ to $T_x^* M$ determined by G must be invertible (see Theorem 16.5). The inverse transformation, denoted by g^{ij}, thus exists, and we have

$$U^i = g^{ij}U_j \quad , \tag{8.11}$$

$$g^{ij}g_{jk} = \delta^i_k \qquad . \tag{8.12}$$

The metric tensor g_{ij} thus provides a one-to-one mapping between a tangent space and its dual, the corresponding cotangent space. This is accomplished by **lowering indices** (from the tangent to the cotangent space) and **raising indices** (from the cotangent space to the tangent space).

An (r, s)-type tensor of the form $T^{i_1,\ldots,i_r}_{j_1,\ldots,j_s}$ (with r upper and s lower indices) is thus an object living in the tensor space

$$\underbrace{T_x M \otimes \cdots \otimes T_x M}_{r \text{ times}} \otimes \underbrace{T_x^* M \otimes \cdots \otimes T_x^* M}_{s \text{ times}}$$

where $\otimes$ denotes the **tensor product**. The concatenation of these tensor spaces, one at each point $x \in M$, is called a **tensor bundle**. One can then work with (r, s)-type tensor fields, which are sections on the corresponding tensor bundle. To summarize, the metric G (also called a **Riemannian metric** when it is positive definite) is a symmetric, non-degenerate (0,2)-type tensor field. A (0,2)-type tensor field is also called a **rank-two covariant tensor field**. A symmetric non-degenerate metric is called **pseudo-Riemannian** if it is not positive definite. For example, the Minkowski metric $\eta_{\mu\nu}$ of (8.3) is not positive definite.

The above discussion of dual spaces has been carried out in a geometrical context in order to be intuitive. We will now give a more abstract (and more general) algebraic characterization of the concept of a dual space. This will be found to be very useful in the mathematical formalism of quantum mechanics.

Let V be a vector space over the field $\mathbb{F}$ (usually the field of real numbers $\mathbb{R}$ or of complex numbers $\mathbb{C}$ in physics applications).

Definition 8.1. *The set of all $\mathbb{F}$-valued linear functions on a vector space V over the field $\mathbb{F}$ is called the **dual space** of V, denoted by V^*.*

Recall that a linear function is simply a special case of a linear map $f : V \longrightarrow W$ (defined by Def. 1.1) where W is the field $\mathbb{F}$. The first important fact concerning V^*, the dual of V, is that $dim(V^*) = n$ if $dim(V) = n$. The proof is quite simple and instructive, and we present it as follows. Suppose $\{e_1, \ldots, e_n\}$ is a basis of V. An arbitrary vector $v \in V$ can then be expressed as

$$v = v^i e_i \quad . \tag{8.13}$$

Let $f \in V^*$, that is, f is a linear function on V. Then linearity implies that

$$f(v) = f(v^i e_i) = v^i f(e_i) \quad . \tag{8.14}$$

Thus any f is completely determined by its values on the basis set $\{e_i\}$ of V. Now choose n linear functions $e^{*\,i}$, $i = 1, \ldots, n$, in V^* such that

$$e^{*\,i}(e_j) = \delta^i_j \quad , \quad 1 \leq j \leq n. \tag{8.15}$$

Then

$$e^{*\,i}(v) = e^{*\,i}(v^j e_j) = v^j e^{*\,i}(e_j) = v^j \delta^i_j = v^i \quad . \tag{8.16}$$

In effect, $e^{*\,i}$ acting on an arbitrary $v \in V$ picks out the i-th component of v. Then, from (8.14),

$$f(v) = v^i f(e_i) = f(e_i) e^{*\,i}(v) \quad . \tag{8.17}$$

Define

$$f_i \equiv f(e_i) \in \mathbb{F} \quad . \tag{8.18}$$

These are numbers when $\mathbb{F}$ is a number field. So

$$f(v) = f_i e^{*\,i}(v) \tag{8.19}$$

for all $v \in V$, and

$$f = f_i e^{*\,i} \quad . \tag{8.20}$$

In other words, an arbitrary $f \in V^*$ can be expressed as a linear combination of the vectors $e^{*i}, i = 1, \ldots, n$, in V^*. We will show that $\{e^{*i}\}$ is a basis of V^* by showing that they are linearly independent as vectors in V^*. Suppose

$$f = f_1 e^{*1} + \cdots + f_n e^{*n} = 0 \quad , \tag{8.21}$$

that is, f is the zero-function in V^*. Thus $f(v) = 0$ for all $v \in V$. Restating this fact, we have, on using (8.16),

$$f_1 v^1 + f_2 v^2 + \cdots + f_n v^n = 0 \quad \text{(the number zero)} \quad , \tag{8.22}$$

for any n-tuplet $\{v^1, \ldots, v^n\} \in \mathbb{F}^n$. This necessarily implies that $f_1 = f_2 = \cdots = f_n = 0$, establishing the linear independence of $\{e^{*i}\}$. Since the basis $\{e^{*i}\}, i = 1, \ldots, n$, of V^* has n elements, V^* is n-dimensional. This completes the proof.

The basis set $\{e^{*i}\}$ of V^* defined by (8.15) is called the **dual basis** of $\{e_i\}$. According to (8.16) they are in fact the **coordinate functions** with respect to the basis $\{e_i\}$ of V. Since V and V^* have the same dimension as vector spaces, they are necessarily **isomorphic** to each other (one-to-one and onto). We can define a **pairing** between a $v \in V$ and a $v^* \in V^*$ by

$$\langle v, v^* \rangle = v^*(v) \quad . \tag{8.23}$$

The bracket way of writing $v^*(v)$, as in the LHS of the above equation, suggests that $v \in V$ can also be viewed as a linear function on V^*. In fact, corresponding to any linear function φ on V^* ($\varphi : V^* \longrightarrow \mathbb{F}$), there is a $v \in V$ such that

$$\langle v, v^* \rangle = \varphi(v^*) \tag{8.24}$$

for all $v^* \in V^*$. Thus

$$(V^*)^* = V \quad . \tag{8.25}$$

Exercise 8.1   For a given function φ on V^* and a choice of basis $\{e_i\}$ of V, express the $v \in V$ satisfying (8.24) in terms of the e_i.

With the above abstract setup, a non-degenerate, symmetric inner product can be defined on V:

$$(v_1, v_2) = (v_2, v_1) \in \mathbb{F} \quad , \quad \text{for all} \quad v_1, v_2 \in V \quad ; \tag{8.26a}$$

if $(v_1, v_2) = 0$ for all $v_2 \in V$, then $v_1 = 0$. This inner product, which is a bilinear map from $V \otimes V$ to $\mathbb{F}$, provides an isomorphism between V and V^* in the following way. For a particular $v \in V$, there exists a unique $v^* \in V^*$ such that

$$\boxed{(u, v) = \langle u, v^* \rangle} \tag{8.26b}$$

for all $u \in V$. Conversely, for a particular $v^* \in V^*$ there exists a unique $v \in V$ such that (8.26b) holds. The proof of the above statement is not difficult but we will not present it here. In Chapter 12, we will in fact present the proof of the infinite-dimensional version of this theorem, known as the Riesz Theorem.

In quantum mechanics, where **complex vector spaces** (often infinite dimensional ones called **Hilbert spaces**) are used, in which the field $\mathbb{F}$ is the set of complex numbers $\mathbb{C}$ rather than the reals, the symmetry property (8.26a) needs to be replaced by the **conjugate (hermitian) symmetry** property:

$$(v_1, v_2) = \overline{(v_2, v_1)} \quad , \tag{8.27}$$

and the bilinearity property replaced by the **conjugate-bilinear** property:

$$(\alpha_1 v_1 + \alpha_2 v_2, v) = \alpha_1(v_1, v) + \alpha_2(v_2, v) \quad , \tag{8.28a}$$

$$(v, \alpha_1 v_1 + \alpha_2 v_2) = \overline{\alpha_1}(v, v_1) + \overline{\alpha_2}(v, v_2) \quad , \tag{8.28b}$$

for all $\alpha_1, \alpha_2 \in \mathbb{C}$. In the above three equations, the overbar represents the complex conjugate. In the physics literature, the **Dirac bracket** notation is almost universally used, in which the order of the entries in $(\,,)$ is reversed and $\langle \,|\, \rangle$ is written instead of parentheses. Thus

$$(v_1, v_2) = \langle v_2 | v_1 \rangle \quad . \tag{8.29}$$

Eq. (8.28b) above is actually a consequence of (8.27) and (8.28a). We will discuss the usefulness of the Dirac notation in Chapter 12.

A complex vector space equipped with an inner product satisfying (8.27) and (8.28) is called a **unitary space**. For a unitary space V, the **conjugate isomorphism** $G : V \longrightarrow V^*$ induced by the inner product $(\,,\,)$ given by (8.26b) satisfies the following. If, under G, we have

$$v \mapsto v^* \quad , \tag{8.30}$$

then

$$\alpha v \mapsto \overline{\alpha} v^* \quad , \quad \alpha \in \mathbb{C} \quad . \tag{8.31}$$

This is required by (8.28b).

Exercise 8.2   Verify the above stated condition.

Once an inner (scalar) product is defined on $V \times V$, the corresponding conjugate isomorphism $G : V \longrightarrow V^*$ induces a scalar product in V^* which satisfies

$$(v_1^*, v_2^*) = (v_2, v_1) = \overline{(v_1, v_2)} \quad . \tag{8.32}$$

Exercise 8.3   Convince yourself that the rule

$$(v_1^*, v_2^*) = (v_1, v_2)$$

would *not* work by verifying that it does not satisfy (8.28a) and (8.31).

Note that since $\bar{\alpha} = \alpha$ for all $\alpha \in \mathbb{R}$, the rules for an inner product in unitary spaces [(8.27), (8.28), (8.31) and (8.32)] will automatically hold for inner products in real vector spaces also.

Let us now re-examine the equations (8.10), (8.11) and (8.12). Consider $V = T_x M$ and $V^* = T_x^* M$. Choose a basis $\{e_i\}$ for $T_x M$, and write $U \in T_x M$ as

$$U = U^i e_i \quad . \tag{8.33}$$

Then the metric tensor G given by $g_{ij}(x)$ specifies an inner product

$$(U, V) = G(U, V) = g_{ij} U^i V^j \quad . \tag{8.34}$$

The induced isomorphism $G : T_x M \longrightarrow T_x^* M$ is then given by

$$U^i e_i \mapsto g_{ij} U^j e^{*i} = U_i e^{*i} \quad , \tag{8.35}$$

where $\{e^{*i}\}$ is the dual basis to $\{e_i\}$. Conversely

$$U_i e^{*i} \mapsto g^{ij} U_j e_i = U^i e_i \quad . \tag{8.36}$$

Eq. (8.36) is just the inverse map of (8.35). The latter equation is valid, since by (8.26),

$$\begin{aligned}
\langle U, V^* \rangle &= \langle U^i e_i, V_j e^{*j} \rangle = \langle U^i e_i, g_{jk} V^k e^{*j} \rangle = g_{jk} U^i V^k \langle e_i, e^{*j} \rangle \\
&= g_{jk} U^i V^k \delta_i^j = g_{ik} U^i V^k = (U, V) \, .
\end{aligned} \tag{8.37}$$

To conclude this chapter, we introduce an important metric on a Lie algebra $\mathcal{G}$ (considered as a vector space), called the Killing form on $\mathcal{G}$. This form is induced by the adjoint representation of $\mathcal{G}$ (see Chapter 7).

Definition 8.2. *The **Killing form** of a Lie algebra $\mathcal{G}$ is the symmetric bilinear form on $\mathcal{G}$ given by*

$$K(A, B) \equiv Tr(Ad_A Ad_B) \, , \tag{8.38}$$

where $A, B \in \mathcal{G}$ and Tr stands for trace.

In the above definition, Ad_A, Ad_B are linear operators on $\mathcal{G}$. Thus the product $Ad_A Ad_B$ is also a linear operator on $\mathcal{G}$, whose trace is given by the trace of some matrix representation of the operator. In Chapter 17 [Theorem 17.2 and Eq. (17.33)] we will see that the trace of a linear operator is independent of the matrix representation.

With respect to a choice of basis $\{e_i\}$ for $\mathcal{G}$, the metric tensor for $\mathcal{G}$ is given by [c.f. (8.6)]

$$g_{ij} = K(e_i, e_j) \, . \tag{8.39}$$

In terms of the structure constants, we see from (7.18) that

$$g_{ij} = Tr(Ad_{e_i} Ad_{e_j}) = (Ad_{e_i})_k^l (Ad_{e_j})_l^k = c_{ik}^l c_{jl}^k \, . \tag{8.40}$$

Exercise 8.4   Show that the metric tensor for $\mathcal{SO}(3)$ is

$$g_{ij} = -2\delta_{ij} \ , \tag{8.41}$$

where δ_{ij} is the Kronecker delta. [Use (2.13).]

    According to the above result, the metric tensor for $\mathcal{SO}(3)$ is negative definite. This is an important property for $\mathcal{SO}(3)$ and its corresponding Lie group $SO(3)$. In fact we have the following useful theorem (stated without proof).

Theorem 8.1. *A Lie algebra $\mathcal{G}$ generates a compact Lie group if and only if its Killing form is negative definite. Negative definiteness means that*

$$g_{ij} A^i B^j < 0 \ , \tag{8.42}$$

for any $A = A^i e_i$, $B = B^i e_i \in \mathcal{G}$, where $\{e_i\}$ is a basis of $\mathcal{G}$.

Exercise 8.5   Show that the Killing form satisfies

$$K([A, B], C) = K([B, C], A) = K([C, A], B) = -K([B, A], C) \ . \tag{8.43}$$

    The Killing form of a Lie algebra $\mathcal{G}$ provides fundamental information on the structure of $\mathcal{G}$. This is exemplified by the following basic theorems (stated without proof) known as Cartan's Criteria.

Theorem 8.2 (Cartan's first Criterion). *A Lie algebra is semisimple if and only if the Killing form is non-degenerate:* $\det(g_{ij}) \neq 0$.

Theorem 8.3 (Cartan's second Criterion). *A Lie algebra is solvable if and only if $K(A, B) = 0$ for all $A \in \mathcal{G}$ and $B \in \mathcal{G}^{(1)}$ [c.f. (7.6)].*

    For a semisimple Lie algebra, $\det(g_{ij}) \neq 0$. Hence the inverse g^{ij} exists and we can define the **Casimir operator** of the adjoint representation by

$$C \equiv g^{ij} Ad_{e_i} Ad_{e_j} \ , \tag{8.44}$$

where $\{e_i\}$ is a basis of $\mathcal{G}$. It is an element of the enveloping algebra $\mathcal{A}_{Ad}(\mathcal{G})$ [c.f. Chapter 7]. We must check that the above definition is independent of the choice of basis of $\mathcal{G}$. Indeed, introduce a change of basis $e_i = a_i^j e_j'$, so that $e_i' = (a^{-1})_i^j e_j$ (c.f. Table 1.1). Then, by (2.1) and (7.15),

$$(g')^{ij} Ad_{e_i'} Ad_{e_j'} (e_p) = a_l^i a_k^j g^{lk} [e_i', [e_j', e_p]] = g^{lk} a_l^i (a^{-1})_i^m a_k^j (a^{-1})_j^n [e_m, [e_n, e_p]]$$

$$= g^{lk} \delta_l^m \delta_k^n [e_m, [e_n, e_p]] = g^{lk} [e_l, [e_k, e_p]] = g^{lk} Ad_{e_l} Ad_{e_k} (e_p) \ ,$$

$$\tag{8.45}$$

for any basis vector e_p. Independence of the definition (8.44) on the choice of basis follows by linearity.

Let us write

$$E_i \equiv Ad_{e_i} . \tag{8.46}$$

Then E_i transforms covariantly and the Casimir operator can be written

$$C = g^{ij} E_i E_j = E^j E_j , \tag{8.47}$$

where

$$E^i = g^{ij} E_j . \tag{8.48}$$

Finally we have

Theorem 8.4. *The Casimir operator C of the adjoint representation of a Lie algebra $\mathcal{G}$ lies in the center of the enveloping algebra $\mathcal{A}_{Ad}(\mathcal{G})$, that is, it commutes with every element in $\mathcal{A}_{Ad}(\mathcal{G})$.*

Proof. It is sufficient to show that C commutes with $Ad_{e_k} = E_k$ for any e_k in a basis of $\mathcal{G}$. We have

$$[C, E_k] = g^{ij}[E_i E_j, E_k] = g^{ij}\{E_i, [E_j, E_k] + [E_i, E_k]E_j\}$$
$$= g^{ij}\{c^l_{jk} E_i E_l + c^l_{ik} E_l E_j\} = g^{ij} c^l_{jk} E_i E_l + g^{ji} c^l_{jk} E_l E_i = g^{ij} c^l_{jk}(E_i E_l + E_l E_i)$$
$$= g^{ij} g^{lm} c_{mjk}(E_i E_l + E_l E_i) = -g^{ij} g^{lm} c_{jmk}(E_l E_i + E_i E_l) = -[C, E_k] . \tag{8.49}$$

Thus $[C, E_k] = 0$, as desired. In the 5-th equality of (8.49), we have used the fact that g^{ij} is symmetric; in the 6-th equality, we have defined the $(0,3)$-tensor c_{ijk} by

$$c_{ijk} = g_{il} c^l_{jk} , \tag{8.50}$$

while in the 7-th equality we have used the fact that c_{ijk} is *completely anti-symmetric*. This last fact can be seen as follows. From the definition of the structure constants [(4.29)] it is obvious that

$$c^l_{jk} = -c^l_{kj} . \tag{8.51}$$

It then follows from (8.50) that

$$c_{ijk} = -c_{ikj} . \tag{8.52}$$

Using (8.50) and (8.40), we have

$$c_{ijk} = c^m_{in} c^n_{lm} c^l_{jk} = -c^m_{in} c^n_{lk} c^l_{mj} - c^m_{in} c^n_{lj} c^l_{km} = c^m_{in} c^n_{jl} c^l_{km} + c^m_{ni} c^l_{mj} c^n_{lk} , \tag{8.53}$$

where in the second equality we have used Jacobi's identity for the structure constants [(4.32)], and in the third we have used (8.51). Thus

$$c_{jik} = c^m_{jn} c^n_{il} c^l_{km} + c^m_{nj} c^l_{mi} c^n_{lk} = -c^l_{im} c^m_{jn} c^n_{kl} - c^n_{li} c^m_{nj} c^l_{mk} = -c_{ijk} . \tag{8.54}$$

Similarly,

$$c_{ijk} = -c_{kji} . \tag{8.55}$$

□

Note that even though technically the Casimir operator is only defined for semisimple Lie algebras, there may exist elements in the enveloping algebra of non-semisimple ones which commute with all the generators (basis vectors) of the algebra. Such is the case, for example, in $\mathcal{E}(3)$, the six-dimensional Lie algebra of Euclidean motions [(7.13)]. $\mathcal{E}(3)$ is not semisimple since it contains the abelian ideal $\mathcal{T}(3)$ (the algebra of three-dimensional translations) [c.f. the commutation relations (7.10) to (7.12)]. But one can verify easily that

$$p^2 \equiv p_1^2 + p_2^2 + p_3^2 , \tag{8.56}$$

and

$$\boldsymbol{p} \cdot \boldsymbol{L} \equiv p_1 L^1 + p_2 L^2 + p_3 L^3 , \tag{8.57}$$

(with a slight abuse of notation) commute with all the p_i and L^i. These are sometimes also called Casimir operators.

Chapter 9

$SO(4)$ and the Hydrogen Atom

The hydrogen atom problem is traditionally treated within the context of the Schrödinger wave equation (a second order partial differential equation). But it also affords an excellent example in the use of group theoretic methods, especially the basic ideas in Lie algebra theory developed in Chapters 4 and 7. We shall see that the spectrum of quantized energy levels (with their degeneracies) of the hydrogen atom is basically a consequence of the rotational symmetry and an additional symmetry – sometimes called **dynamical symmetry** – of the electronic Hamiltonian.

We will consider the so-called **Kepler problem**, in which the Hamiltonian includes only the Coulomb interaction between the electron and the nucleus. This Hamiltonian is given by

$$H = \frac{p^2}{2m} - \frac{Ze^2}{r} \ , \tag{9.1}$$

$$p_j = -i\hbar \frac{\partial}{\partial x^j} \ , \tag{9.2}$$

$$r = \sqrt{(x^1)^2 + (x^2)^2 + (x^3)^2} \ , \tag{9.3}$$

where m is the reduced mass of the electron, e is the magnitude of the electronic charge, and Z is the charge number of the nucleus.

An elegant solution of the classical Kepler problem goes as follows. Hamilton's equations of motion (see Chapter 44)

$$\frac{dq^i}{dt} = \frac{\partial H}{\partial p_i} \ , \qquad \frac{dp_i}{dt} = -\frac{\partial H}{\partial q^i} \ , \tag{9.4}$$

where q^i and p_i are the canonical coordinates and momenta, respectively, lead to [with H given by (9.1)]

$$\frac{d\boldsymbol{r}}{dt} = \frac{\boldsymbol{p}}{m} \ , \qquad \frac{d\boldsymbol{p}}{dt} = -\frac{Ze^2\boldsymbol{r}}{r^3} \ . \tag{9.5}$$

71

The second of the above equations is just Newton's second law applied to the present situation. One tries to look for all the **constants of motion**, quantities whose total derivatives with respect to time vanish. One easily finds that the angular momentum

$$\boldsymbol{L} = \boldsymbol{r} \times \boldsymbol{p} \tag{9.6}$$

is one such quantity, since by (9.5),

$$\frac{\mathrm{d}\boldsymbol{L}}{\mathrm{d}t} = 0 \, . \tag{9.7}$$

Another (much less obvious) one is the so-called **Runge-Lenz vector**:

$$\boldsymbol{M} = \frac{\boldsymbol{p} \times \boldsymbol{L}}{m} - \frac{Ze^2\boldsymbol{r}}{r} \, , \tag{9.8}$$

which also satisfies

$$\frac{\mathrm{d}\boldsymbol{M}}{\mathrm{d}t} = 0 \, . \tag{9.9}$$

Exercise 9.1   Verify (9.7) and (9.9).

It is also clear that

$$\boldsymbol{L} \cdot \boldsymbol{r} = \boldsymbol{L} \cdot \boldsymbol{M} = 0 \, . \tag{9.10}$$

Exercise 9.2   Verify (9.10).

Thus $\boldsymbol{L}$ is perpendicular to the plane of the orbit and $\boldsymbol{M}$ lies on that plane. The latter fact immediately yields the orbit, since (9.8) gives

$$\boldsymbol{M} \cdot \boldsymbol{r} = \frac{L^2}{m} - Ze^2 r \, . \tag{9.11}$$

This, in fact, is the equation of a conic section. To see this more explicitly, we let the $x - y$ plane coincide with the orbital plane with the positive x-axis along the direction of $\boldsymbol{M}$. Then (9.11) can be written in polar coordinates as

$$r = \frac{L^2}{mZe^2} \left(1 + \frac{M}{Ze^2} \cos\phi \right)^{-1} \, , \tag{9.12}$$

where ϕ is the angle between $\boldsymbol{r}$ and the x-axis (the direction of $\boldsymbol{M}$), and $M/(Ze^2)$ is the eccentricity of the conic. A bit more vector algebra will reveal that

$$M^2 = \frac{2HL^2}{m} + Z^2 e^4 \, . \tag{9.13}$$

Thus the constant total energy is given by

$$H = \frac{m}{2L^2}(M^2 - Z^2 e^4) \,. \tag{9.14}$$

Now we do quantum mechanics. r, p, L are all operators in the Hilbert space of electron wave functions [see Chapter 12 below] and do not commute with each other [c.f. (4.12), (7.10) to (7.12)]. In particular $p \times L \neq -L \times p$. Thus we redefine a quantum mechanical Runge-Lenz vector by anti-symmetrizing the product $p \times L$:

$$M \equiv \frac{1}{2m}(p \times L - L \times p) - \frac{Ze^2 r}{r} \,. \tag{9.15}$$

The main strategy from this point on is to exploit the deep connection between symmetry and conservation (as explained in Chapter 3). We have already studied in great detail rotational symmetry in Chapter 4, with the symmetry group $SO(3)$ and its infinitesimal generators L^1, L^2, L^3, which form the basis of the Lie algebra $\mathcal{SO}(3)$. The Hamiltonian H in (9.1) is manifestly rotationally invariant (the Coulomb force is a central force), so we expect conservation of angular momentum classically. Quantum mechanically, this is also established by calculating dL/dt explicitly and using Heisenberg's equation of motion for an observable A:

$$\frac{dA}{dt} = \frac{[A, H]}{i\hbar} + \frac{\partial A}{\partial t} \,. \tag{9.16}$$

Exercise 9.3 Show that quantum mechanically $dL/dt = 0$ by using (9.16) for r and p.

Note that $[A, H] = 0$ implies

$$H = AHA^{-1} \,. \tag{9.17}$$

We say that in this case H is invariant under the infinitesimal transformation A. In Chapter 18 we will study in more detail the significance of (9.17) in the classification of energy eigenvalues of the Hamiltonian in the light of group representation theory.

Now classical mechanics suggests that M is a constant of motion also [(9.9)]. It turns out that quantum mechanically it is indeed true that $dM/dt = 0$.

Exercise 9.4 Verify the above statement by differentiating (9.15) with respect to t and evaluating all the time derivatives in the resulting expression by using the right-hand-side of (9.16). [Warning: This is a tedious exercise. It involves repeated use of the commutation relation

$$[f(r), p] = i\hbar \nabla f(r) \,,$$

where f is an analytic function of the components of $\boldsymbol{r}$. (See W. Pauli 1926.)]

We thus have

$$[\boldsymbol{L}, H] = [\boldsymbol{M}, H] = 0 . \tag{9.18}$$

The fact that M^1, M^2, M^3 are also constants of motion in addition to $L^1, L^2,$ L^3 suggests the Lie group here is larger than $SO(3)$. In fact we will establish that M^i and L^j form the basis of a six-dimensional Lie algebra by working out all the commutators of these objects and show that they satisfy (4.29). The following results (whose tedious but straightforward demonstrations we will not present here) are obtained.

$$[L^i, L^j] = i\hbar\, \varepsilon^{ij}{}_k L^k , \tag{9.19}$$

$$[L^i, M^j] = i\hbar\, \varepsilon^{ij}{}_k M^k , \tag{9.20}$$

$$[M^i, M^j] = -\frac{2i\hbar}{m} H\, \varepsilon^{ij}{}_k L^k . \tag{9.21}$$

Actually the last equation seems to spoil things a bit since H occurs on the right-hand-side. But since H commutes with both $\boldsymbol{M}$ and $\boldsymbol{L}$, Schur's Lemma of group representation theory (see Chapter 18) implies that, within an irreducible representation of our group (on a subspace of the Hilbert space of electronic wave functions), $H = EI$, where $E \in \mathbb{R}$ and I is the identity operator. If we work within an irreducible representation, and view all of the L^i, M^j as linear operators on the representation space, then H on the RHS of (9.21) can be replaced by a real number E, and the closure requirement of (4.29) will be satisfied. Within an irreducible representation let us then define

$$\boldsymbol{M'} \equiv \sqrt{-\frac{m}{2E}}\, \boldsymbol{M} , \tag{9.22}$$

where $E < 0$ (bound states of electrons). Note that $\boldsymbol{M'}$ has the same physical dimensions as the angular momentum $\boldsymbol{L}$. The commutation relations (9.19) to (9.21) can then be written as

$$[L^i, L^j] = i\hbar\, \varepsilon^{ij}{}_k L^k , \tag{9.23}$$

$$[L^i, (M')^j] = i\hbar\, \varepsilon^{ij}{}_k (M')^k , \tag{9.24}$$

$$[(M')^i, (M')^j] = i\hbar\, \varepsilon^{ij}{}_k L^k . \tag{9.25}$$

The 3-dimensional algebra $\mathcal{SO}(3)$ generated by the L^i is clearly a subalgebra of our 6-dimensional algebra. The six operators L^i and $(M')^j$ can be viewed as the six independent components of a 4×4 antisymmetric matrix $-iJ$ (of operators) as follows:

$$- i\hbar J$$

$$= -i\hbar \begin{pmatrix} 0 & J_1^2 & J_1^3 & J_1^4 \\ -J_1^2 & 0 & J_2^3 & J_2^4 \\ -J_1^3 & -J_2^3 & 0 & J_3^4 \\ -J_1^4 & -J_2^4 & -J_3^4 & 0 \end{pmatrix} \equiv \begin{pmatrix} 0 & L^3 & -L^2 & (M')^1 \\ -L^3 & 0 & L^1 & (M')^2 \\ L^2 & -L^1 & 0 & (M')^3 \\ -(M')^1 & -(M')^2 & -(M')^3 & 0 \end{pmatrix}.$$

$$(9.26)$$

In terms of the J_i^j, then, the commutation relations (9.23) to (9.25) appear as the single equation

$$[J_i^j, J_k^l] = \delta_k^j J_i^l - \delta_i^l J_k^j + \delta_i^k J_l^j - \delta_l^j J_i^k, \quad i, j, k, l = 1, 2, 3, 4. \qquad (9.27)$$

Exercise 9.5   Show that (9.27) is equivalent to (9.23) through (9.25).

From our earlier result (4.35), these are precisely the commutation relations for the Lie algebra $SO(4)$! Thus the relevant group for the quantum mechanical Kepler problem is $SO(4)$.

$SO(4)$ can in fact be decomposed into a direct sum of two copies of $SO(3)$:

$$SO(4) = SO(3) \oplus SO(3). \qquad (9.28)$$

This can be achieved by a change of basis from $\{\boldsymbol{L}, \boldsymbol{M'}\}$ to

$$\boldsymbol{L}^{\pm} \equiv \frac{1}{2}(\boldsymbol{L} \pm \boldsymbol{M'}). \qquad (9.29)$$

The structure constants with respect to this new basis are given by

$$[(L^{\pm})^i, (L^{\pm})^j] = i\hbar\,\varepsilon^{ij}{}_k (L^{\pm})^k, \qquad (9.30)$$

$$[(L^{+})^i, (L^{-})^j] = 0, \qquad (9.31)$$

as expected from the result (9.28).

Exercise 9.6   Verify the above two equations by using (9.23) to (9.25).

It is obvious from (9.18) that

$$[\boldsymbol{L}^{+}, H] = [\boldsymbol{L}^{-}, H] = 0, \qquad (9.32)$$

and thus $\boldsymbol{L}^{+}$ and $\boldsymbol{L}^{-}$ are both constants of motion.

As discussed at the end of Chapter 7, each copy of $SO(3)$ in $SO(4)$ has a Casimir operator. These are $L^+ \cdot L^+$ and $L^- \cdot L^-$, with eigenvalues $l^+(l^+ + 1)$ and $l^-(l^- + 1)$, respectively; and so the Casimir operator for $SO(4)$ is simply

$$C \equiv L^+ \cdot L^+ + L^- \cdot L^- \,, \tag{9.33}$$

with

$$[C, (L^{\pm})^i] = 0 \,. \tag{9.34}$$

Now within the same irreducible representation of $SO(4)$ [or $\mathcal{SO}(4)$], $L^+ \cdot L^+$ and $L^- \cdot L^-$ have the same eigenvalue, namely,

$$l^+(l^+ + 1) = l^-(l^- + 1) \,. \tag{9.35}$$

We thus have

$$l^+ = l^- = 0, 1/2, 1, 3/2, 2, \ldots \,. \tag{9.36}$$

This actually leads to the result that

$$L^+ \cdot L^+ - L^- \cdot L^- = L \cdot M' = L \cdot M = 0 \,, \tag{9.37}$$

which can also be proved directly. In fact, using the commutation relations (4.12) through (4.14), we can establish the following as operator equations:

$$r \cdot L = L \cdot r = 0 \,, \tag{9.38}$$
$$p \cdot L = L \cdot p = 0 \,, \tag{9.39}$$
$$M \cdot L = L \cdot M = 0 \,, \tag{9.40}$$

which corroborate the classical results. Manipulating with the expression for M given by (9.15) and using the various commutation relations between r, p and L, one also obtains

$$M \cdot M = \frac{2H}{m}(L \cdot L + \hbar^2) + Z^2 e^4 \,. \tag{9.41}$$

Now, due to (9.40), the Casimir operator C [(9.33)] can be written as

$$C = \frac{1}{2}(L \cdot L + M' \cdot M') \,, \tag{9.42}$$

or, on using (9.22) and (9.41) with H replaced by $E \in \mathbb{R}$,

$$C = - \left(\frac{\hbar^2}{2} + \frac{mZ^2 e^4}{4E} \right) \,. \tag{9.43}$$

Replacing C by $2l^+(l^+ + 1)$ [c.f. (9.33), (9.35) and (9.36)], one obtains

$$E = - \frac{mZ^2 e^4}{2\hbar^2 (2l^+ + 1)^2} \,, \qquad l^+ = 0, 1/2, 1, 3/2, 2, \ldots \,. \tag{9.44}$$

From (9.29), we realize that the total orbital angular momentum $\boldsymbol{L}$ is given by

$$\boldsymbol{L} = \boldsymbol{L}^+ + \boldsymbol{L}^- \ . \tag{9.45}$$

The eigenvalues of $\boldsymbol{L} \cdot \boldsymbol{L}$ are $l(l+1)$, $l = 0, 1, 2, 3, \ldots$. (The orbital angular momentum quantum number can only have integer values.) From the theory of the addition of angular momentum in quantum mechanics [which is actually the theory of product representations of irreducible representations of $SO(3)$, (see Chapter 22)], l can assume the values (given by Clebsch-Gordon decomposition)

$$l = \begin{cases} l^+ + l^- = 2l^+ \ , \\ l^+ + l^- - 1 \ , \\ \ldots \ , \\ \ldots \ , \\ l^+ - l^- = 0 \ . \end{cases} \tag{9.46}$$

Since $2l^+$ is always integral, we can replace $2l^+ + 1$ by n, a positive integer. Finally we have the following expression for the energy eigenvalues.

$$\boxed{E = -\frac{mZ^2 e^4}{2\hbar^2 n^2} \ , \quad n = 1, 2, 3, \ldots} \quad . \tag{9.47}$$

In the above equation n is identified as the **principal quantum number**. For a given n, (9.46) implies that the orbital angular momentum quantum number l can assume values from 0 to $n - 1$, in unit intervals. Also, for a given $l^+ = l^-$, each of $(L^+)^3$ and $(L^-)^3$ has $2l^+ + 1 = n$ values. Thus every energy level is n^2-fold degenerate (ignoring spin). The actual degeneracy is $2n^2$, with spin.

Chapter 10

Adjoints and Unitary Transformations

We saw in Chapter 1 that a linear transformation on a vector space V can be represented by an $n \times n$ matrix with respect to a particular choice of basis $\{e_i\}$ of V. We also studied an important class of linear transformations called orthogonal transformations. These are represented by orthogonal matrices a satisfying the condition $aa^T = a^T a = 1$. For Euclidean spaces, that is, spaces equipped with the Euclidean inner product (8.2), orthogonal transformations preserve the length of vectors. In this chapter, we will study the generalization of an orthogonal transformation to a so-called unitary transformation, which plays the same role in unitary spaces [complex spaces equipped with inner products satisfying (8.27) and (8.28)] as orthogonal transformations do in real Euclidean spaces. Unitary transformations preserve the lengths (norms) of vectors in unitary spaces, and are of fundamental importance in quantum mechanics. First, we need to introduce the notion of an adjoint operator (transformation).

Definition 10.1. *Let A be a linear operator (transformation) on an inner product space V. The unique linear operator $A^\dagger$ defined by the property*

$$\boxed{(Av_1, v_2) = (v_1, A^\dagger v_2)} \quad , \tag{10.1}$$

*for all $v_1, v_2 \in V$ is called the **adjoint** of A.*

Exercise 10.1   Prove that the above definition makes sense, that is, prove that $A^\dagger$ as defined is
(1) unique, and
(2) linear.

The following list of properties of the adjoint follows easily from the definition.

$$1) \quad 0^\dagger = 0, \quad \text{(the zero-operator is self adjoint)} \qquad (10.2)$$

$$2) \quad 1^\dagger = 1, \quad \text{(the identity is self adjoint)} \qquad (10.3)$$

$$3) \quad (A+B)^\dagger = A^\dagger + B^\dagger, \qquad (10.4)$$

$$4) \quad (\alpha A)^\dagger = \bar{\alpha} A^\dagger, \quad \alpha \in \mathbb{C}, \qquad (10.5)$$

$$5) \quad (AB)^\dagger = B^\dagger A^\dagger, \qquad (10.6)$$

$$6) \quad (A^{-1})^\dagger = (A^\dagger)^{-1}, \quad \text{if } A \text{ is invertible}, \qquad (10.7)$$

$$7) \quad (A^\dagger)^\dagger = A. \qquad (10.8)$$

$\boxed{\text{Exercise 10.2}}$ Prove 4) to 7) above.

An operator whose adjoint is equal to itself ($A^\dagger = A$) is called **self-adjoint**, or **hermitian**. Hermitian operators play a crucial role in quantum mechanics. In the mathematical formalism of that theory, every physically observable quantity, called an **observable**, has to be represented by a hermitian operator on a unitary space whose vectors describe the **states** of a physical system. We have seen some important examples already: the linear momentum operators [(4.4)] and the angular momentum operators [(4.22)].

Since linear operators in general do not commute, one should be careful to note that the product of two hermitian operators is not necessarily hermitian. In fact, (10.6) implies that the product of two hermitian operators A and B is hermitian if and only if A and B commute.

Two further related and useful facts of hermitian operators are the following:

i) If $A^\dagger = A$, then $B^\dagger AB$ is hermitian for all B;

ii) If B is invertible and $B^\dagger AB$ is hermitian, then A is hermitian.

An operator A is called **skew-hermitian** if

$$A^\dagger = -A \quad . \qquad (10.9)$$

It is easy to see that every linear operator A can be decomposed uniquely as $A = B + C$ where B is hermitian and C is skew-hermitian. In fact, we have

$$B = \frac{A + A^\dagger}{2} \quad , \qquad (10.10)$$

$$C = \frac{A - A^\dagger}{2} \quad . \qquad (10.11)$$

Every hermitian operator can be made skew-hermitian (and vice-versa) simply by multiplication by $i = \sqrt{-1}$. Thus every linear operator A has a unique decomposition

$$A = B + iC \qquad (10.12)$$

where both B and C are hermitian. This is reminiscent of the fact that every complex number $z \in \mathbb{C}$ can be uniquely decomposed as $z = x + iy$, where both x and y are real. Thus hermitian operators play the role of real numbers, and skew-hermitian operators play the role of imaginary numbers in the algebra of operator space.

Definition 10.2. *A **unitary transformation** A is one which satisfies the condition*

$$A^{-1} = A^\dagger \quad , \tag{10.13a}$$

or equivalently,

$$A^\dagger A = A A^\dagger = 1 \quad . \tag{10.13b}$$

From this definition we see immediately that unitary transformations preserve the length of vectors in unitary spaces. Indeed, the length squared of a vector v in a unitary space V is defined via the inner product to be

$$\|v\|^2 = (v, v) \quad . \tag{10.14}$$

Thus, for a unitary operator A,

$$(Av, Av) = (v, A^\dagger Av) = (v, v) \quad . \tag{10.15}$$

The finite spatial translations [(4.10)] and the finite spatial rotations [(4.21)] introduced earlier are very important examples of unitary operators in quantum mechanics. Another important example is the finite-time translation operator (**time-evolution operator**) in quantum mechanics

$$U(t) = \exp\left(\frac{-iHt}{\hbar}\right) \tag{10.16}$$

where H is the **Hamiltonian** of the system (the hermitian operator representing the energy observable of the system).

We will now find the matrix representation of the adjoint $A^\dagger$ given the matrix representation of A with respect to an **orthonormal basis** $\{e_i\}$ of V. Two vectors v_1, v_2 are said to be **orthogonal** with respect to an inner product (,) if $(v_1, v_2) = (v_2, v_1) = 0$. An orthonormal basis is simply one that satisfies

$$(e_i, e_j) = \delta_{ij} \quad , \tag{10.17}$$

that is, every two distinct basis vectors in the basis set are orthogonal to each other, and they all have unit length. An important fact to take note (although we will not prove it) is that *orthonormal bases are guaranteed to exist in unitary spaces.*

To understand the advantage of using orthonormal bases we need to introduce the notion of **the dual basis in** V **to a given basis** $\{e_i\}$ **in** V (as opposed to the notion of the dual basis $\{e^{*i}\}$ in the dual space V^* to a given basis $\{e_i\}$ in V introduced in Chapter 8). First we recall that for a given basis $\{e_i\}$ in V, the dual basis in V^*, $\{e^{*i}\}$, is given by

$$\langle e_i, e^{*j} \rangle = \delta_i^j \quad . \tag{10.18}$$

The unitary space V comes equipped with an inner product and thus an induced conjugate isomorphism between V and V^*:

$$G : V \longrightarrow V^* \quad , \quad G^{-1} : V^* \longrightarrow V \quad .$$

Let e_i' be the image of e^{*i} under the conjugate isomorphism G^{-1}. The set $\{e_i'\}$ then satisfies, by definition,

$$(e_i, e_j') = \langle e_i, e^{*j} \rangle = \delta_{ij} \quad . \tag{10.19}$$

The set $\{e_i'\}$ is linearly independent. Indeed, suppose $\alpha^i e_i' = 0$, $\alpha^i \in \mathbb{C}$. Then, for all j, $(\alpha^i e_i', e_j) = 0 = \alpha^i (e_i', e_j) = \alpha^i \delta_{ij} = \alpha^j$. Furthermore, since $\{e_i'\}$ and $\{e_i\}$ are related by an invertible linear transformation, every vector $v \in V$, being expressible as a linear combination of $\{e_i\}$, can also be expressed as a linear combination of $\{e_i'\}$. Thus $\{e_i'\}$ is also a basis of V. It is called the dual basis in V to $\{e_i\}$. Now comes the crucial fact. An orthonormal basis $\{e_i\}$ is **self-dual**, that is, $e_i' = e_i$. The proof is easy. We have, for all $v \in V$,

$$(v, e_i - e_i') = (v, e_i) - (v, e_i') = (v^j e_j, e_i) - (v^j e_j, e_i') = v^j \delta_{ij} - v^j \delta_{ij}$$
$$= v^i - v^i = 0 \, , \tag{10.20}$$

which implies $e_i' = e_i$ for all i.

Now suppose the matrix representation of a linear operator A with respect to a basis $\{e_i\}$ of V is a_i^j, i.e. [see (1.16)],

$$A(e_i) = a_i^j e_j \quad , \tag{10.21}$$

and the matrix representation of $A^\dagger$ with respect to the dual basis $\{e_i'\}$ introduced above is $(a^\dagger)_i^j$, i.e.,

$$A^\dagger(e_i') = (a^\dagger)_i^k e_k' \quad . \tag{10.22}$$

We then have

$$(e_j, A^\dagger e_i') = \overline{(a^\dagger)_i^k} \, (e_j, e_k') = \overline{(a^\dagger)_i^k} \, \delta_{jk} = \overline{(a^\dagger)_i^j} \quad . \tag{10.23}$$

On the other hand,

$$(e_j, A^\dagger e_i') = (Ae_j, e_i') = a_j^k \, (e_k, e_i') = a_j^k \delta_{ki} = a_j^i \quad . \tag{10.24}$$

Thus

$$\boxed{(a^\dagger)_i^j = \overline{a_j^i}} \quad . \tag{10.25}$$

The matrix $(a^\dagger)_i^j = \overline{a_j^i}$ is called the **hermitian conjugate** of the matrix (a_i^j).

In the above equation, it is important to remember that a_i^j is the matrix representation of A with respect to a basis $\{e_i\}$, while $(a^\dagger)_i^j$ is the matrix representation of $A^\dagger$ with respect to the dual basis $\{e_i'\}$. In general $\{e_i\}$ is not required to be an orthonormal basis. But if it is, then $e_i = e_i'$ as follows from our earlier discussion, and (10.25) is a statement relating the matrix representation of A and $A^\dagger$ with respect to the same (orthonormal) basis.

Note that the matrix $(a^\dagger)_i^j$, according to (10.25), is the complex conjugate transpose of the matrix a_i^j. For a hermitian operator $(A^\dagger = A)$, the matrix representation with respect to an orthonormal basis $\{e_i\}$ then satisfies the condition

$$a_i^j = \overline{a_j^i} \quad . \tag{10.26}$$

Hence the matrix is equal to its complex conjugate transpose. Such a matrix is called a **hermitian matrix**. Note that real hermitian matrices are just real symmetric matrices $(a_i^j = \overline{a_j^i} = a_j^i)$.

If an operator A is unitary $(A^{-1} = A^\dagger)$, and the matrix representation of A with respect to an orthonormal basis $\{e_i\}$ is (a_i^j), the matrix representation of A^{-1} with respect to the same basis is the hermitian conjugate of the matrix (a_i^j), i.e.,

$$(a^{-1})_i^j = \overline{a_j^i} \quad . \tag{10.27}$$

A matrix that satisfies the above property is called a **unitary matrix**.

| Exercise 10.3 | Show that $\exp(iA)$, where A is hermitian, is a unitary operator.

If the matrix (a_i^j) in (6.27) is real, then

$$(a^{-1})_i^j = \overline{a_j^i} = a_j^i \quad , \tag{10.28}$$

that is,

$$a^{-1} = a^T \quad , \tag{10.29}$$

and the property of unitarity reduces to that of orthogonality.

Chapter 11

The Lorentz Group and $SL(2, \mathbb{C})$

The Lorentz group has a special significance in physics because of its fundamental role in the special theory of relativity, and hence in quantum field theory. The so-called space-time symmetries manifested in the invariance of physical laws under Lorentz transformations are the source of the most basic conservation principles in physics, including the conservation of linear and angular momentum considered earlier in Chapters 3 and 4.

Recall the **Minkowski metric** $\eta_{\mu\nu}$ on $\mathbb{R}^4$ introduced in Chapter 8 [c.f. (8.3)],

$$
\eta_{\mu\nu} =
\begin{pmatrix}
-1 & 0 & 0 & 0 \\
0 & 1 & 0 & 0 \\
0 & 0 & 1 & 0 \\
0 & 0 & 0 & 1
\end{pmatrix},
$$

which determines the **norm** $\|x\|^2$ of a vector $x = (x^0, x^1, x^2, x^3) \in \mathbb{R}^4$ by

$$
\|x\|^2 = \eta_{\mu\nu} x^\mu x^\nu = -(x^0)^2 + (x^1)^2 + (x^2)^2 + (x^3)^2 . \tag{11.1}
$$

(Note: It is customary in the physics literature to use Greek letters for space-time indices; Latin letters are reserved for space indices.) The vector space $\mathbb{R}^4$ equipped with the Minkowski metric $\eta_{\mu\nu}$ is called **Minkowski space**, which we will denote in this chapter by M.

Definition 11.1. *A **Lorentz transformation** Λ is a linear transformation on Minkowski space M which preserves the norm of a vector in M, that is,*

$$
\|\Lambda x\|^2 = \|x\|^2 , \tag{11.2}
$$

for all $x \in M$.

Lorentz transformations can be thought of as 'rotations' in space-time. The action of a Lorentz transformation Λ on a space-time point $x \in M$ is usually written in terms of components as

$$x^\mu \longrightarrow x'^\mu = \Lambda^\mu_\nu x^\nu , \quad \mu, \nu = 0, 1, 2, 3 , \tag{11.3}$$

or

$$(x'^0, x'^1, x'^2, x'^3) = (x^0, x^1, x^2, x^3) \begin{pmatrix} \Lambda^0_0 & \Lambda^1_0 & \Lambda^2_0 & \Lambda^3_0 \\ \Lambda^0_1 & \Lambda^1_1 & \Lambda^2_1 & \Lambda^3_1 \\ \Lambda^0_2 & \Lambda^1_2 & \Lambda^2_2 & \Lambda^3_2 \\ \Lambda^0_3 & \Lambda^1_3 & \Lambda^2_3 & \Lambda^3_3 \end{pmatrix} .$$

The set of all Lorentz transformations on M forms a group $L(4)$ [sometimes denoted by $O(1,3)$], called the **Lorentz group**. This is actually a subgroup of the so-called **Poincare group** P, which includes the abelian subgroup $T(4)$ of all translations in Minkowski space in addition to the Lorentz transformations.

$\boxed{\text{Exercise 11.1}}$ Check that the Lorentz transformations satisfy the group axioms [c.f. Def. 4.1].

Definition 11.2. *Suppose A is an invariant (normal) abelian subgroup of G and B is another subgroup of G such that $A \cap B = \{e\}$ (the identity in G) and every element $g \in G$ can be written in a unique way as $g = ba$, $b \in B, a \in A$, then G is said to be a **semi-direct product** of B and A, written $G = B \otimes_S A$.*

Theorem 11.1. *The Poincare group P is a semi-direct product of the Lorentz group $L(4)$ and the group of all translations $T(4)$ on Minkowski space, that is,*

$$P = L(4) \otimes_S T(4) . \tag{11.4}$$

$\boxed{\text{Exercise 11.2}}$ Prove Theorem 11.1.

A particular $p \in P$ is then usually denoted $p = (\Lambda^\mu_\nu, a^\mu)$. Under p, $x \in M$ transforms as

$$x^\mu \longrightarrow x'^\mu = \Lambda^\mu_\nu x^\nu + a^\mu . \tag{11.5}$$

Note that the translations do not commute with the Lorentz transformations, since the translation vectors a^μ can get Lorentz rotated also.

The defining property of a Lorentz transformation (11.2) implies that

$$\eta_{\lambda\sigma} x'^\lambda x'^\sigma = \eta_{\lambda\sigma} \Lambda^\lambda_\mu \Lambda^\sigma_\nu x^\mu x^\nu = \eta_{\mu\nu} x^\mu x^\nu .$$

Thus

$$\eta_{\mu\nu} = \Lambda^\lambda_\mu \eta_{\lambda\sigma} \Lambda^\sigma_\nu = \sum_\sigma \Lambda^\lambda_\mu \eta_{\lambda\sigma} (\Lambda^T)^\nu_\sigma . \tag{11.6}$$

In matrix notation

$$\eta = \Lambda \eta \Lambda^T , \tag{11.7}$$

which in turn implies that

$$\det(\eta) = \det(\Lambda) \det(\eta) \det(\Lambda^T) = \det(\eta)(\det(\Lambda))^2 .$$

(For a discussion of the properties of determinants, see Chapter 16 below.) Thus we arrive at an important property of Lorentz transformations:

$$\det(\Lambda) = \pm 1 , \quad \Lambda \in L(4) . \tag{11.8}$$

(Compare the situation for orthogonal matrices discussed in Chapter 1.)

All $\Lambda \in L(4)$ for which $\det(\Lambda) = +1(-1)$ are called **proper (improper) Lorentz transformations**. These are analogous to the pure rotations $[\det(a) = +1]$ and rotation-inversion combinations $[\det(a) = -1]$ in $O(3)$. It is interesting to note that η itself is an improper Lorentz transformation. Physically it corresponds to a time inversion.

In any matrix representation on M, the element Λ^0_0 of a Lorentz transformation plays a special role. It follows from (11.6) and (8.3) that

$$\eta_{00} = -1 = \Lambda^\lambda_0 \eta_{\lambda\sigma} \Lambda^\sigma_0 = -(\Lambda^0_0)^2 + \sum_i (\Lambda^i_0)^2 . \tag{11.9}$$

Since all $\Lambda^\nu_\mu \in \mathbb{R}$ (real numbers),

$$(\Lambda^0_0)^2 \geq 1 , \tag{11.10}$$

that is, $\Lambda^0_0 \geq 1$ or ≤ -1.

Based on the properties (11.8) and (11.10), the Lorentz group $L(4)$ is seen to be the union of four disconnected components:

1) L_0 $[SO(1,3)^\uparrow$ or $L^\uparrow_+]$ (proper orthochronous), $\det(\Lambda) = +1$, $\Lambda^0_0 \geq 1$,

2) sL_0 $(L^\uparrow_-)$ (improper orthochronous), $\det(\Lambda) = -1$, $\Lambda^0_0 \geq 1$,

3) tL_0 $(L^\downarrow_-)$ (improper non-orthochronous), $\det(\Lambda) = -1$, $\Lambda^0_0 \leq -1$,

4) stL_0 $(L^\downarrow_+)$ (proper non-orthochronous), $\det(\Lambda) = 1$, $\Lambda^0_0 \leq -1$,

where

$$s = \begin{pmatrix} 1 & 0 & 0 & 0 \\ 0 & -1 & 0 & 0 \\ 0 & 0 & -1 & 0 \\ 0 & 0 & 0 & -1 \end{pmatrix} , \tag{11.11}$$

and

$$t = \eta = \begin{pmatrix} -1 & 0 & 0 & 0 \\ 0 & 1 & 0 & 0 \\ 0 & 0 & 1 & 0 \\ 0 & 0 & 0 & 1 \end{pmatrix} \tag{11.12}$$

are the spatial and time inversions, respectively.

Any $\Lambda \in L(4)$ can be decomposed into a product of transformations of these four types. Since they are all derived from L_0, the group parameters of $L(4)$ [local coordinates of the group manifold (see Chapter 30)] are those of L_0. We will see below that $L(4)$ as a *differentiable manifold is neither connected nor compact*. This topological fact has profound implications on the representation theory of the Lorentz group, a detailed study of which is beyond the scope of the present text.

$\boxed{\text{Exercise 11.3}}$ Show that $SO(1,3)^\uparrow$ is a normal subgroup of $O(1,3)$ and

$$\frac{O(1,3)}{SO(1,3)^\uparrow} = \{SO(1,3)^\uparrow, L_-^\uparrow, L_-^\downarrow, L_+^\downarrow\} \,.$$

It turns out that the properties of the Lorentz group $L(4)$ can be greatly illuminated by its relationship to $SL(2,\mathbb{C})$, the special (unimodular, determinant $= +1$) linear group of 2×2 matrices with complex entries. We will now embark on a study of this relationship.

To begin, we consider the set H_2 of 2×2 hermitian matrices, a general member of which can be written [c.f. (10.25)]

$$h = \begin{pmatrix} a & \bar{c} \\ c & b \end{pmatrix} \,, \quad a,b \in \mathbb{R}, \ c \in \mathbb{C} \,. \tag{11.13}$$

There is an isomorphism between H_2 and Minkowski space M, $f : M \to H_2$, given by

$$f(x) = \begin{pmatrix} x^0 + x^3 & x^1 - ix^2 \\ x^1 + ix^2 & x^0 - x^3 \end{pmatrix} \in H_2 \,, \tag{11.14}$$

where $x = (x^0, x^1, x^2, x^3) \in M$.

$\boxed{\text{Exercise 11.4}}$ Show that $f(x)$ given above is indeed a bijection with the inverse $f^{-1} : H_2 \to M$ given by

$$x^0 = \frac{1}{2}Tr(h) = \frac{1}{2}(a+b) \,, \ x^1 = \frac{1}{2}(c+\bar{c}) \,, \ x^2 = \frac{1}{2i}(c-\bar{c}) \,, \ x^3 = \frac{1}{2}(a-b) \,. \tag{11.15}$$

Under f the unit vectors in M map to the unit (2×2) matrix e and the **Pauli spin matrices** σ_i:

$$e_0 = (1,0,0,0) \longrightarrow \sigma_0 = \begin{pmatrix} 1 & 0 \\ 0 & 1 \end{pmatrix} = e \,,$$

$$e_1 = (0,1,0,0) \longrightarrow \sigma_1 = \begin{pmatrix} 0 & 1 \\ 1 & 0 \end{pmatrix} \,,$$

$$e_2 = (0,0,1,0) \longrightarrow \sigma_2 = \begin{pmatrix} 0 & -i \\ i & 0 \end{pmatrix} \,, \tag{11.16}$$

$$e_3 = (0,0,0,1) \longrightarrow \sigma_3 = \begin{pmatrix} 1 & 0 \\ 0 & -1 \end{pmatrix} \,.$$

Thus H_2 has a real vector space structure (four-dimensional), where $(\sigma_0, \sigma_1, \sigma_2, \sigma_3)$ is a basis and any $h \in H_2$ can be written

$$h = x^\mu \sigma_\mu \,, \quad x^\mu \in \mathbb{R}\,, \quad \mu = 0,1,2,3\,. \tag{11.17}$$

Also,

$$\det(f(x)) = (x^0)^2 - (x^1)^2 - (x^2)^2 - (x^3)^2 = -\|x\|^2 \,. \tag{11.18}$$

$\boxed{\text{Exercise 11.5}}$ Verify the following important properties of the Pauli spin matrices

$$\sigma_i = \sigma_i^\dagger \,,$$
$$\operatorname{tr} \sigma_i = 0 \,,$$
$$\sigma_i \sigma_j + \sigma_j \sigma_i = 2\delta_{ij} \,,$$
$$[\sigma_j \,, \sigma_k] = 2i\varepsilon_{jk}^{\,l} \sigma_l \,,$$

and thus

$$\sigma_j \sigma_k = \delta_{jk} + i\varepsilon_{jk}^{\,l} \sigma_l \,,$$
$$(a \cdot \sigma)(b \cdot \sigma) = a \cdot b + i\,(a \times b) \cdot \sigma \,,$$

and

$$e^{i\theta\sigma_j} = \cos\theta + i\,\sin\theta\,\sigma_j \,.$$

Now define a group homomorphism as follows (see Fig. 11.1 for clarification). For any $g \in SL(2,\mathbb{C})$ and $x \in M$,

$$\phi(g)(x) \equiv f^{-1}(gf(x)g^\dagger) \,. \tag{11.19}$$

$$H_2 \xleftarrow{\;\;f\;\;} M$$

$$SL(2,\,\mathbb{C}) \ni \Big\downarrow g \;\xrightarrow{\;\phi\;}\; \phi(g) \Big\downarrow \in L(4)$$

$$H_2 \xrightarrow[\;f^{-1}\;]{} M$$

FIGURE 11.1

To check that this definition makes sense we have to show that i) for $h \in H_2$ and $g \in SL(2,\mathbb{C})$, $ghg^\dagger \in H_2$, ii) $\phi(g)$ is indeed in $L(4)$, and iii) ϕ is indeed a group homomorphism (preserves group multiplication). The validity of i) is obvious [from (10.6) and (10.8)]. From (11.18) and (11.19), we have

$$\|\phi(g)(x)\|^2 = -\det(f(\phi(g)(x))) = -\det(gf(x)g^\dagger) = -\det(g)\det(f(x))\det(g^\dagger)$$
$$= -\det(g)\,\overline{\det(g)}\,\det(f(x)) = -\det(f(x)) = \|x\|^2 \;,$$
$$(11.20)$$

where we have used some standard properties of determinants (c.f. Chapter 16), and the fact that, for $g \in SL(2,\mathbb{C})$, $\det(g)$ is by definition equal to 1. Thus $\phi(g) \in L(4)$. To establish that ϕ is a homomorphism we have to show that

$$\phi(g_1 g_2) = \phi(g_1)\phi(g_2)\;, \quad \text{for all } g_1,\, g_2 \in SL(2,\mathbb{C})\;. \tag{11.21}$$

Indeed, by (11.19), for all $x \in M$,

$$\phi(g_1 g_2)(x) = f^{-1}(g_1 g_2 f(x) g_2^\dagger g_1^\dagger) = f^{-1}(g_1 f(\phi(g_2)(x)) g_1^\dagger)$$
$$= f^{-1} f(\phi(g_1)\phi(g_2)(x)) = \phi(g_1)\phi(g_2)(x)\;. \tag{11.22}$$

The homomorphism $\phi : SL(2,\mathbb{C}) \longrightarrow L(4)$ has the crucial property that it is two-to-one (see Fig. 11.2):

$$\phi(g) = \phi(-g)\;. \tag{11.23}$$

Indeed, for all $x \in M$,

$$\phi(g)(x) = f^{-1}(gf(x)g^\dagger) = f^{-1}\left((-g)f(x)(-g)^\dagger\right) = \phi(-g)(x)\;, \tag{11.24}$$

Now consider the subgroup $SU(2)$ of $SL(2,\mathbb{C})$. By definition, $g \in SU(2)$ means that g is a complex 2×2 matrix such that $gg^\dagger = g^\dagger g = e$ and $\det(g) = 1$. It is easy to see that, if $g \in SU(2)$, then $\phi(g)(e_0) = e_0$, where $e_0 = (1,0,0,0)$ is the unit vector along the time axis in Minkowski space. Indeed, by (11.19),

$$\phi(g)(e_0) = f^{-1}\left(gf(e_0)g^\dagger\right) = f^{-1}\left(geg^\dagger\right) = f^{-1}\left(gg^{-1}\right) = f^{-1}(e) = e_0\;. \tag{11.25}$$

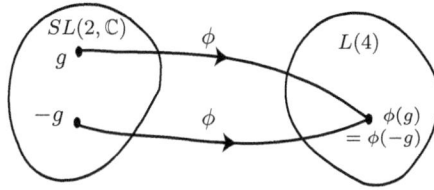

FIGURE 11.2

Thus the Lorentz transformations in $Im\left(\phi\mid_{SU(2)}\right) \subset L(4)$ leave the time axis (subspace in M generated by e_0) invariant. The notation $\phi\mid_{SU(2)}$ means ϕ restricted to $SU(2)$. Let $e_0^\perp$ be the orthogonal subspace to the time axis:

$$e_0^\perp = \{x \in M \mid \eta_{\mu\nu}\, x^\mu (e_0)^\nu = 0\} . \tag{11.26}$$

$e_0^\perp$ consists of vectors of the form $(0, x^1, x^2, x^3)$ (Fig. 11.3). We will show that $Im\left(\phi\mid_{SU(2)}\right)$ also leaves $e_0^\perp$ invariant. Indeed, for $g \in SU(2)$, and $x \in e_0^\perp$,

$$f(x) = x^i \sigma_i , \quad i = 1, 2, 3 , \tag{11.27}$$

as follows from (11.17); so

$$\phi(g)(x) = f^{-1}\left(gx^i\sigma_i g^\dagger\right) = x^i f^{-1}\left(g\sigma_i g^\dagger\right) . \tag{11.28}$$

But

$$Tr\left(g\sigma_i g^\dagger\right) = Tr\left(gg^\dagger \sigma_i\right) = Tr(\sigma_i) = 0 , \quad i = 1, 2, 3 , \tag{11.29}$$

where the last equality follows from the explicit forms for σ_i given in (11.16). The first of equations (11.15) then shows that the first (time) component of the vector $f^{-1}\left(g\sigma_i g^\dagger\right)$ vanishes. Thus $\phi(g)(x) \in e_0^\perp$.

The above development shows that both ae_0 ($a \in \mathbb{R}$) and $e_0^\perp$ are invariant subspaces of the Lorentz transformations in $Im\left(\phi\mid_{SU(2)}\right)$. In fact these transformations act as the identity operator in ae_0 (the time axis) and preserve Euclidean lengths in $e_0^\perp = \mathbb{R}^3$ (with the Euclidean metric $\eta_{ij} = \delta_{ij}$). Thus we have established the important restricted homomorphism of ϕ:

$$\phi\mid_{SU(2)} : SU(2) \longrightarrow O(3) \subset L(4) . \tag{11.30}$$

It is clear that this restricted homomorphism is still two-to-one, since $g \in SU(2)$ implies $-g \in SU(2)$ also (Fig. 11.4).

Let us now consider specific one-parameter subgroups of $SU(2)$. First look at $g \in SU(2)$ given by

$$g(\theta) = \begin{pmatrix} e^{-i\theta} & 0 \\ 0 & e^{i\theta} \end{pmatrix} , \tag{11.31}$$

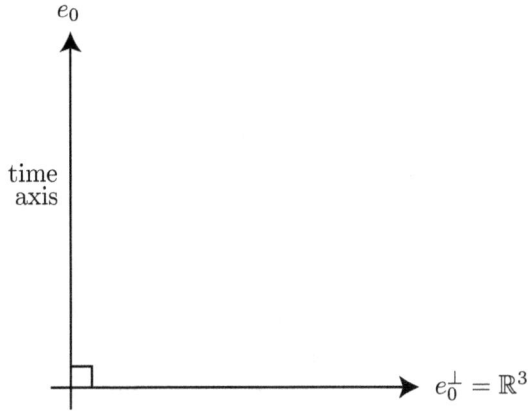

FIGURE 11.3

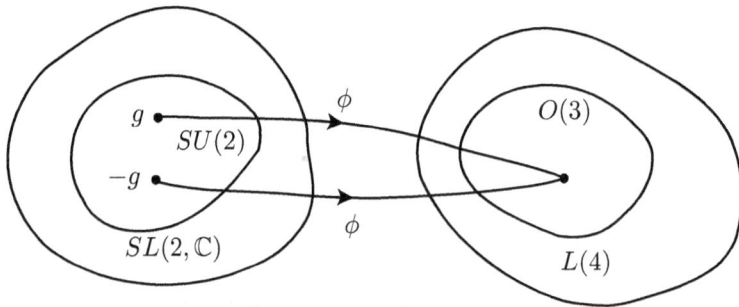

FIGURE 11.4

and compute explicitly the effect of $\phi(g(\theta))$ on $x \in M$. Recalling (11.14) and (11.19) we have

$$
g(\theta)f(x)(g(\theta))^\dagger = \begin{pmatrix} e^{-i\theta} & 0 \\ 0 & e^{i\theta} \end{pmatrix} \begin{pmatrix} x^0 + x^3 & x^1 - ix^2 \\ x^1 + ix^2 & x^0 - x^3 \end{pmatrix} \begin{pmatrix} e^{i\theta} & 0 \\ 0 & e^{-i\theta} \end{pmatrix}
$$

$$
= \begin{pmatrix} x^0 + x^3 & e^{-2i\theta}(x^1 - ix^2) \\ e^{2i\theta}(x^1 + ix^2) & x^0 - x^3 \end{pmatrix}. \tag{11.32}
$$

This shows that $\phi(g(\theta))$ leaves x^0 and x^3 unchanged while a vector in the x^1x^2-plane is rotated in the positive sense about the x^3-axis by an angle 2θ. Thus, in the notation of Chapter 6 [c.f. (6.25)],

$$\phi(g(\theta)) = R_3(2\theta) \, . \tag{11.33}$$

We also note that

$$-g(\theta) = \begin{pmatrix} -e^{-i\theta} & 0 \\ 0 & -e^{i\theta} \end{pmatrix} = \begin{pmatrix} e^{-i(\theta+\pi)} & 0 \\ 0 & e^{i(\theta+\pi)} \end{pmatrix} = g(\theta + \pi) \, . \tag{11.34}$$

Thus

$$\phi(-g(\theta)) = R_3(2\theta + 2\pi) = R_3(2\theta) = \phi(g(\theta)) \, . \tag{11.35}$$

This verifies the two-to-one nature of $\phi \mid_{SU(2)}$ restricted further to the subgroup to which $g(\theta)$ belongs.

Next consider another one-parameter subgroup of $SU(2)$ whose elements have the form

$$g'(\alpha) = \begin{pmatrix} \cos\alpha & -\sin\alpha \\ \sin\alpha & \cos\alpha \end{pmatrix} \, . \tag{11.36}$$

[This is the first matrix we encountered in this book (c.f. (1.6)).] Note that

$$(g'(\alpha))^\dagger = g'(-\alpha) \, . \tag{11.37}$$

The effect of $\phi(g'(\alpha))$ on an arbitrary $x \in M$ is given by

$$g'(\alpha)f(x)\,(g'(\alpha))^\dagger =$$

$$\begin{pmatrix} \cos\alpha & -\sin\alpha \\ \sin\alpha & \cos\alpha \end{pmatrix} \begin{pmatrix} x^0 + x^3 & x^1 - ix^2 \\ x^1 + ix^2 & x^0 - x^3 \end{pmatrix} \begin{pmatrix} \cos\alpha & \sin\alpha \\ -\sin\alpha & \cos\alpha \end{pmatrix} \, . \tag{11.38}$$

Let us try this on $x = e_2 = (0,0,1,0)$. Recalling that [c.f. (11.16)] $f(e_2) = \sigma_2$, Eq. (11.38) yields

$$g'(\alpha)f(e_2)\,(g'(\alpha))^\dagger = \sigma_2 \, . \tag{11.39}$$

Exercise 11.6   Verify the above equation.

Thus

$$\phi(g'(\alpha))(e_2) = f^{-1}(\sigma_2) = e_2 \, , \tag{11.40}$$

implying that $\phi(g'(\alpha))$ must be a rotation about the x^2-axis. To see what the angle of rotation is, we apply (11.38) to $x = e_3 = (0, 0, 0, 1)$. Using $f(e_3) = \sigma_3$, it is straightforward to show that

$$g'(\alpha) f(e_3) (g'(\alpha))^\dagger = g'(\alpha)\sigma_3 (g'(\alpha))^\dagger = \begin{pmatrix} \cos 2\alpha & \sin 2\alpha \\ \sin 2\alpha & -\cos 2\alpha \end{pmatrix} \tag{11.41}$$

$$= f((0, \sin 2\alpha, 0, \cos 2\alpha)) .$$

Exercise 11.7   Verify the above equation.

We conclude from (11.40) and (11.41) that $\phi(g'(\alpha))$ is a rotation about the x^2-axis by an angle 2α:

$$\phi(g'(\alpha)) = R_2(2\alpha) . \tag{11.42}$$

We turn now to consider $g \in SL(2, \mathbb{C})$, but not in $SU(2)$. Specifically we consider

$$g''(\gamma) = \begin{pmatrix} \gamma & 0 \\ 0 & \dfrac{1}{\gamma} \end{pmatrix} = (g''(\gamma))^\dagger , \quad \gamma \in \mathbb{R} . \tag{11.43}$$

To determine its action on $x \in M$, we proceed as before to calculate

$$g''(\gamma) f(x)(g''(\gamma))^\dagger = \begin{pmatrix} \gamma & 0 \\ 0 & \dfrac{1}{\gamma} \end{pmatrix} \begin{pmatrix} x^0 + x^3 & x^1 - ix^2 \\ x^1 + ix^2 & x^0 - x^3 \end{pmatrix} \begin{pmatrix} \gamma & 0 \\ 0 & \dfrac{1}{\gamma} \end{pmatrix}$$

$$= \begin{pmatrix} \gamma^2(x^0 + x^3) & x^1 - ix^2 \\ x^1 + ix^2 & \dfrac{1}{\gamma^2}(x^0 - x^3) \end{pmatrix} . \tag{11.44}$$

Thus $\phi(g''(\gamma))$ leaves x^1 and x^2 unchanged whereas x^0 and x^3 are transformed according to

$$x'^0 + x'^3 = \gamma^2(x^0 + x^3) , \quad x'^0 - x'^3 = \frac{1}{\gamma^2}(x^0 - x^3) . \tag{11.45}$$

Solving these, we have

$$x'^0 = \frac{1}{2}(\gamma^2 + \frac{1}{\gamma^2})x^0 + \frac{1}{2}(\gamma^2 - \frac{1}{\gamma^2})x^3 ,$$

$$x'^3 = \frac{1}{2}(\gamma^2 - \frac{1}{\gamma^2})x^0 + \frac{1}{2}(\gamma^2 + \frac{1}{\gamma^2})x^3 , \tag{11.46}$$

$$x'^1 = x^1 , \quad x'^2 = x^2 .$$

On letting $|\gamma| = e^{\eta}$, the above equations can be written as

$$
\begin{aligned}
x'^0 &= (\cosh 2\eta)x^0 + (\sinh 2\eta)x^3 \;, \\
x'^3 &= (\sinh 2\eta)x^0 + (\cosh 2\eta)x^3 \;, \\
x'^1 &= x^1 \;, \quad x'^2 = x^2 \;.
\end{aligned}
\tag{11.47}
$$

Thus the Lorentz transformation $\phi(g''(\gamma))$ is given by

$$
\phi(g''(\gamma)) = \Lambda^{\nu}_{\mu}(\eta) =
\begin{pmatrix}
\cosh 2\eta & 0 & 0 & \sinh 2\eta \\
0 & 1 & 0 & 0 \\
0 & 0 & 1 & 0 \\
\sinh 2\eta & 0 & 0 & \cosh 2\eta
\end{pmatrix}
,
\tag{11.48}
$$

which is recognized to be a **Lorentz boost** $B_3(v)$ along the x^3-direction with a velocity v determined by

$$
\cosh 2\eta = \frac{1}{\sqrt{1 - \dfrac{v^2}{c^2}}} \;, \qquad
\sinh 2\eta = \frac{-\dfrac{v}{c}}{\sqrt{1 - \dfrac{v^2}{c^2}}} \;,
\tag{11.49}
$$

where c is the speed of light. We thus have

$$
\phi(g''(\gamma)) = B_3(v) \;.
\tag{11.50}
$$

Note that since $\eta \in \mathbb{R}$ lies in the range $(-\infty, \infty)$, the Lorentz group is not compact.

We have seen that the homomorphism $\phi : SL(2,\mathbb{C}) \to L(4)$ is not injective (being two-to-one). It will now be shown that it is not surjective (onto) either. In fact, it is onto only one of the four connected components of $L(4)$, the proper orthochronous Lorentz group, that is,

$$
Im(\phi) = L_0 = L^{\uparrow}_{+} \;.
\tag{11.51}
$$

This will be established by the following sequence of steps (proved in due course).

1) Observe that any element $g \in SL(2,\mathbb{C})$ can be connected to the identity by a continuous curve in $SL(2,\mathbb{C})$.

2) Only elements in L_0 can be connected to the identity in $L(4)$ by a continuous curve, which necessarily lies entirely in L_0.

3) Since $\phi(e) = $ the identity in $L(4)$, the range of ϕ, $Im(\phi)$, must be in a subset of $L(4)$ all of whose elements are connected to the identity [in $L(4)$] by continuous curves, that is, $Im(\phi) \subset L_0$.

4) Make use of the following lemma.

Lemma 11.1. *Any element in the proper orthochronous Lorentz group L_0 can be expressed as*

$$\Lambda = R B_3 R' , \tag{11.52}$$

where R and R' are pure rotations in $\mathbb{R}^3$ and B_3 is a Lorentz boost along the x^3-direction.

5) Recall Euler's theorem (Theorem 6.3) that every rotation R in $\mathbb{R}^3$ can be written as a product

$$R = R_3(\phi) R_2(\theta) R_3(\psi) . \tag{11.53}$$

6) All of R_3, R_2 and B_3 ($\in L_0$) have preimages in $SL(2, \mathbb{C})$ under ϕ (shown earlier in this chapter):

$$\phi^{-1}(B_3(v)) = \pm \begin{pmatrix} \gamma & 0 \\ 0 & \dfrac{1}{\gamma} \end{pmatrix} , \tag{11.54}$$

where v and γ are related by $|\gamma| = e^\eta$ and (11.49),

$$\phi^{-1}(R_2(\theta)) = \pm \begin{pmatrix} \cos\dfrac{\theta}{2} & -\sin\dfrac{\theta}{2} \\ \sin\dfrac{\theta}{2} & \cos\dfrac{\theta}{2} \end{pmatrix} = \pm \exp\left(-i\dfrac{\theta}{2}\sigma_2\right) , \tag{11.55}$$

$$\phi^{-1}(R_3(\psi)) = \pm \begin{pmatrix} \exp\left(-i\dfrac{\psi}{2}\right) & 0 \\ 0 & \exp\left(i\dfrac{\psi}{2}\right) \end{pmatrix} = \pm \exp\left(-i\dfrac{\psi}{2}\sigma_3\right) . \tag{11.56}$$

Remember that the ($\pm$) results from the two-to-one nature of ϕ. Thus $L_0 \subset Im(\phi)$. Together with the result of step 3), we conclude that ϕ indeed maps $SL(2, \mathbb{C})$ onto L_0.

Exercise 11.8   Verify the second equalities in (11.55) and (11.56) by using the results of Exercise 11.5.

Exercise 11.9   Show that

$$\phi^{-1}(R_n(\theta)) = \pm \exp\left(-i\dfrac{\theta}{2}\boldsymbol{n}\cdot\boldsymbol{\sigma}\right) , \tag{11.57}$$

where n is a unit vector in $\mathbb{R}^3$. [Compare with (4.21).]

The above development culminates in the following interesting scheme relating the various groups considered in this chapter.

$$\phi\,|_{SU(2)} : SU(2) \xrightarrow{\text{2-to-1 onto}} SO(3) \quad \subset O(3)$$
$$\cap \qquad\qquad\qquad \cap \qquad\qquad \cap$$
$$\phi : SL(2,\mathbb{C}) \xrightarrow{\text{2-to-1 onto}} SO(1,3)^{\uparrow} \subset O(1,3)$$

Let us now prove the various assertions in the above steps.

Proof of 1). Any $A \in SL(2,\mathbb{C})$ is similar (conjugate) to an upper diagonal matrix in $SL(2,\mathbb{C})$, that is, for some invertible 2×2 matrix B,

$$A = B \begin{pmatrix} a & b \\ 0 & \frac{1}{a} \end{pmatrix} B^{-1}, \quad a, b \in \mathbb{C}. \tag{11.58}$$

This is a special case of a standard result in linear algebra (which will not be proved here). There exists, of course, continuous curves on the complex plane $a(t)$ and $b(t)$, $0 \le t \le 1$, such that $a(0) = 1$, $a(1) = a$ and $b(0) = 0$, $b(1) = b$. (Any complex number can be joined to either 0 or 1 by continuous paths on the complex plane.) Consider the family of matrices in $SL(2,\mathbb{C})$:

$$A(t) = B \begin{pmatrix} a(t) & b(t) \\ 0 & \frac{1}{a(t)} \end{pmatrix} B^{-1}, \quad 0 \le t \le 1. \tag{11.59}$$

They satisfy $A(0) = e$ and $A(1) = A$; and thus describe a continuous curve in $SL(2,\mathbb{C})$ connecting the identity and A. $\qquad\square$

Proof of 2). First note that the identity is in L_0, and since L_0 is a connected component of $L(4)$, all elements in L_0 are connected to the identity. Next we show that all $\Lambda \in L(4)$ with $\det(\Lambda) = -1$ cannot be connected to the identity. This includes all elements in the cosets $sL_0 = L_-^{\uparrow}$ (improper orthochronous) and $tL_0 = L_-^{\downarrow}$ (improper non-orthochronous). Indeed, suppose the contrary, that there does exist a continuous curve $\Lambda(t)$ in $L(4)$ connecting the identity to an element Λ' with $\det(\Lambda') = -1$, such that $\Lambda(0) = e$ and $\Lambda(1) = \Lambda'$. Since $\det(\Lambda(t))$ is a continuous function of t, there must exist a $t', 0 < t' < 1$, such that $\det(\Lambda(t')) = 0$. But this is impossible due to (11.8).

This leaves elements in $stL_0 = L_+^{\downarrow}$, with determinant $+1$ but with $\Lambda^0_0 \le -1$. Recall the distinction between **space-like** ($\eta_{\mu\nu}x^\mu x^\nu > 0$), **light-like** ($\eta_{\mu\nu}x^\mu x^\nu = 0$), and **time-like** ($\eta_{\mu\nu}x^\mu x^\nu < 0$) vectors in Minkowski space M [c.f. discussion following (8.4)]. (See Fig. 11.5.)

We will use the following lemma (proved below).

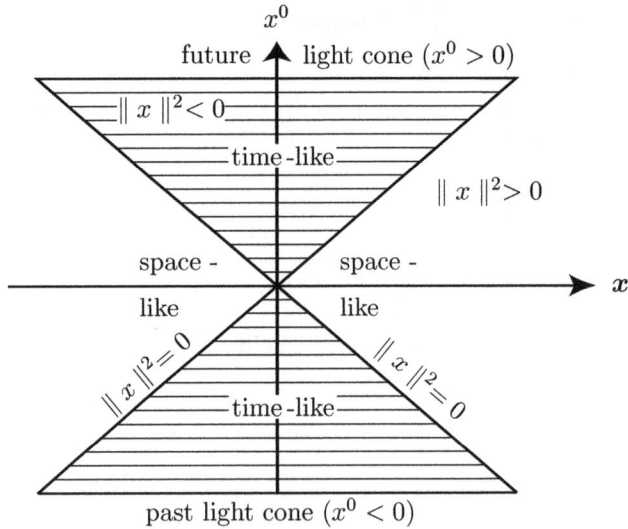

FIGURE 11.5

Lemma 11.2. *Any* $\Lambda \in L(4)$ *that can be connected to the identity by a contin-uous curve preserves each half-cone of time-like vectors, that is, such a Λ maps future light cone vectors to future light cone vectors and past light cone vectors to past light cone vectors.*

An element in the coset stL_0 can be written as $ts\Lambda$, with $\Lambda \in L_0$. Since Λ is connected to the identity, Lemma 11.2 implies that it, as well as $s\Lambda$, maps a vector x in the future light cone, say, to another vector in the same half-cone. Thus $ts\Lambda$ maps x to a vector in the past light cone, since t is the time-inversion. By Lemma 11.2 again, elements in tsL_0 cannot be connected to the identity. $\square$

Proof of 3). This is obvious (see Fig. 11.6). $\square$

We will now provide the proofs for Lemmas 11.1 and 11.2.

Proof of Lemma 11.1. Suppose $\Lambda \in L_0$ and

$$\Lambda e_0 = (x^0, \boldsymbol{x}) \,, \tag{11.60}$$

where $e_0 = (1, 0, 0, 0)$ and $\boldsymbol{x} = (x^1, x^2, x^3) \in \mathbb{R}^3$.
Now $\|e_0\|^2 = -1$. Hence

$$|\boldsymbol{x}|^2 - (x^0)^2 = -1 \,, \tag{11.61}$$

since all Lorentz transformations preserve norms. In the above equation

$$|\boldsymbol{x}|^2 = (x^1)^2 + (x^2)^2 + (x^3)^2 \tag{11.62}$$

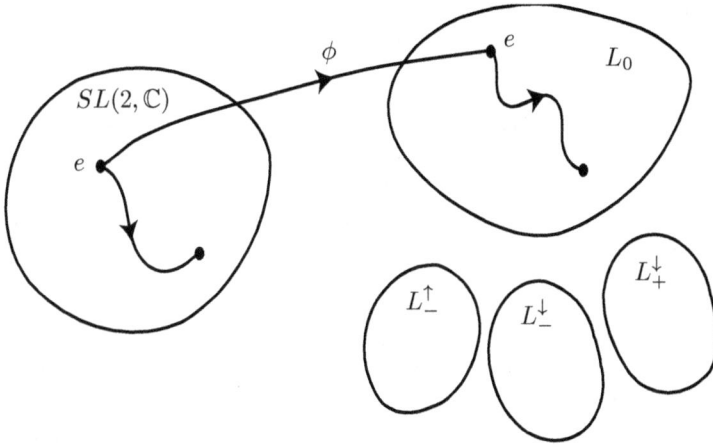

FIGURE 11.6

is the norm of $x \in \mathbb{R}^3$ (with the Euclidean metric). Since the action of $SO(3)$ on S^2 (the 2-sphere) is transitive [c.f. Def. 6.7], there must be a rotation R which rotates x to the direction of the x^3-axis:

$$R\Lambda e_0 = (x^0, 0, 0, |x|) .\tag{11.63}$$

By (11.14), then,

$$f(R\Lambda e_0) = \begin{pmatrix} x^0 + |x| & 0 \\ 0 & x^0 - |x| \end{pmatrix} .\tag{11.64}$$

To this we will apply $g''(\gamma)$ [(11.43)] with

$$\gamma^2 = \frac{1}{x^0 + |x|} = x^0 - |x| ,\tag{11.65}$$

[where the second equality follows from (11.61)]. The quantity $x^0 + |x|$ is guaranteed to be positive since Λe_0 and hence $R\Lambda e_0$ are both in the future light cone, by virtue of the fact that $\Lambda \in L_0$ and thus preserves the future light cone. Eqs. (11.43) and (11.64) then give

$$g''(\gamma) f(R\Lambda e_0)(g''(\gamma))^\dagger = \begin{pmatrix} 1 & 0 \\ 0 & 1 \end{pmatrix} = f(e_0) .\tag{11.66}$$

Thus

$$\phi(g''(\gamma)) R\Lambda(e_0) = e_0 ,\tag{11.67}$$

or, by (11.50),

$$B'_3 R \Lambda(e_0) = e_0 , \tag{11.68}$$

where B'_3 is a Lorentz boost along the x^3-axis. Since $B'_3, R \in L_0$ and, by assumption, $\Lambda \in L_0$, $\det(B'_3 R \Lambda) = +1$. This last fact, together with (11.68), implies that $B'_3 R \Lambda$ is a rotation in $e_0^{\perp} = \mathbb{R}^3$. Denote this rotation by R'. It follows that $B'_3 R \Lambda = R'$, or $\Lambda = R^{-1}(B'_3)^{-1} R'$. Since the inverse of a rotation in $\mathbb{R}^3$ is a rotation in $\mathbb{R}^3$ and the inverse of a Lorentz boost is a Lorentz boost, (11.52) and the Lemma follows. $\square$

Proof of Lemma 11.2. Suppose there existed a $\Lambda \in L(4)$ which maps x in the future light-cone to some x' in the past light-cone, and which is connected to the identity by a continuous curve $\Lambda(t)$, $0 \leq t \leq 1$, such that $\Lambda(0) = e$ and $\Lambda(1) = \Lambda$. By continuity there must exist a t', $0 < t' < 1$, such that $\Lambda(t')x = x''$, where x'' is in the space-like region (Fig. 11.5). But this is impossible, since the defining property of a Lorentz transformation is that it is norm-preserving, and no $\Lambda \in L(4)$ can map a time-like vector to a space-like vector (and vice versa). $\square$

We will now consider in more detail the relationship under ϕ between $SU(2)$ and $SO(3)$ in order to shed light on their topological properties as manifolds. Every $g \in SU(2)$ can be written as

$$g = \begin{pmatrix} a & b \\ -\bar{b} & \bar{a} \end{pmatrix} , \qquad a, b \in \mathbb{C} , \tag{11.69}$$

where the defining property $\det(g) = 1$ requires that $|a|^2 + |b|^2 = 1$. The complex quantities a and b are called the **Cayley-Klein parameters**. Let $a = x^1 - ix^2, b = x^3 - ix^4, x^1, x^2, x^3, x^4 \in \mathbb{R}$. The above condition implies

$$(x^1)^2 + (x^2)^2 + (x^3)^2 + (x^4)^2 = 1 . \tag{11.70}$$

This is the equation of the unit 3-sphere embedded in $\mathbb{R}^4$ (with the Euclidean metric). Hence $SU(2)$ is homeomorphic to S^3:

$$\boxed{SU(2) \sim S^3} . \tag{11.71}$$

In fact, under the map

$$g = \begin{pmatrix} a & b \\ -\bar{b} & \bar{a} \end{pmatrix} \overset{h}{\mapsto} (x^1, x^2, x^3, x^4) , \tag{11.72}$$

g and $-g$ correspond to antipodal points on S^3. Now there is a theorem in topology that S^n for $n \geq 2$ is **simply connected**. [The fundamental group

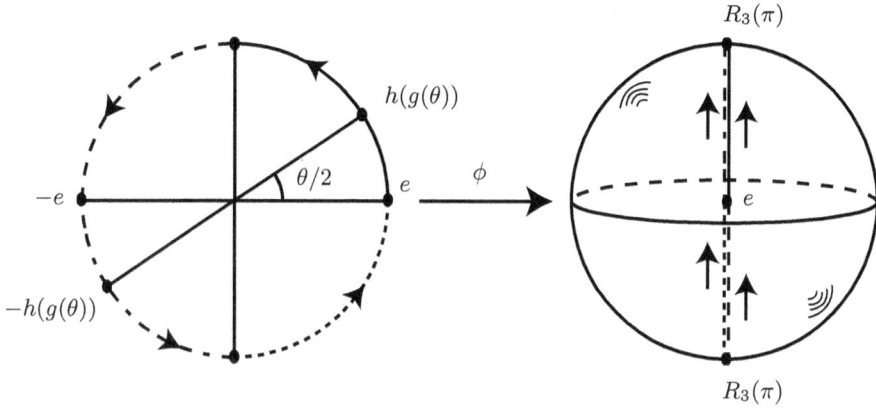

FIGURE 11.7

$\Pi_1(S^n) = e$ for $n \geq 2$ (for a discussion of the fundamental group and the related topological notion of homotopy see Chapter 38).] Hence $SU(2)$ *as a topological manifold is simply connected.* We will show that $SO(3)$ does not share this property.

Consider the closed curve $R_3(\theta), 0 \leq \theta \leq 2\pi$, in $SO(3)$. The preimage of this curve under $\phi \mid_{SU(2)}, g(\theta)$, such that $g(0) = e$, is given by [c.f. (11.56)]

$$g(\theta) = + \begin{pmatrix} \exp\left(-i\dfrac{\theta}{2}\right) & 0 \\ 0 & \exp\left(i\dfrac{\theta}{2}\right) \end{pmatrix}, \tag{11.73}$$

where the overall + sign is required by the above condition on $g(0)$. Under the map given by (11.72),

$$g(\theta) \overset{h}{\mapsto} \left(\cos\frac{\theta}{2}, \sin\frac{\theta}{2}, 0, 0\right). \tag{11.74}$$

As θ varies from 0 to 2π, $h(g(\theta))$ traces out the semicircle $x^3 = x^4 = 0, x^1 = \cos(\theta/2), x^2 = \sin(\theta/2)$ on the projection of the unit 3-sphere S^3 on the $x^1 x^2$-plane of $\mathbb{R}^4$, with the end-point corresponding to $g(2\pi) = -e$ (see Fig. 11.7, which also shows the image under ϕ of $g(\theta)$ in $SO(3)$). This obviously does not correspond to a closed curve in $SU(2)$. Hence, under $\phi \mid_{SU(2)}$, the corresponding closed curve $R_3(\theta)$ in $SO(3), 0 \leq \theta \leq 2\pi$, (Fig. 11.8) cannot be shrunk to a point in $SO(3)$.

The preimage of the double loop in $SO(3)$ under $\phi \mid_{SU(2)}$ (shown in Fig. 11.9), however, is a closed path in $SU(2)$ corresponding to a full circle on the

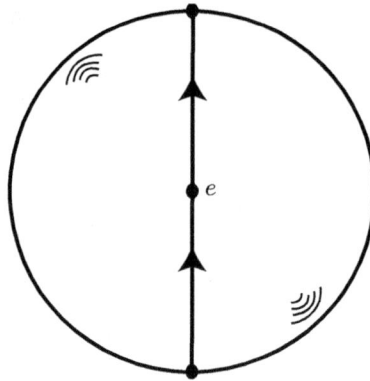

FIGURE 11.8

This closed curve in
$SO(3)$ *cannot* be
shrunk to a point

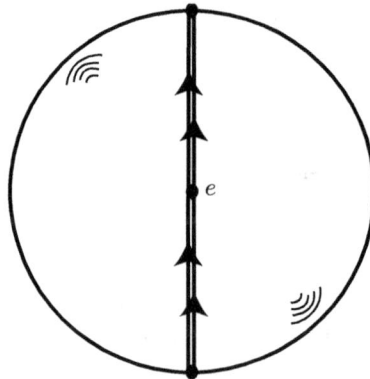

FIGURE 11.9

This closed curve in
$SO(3)$ *can* be
shrunk to a point

$x^1 x^2$-plane (Fig. 11.7) in $\mathbb{R}^4$. Since S^3 is simply connected, this closed path in $SU(2)$ can be shrunk to a point (in $SU(2)$). Thus the double loop of Fig. 11.9 in $SO(3)$ can be shrunk to a point also. This can be seen pictorially in the sequence of deformations of the double loop in $SO(3)$ shown in Fig. 11.10.

$SO(3)$ as a manifold is said to be **doubly connected**. $SU(2)$ is called the universal covering group of $SO(3)$. The general definition of a universal covering group is given as follows.

Definition 11.3. *The **universal covering group** of a connected topological group G is a **simply connected** topological group G_{UC} such that there is a continuous homomorphism $\rho : G_{UC} \rightarrow G$ which is locally one-to-one. (See Fig. 11.11.)*

A universal covering group exists for every connected topological group (and in particular, for every connected Lie group), and is unique up to isomorphism.

Moreover (proof omitted here)

$$\frac{G_{UC}}{ker(\rho)} \sim G \,, \tag{11.75}$$

where $ker(\rho)$ is the kernel of the map ρ given by

$$ker(\rho) = \{g \in G_{UC} \mid \rho(g) = e \in G\} \,. \tag{11.76}$$

Our discussion earlier in this chapter in fact shows that $SL(2, \mathbb{C})$ *is the universal covering group of the proper orthochronous Lorentz group* $SO(1, 3)^\uparrow$ *(or L_0).*

The physical significance of $SL(2, \mathbb{C})$ is that elementary particles transform according to their representations. For particles of spin equal to an odd multiple of $1/2$, such as the electron (spin $1/2$), these are called **spinor representations** in physics.

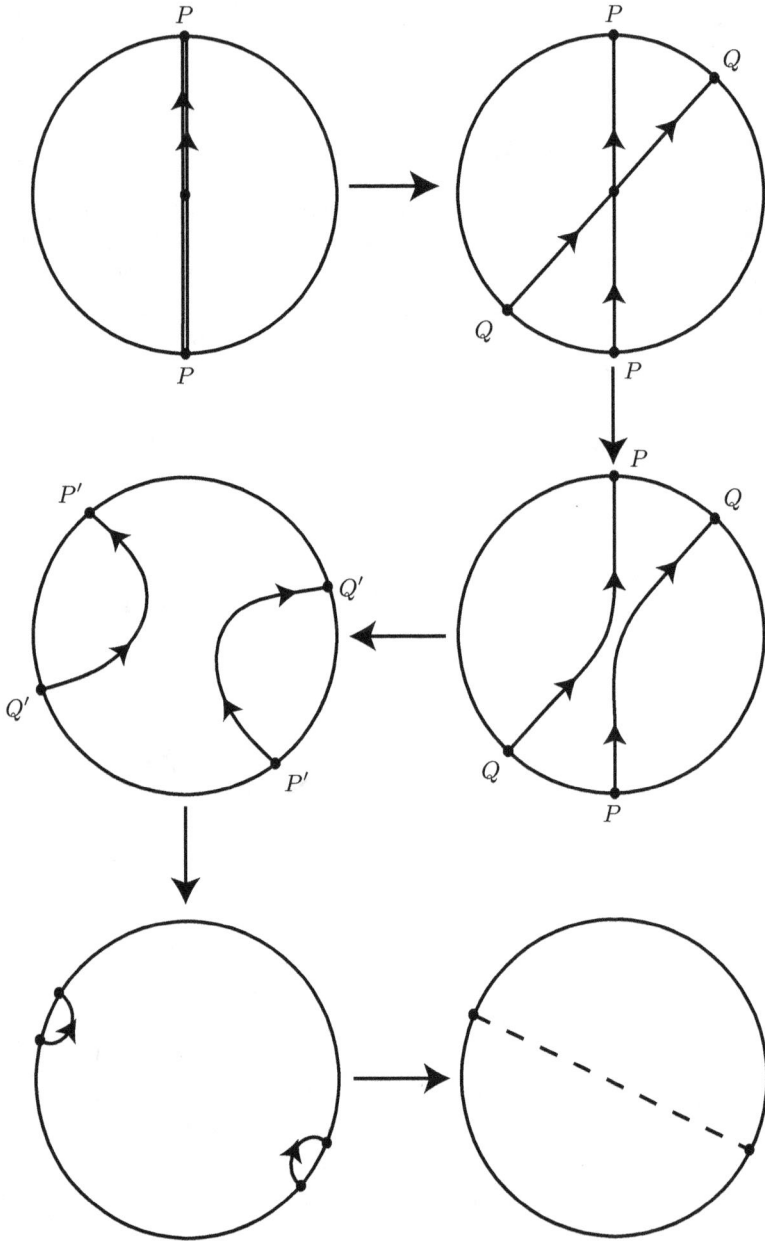

FIGURE 11.10

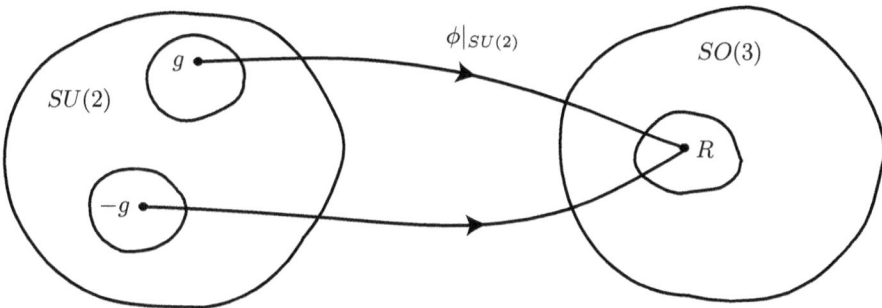

FIGURE 11.11

Chapter 12

The Dirac Bracket Notation in Quantum Theory

The Dirac notation was introduced very briefly in Chapter 8. We would like to elaborate on its use here due to its almost universal occurrence in the physics literature and its great elegance.

In quantum mechanics, the vector spaces of interest are most often infinite-dimensional ones called Hilbert spaces. We will not be concerned here with the technical complications which arise as a result of the infinite dimensionality of such spaces. All vector space notions introduced so far which do not make explicit reference to the dimension of the space, as well as many important facts, generalize to the case of Hilbert spaces.

Given a finite dimensional unitary space V and its dual V^*, the inner product defined in V induces a conjugate isomorphism $G : V \longrightarrow V^*$ [c.f. (8.26b)]. The generalization of this fact to infinite dimensional Hilbert spaces is called the **Riesz Theorem**. The Dirac bracket notation is essentially a very convenient notational device which encodes the mathematical content of this theorem.

Theorem 12.1 (Riesz Theorem). *Let $\mathcal{H}$ be a Hilbert space and $\mathcal{H}^*$ its dual. Then a hermitian inner product (,) [satisfying (8.27) and (8.28)] induces a conjugate isomorphism $G : \mathcal{H} \to \mathcal{H}^*$:*

$$G(\psi) = \psi^* \in \mathcal{H}^* , \tag{12.1}$$

such that

$$\psi^*(\phi) = \langle \phi, \psi^* \rangle = (\phi, \psi) , \quad \text{for all } \phi \in \mathcal{H} , \tag{12.2}$$

and

$$G(\alpha\psi) = \overline{\alpha}\psi^* , \quad \alpha \in \mathbb{C} . \tag{12.3}$$

Proof. We first note that ψ^* as defined by (12.2) is a continuous linear function on $\mathcal{H}$. In fact, for any $\phi_1, \phi_2 \in \mathcal{H}$ and $\alpha, \beta \in \mathbb{C}$,

$$\psi^*(\alpha\phi_1 + \beta\phi_2) = (\alpha\phi_1 + \beta\phi_2, \psi) = \alpha(\phi_1, \psi) + \beta(\phi_2, \psi) = \alpha\psi^*(\phi_1) + \beta\psi^*(\phi_2) ,$$

where the second equality follows from (8.28a). Continuity of ψ^* follows from

$$|\psi^*(\phi)| = |(\phi, \psi)| \le (\phi, \phi)(\psi, \psi) .$$

Now choose an arbitrary non-zero $\psi^* \in \mathcal{H}^*$. We wish to prove the existence of a unique $\psi \in \mathcal{H}$ such that $\psi^*(\phi) = (\phi, \psi)$ for all $\phi \in \mathcal{H}$.

Let $ker(\psi^*) \subset \mathcal{H}$ be the kernel of ψ^*:

$$ker(\psi^*) = \{\phi \in \mathcal{H} \mid \psi^*(\phi) = 0\} .$$

The set $ker(\psi^*)$ is a closed subspace of $\mathcal{H}$. In fact, consider a sequence $\{\phi_n\} \subset ker(\psi^*)$. Thus $\psi^*(\phi_n) = 0$ for all n. If $\lim_{n\to\infty} \phi_n = \phi$, then the continuity of ψ^* implies that $\psi^*(\phi) = 0$.

Now there exists a non-zero $\phi_0 \in (ker(\psi^*))^\perp$ [ϕ_0 is in the orthogonal complement of $ker(\psi^*)$]. Otherwise $(ker(\psi^*))^\perp = \{0\}$, implying that $ker(\psi^*) = \mathcal{H}$. But this is impossible because ψ^* is not identically zero by assumption. Since $\psi^*(\phi_0) \ne 0$, we can, without loss of generality, set $\psi^*(\phi_0) = 1$ [by replacing ϕ_0 by $\phi_0/\psi^*(\phi_0)$]. Then

$$\psi^*(\phi - \psi^*(\phi)\phi_0) = 0 , \quad \text{for all } \phi \in \mathcal{H} ,$$

which implies that

$$\phi' \equiv \phi - \psi^*(\phi)\phi_0 \in ker(\psi^*) .$$

Hence we have the orthogonal decomposition

$$\phi = \phi' + \psi^*(\phi)\phi_0 ; \quad \phi' \in ker(\psi^*) , \quad \phi_0 \in (ker(\psi^*))^\perp ,$$

for all $\phi \in \mathcal{H}$. Taking the hermitian inner product (ϕ, ϕ_0), we have

$$(\phi, \phi_0) = (\phi', \phi_0) + \psi^*(\phi)(\phi_0, \phi_0) = \psi^*(\phi)(\phi_0, \phi_0) .$$

The sought-for $\psi \in \mathcal{H}$ is given by

$$\psi = \frac{\phi_0}{(\phi_0, \phi_0)} .$$

Indeed, (12.2) follows from

$$\psi^*(\phi) = \frac{(\phi, \phi_0)}{(\phi_0, \phi_0)} = (\phi, \psi) .$$

Uniqueness of ψ can be established as follows. Suppose there exists a $\psi' \ne \psi$ such that

$$(\phi, \psi) = (\phi, \psi') \quad \text{for all } \phi \in \mathcal{H} .$$

Then

$$(\phi, \psi - \psi') = 0 \quad \text{for all } \phi \in \mathcal{H} .$$

Choosing $\phi = \psi - \psi'$, we have

$$(\psi - \psi', \psi - \psi') = 0 ,$$

which implies $\psi = \psi'$.

Finally, (8.28b) implies that

$$G(\alpha\psi)(\phi) = (\phi, \alpha\psi) = \overline{\alpha}(\phi, \psi) = \overline{\alpha}\psi^*(\phi) = \overline{\alpha}G(\psi)(\phi) \,,$$

for all $\phi \in \mathcal{H}$. Thus (12.3) follows. $\qquad\qquad\qquad\qquad\qquad\qquad$ □

In the Dirac notation, a vector $\psi \in \mathcal{H}$ is labelled $|\psi\rangle$, called a **ket vector**, or simply a **ket**. Dirac also labels ψ^* by $\langle\psi|$, and calls it a **bra vector**, or simply a **bra**. The inner product $(\phi, \psi) = \langle\phi, \psi^*\rangle$ is then denoted by $\langle\psi|\phi\rangle$, that is,

$$\langle\psi|\phi\rangle \equiv (\phi, \psi) \quad . \tag{12.4}$$

As remarked earlier, the order of the entries in the product is reversed on going from the mathematical notation with ordinary parentheses to the Dirac notation. Note again that $\langle\psi|\phi\rangle$ is in general a complex number, $|\phi\rangle \in \mathcal{H}$, and $\langle\psi| \in \mathcal{H}^*$. Eqs. (8.28a) and (8.28b) can then be written in the Dirac notation as

$$\langle\phi|\alpha_1\psi_1 + \alpha_2\psi_2\rangle = \alpha_1\langle\phi|\psi_1\rangle + \alpha_2\langle\phi|\psi_2\rangle \quad , \tag{12.5a}$$

and

$$\langle\alpha_1\psi_1 + \alpha_2\psi_2|\phi\rangle = \overline{\alpha_1}\langle\psi_1|\phi\rangle + \overline{\alpha_2}\langle\psi_2|\phi\rangle \quad , \tag{12.5b}$$

respectively. These conditions are called linearity conditions of the inner product with respect to the ket and to the bra, respectively. The property of **conjugate linearity** of the conjugate isomorphism G is expressed as

$$|\alpha_1\psi_1 + \alpha_2\psi_2\rangle \mapsto \overline{\alpha_1}\langle\psi_1| + \overline{\alpha_2}\langle\psi_2| \quad . \tag{12.6}$$

The property of **hermitian symmetry** [c.f. (8.27)] of the inner product is written as

$$\boxed{\langle\phi|\psi\rangle = \overline{\langle\psi|\phi\rangle}} \quad . \tag{12.7}$$

If a vector $|\psi\rangle$ represents a **physical state**, the norm $\langle\psi|\psi\rangle$ is interpreted as a probability in quantum theory, and hence must be greater than or equal to zero:

$$\langle\psi|\psi\rangle \geq 0 \qquad \text{for physical states.}$$

However, in quantum field theories, **ghost states** may enter the mathematical formalism. These are states with negative norm:

$$\langle\psi|\psi\rangle < 0 \qquad \text{for ghost states.}$$

Suppose $\{e_i\}, i = 1, 2, \ldots, \infty$, is an orthonormal basis of $\mathcal{H}$. Then any $\psi \in \mathcal{H}$ can be written as

$$\psi = \psi^i e_i \quad , \tag{12.8}$$

where

$$\psi^i = \langle\psi, e^{*i}\rangle = (\psi, e_i) \quad , \tag{12.9}$$

and $\{e^{*i}\}$ is the dual basis to $\{e_i\}$ in $\mathcal{H}^*$. Note that the orthonormality of $\{e_i\}$ guarantees that $G^{-1}(e^{*i}) = e_i$ [see discussion preceding (10.20)]. In the Dirac notation, Eqs. (12.8) and (12.9) can be combined:

$$|\psi\rangle = \sum_i \langle e_i|\psi\rangle \, |e_i\rangle \quad , \tag{12.10}$$

or equivalently

$$|\psi\rangle = \sum_i |e_i\rangle\langle e_i|\psi\rangle \quad . \tag{12.11}$$

Since $|\psi\rangle$ can always be written as $1 \cdot |\psi\rangle$, where 1 represents the identity operator, we can formally write

$$\boxed{\sum_i |e_i\rangle\langle e_i| = 1} \quad , \tag{12.12}$$

or equivalently

$$\sum_i P_i = 1 \quad , \tag{12.13}$$

where

$$P_i \equiv |\,e_i\rangle\langle e_i| \tag{12.14}$$

is the **projection operator** along the subspace spanned by $|\,e_i\rangle$. Eq. (12.12) is called the **completeness condition** for the orthonormal basis $\{e_i\}$. The projection operator P_i is **idempotent**:

$$(P_i)^2 = P_i \quad .$$

Indeed

$$(P_i)^2 = |\,e_i\rangle\langle e_i|\,e_i\rangle\langle e_i| = |\,e_i\rangle\langle e_i| = P_i \quad . \tag{12.15}$$

It is also easily seen to be hermitian [see (12.30) below]. A projection on a finite-dimensional subspace spanned by a finite subset $\{e_1, \dots, e_n\}$ of the orthonormal basis $\{e_i\}$ is given by

$$P = \sum_{i=1}^n |\,e_i\rangle\langle e_i| \quad . \tag{12.16}$$

Eq. (12.12) (the completeness condition) is extremely useful in quantum mechanical calculations. Note that an expression of the form $|\phi\rangle\langle\psi|$, where $|\phi\rangle, |\psi\rangle \in \mathcal{H}$, represents a linear operator, although not a projection operator when $|\phi\rangle \neq |\psi\rangle$.

Now consider an arbitrary linear operator A. Its action on a ket $|\psi\rangle$ is written $A|\psi\rangle$. The inner product $(A\psi, \phi)$ is then, in the Dirac notation, $\langle\phi|A|\psi\rangle$:

$$\langle\phi|A|\psi\rangle \equiv (A\psi, \phi) \quad . \tag{12.17}$$

This notation suggests that in the "sandwiched" position, A can be thought of as acting to the left also. Indeed, by (10.1),

$$\langle\phi|A|\psi\rangle = (A\psi, \phi) = (\psi, A^\dagger\phi) = \langle A^\dagger\phi|\psi\rangle \quad . \tag{12.18}$$

Thus
$$\langle\phi|A = \langle A^\dagger\phi| \tag{12.19}$$
is the bra vector conjugate to $|A^\dagger\phi\rangle$, or the dual vector in $\mathcal{H}^*$ mapped to by the conjugate isomorphism G acting on $|A^\dagger\phi\rangle$. It follows that
$$\langle\phi|A^\dagger = \langle A\phi| \quad, \tag{12.20}$$
or, under G,
$$\boxed{A|\phi\rangle \mapsto \langle\phi|A^\dagger} \quad. \tag{12.21}$$
Now (12.20) implies that, for all $|\psi\rangle$,
$$\langle\phi|A^\dagger|\psi\rangle = \langle A\phi|\psi\rangle \quad. \tag{12.22}$$
Hence, by (12.7),
$$\boxed{\overline{\langle\psi|A|\phi\rangle} = \langle\phi|A^\dagger|\psi\rangle} \quad. \tag{12.23}$$

What is the matrix representation of A with respect to a choice of orthonormal basis $\{e_i\}$? Analogous to (1.16) we write, using the Dirac notation,
$$A|e_i\rangle = a_i^k|e_k\rangle \quad. \tag{12.24}$$
"Multiplying" on the left by $\langle e_j|$, we have
$$\langle e_j|A|e_i\rangle = a_i^k\langle e_j|e_k\rangle = \delta_{jk}\, a_i^k \quad. \tag{12.25}$$
Thus
$$\boxed{a_i^j = \langle e_j|A|e_i\rangle} \quad. \tag{12.26}$$

Exercise 12.1 Obtain the above equation [(12.26)] by using the completeness condition (12.12) for the identity operator to act on $A|e_i\rangle$.

Exercise 12.2 Suppose $A|\psi\rangle = |\phi\rangle$, and expand $|\psi\rangle$ and $|\phi\rangle$ as follows:
$$|\psi\rangle = \psi^i|e_i\rangle \quad, \tag{12.27}$$
$$|\phi\rangle = \phi^i|e_i\rangle \quad. \tag{12.28}$$
Use the completeness condition (12.12) to show that
$$\phi^i = a_j^i\psi^j \quad, \tag{12.29}$$
where a_i^j is given by (12.26).

Exercise 12.3 Use (12.7) and (12.23) to show that
$$(|\phi\rangle\langle\psi|)^\dagger = |\psi\rangle\langle\phi| \tag{12.30}$$
for any $|\phi\rangle, |\psi\rangle \in \mathcal{H}$. Start with the expression $\langle\alpha|(|\phi\rangle\langle\psi|)^\dagger|\beta\rangle$ for any $|\alpha\rangle$, $|\beta\rangle \in \mathcal{H}$.

Chapter 13

The Quantum Mechanical Simple Harmonic Oscillator

The quantum mechanical simple harmonic oscillator is undeniably one of the most fundamental and far-reaching examples in all of theoretical physics. Its solution by Dirac is also marked by extreme elegance.

For a one-dimensional oscillator of mass m and frequency ω, the Hamiltonian (operator) is given by

$$H = \frac{p^2}{2m} + \frac{1}{2}m\omega^2 x^2 \quad , \tag{13.1}$$

where p is the linear momentum (operator) [c.f. (4.4)] and x is the coordinate (operator) of the oscillator. It will be convenient to define the dimensionless coordinate Q and the dimensionless momentum P by

$$\begin{cases} Q \equiv \sqrt{\dfrac{m\omega}{\hbar}}\, x \\[3mm] P \equiv \dfrac{1}{\sqrt{m\hbar\omega}}\, p \quad . \end{cases} \tag{13.2}$$

The Hamiltonian in (13.1) is then given by

$$H = \frac{\hbar\omega}{2}\left(P^2 + Q^2\right) \quad . \tag{13.3}$$

Exercise 13.1   Verify (13.3).

Now define the following pair of operators:

$$a = \frac{1}{\sqrt{2}}\left(Q + iP\right) \tag{13.4}$$

113

and its hermitian adjoint

$$a^\dagger = \frac{1}{\sqrt{2}}(Q - iP) \quad . \tag{13.5}$$

The fact that the RHS of (13.5) is the hermitian adjoint of the RHS of (13.4) follows from the fact that Q and P are both hermitian, and Eqs. (10.4) and (10.5). Solving for Q and P in terms of a and $a^\dagger$, we have

$$Q = \frac{1}{2}(a + a^\dagger) , \quad P = \frac{-i}{\sqrt{2}}(a - a^\dagger) , \tag{13.6}$$

from which it follows that

$$x = \sqrt{\frac{\hbar}{m\omega}} \frac{1}{\sqrt{2}}(a + a^\dagger) , \quad p = \sqrt{m\hbar\omega}\left(-\frac{i}{\sqrt{2}}\right)(a - a^\dagger) . \tag{13.7}$$

The crucial condition is that the operators x and p are required to satisfy the quantum commutation relations (4.12):

$$[x, p] = i\hbar \quad , \tag{13.8}$$

which means that

$$[Q, P] = i \quad . \tag{13.9}$$

Thus

$$[a, a^\dagger] = \frac{1}{2}[Q + iP, Q - iP] = \frac{i}{2}([P, Q] - [Q, P]) = \frac{i}{2}(-i - (-i)) = 1 . \tag{13.10}$$

Either one of the sets spanned by $\{P, Q, \mathbb{I}\}$ or $\{a, a^\dagger, \mathbb{I}\}$, where $\mathbb{I}$ is the identity operator, constitutes a complex Lie algebra called the **Heisenberg algebra**. To the Euclidean position coordinates x^i and momenta p^j, there correspond the operators a^i and $(a^i)^\dagger$. The general commutation relations for these, which follow from (4.12), are:

$$[a^i, (a^j)^\dagger] = \delta^{ij} \quad ,$$
$$[a^i, a^j] = [(a^i)^\dagger, (a^j)^\dagger] = 0 \quad . \tag{13.11}$$

Returning to the one-dimensional oscillator, it is easily seen that

$$\boxed{H = \hbar\omega\left(a^\dagger a + \frac{1}{2}\right)} \quad . \tag{13.12}$$

Exercise 13.2 Verify (13.12) by using (13.3) and the commutation relations (13.9) or (13.11).

Exercise 13.3 Verify that H as given in (13.12) is hermitian.

The solution of our problem then boils down to the mathematical problem of finding the **spectrum** of H, that is, finding all the vectors (in the Hilbert space of vectors describing the oscillator) $|\phi_n\rangle$ and all real numbers E_n such that

$$H\,|\,\phi_n\rangle = E_n\,|\,\phi_n\rangle \quad . \tag{13.13}$$

The set of all E_n is called the spectrum of H. Each E_n is called an **eigenvalue** of the operator H, and $|\,\phi_n\rangle$ is called the **eigenvector**, or eigenstate, belonging to the eigenvalue E_n.

Let us digress on the so-called **eigenvalue problem**, which will be revisited in more detail in Chapter 17.

That all energy eigenvalues of H are real follows from the fact that H is hermitian. In fact, *the eigenvalues of a hermitian operator are always real.* The easy proof is as follows. Suppose $A = A^\dagger$ and $A|\,\phi\rangle = \lambda|\,\phi\rangle$, for some $\lambda \in \mathbb{C}$, $|\,\phi\rangle \in \mathcal{H}$, $(|\,\phi\rangle \neq 0)$. Then

$$\langle\phi|A|\phi\rangle = \overline{\langle\phi|A^\dagger|\phi\rangle} = \overline{\langle\phi|A|\phi\rangle} \,, \tag{13.14}$$

where the first equality follows from (12.23) and the second from the fact that $A = A^\dagger$. This implies that

$$\lambda\langle\phi|\phi\rangle = \overline{\lambda}\,\overline{\langle\phi|\phi\rangle} = \overline{\lambda}\langle\phi|\phi\rangle \quad , \tag{13.15}$$

and thus, since $\langle\phi|\phi\rangle \neq 0$, $\lambda = \overline{\lambda}$.

Another important fact concerning hermitian operators is that *eigenvectors belonging to distinct eigenvalues of a hermitian operator are orthogonal to each other.* The proof of this statement is also quite simple. Suppose $A = A^\dagger$ and

$$A|\,\phi_1\rangle = \lambda_1\,|\,\phi_1\rangle\,, \quad A|\,\phi_2\rangle = \lambda_2\,|\,\phi_2\rangle\,, \quad \lambda_1 \neq \lambda_2 \quad . \tag{13.16}$$

Since A is hermitian, we know from the above that λ_1 and λ_2 are both real. Thus

$$\begin{aligned}
0 &= \langle\phi_1|A|\phi_2\rangle - \overline{\langle\phi_2|A|\phi_1\rangle} = \lambda_2\,\langle\phi_1|\phi_2\rangle - \lambda_1\,\overline{\langle\phi_2|\phi_1\rangle} \\
&= \lambda_2\,\langle\phi_1|\phi_2\rangle - \lambda_1\,\langle\phi_1|\phi_2\rangle = (\lambda_2 - \lambda_1)\,\langle\phi_1|\phi_2\rangle \,.
\end{aligned} \tag{13.17}$$

Since $\lambda_2 \neq \lambda_1$ by assumption, $\langle\phi_1|\phi_2\rangle = 0$, that is, $|\,\phi_1\rangle$ is orthogonal to $|\,\phi_2\rangle$.

In general, different eigenvectors may belong to the same eigenvalue of a linear operator. In such a case we may speak of a **degenerate eigenvalue**.

We will now return to the solution of the simple harmonic oscillator problem. The relevant Hilbert space in this problem is the infinite-dimensional vector space of **square integrable functions** $\psi(x)$:

$$\mathcal{H} = \left\{ \psi(x) \,\middle|\, \int_{-\infty}^{\infty} dx\, |\psi(x)|^2 < \infty \right\} \quad . \tag{13.18}$$

By the celebrated **Fischer-Riesz Theorem** (see J. von Neumann 1955) $\mathcal{H}$ is isomorphic to the set of infinite-tuples of complex numbers $(z_1, z_2, z_3, \ldots)$ such that $\sum_{n=1}^{\infty} |z_n|^2$ is finite. In other words,

$$\mathcal{H} \sim \mathcal{H}' = \left\{ (z_1, z_2, \ldots) \,\middle|\, z_i \in \mathbb{C}, \sum_{i=1}^{\infty} |z_i|^2 < \infty \right\} \quad . \tag{13.19}$$

If we define respective hermitian inner products in $\mathcal{H}$ and $\mathcal{H}'$:

$$\langle \phi(x) | \psi(x) \rangle = \int_{-\infty}^{\infty} dx \, \overline{\phi}(x) \psi(x) \quad , \tag{13.20}$$

$$\langle w | z \rangle = \sum_{n=1}^{\infty} \overline{w}_i \, z_i \quad , \tag{13.21}$$

the isomorphism is in fact an **isometry** (norm-preserving). In other words, if $\psi(x) \leftrightarrow (z_1, z_2, \ldots)$ under the isomorphism, then $\|\psi(x)\|^2 = \|z\|^2$, or

$$\int_{-\infty}^{\infty} |f(x)|^2 dx = \sum_{n=1}^{\infty} |z_i|^2 \quad . \tag{13.22}$$

Instead of working with $\mathcal{H}$ directly, we will work with $\mathcal{H}'$.

Let us choose the **standard basis** for $\mathcal{H}'$:

$$\begin{cases} | e_1 \rangle = (1, 0, 0, \ldots) \equiv | 0 \rangle \quad , \\ | e_2 \rangle = (0, 1, 0, \ldots) \equiv | 1 \rangle \quad , \\ \quad \vdots \\ | e_i \rangle = (0, 0, \ldots, 0, 1, 0, \ldots) \equiv | i - 1 \rangle \quad , \\ \quad \vdots \end{cases} \tag{13.23}$$

where in the equation for $| e_i \rangle$, the number 1 occurs in the i-th entry in the infinite row of numbers. This is clearly an orthonormal basis. We claim that with respect to this basis,

$$a = \begin{pmatrix} 0 & 0 & 0 & 0 & \cdots \\ \sqrt{1} & 0 & 0 & 0 & \cdots \\ 0 & \sqrt{2} & 0 & 0 & \cdots \\ 0 & 0 & \sqrt{3} & 0 & \cdots \\ & & & \vdots & \end{pmatrix} \quad , \tag{13.24}$$

$$a^{\dagger} = \begin{pmatrix} 0 & \sqrt{1} & 0 & 0 & 0 & \cdots \\ 0 & 0 & \sqrt{2} & 0 & 0 & \cdots \\ 0 & 0 & 0 & \sqrt{3} & 0 & \cdots \\ & & & \vdots & & \end{pmatrix} = a^T \quad . \tag{13.25}$$

These two equations imply that

$$a^\dagger a = \begin{pmatrix} 0 & 0 & 0 & 0 & \cdots \\ 0 & 1 & 0 & 0 & \cdots \\ 0 & 0 & 2 & 0 & \cdots \\ 0 & 0 & 0 & 3 & \cdots \\ & & \vdots & & \end{pmatrix} . \tag{13.26}$$

Thus

$$a^\dagger a \, |n\rangle = n \, |n\rangle , \quad n = 0, 1, 2, 3, \ldots \quad , \tag{13.27}$$

where $|n\rangle = |e_{n+1}\rangle$ are the basis states defined in (13.23). Thus $|n\rangle, n = 0, 1, 2, \ldots$, is an eigenstate of $a^\dagger a$ with eigenvalue n. For this reason, $a^\dagger a$ is called the **number operator**. Eq. (13.12) then gives

$$H = \frac{\hbar \omega}{2} \begin{pmatrix} 1 & 0 & 0 & 0 & \cdots \\ 0 & 3 & 0 & 0 & \cdots \\ 0 & 0 & 5 & 0 & \cdots \\ 0 & 0 & 0 & 7 & \cdots \\ & & \vdots & & \end{pmatrix} . \tag{13.28}$$

This implies that

$$H | n\rangle = \hbar \omega \left(n + \frac{1}{2} \right) |n\rangle, \quad n = 0, 1, 2, \ldots \quad , \tag{13.29}$$

so the energy eigenvalues are

$$E_n = \hbar \omega \left(n + \frac{1}{2} \right), \quad n = 0, 1, 2, \ldots \quad . \tag{13.30}$$

The state $|0\rangle$ is called the **ground state**. Note that it is *not* the zero vector in $\mathcal{H}'$, which is the vector $(0, 0, 0, \ldots)$ (with all zeros). Eq. (13.29) shows that all the energy eigenvalues for the simple harmonic oscillator are **non-degenerate**. In the physics literature, the normalized eigenstates $|n\rangle$ are called **number states**, the eigenvalue n of $a^\dagger a$ is called the **occupation number**, and the Hilbert space spanned by $|n\rangle, n = 0, 1, 2, \ldots$, is called a **Fock space**. Clearly, the physical interpretation of n is the number of quanta, or excitations, of a certain kind.

Using (12.29), the effects of a and $a^\dagger$ on $|n\rangle$ can be obtained by matrix multiplication:

$$a|n\rangle = (0,\ldots,0,1,0,\ldots)\begin{pmatrix} 0 & 0 & 0 & \cdots \\ \sqrt{1} & 0 & 0 & \cdots \\ 0 & \sqrt{2} & 0 & \cdots \\ 0 & 0 & \sqrt{3} & \cdots \\ & & & \vdots \end{pmatrix} \tag{13.31}$$

$$= \sqrt{n}\,(0,\ldots,0,1,0,\ldots) \quad,$$

where in the row matrix on the RHS of the first equality the 1 occurs in the $(n+1)$-th entry, and in the row matrix on the RHS of the second equality the 1 occurs in the n-th entry. Thus

$$a|n\rangle = \sqrt{n}\,|n-1\rangle, \quad n = 1, 2, \ldots, \tag{13.32}$$
$$a|0\rangle = 0 \quad. \tag{13.33}$$

Similarly, one can show that

$$a^\dagger|n\rangle = \sqrt{n+1}\,|n+1\rangle, \quad n = 0, 1, 2, \ldots \quad. \tag{13.34}$$

Based on the above three equations, the operators a and $a^\dagger$ are called the **annihilation** and **creation operators**, respectively. Starting with the ground state $|0\rangle$ and using (13.34) repeatedly, one obtains

$$|n\rangle = \frac{1}{\sqrt{n!}}\,(a^\dagger)^n\,|0\rangle, \quad n = 1, 2, 3, \ldots \quad. \tag{13.35}$$

Exercise 13.4   Use (13.23) and (13.25) to verify (13.34).

Exercise 13.5   Verify (13.35) by using (13.34).

It is extremely important to realize that the equations describing the actions of the creation and annihilation operators [(13.32) to (13.34)], and Eq. (13.35) showing how to construct a normalized n-particle state from the ground state (also called the **vacuum state** since its occupation number is zero), are all direct consequences of the commutation relations (13.11). In particular (13.35) shows that it is possible for a quantum state to have occupation number n equal to any non-negative integer, provided that the corresponding creation and annihilation operators generate an algebra satisfying the multiplication rules (the commutation relations) given by (13.11). Particles (or any elementary excitations) whose creation and annihilation operators satisfy the commutation rules of (13.11) are called **bosons**. *It is a consequence of special relativity that bosons are particles of integral spin and satisfy the so-called **Bose-Einstein statistics**.*

The other class of particles are called **fermions**, and it is an amazing fact of nature that, as we know it, the world consists of only bosons and fermions, and nothing else. Again, because of the requirements of special relativity, *fermions are particles of half-integral spin and satisfy the so-called **Fermi-Dirac statistics**.* One of the consequences of the latter is the **Pauli exclusion principle**, which stipulates that the occupation number of a fermion quantum state is either 0 or 1, as opposed to the boson case, where $n = 0, 1, 2, \ldots$. Fermion creation and annihilation operators satisfy a different set of multiplication rules than (13.11). Instead of the commutation rules for bosons, we have the **anti-commutation** rules for fermions:

$$\{a^i, (a^j)^\dagger\} = \delta^{ij} \quad ,$$
$$\{a^i, a^j\} = \{(a^i)^\dagger, (a^j)^\dagger\} = 0 \quad , \tag{13.36}$$

where $\{A, B\} \equiv AB + BA$. The second of these equations immediately implies that $(a^\dagger)^2 = 0$. We still have, for fermions,

$$a|0\rangle = 0 , \quad a|1\rangle = |0\rangle , \quad a^\dagger|0\rangle = |1\rangle . \tag{13.37}$$

Exercise 13.6   Verify that (13.37) is consistent with (13.36).

Eqs. (13.36) are called the **anticommutation relations** for fermion creation and annihilation operators (see Chapter 27).

According to the **standard model**, the world is made up of **quarks** and **leptons**, which are fermions, and the **gauge bosons**, which are responsible for the fundamental forces. For example, the **photon** is the gauge boson responsible for the electromagnetic force, and the most well-known lepton is the electron.

In relativistic quantum electrodynamics, we have the so-called **longitudinal modes** of photons, described by the time-like components of the creation and annihilation operators $(a^0)^\dagger$ and a^0 (remember in special relativity, the superscript 0 refers to the time component). The **transverse modes** are described by $(a^i)^\dagger$ and $a^i, i = 1, 2, 3$. Instead of (13.11) we have

$$[a^\mu, (a^\nu)^\dagger] = \eta^{\mu\nu} \quad ,$$
$$[a^\mu, a^\nu] = [(a^\mu)^\dagger, (a^\nu)^\dagger] = 0 \quad , \tag{13.38}$$

where $\mu, \nu = 0, 1, 2, 3$, and $\eta^{\mu\nu} = $ inverse of the Lorentz metric $\eta_{\mu\nu} = \eta_{\mu\nu}$. [c.f. (2.20) and (8.3)].

There is a peculiarity about the longitudinal photons. Since $\eta^{00} = \eta_{00} = -1$, $[a^0, (a^0)^\dagger] = -1$, which implies

$$a^0(a^0)^\dagger = -1 + (a^0)^\dagger a^0 \quad . \tag{13.39}$$

Thus the norm of the one-photon state $(a^0)^\dagger | 0 \rangle$, which is given by $\langle 0 | a^0 (a^0)^\dagger | 0 \rangle$, is -1. As mentioned before, these non-physical negative-norm states are called **ghost states**. In the quantization of Lorentz invariant field theories, they have to be handled very carefully: these states cannot have any effects on physically observable quantities.

Chapter 14

Fourier Series and Fourier Transforms, the Dirac Delta Function, Green's Functions

A particularly important Hilbert space in theoretical physics is the space of periodic, piecewise-continuous functions equipped with the hermitian inner product

$$\langle g| f \rangle = \int_a^b \mathrm{d}x \, \overline{g}(x) f(x) \quad , \tag{14.1}$$

where $[a, b]$ is one period of the functions in the Hilbert space $\mathcal{H}$:

$$f(x) = f(x + n(b - a)), \quad n = 0, \pm 1, \pm 2, \ldots \quad . \tag{14.2}$$

All $f \in \mathcal{H}$ are supposed to be complex-valued functions of a real variable ($f : \mathbb{R} \longrightarrow \mathbb{C}$). We will henceforth assume that $[a, b] = [0, 2\pi]$, with the general interval regainable by the linear transformation

$$x \in [0, 2\pi] \mapsto a + x \left(\frac{b - a}{2\pi} \right) \in [a, b] \quad . \tag{14.3}$$

The most important examples are $\sin nx$ and $\cos nx$, $n = 0, \pm 1, \pm 2, \ldots$.

Let us consider the orthonormal set

$$| e_n \rangle = \frac{1}{\sqrt{2\pi}} e^{inx}, \quad n = 0, \pm 1, \pm 2, \ldots \quad . \tag{14.4}$$

Orthonormality of this set can be verified quite easily. Indeed, according to (14.1),

$$\langle e_m| e_n \rangle = \frac{1}{2\pi} \int_0^{2\pi} \mathrm{d}x \, e^{-imx} e^{inx} = \delta_{mn} \quad . \tag{14.5}$$

The central fact of the theory of Fourier series (which we will not prove here) is that the orthonormal set specified in (14.4) is complete. In the Dirac notation, this condition is stated as [see (12.12)]:

$$\sum_{n=-\infty}^{\infty} |e_n\rangle\langle e_n| = 1 \quad . \tag{14.6}$$

Thus any $|f\rangle = f(x) \in \mathcal{H}$ can be expanded as:

$$f(x) = |f\rangle = 1 \cdot |f\rangle = \sum_{n=-\infty}^{\infty} |e_n\rangle\langle e_n|f\rangle = \sum_{n=-\infty}^{\infty} \langle e_n|f\rangle\,|e_n\rangle ,$$

or

$$f(x) = \sum_{n=-\infty}^{\infty} f_n \frac{e^{inx}}{\sqrt{2\pi}} \quad , \tag{14.7}$$

where the **Fourier coefficient** f_n is given by

$$f_n \equiv \langle e_n|f\rangle \quad ,$$

or

$$f_n = \frac{1}{\sqrt{2\pi}} \int_0^{2\pi} dx\, e^{-inx} f(x) \quad . \tag{14.8}$$

Strictly speaking, the equality in (14.7) does not quite mean **pointwise equality** (equal at every point x). This is only achieved at points x where $f(x)$ is continuous. In general, we have equality in the sense of **norm convergence**:

$$\lim_{N \to \infty} \|f(x) - S_N(x)\| = 0 \quad , \tag{14.9}$$

where

$$S_N = \sum_{m=-N}^{N} f_m \frac{e^{imx}}{\sqrt{2\pi}} \quad , \tag{14.10}$$

and, from (14.1), the norm $\|g\|$ of a function $g(x) \in \mathcal{H}$ is given by

$$\|g\|^2 = \int_0^{2\pi} dx\, |g(x)|^2 \quad . \tag{14.11}$$

The result for the Fourier series given by Eqs. (14.7) and (14.8) is a special case of the Fischer-Riesz theorem mentioned in the previous chapter.

Using $e^{inx} = \cos nx + i \sin nx$, the complex Fourier series (14.7) can be written in terms of sines and cosines. Let us further suppose that $f(x)$ is real.

Then, from (14.7)

$$\overline{f(x)} = \overline{f_0} + \overline{f_1} \frac{e^{-ix}}{\sqrt{2\pi}} + \cdots + \overline{f_n} \frac{e^{-inx}}{\sqrt{2\pi}} + \cdots$$

$$+ \overline{f_{-1}} \frac{e^{ix}}{\sqrt{2\pi}} + \cdots + \overline{f_{-n}} \frac{e^{inx}}{\sqrt{2\pi}} + \cdots$$

$$= f(x) \tag{14.12}$$

$$= f_0 + f_1 \frac{e^{ix}}{\sqrt{2\pi}} + \cdots + f_n \frac{e^{inx}}{\sqrt{2\pi}} + \cdots$$

$$+ f_{-1} \frac{e^{-ix}}{\sqrt{2\pi}} + \cdots + f_{-n} \frac{e^{-inx}}{\sqrt{2\pi}} + \cdots \quad .$$

Since the set $\{|\, e_n\rangle\} = \frac{1}{\sqrt{2\pi}} e^{inx}$, $n = 0, \pm 1, \pm 2, \ldots$, is linearly independent, we can equate the coefficients of $|\, e_n\rangle$ for each n in (14.12) and thus obtain

$$f_0 = \overline{f_0} \, , \; f_{-1} = \overline{f_1} \, , \; \ldots \, , \; f_{-n} = \overline{f_n} \, , \; \ldots \, . \tag{14.13}$$

Hence a real periodic function with period $[0, 2\pi]$ can be expanded as

$$f(x) = f_0 + \sum_{n=1}^{\infty} \left(f_n \frac{e^{inx}}{\sqrt{2\pi}} + \text{c.c.} \right) \quad , \tag{14.14}$$

where c.c. means complex conjugate. Writing

$$f_n = c_n - id_n \quad , \tag{14.15}$$

where c_n, d_n are real, $f(x)$ can be cast in the classical Fourier expansion

$$\boxed{f(x) = f_0 + \sum_{n=1}^{\infty} (a_n \cos nx + b_n \sin nx)} \quad , \tag{14.16}$$

where $a_n = \frac{2c_n}{\sqrt{2\pi}}, b_n = \frac{2d_n}{\sqrt{2\pi}}$, and f_0 are all real. Using Eqs. (14.8) and (14.15) the Fourier coefficients a_n and b_n are given by

$$a_n = \frac{1}{\pi} \int_0^{2\pi} dx \, f(x) \cos nx \, , \quad b_n = \frac{1}{\pi} \int_0^{2\pi} dx \, f(x) \sin nx \, . \tag{14.17}$$

Exercise 14.1   Find the Fourier expansion (14.16) for the periodic function shown in Fig. 14.1.

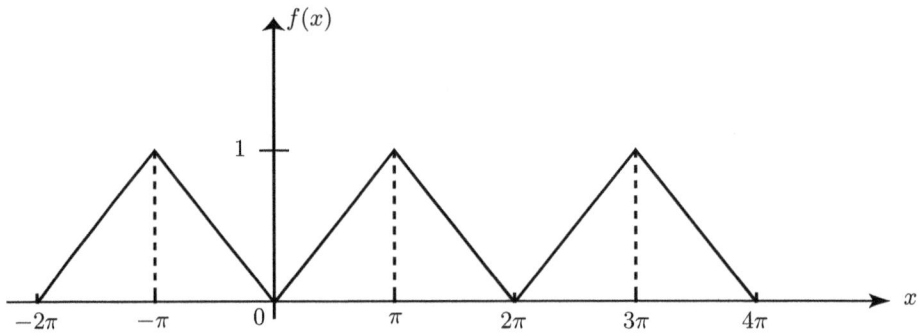

FIGURE 14.1

From (14.17) it is clear that $b_n = 0$ if $f(x)$ is even with respect to $x = \pi$, and $a_n = f_0 = 0$ if $f(x)$ is odd with respect to $x = \pi$. Thus

$$f(x) = \begin{cases} f_0 + \displaystyle\sum_{n=1}^{\infty} a_n \cos nx & , \quad f(x) \text{ even w.r.t. } x = \pi, \\[3mm] \displaystyle\sum_{n=1}^{\infty} b_n \sin nx & , \quad f(x) \text{ odd w.r.t. } x = \pi. \end{cases} \tag{14.18}$$

On passing to the continuous limit, the Fourier series (14.7) is replaced by the **Fourier integral**, and (14.7) and (14.8) are replaced by

$$f(x) = \frac{1}{\sqrt{2\pi}} \int_{-\infty}^{\infty} dk\, F(k)\, e^{ikx} \quad , \tag{14.19}$$

and

$$F(k) = \frac{1}{\sqrt{2\pi}} \int_{-\infty}^{\infty} dx\, f(x)\, e^{-ikx} \quad , \tag{14.20}$$

respectively. The functions $f(x)$ and $F(k)$ are called **Fourier transforms** of each other. Eq. (14.19) is used instead of (14.7) when $f(x)$ is not a periodic function of x, or equivalently, when the period $[0, 2\pi] \to (-\infty, \infty)$. The Fourier transform $F(k)$ of $f(x)$ can be viewed as the amplitude of the harmonic oscillation whose wave number is k.

Fourier transforms are of the greatest importance in the solution of differential equations. Using the transform method on an ordinary differential equation with constant coefficients, for example, the differential equation is transformed into an algebraic equation. To see how this works let us consider a second order equation of the form

$$\frac{d^2 f(x)}{dx^2} + \alpha \frac{df(x)}{dx} + \beta f(x) = h(x) \quad , \tag{14.21}$$

where α and β are constants. Fourier transforming each side of the equation, we obtain

$$\frac{1}{\sqrt{2\pi}} \int_{-\infty}^{\infty} dk\, F(k) \frac{d^2}{dx^2} e^{ikx} + \frac{\alpha}{\sqrt{2\pi}} \int_{-\infty}^{\infty} dk\, F(k) \frac{d}{dx} e^{ikx}$$

$$+ \frac{\beta}{\sqrt{2\pi}} \int_{-\infty}^{\infty} dk\, F(k) e^{ikx} = \frac{1}{\sqrt{2\pi}} \int_{-\infty}^{\infty} dk\, H(k) e^{ikx} \quad , \quad (14.22)$$

where $F(k)$ and $H(k)$ are the Fourier transforms of $f(x)$ and $h(x)$, respectively. Note that we have interchanged the order of the differential and integral operations. (The validity of this step will not be proved here.) It follows that

$$(ik)^2 F(k) + \alpha(ik)F(k) + \beta F(k) = H(k) \quad ,$$

or

$$F(k) = H(k)G(k) \quad , \quad (14.23)$$

where $G(k)$, called the **Green's function** in the k-domain, is given by

$$G(k) = \frac{-1}{k^2 - i\alpha k - \beta} = \frac{-1}{(k - k_1)(k - k_2)} \quad , \quad (14.24)$$

with

$$k_{\frac{1}{2}} \equiv \frac{i\alpha \pm \sqrt{-\alpha^2 + 4\beta}}{2} \quad . \quad (14.25)$$

The so-called particular solution to (14.21) is then obtained by Fourier transforming $F(k)$ back to $f(x)$, that is,

$$f_p(x) = \frac{-1}{\sqrt{2\pi}} \int_{-\infty}^{\infty} dk \frac{e^{ikx} H(k)}{(k - k_1)(k - k_2)} \quad , \quad (14.26)$$

where the subscript p denotes "particular solution". This type of integral can most easily be evaluated by **contour integration** (see Chapter 40).

We will now introduce a very important mathematical device in theoretical physics, called the **Dirac delta function**. Again, we will not be mathematically rigorous, but will use instead the physicist's customary way of combining physical intuition with formal manipulations. Substituting (14.20) formally back into (14.19), we find that

$$f(x) = \frac{1}{2\pi} \int_{-\infty}^{\infty} dk\, e^{ikx} \int_{-\infty}^{\infty} dx'\, f(x') e^{-ikx'}$$

$$= \int_{-\infty}^{\infty} dx'\, f(x') \left(\frac{1}{2\pi} \int_{-\infty}^{\infty} dk\, e^{-ik(x-x')} \right) \quad , \quad (14.27)$$

where we have assumed the legitimacy of the interchange of the order of integrations. On formally defining the Dirac delta function as follows:

$$\delta(x) = \frac{1}{2\pi} \int_{-\infty}^{\infty} dk\, e^{ikx} = \frac{1}{2\pi} \int_{-\infty}^{\infty} dk\, e^{-ikx} = \delta(-x) \quad , \quad (14.28)$$

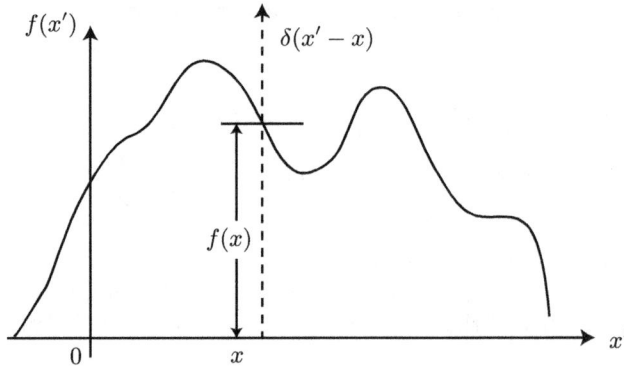

FIGURE 14.2

we can write (14.27) as

$$f(x) = \int_{-\infty}^{\infty} dx'\, f(x')\delta(x' - x) \quad . \tag{14.29}$$

The above equation is to be intuitively understood as follows. The "function" $\delta(x' - x)$ acts like a mask with an infinitely narrow and infinitely tall slit which covers up all values of x' except the point x, and thus filters out all values of the function $f(x')$ except the value $f(x)$. Of course it is assumed that $f(x')$ itself is well-behaved at $x' = x$. (See Fig. 14.2.)

Eq. (14.28) gives the so-called **integral representation** of the Dirac delta function. [There are many other representations for $\delta(x)$, as we shall see later, for example, those given by (14.45) and (14.46).] Putting $f(x) = 1$ in (14.29) and making the interchange $x \leftrightarrow x'$, we have

$$\int_{-\infty}^{\infty} dx\, \delta(x - x') = 1 \quad . \tag{14.30}$$

In fact, changing the variable of integration: $x - x' \to x$, we further have

$$\boxed{\int_{-\infty}^{\infty} dx\, \delta(x) = 1} \quad . \tag{14.31}$$

Another corollary of (14.29) is obtained by putting $x = 0$:

$$f(0) = \int_{-\infty}^{\infty} dx\, f(x)\delta(x) \quad . \tag{14.32}$$

Now, according to (14.28), it is clear that $\delta(x)$ vanishes everywhere except at $x = 0$, at which point $\delta(x) \to \infty$. Intuitively, $\delta(x)$ can be pictured as an infinitely sharp and infinitely tall spike at $x = 0$, such that its integral over all of $\mathbb{R}$, and in fact, over any open interval in $\mathbb{R}$ which covers $x = 0$, is equal to 1! Physicists describe this situation loosely, and with some abuse of mathematical notation, by writing

$$\delta(x) = \begin{cases} 0 & , & x \neq 0 \\ \infty & , & x = 0 \end{cases} . \tag{14.33}$$

$\delta(x)$ is of course not a bona-fide function mathematically. It is, in fact, a map from the space of well-behaved functions $f(x)$ (in some specific sense which we will not spell out) to $\mathbb{R}$ (or $\mathbb{C}$), the set of real (or complex) numbers. Such an object is called a **distribution**, or a **generalized function**.

Let us now return to the Green's function $G(k)$ introduced in (14.24). Aside from being an extremely useful tool in the solution of differential equations, it also has a very important physical interpretation. To see this we rewrite (14.26) as follows:

$$\begin{aligned}
f_p(x) &= \frac{1}{\sqrt{2\pi}} \int_{-\infty}^{\infty} dk \, e^{ikx} H(k) G(k) \\
&= \frac{1}{\sqrt{2\pi}} \int_{-\infty}^{\infty} dk \, e^{ikx} G(k) \frac{1}{\sqrt{2\pi}} \int_{-\infty}^{\infty} dx' \, e^{-ikx'} h(x') \\
&= \frac{1}{\sqrt{2\pi}} \int_{-\infty}^{\infty} dx' \, h(x') \left(\frac{1}{\sqrt{2\pi}} \int_{-\infty}^{\infty} dk \, G(k) e^{ik(x-x')} \right) ,
\end{aligned} \tag{14.34}$$

where in the RHS of the second equality we have used (14.20) for $H(k)$. On recognizing that the quantity in parentheses is in fact the Fourier transform of $G(k)$, which we will denote by $g(x - x')$:

$$g(x) = \frac{1}{\sqrt{2\pi}} \int_{-\infty}^{\infty} dk \, e^{ikx} G(k) , \tag{14.35}$$

we can write

$$f_p(x) = \frac{1}{\sqrt{2\pi}} \int_{-\infty}^{\infty} dx' \, h(x') g(x - x') . \tag{14.36}$$

The quantity $g(x)$ is called the **Green's function** (in the x-domain) associated with the differential equation (14.21). Eq. (14.36) gives the particular solution of (14.21) as a so-called **correlation function**, or a **convolution product**, sometimes written as

$$h(x) * g(x) \equiv \frac{1}{\sqrt{2\pi}} \int_{-\infty}^{\infty} dx' \, h(x') g(x - x') . \tag{14.37}$$

Thus, according to (14.36) and the first equality of (14.34), the Fourier transform of a convolution product $h(x) * g(x)$ is an ordinary product of the Fourier transforms of $h(x)$ and $g(x)$ (and vice versa):

$$h(x) * g(x) \xleftrightarrow{\text{F.T.}} H(k) G(k) . \tag{14.38}$$

Now suppose the **inhomogeneous term** on the RHS of the differential equation (14.21) (the "driving force") is given by the Dirac delta function $\delta(x)$. Then (14.36), with the help of (9.29), implies

$$f_p(x) = \frac{1}{\sqrt{2\pi}} \int_{-\infty}^{\infty} dx'\, \delta(x')g(x-x') = \frac{g(x)}{\sqrt{2\pi}} \quad . \tag{14.39}$$

Thus the Green's function $g(x)$ (divided by $\sqrt{2\pi}$) is the particular solution (or **response**) to the differential equation (14.21) when the inhomogeneous term (or "driving force") is the infinitely sharp $\delta(x)$ (applied at $x = 0$).

If we evaluate $g(x)$ using (14.35) and the expression for $G(k)$ given by (14.24), it will be seen that, for $\alpha > 0$ and $\beta > 0$,

$$g(x) = 0, \quad \text{for } x < 0 \quad . \tag{14.40}$$

The upper limit in the integral expression for the particular solution (or **response function**) can then be capped at x:

$$f_p(x) = \frac{1}{\sqrt{2\pi}} \int_{-\infty}^{x} dx'\, h(x')g(x-x') \quad . \tag{14.41}$$

If x were a time variable, the above equation then says that the response $f_p(x)$ at the time x only depends on the driving force $h(x')$ at times x' earlier than x. This is in fact a statement of the physical **principle of causality**.

It is a most amazing and interesting mathematical fact that the causality condition (14.40) on $g(x)$ is encoded in certain **analytic properties** of its Fourier transform $G(k)$, when the latter is viewed as a function of the complex variable k. We will not explore these properties here.

The integral representation of $\delta(x)$ [(14.28)] also leads to the so-called **Parsevaal's Theorem**:

$$\boxed{\int_{-\infty}^{\infty} dx\, \overline{f(x)}h(x) = \int_{-\infty}^{\infty} dk\, \overline{F(k)}H(k)} \quad , \tag{14.42}$$

where $(f(x), F(k))$ and $(h(x), H(k))$ are Fourier transform pairs.

Exercise 14.2   Verify (14.42).

In particular, we have

$$\int_{-\infty}^{\infty} |f(x)|^2\, dx = \int_{-\infty}^{\infty} |F(k)|^2\, dk \quad . \tag{14.43}$$

Suppose $f(x)$ represents the field strength of some signal as a function of the position x. Then $|f(x)|^2$ is proportional to the signal power as a function of position. Thus (14.43) has the important physical interpretation that the

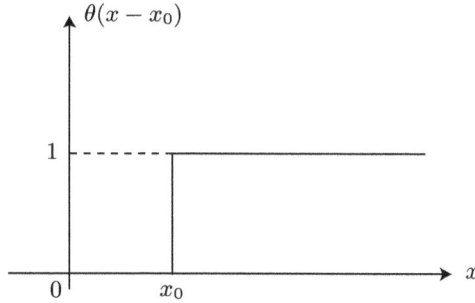

FIGURE 14.3

integrated power over space is the same as the integrated power of the Fourier component $F(k)$ over the wave number k.

We will gather here some of the most frequently used representations of the Dirac delta function:

$$\delta(x - x_0) = \frac{1}{2\pi} \int_{-\infty}^{\infty} dk \, e^{ik(x-x_0)} \quad , \tag{14.44}$$

$$\delta(x - x_0) = \frac{1}{\pi} \lim_{\epsilon \to 0^+} \frac{\epsilon}{(x - x_0)^2 + \epsilon^2} \quad , \tag{14.45}$$

$$\delta(x - x_0) = \lim_{\eta \to 0} \frac{\theta(x - x_0 + \eta) - \theta(x - x_0)}{\eta} = \frac{d}{dx} \theta(x - x_0) \quad , \tag{14.46}$$

where $\theta(x - x_0)$ is the **step function** shown in Fig. 14.3.

We also state the following commonly used properties of the Dirac delta function for reference:

$$\delta(x) = \delta(-x) \quad , \tag{14.47}$$

$$\delta(ax) = \frac{1}{|a|} \delta(x) \quad , \quad a \in \mathbb{R} \quad , \tag{14.48}$$

$$x\delta(x) = 0 \quad , \tag{14.49}$$

$$\delta(f(x)) = \sum_n \frac{1}{|f'(x_n)|} \delta(x - x_n) \quad , \tag{14.50}$$

where x_n are the roots of the equation $f(x) = 0$ with $f'(x_n) \neq 0$,

$$\int_{-\infty}^{\infty} dx \, f(x) \frac{d^n \delta(x)}{dx^n} = (-1)^n \left. \frac{d^n f(x)}{dx^n} \right|_{x=0} \quad , \tag{14.51}$$

$$\sum_{n=-\infty}^{\infty} \delta(x - a - nb) = \frac{1}{b} \sum_{m=-\infty}^{\infty} e^{i2\pi m(x-a)/b} \quad . \tag{14.52}$$

The last formula is known as **Poisson's sum rule**.

Chapter 15

The Continuous Spectrum and Non-Normalizable States

In Chapter 13 we introduced the Hilbert space $\mathcal{H}$ of square-integrable functions [(13.18)]. These functions are also called normalizable functions, since for a $\psi(x) \in \mathcal{H}$ it is usual to define the **norm** (or length squared) of $\psi(x)$ by

$$\|\psi(x)\|^2 = \int_{-\infty}^{\infty} dx \, |\psi(x)|^2 = \langle \psi | \psi \rangle \quad . \tag{15.1}$$

In quantum mechanics normalizable wave functions describe **bound states** of physical systems, and usually arise as eigenstates belonging to discrete eigenvalues of hermitian (self-adjoint) operators. Examples are bound states of atoms and molecules, nuclei, and bound quark states of nucleons. The term "bound" means that the system is spatially localized and hence the integral on the RHS of (15.1) must not diverge. The harmonic oscillator studied in Chapter 13 provides an example of a hermitian operator – the Hamiltonian H – possessing a spectrum which is entirely discrete and non-degenerate, and whose eigenstates form a complete set.

There are many situations where a continuous spectrum arises naturally also. Just think of a freely moving electron whose **configuration space** is an infinite volume. The possible energy eigenvalues then make up the entire interval $[0, \infty)$. We will see that the corresponding eigenstates are necessarily non-normalizable.

Let us now introduce the **position operator** Q in $\mathcal{H}$, whose action on any normalizable wave function $\psi(x) \in \mathcal{H}$ is given by

$$Q\psi(x) = x\psi(x) \quad . \tag{15.2}$$

$\boxed{\text{Exercise 15.1}}$ Show that Q is a hermitian operator, i.e., show that for any

$|\phi\rangle, |\psi\rangle \in \mathcal{H}$,

$$\langle \phi | Q | \psi \rangle = \overline{\langle \psi | Q | \phi \rangle} \quad . \tag{15.3}$$

What are the eigenvalues and eigenfunctions of Q? On writing

$$Q\phi(x) = x\phi(x) = \lambda\phi(x) \quad , \tag{15.4}$$

where $\lambda \in \mathbb{R}$ is a constant, we see that a formal solution is the Dirac delta function

$$\phi(x) = \delta(x - \lambda) \quad . \tag{15.5}$$

Hence we have a situation where the eigenfunction $\phi(x)$ does not belong to $\mathcal{H}$. It is not even a function, let alone a square-integrable function. In the mathematics literature, it is more properly called a **distribution**. Dirac, however, still considers it as a legitimate eigenstate of Q, and labels it as the ket $|\lambda\rangle$: the eigenstate belonging to the eigenvalue λ of the position operator Q. Thus he writes

$$Q|\lambda\rangle = \lambda|\lambda\rangle \quad , \tag{15.6}$$

from which it follows that

$$\langle \lambda' | Q | \lambda \rangle = \lambda \langle \lambda' | \lambda \rangle \quad . \tag{15.7}$$

On the other hand,

$$\langle \lambda' | Q | \lambda \rangle = \int_{-\infty}^{\infty} dx\, \delta(x - \lambda')\lambda\delta(x - \lambda)$$
$$= \lambda\delta(\lambda' - \lambda) \quad . \tag{15.8}$$

Comparing (15.7) and (15.8), we have

$$\boxed{\langle \lambda' | \lambda \rangle = \delta(\lambda' - \lambda)} \quad . \tag{15.9}$$

Thus the states $|\lambda\rangle$ are definitely not normalizable, since $\langle \lambda | \lambda \rangle = \delta(0) = \infty$. We sometimes say that the states $|\lambda\rangle$ belonging to the continuous spectrum normalize to the delta function.

Dirac's approach to the mathematical formulation of quantum mechanics is to admit these vectors of infinite norm into the theory, and thus enlarge the Hilbert space of normalizable functions to what is called **rigged Hilbert space**. Because of the convenience and the elegance of the Dirac bracket notation, his approach is almost universally adopted by physicists. Another approach is to use the **spectral theorem** of hermitian operators in Hilbert space and formulate quantum theory in terms of **projection operators** when dealing with the continuous part of the spectrum. This latter approach was pioneered by von-Neumann (see J. von Neumann 1955); and its rigorous formulation actually preceded that of rigged Hilbert space. We will only consider Dirac's approach.

Making use of the delta function property (14.29), we can calculate the inner product of the position eigenket $|\lambda\rangle$ and an arbitrary $|\psi\rangle$:

$$\langle\lambda|\psi\rangle = \int_{-\infty}^{\infty} dx\, \delta(x - \lambda)\psi(x) = \psi(\lambda) \quad . \tag{15.10}$$

Thus a wave function $\psi(x)$ is given in Dirac notation by

$$\boxed{\psi(x) = \langle x|\psi\rangle} \quad , \tag{15.11}$$

which is called the **coordinate representation** of the state $|\psi\rangle$.

Now consider the formal integral over x of the "projection operators" $|x\rangle\langle x|$, that is,

$$\int_{-\infty}^{\infty} dx\, |x\rangle\langle x| \quad . \tag{15.12}$$

For any $|\phi\rangle, |\psi\rangle \in \mathcal{H}$, we have

$$\langle\phi| \left(\int_{-\infty}^{\infty} dx\, |x\rangle\langle x| \right) |\psi\rangle = \int_{-\infty}^{\infty} dx\, \langle\phi|x\rangle\langle|\psi\rangle = \int_{-\infty}^{\infty} dx\, \overline{\phi(x)}\psi(x) = \langle\phi|\psi\rangle \quad . \tag{15.13}$$

Thus we have the completeness condition for $|x\rangle, x \in \mathbb{R}$:

$$\boxed{\int_{-\infty}^{\infty} dx\, |x\rangle\langle x| = 1} \quad . \tag{15.14}$$

This is to be compared with (12.12), which gives the completeness condition for a countable orthonormal basis set of $\mathcal{H}$.

Eq. (15.10) is consistent with our understanding that the wave function corresponding to $|\lambda\rangle$ is $\delta(x - \lambda)$, since by (15.9) and (15.11),

$$\langle x|\lambda\rangle = \delta(x - \lambda) \quad . \tag{15.15}$$

$|\lambda\rangle$ can then be interpreted as a state describing a particle located with certainty at the position $x = \lambda$.

Now we know that in quantum mechanics a particle of definite momentum $\hbar k$ is described by the **plane-wave** function

$$\psi_k(x) = \frac{1}{\sqrt{2\pi}} e^{ikx} \quad . \tag{15.16}$$

Thus, analogous to the position states $|\lambda\rangle$, $-\infty < \lambda < \infty$, we can introduce a complete set of momentum states $|k\rangle$, $-\infty < k < \infty$, such that

$$\int_{-\infty}^{\infty} dk\, |k\rangle\langle k| = 1 \quad , \tag{15.17}$$

and

$$\psi_k(x) = \langle x | k \rangle = \frac{1}{\sqrt{2\pi}} \, e^{ikx} \quad . \tag{15.18}$$

It is easily verified that, using the expression

$$p = -i\hbar \frac{\partial}{\partial x} \tag{15.19}$$

[c.f. (4.4)] for the linear momentum operator,

$$p \, \psi_k(x) = \hbar k \, \psi_k(x) \quad . \tag{15.20}$$

The above equation is the **coordinate representation** of the abstract statement

$$\boxed{p | k \rangle = \hbar k | k \rangle} \quad , \tag{15.21}$$

which is the analog of (15.6) for the linear momentum operator.

Referring to (15.11), let us now consider the quantity

$$\psi(k) = \langle k | \psi \rangle \quad , \tag{15.22}$$

which is the **momentum representation** of the state $| \psi \rangle$. We have, using the completeness condition (15.14),

$$\psi(k) = \langle k | \psi \rangle = \int_{-\infty}^{\infty} dx \, \langle k | x \rangle \langle x | \psi \rangle = \int_{-\infty}^{\infty} dx \, \overline{\langle x | k \rangle} \langle x | \psi \rangle \ ,$$

or

$$\boxed{\psi(k) = \frac{1}{\sqrt{2\pi}} \int_{-\infty}^{\infty} dx \, e^{-ikx} \psi(x)} \quad . \tag{15.23}$$

Conversely, using (15.17),

$$\psi(x) = \langle x | \psi \rangle = \int_{-\infty}^{\infty} dk \, \langle x | k \rangle \langle k | \psi \rangle \quad ,$$

or

$$\boxed{\psi(x) = \frac{1}{\sqrt{2\pi}} \int_{-\infty}^{\infty} dk \, e^{ikx} \psi(k)} \quad . \tag{15.24}$$

Thus we have the important result that $\langle x | \psi \rangle = \psi(x)$ and $\langle k | \psi \rangle = \psi(k)$, the coordinate and momentum representations of the state $| \psi \rangle$, respectively, are Fourier transforms of each other.

Exercise 15.2 Starting with the fact $\langle x | \lambda \rangle = \delta(x - \lambda)$ and using the integral representation (14.28) for the delta function $\delta(x - \lambda)$:

$$\delta(x - \lambda) = \frac{1}{2\pi} \int_{-\infty}^{\infty} dk \, e^{ik(x - \lambda)} \quad , \tag{15.25}$$

deduce the completeness condition (15.17) for $|k\rangle$.

Exercise 15.3 Start with Eq. (15.21):

$$p\,|\,k\rangle = \hbar k\,|\,k\rangle \quad ,$$

show that

$$\langle x|\,p\,|\,k\rangle = -i\hbar \frac{\mathrm{d}}{\mathrm{d}x}\,\psi_k(x) \quad , \tag{15.26}$$

and hence the validity of (15.20).

Exercise 15.4 Start with Eq. (15.6):

$$Q\,|\,x\rangle = x\,|\,x\rangle \quad , \tag{15.27}$$

show that

$$\langle k|\,Q\,|\,x\rangle = i\frac{\mathrm{d}}{\mathrm{d}k}\,\langle k|\,x\rangle \quad , \tag{15.28}$$

and thus

$$i\frac{\mathrm{d}}{\mathrm{d}k}\,\langle k|\,x\rangle = x\,\langle k|\,x\rangle \quad . \tag{15.29}$$

Since $\langle k|\,x\rangle = \psi_x(k)$ can be interpreted as the momentum representation of a state $|\,x\rangle$ of definite position x, (15.29) implies that the position operator Q defined by (15.6) is replaced by

$$\boxed{Q \longrightarrow i\frac{\mathrm{d}}{\mathrm{d}k}} \tag{15.30}$$

when acting on wave functions in the momentum representation, just as the momentum operator is replaced by

$$\boxed{p \longrightarrow -i\hbar\frac{\mathrm{d}}{\mathrm{d}x}} \tag{15.31}$$

when acting on wave functions in the coordinate representation.

By way of illustrating the power and elegance of the Dirac notation, let us prove once more the fundamental quantum principle (15.31) as applied to any $\psi(x)$ [and not just $\psi_k(x)$]. We have

$$\langle x|\,p\,|\,\psi\rangle = \int_{-\infty}^{\infty} \mathrm{d}k\,\langle x|\,k\rangle\langle k|\,p\,|\,\psi\rangle = \frac{1}{\sqrt{2\pi}}\int_{-\infty}^{\infty} \mathrm{d}k\,e^{ikx}\hbar k\langle k|\,\psi\rangle$$

$$= -i\hbar\frac{\mathrm{d}}{\mathrm{d}x}\left(\frac{1}{\sqrt{2\pi}}\int_{-\infty}^{\infty} \mathrm{d}k\,e^{ikx}\psi(k)\right) = -i\hbar\frac{\mathrm{d}}{\mathrm{d}x}\psi(x) \quad . \tag{15.32}$$

Exercise 15.5   Repeat the above procedure to verify (15.30) by calculating $\langle k|\, Q\, |\psi\rangle$.

The fact that $\psi(x)$ and $\psi(k)$ are Fourier transforms of each other is intimately related to the uncertainty principle in quantum mechanics.

Exercise 15.6   Use the Gaussian integral formula

$$\int_{-\infty}^{\infty} e^{-ax^2}\, dx = \sqrt{\frac{\pi}{a}} \tag{15.33}$$

to calculate the Fourier transform $\psi(k)$ of the normalized Gaussian wave function

$$\psi(x) = \left(\frac{2a}{\pi}\right)^{\frac{1}{4}} e^{-ax^2} \quad, \tag{15.34}$$

where

$$a = \left(\frac{1}{2\Delta x}\right)^2 \quad, \tag{15.35}$$

with Δx being the width of $|\psi(x)|^2$. (See Fig. 15.1.) Show that $\psi(k)$ is also a Gaussian, and the width Δk of $|\psi(k)|^2$ satisfies

$$(\Delta x)(\Delta k) = \frac{1}{2} \quad. \tag{15.36}$$

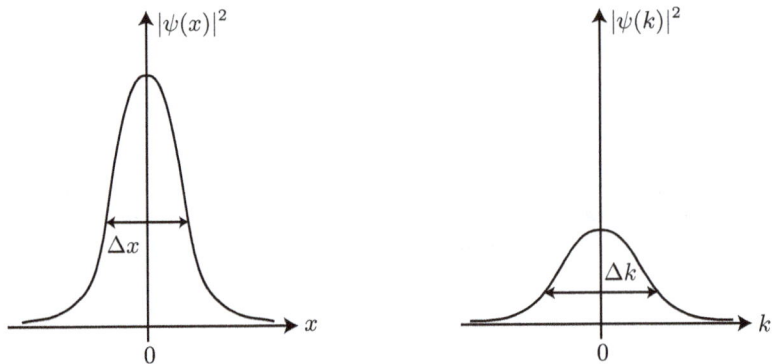

FIGURE 15.1.

The Gaussian $\psi(x)$ is in fact a **minimum uncertainty wave packet**. All other wave functions satisfy

$$\boxed{(\Delta x)(\Delta p) \geq \frac{\hbar}{2}} \quad, \tag{15.37}$$

which is the statement of **Heisenberg's uncertainty principle.**

Chapter 16

Skew-Symmetric Tensors and Determinants

In this chapter we begin the development of some algebraic tools which will be important in the study of the algebra of skew-symmetric tensors – or **exterior algebra**. This subject is the foundation of the so-called **exterior differential calculus of differential forms**, which is an invaluable tool in understanding the topology and geometry of manifolds. Much of the fundamental laws of physics, as it turns out, are best formulated and understood in terms of topological and geometrical concepts; and tensor analysis and differential forms provide powerful and complementary analytic tools for their study. Vector analysis of vector fields living on three-dimensional Euclidean manifolds is but a very special case of exterior calculus; and we will see later (in Chapter 31) how the techniques of differential forms, which are applicable to manifolds of arbitrary type and dimensionality, completely subsume conventional vector calculus.

Let V be a vector space and V^* its dual space. The tensor product

$$\underbrace{V \otimes \cdots \otimes V}_{r \text{ times}} \otimes \underbrace{V^* \otimes \cdots \otimes V^*}_{s \text{ times}} = T_s^r(V) \tag{16.1}$$

is itself a vector space, namely, a space of (r, s)-type tensors. In particular let us consider

$$T^r(V) = \underbrace{V \otimes \cdots \otimes V}_{r \text{ times}} \tag{16.2}$$

and

$$T^r(V^*) = \underbrace{V^* \otimes \cdots \otimes V^*}_{r \text{ times}} \quad . \tag{16.3}$$

These are dual spaces to each other, with the pairing between them given by

$$\langle v_1 \otimes \cdots \otimes v_r \,, v^{*1} \otimes \cdots \otimes v^{*r} \rangle = \langle v_1, v^{*1} \rangle \cdots \langle v_r, v^{*r} \rangle \quad , \tag{16.4}$$

where $v_i \in V$ and $v^{*i} \in V^*$, $i = 1, \ldots, r$.

To characterize the **symmetry properties** of tensors which are not of mixed type, we need to introduce the so-called **symmetric group** (or **permutation group**) $S(r)$. This is simply the group of permutations of the set of r numbers $\{1, 2, \ldots, r\}$. It has $r!$ elements. An element $\sigma \in S(r)$ is usually represented by

$$\begin{pmatrix} 1 & 2 & 3 & \cdots & r \\ \sigma(1) & \sigma(2) & \sigma(3) & \cdots & \sigma(r) \end{pmatrix} . \tag{16.5}$$

For example, in $S(4)$, we have the following permutation:

$$\begin{pmatrix} 1 & 2 & 3 & 4 \\ 4 & 2 & 1 & 3 \end{pmatrix} . \tag{16.6}$$

The identity $e \in S(4)$ is, of course,

$$\begin{pmatrix} 1 & 2 & 3 & 4 \\ 1 & 2 & 3 & 4 \end{pmatrix} .$$

Group multiplication of two elements $\sigma, \rho \in S(r)$ is defined as the composition

$$(\sigma\rho)(i) = (\sigma \circ \rho)(i) = \sigma(\rho(i)), \quad i = 1, \ldots, r . \tag{16.7}$$

For instance, if σ is given by (16.6) and

$$\rho = \begin{pmatrix} 1 & 2 & 3 & 4 \\ 1 & 3 & 2 & 4 \end{pmatrix} , \tag{16.8}$$

then

$$\sigma\rho = \begin{pmatrix} 1 & 2 & 3 & 4 \\ 4 & 1 & 2 & 3 \end{pmatrix} . \tag{16.9}$$

The inverse of σ, σ^{-1}, is simply the permutation which undoes the effects of σ. For σ given by (16.6), σ^{-1} would be

$$\sigma^{-1} = \begin{pmatrix} 1 & 2 & 3 & 4 \\ 3 & 2 & 4 & 1 \end{pmatrix} . \tag{16.10}$$

We note that $S(r)$ is a non-abelian group for any $r > 2$.

| Exercise 16.1 | For σ and ρ given by (16.6) and (16.8), respectively, show that $\sigma\rho \neq \rho\sigma$.

Remark. The group representation theory of $\mathcal{S}(r)$ is an extremely important topic in the quantum mechanical description of systems of r identical particles (see Chapter 20).

Any element $\sigma \in \mathcal{S}(r)$ determines an automorphism of the vector space $T^r(V)$ [and $T^r(V^*)$]. Suppose $x \in T^r(V)$. Then x is an r-linear function on $T^r(V^*)$. The r-linear function $\sigma x \in T^r(V)$ is defined by

$$(\sigma x)(v^{*1}, \ldots, v^{*r}) = x(v^{*\sigma(1)}, \ldots, v^{*\sigma(r)}) \qquad , \tag{16.11}$$

where $v^{*i} \in V^*$.

Exercise 16.2 Show that if x is a monomial, that is, if

$$x = v_1 \otimes \cdots \otimes v_r, \quad v_1, \ldots, v_r \in V \quad , \tag{16.12}$$

then

$$\sigma x = v_{\sigma^{-1}(1)} \otimes \cdots \otimes v_{\sigma^{-1}(r)} \quad . \tag{16.13}$$

Show that this is equivalent to permuting the positions of the vectors $v_1, \ldots, v_r$.

Definition 16.1. *Suppose $x \in T^r(V)$. If for any $\sigma \in \mathcal{S}(r)$ we have*

$$\sigma x = x \quad , \tag{16.14}$$

*then x is called a **symmetric contravariant tensor of rank** r, or a symmetric $(r, 0)$-type tensor. If for any $\sigma \in \mathcal{S}(r)$ we have*

$$\sigma x = sgn\,\sigma \cdot x \quad , \tag{16.15}$$

where $sgn\,\sigma$ denotes the sign of the permutation σ:

$$sgn\,\sigma = \begin{cases} +1 & \text{if } \sigma \text{ is an **even permutation,}} \\ -1 & \text{if } \sigma \text{ is an **odd permutation,}} \end{cases} \tag{16.16}$$

*then x is called a **skew-symmetric** (or **alternating**) **contravariant tensor of rank** r, or a skew-symmetric (or alternating) $(r, 0)$-type tensor.*

Instead of viewing the symmetry properties of tensors through the above definition (which is preferred by mathematicians), physicists are more accustomed to look at the symmetry properties of the components of tensors under permutations of their indices. This is possible due to the following theorem.

Theorem 16.1. *Suppose $x \in T^r(V)$ or $[T^r(V^*)]$. x is a symmetric (skew-symmetric) tensor if and only if all its components are symmetric (skew-symmetric) with respect to any permutation of the indices.*

Proof. Choose a basis $\{e_1, \ldots, e_n\}$ of V and write

$$x = x^{i_1 \cdots i_r} e_{i_1} \otimes \cdots \otimes e_{i_r} \quad , \tag{16.17}$$

so that

$$x^{i_1 \cdots i_r} = x(e^{*i_1}, \ldots, e^{*i_r}) \quad . \tag{16.18}$$

Suppose $x \in T^r(V)$ is symmetric. Then, for any $\sigma \in \mathcal{S}(r)$,

$$x^{i_1 \cdots i_r} = x(e^{*i_1}, \ldots, e^{*i_r}) = \sigma x(e^{*i_1}, \ldots, e^{*i_r})$$
$$= x(e^{*i_{\sigma(1)}}, \ldots, e^{*i_{\sigma(r)}}) = x^{i_{\sigma(1)} \cdots i_{\sigma(r)}} \ . \tag{16.19}$$

On the other hand, if $x \in T^{(V)}$ is skew-symmetric, then for any $\sigma \in \mathcal{S}(r)$,

$$x^{i_1 \cdots i_r} = x(e^{*i_1}, \ldots, e^{*i_r}) = (sgn \, \sigma) \, \sigma x(e^{*i_1}, \ldots, e^{*i_r})$$
$$= (sgn \, \sigma) \, x(e^{*i_{\sigma(1)}}, \ldots, e^{*i_{\sigma(r)}}) = (sgn \, \sigma) \, x^{i_{\sigma(1)} \cdots i_{\sigma(r)}} \ . \tag{16.20}$$

| Exercise 16.3 | Prove the converses, that is,

(i) if $x^{i_{\sigma(1)} \cdots i_{\sigma(r)}} = x^{i_1 \cdots i_r}$ for any $\mathcal{S}(r)$, then x is symmetric; and
(ii) if $x^{i_{\sigma(1)} \cdots i_{\sigma(r)}} = (sgn \, \sigma) \, x^{i_1 \cdots i_r}$, then x is skew-symmetric.

The case for $x \in T^r(V^*)$ is entirely similar.   □

Note that any permutation $\sigma \in \mathcal{S}(r)$ can be expressed as the product of a certain number n of pair interchanges. If n is even(odd), the σ is said to be an even(odd) permutation. Each of the indices i_1 through i_r in (16.19) or (16.20) can range from 1 to n, where $n = dim(V)$ (the dimension of V), and so the values of some of the indices may be repeated. It is clear that if x is skew-symmetric, then $x^{i_1, \ldots, i_r} = 0$ whenever any pair of indices have the same value.

| Exercise 16.4 | Determine $sgn \, \sigma$ and $sgn \, \rho$ for σ and ρ given by (16.6) and (16.8), respectively.

Denote the subset of all symmetric $x \in T^r(V)$ by $P^r(V)$, and the subset of all skew-symmetric $x \in T^r(V)$ by $\Lambda^r(V)$. Since

a) the sum of two symmetric (skew-symmetric) tensors is still symmetric (skew-symmetric);

b) scalar multiplication of $x \in T^r(V)$ by $\alpha \in \mathbb{F}$ (the field of V) does not change the symmetry property of x; and

c) the zero vector $0 \in T^r(V)$ is both symmetric and skew-symmetric,

$P^r(V)$ and $\Lambda^r(V)$ are both vector subspaces of $T^r(V)$. In fact, they can be obtained from $T^r(V)$ by the so-called **symmetrizing map** S_r and the **alternating map** A_r, respectively. These are defined by

$$S_r(x) = \frac{1}{r!} \sum_{\sigma \in \mathcal{S}(r)} \sigma x \quad , \tag{16.21}$$

$$A_r(x) = \frac{1}{r!} \sum_{\sigma \in \mathcal{S}(r)} (sgn\,\sigma)\,\sigma x \quad , \tag{16.22}$$

for all $x \in T^r(V)$. Each sum in the above two equations runs through all permutations in $\mathcal{S}(r)$. We have the following useful theorem, which will be stated without proof.

Theorem 16.2.

$$P^r(V) = S_r(T^r(V)) \quad , \tag{16.23}$$
$$\Lambda^r(V) = A_r(T^r(V)) \quad . \tag{16.24}$$

As a simple but important example, consider a two-dimensional vector space V with a chosen basis $\{e_1, e_2\}$, so that an arbitrary $x \in T^2(V)$ can be written

$$x = x^{i_1 i_2} e_{i_1} \otimes e_{i_2}$$
$$= x^{11} e_1 \otimes e_1 + x^{12} e_1 \otimes e_2 + x^{21} e_2 \otimes e_1 + x^{22} e_2 \otimes e_2 \,. \tag{16.25}$$

In general, the four components x^{11}, x^{12}, x^{21} and x^{22} of x do not bear any relationship to each other. Now the symmetry group $\mathcal{S}(2)$ has just two elements: the identity

$$e = \begin{pmatrix} 1 & 2 \\ 1 & 2 \end{pmatrix} \quad ,$$

and the single interchange

$$\sigma = \begin{pmatrix} 1 & 2 \\ 2 & 1 \end{pmatrix} \quad .$$

The inverse of the latter, σ^{-1}, is again equal to σ. Applying (16.13), we then have

$$S_2(e_1 \otimes e_1) = \frac{1}{2}(e_1 \otimes e_1 + e_1 \otimes e_1) = e_1 \otimes e_1 \quad , \tag{16.26}$$

$$S_2(e_1 \otimes e_2) = \frac{1}{2}(e_1 \otimes e_2 + e_2 \otimes e_1) = S_2(e_2 \otimes e_1) \quad , \tag{16.27}$$

$$S_2(e_2 \otimes e_2) = e_2 \otimes e_2 \quad , \tag{16.28}$$

$$A_2(e_1 \otimes e_1) = \frac{1}{2}(e_1 \otimes e_1 - e_1 \otimes e_1) = 0 = A_2(e_2 \otimes e_2) \quad , \tag{16.29}$$

$$A_2(e_1 \otimes e_2) = \frac{1}{2}(e_1 \otimes e_2 - e_2 \otimes e_1) \quad , \tag{16.30}$$

$$A_2(e_2 \otimes e_1) = \frac{1}{2}(e_2 \otimes e_1 - e_1 \otimes e_2) = -A_2(e_1 \otimes e_2) \quad . \tag{16.31}$$

Hence,

$$S_2(x) = x^{11} e_1 \otimes e_1 + x^{22} e_2 \otimes e_2 + \frac{1}{2}(x^{12} + x^{21})(e_1 \otimes e_2 + e_2 \otimes e_1) , \tag{16.32}$$

$$A_2(x) = \frac{1}{2}(x^{12} - x^{21})(e_1 \otimes e_2 - e_2 \otimes e_1) . \tag{16.33}$$

Thus we see that $P^2(V)$ is 3-dimensional (with a possible basis set $\{e_1 \otimes e_1, e_2 \otimes e_2, e_1 \otimes e_2 + e_2 \otimes e_1\}$), and $\Lambda^2(V)$ is 1-dimensional (with a possible basis $\{e_1 \otimes e_2 - e_2 \otimes e_1\}$). In this case

$$T^2(V) = P^2(V) \oplus \Lambda^2(V) \quad , \tag{16.34}$$

that is, the 4-dimensional $T^2(V)$ is the **direct sum** of the 3-dimensional $P^2(V)$ and the 1-dimensional $\Lambda^2(V)$.

Remark. $T^r(V)$ cannot be decomposed into a direct sum similar to (16.34) when $r > 2$.

What does all this have to do with physics? Let us just consider one example: a system of two identical spin $\frac{1}{2}$ particles, such as two electrons. Suppose e_1 and e_2 represent the so-called "spin up" and "spin down" states, respectively. Using the Dirac notation, we can write

$$e_1 = |\uparrow\rangle \quad , \tag{16.35}$$

$$e_2 = |\downarrow\rangle \quad , \tag{16.36}$$

$$e_1 \otimes e_1 = |\uparrow\uparrow\rangle \quad , \tag{16.37}$$

$$e_2 \otimes e_2 = |\downarrow\downarrow\rangle \quad , \tag{16.38}$$

$$e_1 \otimes e_2 + e_2 \otimes e_1 = |\uparrow\downarrow\rangle + |\downarrow\uparrow\rangle \quad , \tag{16.39}$$

$$e_1 \otimes e_2 - e_2 \otimes e_1 = |\uparrow\downarrow\rangle - |\downarrow\uparrow\rangle \quad . \tag{16.40}$$

The set $\{|\uparrow\uparrow\rangle, |\downarrow\downarrow\rangle, |\uparrow\downarrow\rangle + |\downarrow\uparrow\rangle\}$ is a basis of the 3-dimensioanl space of symmetric two-particle states (under exchange of the particles), while

$$\{|\uparrow\downarrow\rangle - |\downarrow\uparrow\rangle\}$$

is a basis of the 1-dimensional space of antisymmetric two-particle states (under exchange of the particles). The one-particle states $| \uparrow \rangle$ and $| \downarrow \rangle$ are usually assumed to be normalized so that

$$\langle \uparrow | \uparrow \rangle = \langle \downarrow | \downarrow \rangle = 1 \quad . \tag{16.41}$$

Then the **triplet** of normalized symmetric states which span $P^2(V)$ are

$$| \uparrow\uparrow \rangle \quad , \quad | \downarrow\downarrow \rangle \quad , \quad \frac{1}{\sqrt{2}}(| \uparrow\downarrow \rangle + | \downarrow\uparrow \rangle) \; ; \tag{16.42}$$

while the **singlet** of normalized antisymmetric state which spans $\Lambda^2(V)$ is

$$\frac{1}{\sqrt{2}}(| \uparrow\downarrow \rangle - | \downarrow\uparrow \rangle) \quad . \tag{16.43}$$

You will encounter expressions like (16.42) and (16.43) frequently in the physics literature. The spaces $P^2(V)$ and $\Lambda^2(V)$ are each **invariant subspaces** of $T^2(V)$ under the rotation group $SO(3)$. In other words, they provide **irreducible representation spaces** for $SO(3)$. This fact is very important in understanding the addition of angular momentum in quantum mechanics.

Exercise 16.5   Verify that (16.32) and (16.33) can be written as

$$S_2(x) = \left(\frac{x^{ij} + x^{ji}}{2} \right) e_i \otimes e_j \quad , \tag{16.44}$$

$$A_2(x) = \left(\frac{x^{ij} - x^{ji}}{2} \right) e_i \otimes e_j \quad . \tag{16.45}$$

In the above development the action of a $\sigma \in S(r)$ on $x \in T^r(V)$ is defined in terms of the actions of x and σx on some element in $T^r(V^*)$, the dual space of $T^r(V)$. Physicists, however, like to work with components of tensors. Starting with (16.17) again:

$$x = x^{i_1 \cdots i_r} e_{i_1} \otimes \cdots \otimes e_{i_r} \quad .$$

We have, by (16.13),

$$\sigma x = x^{i_1 \cdots i_r} \sigma(e_{i_1} \otimes \cdots \otimes e_{i_r}) = x^{i_1 \cdots i_r} e_{i_{\sigma^{-1}(1)}} \otimes \cdots \otimes e_{i_{\sigma^{-1}(r)}} \quad , \tag{16.46}$$

or

$$\boxed{\sigma x = x^{i_{\sigma(1)} \cdots i_{\sigma(r)}} e_{i_1} \otimes \cdots \otimes e_{i_r}} \quad . \tag{16.47}$$

Thus it follows that

$$S_r(x) = \frac{1}{r!} \sum_{\sigma \in S(r)} x^{i_{\sigma(1)} \cdots i_{\sigma(r)}} e_{i_1} \otimes \cdots \otimes e_{i_r} \quad , \tag{16.48}$$

$$A_r(x) = \frac{1}{r!} \sum_{\sigma \in S(r)} (sgn\,\sigma)\, x^{i_{\sigma(1)} \cdots i_{\sigma(r)}} e_{i_1} \otimes \cdots \otimes e_{i_r} \quad . \tag{16.49}$$

Eqs. (16.44) and (16.45) are merely special cases of (16.48) and (16.49), which give $S_r(x)$ and $A_r(x)$ in component form.

Exercise 16.6 Suppose $\{e_1, e_2\}$ is a basis for a 2-dimensional vector space V, and $x \in T^3(V)$ is given by $x = x^{i_1 i_2 i_3} e_{i_1} \otimes e_{i_2} \otimes e_{i_3}, i_1, i_2, i_3 = 1, 2$. Use (16.48) and (16.49) to write explicit expressions for $S_3(x)$ and $A_3(x)$.

If V is spanned by $\{e_1 = |\uparrow\rangle, e_2 = |\downarrow\rangle\}$, then $S_r(x)$ and $A_r(x)$ represent possible spin states of r identical spin $1/2$ particles, such as the spin states of an r-electron atom.

The spaces of skew-symmetric tensors $\Lambda^r(V)$ and $\Lambda^r(V^*)$ will be used to construct exterior algebras later on (Chapter 28), with a multiplication rule called the **exterior product** (denoted by $\wedge$, and also referred to as the **wedge product**) that is a generalization of the ordinary vector (cross) product. The wedge product is a basic operation on differential forms.

Let us now note two facts whose proofs will be deferred until we study exterior products systematically:

i) $\Lambda^r(V)$ (and also $\Lambda^r(V^*)$) $= 0$ if $r > dim(V)$;

ii) $\Lambda^n(V)$ (and also $\Lambda^n(V^*)$) is a one-dimensional vector space if $dim(V) = n$.

The second fact above leads to a most important invariant of a linear transformation A on V, called the **determinant** of A, and denoted $\det(A)$. The term "invariant" means that $\det(A)$ is independent of the choice of basis for V.

The determinant function comes about in a somewhat circuitous route as follows. Corresponding to a linear transformation $A : V \longrightarrow V$, $dim(V) = n$, there is an induced map, called the **pullback** of A and denoted

$$A^* : \Lambda^n(V^*) \longrightarrow \Lambda^n(V^*) \quad , \tag{16.50}$$

that is defined as follows. For any $x \in \Lambda^n(V^*)$, and any $v_1, \dots, v_n \in V$,

$$\boxed{(A^*x)(v_1, \dots, v_n) = x(Av_1, \dots, Av_n)} \quad . \tag{16.51}$$

Since $\Lambda^n(V^*)$ is one-dimensional, the action of A^* on x must be equal to some scalar multiplied by x. The determinant of A is then defined by

$$A^*x = (\det(A))x \quad , \tag{16.52}$$

where $\det(A)$ is the scalar referred to above. Before making contact with the more elementary definition of the determinant of a matrix representing A, we will use (16.51) to derive some familiar properties of the determinant function.

Theorem 16.3. *If the linear operator A is a scalar multiplication, that is, $Av = \alpha v$, for all $v \in V$, where α is a scalar, and if $dim(V) = n$, then $\det(A) = \alpha^n$.*

Proof. For $x \in \Lambda^n(V^*)$, we have, by (16.51),

$$(A^*x)(v_1, \ldots, v_n) = x(Av_1, \ldots, Av_n) = x(\alpha v_1, \ldots, \alpha v_n) = \alpha^n x(v_1, \ldots, v_n) ,$$
(16.53)

where the last inequality follows from the property of the n-linearity of x. $\square$

As special cases of the above theorem we have the following:

$$\det(0) = 0 \quad , \quad \det(1) = 1 ,$$
(16.54)

where the quantities "0" and "1" on the left hand sides represent the zero operator and the identity operator on V, respectively; and the similarly denoted quantities on the right hand sides mean the numbers zero and one, respectively.

Theorem 16.4. *If A and B are linear operators on a vector space, then*

$$\det(AB) = \det(A)\det(B) \quad .$$
(16.55)

Proof. Let $C = AB$. Then, for any $x \in \Lambda^n(V^*)$ and $v_1, \ldots, v_n \in V$,

$$(C^*x)(v_1, \ldots, v_n) = x(Cv_1, \ldots, Cv_n) = x(ABv_1, \ldots, ABv_n)$$
$$= (A^*x)(Bv_1, \ldots, Bv_n) = (B^*A^*x)(v_1, \ldots, v_n) . \quad (16.56)$$

Hence

$$C^* = B^*A^* \quad .$$
(16.57)

But

$$C^*x = \det(C)x = \det(AB)x \quad ,$$
(16.58)

and

$$B^*A^*x = \det(B)A^*x = \det(A)\det(B)x \quad .$$
(16.59)

The theorem then follows from the above two equations. $\square$

The value of $\det(A)$ provides crucial information on the invertibility of A. This is given by the following theorem.

Theorem 16.5. *A linear operator A on a vector space is **invertible** (that is, A^{-1} exists, or A is **non-singular**) if and only if $\det(A) \neq 0$.*

Proof. If A^{-1} exists, then, from (16.54) and (16.55),

$$1 = \det(1) = \det(AA^{-1}) = \det(A)\det(A^{-1}) \quad .$$
(16.60)

It follows that $\det(A) \neq 0$. The above equation also implies that, if A^{-1} exists,

$$\det(A^{-1}) = (\det(A))^{-1} \quad .$$
(16.61)

The proof of the converse is a bit more difficult. We need to prove that, if $\det(A) \neq 0$, then A^{-1} exists. Let $\{e_1, \ldots, e_n\}$ be a basis of V, and $x \in \Lambda^n(V^*)$, $x \neq 0$. If we write

$$x = x_{i_1 \ldots i_n} e^{*i_1} \otimes \cdots \otimes e^{*i_n} \quad ,$$
(16.62)

then
$$x(e_1, \ldots, e_n) = x_{123\ldots n} \neq 0 \quad . \tag{16.63}$$

Indeed, if $x_{123\ldots n}$ did vanish, we would have, by Theorem 16.1,

$$0 = x_{123\ldots n} = -x_{213\ldots n} \quad , \tag{16.64}$$

which implies

$$x_{213\ldots n} = 0 = x_{123\ldots n} \quad . \tag{16.65}$$

The last statement contradicts the supposition that $x \in \Lambda^n(V^*)$, namely, that x is a skew-symmetric tensor (again by Theorem 16.1). Together with the assumption that $\det(A) \neq 0$, Eq. (16.63) implies

$$\det(A) \cdot x(e_1, \ldots, e_n) \neq 0 \quad . \tag{16.66}$$

By (16.52),
$$(A^*x)(e_1, \ldots, e_n) = x(Ae_1, \ldots, Ae_n) \neq 0 \quad . \tag{16.67}$$

This implies that $\{Ae_1, \ldots, Ae_n\}$ is linearly independent, based on the fact that $x(v_1, \ldots, v_n) \neq 0$ implies $\{v_1, \ldots, v_n\}$ is linearly independent, for any $x \in \Lambda^n(V^*)$ and $v_1, \ldots, v_n \in V$. To see the validity of this last assertion, we suppose $\{v_1, \ldots, v_n\}$ to be linearly dependent. Without loss of generality, assume

$$v_1 = \alpha_2 v_2 + \cdots + \alpha_n v_n \quad , \tag{16.68}$$

where $\alpha_2, \ldots, \alpha_n$ are not all zero. Then

$$x(v_1, v_2, \ldots, v_n) = x(\alpha_2 v_2 + \cdots + \alpha_n v_n, v_2, v_3, \ldots, v_n)$$
$$= \alpha_2 x(v_2, v_2, v_3, \ldots, v_n) + \cdots + \alpha_n x(v_n, v_2, v_3, \ldots, v_n) = 0 \,, \quad (16.69)$$

where the last equality follows from the fact that x is a skew-symmetric tensor. Now since $dim(V) = n$, the set of n vectors, $\{Ae_1, \ldots, Ae_n\}$, having been just established to be linearly independent, must also be a basis of V. Thus an e_i must be expressible as a linear combination of $Ae_1, \ldots, Ae_n$. We write

$$e_i = \beta_i^j A e_j = \beta_i^j a_j^k e_k \quad , \tag{16.70}$$

where (a_i^j) is the matrix representation of the linear operator A with respect to the basis $\{e_1, \ldots, e_n\}$ of V. On the other hand

$$e_i = \delta_i^k e_k \quad . \tag{16.71}$$

Thus
$$\beta_i^j a_j^k = \delta_i^k \quad , \tag{16.72}$$

so the matrix (β_i^j) is in fact the inverse of the matrix (a_i^j). It follows that A is invertible. $\qquad \square$

The following theorem is also very important in practical applications.

Theorem 16.6. *The determinant of a linear operator A is invariant under similarity transformations:*

$$\boxed{\det(A) = \det(B^{-1}AB)} \quad , \tag{16.73}$$

for any invertible linear operator B.

Proof. This is a direct corollary of Theorem 16.4. For any invertible B, we have

$$\det(B^{-1}AB) = \det(B^{-1})\det(A)\det(B) = \det(B^{-1}B)\det(A)$$
$$= \det(1)\det(A) = \det(A) .$$

$\square$

We recall that under a change of basis of V, the matrix representation of a linear transformation A on V is changed according to a similarity transformation [c.f. (1.42)]. Thus, according to Theorem 16.6, $\det(A)$ is independent of the choice of basis for V when calculated with a specific matrix representation of A. [See Eqs. (16.75) and (16.78) below.]

We will now derive an expression for $\det(A)$ in terms of the matrix elements a_i^j of a particular representation of A (with respect to a basis $\{e_1,\ldots,e_n\}$ of V). Let $x \in \Lambda^n(V^*)$. Then

$$\det(A)x(e_1,\ldots,e_n) = (A^*x)(e_1,\ldots,e_n) = x(Ae_1,\ldots,Ae_n)$$
$$= x(a_1^{i_1}e_{i_1},\ldots,a_n^{i_n}e_{i_n}) = a_1^{i_1}a_2^{i_2}\ldots a_n^{i_n}x(e_{i_1},\ldots,e_{i_n})$$
$$= \sum_{\sigma \in S(n)} a_1^{\sigma(1)}a_2^{\sigma(2)}\ldots a_n^{\sigma(n)}x(e_{\sigma(1)},\ldots,e_{\sigma(n)})$$
$$= \sum_{\sigma \in S(n)} a_1^{\sigma(1)}a_2^{\sigma(2)}\ldots a_n^{\sigma(n)}(\sigma x)(e_1,\ldots,e_n) \tag{16.74}$$
$$= \sum_{\sigma \in S(n)} a_1^{\sigma(1)}a_2^{\sigma(2)}\ldots a_n^{\sigma(n)}(sgn\,\sigma)x(e_1,\ldots,e_n) ,$$

where in the fifth equality we have used the fact that $x(e_1,\ldots,e_n) = 0$ whenever any two of the arguments are identical, which in turn follows from the fact that x is a skew-symmetric tensor. Now $x(e_1,\ldots,e_n) \neq 0$ [see the argument immediately following (16.63)]. Eq. (16.74) then gives the following result for the computation of $\det(A)$ in terms of the matrix elements of a specific matrix representation of A.

$$\boxed{\det(A) = \sum_{\sigma \in S(n)} (sgn\,\sigma)\,a_1^{\sigma(1)}\ldots a_n^{\sigma(n)}} \quad . \tag{16.75}$$

This result is actually the same as the elementary definition of the determinant of a matrix. We will verify it for the case $n = 3$. Eq. (16.75) yields

$$\det(A) = a_1^1 a_2^2 a_3^3 + a_1^2 a_2^3 a_3^1 + a_1^3 a_2^1 a_3^2 - a_1^1 a_2^3 a_3^2 - a_1^3 a_2^2 a_3^1 - a_1^2 a_2^1 a_3^3$$

$$= \begin{vmatrix} a_1^1 & a_1^2 & a_1^3 \\ a_2^1 & a_2^2 & a_2^3 \\ a_3^1 & a_3^2 & a_3^3 \end{vmatrix} . \tag{16.76}$$

Notice that in (16.75), the column indices (upper indices) of the matrix elements are permuted. It turns out that a similar expression can be written for $\det(A)$ in which the row indices (lower indices) of a_i^j are permuted. Starting with (16.75) we have

$$\det(A) = \sum_\sigma (sgn\,\sigma) a_1^{\sigma(1)} \dots a_n^{\sigma(n)} = \sum_\sigma (sgn\,\sigma) a_{\sigma^{-1}(1)}^1 a_{\sigma^{-1}(2)}^2 \cdots a_{\sigma^{-1}(n)}^n$$

$$= \sum_{\sigma^{-1}} (sgn\,\sigma^{-1}) a_{\sigma^{-1}(1)}^1 a_{\sigma^{-1}(2)}^2 \cdots a_{\sigma^{-1}(n)}^n ,$$

$$\tag{16.77}$$

or

$$\boxed{\det(A) = \sum_{\sigma \in S(n)} (sgn\,\sigma) a_{\sigma(1)}^1 \cdots a_{\sigma(n)}^n} . \tag{16.78}$$

In deriving the above equation, we have used the facts that $sgn\,\sigma^{-1} = sgn\,\sigma$ and $\sum_\sigma = \sum_{\sigma^{-1}}$.

Exercise 16.7 Verify that for the case $n = 3$, (16.78) gives exactly the same result as (16.76).

On replacing $a_{\sigma(i)}^i$ by $(a^T)_i^{\sigma(i)}$ in (16.78) and then comparing with (16.75), where a^T is the transpose of the matrix a, we see that

$$\det(a) = \det(a^T) ,$$

a very well-known result.

Exercise 16.8 Show that if any two rows or two columns of a square matrix are interchanged, the determinant of the matrix changes sign.

Let A_i^j be the $(n-1) \times (n-1)$ matrix obtained by deleting the i-th row and the j-th column from the $(n \times n)$ matrix (a_i^j). Then the following formula gives an algorithm for finding the determinant of a matrix by expanding along the i-th row of the matrix.

$$\det(a_i^j) = (-1)^{i+1} a_i^1 \det(A_i^1) + \cdots + (-1)^{i+n} a_i^n \det(A_i^n) \qquad . \qquad (16.79)$$

Exercise 16.9   Derive the above formula starting with either (16.75) or (16.78).

Finally we give a formula for a practical algorithm for finding the inverse of a matrix, if it exists. If $\det(a_i^j) \neq 0$, then

$$(a_i^j)^{-1} = \left[\frac{(-1)^{i+j} \det(A_i^j)}{\det(a_i^j)} \right]^T \qquad . \qquad (16.80)$$

Exercise 16.10   Verify (16.80).

Chapter 17

Eigenvalue Problems

One of the most frequently encountered problems in the application of linear algebra (a term that is used here synonymously with the theory of linear transformations on vector spaces) to physics, engineering, and other scientific disciplines is the following: Given a linear transformation A on a vector space V, find all scalars $\lambda \in \mathbb{F}$ (the field for V) and vectors $v \in V$, $v \neq 0$, such that

$$Av = \lambda v \quad . \tag{17.1}$$

This is usually referred to as an **eigenvalue problem**. We have seen an example of this in our discussion of the quantum mechanical harmonic oscillator already [c.f. (13.13)], where V is actually an infinite-dimensional (Hilbert) space. In this chapter, we will elaborate on the theory behind eigenvalue problems for finite-dimensional spaces. This theory is important in its own right, since very often one works with finite-dimensional invariant subspaces (under some symmetry operation) of an infinite-dimensional one.

As mentioned in Chapter 13, the set of all eigenvalues of a linear operator A is called the **spectrum** of A. We begin with the following fact.

Theorem 17.1. *The spectrum of a linear operator A is the set of all scalars λ for which $(A - \lambda)$ is not invertible.*

Proof. Suppose λ is an eigenvalue of A. Then there exists a $v \neq 0$ such that

$$(A - \lambda)v = 0 \quad . \tag{17.2}$$

If $(A - \lambda)$ were invertible, we would have

$$(A - \lambda)^{-1}(A - \lambda)v = 1 \cdot v = v \neq 0 \quad . \tag{17.3}$$

On the other hand, from (17.2),

$$(A - \lambda)^{-1}(A - \lambda)v = (A - \lambda)^{-1} \cdot 0 = 0 \quad . \tag{17.4}$$

Thus we arrive at a contradiction; and $(A - \lambda)$ must be not invertible. □

It follows immediately from Theorem 16.5 that the spectrum of a linear operator A on a finite-dimensional vector space can be obtained by solving the equation

$$\boxed{p(\lambda) \equiv \det(A - \lambda) = 0} \qquad . \tag{17.5}$$

As defined above, $p(\lambda)$ is a polynomial of degree n in λ if $dim(V) = n$. This polynomial is called the **characteristic polynomial** of the linear operator A, and (17.5) is called the **characteristic equation** of A. The eigenvalues λ of A are also called the **characteristic roots** of A.

If V is a real vector space ($\mathbb{F} = \mathbb{R}$), (17.5) does not always have roots (in $\mathbb{R}$). By the **fundamental theorem of algebra**, however, any polynomial with coefficients in $\mathbb{C}$ (the complex number field) will always have roots in $\mathbb{C}$. Thus $\mathbb{C}$ is an algebraically closed field; and a complex vector space has the attractive property that all linear transformations on it necessarily have eigenvalues. In quantum theory, one usually works with complex vector spaces.

Theorem 17.2. *The characteristic polynomial of a linear operator is invariant under similarity transformations.*

Proof. This is a direct consequence of Theorem 16.6. We have, for any invertible operator B,

$$B^{-1}(A - \lambda)B = B^{-1}AB - \lambda \qquad . \tag{17.6}$$

Thus, according to Theorem 16.6,

$$\det(B^{-1}AB - \lambda) = \det(B^{-1}(A - \lambda)B) = \det(A - \lambda) \qquad . \tag{17.7}$$

$$\square$$

The above theorem immediately implies that *the spectrum of a linear operator is invariant under similarity transformations.*

To develop the theory further, we need to introduce some more basic concepts and facts related to linear transformations.

Definition 17.1. *Given a linear operator A on a vector space V, the set*

$$\{v \in V \mid A(v) = 0\} \tag{17.8}$$

*is called the **kernel** of A, and is usually denoted $ker A$. The set*

$$A(v) \subset V \tag{17.9}$$

*is called the **image** of A, and is usually denoted $im A$.*

It is quite obvious that $ker A$ and $im A$ are both vector subspaces of V.

$\boxed{\text{Exercise 17.1}}$ Prove the above statement.

We have the following important theorem.

Theorem 17.3. *If $A : V \longrightarrow W$ is a linear map, where V and W are vector spaces over the same field $\mathbb{F}$, then*

$$\boxed{dim(V) = dim(ker\, A) + dim(im\, A)} \quad . \tag{17.10}$$

Proof. Clearly $ker\, A$ and $im\, A$ are vector subspaces of V and W, respectively. Let $\{v_1, \ldots, v_r\}$ be a basis of $ker\, A$ and $\{w'_1, \ldots, w'_s\}$ be a basis of $im\, A$. For each i, $1 \leq i \leq s$, let v'_i be the vector in V such that $A(v'_i) = w'_i$. Consider the set of vectors $\{v_1, \ldots, v_r, v'_1, \ldots, v'_s\}$. We will show that it forms a basis of V.

Let v be an arbitrary vector in V. Then $A(v) \in im\, A$. Hence we can write

$$A(v) = \sum_{i=1}^{s} c^i w'_i = \sum_{i=1}^{s} c^i A(v'_i), \quad c^i \in \mathbb{F}.$$

From the linearity of A, the above equation implies

$$A\left(v - \sum_{i=1}^{s} c^i v'_i\right) = 0.$$

Thus $v - \sum_{i=1}^{s} c^i v'_i \in ker\, A$, and, by virtue of the fact that $\{v_1, \ldots, v_r\}$ is a basis of $ker\, A$, we can write

$$v - \sum_{i=1}^{s} c^i v'_i = \sum_{i=1}^{r} d^i v_i, \quad d^i \in \mathbb{F},$$

or

$$v = \sum_{i=1}^{r} d^i v_i + \sum_{i=1}^{s} c^i v'_i,$$

that is, v is a linear combination of $\{v_1, \ldots, v_r, v'_1, \ldots, v'_s\}$. Next suppose

$$\sum_{i=1}^{r} a^i v_i + \sum_{i=1}^{s} b^i v'_i = 0,$$

where $a^i, b^j \in \mathbb{F}$, $1 \leq i \leq r$, $1 \leq j \leq s$. Then, since $A(0) = 0$,

$$0 = A(a^i v_i + b^i v'_i) = \sum_{i=1}^{s} b^i A(v'_i) = \sum_{i=1}^{s} b^i w'_i,$$

where the second equality follows from the facts that A is linear and $v_i \in ker\, A$. Since $\{w'_i\}$, being a basis of $im\, A$, is linearly independent, $b^i = 0$, $1 \leq i \leq s$. Thus $\sum_{i=1}^{r} a^i v_i = 0$. The linear independence of the set $\{v_i\}$ (it being a basis of $ker\, A$) then implies that the a^i, $1 \leq i \leq r$, vanish also. This establishes the linear independence of the set $\{v_1, \ldots, v_r, v'_1, \ldots, v'_s\}$. Since we have also shown that it spans V, it must be a basis of V. The theorem then follows from the facts that this basis has $r + s$ elements, $r = dim(ker\, A)$, and $s = dim(im\, A)$. $\square$

The kernel of a linear operator A bears important information on A, as seen in the following theorem.

Theorem 17.4. *A linear operator $A : V \longrightarrow V$ is **injective** (one-to-one) if and only if $\ker A = \{0\}$.*

Proof. Since A is linear,

$$A(0) = A(v - v) = A(v) - A(v) = 0 \quad .$$

Then, A is injective implies that $0 \in V$ is the only element in $\ker A$, that is, $\ker A = \{0\}$.

Conversely, suppose $\ker A = \{0\}$, and

$$A(v) = A(w), \quad v, w \in V \quad .$$

Then, by linearity of A, $A(v - w) = 0$. Since, by assumption, $\ker A = \{0\}$, it follows that $v - w = 0$. Thus $v = w$, and A is injective. $\qquad \square$

Theorems 17.3 and 17.4 lead to the following corollary.

Corollary 17.1. *Let $A : V \longrightarrow V$ be a linear operator. Then the following statements are equivalent:*

1. *$\ker A = \{0\}$,*

2. *$\operatorname{im} A = V$,*

3. *A is an isomorphism (a one-to-one and onto map).*

Exercise 17.2   Prove Corollary 17.1.

Definition 17.2. *The dimension of the subspace $\operatorname{im} A \subset V$ is called the **rank** of the linear operator A:*

$$rank(A) \equiv dim(imA) \quad . \tag{17.11}$$

We know that all n-dimensional complex vector spaces are isomorphic to the vector space $\mathbb{C}^n$ – the set of all n-tuples of complex numbers. Since every $n \times n$ matrix of complex entries is a linear transformation on $\mathbb{C}^n$, every $n \times n$ complex matrix represents some linear transformation on an n-dimensional complex vector space.

Definition 17.3. *The **rank** of an $n \times n$ matrix (a_i^j) is its rank when the matrix is considered as a linear transformation on $\mathbb{C}^n$.*

Consider the **standard basis** for $\mathbb{C}^n$ given as follows:

$$e_1 = (1, 0, 0, \ldots, 0),$$
$$e_2 = (0, 1, 0, \ldots, 0),$$

$$\vdots$$

$$e_i = (0, 0, \ldots, 0, 1, 0, \ldots, 0), \quad 1 \le i \le n, \text{ 1 in the } i\text{-th position},$$

$$\vdots$$

$$e_n = (0, 0, 0, \ldots, 1) \quad .$$

Let (a_i^j) be an arbitrary $n \times n$ complex matrix, and $(z^1, \ldots, z^n) \in \mathbb{C}^n$, $z^i \in \mathbb{C}$, $1 \le i \le n$. Then by

$$z'^i = a_j^i z^j \in \mathbb{C}^n \quad , \tag{17.12}$$

(a_i^j) acts as a linear operator A on $\mathbb{C}^n$. Furthermore

$$A(e_i) = (a_i^1, a_i^2, \ldots, a_i^n) \quad . \tag{17.13}$$

Thus, on recalling (1.16), (a_i^j) represents itself with respect to the standard basis in $\mathbb{C}^n$. Now for any vector space V, any linear operator $A : V \longrightarrow V$, and any basis $\{e_1, \ldots, e_n\}$ of V, $im\, A$ is spanned by a **maximally linearly independent subset** of $\{A(e_1), \ldots, A(e_n)\}$. It follows from Def. 17.3 and Eq. (17.13) that *the rank of an $n \times n$ matrix is equal to the maximum number of linearly independent rows of the matrix.*

The following facts also follow quite obviously from the above discussion.

Theorem 17.5. *The rank of a linear operator A is equal to the rank of any matrix that represents it.*

Theorem 17.6. *The rank of an $n \times n$ matrix is invariant under similarity transformations, that is, for any $n \times n$ invertible matrix B,*

$$rank(B^{-1}AB) = rank(A) \quad . \tag{17.14}$$

Finally, from Theorem 17.3 and Corollary 17.1, we have

Theorem 17.7.   *(a) A linear operator $A : V \longrightarrow V$, with $dim(V) = n$, is invertible if and only if $rank(A) = n$.*

(b) An $n \times n$ matrix (a_i^j) is invertible if and only if $rank((a_i^j)) = n$.

Exercise 17.3   Fill in the details in the proof of Theorem 17.7.

In addition to the above three theorems we have one more important fact concerning the rank of a linear operator.

Theorem 17.8. *Let A be a linear operator on an inner product space V. Then*

$$rank(A) = rank(A^\dagger) \quad , \tag{17.15}$$

where $A^\dagger$ is the adjoint of A.

Proof. Recall that the **orthogonal complement** of a subspace $M \subset V$, denoted $M^\perp$, is the set of all vectors in V that are orthogonal to every vector in M. $M^\perp$ itself is a subspace of V.

Exercise 17.4   Prove the above statement.

We will first show that

$$(imA)^\perp = ker(A^\dagger) \quad . \tag{17.16}$$

Suppose $x \in (imA)^\perp$. Then, for all $y \in V$,

$$(x, Ay) = 0 \quad . \tag{17.17}$$

But $(x, Ay) = (A^\dagger x, y)$ for all $y \in V$. Hence $(A^\dagger x, y) = 0$, for all $y \in V$. This implies that $A^\dagger x = 0$, and hence $x \in ker(A^\dagger)$. Thus

$$(imA)^\perp \subset ker(A^\dagger) \quad . \tag{17.18}$$

Conversely, suppose $x \in ker(A^\dagger)$. Then, for all $y \in V$,

$$(y, A^\dagger x) = 0 = (Ay, x) \quad , \tag{17.19}$$

where the second equality follows from (10.8). This implies that $x \in (imA)^\perp$. Thus

$$ker(A^\dagger) \subset (imA)^\perp \quad . \tag{17.20}$$

Eqs. (17.18) and (17.20) then imply (17.16). From (17.16) and (17.10), we then have

$$dim(ker(A^\dagger)) = dim(imA)^\perp = dim(V) - dim(imA)$$
$$= dim(V) - rank(A) = dim(kerA) . \tag{17.21}$$

The third equality of the above equation can be ritten

$$dim(V) = rank(A) + dim(ker(A^\dagger)) \quad . \tag{17.22}$$

Meanwhile, (17.10) applied to $A^\dagger$ yields

$$dim(V) = rank(A^\dagger) + dim(ker(A^\dagger)) \quad . \tag{17.23}$$

The theorem then follows by comparison of the above two equations.   □

Now let us recall our discussion of the dual basis $\{e_i'\}$ to a given basis $\{e_i\}$ in an inner product space [c.f. the discussion surrounding (10.18)]. We saw that if a matrix (a_i^j) represents some linear operator A with respect to $\{e_i\}$, then the hermitian conjugate matrix $(\overline{a_i^j})^T$ (transpose of the complex-conjugated matrix) represents the adjoint operator $A^\dagger$ with respect to $\{e_i'\}$. Let us call the maximum number of linearly independent rows of a matrix its **row rank** and the maximum number of linearly independent columns of a matrix its **column rank**. By Theorems 17.5 and 17.8,

$$\text{row rank of } (a_i^j) = rank(A) = rank(A^\dagger) = \text{row rank of } (\overline{a_i^j})^T$$

$$= \text{column rank of } (\overline{a_i^j}) = \text{column rank of } (a_i^j) . \quad (17.24)$$

We thus have the following useful fact.

Theorem 17.9. *The row rank of a square matrix is equal to its column rank, and both are referred to as the **rank** of the matrix.*

In all our previous theorems where the term "rank of a matrix" appears, we can thus understand it to mean either the row rank or the column rank.

We now return to the study of eigenvalues of a linear operator.

Definition 17.4. *For an arbitrary scalar λ and linear operator $A : V \longrightarrow V$, the number*

$$dim(ker(A - \lambda))$$

*is called the **geometric multiplicity** of λ. In other words it is the dimension of the subspace of V consisting of eigenvectors belonging to the eigenvalue λ.*

If λ is not an eigenvalue of A, then $ker(A - \lambda) = \{0\}$, and the geometric multiplicity of λ is zero.

Definition 17.5. *For an arbitrary scalar λ and linear operator A, the multiplicity of λ as a root of the characteristic equation*

$$det(A - \lambda) = 0$$

*is called the **algebraic multiplicity** of λ. In other words, the algebraic multiplicity of λ is the number of times λ occurs as a root of the characteristic equation.*

The above two concepts of multiplicity are not exactly the same. The strongest statement one can make concerning their relationship in general is that *the geometrical multiplicity of an eigenvalue is never greater greater than its algebraic multiplicity.* However, we will not be concerned with this somewhat technical difference and will work entirely with cases where the two can be identified. We state without proof the condition as follows.

Theorem 17.10. *Suppose $\lambda_1, \ldots, \lambda_p$ are all the distinct eigenvalues of a linear operator A on a vector space of dimension n, with respective geometric multiplicities $m_1, \ldots, m_p$. If*

$$\sum_{i=1}^{p} m_i = n \quad , \tag{17.25}$$

then the geometric multiplicity m_i is equal to the algebraic multiplicity of λ_i, for each $i = 1, \ldots, p$.

From now on we will refer to both types of multiplicity as just the multiplicity.

Assuming the conditions of Theorem 17.10 are met, we have

$$\det(A - \lambda) = (\lambda_1 - \lambda)^{m_1} \ldots (\lambda_p - \lambda)^{m_p}, \tag{17.26}$$

$$m_1 + \cdots + m_p = dim(V) \quad . \tag{17.27}$$

On setting $\lambda = 0$, we immediately have the useful result

$$\boxed{\det(A) = \prod_{i=1}^{p} \lambda_i^{m_i}} \quad , \tag{17.28}$$

that is, the determinant of a linear operator is equal to the product of its eigenvalues, including multiplicities.

Consider a matrix representation (a_i^j) of a linear operator A. Then a matrix representation of $A - \lambda$ is $(a_i^j - \lambda \delta_i^j)$. Using (16.75) we have

$$\det(A - \lambda) = \sum_{\sigma} (sgn\,\sigma)(a_1^{\sigma(1)} - \lambda \delta_1^{\sigma(1)}) \ldots (a_n^{\sigma(n)} - \lambda \delta_n^{\sigma(n)}) \quad . \tag{17.29}$$

If we also write

$$\det(A - \lambda) = \alpha_0 + \alpha_1 \lambda + \cdots + \alpha_{n-1} \lambda^{n-1} + \alpha_n \lambda^n \quad , \tag{17.30}$$

we see that

$$\alpha_n = (-1)^n \quad , \tag{17.31}$$

$$\alpha_0 = \det(A) \quad , \tag{17.32}$$

and

$$\frac{\alpha_{n-1}}{(-1)^{n-1}} = a_1^1 + a_2^2 + \cdots + a_n^n = Tr(a_i^j) \quad , \tag{17.33}$$

where "Tr" means "trace", that is, sum of the diagonal elements of a matrix.

| Exercise 17.5 | Verify (17.33) for the case $n = 4$.

On the other hand, from (17.26), the coefficient α_{n-1} of λ^{n-1} in $\det(A - \lambda)$ is also given by

$$\alpha_{n-1} = (-1)^{n-1}(m_1 \lambda_1 + \cdots + m_p \lambda_p) \quad . \tag{17.34}$$

Comparing (17.33) with (17.34), we have another useful result:

$$Tr(A) = Tr(a_i^j) = \sum_{i=1}^{p} m_i \lambda_i \qquad . \qquad (17.35)$$

Note that by virtue of Theorem 17.2, *the trace of a matrix is invariant under similarity transformations*, and thus $Tr(a_i^j)$ is a number characteristic of the linear operator A that (a_i^j) represents. The determinant and trace are two of the most important invariants (under similarity transformations) of a linear operator.

An important question concerning a linear operator $A : V \longrightarrow V$ is : Does there exist a basis $\{e_1, \ldots, e_n\}$ of V with respect to which the matrix representation of A is diagonal? If such a basis exists, A is said to be **diagonalizable**, and a glance at (17.26) reveals that the diagonal elements of the diagonal matrix representation are precisely the eigenvalues of A, including multiplicities. With a suitable rearrangement of the basis vectors, the diagonal matrix will appear in the following standard form:

$$(a_i^j) = \begin{pmatrix} \lambda_1 & 0 & 0 & 0 & \cdots \cdots \cdots \\ 0 & \lambda_1 & 0 & 0 & \cdots \cdots \cdots \\ 0 & 0 & \lambda_1 & 0 & \cdots \cdots \cdots \\ \cdots & \cdots & \cdots & \cdots & \cdots \cdots \\ \cdots & \cdots & \cdots & \cdots & \cdots \cdots \\ \cdots \cdots & 0 & \lambda_p & 0 & 0 \\ \cdots & \cdots & \cdots & \cdots & \cdots \\ \cdots \cdots & 0 & 0 & 0 & \lambda_p \end{pmatrix} , \qquad (17.36)$$

where the eigenvalue λ_1 occurs m_1 times, the eigenvalue λ_2 occurs m_2 times, $\ldots$, and the eigenvalue λ_p occurs m_p times along the diagonal.

$\boxed{\text{Exercise 17.6}}$ Show that if A is diagonalizable, then the diagonal elements of the diagonalized matrix are precisely all the eigenvalues of A, including multiplicities.

The following theorem is our first important result on diagonalizability.

Theorem 17.11. *If a linear operator A on an n-dimensional vector space has n distinct non-zero eigenvalues, then A is diagonalizable.*

Proof. Suppose A has n distinct non-zero eigenvalues $\lambda_1, \ldots, \lambda_n$. Then there correspond n distinct non-zero eigenvectors $e_1, \ldots, e_n$ such that

$$A(e_1) = \lambda_1 e_1 , \ldots , A(e_n) = \lambda_n e_n . \tag{17.37}$$

We will prove that $\{e_1, \ldots, e_n\}$ is linearly independent. Indeed, suppose the contrary, that is, $\{e_1, \ldots, e_n\}$ is linearly dependent. We can always rearrange the vectors in this set so that, without loss of generality, we can write

$$e_n = \sum_{i=1}^{m} \alpha_i e_i , \quad m < n, \tag{17.38}$$

where $\{e_1, e_2, \ldots, e_m\}$ is a maximal independent subset. Thus

$$0 = (A - \lambda_n) e_n = \sum_{i=1}^{m} \alpha_i (A - \lambda_n) e_i = \sum_{i=1}^{m} \alpha_i (\lambda_i - \lambda_n) e_i . \tag{17.39}$$

Since $\{e_1, \ldots, e_m\}$ is linearly independent,

$$\alpha_i (\lambda_i - \lambda_n) = 0 , \quad i = 1, \ldots, m .$$

But by hypothesis, $\lambda_i \neq \lambda_n$ (the eigenvalues are distinct). Hence $\alpha_i = 0$, $i = 1, \ldots, m$, implying that $e_n = 0$, by (17.38). This is a contradiction, since e_n is supposed to be a non-zero eigenvector belonging to the non-zero eigenvalue λ_n. Thus $\{e_1, \ldots, e_n\}$ must be linearly independent. It must also then be a basis of V since $dim(V) = n$. Comparing (17.37) with (1.16), we see that the matrix representation of A with respect to the basis $\{e_1, \ldots, e_n\}$ consisting of the eigenvectors belonging to the distinct eigenvalues is given by the diagonal matrix

$$\begin{pmatrix} \lambda_1 & 0 & 0 & \ldots\ldots \\ 0 & \lambda_2 & 0 & \ldots\ldots \\ 0 & 0 & \lambda_3 & \ldots\ldots \\ \ldots\ldots\ldots\ldots\ldots \\ \ldots\ldots & 0 & 0 & \lambda_n \end{pmatrix} .$$

The theorem follows.   □

Remark. Theorem 17.11 gives only a sufficient but not a necessary condition for a linear operator A to be diagonalizable.

We will now give a necessary and sufficient condition for a linear operator to be diagonalizable (as Theorem 17.12). First we give

Definition 17.6. *The subspace* $ker(A-\lambda_k)$ *of* V *corresponding to an eigenvalue* λ_k *of a linear operator* A *is called the **eigenspace** of* A *belonging to* λ_k.

Theorem 17.12. *A linear operator* $A : V \longrightarrow V$ *is diagonalizable if and only if* V *is the direct sum of all the eigenspaces of* A.

Proof. First we recall the meaning of the **direct sum** of two vector subspaces $H \subset V$ and $K \subset V$ of V, written $H \oplus K$. The statement $V = H \oplus K$ means that

i) $H \cap K = \{0\}$ (the zero vector in V);

ii) any $v \in V$ can be expressed uniquely as

$$v = h + k \quad , \quad h \in H, \, k \in K. \tag{17.40}$$

Now suppose the distinct eigenvalues of A are $\lambda_1, \ldots, \lambda_p$ with algebraic multiplicities $m_1, \ldots, m_p$, respectively. Assume that

$$V = E_1 \oplus E_2 \oplus \cdots \oplus E_p \quad , \tag{17.41}$$

where $E_i = ker(A - \lambda_i)$, $i = 1, \ldots, p$, is the eigenspace belonging to the eigenvalue λ_i. By Theorem 17.10, the geometric and algebraic multiplicities are in fact equal, and thus

$$dim(E_i) = m_i , \quad i = 1, \ldots, p. \tag{17.42}$$

Choose p bases:

$$\{e_1^{(1)}, \ldots, e_{m_1}^{(1)}\}, \{e_1^{(2)}, \ldots, e_{m_2}^{(2)}\}, \ldots, \{e_1^{(p)}, \ldots, e_{m_p}^{(p)}\},$$

for the subspaces $E_1, \ldots, E_p$, respectively. Then the union of these p sets:

$$\mathcal{B} = \{e_1^{(1)}, \ldots, e_{m_1}^{(1)}, e_1^{(2)}, \ldots, e_{m_2}^{(2)}, \ldots, e_1^{(p)}, \ldots, e_{m_p}^{(p)}\} \quad ,$$
$$m_1 + \cdots + m_p = n \quad ,$$

constitutes a basis for V, by virtue of (17.41). We have

$$A(e_j^{(i)}) = \lambda_i e_j^{(i)} \quad , \quad i = 1, \ldots, p; \, j = 1, \ldots, m_i \quad . \tag{17.43}$$

Thus, with respect to the basis $\mathcal{B}$, A is diagonalized and its matrix representation is given by (17.36).

Conversely, suppose A is diagonalizable. Then there exists a basis $\{e_1, \ldots, e_n\}$ such that

$$A(e_i) = \gamma_i e_i \quad , \quad i = 1, \ldots, n. \tag{17.44}$$

It follows that the characteristic equation of A is

$$(\gamma_1 - \lambda)(\gamma_2 - \lambda)\ldots(\gamma_n - \lambda) = 0 \quad . \tag{17.45}$$

The same characteristic equation is also given by

$$(\lambda_1 - \lambda)^{m_1}(\lambda_2 - \lambda)^{m_2} \ldots (\lambda_p - \lambda)^{m_p} = 0 \quad . \tag{17.46}$$

We can then arrange the numbers γ_i, $i = 1, \ldots, n$, such that

$$\begin{aligned}
&(\gamma_1, \ldots, \gamma_{m_1}, \gamma_{m_1+1}, \ldots, \gamma_{m_1+m_2}, \ldots \ldots, \gamma_{m_1+\cdots+m_{p-1}+1}, \ldots, \gamma_{m_1+\cdots+m_p}) \\
&= (\underbrace{\lambda_1, \ldots, \lambda_1}_{m_1}, \underbrace{\lambda_2, \ldots, \lambda_2}_{m_2}, \ldots \ldots, \underbrace{\lambda_p, \ldots, \lambda_p}_{m_p}) \quad .
\end{aligned} \tag{17.47}$$

The above equation says that the first m_1 of the γ's are all equal to λ_1, the next m_2 of the γ's are all equal to λ_2, and so on. Obviously,

$$\begin{aligned}
E_1 &= span\{e_1, \ldots, e_{m_1}\} \quad , \\
E_2 &= span\{e_{m_1+1}, \ldots, e_{m_1+m_2}\} \quad , \\
&\vdots \\
E_p &= span\{e_{m_1+\cdots+m_{p-1}+1}, \ldots, e_{m_1+\cdots+m_p}\} \quad ,
\end{aligned} \tag{17.48}$$

and thus

$$V = E_1 \oplus E_2 \oplus \cdots \oplus E_p \quad .$$

$\square$

Now comes a central theorem in a variety of physics applications.

Theorem 17.13. *A hermitian operator on an inner-product space (unitary space) is always diagonalizable.*

Proof. Let $\lambda_1 \in \mathbb{R}$ be a non-zero eigenvalue of a hermitian operator A on an inner product space V, and let e_1 be an eigenvector of A belonging to λ_1. (Remember that a linear operator on a complex vector space is guaranteed to have at least one non-zero eigenvalue.) We construct the orthogonal complement of e_1:

$$E_1^\perp \equiv \{x \in V \,|\, (x, e_1) = 0\} \quad . \tag{17.49}$$

In the above equation, the inner product satisfies (8.27) and (8.28). $E_1^\perp$ is obviously a subspace of V. Furthermore, it is invariant under A. Indeed, for $x \in E_1^\perp$,

$$(Ax, e_1) = (x, A^\dagger e_1) = (x, Ae_1) = (x, \lambda_1 e_1) = \lambda_1(x, e_1) = 0 , \tag{17.50}$$

where the second equality follows from the fact that $A = A^\dagger$ and the fourth from the fact that $\lambda_1 \in \mathbb{R}$. $E_1^\perp$ is then an invariant subspace of V under A. Let the restriction of A to $E_1^\perp$ be denoted A_1. A_1 is then a hermitian operator on $E_1^\perp$. It is guaranteed to have at least one non-zero eigenvalue $\lambda_2 \in \mathbb{R}$ (which may not necessarily be different from λ_1). Let $e_2 \in E_1^\perp$ be an eigenvector of A_1 belonging to the eigenvalue λ_2:

$$A_1 e_2 = \lambda_2 e_2 = A e_2 \quad . \tag{17.51}$$

$\{e_1, e_2\}$ must be linearly independent since the vectors in the set are orthogonal to each other. Next we construct the set

$$E_2^\perp \equiv \{x \in V \,|\, (x, e_1) = (x, e_2) = 0\} \quad . \tag{17.52}$$

By the same argument as above, $E_2^\perp$ is also an invariant subspace of V under A. Let the restriction of A to $E_2^\perp$ be denoted A_2. A_2 is then a hermitian operator on $E_2^\perp$. It is guaranteed to have a real non-zero eigenvalue λ_3, with an eigenvector $e_3 \in E_2^\perp$:

$$A_2 e_3 = \lambda_3 e_3 = A e_3 \quad . \tag{17.53}$$

$\{e_1, e_2, e_3\}$ is an orthogonal and linearly independent set. Continuing in the same way, we can generate a maximal linearly independent orthogonal set of eigenvectors of A, $\mathcal{B} = \{e_1, e_2, \ldots, e_n\}$, where $n = dim(V)$. $\mathcal{B}$ is in fact a basis of V. It is quite obvious that with respect to this basis, A is represented by a diagonal matrix, whose diagonal elements are the eigenvalues of A. $\qquad \square$

In Chapter 13 we saw that the eigenvalues of a hermitian operator are always real. Since physical measurements always yield real numbers, the operators representing physical observables in quantum mechanics, such as energy and momentum, are generally required to be hermitian operators. The fact that these are always diagonalizable means that there are always states of the physical system described by the eigenvectors of the hermitian operator such that, if a measurement corresponding to the operator is performed on the system, one would obtain a value belonging to the real spectrum of the operator.

The basis $\mathcal{B}$ constructed in the proof of Theorem 17.13 may of course be normalized: $(e_i, e_i) = 1$, $i = 1, \ldots, n$. Thus Theorem 17.13 also implies

Corollary 17.2. *A hermitian operator A on a unitary space can always be diagonalized by an orthonormal basis of eigenvectors of A.*

Recall that with respect to an orthonormal basis, a hermitian operator is represented by a hermitian matrix:

$$a_i^j = \overline{a_j^i} \tag{17.54}$$

[c.f. Eq. (10.26)]. Since under a change of basis, a matrix representation of a linear operator undergoes a similarity transformation [c.f. Eq. (1.41)], we have the following result.

Corollary 17.3. *A hermitian matrix can always be diagonalized by a similarity transformation.*

Since a real symmetric matrix is just a special case of a hermitian matrix, we have,

Corollary 17.4. *A real symmetric matrix can always be diagonalized by a similarity transformation.*

Recall the discussion in Chapter 2 of the moment of inertia tensor as an example of a real symmetric matrix. The diagonalization of real symmetric matrices is the most important mathematical tool for the solution of the so-called **normal mode** problems involving small oscillations in classical mechanics and solid state physics.

When a linear transformation A on a unitary space V is not diagonalizable, there is still always the possibility of finding a suitable basis of V with respect to which A appears in a standard (in some sense the simplest) form called the **Jordan canonical form**. The theory behind this will not be presented in detail here; and we will only state the main facts without proofs (in most cases). The reader will find detailed expositions of this topic in books on the theory of matrices, and techniques in the solutions of linear systems of differential equations (see, for example, M. W. Hirsch and S. Smale 1974).

Definition 17.7. *Let A be a linear transformation on a vector space V, and $\lambda_1, \ldots, \lambda_p$ be the eigenvalues of A with algebraic multiplicities $m_1, \ldots, m_p$, respectively. The subspace*

$$\mathcal{E}_k \equiv ker\,(A - \lambda_k)^{m_k}\,, \quad k = 1, \ldots, p\,, \tag{17.55}$$

*is called the **generalized eigenspace** of A belonging to λ_k.*

We note that $\mathcal{E}_k$ is invariant under A. In fact, let $x \in \mathcal{E}_k$, then by definition,

$$(A - \lambda_k)^{m_k}\,x = 0\,. \tag{17.56}$$

Thus

$$(A - \lambda_k)^{m_k}\,Ax = A(A - \lambda_k)^{m_k}\,x = 0\,.$$

We have the following fundamental theorem.

Theorem 17.14 (Primary Decomposition Theorem). *Let A be an operator on a complex space V. Then V is the direct sum of the generalized eigenspaces of A. The dimension of each generalized eigenspace is equal to the algebraic multiplicity of the corresponding eigenvalue.*

Definition 17.8. *An operator N is said to be **nilpotent** if there exists a positive integer m such that $N^m = 0$; the least such integer is called the **index** of the nilpotence.*

The following theorem is the most important consequence of Theorem 17.14.

Theorem 17.15. *Let A be a linear operator on a complex space V. Then A can be uniquely written as*

$$A = S + N\,, \tag{17.57}$$

where i) $SN = NS$, ii) S is diagonalizable, and iii) N is nilpotent.

Recall that for $X \in V$, the formal solution to the system of differential equations

$$\frac{dX}{dt} = AX \tag{17.58}$$

is given by

$$X(t) = X(0) e^{At} . \tag{17.59}$$

Using the above theorem and a basis with respect to which $S = A - N$ is diagonal, the exponential can be calculated relatively simply. Indeed,

$$e^{At} = e^{(S+N)t} = e^{St} e^{Nt} , \tag{17.60}$$

where in the second equality we have used the property i) in Theorem 17.15. The exponential of a diagonal matrix is easily computed:

$$\exp \begin{pmatrix} s_1 t & & 0 \\ & \ddots & \\ 0 & & s_n t \end{pmatrix} = \begin{pmatrix} e^{s_1 t} & & 0 \\ & \ddots & \\ 0 & & e^{s_n t} \end{pmatrix} , \tag{17.61}$$

while the exponential of the nilpotent matrix is given by

$$e^{Nt} = \sum_{k=0}^{m-1} \frac{(Nt)^k}{k!} , \tag{17.62}$$

which is a finite sum.

Exercise 17.7   Prove (17.61).

An interesting consequence of the Primary Decomposition Theorem is the following.

Theorem 17.16 (Cayley-Hamilton Theorem). *Let A be an operator on a vector space V with characteristic polynomial*

$$p(t) = \sum_{k=0}^{n} \alpha_k t^k . \tag{17.63}$$

Then

$$p(A) = 0 , \tag{17.64}$$

that is

$$\sum_{k=0}^{n} \alpha_k A^k v = 0 , \tag{17.65}$$

for all $v \in V$.

Proof. If V is a real space we can form the **complexification** of V, which is a complex space. This process consists of forming all complex linear combinations of a basis of V, which leaves the characteristic polynomial of A invariant. Thus for the purpose of proving this theorem, there is no loss of generality in assuming V to be complex.

It suffices to show that

$$p(A)x_k = 0 , \tag{17.66}$$

where $x_k \in \mathcal{E}_k$ (the k-th generalized eigenspace corresponding to the eigenvalue λ_k). Since $(t - \lambda_k)^{m_p}$ divides $p(t)$ we can write

$$p(t) = q(t)(t - \lambda_k)^{m_p} . \tag{17.67}$$

Thus, for $x_k \in \mathcal{E}_k$, we have

$$p(A)x_k = q(A)(A - \lambda_k)^{m_p} x_k = q(A)(0) = 0 . \tag{17.68}$$

$\square$

We will now discuss how the $S + N$ decomposition of Theorem 17.15 can be carried out. Let $\lambda_1, \ldots, \lambda_p$ be the distinct eigenvalues of an operator A on a complex space V with algebraic multiplicities $m_1, \ldots, m_p$ respectively. Let $\mathcal{E}_1, \ldots, \mathcal{E}_p$ be the corresponding generalized eigenspaces. Then by Theorem 17.4

$$V = \mathcal{E}_1 \oplus \mathcal{E}_2 \oplus \cdots \oplus \mathcal{E}_p . \tag{17.69}$$

Consider the restrictions of A on the generalized eigenspaces:

$$A_k \equiv A \,|\, \mathcal{E}_k , \quad k = 1, \ldots, p . \tag{17.70}$$

The linear operator A_k on the complex space $\mathcal{E}_k$ obviously has only one eigenvalue, that is, λ_k. By definition

$$N_k \equiv A_k - \lambda_k$$

is nilpotent, since $x \in \mathcal{E}_k$ implies $(A_k - \lambda_k)^{m_k} x = 0$. Thus

$$A_k = \lambda_k + N_k , \tag{17.71}$$

where $\lambda_k (= \lambda_k 1)$ is diagonalizable (in fact diagonal with respect to any basis in $\mathcal{E}_k$) and N_k is nilpotent. At this point we need to introduce the notion of the direct sum of operators.

Definition 17.9. *Suppose $A_1, \ldots, A_p$ are linear operators on vector subspaces $V_1, \ldots, V_p$ of V, respectively; and*

$$V = V_1 \oplus \cdots \oplus V_p .$$

Then the direct sum of operators

$$A = A_1 \oplus \cdots \oplus A_p \tag{17.72}$$

acting on V is given by

$$Av = A_1 v_1 + \cdots + A_p v_P \,, \tag{17.73}$$

where $v \in V$ has the unique decomposition

$$v = v_1 + \cdots + v_p \,,$$

with $v_k \in V_k$, $k = 1, \ldots, p$.

The restrictions A_k given by (17.70) then satisfy (17.71) and (17.72), with N_k nilpotent. Now let

$$S = \lambda_1 1 \oplus \cdots \oplus \lambda_p 1 \,, \tag{17.74}$$
$$N = N_1 \oplus \cdots \oplus N_p \,. \tag{17.75}$$

Then

$$A = S + N \,. \tag{17.76}$$

S is obviously diagonal [having the form (17.36) with respect to any basis of V]; and N is nilpotent, since $N^m = 0$, where $m = max\,(m_1, \ldots, m_p)$. It is also obvious that S and N commute. Theorem 17.15 then asserts that the above construction for S and N is unique.

The canonical form (17.36) for S will be denoted by

$$S = diag\,\{\,\lambda_1(m_1), \ldots, \lambda_p(m_p)\,\} \,. \tag{17.77}$$

A canonical form for N can also be constructed.

Definition 17.10. *An **elementary nilpotent block** is an $n \times n$ matrix of the form*

$$\begin{pmatrix} 0 & 1 & & & & \\ & 0 & 1 & & & \\ & & 0 & 1 & & \\ & & & \ddots & \ddots & \\ & & & & 0 & 1 \\ & & & & & 0 \end{pmatrix} \,, \tag{17.78}$$

with 1's just above the diagonal and 0's elsewhere.

Suppose the matrix (17.78) represents an operator N with respect to a basis $\{\,e_1, \ldots, e_n\,\}$. Then

$$N(e_1) = e_2, \; N(e_2) = e_3, \; \ldots, \; N(e_{n-1}) = e_n, \; N(e_n) = 0 \,. \tag{17.79}$$

Thus $N^n(e_k) = 0$, $k = 1, \ldots, n$, which implies $N^n = 0$. Moreover $N^k \neq 0$ for $0 \leq k \leq n$, since $N^k e_1 = e_{k+1} \neq 0$. N is thus nilpotent with index n.

Theorem 17.17. *Suppose N is a nilpotent operator on a vector space V (real or complex). Then there exists a basis of V giving N a matrix representation of the form*

$$N = diag\left\{ N_1, \ldots, N_r \right\},\tag{17.80}$$

where

$$r = dim\left(ker\, N\right),\tag{17.81}$$

N_j, $j = 1, \ldots, r$, *is an elementary nilpotent block, and the size of N_j is a non-increasing function of j. The blocks N_j are uniquely determined by N.*

As an illustration, we may have the following arrangement of nilpotent blocks.

$$N = \begin{pmatrix} \begin{pmatrix} 0 & 1 & 0 \\ 0 & 0 & 1 \\ 0 & 0 & 0 \end{pmatrix} & & & & & \\ & \begin{pmatrix} 0 & 1 \\ 0 & 0 \end{pmatrix} & & & & \\ & & \begin{pmatrix} 0 & 1 \\ 0 & 0 \end{pmatrix} & & & \\ & & & (0) & & \\ & & & & (0) & \\ & & & & & (0) \end{pmatrix},\tag{17.82}$$

in which the empty slots are all filled by 0's. The form (17.80) is called the **nilpotent canonical form** of a nilpotent operator. It is obvious that the rank (maximum number of independent rows or columns) of each $k \times k$ elementary nilpotent block is $k - 1$, and so

$$dim\left(im\, N\right) = n - r,\tag{17.83}$$

where $r = $ total number of elementary nilpotent blocks in N. From theorem 17.3, it follows that

$$n = dim\left(ker\, N\right) + n - r.\tag{17.84}$$

This immediately implies (17.81).

Theorem 17.18. *Two nilpotent operators on the same vector space have the same canonical form if and only if they are related by a similarity transformation.*

We will not address the methods of construction of the nilpotent canonical form, or of the basis leading to this form. The requisite techniques are mainly of importance in numerical work involving matrices, and do not concern us directly.

The last step in the reduction of a matrix representation of A (on a complex vector space) to the Jordan canonical form consists in writing down a **Jordan matrix** belonging to distinct eigenvalue λ_k, $k = 1, \ldots, p$. This is the matrix representation of

$$A_k = A \,|\, \mathcal{E}_k = \lambda_k 1 + N_k \tag{17.85}$$

with respect to the basis $\mathcal{B}_k$ which gives the nilpotent canonical form for N_k. It has the form given by (17.86) below: the matrix has λ_k along the diagonal, 1's and 0's immediately above the diagonal, and 0's everywhere else.

$$C_k = \begin{pmatrix} \begin{pmatrix} \lambda_k & 1 & & \\ & \ddots & 1 & \\ & & & \lambda_k \end{pmatrix} & & & \\ & \begin{pmatrix} \lambda_k & 1 & & \\ & \ddots & 1 & \\ & & & \lambda_k \end{pmatrix} & & \\ & & \ddots & \\ & & & \begin{pmatrix} \lambda_k & 1 & & \\ & \ddots & 1 & \\ & & & \lambda_k \end{pmatrix} \end{pmatrix} \tag{17.86}$$

Each C_k is $(m_k \times m_k)$, $k = 1, \ldots, p$. Each of the blocks making up C_k is called an **elementary Jordan matrix**, or **elementary λ-block**. The basis $\mathcal{B} = \mathcal{B}_1 \cup \cdots \cup \mathcal{B}_p$ spans the entire V according to Theorem 17.14. With respect to $\mathcal{B}$, $A = S + N$ is then represented by the matrix

$$C = diag\,\{\,C_1, \ldots, C_p\,\}\,. \tag{17.87}$$

This matrix is called the **Jordan canonical form** of A. An operator A determines its Jordan canonical form uniquely up to the order in which the Jordan matrices C_k (corresponding to the different eigenvalues) appear.

Finally, according to (17.81), the number of elementary λ-blocks in each Jordan matrix C_k is $dim\,(ker\,(A_k - \lambda_k))$.

Chapter 18

Group Representation Theory

The general definition of a group representation was given in Chapter 4 (Def. 4.3). In this chapter we will present some basic concepts and techniques concerning its theory, especially for finite groups. The significance of group representations in physics, and indeed in mathematics itself, can hardly be overestimated.

Let us first motivate our study by considering in general the relevance of group representation theory in quantum mechanics. Recall that in quantum theory, states of physical systems are represented by vectors in Hilbert space (actually rays, or vectors up to constant multiplicative complex numbers, but we will ignore this detail here), and unitary operators on Hilbert space, such as finite translations, finite rotations, and time evolutions, are of great importance because they are norm-preserving. Let $U : G \to \mathcal{U}(\mathcal{H})$ be a unitary representation of a **symmetry group** G [such as $SO(3)$] on a Hilbert space $\mathcal{H}$, where $\mathcal{U}(\mathcal{H})$ is a group of unitary transformations on $\mathcal{H}$. G is a symmetry group means that it leaves the Hamiltonian H invariant:

$$U^\dagger(g)HU(g) = H , \tag{18.1}$$

or

$$U(g)H = HU(g) , \tag{18.2}$$

for all $g \in G$. Recall that $U(g)$ is unitary implies that $U^{-1}(g) = U^\dagger(g)$. As a result of the symmetry property of G, any $U(g)|\psi\rangle$ is also a solution of the Schrödinger equation

$$H|\psi\rangle = E|\psi\rangle \tag{18.3}$$

with the same eigenvalue E if $|\psi\rangle$ is. Indeed

$$HU(g)|\psi\rangle = U(g)U^\dagger(g)HU(g)|\psi\rangle = U(g)H|\psi\rangle$$
$$= U(g)E|\psi\rangle = E(U(g)|\psi\rangle) . \tag{18.4}$$

Exercise 18.1   Show that both the Schrödinger equation (18.3) and **Heisenberg's equation of motion** for an observable A [(9.16)],

$$\frac{\mathrm{d}A}{\mathrm{d}t} = \frac{[A, H]}{i\hbar} + \frac{\partial A}{\partial t} \ ,$$

are left invariant under $U(g)$ if and only if $[H, U(g)] = 0$ for all $g \in G$.

In general, an energy eigenvalue E may be degenerate. It is said to be r-fold degenerate if a set of r vectors $\{|\psi_1\rangle, \ldots, |\psi_r\rangle\}$ in $\mathcal{H}$ span the eigenspace V_E for a particular E. The above discussion implies that if $|\psi\rangle \in V_E$, then $U(g)|\psi\rangle \in V_E$ also, for all $g \in G$. In other words, any eigensubspace V_E of $\mathcal{H}$ is invariant under $U(g)$, for all $g \in G$. We also say that *the group action of G leaves any eigenspace invariant.*

Suppose G acts on $x \in M$, for example, $M = \mathbb{R}^3$. It is interesting to use the Dirac notation to obtain the action of $U(g)$ on the wave function $\psi(x) = \langle x|\psi\rangle$ (the coordinate representation of $|\psi\rangle$), given its action on $|\psi\rangle$. We generalize slightly and consider a unitary representation of G on rigged Hilbert space $\mathcal{H}_R$ (see Chapter 15) instead of $\mathcal{H}$, $U_R : G \to \mathcal{U}(\mathcal{H}_R)$, where, for normalizable states $|\psi\rangle$, $U_R(g)|\psi\rangle \equiv U(g)|\psi\rangle$, and for non-normalizable states $|x\rangle$, $U_R(g)|x\rangle \equiv |gx\rangle$. The latter condition satisfies unitarity since, for any $x_1, x_2 \in M$, $\langle x_1|gx_2\rangle = \langle g^{-1}x_1|x_2\rangle$.

Suppose

$$U(g)|\psi\rangle = |\psi'\rangle \ . \tag{18.5}$$

Then,

$$U(g)\psi(x) = \psi'(x) = \langle x|\psi'\rangle = \langle x|U(g)|\psi\rangle$$
$$= \langle x|U_R(g)|\psi\rangle = \langle U_R^{-1}x|\psi\rangle = \langle g^{-1}x|\psi\rangle = \psi(g^{-1}x) \ . \tag{18.6}$$

We have already used this result in (4.17).

Exercise 18.2   Verify that the group action

$$U(g)\psi(x) = \psi(g^{-1}x) \tag{18.7}$$

satisfies the group homomorphism requirement

$$U(g_1)U(g_2)\psi(x) = U(g_1 g_2)\psi(x) \ . \tag{18.8}$$

From our earlier discussion, we conclude that if the Hamiltonian H commutes with $U(g)$ for all $g \in G$, then $U(g)$ gives a unitary representation of G on any eigensubspace of H. Thus, if we succeed in finding all the inequivalent, irreducible representations of a group G under which H is invariant, then we will have found a complete classification of all the energy eigenvalues of H, without solving the Schrödinger equation explicitly. (The notions of inequivalent, irreducible representations will be defined shortly.) This kind of classification using group representation theory finds important applications in much of atomic, molecular, condensed matter, and particle physics.

Let us now define the important concepts of equivalence and irreducibility in group representation theory.

Definition 18.1. *Two representations $D_1 : G \to \mathcal{T}(\mathbb{V}_1)$ and $D_2 : G \to \mathcal{T}(\mathbb{V}_2)$ of a group G are said to be **equivalent group representations** if they are related by a similarity transformation [(1.42)]*

$$D_2(g) = S D_1(g) S^{-1} \tag{18.9}$$

for all $g \in G$, where S is an invertible linear transformation from $\mathbb{V}_1$ to $\mathbb{V}_2$.

Definition 18.2. *A subspace $\mathbb{V}_1 \subset \mathbb{V}$ is said to be an **invariant subspace** of $\mathbb{V}$ with respect to a group representation $D : G \to \mathcal{T}(\mathbb{V})$ if $v \in \mathbb{V}_1$ implies $D(g)v \in \mathbb{V}_1$ also, for all $g \in G$. An invariant subspace is said to be **proper** (or **minimal**) if it does not contain any invariant subspace with respect to D other than itself and $\{0\}$, that is, if it contains no non-trivial invariant subspace.*

Definition 18.3. *A representation $D : G \to \mathcal{T}(\mathbb{V})$ of a group G on a vector space $\mathbb{V}$ is said to be **irreducible** if $\mathbb{V}$ does not contain a non-trivial invariant subspace with respect to D; otherwise D is said to be **reducible**. In the latter case, if the orthogonal complement of the non-trivial invariant subspace is also invariant with respect to D, then D is said to be **fully reducible** (or **decomposable**).*

Expressed as a matrix (with respect to some choice of basis vectors in $\mathbb{V}$), an operator $D(g)$ of a reducible representation (up to similarity transformations) looks like:

$$D(g) = \begin{pmatrix} D_1(g) & 0 \\ D'(g) & D_2(g) \end{pmatrix} . \tag{18.10}$$

If the representation space $\mathbb{V}$ is n-dimensional, then $D(g)$ is an $n \times n$ matrix. If the non-trivial invariant subspace $\mathbb{V}_1$ is n_1-dimensional, $n_1 < n$, then the block $D_1(g)$ is $n_1 \times n_1$. To see the form of (18.10), we choose a basis set of $\mathbb{V}$, $\{e_1, \ldots, e_{n_1}, e_{n_1+1}, \ldots, e_n\}$ such that the subset $\{e_1, \ldots, e_{n_1}\}$ is a basis of $\mathbb{V}_1$. The fact that $\mathbb{V}_1$ is invariant implies that

$$D(g)e_i = (D(g))_i^j e_j \in \mathbb{V}_1 , \quad i = 1, \ldots, n_1 . \tag{18.11}$$

Thus

$$(D(g))_i^j = 0 , \quad i = 1, \ldots, n_1 ; j = n_1 + 1, \ldots, n . \tag{18.12}$$

These are precisely elements in the upper right block of the RHS of (18.10). (Recall our convention that in a matrix a_i^j, i is the row index and j is the column index.)

A fully reducible representation matrix looks like, up to similarity transformations,

$$D(g) = \begin{pmatrix} D_1(g) & 0 \\ 0 & D_2(g) \end{pmatrix} . \tag{18.13}$$

In this case, the space $V_2 = V_1^\perp$ spanned by $\{e_{n_1+1}, \ldots, e_n\}$ is also invariant under $D(g)$. Thus

$$D(g)e_i = (D(g))_i^j e_j \in V_2 , \quad i = n_1 + 1, \ldots, n . \tag{18.14}$$

This implies that

$$(D(g))_i^j = 0 , \quad i = n_1 + 1, \ldots, n ; \ j = 1, \ldots, n_1 . \tag{18.15}$$

These are precisely elements in the lower left block of the RHS of (18.13).

Exercise 18.3   Consider

$$(D(\theta))_i^j = \begin{pmatrix} \cos\theta & -\sin\theta \\ \sin\theta & \cos\theta \end{pmatrix} , \quad 0 \le \theta \le 2\pi ,$$

the defining representation of the one-parameter group $SO(2)$ on $\mathbb{R}^2$. Show that this representation is irreducible on $\mathbb{R}^2$. However, there exists a fully reducible representation of $SO(2)$ on complexified $\mathbb{R}^2$, namely $\mathbb{C}^2$, considered as a vector space. Let $\{e_1, e_2\}$ be a basis in $\mathbb{R}^2$. Then it is a basis in $\mathbb{C}^2$ also. Show that under the change of basis $e_i = S_i^j e'_j$ in $\mathbb{C}^2$ given by

$$e_+ = e'_1 = \frac{1}{\sqrt{2}}(e_1 + ie_2) , \qquad e_- = e'_2 = \frac{1}{\sqrt{2}}(e_1 - ie_2) , \tag{18.16}$$

$(D(\theta))_i^j$ can be diagonalized to $D' = S^\dagger D S$ given by

$$(D'(\theta))_i^j = \begin{pmatrix} e^{i\theta} & 0 \\ 0 & e^{-i\theta} \end{pmatrix} , \tag{18.17}$$

which is clearly fully reducible. This representation has very useful applications in physics, for example, in the description of polarization states of photons.

Exercise 18.4   Show that if a unitary representation of a group G is reducible, then it is also fully reducible.

Theorem 18.1. *Every representation of a finite group on an inner product space is equivalent to a unitary representation.*

Note: This important theorem generalizes to compact Lie groups also (see Theorem 22.3).

Proof. Let $D : G \to \mathcal{T}(\mathbb{V})$ be a representation of a finite group G. We will show that there exists an invertible operator $S \in \mathcal{T}(\mathbb{V})$ such that

$$U(g) = S^{-1}D(g)S \tag{18.18}$$

is unitary for all $g \in G$. Define an operator S by

$$Sv = \sum_{g \in G} D(g)v , \tag{18.19}$$

for all $v \in \mathbb{V}$. Denote the inner product in $\mathbb{V}$ by (u, v). Then

$$\begin{aligned}
(U(g)u, U(g)v) &= (S^{-1}D(g)Su, S^{-1}D(g)Sv) \\
&= \sum_{g' g''}(S^{-1}D(g)D(g')u, S^{-1}D(g)D(g'')v) = \sum_{h' h''}(S^{-1}D(h')u, S^{-1}D(h'')v) \\
&= (S^{-1}Su, S^{-1}Sv) = (u, v) ,
\end{aligned}$$

$$\tag{18.20}$$

where $h' = gg'$, $h'' = gg''$, and in the third equality we have used the fact that when g' ranges over the whole group, so does gg' for fixed g. Thus $U(g)$ is unitary [c.f. (10.15)]. □

By the result of Exercise 18.4 and the above theorem, we see that *all reducible representations of a finite group are fully reducible.*

For the remainder of this chapter we will develop some basic and useful facts in the theory of the irreducible representations of finite groups. First we introduce a central theorem, Schur's Lemma, on which all these facts are based. Representations for some important infinite groups [such as $SO(3)$ and $SU(2)$] will be discussed later (see Chapters 21 and 22).

Theorem 18.2 (Schur's Lemma). *Let $D_1 : G \to \mathcal{T}(\mathbb{V}_1)$ and $D_2 : G \to \mathcal{T}(\mathbb{V}_2)$ be two irreducible representations of a group G on vector spaces $\mathbb{V}_1$ and $\mathbb{V}_2$, respectively. Let $A : \mathbb{V}_1 \to \mathbb{V}_2$ be a linear map such that*

$$D_2(g)A = AD_1(g) \quad \text{for all } g \in G . \tag{18.21}$$

Then

(1) If D_1 is not equivalent to D_2, then $A = 0$, and

(2) If $D_1 = D_2$ (thus $\mathbb{V}_1 = \mathbb{V}_2$), then $A = \lambda I$, where λ is a scalar and I is the identity operator (in $\mathbb{V}_1 = \mathbb{V}_2$).

Proof. Consider $ker(A) = \{v_1 \in \mathbb{V}_1 | \, Av_1 = 0 \in \mathbb{V}_2\}$. This is an invariant subspace of $\mathbb{V}_1$ under $D_1(g)$ for all $g \in G$, since, if $v_1 \in ker(A)$, then

$$AD_1(g)v_1 = D_2(g)Av_1 = D_2(g)(0) = 0 \in \mathbb{V}_2 \,. \tag{18.22}$$

Since D_1 is by assumption irreducible, $\mathbb{V}_1$ contains no non-trivial invariant subspaces. Thus $ker(A)$ is either all of $\mathbb{V}_1$ or $\{0\}$. In the former case $A = 0$ (the zero operator), and the entire theorem is proved [$\lambda = 0$ for statement (2)]. Suppose $ker(A) = \{0\}$. Then, by Corollary 17.1, A establishes an isomorphism between $\mathbb{V}_1$ and $\mathbb{V}_2$. Thus the condition (18.21), which can be rewritten as

$$D_2(g) = AD_1(g)A^{-1} \,, \quad \text{for all } g \in G \,, \tag{18.23}$$

implies that D_1 and D_2 are equivalent representations. This proves (1).

To prove statement (2) we assume that $\mathbb{V}_1 = \mathbb{V}_2 = \mathbb{V}$ is a complex vector space. Then A has at least one eigenvalue [see discussion following (17.5)]. Let λ be an eigenvalue of A, and $\mathbb{V}_\lambda \subset \mathbb{V}$ be the eigenspace corresponding to λ. Then, for any $v \in \mathbb{V}_\lambda$,

$$Av = \lambda v \,, \tag{18.24}$$

and

$$AD_1(g)v = D_1(g)Av = \lambda D_1(g)v \,. \tag{18.25}$$

Thus $\mathbb{V}_\lambda$ is an invariant subspace of $\mathbb{V}$; and since D_1 is irreducible by assumption, $\mathbb{V}_\lambda = \mathbb{V}$. It follows that $A = \lambda I$. $\square$

Corollary 18.1. *Irreducible representations of an abelian group are all one-dimensional.*

Proof. Let $D : G \to \mathcal{T}(\mathbb{V})$ be an irreducible representation of an abelian group G. For a fixed $g' \in G$, the abelian property of G implies that

$$D(g)D(g') = D(g')D(g) \,, \quad \text{for all } g' \in G \,. \tag{18.26}$$

Let $D(g')$ play the role of A in the above statement of Schur's Lemma. Then statement (2) in that Lemma implies that

$$D(g') = \lambda I \,. \tag{18.27}$$

Since g' was picked arbitrarily, $D(g)$ is diagonal for all $g \in G$. Since D is irreducible, the matrix $D(g)$ must be 1×1, and the representation thus one-dimensional. $\square$

Before presenting the following results for the representations of finite groups, it is useful to list here the various quantities related by them, and to establish some notation:

$$N_G = \text{number of elements in the group } G \ (\textbf{order of the group}),$$
$$N_{(i)} = \text{number of elements in the } i\text{-th conjugacy (equivalence) class},$$
$$N_C = \text{number of conjugacy classes in the group},$$
$$N_R = \text{number of inequivalent, irreducible representations of } G,$$
$$N_{(\mu)} = \text{dimension of the } \mu\text{-irreducible representation}$$
$$\text{(only inequivalant ones are counted)},$$
$$D_{(\mu)}(g) = \text{the linear operator corresponding to } g \in G \text{ in the}$$
$$\mu\text{-th irreducible representation},$$
$$\chi_{(\mu i)} = \textbf{character} \text{ of } i\text{-th conjugacy class group elements in the}$$
$$\mu\text{-th representation}$$
$$\equiv Tr(D_{(\mu)}(g)), \quad g \in i\text{-th class}.$$

Note that by Theorem 17.2 and Eq. (17.35), the character $\chi_{(\mu i)}$ is independent of the particular representatives in either the representation or the class. We have used Greek letters for representation indices and Latin letters for class indices. Neither of these kinds are tensorial and they have been bracketed to distinguish them from the ones that are. Sums over bracketed indices will always be indicated explicitly, while the Einstein summation convention will be used for repeated tensorial indices only.

Theorem 18.3 (Orthonormality of Group Representations). *For a finite group G*

$$\boxed{\frac{N_{(\mu)}}{N_G} \sum_{g \in G} (D^{-1}_{(\mu)}(g))^k_i (D_{(\nu)}(g))^j_l = \delta_{\mu\nu} \delta^j_i \delta^k_l}. \tag{18.28}$$

Proof. Let $\mathcal{M}$ be an arbitrary $N_{(\mu)} \times N_{(\nu)}$ matrix. Then define the $N_{(\mu)} \times N_{(\nu)}$ matrix $A_{\mathcal{M}}$ by

$$A_{\mathcal{M}} \equiv \sum_{g' \in G} D^{-1}_{(\mu)}(g') \mathcal{M} D_{(\nu)}(g'). \tag{18.29}$$

We have, for all $g \in G$,

$$D^{-1}_{(\mu)}(g) A_{\mathcal{M}} D_{(\nu)}(g) = \sum_{g'} D^{-1}_{(\mu)}(g) D^{-1}_{(\mu)}(g') \mathcal{M} D_{(\nu)}(g') D_{(\nu)}(g)$$

$$= \sum_{g'} \left(D_{(\mu)}(g') D_{(\mu)}(g) \right)^{-1} \mathcal{M} D_{(\nu)}(g') D_{(\nu)}(g) = \sum_{g'} D^{-1}_{(\mu)}(g'g) \mathcal{M} D_{(\nu)}(g'g)$$

$$= \sum_{g'} D^{-1}_{(\mu)}(g') \mathcal{M} D_{(\nu)}(g') = A_{\mathcal{M}}. \tag{18.30}$$

Thus

$$A_{\mathcal{M}} D_{(\nu)}(g) = D_{(\mu)}(g) A_{\mathcal{M}}, \quad \text{for all } g \in G. \tag{18.31}$$

By Schur's lemma, either $\mu \neq \nu$, in which case $A_{\mathcal{M}} = 0$, or $\mu = \nu$, in which case $A_{\mathcal{M}} = \lambda_{\mathcal{M}} I$.

Now let us construct a set of $N_{(\mu)} N_{(\nu)}$ matrices, $\mathcal{M}_l^k$, $l = 1, \ldots, N_{(\mu)}$, $k = 1, \ldots, N_{(\nu)}$. Each is an $N_{(\mu)} \times N_{(\nu)}$ matrix, with the (ij) element given by

$$\left(\mathcal{M}_l^k \right)_i^j = \delta_l^j \, \delta_i^k \, . \tag{18.32}$$

Let the matrix $\mathcal{M}$ in (18.29) be any one of the $\mathcal{M}_l^k$. Correspondingly we have the matrices A_l^k, given according to (18.29) by

$$\left(A_l^k \right)_i^j = \sum_g \left(D_{(\mu)}^{-1}(g) \right)_i^n \, (\mathcal{M}_l^k)_n^p \, \left(D_{(\nu)}(g) \right)_p^j$$

$$= \sum_g \left(D_{(\mu)}^{-1}(g) \right)_i^n \, \delta_l^p \, \delta_n^k \, \left(D_{(\nu)}(g) \right)_p^j = \sum_g \left(D_{(\mu)}^{-1}(g) \right)_i^k \, \left(D_{(\nu)}(g) \right)_l^j \, . \tag{18.33}$$

We know that A_l^k has to vanish if $\mu \neq \nu$. Thus we have proved the theorem when $\mu \neq \nu$.

When $\mu = \nu$, $A_l^k = \lambda_l^k \, I$, that is

$$\left(A_l^k \right)_i^j = \lambda_l^k \, \delta_i^j \, , \tag{18.34}$$

where λ_l^k is a scalar ($\in \mathbb{C}$). Hence (18.33) yields (for $\mu = \nu$)

$$\lambda_l^k \, \delta_i^j = \sum_g (D_{(\mu)}^{-1}(g))_i^k \, (D_{(\mu)}(g))_l^j \, , \tag{18.35}$$

where all the indices run from 1 to $N_{(\mu)}$. Contracting the pair of indices i and j, we have

$$\lambda_l^k \, \delta_i^i = N_{(\mu)} \, \lambda_l^k = \sum_g (D_{(\mu)}(g))_l^i \, (D_{(\mu)}^{-1}(g))_i^k$$

$$= \sum_g \left(D_{(\mu)}(g) \, D_{(\mu)}^{-1}(g) \right)_l^k = \sum_g \delta_l^k = N_G \, \delta_l^k \, , \tag{18.36}$$

or

$$\lambda_l^k = \frac{N_G}{N_{(\mu)}} \, \delta_l^k \, . \tag{18.37}$$

Putting this result in (18.35) the theorem follows.   □

As an example consider the two-element **cyclic group** $C_2 = \{e, a\}$, where $a^2 = e$ (hence $a^{-1} = a$). This group is abelian, so by Corollary 18.1, all the irreducible representations are one-diemnsional. We have, first of all, the trivial representation: $D_{(1)}(e) = 1, D_{(1)}(a) = 1$. With $N_{(1)} = 1$ and $N_G = 2$, this obviously satisfies (18.28). By letting $\mu = 1$, the orthonormality theorem implies that all other representations ($\nu \neq 1$) must satisfy

$$\frac{1}{2} \left(D_{(\nu)}(e) + D_{(\nu)}(a) \right) = 0 \, , \tag{18.38}$$

For any one-dimensional representation, $D_{(\nu)}(e) = 1$. Hence $D_{(\nu)}(a) = -1$. We conclude that, besides the trivial representation $D_{(1)}$, there is only one other: $D_{(2)}(e) = 1, D_{(2)}(a) = -1$.

The following important theorem complements the orthonormality theorem (Theorem 18.3). It will be stated without proof (see W.-K. Tung 1985).

Theorem 18.4 (Completeness of Irreducible Representations).

$$\sum_{\mu} \frac{N_{(\mu)}}{N_G} Tr \left(D_{(\mu)}(g) \, D_{(\mu)}^{-1}(g') \right) = \delta_{gg'} \qquad (18.39)$$

An immediate consequence of the above theorem is, on setting $g = g'$,

$$\sum_{\mu} (N_{(\mu)})^2 = N_G . \qquad (18.40)$$

This shows explicitly that all irreducible representations of a finite group are finite-dimensional.

The following theorems concern the character $\chi_{(\mu i)}$ of irreducible representations of group elements in particular conjugacy classes.

Theorem 18.5.

$$\sum_{g \in C_i} D_{(\mu)}(g) = \frac{N_{(i)}}{N_{(\mu)}} \chi_{(\mu i)} I , \qquad (18.41)$$

where $C_i \subset G$ consists of elements in the i-th conjugacy class and I is the identity operator.

Proof. Let

$$A_{(\mu i)} \equiv \sum_{g' \in C_i} D_{(\mu)}(g') . \qquad (18.42)$$

Then, for all $g \in G$,

$$D_{(\mu)}(g) A_{(\mu i)} D_{(\mu)}^{-1}(g) = \sum_{g' \in C_i} D_{(\mu)}(g) D_{(\mu)}(g') D_{(\mu)}^{-1}(g)$$

$$= \sum_{g' \in C_i} D_{(\mu)}(gg'g^{-1}) = \sum_{g' \in C_i} D_{(\mu)}(g') = A_{(\mu i)} . \qquad (18.43)$$

Thus, by Schur's lemma,

$$A_{(\mu i)} = \lambda_{(\mu i)} I . \qquad (18.44)$$

To calculate the number $\lambda_{(\mu i)}$ we take the trace of both sides of (18.42) to obtain

$$\lambda_{(\mu i)} N_{(\mu)} = N_{(i)} \chi_{(\mu i)} . \qquad (18.45)$$

The theorem follows from putting this result for $\lambda_{(\mu i)}$ in (18.44), and using (18.42) for $A_{(\mu i)}$. □

Theorem 18.6 (Orthonormality and Completeness of Group Characters).

$$\sum_i \frac{N_{(i)}}{N_G} \overline{\chi_{(\mu i)}}\, \chi_{(\nu i)} = \delta_{\mu\nu} \quad \textit{(orthonormality)}, \tag{18.46}$$

$$\frac{N_{(i)}}{N_G} \sum_\mu \overline{\chi_{(\mu i)}}\, \chi_{(\mu j)} = \delta_{ij} \quad \textit{(completeness)}, \tag{18.47}$$

where in (18.46) the sum is over conjugacy classes and in (18.47) it is over inequivalent irreducible representations. (The overbar in both equations means complex conjugation.)

Proof. Contracting the pairs of indices (i, k) and (l, j) in the orthonormality theorem [(18.28)], we have

$$\frac{N_{(\mu)}}{N_G} \sum_{g \in G} (D^{-1}_{(\mu)}(g))^i_i (D_{(\nu)}(g))^j_j = \delta_{\mu\nu}\, \delta^i_i = N_{(\mu)}\, \delta_{\mu\nu}\,. \tag{18.48}$$

Since every representation of a finite group is equivalent to a unitary representation (Theorem 18.1), and since the trace of a matrix is invariant under similarity transformations (Theorem 17.2), the above equation can be rewritten as

$$\frac{1}{N_G} \sum_{g \in G} (U^\dagger_{(\mu)}(g))^i_i\, U_{(\nu)}(g))^j_j = \delta_{\mu\nu}\,, \tag{18.49}$$

where $U_{(\mu)}$ and $U_{(\nu)}$ are unitary representations, and we have used the fact that, for unitary operators, $U^{-1} = U^\dagger$. Furthermore, since

$$\chi_{(\mu i)} = Tr(U_{(\mu)}(g)) = (U_{(\nu)}(g))^j_j \tag{18.50}$$

just depends on the representation and the conjugacy class, and

$$\overline{(U_{(\mu)}(g))^j_j} = (U^\dagger_{(\mu)}(g))^j_j\,, \tag{18.51}$$

the orthonormality condition (18.46) follows directly from (18.49).

   To prove the completeness condition, we take the LHS of (18.39) and manipulate as follows:

$$\sum_{g \in C_i} \sum_{g' \in C_j} \sum_\mu \frac{N_{(\mu)}}{N_G} Tr(D_{(\mu)}(g) D^{-1}_{(\mu)}(g'))$$

$$= \sum_\mu \sum_{g \in C_i} \sum_{g' \in C_j} \frac{N_{(\mu)}}{N_G} Tr(U_{(\mu)}(g) U^\dagger_{(\mu)}(g')) \tag{18.52}$$

$$= \sum_\mu \frac{N_{(\mu)}}{N_G} \frac{N_{(i)}}{N_{(\mu)}} \frac{N_{(j)}}{N_{(\mu)}} \chi_{(\mu i)} \overline{\chi_{(\mu j)}}\, Tr(I) = \frac{N_{(i)} N_{(j)}}{N_G} \sum_\mu \chi_{(\mu i)} \overline{\chi_{(\mu j)}}\,,$$

where in the second equality we have used (18.41). By (18.39), the LHS of the first equality in the above equation is also equal to

$$\sum_{g \in C_i} \sum_{g' \in C_j} \delta_{gg'} = N_{(i)} \, \delta_{ij} \,. \tag{18.53}$$

On equating the RHS of the last equality in (18.52) to $N_{(i)} \delta_{ij}$ and taking the complex conjugate, the completeness condition (18.47) follows. $\qquad \square$

As a corollary of Theorem 18.6, we have the following important fact.

Corollary 18.2. *The number of inequivalent irreducible representations of any finite group G is equal to the number of distinct conjugacy classes of G.*

Proof. Setting $\mu = \nu$ in (18.46) and $i = j$ in (18.47) we have

$$\sum_i N_{(i)} \, \overline{\chi_{(\mu i)}} \, \chi_{(\mu i)} = 1 \,, \tag{18.54}$$

$$\sum_\mu N_{(i)} \, \overline{\chi_{(\mu i)}} \, \chi_{(\mu i)} = 1 \,. \tag{18.55}$$

Summing over μ in (18.54) and i in (18.55) the corollary follows. $\qquad \square$

Due to Corollary 18.2, the characters $\chi_{(\mu i)}$ of a group G can be arranged as entries in an $N_C \times N_C$ matrix, where N_C is the number of distinct conjugacy classes in G. This matrix is often called the **character table** of G. Note that for abelian groups, each group element is in a conjugacy class by itself, and by Corollary 18.1, all irreducible representations are one-diemnsional. Hence $D_{(\mu)}(g) = \chi_{(\mu i)}$, for $g \in C_i$.

Theorem 18.7. *In the reduction of a given representation $D(G)$ into its irreducible components, the number of times $n_{(\mu)}$ that the irreducible representation $D_{(\mu)}$ occurs is given by*

$$n_{(\mu)} = \sum_i \frac{N_{(i)}}{N_G} \, \overline{\chi_{(\mu i)}} \, \chi_{(i)} \,, \tag{18.56}$$

where $\chi_{(i)} \equiv Tr(D(g))$, $g \in C_i$ (the i-th conjugacy class).

Proof. We have, for $g \in C_i$,

$$\chi_{(i)} = Tr(D(g)) = \sum_\nu n_{(\nu)} \chi_{(\nu i)} \,. \tag{18.57}$$

Thus, multiplying both sides by $\dfrac{N_{(i)}}{N_G} \overline{\chi_{(\mu i)}}$ and summing over i, we have

$$\sum_i \frac{N_{(i)}}{N_G} \, \overline{\chi_{(\mu i)}} \, \chi_{(i)} = \sum_\nu n_{(\nu)} \sum_i \frac{N_{(i)}}{N_G} \overline{\chi_{(\mu i)}} \, \chi_{(\nu i)} = \sum_\nu n_{(\nu)} \, \delta_{\mu\nu} = n_{(\mu)} \,,$$

where in the second equality we have used (18.46). $\qquad \square$

Finally, with regard to group characters, we have

Theorem 18.8. *A representation of a group G with characters $\chi_{(i)}$ is irreducible if and only if*

$$\sum_i N_{(i)} |\chi_{(i)}|^2 = N_G . \tag{18.58}$$

Proof. We have

$$\sum_i N_{(i)} |\chi_{(i)}|^2 = \sum_i N_{(i)} \sum_\mu n_{(\mu)} \overline{\chi_{(\mu i)}} \sum_\nu n_{(\nu)} \chi_{(\nu i)}$$

$$= N_G \sum_{\mu,\nu} n_{(\mu)} n_{(\nu)} \delta_{\mu\nu} = N_G \sum_\mu n_{(\mu)}^2 . \tag{18.59}$$

If the representation is irreducible, it must be true that $n_{(\mu)} = 1$ for some μ, and $n_{(\nu)} = 0$ for all $\nu \neq \mu$. Hence (18.58) follows. Conversely, suppose (18.58) holds. Then (18.59) implies that

$$\sum_\mu n_{(\mu)}^2 = 1 . \tag{18.60}$$

This in turn implies that $n_{(\mu)} = 1$ for some μ and $n_{(\nu)} = 0$ for all $\nu \neq \mu$. The representation is thus irreducible.   □

As an illustration of the above discussion on group characters we will work out the character table of the **dihedral group** D_6. This finite group consists of all products of rotations on the plane by $60°$ and products of these rotations and a reflection about the x-axis. Its 12 elements can be listed as

$$D_6 = \{e, a, a^2, a^3, a^4, a^5, b, ba, ba^2, ba^3, ba^4, ba^5\} ,$$

where $a =$ a rotation about the z-axis in the counterclockwise sense by $60°$, and $b =$ a reflection about the x-axis. It decomposes into the following six conjugacy classes:

$$
\begin{aligned}
C_1 &= \{e\} , & &\text{so } N_{(1)} = 1 , \\
C_2 &= \{a^3\} , & &\text{so } N_{(2)} = 1 , \\
C_3 &= \{a, a^5\} , & &\text{so } N_{(3)} = 2 , \\
C_4 &= \{a^2, a^4\} , & &\text{so } N_{(4)} = 2 , \\
C_5 &= \{ab, ba, a^3 b\} , & &\text{so } N_{(5)} = 3 , \\
C_6 &= \{b, ba^2, a^2 b\} , & &\text{so } N_{(6)} = 3 .
\end{aligned}
$$

Exercise 18.5   Verify that the C_i given above are indeed conjugacy classes.

By Corollary 18.2, D_6 then has six inequivalent irreducible representations, which will be labelled by $\Gamma_{(i)}, i = 1, 2, \ldots, 6$. From Theorem (18.4) [(18.40)] we have

$$N_G = 12 = \sum_{\mu=1}^{6} (N_{(\mu)})^2 .$$

This equation has the unique solution: $12 = 1^2 + 1^2 + 1^2 + 1^2 + 2^2 + 2^2$. Thus the $\Gamma_{(i)}, i = 1, \ldots, 6$, have dimensions $1, 1, 1, 1, 2, 2$, respectively. The 36 entries in the character table are $\chi_{(\mu i)}, \mu, i = 1, \ldots, 6$. This will be written as a matrix where μ is the representation index (row index) and i is the class index (column index).

Let $\Gamma_{(1)}$ be the (one-dimensional) identity representation given by $\Gamma_{(1)}(g) = 1$ for all $g \in D_6$. Thus the first row of the character table (matrix) are all 1's. Since $\Gamma_{(\mu)}(e) = I$ (the identity operator), $\chi_{(\mu 1)} = Tr(I_{N_{(\mu)}}) = N_{(\mu)}$, where $I_{N_{(\mu)}}$ is the $N_{(\mu)} \times N_{(\mu)}$ unit matrix. The first column is thus $1, 1, 1, 1, 2, 2$. The rest of the characters for the one-dimensional representations $\Gamma_{(1)}, \Gamma_{(2)}, \Gamma_{(3)}$ and $\Gamma_{(4)}$ are relatively easy to work out since the $\Gamma_{(\mu)}(g), \mu = 1, 2, 3, 4$, are just numbers, and $\Gamma_{(\mu)}(g) = \chi_{(\mu i)}$ if $g \in C_i$. For these $\Gamma_{(\mu)}(g)$ we have

$$\chi_{(\mu i)} \chi_{(\mu j)} = \chi_{(\mu k)} , \quad \mu = 1, 2, 3, 4,$$

where k is the class index for the group element $g_1 g_2$, with $g_1 \in C_i$ and $g_2 \in C_j$. Looking at the explicitly displayed members of C_i, and making use of the multiplication table for D_6, we have, for instance,

$$\mu = 2, 3, 4 \quad \begin{cases} \chi^2_{(\mu 2)} = 1 & \implies \chi_{(\mu 2)} = \pm 1 , \\ \chi_{(\mu 4)} \chi_{(\mu 5)} = \chi_{(\mu 5)} & \implies \chi_{(\mu 4)} = 1 , \\ \chi_{(\mu 3)} \chi_{(\mu 4)} = \chi_{(\mu 2)} & \implies \chi_{(\mu 2)} = \chi_{(\mu 3)} . \end{cases}$$

We also have

$$\mu = 2, 3, 4 \quad \begin{cases} \chi^2_{(\mu 5)} = \chi^2_{(\mu 6)} = 1 & \implies \chi^2_{(\mu 5)} = \pm 1 , \; \chi^2_{(\mu 6)} = \pm 1 , \\ \chi_{(\mu 3)} \chi_{(\mu 5)} = \chi_{(\mu 6)} . \end{cases}$$

To determine all the $\chi_{(\mu i)}$ for the one-dimensional representations, we just need one more linear equation for the $\chi_{(\mu i)}$. This can be obtained from setting $\mu = 1$ and $\nu = 2, 3$ or 4 in the orthonormality relation [(18.46)]:

$$1 + \chi_{(\nu 2)} + 2\chi_{(\nu 3)} + 2\chi_{(\nu 4)} + 3\chi_{(\nu 5)} + 3\chi_{(\nu 6)} = 0 ,$$

or, on recalling $\chi_{(\nu 2)} = \chi_{(\nu 3)}$ and $\chi_{(\nu 4)} = 1$,

$$1 + \chi_{(\nu 2)} + \chi_{(\nu 5)} + \chi_{(\nu 6)} = 0 , \quad \nu = 2, 3, 4 .$$

The above data will determine all the $\chi_{(\mu i)}$ uniquely, for $\mu = 1, 2, 3, 4$.

| Exercise 18.6 | Calculate all the characters $\chi_{(\mu i)}$ for D_6 for the one-dimensional representations. (See Table 18.1.)

One of the two 2-dimensional representations $\Gamma_{(5)}$ and $\Gamma_{(6)}$ is obviously given by the rotation matrix

$$\Gamma_{(5)}(a) = \begin{pmatrix} \cos\dfrac{\pi}{3} & -\sin\dfrac{\pi}{3} \\ \sin\dfrac{\pi}{3} & \cos\dfrac{\pi}{3} \end{pmatrix} = \begin{pmatrix} 1/2 & -\sqrt{3}/2 \\ \sqrt{3}/2 & 1/2 \end{pmatrix},$$

and the reflection matrix (about the x-axis)

$$\Gamma_{(5)}(b) = \begin{pmatrix} 1 & 0 \\ 0 & -1 \end{pmatrix}.$$

The 2×2 representation matrices for all the other group elements for $\Gamma_{(5)}$ can be determined simply by matrix multiplication. Taking traces of the appropriate 2×2 matrices, we find: $\chi_{(51)} = 2$, $\chi_{(52)} = -2$, $\chi_{(53)} = 1$, $\chi_{(54)} = -1$, $\chi_{(55)} = \chi_{(56)} = 0$. While the remaining 2-dimensional representation $\Gamma_{(6)}$ is not obvious, its characters can be determined by the Orthonormality and Completeness Theorem of Group Characters (Theorem 18.6). Setting $i = j = 5$ and then $i = j = 6$ in (18.47), we immediately obtain $\chi_{(65)} = \chi_{(66)} = 0$. The last three characters, $\chi_{(62)}, \chi_{(63)}$ and $\chi_{(64)}$, can be determined from (18.46), for the instances $(\mu = 1, \nu = 6)$, $(\mu = 4, \nu = 6)$ and $(\mu = 5, \nu = 6)$. Thus

$$\chi_{(62)} + 2\chi_{(63)} + 2\chi_{(64)} = -2 \ , \ \ \chi_{(62)} + 2\chi_{(63)} - 2\chi_{(64)} = 2 \ , \ \ \chi_{(62)} - \chi_{(63)} + \chi_{(64)} = 2 \ .$$

These lead to the results $\chi_{(62)} = 2, \chi_{(63)} = -1$, and $\chi_{(64)} = -1$.

We finally present the character table for D_6 as follows (Table 18.1). In the table, the number within parentheses in the class designation C_i is the number of elements in that class, $N_{(i)}$.

	$C_1(1)$	$C_2(1)$	$C_3(2)$	$C_4(2)$	$C_5(3)$	$C_6(3)$
$\Gamma_{(1)}$	1	1	1	1	1	1
$\Gamma_{(2)}$	1	1	1	1	-1	-1
$\Gamma_{(3)}$	1	-1	-1	1	1	-1
$\Gamma_{(4)}$	1	-1	-1	1	-1	1
$\Gamma_{(5)}$	2	-2	1	-1	0	0
$\Gamma_{(6)}$	2	2	-1	-1	0	0

TABLE 18.1

Exercise 18.7 Verify that the character table in Table 18.1 satisfies Theorem 18.6.

Given a representation $D : G \to \mathcal{T}(\mathbb{V})$ that is fully reducible, we would now like to find all the irreducible representations of G that $D(G)$ contains. The objective is to find projection operators $P_{(\mu)} \in \mathcal{T}(\mathbb{V})$ such that, for any $v \in \mathbb{V}$, $P_{(\mu)} v \in \mathbb{V}_{(\mu)}$, where $\mathbb{V}_{(\mu)}$ is the representation space of the μ-th inequivalent irreducible representation.

Recall the notion of an enveloping algebra of a Lie algebra with respect to a particular representation (introduced in Chapter 7). We have a similar notion for groups. In a group G, only group multiplication is defined, so that $g_1 g_2 \in G$ if $g_1 \in G$ and $g_2 \in G$. Neither αg, $\alpha \in \mathbb{C}$, nor $g_1 + g_2$ is defined. However, for a given representation $D(G)$, both $\alpha D(g)$ and $D(g_1) + D(g_2)$ are well-defined, since $D(g)$ is a linear operator for all $g \in G$. The sums and products of all elements in $D(G) \subset \mathcal{T}(\mathbb{V})$ with the customary distributive and associative laws is called the **group algebra** of G with respect to the representation D, denoted by $A_D(G)$. Consider operators $(P_{(\mu)})_i^j \in A_D(G)$ defined by

$$(P_{(\mu)})_i^j \equiv \frac{N_{(\mu)}}{N_G} \sum_{g \in G} (D_{(\mu)}(g^{-1}))_j^i \, D(g), \quad i, j = 1, \dots, N_{(\mu)}. \tag{18.61}$$

We have

$$(P_{(\mu)})_i^j \, (P_{(\nu)})_k^l = \frac{N_{(\mu)} N_{(\nu)}}{N_G^2} \sum_{g_1, g_2 \in G} (D_{(\mu)}(g_1^{-1}))_j^i \, (D_{(\nu)}(g_2^{-1}))_l^k \, D(g_1) D(g_2)$$

$$= \frac{N_{(\mu)} N_{(\nu)}}{N_G^2} \sum_{g_1, g \in G} (D_{(\mu)}(g_1^{-1}))_j^i \, (D_{(\nu)}(g^{-1} g_1))_l^k \, D(g)$$

$$= \frac{N_{(\mu)} N_{(\nu)}}{N_G^2} \sum_{g_1, g \in G} (D_{(\mu)}(g_1^{-1}))_j^i \sum_{m=1}^{N_{(\nu)}} (D_{(\nu)}(g^{-1}))_l^m \, (D_{(\nu)}(g_1))_m^k \, D(g)$$

$$= \frac{N_{(\nu)}}{N_G} \sum_{g \in G} \sum_{m=1}^{N_{(\nu)}} (D_{(\nu)}(g^{-1}))_l^m \left(\frac{N_{(\mu)}}{N_G} \sum_{g_1 \in G} (D_{(\mu)}^{-1}(g_1))_j^i \, (D_{(\nu)}(g_1))_m^k \right) D(g)$$

$$= \frac{N_{(\nu)}}{N_G} \sum_{g \in G} \sum_{m=1}^{N_{(\nu)}} (D_{(\nu)}(g^{-1}))_l^m \, \delta_{\mu\nu} \, \delta_j^k \, \delta_m^i \, D(g) = \delta_{\mu\nu} \, \delta_j^k \sum_{g \in G} (D_{(\mu)}(g^{-1}))_l^i \, D(g),$$

$$\tag{18.62}$$

where in the second equality we have let $g = g_1 g_2$, and in the fifth we have used the orthonormality theorem [(18.28) in Theorem 18.3]. Thus, by (18.61),

$$(P_{(\mu)})_i^j \, (P_{(\nu)})_k^l = \delta_{\mu\nu} \, \delta_j^k \, (P_{(\mu)})_i^l. \tag{18.63}$$

Let us now define

$$P_{(\mu)} \equiv (P_{(\mu)})_i^i = \frac{N_{(\mu)}}{N_G} \sum_{g \in G} \chi_{(\mu)}(g^{-1}) D(g) = \frac{N_{(\mu)}}{N_G} \sum_{i=1}^{N_C} \overline{\chi(\mu i)} \sum_{g \in C_i} D(g), \tag{18.64}$$

where $\chi_{(\mu)}$ is the character in the μ-th representation. We will show that this is our sought-for projection operator.

First of all, (18.63) implies that $P_{(\mu)}$ is **idempotent** (a property that any projection operator has to satisfy):

$$P_{(\mu)}P_{(\nu)} = \delta_{\mu\nu}\,P_{(\mu)}\,.\tag{18.65}$$

We also need to prove completeness:

$$\sum_{\mu=1}^{N_R} P_{(\mu)} = I\,,\tag{18.66}$$

where I is the identity operator.

Since $D(G)$ is always fully reducible, there exists an invertible operator T (on $\mathbb{V}$) such that, for all $g \in G$,

$$T^{-1}D(g)T = \begin{pmatrix} \boxed{D_{(1)}(g)} & & 0 \\ & \ddots & \\ 0 & & \boxed{D_{(N_R)}(g)} \end{pmatrix}\,,\tag{18.67}$$

where the μ-th representation $D_{(\mu)}(g)$ occurs $n_{(\mu)}$ times [(18.56)] ($n_{(\mu)}$ may be zero). Thus

$$T^{-1}A_{(i)}T = \begin{pmatrix} \boxed{A_{(1i)}} & & 0 \\ & \ddots & \\ 0 & & \boxed{A_{(N_R i)}} \end{pmatrix}\,,\tag{18.68}$$

where

$$A_{(i)} \equiv \sum_{g \in C_i} D(g)\,,\tag{18.69}$$

[recall (18.42)]. From Theorem 18.5,

$$A_{(\mu i)} = \frac{N_{(i)}}{N_{(\mu)}}\chi_{(\mu i)}I_{N_{(\mu)}}\,,\tag{18.70}$$

where $I_{N_{(\mu)}}$ is the $N_{(\mu)} \times N_{(\mu)}$ unit matrix. Thus, by the orthonormality group characters [(18.46) in Theorem 18.6],

$$\frac{N_{(\mu)}}{N_G}\sum_{i=1}^{N_C}\overline{\chi_{(\mu i)}}A_{(\nu i)} = \delta_{\mu\nu}I_{N_{(\mu)}}\,.\tag{18.71}$$

It follows from (18.64) that

$$T^{-1} \left(\sum_{\mu} P_{(\mu)} \right) T = \sum_{\mu} T^{-1} P_{(\mu)} T = \sum_{\mu} \frac{N_{(\mu)}}{N_G} \sum_{i=1}^{N_C} \overline{X_{(\mu i)}} \sum_{g \in C_i} T^{-1} D(g) T$$

$$= \sum_{\mu} \frac{N_{(\mu)}}{N_G} \sum_{i=1}^{N_C} \overline{X_{(\mu i)}} \, T^{-1} A_{(i)} T = \sum_{\mu} \frac{N_{(\mu)}}{N_G} \sum_{i=1}^{N_C} \overline{X_{(\mu i)}} \begin{pmatrix} \boxed{A_{(1i)}} & & 0 \\ & \ddots & \\ 0 & & \boxed{A_{(N_R i)}} \end{pmatrix}$$

$$= \sum_{\mu} \begin{pmatrix} \boxed{\begin{matrix} \delta_{1\mu} & 0 \\ \cdots\cdots \\ 0 & \delta_{1\mu} \end{matrix}} & & 0 \\ \cdots\cdots\cdots\cdots\cdots\cdots\cdots \\ \cdots\cdots\cdots\cdots\cdots\cdots\cdots \\ 0 & & \boxed{\begin{matrix} \delta_{N_R\mu} & 0 \\ \cdots\cdots \\ 0 & \delta_{N_R\mu} \end{matrix}} \end{pmatrix} = I$$

(18.72)

where, on the RHS of the next to last equality, the top left box contains an $N_{(1)} \times N_{(1)}$ submatrix and the bottom right box contains an $N_{(N_R)} \times N_{(N_R)}$ submatrix. Hence $\sum_{\mu} P_{(\mu)} = I$ and completeness is established.

The projection operators $P_{(\mu)}$ also satisfy the following properties:

1)
$$[P_{(\mu)}, D(g)] = 0 \quad \text{for all } g \in G,$$
(18.73)

2) $P_{(\mu)}$ is hermitian if D is unitary.

Exercise 18.8 Prove the above two properties.

In the next chapter we will see an application of many of the results developed in the present one.

We will conclude this chapter with a general consideration of the problem of the reduction of the product of irreducible representations. This topic has important applications in the physics of composite systems.

Let $D_{(\mu)} : G \to \mathcal{T}(\mathbb{V}_{(\mu)})$ and $D_{(\nu)} : G \to \mathcal{T}(\mathbb{V}_{(\nu)})$ be two irreducible representations of a group G. The **product representation** $D_{(\mu)} \times D_{(\nu)} : G \to \mathcal{T}(\mathbb{V}_{(\mu)} \otimes \mathbb{V}_{(\nu)})$, of dimension $N_{(\mu)} N_{(\nu)}$, is given by

$$((D_{(\mu)} \times D_{(\nu)})(g)) \left(e_i^{(\mu)} \otimes e_j^{(\nu)} \right)$$
$$= \left(D_{(\mu)}(g) e_i^{(\mu)} \right) \otimes \left(D_{(\nu)}(g) e_j^{(\nu)} \right) . \quad (18.74)$$

This representation is in general reducible. The formal reduction is written as the so-called **Clebsch-Gordan series**:

$$D_{(\mu)} \times D_{(\nu)} = \sum_{\otimes \sigma} n_{(\mu\nu, \sigma)} D_{(\sigma)} , \quad (18.75)$$

where $n_{(\mu\nu, \sigma)}$ is the number of times that the σ-th irreducible representation occurs in the product representation [c.f. (18.56)]. Eq. (18.74) implies that the characters of the product representation and those of the irreducible representations are related by

$$\chi_{(\mu \times \nu)}(g) = \chi_{(\mu)}(g) \chi_{(\nu)}(g) . \quad (18.76)$$

Thus by (18.56)

$$n_{(\mu\nu, \sigma)} = \frac{1}{N_G} \sum_{g \in G} \overline{\chi_{(\sigma)}(g)} \, \chi_{(\mu)}(g) \chi_{(\nu)}(g) . \quad (18.77)$$

Suppose $\{e_i^{(\mu)}\}$ and $\{e_j^{(\nu)}\}$ are orthonormal bases of the irreducible representation spaces $\mathbb{V}_{(\mu)}$ and $\mathbb{V}_{(\nu)}$ respectively. We shall use the Dirac notation to write the basis vectors as $|(\mu)i\rangle$ and $|(\nu)j\rangle$. Basis vectors of the product representation space can thus be written as

$$|(\mu)i\,(\nu)j\rangle \equiv |(\mu)i\rangle \otimes |(\nu)j\rangle . \quad (18.78)$$

We have the completeness condition [c.f. (12.12)]

$$\sum_{ij} |(\mu)i\,(\nu)j\,\rangle\langle\,(\mu)i\,(\nu)j\,| = 1 . \quad (18.79)$$

The orthonormal basis states on the RHS of the Clebsch-Gordan series (18.75) can be denoted by $|(\sigma)\alpha\,l\,\rangle$, where $\alpha = 1, \dots, n_{(\mu\nu, \sigma)}$ labels the (possibly) multiple occurrence of the σ-th irreducible representation, and $l = 1, \dots, N_{(\sigma)}$. It follows from (18.79) that

$$|(\sigma)\alpha\,l\,\rangle = \sum_{ij} |(\mu)i\,(\nu)j\,\rangle\langle\,(\mu)i\,(\nu)j\,|(\sigma)\alpha\,l\,\rangle . \quad (18.80)$$

The matrix elements $\langle (\mu)i\,(\nu)j \,|\, (\sigma)\alpha\,l \rangle$ are called **Clebsch-Gordan coeffi-cients**. They obviously satisfy the following pair of orthonormality and com-pleteness conditions:

$$\sum_{ij} \langle (\sigma')\alpha'\,l' \,|\, (\mu)i\,(\nu)j \rangle \langle (\mu)i\,(\nu)j \,|\, (\sigma)\alpha\,l \rangle = \delta_{\sigma'\sigma}\delta_{\alpha'\alpha}\delta_{l'l} \,, \tag{18.81}$$

$$\sum_{\sigma\alpha l} \langle (\mu)i'\,(\nu)j' \,|\, (\sigma)\alpha\,l \rangle \langle (\sigma)\alpha\,l \,|\, (\mu)i\,(\nu)j \rangle = \delta_{i'i}\delta_{j'j} \,. \tag{18.82}$$

Eq. (18.81) is a consequence of the completeness of $\{|\,(\mu)i\,(\nu)j \rangle\}$ (for fixed μ and ν) and the orthonormality of $\{|\,(\sigma)\alpha\,l \rangle\}$, while (18.82) is a consequence of the completeness of $\{|\,(\sigma)\alpha\,l \rangle\}$ and the orthonormality of $\{|\,(\mu)i\,(\nu)j \rangle\}$ (for fixed μ and ν).

The following reciprocal relations, which are consequences of the Clebsch-Gordon series (18.75), are also very useful in applications. We have, for all $g \in G$,

$$(D_{(\mu)}(g))_i^k \,(D_{(\nu)}(g))_j^m$$
$$= \sum_{\sigma\alpha l\,l'} \langle (\mu)k\,(\nu)m \,|\, (\sigma)\alpha\,l' \rangle (D_{(\sigma)}(g))_{l'}^l \langle (\sigma)\alpha l \,|\, (\mu)i\,(\nu)j \rangle \,, \tag{18.83}$$

$$\delta_{\sigma'\sigma}\,\delta_{\alpha'\alpha}(D_{(\sigma)}(g))_l^{l'}$$
$$= \sum_{ijkm} \langle (\sigma')\alpha'\,l' \,|\, (\mu)k\,(\nu)m \rangle\, (D_{(\mu)}(g))_i^k (D_{(\nu)}(g))_j^m \langle (\mu)i\,(\nu)j \,|\, (\sigma)\alpha\,l \rangle \,.$$
$$\tag{18.84}$$

⎍Exercise 18.9⎍ Prove the above two equations.

Chapter 19

The Dihedral Group D_6 and the Benzene Molecule

In this chapter we will demonstrate an interesting quantum mechanical application of the theory of representations of finite groups. We will study the electronic energy levels of the benzene molecule $C_6 H_6$, which has a hexagonal ring structure shown in Fig. 19.1. The atomic number Z of carbon is 6. Of the

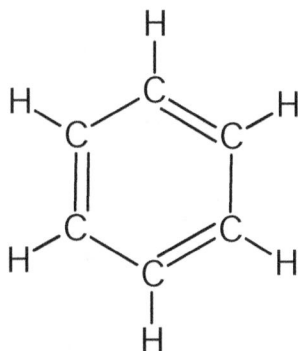

FIGURE 19.1

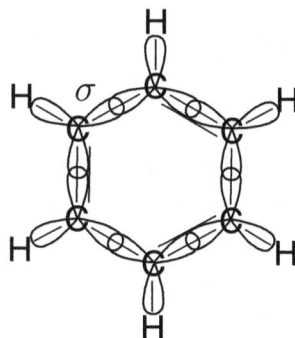

FIGURE 19.2

6 electrons associated with each carbon atom, 2 are in the K-shell (principal quantum number $n = 1$) and 4 are in the L-shell ($n = 2$). Three of the four valence electrons form hybridized σ-bonds with the axially symmetric orbitals (along the $C - C$ and the $C - H$ axes) in the plane of the hexagonal ring (Fig. 19.2). These are relatively tightly bound to the carbon nuclei. The fourth valence electron is a π-electron with orbitals perpendicular to the plane of the ring (p_z-orbitals) (Fig. 19.3).

Each π-electron can be pictured to "roam" around the ring, so its wave function (molecular orbital) is a linear superposition of localized p_z-orbitals

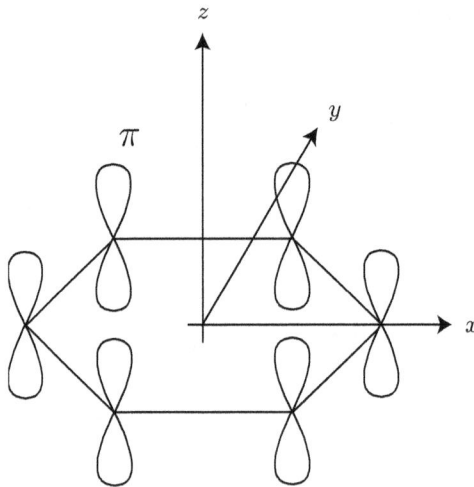

FIGURE 19.3

given by

$$\psi(\boldsymbol{r}) = \sum_{i=1}^{6} a^i\, \phi_i(\boldsymbol{r}) \,, \tag{19.1}$$

where $\phi_i(\boldsymbol{r})$ is a localized p_z-orbital centered at the i-th carbon nuclear site ($\boldsymbol{R}_i$) (see Fig. 19.4), and is given by

$$\phi_i(\boldsymbol{r}) \equiv \phi(\boldsymbol{r} - \boldsymbol{R}_i) = \beta|\,\boldsymbol{r} - \boldsymbol{R}_i\,|\cos\theta\,\exp\{-\alpha|\boldsymbol{r} - \boldsymbol{R}_i|\} \,, \tag{19.2}$$

where α and β are constants.

We have to solve the Schrödinger equation

$$H\psi = E\psi \tag{19.3}$$

with the one-electron Hamiltonian given by

$$H = -\frac{\hbar^2}{2m}\nabla^2 - \sum_{i=1}^{6} \frac{Ze^2}{|\boldsymbol{r} - \boldsymbol{R}_i|} \,, \tag{19.4}$$

which includes only the Coulomb interaction between the electron and the carbon nuclei. Assuming that

$$\langle\, \phi_i \,|\, \phi_j \,\rangle \equiv \int \mathrm{d}^3 r\, \overline{\phi_i(\boldsymbol{r})}\, \phi_j(\boldsymbol{r}) = \delta_{ij} \,, \tag{19.5}$$

(that is, the localized atomic orbitals at all the nuclear sites are normalized and the overlap integrals for $i \neq j$ all vanish), the Schrödinger equation (19.3) is equivalent to

$$H^i_j a^j = E a^i \,, \qquad i = 1,\ldots,6 \,, \tag{19.6}$$

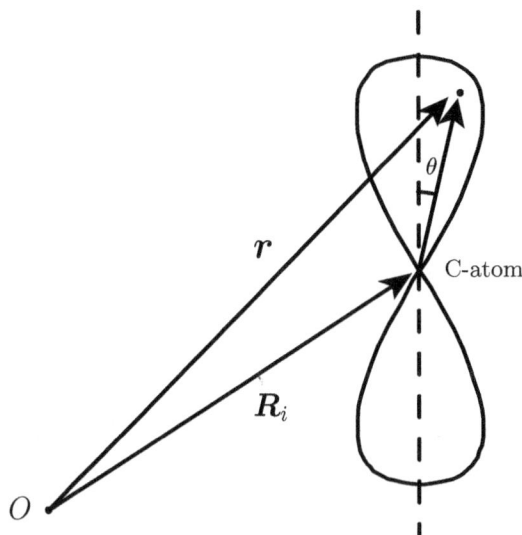

FIGURE 19.4

where [c.f. (12.26)]

$$H_j^i \equiv \langle \phi_i | H | \phi_j \rangle = \int d^3 r \, \overline{\phi_i(r)} \, H \, \phi_j(r) \, . \tag{19.7}$$

Exercise 19.1 Derive (19.6) from the Schrödinger equation (19.3) and the assumption of non-overlapping orbitals (19.5).

Eq. (19.6) leads to the secular equation

$$\det(H - E) = 0 \, , \tag{19.8}$$

whose solutions yield the energy eigenvalues [c.f. (17.5)].

Instead of tackling the problem directly, we will use group theory to obtain important information on the classification of the spectrum of the Hamiltonian H given by (19.4). In the present problem, the relevant symmetry group is apparently the dihedral group D_6 introduced in the last chapter. The Hamiltonian H is quite obviously invariant under the group action of D_6 on the nuclear coordinates $R_i, i = 1, \ldots, 6$. Thus there exists a unitary representation of D_6 on the space spanned by $\phi_i(r), i = 1, \ldots, 6$, a general vector of which, $\psi(r)$, is of the form (19.1). This representation is given by [c.f. (19.2) and (18.6)]

$$U(g)\phi_i(r) = \phi(r - g^{-1}R_i) \, , \quad g \in D_6 \, . \tag{19.9}$$

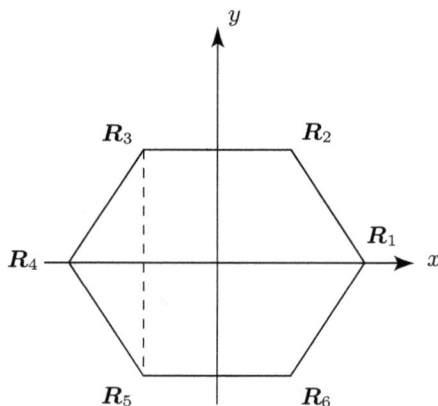

FIGURE 19.5

Choosing a coordinate system specifying the nuclear positions R_i as shown in Fig. 19.5, we thus have, under $U(a)$ (counterclockwise rotation by $60°$):

$$\phi_1 \to \phi_6 \to \phi_5 \to \phi_4 \to \phi_3 \to \phi_2 \to \phi_1 \ ,$$

and under $U(b)$ (reflection about the x-axis):

$$\phi_1 \to \phi_1 \ , \ \phi_2 \leftrightarrow \phi_6 \ , \ \phi_3 \leftrightarrow \phi_5 \ , \ \phi_4 \to \phi_4 \ .$$

$\boxed{\text{Exercise 19.2}}$ Verify directly that the representation of D_6 given by (19.9) is unitary: $\langle\, U(g)\phi_i\,|\,U(g)\phi_j\,\rangle = \langle\,\phi_i\,|\,\phi_j\,\rangle$.

Defining the representation matrix elements by

$$U(g)\phi_i = (U(g))_i^j\,\phi_j \ , \tag{19.10}$$

we have

$$U(a) = \begin{pmatrix} 0 & 0 & 0 & 0 & 0 & 1 \\ 1 & 0 & 0 & 0 & 0 & 0 \\ 0 & 1 & 0 & 0 & 0 & 0 \\ 0 & 0 & 1 & 0 & 0 & 0 \\ 0 & 0 & 0 & 1 & 0 & 0 \\ 0 & 0 & 0 & 0 & 0 & 1 \end{pmatrix} \ , \quad U(b) = \begin{pmatrix} 1 & 0 & 0 & 0 & 0 & 0 \\ 0 & 0 & 0 & 0 & 0 & 1 \\ 0 & 0 & 0 & 0 & 1 & 0 \\ 0 & 0 & 0 & 1 & 0 & 0 \\ 0 & 0 & 1 & 0 & 0 & 0 \\ 0 & 1 & 0 & 0 & 0 & 0 \end{pmatrix} \ . \tag{19.11}$$

All the other $U(g)$ can be obtained easily by matrix multiplication of the above two matrices. Note that the representation U is reducible.

Our discussion at the beginning of the last chapter indicated that the energy eigenvalues of H are classified by the inequivalent, irreducible representations of D_6. To find these let us begin by using the results on D_6 obtained in the last chapter. We found that D_6 has six inequivalent, irreducible representations with dimensions $1, 1, 1, 1, 2, 2$. The character table for D_6 (Table 18.1) and Theorem 18.7 can then be used to find the number of times, $n_{(\mu)}$, that the μ-th irreducible representation occurs in our representation $U(D_6)$. Observing that

$$\chi_{(1)} = Tr(U(e)) = 6 , \quad \chi_{(2)} = Tr(U(a^3)) = 0 , \quad \chi_{(3)} = Tr(U(a)) = 0 ,$$
$$\chi_{(4)} = Tr(U(a^2)) = 0 , \quad \chi_{(5)} = Tr(U(ab)) = 0 , \quad \chi_{(6)} = Tr(U(b)) = 2 ,$$
$$\tag{19.12}$$

we have, from (18.56),

$$n_{(\mu)} = \frac{1}{2}(\chi_{(\mu 1)} + \chi_{(\mu 6)}) . \tag{19.13}$$

The character table then reveals that

$$n_{(1)} = \frac{1}{2}(1 + 1) = 1 , \quad n_{(2)} = \frac{1}{2}(1 - 1) = 0 , \quad n_{(3)} = \frac{1}{2}(1 - 1) = 0 ,$$
$$n_{(4)} = \frac{1}{2}(1 + 1) = 1 , \quad n_{(5)} = \frac{1}{2}(2 + 0) = 1 , \quad n_{(6)} = \frac{1}{2}(2 + 0) = 1 . \tag{19.14}$$

We see that the representation U decomposes into the following irreducible representations:

$$U = \Gamma_{(1)} \oplus \Gamma_{(4)} \oplus \Gamma_{(5)} \oplus \Gamma_{(6)} . \tag{19.15}$$

Our next task is to project out the irreducible representation spaces $V_{(1)}, V_{(4)}, V_{(5)}$ and $V_{(6)}$ from the representation space of U. We need to calculate the projection operators $P_{(\mu)}$, $\mu = 1, 4, 5, 6$, as given by (18.64). Writing

$$S_i \equiv \sum_{g \in C_i} U(g) , \tag{19.16}$$

these are given by

$$P_\mu = \frac{N_{(\mu)}}{12} \sum_{i=1}^{6} \chi_{(\mu i)} S_i . \tag{19.17}$$

The calculations are long but quite straightforward. We give the steps as follows. Eq. (19.17) gives

$$P_{(1)} = \frac{1}{12} \sum_{i=1}^{6} S_i ,$$

$$P_{(4)} = \frac{1}{12} (S_1 - S_2 - S_3 + S_4 - S_5 + S_6) , \tag{19.18}$$

$$P_{(5)} = \frac{1}{6} (2S_1 - 2S_2 + S_3 - S_4) ,$$

$$P_{(6)} = \frac{1}{6} (2S_1 + 2S_2 - S_3 - S_4) ,$$

where

$$S_1 = U(e) = I_6 \quad (6 \times 6 \text{ unit matrix}) ,$$

$$S_2 = U(a^3) = \begin{pmatrix} 0 & I_3 \\ I_3 & 0 \end{pmatrix} , \qquad (19.19)$$

(with the blocks being 3×3 matrices and I_3 the 3×3 identity matrix),

$$S_3 = U(a) + U(a^5) = \begin{pmatrix} A & B \\ B & A \end{pmatrix} , \qquad (19.20\text{a})$$

$$S_4 = U(a^2) + U(a^4) = \begin{pmatrix} B & A \\ A & B \end{pmatrix} , \qquad (19.20\text{b})$$

$$S_5 = U(ab) + U(ba) + U(a^3b) = \begin{pmatrix} A & C \\ C & A \end{pmatrix} , \qquad (19.20\text{c})$$

$$S_6 = U(b) + U(ba^2) + U(a^2b) = \begin{pmatrix} C & A \\ A & C \end{pmatrix} . \qquad (19.20\text{d})$$

In the above equations,

$$A \equiv \begin{pmatrix} 0 & 1 & 0 \\ 1 & 0 & 1 \\ 0 & 1 & 0 \end{pmatrix} , \quad B \equiv \begin{pmatrix} 0 & 0 & 1 \\ 0 & 0 & 0 \\ 1 & 0 & 0 \end{pmatrix} , \quad C \equiv \begin{pmatrix} 1 & 0 & 1 \\ 0 & 1 & 0 \\ 1 & 0 & 1 \end{pmatrix} . \qquad (19.21)$$

Using (19.19), (19.20) and (19.21), the projection operators of (19.18) can be written explicitly in matrix form (with respect to the basis $\{\phi_i\}$) as follows:

$$P_{(1)} = \frac{1}{6} \begin{pmatrix} 1 & 1 & 1 & 1 & 1 & 1 \\ 1 & 1 & 1 & 1 & 1 & 1 \\ 1 & 1 & 1 & 1 & 1 & 1 \\ 1 & 1 & 1 & 1 & 1 & 1 \\ 1 & 1 & 1 & 1 & 1 & 1 \\ 1 & 1 & 1 & 1 & 1 & 1 \end{pmatrix} , \quad P_{(4)} = \frac{1}{6} \begin{pmatrix} 1 & -1 & 1 & -1 & 1 & -1 \\ -1 & 1 & -1 & 1 & -1 & 1 \\ 1 & -1 & 1 & -1 & 1 & -1 \\ -1 & 1 & -1 & 1 & -1 & 1 \\ 1 & -1 & 1 & -1 & 1 & -1 \\ -1 & 1 & -1 & 1 & -1 & 1 \end{pmatrix} ,$$

$$(19.22\text{a})$$

$$P_{(5)} = \frac{1}{6} \begin{pmatrix} 2 & 1 & -1 & -2 & -1 & 1 \\ 1 & 2 & 1 & -1 & -2 & -1 \\ -1 & 1 & 2 & 1 & -1 & -2 \\ -2 & -1 & 1 & 2 & 1 & -1 \\ -1 & -2 & -1 & 1 & 2 & 1 \\ 1 & -1 & -2 & -1 & 1 & 2 \end{pmatrix}, \qquad (19.22\text{b})$$

$$P_{(6)} = \frac{1}{6} \begin{pmatrix} 2 & -1 & -1 & 2 & -1 & -1 \\ -1 & 2 & -1 & -1 & 2 & -1 \\ -1 & -1 & 2 & -1 & -1 & 2 \\ 2 & -1 & -1 & 2 & -1 & -1 \\ -1 & 2 & -1 & -1 & 2 & -1 \\ -1 & -1 & 2 & -1 & -1 & 2 \end{pmatrix}. \qquad (19.22\text{c})$$

Exercise 19.3   Verify the above expressions for $P_{(\mu)}$.

The above projection operators can then be used to pick up $N_{(\mu)}$ linearly independent basis vectors (which can be normalized): $\psi_{(\mu)i}$, $i = 1, \ldots, N_{(\mu)}$, in each $\mathbb{V}_{(\mu)}$. Thus

$$\psi_{(1)1} = \sqrt{6}\,(P_{(1)})_1^j\,\phi_j = \frac{1}{\sqrt{6}}\,(\phi_1 + \phi_2 + \phi_3 + \phi_4 + \phi_5 + \phi_6), \qquad (19.23)$$

$$\psi_{(4)1} = \sqrt{6}\,(P_{(4)})_1^j\,\phi_j = \frac{1}{\sqrt{6}}\,(\phi_1 - \phi_2 + \phi_3 - \phi_4 + \phi_5 - \phi_6), \qquad (19.24)$$

$$\psi_{(5)1} = \sqrt{3}\,(P_{(5)})_1^j\,\phi_j = \frac{1}{\sqrt{12}}\,(2\phi_1 + \phi_2 - \phi_3 - 2\phi_4 - \phi_5 + \phi_6), \qquad (19.25\text{a})$$

$$\psi_{(5)2} = \sqrt{3}\,(P_{(5)})_2^j\,\phi_j = \frac{1}{\sqrt{12}}\,(\phi_1 + 2\phi_2 + \phi_3 - \phi_4 - 2\phi_5 - \phi_6), \qquad (19.25\text{b})$$

$$\psi_{(6)1} = \sqrt{3}\,(P_{(6)})_1^j\,\phi_j = \frac{1}{\sqrt{12}}\,(2\phi_1 - \phi_2 - \phi_3 + 2\phi_4 - \phi_5 - \phi_6), \qquad (19.26\text{a})$$

$$\psi_{(6)2} = \sqrt{3}\,(P_{(6)})_2^j\,\phi_j = \frac{1}{\sqrt{12}}\,(-\phi_1 + 2\phi_2 - \phi_3 - \phi_4 + 2\phi_5 - \phi_6). \qquad (19.26\text{b})$$

Note that
$$\langle \psi_{(5)1} \,|\, \psi_{(5)2} \rangle \neq 0 \,, \quad \langle \psi_{(6)1} \,|\, \psi_{(6)2} \rangle \neq 0 \,; \tag{19.27}$$
but we can use the **Schmidt orthogonalization** procedure to construct orthonormal bases for $\mathbb{V}_{(5)}$ and $\mathbb{V}_{(6)}$, using $\psi_{(5)1}, \psi_{(5)2}, \psi_{(6)1}$ and $\psi_{(6)2}$. Thus we define the following normalized vectors:
$$\psi'_{(\mu)2} = (\psi_{(\mu)2} - \langle \psi_{(\mu)1} \,|\, \psi_{(\mu)2} \rangle \, \psi_{(\mu)1})/(length) \,, \quad \mu = 5, 6 \,. \tag{19.28}$$
The sets $(\psi_{(5)1}, \psi'_{(5)2})$ and $(\psi_{(6)1}, \psi'_{(6)2})$ will then form orthonormal bases for $\mathbb{V}_{(5)}$ and $\mathbb{V}_{(6)}$, respectively. We find that these are given by

$$\psi_{(5)1} = \frac{1}{\sqrt{12}} \, (2\phi_1 + \phi_2 - \phi_3 - 2\phi_4 - \phi_5 + \phi_6) \,, \tag{19.29a}$$

$$\psi'_{(5)2} = \frac{1}{2} \, (\phi_2 + \phi_3 - \phi_5 - \phi_6) \,, \tag{19.29b}$$

$$\psi_{(6)1} = \frac{1}{\sqrt{12}} \, (2\phi_1 - \phi_2 - \phi_3 + 2\phi_4 - \phi_5 - \phi_6) \,, \tag{19.30a}$$

$$\psi'_{(6)2} = \frac{1}{2} \, (\phi_2 - \phi_3 + \phi_5 - \phi_6) \,. \tag{19.30b}$$

$\boxed{\text{Exercise 19.4}}$ Use (19.28) to verify the above expressions for $\psi'_{(5)2}$ and $\psi'_{(6)2}$.

To summarize

$$\begin{aligned} \psi_{(1)1} \text{ spans } \mathbb{V}_{(1)} \,, &\quad (1-d) \,, \\ \psi_{(4)1} \text{ spans } \mathbb{V}_{(4)} \,, &\quad (1-d) \,, \\ \psi_{(5)1} \text{ and } \psi'_{(5)2} \text{ span } \mathbb{V}_{(5)} \,, &\quad (2-d) \,, \\ \psi_{(6)1} \text{ and } \psi'_{(6)2} \text{ span } \mathbb{V}_{(6)} \,, &\quad (2-d) \,. \end{aligned}$$

The orthonormal set of vectors $\{\psi_{(1)1}, \psi_{(4)1}, \psi_{(5)1}, \psi'_{(5)2}, \psi_{(6)1}, \psi'_{(6)2}\}$ spans

$$\mathbb{V} = \mathbb{V}_{(1)} \oplus \mathbb{V}_{(4)} \oplus \mathbb{V}_{(5)} \oplus \mathbb{V}_{(6)} \,. \tag{19.31}$$

With respect to this set, the 6×6 representation matrices of D_6 are fully reduced to the form

$$S^{-1}U(g)S = \begin{pmatrix} \square & 0 & 0 & 0 & 0 & 0 \\ 0 & \square & 0 & 0 & 0 & 0 \\ 0 & 0 & \square & \square & 0 & 0 \\ 0 & 0 & \square & \square & 0 & 0 \\ 0 & 0 & 0 & 0 & \square & \square \\ 0 & 0 & 0 & 0 & \square & \square \end{pmatrix} \,, \tag{19.32}$$

where the boxes denote entries that are not necessarily zero,

$$S = \begin{pmatrix} \dfrac{1}{\sqrt{6}} & \dfrac{1}{\sqrt{6}} & \dfrac{1}{\sqrt{3}} & 0 & \dfrac{1}{\sqrt{3}} & 0 \\[2mm] \dfrac{1}{\sqrt{6}} & -\dfrac{1}{\sqrt{6}} & \dfrac{1}{\sqrt{12}} & \dfrac{1}{2} & -\dfrac{1}{\sqrt{12}} & \dfrac{1}{2} \\[2mm] \dfrac{1}{\sqrt{6}} & \dfrac{1}{\sqrt{6}} & -\dfrac{1}{\sqrt{12}} & \dfrac{1}{2} & -\dfrac{1}{\sqrt{12}} & -\dfrac{1}{2} \\[2mm] \dfrac{1}{\sqrt{6}} & -\dfrac{1}{\sqrt{6}} & -\dfrac{1}{\sqrt{3}} & 0 & \dfrac{1}{\sqrt{3}} & 0 \\[2mm] \dfrac{1}{\sqrt{6}} & \dfrac{1}{\sqrt{6}} & -\dfrac{1}{\sqrt{12}} & -\dfrac{1}{2} & -\dfrac{1}{\sqrt{12}} & \dfrac{1}{2} \\[2mm] \dfrac{1}{\sqrt{6}} & -\dfrac{1}{\sqrt{6}} & \dfrac{1}{\sqrt{12}} & -\dfrac{1}{2} & -\dfrac{1}{\sqrt{12}} & -\dfrac{1}{2} \end{pmatrix}, \qquad (19.33)$$

and $S^{-1} = S^T$.

Exercise 19.5   Use (19.23), (19.24), (19.29) and (19.30) in conjunction with (1.36) to read off the entries for S as given in (19.33), and show explicitly that $S^{-1} = S^T$.

Let us now recall our discussion at the beginning of the last chapter concerning the relevance of group theory in quantum mechanics. Since our Hamiltonian H [of (19.4)] is invariant under the group D_6, it commutes with $U(g)$ for all $g \in D_6$. The irreducible representation spaces $V_{(1)}, V_{(4)}, V_{(5)}$ and $V_{(6)}$ of D_6 are then also eigensubspaces of H, the dimension of each of these subspaces being the degeneracy of the corresponding eigenvalue. Let these eigenvalues be E_1, E_4, E_5 and E_6. Then the 6×6 matrix H^i_j [c.f. (19.7)] can be diagonalized (by a non-singular matrix S):

$$S^{-1}HS = \begin{pmatrix} E_1 & 0 & 0 & 0 & 0 & 0 \\ 0 & E_4 & 0 & 0 & 0 & 0 \\ 0 & 0 & E_5 & 0 & 0 & 0 \\ 0 & 0 & 0 & E_5 & 0 & 0 \\ 0 & 0 & 0 & 0 & E_6 & 0 \\ 0 & 0 & 0 & 0 & 0 & E_6 \end{pmatrix}, \qquad (19.34)$$

where

$$E_1 = \langle \psi_{(1)1} | H | \psi_{(1)1} \rangle, \qquad (19.35a)$$
$$E_4 = \langle \psi_{(4)1} | H | \psi_{(4)1} \rangle, \qquad (19.35b)$$
$$E_5 = \langle \psi_{(5)1} | H | \psi_{(5)1} \rangle = \langle \psi'_{(5)2} | H | \psi'_{(5)2} \rangle, \qquad (19.35c)$$
$$E_6 = \langle \psi_{(6)1} | H | \psi_{(6)1} \rangle = \langle \psi'_{(6)2} | H | \psi'_{(6)2} \rangle. \qquad (19.35d)$$

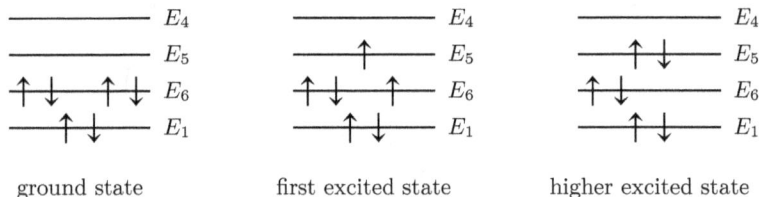

| ground state | first excited state | higher excited state |

FIGURE 19.6

Exercise 19.6   Use the above four equations to give expressions for the energy eigenvalues E_1, E_4, E_5 and E_6 in terms of H_j^i .

An explicit calculation using (19.4) for H, (19.7) for H_i^j and (19.2) for the ϕ_i shows that $E_1 < E_6 < E_5 < E_4$. The electronic states are then described by using the Aufbau principle in atomic physics to accommodate the six π-electrons (one from each carbon atom of the benzene ring) in these four energy levels, keeping in mind that E_1 and E_4 are non-degenerate, while E_5 and E_6 are doubly degenerate, and that due to Pauli's exclusion principle, each non-degenerate level can accommodate two electrons of opposite spin. Thus we have, for example, the three electronic states pictured in Fig. 19.6.

Chapter 20

Representations of the Symmetric Groups and the General Linear Groups, Young Diagrams

The symmetric (permutation) group S_n was introduced in Chapter 16 for the purpose of the construction of symmetric and skew-symmetric tensors. Besides these two symmetry classes, there may be many others that can be obtained by the action of S_n on the tensor space $T^n(V)$ [or $T^n(V^*)$]. As we shall see, the different symmetry classes correspond to inequivalent irreducible representations of the symmetric group, and indeed, of the general linear group $GL(n)$, if $dim(V) = n$. The symmetric group S_n is also of fundamental importance in quantum mechanics due to the fact that the Hamiltonian H describing a system of identical particles (such as an n-electron atom) is obviously invariant under S_n. Thus the irreducible representations of S_n also serve to classify the spectrum of H.

Let us first review some basic facts of S_n. Each element $\sigma \in S_n$ can be written as a **cycle** or a product of cycles. A cycle of length r, or an r-cycle, is written $(i_1 \, i_2 \ldots i_r)$, where $i_j \in \{1, 2, \ldots, n\}, j = 1, \ldots, r$. Its action is given by

$$i_1 \to i_2 \to i_3 \to \cdots \to i_{r-1} \to i_r \to i_1 \,.$$

Thus, for S_3, the 2-cycle (13) gives the permutation [c.f. (16.5)]

$$\begin{pmatrix} 1 & 2 & 3 \\ 3 & 2 & 1 \end{pmatrix},$$

while the 3-cycle (123) gives

$$\begin{pmatrix} 1 & 2 & 3 \\ 2 & 3 & 1 \end{pmatrix}.$$

Note that all 1-cycles are equal to the identity $e \in S_n$. It is evident that all $g \in S_n$ except the identity e can be written as a product of 2-cycles, or **transpositions**. In fact, an even permutation σ (with $sgn\,\sigma = 1$) is one which can be written as a product of an even number of transpositions, while an odd permutation $(sgn\,\sigma = -1)$ is one expressible as the product of an odd number of transpositions. Thus

$$(123) = (13)(12) = (23)(13) \tag{20.1}$$

is even while

$$(13) = (12)(23)(12) = (12)(13)(23) \tag{20.2}$$

is odd. Note that the decomposition into transpositions is not unique. As for every group, elements of S_n can be partitioned into conjugacy classes (c.f. Def. 6.2). For example, S_3 can be partitioned into the three classes

$$C_1 = \{e\}, \; C_2 = \{(12),(13),(23)\}, \; C_3 = \{(123),(321)\},$$

since

$$(13) = (23)(12)(23)^{-1} = (12)(23)(12)^{-1}, \tag{20.3}$$

and

$$(321) = (12)(123)(12)^{-1}. \tag{20.4}$$

| Exercise 20.1 | Verify (20.3) and (20.4).

The representation theory of S_n begins with the following theorem, which will be stated without proof.

Theorem 20.1. *Each conjugacy class C_i of the symmetric group S_n is characterized by the common cycle-structure of elements $g \in C_i$. By a cycle-structure of an element $\sigma \in S_n$ we mean an ordered set of non-negative integers $\{\alpha_1, \alpha_2, \ldots, \alpha_n\}$ such that σ can be expressed as a product of α_1 1-cycles, α_2 2-cycles,...., and α_n n-cycles; and*

$$\alpha_1 + 2\alpha_2 + \cdots + n\alpha_n = n. \tag{20.5}$$

*The set $\{\alpha_1, \ldots, \alpha_n\}$ satisfying (20.5) is called a **partition** of the integer n.*

Let us now use S_3 again to illustrate Theorem 20.1. The three classes C_1, C_2 and C_3 have the following cycle-structures:

$$C_1 = \{e\} = \{(1)(2)(3)\}\,, \qquad\qquad \alpha_1 = 3\,, \alpha_2 = 0\,, \alpha_3 = 0\,,$$
$$C_2 = \{(3)(12), (2)(13), (1)(23)\}\,, \qquad \alpha_1 = 1\,, \alpha_2 = 1\,, \alpha_3 = 0\,,$$
$$C_3 = \{(123), (321)\}\,, \qquad\qquad \alpha_1 = 0\,, \alpha_2 = 0\,, \alpha_3 = 1\,.$$

Note that if an element has more than one type of cycle, the numbers that occur in one type must all be different from those that occur in another type.

Using (20.5), the partition $\{\alpha_1, \ldots, \alpha_n\}$ can also be labeled by the set of integers $\{\mu_1, \ldots, \mu_n\}$ given by

$$\alpha_1 + \alpha_2 + \alpha_3 + \cdots + \alpha_n = \mu_1\,,$$
$$\alpha_2 + \alpha_3 + \cdots + \alpha_n = \mu_2\,,$$
$$\alpha_3 + \cdots + \alpha_n = \mu_3\,, \tag{20.6}$$
$$\vdots$$
$$\alpha_n = \mu_n\,.$$

It is clear that

$$\mu_1 + \mu_2 + \cdots + \mu_n = n\,, \tag{20.7}$$

and

$$\mu_1 \geq \mu_2 \geq \cdots \geq \mu_n \geq 0\,. \tag{20.8}$$

The set $\{\mu_1, \ldots, \mu_n\}$ can also be used to label a partition, and thus a conjugacy class. Using the μ-labeling for the conjugacy classes of S_3, we have

$$C_1 : \mu_1 = 3, \mu_2 = \mu_3 = 0\,, \; C_2 : \mu_1 = 2, \mu_2 = 1, \mu_3 = 0\,, \; C_3 : \mu_1 = \mu_2 = \mu_3 = 1\,.$$

Each partition $\{\mu_1, \ldots, \mu_n\}$ of n can be conveniently represented by what is called a **Young diagram**, each of which is an array of exactly n boxes, with at most n rows. The first row contains μ_1 boxes, the second row μ_2 ($\leq \mu_1$) boxes, and so on. Thus for S_3 there are three Young diagrams:

$$\{\mu_i\} = \{3, 0, 0\} \qquad \square\square\square \,,$$

$$\{\mu_i\} = \{2, 1, 0\} \qquad \square\square \atop \square \,,$$

$$\{\mu_i\} = \{1, 1, 1\} \qquad {\square \atop \square \atop \square}\,.$$

For S_2 there are only two Young diagrams:

$$\{\mu_i\} = \{2, 0\} \;\; \square\square \,, \qquad \{\mu_i\} = \{1, 1\} \;\; {\square \atop \square}\,;$$

while for S_4 there are five:

$\{\mu_i\} = \{4, 0, 0, 0\}$, $\{\mu_i\} = \{3, 1, 0, 0\}$,

$\{\mu_i\} = \{2, 2, 0, 0\}$, $\{\mu_i\} = \{2, 1, 1, 0\}$,

$\{\mu_i\} = \{1, 1, 1, 1\}$.

By Corollary 18.2, the inequivalent irreducible representations of S_n are also characterized by the Young diagrams. Thus S_2 has two, S_3 has three, and S_4 has five inequivalent representations. The Young diagrams corresponding to the seven inequivalent irreducible representations of S_5 are also shown below:

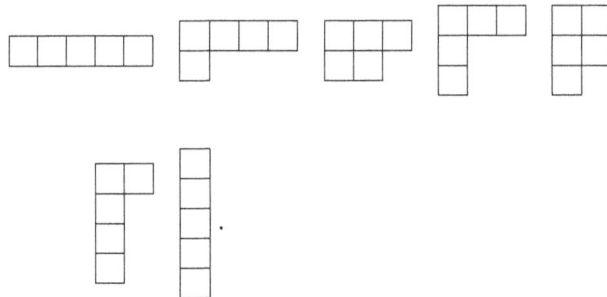

Next we present a remarkable formula for the characters $\chi_{(\mu i)}$ of S_n:

Theorem 20.2 (Frobenius Character Formula). *Let C_i be the i-th conjugacy class of S_n characterized by the cyclic structure $\{\alpha_1, \ldots, \alpha_n\}$ and $\Gamma_{(\mu)}$ be the μ-th irreducible representation of S_n characterized by the Young diagram $\{\mu_1, \ldots, \mu_n\}$. [The μ_i need not be related to the α_j by (20.6).] Introduce p independent variables $z_1, \ldots, z_p$, where the integer $p \leq n$ is determined by*

$$\mu_1 \geq \mu_2 \geq \cdots \geq \mu_p > 0, \quad \mu_{p+1} = \mu_{p+2} = \cdots = \mu_n = 0.$$

Define the power sums $P_j(z_1, \ldots, z_p)$, $1 \leq j \leq n$, and the **discriminant** *$\Delta(z_1, \ldots, z_p)$ by*

$$P_j(z_1, \ldots, z_p) \equiv z_1^j + z_2^j + \cdots + z_p^j, \tag{20.9}$$

$$\Delta(z_1, \ldots, z_p) \equiv \prod_{i < k}^{p} (z_i - z_k). \tag{20.10}$$

Set

$$l_1 \equiv \mu_1 + p - 1 \ , \ l_2 \equiv \mu_2 + p - 2 \ , \ \dots \ , \ l_p \equiv \mu_p \ ; \tag{20.11}$$

and, for a formal power series $f(z_1, \dots, z_p)$, write

$$[f(z_1, \dots, z_p)]_{(l_1 \dots l_p)} \equiv \text{ coefficient of } z_1^{l_1} \dots z_p^{l_p} \text{ in } f(z_1, \dots, z_p) \ . \tag{20.12}$$

Then

$$\chi_{(\mu i)} = \left[\Delta(z_1, \dots, z_p) \prod_{j=1}^n (P_j(z_1, \dots, z_p))^{\alpha_j} \right]_{(l_1 \dots l_p)} \ . \tag{20.13}$$

The proof of the Frobenius character formula is quite involved, and will not be presented here. For details, see W. Fulton and J. Harris 1991.

Exercise 20.2 Use the Frobenius character formula (20.13) to verify the following character table for S_3. In the table the row designations are for $\{\mu_i\}$, while the column designations are for $\{\alpha_i\}$.

	$\{3,0,0\}$	$\{1,1,0\}$	$\{0,0,1\}$
$\{3,0,0\}$	1	1	1
$\{2,1,0\}$	2	0	-1
$\{1,1,1\}$	1	-1	1

Now the dimension of the representation space $\mathbb{V}_{(\mu)}$ of $\Gamma_{(\mu)}$ is given by $\chi_{(\mu 1)}$, where $C_1 = \{e\}$ and the identity e can be written $e = (1)(2)\dots(n)$. Thus for C_1, $\alpha_1 = n, \alpha_2 = \dots = \alpha_n = 0$, and

$$dim \, \mathbb{V}_{(\mu)} = \chi_{(\mu 1)} = [\Delta(z_1, \dots, z_p)(z_1 + \dots + z_p)^n]_{(l_1 \dots l_p)} \ . \tag{20.14}$$

This equation yields the following useful formula for the dimension of the irreducible representation spaces of S_n. (For a derivation of this formula see W. Fulton and J. Harris 1991.)

Corollary 20.1. *Let* $\Gamma_{(\mu)}$ *be the* μ*-th irreducible representation of* S_n, *characterized by the Young diagram* $\{\mu_1, \dots, \mu_n\}$, *with* $\mu_1 \geq \mu_2 \geq \dots \geq \mu_p >$

0, $\mu_{p+1} = \mu_{p+2} = \cdots = \mu_n = 0$. *Then the dimension of the representation space* $\mathbb{V}_{(\mu)}$ *is given by*

$$\boxed{dim\, \mathbb{V}_{(\mu)} = \frac{n!\,\Delta(l_1,\ldots,l_p)}{l_1!\,l_2!\,\ldots l_p!}}\ ,\qquad (20.15)$$

where

$$\Delta(l_1,\ldots,l_p) = \prod_{i<k}^{p}(l_i - l_k)\ .\qquad (20.16)$$

Exercise 20.3 Verify the first column of the character table for S_3 (given in Exercise 20.2) by using (20.15).

To further develop the representation theory of S_n we will introduce some notions related to a Young diagram.

Definition 20.1. *A **Young tableau** on a given Young diagram for* S_n *is a numbering of the boxes of the diagram by the integers* $1,\ldots,n$ *in any order, each number being used only once. For example*

7	8	2
6	4	
5	1	
3		

Definition 20.2. *A **standard Young tableau** is a Young tableau in which the numbers in each row increase to the right (not necessarily in strict order) and those in each column increase to the bottom. For example,*

1	3	4
2	5	
6	8	
7		

Definition 20.3. *The **hook length** of a box in a Young diagram is the number of boxes directly to the right and directly below, including the box itself. In the following Young diagram, for example, each box is labeled by its hooked length.*

6	4	1
4	2	
3	1	
1		

We have the following easy methods (presented without proofs) to calculate the dimensions of irreducible representation spaces for S_n.

Theorem 20.3. *The dimension of the (irreducible) representation space* $\mathbb{V}_{(\mu)}$ *corresponding to a given Young diagram of* S_n *is given by*

$$N_{(\mu)} = dim\,\mathbb{V}_{(\mu)} = \frac{n!}{\prod(hook\ lengths)} \tag{20.17}$$
$$= number\ of\ distinct\ standard\ Young\ tableau\,.$$

As an illustration consider the Young diagram $\{\mu_i\} = \{2,1,1,1,0\}$ for S_5. The possible standard Young tableau are

1 2	1 3	1 4	1 5
3	2	2	2
4	4	3	3
5	5	5	4

Hence $dim\,\mathbb{V}_{(\mu)} = 4$. On the other hand, the hook length tableau is

5	1
3	
2	
1	

Thus,

$$dim\,\mathbb{V}_{(\mu)} = \frac{5!}{5.3.2.1.1} = 4\,.$$

For the rest of this chapter, we will use S_3 as an example to illustrate the representation theory of S_n. The general statements for S_n will be presented without proofs.

We will first discuss the so-called **standard representation** of S_n, which is a unitary representation. For a given standard Young tableau we define the **Yamanouchi row-lattice vector** $|\,r_1,\ldots,r_n\,\rangle$, $1 \le r_1,\ldots,r_n \le n$, where r_i is the row in which the integer i appears in the given tableau; similarly the **Yamanouchi column-lattice vector** $|\,c_1,\ldots,c_n\,\rangle$, where c_i is the column in which i appears. The numbers r_i and c_i are called **Yamanouchi symbols**. For example, the Yamanouchi symbols for the standard Young tableau for S_3 are listed as follows:

1 2 3 : $r_1 = 1, r_2 = 1, r_3 = 1$ ($|\,1,1,1\,\rangle$), $c_1 = 1, c_2 = 2, c_3 = 3$;

1 / 2 / 3 : $r_1 = 1, r_2 = 2, r_3 = 3$ ($|\,1,2,3\,\rangle$), $c_1 = 1, c_2 = 1, c_3 = 1$;

1 2 / 3 : $r_1 = 1, r_2 = 1, r_3 = 2$ ($|\,1,1,2\,\rangle$), $c_1 = 1, c_2 = 2, c_3 = 1$;

1 3 / 2 : $r_1 = 1, r_2 = 2, r_3 = 1$ ($|\,1,2,1\,\rangle$), $c_1 = 1, c_2 = 1, c_3 = 2$.

Notice that $r_1 = c_1 = 1$ always. Define the following group action on the lattice vectors $|r_1, \ldots, r_n\rangle$:

$$(i-1, i) \, |\, 1, r_2, \ldots, r_n\rangle$$

$$= \frac{1}{\tau_{i-1, i}} |\, 1, r_2, \ldots, r_{i-1}, r_i, \ldots, r_n\rangle \tag{20.18}$$

$$+ \sqrt{1 - \left(\frac{1}{\tau_{i-1, i}}\right)^2} \, |\, 1, r_2, \ldots, r_i, r_{i-1}, \ldots, r_n\rangle \,,$$

where $i = 2, \ldots, n$, and

$$\tau_{i-1, i} = \begin{cases} 1, & \text{if } i-1 \text{ and } i \text{ are in the same row,} \\ -1, & \text{if } i-1 \text{ and } i \text{ are in the same column,} \\ c_i - c_{i-1} - (r_i - r_{i-1}), & \text{if } i-1 \text{ and } i \text{ do not lie in the same row} \\ & \text{or column.} \end{cases}$$

To determine the group actions for all $g \in S_3$ we need only determine the actions of the 2-cycles (12) and (23). We have

$$(1, 2) \, |\, 1, 1, 1\rangle = |\, 1, 1, 1\rangle \,, \tag{20.19}$$
$$(2, 3) \, |\, 1, 1, 1\rangle = |\, 1, 1, 1\rangle \,, \tag{20.20}$$
$$(1, 2) \, |\, 1, 2, 3\rangle = -|\, 1, 2, 3\rangle \,, \tag{20.21}$$
$$(2, 3) \, |\, 1, 2, 3\rangle = -|\, 1, 2, 3\rangle \,, \tag{20.22}$$
$$(1, 2) \, |\, 1, 1, 2\rangle = |\, 1, 1, 2\rangle \,, \tag{20.23}$$
$$(1, 2) \, |\, 1, 2, 1\rangle = -|\, 1, 2, 1\rangle \,, \tag{20.24}$$

$$(2, 3) \, |\, 1, 1, 2\rangle = -\frac{1}{2} |\, 1, 1, 2\rangle + \frac{\sqrt{3}}{2} |\, 1, 2, 1\rangle \,, \tag{20.25}$$

$$(2, 3) \, |\, 1, 2, 1\rangle = \frac{1}{2} |\, 1, 2, 1\rangle + \frac{\sqrt{3}}{2} |\, 1, 1, 2\rangle \,. \tag{20.26}$$

Thus $|\, 1, 1, 1\rangle$ and $|\, 1, 2, 3\rangle$ are both basis vectors for one-dimensional representations, while $|\, 1, 1, 2\rangle$ and $|\, 1, 2, 1\rangle$ are basis vectors for a 2-dimensional representation. The representation matrices for $(1, 2)$ and $(2, 3)$ are given by [c.f. (1.16)]:

$$\boxed{} \;:\; (1, 2) = 1, \quad (2, 3) = 1 \,; \tag{20.27}$$

$$\;:\; (1, 2) = -1, \quad (2, 3) = -1 \,; \tag{20.28}$$

$$\;:\; (1, 2) = \begin{pmatrix} 1 & 0 \\ 0 & -1 \end{pmatrix}, \quad (2, 3) = \begin{pmatrix} -\dfrac{1}{2} & \dfrac{\sqrt{3}}{2} \\ \dfrac{\sqrt{3}}{2} & \dfrac{1}{2} \end{pmatrix} \,. \tag{20.29}$$

Exercise 20.4 Verify that the above representations of S_3 are unitary.

We will now represent S_n on the tensor space

$$T^n(V) = \underbrace{V \otimes \cdots \otimes V}_{n \text{ times}} \ .$$

Recall the action of a particular $\sigma \in S_n$ on $T^n(V)$ [c.f.(16.48)]:

$$\sigma \, x^{i_1 \cdots i_n} \, e_{i_1} \otimes \cdots \otimes e_{i_n} = x^{i_{\sigma(1)} \cdots i_{\sigma(n)}} \, e_{i_1} \otimes \cdots \otimes e_{i_n}$$

$$= x^{i_1 \cdots i_n} \, e_{i_{\sigma^{-1}(1)}} \otimes \cdots \otimes e_{i_{\sigma^{-1}(n)}} \ .$$

This induces the action of the group algebra $\mathcal{S}_n$ of S_n on $T^n(V)$ [c.f. the discussion immediately above (18.61)]. Corresponding to each standard Young tableau we define the so-called **Young symmetrizer** ($\in \mathcal{S}_n$) as follows:

$$Y_n = \frac{N_{(\mu)}}{n!} \sum_{\sigma_{(r)}, \, \sigma_{(c)}} \left(sgn \, \sigma_{(c)} \right) \sigma_{(c)} \sigma_{(r)} \ , \tag{20.30}$$

where $N_{(\mu)}$ = dimension of the irreducible representation (which has been labeled as the μ-th) corresponding to the Young diagram of the given Young tableau, $\sigma_{(r)} \in S_n$ is a permutation without changing the numbers in a row, and $\sigma_{(c)} \in S_n$ is a permutation without changing the numbers in a column. We have seen two important Young symmetrizers already [c.f. (16.21) and (16.22)]:

$$Y_n \left(\boxed{1\,2\,3\,\cdot\,\cdot\,\cdot\,n} \right) = \text{ the symmetrizing map }, \tag{20.31a}$$

$$Y_n \left(\begin{array}{c} \boxed{1} \\ \boxed{2} \\ \boxed{3} \\ \cdot \\ \cdot \\ \boxed{n} \end{array} \right) = \text{ the alternating map .} \tag{20.31b}$$

As examples of mixed symmetry, let us consider the standard tableau $\begin{array}{cc}\boxed{1\,2}\\ \boxed{3}\end{array}$ and $\begin{array}{cc}\boxed{1\,3}\\ \boxed{2}\end{array}$. We have

$$Y_3 \left(\begin{array}{c}\boxed{1\,2}\\ \boxed{3}\end{array} \right) = \frac{2}{3!} \left(e - (1,3)\right)\left(e + (1,2)\right) = \frac{1}{3} \left(e - (1,3) + (1,2) - (123)\right) , \tag{20.32}$$

$$Y_3 \left(\begin{array}{c}\boxed{1\,3}\\ \boxed{2}\end{array} \right) = \frac{2}{3!} \left(e - (1,2)\right)\left(e + (1,3)\right) = \frac{1}{3} \left(e + (1,3) - (1,2) - (321)\right) . \tag{20.33}$$

$\boxed{\text{Exercise 20.5}}$ Calculate Y_4 for the standard tableau $\begin{array}{|c|c|}\hline 1 & 2 \\ \hline 3 & 4 \\ \hline\end{array}$ and $\begin{array}{|c|c|}\hline 1 & 3 \\ \hline 2 & 4 \\ \hline\end{array}$.

Important facts concerning Young symmetrizers are stated in the following theorems (without proofs).

Theorem 20.4. *All Young symmetrizers Y_n are **primitive idempotents** of the group algebra $\mathcal{S}_n$, and are projection operators for irreducible representation spaces. Idempotency of an element $s \in \mathcal{S}_n$ means that $s^2 = s$, and primitivity means that if $s = s_1 + s_2$, $s, s_1, s_2 \in \mathcal{S}_n$, such that $s_1^2 = s_1, s_2^2 = s_2$, and $s_1 s_2 = s_2 s_1 = 0$, then either $s_1 = 0$ or $s_2 = 0$.*

In Chapter 7 (Def. 7.1) we defined an ideal of a Lie algebra (with the Lie bracket for internal multiplication). For the group algebra $\mathcal{S}_n$, with ordinary multiplication, we can similarly define left (right) ideals.

Definition 20.4. *A subalgebra $\mathcal{S}'$ of $\mathcal{S}$ is a **left ideal** of $\mathcal{S}$ if, for all $s \in \mathcal{S}$,*

$$s\mathcal{S}' \subset \mathcal{S}' . \tag{20.34}$$

*$\mathcal{S}'$ is a **right ideal** of $\mathcal{S}$ if, for all $s \in \mathcal{S}$,*

$$\mathcal{S}'s \subset \mathcal{S}' . \tag{20.35}$$

Theorem 20.5. *Let $Y_n^{(i)}$ be all the Young symmetrizers corresponding to the standard Young tableau of the symmetric group $\mathcal{S}_n$. Each $Y_n^{(i)}$ generates a left ideal $L^{(i)}$ of the group algebra $\mathcal{S}_n$ given by*

$$L^{(i)} = \{ sY_n^{(i)} \,|\, s \in \mathcal{S}_n \} , \tag{20.36}$$

and $\mathcal{S}_n$ (as a vector space) can be decomposed as the direct sum

$$\mathcal{S}_n = \sum_{\oplus i} L^{(i)} . \tag{20.37}$$

For example, $\mathcal{S}_3$ is a $6 \,(= 3!)$-dimensional vector space which can be decomposed as

$$\mathcal{S}_3 = L\left(\begin{array}{|c|c|c|}\hline 1 & 2 & 3 \\ \hline\end{array}\right) \oplus L\left(\begin{array}{|c|}\hline 1 \\ \hline 2 \\ \hline 3 \\ \hline\end{array}\right) \oplus L\left(\begin{array}{|c|c|}\hline 1 & 2 \\ \hline 3 \\ \cline{1-1}\end{array}\right) \oplus L\left(\begin{array}{|c|c|}\hline 1 & 3 \\ \hline 2 \\ \cline{1-1}\end{array}\right) . \tag{20.38}$$

As vector spaces, $L\left(\begin{array}{|c|c|c|}\hline 1 & 2 & 3 \\ \hline\end{array}\right)$ and $L\left(\begin{array}{|c|}\hline 1 \\ \hline 2 \\ \hline 3 \\ \hline\end{array}\right)$ are both one-dimensional, while

$L\left(\begin{array}{|c|c|}\hline 1 & 2 \\ \hline 3 \\ \cline{1-1}\end{array}\right)$ and $L\left(\begin{array}{|c|c|}\hline 1 & 3 \\ \hline 2 \\ \cline{1-1}\end{array}\right)$ are both two-dimensional.

Since Y_3 ($\boxed{1|2|3}$) is the symmetrizing map and Y_3 $\left(\boxed{\begin{smallmatrix}1\\2\\3\end{smallmatrix}}\right)$ is the alternating map, Theorem 16.2 states that

$$P^3(V) = Y_3 \left(\boxed{1|2|3}\right) \left(T^3(V)\right) , \quad \Lambda^3(V) = Y_3 \left(\boxed{\begin{smallmatrix}1\\2\\3\end{smallmatrix}}\right) \left(T^3(V)\right) , \quad (20.39)$$

where $P^3(V)$ $[\Lambda^3(V)]$ is the subspace of $T^3(V)$ consisting of all symmetric (skew-symmetric) tensors $x \in T^3(V)$. Note that if $dim\, V < 3$, $\Lambda^3(V) = 0$. One-dimensional subspaces generated by any vector in either $P^3(V)$ or $\Lambda^3(V)$ clearly form one-dimensional irreducible representations of S_3.

Tensors of mixed symmetry can be obtained by the actions of Y_3 $\left(\boxed{\begin{smallmatrix}1&2\\3\end{smallmatrix}}\right)$ and Y_3 $\left(\boxed{\begin{smallmatrix}1&3\\2\end{smallmatrix}}\right)$ on $T^3(V)$. Consider a general basis vector in $T^3(V)$ of the form

$$\varphi_{\mu\nu\lambda} \equiv e_\mu \otimes e_\nu \otimes e_\lambda ,$$

where $\mu, \nu, \lambda = 1, \ldots, dim(V)$, and $\{e_1, \ldots, e_{dim(V)}\}$ is a basis of V. We have, from (20.32) and (16.13),

$$M_{\mu\nu\lambda} \equiv Y_3 \left(\boxed{\begin{smallmatrix}1&2\\3\end{smallmatrix}}\right) e_\mu \otimes e_\nu \otimes e_\lambda \equiv \boxed{\begin{smallmatrix}\mu&\nu\\\lambda\end{smallmatrix}}$$

$$= \frac{1}{3} (e - (1,3))(e + (1,2))\, e_\mu \otimes e_\nu \otimes e_\lambda$$

$$= \frac{1}{3} (e - (1,3)) \left(e_\mu \otimes e_\nu \otimes e_\lambda + e_\nu \otimes e_\mu \otimes e_\lambda\right)$$

$$= \frac{1}{3} \left(e_\mu \otimes e_\nu \otimes e_\lambda + e_\nu \otimes e_\mu \otimes e_\lambda - e_\lambda \otimes e_\nu \otimes e_\mu - e_\lambda \otimes e_\mu \otimes e_\nu\right) .$$

$$(20.40)$$

Thus

$$M_{\mu\nu\lambda} = \frac{1}{3} \left(\varphi_{\mu\nu\lambda} + \varphi_{\nu\mu\lambda} - \varphi_{\lambda\nu\mu} - \varphi_{\lambda\mu\nu}\right) . \quad (20.41)$$

Similarly, we can define

$$\tilde{M}_{\mu\nu\lambda} \equiv Y_3 \left(\boxed{\begin{smallmatrix}1&3\\2\end{smallmatrix}}\right) \varphi_{\mu\nu\lambda} \equiv \boxed{\begin{smallmatrix}\mu&\lambda\\\nu\end{smallmatrix}} , \quad (20.42)$$

and obtain

$$\tilde{M}_{\mu\nu\lambda} = \frac{1}{3} \left(\varphi_{\mu\nu\lambda} + \varphi_{\lambda\nu\mu} - \varphi_{\nu\mu\lambda} - \varphi_{\nu\lambda\mu}\right) . \quad (20.43)$$

It is not hard to verify that the tensors $M_{\mu\nu\lambda}$ and $\tilde{M}_{\mu\nu\lambda}$ have the following symmetries:

$$M_{\mu\nu\lambda} = M_{\nu\mu\lambda} , \tag{20.44}$$
$$M_{\mu\nu\lambda} + M_{\nu\lambda\mu} + M_{\lambda\mu\nu} = 0 , \tag{20.45}$$
$$\tilde{M}_{\mu\nu\lambda} = \tilde{M}_{\lambda\nu\mu} , \tag{20.46}$$
$$\tilde{M}_{\mu\nu\lambda} + \tilde{M}_{\nu\lambda\mu} + \tilde{M}_{\lambda\mu\nu} = 0 . \tag{20.47}$$

Exercise 20.6   Verify (20.44) to (20.47).

Now consider the tensor

$$\xi_{\mu\nu\lambda} \equiv (1,2)\, M_{\mu\nu\lambda} = \frac{1}{3}\left(\varphi_{\nu\mu\lambda} + \varphi_{\mu\nu\lambda} - \varphi_{\nu\lambda\mu} - \varphi_{\mu\lambda\nu}\right) . \tag{20.48}$$

Obviously,

$$(1,2)\xi_{\mu\nu\lambda} = M_{\mu\nu\lambda} . \tag{20.49}$$

Also,

$$(1,3)\, M_{\mu\nu\lambda} = -M_{\mu\nu\lambda} , \quad (1,3)\,\xi_{\mu\nu\lambda} = \xi_{\mu\nu\lambda} - M_{\mu\nu\lambda} , \tag{20.50}$$
$$(2,3)\, M_{\mu\nu\lambda} = M_{\mu\nu\lambda} - \xi_{\mu\nu\lambda} , \quad (2,3)\,\xi_{\mu\nu\lambda} = -\xi_{\mu\nu\lambda} . \tag{20.51}$$

Exercise 20.7   Verify (20.50) and (20.51).

Since all elements of S_3 can be written as products of $e, (1,2), (1,3)$ and $(2,3)$, we have a two-dimensional irreducible representation of S_3 with the representation space spanned by $e_1 = M_{\mu\nu\lambda}$ and $e_2 = \xi_{\mu\nu\lambda}$. With respect to these basis vectors, the matrix representations of $(1,2), (1,3)$ and $(2,3)$ are given by

$$(1,2) = \begin{pmatrix} 0 & 1 \\ 1 & 0 \end{pmatrix} , \tag{20.52}$$

$$(1,3) = \begin{pmatrix} -1 & 0 \\ -1 & 1 \end{pmatrix} , \tag{20.53}$$

$$(2,3) = \begin{pmatrix} 1 & -1 \\ 0 & -1 \end{pmatrix} . \tag{20.54}$$

Exercise 20.8   Obtain the matrix representations of (123) and (321) with respect to $M_{\mu\nu\lambda}$ and $\xi_{\mu\nu\lambda}$.

Note that the above representation is not unitary. But according to Theorem 18.1, it must be equivalent to a unitary representation. Indeed, consider the linear transformations

$$M^{(+)}_{\mu\nu\lambda} \equiv (M_{\mu\nu\lambda} + \xi_{\mu\nu\lambda}), \quad M^{(-)}_{\mu\nu\lambda} \equiv \sqrt{3}\,(M_{\mu\nu\lambda} - \xi_{\mu\nu\lambda}). \tag{20.55}$$

Then

$$(1,2)\,M^{(+)}_{\mu\nu\lambda} = M^{(+)}_{\mu\nu\lambda}, \quad (1,2)\,M^{(-)}_{\mu\nu\lambda} = -M^{(-)}_{\mu\nu\lambda}, \tag{20.56}$$

$$(1,3)\,M^{(+)}_{\mu\nu\lambda} = -\frac{1}{2}M^{(+)}_{\mu\nu\lambda} - \frac{\sqrt{3}}{2}M^{(-)}_{\mu\nu\lambda}, \tag{20.57a}$$

$$(1,3)\,M^{(-)}_{\mu\nu\lambda} = -\frac{\sqrt{3}}{2}M^{(+)}_{\mu\nu\lambda} + \frac{1}{2}M^{(-)}_{\mu\nu\lambda}, \tag{20.57b}$$

$$(2,3)\,M^{(+)}_{\mu\nu\lambda} = -\frac{1}{2}M^{(+)}_{\mu\nu\lambda} + \frac{\sqrt{3}}{2}M^{(-)}_{\mu\nu\lambda}, \tag{20.58a}$$

$$(2,3)\,M^{(-)}_{\mu\nu\lambda} = \frac{\sqrt{3}}{2}M^{(+)}_{\mu\nu\lambda} + \frac{1}{2}M^{(-)}_{\mu\nu\lambda}. \tag{20.58b}$$

Hence, with respect to the basis $\{M^{(+)}_{\mu\nu\lambda}, M^{(-)}_{\mu\nu\lambda}\}$,

$$(1,2) = \begin{pmatrix} 1 & 0 \\ 0 & -1 \end{pmatrix}, \tag{20.59}$$

$$(1,3) = \begin{pmatrix} -\dfrac{1}{2} & -\dfrac{\sqrt{3}}{2} \\ -\dfrac{\sqrt{3}}{2} & \dfrac{1}{2} \end{pmatrix}, \tag{20.60}$$

$$(2,3) = \begin{pmatrix} -\dfrac{1}{2} & \dfrac{\sqrt{3}}{2} \\ \dfrac{\sqrt{3}}{2} & \dfrac{1}{2} \end{pmatrix}. \tag{20.61}$$

This irreducible two-dimensional representation is unitary and is in fact the same as the standard (Yamanouchi) representation.

We have thus found all the inequivalent irreducible representations of S_3. Note that the tensors $M_{\mu\nu\lambda}$ and $\xi_{\mu\nu\lambda}$, for fixed values of μ, ν, and λ, span a two-dimensional vector subspace of $T^3(V)$ which is isomorphic to the ideal $L\left(\begin{smallmatrix}\boxed{1\ 2}\\\boxed{3}\end{smallmatrix}\right)$ of the group algebra S_3. One can in fact represent S_3 on the six-dimensional group algebra S_3 (regarded as a vector space). This representation is reducible according to the decomposition (20.38). Both $L(\boxed{1\,2\,3})$ and $L\left(\begin{smallmatrix}\boxed{1}\\\boxed{2}\\\boxed{3}\end{smallmatrix}\right)$ are one-dimensional representation spaces, while $L\left(\begin{smallmatrix}\boxed{1\ 2}\\\boxed{3}\end{smallmatrix}\right)$ and

$L\left(\begin{array}{|c|c|}\hline 1 & 3 \\\hline 2 \\\hline\end{array}\right)$ each furnishes a two-dimensional irreducible representation space.
The latter two (two-dimensional) representations are equivalent to each other.

We will now turn our attention to the representation of $GL(n)$ (the general linear group) on tensor spaces $T^m(V)$, $dim\, V = n$ (in general $m \neq n$). That this is possible is evident from (2.1). The representations are in general reducible, but the remarkable fact is that Young diagrams can again be employed to find all the irreducible representations. The following crucial theorem (stated without proof) shows how this is done.

Theorem 20.6. *In the representation of $GL(n)$ on the tensor space $T^m(V)$, $dim\, V = n$, each standard Young tableau for the Young diagram labeled by $\{\mu_1, \ldots, \mu_n\}$, where $\mu_1 \geq \mu_2 \geq \cdots \geq \mu_n \geq 0$ and*

$$\mu_1 + \mu_2 + \cdots + \mu_n = m \,, \tag{20.62}$$

generates an irreducible representation subspace by the corresponding Young symmetrizer. Furthermore $T^m(V)$ can be decomposed uniquely as a direct sum of all such irreducible representation subspaces.

Note the difference between (20.62) and (20.7).

Let us illustrate this theorem by working out the representation of $GL(3)$ on $T^3(V)$, $dim\, V = 3$. In this case, $m = n = 3$. The Young symmetrizers corresponding to the four standard tableau yield (with $\varphi_{\mu\nu\lambda} = e_\mu \otimes e_\nu \otimes e_\lambda$)

$$Y_3(\begin{array}{|c|c|c|}\hline 1 & 2 & 3 \\\hline\end{array})\,\varphi_{\mu\nu\lambda} \equiv S_{\mu\nu\lambda} \equiv \begin{array}{|c|c|c|}\hline \mu & \nu & \lambda \\\hline\end{array}$$
$$= \frac{1}{6}\left(\varphi_{\mu\nu\lambda} + \varphi_{\nu\lambda\mu} + \varphi_{\lambda\mu\nu} + \varphi_{\mu\lambda\nu} + \varphi_{\lambda\nu\mu} + \varphi_{\nu\mu\lambda}\right), \tag{20.63}$$

$$Y_3\left(\begin{array}{|c|}\hline 1 \\\hline 2 \\\hline 3 \\\hline\end{array}\right)\,\varphi_{\mu\nu\lambda} \equiv A_{\mu\nu\lambda} \equiv \begin{array}{|c|}\hline \mu \\\hline \nu \\\hline \lambda \\\hline\end{array}$$
$$= \frac{1}{6}\left(\varphi_{\mu\nu\lambda} + \varphi_{\nu\lambda\mu} + \varphi_{\lambda\mu\nu} - \varphi_{\mu\lambda\nu} - \varphi_{\lambda\nu\mu} - \varphi_{\nu\mu\lambda}\right), \tag{20.64}$$

$$Y_3\left(\begin{array}{|c|c|}\hline 1 & 2 \\\hline 3 \\\hline\end{array}\right)\,\varphi_{\mu\nu\lambda} \equiv M_{\mu\nu\lambda} \equiv \begin{array}{|c|c|}\hline \mu & \nu \\\hline \lambda \\\hline\end{array}, \tag{20.65}$$

$$Y_3\left(\begin{array}{|c|c|}\hline 1 & 3 \\\hline 2 \\\hline\end{array}\right)\,\varphi_{\mu\nu\lambda} \equiv \tilde{M}_{\mu\nu\lambda} \equiv \begin{array}{|c|c|}\hline \mu & \lambda \\\hline \nu \\\hline\end{array}, \tag{20.66}$$

where $M_{\mu\nu\lambda}$ and $\tilde{M}_{\mu\nu\lambda}$ have been given by (20.41) and (20.43). Since $dim\, V = 3$, the indices μ, ν, λ can each assume the values $1, 2$ and 3; and the space $T^3(V)$ is 27-dimensional.

The irreducible subspace generated by $S_{\mu\nu\lambda}$ is 10-dimensional, with a possible basis given by

$$\{S_{111}, S_{222}, S_{333}, S_{112}, S_{113}, S_{221}, S_{223}, S_{331}, S_{332}, S_{123}\} \,. \tag{20.67}$$

The irreducible subspace generated by $A_{\mu\nu\lambda}$ is one-dimensional, with A_{123} as the only basis vector. Due to the symmetry conditions (20.44) to (20.47) for $M_{\mu\nu\lambda}$ and $\tilde{M}_{\mu\nu\lambda}$, each of the irreducible subspaces generated by $M_{\mu\nu\lambda}$ and $\tilde{M}_{\mu\nu\lambda}$ is 8-dimensional. Two respective possible bases are given by

$$\{M_{112}, M_{113}, M_{221}, M_{223}, M_{331}, M_{332}, M_{123}, M_{231}\}\,, \tag{20.68}$$

$$\{\tilde{M}_{112}, \tilde{M}_{113}, \tilde{M}_{221}, \tilde{M}_{223}, \tilde{M}_{331}, \tilde{M}_{332}, \tilde{M}_{123}, \tilde{M}_{231}\}\,. \tag{20.69}$$

$\boxed{\text{Exercise 20.9}}$ Use the conditions (20.44) and (20.45) to verify that, of the 27 quantities $M_{\mu\nu\lambda}$, $\mu, \nu, \lambda = 1, 2, 3$, there are only eight independent ones, a possible set of which is given by (20.68). Show that (20.46) and (20.47) lead to a similar conclusion for $\tilde{M}_{\mu\nu\lambda}$.

We can represent the decomposition of $T^3(V)$, $dim\,V = 3$, into the irreducible subspaces of $GL(3)$ by

$$\boxed{3} \otimes \boxed{3} \otimes \boxed{3} \;=\; \boxed{0} \;\oplus\; \boxed{1} \;\oplus\; \boxed{8} \;\oplus\; \boxed{8}.$$

$$\tag{20.70}$$

In fact, a basis tensor $\varphi_{\mu\nu\lambda} = e_\mu \otimes e_\nu \otimes e_\lambda \in T^3(V)$ can be expressed as

$$\varphi_{\mu\nu\lambda} = S_{\mu\nu\lambda} + A_{\mu\nu\lambda} + M_{\mu\nu\lambda} + \tilde{M}_{\mu\nu\lambda}\,. \tag{20.71}$$

$\boxed{\text{Exercise 20.10}}$ Verify the above result.

The RHS of (20.70) can be represented pictorially as

In this diagram, reading horizontally, we have the irreducible representations of $GL(3)$; reading vertically, we have 10 one-dimensional equivalent representations of S_3, 1 one-dimensional representation of S_3, and 8 equivalent two-dimensional representations of S_3.

The decomposition (20.70) has interesting applications in, for example, atomic physics [where the relevant symmetry group is $SO(3)$], and the quark model, [where the relevant symmetry group is $SU(3)$] (see Chapter 26).

Exercise 20.11   Consider the product representation of two 2-dimensional irreducible representations of S_3, each belonging to the set of eight equivalent ones generated by the Young diagram ⊞ [see (20.70)]. Denote the basis sets of these two representations by $\{M_{\mu\nu\lambda}^{(o)}, \xi_{\mu\nu\lambda}^{(o)}\}$ and $\{M_{\mu\nu\lambda}^{(s)}, \xi_{\mu\nu\lambda}^{(s)}\}$ [where, in applications in atomic physics for example, (o) may indicate the orbital part and (s) the spin part of electronic states, respectively]. Show that this product representation is reducible, and

$$\zeta \equiv M^{(o)} \otimes \xi^{(s)} - \xi^{(o)} \otimes M^{(s)}$$

generates a 1-dimensional antisymmetric irreducible representation of S_3. In other words, show that if $\sigma = (i, j) \in S_3$ (exchange of a pair), then $\sigma\zeta = -\zeta$. Hint: Write

$$\sigma M = \sigma_1^1 M + \sigma_1^2 \xi, \quad \sigma\xi = \sigma_2^1 M + \sigma_2^2 \xi,$$

for both the (o) and (s) representations.

There is a very useful, general formula for the dimension of an irreducible representation of $GL(n)$ on $T^m(V)$ corresponding to a particular Young diagram. We will present it here without proof. The reader should check that it yields the special results of (20.70) for $n = m = 3$.

Theorem 20.7. *The dimension of the irreducible representation of $GL(n)$ on $T^m(V)$ corresponding to a Young diagram specified by $\{\mu_1, \ldots, \mu_n\}$, $\mu_1 + \cdots + \mu_n = m$, is given by*

$$d(\mu_1, \ldots, \mu_n) = \frac{\prod\limits_{i<j}^{n} (l_i - l_j)}{1!\, 2!\, 3! \ldots (n-1)!}, \tag{20.72}$$

where $l_1 > l_2 > \cdots > l_n$ are given by

$$l_1 \equiv \mu_1 + (n-1), \; l_2 \equiv \mu_2 + (n-2), \ldots, l_j \equiv \mu_j + (n-j), \ldots, l_n \equiv \mu_n.$$

Exercise 20.12   Find the dimensions of the irreducible representations of $GL(2)$ on $T^3(V)$. Show that

$$⑧ \;=\; ④ \;\oplus\; ② \;\oplus\; ② .$$

$$\boxed{1\,|\,2\,|\,3} \qquad \begin{array}{|c|c|}\hline 1 & 2 \\ \hline 3 \\ \hline \end{array} \qquad \begin{array}{|c|c|}\hline 1 & 3 \\ \hline 2 \\ \hline \end{array}$$

Exercise 20.13   Consider the ground electronic state of the nitrogen (N) atom [Z (atomic number) $= 7$, shell structure $(1s)^2 (2s)^2 (2p)^3$]. All three valence electrons are in the $2p$ orbital, and the one-electron orbital wave function for each can be written $|lm\rangle = |1m\rangle$, where $l = 1$ and $m = 1, 0$, or -1 [basis states of the $l = 1$ irreducible representation of $SO(3)$ (see Chapter 22 for more details)]. Therefore the 3-electron orbital wave functions can be represented by $|m_1, m_2, m_3\rangle$, where $m_i = 1, 0, -1$ and $m_1 + m_2 + m_3 = $ z-component of the total orbital angular momentum (we omit the label $l = 1$, since it is the same for all three electrons). These span the 27-dimensional space $T^3(V)$, $dim(V) = 3$. This is a representation space of $GL(3)$, and it partitions into irreducible subspaces according to (20.70). Each of these irreducible representations further partitions into irreducible representations of $SO(3)$ [which is a symmetry group of the electronic Hamiltonian (recall the discussion at the beginning of Chapter 18)] as follows:

$$\boxed{10} = \boxed{7} \oplus \boxed{3}, \quad \boxed{8} = \boxed{5} \oplus \boxed{3} \ \ (\text{twice}), \quad \boxed{1} = \boxed{1}.$$

For example, $\boxed{7}$ is the irreducible representation of $SO(3)$ with total orbital angular momentum $L = 3$ and multiplicity $2L + 1 = 7$. Since the electron has spin $1/2$, the total electronic wave function must be given by a product representation of orbital and spin parts. The 3-electron spin wave function can be written as $|m_{s1}m_{s2}m_{s3}\rangle$ where $m_{si} = \pm 1/2$. Thus the spin representation space is 8-dimensional and partitions into irreducible subspaces of $GL(2)$ according to the result of Exercise 20.12. By **Pauli's exclusion principle** the total electronic wave function is required to be antisymmetric under the exchange of any pair of electrons. The total wave function is usually labeled by the **spectroscopic notation**

$$|^{2S+1}L_J; Jm_J\rangle, \quad m_J = J, J - 1, \ldots, -J + 1, -J,$$

where $L = $ total orbital angular momentum, $S = $ total spin angular momentum, $J = $ total angular momentum, and $m_J = $ z-component of total angular momentum. The possible values of J are $L + S, L + S - 1, \ldots, |L - S|$ (see Chapter 22). Putting all the above information together, and using the result of Exercise 20.11, show that the only possible spectroscopic terms for the $(2p)^3$ configuration are

$$^4S_{3/2}, \ ^2D_{5/2}, \ ^2D_{3/2}, \ ^2P_{3/2} \text{ and } ^2P_{1/2},$$

a total of 20 states, where S means $L = 0$, P means $L = 1$, and D means $L = 2$.

The above discussion on the representation of $GL(n)$ on $T^m(V)$, $dim\, V = n$, or $(m, 0)$-type tensor space, applies analogously to the dual tensor space $T^m(V^*)$, or $(0, m)$-type tensor space. Indeed, one can represent $GL(n)$ on (r, s)-type mixed tensor spaces. First consider a $(1, 1)$-type tensor space, with basis tensors φ^ν_μ. If the trace $\varphi^\mu_\mu \neq 0$, we can decompose φ^ν_μ as

$$\varphi^\nu_\mu = \psi^\nu_\mu + \delta^\nu_\mu S, \tag{20.73}$$

where

$$S \equiv \frac{1}{n} \varphi^\mu_\mu, \tag{20.74}$$

and ψ_μ^ν is the traceless part: $\psi_\mu^\mu = 0$. Obviously S spans a one-dimensional space that is invariant under $GL(n)$, while ψ_μ^ν forms the basis of an irreducible representation. Thus the $(1,1)$-type tensor space spanned by φ_μ^ν decomposes into a direct sum of the singlet space spanned by S and an $(n^2 - 1)$-dimensional irreducible representation space spanned by the traceless tensors ψ_μ^ν. This example shows that if a mixed (s,r)-type tensor $\psi_{\mu_1,\,...,\,\mu_r}^{\nu_1,\,...,\,\nu_s}$ is a tensor of an irreducible representation space, then it must satisfy the traceless conditions

$$\psi_{\mu_1,\,...,\,\mu_{p-1},\,\lambda,\,\mu_{p+1},\,...,\,\mu_r}^{\nu_1,\,...,\,\nu_{q-1},\,\lambda,\,\nu_{q+1},\,...,\,\nu_s} = 0 \ , \tag{20.75}$$

where $1 \leq p \leq r$, $1 \leq q \leq s$. In these irreducible representations, separate standard Young tableau Y_r and Y_s are assigned to the lower and upper indices, respectively:

$$Y_r = \{\mu_1, \ldots, \mu_l\} \ , \quad Y_s = \{\mu_1', \ldots, \mu_{n-l}'\} \ , \tag{20.76}$$

such that [in analogy to (20.62)]

$$\mu_1 + \mu_2 + \cdots + \mu_l = r \ , \tag{20.77}$$
$$\mu_1' + \mu_2' + \cdots + \mu_{n-l}' = s \ , \tag{20.78}$$

and

$$\mu_1 \geq \mu_2 \geq \cdots \geq \mu_l \geq 0 \geq \mu_{l+1} \geq \mu_{l+2} \geq \cdots \geq \mu_n \ , \tag{20.79}$$

where we have set

$$\mu_1' = -\mu_n, \ \mu_2' = -\mu_{n-1}, \ldots, \mu_{n-l}' = -\mu_{l+1} \tag{20.80}$$

to fulfill the condition

$$\mu_1' \geq \mu_2' \geq \cdots \geq \mu_{n-l}' \geq 0 \ . \tag{20.81}$$

Chapter 21

Irreducible Representations of $U(n), SL(n), SU(n)$ and $O(n)$

We wish to inquire whether irreducible representations of $GL(n)$ become reducible when restricted to its subgroups $U(n), SL(n), SU(n)$ and $O(n)$.

First consider $U(n)$. Suppose $D(g)$ is a matrix representation of $U(n)$ on $T^1(V) = V$ (a unitary space) with respect to a basis $\{e_i\}$ of V, so that, under the action of $g \in U(n)$ [recall Table 1.1],

$$e_i \longrightarrow e_i' = (D^{-1}(g))_i^j\, e_j \ . \tag{21.1}$$

By unitarity, $D^{-1}(g) = D^\dagger(g)$, for all $g \in U(n)$. Thus

$$e_i' = (D^\dagger(g))_i^j\, e_j = \sum_j \overline{(D(g))_j^i}\, e_j \ . \tag{21.2}$$

On the other hand, the same matrix representation $D(g)$ of $U(n)$ on $T^*(V)$ yields

$$e'^i = (D(g))_j^i\, e^j \ , \tag{21.3}$$

where $\{e^i\}$ is a basis in V^* dual to $\{e_i\}$. On complex conjugating the above equation, we see that $\overline{e^i}$ transforms under $U(n)$ in the same way as e_i. In fact, it is quite obvious that under $U(n)$, we have the following equivalence of mixed tensors:

$$\varphi_{\mu_1,\ldots,\,\mu_r}^{\nu_1,\ldots,\,\nu_s} \sim \overline{\varphi_{\nu_1,\ldots,\,\nu_s}^{\mu_1,\ldots,\,\mu_r}} \ . \tag{21.4}$$

Apart from this equivalence, *irreducible representations of $GL(n)$, characterized by distinct Young diagrams appropriate to the ranks of tensors, remain irreducible representations of $U(n)$.*

We will next consider the special linear (unimodular) group $SL(n)$:

$$SL(n) = \{g \,|\, g \in GL(n)\,,\ \det(g) = 1\} \ .$$

First note that any $g \in GL(n)$ can be expressed as

$$g = (\det(g))^{\frac{1}{n}} g' \,, \tag{21.5}$$

where $g' \in SL(n)$.

| Exercise 21.1 | Verify the above statement (with the help of Theorems 16.3 and 16.4).

Now suppose $D : SU(n) \longrightarrow \mathcal{T}(T^r(V))$ is a reducible representation of $SU(n)$ on some rank-r tensor space. It follows from (21.5) that $D : GL(n) \longrightarrow \mathcal{T}(T^r(V))$ given by

$$D(g) = (\det(g))^{\frac{r}{n}} D(g') \tag{21.6}$$

is also reducible.

| Exercise 21.2 | Verify the above statement.

Hence irreducible representations of $GL(n)$ remain irreducible representations of $SL(n)$, and thus of $SU(n)$.

Certain inequivalent irreducible representations of $GL(n)$, however, may become equivalent for $SL(n)$ [see (21.20) below]. Consider the completely antisymmetric rank-n tensor corresponding to the Young tableau

$$\boxed{\begin{array}{c} 1 \\ \hline 2 \\ \hline \cdot \\ \hline \cdot \\ \hline n \end{array}}$$

. Its components

are necessarily of the form

$$A^{i_1 \cdots i_n} = c\, \varepsilon^{i_1 \cdots i_n} \,, \tag{21.7}$$

where c is a constant and $\varepsilon^{i_1 \cdots i_n}$ is the Levi-Civita tensor [c.f. (2.8)]. Under $g \in GL(n)$,

$$\varepsilon^{i_1 \cdots i_n} \longrightarrow \varepsilon'^{i_1 \cdots i_n} = g^{i_1}_{j_1} \cdots g^{i_n}_{j_n} \varepsilon^{j_1 \cdots j_n}$$
$$= \alpha\, \varepsilon^{i_1 \cdots i_n} \,. \tag{21.8}$$

The constant α can be determined by setting $i_1 = 1, i_2 = 2, \ldots, i_n = n$ in the above equation. Thus

$$\alpha = g^1_{j_1} \cdots g^n_{j_n} \varepsilon^{j_1 \cdots j_n} \,, \tag{21.9}$$

which, according to (16.78), is equal to $\det(g)$. We then have

$$\boxed{\varepsilon^{i_1 \cdots i_n} \longrightarrow \varepsilon'^{i_1 \cdots i_n} = \det(g)\, \varepsilon^{i_1 \cdots i_n}} \,. \tag{21.10}$$

The following theorem can now be stated (without proof).

Theorem 21.1. *Let $\psi_{i_1 \ldots i_N}$, $N = mn$, be the tensor obtained from the (rectangular) Young tableau (corresponding to the Young diagram $\{\mu_1 = m, \mu_2 = m, \ldots, \mu_n = m\}$ with n rows and m columns):*

1	$n+1$	$2n+1$	$\cdots\cdots$	$(m-2)n+1$	$(m-1)n+1$
2	$n+2$	$2n+2$	$\cdots\cdots$		
$\vdots$	$\vdots$	$\vdots$	$\cdots\cdots$		
$n-1$					
n	$2n$	$3n$	$\cdots\cdots$	$mn-n$	mn

with the corresponding Young symmetrizer Y, where each of the indices i_1 to i_N can assume values $1, 2, \ldots, n$, that is,

$$\psi_{i_1 \ldots i_N} = Y\, e_{i_1} \otimes \cdots \otimes e_{i_N} .$$

Then

$$\psi_{i_1 \ldots i_N} = c\,\varepsilon^{i_1 \ldots i_n}\,\varepsilon^{i_{n+1} \ldots i_{2n}} \ldots \varepsilon^{i_{(m-1)n+1} \ldots i_{mn}}$$
$$e_{i_1} \otimes \cdots \otimes e_{i_n} \otimes \cdots \otimes e_{i_{(m-1)n+1}} \otimes \cdots \otimes e_{i_{mn}} , \quad (21.11)$$

where c is a constant, and each ε is a Levi-Civita symbol.

Exercise 21.3 Verify the validity of the above theorem for the Young tableau

$$\begin{array}{|c|c|} \hline 1 & 3 \\ \hline 2 & 4 \\ \hline \end{array}.$$

Show that for this case, with $i_1 = i_3 = 1$, $i_2 = i_4 = 2$,

$$\psi_{1212} = \frac{1}{3}\,(e_1 \otimes e_2 - e_2 \otimes e_1) \otimes (e_1 \otimes e_2 - e_2 \otimes e_1) ,$$

which is the tensor product of two antisymmetric tensors. (Consult Exercise 20.5.)

An immediate consequence of Theorem 21.1 and Eq. (21.10) is that, under any $g \in GL(n)$,

$$\psi_{i_1 \ldots i_N} \longrightarrow \psi'_{i_1 \ldots i_N} = (\det(g))^m\, \psi_{i_1 \ldots i_N} , \quad N = mn . \quad (21.12)$$

Thus $\psi_{i_1 \ldots i_N}$ is a one-dimensional representation of $GL(n)$; in particular, it is invariant under $SL(n)$ [for which $\det(g) = 1$ by definition].

Let us now state a rule (without proof) for the **Clebsch-Gordan decomposition** of product representations [c.f. (18.75)] in terms of allowable Young diagrams. Let $[\mu]$, $[\nu]$ be two irreducible representations of $GL(n)$ specified by two Young diagrams. For example,

$$[\mu] = \quad\boxed{} \qquad\qquad [\nu] = \begin{array}{|c|c|}\hline 1 & 1 \\\hline 2 \\\cline{1-1}\end{array} \quad .$$

Put ν_1 1's, ν_2 2's, etc. in the $[\nu]$ diagram (as shown in the $[\nu]$ diagram above). Add the 1-boxes to the $[\mu]$ diagram to make other legitimate diagrams such that no two 1-boxes are in the same column:

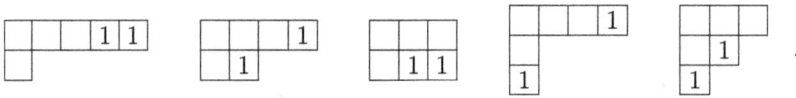

Next add the 2-boxes to the above obtained diagrams to make further legitimate diagrams, such that no two 2-boxes are in the same column:

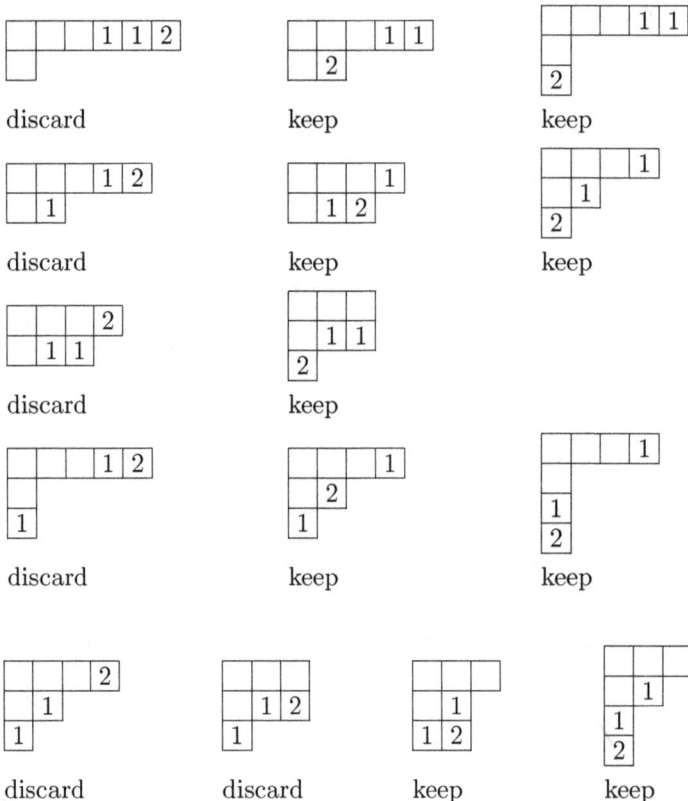

discard	keep	keep
discard	keep	keep
discard	keep	
discard	keep	keep
discard	discard	keep	keep

From the diagrams obtained in the last step, discard all the ones which, when read from right to left in successive rows, have the i-th 2 appearing before the i-th 1, the i-th 3 appearing before the i-th 2, etc., for any positive integer *i*. The remaining diagrams, with all numbers erased from the added boxes and counted with multiplicity, give the Clebsch-Gordan decomposition. Thus in the above example, we have

(21.13)

Exercise 21.4 Use the above rules for Clebsch-Gordan decomposition to verify that

i)

(21.14)

ii)

(21.15)

iii)

(21.16)

iv)

(21.17)

The result iv) in the above exercise reproduces (20.70).

The Clebsch-Gordan decomposition rule introduced above implies the following rule.

Theorem 21.2. *Let the irreducible representations of $GL(n)$ be specified by Young diagrams $[\mu_1, \ldots, \mu_n]$. Then, for $l < n$,*

$$[\underbrace{m, m, \ldots, m}_{n \ times}] \otimes [\mu_1, \ldots, \mu_l] = [\mu_1 + m, \ldots, \mu_l + m, \underbrace{m, \ldots, m}_{(n-l) \ times}]. \quad (21.18)$$

For example, for $GL(3)$,

(21.19)

$[m, m, \ldots, m]$ in (21.18) is a rectangular Young diagram with n rows and m columns. According to Theorem 21.1, it is a one-dimensional representation which transforms with a multiplicative factor of $(\det(g))^m$, $g \in GL(n)$. It is then immediately clear that for $SL(n)$, we have the following equivalence of representations.

$$\boxed{[\mu_1 + m, \ldots, \mu_l + m, \underbrace{m, \ldots, m}_{(n-l)\text{ times}}] \approx [\mu_1, \ldots, \mu_l] \,, \quad \text{for } SL(n) \,.} \qquad (21.20)$$

Thus, to find the irreducible representations for $SL(n)$, we only need to consider Young diagrams with fewer than n rows, specified by at most $n - 1$ integers, $\mu_1, \ldots, \mu_{n-1}$. Eq. (21.19) for $GL(3)$ becomes, for $SL(3)$,

$$\qquad (21.21)$$

Exercise 21.5   Show that for $SL(n)$,

$$[1^{n-1}] \approx [1] \,. \qquad (21.22)$$

The generalization of (21.22) for $SL(n)$, which is an equivalence distinct from (21.20), is given by the following theorem (stated without proof).

Theorem 21.3. *For $SL(n)$,*

$$[\mu_1, \mu_2, \ldots, \mu_{n-1}] \approx [\mu_1, \mu_1 - \mu_{n-1}, \ldots, \mu_1 - \mu_2] \,. \qquad (21.23)$$

In Fig. 21.1, the dotted Young diagram [RHS of (21.3)] completes an $n \times \mu_1$ rectangle when adjoined to the solid Young diagram [LHS of (21.23)].

Note that (21.20) can be written

$$[\mu_1, \mu_2, \ldots, \mu_n] \approx [\mu_1 - m, \mu_2 - m, \ldots, \mu_n - m] \,, \quad \text{for } SL(n) \,, \qquad (21.24)$$

where m is any integer, positive or negative. Starting with $\mu_1 \geq \mu_2 \geq \cdots \geq \mu_n \geq 0$, one may then produce equivalent tensor representations of mixed rank [c.f. (20.77)]. For example, $[0, 0, \ldots, 0, -1]$ corresponds to the tensor φ^μ (an n-dimensional representation).

Setting $m = -1$ in (21.24) we have

$$[0, 0, \ldots, 0, -1] \approx [1, 1, \ldots, 1, 0] \,.$$

By Theorem 21.1, the RHS corresponds to the tensor $c\,\varepsilon^{i_1 \cdots i_{n-1}} e_{i_1} \otimes \cdots \otimes e_{i_{n-1}}$, which is n-dimensional (since each of $i_1, \ldots, i_{n-1}$ can take values $1, 2, \ldots, n$). φ^μ is thus equivalent to

$$\varepsilon_{i_1 \ldots i_{n-1} \nu} \, \varphi^\nu = \varphi_{i_1 \ldots i_{n-1}} \,, \qquad (21.25)$$

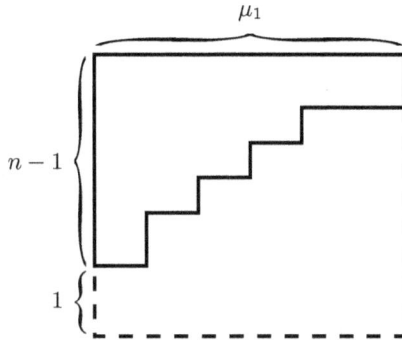

FIGURE 21.1

where $\varepsilon_{i_1,\ldots i_{n-1}\nu}$ is the Levi-Civita symbol. We conclude that *the Levi-Civita symbol can be used to raise or lower indices to obtain equivalent irreducible representations of $SL(n)$.*

Consider $SL(3)$ [or $SU(3)$]. The mixed tensor φ^ν_μ corresponding to $[1, 0, -1]$ is equivalent to $\varphi_{\mu\nu\lambda}$, corresponding to $[2, 1, 0]$ ($\boxplus$) via

$$\varphi_{\mu\nu\lambda} = \varepsilon_{\mu\nu\alpha}\,\varphi^\alpha_\lambda, \qquad \varphi^\alpha_\lambda = \frac{1}{2}\varepsilon^{\alpha\mu\nu}\,\varphi_{\mu\nu\lambda}. \tag{21.26}$$

Exercise 21.6 Use (2.12) to derive the second equation of (21.26) from the first.

For φ^α_λ to be irreducible, it is required to be traceless: $\varphi^\lambda_\lambda = 0$ [c.f. (20.73)]. This implies

$$\varphi_{\mu\nu\lambda} + \varphi_{\nu\lambda\mu} + \varphi_{\lambda\mu\nu} = 0. \tag{21.27}$$

The first equation of (21.26) also implies

$$\varphi_{\mu\nu\lambda} + \varphi_{\nu\mu\lambda} = 0. \tag{21.28}$$

Conversely, if $\varphi_{\mu\nu\lambda}$ satisfies the above two conditions, then $\varphi^\alpha_\lambda = \frac{1}{2}\varepsilon^{\alpha\mu\nu}\varphi_{\mu\nu\lambda}$ satisfies the traceless condition $\varphi^\alpha_\alpha = 0$.

Finally, we consider $O(n)$ and its subgroup $SO(n)$. Recall that for the matrix $(a^j_i) \in O(n)$, the orthogonality condition implies that

$$\sum_{k=1}^{n} a^i_k a^j_k = \sum_{k=1}^{n} a^k_i a^k_j = \delta_{ij}, \tag{21.29}$$

[c.f. (1.29)]. Thus in a representation space $T^r(V)$, in which a tensor with components $T^{i_1\ldots i_r}$ transforms under $O(n)$ as

$$(T')^{i_1\ldots i_r} = a^{i_1}_{j_1} a^{i_2}_{j_2} \ldots a^{i_r}_{j_r}\, T^{j_1\ldots j_r}, \tag{21.30}$$

contraction of indices is an invariant operation under $O(n)$ transformations. In other words, *contractions commute with $O(n)$ transformations*. Indeed, considering the (12) contraction for example (contracting the first two indices), (21.29) and (21.30) imply

$$\sum_k (T')^{kki_3\cdots i_r} = \delta_{j_1j_2}\, a^{i_3}_{j_3}\cdots a^{i_r}_{j_r}\, T^{j_1j_2j_3\cdots j_r} = a^{i_3}_{j_3}\cdots a^{i_r}_{j_r} \sum_k T^{kkj_3\cdots j_r}\,. \quad (21.31)$$

For a rank-r tensor there are $\binom{r}{2} = \frac{r(r-1)}{2}$ possible contractions:

$$\left(T_{(12)}\right)^{i_3\cdots i_r} \equiv \sum_{k=1}^{n} T^{kki_3\cdots i_r}\,,$$

$$\left(T_{(13)}\right)^{i_2i_4\cdots i_r} \equiv \sum_{k=1}^{n} T^{ki_2ki_4\cdots i_r}\,,$$

$$\vdots$$

$$\left(T_{(\alpha\beta)}\right)^{i_1\cdots i_{\alpha-1}i_{\alpha+1}\cdots i_{\beta-1}i_{\beta+1}\cdots i_r} \equiv \sum_{k=1}^{n} T^{i_1\cdots i_{\alpha-1}ki_{\alpha+1}\cdots i_{\beta-1}ki_{\beta+1}\cdots i_r}\,,$$

$$\vdots$$

where each is a rank-$(r-2)$ tensor. The subscripts inside the parentheses indicate the positions of the tensor indices contracted. Then we can construct a traceless tensor T_0 from T as follows. Write

$$T^{i_1\cdots i_r} = (T_0)^{i_1\cdots i_r} + \delta^{i_1i_2}\left(V_{(12)}\right)^{i_3\cdots i_r} + \cdots$$
$$+ \delta^{i_\alpha i_\beta}\left(V_{(\alpha\beta)}\right)^{i_1\cdots i_{\alpha-1}i_{\alpha+1}\cdots i_{\beta-1}i_{\beta+1}\cdots i_r} + \cdots \quad . \qquad (21.32)$$

Requiring T_0 to be traceless (with respect to all pairs of indices), we obtain $\binom{r}{2}$ equations from which the rank-$(r-2)$ contracted tensors $V_{(\alpha\beta)}$ can be solved in terms of the rank-$(r-2)$ contracted tensors $T_{(\alpha\beta)}$.

$\boxed{\text{Exercise 21.7}}$ Perform the decomposition of a general rank-3 tensor $T^{i_1i_2i_3}$ into a traceless part T_0 and its orthogonal complement:

$$T^{i_1i_2i_3} = (T_0)^{i_1i_2i_3} + \delta^{i_1i_2}\left(V_{(12)}\right)^{i_3} + \delta^{i_1i_3}\left(V_{(13)}\right)^{i_2} + \delta^{i_2i_3}\left(V_{(23)}\right)^{i_1}\,. \quad (21.33)$$

Solve for $V_{(\alpha\beta)}$ in terms of $T_{(\alpha\beta)}$.

We will write the decomposition (21.32) of a tensor T into its traceless part T_0 and the orthogonal complement V as the direct sum

$$T = T_0 \oplus V\,, \qquad (21.34)$$

where $V^{i_1\cdots i_r}$ has the general form given by the last $\binom{r}{2}$ terms on the RHS of (21.32). Our discussion immediately preceding (21.31) suggests the following theorem, which will be stated without proof.

Theorem 21.4. *The decomposition of a tensor into its traceless part and the orthogonal complement [(21.34)] is invariant under $O(n)$ transformations.*

We further note that the subspace of traceless tensors $T_0^r(V) \subset T^r(V)$, $dim(V) = n$, is invariant under the symmetric group S_n. Thus one can construct all the irreducible representations of $O(n)$ on $T^r(V)$ by starting with $T_0^r(V)$, and then by applying Young symmetrizers to obtain traceless tensors of given symmetry types. However, not all possible Young diagrams will yield non-vanishing traceless tensors, as the following example shows.

Consider the group $O(2)$ and its representation on $T^3(V)$, $dim(V) = 2$. Let a traceless tensor $T_0 \in T^3(V)$ be written as

$$T_0 = T^{i_1 i_2 i_3} \varphi_{i_1 i_2 i_3} , \quad i_1, i_2, i_3 = 1, 2 , \tag{21.35}$$

where $\varphi_{i_1 i_2 i_3} = e_{i_1} \otimes e_{i_2} \otimes e_{i_3}$. From (20.40) we have

$$Y_3 \left(\begin{array}{|c|c|} \hline 1 & 2 \\ \hline 3 \\ \cline{1-1} \end{array} \right) T_0 = T^{i_1 i_2 i_3} M_{i_1 i_2 i_3} , \tag{21.36}$$

where

$$M_{i_1 i_2 i_3} = \frac{1}{3} \left(\varphi_{i_1 i_2 i_3} + \varphi_{i_2 i_1 i_3} - \varphi_{i_3 i_2 i_1} - \varphi_{i_3 i_1 i_2} \right) , \tag{21.37}$$

[c.f. (20.41)]. The only independent $M_{i_1 i_2 i_3}$ are

$$M_{112} = \frac{2}{3} (\varphi_{112} - \varphi_{211}) = -2M_{211} = -2M_{121} ,$$

$$M_{221} = \frac{2}{3} (\varphi_{221} - \varphi_{122}) = -2M_{122} = -2M_{212} .$$

Thus we have

$$Y_3 \left(\begin{array}{|c|c|} \hline 1 & 2 \\ \hline 3 \\ \cline{1-1} \end{array} \right) T_0 = (T^{112} - \frac{1}{2} (T^{211} + T^{121})) M_{112}$$
$$+ (T^{221} - \frac{1}{2} (T^{122} + T^{212})) M_{221} . \tag{21.38}$$

Since T_0 is traceless, its components satisfy

$$\sum_i T^{iij} = \sum_i T^{iji} = \sum_i T^{jii} = 0, \quad j = 1, 2 . \tag{21.39}$$

These conditions imply

$$T^{112} = T^{211} = T^{121} , \quad T^{221} = T^{122} = T^{212} . \tag{21.40}$$

We conclude from (21.38), then, that

$$Y_3 \left(\begin{array}{|c|c|} \hline 1 & 2 \\ \hline 3 \\ \cline{1-1} \end{array} \right) T_0 = 0 . \tag{21.41}$$

Similarly, one can show that

$$Y_3 \left(\begin{array}{|c|c|} \hline 1 & 3 \\ \hline 2 \\ \hline \end{array} \right) T_0 = 0 . \tag{21.42}$$

Exercise 21.8   Verify the above equation.

Exercise 21.9   Show that the traceless tensor $Y_3 \left(\begin{array}{|c|} \hline 1 \\ \hline 2 \\ \hline 3 \\ \hline \end{array} \right) T_0$ vanishes identically for $n = 2$.

Exercise 21.10   Calculate the traceless tensor $Y_2 \left(\begin{array}{|c|} \hline 1 \\ \hline 2 \\ \hline \end{array} \right) T_0$, where T_0 is a rank-2 traceless tensor $T_0 = T^{i_1 i_2} \varphi_{i_1 i_2}$, $i_1, i_2 = 1, 2$; and show that it does not vanish identically.

We will state the following important theorem without proof.

Theorem 21.5. *Traceless tensors in $T^r(V)$, $\dim V = n$, corresponding to Young diagrams in which the sum of the lengths of the first two columns is greater than n must vanish identically.*

This general result corroborates the results of the special examples studied earlier.

Corollary 21.1. *The irreducible representations of $O(n)$ can be constructed from traceless tensors corresponding to Young diagrams in which the sum of the lengths of the first two columns is less than or equal to n.*

For example, for $O(3)$, we have the traceless tensors

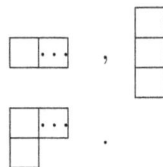

For $O(4)$, we have

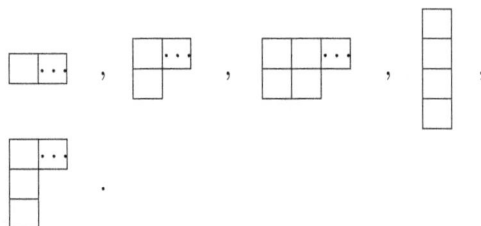

For $O(5)$, there are the following:

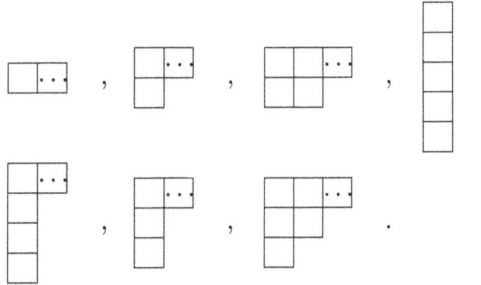

In the above Young diagrams, a box with dots inside represents an arbitrary number of boxes in the same row.

Some of the possible diagrams may pair up in the following way: a diagram with c_1 boxes in the first column paired with one with $n - c_1$ boxes in the first column, and the same number of boxes in each of the other columns. (These pairs are shown in the same column in the above diagrams.) When n is even, such as in $O(4)$, the diagrams with $c_1 = n/2$ are not paired. Each completely antisymmetric diagram $[1^n], (c_1 = n)$ can be thought of as being paired with a diagram with no boxes at all, representing a scalar.

Let us now focus on $SO(n)$. It turns out that for this case, the irreducible representations for the paired diagrams are equivalent, via contractions with the Levi-Civita symbol. We will not prove this fact in general, but instead illustrate it by some examples. First note that for $SO(n)$, we do not have to consider the completely antisymmetric diagrams $[1^n]$.

Consider $SO(3)$ and the irreducible representation corresponding to the Young diagram $\square$, with tensors φ_μ. Construct the antisymmetric tensor

$$\psi_{\mu\nu} \equiv \varepsilon_{\mu\nu}{}^\lambda \varphi_\lambda = -\psi_{\nu\mu} , \tag{21.43}$$

where $\varepsilon_{\mu\nu}{}^\lambda$ is the Levi-Civita symbol. Suppose φ_λ transforms, under $SO(3)$, according to

$$\varphi_\lambda \to \varphi'_\lambda = a^\rho_\lambda \varphi_\rho . \tag{21.44}$$

Then it follows from (21.10) that

$$\psi_{\mu\nu} \to \psi'_{\mu\nu} = \varepsilon'_{\mu\nu}{}^\lambda \varphi'_\lambda = \varepsilon_{\mu\nu}{}^\lambda a^\rho_\lambda \varphi_\rho . \tag{21.45}$$

Thus

$$\psi'_{12} = \varepsilon_{12}{}^3 a^\rho_3 \varphi_\rho = a^\rho_3 \varphi_\rho = -\psi'_{21} , \tag{21.46a}$$

$$\psi'_{23} = \varepsilon_{23}{}^1 a^\rho_1 \varphi_\rho = a^\rho_1 \varphi_\rho = -\psi'_{32} , \tag{21.46b}$$

$$\psi'_{31} = \varepsilon_{31}{}^2 a^\rho_2 \varphi_\rho = a^\rho_2 \varphi_\rho = -\psi'_{13} . \tag{21.46c}$$

$\psi_{12}, \psi_{23}, \psi_{31}$ form the basis of the 3-dimensional irreducible representation corresponding to the Young diagram $\boxminus$ and transform in exactly the same way as

$\varphi_3, \varphi_1, \varphi_2$, respectively. We can represent the equivalence between $\square$ and $\underset{\square}{\square}$ for $SO(3)$ as follows:

$$\boxed{\lambda} \approx \begin{array}{|c|c|}\hline \mu & \lambda \\\hline \nu \\\cline{1-1} \lambda \\\cline{1-1}\end{array} = \begin{array}{|c|}\hline \mu \\\hline \nu \\\hline\end{array}, \tag{21.47}$$

where the repeated index λ in the middle diagram is summed over (from 1 to n). Similarly,

$$\varphi_{\mu\nu} = \begin{array}{|c|c|}\hline \mu & \nu \\\hline\end{array} \approx \begin{array}{|c|c|c|}\hline \mu & \rho & \lambda \\\hline \nu \\\cline{1-1} \rho \\\cline{1-1}\end{array} = \varepsilon_{\mu\nu}{}^{\rho}\,\varphi_{\rho\lambda} \equiv \psi_{\mu\nu\lambda} = \begin{array}{|c|c|}\hline \mu & \lambda \\\hline \nu \\\cline{1-1}\end{array} . \tag{21.48}$$

In fact, all the tensorial inequivalent irreducible representations of $SO(3)$ can be specified by the diagrams

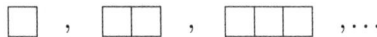

$$\square \quad , \quad \square\square \quad , \quad \square\square\square \quad , \ldots$$

where each diagram is characterized by the single positive integer μ_1, assuming the values $0, 1, 2, 3, \ldots$.

As a representation for $GL(3)$, the diagram $[\mu_1]$ has, according to (20.71), a dimension of

$$\frac{(\mu_1 + 2)(\mu_1 + 1)}{2} = \binom{\mu_1 + 2}{2} .$$

The traceless conditions on the totally symmetric tensors of $[\mu_1]$ is given by

$$S_{11 i_3 \ldots i_{\mu_1}} + S_{22 i_3 \ldots i_{\mu_1}} + S_{33 i_3 \ldots i_{\mu_1}} = 0 , \tag{21.49}$$

which comprises exactly as many equations as the dimensionality of $[\mu_1 - 2]$, given by (20.72). For $n = 3$, this number is $\binom{\mu_1}{2}$. Thus, as a representation for $SO(3)$, the dimensionality of $[\mu_1]$ is

$$\binom{\mu_1 + 2}{2} - \binom{\mu_1}{2} = \frac{(\mu_1 + 2)(\mu_1 + 1)}{2} - \frac{\mu_1(\mu_1 - 1)}{2} = 2\mu_1 + 1 . \tag{21.50}$$

This is a familiar result in the theory of orbital angular momentum in quantum mechanics. Thus the Completely symmetric traceless tensor

$$\overset{\mu_1}{\overbrace{\begin{array}{|c|c|c|c|}\hline & & & \cdots \\\hline\end{array}}}$$

describes a spin μ_1 ($\mu_1 = 0, 1, 2, \ldots$) particle with the z-component of the spin having $2\mu_1 + 1$ independent values. Writing $\mu_1 = l$ (which is the symbol usually used for orbital angular momentum in quantum theory), the basis states in this representation are usually written (in the Dirac notation) as $|\,lm\,\rangle$, where the $2l + 1$ values of m are $l, l - 1, \ldots, -l$. The half-integral values for angular momentum arises from the so-called **spinor representation** of $SO(3)$, which will be considered in the next chapter.

For $SO(4)$, contraction with the Levi-Civita symbol yields

$$\varphi_{\mu\nu} = \boxed{\mu\,\nu} \approx \ytableau{\lambda & \mu & \nu \\ \rho \\ \delta \\ \nu} = \varepsilon_{\lambda\rho\delta}{}^{\nu}\varphi_{\mu\nu} \equiv \psi_{\lambda\rho\delta\mu} = \ytableau{\lambda & \mu \\ \rho \\ \delta} \,. \qquad (21.51)$$

For $SO(5)$, the following equivalences are obtained by contraction:

$$\varphi_{\mu\nu} = \boxed{\mu\,\nu} \approx \ytableau{\alpha & \mu & \nu \\ \beta \\ \gamma \\ \delta \\ \mu} = \varepsilon_{\alpha\beta\gamma\delta}{}^{\mu}\varphi_{\mu\nu} \equiv \psi_{\alpha\beta\gamma\delta\nu} = \ytableau{\alpha & \nu \\ \beta \\ \gamma \\ \delta} \,, \qquad (21.52)$$

$$\varphi_{\mu\nu\lambda} = \ytableau{\mu & \nu \\ \lambda} \approx \ytableau{\alpha & \mu & \nu \\ \beta & \lambda \\ \gamma \\ \mu \\ \lambda} = \varepsilon_{\alpha\beta\gamma}{}^{\mu\nu}\varphi_{\mu\nu\lambda} \equiv \psi_{\alpha\beta\gamma\nu} = \ytableau{\alpha & \nu \\ \beta \\ \gamma} \,,$$

$$\qquad (21.53)$$

$$\varphi_{\mu\nu\lambda\rho\delta} = \ytableau{\mu & \nu & \lambda \\ \rho & \delta} \approx \ytableau{\alpha & \mu & \nu & \lambda \\ \beta & \rho & \delta \\ \gamma \\ \mu \\ \rho} = \varepsilon_{\alpha\beta\gamma}{}^{\mu\rho}\varphi_{\mu\nu\lambda\rho\delta} \equiv \psi_{\alpha\beta\gamma\nu\lambda\delta} = \ytableau{\alpha & \nu & \lambda \\ \beta & \delta \\ \gamma} \,.$$

$$\qquad (21.54)$$

Equivalences for $SO(n)$, $n > 5$, are obtained similarly.

Chapter 22

Irreducible Representations of $SU(2)$ and $SO(3)$

Various aspects of $SO(3)$, the rotation group in three dimensions, have been discussed previously (Chapters 1, 4, 9 and 11). Its fundamental role in physics applications, especially problems involving rotational invariance in quantum theory, has also been stressed. In this chapter we will finally work out explicitly the irreducible representations of this group. Our approach here complements that of Chapter 4, which focused on the Lie algebra (the infinitesimal generators) of $SO(3)$.

We begin by constructing the irreducible representations of $SU(2)$, the universal covering group of $SO(3)$ (c.f. Chapter 11). Recall that [c.f. (11.69)] every $g \in SU(2)$ can be written as

$$g = \begin{pmatrix} a & b \\ -\bar{b} & \bar{a} \end{pmatrix} , \quad a, b \in \mathbb{C} , \tag{22.1}$$

with $|a|^2 + |b|^2 = 1$. Let $\{e_1, e_2\}$ be a basis of the complex 2-dimensional space $\mathbb{C}^2$, on which g acts by

$$g(e_i) = g_i^j \, e_j . \tag{22.2}$$

For $j = n/2, n = 0, 1, 2, 3, \ldots$, construct the following basis for a $(2j + 1)$-

dimensional complex vector space $\mathbb{V}^{(j)}$:

$$\{e_0^{(j)} = \sqrt{(2j)!}\, S_{2j}(e_1^{(2j)}) \equiv \sqrt{(2j)!}\, S_{2j}(\overbrace{e_1 \otimes \cdots \otimes e_1}^{2j \text{ times}}),$$

$$e_1^{(j)} = \sqrt{(2j)!}\, S_{2j}(e_1^{2j-1} \otimes e_2) \equiv \sqrt{(2j)!}\, S_{2j}(\overbrace{e_1 \otimes \cdots \otimes e_1}^{(2j-1) \text{ times}} \otimes e_2),$$

$$e_2^{(j)} = \sqrt{(2j)!}\, S_{2j}(e_1^{2j-2} \otimes e_2^2),$$

$$\vdots \tag{22.3}$$

$$e_k^{(j)} = \sqrt{(2j)!}\, S_{2j}(e_1^{2j-k} \otimes e_2^k),$$

$$\vdots$$

$$e_{2j}^{(j)} = \sqrt{(2j)!}\, S_{2j}(e_2^{2j})\},$$

where S_{2j} is the symmetrizing map $[(16.21)]$. The above basis vectors can be regarded as homogeneous polynomials of order $2j$ in e_1 and e_2. For example, for $j = 2$, we have, in an obvious notation,

$$e_0^{(2)} = e_1^4, \; e_1^{(2)} = e_1^3 e_2, \; e_2^{(2)} = e_1^2 e_2^2, \; e_3^{(2)} = e_1 e_2^3, \; e_4^{(2)} = e_2^4. \tag{22.4}$$

According to (22.1) and (22.2), $g \in SU(2)$ acts on $\mathbb{V}^{(j)}$ as follows.

$$g(e_k^{(j)}) = (g(e_1))^{2j-k}\,(g(e_2))^k = (ae_1 + be_2)^{2j-k}\,(-\bar{b}e_1 + \bar{a}e_2)^k$$

$$= \sum_{m=0}^{2j-k} \binom{2j-k}{m}(ae_1)^{2j-k-m}(be_2)^m \sum_{l=0}^{k}\binom{k}{l}(-\bar{b}e_1)^{k-l}(\bar{a}e_2)^l \tag{22.5}$$

$$= \sum_{m=0}^{2j-k}\sum_{l=0}^{k}\binom{2j-k}{m}\binom{k}{l}a^{2j-k-m}b^m(-\bar{b})^{k-l}(\bar{a})^l e_1^{2j-(m+l)} e_2^{m+l}.$$

Letting $m+l = p$, we see that in the above double sum, p can assume the values $0, 1, 2, \ldots, 2j$. Thus we can write

$$(e_k^{(j)})' = g(e_k^{(j)}) = \left(\mathcal{D}^{(j)}\right)_k^p e_p^{(j)}, \tag{22.6}$$

where

$$\left(\mathcal{D}^{(j)}\right)_k^p = \sum_{m+l=p}\binom{2j-k}{m}\binom{k}{l}a^{2j-k-m}b^m(-\bar{b})^{k-l}(\bar{a})^l \; ; \; k, p = 0, 1, \ldots, 2j,$$
$$\tag{22.7}$$

and the space $\mathbb{V}^{(j)}$ is seen to be an irreducible representation space of $SU(2)$. The $(2j+1)$ basis vectors $e_k^{(j)}$, $k = 0, 1, 2, \ldots, 2j$, are called **spinors** of rank $2j$. As an example, we have the following result:

$$(e_1^{(2)})' = (-a^3\bar{b})\,e_0^{(2)} + a^2(|a|^2 - 3|b|^2)\,e_1^{(2)}$$
$$+ 3ab(|a|^2 - |b|^2)\,e_2^{(2)} - b^2(|b|^2 - 3|a|^2)\,e_3^{(2)} + (b^3\bar{a})\,e_4^{(2)}. \tag{22.8}$$

Exercise 22.1 Verify the above equation by using (22.5).

The above representation can be rendered unitary by rescaling the basis vectors in (22.3). Define the normalized states

$$|jm\rangle \equiv \epsilon_m^{(j)} \equiv \frac{e_k^{(j)}}{\sqrt{(2j-k)!k!}} \equiv \frac{e_1^{j+m} \, e_2^{j-m}}{\sqrt{(j+m)!(j-m)!}} \,, \qquad (22.9)$$

where $m \equiv j - k$. Since k runs through $0, 1, \ldots, 2j$ for a given value of j, the possible values of m are $j, j-1, \ldots, -j+1, -j$. Note that in the above equation $e_1^{j+m} e_2^{j-m} \equiv e_k^{(j)}$ means a symmetrized product of $j+m$ factors of e_1 and $j-m$ factors of e_2 [as defined by (22.3)]. Unitarity is demonstrated as follows. Let $e_1 = (1,0), e_2 = (0,1)$ be the standard basis in $\mathbb{C}^2$. Then $u = (u^1, u^2) \in \mathbb{C}^2$ can be written $u = u^1 e_1 + u^2 e_2$. Define a hermitian inner product in $\mathbb{C}^2$ by

$$(u, v) = u^1 \, \overline{v^1} + u^2 \, \overline{v^2} \,. \qquad (22.10)$$

The vector $(1/\sqrt{(2j)!}) \, \overbrace{u \otimes \cdots \otimes u}^{2j \text{ times}}$ in the tensor space $\overbrace{\mathbb{C}^2 \otimes \cdots \otimes \mathbb{C}^2}^{2j \text{ times}}$ then projects on $V^{(j)}$, giving a vector $u^{(j)} \in V^{(j)}$ with components along $\epsilon_j^{(j)}, \ldots, \epsilon_{-j}^{(j)}$ given by, respectively,

$$(u^{(j)})^0 = \frac{(u^1)^{2j}}{\sqrt{(2j)!}}, \quad (u^{(j)})^1 = \frac{(u^1)^{2j-1} (u^2)}{\sqrt{(2j-1)!}}, \ldots, (u^{(j)})^k = \frac{(u^1)^{2j-k} (u^2)^k}{\sqrt{(2j-k)!k!}}, \ldots,$$

$$(u^{(j)})^{2j} = \frac{(u^2)^{2j}}{\sqrt{(2j)!}} \,.$$

$$(22.11)$$

The inner product (22.10) induces in $V^{(j)}$ the following hermitian inner product:

$$(u^{(j)}, v^{(j)}) = \sum_{k=0}^{2j} (u^{(j)})^k \, \overline{(v^{(j)})^k} \,. \qquad (22.12)$$

Under $g \in SU(2)$ given by (22.1), $u^{(j)}$ transforms as [compare with (22.5)]:

$$(u^{(j)})^k \longrightarrow ((u^{(j)})')^k = \frac{1}{\sqrt{(2j-k)!k!}} ((u')^1)^{2j-k} ((u')^2)^k$$

$$= \frac{1}{\sqrt{(2j-k)!k!}} (au^1 - \bar{b}u^2)^{2j-k} (bu^1 + \bar{a}u^2)^k \,, \qquad (22.13)$$

Then,

$$(2j)! \left((u^{(j)})', (v^{(j)})' \right) = \sum_{k=0}^{2j} \frac{(2j)!}{(2j-k)!k!} ((u')^1)^{2j-k}((u')^2)^k (\overline{(v')^1})^{2j-k}(\overline{(v')^2})^k$$

$$= \sum_{k=0}^{2j} \binom{2j}{k} \left((u')^1 \overline{(v')^1} \right)^{2j-k} \left((u')^2 \overline{(v')^2} \right)^k = \left((u')^1 \overline{(v')^1} + (u')^2 \overline{(v')^2} \right)^{2j}$$

$$= \left(u^1 \overline{v^1} + u^2 \overline{v^2} \right)^{2j} = (2j)! \left(u^{(j)}, v^{(j)} \right) ,$$

$$\tag{22.14}$$

where the fourth equality follows from the unitarity of g. Thus $((u^{(j)})', (v^{(j)})') = (u^{(j)}, v^{(j)})$, and the irreducible $(2j+1)$-dimensional representation of $SU(2)$ on $\mathbb{V}^{(j)}$ specified by the basis (22.9) is unitary. In fact, the collection of all these irreducible representations, labelled $D^{(j)}$ (for $j = n/2, n = 0, 1, 2, \ldots$), exhaust all the irreducible representations of $SU(2)$. (We will omit the proof of this central fact here). Analogous to (22.6) we can write

$$((u^{(j)})')^m = \left(D^{(j)} \right)^m_{m'} (u^{(j)})^{m'} ,$$

$$\tag{22.15}$$

where

$$(u^{(j)})^m = \frac{(u^1)^{j+m} (u^2)^{j-m}}{\sqrt{(j+m)!(j-m)!}} ,$$

$$\tag{22.16}$$

with $m, m' = j, j-1, \ldots, -j+1, -j$. [These are the same as the quantities $(u^{(j)})^k$ in (22.11), with $m = j-k$.] Note that a vector with components $(u^{(j)})^m$ can be regarded as a vector in the linear space of homogeneous polynomials in $\mathbb{C}^2$ of degree $2j$. This space is isomorphic to $\mathbb{V}^{(j)}$ [(22.3)] and can also be regarded as the $(2j+1)$-dimensional irreducible representation space for $SO(3)$. It follows from (22.13) that

$$((u^{(j)})')^m = \frac{((u')^1)^{j+m} ((u')^2)^{j-m}}{\sqrt{(j+m)!(j-m)!}} = \frac{(au^1 - \bar{b}u^2)^{j+m} (bu^1 + \bar{a}u^2)^{j-m}}{\sqrt{(j+m)!(j-m)!}}$$

$$= \frac{1}{\sqrt{(j+m)!(j-m)!}} \sum_{p=0}^{j+m} \binom{j+m}{p} (au^1)^{j+m-p}(-\bar{b}u^2)^p$$

$$\times \sum_{q=0}^{j-m} \binom{j-m}{q} (bu^1)^{j-m-q} (\bar{a}u^2)^q$$

$$= \sum_{p=0}^{j+m} \sum_{q=0}^{j-m} \frac{\sqrt{(j+m)!(j-m)!}}{(p!)(j+m-p)!q!(j-m-q)!} (-1)^p a^{j+m-p} (\bar{b})^p b^{j-m-q} (\bar{a})^q$$

$$\times (u^1)^{2j-(p+q)} (u^2)^{p+q}$$

$$= \sum_{p=0}^{j+m} \sum_{q=0}^{j-m} \frac{\sqrt{(j+m)!(j-m)!(j+m')(j-m')}}{(p!)(j+m-p)!(j-m'-p)!(p+m'-m)!} (-1)^p$$

$$\times a^{j+m-p} (\bar{a})^{j-m'-p} (\bar{b})^p b^{p+m'-m} \frac{(u^1)^{j+m'}(u^2)^{j-m'}}{\sqrt{(j+m')!(j-m')!}} , \qquad (22.17)$$

where in the last equality we have set $m' \equiv j - (p+q)$. As p and q assume the various allowed values (indicated by limits on the double sum above), m' ranges over the $2j+1$ values $j, j-1, \ldots, -j+1, -j$. Table 22.1 shows the situation for $j = 2, m = 1$. For this case the double sum $\sum_{p=0}^{3} \sum_{q=0}^{1}$ can be rewritten as the double sum over m' and p, where the possible combination of values are:

$$m' = -2, \ p = 3 \ ; m' = -1, \ p = 2, 3 \ ; m' = 0, \ p = 1, 2 \ ;$$
$$m' = 1, \ p = 0, 1 \ ; m' = 2, \ p = 0 \ .$$

p	0	0	1	1	2	2	3	3
q	0	1	0	1	0	1	0	1
$p+q$	0	1	1	2	2	3	3	4
m'	2	1	1	0	0	-1	-1	-2

TABLE 22.1

In general, the double sum in (22.17) can be rewritten as

$$\sum_{m'=-j}^{j} \sum_{p=\underline{p}}^{\bar{p}} ,$$

where for a given m', the range of p is given by

$$\underline{p} = max\{0, m - m'\} , \quad \bar{p} = min\{j - m', j + m\} . \qquad (22.18)$$

$\boxed{\text{Exercise 22.2}}$ Verify the above assertion for the case $j = 2, m = 1$; and in general.

Eqs. (22.25), (22.16) and (22.17) imply that the irreducible unitary representation $D^{(j)}(g)$ has matrix elements given by

$$(D^{(j)})^m_{m'} = \sum_{p=\underline{p}}^{\bar{p}} \frac{(-1)^p \sqrt{(j+m)!(j-m)!(j+m')(j-m')}}{p! \, (j+m-p)!(j-m'-p)!(p+m'-m)!}$$

$$\times a^{j+m-p} (\bar{a})^{j-m'-p} (\bar{b})^p b^{p+m'-m} , \qquad (22.19)$$

where p and $\bar{p}$ are given by (22.18).

Let us immediately apply this general formula to the case $j = 1/2$. One sees that

$$(D^{(\frac{1}{2})})^{\frac{1}{2}}_{\frac{1}{2}} = a, \ (D^{(\frac{1}{2})})^{-\frac{1}{2}}_{\frac{1}{2}} = b, \ (D^{(\frac{1}{2})})^{\frac{1}{2}}_{-\frac{1}{2}} = -\bar{b}, \ (D^{(\frac{1}{2})})^{-\frac{1}{2}}_{-\frac{1}{2}} = \bar{a} \ .$$

Thus,

$$D^{(\frac{1}{2})} = \begin{pmatrix} a & b \\ -\bar{b} & \bar{a} \end{pmatrix} = g \ , \tag{22.20}$$

as expected.

Exercise 22.3   Use the result of (22.19) to show that

$$D^{(1)} = \begin{pmatrix} (D^{(1)})^1_1 & (D^{(1)})^0_1 & (D^{(1)})^{-1}_1 \\ (D^{(1)})^1_0 & (D^{(1)})^0_0 & (D^{(1)})^{-1}_0 \\ (D^{(1)})^1_{-1} & (D^{(1)})^0_{-1} & (D^{(1)})^{-1}_{-1} \end{pmatrix} = \begin{pmatrix} a^2 & \sqrt{2}\,ab & b^2 \\ -\sqrt{2}\,a\bar{b} & |a|^2 - |b|^2 & \sqrt{2}\,\bar{a}b \\ \bar{b}^2 & -\sqrt{2}\,\bar{a}\bar{b} & \bar{a}^2 \end{pmatrix} \ . \tag{22.21}$$

Verify that $\det(D^{(1)}) = 1$ and $(D^{(1)})^{-1} = (D^{(1)})^\dagger$.

Having determined the irreducible representations of $SU(2)$, those of $SO(3)$ follow on recalling that the group homomorphism $\phi|_{SU(2)}$ given by (11.19) is a two-to-one and onto map from $SU(2)$ to $SO(3)$. Suppose for a certain $g \in SU(2)$, $\phi(g) = \phi(-g) = R \in SO(3)$. For $(u^1, u^2) \in \mathbb{C}^2$, let

$$((u')^1, (u')^2) = (u^1, u^2)\,g \ , \tag{22.22}$$
$$((u'')^1, (u'')^2) = (u^1, u^2)(-g) = (-u^1, -u^2)\,g \ . \tag{22.23}$$

Then, from (22.15) and (22.16),

$$\left((u^{(j)})'\right)^m = \left(D^{(j)}(g)\right)^m_{m'} \left(u^{(j)}\right)^{m'} = \sum_{m'} \left(D^{(j)}(g)\right)^m_{m'} \frac{(u^1)^{j+m'}(u^2)^{j-m'}}{\sqrt{(j+m')!(j-m')!}} \ , \tag{22.24}$$

$$\left((u^{(j)})''\right)^m = \left(D^{(j)}(-g)\right)^m_{m'} \left(u^{(j)}\right)^{m'}$$
$$= \sum_{m'} \left(D^{(j)}(g)\right)^m_{m'} \frac{(-u^1)^{j+m'}(-u^2)^{j-m'}}{\sqrt{(j+m')!(j-m')!}} \tag{22.25}$$
$$= (-1)^{2j} \sum_{m'} \left(D^{(j)}(g)\right)^m_{m'} \left(u^{(j)}\right)^{m'} \ .$$

Hence, for $2j =$ an even integer, $D^{(j)}$ can be considered as a single-valued representation for $SO(3)$. This representation is none other than the $(2j + 1)$-dimensional tensorial representation characterized by the Young diagram

j boxes

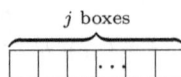

[see discussion preceding (21.50)]. For $2j$ = an odd integer, $D^{(j)}$ cannot be reduced to a tensorial representation. In this case, it is called a **spinor representation**, and is doubled-valued. In physics spinor representations describe elementary particles of half-integer spins, such as the electron, with spin $j = 1/2$.

Remark: The terminology "spinor representation" of $SO(3)$, as well as the "single-valuedness" and "double-valuedness" of representations, is more frequently used in the physics literature. Strictly speaking, all $D^{(j)}$ are representations of $SU(2)$, while the $D^{(j)}$ for $j = 0, 1, 2, \ldots$ are representations of $SO(3)$.

Let us work out the spinor representation $D^{(\frac{1}{2})}(R)$ for an arbitrary rotation $R \in SO(3)$. According to Euler's theorem (Theorem 6.3), every rotation R in $\mathbb{R}^3$ can be written as $R = R_3(\phi)R_2(\theta)R_3(\psi)$. It then follows from (11.55), (11.56) and (22.20) that the two values for the spinor representation for $j = 1/2$ are given by $\pm D^{(\frac{1}{2})}(R_3(\phi)R_2(\theta)R_3(\psi))$, where

$$D^{(\frac{1}{2})}(R_3(\phi)R_2(\theta)R_3(\psi)) = \exp\left(-i\frac{\phi}{2}\sigma_3\right)\exp\left(-i\frac{\theta}{2}\sigma_2\right)\exp\left(-i\frac{\psi}{2}\sigma_3\right)$$

$$= \begin{pmatrix} \exp(-i\frac{\phi}{2}) & 0 \\ 0 & \exp(i\frac{\phi}{2}) \end{pmatrix} \begin{pmatrix} \cos\frac{\theta}{2} & -\sin\frac{\theta}{2} \\ \sin\frac{\theta}{2} & \cos\frac{\theta}{2} \end{pmatrix} \begin{pmatrix} \exp(-i\frac{\psi}{2}) & 0 \\ 0 & \exp(i\frac{\psi}{2}) \end{pmatrix},$$

$$(22.26)$$

or, on multiplication of the three matrices above,

$$D^{(\frac{1}{2})}(R_3(\phi)R_2(\theta)R_3(\psi))$$

$$= \begin{pmatrix} \exp(-i\frac{\phi}{2})\cos\frac{\theta}{2}\exp(-i\frac{\psi}{2}) & -\exp(-i\frac{\phi}{2})\sin\frac{\theta}{2}\exp(i\frac{\psi}{2}) \\ \exp(i\frac{\phi}{2})\sin\frac{\theta}{2}\exp(-i\frac{\psi}{2}) & \exp(i\frac{\phi}{2})\cos\frac{\theta}{2}\exp(i\frac{\psi}{2}) \end{pmatrix}. \quad (22.27)$$

Note that this is of the form (22.20), as it should be.

A review of the proof of Euler's theorem (Theorem 6.3) shows that $D^{(\frac{1}{2})}(\phi, \theta, \psi)$ represents the rotation $R = g'g$, $g', g \in SO(3)$, where g' is the rotation about the unit vector axis $\hat{n}$ through the point (θ, ϕ) on the unit sphere by an angle ψ, and $g = R_3(\phi)R_2(\theta)$. Thus

$$R_{\hat{n}}(\psi) = Rg^{-1} = gR_3(\psi)g^{-1}. \quad (22.28)$$

Setting $\psi = 2\pi$ and 4π, and using (11.56), we see that

$$D^{(\frac{1}{2})}(R_{\hat{n}}(2\pi)) = -e, \quad D^{(\frac{1}{2})}(R_{\hat{n}}(4\pi)) = e, \quad (22.29)$$

where e is the identity in $SU(2)$. Thus 2-component spinors (describing spin $1/2$ particles) change sign under a rotation of 2π about any axis in $\mathbb{R}^3$, but is invariant under a rotation of 4π. This is an important feature in the quantum mechanical description of spin $1/2$ particles.

The $j = 1$ (vector) representation can be similarly worked out [by using (22.21)]. We have, on recalling (11.55) and (11.56),

$$D^{(1)}(R_3(\phi)) = \begin{pmatrix} e^{-i\phi} & 0 & 0 \\ 0 & 1 & 0 \\ 0 & 0 & e^{i\phi} \end{pmatrix} , \tag{22.30}$$

[and a similar expression for $D^{(1)}(R_3(\psi))$],

$$D^{(1)}(R_2(\theta)) = \begin{pmatrix} \frac{1}{2}(1+\cos\theta) & -\frac{\sin\theta}{\sqrt{2}} & \frac{1}{2}(1-\cos\theta) \\ \frac{\sin\theta}{\sqrt{2}} & \cos\theta & -\frac{\sin\theta}{\sqrt{2}} \\ \frac{1}{2}(1-\cos\theta) & \frac{\sin\theta}{\sqrt{2}} & \frac{1}{2}(1+\cos\theta) \end{pmatrix} . \tag{22.31}$$

By multiplying the above 3×3 matrices we can obtain, for a general rotation in $\mathbb{R}^3$,

$$D^{(1)}(R_3(\phi)R_2(\theta)R_3(\psi)) = D^{(1)}(R_3(\phi))D^{(1)}(R_2(\theta))D^{(1)}(R_3(\psi)) . \tag{22.32}$$

Exercise 22.4   Verify (22.30) and (22.31).

Higher order $(j > 1)$ representations are given directly by using (22.19) in

$$D^{(j)}(R_3(\phi)R_2(\theta)R_3(\psi)) = D^{(j)}(R_3(\phi))D^{(j)}(R_2(\theta))D^{(j)}(R_3(\psi)) , \tag{22.33}$$

where the Cayley-Klein parameters (a, b) to be used in (22.19) are

$$\begin{aligned}
D^{(j)}(R_3(\phi)): & \quad a = e^{-i\phi/2}, \ b = 0 ; \\
D^{(j)}(R_2(\theta)): & \quad a = \cos(\theta/2), \ b = -\sin(\theta/2) ; \\
D^{(j)}(R_3(\psi)): & \quad a = e^{-i\psi/2}, \ b = 0 .
\end{aligned} \tag{22.34}$$

Since $D^{(1)}(R_3(2\pi)) = e \in SO(3)$, (22.28) implies that a vector in the representation space of $D^{(1)}$, unlike spinors, is invariant under a rotation of 2π about any axis, as expected.

Exercise 22.5   Use (22.19) to verify directly that

$$\left(D^{(j)}(R_3(\phi)) \right)^m_{m'} = \delta_{m'm} \, e^{-im\phi} , \tag{22.35}$$

where $m, m' = j, j-1, \ldots, -j+1, -j$.

Exercise 22.6 Use the result in the above exercise to show directly that all $D^{(j)}$ matrices are unimodular, that is

$$\det[D^{(j)}(R_{\hat{n}})] = 1 . \tag{22.36}$$

Since $D^{(j)}$ is unitary, we have

$$\left(D^{(j)}(R_3(\phi)R_2(\theta)R_3(\psi))\right)^\dagger = \left(D^{(j)}(R_3(\phi)R_2(\theta)R_3(\psi))\right)^{-1}$$
$$= D^{(j)}(R_3(-\psi)R_2(-\theta)R_3(-\phi)) . \tag{22.37}$$

The result of Exercise 22.5 [(22.35)] indicates that $R_3(\phi)$ defines a hermitian operator J^3 given by

$$R_3(\phi) = \exp(-i\phi J^3) , \tag{22.38}$$

such that, in the representation $D^{(j)}$, J^3 is diagonal and is given by

$$(J^3)_{m'}^m = \delta_{m'm} \, m .$$

Equivalently,

$$J^3 \,|\, jm \rangle = m \,|\, jm \rangle , \tag{22.39}$$

where $|\, jm \rangle$ is given by (22.9).

We recall from (4.39) that J^3 is related to the basis $\{J_i^j\}$ of the Lie algebra $\mathcal{SO}(3)$. In fact,

$$J^3 = -iJ_1^2 , \quad J^1 = -iJ_2^3 , \quad J^2 = -iJ_3^1 . \tag{22.40}$$

These are precisely the angular momentum operators studied in Chapter 4. We can now write [in analogy to (4.21)]

$$\boxed{R_{\hat{n}}(\theta) = \exp(-i\theta \hat{n} \cdot \boldsymbol{J})} , \tag{22.41}$$

where $\boldsymbol{J} = (J^1, J^2, J^3)$ are hermitian operators acting on all the irreducible representation spaces $D^{(j)}$ of $SO(3)$ (both tensor and spinor). (In the physics literature, the symbol $\boldsymbol{L}$ is usually reserved for the orbital angular momentum, while $\boldsymbol{J}$ denotes the total angular momentum, orbital and spin.)

The complex-conjugate representation $\overline{D^{(j)}}$ can be shown to be equivalent to $D^{(j)}$. First note that

$$\overline{D^{(j)}}(R_3(\psi)) = \left((D^{(j)})^\dagger\right)^T (R_3(\psi)) = \left((D^{(j)})^{-1}(R_3(\psi))\right)^T$$

$$= \left(D^{(j)}(R_3(-\psi))\right)^T = D^{(j)}(R_3(-\psi)) = D^{(j)}(\exp(-i(-\psi)e_3 \cdot \boldsymbol{J}))$$
$$= D^{(j)}(\exp(-i\psi(-e_3) \cdot \boldsymbol{J})) = D^{(j)}(R_{-e_3}(\psi)) = D^{(j)}(R_2(\pi)R_3(\psi)(R_2(\pi))^{-1})$$
$$= D^{(j)}(R_2(\pi))D^{(j)}(R_3(\psi))(D^{(j)}(R_2(\pi)))^{-1} , \tag{22.42}$$

where the fourth equality follows from the fact that J^3 is diagonal [c.f. (22.39)] and the eighth from Lemma 6.1. From (22.19) and (22.34) we see that $D^{(j)}(R_2(\theta))$ is real (since both a and b are real for this rotation). Hence, similar to (22.42),

$$\overline{D^{(j)}}(R_2(\theta)) = D^{(j)}(R_2(\theta)) = D^{(j)}(R_2(\pi))D^{(j)}(R_2(\theta))(D^{(j)}(R_2(\pi)))^{-1}, \tag{22.43}$$

where the last equality follows from the fact that $R_2(\pi)$ does not rotate e_2. Then it follows from (22.42) and (22.43) that

$$\overline{D^{(j)}}(R_3(\phi)R_2(\theta)R_3(\psi)) = \overline{D^{(j)}}(R_3(\phi))\overline{D^{(j)}}(R_2(\theta))\overline{D^{(j)}}(R_3(\psi))$$
$$= D^{(j)}(R_2(\pi))D^{(j)}(R_3(\phi))D^{(j)}(R_2(\theta))D^{(j)}(R_3(\psi))(D^{(j)}(R_2(\pi)))^{-1} \tag{22.44}$$
$$= D^{(j)}(R_2(\pi))D^{(j)}(R_3(\phi)R_2(\theta)R_3(\psi))(D^{(j)}(R_2(\pi)))^{-1}.$$

In the physics literature the following notation is often used:

$$d^{(j)}(\theta) \equiv D^{(j)}(R_2(\theta)), \tag{22.45}$$

and the matrices $D^{(j)}$ are called **rotation matrices**. Using (22.35), $D^{(j)}$ can be written in terms of $d^{(j)}(\theta)$ as

$$\boxed{\begin{array}{c}(D^{(j)}(R_3(\phi)R_2(\theta)R_3(\psi)))^m_{m'} = \langle jm | D^{(j)}(R) | jm' \rangle \\ = e^{-im\phi} \left(d^{(j)}(\theta)\right)^m_{m'} e^{-im'\psi}\end{array}}, \tag{22.46}$$

where, according to (22.19),

$$\boxed{\begin{array}{c}\left(d^{(j)}(\theta)\right)^m_{m'} = \sum_{p=\underline{p}}^{\overline{p}} \dfrac{(-1)^p \sqrt{(j+m)!(j-m)!(j+m')!(j-m')!}}{p!\,(j+m-p)!(j-m'-p)!(p+m'-m)!} \\ \times (\cos(\theta/2))^{2j+m-m'-2p}\,(\sin(\theta/2))^{2p+m'-m}\end{array}}, \tag{22.47}$$

and the range of p in the sum is given by (22.18), (which is equivalent to the requirement that the factorials in the equation have meaning as positive integer factorials).

Exercise 22.7 Prove the following symmetry relations for $d^{(j)}(\theta)$:

$$\left(d^{(j)}(\theta)\right)^m_{m'} = \left(d^{(j)}(-\theta)\right)^{m'}_{m} = \left(d^{(j)}(\theta)\right)^{-m}_{-m'}(-1)^{m-m'}. \tag{22.48}$$

We would now like to use the properties of group characters to further study the irreducible representations of $SO(3)$. Recall the orthonormality theorem for group characters (Theorem 18.6), which is stated with reference to finite groups.

For infinite compact groups [of which $SU(2)$ is an example], one needs to define an integral over the group manifold in place of the sum over group elements. The required integration measure has to generalize the following finite-group invariance requirement:

$$\sum_{g \in G} f(g) = \sum_{g \in G} f(h^{-1}g) \tag{22.49}$$

for any $h \in G$, where $f : G \to \mathbb{C}$ is a continuous function on G. Thus

$$\int_G dg\, f(g) = \int_G dg\, f(h^{-1}g) . \tag{22.50}$$

This measure is called a **Haar measure**. Its definition is given most succinctly in terms of differential forms (see Chapter 30): a Haar measure for an n-dimensional group manifold is an invariant n-form Ω (a volume form) on G. A form Ω is said to be invariant if it satisfies

$$(L_{h^{-1}})^*\Omega = \Omega , \tag{22.51}$$

for all $h \in G$, where $L_{h^{-1}} : G \to G$ is the left multiplication by h^{-1}, and $(L_h)^*$ is the pullback map induced by L_h. These differential geometric concepts will be explained in detail in later chapters. For now we will just give the result for $SO(3)$

$$\Omega = \frac{1}{8\pi^2} \sin\theta d\theta \wedge d\phi \wedge d\psi , \tag{22.52}$$

where $\wedge$ is the exterior product (see Chapter 28), and $0 \le \theta \le \pi$, $0 \le \phi, \psi \le 2\pi$, are the angles in (22.27). The constant $1/(8\pi^2)$ is chosen so that the group volume is normalized to unity:

$$\int_G \Omega = 1 . \tag{22.53}$$

The generalization of Theorem 18.6 to the case of compact groups is then given by

Theorem 22.1 (Orthonormality of Characters of a Compact Group).

$$\int_G \Omega\, \overline{\chi_{(\mu)}(g)}\, \chi_{(\nu)}(g) = \delta_{\mu\nu} . \tag{22.54}$$

We will not prove this theorem here. In the above equation μ and ν are representation indices. Thus the character $\chi_{(\mu)}(g)$ for any representation is a square-integrable class function (one which depends only on the equivalence class):

$$\int_G \Omega |\chi_{(\mu)}(g)|^2 = 1 . \tag{22.55}$$

In fact, we have the following important result (stated without proof):

Theorem 22.2. *The characters* $\chi_{(\mu)}(g)$ *for the irreducible representations of a compact group G form an orthonormal basis for the Hilbert space of square-integrable class functions.*

At this point, it is worthwhile to state several basic and useful results on the irreducible representations of compact Lie groups. These are all generalizations on the corresponding theorems for finite groups and depend on the notion of the Haar measure. (The proofs of these theorems are beyond the scope of this book.)

Theorem 22.3. *Every continuous, bounded representation of a compact Lie group is equivalent to a unitary representation.*

Theorem 22.4. *All continuous unitary irreducible representations of a compact Lie group are finite dimensional.*

Theorem 22.5 (Orthonormality of Irreducible Representations of a Compact Group).

$$N_{(\mu)} \int_G \Omega \, (D_{(\mu)}^\dagger(g))_i^k \, (D_{(\nu)}(g))_l^j = \delta_{\mu\nu} \, \delta_i^j \, \delta_l^k \, . \tag{22.56}$$

Theorem 22.6 (Peter-Weyl Theorem). *The Hilbert space of square-integrable functions on a compact group G, denoted by $L^2(G)$, decomposes into a direct sum of irreducible representations of G, each of which is finite dimensional. Equivalently, any $f(g) \in L^2(G)$ can be expanded as*

$$f(g) = \sum_{\mu, \, i, j} (C_{(\mu)})_i^j \, (D_{(\mu)}(g))_i^j \, , \tag{22.57}$$

where μ labels the irreducible representations and $(C_{(\mu)})_i^j \in \mathbb{C}$; and

$$\int_G \Omega \, | \, f(g) \, |^2 = \sum_{\mu, \, i, j} | \, (C_{(\mu)})_i^j \, |^2 \, . \tag{22.58}$$

As an example on the use of the above theorems, consider the group $U(1)$ (the circle group):

$$U(1) = \{ e^{i\theta} \, | \, 0 \le \theta \le 2\pi \} \, , \tag{22.59}$$

whose action on $z \in \mathbb{C}$ is given by

$$z \to z' = e^{i\theta} \, z \, . \tag{22.60}$$

This group is obviously compact, and is also abelian. Hence, by Corollary 18.1, its irreducible representations are all one-dimensional. These are given by

$$D_{(\mu)}(\theta) = e^{i\mu\theta} \, , \tag{22.61}$$

where $\mu = 0, \pm 1, \pm 2, \ldots$. The integral values for μ are required by the fact that

$$D_{(\mu)}(\theta + 2\pi) = D_{(\mu)}(\theta) \, .$$

The invariant volume element (Haar measure) is simply

$$\Omega = \frac{1}{2\pi}\, d\theta \; . \tag{22.62}$$

Thus the orthonormality of representations theorem (Theorem 22.5) applied to $U(1)$ gives

$$\frac{1}{2\pi} \int_0^{2\pi} d\theta \, e^{-i\mu\theta} e^{i\nu\theta} = \delta_{\mu\nu} \; , \tag{22.63}$$

where $\mu, \nu = 0, \pm 1, \pm 2, \dots$. This is a familiar result in the theory of Fourier series [c.f. (14.5)]. Since all $D_{(\mu)}(\theta)$ are one-dimensional, $\chi_{(\mu)}(\theta) = D_{(\mu)}(\theta)$, and the orthonormality of characters theorem yields exactly the same result. Finally, the Peter-Weyl Theorem applied to $U(1)$ gives the central result of Fourier series theory:

$$f(\theta) = \sum_{n=-\infty}^{\infty} f_n \frac{e^{in\theta}}{\sqrt{2\pi}} \; ,$$

[c.f. (14.7)], where $f(\theta) \in L^2(U(1))$.

Let us now compute the irreducible characters $\chi_{(\mu)}$ of $SO(3)$. We first note that (c.f. Exercise 6.3) all rotations by the same angle ϕ (about any axis) belong to the same equivalence class in $SO(3)$. Thus $\chi_{(\mu)}$ is a function of ϕ only. The representation index μ is given by $j = 0, 1/2, 1, 3/2, \dots$. Let $\mathbb{V}^{(j)}$, then, be the irreducible representation space, and $D^{(j)}$ be the unitary representation (of dimension $2j + 1$), corresponding to a particular j. By our earlier results, $\mathbb{V}^{(j)}$ decomposes into a direct sum of $(2j + 1)$ one-dimensional eignespaces of $D^{(j)}$. Indeed, by (22.38), the $2j + 1$ eigenvalues are given by

$$e^{-ij\phi}, e^{-i(j-1)\phi}, \dots, e^{-i(-j+1)\phi}, e^{-i(-j)\phi},$$

for some $0 \le \phi \le \pi$. Hence

$$\chi^{(j)}(\phi) = \sum_{m=-j}^{j} e^{-im\phi} \; . \tag{22.64}$$

For example,

$$\begin{aligned}
\chi^{(0)}(\phi) &= 1 \; , \\
\chi^{(1/2)}(\phi) &= e^{-i\phi/2} + e^{i\phi/2} = 2\cos(\phi/2) \; , \\
\chi^{(1)}(\phi) &= e^{-i\phi} + e^{i\phi} + 1 = 1 + 2\cos\phi \; .
\end{aligned} \tag{22.65}$$

Exercise 22.8 Show that

$$\chi^{(j)}(\phi) = \frac{\sin(j + \frac{1}{2})\phi}{\sin\frac{\phi}{2}} \; , \tag{22.66}$$

where $j = 0, 1/2, 1, 3/2, \ldots$.

Exercise 22.9   Verify Theorem 22.4 directly for $j = 1/2$ using the explicit expressions for $D^{(1/2)}(\phi, \theta, \psi)$ given by (22.27).

Exercise 22.10   Verify Theorem 22.5 directly for $j = 1/2$ using (22.27).

We can use our formulas for the characters χ_j to determine the Clebsch-Gordon decomposition [c.f. (18.75)] of the product representation $D^{(j_1)} \times D^{(j_2)}$. By (18.76),

$$\chi^{(j_1 \times j_2)} = \chi^{(j_1)} \chi^{(j_2)} . \tag{22.67}$$

Writing $z = e^{i\phi}$, it follows from (22.64) that

$$\chi^{(j)} = z^{-j} + z^{-j+1} + \cdots + z^{j} = \frac{z^{-j}(1 - z^{2j+1})}{1 - z} = \frac{z^{j+1} - z^{-j}}{z - 1} . \tag{22.68}$$

For $j_1 \leq j_2$, we then have

$$\chi^{(j_1)}(z)\chi^{(j_2)}(z) = (z^{-j_1} + \cdots + z^{j_1}) \left(\frac{z^{j_2+1} - z^{-j_2}}{z - 1} \right)$$

$$= \frac{z^{j_1+j_2+1} - z^{-(j_1+j_2)}}{z - 1} + \cdots + \frac{z^{j_2-j_1+1} - z^{-(j_2-j_1)}}{z - 1} \tag{22.69}$$

$$= \chi^{(j_1+j_2)} + \chi^{(j_1+j_2-1)} + \cdots + \chi^{(j_2-j_1)} .$$

For any values of j_1 and j_2, we thus have

$$\boxed{D^{(j_1)} \times D^{(j_2)} = \sum_{j=|j_1-j_2|}^{j_1+j_2} D^{(j)}} . \tag{22.70}$$

For example

$$D^{(\frac{1}{2})} \times D^{(\frac{1}{2})} = D^{(0)} + D^{(1)} ,$$

$$D^{(\frac{1}{2})} \times D^{(1)} = D^{(\frac{1}{2})} + D^{(\frac{3}{2})} ,$$

$$D^{(1)} \times D^{(1)} = D^{(0)} + D^{(1)} + D^{(2)} , \tag{22.71}$$

$$D^{(1)} \times D^{(2)} = D^{(1)} + D^{(2)} + D^{(3)} , \text{ etc.}$$

Eq. (22.70) is the central result in the theory of the addition of angular momentum in quantum mechanics.

Chapter 23

The Spherical Harmonics

The spherical harmonics are functions defined on the unit sphere (in $\mathbb{R}^3$), and arise as solutions to the **Laplace's equation**:

$$\nabla^2 \psi(\boldsymbol{r}) \equiv \left(\frac{\partial^2}{\partial x^2} + \frac{\partial^2}{\partial y^2} + \frac{\partial^2}{\partial z^2} \right) \psi(\boldsymbol{r}) = 0 . \tag{23.1}$$

We will see in this chapter that they are also basis vectors in certain irreducible representation spaces of $SO(3)$, and in fact constitute a basis for square-integrable functions defined on the unit sphere. We begin with the irreducible representations of $SU(2)$.

Consider

$$g(\theta) = \begin{pmatrix} e^{-i\theta} & 0 \\ 0 & e^{i\theta} \end{pmatrix} \in SU(2) .$$

Our previous results [(22.35)] show that

$$\left(D_{(j)}(g(\theta)) \right)^m_{m'} = \delta_{mm'} \, e^{-2im\theta} , \tag{23.2}$$

where $m, m' = j, j-1, \ldots, -j$. The factor 2θ occurs in the exponent because under the group homomorphism $\phi : SU(2) \to SO(3)$ given by (11.19), $g(\theta)$ corresponds to the rotation $R_3(2\theta)$ [c.f. (11.33)]. It is clear that $\exp(2ij\theta)$ is an eigenvalue of $D_{(j)}(g(\theta))$, which is a $(2j+1)$-dimensional representation of $SU(2)$. One can in fact establish the following fact, which will be stated without proof (S. Sternberg 1995).

Lemma 23.1. *If D is a $(2j+1)$-dimensional representation of $SU(2)$ such that $\exp(2ij\theta)$ occurs as an eigenvalue of $D(g(\theta))$, where $j = 0, 1/2, 1, 3/2, \ldots$, and $g(\theta) \in SU(2)$ is given above, then D is equivalent to the irreducible representation $D_{(j)}$.*

Now let us recall, by (22.25) and the discussion immediately following, that if j is an integer ≥ 0, then for any $g \in SU(2)$,

$$D_{(j)}(g) = D_{(j)}(-g) . \tag{23.3}$$

247

Since $\phi(g) = \phi(-g) \in SO(3)$, where $\phi : SU(2) \to SO(3)$ is the homomorphism of (11.9), the representations for $SO(3)$ defined by

$$(D')_{(j)}(R) \equiv D_{(j)}(\phi^{-1}(R)) \,, \tag{23.4}$$

for $j = 0, 1, 2, \ldots, R \in SO(3)$, furnish all the irreducible representations of $SO(3)$. Following the practice in the physics literature, we will still label the $(D')_{(j)}$ by $D_{(j)}$ for non-negative integral values of j in our subsequent developments.

It is easily seen that the Laplacian operator ∇^2 commutes with all Euclidean motions (rotations plus translations), so that it commutes with all rotations in particular. Thus, for a representation $D : SO(3) \to \mathcal{U}(L^2(S^2))$, [where $L^2(S^2)$ is the space of all square-integrable functions on the unit sphere S^2, and $\mathcal{U}(\mathcal{H})$ is a group of unitary transformations on the Hilbert space $\mathcal{H}$], such that, for all $R \in SO(3)$,

$$D(R)\psi(\boldsymbol{r}) = \psi(R^{-1}\boldsymbol{r}) \,, \tag{23.5}$$

[c.f. (18.6)], we have

$$(\nabla^2 D(R))\psi(\boldsymbol{r}) = D(R)\nabla^2\psi(\boldsymbol{r}) \,. \tag{23.6}$$

Writing ∇^2 in terms of spherical polar coordinates [see (31.45)]:

$$\nabla^2 = \frac{\partial^2}{\partial r^2} + \frac{2}{r}\frac{\partial}{\partial r} + \frac{1}{r^2}\nabla_S^2 \,, \tag{23.7}$$

where

$$\nabla_S^2 \equiv \frac{1}{\sin\theta}\frac{\partial}{\partial\theta}\sin\theta\frac{\partial}{\partial\theta} + \frac{1}{\sin^2\theta}\frac{\partial^2}{\partial\phi^2} \,, \tag{23.8}$$

we further see that each of the three terms in (23.7) commutes with $D(R)$. In particular, the **spherical Laplacian** ∇_S^2 commutes with $D(R)$, for all $R \in SO(3)$:

$$(\nabla_S^2 D(R))\psi(\boldsymbol{r}) = D(R)\nabla_S^2\psi(\boldsymbol{r}) \,. \tag{23.9}$$

Exercise 23.1   Show explicitly that $\nabla^2, \partial^2/\partial r^2, (1/r)(\partial/\partial r)$ and ∇_S^2 all commute with $D(R)$, for all $R \in SO(3)$, where the action of $D(R)$ is given by (23.5).

Let P^k denote the space of homogeneous polynomials of degree k with complex coefficients on $\mathbb{R}^3$. For example,

$$P^0 = span\,\{1\} \,, \quad P^1 = span\,\{x, y, z\} \,, \quad P^2 = span\,\{x^2, y^2, z^2, xy, yz, xz\} \,.$$

In general, a $p^k(x, y, z) \in P^k$ can be written

$$p^k(x, y, z) = \sum_{i=0}^{k} p^i(x, y)z^{k-i} \,, \tag{23.10}$$

where $p^i(x, y)$ is a complex-valued homogeneous polynomial of degree i on $\mathbb{R}^2$. It is quite obvious from the binomial expansion

$$(x + y)^i = \sum_{r=0}^{i} \binom{i}{r} x^r y^{i-r} \tag{23.11}$$

that p^i belongs to a space of dimension $i + 1$. Thus (23.10) implies that

$$dim\,(P^k) = 1 + 2 + \cdots + k + (k+1) = \frac{(k+1)(k+2)}{2}\, . \tag{23.12}$$

The space P^k, being invariant under $SO(3)$, constitutes a representation space of $SO(3)$. It is, however, not irreducible. In fact, consider a subspace $H^k \subset P^k$ defined by

$$H^k \equiv ker(\nabla^2)\, . \tag{23.13}$$

H^k is called the space of **harmonic polynomials** of degree k. It is of fundamental importance in analysis and geometry, in addition to its frequent occurrence in physics applications. We have the following basic theorem.

Theorem 23.1. *The space H^k of complex-valued harmonic polynomials of degree k on $\mathbb{R}^3$ furnishes an irreducible representation of $SO(3)$ which is equivalent to $D_{(k)}$ [c.f. (23.4) and the remark following it].*

Proof. First note that, by (23.6) and (23.1), H^k is clearly invariant under $SO(3)$. So it furnishes a representation space for $SO(3)$. Denote this representation by $D\,:\,SO(3) \rightarrow \mathcal{U}(H^k)$. We then note that the Laplacian $\nabla^2\,:\,P^k \rightarrow P^{k-2}$ is a surjective map, that is, $P^{k-2} = Im(\nabla^2)$. It follows from Theorem 17.3 [Eq. (17.10)] and (23.12) that

$$dim(H^k) = dim(ker(\nabla^2)) = dim(P^k) - dim(P^{k-2}) = 2k + 1\, . \tag{23.14}$$

Now

$$\frac{\partial^2}{\partial x^2} + \frac{\partial^2}{\partial y^2} = \left(\frac{\partial}{\partial x} - i\frac{\partial}{\partial y}\right)\left(\frac{\partial}{\partial x} + i\frac{\partial}{\partial y}\right)\, , \tag{23.15}$$

and

$$\left(\frac{\partial}{\partial x} + i\frac{\partial}{\partial y}\right)(x + iy)^k = k(x + iy)^{k-1}\left(\frac{\partial}{\partial x} + i\frac{\partial}{\partial y}\right)(x + iy) = 0\, . \tag{23.16}$$

Hence

$$\nabla^2(x + iy)^k = 0\, , \tag{23.17}$$

implying

$$(x + iy)^k \in H^k\, . \tag{23.18}$$

On the other hand, from (11.33) and the discussion following (11.32), $(x + iy)^k$ is an eigenvector of $D(g(\theta))$ with eigenvalue $\exp(2ik\theta)$. It follows from Lemma 23.1 that D is equivalant to the irreducible representation $D_{(k)}$. $\quad\square$

To further reduce P^k into irreducible components, we need the following fact, which will be stated without proof.

Lemma 23.2. For $l = 0, 1, 2, \ldots,$

$$P^l = H^l \oplus r^2 P^{l-2} , \tag{23.19}$$

where $r^2 = x^2 + y^2 + z^2$.

Special cases of the above Lemma are:

$$P^0 = H^0 = \text{constants} \in \mathbb{C} , \quad P^1 = H^1 , \quad P^2 = H^2 \oplus \{r^2\} . \tag{23.20}$$

The last equation follows from the fact that a general polynomial in P^2 can always be expressed as

$$ax^2 + by^2 + cz^2 + dxy + exz + fyz = \left(\frac{a+b+c}{3}\right)(x^2 + y^2 + z^2)$$
$$+ \left\{a - \left(\frac{a+b+c}{3}\right)\right\}x^2 + \left\{b - \left(\frac{a+b+c}{3}\right)\right\}y^2 + \left\{c - \left(\frac{a+b+c}{3}\right)\right\}z^2$$
$$+ dxy + exz + fyz ,$$
$$\tag{23.21}$$

where $a, b, c, d, e, f \in \mathbb{C}$, and

$$H^2 = \{ax^2 + by^2 + cz^2 + dxy + exz + fyz \mid a + b + c = 0\} . \tag{23.22}$$

As a result of Lemma 23.1, P^l decomposes into irreducible subspaces as follows:

$$P^l = H^l \oplus r^2 H^{l-2} \oplus r^4 H^{l-4} \oplus \cdots . \tag{23.23}$$

Since the harmonic functions H^l are all homogeneous polynomials of degree l, any function $f^{(l)}(x, y, z) \in H^l$ can be written in terms of spherical polar coordinates as

$$f^{(l)}(x, y, z) = r^l \, Y_l(\theta, \phi) . \tag{23.24}$$

The functions $Y_l(\theta, \phi)$ are called **spherical harmonics** of degree l, and are defined on the unit sphere. Using (23.7) and (23.8), we have

$$0 = \nabla^2 f^{(l)} = r^{l-2}\{\nabla_S^2 Y_l + l(l-1)Y_l + 2lY_l\} , \tag{23.25}$$

which implies

$$\nabla_S^2 Y_l = -l(l+1)Y_l . \tag{23.26}$$

Let $\tilde{H}^l$ denote the space generated by the spherical harmonics Y_l. From (23.14),

$$dim(\tilde{H}^l) = 2l + 1 . \tag{23.27}$$

A standard basis of $\tilde{H}^l$ consists of eigenvectors of $D_{(l)}(g(\phi/2))$, with $g(\phi)$ given by (23.1), or $D_{(l)}(R_3(\phi))$ [c.f. (22.38)]. For $l = 1$, these are easily seen to be proportional to $x + iy$, $x - iy$ and z, with eigenvalues $\exp(2i\theta)$, $\exp(-2i\theta)$ and 1, respectively. We choose the following functions, Y_1^m, $m = 1, 0, -1$, that are L^2-normalized to 1 when integrated over the unit sphere:

$$Y_1^1 = -\sqrt{\frac{3}{8\pi}} \left(\frac{x + iy}{r}\right) , \qquad Y_1^0 = \sqrt{\frac{3}{4\pi}} \frac{z}{r} , \qquad Y_1^{-1} = \sqrt{\frac{3}{8\pi}} \left(\frac{x - iy}{r}\right) . \tag{23.28}$$

These form a basis for $\tilde{H}^1$, the space of spherical harmonics of degree 1. For arbitrary l $(= 0, 1, 2, \dots)$, $\tilde{H}^l$ is generated by the normalized functions Y_l^m, $m = l, l-1, \dots, -l$, which are proportional to

$$\frac{(x + iy)^l}{r^l} , \quad \frac{(x - iy)^l}{r^l} , \quad \frac{z^l}{r^l} , \quad \frac{(x + iy)^{l-1}(x - iy)}{r^l} , \quad \dots .$$

Recall the rotation matrices $D_{(l)}(\phi, \theta, \psi)$ given by (22.46). The spherical harmonics $Y_l^m(\theta, \phi)$ can in fact be related to them by the following:

$$\boxed{Y_l^m(\theta, \phi) = (-1)^m \sqrt{\frac{2l + 1}{4\pi}} \overline{(D_{(l)}(\phi, \theta, 0))_0^m}} , \tag{23.29}$$

where the overbar means complex conjugation. Related functions are the **associated Legendre polynomials** $P_l^m(\cos\theta)$ and the **Legendre polynomials** $P_l(\cos\theta)$ given by

$$P_l^m(\cos\theta) = \sqrt{\frac{(l + |m|)!}{(l - |m|)!}} \, (d_{(l)}(\theta))_0^m , \tag{23.30}$$

$$P_l(\cos\theta) = P_l^0(\cos\theta) = (d_{(l)}(\theta))_0^0 , \tag{23.31}$$

where $(d_{(l)}(\theta))_{m'}^m$ is given by (22.47). It follows from (22.46) that

$$Y_l^m(\theta, \phi) = (-1)^m \sqrt{\frac{(2l + 1)(l - |m|)!}{4\pi(l + |m|)!}} \, P_l^m(\cos\theta) \, e^{im\phi} . \tag{23.32}$$

The P_l^m and P_l are also related as follows:

$$\begin{aligned}
P_l^m(\cos\theta) &= (1 - \cos^2\theta)^{|m|/2} \frac{d^{|m|}}{d(\cos\theta)^{|m|}} \, P_l(\cos\theta) \\
&= \frac{(1 - \cos^2\theta)^{|m|/2}}{2^l \, l!} \frac{d^{l+|m|}}{d(\cos\theta)^{l+|m|}} \, (\cos^2\theta - 1)^l .
\end{aligned} \tag{23.33}$$

It is easy to show that

$$Y_l^{-m}(\theta, \phi) = (-1)^m \overline{Y_l^m(\theta, \phi)} , \tag{23.34}$$

and

$$Y_l^m(\pi - \theta, \phi + \pi) = (-1)^l \, Y_l^m(\theta, \phi) ,$$

that is,

$$Y_l^m(-\hat{n}) = (-1)^l\, Y_l^m(\hat{n}) . \tag{23.35}$$

In physics terminology, the wave function Y_l^m is said to have a **parity** of $(-1)^l$ under inversion of coordinates.

We list below the first few (low-order) Y_l^m's for reference.

$$Y_0^0 = \frac{1}{\sqrt{4\pi}} ,$$

$$Y_1^0 = \sqrt{\frac{3}{4\pi}}\, \cos\theta , \qquad\qquad Y_1^1 = -\sqrt{\frac{3}{8\pi}}\, \sin\theta\, e^{i\phi} ,$$

$$Y_2^0 = \sqrt{\frac{5}{16\pi}}\,(3\cos^2\theta - 1) , \qquad\qquad Y_2^1 = -\sqrt{\frac{15}{8\pi}}\, \sin\theta\,\cos\theta\, e^{i\phi} ,$$

$$Y_2^2 = \sqrt{\frac{15}{32\pi}}\, \sin^2\theta\, e^{2i\phi} ,$$

$$Y_3^0 = \sqrt{\frac{7}{16\pi}}\,(5\cos^3\theta - 3\cos\theta) ,$$

$$Y_3^1 = -\sqrt{\frac{21}{64\pi}}\, \sin\theta\,(5\cos^2\theta - 1)\, e^{i\phi} ,$$

$$Y_3^2 = \sqrt{\frac{105}{32\pi}}\, \sin^2\theta\, \cos\theta\, e^{i\phi} ,$$

$$Y_3^3 = -\sqrt{\frac{35}{64\pi}}\, \sin^3\theta\, e^{3i\phi} . \tag{23.36}$$

Exercise 23.2   Show that Y_2^m and Y_3^m as given above can be constructed from homogeneous polynomials arising from powers of $x + iy, x - iy$ and z.

The spherical harmonics $Y_l^m(\theta,\phi)$ form the angular part of wave functions of quantum mechanical systems that are eigenfunctions of rotationally invariant Hamiltonians. Let us now consider a unitary representation of $SO(3)$ on a rigged Hilbert space $\mathcal{H}_R$ (consisting of normalizable as well as non-normalizable states), $U_R : SO(3) \to \mathcal{U}(\mathcal{H}_R)$ (c.f. discussion in Chapter 18 following Exercise 18.1), and a rotationally invariant Hamiltonian H acting on states in $\mathcal{H}_R$. Rotational invariance of H implies

$$[H, U_R(g)] = 0, \qquad \text{for all } g \in SO(3) . \tag{23.37}$$

Equivalently,

$$[H, J^i] = 0, \qquad i = 1, 2, 3, \tag{23.38}$$

where J^i is the i-th component of the angular momentum operator [recall (22.41)]. The set of orthonormal eigenstates of H are labeled in the Dirac notation by $|\,Elm\,\rangle$, assuming that the complete set of commuting observables is $\{H, \boldsymbol{J} \cdot \boldsymbol{J}, J^3\}$. They satisfy

$$H\,|\,Elm\,\rangle = E\,|\,Elm\,\rangle\,, \tag{23.39}$$

$$\boldsymbol{J} \cdot \boldsymbol{J}\,|\,Elm\,\rangle = l(l+1)\,|\,Elm\,\rangle\,, \tag{23.40}$$

$$J^3\,|\,Elm\,\rangle = m\,|\,Elm\,\rangle\,, \tag{23.41}$$

where $l = 0, 1, 2, \dots\,; m = l, l-1, \dots, -l$. For a given energy eigenvalue E, these eigenstates form an irreducible representation space of $SO(3)$, namely, the $D_{(l)}$ representation.

The Schrödinger wave function for $|\,Elm\,\rangle$ is [c.f. (15.11)]

$$\psi_{Elm}(\boldsymbol{x}) = \langle\,\boldsymbol{x}\,|\,Elm\,\rangle = \langle\,r, \theta, \phi\,|\,Elm\,\rangle\,. \tag{23.42}$$

Now

$$\begin{aligned}
|\,\boldsymbol{x}\,\rangle \equiv |\,r, \theta, \phi\,\rangle &= |\,R_3(\phi)R_2(\theta)r\hat{z}\,\rangle \\
&= \exp(-i\phi J^3)\exp(-i\theta J^2)\,|\,r, 0, 0\,\rangle = U_R(\phi, \theta, 0)\,|\,r, 0, 0\,\rangle\,. \tag{23.43}
\end{aligned}$$

Hence

$$\psi_{Elm}(\boldsymbol{x}) = \langle\,r, 0, 0\,|\,U_R^\dagger(\phi, \theta, 0)\,|\,Elm\,\rangle = (D_{(l)}^\dagger(\phi, \theta, 0))_m^{m'}\,\langle\,r, 0, 0\,|\,Elm'\,\rangle\,. \tag{23.44}$$

Rotation about the $\hat{z}$-axis must leave $|\,r\hat{z}\,\rangle$ invariant. Thus

$$\exp(-i\alpha J^3)\,|\,r, 0, 0\,\rangle = |\,r, 0, 0\,\rangle\,, \tag{23.45}$$

or

$$J^3\,|\,r, 0, 0\,\rangle = 0\,. \tag{23.46}$$

Eq. (22.42) then implies

$$|\,r, 0, 0\,\rangle = f_{El}(r)\,|\,El0\,\rangle\,, \tag{23.47}$$

where $f_{El}(r)$ is some (normalizable) function of r. We thus have

$$\langle\,r, 0, 0\,|\,Elm\,\rangle = f_{El}(r)\,\delta_{m0}\,, \tag{23.48}$$

and from (23.44),

$$\psi_{Elm}(\boldsymbol{x}) = f_{El}(r)\overline{(D_{(l)}(\phi, \theta, 0))_0^m}\,. \tag{23.49}$$

This translates, in view of (23.29), to

$$\boxed{\psi_{Elm}(\boldsymbol{x}) = \psi_{El}(r)(-1)^m Y_l^m(\theta, \phi)}\,, \tag{23.50}$$

where

$$\psi_{El}(r) \equiv \sqrt{\frac{4\pi}{2l+1}}\,f_{El}(r)\,. \tag{23.51}$$

The above general form of $\psi_{Elm}(\boldsymbol{x})$ is a direct consequence of the rotational symmetry of the Hamiltonian, and is independent of the details of the potential function $V(r)$,

Under a rotation $R \in SO(3); |\, Elm \,\rangle$ transforms according to

$$| \psi' \,\rangle = U(R) \,|\, Elm \,\rangle = \sum_{m'} (D_{(l)}(R))^{m'}_m \,|\, Elm' \,\rangle \,. \qquad (23.52)$$

Hence

$$\psi'(\boldsymbol{x}) = \langle\, \boldsymbol{x} \,|\, \psi' \,\rangle = \sum_{m'} (D_{(l)}(R))^{m'}_m \, \langle\, \boldsymbol{x} \,|\, Elm' \,\rangle$$

$$= \sum_{m'} (D_{(l)}(R))^{m'}_m \, \psi_{Elm'}(\boldsymbol{x}) = \sum_{m'} (D_{(l)}(R))^{m'}_m \, \psi_{El}(r)(-1)^{m'} \, Y_l^{m'}(\theta, \phi) \,.$$

$$(23.53)$$

On the other hand,

$$\psi'(\boldsymbol{x}) = \psi(R^{-1}\boldsymbol{x}) = \psi_{El}(r)(-1)^m Y_l^m (R^{-1}\hat{x}) \,, \qquad (23.54)$$

whence

$$Y_l^m (R^{-1}\hat{x}) = \sum_{m'} (-1)^{m+m'} \, (D_{(l)}(R))^{m'}_m \, Y_l^{m'}(\hat{x}) \,. \qquad (23.55)$$

The wave function $\psi_{Elm}(\boldsymbol{x})$ describes the state of a spinless particle ($j = 0$). For a particle with spin ($j = 1/2, 1, 3/2, 2, \dots$), the rigged Hilbert space $(\mathcal{H}')_R$ consists of states of the form $|\, \boldsymbol{x}, \sigma \,\rangle, \sigma = j, j-1, \dots, -j$, satisfying the completeness condition

$$\sum_{\sigma=-j}^{j} \int d^3x \,|\, \boldsymbol{x}, \sigma \,\rangle\langle\, \boldsymbol{x}, \sigma \,| = 1 \,. \qquad (23.56)$$

The unitary representation of $SO(3)$ on this space is given by

$$U(R) \,|\, \boldsymbol{x}, \sigma \,\rangle = \sum_{\lambda} (D_{(j)}(R))^{\lambda}_{\sigma} \,|\, R\boldsymbol{x}, \lambda \,\rangle \,. \qquad (23.57)$$

An arbitrary $|\, \psi \,\rangle \in (\mathcal{H}')_R$ can be expressed, on using (23.56), as

$$|\, \psi \,\rangle = \sum_{\sigma} \int d^3x \,|\, \boldsymbol{x}, \sigma \,\rangle\langle\, \boldsymbol{x}, \sigma \,|\, \psi \,\rangle = \sum_{\sigma} \int d^3x \,|\, \boldsymbol{x}, \sigma \,\rangle \, \psi^{\sigma}(\boldsymbol{x}) \,, \qquad (23.58)$$

where

$$\psi^{\sigma}(\boldsymbol{x}) \equiv \langle\, \boldsymbol{x}, \sigma \,|\, \psi \,\rangle \,, \qquad \sigma = j, j-1, \dots, -j, \qquad (23.59)$$

are called the **irreducible field components** of spin j. For $j = 1/2$, the two-component object $(\psi^{\frac{1}{2}}(\boldsymbol{x}), \psi^{-\frac{1}{2}}(\boldsymbol{x}))$ is called the **Pauli spin wave function** in non-relativistic quantum mechanics. For arbitrary j, we have, under a rotation $R \in SO(3)$,

$$| \psi' \rangle = U(R) | \psi \rangle = \sum_\sigma \int d^3x \, U(R) | \boldsymbol{x}, \sigma \rangle \, \psi^\sigma (\boldsymbol{x})$$

$$= \sum_{\sigma\lambda} \int d^3x \, | R\boldsymbol{x}, \lambda \rangle (D_{(j)}(R))_\sigma^\lambda \, \psi^\sigma (\boldsymbol{x}) \qquad (23.60)$$

$$= \sum_\lambda \int d^3x \, | \boldsymbol{x}, \lambda \rangle \sum_\sigma (D_{(j)}(R))_\sigma^\lambda \, \psi^\sigma (R^{-1}\boldsymbol{x}) \, .$$

Thus the transformation rule under a rotation $R \in SO(3)$ for the irreducible field components of spin j is given by

$$\boxed{(\psi')^\lambda (\boldsymbol{x}) = \sum_{\sigma=-j}^{j} (D_{(j)}(R))_\sigma^\lambda \, \psi^\sigma (R^{-1}\boldsymbol{x})} \quad , \qquad (23.61)$$

where $D_{(j)}(R)$ is given by (22.46).

Chapter 24

The Structure of
Semisimple Lie Algebras

The semisimple Lie algebras (Def. 7.11) play a fundamental role in the classification of Lie algebras [c.f. Theorem 7.1 (Levi's Decomposition Theorem)]. Their representations, as those of Lie groups, are also of great importance in physics applications.

Let $\mathcal{G}$ be an n-dimensional semisimple Lie algebra, with a basis $\{e_i\}, i = 1, \ldots, n$. Suppose $A, X \in \mathcal{G}, X \neq 0$, satisfy

$$Ad_A X = [A, X] = \alpha X, \quad \alpha \in \mathbb{C}. \tag{24.1}$$

X is then an eigenvector of Ad_A (the adjoint representation of A [c.f. (7.15)]) with eigenvalue α. If we write $A = a^i e_i, X = x^i e_i$, then (24.1) becomes

$$a^i x^j [e_i, e_j] = \alpha x^k e_k, \tag{24.2}$$

or, in terms of the structure constants c^i_{jk} with respect to $\{e_i\}$,

$$(a^i c^k_{ij} - \alpha \delta^k_j) x^j = 0. \tag{24.3}$$

Since $X \neq 0$ by assumption, the x^j are not all zero, and we obtain the secular equation

$$\det |a^i c^k_{ij} - \alpha \delta^k_j| = 0. \tag{24.4}$$

To proceed, we need the following lemma (which will be stated without proof).

Lemma 24.1. *Let $\mathcal{G}$ be a semisimple Lie algebra. If $Ad_A, A \in \mathcal{G}$, has the largest possible number of distinct eigenvalues (of all $Ad_Y, Y \in \mathcal{G}$), then among these eigenvalues, only the zero-eigenvalue can be degenerate.*

Definition 24.1. *Suppose $\mathcal{G}$ is a semisimple Lie algebra and $Ad_A, A \in \mathcal{G}$ has the largest possible number of distinct eigenvalues, then the degree of degeneracy m (multiplicity) of the eigenvalue 0 of Ad_A is called the **rank** of the Lie algebra $\mathcal{G}$. We write $rank(\mathcal{G}) = m$. The element A is called a **regular element** of $\mathcal{G}$.*

Let $\mathcal{G}$ be a semisimple Lie algebra and $rank(\mathcal{G}) = m$, $dim(\mathcal{G}) = n$. Then the 0-eigenspace of Ad_A, where A is a regular element of $\mathcal{G}$, is of dimension m. Let $\{H_i\}, i = 1, \ldots, m$, be a basis of this eigenspace, denoted by $\mathcal{G}_0$. Then

$$Ad_A \, H_i = [A, H_i] = 0, \quad i = 1, \ldots, m \, . \tag{24.5}$$

The subspace $\mathcal{G}_0$ forms an m-dimensional subalgebra of $\mathcal{G}$, called the **Cartan subalgebra** of $\mathcal{G}$. As will be seen below, $\mathcal{G}_0$ is the maximal abelian subalgebra in $\mathcal{G}$ such that Ad_H is diagonalizable if $H \in \mathcal{G}_0$. Indeed,

$$[A, [H_i, H_j]] = [A, H_i H_j] - [A, H_j H_i] = [A, H_i]H_j + H_i[A, H_j]$$
$$+ [H_j, A]H_i + H_j[H_i, A] = 0 , \quad (24.6)$$

which implies $[H_i, H_j] \in \mathcal{G}_0$.

Now let α be a non-zero, and thus according to Lemma 24.1, a non-degenerate eigenvalue of Ad_A, with eigenvector E_α:

$$[A, E_\alpha] = \alpha E_\alpha \, . \tag{24.7}$$

Consider an element $[H_i, E_\alpha] \in \mathcal{G}$. We have

$$[A, [H_i, E_\alpha]] = [A, H_i E_\alpha - E_\alpha H_i] = [A, H_i E_\alpha] - [A, E_\alpha H_i]$$
$$= [A, H_i]E_\alpha + H_i[A, E_\alpha] - [A, E_\alpha]H_i - E_\alpha[A, H_i] \tag{24.8}$$
$$= \alpha H_i E_\alpha - \alpha E_\alpha H_i = \alpha[H_i, E_\alpha] \, .$$

Thus $[H_i, E_\alpha]$ is also an eigenvector of Ad_A with eigenvalue α. Since α is non-degenerate, $[H_i, E_\alpha]$ must be proportional to E_α:

$$[H_i, E_\alpha] = \alpha_i E_\alpha, \quad i = 1, \ldots, m \, . \tag{24.9}$$

By Theorem 17.14 (the Primary Decomposition Theorem), $\mathcal{G}$ is the (vector space) direct sum of the eigenspaces of Ad_A. Thus we can write

$$A = a^i H_i + \sum_\alpha a^\alpha E_\alpha \, , \tag{24.10}$$

where the second sum is over all non-zero eigenvalues α; and so, by (24.9),

$$0 = [H_j, A] = a^i[H_j, H_i] + \sum_\alpha a^\alpha[H_j, E_\alpha] = a^i[H_j, H_i] + \sum_\alpha \alpha_j a^\alpha E_\alpha \, . \tag{24.11}$$

Since $[H_j, H_i] \in \mathcal{G}_0$ and $\alpha_j E_\alpha \neq 0$ for each j and α, $a^\alpha = 0$ for all α. Thus

$$A = a^i H_i \in \mathcal{G}_0 \, . \tag{24.12}$$

Furthermore since the a^i can obviously not all be zero, (24.11) also implies that

$$[H_i, H_j] = 0 \, . \tag{24.13}$$

Eqs. (24.7), (24.9) and (24.12) imply that the non-zero eigenvalues of Ad_A are given by

$$\alpha = a^i \alpha_i , \quad i = 1, \ldots, m (= dim(\mathcal{G}_0)) . \tag{24.14}$$

Note that α is in general complex, and is in fact equal to $\alpha^*(a^i H_i)$, where $\alpha^* = \alpha_i (H^*)^i \in \mathcal{G}_0^*$ is an element in the dual space $\mathcal{G}_0^*$ of $\mathcal{G}_0$, and $(H^*)^i \in \mathcal{G}_0^*$ are the dual basis vectors to H_i [given by $(H^*)^i(H_j) = \delta_j^i$]. Each α is called a **root** of the Lie algebra $\mathcal{G}$. (The functionals α^* are also referred to as the roots of $\mathcal{G}$.) From Jacobi's identity [(4.26)] we have

$$[A, [E_\alpha, E_\beta]] + [E_\alpha, [E_\beta, A]] + [E_\beta, [A, E_\alpha]] = 0 . \tag{24.15}$$

It follows from (24.7) that

$$[A, [E_\alpha, E_\beta]] - \beta[E_\alpha, E_\beta] + \alpha[E_\beta, E_\alpha] = 0 ,$$

or

$$[A, [E_\alpha, E_\beta]] = (\alpha + \beta)[E_\alpha, E_\beta] . \tag{24.16}$$

Thus, if E_α and E_β are eigenvectors of Ad_A with eigenvalues α and β respectively, then $[E_\alpha, E_\beta]$ is either the zero vector (in which case $\alpha + \beta$ is not a root), or an eigenvector of Ad_A with eigenvector $\alpha + \beta$. Then, if $\alpha + \beta = 0$, that is, if $\beta = -\alpha$, (24.5) implies that

$$[E_\alpha, E_{-\alpha}] = c_{\alpha, -\alpha}^i H_i , \tag{24.17}$$

where the subscript α in the structure constant refers to a particular value of a root and not to a numerical tensorial index. If $\alpha + \beta$ is a non-zero root, we have

$$[E_\alpha, E_\beta] = N_{\alpha\beta} E_{\alpha+\beta} , \tag{24.18}$$

where the constant $N_{\alpha\beta}$ can be written in terms of the structure constants with respect to the basis $\{H_i, E_\alpha\}$ of $\mathcal{G}$ as

$$N_{\alpha\beta} = c_{\alpha\beta}^{\alpha+\beta} . \tag{24.19}$$

We will next show that *if α is a root of a semisimple Lie algebra $\mathcal{G}$, then $-\alpha$ is also a root of $\mathcal{G}$.* Indeed consider the symmetric metric tensor g_{ij} on $\mathcal{G}$ given by the Killing form [(8.40)]:

$$g_{ij} = c_{ik}^l c_{jl}^k . \tag{24.20}$$

We will consider g_{ij} and the structure constants with respect to the basis $\{H_i, E_\alpha\}, i = 1, \ldots, m$, of $\mathcal{G}$, with α referring to a root. Our results so far indicate that (with Latin letters running from 1 to m and Greek letters referring to distinct non-zero roots of $\mathcal{G}$)

$$c_{i\alpha}^\sigma = \alpha_i \delta_\alpha^\sigma , \quad c_{i\alpha}^j = 0 ,$$
$$c_{ij}^k = c_{ij}^\alpha = 0 ,$$
$$c_{\alpha\beta}^i = \begin{cases} c_{\alpha, -\alpha}^i , & \text{if } \alpha + \beta = 0 , \\ 0 , & \text{if } \alpha + \beta \neq 0 , \end{cases} \tag{24.21}$$
$$c_{\alpha\beta}^\sigma = \begin{cases} c_{\alpha\beta}^{\alpha+\beta} , & \text{if } \sigma = \alpha + \beta , \text{ and } \alpha + \beta \text{ is a root} , \\ 0 , & \text{if } \alpha + \beta \text{ is not a root} . \end{cases}$$

The metric tensor can be blocked off as follows:

$$g = \begin{pmatrix} g_{ij} & g_{i\alpha} \\ g_{\alpha i} & g_{\alpha\beta} \end{pmatrix} . \tag{24.22}$$

We have

$$g_{i\alpha} = \sum_{k=1}^{m} c_{ik}^{l} c_{\alpha l}^{k} + \sum_{\beta} c_{i\beta}^{l} c_{\alpha l}^{\beta}$$

$$= \sum_{k=1}^{m} \left(\sum_{l=1}^{m} \underbrace{c_{ik}^{l}}_{=0} c_{\alpha l}^{k} + \sum_{\beta} \underbrace{c_{ik}^{\beta}}_{=0} c_{\alpha\beta}^{k} \right) + \sum_{\beta} \left(\sum_{l=1}^{m} \underbrace{c_{i\beta}^{l}}_{=0} c_{\alpha l}^{\beta} + \sum_{\gamma} c_{i\beta}^{\gamma} c_{\alpha\gamma}^{\beta} \right)$$

$$= \sum_{\beta\gamma} c_{i\beta}^{\gamma} c_{\alpha\gamma}^{\beta} = \sum_{\beta} c_{i\beta}^{\beta} c_{\alpha\beta}^{\beta} = \sum_{\beta} \beta_{i} \underbrace{c_{\alpha\beta}^{\beta}}_{=0} = 0 ,$$

$$\tag{24.23}$$

$$g_{\alpha\beta} = \sum_{k=1}^{m} c_{\alpha k}^{l} c_{\beta l}^{k} + \sum_{\gamma} c_{\alpha\gamma}^{l} c_{\beta l}^{\gamma}$$

$$= \sum_{k=1}^{m} \left(\sum_{l=1}^{m} c_{\alpha k}^{l} \underbrace{c_{\beta l}^{k}}_{=0} + \sum_{\sigma} c_{\alpha k}^{\sigma} \underbrace{c_{\beta\sigma}^{k}}_{=0} \right) + \sum_{\gamma} \left(\sum_{l=1}^{m} c_{\alpha\gamma}^{l} c_{\beta l}^{\gamma} + \sum_{\tau} c_{\alpha\gamma}^{\tau} c_{\beta\tau}^{\gamma} \right)$$

$$= \sum_{k=1}^{m} (-\alpha_k) c_{\beta\alpha}^{k} + \sum_{l=1}^{m} c_{\alpha\beta}^{l} (-\beta_l) + \sum_{\gamma} c_{\alpha\gamma}^{\alpha+\gamma} c_{\beta, \alpha+\gamma}^{\gamma} .$$

$$\tag{24.24}$$

Each term on the RHS of the last equality vanishes unless $\beta = -\alpha$. Thus $g_{\alpha\beta}$ vanishes unless $\alpha + \beta = 0$. Putting the results of the above two equations together, we see that $\det(g) = 0$ unless $\alpha + \beta = 0$. The assertion that α is a root of a semisimple Lie algebra implies that $-\alpha$ is also a root then follows from Theorem 8.2 (Cartan's first criterion).

Exercise 24.1   Show that in the blocked matrix form of the metric tensor g in (24.22), $\det(g) = 0$ if $g_{i\alpha} = g_{\alpha\beta} = 0$.

Since the non-zero roots come in pairs, $\pm\alpha$, and since $g_{\alpha\beta} = 0$ unless $\alpha + \beta =$

0, the metric tensor (24.22) appears as

$$g = \begin{pmatrix} g_{ij} & 0 & 0 & \cdots & \cdots & 0 & 0 \\ 0 & 0 & A & 0 & \cdots & 0 & 0 \\ 0 & A & 0 & 0 & \cdots & 0 & 0 \\ \vdots & 0 & 0 & \cdots & \cdots & 0 & 0 \\ \vdots & \cdots & \cdots & \cdots & \cdots & 0 & 0 \\ 0 & 0 & 0 & \cdots & 0 & 0 & B \\ 0 & 0 & 0 & \cdots & 0 & B & 0 \end{pmatrix} . \tag{24.25}$$

In this form it is readily seen that $\det(g) \neq 0$ implies $\det(g_{ij}) \neq 0$.

Exercise 24.2 Verify the above statement.

Now

$$g_{ij} = \sum_{k=1}^{m} c_{ik}^{l}{}_{=0} \, c_{jl}^{k}{}_{=0} + \sum_{\alpha} c_{i\alpha}^{l} \, c_{jl}^{\alpha} = \sum_{\alpha} \left(\sum_{l=1}^{m} c_{i\alpha}^{l}{}_{=0} \, c_{jl}^{\alpha} + \sum_{\beta} c_{i\alpha}^{\beta} \, c_{j\beta}^{\alpha} \right) \tag{24.26}$$

$$= \sum_{\alpha} \sum_{\beta} c_{i\alpha}^{\beta} \, c_{j\beta}^{\alpha} = \sum_{\alpha} c_{i\alpha}^{\alpha} \, c_{j\alpha}^{\alpha} = \sum_{\alpha} \alpha_i \alpha_j = g_{ji} .$$

Thus g_{ij} $(i, j = 1, \ldots, m)$, (the restriction of g to $\mathcal{G}_0$), also serves as a (symmetric and non-degenerate) metric tensor for the real m-dimensional space P spanned by the roots, called the **root space**. The fact that P is m-dimensional will be proved in Theorem 24.5 below. Using g_{ij}, the structure constants $c_{\alpha,-\alpha}^{i}$ [of (24.17)] are seen to be dual to α_i. Indeed,

$$c_{\alpha,-\alpha}^{i} = \delta_{l}^{i} \, c_{\alpha,-\alpha}^{l} = g^{ik} \, g_{kl} \, c_{\alpha,-\alpha}^{l} = g^{ik} \, c_{k, \alpha,-\alpha}$$

$$= g^{ik} \, c_{-\alpha, k, \alpha} = g^{ik} \, g_{-\alpha\beta} \, c_{k\alpha}^{\beta} = g^{ik} \, g_{-\alpha,\alpha} \, c_{k\alpha}^{\alpha} \tag{24.27}$$

$$= g^{ik} \, g_{-\alpha,\alpha} \, \alpha_k = g_{-\alpha,\alpha} \, \alpha^i .$$

We can then normalize the E_α in (24.7) so that, for all roots α, $g_{\alpha,-\alpha} = 1$. With the normalized E_α,

$$c_{\alpha,-\alpha}^{i} = \alpha^i , \tag{24.28}$$

and thus (24.17) becomes

$$[E_\alpha, E_{-\alpha}] = \alpha^i H_i . \tag{24.29}$$

From now on we will assume that the E_α are normalized unless specified otherwise. The scalar product between two roots α and β is defined by

$$\alpha \cdot \beta \equiv (\alpha^*, \beta^*) = g_{ij}\, \alpha^i \beta^j = \alpha_j \beta^j = \alpha_j \beta_i\, g^{ij} . \tag{24.30}$$

Note that this scalar product is defined on $\mathcal{G}_0^*$.

Gathering the above several results, we can identify the so-called **Cartan-Weyl-Chevalley normal form** of a complex semisimple Lie algebra as follows.

Theorem 24.1. *Let $\mathcal{G}$ be a complex semisimple Lie algebra and*

$$\mathcal{G}_0 = span\{H_1, \ldots, H_m\}$$

be its Cartan subalgebra. $\mathcal{G}$ decomposes into a direct sum of $\mathcal{G}_0$ and its one-dimensional root spaces $\mathcal{G}_\alpha = span\{E_\alpha\}$, such that

$$[H_i, H_j] = 0, \quad i,j = 1, \ldots, m , \tag{24.31}$$

$$[H_i, E_\alpha] = \alpha_i E_\alpha , \tag{24.32}$$

$$[E_\alpha, E_{-\alpha}] = \alpha^i H_i , \qquad \alpha^i = g^{ij}\,\alpha_j , \tag{24.33}$$

$$[E_\alpha, E_\beta] = N_{\alpha\beta}\, E_{\alpha+\beta} , \tag{24.34}$$

where $N_{\alpha\beta} = 0$ unless $\alpha + \beta$ is a root.

We will now study some further important properties of roots.

Theorem 24.2. *If α and β are roots of a semisimple Lie algebra $\mathcal{G}$, then $2(\alpha \cdot \beta)/(\alpha \cdot \alpha)$ is an integer, and $\beta - 2\alpha(\alpha \cdot \beta)/(\alpha \cdot \alpha)$ is also a root of $\mathcal{G}$.*

Proof. Suppose $\alpha + \beta$ is not a root. From (24.34),

$$[E_{-\alpha}, E_\beta] = N_{-\alpha, \beta}E_{\beta-\alpha} \equiv E'_{\beta-\alpha} , \quad [E_{-\alpha}, E'_{\beta-\alpha}] \equiv E'_{\beta-2\alpha} , \ldots \ldots$$
$$[E_{-\alpha}, E'_{\beta-j\alpha}] \equiv E'_{\beta-(j+1)\alpha} , \ldots \ldots \tag{24.35}$$

where $E'_{\beta-j\alpha}$ $(j > 0)$ are non-normalized elements in the corresponding root spaces. Since there are only a finite number of root spaces, there exists a positive integer h such that, after h iterations,

$$[E_{-\alpha}, E'_{\beta-h\alpha}] = E'_{\beta-(h+1)\alpha} = 0 . \tag{24.36}$$

From (24.34) we also see that

$$[E_\alpha, E'_{\beta-(j+1)\alpha}] = N_{j+1}E'_{\beta-j\alpha} , \tag{24.37}$$

where the constant N_{j+1} is defined by the equation. It can be obtained by eliminating $E'_{\beta-(j+1)\alpha}$ between (24.37) and the last displayed equation of (24.35):

$$N_{j+1}E'_{\beta-j\alpha} = [E_\alpha, [E_{-\alpha}, E'_{\beta-j\alpha}]]$$
$$= -[E'_{\beta-j\alpha}, [E_\alpha, E_{-\alpha}]] - [E_{-\alpha}, [E'_{\beta-j\alpha}, E_\alpha]]$$
$$= \alpha^i [H_i, E'_{\beta-j\alpha}] + N_j[E_{-\alpha}, E'_{\beta-(j-1)\alpha}] = \alpha^i(\beta - j\alpha)_i\, E'_{\beta-j\alpha} + N_j E'_{\beta-j\alpha}$$
$$= \{\alpha^i(\beta - j\alpha)_i + N_j\}\, E'_{\beta-j\alpha} ,$$

$$\tag{24.38}$$

where the second equality follows from the Jacobi identity. Thus we obtain the iteration formula

$$N_{j+1} = \alpha^i(\beta - j\alpha)_i + N_j = \alpha \cdot \beta - j(\alpha \cdot \alpha) + N_j , \quad j \geq 1 . \tag{24.39}$$

If we define $N_0 \equiv 0$, (24.39) yields

$$\begin{aligned}
&N_0 = 0 , N_1 = (\alpha \cdot \beta) , N_2 = (\alpha \cdot \beta) - (\alpha \cdot \alpha) + N_1 , \\
&N_3 = (\alpha \cdot \beta) - 2(\alpha \cdot \alpha) + N_2 , \dots , \\
&N_j = (\alpha \cdot \beta) - (j-1)(\alpha \cdot \alpha) + N_{j-1} ,
\end{aligned} \tag{24.40}$$

or,

$$\begin{aligned}
N_j &= j(\alpha \cdot \beta) - (\alpha \cdot \alpha)(1 + 2 + \cdots + (j-1)) \\
&= j(\alpha \cdot \beta) - \left(\frac{j(j-1)}{2} \right) (\alpha \cdot \alpha) .
\end{aligned} \tag{24.41}$$

Now (24.36) and (24.37) imply that

$$N_{h+1} = 0 . \tag{24.42}$$

Thus it follows from (24.41) (on setting $j = h + 1$) that

$$(\alpha \cdot \beta) = \frac{h}{2}(\alpha \cdot \alpha) . \tag{24.43}$$

Substituting into (24.41), we have

$$N_j = \frac{j}{2}(h - j + 1)(\alpha \cdot \alpha) . \tag{24.44}$$

Note that $(\alpha \cdot \alpha) \neq 0$. Indeed, suppose $(\alpha \cdot \alpha) = 0$. Then (24.41) implies that $N_j = j(\alpha \cdot \beta)$, $j \geq 1$, which contradicts (24.42) if $\alpha \cdot \beta \neq 0$. Eq. (24.43) shows that $2(\alpha \cdot \beta)/(\alpha \cdot \alpha)$ must be a positive integer. Also, (24.36), (24.37), and the values of N_j given by (24.44) show that, if α and β are roots and $\alpha + \beta$ is not a root, then

$$\beta, \beta - \alpha, \beta - 2\alpha, \dots, \beta - h\alpha$$

are all roots, where $h = 2(\alpha \cdot \beta)/(\alpha \cdot \alpha)$ is a positive integer. The above sequence of roots is called the α-**root string** of β.

Now suppose $\alpha + \beta$ is a root. Then there exists a positive integer k such that

$$\begin{aligned}
&[E_\alpha, E_\beta] = E'_{\alpha+\beta} , \dots , [E_\alpha, E'_{\beta+j\alpha}] = E'_{\beta+(j+1)\alpha} , \dots , \\
&[E_\alpha, E'_{\beta+(k-1)\alpha}] = E'_{\beta+k\alpha} , [E_\alpha, E'_{\beta+k\alpha}] = E'_{\beta+(k+1)\alpha} = 0 .
\end{aligned} \tag{24.45}$$

On defining the constants $\mathcal{N}_j$ by

$$[E_{-\alpha}, E'_{\beta+(j+1)\alpha}] = \mathcal{N}_{j+1} E'_{\beta+j\alpha} , \tag{24.46}$$

we can obtain, by steps entirely analogous to those leading to (24.41),

$$\mathcal{N}_j = -j(\alpha \cdot \beta) - \left(\frac{j(j-1)}{2}\right)(\alpha \cdot \alpha) \ . \tag{24.47}$$

On setting $\mathcal{N}_{k+1} = 0$, we conclude that

$$\frac{2(\alpha \cdot \beta)}{(\alpha \cdot \alpha)} = -k \tag{24.48}$$

is a negative integer. Then we have the following root-string:

$$\beta, \beta + \alpha, \dots, \beta + k\alpha \ ,$$

the last root of which is $\beta - \dfrac{2(\alpha \cdot \beta)}{(\alpha \cdot \alpha)}\alpha$. The integers $2(\alpha \cdot \beta)/(\alpha \cdot \alpha)$ of a semisimple Lie algebra are called the **Cartan integers**. ☐

Theorem 24.3. *If α is a root, then, of all the integral multiples $n\alpha$ of α, only $\alpha, 0$, and $-\alpha$ are roots.*

Proof. Since $[E_\alpha, E_\alpha] = 0$, it follows from (24.34) that 2α is not a root. Now suppose that $n\alpha$ is a root for a certain $n > 1$. Then, by the previous theorem, since

$$h = \frac{2(\alpha \cdot n\alpha)}{\alpha \cdot \alpha} = 2n \ , \tag{24.49}$$

$n\alpha$ has the α-root string

$$n\alpha, (n-1)\alpha, \dots, -n\alpha \ ,$$

which includes 2α. This contradicts the fact that 2α is not a root. Thus $n\alpha$ cannot be a root for all $n > 1$. Similarly, suppose $-n\alpha$ is a root for some $n > 1$. Then

$$k = -\frac{2(\alpha \cdot (-n\alpha))}{(\alpha \cdot \alpha)} = 2n \ , \tag{24.50}$$

and $-n\alpha$ has the α-root string

$$-n\alpha, -(n-1)\alpha, \dots, n\alpha \ ,$$

which again includes 2α. This proves the theorem. ☐

Theorem 24.4. *Suppose α and β are both non-zero roots, then the α-root string of β contains at most four roots. Thus*

$$\frac{2(\alpha \cdot \beta)}{(\alpha \cdot \alpha)} = 0, \pm 1, \pm 2, \ or \ \pm 3 \ . \tag{24.51}$$

Proof. We suppose $\beta \neq \pm\alpha$; for if $\beta = \pm\alpha$, then the α-root string of β consists of just the three roots $\alpha, 0, -\alpha$, in which case the theorem is proved.

Suppose the α-root string of β has five roots. Let these be

$$\beta, \beta - \alpha, \beta - 2\alpha, \beta - 3\alpha, \beta - 4\alpha .$$

Neither $\beta - (\beta - 2\alpha) = 2\alpha$ nor $\beta + (\beta - 2\alpha) = 2\beta - 2\alpha$ are roots. Hence the $(\beta - 2\alpha)$-root string of β consists of only one root, namely, β. Eq. (24.43) then implies that

$$\beta \cdot (\beta - 2\alpha) = \beta \cdot \beta - 2\alpha \cdot \beta = 0 . \tag{24.52}$$

Similarly, the $(\beta - 3\alpha)$-string of $\beta - \alpha$ consists of only one root, $\beta - \alpha$; and the $(\beta - 4\alpha)$-string of $\beta - 2\alpha$ consists of only the root $\beta - 2\alpha$. Again, by (24.43),

$$(\beta - \alpha) \cdot (\beta - 3\alpha) = \beta \cdot \beta - 4\alpha \cdot \beta + 3\alpha \cdot \alpha = 0 , \tag{24.53}$$
$$(\beta - 2\alpha) \cdot (\beta - 4\alpha) = \beta \cdot \beta - 6\alpha \cdot \beta + 8\alpha \cdot \alpha = 0 . \tag{24.54}$$

The last three equations together can be written in matrix form:

$$\begin{pmatrix} 1 & -2 & 0 \\ 1 & -4 & 3 \\ 1 & -6 & 8 \end{pmatrix} \begin{pmatrix} \beta \cdot \beta \\ \alpha \cdot \beta \\ \alpha \cdot \alpha \end{pmatrix} = 0 . \tag{24.55}$$

Since the determinant of the coefficient matrix does not vanish (being equal to -4), the only solution is the trivial one:

$$\alpha \cdot \alpha = \beta \cdot \beta = \alpha \cdot \beta = 0 . \tag{24.56}$$

This obviously contradicts the assumption that α and β are both non-zero roots. Hence the supposition that the α-string of β has five roots must be false. Similarly the root string

$$\beta, \beta + \alpha, \beta + 2\alpha, \beta + 3\alpha, \beta + 4\alpha$$

cannot exist. On the other hand, consider the α-root string of β

$$\beta, \beta - \alpha, \beta - 2\alpha, \beta - 3\alpha .$$

Eqs. (24.52) and (24.53) are still valid, which together yield

$$\frac{2(\alpha \cdot \beta)}{(\alpha \cdot \alpha)} = 3 , \tag{24.57}$$

confirming (24.43). For the root string

$$\beta, \beta + \alpha, \beta + 2\alpha, \beta + 3\alpha ,$$

we have, instead,

$$\frac{2(\alpha \cdot \beta)}{(\alpha \cdot \alpha)} = -3 . \tag{24.58}$$

Eq. (24.52) follows, and the theorem is proved. $\qquad\square$

Let us now calculate $N_{\alpha\beta}$ in (24.34) when $\alpha+\beta$ is a root (otherwise $N_{\alpha\beta} = 0$). Consider the following α-root string of β:

$$\beta + j\alpha, \beta + (j-1)\alpha, \ldots, \beta + \alpha, \beta, \beta - \alpha, \ldots, \beta - k\alpha . \tag{24.59}$$

Letting

$$\gamma \equiv \beta + j\alpha , \tag{24.60}$$

this can be written equivalently as

$$\gamma, \gamma - \alpha, \ldots, \gamma - h\alpha , \tag{24.61}$$

where

$$h = j + k . \tag{24.62}$$

From (24.37), we have (on replacing β by γ and $j+1$ by j in that equation),

$$[E_\alpha, E'_{\gamma-j\alpha}] = N_j E'_{\gamma-(j-1)\alpha} , \tag{24.63}$$

or, from the last of Eqs. (24.35),

$$[E_\alpha, [E_{-\alpha}, E'_{\gamma-(j-1)\alpha}]] = N_j E'_{\gamma-(j-1)\alpha} . \tag{24.64}$$

Since, according to (24.35), $E'_{\gamma-(j-1)\alpha}$ differs from $E_{\gamma-(j-1)\alpha}$ by a constant, the above equation and (24.60) together imply

$$N_j E_{\alpha+\beta} = [E_\alpha, [E_{-\alpha}, E_{\alpha+\beta}]] = N_{-\alpha, \alpha+\beta} [E_\alpha, E_\beta] = N_{\alpha\beta} N_{-\alpha, \alpha+\beta} E_{\alpha+\beta} . \tag{24.65}$$

Thus, on recalling (24.44), and substituting $j+k$ for h [(24.62)] in that equation, we have

$$N_{\alpha\beta} N_{-\alpha, \alpha+\beta} = \frac{j(k+1)(\alpha \cdot \alpha)}{2} . \tag{24.66}$$

Next we use the Jacobi identity [(4.26)] to obtain

$$[[E_\alpha, E_\beta], E_{-\alpha-\beta}] = [E_\alpha, [E_\beta, E_{-\alpha-\beta}]] + [E_\beta, [E_{-\alpha-\beta}, E_\alpha]] , \tag{24.67}$$

and a similar equation on replacing α by $-\alpha$ and β by $-\beta$. (This is allowed since α is a root implies $-\alpha$ is also a root.) These equations yield, on using the Cartan-Weyl-Chevalley form [(24.31) to (24.34)],

$$N_{\alpha\beta} (\alpha + \beta)^i H_i = (N_{\beta, -\alpha-\beta} \alpha^i + N_{-\alpha-\beta, \alpha} \beta^i) H_i , \tag{24.68}$$

$$N_{-\alpha, -\beta} (\alpha + \beta)^i H_i = (N_{-\beta, \alpha+\beta} \alpha^i + N_{\alpha+\beta, -\alpha} \beta^i) H_i . \tag{24.69}$$

We thus have

$$N_{\alpha\beta} = N_{-\beta, -\alpha-\beta} = N_{-\alpha-\beta, \alpha} , \tag{24.70}$$

$$N_{-\alpha, -\beta} = N_{-\beta, \alpha+\beta} = N_{\alpha+\beta, -\alpha} . \tag{24.71}$$

Eq. (24.66) can then be rewritten

$$-N_{\alpha\beta} N_{-\alpha,-\beta} = \frac{j(k+1)(\alpha \cdot \alpha)}{2} , \qquad (24.72)$$

which implies

$$N_{\alpha\beta} N_{-\alpha,-\beta} < 0 . \qquad (24.73)$$

We now observe that under the rescalings

$$E_\alpha \rightarrow \tilde{E}_\alpha = C_\alpha E_\alpha , \qquad (24.74)$$

$$N_{\alpha\beta} \rightarrow \tilde{N}_{\alpha\beta} = \frac{C_\alpha C_\beta}{C_{\alpha+\beta}} N_{\alpha\beta} , \qquad (24.75)$$

the Cartan-Weyl-Chevalley form [(24.32) to (24.34)] remains invariant provided

$$C_\alpha C_{-\alpha} = 1 . \qquad (24.76)$$

Exercise 24.3 Verify the above statement.

With this rescaling freedom, one can always set

$$N_{-\alpha,-\beta} = -N_{\alpha\beta} . \qquad (24.77)$$

Eq. (24.72) then gives the following result for $N_{\alpha\beta}$:

$$\boxed{N_{\alpha\beta} = \sqrt{\frac{j(k+1)(\alpha \cdot \alpha)}{2}}} . \qquad (24.78)$$

In the above equation, the positive integers j and k are determined from the root string (24.59) relating the roots α and β, with $\alpha + \beta$ being a root also.

Before we proceed, we illustrate the results obtained thus far with the simple but important case of $SU(2) \sim SO(3)$, which, as we have seen many times, is the Lie algebra underlying the quantum theory of angular momentum.

Recall from Chapter 4 that $SO(3)$ is a three-dimensional Lie algebra with generators J_2^3, J_3^1 and J_1^2. Writing

$$\boldsymbol{L} = (L^1, L^2, L^3)$$

(the three components of the angular momentum operator) with [c.f. (9.26)]

$$L^1 = -i\hbar J_2^3, \quad L^2 = -i\hbar J_3^1, \quad L^3 = -i\hbar J_1^2 , \qquad (24.79)$$

and defining the dimensionless angular momentum operator $\boldsymbol{J}$ by

$$\boldsymbol{L} \equiv \hbar \boldsymbol{J} = \hbar (J_1, J_2, J_3) , \qquad (24.80)$$

the structure equation for $\mathcal{SO}(n)$ [(4.35)] applied to $\mathcal{SO}(3)$ gives

$$[\, J_j, J_k \,] = i \sum_l \varepsilon_{jkl}\, J_l \tag{24.81}$$

as the structure equation for $\mathcal{SO}(3)$ [c.f. (9.23)]. If we define

$$J_+ \equiv \frac{J_1 + iJ_2}{\sqrt{2}}, \tag{24.82}$$

$$J_- \equiv \frac{J_1 - iJ_2}{\sqrt{2}} = J_+^\dagger, \tag{24.83}$$

it follows from (24.81) that

$$[\, J_3, J_\pm \,] = \pm J_\pm, \tag{24.84}$$
$$[\, J_+, J_- \,] = J_3. \tag{24.85}$$

Thus the Cartan subalgebra of $\mathcal{SO}(3)$ is one-dimensional:

$$(\mathcal{SO}(3))_0 = \{aJ_3\}, \quad a \in \mathbb{C}.$$

There are only two roots: $\alpha_+ = 1$ and $\alpha_- = -\alpha_+ = -1$, and the metric tensor is given by $g_{ij} = g^{ij} = 1$. The corresponding eignevectors of Ad_{J_3} are J_+ and J_-. (The structure constants $N_{\alpha\beta}$ do not enter into this algebra.)

As discussed before, the quantities α_i can be represented as covariant components of a root vector $\alpha^* = \alpha_i (H^*)^i \in \mathcal{G}_0^*$, where $(H^*)^i(H_j) = \delta_j^i$. Following our slight abuse of notation, we will still use the symbol α to denote α^*, remembering that the symmetric, positive-definite inner product in $\mathcal{G}_0^*$ is given by

$$\alpha \cdot \beta = (\alpha^*, \beta^*) = g^{ij}\alpha_i\beta_j = \alpha^j\beta_j, \quad i,j = 1,\ldots,m = \text{rank}. \tag{24.86}$$

Based on this inner product, we can define the angle $\theta_{\alpha\beta}$ between the root vectors α and β by

$$\cos\theta_{\alpha\beta} \equiv \frac{(\alpha \cdot \beta)}{\sqrt{(\alpha \cdot \alpha)(\beta \cdot \beta)}}, \tag{24.87}$$

or

$$\cos^2\theta_{\alpha\beta} = \frac{(\alpha \cdot \beta)^2}{(\alpha \cdot \alpha)(\beta \cdot \beta)}. \tag{24.88}$$

One can also specify the ratio $r_{\alpha\beta}$ of the lengths of the root vectors α and β:

$$r_{\alpha\beta} \equiv \sqrt{\frac{(\alpha \cdot \alpha)}{(\beta \cdot \beta)}} = \sqrt{\frac{2(\alpha \cdot \beta)/(\beta \cdot \beta)}{2(\alpha \cdot \beta)/(\alpha \cdot \alpha)}}. \tag{24.89}$$

It follows from Theorem 24.4 that the possible values for $\theta_{\alpha\beta}$ are given by

$$\cos^2\theta_{\alpha\beta} = 0, \frac{1}{4}, \frac{1}{2}, \frac{3}{4}, 1, \tag{24.90}$$

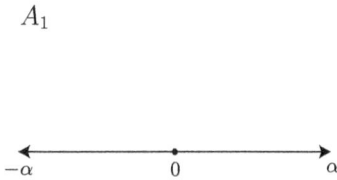

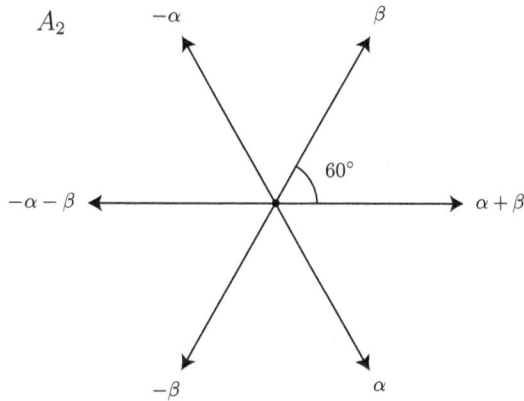

FIGURE 24.1   FIGURE 24.2

where the last value is for $\alpha = \beta$. Thus the possible acute angles are

$$\theta_{\alpha\beta} = 0,\ 30°,\ 45°,\ 60°,\ 90°\ ; \tag{24.91}$$

and correspondingly, by (24.89), (24.43), and (24.48),

$$\begin{aligned}
\theta_{\alpha\beta} &= 30°, \quad r_{\alpha\beta} = \sqrt{3}\ , \\
\theta_{\alpha\beta} &= 45°, \quad r_{\alpha\beta} = \sqrt{2}\ , \\
\theta_{\alpha\beta} &= 60°, \quad r_{\alpha\beta} = 1\ , \\
\theta_{\alpha\beta} &= 90°, \quad r_{\alpha\beta}\ \text{indeterminate}\ .
\end{aligned} \tag{24.92}$$

For rank-one and rank-two ($m = 1, 2$) semisimple Lie algebras, one can represent the collection of roots for each algebra very conveniently by a set of vectors on a plane with the ordinary Euclidean metric and satisfying (24.92). These are called **root diagrams**, each of which specifies a particular semisimple Lie algebra.

For $m = 1$, there is only the classical algebra $A_1\ [\sim \mathcal{SO}(3) \sim \mathcal{SU}(2)]$. Its root diagram is shown in Fig. 24.1.

For $m = 2$, we have the classical algebras $A_2\ [\sim \mathcal{SU}(3)]$, $B_2\ [\sim \mathcal{SO}(5)]$ and the exceptional algebra G_2. The root diagrams for these are shown in Figures 24.2 through 24.4; and scalar products can be calculated by elementary means with these diagrams using the Euclidean metric.

For semisimple Lie algebras of higher rank than two, it is not possible to represent all the roots by planar root diagrams. E. B. Dynkhin showed, however, that all semisimple Lie algebras can still be represented, and indeed classified, by a class of planar diagrams, known as **Dynkin diagrams**. In the remainder of this chapter, we will discuss (but omit the proof) of this procedure. The final result will be given in Theorem 24.7. First, we have to define an ordering of the non-zero roots.

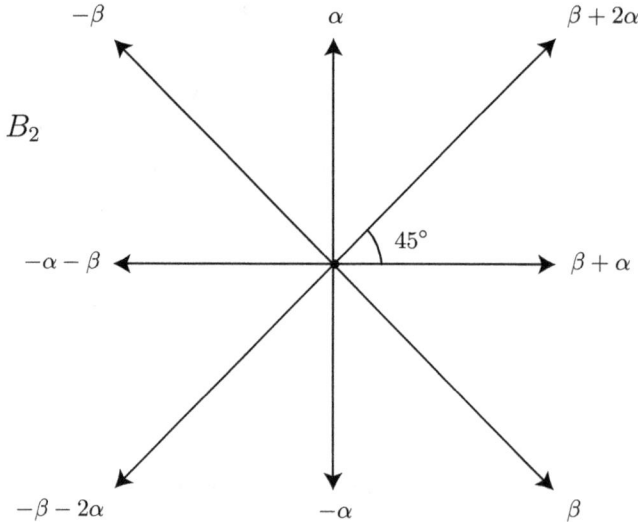

FIGURE 24.3

Definition 24.2. *A root α is said to be **positive** if the first non-zero component of α is a positive number. If two roots α' and α are such that $\alpha' - \alpha$ is positive, then α' is said to be larger than α.*

For example, α, β and $\alpha + \beta$ in A_2; α, β, $\beta + \alpha$, $\beta + 2\alpha$ in B_2; and α, β, $\beta + \alpha$, $\beta + 2\alpha$, $\beta + 3\alpha$, and $2\beta + 3\alpha$ in G_2 are all positive roots (refer to Figures 24.2, 24.3 and 24.4). These examples show that an ordering is simply a way of dividing the root space P into two halves.

Let us denote the set of positive roots by P^+.

Definition 24.3. *A positive root $\gamma \in P^+$ is said to be **simple**, or **fundamental**, if it is not the sum of two other positive roots.*

We will denote the set of all simple roots by Π, and the set of all roots by Σ.

Theorem 24.5. *The elements of Π (the set of all simple roots) form a basis of the m-dimensional real vector space $P = $ span of Σ (where Σ is the set of all roots). If $\alpha_I \neq \alpha_J$ belong to Π (that is, both are simple), then $\alpha_I \cdot \alpha_J \leq 0$. Furthermore, every positive root is a linear combination of simple roots with positive integer coefficients.*

Proof. If $\alpha_I \cdot \alpha_J > 0$, then (24.43) implies that $\alpha_I - \alpha_J$ and $\alpha_J - \alpha_I$ are both roots, one of which is positive. Suppose $\alpha_I - \alpha_J$ is positive, then $\alpha_I = \alpha_J + (\alpha_I - \alpha_J)$ is not simple, contradicting the hypothesis that both α_I and α_J are simple. It follows that $\alpha_I \cdot \alpha_J \leq 0$ (which also implies $\theta_{\alpha_I \alpha_J} \geq \pi/2$).

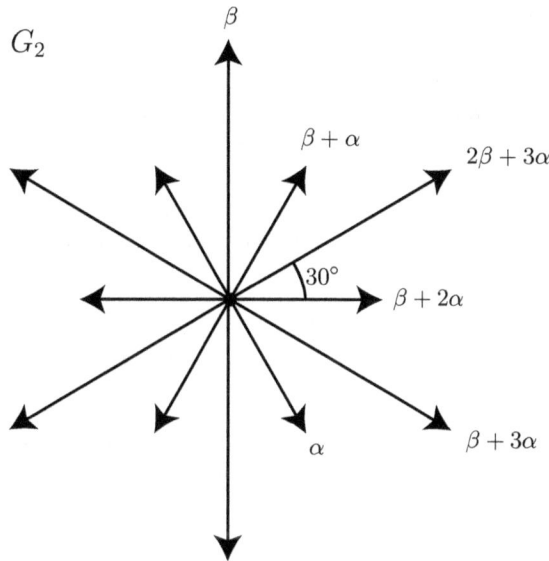

FIGURE 24.4

Let us show next that the simple roots are linearly independent. Let the set of simple roots be $\Pi = \{\alpha_I\}$. If $\{\alpha_I\}$ is linearly dependent, then there exists a set of $x^I \in \mathbb{R}$, not all zero, such that $x^I \alpha_I = 0$. Divide the non-zero terms in this sum into two groups, one with the $x^I > 0$ (call these p^I), and the other with the $x^I < 0$ (call these $-n^J$), and write

$$p^I \alpha_I - n^J \alpha_J = 0 , \qquad (24.93)$$

where $p^I, n^J > 0$. Consider the vector

$$\alpha \equiv p^I \alpha_I = n^J \alpha_J . \qquad (24.94)$$

We have

$$\alpha \cdot \alpha = p^I n^J (\alpha_I \cdot \alpha_J) \leq 0 , \qquad (24.95)$$

where the inequality follows from the first part of the theorem (just proved). But α is a positive root since it is a sum of positive roots. Therefore $\alpha \cdot \alpha \geq 0$ also. Hence $\alpha \cdot \alpha = 0$, which implies $\alpha = 0$. But this is a contradiction, since the p^I are all supposed to be > 0.

Now suppose $\alpha \in P^+$ is not simple. Then it can be written $\alpha = \alpha_1 + \alpha_2$, where $\alpha_1, \alpha_2 \in P^+$. Since $\alpha_1 = \alpha - \alpha_2$ and $\alpha_2 = \alpha - \alpha_1$ are both positive, α is larger than both α_1 and α_2. If either α_1 or α_2 is not simple, the non-simple root(s) can be reduced further to a sum (sums) of positive roots. The process must eventually terminate. So the last statement of the theorem is proved.

We have so far shown that the set of simple roots Π is linearly independent and that it spans P^+. It must then also span Σ (the set of all roots, both positive and negative) and thus P (the root space). Therefore Π forms a basis for P. We will finally show that *there are exactly m simple roots for a rank-m semisimple Lie algebra*, which will imply that P is m-dimensional.

Obviously there cannot be more than m simple roots, since they are linearly independent m-vectors (having components α_i, $i = 1, \ldots, m$). Suppose there are fewer than m simple roots. Let these be $\alpha_{(1)}, \ldots, \alpha_{(k)}$, $k < m$. Then every root α can be expressed as

$$\alpha = \sum_{i=1}^{k} x^i \alpha_{(i)}, \quad k < m. \tag{24.96}$$

Thus P is at most k-dimensional. It follows that there exists a basis $\{H_i, i = 1, \ldots, m\}$ of the Cartan subalgebra $\mathcal{G}_0$ with respect to which the first, say, of the m (covariant) components of all the roots vanish:

$$\alpha = \{0, \alpha_2, \ldots, \alpha_m\}. \tag{24.97}$$

Thus, for all roots α,

$$[H_1, E_\alpha] = 0. \tag{24.98}$$

But at the same time,

$$[H_1, H_j] = 0, \quad j = 1, \ldots, m. \tag{24.99}$$

Therefore, $\{H_1, 0\}$ generates an abelian ideal of $\mathcal{G}$, which contradicts the fact that $\mathcal{G}$ is semisimple. $\qquad\square$

The fact that $\alpha \cdot \beta \leq 0$ if both α and β are simple implies that the angle $\theta_{\alpha\beta}$ between the two simple roots must satisfy

$$\pi/2 \leq \theta_{\alpha\beta} \leq \pi. \tag{24.100}$$

We then have the following straightforward consequence of Theorem 24.4.

Theorem 24.6. *If α and β are two simple roots, then $\theta_{\alpha\beta}$ is either $\pi/2, (2\pi)/3$, $(3\pi)/4$, or $(5\pi)/6$. If, in addition, $\alpha \cdot \alpha \leq \beta \cdot \beta$, then*

$$\frac{\beta \cdot \beta}{\alpha \cdot \alpha} = \begin{cases} indeterminate, & \theta_{\alpha\beta} = \pi/2, \\ 1, & \theta_{\alpha\beta} = (2\pi)/3, \\ 2, & \theta_{\alpha\beta} = (3\pi)/4, \\ 3, & \theta_{\alpha\beta} = (5\pi)/6. \end{cases} \tag{24.101}$$

Exercise 24.4   Prove the above theorem using the result of Theorem 24.4.

The above two theorems reveal that knowledge of the simple roots of a semisimple Lie algebra is sufficient to determine the structure of the algebra. A Dynkin diagram is just a shorthand, diagrammatic notation for representing the simple roots. The rules for the construction of a Dynkin diagram are as follows:

(1) Every simple root is represented by an open circle.

(2) Pairs of circles are connected by n lines, where $n = 0, 1, 2,$ or 3 for $\theta_{\alpha\beta} = \pi/2, (2\pi)/3, (3\pi)/4$ and $(5\pi)/6$, respectively [as in (24.101)].

(3) When circles are connected by two or more lines, an arrow is drawn from the shorter to the longer root.

Let $e_{(i)}$ be the i-th normailzed root:

$$e_{(i)} = \frac{\alpha_{(i)}}{\sqrt{\alpha_{(i)} \cdot \alpha_{(i)}}} \, . \tag{24.102}$$

Then, for $i \neq j$,

$$e_{(i)} \cdot e_{(j)} = \frac{\alpha_{(i)} \cdot \alpha_{(j)}}{\sqrt{(\alpha_{(i)} \cdot \alpha_{(i)})(\alpha_{(j)} \cdot \alpha_{(j)})}} = \cos\theta_{\alpha_{(i)}\alpha_{(j)}} = -\frac{1}{2}\sqrt{n_{ij}} \, , \tag{24.103}$$

where n_{ij} is the number of lines joining the circles representing the i-th and j-th roots in a Dynkin diagram. To each diagram one can then associate a real, positive-definite, quadratic form $x \cdot x$ on the root space P, where $x \in P$ is given by

$$x = x^i e_{(i)} \, . \tag{24.104}$$

We have, from (24.103),

$$x \cdot x = \sum_i (x^i)^2 + 2 \sum_{i<j} x^i x^j \, e_{(i)} \cdot e_{(j)} = \sum_i (x^i)^2 - \sum_{i<j} x^i x^j \, \sqrt{n_{ij}} \, . \tag{24.105}$$

The requirement of positive definiteness of the RHS of the above equation determines all the admissable Dynkin diagrams constructed by the above rules. We state without proofs the following facts.

Lemma 24.2. *Admissable Dynkin diagrams have no loops, and each circle meets at most three lines. Moreover, any pair of circles joined by a single line can be merged into one circle, and the resulting diagram is still admissable.*

Using (admissable) Dynkin diagrams we can represent the so-called **Cartan classification** of simple Lie algebras according to the following theorem. (The classification of semisimple Lie algebras can be reduced to that of simple Lie algebras.)

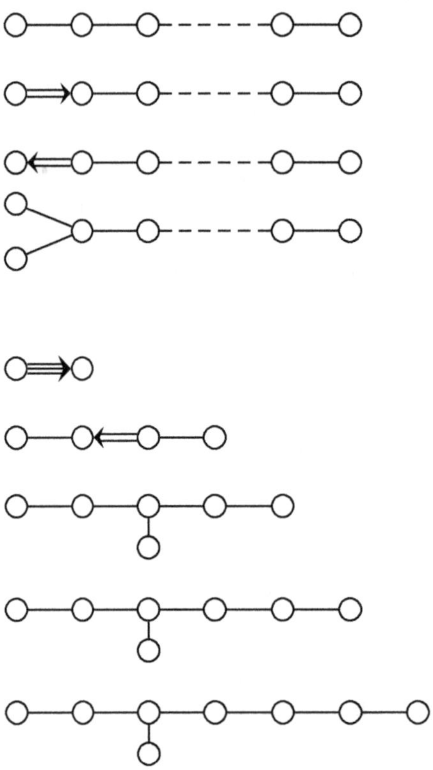

	rank	number of roots
A_m	$m \geq 1$	$m^2 + m$
B_m	$m \geq 2$	$2m^2$
C_m	$m \geq 3$	$2m^2$
D_m	$m \geq 4$	$2(m^2 - m)$
G_2	2	12
F_4	4	48
E_6	6	72
E_7	7	126
E_8	8	240

FIGURE 24.5

Theorem 24.7. *The complex simple Lie algebras are exhausted, up to equivalence, by the following classification: the four series of* ***classical Lie algebras***

$$A_m\,(m \geq 1),\; B_m\,(m \geq 1),\; C_m\,(m \geq 3),\; D_m\,(m \geq 4),$$

where m is the rank of the Lie algebra, and the five ***exceptional Lie algebras***

$$G_2,\; F_4,\; E_6,\; E_7,\; E_8,$$

where the subscript in each case is equal to the rank of the algebra. These algebras have corresponding Dynkin diagrams shown below (Figure 24.5).

The number of circles in each diagram is equal to the number of roots (or the rank) of the algebra.

Exercise 24.5 Show that $\bigcirc\!\!-\!\!\bigcirc\!\!=\!\!\bigcirc\!\!-\!\!\bigcirc\!\!-\!\!\bigcirc$ is not an admissable Dynkin diagram.

Exercise 24.6 Show that $\mathcal{SO}(4) \sim \mathcal{SO}(3) \oplus \mathcal{SO}(3)$ [c.f. (9.28)] is semisimple but not simple.

The four infinite series of complex semisimple Lie algebras can be identified with the complexifications of the Lie algebras of the corresponding classical Lie groups as shown in the following table.

$\mathcal{G}$	Constraint		Dimension
$\mathcal{SL}(n) = A_{n-1}$ (real form $= \mathcal{SU}(n)$)	$\mathrm{Tr}\,X = 0$		$n^2 - 1$
$\mathcal{SO}(2n+1) = B_n$	$X^T + X = 0$		$n(2n+1)$
$\mathcal{SP}(2n) = C_n$	$\begin{pmatrix} A & B \\ C & -A^T \end{pmatrix},$	$\begin{matrix} B^T = B, \\ C^T = C \end{matrix}$	$n(2n+1)$
$\mathcal{SO}(2n) = D_n$	$X^T + X = 0$		$n(2n-1)$

TABLE 24.1

$\mathcal{SP}(2n)$ is the Lie algebra of the symplectic group $Sp\,(2n)$, and X, A, B, C are complex matrices satisfying the listed constraints. Note that $\mathcal{SU}(n)$ is a real Lie algebra (of traceless skew-hermitian $n \times n$ matrices) whose complexification is $\mathcal{SL}(n)$. In other words, it is a **real form** of $\mathcal{SL}(n)$.

Chapter 25

The Representations of Semisimple Lie Algebras

The general definition for the representation of a Lie algebra has been given in Definition 7.12. In particular, the basic concepts of reducible and irreducible representations, and equivalence of representations, are entirely analogous to those for group representations (c.f. Definitions 18.1 and 18.3). In this chapter we will develop the basic concepts, facts, and techniques that are specific to the representation theory of semisimple Lie algebras, a complete classification of which has been given at the end of the last chapter. These are algebras corresponding to the compact Lie groups of importance in physics.

We begin with the following theorem (stated without proof), which is a consequence of Theorem 22.3.

Theorem 25.1. *Every representation of a compact Lie group is equivalent to a representation by hermitian operators on a unitary space (one endowed with a hermitian inner product); and every irreducible representation is equivalent to a representation by finite hermitian matrices.*

Note that the adjoint representation, in terms of which the theory in the last chapter has been worked out, is not a representation by hermitian matrices, since the matrix elements of the adjoint representation are the structure constants [c.f. (7.18)], which do not in general satisfy the requirements of hermiticity. This fact can also be seen explicitly as follows. Elements H_i in the Cartan subalgebra are necessarily hermitian [with respect to the metric of (24.20)], so that the α_i [c.f. (24.32)] are all real. But on taking the hermitian conjugate of (24.32) we have

$$[H_i, E_\alpha]^\dagger = [H_i E_\alpha - E_\alpha H_i]^\dagger = E_\alpha^\dagger H_i - H_i E_\alpha^\dagger = -[H_i, E_\alpha^\dagger] = \alpha_i E_\alpha^\dagger . \quad (25.1)$$

Thus

$$[H_i, E_\alpha^\dagger] = -\alpha_i E_\alpha^\dagger , \quad (25.2)$$

which implies

$$E_\alpha^\dagger = -E_{-\alpha} \ . \tag{25.3}$$

Since elements in a basis $\{H_i\}$ of the Cartan subalgebra commute with each other, there exist simultaneous eigenvectors of the linear operators $D'(H_i)$ in any representation D.

Definition 25.1. *In a Lie algebra representation $D : \mathcal{G} \to \mathcal{T}(V)$, where V is a unitary space, a set of simultaneous eigenvalues $\lambda_1, \ldots, \lambda_m$ of $D(H_i)$, $i = 1, \ldots, m (= \text{rank of } \mathcal{G})$, where $\{H_i\}$ is a basis of the Cartan subalgebra $\mathcal{G}_0$, forms the covariant components of a vector in an m-dimensional linear space. Such a vector is called a* **weight vector** *of the representation D. The corresponding m-dimensional space Δ_D spanned by all the weight vectors is called the* **weight space** *of the representation D.*

Writing a simultaneous eigenvector corresponding to the eigenvalues $\lambda_1, \ldots, \lambda_m$ as

$$| \boldsymbol{\lambda} (\nu) \rangle = | \lambda_1, \ldots, \lambda_m (\nu) \rangle \ , \tag{25.4}$$

we have the following defining equation for the weights:

$$H_i | \boldsymbol{\lambda} (\nu) \rangle = \lambda_i | \boldsymbol{\lambda} (\nu) \rangle \ , \quad i = 1, \ldots, m \ , \tag{25.5}$$

where, with a slight abuse of notation, we have written H_i instead of $D(H_i)$, and ν is a degeneracy index. The roots introduced in the last chapter are seen to be the weights in the adjoint representation.

For a root α, E_α and $E_{-\alpha}$ are raising and lowering operators for the eigenstates $| \boldsymbol{\lambda} \rangle$ of H_i. Indeed, it follows from (24.32) that

$$H_i E_{\pm\alpha} | \boldsymbol{\lambda} \rangle = ([\, H_i, E_{\pm\alpha} \,] + E_{\pm\alpha} H_i) | \boldsymbol{\lambda} \rangle = (\lambda_i \pm \alpha_i) E_{\pm\alpha} | \boldsymbol{\lambda} \rangle \ . \tag{25.6}$$

Thus $E_{\pm\alpha} | \boldsymbol{\lambda} \rangle$ is an eigenstate of H_i with eigenvalue $\lambda_i \pm \alpha_i$, if $\lambda_i \pm \alpha_i$ is a weight; it is equal to the zero vector if $\lambda_i \pm \alpha_i$ is not a weight. As in the case of roots, the weights λ_i depend on the choice of the basis set $\{H_i\}$ in the Cartan subalgebra $\mathcal{G}_0$. Similar to Def. 24.2, we have the following.

Definition 25.2. *A weight $\boldsymbol{\lambda}$ is said to be positive if the first non-vanishing component of $\boldsymbol{\lambda}$ is positive. If two weights $\boldsymbol{\lambda}'$ and $\boldsymbol{\lambda}$ are such that $\boldsymbol{\lambda}' - \boldsymbol{\lambda}$ is positive, then $\boldsymbol{\lambda}'$ is said to be a higher weight than $\boldsymbol{\lambda}$. If, in a representation D, a weight $\boldsymbol{\lambda}$ is a higher weight than all other weights, then $\boldsymbol{\lambda}$ is said to be the* **highest weight** *in the representation D.*

Let us define an inner product of two weight vectors $\boldsymbol{\lambda}$ and $\boldsymbol{\gamma}$ in the same weight space by

$$\boldsymbol{\lambda} \cdot \boldsymbol{\gamma} = \sum_i \lambda_i \gamma_i \ , \tag{25.7}$$

and also write

$$(\boldsymbol{\lambda} + \alpha)_i = \lambda_i + \alpha_i \ , \quad \boldsymbol{\lambda} \cdot \alpha \equiv \sum_i \lambda_i \alpha_i \ , \tag{25.8}$$

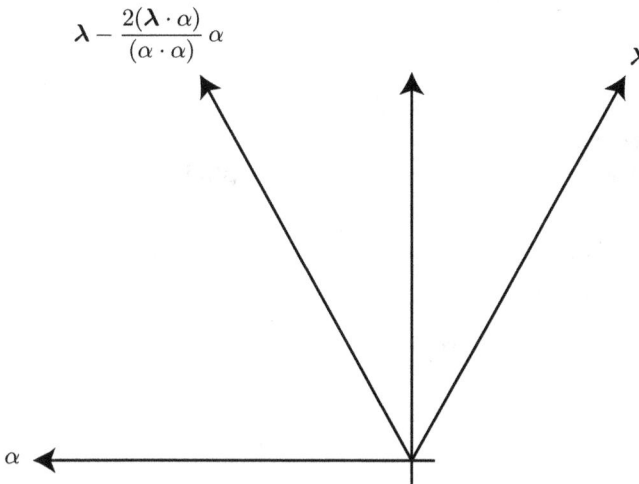

$$\lambda - \frac{2(\boldsymbol{\lambda} \cdot \alpha)}{(\alpha \cdot \alpha)}\,\alpha$$

FIGURE 25.1

where α is a root. We then have the following useful theorems (stated without proofs) concerning weight systems, some of which are close analogs of those developed in the last chapter for roots.

Theorem 25.2. *If λ is any weight (of some representation) and α is any root, then $2(\boldsymbol{\lambda} \cdot \alpha)/(\alpha \cdot \alpha)$ is an integer, and $\boldsymbol{\lambda} - 2\alpha(\boldsymbol{\lambda} \cdot \alpha)/(\alpha \cdot \alpha)$ is also a weight of the same representation. This weight is the reflection of the weight $\boldsymbol{\lambda}$ in the hyperplane (in the weight space Δ) perpendicular to α (see Figure 25.1). The set of all such reflections for all roots forms a group called the* **Weyl group** *of the algebra. (Compare with Theorem 24.2.)*

Exercise 25.1   Show that the Weyl group as defined above is indeed a group.

Exercise 25.2   Use elementary vector algebra (and with the help of Fig. 25.1) to verify that $\boldsymbol{\lambda} - 2\alpha(\boldsymbol{\lambda} \cdot \alpha)/(\alpha \cdot \alpha)$ is indeed the reflection as claimed in the above theorem.

Theorem 25.3. *In any representation space there is at least one weight.*

Theorem 25.4. *If $|\,\boldsymbol{\lambda}\,\rangle$ is an eigenvector belonging to a weight $\boldsymbol{\lambda}$, and can be expressed as a linear superposition of a set of eigenvectors $|\,\boldsymbol{\lambda}_{(k)}\,\rangle$ (belonging to the weights $\boldsymbol{\lambda}_{(k)}$) where none of the weights $\boldsymbol{\lambda}_{(k)}$ is equal to $\boldsymbol{\lambda}$, then $|\,\boldsymbol{\lambda}\,\rangle = 0$.*

As a direct consequence of this theorem we have the following.

Theorem 25.5. *Eigenvectors of H_i belonging to distinct weights are linearly independent. In a representation space of dimension N, there can be at most N distinct weights.*

As mentioned before [c.f. (25.4)], a particular weight may be degenerate. If the eigenspace corresponding to a degenerate weight λ is l-dimensional, then we say that λ is l-fold degenerate, or the **multiplicity** of λ is l.

Theorem 25.6. *The highest weight of an irreducible representation is non-degenerate. If two irreducible representations have identical highest weights, then these two representations are equivalent.*

Theorem 25.7. *A weight vector λ is the highest weight of an irreducible representation if and only if*

$$\lambda_{(\alpha)} \equiv \frac{2(\lambda \cdot \alpha)}{(\alpha \cdot \alpha)} \tag{25.9}$$

is a non-negative integer (including 0) for any simple root α. If $|\Lambda\rangle$ is an eigenvector corresponding to the highest weight Λ, then

$$(E_{-\alpha})^k |\Lambda\rangle \begin{cases} \neq 0 & if \quad k \leq \Lambda_{(\alpha)}, \\ = 0 & if \quad k > \Lambda_{(\alpha)}. \end{cases} \tag{25.10}$$

[Recall (24.43).]

As a consequence of the above two theorems, we have the following important and useful fact.

Theorem 25.8. *An irreducible representation of a semisimple Lie algebra $\mathcal{G}$ of rank m is uniquely determined by its highest weight Λ, and specified completely by the m non-negative integers*

$$\Lambda_{\alpha_{(i)}} = \frac{2\Lambda \cdot \alpha_{(i)}}{\alpha_{(i)} \cdot \alpha_{(i)}}, \qquad i = 1, \dots, m, \tag{25.11}$$

one for each simple root, where $\alpha_{(i)}, i = 1, \dots, m$, are the m simple roots of the algebra.

One usually writes the non-negative integers $\Lambda_{(\alpha_i)}$ on top of each circle in the Dynkin diagram to specify the irreducible representation. For example, for the algebra $\mathcal{SU}(3)$, one has

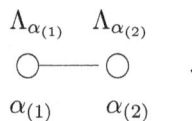

$$\overset{\Lambda_{\alpha_{(1)}} \quad \Lambda_{\alpha_{(2)}}}{\underset{\alpha_{(1)} \qquad \alpha_{(2)}}{\circ\!\!-\!\!\circ}} \quad .$$

Given $\Lambda_{\alpha_{(i)}}, i = 1, \dots, m$, one can compute the highest weight Λ from (25.11). Suppose, for $\mathcal{SU}(3)$,

$$\Lambda = \zeta_1 \alpha_{(1)} + \zeta_2 \alpha_{(2)}, \tag{25.12}$$

where

$$\alpha_{(1)} = \beta, \quad \alpha_{(2)} = \alpha \tag{25.13}$$

(c.f. the root diagram in Figure 24.2). Since

$$\frac{\alpha_{(1)} \cdot \alpha_{(2)}}{\alpha_{(1)} \cdot \alpha_{(1)}} = \frac{\alpha_{(1)} \cdot \alpha_{(2)}}{\alpha_{(2)} \cdot \alpha_{(2)}} = -1/2 , \tag{25.14}$$

we have

$$\frac{\Lambda \cdot \alpha_{(1)}}{\alpha_{(1)} \cdot \alpha_{(1)}} = \zeta_1 \frac{\alpha_{(1)} \cdot \alpha_{(1)}}{\alpha_{(1)} \cdot \alpha_{(1)}} + \zeta_2 \frac{\alpha_{(2)} \cdot \alpha_{(1)}}{\alpha_{(1)} \cdot \alpha_{(1)}} = \zeta_1 - \zeta_2/2 , \tag{25.15}$$

and, by (25.9),

$$\Lambda_{\alpha_{(1)}} = \frac{2\Lambda \cdot \alpha_{(1)}}{\alpha_{(1)} \cdot \alpha_{(1)}} = 2\zeta_1 - \zeta_2 . \tag{25.16}$$

Similarly,

$$\Lambda_{\alpha_{(2)}} = -\zeta_1 + 2\zeta_2 . \tag{25.17}$$

The above two equations give the following expression for the highest weight of $\mathcal{SU}(3)$:

$$\Lambda = \left(\frac{2\Lambda_{\alpha_{(1)}} + \Lambda_{\alpha_{(2)}}}{3} \right) \alpha_{(1)} + \left(\frac{\Lambda_{\alpha_{(1)}} + 2\Lambda_{\alpha_{(2)}}}{3} \right) \alpha_{(2)} . \tag{25.18}$$

Thus, for example,

$$\Lambda_{\underset{\circ-\circ}{1\ \ 0}} = \frac{2}{3} \alpha_{(1)} + \frac{1}{3} \alpha_{(2)} . \tag{25.19}$$

From the highest weight Λ, one can calculate all the weights of the irreducible representation, and obtain the so-called **weight system** of the representation (see Theorem 25.13 below).

$\boxed{\text{Exercise 25.3}}$ Show that for the irreducible representation for $\mathcal{SU}(2) \sim A_1$ with the Dynkin diagram

$$\Lambda_{(\alpha_+)} = l$$
$$\bigcirc$$
$$\alpha_+ \quad ,$$

where l is a non-negative integer and $\alpha_+ = 1$ is the only simple root, the highest weight (of the so-called spin-j representation) is given by

$$\Lambda_{\underset{\circ}{l}} \equiv j = l/2 . \tag{25.20}$$

This is a familiar result in the quantum theory of angular momenta.

Let us first work out some specifics of the spin-j representation of $\mathcal{SU}(2)$ [the angular momentum algebra given by (24.84) and (24.85)] by elementary means. Eq. (25.5) appears as

$$J_3 \,|\, m(\nu)\rangle = m \,|\, m(\nu)\rangle \,, \qquad (25.21)$$

where m is an eigenvalue of J_3 and ν is a degeneracy index. Then, by (25.6),

$$J_3 J_\pm \,|\, m(\nu)\rangle = (J_\pm J_3 \pm J_\pm)\,|\, m(\nu)\rangle = (m \pm 1)J_\pm \,|\, m(\nu)\rangle \,, \qquad (25.22)$$

demonstrating that J_+ and J_- are raising and lowering operators respectively. Since j is the highest weight (highest J_3 eigenvalue), we must have

$$J_+ \,|\, j(\nu)\rangle = 0 \,. \qquad (25.23)$$

Define a normalization constant $N_j(\nu)$ by

$$J_- \,|\, j(\nu)\rangle \equiv N_j(\nu)\,|\, j - 1(\nu)\rangle \,. \qquad (25.24)$$

Then, for arbitrary degeneracy indices μ and ν,

$$\begin{aligned}
\overline{N_j(\mu)} N_j(\nu)\,\langle\, j - 1(\mu)\,|\, j - 1(\nu)\rangle &= \langle\, j(\mu)\,|\, J_+ J_- \,|\, j(\nu)\rangle \\
&= \langle\, j(\mu)\,|\, [\,J_+, J_-\,]\,|\, j(\nu)\rangle = \langle\, j(\mu)\,|\, J_3 \,|\, j(\nu)\rangle = j\,\langle\, j(\mu)\,|\, j(\nu)\rangle \,. \quad (25.25)
\end{aligned}$$

On choosing $|\, j(\mu)\rangle$ to be an orthonormal set:

$$\langle\, j(\mu)\,|\, j(\nu)\rangle = \delta_{\mu\nu} \,, \qquad (25.26)$$

the $|\, j - 1(\mu)\rangle$ can be orthonormalized by setting $|\, N_j(\nu)\,|^2 = j$. We choose a phase so that

$$N_j(\nu) = N_j = \sqrt{j} \,. \qquad (25.27)$$

Then,

$$\begin{aligned}
J_+ \,|\, j - 1\,(\nu)\rangle &= \frac{1}{N_j}\, J_+ J_- \,|\, j\,(\nu)\rangle = \frac{1}{N_j}\,[\,J_+, J_-\,]\,|\, j\,(\nu)\rangle \\
&= \frac{1}{N_j}\, J_3 \,|\, j\,(\nu)\rangle = \frac{j}{N_j}\,|\, j\,(\nu)\rangle = N_j \,|\, j\,(\nu)\rangle \,. \quad (25.28)
\end{aligned}$$

The above equation and (25.24) show that $J_\pm$ do not change the degeneracy index ν. Repeating the action of J_-, we find that there are orthonormal states $|\, j - k\,(\nu)\rangle$ such that [analogous to (25.24) and (25.28)]:

$$J_- \,|\, j - k\,(\nu)\rangle = N_{j-k}\,|\, j - k - 1\,(\nu)\rangle \,, \qquad (25.29a)$$
$$J_+ \,|\, j - k - 1\,(\nu)\rangle = N_{j-k}\,|\, j - k\,(\nu)\rangle \,. \qquad (25.29b)$$

The normalization constants N_{j-k} (chosen real) thus satisfy

$$\begin{aligned}
N_{j-k}^2 &= \langle\, j - k\,(\nu)\,|\, J_+ J_- \,|\, j - k\,(\nu)\rangle \\
&= \langle\, j - k\,(\nu)\,|\, [\,J_+, J_-\,]\,|\, j - k\,(\nu)\rangle + \langle\, j - k\,(\nu)\,|\, J_- J_+ \,|\, j - k\,(\nu)\rangle \\
&= N_{j-k+1}^2 + j - k \,.
\end{aligned}$$

$$(25.30)$$

Incorporating (25.27), this recursion relation can be displayed explicitly as follows:

$$N_j^2 = j \,, \quad N_{j-1}^2 - N_j^2 = j - 1 \,, \ldots \,, \quad N_{j-k}^2 - N_{j-k+1}^2 = j - k \,. \quad (25.31)$$

Adding the above, we obtain

$$N_{j-k}^2 = (k+1)j - \frac{k(k+1)}{2} = \frac{1}{2}(k+1)(2j-k) \,. \quad (25.32)$$

Letting

$$j - k = m \,, \quad (25.33)$$

we have

$$\boxed{N_m = \frac{1}{\sqrt{2}} \sqrt{(j+m)(j-m+1)}} \,, \quad (25.34)$$

which is also a standard result in the quantum theory of angular momenta. Eq. (25.32) implies that

$$N_{j-2j} = N_{-j} = 0 \,. \quad (25.35)$$

Thus, after $2j$ applications of J_- on $|j(\nu)\rangle$, where j is the highest weight, the series (25.29a) terminates, and the representation decomposes into n irreducible representations ($n = $ degree of degeneracy), each of dimension $2j+1$. So a spin-j irreducible representation of $\mathcal{SU}(2)$ has a dimension of $2j+1$, in which the eigenvalues of J_3 are $j, j-1, \ldots, -j$, with the corresponding eigenstates labeled as $|jm\rangle$. These eigenstates are orthonormal:

$$\langle jm' | jm \rangle = \delta_{m'm} \,,$$

since J_3 is hermitian [c.f. discussion prior to (13.17)]. Eqs. (25.29) and (25.34) imply that

$$J_+ |jm\rangle = \frac{1}{\sqrt{2}} \sqrt{j(j+1) - m(m+1)} \, |j, m+1\rangle \,, \quad (25.36a)$$

$$J_- |j, m+1\rangle = \frac{1}{\sqrt{2}} \sqrt{j(j+1) - m(m+1)} \, |j, m\rangle \,. \quad (25.36b)$$

Exercise 25.4   Show that $\sigma_i/2$, $i = 1, 2, 3$, where σ_i are the Pauli spin matrices given by (11.16), give a spin-$\frac{1}{2}$ representation of $\mathcal{SU}(2)$.

For the general calculation of the weight system of an irreducible representation starting with the highest weight $\mathbf{\Lambda}$, it is useful to introduce some additional terminology.

Definition 25.3. *In an irreducible representation D, if a weight $\lambda \in \Delta_D$ can be obtained from the highest weight Λ by the subtraction from Λ of k simple roots, then λ is said to belong to the k-th **level** $\Delta_D^{(k)}$ of the weight space Δ_D. Thus the highest weight Λ belongs to the 0-th level $\Delta_D^{(0)}$. The level of the lowest weight, denoted by T_D, is called the **height** of the irreducible representation D. Thus*

$$\Delta_D = \Delta_D^{(0)} \cup \Delta_D^{(1)} \cup \cdots \cup \Delta_D^{(T_D)} . \tag{25.37}$$

Let us denote the number of weights (including multiplicities) belonging to the k-th level of Δ_D by $S_k(D)$. Then the dimension N of the irreducible representation D is given by

$$N = 1 + S_1(D) + S_2(D) + \cdots + S_{T_D}(D) . \tag{25.38}$$

Definition 25.4. *The integer*

$$m_D \equiv max \ S_k(D) \tag{25.39}$$

*is called the **width** of the irreducible representation D.*

We have the following theorems (given without proofs) concerning multiplicities of weights.

Theorem 25.9. *The multiplicity ν_λ of a weight λ of a representation D with highest weight Λ is given by the following iteration formula:*

$$\{(\Lambda + g) \cdot (\Lambda + g) - (\lambda + g) \cdot (\lambda + g)\} \nu_\lambda$$

$$= 2 \sum_{\alpha \in P^+} \sum_{k=0}^{\infty} \{(\lambda + k\alpha) \cdot \alpha\} \nu_{\lambda + k\alpha} , \tag{25.40}$$

where P^+ is the set of positive roots, and

$$g \equiv \frac{1}{2} \sum_{\alpha \in P^+} \alpha . \tag{25.41}$$

Theorem 25.10. *The multiplicities of the various levels of the weight space Δ_D of an irreducible representation D satisfies the following conditions:*

$$S_k(D) = S_{T_D - k}(D) , \tag{25.42}$$
$$S_h(D) \geq S_{h-1}(D) \geq \cdots \geq S_1(D) \geq 1 , \tag{25.43}$$

where

$$h \equiv \begin{cases} T_D/2 & , \ T_D \ even , \\ (T_D - 1)/2 & , \ T_D \ odd . \end{cases} \tag{25.44}$$

Corollary 25.1. *The width of the weight system of an irreducible representation D is given by*

$$m_D = \begin{cases} S_h(D) & , \ T_D = 2h , \\ S_{h+1}(D) & , \ T_D = 2h + 1 . \end{cases} \tag{25.45}$$

Explicit results for the heights and dimensions of irreducible representations are given by the following two theorems

Theorem 25.11. *Suppose Λ is the highest weight of an irreducible representation D of a semisimple Lie algebra, then the height T_D of the weight system of the irreducible representation D is given by*

$$T_D = \sum_{i,\,\alpha_{(i)} \in \Pi} r_i \Lambda_{\alpha_{(i)}} , \tag{25.46}$$

where the sum is over the simple roots $\alpha_{(i)}$, $\Lambda_{\alpha_{(i)}}$ is given by (25.11), and the coefficients r_i are completely given according to the Cartan classification of complex simple Lie algebras (c.f. Theorem 24.7) in Table 25.1.

Theorem 25.12. *The dimension N of an irreducible representation D of a semisimple Lie algebra with highest weight Λ is given by*

$$N = \prod_{\alpha \in P^+} \left(\frac{\mathbf{\Lambda} \cdot \alpha}{g \cdot \alpha} + 1 \right) , \tag{25.47}$$

where the product is over the positive roots, and g is defined in (25.41).

The following theorem gives a procedure for the determination of the weight system of an irreducible representation when the highest weight is known.

Theorem 25.13. *Suppose that in an irreducible representation D of a semisimple Lie algebra, the weight λ belongs to the level $\Delta_D^{(k-1)}$, then*

$$\lambda - \alpha \in \Delta_D^{(k)} , \quad \alpha \in \Pi \text{ (α is a simple root)} \tag{25.48}$$

if and only if

$$\frac{2(\lambda \cdot \alpha)}{\alpha \cdot \alpha} + q \geq 1 , \tag{25.49}$$

where the integer q is determined by the following condition:

$$\lambda + \alpha, \lambda + 2\alpha, \ldots, \lambda + q\alpha$$

are weights, but $\lambda + (q+1)\alpha$ is not a weight.

As an example, let us calculate the weight system of the representation

$$\overset{0}{\underset{}{\bigcirc}}\!\!-\!\!\overset{1}{\underset{}{\bigcirc}}$$

of the algebra $A_2 \sim \mathcal{SU}(3)$. For this case $\Lambda_{\alpha_{(1)}} = 0$ and $\Lambda_{\alpha_{(2)}} = 1$, and (25.18) implies that the highest weight is given by

$$\Lambda = \frac{1}{3} \alpha_{(1)} + \frac{2}{3} \alpha_{(2)} = \frac{1}{3}(\alpha + 2\beta) , \tag{25.50}$$

where α and β are the simple roots in Fig. 24.2. Eq. (25.46), together with Table 25.1, show that the height of this representation is

$$T_D = 2(0) + 2(1) = 2 \,,$$

so that there are three levels. To find the weights in the first level $\Delta_D^{(1)}$, we consider the quantities $\Lambda - \alpha$ and $\Lambda - \beta$. We have

$$\frac{2(\Lambda \cdot \alpha)}{\alpha \cdot \alpha} = \frac{2}{3} \frac{(\alpha + 2\beta) \cdot \alpha}{\alpha \cdot \alpha} = \frac{2}{3} \left(1 + \frac{2\alpha \cdot \beta}{\alpha \cdot \alpha} \right) = \frac{2}{3} \left(1 + 2(-\frac{1}{2}) \right) = 0 \,, \quad (25.51)$$

$$\frac{2(\Lambda \cdot \beta)}{\beta \cdot \beta} = \frac{2}{3} \frac{(\alpha + 2\beta) \cdot \beta}{\beta \cdot \beta} = \frac{2}{3} \left(\frac{\alpha \cdot \beta}{\beta \cdot \beta} + 2 \right) = \frac{2}{3} \left(-\frac{1}{2} + 2 \right) = 1 \,. \quad (25.52)$$

Since Λ is the highest weight, neither $\Lambda + \alpha$ nor $\Lambda + \beta$ is a weight. So $q = 0$ in (25.49) for both simple roots α and β. It follows then that

$$\frac{2(\Lambda \cdot \alpha)}{\alpha \cdot \alpha} + q = \frac{2(\Lambda \cdot \alpha)}{\alpha \cdot \alpha} = 0 < 1 \,, \quad (25.53)$$

$$\frac{2(\Lambda \cdot \beta)}{\beta \cdot \beta} + q = \frac{2(\Lambda \cdot \beta)}{\beta \cdot \beta} = 1 \,, \quad (25.54)$$

and condition (25.49) is only satisfied for the simple root β. Thus $\Lambda - \alpha$ is not a weight and $\Lambda - \beta = \frac{1}{3}(\alpha - \beta)$ is the only weight in the first level $\Delta_D^{(1)}$. To continue with the second level we compute

$$\frac{2(\Lambda - \beta) \cdot \alpha}{\alpha \cdot \alpha} = \frac{2}{3} \frac{(\alpha - \beta) \cdot \alpha}{\alpha \cdot \alpha} = \frac{2}{3} \left(1 - \frac{\alpha \cdot \beta}{\alpha \cdot \alpha} \right) = \frac{2}{3} \left(1 - (-\frac{1}{2}) \right) = 1 \,,$$
$$(25.55)$$

$$\frac{2(\Lambda - \beta) \cdot \beta}{\beta \cdot \beta} = \frac{2}{3} \frac{(\alpha - \beta) \cdot \beta}{\beta \cdot \beta} = \frac{2}{3} \left(\frac{\alpha \cdot \beta}{\beta \cdot \beta} - 1 \right) = \frac{2}{3} \left(-\frac{1}{2} - 1 \right) = -1 \,.$$
$$(25.56)$$

Writing out the series

$$\Lambda - \beta, \, (\Lambda - \beta) + \alpha, \, \ldots \,, \qquad \Lambda - \beta, \, (\Lambda - \beta) + \beta, \, \ldots \,, \quad (25.57)$$

we see that $q = 0$ for the first case ($\Lambda + \alpha - \beta$ is not a weight) and $q = 1$ for the second (since $\Lambda - \beta + \beta = \Lambda$ is the highest weight). Thus

$$\frac{2(\Lambda - \beta) \cdot \alpha}{\alpha \cdot \alpha} + q = 1 + 0 = 1 \,, \quad (25.58)$$

$$\frac{2(\Lambda - \beta) \cdot \beta}{\beta \cdot \beta} + q = -1 + 0 < 1 \,. \quad (25.59)$$

It follows from the condition (25.49) that $\Lambda - \beta - \alpha = -\frac{1}{3}(2\alpha + \beta)$ is the only weight in the second level $\Delta_D^{(2)}$, whereas $(\Lambda - \beta) - \beta$ is not a weight. We need go no further since it has already been established that there are only three levels.

The entire weight system of the irreducible representation $(\Lambda_\alpha = 0, \Lambda_\beta = 1)$ of $A_2 \sim \mathcal{SU}(3)$ is given by

$$\Lambda = \frac{1}{3}(\alpha + 2\beta) \in \Delta_D^{(0)},$$

$$\Lambda - \beta = \frac{1}{3}(\alpha - \beta) \in \Delta_D^{(1)}, \qquad (25.60)$$

$$\Lambda - \beta - \alpha = -\frac{1}{3}(2\alpha + \beta) \in \Delta_D^{(2)}.$$

We have $S_1(D) = 1, S_2(D) = 1$. Thus the width of this representation is $m_D = S_1(D) = 1$, and the dimension of this representation is 3 [by (25.38)].

Exercise 25.5 Find the weight system of the irreducible representation

0   2

◯—◯

of A_2.

Among all the irreducible representations of a semisimple Lie algebra, the so-called fundamental and primary representations are especially important.

Definition 25.5. *An irreducible representation in which all the $\Lambda_{\alpha_{(i)}}$, ($i = 1, \ldots, m = $ rank of the algebra) are 0 except one with the value 1 is called a* ***fundamental representation.***

A rank-m algebra has m fundamental representations. For example, the three fundamental representations of C_3 are

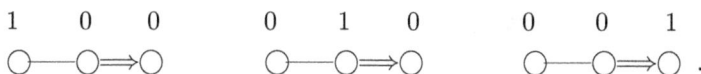

1   0   0   0   1   0   0   0   1

◯—◯⇒◯   ◯—◯⇒◯   ◯—◯⇒◯ .

Definition 25.6. *A fundamental representation with the value 1 (for $\Lambda_{\alpha_{(i)}}$) attached to an end circle of the Dynkin diagram is called a* ***primary representation.***

For example, the two primary representations of C_3 are

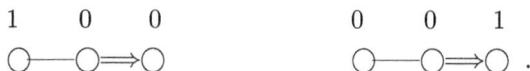

1   0   0   0   0   1

◯—◯⇒◯   ◯—◯⇒◯ .

The importance of the primary representations rests on the following facts (stated without proofs).

Theorem 25.14. *All roots of a semisimple Lie algebra can be expressed in terms of the weights of a primary representation of the algebra.*

For example, the representation

$$
\begin{array}{cc}
0 & 1 \\
\circ\!\!-\!\!\!-\!\!\circ
\end{array}
$$

of A_2 is clearly a primary representation, whose weights have been worked out and given in (25.60). Denote these weights by

$$\boldsymbol{\lambda}_{(1)} \equiv \boldsymbol{\Lambda}, \quad \boldsymbol{\lambda}_{(2)} \equiv \boldsymbol{\Lambda} - \beta, \quad \boldsymbol{\lambda}_{(3)} \equiv \boldsymbol{\Lambda} - (\alpha + \beta).$$

Then the equations (25.60) can be solved for the simple roots α and β, and thus for all roots of A_2, in terms of $\boldsymbol{\lambda}_{(1)}, \boldsymbol{\lambda}_{(2)}$ and $\boldsymbol{\lambda}_{(3)}$.

Theorem 25.15. *The highest weight $\boldsymbol{\Lambda}$ of any irreducible representation of a semisimple Lie algebra can be written as a linear combination of the weights $\boldsymbol{\lambda}_{(k)}$ of a primary representation of the algebra:*

$$\boldsymbol{\Lambda} = \sum_k c_k \boldsymbol{\lambda}_{(k)} , \tag{25.61}$$

where the coefficients c_k depend only on the m quantities [c.f. (25.11)]

$$\Lambda_{\alpha_{(i)}} = \frac{2\boldsymbol{\Lambda} \cdot \alpha_{(i)}}{\alpha_{(i)} \cdot \alpha_{(i)}} , \quad i = 1, \ldots, m ,$$

with $\alpha_{(i)}$ being the simple roots and m the rank of the algebra.

Theorem 25.16. *Suppose [c.f. (25.41)]*

$$g \equiv \frac{1}{2} \sum_{\alpha \in P^+} \alpha ,$$

(one-half the sum of all the positive roots), then,

i) for a simple root $\alpha \, (\in \Pi)$,

$$g_\alpha \equiv \frac{2(g \cdot \alpha)}{\alpha \cdot \alpha} = 1 ; \tag{25.62}$$

ii) g can be written as a linear superposition of the weights $\boldsymbol{\lambda}_{(k)}$ of a primary representation:

$$g = \sum_k g_k \boldsymbol{\lambda}_{(k)} , \tag{25.63}$$

where the g_k are numerical constants characteristic of the algebra.

The specific details of how the above three theorems can be implemented according to the Cartan classification can be found in standard reference works on Lie algebras (see, for example, W. Foulton and J. Harris 1991), and will not be given here.

Using the result of (25.47) of Theorem 25.12, in conjunction with (25.61) and (25.63), one can express the dimension N of an irreducible representation in terms of the c_k and g_k, and thus ultimately in terms of only the $\Lambda_{\alpha_{(i)}}$ [c.f. (25.11)]. For example, for the algebra $A_2 \sim \mathcal{SU}(3)$, the dimensions of the various irreducible representations are given by

$$N = (1 + \Lambda_{\alpha_{(1)}})(1 + \Lambda_{\alpha_{(2)}}) \left(1 + \frac{\Lambda_{\alpha_{(1)}} + \Lambda_{\alpha_{(2)}}}{2} \right) . \tag{25.64}$$

Denoting the dimensions by $N(\Lambda_{\alpha_{(1)}}, \Lambda_{\alpha_{(2)}})$ we thus have,

$$\begin{aligned}
N(0,0) &= 1 , \\
N(1,0) &= N(0,1) = 3 , \\
N(1,1) &= 8 , \\
N(2,0) &= N(0,2) = 6 , \\
N(1,2) &= N(2,1) = 15 , \\
N(2,2) &= 27 , \\
N(3,0) &= N(0,3) = 10 , \text{ etc.}
\end{aligned} \tag{25.65}$$

The representation (j, i) is in fact the complex conjugate representation of (i, j). This can be seen as follows. Recall the commutation relations (24.31) to (24.34), which define the algebra. Using the same symbols H_i and E_α to denote the respective representation matrices in a particular representation, and the facts that, for a general matrix A, the complex conjugate matrix $\overline{A}$ is given by $(A^\dagger)^T$ (the transpose of the hermitian conjugate matrix), $(H_i)^\dagger = H_i$, $E_\alpha^\dagger = E_{-\alpha}$, and (24.77), the equations (24.31) to (24.34), regarded as matrix equations, are equivalent to

$$[-\overline{H_i}, -\overline{H_j}] = 0 , \quad i, j = 1, \ldots, m , \tag{25.66}$$
$$[-\overline{H_i}, -\overline{E_{-\alpha}}] = \alpha_i \left(-\overline{E_{-\alpha}} \right) , \tag{25.67}$$
$$[-\overline{E_{-\alpha}}, -\overline{E_\alpha}] = \alpha^i \left(-\overline{H_i} \right) , \tag{25.68}$$
$$[-\overline{E_{-\alpha}}, -\overline{E_{-\beta}}] = N_{\alpha\beta} \left(-\overline{E_{-\alpha-\beta}} \right) . \tag{25.69}$$

Thus, as representation matrices,

$$H_i \longrightarrow -\overline{H_i} , \qquad E_\alpha \longrightarrow -\overline{E_{-\alpha}} , \tag{25.70}$$

is also a representation, called the complex conjugate representation, denoted by $\overline{D}$, of the original one, D. $\overline{D}$ has the same dimension as D, and the set of weights of $\overline{D}$ are the negatives of those of D. It follows that the highest weight of $\overline{D}$ is the negative of the lowest weight of D. A representation D is said to be **real** if $\overline{D}$ is equivalent to D, otherwise it is said to be **complex**. Thus, if the lowest weight of D is the negative of the highest weight of D, then D is real. For $\mathcal{SU}(3)$, for example, it is easily seen that the highest weight of $(0, 1)$ is the negative of the lowest weight of $(1, 0)$. Thus, $(1, 0)$ and $(0, 1)$ are complex conjugate representations of each other. In the physics literature, it is quite common to designate a representation by its dimension. Thus we write $(1,0) = \boxed{3}$ and $(0,1) = \overline{\boxed{3}}$. Now consider a representation (i,j) of $\mathcal{SU}(3)$. It can be thought of as constructed from the primary representations $(1,0)$ and $(0,1)$. In fact, the highest weight of (i,j) is given by

$$\Lambda_{(i,j)} = i\Lambda_{(1,0)} + j\Lambda_{(0,1)} , \tag{25.71}$$

where $\Lambda_{(1,0)}$ and $\Lambda_{(0,1)}$ are the highest weights of $(1,0)$ and $(0,1)$, respectively. But the lowest weight of (j,i) is given by j times the lowest weight of $(1,0)$ plus i times the lowest weight of $(0,1)$, or

$$-j\Lambda_{(0,1)} - i\Lambda_{(1,0)} ,$$

which is just the negative of the highest weight $\Lambda_{(i,j)}$ of (i,j). Thus, (j,i) is the complex conjugate representation of (i,j). A representation of the form (i,i) is necessarily real.

We conclude this chapter with the description of a procedure (without proof) to use Young diagrams to determine the Clebsch-Gordon decomposition of direct products of irreducible representations of Lie algebras into sums of irreducible representations. This procedure closely resembles that for group representations described in Chapter 21 [c.f. the discussion leading up to (21.13)].

Given an irreducible representation with highest weight $\mathbf{\Lambda}$, and characterized by the m non-negative integers $\Lambda_{\alpha(1)},\dots,\Lambda_{\alpha(m)}$, where $\alpha_{(1)},\dots,\alpha_{(m)}$ are the simple roots of the algebra, there corresponds a Young diagram of the form

$$(25.72)$$

where there are at most m rows, the first row has

$$\Lambda_{\alpha(m)} + \Lambda_{\alpha(m-1)} + \cdots + \Lambda_{\alpha(3)} + \Lambda_{\alpha(2)} + \Lambda_{\alpha(1)} \text{ boxes} ,$$

the second row has

$$\Lambda_{\alpha(m)} + \Lambda_{\alpha(m-1)} + \cdots + \Lambda_{\alpha(3)} + \Lambda_{\alpha(2)} \text{ boxes} ,$$

etc., and the m-th row has $\Lambda_{\alpha(m)}$ boxes. Thus there are a total of

$$\Lambda_{\alpha(1)} + 2\Lambda_{\alpha(2)} + \cdots + m\Lambda_{\alpha(m)}$$

boxes. For example, the Young diagram corresponding to the representation $(\Lambda_{\alpha_{(1)}} = 1, \Lambda_{\alpha_{(2)}} = 2)$ is

A diagram with no boxes at all, or only with columns consisting of more than m boxes, are all taken to correspond to the representation

$$\Lambda_{\alpha_{(1)}} = \Lambda_{\alpha_{(2)}} = \cdots = \Lambda_{\alpha_{(m)}} = 0 ,$$

which is one-dimensional. The rules for the Clebsch-Gordon decomposition are exactly the same as those described in Chapter 21 [in the paragraph ending with (21.13)], with the additional step at the end that any column with more than m (= rank of algebra) boxes can be deleted from the diagram. If after such a deletion from a diagram no boxes are left, then the diagram corresponds to a 1-dimensional representation. We illustrate with the following examples for $SU(3)$:

$$\square \otimes \boxed{1} = \boxed{1} \oplus \boxed{\begin{smallmatrix}\\1\end{smallmatrix}} = (2,0) \oplus (0,1) , \tag{25.73}$$

or $\quad ③ \otimes ③ = ⑥ \oplus \overline{③} ;$ (25.74)

$$\boxed{\begin{smallmatrix}\\ \end{smallmatrix}} \otimes \boxed{1} = \boxed{\begin{smallmatrix}1\\ \end{smallmatrix}} \oplus \boxed{\begin{smallmatrix}\\ \\1\end{smallmatrix}} = (1,1) \oplus (0,0) , \tag{25.75}$$

or $\quad \overline{③} \otimes ③ = ⑧ \oplus ① ;$ (25.76)

$$(25.77)$$

or $\quad ⑧ \otimes ⑧ = ㉗ \oplus ⑩ \oplus \overline{⑩} \oplus ⑧ \oplus ⑧ \oplus ① . \tag{25.78}$$

$\boxed{\text{Exercise 25.6}}$ Find the Clebsch-Gordon decomposition of $\begin{smallmatrix}\square\square\\\square\end{smallmatrix} \otimes \square\square$ for $SU(3)$.

Finally we give a convenient formula to calculate the dimension of an irreducible representation of $SU(n)$ specified by a particular Young diagram.

Theorem 25.17. *Suppose the Young diagram corresponding to a representation D of $SU(n)$ has r_1 boxes in the first row, r_2 boxes in the second row,..., and r_{n-1} boxes in the $(n-1)$-th row $[n-1 = \text{rank of } SU(n)]$. Then the dimension of D, $N_{(D)}$, is obtained by the following procedure:*

(1) *Put the numbers $n, n+1, n+2, \ldots, n+r_1-1$ in succession from left to right in the boxes in the first row; put the numbers $n-1, n, n+1, \ldots, (n-1)+(r_2-1)$ in succession from left to right in the boxes in the second row; put the numbers $n-2, n-1, \ldots, n-2+(r_3-1)$ in succession from left to right in the boxes in the third row, etc., until all the boxes are filled.*

(2) *Multiply all the numbers in the boxes, call the result M.*

(3) *Calculate the product of all the hook lengths of the Young diagram (c.f. Def. 20.3), call the result h.*

(4)

$$N_{(D)} = \frac{M}{h} \, . \tag{25.79}$$

As examples, consider $D_1 = \begin{array}{|c|c|c|}\hline & & \\\hline\end{array}$ of $SU(3)$ and $D_2 = \begin{array}{|c|}\hline \\\hline \\\hline\end{array}$ of $SU(4)$. We have

$$N_{(D_1)} = \left(\begin{array}{|c|c|c|c|}\hline 3 & 4 & 5 & 6 \\\hline 2 & 3 & & \\\hline\end{array}\right) \div \left(\begin{array}{|c|c|c|c|}\hline 5 & 4 & 2 & 1 \\\hline 2 & 1 & & \\\hline\end{array}\right) = \frac{3.4.5.6.2.3}{5.4.2.1.2.1} = 27 \, , \tag{25.80}$$

$$N_{(D_2)} = \left(\begin{array}{|c|c|}\hline 4 & 5 \\\hline 3 & \\\hline 2 & \\\hline\end{array}\right) \div \left(\begin{array}{|c|c|}\hline 4 & 1 \\\hline 2 & \\\hline 1 & \\\hline\end{array}\right) = \frac{4.5.3.2}{4.1.2.1} = 15 \, . \tag{25.81}$$

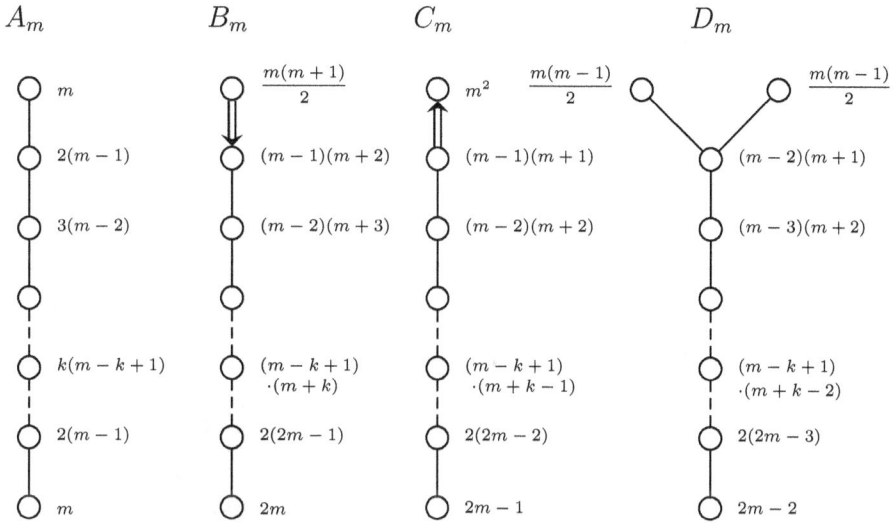

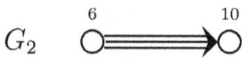

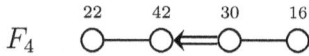

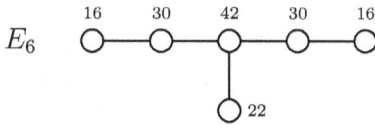

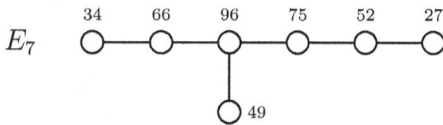

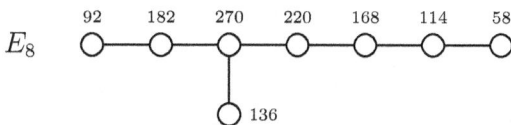

A_m $\qquad$ B_m $\qquad$ C_m $\qquad$ D_m

A_m	B_m	C_m	D_m
m	$\dfrac{m(m+1)}{2}$	m^2	$\dfrac{m(m-1)}{2}$ $\qquad$ $\dfrac{m(m-1)}{2}$
$2(m-1)$	$(m-1)(m+2)$	$(m-1)(m+1)$	$(m-2)(m+1)$
$3(m-2)$	$(m-2)(m+3)$	$(m-2)(m+2)$	$(m-3)(m+2)$
$k(m-k+1)$	$(m-k+1)$ $\cdot(m+k)$	$(m-k+1)$ $\cdot(m+k-1)$	$(m-k+1)$ $\cdot(m+k-2)$
$2(m-1)$	$2(2m-1)$	$2(2m-2)$	$2(2m-3)$
m	$2m$	$2m-1$	$2m-2$

G_2 $\quad$ $\overset{6}{\bigcirc}\!\!\Longrightarrow\!\!\overset{10}{\bigcirc}$

F_4 $\quad$ $\overset{22}{\bigcirc}\!-\!\overset{42}{\bigcirc}\!\Longleftarrow\!\overset{30}{\bigcirc}\!-\!\overset{16}{\bigcirc}$

E_6 $\quad$ $\overset{16}{\bigcirc}\!-\!\overset{30}{\bigcirc}\!-\!\overset{42}{\bigcirc}\!-\!\overset{30}{\bigcirc}\!-\!\overset{16}{\bigcirc}$
$\underset{22}{\bigcirc}$

E_7 $\quad$ $\overset{34}{\bigcirc}\!-\!\overset{66}{\bigcirc}\!-\!\overset{96}{\bigcirc}\!-\!\overset{75}{\bigcirc}\!-\!\overset{52}{\bigcirc}\!-\!\overset{27}{\bigcirc}$
$\underset{49}{\bigcirc}$

E_8 $\quad$ $\overset{92}{\bigcirc}\!-\!\overset{182}{\bigcirc}\!-\!\overset{270}{\bigcirc}\!-\!\overset{220}{\bigcirc}\!-\!\overset{168}{\bigcirc}\!-\!\overset{114}{\bigcirc}\!-\!\overset{58}{\bigcirc}$
$\underset{136}{\bigcirc}$

TABLE 25.1

Chapter 26

$SU(3)$ and the Strong Interaction

The irreducible representations of $\mathcal{SU}(3)$, the Lie algebra of the group of 3×3 unitary matrices of determinant $1, SU(3)$, have found numerous interesting applications in strong interaction physics. Among these we will focus on an early one, called the **flavor-$SU(3)$** quark model of hadrons (baryons and mesons).

We will begin with a standard set of generators [for $SU(3)$] commonly used in the physics literature. These are given in terms of the so-called **Gell-Mann matrices**, $G_i, i = 1, \ldots, 8$, which are generalizations of the Pauli spin matrices, and correspond to the 3-dimensional $(1,0)$ representation of $\mathcal{SU}(3)$. For any $g \in SU(3)$, $g = \exp\left(\frac{i}{2} \epsilon^i G_i\right)$, $\quad i = 1, \ldots, 8$,

$$G_1 = \begin{pmatrix} 0 & 1 & 0 \\ 1 & 0 & 0 \\ 0 & 0 & 0 \end{pmatrix}, \quad G_2 = \begin{pmatrix} 0 & -i & 0 \\ i & 0 & 0 \\ 0 & 0 & 0 \end{pmatrix}, \quad G_3 = \begin{pmatrix} 1 & 0 & 0 \\ 0 & -1 & 0 \\ 0 & 0 & 0 \end{pmatrix},$$

$$G_4 = \begin{pmatrix} 0 & 0 & 1 \\ 0 & 0 & 0 \\ 1 & 0 & 0 \end{pmatrix}, \quad G_5 = \begin{pmatrix} 0 & 0 & -i \\ 0 & 0 & 0 \\ i & 0 & 0 \end{pmatrix}, \quad G_6 = \begin{pmatrix} 0 & 0 & 0 \\ 0 & 0 & 1 \\ 0 & 1 & 0 \end{pmatrix},$$

$$G_7 = \begin{pmatrix} 0 & 0 & 0 \\ 0 & 0 & -i \\ 0 & i & 0 \end{pmatrix}, \quad G_8 = \frac{1}{\sqrt{3}} \begin{pmatrix} 1 & 0 & 0 \\ 0 & 1 & 0 \\ 0 & 0 & -2 \end{pmatrix}. \tag{26.1}$$

The generators are

$$T_i = G_i/2 , \quad i = 1, \ldots, 8 . \tag{26.2}$$

Note that all the generators are traceless hermitian matrices. These properties are required of any element in the Lie algebra of $SU(n)$; tracelessness being required by unimodularity (determinant $= 1$), and hermiticity being required by unitarity. Note also that

$$Tr\, (T_i T_j) = \frac{1}{2} \delta_{ij} . \tag{26.3}$$

T_1, T_2 and T_3 generate a subgroup of $SU(3)$ isomorphic to $SU(2)$, called the **isospin subgroup**. In fact, the upper-left 2×2 blocks of G_1, G_2, G_3 are precisely the Pauli spin matrices σ_1, σ_2 and σ_3 [c.f. (11.16)]. Since G_3 is diagonal, we take T_3 to be an element of the Cartan subalgebra. It is easily seen that the only other generator that commutes with T_3 is T_8. Thus the Cartan subalgebra is 2-dimensional, and we set

$$H_1 = T_3 , \qquad H_2 = T_8 . \tag{26.4}$$

Since T_3 and T_8 are already diagonal, the simultaneous eigenvectors are

$$\begin{pmatrix} 1 \\ 0 \\ 0 \end{pmatrix} = \left| \frac{1}{2}, \frac{1}{2\sqrt{3}} \right\rangle , \quad \begin{pmatrix} 0 \\ 1 \\ 0 \end{pmatrix} = \left| -\frac{1}{2}, \frac{1}{2\sqrt{3}} \right\rangle , \quad \begin{pmatrix} 0 \\ 0 \\ 1 \end{pmatrix} = \left| 0, -\frac{1}{\sqrt{3}} \right\rangle , \tag{26.5}$$

where $|\lambda_1, \lambda_2\rangle$ [c.f. (25.5)] are non-degenerate states satisfying

$$H_1 |\lambda_1 \lambda_2\rangle = \lambda_1 |\lambda_1 \lambda_2\rangle , \qquad H_2 |\lambda_1 \lambda_2\rangle = \lambda_2 |\lambda_1 \lambda_2\rangle . \tag{26.6}$$

The weight vectors of this representation are

$$\left(\frac{1}{2}, \frac{1}{2\sqrt{3}} \right) , \quad \left(0, -\frac{1}{\sqrt{3}} \right) , \quad \left(-\frac{1}{2}, \frac{1}{2\sqrt{3}} \right) ,$$

with the first one being the highest weight. We can plot these weight vectors λ in the $\lambda_1 - \lambda_2$ plane (Fig. 26.1).

For comparison, it is also useful to plot the root vectors (c.f. Fig. 24.2) $\pm\alpha, \pm\beta, \alpha + \beta, -\alpha - \beta$ on the same plane (Fig. 26.2). Note that in Fig. 26.2 we have also indicated the two zero-roots with the two crosses at the origin.

It is not hard to work out the raising and lowering operators E_α in terms of the generators T_i. We have

$$E_{(\pm 1, 0)} = \frac{1}{\sqrt{2}} (T_1 \pm iT_2) , \quad E_{(\pm 1/2, \pm\sqrt{3}/2)} = \frac{1}{\sqrt{2}} (T_4 \pm iT_5) ,$$

$$E_{(\mp 1/2, \pm\sqrt{3}/2)} = \frac{1}{\sqrt{2}} (T_6 \pm iT_7) . \tag{26.7}$$

$(0, 1) = \overline{③}$ representation

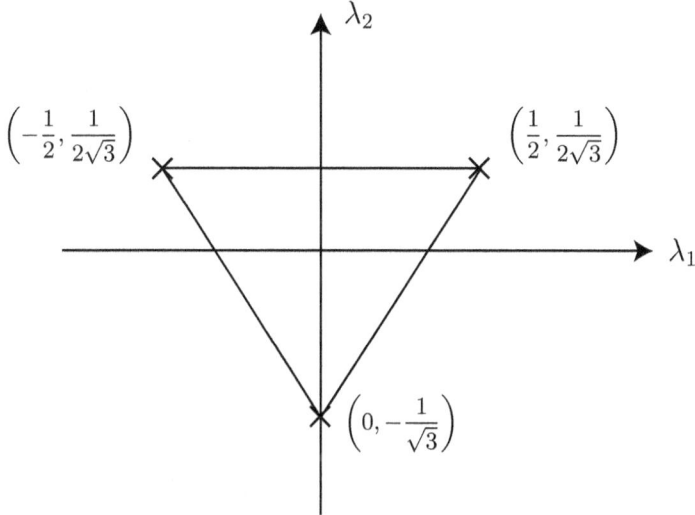

FIGURE 26.1

Exercise 26.1 | Verify the three equations in (26.7).

Starting with $|1/2, 1/(2\sqrt{3})\rangle$, the eigenvector corresponding to the highest weight, the two others in (26.5) are obtained by

$$\left| 0, -\frac{1}{\sqrt{3}} \right\rangle = E_{(-1/2, -\sqrt{3}/2)} \left| \frac{1}{2}, \frac{1}{2\sqrt{3}} \right\rangle = E_{-\beta} \left| \frac{1}{2}, \frac{1}{2\sqrt{3}} \right\rangle, \qquad (26.8)$$

$$\left| -\frac{1}{2}, \frac{1}{2\sqrt{3}} \right\rangle = E_{(-1/2, \sqrt{3}/2)} \left| 0, -\frac{1}{\sqrt{3}} \right\rangle = E_{-\alpha} \left| 0, -\frac{1}{\sqrt{3}} \right\rangle$$
$$= E_{-\alpha} E_{-\beta} \left| \frac{1}{2}, \frac{1}{2\sqrt{3}} \right\rangle. \qquad (26.9)$$

Each is obtained in a unique way by applications of lowering operators. Since the highest-weight eigenvector is non-degenerate, the two other eigenvectors are non-degenerate also. In fact, a similar and more general statement is true for the eigenvectors of any representation of a semisimple Lie algebra: *Any state obtained in a unique way by applications of lowering operators on the highest-weight eigenvector is non-degenerate.*

From our discussion in the last chapter [following (25.77)], the weights of the other primary representation, the complex conjugate representation of $(1, 0) = ③$, namely, $(0, 1) = \overline{③}$, are just the negatives of those of $③$. These can be

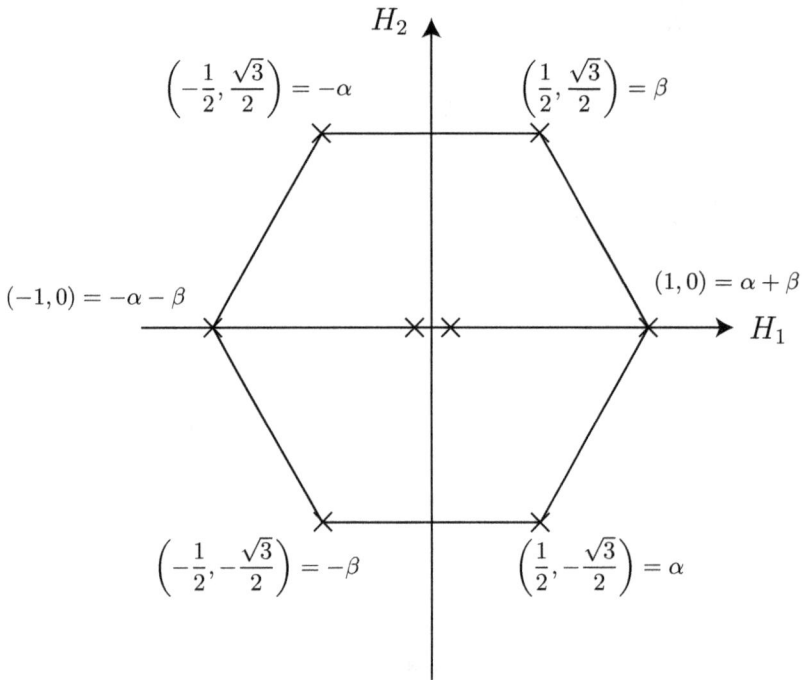

$$
\begin{array}{cc}
\left(-\dfrac{1}{2}, \dfrac{\sqrt{3}}{2}\right) = -\alpha & \left(\dfrac{1}{2}, \dfrac{\sqrt{3}}{2}\right) = \beta
\end{array}
$$

$(-1,0) = -\alpha - \beta$

$(1,0) = \alpha + \beta$

$\left(-\dfrac{1}{2}, -\dfrac{\sqrt{3}}{2}\right) = -\beta$ $\left(\dfrac{1}{2}, -\dfrac{\sqrt{3}}{2}\right) = \alpha$

FIGURE 26.2

represented on the $\lambda_1 - \lambda_2$ plane as in Fig. 26.3. The highest weight of $\overline{\mathbf{3}}$ is $(1/2, -1/(2\sqrt{3}))$, which, as seen before, is the negative of the lowest weight $(-1/2, 1/(2\sqrt{3}))$ of $\mathbf{3}$.

Recall that the highest weight for any irreducible representation of $\mathcal{SU}(3)$ is given by (25.18), with the simple roots $\alpha_{(1)}$ and $\alpha_{(2)}$ given by

$$
\alpha_{(1)} = \beta = \left(\frac{1}{2}, \frac{\sqrt{3}}{2}\right), \quad \alpha_{(2)} = \alpha = \left(\frac{1}{2}, -\frac{\sqrt{3}}{2}\right). \tag{26.10}
$$

Once the highest weight is known, the general procedure for determining the complete weight system has already been given by Theorem 25.13. Without showing explicit calculations, we will present the results for the weight systems of the representations

$$
(2,0) = \textcircled{6}, \quad (3,0) = \textcircled{10}, \text{ and } (2,1) = \textcircled{15}
$$

in Figures 26.4, 26.5 and 26.6, respectively. In these figures,

$$
\Lambda_{(1)} \equiv \Lambda(1,0) = \left(\frac{1}{2}, \frac{1}{2\sqrt{3}}\right), \quad \Lambda_{(2)} \equiv \Lambda(0,1) = \left(\frac{1}{2}, -\frac{1}{2\sqrt{3}}\right), \tag{26.11}
$$

$(0,1) = \overline{③}$ representation

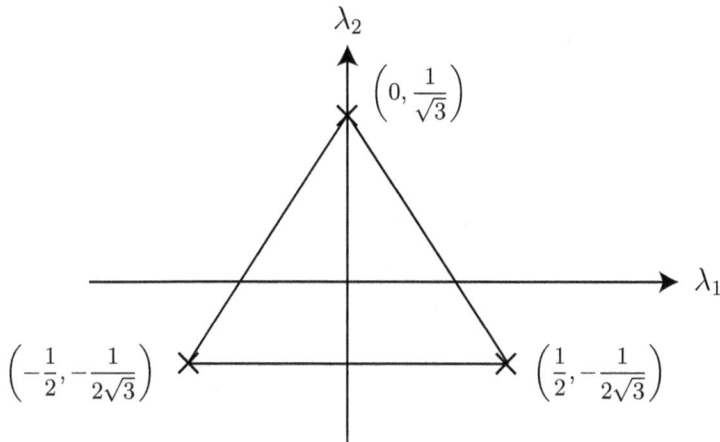

FIGURE 26.3

the weights are indicated by crosses, and degenerate weights by multiple crosses (with the number of crosses equal to the degree of degeneracy).

Figures 26.4, 26.5 and 26.6 indicate that the weight diagrams of any irreducible representation of $\mathcal{SU}(3)$ are either triangles or hexagons. Any weight in the outest layer of weights is always non-degenerate. The degree of degeneracy increases by one on going one layer inwards from any hexagonal layer, until a triangular layer is reached, when the degree of degeneracy stays constant on moving further in.

The original quark model assumes that all hadrons are composed of three flavors of quarks – up, down, strange – and their respective antiparticles, each of which is a charged fermion with an intrinsic spin $J = 1/2$. The up, down, and strange quarks are represented by orthonormal basis vectors in a 3-dimensional irreducible representation space of $SU(3)$ [or of $\mathcal{SU}(3)$], and will be labeled by $|u\rangle, |d\rangle$ and $|s\rangle$, respectively. This representation space is taken to be $(1,0) = ③$ (c.f. Fig. 26.1). The antiparticle states are represented similarly by the conjugate representation of $③$, namely, $(0,1) = \overline{③}$; and the corresponding basis states are labeled by $|\bar{u}\rangle, |\bar{d}\rangle$, and $|\bar{s}\rangle$. Under $SU(3)$, the states in $③$ transform according to the generators T_i [(26.2)] given in terms of the Gell-Mann matrices. Note that the isospin $SU(2)$ is a subgroup of the flavor $SU(3)$, with $|u\rangle = |\uparrow\rangle$ and $|d\rangle = |\downarrow\rangle$ forming an **isospin singlet** [recall the discussion following (26.3)]. Composite quark-antiquark states can thus be obtained from the product representation $③ \otimes \overline{③}$; and 3-quark states from $③ \otimes ③ \otimes ③$.

$(2,0) = \textcircled{6}$

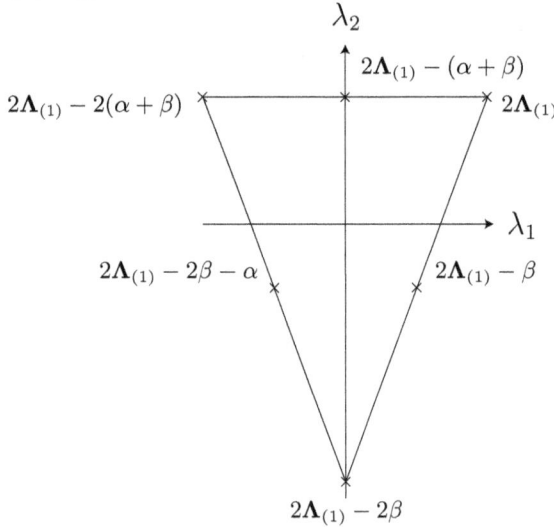

FIGURE 26.4

We already know from (25.70) that

$$\textcircled{3} \otimes \overline{\textcircled{3}} = \textcircled{8} \oplus \textcircled{1}. \tag{26.12}$$

Using this result with (25.68) we obtain

$$\textcircled{3} \otimes \textcircled{3} \otimes \textcircled{3} = (\textcircled{6} \oplus \overline{\textcircled{3}}) \otimes \textcircled{3} = (\textcircled{6} \otimes \textcircled{3}) \oplus (\overline{\textcircled{3}} \otimes \textcircled{3}). \tag{26.13}$$

We can then use the Young-diagrams method to reduce the direct product of irreducible representations [similar to the procedures leading to (25.67) to (25.71)] to get

$$\textcircled{6} \otimes \textcircled{3} = \boxed{} \otimes \boxed{1} = \boxed{1} \oplus \boxed{}\boxed{1} \tag{26.14}$$

$$= (3,0) \oplus (1,1) = \textcircled{10} \oplus \textcircled{8} .$$

Hence

$$\textcircled{3} \otimes \textcircled{3} \otimes \textcircled{3} = \textcircled{10} \oplus \textcircled{8} \oplus \textcircled{8} \oplus \textcircled{1} . \tag{26.15}$$

The quark-antiquark states, known as **mesons**, are realized in nature as **octets** in $\textcircled{8}$, while the 3-quark (or 3-antiquark) states, the **baryons** and their antiparticles, occur as octets in $\textcircled{8}$ and **decaplets** in $\textcircled{10}$. The mesons and baryons together constitute the so-called **hadrons**. The masses in any of these multiplets are actually not exactly the same, so the $SU(3)$ symmetry is only an approximate one. We will not go into the details of the interactions which break this symmetry.

$(3,0) =$ ⑩

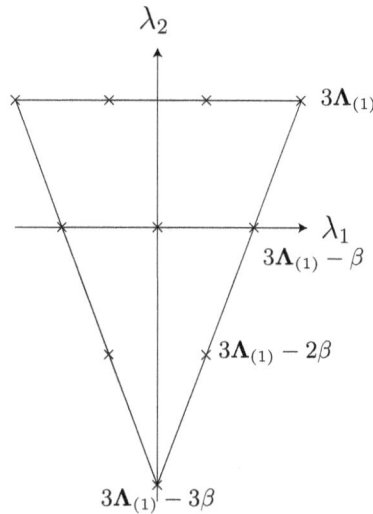

FIGURE 26.5

Among the hadrons, we have the internal symmetries I (**isospin**), S (**strangeness**), B (**baryon number**) and Q (**electric charge**). S, B and Q label one-dimensional representations of three mutually commuting gauge groups isomorphic to $U(1)$, thus these quantum numbers are additive. We have, under these $U(1)$ transformations [analogous to (4.10) and (4.21)],

$$| \; \rangle \to \exp(-i\alpha B)| \; \rangle \,, \; | \; \rangle \to \exp(-i\beta Q)| \; \rangle \,, \; | \; \rangle \to \exp(-i\gamma S)| \; \rangle \,, \quad (26.16)$$

where $\alpha, \beta, \gamma \in \mathbb{R}$ and $| \; \rangle$ represents an arbitrary hadron state. We assign to quarks $B = 1/3$ and to antiquarks $B = -1/3$. Thus mesons have $B = 0$ and baryons have $B = \pm 1$.

I^2 [with eigenvalues $I(I+1)$] is the Casimir operator and I_3 (the third component of the isospin, with eigenvalues $I, I-1, \ldots, -I$), is the sole element generating the Cartan subalgebra of the isospin subalgebra $\mathcal{SU}(2)$. If we define the **hypercharge** Y by

$$Y \equiv B + S \,, \qquad (26.17)$$

then we have, for all hadrons, the empirical rule

$$Q = I_3 + Y/2 \,. \qquad (26.18)$$

Furthermore, it is observed that on plotting the positions (I_3, Y) of the known hadrons on a planar diagram, they remarkably occur as octets and decaplets. Thus we have, for example, the Π-octet of spin 0 mesons (Fig. 26.7), the N-octet of spin 1/2 baryons (Fig. 26.8), and the Δ-decaplet of spin 3/2 resonances (Fig. 26.9).

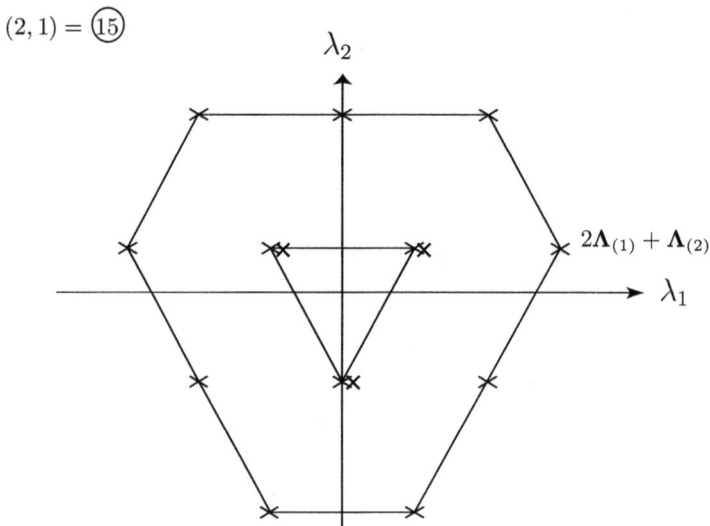

$(2,1) = \textcircled{15}$

FIGURE 26.6

Comparing Figs. 26.7, 26.8 and 26.9 with the diagrams in Figs. 26.2 and 26.5, it is seen that the hypercharge operator (also written Y) is related to T_8 [(26.2)] by

$$Y = \frac{2}{\sqrt{3}} T_8 , \qquad (26.19)$$

while

$$I_3 = T_3 . \qquad (26.20)$$

Exercise 26.2   Verify that (26.18), which relates Q, I_3 and Y, is satisfied for all the hadrons shown in Figs. 26.7, 26.8 and 26.9.

As mentioned before, the three flavors of quarks, u, d and s, appear in a Y-I_3 plot as shown in Fig. 26.10.
Eq. (26.18) implies that the up(u) quark has charge $2/3$, while the down and strange quarks each has charge $-1/3$ (in units of e, the proton charge).

Exercise 26.3   Show that, with respect to the quark basis [c.f. (26.5)]:

$$|u\rangle = |\lambda_1, \lambda_2\rangle = |1/2, 1/(2\sqrt{3})\rangle ,$$
$$|d\rangle = |\lambda_1, \lambda_2\rangle = |-1/2, 1/(2\sqrt{3})\rangle , \qquad (26.21)$$
$$|s\rangle = |\lambda_1, \lambda_2\rangle = |0, -1/\sqrt{3}\rangle ,$$

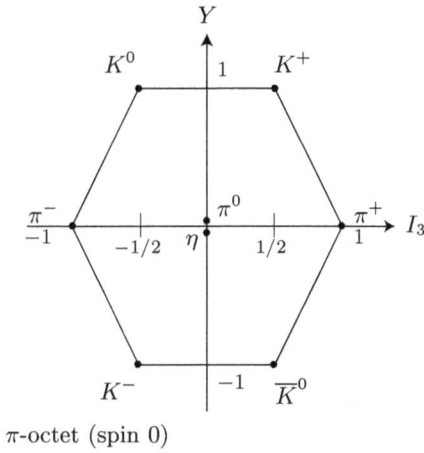

FIGURE 26.7

π-octet (spin 0)

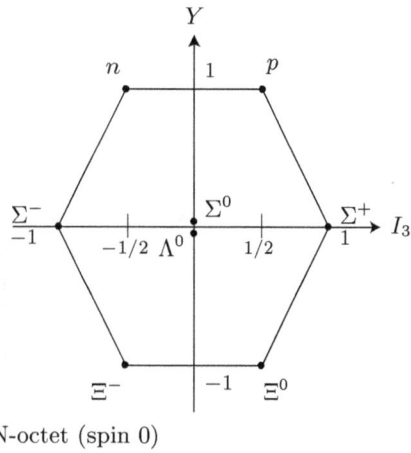

FIGURE 26.8

N-octet (spin 0)

where λ_1 and λ_2 are the eigenvalues of T_3 and T_8, respectively, the matrix representations of I_3, Y, S and Q are given by

$$
I_3 = \begin{pmatrix} \frac{1}{2} & 0 & 0 \\ 0 & -\frac{1}{2} & 0 \\ 0 & 0 & 0 \end{pmatrix}, \quad Y = \begin{pmatrix} \frac{1}{3} & 0 & 0 \\ 0 & \frac{1}{3} & 0 \\ 0 & 0 & -\frac{2}{3} \end{pmatrix}, \tag{26.22}
$$

$$
S = Y - B = \begin{pmatrix} 0 & 0 & 0 \\ 0 & 0 & 0 \\ 0 & 0 & -1 \end{pmatrix}, \quad Q = I_3 + Y/2 = \begin{pmatrix} \frac{2}{3} & 0 & 0 \\ 0 & -\frac{1}{3} & 0 \\ 0 & 0 & -\frac{1}{3} \end{pmatrix}. \tag{26.23}
$$

As mentioned before, flavor-$SU(3)$ is only a very approximate symmetry. In the current picture presented by the **standard model**, there are six flavors of quarks, divided into three families of two flavors each: $(u, d), (c, s)$ and (t, b), where, besides the u, d, s already encountered, the c, t, b stand for **charm, top,** and **bottom** quarks, respectively. Because of the connection between spin and statistics as required by special relativity, the quarks are required to have another internal quantum number, called **color**, which can assume three values. Color is the analog of the electric charge in electrodynamics, and is understood to be the source of the gauge fields responsible for the strong interaction, called **gluon fields.** The gauge group (see Chapter 37) of the strong interaction is $SU(3)$, called **color-**$SU(3)$ [as opposed to flavor-$SU(3)$] in the physics literature. In analogy to the term "electrodynamics", the dynamical theory of the strong interaction is called **chromodynamics.**

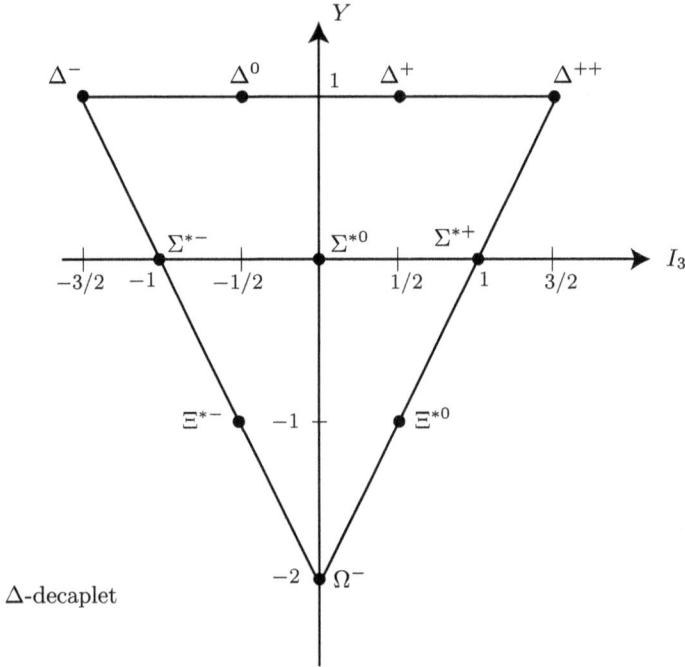

FIGURE 26.9

In closing this chapter, we will specify the Young diagrams of the various irreducible representations of $SU(3)$, and the corresponding tensorial characters. Recalling (25.72), only Young diagrams of at most two rows need to be considered, since the rank of $SU(3)$ is 2. Consider a general example of such a diagram:

$$[m+n, m, 0] = \boxed{} . \qquad (26.24)$$

By Theorem 21.2, this is equivalent to $[n, 0, -m]$, which corresponds to a tensor of the form

$$x^{j_1 \ldots j_m}_{i_1 \ldots i_n} ,$$

obtained from $[m+n, m, 0]$ by raising indices with the Levi-Civita tensors:

$$x^{j_1 \ldots j_m}_{i_1 \ldots i_n} = x_{k_1 \ldots k_m i_1 \ldots i_n l_1 \ldots l_m} \, \varepsilon^{j_1 k_1 l_1} \ldots \varepsilon^{j_m k_m l_m} . \qquad (26.25)$$

Because of the symmetry character of the Young diagram in (26.24) and the skew-symmetry of the Levi-Civita tensor, $x^{j_1 \ldots j_m}_{i_1 \ldots i_n}$ must be completely symmetric in $\{j_1, \ldots, j_m\}$ and $\{i_1, \ldots, i_n\}$. Furthermore, since $x^{j_1 \ldots j_m}_{i_1 \ldots i_n}$ is irreducible, it must satisfy the traceless condition

$$x^{j j_2 \ldots j_m}_{j i_2 \ldots i_n} = 0 . \qquad (26.26)$$

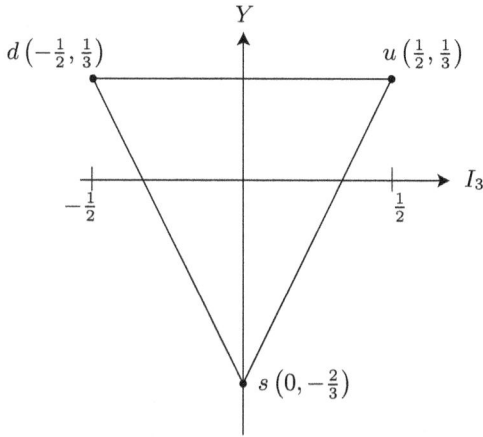

FIGURE 26.10

We summarize in the following table the Young diagrams and the corresponding tensors for some low-ranked irreducible representations of $SU(3)$.

Representation	Young Tableau	Tensor
$(0,0) = \mathbf{1}$		1
$(1,0) = \mathbf{3}$		$x_i,\ i = 1,2,3$
$(0,1) = \overline{\mathbf{3}}$		x^i
$(2,0) = \mathbf{6}$		x_{ij}
$(0,2) = \overline{\mathbf{6}}$		x^{ij}
$(1,1) = \mathbf{8} = \overline{\mathbf{8}}$		$x^i_j\ (x^i_i = 0)$
$(3,0) = \mathbf{10}$		x_{ijk}
$(0,3) = \overline{\mathbf{10}}$		x^{ijk}
$(2,1) = \mathbf{15}$		$x^i_{jk}\ (x^i_{ik} = 0)$
$(1,2) = \overline{\mathbf{15}}$		$x^{ij}_k\ (x^{ij}_i = 0)$
$(2,2) = \mathbf{27} = \overline{\mathbf{27}}$		$x^{ij}_{kl}\ (x^{ij}_{il} = 0)$

TABLE 26.1 – Irreducible Representations of $SU(3)$

Chapter 27

Clifford Algebras

In physics, the use of Clifford algebras began with Dirac's attempt to factor the relativistic (Lorentz-invariant) second order **Klein-Gordon equation**

$$\nabla^2\psi - \frac{\partial^2\psi}{\partial t^2} = m^2\psi \tag{27.1}$$

($\hbar = c = 1, m = $ mass of particle) into a first-order system of relativistic equations. To do so, he proposed a Lorentz-covariant first-order operator of the form $\gamma^\mu\,\partial_\mu$ which would satisfy

$$\left(\sum_\mu \gamma_\mu\partial_\mu\right)^2 = \partial_t^2 - \nabla^2 , \tag{27.2}$$

where ∇^2 is the Laplacian operator on Euclidean $\mathbb{R}^3$, and γ^μ, $\mu = 0, 1, 2, 3$, are unspecified algebraic objects. On computing the LHS of (27.2) formally, we see that the equation requires

$$\sum_{\mu,\nu} \gamma_\mu\gamma_\nu\partial_\mu\partial_\nu = \partial_0^2 - \partial_i^2 , \tag{27.3}$$

where $\mu, \nu = 0, 1, 2, 3$, and $i = 1, 2, 3$. Using the Minkowski metric

$$\eta_{\mu\nu} = \begin{pmatrix} 1 & 0 & 0 & 0 \\ 0 & -1 & 0 & 0 \\ 0 & 0 & -1 & 0 \\ 0 & 0 & 0 & -1 \end{pmatrix} , \tag{27.4}$$

[which is equivalent to the one given by (8.3)], (27.3) is equivalent to

$$\boxed{\{\gamma_\mu, \gamma_\nu\} = 2\eta_{\mu\nu}} , \tag{27.5}$$

where $\{\gamma_\mu, \gamma_\nu\}$ is the anticommutator [c.f. (13.36)]. Dirac found a set of four 4×4 matrices which satisfied (27.5) and thus constructed the equation

$$\sum_\mu \gamma_\mu \partial_\mu \psi - im\psi = 0 , \tag{27.6}$$

known as the **Dirac equation**. The fact that the γ_μ are 4×4 matrices requires that the wave function ψ be 4-component spinors. This led Dirac to predict the existence of anti-matter.

Recall the isomorphism $f : M \to H_2$ from Minkowski's space M to the set of 2×2 hermitian matrices H_2 given by (11.14):

$$f(x) = \begin{pmatrix} x^0 + x^3 & x^1 - ix^2 \\ x^1 + ix^2 & x^0 - x^3 \end{pmatrix} . \tag{27.7}$$

There is a related isomorphism $f' : M \to H_2$ given by

$$f'(x) = \begin{pmatrix} x^0 - x^3 & -x^1 + ix^2 \\ -x^1 - ix^2 & x^0 + x^3 \end{pmatrix} . \tag{27.8}$$

Exercise 27.1 Show that f' as defined by (27.8) is an isomorphism. Note that [c.f. (11.18)],

$$\det(f'(x)) = (x^0)^2 - (x^1)^2 - (x^2)^2 - (x^3)^2 = \det(f(x)) . \tag{27.9}$$

Consider a linear map $\gamma : M \to \mathcal{GL}(4, \mathbb{C})$ defined by

$$\gamma(x) \equiv \begin{pmatrix} 0 & f(x) \\ f'(x) & 0 \end{pmatrix} , \tag{27.10}$$

where the entries on the RHS are 2×2 matrices. For the standard basis of $M, \{e_0, e_1, e_2, e_3\}$, given by (11.16), we have

$$\gamma_0 \equiv \gamma(e_0) = \begin{pmatrix} 0 & 1 \\ 1 & 0 \end{pmatrix} , \qquad \gamma_1 \equiv \gamma(e_1) = \begin{pmatrix} 0 & \sigma_1 \\ -\sigma_1 & 0 \end{pmatrix} , \tag{27.11a}$$

$$\gamma_2 \equiv \gamma(e_2) = \begin{pmatrix} 0 & \sigma_2 \\ -\sigma_2 & 0 \end{pmatrix} , \qquad \gamma_3 \equiv \gamma(e_3) = \begin{pmatrix} 0 & \sigma_3 \\ -\sigma_3 & 0 \end{pmatrix} , \tag{27.11b}$$

where, again, the entries are 2×2 matrices, and $\sigma_1, \sigma_2, \sigma_3$ are the Pauli spin matrices given by (11.16). It is easily seen that the matrices $\gamma_0, \gamma_1, \gamma_2, \gamma_3$ satisfy the condition (27.5). They are called the **Dirac matrices**.

| **Exercise 27.2** | Verify directly that the Dirac matrices $\gamma_\mu, \mu = 0, 1, 2, 3$, satisfy (27.5). Use the properties of the Pauli spin matrices given in Exercise 11.5.

The Dirac matrices $\gamma_\mu, \mu = 0, 1, 2, 3$, can be viewed as basis vectors of a four-dimensional vector space endowed with an inner product

$$(\gamma_\mu, \gamma_\nu) = \eta_{\mu\nu} . \tag{27.12}$$

With matrix multiplication as the internal product, the γ_μ also generate an algebra with the unit equal to the 4×4 unit matrix (denoted 1), called the **Dirac algebra**. It is a special example of what is known as a Clifford algebra. In fact, because of (27.5), this algebra can itself be viewed as a 16-dimensional linear vector space with the basis

$$\{1, \gamma_\mu, \gamma_{\mu_1}\gamma_{\mu_2}, \gamma_{\mu_1}\gamma_{\mu_2}\gamma_{\mu_3}, \gamma_{\mu_1}\gamma_{\mu_2}\gamma_{\mu_3}\gamma_{\mu_4}\} ,$$

where $\mu_i = 0, 1, 2, 3$, and $\mu_i < \mu_{i+1}$. We now give the general definition of a Clifford algebra.

Definition 27.1. *Let $V_{(s)}^n$, $s \in \mathbb{Z}^+$ (the non-negative integers), $s \leq n$, be an n-dimensional real vector space with inner product $(,)$ and a basis $\{e_i\}$, $i = 1, \ldots, n$, such that*

$$(e_i, e_j) = 0 , \; i \neq j ,$$
$$(e_i, e_i) = 1 , \; i = 1, \ldots, s , \tag{27.13}$$
$$(e_i, e_i) = -1 , \; i = s+1, \ldots, n .$$

Introduce an internal product vw, $v, w \in V_{(s)}^n$, which is associative and distributive with respect to vector addition, and which satisfies the condition

$$\{v, w\} = 2 (v, w) , \tag{27.14}$$

where

$$\{v, w\} \equiv vw + wv . \tag{27.15}$$

*The algebra generated by all possible sums and (internal) products of vectors in $V_{(s)}^n$ is called the **Clifford algebra** $C(V_{(s)}^n)$ of $V_{(s)}^n$.*

Eqs. (27.13) and (27.14) imply that

$$\{e_i, e_j\} = 0 , \quad i \neq j , \tag{27.16}$$

$$(e_i)^2 = \begin{cases} 1 , & i = 1, \ldots, s , \\ -1 , & i = s+1, \ldots, n , \end{cases} \tag{27.17}$$

where 1 is the unit of the algebra. Thus $C(V_{(s)}^n)$ is itself a linear vector space of dimension $\sum_{r=0}^n \binom{n}{r} = 2^n$ with basis

$$\{\, 1, \, e_{i_1}, \, e_{i_1} e_{i_2}, \, \ldots \ldots, \, e_{i_1} e_{i_2} \ldots e_{i_n} \,\} \,,$$

where $i_k = 1, \ldots, n$, $1 \le k \le n$ and $i_j < i_{j+1}$.

A familiar Clifford algebra is

$$C(V_{(0)}^1) = \mathbb{C} \,, \tag{27.18}$$

the algebra of complex numbers. Indeed, a basis of $V_{(0)}^1$ consists of the elements 1 and i, with $i^2 = -1$ [c.f. (27.17)]. A general element of $C(V_{(0)}^1)$ is then of the form $x + yi$, $x, y \in \mathbb{R}$. Similarly, $C(V_{(0)}^2)$ is the algebra of **quaternions**. Let $\{\, 1, e_1, e_2 \,\}$ be a basis of $V_{(0)}^2$ satisfying (27.13). A general element of $C(V_{(0)}^2)$ is then of the form

$$a + be_1 + ce_2 + de_1 e_2 \,,$$

where $a, b, c, d \in \mathbb{R}$. Rename the basis vectors of $C(V_{(0)}^2)$ by

$$i = e_1, \quad j = e_2, \quad k = e_1 e_2 \,. \tag{27.19}$$

Then

$$i^2 = j^2 = k^2 = -1 \,, \tag{27.20}$$

$$ij = -ji = k, \quad jk = -kj = i, \quad ki = -ik = j \,. \tag{27.21}$$

Exercise 27.3   Verify that the Dirac algebra is $C(V_{(1)}^4)$, and the algebra generated by the Pauli spin matrices, called the **Pauli algebra**, is $C(V_{(3)}^3)$.

Definition 27.2. *Denote the linear subspace of $C(V_{(s)}^n)$ spanned by the $\binom{n}{p}$ products $(e_{i_1} e_{i_2} \ldots e_{i_p})$ by $C_p(V_{(s)}^n)$. The linear subspaces*

$$C_+ \equiv \sum_{\oplus p \text{ even } \le n} C_p \quad \text{and} \quad C_- \equiv \sum_{\oplus p \text{ odd } \le n} C_p \tag{27.22}$$

*are called the **even subspace** and **odd subspace** of $C(V_{(s)}^n)$, respectively.*

Note that C_+ is also a subalgebra of $C(V_{(s)}^n)$, since it contains the unit of the algebra.

Exercise 27.4   Show that the dimension of both C_+ and C_- is 2^{n-1}.

We state without proof the following fact.

Theorem 27.1. *The even subalgebra C_+ of the Clifford algebra $C(V_{(s)}^n)$ is isomorphic to the Clifford algebra $C(V_{(t)}^{n-1})$ for certain values of t.*

Starting with the Dirac algebra $C(V_{(1)}^4)$, one can successively extract even subalgebras and obtain the following interesting hierarchy of familiar Clifford algebras:

$$(C(V_{(1)}^4))_+ \sim C(V_{(3)}^3) \quad \text{(the Pauli algebra)} , \tag{27.23}$$

$$(C(V_{(3)}^3))_+ \sim C(V_{(0)}^2) \quad \text{(the quaternions)} , \tag{27.24}$$

$$(C(V_{(0)}^2))_+ \sim C(V_{(0)}^1) \quad (= \mathbb{C}, \text{ the complex numbers}) , \tag{27.25}$$

$$(C(V_{(0)}^1))_+ \sim C(V_{(0)}^0) \quad (= \mathbb{R}, \text{ the real numbers}) . \tag{27.26}$$

Eq. (27.23), for example, can be verified as follows. A basis for the even subalgebra of the Dirac algebra is [c.f. (27.11)]

$$1, \ \gamma_0\gamma_1, \ \gamma_0\gamma_2, \ \gamma_0\gamma_3, \ \gamma_1\gamma_2, \ \gamma_1\gamma_3, \ \gamma_2\gamma_3, \ \gamma_0\gamma_1\gamma_2\gamma_3 . \tag{27.27}$$

The isomorphism is given by

$$\gamma_0\gamma_1 \mapsto \sigma_1, \quad \gamma_0\gamma_2 \mapsto \sigma_2, \quad \gamma_0\gamma_2 \mapsto \sigma_3 . \tag{27.28}$$

$\boxed{\text{Exercise 27.5}}$ Verify (27.24), (27.25) and (27.26).

Chapter 28

Exterior Products

Exterior products are generalizations of vector cross products in 3-dimensional vector algebra to vector spaces of arbitrary dimension. The study of these products and the algebraic structures arising from them has led to very fundamental results which played a crucial role in the development of the so-called **exterior differential calculus**, largely by the geometer Elie Cartan. Together with tensor analysis, the use of exterior differentiation on **differential forms** has revolutionized modern differential geometry, a discipline that has become increasingly important to physics. Using the powerful techniques of differential forms, we will also see how to do vector and tensor calculus in a much more elegant fashion than that presented in the conventional physics literature. This chapter will present the algebraic background for the study of differential forms.

Recall the spaces of alternating tensors $\Lambda^r(V)$ and $\Lambda^r(V^*)$ introduced in Chapter 16. An alternating contravariant tensor of order r, $x \in \Lambda^r(V)$, is called an **exterior vector of degree** r, or an **exterior r-vector**. $\Lambda^r(V)$ is called the **exterior space** of V of degree r. [A similar nomenclature applies to alternating covariant tensors in $\Lambda^r(V^*)$.] Note, of course, that $x \in \Lambda^r(V)$ is an $(r, 0)$ tensor, while $x \in \Lambda^r(V^*)$ is a $(0, r)$ tensor. For convenience we set

$$\Lambda^0(V) = \mathbb{F} \quad , \tag{28.1}$$

where $\mathbb{F}$ is the field associated with the vector space V, and

$$\Lambda^1(V) = V \quad . \tag{28.2}$$

The crucial fact concerning exterior vectors is that there exists an operation, called the **exterior** or **wedge product**, such that the exterior product of two exterior vectors is another exterior vector.

Definition 28.1. *Let ξ be an exterior k-vector and η an exterior l-vector. The exterior product of ξ and η, written $\xi \wedge \eta$, is the exterior $(k + l)$-vector defined by*

$$\xi \wedge \eta = A_{k+l}(\xi \otimes \eta) \quad , \tag{28.3}$$

where A_{k+l} is the alternating map given by (16.22).

Note that even though ξ and η are individually alternating in the above definition, $\xi \otimes \eta$, their tensor product, is not necessarily alternating, and the alternating map A_{k+l} must be applied to make it so.

We will state the following theorem without proof, which gives the algebraic rules of operation for the exterior product.

Theorem 28.1. *The exterior product satisfies the following rules. Suppose* $\xi, \xi_1, \xi_2 \in \Lambda^k(V)$, $\eta, \eta_1, \eta_2 \in \Lambda^l(V)$, *and* $\zeta \in \Lambda^k(V)$. *Then we have*

1) *the distributive rule:*

$$\begin{cases} (\xi_1 + \xi_2) \wedge \eta &= \xi_1 \wedge \eta + \xi_2 \wedge \eta \quad, \\ \xi \wedge (\eta_1 + \eta_2) &= \xi \wedge \eta_1 + \xi \wedge \eta_2 \quad; \end{cases} \tag{28.4}$$

2) *the anti-commutative rule:*

$$\xi \wedge \eta = (-1)^{kl} \eta \wedge \xi \quad; \tag{28.5}$$

3) *the associative rule:*

$$(\xi \wedge \eta) \wedge \zeta = \xi \wedge (\eta \wedge \zeta) \quad. \tag{28.6}$$

$\boxed{\text{Exercise 28.1}}$ Prove (28.4).

The anti-commutative rule immediately implies that, if $\xi, \eta \in \Lambda^1(V) = V$, then

$$\xi \wedge \eta = -\eta \wedge \xi \quad.$$

Thus, if $\xi \in \Lambda^1(V)$, then $\xi \wedge \xi = 0$. In general, if there are repeated exterior 1-vectors in a wedge product, then the product vanishes.

Now suppose $\{e_1, \ldots, e_n\}$ is a basis of V. According to the associative rule

$$e_{i_1} \wedge \cdots \wedge e_{i_r} = A_r(e_{i_1} \otimes \cdots \otimes e_{i_r}), \quad 1 \le i_1, \ldots, i_r \le n. \tag{28.7}$$

This exterior product is non-zero only if $i_1, \ldots, i_r$ are distinct. If $r > n$, there must be repeated indices among the $i_1, \ldots, i_r$, and the exterior product (28.7) vanishes, that is,

$$e_{i_1} \wedge \cdots \wedge e_{i_r} = 0 \quad \text{if} \quad r > n = dim(V) \quad. \tag{28.8}$$

Suppose an exterior r-vector $\xi \in \Lambda^r(V)$ is written in component form:

$$\xi = \xi^{i_1 \cdots i_r} e_{i_1} \otimes \cdots \otimes e_{i_r} \quad. \tag{28.9}$$

Since any alternating map is linear, we have

$$\xi = A_r \xi = \xi^{i_1 \cdots i_r} A_r(e_{i_1} \otimes \cdots \otimes e_{i_r}) = \xi^{i_1 \cdots i_r} e_{i_1} \wedge \cdots \wedge e_{i_r} \quad. \tag{28.10}$$

By virtue of (28.8), this implies that

$$\Lambda^r(V) = 0 \quad \text{for} \quad r > n = dim(V).$$ (28.11)

For $r \leq n$, Theorem 16.1 asserts that the components $\xi^{i_1 \cdots i_r}$ are skew-symmetric with respect to any permutation of the indices. The same is true of the exterior product $e_{i_1} \wedge \cdots \wedge e_{i_r}$ by the discussion immediately following Theorem 28.1. Since there are $r!$ permutations of the r indices $i_1, \ldots, i_r$, ξ can be expressed as

$$\xi = r! \sum_{i_1 < \cdots < i_r} \xi^{i_1 \cdots i_r} e_{i_1} \wedge \cdots \wedge e_{i_r} \,,$$ (28.12)

where the summation only includes terms for which $i_1 < i_2 < \cdots < i_r$.

We will now show that, for $r \leq n$, the set $\{e_{i_1} \wedge \cdots \wedge e_{i_r} \,|\, 1 \leq i_1 < \cdots < i_r \leq n\}$ forms a basis of $\Lambda^r(V)$. Eq. (28.12) already shows that any $\xi \in \Lambda^r(V)$ can be expressed as a linear combination of vectors in this set. Thus we only need to show that the

$$\binom{n}{r} = \frac{n!}{r!(n-r)!}$$

vectors in the set are linearly independent. First we derive a formula for evaluating $e_{i_1} \wedge \cdots \wedge e_{i_r}$. Let $v^{*1}, \ldots, v^{*r}$ be r arbitrary vectors in V^*. Then, by definition,

$$(e_{i_1} \wedge \cdots \wedge e_{i_r})(v^{*1}, \ldots, v^{*r})$$

$$= \frac{1}{r!} \sum_{\sigma \in S(r)} (sgn\,\sigma)\sigma(e_{i_1} \otimes \cdots \otimes e_{i_r})(v^{*1}, \ldots, v^{*r})$$

$$= \frac{1}{r!} \sum_{\sigma \in S(r)} (sgn\,\sigma)(e_{i_1} \otimes \cdots \otimes e_{i_r})(v^{*\sigma(1)}, \ldots, v^{*\sigma(r)})$$ (28.13)

$$= \frac{1}{r!} \sum_{\sigma \in S(r)} (sgn\,\sigma)\langle e_{i_1}, v^{*\sigma(1)} \rangle \ldots \langle e_{i_r}, v^{*\sigma(r)} \rangle \,.$$

By the expression for the determinant given in (16.77), we see that

$$(e_{i_1} \wedge \cdots \wedge e_{i_r})(v^{*1}, \ldots, v^{*r}) = \frac{1}{r!} \begin{vmatrix} \langle e_{i_1}, v^{*1} \rangle & \cdots & \cdots & \langle e_{i_1}, v^{*r} \rangle \\ \langle e_{i_2}, v^{*1} \rangle & \cdots & \cdots & \langle e_{i_2}, v^{*r} \rangle \\ \vdots & & & \\ \langle e_{i_r}, v^{*1} \rangle & \cdots & \cdots & \langle e_{i_r}, v^{*r} \rangle \end{vmatrix} \,.$$ (28.14)

This formula is called the **evaluation formula** for the exterior product $e_{i_1} \wedge \cdots \wedge e_{i_r}$. In particular

$$(e_{i_1} \wedge \cdots \wedge e_{i_r})(e^{*j_1}, \ldots, e^{*j_r}) = \frac{1}{r!} \det(\langle e_{i_\alpha}, e^{*j_\beta} \rangle) = \frac{1}{r!} \delta^{j_1 \cdots j_r}_{i_1 \cdots i_r} \,,$$ (28.15)

where

$$
\delta^{j_1 \dots j_r}_{i_1 \dots i_r} =
\begin{cases}
1 & \text{if } i_1, \dots, i_r \text{ are distinct, and } \{j_1, \dots, j_r\} \\
  & \text{is an even permutation of } \{i_1, \dots, i_r\}; \\
-1 & \text{if } i_1, \dots, i_r \text{ are distinct, and } \{j_1, \dots, j_r\} \\
  & \text{is an odd permutation of } \{i_1, \dots, i_r\}; \\
0 & \text{otherwise}
\end{cases}
\tag{28.16}
$$

is called the **generalized Kronecker δ-symbol**. We see immediately from (28.5) that

$$
(e_1 \wedge \cdots \wedge e_n)(e^{*1}, \dots, e^{*n}) = \frac{1}{n!} \quad .
\tag{28.17}
$$

Hence $e_1 \wedge \cdots \wedge e_n \neq 0$. (Later on we will see that this is the necessary and sufficient condition for $\{e_1, \dots, e_n\}$ to be linearly independent). Now we will show that the set

$$
\{ e_{i_1} \wedge \cdots \wedge e_{i_r} \,|\, 1 \leq i_1 < \cdots < i_r \leq n \}
$$

is linearly independent. Suppose the contrary, that is, that it is linearly dependent. Then there exist scalars $a^{i_1 \dots i_r} \in \mathbb{F}$ (the field associated with V), not all zero, such that

$$
\sum_{1 \leq i_1 < \cdots < i_r \leq n} a^{i_1 \dots i_r} e_{i_1} \wedge \cdots \wedge e_{i_r} = 0 \quad .
\tag{28.18}
$$

Assume one of the non-zero scalars to be $a^{j_1 \dots j_r}$, $1 \leq j_1 < \cdots < j_r \leq n$, with the remaining index set labelled $\{k_1, k_2, \dots, k_{n-r}\}$ where $k_1 < \cdots < k_{n-r}$. Thus $(j_1, \dots, j_r, k_1, \dots, k_{n-r})$ is a permutation of $(1, \dots, n)$. Wedge multiplying both sides of (28.18) by $e_{k_1} \wedge \cdots \wedge e_{k_{n-r}}$, we get

$$
a^{j_1 \dots j_r} e_{j_1} \wedge \cdots \wedge e_{j_r} \wedge e_{k_1} \wedge \cdots \wedge e_{k_{n-r}} \ \ (\text{no sum}) = \pm a^{j_1 \dots j_r} e_1 \wedge \cdots \wedge e_n = 0 \quad .
\tag{28.19}
$$

Only one term in the sum in the LHS of (28.18) is picked out because all other terms involve repeated factors of e_i in the exterior product and thus vanish. Since $e_1 \wedge \cdots \wedge e_n \neq 0$, as shown earlier, it follows that $a^{j_1 \dots j_r} = 0$, which is a contradiction. We conclude that

$$
\{ e_{i_1} \wedge \cdots \wedge e_{i_r} \,|\, 1 \leq i_1 < \cdots < i_r \leq n \}
$$

is linearly independent, and forms a basis of $\Lambda^r(V)$. This implies that

$$
dim(\Lambda^r(V)) = \binom{n}{r} \quad .
\tag{28.20}
$$

In particular, $dim(\Lambda^n(V)) = 1$ if $dim(V) = n$, as was claimed in Chapter 16.

| Exercise 28.2 | Write down explicitly a basis for $\Lambda^3(V)$ when $dim(V) = 4$.

From the exterior spaces $\Lambda^r(V)$ we can build the so-called exterior or **Grassmann algebra**. Suppose $dim(V) = n$. Consider the formal sum

$$\Lambda(V) = \sum_{r=0}^{n} \Lambda^r(V) \quad . \tag{28.21}$$

Let $\xi, \eta \in \Lambda(V)$ be written as

$$\xi = \sum_{r=0}^{n} \xi^r , \quad \eta = \sum_{s=0}^{n} \eta^s , \tag{28.22}$$

where $\xi^r \in \Lambda^r(V)$ and $\eta^s \in \Lambda^s(V)$. Define the exterior product of ξ and η by

$$\xi \wedge \eta = \sum_{r, s=0}^{n} \xi^r \wedge \eta^s \quad . \tag{28.23}$$

Then $\Lambda(V)$ becomes an algebra with respect to this exterior product, called the **exterior algebra** or **Grassmann algebra** of V.

If $\{e_1, \ldots, e_n\}$ is a basis of V, it follows from our previous discussion that the following set is a basis of $\Lambda(V)$:

$$\{1, \; e_i \; (1 \leq i \leq n) , \; e_{i_1} \wedge e_{i_2} \; (1 \leq i_1 < i_2 \leq n) , \ldots ,$$
$$e_{i_1} \wedge e_{i_2} \wedge \cdots \wedge e_{i_r} \; (1 \leq i_1 < \cdots < i_r \leq n) , \; \ldots , \; e_1 \wedge \cdots \wedge e_n\} \quad . \tag{28.24}$$

The total number of elements in this set is $\sum_{r=0}^{n} \binom{n}{r}$, which, on setting $x = 1$ in the binomial theorem

$$(1 + x)^n = \sum_{r=0}^{n} \binom{n}{r} x^r , \tag{28.25}$$

is seen to be equal to 2^n. Thus

$$dim(\Lambda(V)) = 2^n , \quad \text{if } dim(V) = n . \tag{28.26}$$

Similarly, we have the Grassman algebra of the dual space V^*:

$$\Lambda(V^*) = \sum_{r=0}^{n} \Lambda^r(V^*) \quad . \tag{28.27}$$

An element $x \in \Lambda^r(V^*)$ is called an **exterior form** of degree r, or an **exterior r-form**. It is an alternating $\mathbb{F}$-valued r-linear function on V.

The vector spaces $\Lambda^r(V)$ and $\Lambda^r(V^*)$ are subspaces of $T^r(V)$ and $T^r(V^*)$, respectively. The pairing between $T^r(V)$ and $T^r(V^*)$, as spaces dual to each other, has already been given by (16.4). Hence $\Lambda^r(V)$ and $\Lambda^r(V^*)$ are dual to each other and inherits the pairing from $T^r(V)$ and $T^r(V^*)$. We will, however, define the pairing between $\Lambda^r(V)$ and $\Lambda^r(V^*)$ by

$$\boxed{\langle v_1 \wedge \cdots \wedge v_r, v^{*1} \wedge \cdots \wedge v^{*r} \rangle = \det(\langle v_\alpha, v^{*\beta} \rangle)} \quad , \tag{28.28}$$

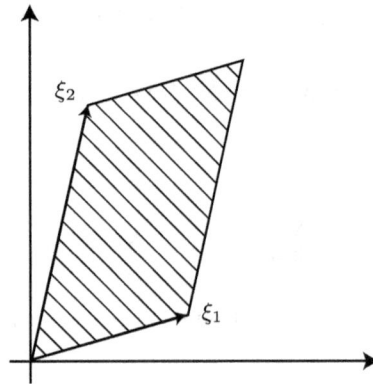

FIGURE 28.1

which differs from the inherited pairing by just a factor of $r!$. Although we will still use the same notation $\langle\,,\,\rangle$ to denote the two different pairings, no confusion should arise if we keep in mind which spaces we are working with.

Exercise 28.3 Show explicitly, using the evaluation formula (28.14) and the properties of determinants, that

$$\langle\quad,\quad\rangle_{T^r(V),\,T^r(V^*)} = \frac{1}{r!}\langle\quad,\quad\rangle_{\Lambda^r(V),\,\Lambda^r(V^*)}\quad. \tag{28.29}$$

The following exercises will illustrate the geometric (and intuitive) meaning of exterior products.

Exercise 28.4 In $V = \mathbb{R}^2$, choose an orthonormal basis $\{e_1, e_2\}$ with orientation $e_1 \wedge e_2$. Show that the **oriented area** of a parallelogram with sides $\xi_1, \xi_2 \in \mathbb{R}^2$ (Fig. 28.1) is given by

$$\langle \xi_1 \wedge \xi_2, e^{*1} \wedge e^{*2}\rangle = (\xi_1 \wedge \xi_2, e_1 \wedge e_2)\quad. \tag{28.30}$$

Equivalently,

$$\text{oriented area} = \begin{vmatrix} \xi_1^1 & \xi_1^2 \\ \xi_2^1 & \xi_2^2 \end{vmatrix}\quad, \tag{28.31}$$

where

$$\xi_1 = \xi_1^1 e_1 + \xi_1^2 e_2, \quad \xi_2 = \xi_2^1 e_1 + \xi_2^2 e_2. \tag{28.32}$$

The form $e^{*1} \wedge e^{*2}$ is called the **area form**.

Exercise 28.5 In $V = \mathbb{R}^3$, choose an orthonormal basis $\{e_1, e_2, e_3\}$ with orientation $e_1 \wedge e_2 \wedge e_3$. The 3-form $e^{*1} \wedge e^{*2} \wedge e^{*3}$ is called the **volume form**.

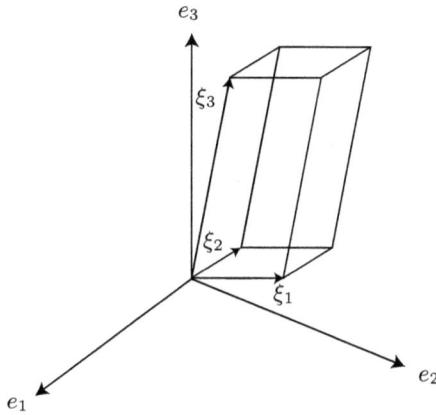

FIGURE 28.2

Show that the **oriented volume** of a parallelopiped with sides $\xi_1, \xi_2, \xi_3 \in \mathbb{R}^3$ (Fig. 28.2) is given by

$$\langle \xi_1 \wedge \xi_2 \wedge \xi_3, e^{*1} \wedge e^{*2} \wedge e^{*3} \rangle = (\xi_1 \wedge \xi_2 \wedge \xi_3, e_1 \wedge e_2 \wedge e_3), \qquad (28.33)$$

Equivalently,

$$\text{oriented volume} = \begin{vmatrix} \xi_1^1 & \xi_1^2 & \xi_1^3 \\ \xi_2^1 & \xi_2^2 & \xi_2^3 \\ \xi_3^1 & \xi_3^2 & \xi_3^3 \end{vmatrix}, \qquad (28.34)$$

where

$$\xi_1 = \xi_1^1 e_1 + \xi_1^2 e_2 + \xi_1^3 e_3 \ , \ \xi_2 = \xi_2^1 e_1 + \xi_2^2 e_2 + \xi_2^3 e_3 \ , \ \xi_3 = \xi_3^1 e_1 + \xi_3^2 e_2 + \xi_3^3 e_3 \ . \tag{28.35}$$

We can easily generalize the expressions for the oriented area and oriented volume given in the above two exercises to give the oriented volume of a parallelopiped in the Euclidean space $\mathbb{R}^n$, with edges $\xi_1, \ldots, \xi_n$:

$$V(\xi_1, \ldots, \xi_n) = \langle \xi_1 \wedge \cdots \wedge \xi_n, e^{*1} \wedge \cdots \wedge e^{*n} \rangle = (\xi_1 \wedge \cdots \wedge \xi_n, e_1 \wedge \cdots \wedge e_n)$$

$$= \begin{vmatrix} \xi_1^1 & \cdots & \xi_1^n \\ \vdots & & \vdots \\ \xi_n^1 & \cdots & \xi_n^n \end{vmatrix}, $$

$$(28.36)$$

where

$$\xi_i = \xi_i^j e_j, \quad i, j = 1, \ldots, n. \tag{28.37}$$

The algebraic structure of a Grassman algebra arising from the exterior product is preserved by linear maps, in the following sense. Suppose $f : V \longrightarrow W$ is a linear map from a vector space V to another vector space W, over the same field. Then f induces a linear map $f^* : \Lambda^r(W^*) \longrightarrow \Lambda^r(V^*)$ defined as follows. Let $\varphi \in \Lambda^r(W^*)$. For arbitrary $v_1, \ldots, v_r \in V$, we have

$$\boxed{(f^*\varphi)(v_1, \ldots, v_r) = \varphi(f(v_1), \ldots, f(v_r))} \tag{28.38}$$

f^* is called the **pullback map** from $\Lambda^r(W^*)$ to $\Lambda^r(V^*)$, and is uniquely determined by f. It is obviously linear. (The notion of a pullback map has already been introduced in Chapter 16, as a prerequisite for the definition of the determinant of a linear operator.)

Exercise 28.6 Prove that f^* is linear, that is, for $a, b \in \mathbb{F}$ and $\varphi_1, \varphi_2 \in \Lambda^r(W^*)$,

$$f^*(a\varphi_1 + b\varphi_2) = af^*(\varphi_1) + bf^*(\varphi_2) \tag{28.39}$$

Theorem 28.2. *Suppose $f : V \longrightarrow W$ is a linear map. Then f^* commutes with the exterior product, that is, for any $\varphi \in \Lambda^r(W^*)$ and $\psi \in \Lambda^s(W^*)$,*

$$f^*(\varphi \wedge \psi) = (f^*\varphi) \wedge (f^*\psi) \tag{28.40}$$

Proof. Choose any $v_1, \ldots, v_{r+s} \in V$. Then

$$f^*(\varphi \wedge \psi)(v_1, \ldots, v_{r+s}) = (\varphi \wedge \psi)(f(v_1), \ldots, f(v_{r+s}))$$

$$= \frac{1}{(r+s)!} \sum_{\sigma \in S(r+s)} (sgn\,\sigma)\varphi(f(v_{\sigma(1)}), \ldots, f(v_{\sigma(r)}))\psi(f(v_{\sigma(r+1)}), \ldots, f(v_{\sigma(r+s)}))$$

$$= \frac{1}{(r+s)!} \sum_{\sigma \in S(r+s)} (sgn\,\sigma)(f^*\varphi)(v_{\sigma(1)}, \ldots, v_{\sigma(r)})(f^*\psi)(v_{\sigma(r+1)}, \ldots, v_{\sigma(r+s)})$$

$$= ((f^*\varphi) \wedge (f^*\psi))(v_1, \ldots, v_{r+s}) \tag{28.41}$$

Eq. (28.40) follows. $\qquad \square$

From the above theorem, it is clear that f^* is a homomorphism from the exterior algebra $\Lambda(W^*)$ to the exterior algebra $\Lambda(V^*)$, that is

$$f^* : \Lambda(W^*) \longrightarrow \Lambda(V^*)$$

preserves the algebraic structure of the two exterior (Grassman) algebras.

The concept of an exterior algebra was originally introduced by H. Grassman for the purpose of studying linear subspaces of vector spaces. In the hands of the great geometer Elie Cartan, it was subsequently developed into the theory of exterior differentiation, and applied to the study of differential equations and differential geometry. At present, exterior products and exterior differentiation of differential forms have become powerful and irreplaceable tools in modern differential geometry.

We will conclude this chapter with five useful theorems on exterior products, which will be stated without proof.

Theorem 28.3. *A set of vectors $v_1, \ldots, v_r \in V$ is linearly dependent if and only if*

$$v_1 \wedge \cdots \wedge v_r = 0 \quad . \tag{28.42}$$

Theorem 28.4 (Cartan's Lemma). *Suppose $\{v_1, \ldots, v_r\}$ and $\{w_1, \ldots, w_r\}$ are two sets of vectors in V such that*

$$\sum_{\alpha=1}^{r} v_\alpha \wedge w_\alpha = 0 \quad . \tag{28.43}$$

If $v_1, \ldots, v_r$ are linearly independent, then each of the w_α can be expressed as a linear combination of the v_β:

$$w_\alpha = a_\alpha^\beta v_\beta, \quad 1 \le \alpha \le r, \tag{28.44}$$

with $a_\alpha^\beta = a_\beta^\alpha$. [The matrix (a_α^β) is symmetric.]

The next theorem gives a necessary and sufficient condition for an arbitrary $\omega \in \Lambda^s(V)$ to be expressible as a linear combination of linearly independent vectors with exterior vectors as "coefficients".

Theorem 28.5. *Let $\omega \in \Lambda^s(V)$, and $v_1, \ldots, v_r$ be r linearly independent vectors in V. ω can be expressed in the form*

$$\omega = v_1 \wedge \psi_1 + \cdots + v_r \wedge \psi_r \quad , \tag{28.45}$$

where $\psi_1, \ldots, \psi_r \in \Lambda^{s-1}(V)$, if and only if

$$v_1 \wedge \cdots \wedge v_r \wedge \omega = 0 \quad . \tag{28.46}$$

Usually we represent (28.45) by the statement

$$\omega = 0 \, mod(v_1, \ldots, v_r) \quad . \tag{28.47}$$

Theorem 28.6. *Suppose $\{v_1, \ldots, v_k, w_1, \ldots, w_k\}$ and $\{v_1', \ldots, v_k', w_1', \ldots, w_k'\}$ are two sets of vectors in V such that $\{v_1, \ldots, v_k, w_1, \ldots, w_k\}$ is linearly independent and*

$$\sum_{\alpha=1}^{k} v_\alpha \wedge w_\alpha = \sum_{\alpha=1}^{k} v_\alpha' \wedge w_\alpha' \quad . \tag{28.48}$$

Then the set $\{v_1', \ldots, v_k', w_1', \ldots, w_k'\}$ is also linearly independent and any vector in the set can be expressed as a linear combination of $v_1, \ldots, v_k, w_1, \ldots, w_k$.

Besides the evaluation formula (28.14) and Eq. (28.28) giving the pairing between an exterior vector space and its dual, the following theorem also illustrates the natural occurrence of the determinant in exterior algebra.

Theorem 28.7. *Suppose $v_1, \ldots, v_k \in V$, and $w_1, \ldots, w_k$ are linear combinations of them given by*

$$w_\alpha = t_\alpha^\beta v_\beta \quad .$$

Then

$$w_1 \wedge \cdots \wedge w_k = \det(t_\alpha^\beta)\, v_1 \wedge \cdots \wedge v_k \quad . \tag{28.49}$$

$\boxed{\text{Exercise 28.7}}$ Prove the above theorem.

$\boxed{\text{Exercise 28.8}}$ Prove the following.

Suppose $\{e_1, \ldots, e_n\}$ is a basis of an n-dimensional vector space V, with the dual basis $\{e^{*1}, \ldots, e^{*n}\}$ of V^*. Then $\{e_{i_1} \wedge \cdots \wedge e_{i_r} \,|\, 1 \leq i_1 < \cdots < i_r \leq n\}$ and $\{e^{*j_1} \wedge \cdots \wedge e^{*j_r} \,|\, 1 \leq j_1 < \cdots < j_r \leq n\}$ are bases of $\Lambda^r(V)$ and $\Lambda^r(V^*)$, respectively, and are dual bases to each other. The pairing between them is given by

$$\langle e_{i_1} \wedge \cdots \wedge e_{i_r}, e^{*j_1} \wedge \cdots \wedge e^{*j_r} \rangle = \det(\langle e_{i_\alpha}, e^{*j_\beta} \rangle) = \delta_{i_1 \ldots i_r}^{j_1 \ldots j_r} \quad , \tag{28.50}$$

where $\delta_{i_1 \ldots i_r}^{j_1 \ldots j_r}$ is the generalized Kronecker δ-symbol given by (28.16). In this case

$$\delta_{i_1 \ldots i_r}^{j_1 \ldots j_r} = \begin{cases} 1 & \text{if } \{j_1, \ldots, j_r\} = \{i_1, \ldots, i_r\}, \\ 0 & \text{if } \{j_1, \ldots, j_r\} \neq \{i_1, \ldots, i_r\}. \end{cases} \tag{28.51}$$

Chapter 29

The Hodge-Star Operator

The pairing between the dual spaces $\Lambda^r(V)$ and $\Lambda^r(V^*)$ given by (28.28) induces an inner product in $\Lambda^r(V)$ given by

$$(v_1 \wedge \cdots \wedge v_r, w_1 \wedge \cdots \wedge w_r) \equiv \langle v_1, \wedge \cdots \wedge v_r, w^{*1} \wedge \cdots \wedge w^{*r} \rangle$$
$$= \det(\langle v_\alpha, w^{*\beta} \rangle) \quad , \tag{29.1}$$

with

$$w^{*\alpha} = g(w_\alpha) \quad , \tag{29.2}$$

where $g : V \longrightarrow V^*$ is the canonical isomorphism induced by a nondegenerate inner product in V [c.f. discussion in Chapter 8 and (8.26b)]. The induced canonical isomorphism $g : \Lambda^r(V) \longrightarrow \Lambda^r(V^*)$ is thus given by

$$g(w_1 \wedge \cdots \wedge w_r) = w^{*1} \wedge \cdots \wedge w^{*r} \quad . \tag{29.3}$$

The inner product (29.1) is nondegenerate if and only if the original inner product $(\,,\,)$ in V, from which it is induced, is nondegenerate.

> Exercise 29.1 Prove the above statement.

A nondegenerate inner product in V and the notion of the orientation of V give rise to a linear mapping

$$\star : \Lambda^r(V) \longrightarrow \Lambda^{n-r}(V) \quad ,$$

where $r \leq n$ and $dim(V) = n$, called the **Hodge star operator**, or **star operator**, for short. We will first explain the notion of the orientation of a vector space.

Definition 29.1. *Suppose* $\{e_1, \ldots, e_n\}$ *and* $\{e_1', \ldots, e_n'\}$ *are two bases of* V *related by a nonsingular matrix* (a_i^j):

$$e_i' = a_i^j e_j \quad . \tag{29.4}$$

Then the two bases are said to be equivalent, or determine the same orientation, if $\det(a_i^j) > 0$. *The set of all bases then partitions into two equivalence classes, each of which is called an* **orientation** *of* V.

According to Theorem 28.7, each orientation of V is specified by a non-zero vector in $\Lambda^n(V)$, if $dim(V) = n$, such as $e_1 \wedge \cdots \wedge e_n$.

Now suppose $\lambda \in \Lambda^r(V)$, where $r \leq n$. Then for any $\omega \in \Lambda^{n-r}(V)$,

$$\lambda \wedge \omega \in \Lambda^n(V) \quad . \tag{29.5}$$

Since $\Lambda^n(V)$ is one-dimensional and $e_1 \wedge \cdots \wedge e_n$ is a basis of $\Lambda^n(V)$, we can write

$$\lambda \wedge \omega = f_\lambda(\omega) e_1 \wedge \cdots \wedge e_n \quad , \tag{29.6}$$

where $f_\lambda : \Lambda^{n-r}(V) \longrightarrow \mathbb{R}$ is a linear function on $\Lambda^{n-r}(V)$ characterized by λ, and is hence an element in the dual space of $\Lambda^{n-r}(V)$, i.e.,

$$f_\lambda \in (\Lambda^{n-r}(V))^* = \Lambda^{n-r}(V^*) \quad . \tag{29.7}$$

The inner product in $\Lambda^{n-r}(V)$ [given by (29.1)] then induces a canonical isomorphism

$$g : \Lambda^{n-r}(V) \longrightarrow \Lambda^{n-r}(V^*) \quad , \tag{29.8}$$

and hence a unique element in $\Lambda^{n-r}(V)$, called the **Hodge-star** of λ, such that

$$g(\star\lambda) = f_\lambda \quad , \tag{29.9}$$

and, for all $\omega \in \Lambda^{n-r}(V)$,

$$(\omega, \star\lambda) = \langle \omega, f_\lambda \rangle = f_\lambda(\omega) \quad . \tag{29.10}$$

Eq. (29.6) can be rewritten as follows. If $\lambda \in \Lambda^r(V)$, then

$$\boxed{\lambda \wedge \omega = (\star\lambda, \omega)\,\sigma} \quad , \tag{29.11}$$

for all $\omega \in \Lambda^{n-r}(V)$. In this equation

$$\sigma = e_1 \wedge \cdots \wedge e_n \tag{29.12}$$

specifies an orientation of V. In fact, the above two equations can be regarded also as the defining equations of the Hodge-star operator. We note that the definition is predicated on two things:

1) a given inner product $(\,,\,)$ in V, which induces an inner product in $\Lambda^r(V)$; and

2) a given orientation σ in V.

Before giving some general results on the star operator, let us define the concept of the signature of a scalar product (which may be non positive-definite, such as that associated with the Lorentz metric in Minkowski space).

Definition 29.2. *Suppose* $\{e_1, \ldots, e_n\}$ *is an orthonormal basis of a vector space equipped with an inner product which may be non positive-definite, so that*

$$(e_i, e_j) = \pm \delta_{ij}, \quad i, j = 1, \ldots, n \quad . \tag{29.13}$$

If the number of $+1$*'s appear* P *times and the number of* -1*'s appear* N *times in the above group of equations, then the integer*

$$S = P - N \tag{29.14}$$

is called the **signature** *of the inner product.*

Exercise 29.2 Show that the signature of an inner product does not depend on the choice of the orthonormal basis.

We are now ready to state (without proof) a number of useful results on the $\star$ operator. None of these are very difficult to prove.

Suppose $dim(V) = n$, $\sigma = e_1 \wedge \cdots \wedge e_n$ specifies an orientation of V, V is endowed with an inner product $(\, , \,)$ with signature S, and λ, μ are arbitrary exterior vectors in $\Lambda^r(V)$. Then

$$\star \sigma = 1 \quad \text{(the scalar 1)} \quad , \tag{29.15}$$
$$\star 1 = (-1)^{(n-S)/2} \sigma \quad , \tag{29.16}$$
$$(\sigma, \sigma) = (-1)^{(n-S)/2} \quad , \tag{29.17}$$
$$\star \star \lambda = (-1)^{r(n-r)+(n-S)/2} \lambda \quad , \tag{29.18}$$
$$(\lambda, \mu) = \star \star \star (\mu \wedge \star \lambda) = \star \star \star (\lambda \wedge \star \mu) \quad , \tag{29.19}$$
$$\mu \wedge \star \lambda = \lambda \wedge \star \mu = (-1)^{(n-S)/2}(\lambda, \mu)\sigma \quad , \tag{29.20}$$
$$(\star \lambda, \star \mu) = \star(\lambda \wedge \star \mu) = \star(\mu \wedge \star \lambda) \quad . \tag{29.21}$$

For the rest of this chapter we will illustrate the use of the $\star$ operator with two frequently encountered examples.

Consider $V = \mathbb{R}^3$ with the Euclidean scalar product. Choose an orthonormal basis $\{e_1, e_2, e_3\}$ and the orientation $\sigma = e_1 \wedge e_2 \wedge e_3$, represented by the right-handed reference frame illustrated in Fig. 29.1.

The scalar product is given by

$$(e_i, e_j) = \delta_{ij}, \quad i, j = 1, 2, 3 \quad . \tag{29.22}$$

Since $S = 3 = n$,

$$(\sigma, \sigma) = 1 \quad . \tag{29.23}$$

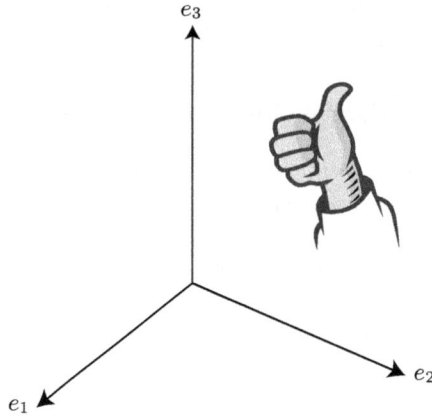

FIGURE 29.1

Applying (29.11), we have

$$\sigma = (e_1 \wedge e_2) \wedge e_3 = (\star(e_1 \wedge e_2), e_3)\sigma\,, \quad \text{so } \star(e_1 \wedge e_2) = e_3\,; \tag{29.24}$$
$$\sigma = (e_2 \wedge e_3) \wedge e_1 = (\star(e_2 \wedge e_3), e_1)\sigma\,, \quad \text{so } \star(e_2 \wedge e_3) = e_1\,; \tag{29.25}$$
$$\sigma = (e_3 \wedge e_1) \wedge e_2 = (\star(e_3 \wedge e_1), e_2)\sigma\,, \quad \text{so } \star(e_3 \wedge e_1) = e_2\,; \tag{29.26}$$
$$-\sigma = (e_1 \wedge e_3) \wedge e_2 = (\star(e_1 \wedge e_3), e_2)\sigma\,, \quad \text{so } \star(e_1 \wedge e_3) = -e_2\,; \tag{29.27}$$
$$-\sigma = (e_3 \wedge e_2) \wedge e_1 = (\star(e_3 \wedge e_2), e_1)\sigma\,, \quad \text{so } \star(e_3 \wedge e_2) = -e_1\,; \tag{29.28}$$
$$-\sigma = (e_2 \wedge e_1) \wedge e_3 = (\star(e_2 \wedge e_1), e_3)\sigma\,, \quad \text{so } \star(e_2 \wedge e_1) = -e_3\,; \tag{29.29}$$
$$\sigma = e_1 \wedge (e_2 \wedge e_3) = (\star e_1, e_2 \wedge e_3)\sigma\,, \quad\quad \text{so } \star e_1 = e_2 \wedge e_3\,; \tag{29.30}$$
$$\sigma = e_2 \wedge (e_3 \wedge e_1) = (\star e_2, e_3 \wedge e_1)\sigma\,, \quad\quad \text{so } \star e_2 = e_3 \wedge e_1\,; \tag{29.31}$$
$$\sigma = e_3 \wedge (e_1 \wedge e_2) = (\star e_3, e_1 \wedge e_2)\sigma\,, \quad\quad \text{so } \star e_3 = e_1 \wedge e_2\,. \tag{29.32}$$

Furthermore,

$$\star(e_1 \wedge e_2 \wedge e_3) = 1, \tag{29.33}$$
$$\star 1 = e_1 \wedge e_2 \wedge e_3\quad. \tag{29.34}$$

Eqs. (29.24) to (29.29) justify the claim, made at the beginning of Chapter 28, that the exterior product is the generalization of the ordinary cross product in 3-dimensional vector algebra. This association is actually made via the $\star$-operator, and the accidental fact that $dim(\mathbb{R}^3) = 3$.

Now let us look at the case where V is the 4-dimensional Minkowski space-time with the scalar product determined by the Lorentz metric $\eta_{\mu\nu}$ [c.f. (2.20)]. Let e_1 be the time axis and e_2, e_3 and e_4 be the spatial x, y and z axes, respectively. Choose $\{e_1, e_2, e_3, e_4\}$ as the basis and $\sigma = e_1 \wedge e_2 \wedge e_3 \wedge e_4$ as the orientation. The Lorentz scalar product is given by

$$\begin{cases} (e_1, e_1) = -1\,, \\ (e_i, e_i) = +1\,, & i = 2, 3, 4\,, \\ (e_i, e_j) = 0\,, & i \neq j \quad. \end{cases} \tag{29.35}$$

The signature S is 2. Thus, from (29.17),

$$(\sigma, \sigma) = (-1)^{(4-2)/2} = -1 \quad. \tag{29.36}$$

As always, from (29.15),

$$\star\sigma = \star(e_1 \wedge e_2 \wedge e_3 \wedge e_4) = 1 \quad. \tag{29.37}$$

We can again apply (29.11) to determine the $\star$ of r-vectors, where $r = 1, 2, 3$. Thus

$$\begin{aligned}
\sigma &= (e_1 \wedge e_2 \wedge e_3) \wedge e_4 \\
&= (\star(e_1 \wedge e_2 \wedge e_3), e_4)\sigma\,, & \text{so } \star(e_1 \wedge e_2 \wedge e_3) = e_4\,; \tag{29.38}
\end{aligned}$$

$$\begin{aligned}
-\sigma &= (e_1 \wedge e_2 \wedge e_4) \wedge e_3 \\
&= (\star(e_1 \wedge e_2 \wedge e_4), e_3)\sigma\,, & \text{so } \star(e_1 \wedge e_2 \wedge e_4) = -e_3\,; \tag{29.39}
\end{aligned}$$

$$\begin{aligned}
\sigma &= (e_1 \wedge e_3 \wedge e_4) \wedge e_2 \\
&= (\star(e_1 \wedge e_3 \wedge e_4), e_2)\sigma\,, & \text{so } \star(e_1 \wedge e_3 \wedge e_4) = e_2\,; \tag{29.40}
\end{aligned}$$

$$\begin{aligned}
-\sigma &= (e_2 \wedge e_3 \wedge e_4) \wedge e_1 \\
&= (\star(e_2 \wedge e_3 \wedge e_4), e_1)\sigma\,, & \text{so } \star(e_2 \wedge e_3 \wedge e_4) = e_1\,; \tag{29.41}
\end{aligned}$$

$$\begin{aligned}
\sigma &= (e_1 \wedge e_2) \wedge (e_3 \wedge e_4) \\
&= (\star(e_1 \wedge e_2), e_3 \wedge e_4)\sigma\,, & \text{so } \star(e_1 \wedge e_2) = e_3 \wedge e_4\,; \tag{29.42}
\end{aligned}$$

$$\begin{aligned}
-\sigma &= (e_1 \wedge e_3) \wedge (e_2 \wedge e_4) \\
&= (\star(e_1 \wedge e_3), e_2 \wedge e_4)\sigma\,, & \text{so } \star(e_1 \wedge e_3) = -e_2 \wedge e_4\,; \tag{29.43}
\end{aligned}$$

$$\begin{aligned}
\sigma &= (e_1 \wedge e_4) \wedge (e_2 \wedge e_3) \\
&= (\star(e_1 \wedge e_4), e_2 \wedge e_3)\sigma\,, & \text{so } \star(e_1 \wedge e_4) = e_2 \wedge e_3\,; \tag{29.44}
\end{aligned}$$

$$\begin{aligned}
\sigma &= (e_2 \wedge e_3) \wedge (e_1 \wedge e_4) \\
&= (\star(e_2 \wedge e_3), e_1 \wedge e_4)\sigma\,, & \text{so } \star(e_2 \wedge e_3) = -e_1 \wedge e_4\,; \tag{29.45}
\end{aligned}$$

$$\begin{aligned}
-\sigma &= (e_2 \wedge e_4) \wedge (e_1 \wedge e_3) \\
&= (\star(e_2 \wedge e_4), e_1 \wedge e_3)\sigma\,, & \text{so } \star(e_2 \wedge e_4) = e_1 \wedge e_3\,; \tag{29.46}
\end{aligned}$$

$$\begin{aligned}
\sigma &= (e_3 \wedge e_4) \wedge (e_1 \wedge e_2) \\
&= (\star(e_3 \wedge e_4), e_1 \wedge e_2)\sigma\,, & \text{so } \star(e_3 \wedge e_4) = -e_1 \wedge e_2\,; \tag{29.47}
\end{aligned}$$

$$\begin{aligned}
\sigma &= e_1 \wedge (e_2 \wedge e_3 \wedge e_4) \\
&= (\star e_1, e_2 \wedge e_3 \wedge e_4)\sigma\,, & \text{so } \star e_1 = e_2 \wedge e_3 \wedge e_4\,; \tag{29.48}
\end{aligned}$$

$$\begin{aligned}
-\sigma &= e_2 \wedge (e_1 \wedge e_3 \wedge e_4) \\
&= (\star e_2, e_1 \wedge e_3 \wedge e_4)\sigma\,, & \text{so } \star e_2 = e_1 \wedge e_3 \wedge e_4\,; \tag{29.49}
\end{aligned}$$

$$\sigma = e_3 \wedge (e_1 \wedge e_2 \wedge e_4)$$
$$= (\star e_3, e_1 \wedge e_2 \wedge e_4)\sigma, \qquad \text{so } \star e_3 = -e_1 \wedge e_2 \wedge e_4; \qquad (29.50)$$
$$-\sigma = e_4 \wedge (e_1 \wedge e_2 \wedge e_3)$$
$$= (\star e_4, e_1 \wedge e_2 \wedge e_3)\sigma, \qquad \text{so } \star e_4 = e_1 \wedge e_2 \wedge e_3. \qquad (29.51)$$

Finally,

$$\star 1 = -e_1 \wedge e_2 \wedge e_3 \wedge e_4. \qquad (29.52)$$

Chapter 30

Differential Forms and Exterior Differentiation

In Chapter 8 we gave a rudimentary discussion of the notion of a tangent space to a differentiable manifold at a particular point in the manifold. This is a very basic and intuitive notion in differential geometry, and forms the starting point to the introduction of differential forms. In our development here we will aim at neither completeness nor mathematical rigor, but will instead rely on an intuitive approach.

A finite-dimensional differentiable manifold M behaves locally like a neighborhood of a finite-dimensional Euclidean space, so that locally, a point $x \in M$ can be characterized by a set of n coordinates $\{x^1, \ldots, x^n\}$, where n is the dimension of the manifold. Such a characterization cannot in general be applied globally, and one requires different "patches" of Euclidean space to cover a manifold. Think of a two-dimensional sphere for example, where locally, a point can be described by two coordinates (latitude and longitude); but the same set of coordinates cannot be used throughout the whole surface. (What is the longitude of the North Pole?) In obvious reference to map-making, we speak of a differentiable manifold as being covered by **charts**, each of which, denoted (U, u^i), consists of a neighborhood $U \subset M$ and a set of coordinates $u^1, \ldots, u^n$ valid in U. If two charts (U, u^i) and (W, w^i) are such that U and W overlap, that is, $U \cap W \neq \emptyset$, then all the w's must be smooth functions of all the u's, and vice versa. Thus one and the same point in M may have different sets of coordinates for its description. A set of charts $(U, u^i), (V, v^i), (W, w^i), \ldots$ such that $\{U, V, W, \ldots\}$ **cover** M, and satisfying certain additional conditions (which will not be spelled out in detail here) besides the differentiability condition mentioned above, is called an **atlas** of the differentiable manifold M, and specifies the so-called **differentiable structure** of the manifold.

Let us now define the concept of a tangent vector at a point $x \in M$. Consider a curve $\gamma : \mathbb{R} \longrightarrow M$ passing through x such that $\gamma(0) = x$. (Fig. 30.1) The **tangent vector** X to the curve $\gamma(t)$ at the point x is the vector whose i-th

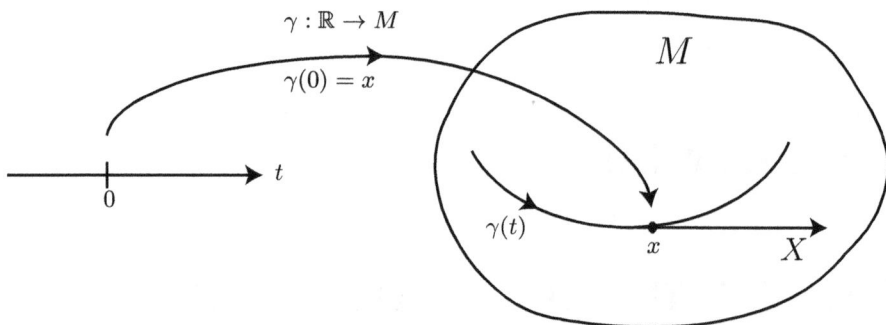

FIGURE 30.1

component ξ^i describes how "fast" the i-th coordinate x^i changes along $\gamma(t)$:

$$\xi^i \equiv \frac{d}{dt}(x^i \circ \gamma)\Big|_{t=0} \quad . \tag{30.1}$$

Now consider a function $f : M \longrightarrow \mathbb{R}$. We may ask: how "fast" does f change along γ at x? The answer is

$$\frac{\partial f}{\partial x^i} \frac{d(x^i \circ \gamma)}{dt}\Big|_{t=0} = (\partial_i f)\xi^i \quad . \tag{30.2}$$

This quantity is called the **directional derivative** of f along the direction X at x. The RHS of (30.2) suggests that $(\partial_i f)\xi^i$ (i summed over) is the pairing between a contravariant vector with components ξ^i and a covariant vector with components $\partial_i f$. In fact, this covariant vector is the ordinary differential of the function f:

$$df = \partial_i f \, dx^i \quad , \tag{30.3}$$

and the pairing in (30.2) is denoted

$$\langle X, df \rangle = (\partial_i f)\xi^i \quad . \tag{30.4}$$

Thus we can write

$$X = \xi^i e_i \quad , \tag{30.5}$$

where $\{e_i\}$ and $\{dx^i\}$ are dual basis sets. How can we write e_i in terms of the local coordinates x^i? The answer is given by the duality requirement that

$$\langle e_i, dx^j \rangle = \delta_i^j = \frac{\partial x^j}{\partial x^i} \quad . \tag{30.6}$$

In other words,

$$e_i = \frac{\partial}{\partial x^i} \quad, \tag{30.7}$$

and we have

$$\boxed{X = \xi^i \frac{\partial}{\partial x^i} = \xi^i \partial_i} \quad. \tag{30.8}$$

The **tangent space** at $x \in M$, denoted $T_x M$, is the linear vector space spanned by the basis set $\left\{ \frac{\partial}{\partial x^1}, \ldots, \frac{\partial}{\partial x^n} \right\}$, whereas the dual space to $T_x M$, denoted $T_x^* M$ and called the **cotangent space** at $x \in M$, is spanned by the basis $\{dx^1, \ldots, d^n\}$. The two bases $\left\{ \frac{\partial}{\partial x^i} \right\}$ and $\{dx^j\}$ are dual bases to each other [compare (8.15)]:

$$\boxed{\left\langle \frac{\partial}{\partial x^i}, dx^j \right\rangle = \delta_i^j} \quad. \tag{30.9}$$

For obvious reasons, $\left\{ \frac{\partial}{\partial x^i} \right\}$ and $\{dx^j\}$ are called the **coordinate bases** (or **natural bases**) of $T_x M$ and $T_x^* M$, respectively.

Remark. The fact that the coordinate basis vectors in $T_x M$ are expressed in terms of differential operators may appear strange at first. But this is perfectly reasonable if we remember that $T_x M$, being dual to $T_x^* M$, is in fact the space of linear functions on $T_x^* M$. Thus $\frac{\partial}{\partial x^i} \in T_x M$ acts on $df \in T_x^* M$ to yield $\frac{\partial f}{\partial x^i}$, a scalar.

Under a change of coordinates

$$x^i \longrightarrow x'^i = x'^i(x^1, \ldots, x^n), \quad i = 1, \ldots, n,$$

we have, by the chain rule,

$$\partial_i = \frac{\partial}{\partial x^i} = \frac{\partial}{\partial x'^j} \frac{\partial x'^j}{\partial x^i} = \frac{\partial x'^j}{\partial x^i} \partial'_j \quad, \tag{30.10}$$

and

$$dx^i = \frac{\partial x^i}{\partial x'^j} dx'^j \quad. \tag{30.11}$$

We can rewrite the above two equations as

$$\partial_i = a_i^j \partial'_j \quad, \tag{30.12}$$

and

$$dx^i = (a^{-1})_j^i dx'^j \quad, \tag{30.13}$$

where

$$a^j_i \equiv \frac{\partial x'^j}{\partial x^i} \quad , \tag{30.14}$$

is the **Jacobian matrix** for a local change of coordinates. The transformation rules (30.12) and (30.13) are in conformity with those given in Table 1.1 in Chapter 1 for covariant (lower indexed) objects and contravariant (upper indexed) objects, respectively.

Based on the notion of a tangent space $T_x M$ and a cotangent space $T_x^* M$ at a particular point $x \in M$, we can first construct the (r, s)-type tensor space

$$T^r_s(x) = \underbrace{T_x \otimes \cdots \otimes T_x}_{r \text{ times}} \otimes \underbrace{T_x^* \otimes \cdots \otimes T_x^*}_{s \text{ times}} \tag{30.15}$$

at $x \in M$, and then the (r, s)-type **tensor bundle**

$$T^r_s = \bigcup_{x \in M} T^r_s(x) \quad . \tag{30.16}$$

Loosely speaking, T^r_s is a "concatenation" of (r, s)-type tensor spaces at all points of the manifold M. Each $T^r_s(x)$ is called the **fiber** of the bundle T^r_s at the point x. The natural projection

$$\pi : T^r_s \longrightarrow M , \tag{30.17}$$

which maps all points in $T^r_s(x)$ to the point $x \in M$, is a smooth surjective map. It is called the **bundle projection** (Fig. 30.2).

The construction of the tensor bundle T^r_s is a sophisticated procedure, a detailed treatment of which is beyond the scope of this book (see Chern, Chen and Lam 1999). We will only provide a very brief and intuitive description here. It consists mainly of the following two steps:

1) specify a topological structure and a differentiable structure such that *locally the tensor bundle is a product*:

$$\varphi_U : U \times V^r_s \longrightarrow \bigcup_{x \in U} T^r_s(x) \quad ; \tag{30.18}$$

2) specify a way whereby fibers "sticking out" of points in overlapping charts of M are "glued" together so that the linear structure of the fiber is preserved. (See Fig. 30.3.)

In Fig. 30.3, the functions

$$g_{UW} : U \cap W \longrightarrow GL(V^r_s) \quad , \tag{30.19}$$

where $GL(V^r_s)$ is the **general linear group** on V^r_s (the group of all linear transformations on V^r_s), are called **transition functions** ("glueing" functions) of T^r_s. They determine the structure of the bundle.

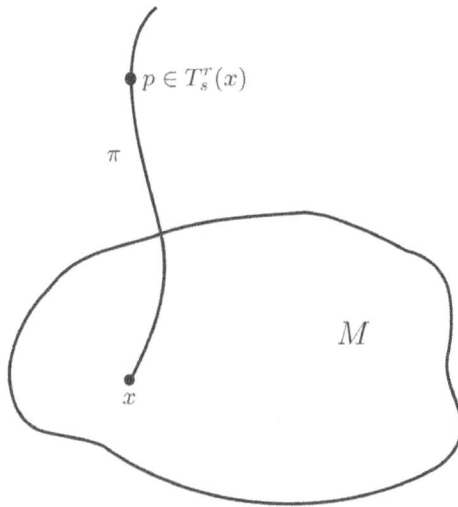

$p \in T_s^r(x)$

π

M

x

FIGURE 30.2

Among the T_s^r we have the **tangent bundle** T_0^1 [also denoted $T(M)$] and the **cotangent bundle** T_1^0 [also denoted $T^*(M)$]. Furthermore, we can construct the so-called **exterior vector bundles** and **exterior form bundles** on M, given respectively by

$$\Lambda^r(M) \equiv \bigcup_{x \in M} \Lambda^r(T_x) \quad \text{and} \quad \Lambda^r(M^*) \equiv \bigcup_{x \in M} \Lambda^r(T_x^*). \tag{30.20}$$

$T_s^r, \Lambda^r(M)$ and $\Lambda^r(M^*)$ are special cases of **vector bundles** (fiber bundles whose fibers are vector spaces). These are usually denoted $\pi : E \longrightarrow M$, where π is the natural projection [c.f. (30.17)], E is called the **total space** (or **bundle space**), and M is called the **base manifold**.

Note that the total space E should be thought of as a manifold in its own right (with its own differentiable structure), and not necessarily as a product of the base manifold and the fiber. In fact, a fiber bundle is only locally a product [see (30.18)]; and in general cannot be represented as such globally. A bundle which is globally a product is called a **trivial bundle**, and much of the theory of fiber bundles is devoted to characterizing various kinds of non-trivialities (see Chapter 41). Non-trivial bundles play a most important role in modern theoretical physics: they form the mathematical basis for the description of many interesting physical objects, such as magnetic monopoles, instantons, and even molecular collision systems.

A smooth map $s : M \longrightarrow E$ such that

$$\pi \circ s = id : M \longrightarrow M \quad, \tag{30.21}$$

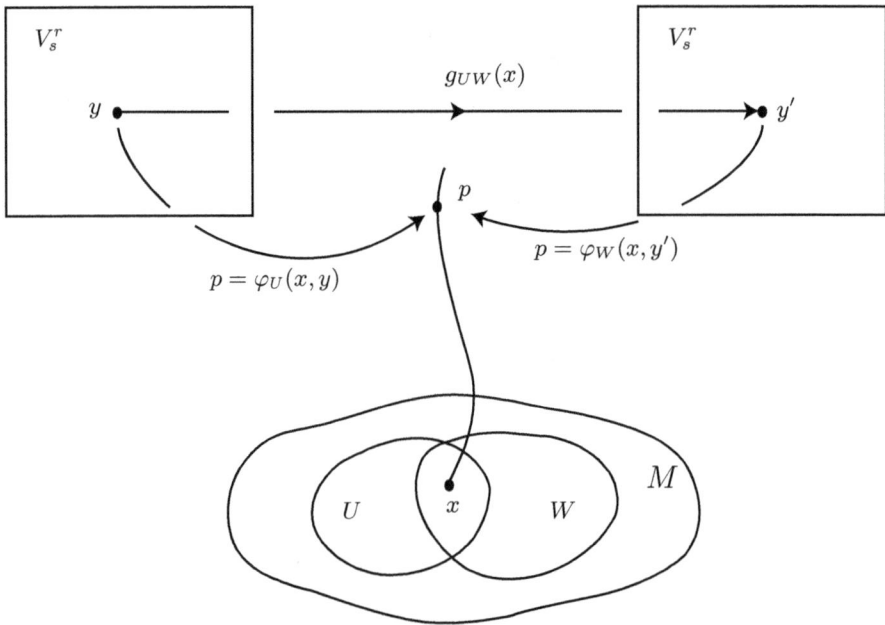

FIGURE 30.3

where id is the identity map, is called a **smooth section** of the vector bundle $\pi : E \longrightarrow M$. The set of all smooth sections of $\pi : E \longrightarrow M$, denoted $\Gamma(E)$, is an important vector space. It is also a $C^\infty(M)$ **module**, that is, a module over the ring of smooth functions on M. A section of a vector bundle $\pi : E \longrightarrow M$ is also called a **vector field** on the manifold M.

We are now finally ready to give a definition of differential forms, or more specifically, exterior differential forms, on which one can perform the all-important operation of exterior differentiation.

Start with an n-dimensional smooth manifold M, and construct the bundle of exterior r-forms on M [c.f. (30.20)]:

$$\Lambda^r(M^*) = \bigcup_{x \in M} \Lambda^r(T_x^*) \quad .$$

This is a vector bundle on M, and so it is meaningful to conceive of the space of smooth sections on it. We denote this vector space by $A^r(M)$:

$$A^r(M) \equiv \Gamma(\Lambda^r(M^*)) \quad . \tag{30.22}$$

As mentioned earlier, $A^r(M)$ is also a $C^\infty(M)$ module.

Definition 30.1. *The elements of the $C^\infty(M)$ module $A^r(M)$ are called the **exterior differential r-forms** on M. In other words, an exterior differential* *r-form on M is a smooth skew-symmetric covariant tensor field of order r [a (0, r)-type tensor field] on M.*

From $A^r(M)$ we can construct the direct sum

$$A(M) = \oplus \sum_{r=0}^{n} A^r(M) = A^0(M) \oplus \cdots \oplus A^n(M) \quad , \tag{30.23}$$

which is in fact the space of all smooth sections of the exterior form bundle

$$\Lambda(M^*) = \bigcup_{x \in M} \Lambda(T_x^*) \quad . \tag{30.24}$$

Elements of $A(M)$ are called **exterior differential forms** on M. Thus every exterior differential form $\omega \in A(M)$ can be written

$$\omega = \omega^0 + \omega^1 + \cdots + \omega^n \quad , \tag{30.25}$$

where $\omega^r \in A^r(M)$, $0 \le r \le n$.

Remark. Technically a differential form need not be an exterior form. The term "exterior" refers to the skew-symmetry of the form in question. When no confusion arises, or when it is clear from the context, we will occasionally speak of just differential forms instead of exterior differential forms.

Now we impose a multiplication rule on $A(M)$ using the exterior (wedge) product (c.f. Def. 28.1). The wedge product of exterior forms at a single point $x \in M$ can be easily extended to the space of exterior differential forms $A(M)$ by a pointwise association: for any $x \in M$, $\omega_1, \omega_2 \in A(M)$, define

$$(\omega_1 \wedge \omega_2)(x) = \omega_1(x) \wedge \omega_2(x) \quad , \tag{30.26}$$

where the RHS is a wedge product of two exterior forms. With this multiplication rule, the space $A(M)$ then becomes a **graded algebra** with respect to addition, scalar multiplication, and the wedge product. The wedge product defines a map

$$\wedge : A^r(M) \times A^s(M) \longrightarrow A^{r+s}(M) \quad , \tag{30.27}$$

where $A^{r+s}(M) = 0$ when $r + s > n$.

Let (U, x^i) be a chart of M. An exterior differential r-form ω can be expressed locally in the neighborhood U in terms of the local coordinates $x^1, \ldots, x^n$ of M as

$$\omega = a_{i_1 \ldots i_r}(x^1, \ldots, x^r) dx^{i_1} \wedge \cdots \wedge dx^{i_r} \quad , \tag{30.28}$$

where each of the $a_{i_1 \ldots i_r}$ is a smooth function on U (i.e., a smooth function of the local coordinates $x^1, \ldots, x^n$), and is required to be skew-symmetric with respect to any permutation of its indices (c.f. Theorem 16.1). Note that in (30.28) the Einstein summation convention has been used.

Since the fibers of $\Lambda^r(M)$ and $\Lambda^r(M^*)$ at a point $x \in M$ are dual spaces to each other, we can introduce a pairing between them according to (28.28). In terms of the natural bases $\left\{ \dfrac{\partial}{\partial x^{i_1}} \wedge \cdots \wedge \dfrac{\partial}{\partial x^{i_r}} \right\}$, $i_1 \leq \cdots \leq i_r$, of $\Lambda^r(T_x)$ and $\{dx^{j_1} \wedge \cdots \wedge dx^{j_r}\}$, $j_1 \leq \cdots \leq j_r$, of $\Lambda^r(T_x^*)$, we have

$$\left\langle \frac{\partial}{\partial x^{i_1}} \wedge \cdots \wedge \frac{\partial}{\partial x^{i_r}}, \ dx^{j_1} \wedge \cdots \wedge dx^{j_r} \right\rangle = \delta_{i_1 \dots i_r}^{j_1 \dots j_r} \qquad , \qquad (30.29)$$

where the Kronecker delta symbol has been defined in (28.51). Thus the components of ω in terms of local coordinates can be expressed as

$$a_{i_1 \dots i_r} = \frac{1}{r!} \left\langle \frac{\partial}{\partial x^{i_1}} \wedge \cdots \wedge \frac{\partial}{\partial x^{i_r}}, \ \omega \right\rangle \ . \qquad (30.30)$$

Exercise 30.1   Verify (30.30).

The space $A(M)$ of exterior differential forms plays a crucial role in the study of differentiable manifolds, due to the existence of a most useful operator, the so-called exterior derivative operator d on $A(M)$, with the property that $d^2 = 0$. We state without proof the following theorem.

Theorem 30.1. *Let M be an n-dimensional smooth manifold. Then there exists a unique map*

$$d : A(M) \longrightarrow A(M), \quad d(A^r(M)) \subset A^{r+1}(M),$$

which satisfies the following properties:

1) *For any $\omega_1, \omega_2 \in A(M)$, $d(\omega_1 + \omega_2) = d\omega_1 + d\omega_2$.*

2) *Suppose $\omega_1 \in A^r(M)$. Then, for any $\omega_2 \in A(M)$,*

$$d(\omega_1 \wedge \omega_2) = d\omega_1 \wedge \omega_2 + (-1)^r \omega_1 \wedge d\omega_2 \qquad . \qquad (30.31)$$

3) *If f is a smooth function on M, that is, if $f \in A^0(M)$, then df is precisely the differential of f.*

4) *If $f \in A^0(M)$, then $d^2 f = 0$.*

The map d is called the **exterior derivative**.

Eq. (30.31) is the "product rule" of exterior differentiation. We note that d is a local operator, which means that, if $\omega_1, \omega_2 \in A(M)$ and their restrictions to a neighborhood U of M are equal $(\omega_1|_U = \omega_2|_U)$, then $d\omega_1|_U = d\omega_2|_U$.

In order to learn how to calculate exterior derivatives, let us begin with applying the operator d to $\omega = a_{i_1\ldots i_r}\,dx^{i_1}\wedge\cdots\wedge dx^{i_r}\in A^r(M)$. Using properties 2) [(30.31)] and 4) in Theorem 30.1, we immediately have

$$d\omega = (da_{i_1\ldots i_r})\wedge dx^{i_1}\wedge\cdots\wedge dx^{i_r}\quad. \tag{30.32}$$

On using (30.31), the second term on its RHS drops out because x^i is a smooth function (the coordinate function) on U and thus property 4) applies to it.

It is instructive to use (30.32) to check the validity of property 2) when ω_1 and ω_2 are both monomials [property 1) takes care of situations when ω_2 is not a monomial]:

$$\omega_1 = a\,dx^{i_1}\wedge\cdots\wedge dx^{i_r}\in A^r(M)\,,\qquad \omega_2 = b\,dx^{j_1}\wedge\cdots\wedge dx^{j_s}\in A^s(M)\,,$$

so that

$$\omega_1\wedge\omega_2 = ab\,dx^{i_1}\wedge\cdots\wedge dx^{i_r}\wedge dx^{j_1}\wedge\cdots\wedge dx^{j_s}\quad.$$

Then, by (30.32),

$$\begin{aligned}
d(\omega_1\wedge\omega_2) &= d(ab)\wedge dx^{i_1}\wedge\cdots\wedge dx^{i_r}\wedge dx^{j_1}\wedge\cdots\wedge dx^{j_s}\\
&= (b\,da + a\,db)\wedge dx^{i_1}\wedge\cdots\wedge dx^{i_r}\wedge dx^{j_1}\wedge\cdots\wedge dx^{j_s}\\
&= (da\wedge dx^{i_1}\wedge\cdots\wedge dx^{i_r})\wedge(b\,dx^{j_1}\wedge\cdots\wedge dx^{j_s})\\
&\quad + (-1)^r\,(a\,dx^{i_1}\wedge\cdots\wedge dx^{i_r})\wedge(db\wedge dx^{j_1}\wedge\cdots\wedge dx^{j_s})\\
&= d\omega_1\wedge\omega_2 + (-1)^r\,\omega_1\wedge d\omega_2\quad,
\end{aligned} \tag{30.33}$$

where, in the third equality, the term db has been commuted to the right by r "slots" and thus a factor of $(-1)^r$ is acquired according to the anticommutative rule [(28.5)] of exterior products.

Eq. (30.32) can also be used to check property 4) of Theorem 30.1. Let f be a smooth function on M. Then by 3) the local expression for df (in U) is

$$df = \frac{\partial f}{\partial x^i}\,dx^i\quad. \tag{30.34}$$

It follows from (30.32) that

$$\begin{aligned}
d(df) &= d\left(\frac{\partial f}{\partial x^i}\right)\wedge dx^i = \frac{\partial^2 f}{\partial x^j\partial x^i}\,dx^j\wedge dx^i\\
&= \frac{1}{2}\frac{\partial^2 f}{\partial x^j\partial x^i}\,dx^j\wedge dx^i + \frac{1}{2}\frac{\partial^2 f}{\partial x^j\partial x^i}\,dx^j\wedge dx^i\\
&= \frac{1}{2}\frac{\partial^2 f}{\partial x^j\partial x^i}\,dx^j\wedge dx^i + \frac{1}{2}\frac{\partial^2 f}{\partial x^i\partial x^j}\,dx^i\wedge dx^j\\
&= \frac{1}{2}\left(\frac{\partial^2 f}{\partial x^j\partial x^i} - \frac{\partial^2 f}{\partial x^i\partial x^j}\right)dx^j\wedge dx^i = 0\quad,
\end{aligned} \tag{30.35}$$

where in the second term on the RHS of the third equality, we have made the interchange $i\leftrightarrow j$ to obtain the second term on the RHS of the fourth equality, and the last equality follows from the fact that

$$\frac{\partial^2 f}{\partial x^i \partial x^j} = \frac{\partial^2 f}{\partial x^j \partial x^i} \quad , \tag{30.36}$$

which in turn follows from the C^∞ (smoothness) nature of f.

From the properties of the exterior derivative specified in Theorem 30.1, we immediately have the following important and useful result:

Theorem 30.2 (Poincare's Lemma).

$$d^2 = 0 \quad , \tag{30.37}$$

that is, for every $w \in A(M)$, $d(dw) = 0$.

Proof. We need only prove the lemma for a monomial $w \in A^r(M)$ since d is linear by property 1) of Theorem 30.1. Also, since d is a local operator, we need only consider

$$w = a \, dx^1 \wedge \cdots \wedge dx^r \, .$$

By (30.32),

$$dw = da \wedge dx^1 \wedge \cdots \wedge dx^r \, .$$

Differentiating once more and applying properties 2) and 4) of Theorem 30.1, we have

$$d(dw) = d(da) \wedge dx^1 \wedge \cdots \wedge dx^r - da \wedge d(dx^1 \wedge \cdots \wedge dx^r)$$
$$= -da \wedge d(dx^1) \wedge \cdots \wedge dx^r + \ldots = 0 \, ,$$

since each of the x^i is a local coordinate function on U (a neighborhood of M), and so, by property 4), $d^2 x^i = 0$. $\qquad \square$

We will rewrite (30.32) in a form most often used in computations of exterior derivatives:

$$\boxed{\begin{aligned} &\text{If} \quad w|_U = a_{i_1 \ldots i_r} \, dx^{i_1} \wedge \cdots \wedge dx^{i_r} \in A^r(M) \, , \\ &\text{then} \quad dw|_U = \frac{\partial a_{i_1 \ldots i_r}}{\partial x^j} \, dx^j \wedge dx^{i_1} \wedge \cdots \wedge dx^{i_r} \, . \end{aligned}} \tag{30.38}$$

The following is a useful fact relating differential forms and tangent vector fields.

Theorem 30.3. *Let w be a differential 1-form on a smooth manifold M; and X and Y are smooth tangent vector fields on M. Then*

$$\boxed{\langle X \wedge Y, dw \rangle = X \langle Y, w \rangle - Y \langle X, w \rangle - \langle [X, Y], w \rangle} \quad , \tag{30.39}$$

where

$$[X, Y] = XY - YX \, . \tag{30.40}$$

Note : The notation $X \langle Y, \omega \rangle$ means the action of the smooth tangent vector field X on the smooth function $\langle Y, \omega \rangle$, which yields another smooth function.

Proof. Since the pairing $\langle \, , \, \rangle$ is a bilinear function, it is sufficient to prove the theorem for $\omega = g \, df$, where g and f are both smooth functions on M. Thus

$$d\omega = dg \wedge df .$$

By (28.28) and (30.4),

$$\langle X \wedge Y, \, dg \wedge df \rangle = \begin{vmatrix} \langle X, dg \rangle & \langle X, df \rangle \\ \langle Y, dg \rangle & \langle Y, df \rangle \end{vmatrix} = X(g)Y(f) - X(f)Y(g) . \quad (30.41)$$

Since

$$\langle X, \omega \rangle = \langle X, gdf \rangle = g \cdot X(f) , \quad (30.42)$$

we have

$$Y \langle X, \omega \rangle = Y(g)X(f) + g \cdot Y(X(f)) . \quad (30.43)$$

Similarly,

$$X \langle Y, \omega \rangle = X(g)Y(f) + g \cdot X(Y(f)) . \quad (30.44)$$

It follows that the RHS of (30.39) is

$$\begin{aligned} & X \langle Y, \omega \rangle - Y \langle X, \omega \rangle - \langle [X,Y], \omega \rangle \\ & = X(g)Y(f) - Y(g)X(f) + g[X(Y(f)) - Y(X(f))] - g \langle [X,Y], df \rangle \\ & = X(g)Y(f) - Y(g)X(f) , \end{aligned}$$
$$(30.45)$$

which is precisely the RHS of the second equality in (30.41). $\qquad \square$

The above theorem can be generalized quite straightforwardly as follows.

Corollary 30.1. *Let $\omega \in A^r(M)$ be a smooth exterior differential r-form on M, and $X_1, \ldots, X_{r+1}$ be smooth tangent vector fields on M. Then*

$$\begin{aligned} \langle X_1 \wedge \cdots \wedge X_{r+1}, \, d\omega \rangle &= \sum_{i=1}^{r+1} (-1)^{i+1} X_i \langle X_1 \wedge \cdots \wedge \hat{X}_i \wedge \cdots \wedge X_{r+1}, \, \omega \rangle \\ &+ \sum_{1 \le i < j \le r+1} (-1)^{i+j} \langle [X_i, X_j] \wedge \cdots \wedge \hat{X}_i \wedge \cdots \wedge \hat{X}_j \wedge \cdots \wedge X_{r+1}, \, \omega \rangle , \end{aligned}$$
$$(30.46)$$

where the $\hat{X}_i$ designates an omitted term.

Exercise 30.2   Prove the above formula by induction (or other means).

For the remainder of this chapter, we will demonstrate the power and elegance of differential forms and exterior differentiation by applying it to 3-dimensional vector calculus. We will see that the operator d completely subsumes the vector calculus operators of "div, grad, curl and all that". Temporarily, we will revert to the use of the Cartesian coordinates (x, y, z).

Let f be a smooth function (a zero-form) on $\mathbb{R}^3$. Then

$$df = \frac{\partial f}{\partial x}\, dx + \frac{\partial f}{\partial y}\, dy + \frac{\partial f}{\partial z}\, dz \,. \tag{30.47}$$

In vector calculus, we call the vector field with components $\left(\dfrac{\partial f}{\partial x}, \dfrac{\partial f}{\partial y}, \dfrac{\partial f}{\partial z}\right)$ the **gradient** of f, denoted by ∇f, or *grad f*.

Next consider a 1-form

$$\omega = A\, dx + B\, dy + C\, dz \,, \tag{30.48}$$

where A, B, C are smooth functions on $\mathbb{R}^3$. Then, according to (30.38),

$$
\begin{aligned}
d\omega = dA \wedge dx + dB \wedge dy + dC \wedge dz = {} & \frac{\partial A}{\partial y}\, dy \wedge dx + \frac{\partial A}{\partial z}\, dz \wedge dx \\
& + \frac{\partial B}{\partial x}\, dx \wedge dy + \frac{\partial B}{\partial z}\, dz \wedge dy + \frac{\partial C}{\partial x}\, dx \wedge dz + \frac{\partial C}{\partial y}\, dy \wedge dz \\
= {} & \left(\frac{\partial C}{\partial y} - \frac{\partial B}{\partial z}\right) dy \wedge dz + \left(\frac{\partial A}{\partial z} - \frac{\partial C}{\partial x}\right) dz \wedge dx + \left(\frac{\partial B}{\partial x} - \frac{\partial A}{\partial y}\right) dx \wedge dy \,.
\end{aligned}
\tag{30.49}
$$

If we let $\boldsymbol{X}$ be the vector field (A, B, C), then the vector field with components

$$\left(\frac{\partial C}{\partial y} - \frac{\partial B}{\partial z},\; \frac{\partial A}{\partial z} - \frac{\partial C}{\partial x},\; \frac{\partial B}{\partial x} - \frac{\partial A}{\partial y}\right)$$

is precisely the **curl** of the vector field $\boldsymbol{X}$ in vector calculus, denoted therein by $\nabla \times \boldsymbol{X}$.

Finally consider the 2-form

$$\psi = A\, dy \wedge dz + B\, dz \wedge dx + C\, dx \wedge dy \,, \tag{30.50}$$

where, again, A, B, C are smooth functions on $\mathbb{R}^3$. We have, on applying (30.38),

$$d\psi = \left(\frac{\partial A}{\partial x} + \frac{\partial B}{\partial y} + \frac{\partial C}{\partial z}\right) dx \wedge dy \wedge dz \,. \tag{30.51}$$

The quantity inside the parenthesis in the above equation is precisely the **divergence** of the vector field $\boldsymbol{X} = (A, B, C)$, denoted in vector calculus by $\nabla \cdot \boldsymbol{X}$ or *div* $\boldsymbol{X}$.

Thus, *div, grad,* and *curl* are all different manifestations of one and the same operator, namely, the exterior derivative d.

From Poincare's lemma ($d^2 = 0$), two fundamental formulas in vector calculus follow immediately. Let f be a smooth function on $\mathbb{R}^3$ and $\boldsymbol{X}$ a smooth tangent vector field on $\mathbb{R}^3$. Then we have

$$\nabla \cdot (\nabla \times \boldsymbol{X}) = 0 \quad , \tag{30.52}$$
$$\nabla \times (\nabla f) = 0 \quad . \tag{30.53}$$

Stated in words, *the divergence of a curl and the curl of a gradient are always equal to zero.*

Eq. (30.53) is simply equivalent to $d^2 f = 0$, where f, as a smooth function on $\mathbb{R}^3$, is considered to be an element of $A^0(\mathbb{R}^3)$, that is, a zero-form on $\mathbb{R}^3$. To see how (30.52) arises from Poincare's lemma, we first introduce the idea of a **local frame field**. According to (30.9) and the discussion immediately preceding that equation, at each point $p \in \mathbb{R}^3$, $\{(dx)_p, (dy)_p, (dz)_p\}$ is an orthonormal frame (basis set) of $T_p^*(\mathbb{R}^3) = \Lambda^1(T_p^*(\mathbb{R}^3))$, $\{\frac{\partial}{\partial x}|_p, \frac{\partial}{\partial y}|_p, \frac{\partial}{\partial z}|_p\}$ is an orthonormal frame of $T_p(\mathbb{R}^3) = \Lambda^1(T_p(\mathbb{R}^3))$, and the two bases are dual to each other. If we let the point p vary over a local coordinate neighborhood U (in $\mathbb{R}^3$, U can actually be the whole space $\mathbb{R}^3$), we obtain the so-called local frame fields $\{dx, dy, dz\}$ of $T^*(\mathbb{R}^3)$ and $\{\frac{\partial}{\partial x}, \frac{\partial}{\partial y}, \frac{\partial}{\partial z}\}$ of $T(\mathbb{R}^3)$. Thus, an arbitrary vector field on $\mathbb{R}^3$, or equivalently, a section of the tangent bundle $T(\mathbb{R}^3)$, can be written as

$$\boldsymbol{X} = A\frac{\partial}{\partial x} + B\frac{\partial}{\partial y} + C\frac{\partial}{\partial z} \quad , \tag{30.54}$$

where A, B, C are smooth functions on $\mathbb{R}^3$. Correspondingly, we have the 1-form

$$\omega = A dx + B dy + C dz \quad ,$$

which is a section of the cotangent bundle $T^*(\mathbb{R}^3)$, or an element of $A^1(\mathbb{R}^3) = \Gamma(\Lambda^1(\mathbb{R}^{3*}))$. According to (30.49), $d\omega$ is then a 2-form whose components are the components of the vector field $\nabla \times \boldsymbol{X}$. We then have, from (30.51),

$$0 = d^2\omega = \nabla \cdot (\nabla \times \boldsymbol{X}) dx \wedge dy \wedge dz \quad . \tag{30.55}$$

Thus (30.52) follows.

The rules of exterior differentiation given in Theorem 30.1, in particular the "product rule", can be used to derive quite easily many useful identities in vector calculus. We will derive the following four basic ones:

(1)	$\nabla(fg) = f\nabla g + g\nabla f \quad ,$	(30.56)
(2)	$\nabla \times (f\boldsymbol{v}) = \nabla f \times \boldsymbol{v} + f\nabla \times \boldsymbol{v} \quad ,$	(30.57)
(3)	$\nabla \cdot (f\boldsymbol{v}) = (\nabla f) \cdot \boldsymbol{v} + f\nabla \cdot \boldsymbol{v} \quad ,$	(30.58)
(4)	$\nabla \cdot (\boldsymbol{v} \times \boldsymbol{u}) = \boldsymbol{u} \cdot (\nabla \times \boldsymbol{v}) - \boldsymbol{v} \cdot (\nabla \times \boldsymbol{u}) \quad .$	(30.59)

Proof of (1): Suppose $f, g \in A^0(\mathbb{R}^3)$, i.e., f and g are both zero-forms, or smooth functions, on $\mathbb{R}^3$. Then

$$f \wedge g = fg \tag{30.60}$$

is also a zero-form on $\mathbb{R}^3$. Now, by (30.31),

$$d(fg) = d(f \wedge g) = df \wedge g + (-1)^0 f \wedge dg = gdf + fdg, \tag{30.61}$$

which is a 1-form whose components are those of $\nabla(fg)$. Thus

$$(\nabla(fg))^1 dx + (\nabla(fg))^2 dy + (\nabla(fg))^3 dz$$
$$= \left(g\frac{\partial f}{\partial x} + f\frac{\partial g}{\partial x} \right) dx + \left(g\frac{\partial f}{\partial y} + f\frac{\partial g}{\partial y} \right) dy + \left(g\frac{\partial f}{\partial z} + f\frac{\partial g}{\partial z} \right) dz \quad . \tag{30.62}$$

On comparing coefficients of dx, dy, dz, (30.56) follows.

Proof of (2): Let

$$\boldsymbol{v} = v^1 \frac{\partial}{\partial x} + v^2 \frac{\partial}{\partial y} + v^3 \frac{\partial}{\partial z} \quad , \tag{30.63}$$

and ω be the 1-form given by

$$\omega = v^1 dx + v^2 dy + v^3 dz \quad . \tag{30.64}$$

Then

$$f\omega = f \wedge \omega \quad , \tag{30.65}$$

where $f \in A^0(\mathbb{R}^3)$ (a zero-form), is a 1-form. We have

$$d(f \wedge \omega) = df \wedge \omega + fd\omega \quad . \tag{30.66}$$

The LHS can be written, by (30.49), as

$$d(f\omega) = (\nabla \times (f\boldsymbol{v}))^1 dy \wedge dz + (\nabla \times (f\boldsymbol{v}))^2 dz \wedge dx + (\nabla \times (f\boldsymbol{v}))^3 dx \wedge dy, \tag{30.67}$$

while the RHS assumes the form

$$\begin{aligned}
& df \wedge \omega + fd\omega \\
&= \left(\frac{\partial f}{\partial x} dx + \frac{\partial f}{\partial y} dy + \frac{\partial f}{\partial z} dz \right) \wedge (v^1 dx + v^2 dy + v^3 dz) \\
& \quad + f(\nabla \times \boldsymbol{v})^1 dy \wedge dz + f(\nabla \times \boldsymbol{v})^2 dz \wedge dx + f(\nabla \times \boldsymbol{v})^3 dx \wedge dy \\
&= \left(\frac{\partial f}{\partial y} v^3 - \frac{\partial f}{\partial z} v^2 + f(\nabla \times \boldsymbol{v})^1 \right) dy \wedge dz \\
& \quad + \left(\frac{\partial f}{\partial z} v^1 - \frac{\partial f}{\partial x} v^3 + f(\nabla \times \boldsymbol{v})^2 \right) dz \wedge dx \\
& \quad + \left(\frac{\partial f}{\partial x} v^2 - \frac{\partial f}{\partial y} v^1 + f(\nabla \times \boldsymbol{v})^3 \right) dx \wedge dy \\
&= \{((\nabla f) \times \boldsymbol{v})^1 + f(\nabla \times \boldsymbol{v})^1\} dy \wedge dz \\
& \quad + \{((\nabla f) \times \boldsymbol{v})^2 + f(\nabla \times \boldsymbol{v})^2\} dz \wedge dx \\
& \quad + \{((\nabla f) \times \boldsymbol{v})^3 + f(\nabla \times \boldsymbol{v})^3\} dx \wedge dy \quad . \tag{30.68}
\end{aligned}$$

Comparing the RHS of (30.67) to the last expression on the RHS of (30.68), we arrive at (30.57).

Proof of (3): Let

$$v = v^1 \frac{\partial}{\partial x} + v^2 \frac{\partial}{\partial y} + v^3 \frac{\partial}{\partial z}$$

$$\text{and} \quad \psi = v^1 dy \wedge dz + v^2 dz \wedge dx + v^3 dx \wedge dy \quad . \tag{30.69}$$

Then, for $f \in A^0(\mathbb{R}^3)$, i.e., a smooth function on $\mathbb{R}^3$,

$$d(f\psi) = df \wedge \psi + f d\psi \quad . \tag{30.70}$$

By (30.51), since $f\psi$ is a 2-form,

$$d(f\psi) = \nabla \cdot (f v) \, dx \wedge dy \wedge dz \quad , \tag{30.71}$$

while

$$df \wedge \psi = \left(\frac{\partial f}{\partial x} dx + \frac{\partial f}{\partial y} dy + \frac{\partial f}{\partial z} dz \right) \wedge$$

$$(v^1 dy \wedge dz + v^2 dz \wedge dx + v^3 dx \wedge dy) = ((\nabla f) \cdot v) \, dx \wedge dy \wedge dz \,, \tag{30.72}$$

and

$$f d\psi = f(\nabla \cdot v) \, dx \wedge dy \wedge dz \quad . \tag{30.73}$$

Eq. (30.58) then follows from the last four equations.

Proof of (4): Let

$$u = u^1 \frac{\partial}{\partial x} + u^2 \frac{\partial}{\partial y} + u^3 \frac{\partial}{\partial z} \,, \tag{30.74}$$

$$v = v^1 \frac{\partial}{\partial x} + v^2 \frac{\partial}{\partial y} + v^3 \frac{\partial}{\partial z} \,, \tag{30.75}$$

be two tangent vector fields on $\mathbb{R}^3$. Then

$$v \wedge u = (v^2 u^3 - v^3 u^2) \frac{\partial}{\partial y} \wedge \frac{\partial}{\partial z} + (v^3 u^1 - v^1 u^3) \frac{\partial}{\partial z} \wedge \frac{\partial}{\partial x}$$

$$+ (v^1 u^2 - v^2 u^1) \frac{\partial}{\partial x} \wedge \frac{\partial}{\partial y} \quad . \tag{30.76}$$

Now let

$$\psi = (v^2 u^3 - v^3 u^2) dy \wedge dz + (v^3 u^1 - v^1 u^3) dz \wedge dx + (v^1 u^2 - v^2 u^1) dx \wedge dy \,, \tag{30.77}$$

$$\theta = v^1 dx + v^2 dy + v^3 dz \,, \tag{30.78}$$

$$\phi = u^1 dx + u^2 dy + u^3 dz \quad . \tag{30.79}$$

Then

$$\psi = \theta \wedge \phi \quad , \tag{30.80}$$

and

$$d\psi = d\theta \wedge \phi + (-1)^1 \theta \wedge d\phi \quad . \tag{30.81}$$

On the other hand, by (30.51),

$$d\psi = \nabla \cdot (\boldsymbol{v} \times \boldsymbol{u}) \, dx \wedge dy \wedge dz \quad , \tag{30.82}$$

while

$$\begin{aligned} d\theta \wedge \phi &= ((\nabla \times \boldsymbol{v})^1 dy \wedge dz + (\nabla \times \boldsymbol{v})^2 dz \wedge dx + (\nabla \times \boldsymbol{v})^3 dx \wedge dy) \\ &\wedge (u^1 dx + u^2 dy + u^3 dz) = (\boldsymbol{u} \cdot (\nabla \times \boldsymbol{v})) \, dx \wedge dy \wedge dz \quad , \end{aligned} \tag{30.83}$$

and similarly,

$$\theta \wedge d\phi = (\boldsymbol{v} \cdot (\nabla \times \boldsymbol{u})) \, dx \wedge dy \wedge dz \quad . \tag{30.84}$$

Eq. (30.59) then follows from the last four equations.

Chapter 31

Moving Frames and Curvilinear Coordinates in $\mathbb{R}^3$

In this chapter we will use the techniques of exterior differential calculus developed in the last to do some geometry on $\mathbb{R}^3$. Several important geometrical concepts will be introduced, and we will present a unified method to derive expressions for the vector calculus operators in terms of curvilinear coordinates. Useful as these are in the limited context of vector calculus, the real importance of them lies in their general applicability to arbitrary manifolds.

Let $(O; \delta_1, \delta_2, \delta_3)$ be a fixed orthonormal frame in $\mathbb{R}^3$ with origin at the point O, and $(p; e_1, e_2, e_3)$ be a moving orthonormal frame with origin at the moving point p and having the same orientation as the fixed frame. (See Fig. 31.1, which shows a moving frame based on the spherical coordinates.) In general, we can write

$$\overrightarrow{Op} = a^i \delta_i \quad , \tag{31.1}$$

$$\text{and} \qquad e_i = a_i^j \delta_j \quad , \tag{31.2}$$

where a^i and the matrix elements a_i^j are smooth functions on $\mathbb{R}^3$. For the case of spherical coordinates, a^i and a_i^j are given explicitly by

$$\overrightarrow{Op} = (r \sin \theta \cos \phi) \delta_1 + (r \sin \theta \sin \phi) \delta_2 + (r \cos \theta) \delta_3 , \tag{31.3}$$

and

$$
\begin{aligned}
e_1 &= e_r = (\sin \theta \cos \phi) \delta_1 + (\sin \theta \sin \phi) \delta_2 + (\cos \theta) \delta_3 , \\
e_2 &= e_\theta = (\cos \theta \cos \phi) \delta_1 + (\cos \theta \sin \phi) \delta_2 - (\sin \theta) \delta_3 , \\
e_3 &= e_\phi = -(\sin \phi) \delta_1 + (\cos \phi) \delta_2 \quad .
\end{aligned}
\tag{31.4}
$$

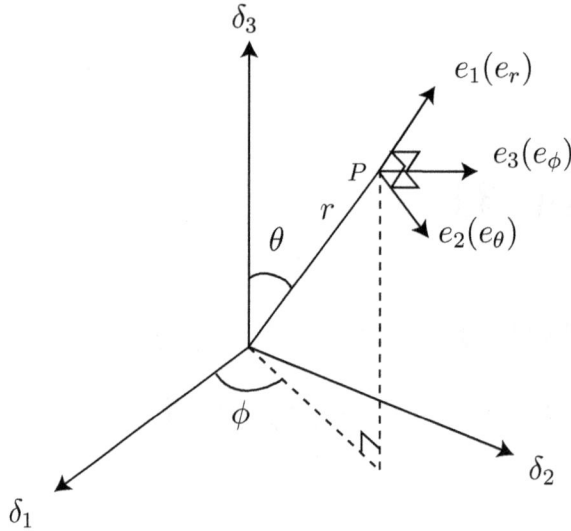

FIGURE 31.1

Note that a^i, a^j_i are all 0-forms on $\mathbb{R}^3$. If we denote

$$\delta = (\delta_1, \delta_2, \delta_3)^T \quad , \tag{31.5}$$

$$e = (e_1, e_2, e_3)^T \quad , \tag{31.6}$$

$$a = (a^1, a^2, a^3) \quad , \tag{31.7}$$

$$A = (a^j_i) \quad , \tag{31.8}$$

where T means the transpose of a matrix, then (31.1) and (31.2) can be written in matrix notation as

$$\overrightarrow{Op} = a \cdot \delta \quad , \tag{31.9}$$

$$\text{and} \quad e = A \cdot \delta \quad , \tag{31.10}$$

respectively, where the dot means matrix multiplication. Suppose now we move the frame at p to a slightly neighboring one $(p + dp \, ; e_i + de_i)$. The vectors dp and de_i can still be expressed in terms of the frame $(p \, ; e_1, e_2, e_3)$:

$$dp = w^i e_i \quad , \tag{31.11}$$

$$de_i = w^j_i e_j \quad , \tag{31.12}$$

where w^i, w^j_i are all 1-forms on $\mathbb{R}^3$. In fact, dp and de_i are so-called **vector-**

valued 1-forms. If we denote

$$\theta = (\omega^1, \omega^2, \omega^3) \quad , \tag{31.13}$$

$$\omega = (\omega_i^j) \quad , \tag{31.14}$$

the we can write, in matrix notation,

$$dp = \theta \cdot e \quad , \tag{31.15}$$

$$de = \omega \cdot e \quad . \tag{31.16}$$

The vector-valued 1-form θ and the **matrix-valued 1-form** ω can be found easily by exterior differentiation of (31.1) and (31.2), or equivalently, (31.9) and (31.10). Thus

$$dp = da \cdot \delta = da \cdot A^{-1} \cdot e \quad , \tag{31.17}$$

$$de = dA \cdot \delta = dA \cdot A^{-1} \cdot e \quad , \tag{31.18}$$

where we have used the fact that $d\delta = 0$, since $(O; \delta_1, \delta_2, \delta_3)$ is a fixed frame. It follows from (31.15) and (31.16) that

$$\theta = da \cdot A^{-1} \quad , \tag{31.19}$$

$$\omega = dA \cdot A^{-1} \quad . \tag{31.20}$$

Now, differentiating exteriorly the equation $A \cdot A^{-1} = 1$, we have

$$dA \cdot A^{-1} + A \cdot d(A^{-1}) = dA \cdot A^{-1} + (dA \cdot A^{-1})^T = 0 \quad , \tag{31.21}$$

where we have used the fact that $A^{-1} = A^T$ (since A is an orthogonal matrix). It follows that ω is an antisymmetric matrix of 1-forms:

$$\boxed{\omega_i^j + \omega_j^i = 0} \quad . \tag{31.22}$$

Now dp can also be expressed, using local (rectangular) coordinates $x^1 = x, x^2 = y, x^3 = z$ with respect to the fixed frame $(O; \delta_1, \delta_2, \delta_3)$ as

$$dp = dx^i \delta_i = \omega^i e_i \quad , \tag{31.23}$$

$$\text{or} \quad dp = dx \cdot \delta = \theta \cdot e = \theta \cdot A \cdot \delta \quad , \tag{31.24}$$

where the last equality follows from (31.10). Thus

$$dx = \theta \cdot A \quad , \tag{31.25}$$

$$\text{or} \quad dx^i = \omega^j a_j^i \quad . \tag{31.26}$$

From Theorem 28.7, the volume form (c.f. Exercise 28.5) $dx^1 \wedge dx^2 \wedge dx^3$ can be written as

$$dx^1 \wedge dx^2 \wedge dx^3 = \det(a_i^j) \, \omega^1 \wedge \omega^2 \wedge \omega^3 \quad . \tag{31.27}$$

Since the matrix (a_i^j) is orthogonal, $\det(a_i^j) = \pm 1$. The requirement that $(p; e_1, e_2, e_3)$ has the same orientation as $(O; \delta_1, \delta_2, \delta_3)$, however, dictates that $\det(a_i^j) = +1$ (c.f. Def. 29.1). Hence

$$dx^1 \wedge dx^2 \wedge dx^3 = \omega^1 \wedge \omega^2 \wedge \omega^3 \quad , \tag{31.28}$$

which is a statement of the invariance of the volume form under rotations (orientation-preserving orthogonal transformations).

Two very important equations can be obtained simply by exteriorly differentiating (31.15) and (31.16), and using Poincare's lemma $(d^2 = 0)$ on the LHS of these equations. We have, on using the product rule of exterior differentiation [(30.31)],

$$d\theta \cdot e - \theta \wedge de = 0 \quad , \tag{31.29}$$
$$d\omega \cdot e - \omega \wedge de = 0 \quad . \tag{31.30}$$

Using (31.16) for de, the above equations reduce to

$$d\theta = \theta \wedge \omega \quad , \tag{31.31}$$
$$d\omega = \omega \wedge \omega \quad , \tag{31.32}$$

or, in terms of components,

$$d\omega^i = \omega^j \wedge \omega_j^i \quad , \tag{31.33}$$
$$d\omega_i^j = \omega_i^k \wedge \omega_k^j \quad . \tag{31.34}$$

Note that (31.32) is a matrix equation, with ω being a matrix of 1-forms. Hence in that equation, $\omega \wedge \omega$ is not necessarily zero (as it would be if ω were an individual 1-form). Eqs. (31.31) and (31.32) are called **structure equations**. The latter is also called the **Maurer-Cartan equation**. Together with (31.22), they determine completely the geometry of $\mathbb{R}^3$.

The matrix of 1-forms (ω_i^j) is called a **connection matrix** of 1-forms. Eq. (31.22) is the condition for the **metric-compatibility** of the connection ω, (31.31) the condition for **torsion freeness** of the connection ω, and (31.32) the condition for **zero-curvature** of the connection ω. These important geometrical concepts will be developed and explained in much greater detail in a later part of this book, when we introduce formally the geometrical notion of the **connection on a vector bundle**, a notion at the heart of the mathematical framework behind gauge field theories in physics. (See Chapters 35 and 39.)

| Exercise 31.1 | According to (31.4), for spherical coordinates,

$$A = (a_i^j) = \begin{pmatrix} \sin\theta\cos\phi & \sin\theta\sin\phi & \cos\theta \\ \cos\theta\cos\phi & \cos\theta\sin\phi & -\sin\theta \\ -\sin\phi & \cos\phi & 0 \end{pmatrix} \quad . \tag{31.35}$$

Show explicitly that $A^{-1} = A^T$.

Exercise 31.2 From (31.4), show by exterior differentiation, that is, by calculating de_1, de_2 and de_3, that in spherical coordinates, the connection matrix of 1-forms (ω_i^j) is given by

$$(\omega_i^j) = \begin{pmatrix} 0 & d\theta & \sin\theta d\phi \\ -d\theta & 0 & \cos\theta d\phi \\ -\sin\theta d\phi & -\cos\theta d\phi & 0 \end{pmatrix} \quad . \tag{31.36}$$

Note that, as anticipated by (31.22), (ω_i^j) is antisymmetric.

Exercise 31.3 Show explicitly that (31.35) and (31.36) satisfy (31.19), that is,

$$\omega = (dA) \cdot A^{-1} \quad .$$

Exercise 31.4 By exteriorly differentiating (31.3) and comparing the result with (31.11), show that for spherical coordinates

$$\omega^1 = dr , \quad \omega^2 = rd\theta , \quad \omega^3 = r\sin\theta d\phi . \tag{31.37}$$

Thus the volume form (a 3-form) is given in spherical coordinates by the familiar result

$$dx \wedge dy \wedge dz = r^2 \sin\theta \, dr \wedge d\theta \wedge d\phi \quad . \tag{31.38}$$

We can write
$$\omega^1 = \lambda_1 dr , \quad \omega^2 = \lambda_2 d\theta , \quad \omega^3 = \lambda_3 d\phi , \tag{31.39}$$

where $\lambda_1, \lambda_2, \lambda_3$ are smooth functions on $\mathbb{R}^3$, given by

$$\lambda_1 = 1 , \quad \lambda_2 = r , \quad \lambda_3 = r\sin\theta . \tag{31.40}$$

Eq. (31.11) [or (31.15)] implies that the ω^i are dual to the e_j, that is,

$$\langle e_i, \omega^j \rangle = \delta_i^j \quad . \tag{31.41}$$

Exercise 31.5 By writing

$$\delta_1 = \frac{\partial}{\partial x} , \quad \delta_2 = \frac{\partial}{\partial y} , \quad \delta_3 = \frac{\partial}{\partial z} ,$$

and using the familiar transformation equations

$$x = r\sin\theta\cos\phi , \quad y = r\sin\theta\sin\phi , \quad z = r\cos\theta , \tag{31.42}$$

show explicitly that the expressions for e_i and ω^j given by (31.4) and (31.37), respectively, satisfy the duality condition (31.41).

The totality of moving frames $\{e_i\}$ is called a **frame field**, and thus $\{\omega^i\}$ is called a **coframe field**. The unit vectors in spherical coordinates $e_r = e_1, e_\theta = e_2$ and $e_\phi = e_3$ are then given, according to (31.37) and the duality condition (31.41), by

$$e_r = \frac{\partial}{\partial r} , \quad e_\theta = \frac{1}{r} \frac{\partial}{\partial \theta} , \quad e_\phi = \frac{1}{r \sin \theta} \frac{\partial}{\partial \phi} . \tag{31.43}$$

Note that $\dfrac{\partial}{\partial \theta}$ and $\dfrac{\partial}{\partial \phi}$ are *not* unit vectors.

We will now use exterior differentiation to obtain familiar formulas in spherical coordinates for the vector calculus operators of the gradient, the Laplacian, the divergence, and the curl. These are given by:

$$\nabla f = \frac{\partial f}{\partial r} e_r + \frac{1}{r} \frac{\partial f}{\partial \theta} e_\theta + \frac{1}{r \sin \theta} \frac{\partial f}{\partial \phi} e_\phi , \tag{31.44}$$

$$\nabla^2 f = \frac{1}{r^2 \sin \theta} \left\{ \frac{\partial}{\partial r} \left(r^2 \sin \theta \frac{\partial f}{\partial r} \right) + \frac{\partial}{\partial \theta} \left(\sin \theta \frac{\partial f}{\partial \theta} \right) + \frac{\partial}{\partial \phi} \left(\frac{1}{\sin \theta} \frac{\partial f}{\partial \phi} \right) \right\} , \tag{31.45}$$

$$\nabla \cdot A = \frac{1}{r^2} \frac{\partial}{\partial r} \left(r^2 A_r \right) + \frac{1}{r \sin \theta} \frac{\partial}{\partial \theta} \left(\sin \theta \, A_\theta \right) + \frac{1}{r \sin \theta} \frac{\partial A_\phi}{\partial \phi} , \tag{31.46}$$

$$\nabla \times A = \frac{1}{r \sin \theta} \left\{ \frac{\partial}{\partial \theta} \left(A_\phi \sin \theta \right) - \frac{\partial A_\theta}{\partial \phi} \right\} e_r$$
$$+ \frac{1}{r \sin \theta} \left\{ \frac{\partial A_r}{\partial \phi} - \sin \theta \frac{\partial}{\partial r} \left(r A_\phi \right) \right\} e_\theta + \frac{1}{r} \left\{ \frac{\partial}{\partial r} \left(r A_\theta \right) - \frac{\partial A_r}{\partial \theta} \right\} e_\phi , \tag{31.47}$$

where f is a smooth function on $\mathbb{R}^3$ and

$$A = A_x \delta_1 + A_y \delta_2 + A_z \delta_3 = A_r e_r + A_\theta e_\theta + A_\phi e_\phi \tag{31.48}$$

is a smooth vector field on $\mathbb{R}^3$. The operators $\nabla, \nabla^2, \nabla \cdot$ and $\nabla \times$ are the gradient, the Laplacian, the divergence, and the curl, respectively, the first two operating on a scalar field and the last two on a vector field.

To derive the above four formulas, it is actually more convenient to consider directly an arbitrary curvilinear coordinate system specified by the local coordinates x^1, x^2, x^3 such that

$$dp = \omega^1 e_1 + \omega^2 e_2 + \omega^3 e_3 , \tag{31.49}$$

$$\omega^1 = \lambda_1 dx^1 , \quad \omega^2 = \lambda_2 dx^2 , \quad \omega^3 = \lambda_3 dx^3 , \tag{31.50}$$

$$\langle e_i, \omega^j \rangle = \delta_i^j , \quad i, j = 1, 2, 3, \tag{31.51}$$

and the volume element is given by

$$dx \wedge dy \wedge dz = \omega^1 \wedge \omega^2 \wedge \omega^3 = \lambda_1 \lambda_2 \lambda_3 \, dx^1 \wedge dx^2 \wedge dx^3 . \tag{31.52}$$

For spherical coordinates $x^1 = r, x^2 = \theta, x^3 = \phi$, (31.50) and (31.51) imply

$$e_i = \frac{1}{\lambda_i} \frac{\partial}{\partial x^i}, \quad i = 1, 2, 3, \tag{31.53}$$

where we have used [c.f. (30.9)]

$$\left\langle \frac{\partial}{\partial x^i}, dx^j \right\rangle = \delta_i^j, \quad i, j = 1, 2, 3 \quad . \tag{31.54}$$

With the above set-up, we can now proceed to derive the generalizations of (31.44) through (31.47) to arbitrary curvilinear coordinates.

First consider the **gradient**. We have

$$df = \frac{\partial f}{\partial x^1} dx^1 + \frac{\partial f}{\partial x^2} dx^2 + \frac{\partial f}{\partial x^3} dx^3 = \frac{1}{\lambda_1} \frac{\partial f}{\partial x^1} \omega^1 + \frac{1}{\lambda_2} \frac{\partial f}{\partial x^2} \omega^2 + \frac{1}{\lambda_3} \frac{\partial f}{\partial x^3} \omega^3 \quad , \tag{31.55}$$

where the second equality follows from (31.50). In view of (31.51) (duality between e_i and ω^j), we immediately have, on recalling the discussion following (30.39),

$$\boxed{\nabla f = \frac{1}{\lambda_1} \frac{\partial f}{\partial x^1} e_1 + \frac{1}{\lambda_2} \frac{\partial f}{\partial x^2} e_2 + \frac{1}{\lambda_3} \frac{\partial f}{\partial x^3} e_3} \quad . \tag{31.56}$$

This yields (31.44) on using (31.40).

Next consider the **Laplacian**. In rectangular coordinates (x, y, z) we have

$$\nabla^2 f = \frac{\partial^2 f}{\partial x^2} + \frac{\partial^2 f}{\partial y^2} + \frac{\partial^2 f}{\partial z^2} \quad . \tag{31.57}$$

We will show that

$$\nabla^2 f \, dx \wedge dy \wedge dz = d(\star \, df) \quad , \tag{31.58}$$

where $\star$ is the Hodge-star operator. Starting with

$$df = \frac{\partial f}{\partial x} dx + \frac{\partial f}{\partial y} dy + \frac{\partial f}{\partial z} dz \quad , \tag{31.59}$$

we have, from the results in Chapter 29,

$$\star \, df = \frac{\partial f}{\partial x} dy \wedge dz + \frac{\partial f}{\partial y} dz \wedge dx + \frac{\partial f}{\partial z} dx \wedge dy \quad . \tag{31.60}$$

On exterior differentiation [using (30.38)] it follows that

$$d(\star \, df) = \left(\frac{\partial^2 f}{\partial x^2} + \frac{\partial^2 f}{\partial y^2} + \frac{\partial^2 f}{\partial z^2} \right) dx \wedge dy \wedge dz \quad . \tag{31.61}$$

Now use curvilinear coordinates. Eqs. (31.52) and (31.58) imply that

$$d(\star \, df) = \nabla^2 f \, \omega^1 \wedge \omega^2 \wedge \omega^3 \quad . \tag{31.62}$$

Using (31.55) and (31.50) we have

$$
\begin{aligned}
\star \, df &= \frac{1}{\lambda_1} \frac{\partial f}{\partial x^1} \, \omega^2 \wedge \omega^3 + \frac{1}{\lambda_2} \frac{\partial f}{\partial x^2} \, \omega^3 \wedge \omega^1 + \frac{1}{\lambda_3} \frac{\partial f}{\partial x^3} \, \omega^1 \wedge \omega^2 \\
&= \frac{\lambda_2 \lambda_3}{\lambda_1} \left(\frac{\partial f}{\partial x^1} \right) dx^2 \wedge dx^3 + \frac{\lambda_1 \lambda_3}{\lambda_2} \left(\frac{\partial f}{\partial x^2} \right) dx^3 \wedge dx^1 \\
&\quad + \frac{\lambda_1 \lambda_2}{\lambda_3} \left(\frac{\partial f}{\partial x^3} \right) dx^1 \wedge dx^2 \quad .
\end{aligned}
\tag{31.63}
$$

Thus, on exterior differentiation of the above equation again [using (30.38)],we obtain

$$
\begin{aligned}
d(\star \, df) &= \left\{ \frac{\partial}{\partial x^1} \left(\frac{\lambda_2 \lambda_3}{\lambda_1} \frac{\partial f}{\partial x^1} \right) + \frac{\partial}{\partial x^2} \left(\frac{\lambda_1 \lambda_3}{\lambda_2} \frac{\partial f}{\partial x^2} \right) + \frac{\partial}{\partial x^3} \left(\frac{\lambda_1 \lambda_2}{\lambda_3} \frac{\partial f}{\partial x^3} \right) \right\} \\
&\quad dx^1 \wedge dx^2 \wedge dx^3 \quad .
\end{aligned}
\tag{31.64}
$$

But by (31.50),

$$
dx^1 \wedge dx^2 \wedge dx^3 = \frac{\omega^1 \wedge \omega^2 \wedge \omega^3}{\lambda_1 \lambda_2 \lambda_3} \quad .
\tag{31.65}
$$

Hence we finally have, from (31.62), the following formula for the Laplacian of a function f in the curvilinear coordinates x^1, x^2, x^3:

$$
\boxed{\nabla^2 f = \frac{1}{\lambda_1 \lambda_2 \lambda_3} \left\{ \frac{\partial}{\partial x^1} \left(\frac{\lambda_2 \lambda_3}{\lambda_1} \frac{\partial f}{\partial x^1} \right) + \frac{\partial}{\partial x^2} \left(\frac{\lambda_1 \lambda_3}{\lambda_2} \frac{\partial f}{\partial x^2} \right) + \frac{\partial}{\partial x^3} \left(\frac{\lambda_1 \lambda_2}{\lambda_3} \frac{\partial f}{\partial x^3} \right) \right\}}
$$

$$
\tag{31.66}
$$

This immediately gives (31.45) for spherical coordinates on using (31.40) for λ_1, λ_2 and λ_3.

Let us now consider the **divergence**. Suppose

$$
\boldsymbol{A} = A^1 \boldsymbol{e}_1 + A^2 \boldsymbol{e}_2 + A^3 \boldsymbol{e}_3
\tag{31.67}
$$

is a vector field on $\mathbb{R}^3$, where A^1, A^2, A^3 are smooth functions on $\mathbb{R}^3$. With the components of $\boldsymbol{A}$ construct the 2-form

$$
\psi = A^1 \omega^2 \wedge \omega^3 + A^2 \omega^3 \wedge \omega^1 + A^3 \omega^1 \wedge \omega^2 \quad .
\tag{31.68}
$$

We then have, from (30.43),

$$
d\psi = (\nabla \cdot \boldsymbol{A}) \, \omega^1 \wedge \omega^2 \wedge \omega^3 \quad .
\tag{31.69}
$$

Rewriting (31.68) as

$$
\psi = \lambda_2 \lambda_3 A^1 dx^2 \wedge dx^3 + \lambda_1 \lambda_3 A^2 dx^3 \wedge dx^1 + \lambda_1 \lambda_2 A^3 dx^1 \wedge dx^2 \quad ,
\tag{31.70}
$$

we can "hit ψ with a d" (exteriorly differentiate) to obtain

$$d\psi = \left\{ \frac{\partial}{\partial x^1} (\lambda_2 \lambda_3 A^1) + \frac{\partial}{\partial x^2} (\lambda_1 \lambda_3 A^2) + \frac{\partial}{\partial x^3} (\lambda_1 \lambda_2 A^3) \right\} dx^1 \wedge dx^2 \wedge dx^3 \ .$$

(31.71)

Using (31.65) again yields the following expression for the divergence of a vector field in curvilinear coordinates:

$$\boxed{ \nabla \cdot \boldsymbol{A} = \frac{1}{\lambda_1 \lambda_2 \lambda_3} \left\{ \frac{\partial}{\partial x^1} (\lambda_2 \lambda_3 A^1) + \frac{\partial}{\partial x^2} (\lambda_1 \lambda_3 A^2) + \frac{\partial}{\partial x^3} (\lambda_1 \lambda_2 A^3) \right\} } \ ,$$

(31.72)

from which the special formula (31.46) for spherical coordinates follows immediately, with the aid of (31.40).

Finally we consider the **curl**. Let $\boldsymbol{A}$ be a vector field as given by (31.67). Its dual 1-form is

$$\alpha = A^1 \omega^1 + A^2 \omega^2 + A^3 \omega^3 \ . \tag{31.73}$$

According to the discussion following (30.40), $\nabla \times \boldsymbol{A}$ is obtained from α through the relation

$$\star (d\alpha) = (\nabla \times \boldsymbol{A})^1 \omega^1 + (\nabla \times \boldsymbol{A})^2 \omega^2 + (\nabla \times \boldsymbol{A})^3 \omega^3 \ . \tag{31.74}$$

Writing

$$\alpha = \lambda_1 A^1 dx^1 + \lambda_2 A^2 dx^2 + \lambda_3 A^3 dx^3 \ , \tag{31.75}$$

we can exteriorly differentiate to obtain

$$\begin{aligned} d\alpha = \ & \frac{\partial}{\partial x^2}(\lambda_1 A^1) dx^2 \wedge dx^1 + \frac{\partial}{\partial x^3}(\lambda_1 A^1) dx^3 \wedge dx^1 \\ & + \frac{\partial}{\partial x^1}(\lambda_2 A^2) dx^1 \wedge dx^2 + \frac{\partial}{\partial x^3}(\lambda_2 A^2) dx^3 \wedge dx^2 \\ & + \frac{\partial}{\partial x^1}(\lambda_3 A^3) dx^1 \wedge dx^3 + \frac{\partial}{\partial x^2}(\lambda_3 A^3) dx^2 \wedge dx^3 \\ = \ & \left\{ \frac{\partial}{\partial x^2}(\lambda_3 A^3) - \frac{\partial}{\partial x^3}(\lambda_2 A^2) \right\} dx^2 \wedge dx^3 \\ & + \left\{ \frac{\partial}{\partial x^3}(\lambda_1 A^1) - \frac{\partial}{\partial x^1}(\lambda_3 A^3) \right\} dx^3 \wedge dx^1 \\ & + \left\{ \frac{\partial}{\partial x^1}(\lambda_2 A^2) - \frac{\partial}{\partial x^2}(\lambda_1 A^1) \right\} dx^1 \wedge dx^2 \\ = \ & \frac{1}{\lambda_2 \lambda_3} \left\{ \frac{\partial}{\partial x^2}(\lambda_3 A^3) - \frac{\partial}{\partial x^3}(\lambda_2 A^2) \right\} \omega^2 \wedge \omega^3 \\ & + \frac{1}{\lambda_1 \lambda_3} \left\{ \frac{\partial}{\partial x^3}(\lambda_1 A^1) - \frac{\partial}{\partial x^1}(\lambda_3 A^3) \right\} \omega^3 \wedge \omega^1 \\ & + \frac{1}{\lambda_1 \lambda_2} \left\{ \frac{\partial}{\partial x^1}(\lambda_2 A^2) - \frac{\partial}{\partial x^2}(\lambda_1 A^1) \right\} \omega^1 \wedge \omega^2 \ . \end{aligned}$$

(31.76)

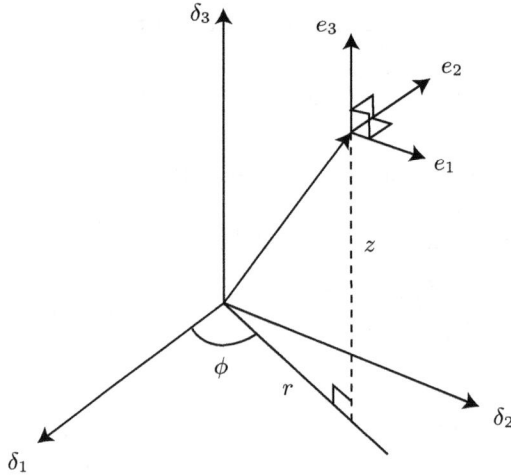

FIGURE 31.2

Noting that

$$\star\,(\omega^2 \wedge \omega^3) = \omega^1 \,, \quad \star\,(\omega^3 \wedge \omega^1) = \omega^2 \,, \quad \star\,(\omega^1 \wedge \omega^2) = \omega^3 \,, \tag{31.77}$$

Eq. (31.74) then allows us to write

$$\nabla \times \boldsymbol{A} = \frac{1}{\lambda_1 \lambda_2 \lambda_3} \begin{vmatrix} \lambda_1 e_1 & \lambda_2 e_2 & \lambda_3 e_3 \\ \dfrac{\partial}{\partial x^1} & \dfrac{\partial}{\partial x^2} & \dfrac{\partial}{\partial x^3} \\ \lambda_1 A^1 & \lambda_2 A^2 & \lambda_3 A^3 \end{vmatrix} \,. \tag{31.78}$$

Eq. (31.40) together with the above equation immediately yields (31.47) for $\nabla \times \boldsymbol{A}$ in spherical coordinates.

Exercise 31.6 Show that for cylindrical coordinates, where $x^1 = r, x^2 = \phi, x^3 = z$ (see Fig. 31.2),

$$\omega^1 = dr \,, \quad \omega^2 = r d\phi \,, \quad \omega^3 = dz \,, \tag{31.79}$$

or

$$\lambda_1 = 1 \,, \quad \lambda_2 = r \,, \quad \lambda_3 = 1 \,. \tag{31.80}$$

Chapter 32

Integrals of Differential Forms and the Stokes Theorem

The pullback map defined by (28.38) on exterior vectors can be readily extended to pullback maps on differential forms. This extension plays a crucial role in the theory of integration of differential forms, which is the generalization to manifolds of multiple integrals over certain domains in $\mathbb{R}^n$.

Suppose $f : M \longrightarrow N$ is a smooth map from a smooth manifold M to a smooth manifold N. Then it induces a linear map between the corresponding spaces of exterior differential forms:

$$f^* : A(N) \longrightarrow A(M) \quad , \tag{32.1}$$

known as the **pullback** of f, defined successively by the following steps.

Definition 32.1. *Suppose $f : M \longrightarrow N$ is a smooth map between smooth manifolds M and N, $x \in M$, and $y = f(x) \in N$. The pullback map*

$$f^* : T_y^*(N) \longrightarrow T_x^*(M) \tag{32.2}$$

is given by

$$\boxed{f^*(d\phi)_y = (d(\phi \circ f))_x} \quad , \tag{32.3}$$

where $(d\phi)_y \in T_y^*(N)$ and ϕ is a function on N.

The map f^* in (32.2) is sometimes known as the **differential** of the map f. From f^* we can immediately obtain its adjoint f_*, the **tangent map**, or the **derivative map**, induced by f.

Definition 32.2. *Let $f : M \longrightarrow N$ be a smooth map, $X \in T_x(M)$, and $d\phi \in T_y^*(N)$. Then the tangent map*

$$f_* : T_x(M) \longrightarrow T_y(N) \tag{32.4}$$

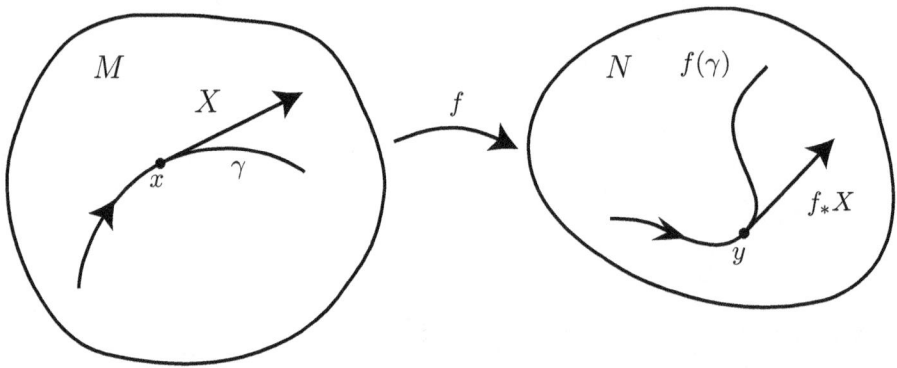

FIGURE 32.1

is given by

$$\langle f_* X, d\phi \rangle_y = \langle X, f^*(d\phi) \rangle_x \qquad . \tag{32.5}$$

Notice the adjoint relationship between the "upper star" and the "lower star" maps f_* and f^* in the above equation. These maps are obviously linear. The action of the tangent map f_* is illustrated intuitively in Fig. 32.1, where X is a tangent vector to M at $x \in M$ and $f_* X$ is a tangent vector to N at $y = f(x) \in N$. The curve γ lying in M is mapped by f to the curve $f(\gamma)$ in N.

Exercise 32.1   Suppose $dim(M) = m$ and $dim(N) = n$. Let $(x^1, \ldots, x^m)$ and $(y^1, \ldots, y^n)$ be local coordinates near $x \in M$ and $y \in N$, respectively. A map $f : M \longrightarrow N$ can be represented near the point x by the n functions

$$y^\alpha = f^\alpha(x^1, \ldots, x^m), \quad 1 \le \alpha \le n. \tag{32.6}$$

Show that the actions of f^* and f_* on the natural bases $\{dy^\alpha, 1 \le \alpha \le n\}$ and $\{\frac{\partial}{\partial x^i}, 1 \le i \le m\}$ are given by:

$$f^*(dy^\alpha) = \sum_{i=1}^m \left(\frac{\partial f^\alpha}{\partial x^i} \right)_x dx^i \quad , \tag{32.7}$$

$$f_* \left(\frac{\partial}{\partial x^i} \right) = \sum_{\beta=1}^n \left(\frac{\partial f^\beta}{\partial x^i} \right)_x \frac{\partial}{\partial y^\beta} \quad . \tag{32.8}$$

The above results show that f_* and f^* have the same matrix representation under the natural (coordinate) bases.

We are now ready to provide the definition of the pullback map in (32.1) promised earlier.

Definition 32.3. *Suppose* $f : M \longrightarrow N$ *is a smooth map,* $X_1, \ldots, X_r$ *are* r *arbitrary smooth tangent vector fields on* M, *and* $\beta \in A^r(N), r \geq 1$, *is a differential* r-*form on* N. *The pullback of* β, $f^*\beta$, *is given in a pointwise fashion by*

$$\langle X_1 \wedge \cdots \wedge X_r, f^*\beta \rangle_x = \langle f_*X_1 \wedge \cdots \wedge f_*X_r, \beta \rangle_{f(x)} \quad , \tag{32.9}$$

where x *is an arbitrary point in* M. *If* $\beta \in A^0(N)$, *then*

$$f^*\beta = \beta \circ f \in A^0(M) \quad . \tag{32.10}$$

The above pairing can be computed by (30.29) when the differential forms and tangent vector fields are expressed in natural coordinates.

By Theorem 28.2, we immediately have

Theorem 32.1. *For any* $\omega, \eta \in A(N)$ *(exterior differential forms on a manifold* N*) and a given smooth function* $f : M \longrightarrow N$, f^* *distributes over the exterior product, that is,*

$$f^*(\omega \wedge \eta) = f^*\omega \wedge f^*\eta \quad . \tag{32.11}$$

The importance of the pullback map f^* also rests on the fact that it commutes with the exterior derivative. More specifically, we have the following useful theorem, which will be stated without proof.

Theorem 32.2. *Let* $f : M \longrightarrow N$ *be a smooth map. Then*

$$f^* \circ d = d \circ f^* : A(N) \longrightarrow A(M) \quad . \tag{32.12}$$

Equivalently, we have the following commuting diagram.

$$
\begin{array}{ccc}
A(N) & \xrightarrow{\ d\ } & A(N) \\
{\scriptstyle f^*}\big\downarrow & & \big\downarrow{\scriptstyle f^*} \\
A(M) & \xrightarrow{\ d\ } & A(M)
\end{array}
$$

The next step in the definition of an integral of a differential form is to introduce the notion of an **orientable manifold**.

Definition 32.4. *An* n-*dimensional smooth manifold* M *is said to be* **orientable** *if there exists a continuous and non-vanishing differential* n-*form* ω *on* M. *Given such an* ω, M *is said to be* **oriented** *(by* ω*). Two such forms differing only by a positive function factor are said to assign the same* **orientation** *to* M.

It follows immediately from the above definition that if a connected manifold M is orientable, then there exist exactly two orientations on M.

Exercise 32.2 Explain why the above statement is true.

Suppose an n-dimensional orientable manifold M is oriented by an exterior form ω, and $(U; x^i)$ is a local coordinate system in M. Then $dx^1 \wedge \cdots \wedge dx^n$ and

$\omega|_U$ (ω restricted to U) are the same up to a non-zero function factor. If the factor is positive, then we say that $(U; x^i)$ is a coordinate system **consistent** with the orientation of M. Obviously, for any oriented manifold, there exists a coordinate covering which is consistent with the orientation of the manifold (this means that every coordinate system in the covering is consistent with the orientation of M, and the Jacobian of the change of coordinates between any two coordinate neighborhoods with non-empty intersection is always positive). Conversely, it is also true that if there exists a compatible coordinate covering such that the Jacobian of the change of coordinates in the non-empty intersection of any two coordinate neighborhoods in the covering is always positive, then M is orientable. The proof of this last statement will not be given here.

Next we introduce the concepts of the **support of a function** and the **support of an exterior differential form**.

Definition 32.5. *Suppose $f : M \longrightarrow \mathbb{R}$ is a real function on M. The **support** of f is the closure of the set of points in M at which f is non-zero:*

$$supp\, f = \overline{\{x \in M \mid f(x) \neq 0\}} \ , \tag{32.13}$$

where $\overline{\{\}}$ denotes the closure of the set $\{\}$.

Definition 32.6. *Suppose $\omega \in A(M)$ (ω is an exterior differential form on M). The **support** of ω, denoted by $supp\, \omega$, is given by*

$$supp\, \omega = \overline{\{x \in M \mid \omega_x \neq 0\}} \ . \tag{32.14}$$

The following theorem provides the crucial technical tool for the definition of an integral of a differential form. The proof of this theorem requires the use of some basic topology, and will not be given here (see Chern, Chen and Lam 1999).

Theorem 32.3 (Partition of Unity). *Suppose Σ is an open covering of a smooth manifold M. Then there exists a family of smooth functions $\{g_\alpha\}$ on M satisfying the following conditions:*

1) *$0 \leq g_\alpha \leq 1$, and $supp\, g_\alpha$ is compact for each α. Moreover, for each α, there exists an open set $W_i \in \Sigma$ such that $supp\, g_\alpha \subset W_i$.*

2) *For each $x \in M$, there exists a neighborhood U of x that intersects $supp\, g_\alpha$ for only finitely many α.*

3)

$$\sum_\alpha g_\alpha = 1 \ .$$

Condition 2) ensures that, for any $x \in M$, there are only finitely many non-zero terms in the sum of condition 3). The family of functions $\{g_\alpha\}$ is called a **partition of unity** subordinate to the coordinate covering Σ.

With the above background, we can now define the integral of an exterior differential form with compact support on M. In order to clarify the procedure this will be done with steps (a) to (d) below. Throughout these steps we will assume that M is an oriented manifold and $dim(M) = n$.

(a) First define the integral of an n-form φ on M. Choose a coordinate covering $\Sigma = \{W_i\}$ which is consistent with the orientation of M. If $supp\ \varphi$ happens to be inside a coordinate neighborhood $U \subset \Sigma$ with local coordinates x^i consistent with the orientation of M, then φ can be expressed as

$$\varphi = f(x^1, \ldots, x^n)\, dx^1 \wedge \cdots \wedge dx^n \quad . \tag{32.15}$$

In this case we define

$$\int_M \varphi \equiv \int_U f(x^1, \ldots, x^n)\, dx^1 \ldots dx^n \quad , \tag{32.16}$$

where the integral on the RHS is the usual Riemann integral.

(b) If $supp\ \varphi$ does not fall inside a particular $U \subset \Sigma$, we make use of the partition of unity $\{g_\alpha\}$ subordinate to Σ as follows.

First write

$$\varphi = \left(\sum_\alpha g_\alpha\right)\varphi = \sum_\alpha (g_\alpha \cdot \varphi) \quad . \tag{32.17}$$

Clearly, $supp\,(g_\alpha \cdot \varphi) \subset supp\ g_\alpha$. By condition 1) of Theorem 32.3, there exists some coordinate neighborhood $W_i \subset \Sigma$ such that

$$supp\,(g_\alpha \cdot \varphi) \subset supp\ g_\alpha \subset W_i \quad .$$

Hence we can define

$$\int_M g_\alpha \cdot \varphi \equiv \int_{W_i} g_\alpha \cdot \varphi \quad , \tag{32.18}$$

where the integral on the RHS is the usual Riemann integral. This means that, if $g_\alpha \cdot \varphi$ with respect to the coordinate system $x^1, \ldots, x^n$ in W_i is expressed as

$$g_\alpha \cdot \varphi = h(x^1, \ldots, x^n)\, dx^1 \wedge \cdots \wedge dx^n \quad , \tag{32.19}$$

then the integral on the RHS of (32.18) is given by

$$\int_{W_i} g_\alpha \cdot \varphi = \int_{W_i} h(x^1, \ldots, x^n)\, dx^1 \ldots dx^n \quad . \tag{32.20}$$

The definition given by (32.18) makes sense because the RHS of that equation is independent of the choice of W_i. Indeed, if $supp\,(g_\alpha \cdot \varphi)$ is contained in two coordinate neighborhoods W_i and W_j with coordinate systems $(x^1, \ldots, x^n)$ and $(y^1, \ldots, y^n)$, respectively, then the Jacobian of the change of coordinates satisfies

$$J = \frac{\partial(y^1, \ldots, y^n)}{\partial(x^1, \ldots, x^n)} > 0 \quad , \tag{32.21}$$

and

$$g_\alpha \cdot \varphi = h(x^1, \ldots, x^n)\, dx^1 \wedge \cdots \wedge dx^n \quad \text{(in } W_i) \quad , \tag{32.22}$$
$$g_\alpha \cdot \varphi = h'(y^1, \ldots, y^n)\, dy^1 \wedge \cdots \wedge dy^n \quad \text{(in } W_j) \quad . \tag{32.23}$$

Theorem 28.7 and Eq. (32.21) imply that

$$h' \cdot J = h'|J| = h \quad . \tag{32.24}$$

On the other hand, by the rules for the change of variables in a Riemann integral, we have

$$\int_{W_i \cap W_j} h'\, dy^1 \ldots dy^n = \int_{W_i \cap W_j} h'|J|\, dx^1 \ldots dx^n = \int_{W_i \cap W_j} h\, dx^1 \ldots dx^n . \tag{32.25}$$

Thus

$$\int_{W_i} g_\alpha \cdot \varphi = \int_{W_j} g_\alpha \cdot \varphi \quad . \tag{32.26}$$

(c) Since *supp* φ is compact by assumption, it only intersects finitely many *supp* g_α by condition 2) of the Partition of Unity Theorem (Theorem 32.3). Hence the RHS of (32.17) is a sum (over α) of finitely many terms. Using (32.18) we now define

$$\boxed{\int_M \varphi \equiv \sum_\alpha \int_M g_\alpha \cdot \varphi} \quad . \tag{32.27}$$

The RHS of the above equation appears to depend on a particular choice of partition of unity, but in fact it does not. Indeed, suppose $\{g'_\beta\}$ is another partition of unity subordinate to the coordinate covering Σ. Then by condition 3) of Theorem 32.3,

$$\sum_\beta \int_M g'_\beta \cdot \varphi = \sum_\beta \sum_\alpha \int_M g_\alpha \cdot g'_\beta \cdot \varphi$$

$$= \sum_\alpha \int_M \sum_\beta g'_\beta \cdot g_\alpha \cdot \varphi = \sum_\alpha \int_M g_\alpha \cdot \varphi \quad . \tag{32.28}$$

(d) Having defined the integral of an n-form with compact support over an n-dimensional manifold M, we can now define the integral of an r-form with compact support, for $r < n$, over any r-dimensional submanifold N of M. (At this time we will just appeal to our intuitive understanding of the notion of an **embedded submanifold**, such as, for example, a two-dimensional surface embedded in three-dimensional Euclidean space.) Let

$$h : N \longrightarrow M$$

be an embedding of an r-dimensional manifold N into an n-dimensional manifold M, $r < n$; and φ be a differential r-form on M with compact support. Then, from Def. 32.3, the pullback $h^*\varphi$ is a differential r-form on N with compact support. So the integral

$$\int_N h^*\varphi$$

is well-defined according to steps (a) to (c) above [c.f. (32.27)]. We then define the integral of φ on the submanifold $h(N)$ by

$$\boxed{\int_{h(N)} \varphi \equiv \int_N h^*\varphi} \quad . \tag{32.29}$$

This completes the definition of the integral of a differential form with compact support.

Let $\varphi, \varphi_1, \varphi_2$ be exterior differential n-forms on M (of dimension n) with compact support. Then $\varphi_1 + \varphi_2$ has compact support, as has $c\varphi$, for any real number c. By the definition (32.27), it is obvious that

$$\int_M (\varphi_1 + \varphi_2) = \int_M \varphi_1 + \int_M \varphi_2 \quad , \tag{32.30}$$

$$\int_M c\varphi = c \int_M \varphi \quad . \tag{32.31}$$

Thus the integration $\int_M$ can be viewed as a **linear functional** on the set of all n-forms on M with compact support. Eq. (32.27) is in fact a generalization of the Riemann (multiple) integral to the case of manifolds.

The central theorem on the integration of differential forms is the **Stokes Theorem**, which, as we will see, is a generalization of several familiar theorems on intrgrals over domains in $\mathbb{R}, \mathbb{R}^2$ and $\mathbb{R}^3$ (including the vector calculus theorems) to the case of integrals over domains in manifolds of arbitrary dimensions.

Before stating the theorem, we will first see how the more familiar integral theorems appear as just different guises of it.

Example 32.1 Suppose $D = [a, b]$ is a closed interval in $\mathbb{R}$ and f is a C^1 function (continuously differentiable function) on D. The **Fundamental Theorem of Calculus** then says:

$$\int_D df = f(b) - f(a) = \int_{\partial D} f \quad , \tag{32.32}$$

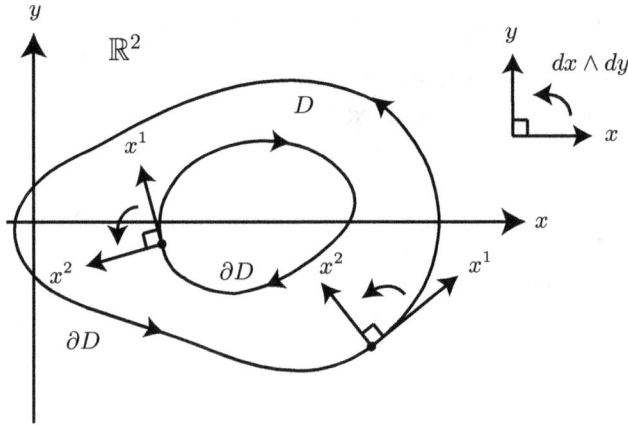

FIGURE 32.2

where the **oriented boundary** of D, denoted by ∂D, is given by $\{b\} - \{a\}$. Note that the RHS of the second equality in the above equation is the integral of a zero-form over a zero-dimensional space, while the LHS of the first equality is the integral of a 1-form over a 1-dimensional space. Both of these integrals are special cases of (32.27).

Example 32.2 Suppose D is a bounded domain in $\mathbb{R}^2$ whose orientation is consistent with a given orientation of $\mathbb{R}^2$. Denote the oriented boundary of D, with orientation induced by that of D, by ∂D. This means that the orientation of ∂D together with the normal vector pointing towards the interior of D form a coordinate system consistent with the given orientation of $\mathbb{R}^2$ (see Fig. 32.2).

In Fig. 32.2, D is the annulus region. The orientation of ∂D is $(-1)^2 dx^1$ [we will explain the significance of the $(-1)^2$ factor later, when we give a more general definition of the induced orientation on a boundary for arbitrary dimensions]. The coordinate system (x^1, x^2) then has an orientation $dx^1 \wedge dx^2$, which is consistent with the given orientation $dx \wedge dy$ of $\mathbb{R}^2$.

Now suppose P and Q are both C^1 functions on D. Then the **Green's Theorem** states that

$$\int_{\partial D} (P dx + Q dy) = \int_D \left(\frac{\partial Q}{\partial x} - \frac{\partial P}{\partial y} \right) dx dy \quad . \tag{32.33}$$

The integrands on both sides of the above equation are simply related. If we let ω be the 1-form

$$\omega = P dx + Q dy \quad , \tag{32.34}$$

then

$$d\omega = \left(\frac{\partial Q}{\partial x} - \frac{\partial P}{\partial y} \right) dx \wedge dy \quad . \tag{32.35}$$

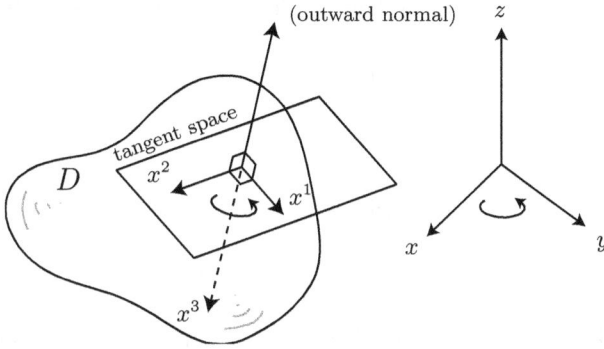

FIGURE 32.3

Thus the Green's theorem (32.33) can also be written as

$$\int_{\partial D} \omega = \int_D d\omega \quad . \tag{32.36}$$

Note the similarity between the above equation and the Fundamental Theorem of Calculus as stated in (32.32).

Example 32.3 Suppose D is a bounded domain in $\mathbb{R}^3$ whose orientation is consistent with a given one in $\mathbb{R}^3$. This time we use the outward normal as the positive direction to induce an orientation on the boundary of D, denoted again by ∂D (see Fig. 32.3). In Fig. 32.3, the induced orientation of ∂D is given by $(-1)^3 dx^1 \wedge dx^2 = dx^2 \wedge dx^1$. [Again, the factor $(-1)^3$ will be explained later.] Together with the outward normal, we have the orientation $dx^2 \wedge dx^1 \wedge (-dx^3) = dx^1 \wedge dx^2 \wedge dx^3$, which is consistent with $dx \wedge dy \wedge dz$, the given one in $\mathbb{R}^3$.

Suppose P, Q, R are C^1 functions on D. Then **Gauss' Theorem** states that

$$\int_{\partial D} (P dy dz + Q dz dx + R dx dy) = \int_D \left(\frac{\partial P}{\partial x} + \frac{\partial Q}{\partial y} + \frac{\partial R}{\partial z} \right) dx dy dz \quad . \tag{32.37}$$

Again, the integrands are simply related through exterior differentiation. Let φ be the 2-form

$$\varphi = P dy \wedge dz + Q dz \wedge dx + R dx \wedge dy \quad . \tag{32.38}$$

Then (32.37) can be written

$$\int_{\partial D} \varphi = \int_D d\varphi \quad . \tag{32.39}$$

In ordinary vector notation, the Gauss' Theorem appears as

$$\int_{\partial D} \boldsymbol{A} \cdot \mathrm{d}\boldsymbol{a} = \int_D \nabla \cdot \boldsymbol{A} \, \mathrm{d}^3 r \quad , \tag{32.40}$$

where $\boldsymbol{A}$ is the vector field

$$\boldsymbol{A} = P e_x + Q e_y + R e_z \quad , \tag{32.41}$$

and $\mathrm{d}\boldsymbol{a}$ is an area element in ∂D along the outward normal.

Exercise 32.3   Consider Fig. 32.4, where (e_1, e_2, e_3) is an orthonormal frame at a boundary point p of D, with e_3 being the outward normal to the boundary ∂D at p. If we let p move to an infinitesimally near position on ∂D, then

$$dp = d\boldsymbol{r} = dx\, e_x + dy\, e_y + dz\, e_z = \omega^1 e_1 + \omega^2 e_2 \quad , \tag{32.42}$$

where $\omega^1 = \lambda_1 dx^1, \omega^2 = \lambda_2 dx^2$ [c.f. (31.39)] are 1-forms on ∂D. Show that the area element $\omega^1 \wedge \omega^2 e_3$ is given by

$$\omega^1 \wedge \omega^2 e_3 = (dy \wedge dz)\, e_x + (dz \wedge dx)\, e_y + (dx \wedge dy)\, e_z \quad . \tag{32.43}$$

This equation allows us to equate the LHS of (32.40) to the LHS of (32.37).

Example 32.4   Suppose Σ is an oriented surface in $\mathbb{R}^3$ whose boundary $\partial \Sigma$ is an oriented closed curve, such that the positive orientation of Σ together with an outward normal vector to Σ form a right-handed coordinate system. We assume, of course, that $\mathbb{R}^3$, in which Σ is embedded, is oriented by such a system (Fig. 32.5).

In Fig. 32.5, the orientation of Σ is given by $dx^1 \wedge dx^2$, the induced orientation on $\partial \Sigma$ is $(-1)^2 dx^1 = dx^1$, and $dx^1 \wedge dx^2 \wedge dx^3 \sim dx \wedge dy \wedge dz$, where $\sim$ denotes equivalence of orientations.

Suppose P, Q, R are C^1 functions on a domain containing Σ. Then the **Stokes Theorem** of 3-dimensional vector calculus states that

$$\int_{\partial \Sigma} (Pdx + Qdy + Rdz) = \iint_\Sigma \left\{ \left(\frac{\partial R}{\partial y} - \frac{\partial Q}{\partial z} \right) dydz \right.$$
$$\left. + \left(\frac{\partial P}{\partial z} - \frac{\partial R}{\partial x} \right) dzdx + \left(\frac{\partial Q}{\partial x} - \frac{\partial P}{\partial y} \right) dxdy \right\} \quad . \tag{32.44}$$

If we define the 1-form ω by

$$\omega = Pdx + Qdy + Rdz \quad , \tag{32.45}$$

Eq. (32.44) reduces to

$$\int_{\partial \Sigma} \omega = \int_\Sigma d\omega \quad . \tag{32.46}$$

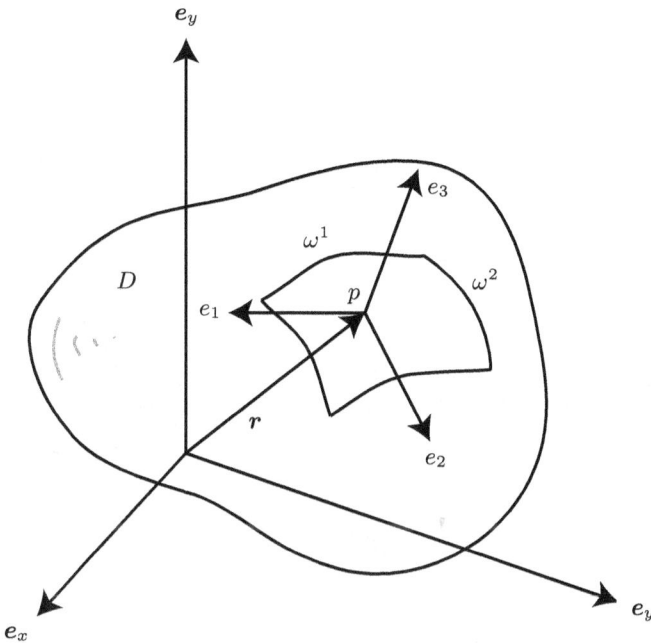

FIGURE 32.4

In ordinary vector notation, the above Stokes Theorem appears as

$$\int_{\partial \Sigma} \boldsymbol{A} \cdot d\boldsymbol{l} = \int_{\Sigma} (\nabla \times \boldsymbol{A}) \cdot d\boldsymbol{a} \quad , \tag{32.47}$$

where $\boldsymbol{A}$ is the vector field given by (32.41), $d\boldsymbol{l}$ is an infinitesimal displacement vector along the positive direction of $\partial \Sigma$, and $d\boldsymbol{a}$ is an area element in Σ (along the outward normal).

We note that the end results of the above four examples, (32.32),(32.36), (32.39) and (32.46), all have the same general form. This suggests that they are special cases of a more general theorem, which is indeed the case. The general theorem is also known as the **Stokes Theorem**, and applies to regions in manifolds of arbitrary dimension. We will now state without proof this important theorem. (For a proof see Chern, Chen and Lam 1999.)

Theorem 32.4 (Stokes Theorem). *Suppose D is a region with boundary in an n-dimensional oriented manifold M, and ω is an exterior differential $(n-1)$-form on M with compact support. Then*

$$\boxed{\int_{D} d\omega = \int_{\partial D} \omega} \quad , \tag{32.48}$$

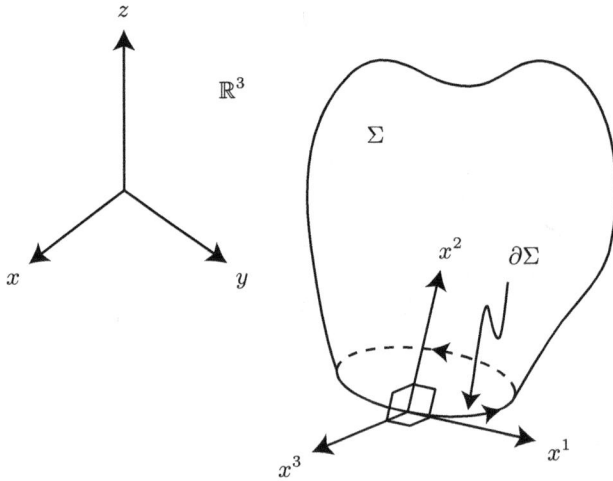

FIGURE 32.5

where ∂D is the boundary of D with the induced orientation. If $\partial D = \emptyset$ (the empty set), then the integral on the RHS is zero.

The notion of the induced orientation of ∂D, where D is an embedded submanifold of an n-dimensional smooth manifold M, requires some explanation. First we state

Definition 32.7. *Suppose M is an n-dimensional smooth manifold. A **region with boundary** D is a subset of M with two kinds of points:*

*1) **interior points**, each of which has a neighborhood in M contained in D;*

*2) **boundary points** p, for each of which there exists a coordinate chart (U, x^i) such that $x^i(p) = 0, i = 1, \ldots, n$, and*

$$U \cap D = \{q \in U \,|\, x^n(q) \geq 0\} \quad.$$

(See Fig. 32.6 for the case $n = 3$.)

*A coordinate system x^i with the above property is called an **adapted coordinate system** for the boundary point p. The set of all boundary points of D is called the boundary of D, denoted by ∂D.*

Let M be an oriented n-dimensional manifold, and D be a region with boundary in M. At a point $p \in \partial D$, choose an adapted coordinate system x^i such that $dx^1 \wedge \cdots \wedge dx^n$ is consistent with the given orientation of M. Then $(x^1, \ldots, x^{n-1})$ is a local coordinate system of ∂D at p. The orientation of ∂D specified by

$$(-1)^n dx^1 \wedge \cdots \wedge dx^{n-1} \tag{32.49}$$

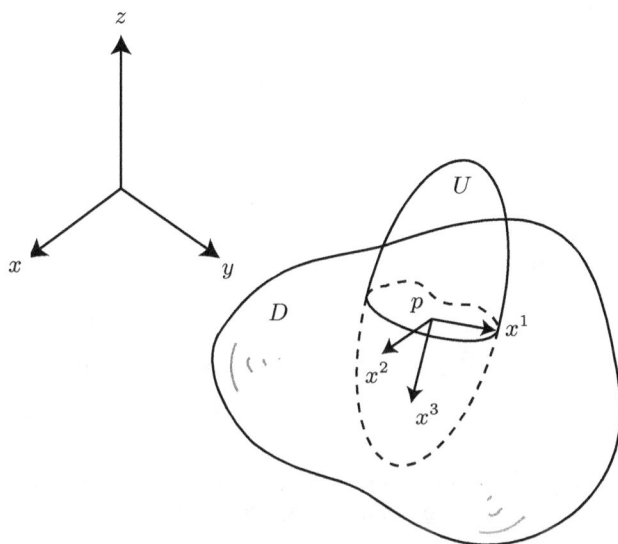

FIGURE 32.6

is called the **induced orientation** on the boundary ∂D. It is easily seen that the induced orientations on the boundaries in Examples 32.1 through 32.4 above all satisfy (32.49).

Writing

$$\int_{\partial D} \omega = (\partial D, \omega) \quad , \tag{32.50}$$

$$\int_{D} d\omega = (D, d\omega) \quad , \tag{32.51}$$

where either entry in each bracket of the RHS's can be regarded as a linear functional acting on the other, we see that the Stokes Theorem describes a duality between the **boundary operator** ∂ and the exterior derivative operator d. Hence d is also called the **coboundary operator**. ∂ and d can also be regarded as adjoint operators of each other.

We will conclude this chapter with a discussion of how the boundary operator ∂ acts on certain standard shapes, called simplexes, in Euclidean spaces $\mathbb{R}^n$. The cases $n = 1, 2, 3$ will provide an intuitive understanding of the formal rules.

Definition 32.8. *An r-simplex in the Euclidean space $\mathbb{R}^n, r \leq n$, denoted by the ordered symbol*

$$S_r = (p_0 p_1 \ldots p_r), \quad r \leq n, \tag{32.52}$$

is an oriented region in $\mathbb{R}^n$ consisting of the point set

$$S_r = \left\{ x \in \mathbb{R}^n \;\middle|\; x = m_0 p_0 + \cdots + m_r p_r, \; m_i \geq 0, \; \sum_{i=0}^{r} m_i = 1 \right\} \quad , \qquad (32.53)$$

where

$$p_0 = (0, \ldots, 0) \quad ,$$
$$p_1 = (1, 0, \ldots, 0) \quad ,$$
$$\vdots$$
$$p_r = (0, 0, \ldots, 0, 1, 0, \ldots, 0) \quad (\text{``1'' in r-th position}) \quad ,$$

are all points in $\mathbb{R}^n$.

The point x in (32.53) can be interpreted as the "center of mass" of a system of point masses, of masses $m_0, \ldots, m_r$, located at the points $p_0, \ldots, p_r$, respectively.

For example, S_0, S_1, S_2, S_3 are shown in Fig. 32.7.

Intuitively we can write

$$\partial S_0 = \partial(p_0) = 0 \quad , \qquad (32.54)$$
$$\partial S_1 = \partial(p_0 p_1) = p_1 - p_0 \quad , \qquad (32.55)$$
$$\partial S_2 = \partial(p_0 p_1 p_2) = p_1 p_2 - p_0 p_2 + p_0 p_1$$
$$= p_1 p_2 + p_2 p_0 + p_0 p_1 \quad , \qquad (32.56)$$
$$\partial S_3 = \partial(p_0 p_1 p_2 p_3) = p_1 p_2 p_3 - p_0 p_2 p_3 + p_0 p_1 p_3 - p_0 p_1 p_2 \quad . \qquad (32.57)$$

In the last equation S_3 is the pyramid in Fig. 32.7 and each term (with the sign) on the RHS is an oriented face of the pyramid, with an orientation consistent with the induced orientation defined by (32.49). Going around the corners of the face according to the induced orientation gives an outward normal according to the right-hand rule.

We can generalize from the above equations. The boundary ∂S_r, with induced orientation, is defined by

$$\partial S_r = \sum_{i=0}^{r} (-1)^i (p_0 p_1 \ldots \hat{p}_i \ldots p_r) \quad , \qquad (32.58)$$

where $\hat{p}_i$ means that p_i is to be omitted from the expression. Each term in the above sum is called an **oriented face** of S_r.

| Exercise 32.4 | Show that Eqs. (32.54) to (32.57) conform with (32.58). |

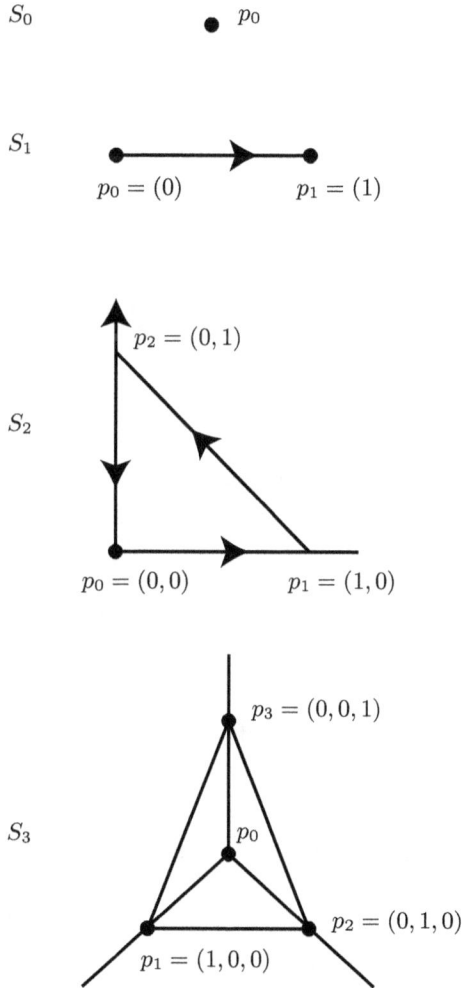

FIGURE 32.7

Chapter 33

Homology and De Rham Cohomology

In this chapter we will provide a brief introduction to the notion of de Rham cohomology, which arises from the operation of exterior differentiation, and plays a crucial role in relating the differentiable structure of a manifold to its topological structure. The associated notion of homology, a purely topological construct, will also be discussed.

Definition 33.1. *A form $\omega \in A^r(M)$ is said to be **closed** if $d\omega = 0$; it is said to be **exact** if $\omega = d\alpha$ for some $\alpha \in A^{r-1}(M)$.*

Since $d^2 = 0$ [by Poincare's Lemma (Theorem 30.2)], every exact form is closed; but a closed form is not necessarily exact.

Definition 33.2. *Let $Z^r(M)$ and $B^r(M)$ be the vector spaces of closed and exact differential r-forms on M, respectively. Then the quotient space*

$$\boxed{H^r(M) \equiv Z^r(M)/B^r(M)} \tag{33.1}$$

*is called the r-th **de Rham cohomology group** of the manifold M.*

If $\alpha \in Z^r(M)$, then $[\alpha] \in H^r(M)$ denotes the class determined by the representative α. Two elements $\alpha, \alpha' \in Z^r(M)$ belong to the same class (are cohomologous to each other) if $\alpha - \alpha' \in B^r(M)$, that is, if $\alpha - \alpha' = d\beta$, for some $\beta \in A^{r-1}(M)$. The group operation of $H^r(M)$ (an addition) is defined by

$$[\alpha] + [\beta] \equiv [\alpha + \beta] , \tag{33.2}$$

which makes $H^r(M)$ an abelian group. If $[\alpha] \in H^r(M)$ and $[\beta] \in H^s(M)$, one can also define an exterior product by

$$[\alpha] \wedge [\beta] \equiv [\alpha \wedge \beta] , \tag{33.3}$$

which is an element in $H^{r+s}(M)$. For this definition to make sense, we must check that 1) $d(\alpha \wedge \beta) = 0$; and 2) If $[\alpha] = [\alpha']$ and $[\beta] = [\beta']$, then $[\alpha \wedge \beta] = [\alpha' \wedge \beta']$. In fact, by (30.31),

$$d(\alpha \wedge \beta) = d\alpha \wedge \beta + (-1)^r \alpha \wedge d\beta = 0 , \qquad (33.4)$$

since $d\alpha = d\beta = 0$. Let $\alpha' - \alpha = da$ and $\beta' - \beta = db$. Then

$$\begin{aligned}
\alpha' \wedge \beta' - \alpha \wedge \beta &= (\alpha + da) \wedge (\beta + db) - \alpha \wedge \beta \\
&= da \wedge \beta + \alpha \wedge db + da \wedge db = d(a \wedge \beta) + (-1)^r d(\alpha \wedge b) + d(a \wedge db) \quad (33.5) \\
&= d(a \wedge \beta + (-1)^r \alpha \wedge b + a \wedge db) ,
\end{aligned}$$

whence $[\alpha' \wedge \beta'] = [\alpha \wedge \beta]$. Note also that

$$[\alpha] \wedge [\beta] = (-1)^{rs} [\beta] \wedge [\alpha] . \qquad (33.6)$$

The exterior product then extends (by linearity) to the direct sum [c.f. (28.21)]

$$H^*(M) = \sum_{\oplus r} H^r(M) . \qquad (33.7)$$

This makes $H^*(M)$ a ring, called the **cohomology ring** of M.

The de Rham cohomology group $H^r(M)$ is so named because it is the dual space to the so-called r-th homology group $H_r(M)$ of the manifold M. This is the content of a central theorem in differential geometry – de Rham's theorem. Before stating this theorem let us introduce the notion of a homology group.

Recall our discussion of r-simplexes in $\mathbb{R}^n, r \leq n$, at the end of the last chapter. Suppose S_r is one such simplex. Consider a smooth map $h : S_r \to M$, which is not necessarily one-to-one. The image of S_r under h in M, denoted by s_r, is called a **singular r-simplex** in M. Such simplexes are called singular due to the fact that h may not be invertible, and hence the s_r do not necessarily provide a **triangulation** of M (roughly speaking, carving M up into the topological equivalents of triangles on a plane).

Definition 33.3. *Denote a set of singular r-simplexes in M by $\{s_{r,i}\}$. An r-**chain** in M is a formal sum of the form*

$$c = \sum_i a_i s_{r,i} , \qquad a_i \in \mathbb{R} . \qquad (33.8)$$

The set of r-chains in M forms an abelian group $C_r(M)$ (under addition), called the r-**chain group** on M. The notion of the boundary of an r-simplex in $\mathbb{R}^n$ also generalizes to that of the boundary ∂s_r of a singular r-simplex, with an induced orientation:

$$\partial s_r \equiv h(\partial S_r) , \qquad (33.9)$$

where S_r is any r-simplex in $\mathbb{R}^n$ such that $h(S_r) = s_r$. By linearity, the domain of ∂ extends to $C_r(M)$. We thus have the boundary map

$$\partial : C_r(M) \longrightarrow C_{r-1}(M) .$$

As in the case of r-simplexes in $\mathbb{R}^n$, the boundary operator acting on C_r is also nilpotent: $\partial^2 = 0$ (the boundary of a boundary is zero).

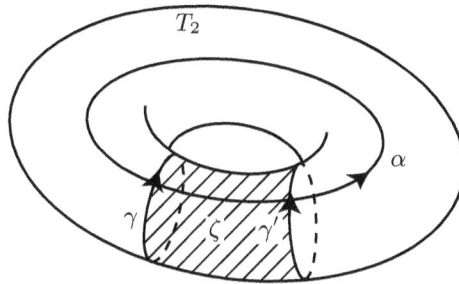

FIGURE 33.1

Definition 33.4. *An r-chain c_r such that $\partial c_r = 0$ is called an r-**cycle**; an r-chain c_r such that $c_r = \partial c_{r+1}$, where c_{r+1} is some $(r+1)$-chain is called an r-**boundary**.*

Due to the fact that $\partial^2 = 0$, an r-boundary is necessarily an r-cycle, but an r-cycle need not be an r-boundary.

Definition 33.5. *Let $Z_r(M)$ and $B_r(M)$ be the groups of r-cycles and r-boundaries in the manifold M, respectively. The quotient group $H_r(M)$ defined by*

$$\boxed{H_r(M) \equiv Z_r(M)/B_r(M)} \qquad (33.10)$$

*is called the r-th **singular homology group** on M.*

Two elements $\gamma, \lambda \in Z_r$ belong to the same class: $[\gamma] = [\lambda] \in H_r(M)$, if there exists some $(r+1)$-chain ζ such that $\gamma - \lambda = \partial \zeta$. The group operation of $H_r(M)$ (an addition) is defined by

$$[\gamma] + [\lambda] = [\gamma + \lambda] . \qquad (33.11)$$

For example, consider the two 1-cycles γ and γ' in the 2-dimensional torus T_2 (see Fig. 33.1). It is clear that $\gamma - \gamma' = \partial \zeta$, where ζ is the shaded region, and so $[\gamma] = [\gamma']$. We also say that γ is homologous to γ'. On the other hand, α and γ belong to different classes. In fact $H_1(T_2)$ is generated by those loops which are not boundaries; and there are only two such classes, $[\gamma]$ and $[\alpha]$. Thus

$$H_1(T_2) \sim \mathbb{R} \oplus \mathbb{R} . \qquad (33.12)$$

In general, for a $2g$-torus (g being the **genus** or the number of holes),

$$\boxed{H_1(T_{2g}) \sim \underbrace{\mathbb{R} \oplus \mathbb{R} \oplus \cdots \oplus \mathbb{R}}_{2g}} . \qquad (33.13)$$

Exercise 33.1 Justify the following equation:

$$H_0(T_{2g}) \sim H_2(T_{2g}) \sim \mathbb{R} \quad . \tag{33.14}$$

Suppose M is a **triangulable** topological space, that is, M is homeomorphic to a **simplicial complex** K of dimension n, where n is a positive integer, (a collection of simplexes "nicely" fitted together). Then one can define an important topological invariant of M, the **Euler characteristic** $\chi(M)$, by

$$\chi(M) \equiv \sum_{r=0}^{n} (-1)^r I_r \quad , \tag{33.15}$$

where I_r is the number of r-simplexes in K. For a polyhedron in $\mathbb{R}^3$, for example,

$$\chi = (\ \# \text{ of vertices }) - (\ \# \text{ of edges }) + (\ \# \text{ of faces }) = 2 \ .$$

The homology groups $H_r(M)$ provide important information on the topology of M by relating its so-called Betti numbers to the Euler characteristic through the Euler-Poincare Theorem.

Definition 33.6. *The r-th* **Betti number** *$b_r(M)$ of a manifold M is defined by*

$$b_r(M) \equiv dim(H_r(M)) \ . \tag{33.16}$$

Theorem 33.1 (Euler–Poincare). *Let M be a triangulable manifold which is homeomorphic to an n-dimensional simplicial complex K. Then the Euler characteristic of M is given by*

$$\chi(M) = \sum_{r=0}^{n} (-1)^r b_r(M) \quad , \tag{33.17}$$

where $b_r(M)$ is the r-th Betti number of M.

Proof. Consider the boundary homomorphism $\partial : C_r(M) \to C_{r-1}(M)$ between chain groups on M. Define $C_{-1}(M) = 0$. Viewing $C_r(M)$ and $C_{r-1}(M)$ as vector spaces, it follows from Theorem 17.3 that

$$\begin{aligned} I_r &= dim(C_r(M)) = dim(ker\,\partial) + dim(im\,\partial) \\ &= dim(Z_r(M)) + dim(B_{r-1}(M)) \ , \end{aligned} \tag{33.18}$$

where $B_{-1}(M)$ is defined to be 0. On the other hand,

$$b_r(M) = dim(Z_r(M)/B_r(M)) = dim(Z_r(M)) - dim(B_r(M)) \ . \tag{33.19}$$

Exercise 33.2 Prove the last equality of the above equation.

Eqs. (33.15), (33.18) and (33.19) then imply that

$$\chi(M) = \sum_{r=0}^{n} (-1)^r \left[dim(Z_r(M)) + dim(B_{r-1}(M)) \right]$$

$$= \sum_{r=0}^{n} \left[(-1)^r \, dim(Z_r(M)) - (-1)^r \, dim(B_r(M)) \right] = \sum_{r=0}^{n} (-1)^r \, b_r(M) , \quad (33.20)$$

where in the second equality we have used the fact that $B_{-1}(M) = B_n(M) = 0$. $\qquad\square$

The integral of an r-form over an r-dimensional submanifold of M [defined by (32.29)] can be extended to that of an r-form over an r-chain in M. In fact, using the notation of (32.50) and (32.51), a pairing

$$(\ , \ ) : C_r(M) \times A^r(M) \longrightarrow \mathbb{R} \qquad (33.21)$$

can be defined by

$$(c, \omega) \equiv \int_c \omega , \qquad c \in C_r(M), \ \omega \in A^r(M) . \qquad (33.22)$$

This pairing induces a pairing between $H_r(M)$ and $H^r(M)$:

$$\langle \, [c], [\omega] \, \rangle \equiv (c, \omega) = \int_c \omega , \qquad (33.23)$$

for $c \in Z_r(M)$ and $\omega \in Z^r(M)$. The soundness of this definition is guaranteed by Stokes Theorem (Theorem 32.4). Indeed, suppose $[c] = [c']$ and $c - c' = \partial c''$. Then

$$\langle \, [c'], [\omega] \, \rangle = (c', \omega) = (c - \partial c'', \omega) = (c, \omega) - (\partial c'', \omega)$$
$$= (c, \omega) - (c'', d\omega) = (c, \omega) = \langle \, [c], [\omega] \, \rangle , \quad (33.24)$$

where the fourth equality follows from Stokes Theorem and the fifth from the fact that $d\omega = 0$. Likewise, suppose $[\omega] = [\omega'']$, so that $\omega - \omega' = d\omega''$. Then

$$\langle \, [c], [\omega'] \, \rangle = (c, \omega') = (c, \omega - d\omega'') = (c, \omega) - (c, d\omega'')$$
$$= (c, \omega) - (\partial c, \omega'') = (c, \omega) = \langle \, [c], [\omega] \, \rangle . \quad (33.25)$$

Thus the pairing $\langle \, [c], [\omega] \, \rangle$ only depends on the homology and cohomology classes and does not depend on the particular representatives of these classes. The pairing is also clearly bilinear.

We can now state de Rham's Theorem. (The proof of this theorem is highly non-trivial and beyond the scope of this book; see, for example, I. M. Singer and J. A. Thorpe 1976.)

Theorem 33.2 (de Rham). *Suppose M is a compact manifold. Then the r-th homology group and the r-th de Rham cohomology group (both with real coefficients) are dual vector spaces to each other, with the pairing between them given by (33.23).*

Corollary 33.1. *The Betti numbers $b_r(M)$ [as defined by (33.16)] are given by*

$$b_r(M) = dim\,(H^r(M))\,. \tag{33.26}$$

If, for a particular r, all closed r-forms are exact, then $H^r(M)$ is trivial and $b_r(M) = 0$. Since a closed form is always exact locally, the de Rham cohomology groups $H^r(M)$ measure obstructions to the global exactness of closed forms.

The de Rham cohomology groups derive from the differentiable structure of a manifold, while the Betti numbers are a purely topological invariants of the manifold. Thus the de Rham Theorem establishes a connection between the local and global properties of a manifold. This is a recurrent and central theme in modern differential geometry and its applications to physics.

$\boxed{\text{Exercise 33.3}}$ Show that for a connected manifold M with a single component,

$$H^0(M) = \mathbb{R}\,, \tag{33.27}$$

and if M has m connected components,

$$H^0(M) = \underbrace{\mathbb{R} \oplus \mathbb{R} \oplus \cdots \oplus \mathbb{R}}_{m \text{ times}}\,. \tag{33.28}$$

There is an interesting duality between $H^r(M)$ and $H^{n-r}(M)$, where $n = dim\,M$ and M is compact. We have the following bilinear, non-degenerate map

$$\langle\,,\,\rangle : H^r(M) \times H^{n-r}(M) \longrightarrow \mathbb{R}\,,$$

given by

$$\langle\,[\omega], [\eta]\,\rangle \equiv \int_M \omega \wedge \eta\,, \tag{33.29}$$

for $\omega \in Z^r(M)$ and $\eta \in X^{n-r}(M)$. The above is well-defined since $\omega \wedge \eta$ is an n-form (a top or "volume" form), so that its integral over the n-dimensional manifold M makes sense. Furthermore, suppose $\omega' - \omega = d\omega''$, where $\omega'' \in A^{r-1}(M)$. Then

$$\langle\,[\omega'], [\eta]\,\rangle = \int_M \omega' \wedge \eta = \int_M (\omega + d\omega'') \wedge \eta = \int_M (\omega \wedge \eta + d\omega'' \wedge \eta)$$

$$= \int_M (\omega \wedge \eta + d(\omega'' \wedge \eta)) = \int_M \omega \wedge \eta + \int_{\partial M} \omega'' \wedge \eta = \int_M \omega \wedge \eta = \langle\,[\omega], [\eta]\,\rangle\,,$$

$$\tag{33.30}$$

where the fourth equality follows from the fact that $d\eta = 0$ [since $\eta \in Z^r(M)$], the fifth from Stokes Theorem, and the sixth from the fact that $\partial M = 0$. A similar argument shows that

$$\langle [\omega], [\eta'] \rangle = \langle [\omega], [\eta] \rangle$$

if $\eta' - \eta = d\eta''$. Non-degeneracy of the map (33.29) means that if neither $[\omega]$ nor $[\eta]$ vanishes, then $\langle [\omega], [\eta] \rangle$ cannot vanish identically. This map establishes an isomorphism between $H^r(M)$ and $H^{n-r}(M)$ as vector spaces, called the **Poincare duality** (the proof will be omitted here):

$$\boxed{H^r(M) \sim H^{n-r}(M) , \quad M \text{ compact}} . \tag{33.31}$$

As a consequence, the Betti numbers satisfy

$$b_r(M) = b_{n-r}(M) . \tag{33.32}$$

It follows that *the Euler characteristic of an odd-dimensional manifold vanishes.*

Exercise 33.4   Prove the above statement.

The structure of cohomology groups is preserved by smooth mappings. Suppose $f : M \to N$ is a smooth map from a manifold M to a manifold N. Then the pullback of f, $f^* : \Lambda^r(N) \to \Lambda^r(M)$, is a homomorphism which commutes with the exterior derivative d (Theorem 32.2). Thus, for $\omega \in Z^r(N)$,

$$d(f^*\omega) = f^*(d\omega) = 0 , \tag{33.33}$$

which implies $f^*\omega \in Z^r(M)$. This makes

$$f^* : Z^r(N) \to Z^r(M)$$

a homomorphism. Similarly, for $\eta \in B^r(M)$ [$\eta = d\xi$, $\xi \in \Lambda^{r-1}(M)$],

$$f^*\eta = f^*(d\xi) = d(f^*\xi) , \tag{33.34}$$

which implies $f^*\eta \in B^r(M)$. Consequently,

$$f^* : B^r(N) \longrightarrow B^r(M)$$

is also a homomorphism. As a result, a smooth map $f : M \to N$ induces, through the pullback map $f^* : H^r(N) \to H^r(M)$, a homomorphism between de Rham groups. This pullback is given by

$$f^*[\omega] \equiv [f^*(\omega)] , \tag{33.35}$$

where $\omega \in Z^r(N)$. Indeed, for $[\omega] \in H^r(N)$ and $[\eta] \in H^s(N)$,

$$f^*([\omega] \wedge [\eta]) = f^*([\omega \wedge \eta]) = [f^*(\omega \wedge \eta)] = [f^*\omega \wedge f^*\eta] = [f^*\omega] \wedge [f^*\eta] . \tag{33.36}$$

This shows that f^* is a homomorphism.

To conclude, we collect some simple and useful examples of de Rham groups as follows.

$$H^r(M) = H^{n-r}(M), \qquad M \text{ compact }, n = dim\, M$$
$$\text{(Poincare duality)}, \qquad\qquad (33.37)$$

$$H^0(M) = \mathbb{R}, \qquad M \text{ connected }, \qquad (33.38)$$

$$H^0(M) = \underbrace{\mathbb{R} \oplus \cdots \oplus \mathbb{R}}_{m \text{ times}}, \qquad M \text{ has } m \text{ connected components }, \qquad (33.39)$$

$$H^1(M) = 0, \qquad M \text{ simply connected }, \qquad (33.40)$$

$$H^r(\mathbb{R}^n) = 0, \qquad 1 \le r \le n, \qquad (33.41)$$

$$H^1(S^1) = \mathbb{R}, \qquad\qquad (33.42)$$

$$H^r(S^n) = 0, \qquad n \ge 2, 1 \le r \le n-1,$$
$$(S^n \text{ is the } n\text{-sphere}), \qquad (33.43)$$

$$H^1(T_{2g}) = \underbrace{\mathbb{R} \oplus \cdots \oplus \mathbb{R}}_{2g}, \qquad (T_{2g} \text{ is the } 2g\text{-torus}). \qquad (33.44)$$

Note that $H^r(\mathbb{R}^n)$ does not satisfy the Poincare duality since $\mathbb{R}^n$ is not compact. Also, (33.44) follows from (33.13) and de Rham's Theorem (Theorem 33.2).

Exercise 33.5   Verify (33.38) and (33.42).

Chapter 34

The Geometry of Lie Groups

Our interests in Lie groups have heretofore been focused on their algebraic structure (in relation to their Lie algebras). There is, however, also a differentiable structure which is equally important, especially in the characterization of the geometry of fiber bundles, which is the basic mathematical setting for numerous physics applications.

We start with the proper definition of a Lie group.

Definition 34.1. *Let G be a group. If G is also a finite-dimensional differentiable manifold such that both the group multiplication map*

$$\varphi : G \times G \to G , \quad \varphi(g_1, g_2) = g_1 g_2$$

and the inverse map

$$\tau : G \to G , \quad \tau(g) = g^{-1}$$

are smooth, then G is called an r-dimensional Lie group, where r is the dimension of G considered as a manifold.

Besides the inverse map, a Lie group G possesses two other sets of diffeomorphisms, called the right and left translations, defined as follows. (*A **diffeomorphism** $f : M \to N$ from a manifold M to a manifold N is a smooth map whose inverse is also smooth. The existence of a diffeomorphism between two manifolds signifies their common **differentiable structure**.*)

Definition 34.2. *Let G be a Lie group. For $g \in G$, the **right translation** by g on G is a map $R_g : G \to G$ defined by*

$$R_g(g') = g'g , \quad \text{for all } g' \in G ; \tag{34.1}$$

*and the **left translation** by g on G is a map $L_g : G \to G$ defined by*

$$L_g(g') = gg' , \quad \text{for all } g' \in G . \tag{34.2}$$

Consider an r-dimensional Lie group G with identity e. Then, for every $g \in G$, $R_{g^{-1}}(g) = gg^{-1} = e$ and the tangent map

$$(R_{g^{-1}})_* : T_g(G) \to T_e(G) \tag{34.3}$$

is a linear isomorphism, where $T_g(G)$ denotes the tangent space to G at g.

Definition 34.3. *Let $X \in T_g(G)$. Then the differential 1-form ω with values in $T_e(G)$ defined by*

$$\boxed{\omega(X) \equiv (R_{g^{-1}})_* X} \tag{34.4}$$

is called the **right fundamental form** *or the* **Maurer-Cartan form** *of the Lie group G.*

Remark. A **left fundamental form** (also called the Maurer-Cartan form) can be similarly defined. From now on we will develop the theory in terms of the right translation.

Note that up to now we have only considered real-valued forms, in which the action of some cotangent vector on a tangent vector produces a real number. The above definition represents a generalization to the so-called **vector-valued forms**. To clarify the concept let us calculate the matrix representation of the right fundamental form ω with respect to a choice of local coordinates. Let $\{\delta_i\}$, $1 \leq i \leq r$, be a basis of $T_e(G)$, then we can write

$$\omega = \omega^i \, \delta_i \,, \tag{34.5}$$

where the ω^i, $1 \leq i \leq r$, are r real-valued 1-forms on G that are linearly independent everywhere. (They are also called the **fundamental forms** of G.) Choose local coordinates $(U; x^i)$ and $(W; y^i)$ at the points e and g, respectively. Since the group multiplication map φ is smooth (and hence continuous), for sufficiently small U, there exists a $W_1 \subset W$ such that

$$\varphi(U \times W_1) \subset W \,.$$

Choose $\delta_i = \left. \dfrac{\partial}{\partial x^i} \right|_e$ and let

$$\varphi^i(x,y) \equiv y^i \circ \varphi(e,g) \,, \quad (e,g) \in (U \times W_1) \,.$$

Then the matrix representation of $(R_g)_* : T_e(G) \to T_g(G)$ is given by [c.f. (32.8)]

$$(R_g)_* \, \delta_i = \left(\frac{\partial \varphi^j(x,y)}{\partial x^i} \right)_{x=e} \left(\frac{\partial}{\partial y^j} \right)_{y=g} \,. \tag{34.6}$$

But $(R_{g^{-1}})_* \circ (R_g)_* = id : T_e(G) \to T_e(G)$. Thus $(R_{g^{-1}})_* = (R_g)_*^{-1}$ and

$$(R_{g^{-1}})_* \left(\frac{\partial}{\partial y^i} \right)_{y=g} = \Lambda_i^j(g) \delta_j \,, \tag{34.7}$$

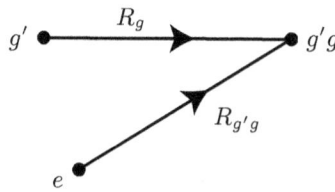

FIGURE 34.1

where $\Lambda_i^j(g)$ is the inverse matrix of $\left(\dfrac{\partial \varphi^j(x, y)}{\partial x^i}\right)_{x=e}$. It follows that

$$\omega^i = \Lambda_j^i(g)\, dy^j , \qquad (34.8)$$

which indeed is a smooth 1-form on G.

Definition 34.4. *Let X be a smooth tangent vector field on a Lie group G [alternatively, we say that $X \in \Gamma(TG)$ is a **smooth section** of TG, $\Gamma(TG)$ being the space of smooth sections of TG]. If, for any $g \in G$,*

$$(R_g)_* X = X , \qquad (34.9)$$

*then X is said to be a **right-invariant vector field** on G. Similarly, let $\omega \in \Gamma(T^*G)$ be a 1-form on G. If*

$$(R_g)^* \omega = \omega , \qquad (34.10)$$

*then ω is said to be a **right-invariant form** on G.*

Theorem 34.1. *The Maurer-Cartan form ω [c.f. Eq. (34.4) in Def. 34.3] of a Lie group G is a right-invariant form on G.*

Proof. Let $g \in G$ be an element of G and $X_{g'} \in T_{g'}(G)$ be a tangent vector at $g' \in G$. Then (see Fig. 34.1)

$$\begin{aligned}
((R_g)^* \omega)(X_{g'}) &= \omega((R_g)_* X_{g'}) = (R_{(g'g)^{-1}})_*(R_g)_* X_{g'} \\
&= (R_{(g')^{-1}})_* X_{g'} = \omega(X_{g'}) .
\end{aligned} \qquad (34.11)$$

Since g' is an arbitrary element of G, (34.10) is satisfied and the theorem is proved. $\qquad \square$

More generally we have the following theorem (given without proof).

Theorem 34.2. *Let $\sigma : G \rightarrow G$ be a smooth map on an r-dimensional Lie group. Then σ is a right translation if and only if*

$$\sigma^* \omega^i = \omega^i , \quad 1 \leq i \leq r , \tag{34.12}$$

where ω^i are the fundamental forms [with respect to a particular choice of basis $\{\delta_i\}$ of $T_e(G)$] corresponding to the Maurer-Cartan form of G [c.f. (34.5)].

Consider an arbitrary tangent vector $A \in T_e(G)$, and let

$$A_g = (R_g)_* A . \tag{34.13}$$

If we let g range over all of G, A_g then gives a smooth tangent vector field $\overline{A}$ on G. Obviously, the Maurer-Cartan form ω pulls A_g back to A:

$$\omega(A_g) = (R_{g^{-1}})_* A_g = (R_{g^{-1}})_* (R_g)_* A = A . \tag{34.14}$$

Now, by Theorem 34.1,

$$A = \omega(A_{g'}) = ((R_g)^* \omega)(A_{g'}) = \omega((R_g)_* A_{g'}) . \tag{34.15}$$

On the other hand,

$$A = \omega(A_{g'g}) . \tag{34.16}$$

Therefore

$$(R_g)_* \overline{A} = \overline{A} , \tag{34.17}$$

that is, $\overline{A}$ is a right-invariant vector field.

Let X_i be the right-invariant vector field obtained by the right translation of $\delta_i \in T_e(G)$, where $\{\delta_i\}$, $1 \leq i \leq r$, is a basis of $T_e(G)$:

$$(R_g)_* \delta_i = (X_i)_g . \tag{34.18}$$

Then the X_i, $1 \leq i \leq r$, constitute r right-invariant vector fields which are everywhere linearly independent on G, and any right-invariant vector field on G is a linear combination of the X_i with constant coefficients. In other words, $\{X_i\}$, $1 \leq i \leq r$, constitute a basis of the vector space $\mathcal{G}$ of right-invariant vector fields on G. Furthermore, this vector space is isomorphic to $T_e(G)$.

Since

$$\omega(X_i) = \delta_i , \tag{34.19}$$

Eq. (34.5) implies that

$$\omega^i \delta_i (X_j) = \delta_j , \tag{34.20}$$

or

$$\omega^i(X_j) = \langle X_j, \omega^i \rangle = \delta^i_j . \tag{34.21}$$

Thus the (right-invariant) fundamental forms ω^i and the right-invariant vector fields X_i constitute sets of naturally dual **coframe** and **frame fields**, respectively, on the Lie group G. We also see that *a vector field X on G is right-invariant if and only if $\omega(X) =$ a constant vector $\in T_e(G)$.*

In addition to being a vector space, $\mathcal{G}$ also admits an internal multiplication, due to the following theorem.

Theorem 34.3. *If X and Y are right-invariant vector fields on a Lie group G, then*

$$[X, Y] \equiv XY - YX \tag{34.22}$$

is also a right-invariant vector field on G.

Before proving this theorem we have to develop some more properties of the fundamental forms ω^i. Since the set $\{\omega^i\}$ forms a coframe field on G, and each ω^i is right-invariant, we can define constants c^i_{jk} such that

$$d\omega^i = -\frac{1}{2} c^i_{jk} \, \omega^j \wedge \omega^k \, , \tag{34.23a}$$

$$c^i_{jk} + c^i_{kj} = 0 \, . \tag{34.23b}$$

Eq. (34.23a) is called the **structure equation** or the **Maurer-Cartan equation** of the Lie group G, and the constants c^i_{jk} are called the **structure constants** of G. It is not hard to see that the structure constants satisfy the Jacobi identity:

$$c^i_{jk} c^j_{hl} + c^i_{jh} c^j_{lk} + c^i_{jl} c^j_{kh} = 0 \, , \tag{34.24}$$

which is the same as (4.32). Indeed, exteriorly differentiating (34.23a), we have

$$\begin{aligned}
0 &= -\frac{1}{2} c^i_{jk} (d\omega^j \wedge \omega^k - \omega^j \wedge d\omega^k) \\
&= \left(\frac{1}{2}\right)\left(\frac{1}{2}\right) c^i_{jk} \, (c^j_{hl} \, \omega^h \wedge \omega^l \wedge \omega^k - c^k_{hl} \, \omega^j \wedge \omega^h \wedge \omega^l) \\
&= \frac{1}{4} c^i_{jk} c^j_{hl} \, \omega^h \wedge \omega^l \wedge \omega^k - \frac{1}{4} c^i_{jk} c^k_{hl} \, \omega^j \wedge \omega^h \wedge \omega^l \, .
\end{aligned} \tag{34.25}$$

The last term can be written as

$$\frac{1}{4} c^i_{kj} c^k_{hl} \, \omega^j \wedge \omega^h \wedge \omega^l = \frac{1}{4} c^i_{jk} c^j_{hl} \, \omega^k \wedge \omega^h \wedge \omega^l = \frac{1}{4} c^i_{jk} c^j_{hl} \, \omega^h \wedge \omega^l \wedge \omega^k \, , \tag{34.26}$$

where the first equality follows from interchanging the dummy indices k and j. Eq. (34.25) then gives

$$\begin{aligned}
0 &= \frac{1}{2} c^i_{jk} c^j_{hl} \, \omega^h \wedge \omega^l \wedge \omega^k \\
&= \frac{1}{6} (c^i_{jk} c^j_{hl} \, \omega^k \wedge \omega^h \wedge \omega^l + c^i_{jh} c^j_{lk} \, \omega^h \wedge \omega^l \wedge \omega^k + c^i_{jl} c^j_{kh} \, \omega^l \wedge \omega^k \wedge \omega^h) \\
&= \frac{1}{6} (c^i_{jk} c^j_{hl} + c^i_{jh} c^j_{lk} + c^i_{jl} c^j_{kh}) \, \omega^k \wedge \omega^h \wedge \omega^l \, .
\end{aligned} \tag{34.27}$$

Since the terms inside the parentheses on the RHS of the last equality are skew-symmetric with respect to the indices k, h and l, the Jacobi identity (34.24) follows.

Let us now prove Theorem 34.3.

Proof. (of Theorem 34.3) By Theorem 30.3,

$$\langle X \wedge Y, d\omega^i \rangle = X \langle Y, \omega^i \rangle - Y \langle X, \omega^i \rangle - \langle [X,Y], \omega^i \rangle . \tag{34.28}$$

On the other hand, from the structure equation (34.23a),

$$\langle X \wedge Y, d\omega^i \rangle = -\frac{1}{2} c^i_{jk} \langle X \wedge Y, \omega^j \wedge \omega^k \rangle$$

$$= -\frac{1}{2} c^i_{jk} \left(\omega^j(X)\omega^k(Y) - \omega^j(Y)\omega^k(X) \right) = -c^i_{jk} \omega^j(X)\omega^k(Y) , \tag{34.29}$$

where the second equality follows from (28.28) and the third from the fact that $c^i_{jk} = -c^i_{kj}$. Since, by assumption, X and Y are both right-invariant, $\omega^j(X)$ and $\omega^k(Y)$ are all constants. It follows from (34.28) that $\omega^i([X,Y])$ must also be a constant. In other words, $[X,Y]$ is a right-invariant vector field. This proves the theorem. $\qquad\square$

It is easily seen that the product $[X,Y]$ satisfies the usual distributive rule, the skew-symmetry property, and the Jacobi identity. We are finally in a position to define the Lie algebra $\mathcal{G}$ of a Lie group G.

Definition 34.5. *The linear space $\mathcal{G}$ of right-invariant vector fields on a Lie group G (which is isomorphic to $T_e(G)$, the tangent space at the identity $e \in G$), together with the internal multiplication $[X,Y] \equiv XY - YX$, constitutes a Lie algebra, called the Lie algebra $\mathcal{G}$ of the Lie group G. $T_e(G)$ is also a Lie algebra isomorphic to $\mathcal{G}$ if we define the following internal multiplication in $T_e(G)$:*

$$[\delta_j, \delta_k] = c^i_{jk} \delta_i , \tag{34.30}$$

where $\{\delta_i\}$ is a basis of $T_e(G)$.

Exercise 34.1   Let ω^i, $i = 1, \ldots, r$, be the fundamental forms of an r-dimensional Lie group G, and X_i, $i = 1, \ldots, r$, be the dual vector fields. Use (34.28) and (34.29) to show that

$$\omega^i([X_j, X_k]) = c^i_{jk} , \tag{34.31}$$

and hence

$$[X_j, X_k] = c^i_{jk} X_i . \tag{34.32}$$

Eqs. (34.31) and (34.32) identify the structure constants c^i_{jk} introduced in the present chapter with those introduced in Chapter 4 [c.f. (4.29)].

In Chapter 4 we calculated the structure constants of the Lie groups $SO(n)$. We will now calculate those of the general linear group $GL(n, \mathbb{R})$, where $g \in GL(n, \mathbb{R})$ is an $n \times n$ invertible real matrix. Let $g = (g^j_i) \in GL(n, \mathbb{R})$. Then the matrix elements g^j_i, $1 \le i, j \le n$, can be regarded as local coordinates on the n^2-dimensional manifold $GL(n, \mathbb{R})$, and $\{dg_i^j\}$ gives a local coframe field on $GL(n, \mathbb{R})$.

For any $g' \in G$, we represent the right translation R_g by

$$\varphi^{(i,j)}(g', g) = (g' g)_i^j = (g')_i^m g_m^j . \tag{34.33}$$

Thus, recalling (34.6),

$$(\Lambda^{-1}(g))_{(k,l)}^{(i,j)} = \left.\frac{\partial \varphi^{(i,j)}(g', g)}{\partial(g')_k^l}\right|_{g' = e} = \frac{\partial((g')_i^l g_l^j)}{\partial(g')_k^l} \quad \text{(no sum on } l) = \delta_k^i \, g_l^j . \tag{34.34}$$

Note that $\Lambda^{-1}(g)$ is an $(n^2 \times n^2)$ matrix. If we order the composite index (i, j) in the following manner:

$$(1, 1) \to (1, 2) \to \cdots \to (1, n)$$
$$\to (2, 1) \to (2, 2) \to \cdots \to (2, n)$$

$$\vdots$$

$$\to (n, 1) \to (n, 2) \to \cdots \to (n, n) ,$$

then the matrix $\Lambda^{-1}(g)$ appears in the block diagonal form

$$\begin{pmatrix} (g_i^j) & 0 & \cdots & 0 \\ 0 & (g_i^j) & \cdots & 0 \\ & \vdots & & \\ 0 & 0 & \cdots & (g_i^j) \end{pmatrix} ,$$

where each block is an $n \times n$ matrix. Thus

$$(\Lambda(g))_{(k,l)}^{(i,j)} = \delta_k^i \, (g^{-1})_l^j , \tag{34.35}$$

and (34.8) implies

$$w_i^j = \sum_{k, l} \Lambda_{(k,l)}^{(i,j)}(g) \, dg_k^l = \sum_{k, l} \delta_k^i \, (g^{-1})_l^j \, dg_k^l = dg_k^l \, (g^{-1})_l^j . \tag{34.36}$$

In matrix notation, the matrix of right fundamental one-forms (w_i^j) is thus given by

$$w = dg \cdot g^{-1} . \tag{34.37}$$

Exteriorly differentiating the above, we get [c.f. (31.32)]

$$dw = -dg \wedge d(g^{-1}) = -dg \cdot g^{-1} \wedge g \cdot d(g^{-1})$$
$$= dg \cdot g^{-1} \wedge dg \cdot g^{-1} = w \wedge w , \tag{34.38}$$

where in the third equality we have used the fact that $g \cdot d(g^{-1}) = -dg \cdot g^{-1}$, which is a consequence of exteriorly differentiating the equation $g \cdot g^{-1} = 1$. Eq. (34.38) can also be written in the component form [c.f. (31.34)]

$$d\omega_i^j = \omega_i^k \wedge \omega_k^j . \tag{34.39}$$

Thus

$$d\omega_i^j = \frac{1}{2} \sum_{p,q,r,s=1}^{n} (\delta_i^p \delta_q^j \delta_s^r - \delta_i^r \delta_s^j \delta_q^p)\, \omega_p^s \wedge \omega_r^q . \tag{34.40}$$

It follows from (34.23a) that the structure constants of $GL(n, \mathbb{R})$ are given by

$$\boxed{c^{(i,j)}_{(p,s)(r,q)} = -\delta_i^p \delta_q^j \delta_s^r + \delta_i^r \delta_s^j \delta_q^p} \quad . \tag{34.41}$$

As discussed earlier, the Lie algebra $\mathcal{GL}(n, \mathbb{R})$ of the Lie group $GL(n, \mathbb{R})$ can be identified with $T_e(GL(n, \mathbb{R}))$, with internal multiplication given by (34.30). On the other hand $T_e(GL(n, \mathbb{R}))$ is the space of $n \times n$ real matrices, which is isomorphic to the n^2-dimensional vector space $\mathbb{R}^{n^2}$. Thus a basis of $\mathcal{GL}(n, \mathbb{R})$ is the set of matrices $\{E_i^j\}$, $i, j = 1, \dots, n$, where E_i^j denotes an $n \times n$ matrix with the value 1 for the element in the j-th row and i-th column, and 0 for all other elements. (Note carefully the row and column positions.) For example, with $n = 2$,

$$E_2^1 = \begin{pmatrix} 0 & 1 \\ 0 & 0 \end{pmatrix} , \quad E_1^2 = \begin{pmatrix} 0 & 0 \\ 1 & 0 \end{pmatrix} .$$

The right fundamental form ω is written

$$\omega = \omega_i^j E_j^i . \tag{34.42}$$

From (34.30) and (34.41), the multiplication table for $\mathcal{GL}(n, \mathbb{R})$ is then given by

$$[E_s^p, E_q^r] = \sum_{i,j} c^{(i,j)}_{(p,s)(r,q)}\, E_j^i = \sum_{i,j} (-\delta_i^p \delta_q^j \delta_s^r + \delta_i^r \delta_s^j \delta_q^p)\, E_j^i = \delta_q^p E_s^r - \delta_s^r E_q^p .$$

$$\tag{34.43}$$

Suppose $A, B \in \mathcal{GL}(n, \mathbb{R})$, written in terms of matrix elements as

$$A = (A_p^s) = A_p^s\, E_s^p , \qquad B = (B_r^q) = B_r^q\, E_q^r .$$

Then, by (34.43),

$$[A, B] = A_p^s B_r^q\, [E_s^p, E_q^r] = A_p^s B_r^q\, (\delta_q^p E_s^r - \delta_s^r E_q^p)$$
$$= B_r^q A_q^s E_s^r - A_p^s B_s^q E_q^p = (BA)_r^s E_s^r - (AB)_p^q E_q^p = BA - AB , \tag{34.44}$$

(where the RHS is interpreted in the sense of matrix products). Note that this differs from the usual definition of a commutator by a sign [c.f. (4.25), which in fact defines $[\,,\,]_{\text{left}}$, derived from the left translation L_g, see Ex. 34.2 below].

Exercise 34.2 The left-fundamental (Maurer-Cartan) form $\tilde{\omega}$ and the associated left-invariant fields $\tilde{X}_i$ on G are given by the following equations:

$$\tilde{\omega} = \tilde{\omega}^i \delta_i , \tag{34.45}$$

$$(L_g)_* \, \delta_i = (\tilde{X}_i)_g , \tag{34.46}$$

$$\tilde{\omega}^i(\tilde{X}_j) = \langle \tilde{X}_j, \tilde{\omega}^i \rangle = \delta_j^i , \tag{34.47}$$

$$d\tilde{\omega}^i = -\frac{1}{2} \tilde{c}_{jk}^i \, \tilde{\omega}^j \wedge \tilde{\omega}^k , \tag{34.48}$$

$$[\tilde{X}_j, \tilde{X}_k] = \tilde{c}_{jk}^i \, \tilde{X}_i . \tag{34.49}$$

Show that

$$\tilde{c}_{jk}^i = -c_{jk}^i . \tag{34.50}$$

Thus we have two different internal multiplications in $\mathcal{G}$, denoted (in this problem) by $[\, , \,]_{\text{right}}$ and $[\, , \,]_{\text{left}}$, corresponding to the sets of structure constants c_{jk}^i and $\tilde{c}_{jk}^i$, respectively. Show that these multiplications just differ by a sign:

$$[\delta_i, \delta_j]_{\text{right}} = -[\delta_i, \delta_j]_{\text{left}} . \tag{34.51}$$

We digress to consider the action of Lie groups on manifolds. This is important in the study of dynamical systems.

Definition 34.6. *Let M be a manifold and $X \in \Gamma(TM)$ be a section in the tangent bundle of M, or a tangent vector field on M. Suppose for each $p \in M$ there is a curve $\gamma_p : \mathbb{R} \to M$ [$\gamma_p(t), t \in \mathbb{R}$] through p such that $\gamma_p(0) = p$ and, for all $t \in \mathbb{R}$,*

$$\frac{d}{dt} \gamma_p(t) = X_{\gamma_p(t)} . \tag{34.52}$$

*(See Fig. 34.2.) Then X is said to be a **complete tangent vector field**.*

For a complete vector field X and any $t \in \mathbb{R}$, we can define a map $\varphi_t : M \to M$ by

$$\varphi_t(p) \equiv \gamma_p(t) . \tag{34.53}$$

One can show that (proof omitted here) φ_t is a diffeomorphism on M, and that

$$\varphi_s \circ \varphi_t = \varphi_{s+t} , \quad t, s \in \mathbb{R} . \tag{34.54}$$

The set $\{\varphi_t \,|\, t \in \mathbb{R}\}$ is called a **one-parameter group of diffeomorphisms** generated by the complete tangent vector field X. The parametrized curve $\gamma_p(t)$ is called the **orbit** of φ_t through p. Note that for each $t \in \mathbb{R}$, φ_t is an element of some group. The most important examples of one-parameter groups of diffeomorphisms are abelian subgroups of Lie groups.

Definition 34.7. *Let $X \in \Gamma(TM)$ and $\psi : M \to M$ be a diffeomorphism on M. X is said to be **invariant** under ψ if*

$$\psi_* X = X . \tag{34.55}$$

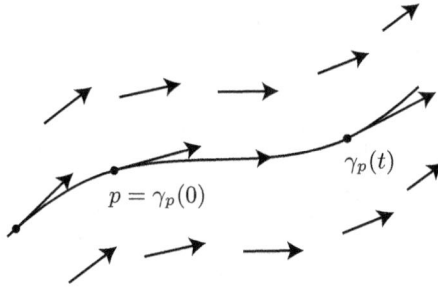

FIGURE 34.2

Theorem 34.4. *Let φ_t be the one-parameter group of diffeomorphisms generated by a complete vector field X on a smooth manifold M, and $\psi : M \to M$ is a diffeomorphism on M. Then $\psi \circ \varphi_t \circ \psi^{-1}$ is the one-parameter group of diffeomorphisms generated by $\psi_* X$.*

Proof. We need to show that the tangent to the curve $\rho_p(t) \equiv (\psi \circ \varphi_t \circ \psi^{-1})(p)$ at the point $\rho_p(t)$ is equal to $(\psi_* X)_p$. Suppose $f \in C^\infty(M)$ is a smooth function on M. Then, if $\psi(q) = p$,

$$(\psi_* X)_p f = X_q (f \circ \psi) = \frac{d}{dt} \, f \circ \psi(\varphi_t(q))|_{t=0}$$

$$= \frac{d}{dt} \, f(\psi \circ \varphi_t \circ \psi^{-1}(\psi(q)))|_{t=0} = \frac{d}{dt} \, f(\psi \circ \varphi_t \circ \psi^{-1}(p))|_{t=0} \quad (34.56)$$

$$= \frac{d}{dt} \, f(\rho_p(t))|_{t=0} \ .$$

This proves the theorem. $\qquad\qquad\qquad\qquad\qquad\qquad\qquad\qquad\qquad\qquad\square$

Corollary 34.1. *A complete tangent vector field $X \in \Gamma(TM)$ is invariant under a diffeomorphism $\psi : M \to M$ if and only if the one-parameter group of diffeomorphisms φ_t generated by X commutes with ψ.*

| Exercise 34.3 | Show that the above corollary is a direct consequence of Theorem 34.4.

Definition 34.8. *Let X and Y be two smooth tangent vector fields on a manifold M, and φ_t be the one-parameter group of diffeomorphisms generated by X. The **Lie derivative** of the vector field Y along the vector field X is the vector field $L_X Y$ defined by*

$$\boxed{ L_X Y \equiv \frac{d}{dt} \, (\varphi_t)_*^{-1}(Y) \Big|_{t=0} } \ , \qquad\qquad (34.57)$$

or, written out in more detail,

$$(L_X Y)_p = \frac{d}{dt}\left(\left(\left((\varphi_t)_*^{-1}\right)_{\varphi_t(p)}\left(Y_{\varphi_t(p)}\right)\right)\right)\Bigg|_{t=0} = \lim_{t\to 0}\frac{(\varphi_t^{-1})_* Y_{\varphi_t(p)} - Y_p}{t} . \quad (34.58)$$

Theorem 34.5. *Let X and Y be two smooth tangent vector fields on M. Then*

$$\boxed{L_X Y = [X,Y] \equiv XY - YX} . \quad (34.59)$$

Proof. Eq. (34.58) can also be written as

$$L_X Y = \lim_{t\to 0}\frac{Y_p - (\varphi_t)_* Y_{\varphi_t^{-1}(p)}}{t} . \quad (34.60)$$

(See Fig. 34.3.) Suppose f is a smooth function on M. Let

$$F(t) \equiv f(\varphi_t(p)) . \quad (34.61)$$

Then

$$F(t) - F(0) = \int_0^1 \frac{dF(st)}{ds}\, ds = t\int_0^1 F'(u)|_{u=st}\, ds . \quad (34.62)$$

It follows from (34.61) that

$$f(\varphi_t(p)) = f(p) + t g_t(p) , \quad (34.63)$$

where

$$g_t(p) \equiv \int_0^1 F'(u)|_{u=st}\, ds = \int_0^1 \frac{df(\varphi_u(p))}{du}\bigg|_{u=st}\, ds , \quad (34.64)$$

and

$$g_0(p) = \int_0^1 \frac{df(\varphi_u(p))}{du}\bigg|_{u=0}\, ds = \frac{df(\varphi_u(p))}{du}\bigg|_{u=0} = X_p f . \quad (34.65)$$

Applying the RHS of (34.60) to f, we have

$$\left(\lim_{t\to 0}\frac{Y_p - (\varphi_t)_* Y_{\varphi_t^{-1}(p)}}{t}\right) f = \lim_{t\to 0}\frac{Y_p f - Y_{\varphi_t^{-1}(p)}(f\circ\varphi_t)}{t}$$

$$= \lim_{t\to 0}\frac{Y_p f - Y_{\varphi_t^{-1}(p)}(f + t g_t)}{t} = \lim_{t\to 0}\frac{Y_p f - Y_{\varphi_t^{-1}(p)} f}{t} - \lim_{t\to 0} Y_{\varphi_t^{-1}(p)} g_t$$

$$= \lim_{t\to 0}\frac{Y_{\varphi_t(p)} f - Y_p f}{t} - Y_p g_0(p) = \lim_{t\to 0}\frac{Yf\circ\varphi_t(p) - Yf\circ\varphi_0(p)}{t} - Y_p X_p f$$

$$= \frac{d}{dt}(Yf)\circ\varphi_t(p)\bigg|_{t=0} - (YX)_p f = X_p(Yf) - (YX)_p f = [X,Y]_p f .$$

$$(34.66)$$

Since f is arbitrary, the theorem is proved. $\qquad\square$

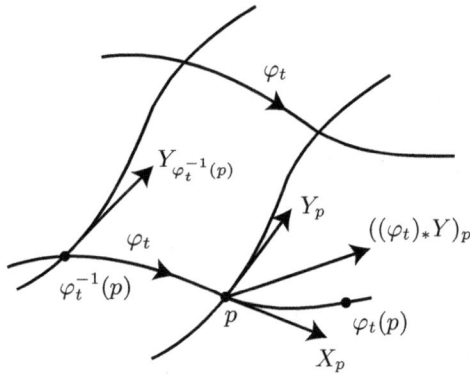

FIGURE 34.3

In a coordinate neighborhood $(U; x^i)$ let

$$X = X^i(x^j)\frac{\partial}{\partial x^i} = X^i \partial_i , \tag{34.67}$$

$$\text{and} \quad Y = Y^i(x^j)\frac{\partial}{\partial x^i} = Y^i \partial_i . \tag{34.68}$$

Then

$$[X,Y]f = X(Yf) - Y(Xf) = X^i\partial_i(Y^j\partial_j f) - Y^i\partial_i(X^j\partial_j f)$$
$$= X^i(\partial_i Y^j \partial_j f + Y^j\partial_i\partial_j f) - Y^i(\partial_i X^j \partial_j f + X^j\partial_i\partial_j f) \tag{34.69}$$
$$= \{X^i(\partial_i Y^j)\partial_j - Y^i(\partial_i X^j)\partial_j\}f .$$

Thus, locally,

$$\boxed{[X,Y] = \{X^i(\partial_i Y^j) - Y^i(\partial_i X^j)\} \partial_j} \quad . \tag{34.70}$$

For $X_e \in T_e(G) \sim \mathcal{G}$, we have the right-invariant field $X = (R_g)_* X_e$. This field can be shown to be complete (proof omitted here). Let $\{\varphi_t\}$ be the one-parameter group of doffeomorphisms generated by X, and the corresponding curve through $e \in G$ be given by

$$\gamma_e(t) = \varphi_t(e) . \tag{34.71}$$

Exercise 34.4   Show that

$$\gamma_e(s+t) = \gamma_e(s)\gamma_e(t) . \tag{34.72}$$

In fact it can be shown that (proof omitted) there is a one-to-one correspondence $X \leftrightarrow \gamma_e(t)$.

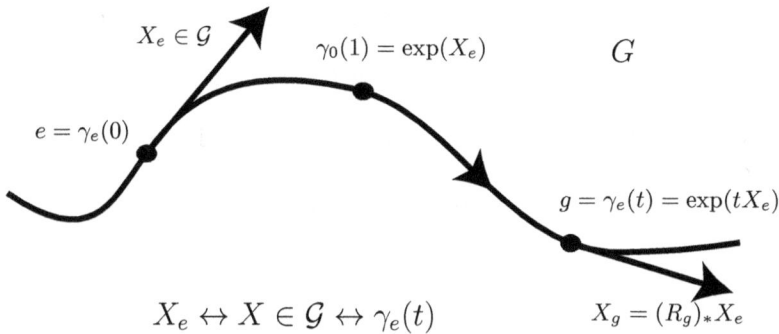

$$X_e \in \mathcal{G}$$
$$\gamma_0(1) = \exp(X_e)$$
$$G$$
$$e = \gamma_e(0)$$
$$g = \gamma_e(t) = \exp(tX_e)$$
$$X_e \leftrightarrow X \in \mathcal{G} \leftrightarrow \gamma_e(t)$$
$$X_g = (R_g)_* X_e$$

FIGURE 34.4

Definition 34.9. *The **exponential map** exp : $\mathcal{G} \to G$ is defined by*

$$\exp(X_e) \equiv \gamma_e(1) , \tag{34.73}$$

for all $X_e \in T_e(G)$.

Note that

$$\gamma_e(t) = \exp(tX_e) , \tag{34.74}$$

and

$$\varphi_t(g) = \gamma_g(t) = \gamma_e(t)g = \exp(tX_e)g . \tag{34.75}$$

The correspondence $X_e \leftrightarrow \gamma_e(t)$ is illustrated in Fig. 34.4.

We recall that $\mathcal{GL}(n, \mathbb{R})$ can be identified with the space of all $n \times n$ real matrices. For a matrix $A \in \mathcal{GL}(n, \mathbb{R})$, we define

$$e^A \equiv 1 + A + \frac{1}{2!} A^2 + \frac{1}{3!} A^3 \ldots . \tag{34.76}$$

Exercise 34.5 Show that the series in (34.76) converges.

Exercise 34.6 Show that

(a) $$e^{(t+s)A} = e^{tA} e^{sA} , \tag{34.77}$$

(b) $$e^{tA} e^{-tA} = 1 \quad \text{(the identity matrix)} . \tag{34.78}$$

We can identify e^{tA} with $\gamma_e(t)$ and thus have $e^A = \exp(A)$, where the exponential map exp is defined according to Def. 34.9 and e^A according to (34.76). The curve $\gamma_e(t)$ in G is given by

$$\left. \frac{d\gamma_e(t)}{dt} \right|_{t=0} = \left. \frac{d}{dt} \exp(tA) \right|_{t=0} = A . \tag{34.79}$$

For $g \in G$, we define the **adjoint isomorphism** $ad_g : G \to G$ by

$$ad_g(g') = gg'g^{-1} = L_g \circ R_{g^{-1}} g' , \qquad (34.80)$$

for all $g' \in G$. The adjoint isomorphism ad_g on the Lie group G induces an isomorphism

$$Ad(g) \equiv (ad_g)_* : T_e(G) \to T_e(G) ,$$

or

$$Ad(g) : \mathcal{G} \to \mathcal{G} ,$$

on the Lie algebra $\mathcal{G}$. This is so since $ad_g(e) = geg^{-1} = e$. We thus have the map

$$Ad : G \to GL(r, \mathbb{R}) , \qquad (34.81)$$

where $r = dim\,\mathcal{G}$, called the **adjoint representation of the Lie group** G. This is obviously a homomorphism between groups since

$$\begin{aligned}
Ad(g_1\,g_2) &= (ad_{g_1 g_2})_* = (ad_{g_1} \circ ad_{g_2})_* \\
&= (ad_{g_1})_* \circ (ad_{g_2})_* = Ad(g_1) \circ Ad(g_2) .
\end{aligned} \qquad (34.82)$$

This homomorphism in turn induces a homomorphism between Lie algebras

$$Ad \equiv Ad_* : \mathcal{G} \to \mathcal{GL}(r, \mathbb{R}), \quad r = dim\mathcal{G} , \qquad (34.83)$$

called the **adjoint representation of the Lie algebra** $\mathcal{G}$. [An element in $\mathcal{GL}(r, \mathbb{R})$ can be viewed as a linear transformation on the r-dimensional space $\mathcal{G}$]. Thus for $X \in \mathcal{G}$, $Ad(X) = Ad_X$ is a linear transformation on $\mathcal{G}$.

Theorem 34.6. *Let X and Y be right-invariant vector fields on a Lie group G (considered as elements of the Lie algebra $\mathcal{G}$ of G), then the adjoint representation of $\mathcal{G}$ is given by*

$$\boxed{Ad_X\, Y = -[X, Y]_r} \quad , \qquad (34.84)$$

where the Lie bracket product $[\,,\,]_r$ is that derived from the right translation R_g on G.

Remark. Note that the Lie bracket product given earlier in (7.15) is actually $[\,,\,]_{\text{left}}$, derived from the left translation L_g, hence a difference in sign between (34.84) and (7.15), as a consequence of (34.51). This latter equation then implies that (34.84) is equivalent to (7.15).

Proof. Consider $X_e, Y_e \in T_e(G) \sim \mathcal{G}$. Let X and Y be the right-invariant fields on G corresponding to X_e and Y_e, respectively, and φ_t be the one-parameter group of diffeomorphisms generated by X. Since X is right-invariant, Corollary 34.1 implies that the right translation R_g commutes with φ_t:

$$R_g \circ \varphi_t = \varphi_t \circ R_g , \qquad \text{for all } g \in G . \qquad (34.85)$$

Hence

$$\varphi_t(g) = (\varphi_t \circ R_g)(e) = (R_g \circ \varphi_t)(e) = R_g \, \varphi_t(e) = \varphi_t(e) \, g \, . \qquad (34.86)$$

Thus

$$\varphi_t = L_{\varphi_t(e)} \, . \qquad (34.87)$$

Now, by definition [(34.80)], the adjoint isomorphism $ad_{\varphi_t(e)}$ is given by

$$ad_{\varphi_t(e)} = L_{\varphi_t(e)} \circ R_{(\varphi_t(e))^{-1}} \, . \qquad (34.88)$$

It follows that

$$Ad(\varphi_t(e)) \equiv (ad_{\varphi_t(e)})_* = (L_{\varphi_t(e)})_* \circ (R_{(\varphi_t(e))^{-1}})_* \, . \qquad (34.89)$$

Furthermore,

$$Ad(X_e) = Ad_* (X_e) = \lim_{t \to 0} \frac{Ad(\varphi_t(e)) - Ad(e)}{t} \, . \qquad (34.90)$$

Since Ad maps the identity $e \in G$ to the identity in $GL(r, \mathbb{R})$,

$$(Ad(e)) \, (Y_e) = Y_e \, , \qquad (34.91)$$

for any $Y_e \in T_e(G)$. On applying the RHS of (34.90) to Y_e and using (34.89) we then have

$$
\begin{aligned}
(Ad(X_e)) \, (Y_e) &= \lim_{t \to 0} \frac{(L_{\varphi_t(e)})_* \circ (R_{(\varphi_t(e))^{-1}})_* \, Y_e - Y_e}{t} \\
&= \lim_{t \to 0} \frac{(\varphi_t)_* \circ Y_{(\varphi_t(e))^{-1}} - Y_e}{t} = -[X_e, Y_e] \, ,
\end{aligned}
\qquad (34.92)
$$

where the second equality follows from (34.87) and the last from Theorem 34.5 [(34.59)]. Since there is a one-to-one correspondence between any $X_e \in T_e(G)$ and its corresponding right-invariant vector field $X \in \mathcal{G}$, the theorem follows. $\qquad \square$

We note again that if $\mathcal{G}$ is considered as the algebra of left-invariant vector fields on G, then we would have

$$Ad_X \, Y = [X, Y]_l \, , \qquad (34.93)$$

where $[\, , \,]_l$ indicates the Lie bracket product for left-invariant fields.

To conclude this chapter, we will compute the Lie algebra $\mathcal{SU}(n)$ of the special unitary group $SU(n)$, using the geometric concepts developed above. Recall that

$$U(n) = \{g \in GL(n) \, | \, gg^\dagger = 1\} \, , \qquad (34.94)$$
$$SU(n) = \{g \in U(n) \, | \, \det(g) = 1\} \, , \qquad (34.95)$$

where

$$(g^\dagger)^j_i = \overline{g^i_j} \; . \tag{34.96}$$

Thus

$$\mathcal{U}(n) \sim T_e(U(n)) \; . \tag{34.97}$$

Consider a curve $g(t)$ through the identity $e \in U(n)$ such that $g(0) = e = g^\dagger(0)$. Then

$$0 = \left.\frac{d}{dt} g(t)g^\dagger(t)\right|_{t=0} = \left.\frac{dg(t)}{dt}\right|_{t=0} g^\dagger(0) + g(0) \left.\frac{dg^\dagger(t)}{dt}\right|_{t=0} \tag{34.98}$$

$$= g'(0) + (g^\dagger)'(0) \; .$$

Hence

$$\mathcal{U}(n) \subset \mathcal{S}(n) \; , \tag{34.99}$$

where

$$\mathcal{S}(n) \equiv \{A \in \mathcal{GL}(n, \mathbb{C}) \,|\, A + A^\dagger = 0\} \; . \tag{34.100}$$

Conversely, suppose $A \in \mathcal{S}(n)$. Then

$$\exp(A)(\exp(A))^\dagger = \exp(A)\exp(A^\dagger) = \exp(A)\exp(-A) = 1 \; . \tag{34.101}$$

Hence $\exp(A) \in U(n)$, and·

$$A = \left.\frac{d}{dt} \exp(tA)\right|_{t=0} \in \mathcal{U}(n) \; . \tag{34.102}$$

Thus

$$\mathcal{S}(n) \subset \mathcal{U}(n) \; . \tag{34.103}$$

Together with (34.99), this implies

$$\mathcal{U}(n) = \mathcal{S}(n) \; .$$

We assert that

$$\mathcal{SU}(n) = \{B \in \mathcal{U}(n) \,|\, tr\, B = 0\} \; , \tag{34.104}$$

or

$$\boxed{\mathcal{SU}(n) = \{B \in \mathcal{GL}(n, \mathbb{C}) \,|\, B + B^\dagger = 0, \; tr\, B = 0\}} \; . \tag{34.105}$$

This is a direct consequence of the following lemma.

Lemma 34.1. *For any $n \times n$ matrix B,*

$$\boxed{\det(e^B) = e^{tr\, B}} \; , \tag{34.106}$$

where e^B is defined by (34.76).

Proof. Let
$$f(t) \equiv \det(e^{tB}) . \tag{34.107}$$

We have

$$
\begin{aligned}
\frac{df(t)}{dt} &= \frac{d}{dh} f(t+h)\Big|_{h=0} = \frac{d}{dh} \left(\det \left(e^{(t+h)B} \right) \right) \Big|_{h=0} \\
&= \frac{d}{dh} \left(\det(e^{tB}) \det(e^{hB}) \right) \Big|_{h=0} \\
&= \det(e^{tB}) \frac{d}{dh} \left(\det \left(1 + hB + \frac{h^2 B^2}{2!} + \dots \right) \right) \Big|_{h=0} \tag{34.108} \\
&= \det(e^{tB}) \frac{d}{dh} \left(\det(1 + hB) \right) \Big|_{h=0} \\
&= \det(e^{tB}) \, tr \, B = tr \, B \, f(t) ,
\end{aligned}
$$

where in the fifth equality we have used the fact that, to first order in h, $\det(e^{hB}) \sim \det(1 + hB)$, and in the sixth, the fact that

$$\det(1 + B) = 1 + tr \, B - \frac{1}{2} tr \, (B^2) + \frac{1}{2} (tr \, B)^2 + \dots . \tag{34.109}$$

Thus
$$f(t) = f(0) \, e^{(tr \, B)t} = e^{(tr \, B)t} . \tag{34.110}$$

Our result is obtained on setting $t = 1$. $\qquad\square$

Chapter 35

Connections and Curvatures on a Vector Bundle

The notion of a vector bundle has been introduced in Chapter 30. In this chapter we will introduce a central concept in the theory of fiber bundles – the connection – which forms the mathematical basis of gauge field theories, as well as many other applications in physics. As we shall see, the curvature is a concept derived from the connection.

Basically, a connection on a vector bundle $\pi : E \to M$ is a specific way to differentiate vector fields on M, or sections of E. Since there is no canonical relationship between points on different fibers, one needs to impose a structure on E, beyond its differentiable structure, so that points on "neighboring" fibers can be "connected".

Recall that the space of smooth sections of E on M, denoted by $\Gamma(E)$, is a real vector space, as well as a $C^\infty(M)$-module.

Definition 35.1. *A **connection** on a vector bundle $\pi : E \to M$ is a map*

$$D : \Gamma(E) \to \Gamma(T^*(M) \otimes E)$$

which satisfies the following conditions:

1) For any $s_1, s_2 \in \Gamma(E)$,

$$D(s_1 + s_2) = Ds_1 + Ds_2 . \tag{35.1}$$

2) For any $s \in \Gamma(E)$ and $\alpha \in C^\infty(M)$,

$$D(\alpha s) = d\alpha \otimes s + \alpha Ds . \tag{35.2}$$

*Ds is also called the **covariant derivative** of the section s.*

Exercise 35.1 Show that D is a linear map.

Definition 35.2. *Suppose X is a smooth tangent field on M and $s \in \Gamma(E)$. The **directional covariant derivative** of s along X is the section given by*

$$D_X s = \langle X, Ds \rangle \in \Gamma(E) . \tag{35.3}$$

The directional covariant derivative satisfies the following properties. Let X and Y be any two smooth tangent vector fields on M, $s, s_1, s_2 \in \Gamma(E)$, and $\alpha \in C^\infty(M)$. Then

$$
\begin{array}{llr}
1) & D_{X+Y}\, s = D_X\, s + D_Y\, s \,; & (35.4) \\
2) & D_{\alpha X}\, s = \alpha D_X\, s \,; & (35.5) \\
3) & D_X\, (s_1 + s_2) = D_X\, s_1 + D_X\, s_2 \,; & (35.6) \\
4) & D_X\, (\alpha s) = (X\alpha)s + \alpha D_X\, s \,. & (35.7)
\end{array}
$$

Exercise 35.2 Prove the above four properties.

In order to understand the concept of a connection more concretely, we will use local coordinates. Suppose the typical fiber of the vector bundle $\pi : E \to M$ is an m-dimensional vector space and M is an n-dimensional manifold. Consider a coordinate neighborhood $(U; x^i)$ with local coordinates x^i, $i = 1, \ldots, n$. Choose m smooth sections s_α, $\alpha = 1, \ldots, m$, which are linearly independent everywhere in U. Such a set of m sections is called a **local frame field** of E on U. With this setup, the set $\{dx^i \otimes s_\alpha, 1 \le i \le n, 1 \le \alpha \le m\}$ forms a basis of $T_p^* \otimes E_p$ for all points $p \in U$, E_p being the fiber above the point $p \in U$. We can thus write

$$Ds_\alpha = A^\beta_{\alpha i}\, dx^i \otimes s_\beta , \tag{35.8}$$

where $A^\beta_{\alpha i} \in C^\infty(U)$ are smooth functions on U. Denote

$$\omega^\beta_\alpha = A^\beta_{\alpha i}\, dx^i , \quad 1 \le \alpha, \beta \le m . \tag{35.9}$$

Then (35.8) becomes

$$Ds_\alpha = \omega^\beta_\alpha \otimes s_\beta , \tag{35.10}$$

where ω^β_α is a matrix of 1-forms, or equivalently, a matrix-valued 1-form. The matrix (ω^β_α) is called the **connection matrix**, and depends on the choice of the local frame field. We can also write (35.10) as the matrix equation

$$DS = \omega \otimes S , \tag{35.11}$$

where

$$
S \equiv \begin{pmatrix} s_1 \\ \vdots \\ s_m \end{pmatrix} , \quad \text{and} \quad \omega = \begin{pmatrix} \omega^1_1 & \omega^2_1 & \cdots & \omega^m_1 \\ & \vdots & & \\ \omega^1_m & \omega^2_m & \cdots & \omega^m_m \end{pmatrix} . \tag{35.12}
$$

Since (ω_α^β) depends on the choice of the local frame field, it is important to establish its transformation properties under a change of local frame fields. Suppose $S' = ((s')_1, \ldots, (s')_m)^t$ (where t denotes the transpose) is related to S by the **gauge transformation**

$$S' = gS, \qquad \text{or } (s')_\alpha = g_\alpha^\beta s_\beta, \tag{35.13}$$

where

$$g = \begin{pmatrix} g_1^1 & \cdots & g_1^m \\ & \vdots & \\ g_m^1 & \cdots & g_m^m \end{pmatrix} \tag{35.14}$$

is an invertible matrix of smooth functions on U $[g_i^j(x) \in C^\infty(U)$ and $\det(g) \neq 0]$. Suppose the connection matrix of 1-forms with respect to S' is $((\omega')_\alpha^\beta)$:

$$DS' = \omega' \otimes S'. \tag{35.15}$$

On the other hand,

$$\begin{aligned} DS' = D(gS) &= dg \otimes S + g DS \\ &= (dg + g \cdot \omega) \otimes S = (dg \cdot g^{-1} + g \cdot \omega \cdot g^{-1}) \otimes S'. \end{aligned} \tag{35.16}$$

Comparing with (35.15), we obtain the following transformation rule for a connection matrix:

$$\boxed{\omega' = dg\, g^{-1} + g\omega\, g^{-1}} \quad . \tag{35.17}$$

Because of the term $dg\, g^{-1}$, which is in general non-vanishing, we see that (ω_α^β) does not transform like a $(1,1)$-type tensor [whose transformation rule only involves the second term on the RHS of (35.17)].

We have the following basic theorem (stated without proof).

Theorem 35.1. *A connection always exists on a vector bundle.*

Given a connection D, the directional covariant derivatives D_X and D_Y for arbitrary tangent vector fields X and Y on M in general fail to commute. To measure the degree of non-commutativity of covariant derivatives, one introduces a matrix-valued 2-form on M, called the curvature matrix, defined as follows.

Definition 35.3. *Given a connection D on a vector bundle $\pi : E \to M$ with connection matrix ω (a matrix-valued 1-form), the **curvature matrix** of two-forms Ω is defined by*

$$\boxed{\Omega \equiv d\omega - \omega \wedge \omega} \quad . \tag{35.18}$$

Note that the above exterior product is also a matrix multiplication, that is,

$$(\omega \wedge \omega)^\alpha_\beta = \omega^\gamma_\beta \wedge \omega^\alpha_\gamma . \tag{35.19}$$

Let us first work out the transformation properties of Ω under a change of local frame fields [given by (35.13)]. Under such a change, ω transforms as (35.17), so that

$$\omega' g = dg + g\omega . \tag{35.20}$$

On exteriorly differentiating this equation, we have

$$d\omega' g - \omega' \wedge dg = dg \wedge \omega + g \, d\omega . \tag{35.21}$$

From (35.20),

$$dg = \omega' g - g\omega . \tag{35.22}$$

On plugging this result into (35.21), we obtain

$$d\omega' g - \omega' \wedge (\omega' g - g\omega) = (\omega' g - g\omega) \wedge \omega + g \, d\omega , \tag{35.23}$$

or

$$(d\omega' - \omega' \wedge \omega') g = g (d\omega - \omega \wedge \omega) . \tag{35.24}$$

Thus, under a change of local frame fields,

$$\Omega' = g\Omega g^{-1} . \tag{35.25}$$

It is seen that the curvature matrix of 2-forms transforms like a type $(1, 1)$-tensor [similarity transformation of matrices [c.f. (1.42)]].

At this point, it is worthwhile to point out a subtlety in our notation which is potentially confusing. Recall that the matrix g determining the change of local frame fields is given [(35.13)] by $(s')_\alpha = g^\beta_\alpha s_\beta$. We also write this as a left multiplication of S by g ($S' = gS$). But we could also have written the same equation as a right multiplication ($S' = Sg$) had we interpreted g^β_α as the element in the β-th row and α-th column of the matrix g, and at the same time written S and S' as row matrices (the reader should verify this). Eq. (35.11) would also have appeared as $DS = S \otimes \omega$. In this case, a similar procedure to the derivation of (35.17) would have yielded $\omega' = g^{-1}\omega g + g^{-1} dg$. An analogous procedure to that leading to (35.24) would then give $g (d\omega' + \omega' \wedge \omega') = (d\omega + \omega \wedge \omega) g$, forcing us to identify the curvature matrix as $\Omega = d\omega + \omega \wedge \omega$ [compare with (35.18)] in order for it to satisfy the expected tensorial transformation property: $\Omega' = g^{-1}\Omega g$. The convention with Ω defined with the plus sign on the RHS of (35.18) is also used quite commonly in the literature.

Given two arbitrary tangent vector fields X and Y on M, and a connection D on $\Gamma(E)$, one can define a **curvature operator**

$$F(X, Y) : \Gamma(E) \to \Gamma(E)$$

[based on the curvature 2-form Ω (35.18)] as follows. Consider a coordinate neighborhood $(U; x^i)$ and $x \in U$. Let $s_x \in \pi^{-1}(x)$ be a point in the fiber above

x, which is an m-dimensional vector space. With respect to a local frame field $S_U = (s_1, \ldots, s_m)^t$ in U, write

$$s_x = \lambda^\alpha \, s_\alpha|_x \,, \qquad \lambda^\alpha \in \mathbb{R} \,. \tag{35.26}$$

Define the map $F(X_x, Y_x) : \pi^{-1}(x) \to \pi^{-1}(x)$ by

$$F(X_x, Y_x) \, s_x \equiv \lambda^\alpha \left(\langle X \wedge Y, \Omega_\alpha^\beta \rangle \, s_\beta \right)\big|_x \,, \tag{35.27a}$$

where the pairing $\langle \, , \, \rangle$ has been given in Theorem 30.3 [Eq. (30.39)] and (28.28). We then define, in a pointwise fashion,

$$(F(X, Y) \, s)(x) = F(X_x, Y_x) \, s_x \,. \tag{35.27b}$$

The curvature operator $F(X, Y)$ has the following properties:

$$
\begin{aligned}
1) \quad & F(X, Y) = -F(Y, X) \,, & (35.28) \\
2) \quad & F(fX, Y) = fF(X, Y) \,, & (35.29) \\
3) \quad & F(X, Y)(fs) = f(F(X, Y)s) \,, & (35.30)
\end{aligned}
$$

where $X, Y \in \Gamma(T(M))$, $f \in C^\infty(M)$ and $s \in \Gamma(E)$.

Exercise 35.3   Prove the above three properties.

Exercise 35.4   Show that for $f, g \in C^\infty(M)$,

$$F(fX, gY) = (fg)F(X, Y) \,. \tag{35.31}$$

We say that $F(X, Y)$ is $C^\infty(M)$-bilinear in X and Y.

We now establish the fact which shows explicitly that the curvature of a connection measures the degree of non-commutativity of directional covariant derivatives.

Theorem 35.2. *Let X and Y be two arbitrary smooth tangent vector fields on a manifold M, and D be a given connection on the vector bundle $\pi : E \to M$. Then*

$$\boxed{F(X, Y) = D_X D_Y - D_Y D_X - D_{[X,Y]}} \,. \tag{35.32}$$

Proof. We need only prove the above equation in a local coordinate neighborhood $(U; x^i)$. Let $s \in \Gamma(E)$ be written as $s = \lambda^\alpha s_\alpha$ under a local choice of frame fields. Then, by (35.3), (noting that λ^α are functions on U),

$$D_X s = \langle X, Ds \rangle = \langle X, (d\lambda^\alpha)s_\alpha + \lambda^\alpha \, \omega_\alpha^\beta s_\beta \rangle = (X\lambda^\alpha + \lambda^\beta \langle X, \omega_\beta^\alpha \rangle) s_\alpha \,, \tag{35.33}$$

and

$$
\begin{aligned}
D_Y D_X s &= \{Y(X\lambda^\alpha) + Y(\lambda^\beta \langle X, \omega_\beta^\alpha \rangle) + (X\lambda^\gamma + \lambda^\beta \langle X, \omega_\beta^\gamma \rangle)\langle Y, \omega_\gamma^\alpha \rangle\} s_\alpha \\
&= \{Y(X\lambda^\alpha) + ((X\lambda^\beta)\langle Y, \omega_\beta^\alpha \rangle + (Y\lambda^\beta)\langle X, \omega_\beta^\alpha \rangle) \\
&\quad + \lambda^\beta (Y\langle X, \omega_\beta^\alpha \rangle + \langle X, \omega_\beta^\gamma \rangle\langle Y, \omega_\gamma^\alpha \rangle)\} s_\alpha \,.
\end{aligned}
\tag{35.34}
$$

Hence

$$(D_X D_Y - D_Y D_X) s = \{ [X,Y] \lambda^\alpha + \lambda^\beta (\langle [X,Y], \omega_\beta^\alpha \rangle$$
$$+ \langle X \wedge Y, d\omega_\beta^\alpha - \omega_\beta^\gamma \wedge \omega_\gamma^\alpha \rangle) \} s_\alpha = D_{[X,Y]} s + \lambda^\beta \langle X \wedge Y, \Omega_\beta^\alpha \rangle s_\alpha , \quad (35.35)$$

where (30.39) and (28.28) have been used. The theorem follows from the definition given by (35.27).   □

Note that even if the curvature $\Omega = 0$ (flat space), D_X and D_Y do not necessarily commute, unless X and Y do.

Any curvature matrix of 2-forms Ω satisfies an important identity, called the **Bianchi identity**:

$$\boxed{d\Omega = \omega \wedge \Omega - \Omega \wedge \omega} . \qquad (35.36)$$

This is easily established by exterior differentiation of both sides of the defining equation for Ω [(35.18)]. Indeed,

$$d\Omega = d^2\omega - (d\omega \wedge \omega - \omega \wedge d\omega)$$
$$= -(\Omega + \omega \wedge \omega) \wedge \omega + \omega \wedge (\Omega + \omega \wedge \omega) = \omega \wedge \Omega - \Omega \wedge \omega .$$

In the context of gauge field theories in physics, the connection is referred to as the **vector potential** and the curvature as the **force**. The relationship between these, given abstractly by (35.18) and a bit less so by (35.32), can be exhibited more concretely in terms of local coordinates, which is the form usually presented in the physics literature. Let us use the (physics) convention that the Greek indices $(\mu, \nu, \dots)$ label the base manifold (spacetime) coordinates, while the Latin indices $(i, j, \dots)$ label the fiber (internal degrees of freedom) coordinates. Suppose $X = X^\mu \partial_\mu$ and $Y = Y^\nu \partial_\nu$ in a coordinate neighborhood $(U; x^i)$. Then, by the bilinear property of F [(35.31)], we can write

$$F(X,Y) = X^\mu Y^\nu F(\partial_\mu, \partial_\nu) \equiv X^\mu Y^\nu F_{\mu\nu} , \qquad (35.37)$$

where

$$F_{\mu\nu} \equiv F(\partial_\mu, \partial_\nu) = D_{\partial_\mu} D_{\partial_\nu} - D_{\partial_\nu} D_{\partial_\mu} - D_{[\partial_\mu, \partial_\nu]} = D_{\partial_\mu} D_{\partial_\nu} - D_{\partial_\nu} D_{\partial_\mu} , \quad (35.38)$$

since $[\partial_\mu, \partial_\nu] = 0$. On the other hand, with a choice of local frame fields in U: $\{ s_i, i = 1, \dots, m \}$, the directional covariant derivative D_{∂_μ} is given by [c.f. (35.8)]

$$D_{\partial_\mu} s_i = A_{i\mu}^j s_j , \qquad (35.39)$$

where $A_{i\mu}^j \in C^\infty(U)$ are smooth functions on U. Eq. (35.39) then implies

$$F_{\mu\nu} s_i = D_{\partial_\mu} D_{\partial_\nu} s_i - D_{\partial_\nu} D_{\partial_\mu} s_i = D_{\partial_\mu} (A_{i\nu}^j s_j) - D_{\partial_\nu} (A_{i\mu}^j s_j)$$
$$= (\partial_\mu A_{i\nu}^j) s_j + A_{i\nu}^j A_{j\mu}^k s_k - (\partial_\nu A_{i\mu}^j) s_j - A_{i\mu}^j A_{j\nu}^k s_k \qquad (35.40)$$
$$= ((\partial_\mu A_{i\nu}^j - \partial_\nu A_{i\mu}^j) + (A_{i\nu}^k A_{k\mu}^j - A_{i\mu}^k A_{k\nu}^j)) s_j ,$$

where in the fourth equality we have relabeled indices $(j \leftrightarrow k)$ in the quadratic terms in A. In the above equation $F_{\mu\nu}$ appears as a linear operator on s_i. With respect to the choice of frame fields $\{s_i\}$, we can then write

$$F_{\mu\nu} s_i = F_{\mu\nu i}{}^j s_j , \tag{35.41}$$

where $F_{\mu\nu i}{}^j \in C^\infty(U)$ are smooth functions on U, and are called the components of the curvature with respect to the local coordinates x^μ and the local frame field $\{s_i\}$. Comparing (35.40) with (35.41) we have

$$F_{\mu\nu i}{}^j = \partial_\mu A_{i\nu}^j - \partial_\nu A_{i\mu}^j + A_{i\nu}^k A_{k\mu}^j - A_{i\mu}^k A_{k\nu}^j . \tag{35.42}$$

Equivalently, in matrix form,

$$\boxed{F_{\mu\nu} = \partial_\mu A_\nu - \partial_\nu A_\mu - [\, A_\mu, A_\nu \,]} \quad . \tag{35.43}$$

This is the equation that one most frequently encounters in the physics literature, in which F represents the **field strength** and A the **vector potential**, in the context of **non-abelian gauge field theories**. In abelian cases $([\, A_\mu, A_\nu \,] = 0)$, the above equation reduces to

$$F_{\mu\nu} = \partial_\mu A_\nu - \partial_\nu A_\mu , \tag{35.44}$$

which is precisely the relationship between the field tensor and the vector potential in Maxwell's electrodynamics. Note that $F_{\mu\nu}$ in (35.38) is related to the original definition of the curvature matrix of 2-forms Ω [(35.18)] by

$$\Omega = \frac{1}{2} F_{\mu\nu} \, dx^\mu \wedge dx^\nu , \tag{35.45}$$

where the factor $1/2$ takes care of the skew-symmetry of $F_{\mu\nu}$ and the wedge product.

$\boxed{\text{Exercise 35.5}}$ Check that (35.45) agrees with (35.27a) and (35.27b).

Given a connection D on a vector bundle $\pi : E \to M$, one can construct the so-called **induced connection** on the **dual bundle** $\pi : E^* \to M$ (also denoted by D). Suppose $s \in \Gamma(E)$ and $s^* \in \Gamma(E^*)$ are smooth sections on the respective bundles, so that the pairing $\langle\, s, s^* \,\rangle$ is a smooth function on M. Then the induced connection D on E^* is determined by the equation

$$d\langle\, s, s^* \,\rangle = \langle\, Ds, s^* \,\rangle + \langle\, s, Ds^* \,\rangle . \tag{35.46}$$

Notice that each term in the above equation is a 1-form on M. Suppose we choose a local frame field $\{s_i\}$ with dual frame field $\{(s^*)^j\}$, $i, j = 1, \ldots, m$, so that

$$\langle\, s_i, (s^*)^j \,\rangle = \delta_i^j . \tag{35.47}$$

Let [c.f. (35.10)]

$$Ds_i = \omega_i^j \otimes s_j ,$$ (35.48)

and

$$D(s^*)^j = (\omega^*)_i^j \otimes (s^*)^i .$$ (35.49)

Then (35.46) implies

$$d(\delta_i^j) = 0 = \langle\, Ds_i, (s^*)^j \,\rangle + \langle\, s_i, D(s^*)^j \,\rangle$$
$$= \omega_i^k \langle\, s_k, (s^*)^j \,\rangle + (\omega^*)_k^j \langle\, s_i, (s^*)^k \,\rangle = \omega_i^j + (\omega^*)_i^j . \quad (35.50)$$

Thus

$$(\omega^*)_i^j = -\omega_i^j ,$$ (35.51)

or

$$D(s^*)^j = -\omega_i^j \otimes (s^*)^i .$$ (35.52)

Given two vector bundles $\pi : E_1 \to M$ and $\pi : E_2 \to M$, one can construct the bundles $\pi : E_1 \oplus E_2 \to M$ and $\pi : E_1 \otimes E_2 \to M$. Then, for $s_1 \in \Gamma(E_1)$, $s_2 \in \Gamma(E_2)$, $s_1 \oplus s_2$ and $s_1 \otimes s_2$ are sections on $E_1 \oplus E_2$ and $E_1 \otimes E_2$, respectively. If connections are separately given on E_1 and E_2 (denoted by the same symbol D), then one has the induced connections on $E_1 \oplus E_2$ and $E_1 \otimes E_2$ given by

$$D(s_1 \oplus s_2) \equiv Ds_1 \oplus Ds_2 ,$$ (35.53)
$$D(s_1 \otimes s_2) \equiv Ds_1 \otimes s_2 + s_1 \otimes Ds_2 .$$ (35.54)

Exercise 35.6   Check that the above definitions for the induced connections $E_1 \oplus E_2$ and $E_1 \otimes E_2$ satisfy the conditions 1) and 2) for any connection (given in Def. 35.1).

Given a vector bundle $\pi : E \to M$ and a connection D on E, let us now define the covariant derivatives of section-valued [$\Gamma(E)$-valued] r-forms [$\in \Gamma(E) \otimes A^r(M)$] and matrix-valued [$\Gamma(E \otimes E^*)$-valued] r-forms [$\in \Gamma(E \otimes E^*) \otimes A^r(M)$]. [A matrix can be regarded as the matrix representation (with respect to a particular choice of local frame fields) of a linear operator on $\Gamma(E)$, and is thus an element of $\Gamma(E \otimes E^*)$.] Let $\lambda \in \Gamma(E) \otimes A^r(M)$, and write

$$\lambda = \lambda_{\mu_1 \dots \mu_r} \otimes dx^{\mu_1} \wedge \dots \wedge dx^{\mu_r} ,$$ (35.55)

where $\lambda_{\mu_1 \dots \mu_r} \in \Gamma(E)$. Then we define

$$D\lambda \equiv D(\lambda_{\mu_1 \dots \mu_r}) \wedge dx^{\mu_1} \wedge \dots \wedge dx^{\mu_r} .$$ (35.56)

Let

$$\lambda_{\mu_1 \dots \mu_r} = \alpha^i_{\{\mu_1 \dots \mu_r\}} s_i ,$$ (35.57)

Then, by (35.3),

$$D\lambda_{\mu_1 \dots \mu_r} = d\alpha^i_{\{\mu_1 \dots \mu_r\}} s_i + \alpha^i_{\{\mu_1 \dots \mu_r\}} Ds_i = d\alpha^i_{\{\mu_1 \dots \mu_r\}} s_i + \alpha^i_{\{u_1 \dots \mu_r\}} \omega_i^j \otimes s_j .$$

Thus

$$
\begin{aligned}
D\lambda &= d\alpha^i_{\{\mu_1\dots\,\mu_r\}} \wedge dx^{\mu_1} \wedge \cdots \wedge dx^{\mu_r} \otimes s_i \\
&\quad + \omega^j_i \wedge (\alpha^i_{\{\mu_1\dots\,\mu_r\}} s_j \otimes dx^{\mu_1} \wedge \cdots \wedge dx^{\mu_r}) \\
&= d\alpha^i_{\{\mu_1\dots\,\mu_r\}} \wedge dx^{\mu_1} \wedge \cdots \wedge dx^{\mu_r} \otimes s_i \\
&\quad + (-1)^r \, \alpha^i_{\{\mu_1\dots\,\mu_r\}} (dx^{\mu_1} \wedge \cdots \wedge dx^{\mu_r}) \wedge \omega^j_i \otimes s_j \ .
\end{aligned}
\tag{35.58}
$$

We will abbreviate the above by writing the matrix equation

$$
\boxed{\,D\lambda = d\lambda + (-1)^r \lambda \wedge \omega \ , \quad \lambda \in \Gamma(E) \otimes A^r(M)\,}
\tag{35.59}
$$

In the above equation λ is a row matrix of r-forms:

$$
\lambda \equiv (\alpha^1_{\{\mu_1\dots\,\mu_r\}}, \dots, \alpha^m_{\{\mu_1\dots\,\mu_r\}}) \, dx^{\mu_1} \wedge \cdots \wedge dx^{\mu_r} \ ,
$$

and $\lambda \wedge \omega$ means both the wedge product as well as matrix multiplication.

Now let $\eta \in \Gamma(E \otimes E^*) \otimes A^r(M)$, and write

$$
\eta = \eta_{\mu_1\dots\,\mu_r} \otimes dx^{\mu_1} \wedge \cdots \wedge dx^{\mu_r} \ ,
\tag{35.60}
$$

where $\eta_{\mu_1\dots\,\mu_r}$ is an element in $\Gamma(E \otimes E^*)$ given by

$$
\eta_{\mu_1\dots\,\mu_r} = \alpha^i_{\{\mu_1\dots\,\mu_r\}} s_i \otimes \beta_{j\,\{\mu_1\dots\,\mu_r\}} (s^*)^j \ ,
\tag{35.61}
$$

where $\alpha^i_{\{\mu_1\dots\,\mu_r\}}$ and $\beta_{j\,\{\mu_1\dots\,\mu_r\}}$ are all smooth functions on M. We then define

$$
D\eta \equiv D(\eta_{\mu_1\dots\,\mu_r}) \wedge dx^{\mu_1} \wedge \cdots \wedge dx^{\mu_r} \ .
\tag{35.62}
$$

By (35.54),

$$
\begin{aligned}
D(\eta_{\mu_1\dots\,\mu_r}) &= D(\alpha^i_{\{\mu_1\dots\,\mu_r\}} s_i) \otimes \beta_{j\,\{\mu_1\dots\,\mu_r\}} (s^*)^j \\
&\quad + \alpha^i_{\{\mu_1\dots\,\mu_r\}} s_i \otimes D(\beta_{j\,\{\mu_1\dots\,\mu_r\}} (s^*)^j) \\
&= \{(d\alpha^i_{\{\mu_1\dots\,\mu_r\}}) \otimes s_i + \alpha^i_{\{\mu_1\dots\,\mu_r\}} \omega^k_i \otimes s_k\} \otimes \beta_{j\,\{\mu_1\dots\,\mu_r\}} (s^*)^j \\
&\quad + \alpha^i_{\{\mu_1\dots\,\mu_r\}} s_i \otimes \{d\beta_{j\,\{\mu_1\dots\,\mu_r\}} \otimes (s^*)^j - \beta_{j\,\{\mu_1\dots\,\mu_r\}} \omega^j_k (s^*)^k\} \\
&= d(\alpha^i_{\{\mu_1\dots\,\mu_r\}} \beta_{j\,\{\mu_1\dots\,\mu_r\}}) s_i \otimes (s^*)^j \\
&\quad + \alpha^i_{\{\mu_1\dots\,\mu_r\}} \beta_{j\,\{\mu_1\dots\,\mu_r\}} \{\omega^k_i \otimes s_k \otimes (s^*)^j - s_i \otimes (s^*)^k \otimes \omega^j_k\} \ .
\end{aligned}
\tag{35.63}
$$

Thus

$$
\begin{aligned}
D\eta &= d(\alpha^i_{\{\mu_1\dots\,\mu_r\}} \beta_{j\,\{\mu_1\dots\,\mu_r\}}) \, s_i \otimes (s^*)^j \wedge dx^{\mu_1} \wedge \cdots \wedge dx^{\mu_r} \\
&\quad + \alpha^i_{\{\mu_1\dots\,\mu_r\}} \beta_{j\,\{\mu_1\dots\,\mu_r\}} \{(-1)^r (s_k \otimes (s^*)^j) \wedge dx^{\mu_1} \wedge \cdots \wedge dx^{\mu_r} \wedge \omega^k_i \\
&\quad - \omega^j_k \wedge (s_i \otimes (s^*)^k) \otimes dx^{\mu_1} \wedge \cdots \wedge dx^{\mu_r}\} \ ,
\end{aligned}
\tag{35.64}
$$

where the factor $(-1)^r$ comes from commuting $(\omega)^k_i$ (a 1-form) and the r-form $dx^{\mu_1} \wedge \cdots \wedge dx^{\mu_r}$.

Let us define the **graded commutator** of two matrix-valued forms η_1 and η_2 by

$$\boxed{[\,\eta_1, \eta_2\,]_{\mathrm{gr}} \equiv \eta_1 \wedge \eta_2 - (-1)^{r_1 r_2}\, \eta_2 \wedge \eta_1} \quad , \qquad (35.65)$$

where η_1 is an r_1-form and η_2 is an r_2-form. Then, similar to (35.59), we can abbreviate (35.64) by writing

$$\boxed{D\eta = d\eta - [\,\omega, \eta\,]_{\mathrm{gr}}\,, \qquad \eta \in \Gamma(E \otimes E^*) \otimes A^r(M)}\,, \qquad (35.66)$$

where, again, $\omega \wedge \eta$ and $\eta \wedge \omega$ involve matrix multiplication as well as wedge multiplication.

The order of ω and η in the graded commutator requires some explanation. Consider the term $\alpha^i_{\{\mu_1\dots\,\mu_r\}} \beta_{j\,\{\mu_1\dots\,\mu_r\}}\, \omega^j_k \wedge (s_i \otimes (s^*)^k) \otimes dx^{\mu_1} \wedge \cdots \wedge dx^{\mu_r}$ in the RHS of (35.64). We can write this as

$$\mathcal{A} = \alpha^i \beta_j\, s_i \otimes (s^*)^k\, \omega^j_k \wedge \xi\,,$$

where ξ is the r-form $dx^{\mu_1} \wedge \cdots \wedge dx^{\mu_r}$, and the designation $\{\mu_1 \dots \mu_r\}$ in α^i and β_j has been neglected. $\mathcal{A}$ can be considered as a map $A : \Gamma(E) \to \Gamma(E) \otimes A^{r+1}(M)$. Thus

$$\begin{aligned}
\mathcal{A}s_l = A^i_l s_i &= \alpha^i \beta_j s_i \langle\, s_l, (s^*)^k \,\rangle\, \omega^j_k \wedge \xi = \alpha^i \beta_j s_i \delta^k_l\, \omega^j_k \wedge \xi \\
&= \omega^j_l \wedge (\beta_j \alpha^i \xi) \otimes s_i = (\omega^j_l \wedge \eta^i_j) \otimes s_i = (\omega \wedge \eta)^i_l s_i\,,
\end{aligned}$$

where

$$\eta^i_j \equiv \beta_j \alpha^i \xi = \beta_{j\,\{\mu_1\dots\,\mu_r\}} \alpha^i_{\{\mu_1\dots\,\mu_r\}}\, dx^{\mu_1} \wedge \cdots \wedge dx^{\mu_r}\,.$$

A similar consideration applies to the term involving $(-1)^r$ in (35.64).

On applying (35.66) to the curvature matrix of 2-forms Ω, we see that

$$\begin{aligned}
D\Omega = d\Omega - [\,\omega, \Omega\,]_{\mathrm{gr}} &= d\Omega - (\omega \wedge \Omega - (-1)^2 \Omega \wedge \omega) \\
&= d\Omega - (\omega \wedge \Omega - \Omega \wedge \omega)\,.
\end{aligned} \qquad (35.67)$$

Thus the Bianchi identity (35.36) can be stated more succinctly as

$$\boxed{D\Omega = 0} \quad . \qquad (35.68)$$

Note that using the graded commutator, the curvature Ω can also be written as

$$\boxed{\Omega = d\omega - \frac{1}{2}[\,\omega, \omega\,]_{\mathrm{gr}}} \quad . \qquad (35.69)$$

To continue with the calculus of the covariant derivative, we will use (35.59) to calculate $D^2\lambda$, where $\lambda \in \Gamma(E) \otimes A^r(M)$. We have

$$D(D\lambda) = d(D\lambda) + (-1)^{r+1}(D\lambda) \wedge \omega$$
$$= d(d\lambda + (-1)^r \lambda \wedge \omega) + (-1)^{r+1}(d\lambda + (-1)^r \lambda \wedge \omega) \wedge \omega$$
$$= (-1)^r((d\lambda) \wedge \omega + (-1)^r \lambda \wedge d\omega) + (-1)^{r+1}(d\lambda) \wedge \omega - \lambda \wedge (\omega \wedge \omega)$$
$$= \lambda \wedge (d\omega - \omega \wedge \omega) .$$

$$(35.70)$$

It follows from (35.18) that

$$\boxed{D^2\lambda = \lambda \wedge \Omega , \qquad \lambda \in \Gamma(E) \otimes A^r(M)} \quad . \qquad (35.71)$$

Exercise 35.7 Obtain the result (35.71) again by starting with

$$D^2\lambda = D(d\lambda + (-1)^r \lambda \wedge \omega) = D(d\lambda) + (-1)^r D(\lambda \wedge \omega) ,$$

and realizing that $d\lambda$ and $\lambda \wedge \omega$ are both elements in $\Gamma(E) \otimes A^{r+1}(M)$, that is, section-valued $(r+1)$-forms.

Covariantly differentiating (35.66) we have

$$D^2\eta = D(d\eta - \omega \wedge \eta + (-1)^r \eta \wedge \omega) = D(d\eta) - D(\omega \wedge \eta) + (-1)^r D(\eta \wedge \omega)$$
$$= -[\omega, d\eta]_{\text{gr}} - (d(\omega \wedge \eta) - [\omega, \omega \wedge \eta]_{\text{gr}}) + (-1)^r (d(\eta \wedge \omega) - [\omega, \eta \wedge \omega]_{\text{gr}})$$
$$= \eta \wedge (d\omega - \omega \wedge \omega) - (d\omega - \omega \wedge \omega) \wedge \eta .$$

$$(35.72)$$

Hence

$$\boxed{D^2\eta = [\eta, \Omega] , \qquad \eta \in \Gamma(E \otimes E^*) \otimes A^r(M)} \quad . \qquad (35.73)$$

Note that since Ω is a 2-form, the commutator in the above equation is equal to the graded commutator.

The connection on a vector bundle $\pi : E \to M$ leads to the useful notion of the parallel displacement of a vector field along a curve C in M.

Definition 35.4. *Let D be a connection on $\pi : E \to M$ and C be a parametrized curve in M. If a section $s \in \Gamma(E)$ satisfies*

$$D_X s = 0 \qquad (35.74)$$

*at every point on C, where X is the tangent vector field along C, then s is said to be **parallel** along the curve C.*

Let the curve C be given in a local coordinate neighborhood $(U; x^\mu)$ by

$$x^\mu = x^\mu(t), \qquad 1 \le \mu \le n. \tag{35.75}$$

Then the tangent vector field X along C is

$$X = \frac{dx^\mu}{dt} \frac{\partial}{\partial x^\mu}. \tag{35.76}$$

Choose a local frame field $\{s_i\}$ on U such that $s = \lambda^i s_i$. Then (35.74) is equivalent to

$$D_X s = \langle X, Ds \rangle = \left\langle \frac{dx^\mu}{dt} \partial_\mu, (d\lambda^i)s_i + \lambda^i A^j_{i\nu} dx^\nu \otimes s_j \right\rangle$$

$$= \frac{\partial \lambda^i}{\partial x^\mu} \frac{dx^\mu}{dt} s_i + A^i_{j\nu} \delta^\nu_\mu \frac{dx^\mu}{dt} \lambda^j \otimes s_i = \left(\frac{d\lambda^i}{dt} + A^i_{j\mu} \frac{dx^\mu}{dt} \lambda^j \right) s_i = 0. \tag{35.77}$$

Since the s_i are independent, the above equation implies the following system of equations:

$$\frac{d\lambda^i}{dt} + A^i_{j\mu} \frac{dx^\mu}{dt} \lambda^j = 0, \qquad 1 \le i \le m. \tag{35.78}$$

Given a set of initial conditions $\lambda^i(0)$, that is, $v \in E_{x(0)} = \pi^{-1}(x(0))$, a unique solution exists (based on the existence theorems of ordinary differential equations). The vector field $s(t) = \lambda^i(t)s_i$ along the curve C is called the **parallel displacement** of the vector $v = \lambda^i(0)s_i$ along the curve C. This notion has many specific applications in physics. In particular, it leads to the notion of holonomy.

Definition 35.5. *Let C be a smooth parametrized curve in a manifold M, and D be a given connection on the vector bundle $\pi : E \to M$. Suppose $C(t_0) = x$ and $C(t_1) = y$ are points on the curve C in M. Then the linear map $H(C; D) : \pi^{-1}(x) \to \pi^{-1}(y)$ given by*

$$H(C; D)(v) = w, \tag{35.79}$$

*where v is some vector in $\pi^{-1}(x)$, and w is the result of parallely displacing v along the curve C from x to y, is called the **holonomy map** from x to y along the curve C.*

Exercise 35.8   Show that the holonomy map $H(C; D)$ is independent of the parametrization of the curve C.

Theorem 35.3. *The parallel displacement, and hence the holonomy map, is invariant under a gauge transformation of local frame fields.*

Proof. Eq. (35.78) describing the parallel displacement of $v(t) = v^i(t)s_i$ (with respect to the local frame field $\{s_i\}$) can be written as

$$\frac{d}{dt} v^i = -\frac{dx^\mu}{dt} A^i_{j\mu} v^j , \qquad (35.80)$$

with $v(t_0) = v$ and $v(t_1) = w$. Now apply the gauge transformation $(s')_i = g^j_i s_j$, so that

$$v^i(t)s_i = (v')^i(t)(s')_i ,$$

where $(v')^i(t) = (g^{-1})^i_j(C(t))v^j(t)$. [Consult Table 1.1.] We then have

$$
\begin{aligned}
\frac{d}{dt}(v')^i(t) &= \left(\frac{d}{dt}(g^{-1})^i_j(C(t))\right)v^j(t) + (g^{-1})^i_j(C(t))\frac{dv^j(t)}{dt} \\
&= \frac{\partial(g^{-1})^i_j}{\partial x^\mu}\frac{dx^\mu}{dt} v^j - (g^{-1})^i_j\frac{dx^\mu}{dt} A^j_{k\mu} v^k \\
&= \frac{\partial(g^{-1})^i_j}{\partial x^\mu}\frac{dx^\mu}{dt} g^j_l(v')^l - (g^{-1})^i_j\frac{dx^\mu}{dt} A^j_{k\mu} g^k_l(v')^l \\
&= -\frac{dx^\mu}{dt}\left(g^k_l A^j_{k\mu}(g^{-1})^i_j - \frac{\partial(g^{-1})^i_j}{\partial x^\mu} g^j_l\right)(v')^l .
\end{aligned}
\qquad (35.81)
$$

Since $g^j_l(g^{-1})^i_j = \delta^i_l$,

$$\frac{\partial(g^{-1})^i_j}{\partial x^\mu} g^j_l = -\frac{\partial g^j_l}{\partial x^\mu}(g^{-1})^i_j .$$

Hence (35.81) implies

$$\frac{d(v')^i}{dt} = -\frac{dx^\mu}{dt}(A')^i_{l\mu}(v')^l , \qquad (35.82)$$

where

$$(A')^i_{l\mu} \equiv \frac{\partial g^j_l}{\partial x^\mu}(g^{-1})^i_j + g^k_l A^j_{k\mu}(g^{-1})^i_j , \qquad (35.83)$$

which agrees precisely with the gauge transformation rule for the connection matrix of 1-forms given by (35.17). Thus (35.80), that is, the process of parallel displacement, is invariant with respect to a gauge transformation. $\qquad\square$

With respect to a local frame field $\{s_i\}$, let us write the matrix representation of the holonomy map $H(C; D) : \pi^{-1}(C(t_0)) \to \pi^{-1}(C(t_1))$ by

$$H s_i = H^j_i s_j . \qquad (35.84)$$

With respect to the gauge-transformed frame field $(s')_i = g^j_i(C(t))s_j$, we have

$$H(s')_i = (H')^j_i(s')_j . \qquad (35.85)$$

Thus

$$H g^j_i(C(t_0))s_j = (H')^j_i g^k_j(C(t_1))s_k ,$$

or

$$g_i^j(C(t_0))H_j^k s_k = (H')_i^j g_j^k(C(t_1))s_k .$$

This implies the matrix equation

$$g(C(t_0))H = H'g(C(t_1)) ,$$

or

$$\boxed{H' = g(C(t_0))Hg^{-1}(C(t_1))} , \qquad (35.86)$$

which is the rule for the gauge transformation of the matrix representation of the holonomy map.

An important situation arises when we consider the holonomy map for closed loops: $x = C(t_1) = C(t_0)$. In this case $H(C; D)$ is a linear operator on the vector space $E_x = \pi^{-1}(x)$. The set

$$H_x \equiv \{ H(C; D) \,|\, C(t_0) = C(t_1) = x \} \qquad (35.87)$$

corresponding to all loops at x forms a subgroup of $GL(m)$, where m is the dimension of E_x, called the **holonomy group** of the connection D with reference to the point $x \in M$.

Suppose $H \in H_x$. Then, under a gauge transformation, (35.86) implies that the matrix representation of H transforms according to

$$H' = g(x)Hg^{-1}(x) . \qquad (35.88)$$

We thus obtain the gauge invariant quantity

$$W \equiv Tr(H) , \qquad H \in H_x , \qquad (35.89)$$

called the **Wilson loop** in the physics literature.

Chapter 36

Yang-Mills Equations

Maxwell's electrodynamics provides the simplest example of a gauge field theory. In this chapter we will first rewrite Maxwell's equations using differential forms, and show that they can be obtained from the covariant derivatives of a curvature 2-form. The generalizations of Maxwell's equations (which describe an abelian gauge field theory) to the non-abelian case are called Yang-Mills equations [given in (36.34)].

The standard form of Maxwell's equations is given by

$$\nabla \cdot \boldsymbol{B} = 0 , \tag{36.1}$$

$$\nabla \times \boldsymbol{E} + \frac{\partial \boldsymbol{B}}{\partial t} = 0 , \tag{36.2}$$

$$\nabla \cdot \boldsymbol{E} = \rho , \tag{36.3}$$

$$\nabla \times \boldsymbol{B} - \frac{\partial \boldsymbol{E}}{\partial t} = \boldsymbol{j} . \tag{36.4}$$

Recall the electromagnetic field tensor $F_{\mu\nu}$, $\mu, \nu = 0, 1, 2, 3$ [c.f. (2.31)]:

$$F_{\mu\nu} = \begin{pmatrix} 0 & -E^1 & -E^2 & -E^3 \\ E^1 & 0 & B^3 & -B^2 \\ E^2 & -B^3 & 0 & B^1 \\ E^3 & B^2 & -B^1 & 0 \end{pmatrix} . \tag{36.5}$$

Introduce the 2-form [c.f. (35.45)]

$$\Omega = \frac{1}{2} F_{\mu\nu} \, dx^\mu \wedge dx^\nu . \tag{36.6}$$

(Note that in this case Ω is not matrix-valued.) Using (36.5) we can write Ω

411

explicitly as

$$\Omega = E^1 \, dx \wedge dt + E^2 \, dy \wedge dt + E^3 \, dz \wedge dt$$
$$+ B^1 \, dy \wedge dz + B^2 \, dz \wedge dx + B^3 \, dx \wedge dy \,. \quad (36.7)$$

Using the notation

$$d\boldsymbol{r} = (dx,\, dy,\, dz) \,, \quad\quad\quad\quad (36.8)$$
$$d\boldsymbol{a} = (dy \wedge dz,\, dz \wedge dx,\, dx \wedge dy) \,, \quad\quad\quad\quad (36.9)$$

Eq. (36.7) can also be written

$$\Omega = \boldsymbol{E} \cdot d\boldsymbol{r} \wedge dt + \boldsymbol{B} \cdot d\boldsymbol{a} \,. \quad\quad\quad\quad (36.10)$$

Exteriorly differentiating (36.7) we obtain

$$
d\Omega = \left(\frac{\partial E^3}{\partial y} - \frac{\partial E^2}{\partial z} \right) dy \wedge dz \wedge dt + \frac{\partial B^1}{\partial t} dy \wedge dz \wedge dt
$$
$$
+ \left(\frac{\partial E^1}{\partial z} - \frac{\partial E^3}{\partial x} \right) dz \wedge dx \wedge dt + \frac{\partial B^2}{\partial t} dz \wedge dx \wedge dt
$$
$$
+ \left(\frac{\partial E^2}{\partial x} - \frac{\partial E^1}{\partial y} \right) dx \wedge dy \wedge dt + \frac{\partial B^3}{\partial t} dx \wedge dy \wedge dt
$$
$$
+ \left(\frac{\partial B^1}{\partial x} + \frac{\partial B^2}{\partial y} + \frac{\partial B^3}{\partial z} \right) dx \wedge dy \wedge dz \,,
$$

$$(36.11)$$

or

$$d\Omega = \left(\nabla \times \boldsymbol{E} + \frac{\partial \boldsymbol{B}}{\partial t} \right) \cdot d\boldsymbol{a} \wedge dt + (\nabla \cdot \boldsymbol{B}) \, d\tau \,, \quad\quad (36.12)$$

where $d\tau \equiv dx \wedge dy \wedge dz$ is the volume element in space.

$\boxed{\text{Exercise 36.1}}$ Use the rules of exterior differentiation to verify (36.11).

It is clearly seen that the first pair of Maxwell's equations [(36.1) and (36.2)] are equivalent to the single equation

$$d\Omega = 0 \,. \quad\quad\quad\quad (36.13)$$

Let us now calculate $\star\Omega$, where $\star$ is the Hodge-star operator defined in (29.11). In the present case, $\star$ is defined in the exterior-form bundle $\Lambda(M^*)$ [c.f. (30.22)], where M is Minkowski spacetime with the metric $\eta_{\mu\nu}$ given by (2.20), and the orientation is chosen to be $\sigma = dt \wedge dx \wedge dy \wedge dz$. We see that, on using the results of Eqs. (29.42) to (29.47),

$$\star\Omega = -E^1 \, dy \wedge dz - E^2 \, dz \wedge dx - E^3 \, dx \wedge dy$$
$$- B^1 \, dt \wedge dx - B^2 \, dt \wedge dy - B^3 \, dt \wedge dz \,. \quad (36.14)$$

Hence

$$
\begin{aligned}
d \star \Omega = \; & -\frac{\partial E^1}{\partial x} \, dx \wedge dy \wedge dz - \frac{\partial E^1}{\partial t} \, dt \wedge dy \wedge dz - \frac{\partial E^2}{\partial y} \, dy \wedge dz \wedge dx \\
& - \frac{\partial E^2}{\partial t} \, dt \wedge dz \wedge dx - \frac{\partial E^3}{\partial z} \, dz \wedge dx \wedge dy - \frac{\partial E^3}{\partial t} \, dt \wedge dx \wedge dy \\
& - \frac{\partial B^1}{\partial y} \, dy \wedge dt \wedge dx - \frac{\partial B^1}{\partial z} \, dz \wedge dt \wedge dx - \frac{\partial B^2}{\partial x} \, dx \wedge dt \wedge dy \\
& - \frac{\partial B^2}{\partial z} \, dz \wedge dt \wedge dy - \frac{\partial B^3}{\partial x} \, dx \wedge dt \wedge dz - \frac{\partial B^3}{\partial y} \, dy \wedge dt \wedge dz \; ,
\end{aligned}
\tag{36.15}
$$

and

$$
\begin{aligned}
\star d \star \Omega = \; & -\left(\frac{\partial E^1}{\partial x} + \frac{\partial E^2}{\partial y} + \frac{\partial E^3}{\partial z} \right) dt + \left(\frac{\partial B^3}{\partial y} - \frac{\partial B^2}{\partial z} - \frac{\partial E^1}{\partial t} \right) dx \\
& + \left(\frac{\partial B^1}{\partial z} - \frac{\partial B^3}{\partial x} - \frac{\partial E^2}{\partial t} \right) dy + \left(\frac{\partial B^2}{\partial x} - \frac{\partial B^1}{\partial y} - \frac{\partial E^3}{\partial t} \right) dz \; .
\end{aligned}
\tag{36.16}
$$

Equivalently,

$$
\star d \star \Omega = -\nabla \cdot \boldsymbol{E} \, dt + \left(\nabla \times \boldsymbol{B} - \frac{\partial \boldsymbol{E}}{\partial t} \right) \cdot d\boldsymbol{r} \; .
\tag{36.17}
$$

The operator $\star d\star$ is sometimes called the **codifferential**. Now we can define the vector field $\boldsymbol{J}$ on M:

$$
\boldsymbol{J} \equiv j^0 \frac{\partial}{\partial t} + j^1 \frac{\partial}{\partial x} + j^2 \frac{\partial}{\partial y} + j^3 \frac{\partial}{\partial z} \; ,
\tag{36.18}
$$

where $j^0 = \rho$ (the charge density), and $\boldsymbol{j} = (j^1, j^2, j^3)$ (the current density). The dual one-form of $\boldsymbol{J}$ is

$$
J = j_0 dx^0 + j_1 dx^1 + j_2 dx^2 + j_3 dx^3 = j_0 dt + j_1 dx + j_2 dy + j_3 dz \; ,
\tag{36.19}
$$

where

$$
j_0 = \eta_{00} j^0 = -j^0 = -\rho \; ,
\tag{36.20}
$$

$$
j_1 = \eta_{11} j^1 = j^1 \; , \quad j_2 = \eta_{22} j^2 = j^2 \; , \quad j_3 = \eta_{33} j^3 = j^3 \; .
\tag{36.21}
$$

Thus

$$
J = -\rho dt + \boldsymbol{j} \cdot d\boldsymbol{r} \; .
\tag{36.22}
$$

On comparison with (36.17), we see that the second pair of Maxwell's equations (36.3) and (36.4) are equivalent to

$$
\star d \star \Omega = J \; .
\tag{36.23}
$$

From (2.18) we see that

$$
F_{\mu\nu} = \partial_\mu A_\nu - \partial_\nu A_\mu \; ,
\tag{36.24}
$$

where $A_0 = \eta_{00} A^0 = -A^0 = -\phi$, $A_i = A^i$ $(i = 1, 2, 3)$, and ϕ, $\boldsymbol{A} = (A^1, A^2, A^3)$ are the scalar and vector potential of electromagnetism, respectively. Since A_μ is not matrix-valued, $[A_\mu, A_\nu] = 0$, and (36.24) becomes a special case of (35.43). Let us define the 1-form

$$w = A_\mu \, dx^\mu \quad A_\mu(x) \in C^\infty(M) . \tag{36.25}$$

Then it follows from (36.6) that

$$\Omega = dw . \tag{36.26}$$

Also

$$w \wedge w = 0 , \tag{36.27}$$

and

$$w \wedge \Omega = \Omega \wedge w . \tag{36.28}$$

$\boxed{\text{Exercise 36.2}}$ For w defined by (36.25), verify (36.27) and (26.28).

From (35.18), it follows that Ω can be identified as the curvature 2-form corresponding to the connection 1-form w, defined on a (hermitian) line bundle $\pi : E \to M$ with structure group $U(1)$, where each fiber E_x is a one-dimensional complex vector space. Furthermore, the equation $d\Omega = 0$ [(36.13)] is just a special case of the Bianchi identity [(35.36)]. Thus, of the Maxwell's equations in the form

$$d\Omega = 0 , \tag{36.29}$$
$$\star d \star \Omega = J , \tag{36.30}$$

the first one is entirely geometric in nature, being just the Bianchi identity. The second one, as it turns out, leads to the physical requirement of the conservation of charge, as expressed by the **equation of continuity**

$$\nabla \cdot \boldsymbol{j} + \frac{\partial \rho}{\partial t} = 0 . \tag{36.31}$$

Indeed, on performing the Hodge star on both sides of (36.30), we get

$$\star J = \star \star d \star \Omega = d \star \Omega , \tag{36.32}$$

since $\star\star = 1$ in the present case [c.f. (29.18)]. Consequently, since $d^2 = 0$,

$$\boxed{d \star J = 0} \quad . \tag{36.33}$$

It is easily verified from the definition of the current density 1-form J [(36.22)] that (36.33) is the same as the equation of continuity (36.31).

$\boxed{\text{Exercise 36.3}}$ Show that (36.33) is the same as (36.31).

The generalization of the Maxwell's equations [(36.29) and (36.30)] to the non-abelian case gives the so-called **Yang-Mills equations** as follows:

$$\boxed{D\Omega = 0 \,, \qquad \star D \star \Omega = J \,,}$$ (36.34)

where D is the covariant derivative corresponding to a given connection on a hermitian vector bundle $\pi : E \to M$, and J is a generalized "current" one-form. [Recall the action of D on a matrix-valued 2-form given by (35.66).] The first equation is again just Bianchi's identity [(35.68)].

The Yang-Mills equations are in general non-linear partial differential equations to be solved for the connection ω whose curvature is Ω. Restricting our attention to the free-field situation ($J = 0$), the second equation in (36.34) appears as

$$D \star \Omega = 0 \,,$$ (36.35)

and actually arises on extremizing the **Yang-Mills action functional**

$$S_{YM} = -\int_M Tr(\Omega \wedge \star\Omega) \,.$$ (36.36)

Suppose one can find an ω such that $\Omega = d\omega - \omega \wedge \omega$ satisfies

$$\lambda \star \Omega = \Omega \,,$$ (36.37)

where λ is some constant. Ω is then automatically a solution of (36.35) because of Bianchi's identity $D\Omega = 0$. On Hodge-starring (36.37), one finds

$$\lambda \star \star\Omega = \star\Omega \,,$$ (36.38)

or

$$\lambda \star \Omega = \lambda^2 \star \star\Omega = \Omega \,,$$ (36.39)

where the last equality follows from (36.37). From (29.18),

$$\star \star \Omega = (-1)^{2(n-2)+(n-S)/2} \, \Omega = \pm\Omega \,,$$ (36.40)

where n is the dimension of the base manifold M and S is the signature of the metric on M [c.f. (29.14)]. It follows from (36.39) that $\lambda^2 = \pm 1$. Eq. (36.37) then divides into two cases:

$$\star\Omega = \pm\Omega \,,$$ (36.41)
$$\star\Omega = \pm i\Omega \,.$$ (36.42)

If $\star\Omega = \Omega$, then the curvature Ω is called **self-dual**; if $\star\Omega = -\Omega$, then Ω is called **anti-self-dual**. Self-dual and anti-self-dual solutions to the Yang-Mills equations are called **instantons**. They play an important role in field theories in physics as well as in the elucidation of differentiable structures in $\mathbb{R}^4$ (see Donaldson and Kronheimer 1991).

From (36.40) it is seen that $\star\Omega = \pm\Omega$ if $T_x M$ has a Euclidean metric for all $x \in M$ ($S = n$); while $\star\Omega = \pm i\Omega$ if $T_x M$ has a Lorentz metric ($S = n - 2$). Equivalently, (36.41) arises from Riemannian manifolds while (36.42) arises from semi-Riemannian manifolds (manifolds with a pseudo-Riemannian metric) (see Chapter 39). This difference is significant, since it places a strong constraint on the types of structure groups (gauge groups) that the vector bundle $\pi : E \to M$ can have. We will only briefly indicate the reason here without proof. Since Ω is Lie-algebra ($\mathcal{G}$)-valued, so is $\star\Omega$. Eq. (36.42) then requires that $i\mathcal{G} = \mathcal{G}$. But this condition is not satisfied for any Lie algebra of a compact Lie group. On the other hand, there is no such restriction on the gauge group when (36.41) applies. Since many of the most interesting gauge groups in physics are compact Lie groups [such as $G = U(n)$], the instanton solutions are of intrinsic interest in physics.

Chapter 37

Connections on a Principal Bundle

To introduce the notion of a principal fiber bundle we will first discuss a concrete example of such a bundle, namely, a frame bundle.

Definition 37.1. *Let M be an n-dimensional manifold. A **frame** on M at $x \in M$ is a linear isomorphism $f : \mathbb{R}^n \to T_x M$. Suppose*

$$f(1, 0, \ldots, 0) = e_1 \,,$$

$$\vdots \tag{37.1}$$

$$f(0, 0, \ldots, 1) = e_n \,,$$

then $\{e_1, \ldots, e_n\}$ is a basis of $T_x M$. Thus a frame can also be written as a combination of the form $(x; e_1, \ldots, e_n)$, where $x \in M$ and $(e_1, \ldots, e_n)$ are n linearly independent tangent vectors at $x \in M$.

Denote the set of all frames on M by P. We can introduce a differentiable structure on P so that it becomes a smooth manifold, and the natural projection $\pi : P \to M$ defined by

$$\pi(x; e_1, \ldots, e_n) = x \tag{37.2}$$

is a smooth map. The triplet (P, M, π) is then called the **frame bundle** on M.

The differentiable structure is inherited from that of the base manifold M, and is constructed as follows. Suppose $(U; x^i)$ is a coordinate neighborhood of M. Then there is a natural frame field $\left(\dfrac{\partial}{\partial x^1}, \ldots, \dfrac{\partial}{\partial x^n} \right)$ on U. Hence any frame $(x; e_1, \ldots, e_n)$ on U can be expressed as

$$e_i = X_i^k \left(\frac{\partial}{\partial x^k} \right)_x \,, \quad 1 \le i \le n \,, \tag{37.3}$$

where (X_i^k) is a non-singular $n \times n$ matrix. Then we can define a map

$$\varphi_U : U \times GL(n, \mathbb{R}) \to \pi^{-1}(U) \,,$$

called a **local trivialization** (or local product structure), such that for any $x \in U$ and $(X_i^k) \in GL(n, \mathbb{R})$,

$$\varphi_U(x, (X_i^k)) = (x; e_1, \ldots, e_n) \,, \qquad (37.4)$$

where the e_i are given by (37.3). The numbers (x^i, X_i^k) can be regarded as local coordinates for the frame $p = (x; e_1, \ldots, e_n) \in P$. Hence P as a manifold is $(n + n^2)$-dimensional.

Let $(V; y^i)$ be another coordinate neighborhood of M such that $U \cap V \neq \emptyset$. We have the local change of coordinates

$$y^i = y^i(x^1, \ldots, x^n) \,, \quad i = 1, \ldots, n \,; \qquad (37.5)$$

and the following transformation for the natural bases:

$$\frac{\partial}{\partial x^i} = \frac{\partial y^j}{\partial x^i} \frac{\partial}{\partial y^j} \,. \qquad (37.6)$$

Let (y^i, Y_i^k) be the local coordinates of $p = (x; e_1, \ldots, e_n)$, corresponding to the local trivialization $\varphi_V : V \times GL(n, \mathbb{R}) \to \pi^{-1}(V)$, where $x \in U \cap V$. We thus have, from (37.3),

$$e_i = X_i^k \frac{\partial}{\partial x^k} = Y_i^k \frac{\partial}{\partial y^k} \,, \qquad (37.7)$$

or

$$Y_i^k = X_i^j \frac{\partial y^k}{\partial x^j} \,. \qquad (37.8)$$

Eqs. (37.5) and (37.8) together constitute the coordinate transformation rules for any two coordinate neighborhoods $\pi^{-1}(U)$ and $\pi^{-1}(V)$ in P, where $U \cap V \neq \emptyset$, and thus $\pi^{-1}(U) \cap \pi^{-1}(V) \neq \emptyset$. These determine the differentiable structure on P.

Let us write $X \equiv (X_i^j) \in GL(n, \mathbb{R})$. For $x \in U$, let

$$\varphi_{U,x}(X) \equiv \varphi_U(x, X) \,. \qquad (37.9)$$

Then $\varphi_{U,x} : G(n, \mathbb{R}) \to \pi^{-1}(x)$ is a homeomorphism, and the typical fiber of the frame bundle P is $GL(n, \mathbb{R})$.

As in the case of vector bundles, one can identify a family of **transition functions**

$$g_{UV} : U \cap V \to GL(n, \mathbb{R}) \qquad (37.10)$$

given by

$$g_{UV}(x) = \varphi_{V,x}^{-1} \circ \varphi_{U,x} \,. \qquad (37.11)$$

In fact, by (37.8), $\varphi_{V,x}^{-1} \circ \varphi_{U,x}$ is precisely the right translation of X by the Jacobian matrix $\left(\dfrac{\partial y^k}{\partial x^j} \right)_x \in GL(n, \mathbb{R})$. These transition functions specify how the typical fibers are "glued" together in overlapping coordinate neighborhoods $(U \cap V \neq \emptyset)$ in the base manifold M, and determine the differentiable structure of the frame bundle P. Note that $g_{UV}(x)$ only depends on $x \in M$ and does not depend on the fiber element $X \in GL(n, \mathbb{R})$. Since the Jacobian $\partial y^k / \partial x^j$ also determines the transition functions of the tangent bundle $T(M)$, we say that the frame bundle (P, M, π) is the principal bundle associated with the tangent bundle $T(M)$. The typical fiber as well as the structure group of the frame bundle are both $GL(n, \mathbb{R})$. This is a general feature of principal fiber bundles. Note also that the frame bundle, and principal bundles in general, are not vector bundles.

The structure group $GL(n, \mathbb{R})$ acts naturally on the frame bundle P by a left action L_g, $g \in GL(n, \mathbb{R})$, which preserves the fibers. L_g is defined by

$$L_g(x; e_1, \ldots, e_n) = (x; (e')_1, \ldots, (e')_n), \tag{37.12}$$

where

$$(e')_i = g_i^j \, e_j . \tag{37.13}$$

It is clear that L_g is fiber-preserving (mapping a fiber onto itself):

$$\pi \circ L_g = \pi : P \to M . \tag{37.14}$$

Exercise 37.1   Prove that

$$L_{g_1 g_2} = L_{g_1} \circ L_{g_2} . \tag{37.15}$$

We will now construct a set of differential one-forms (θ^i, θ_j^k), $i, j, k = 1, \ldots, n$, on P which will allow us to define a connection on P. Suppose $(U; x^i)$ and $(V; y^i)$ are two coordinate neighborhoods on M, with the corresponding coordinate systems $(x^i, X_j^k), (y^i, Y_j^k)$ on P. If $U \cap V \neq \emptyset$ we have, on $U \cap V$,

$$dy^i = \frac{\partial y^i}{\partial x^j} \, dx^j . \tag{37.16}$$

By (37.8), on the other hand,

$$(X^{-1})_i^j = \frac{\partial y^k}{\partial x^i} \, (Y^{-1})_k^j , \tag{37.17}$$

where X^{-1} and Y^{-1} are the inverse matrices of X and Y, respectively. From the above two equations it follows that

$$(X^{-1})_i^j \, dx^i = (Y^{-1})_i^j \, dy^i . \tag{37.18}$$

This implies that the 1-form

$$\boxed{\theta^i \equiv (X^{-1})_j^i \, dx^j} \tag{37.19}$$

is independent of local coordinates and is a 1-form on P.

Exercise 37.2   Verify (37.18) by using (37.16) and (37.17) in their matrix forms.

The **Pfaffian** system of equations

$$\theta^i = 0 , \qquad i = 1, \ldots, n , \tag{37.20}$$

then determines an n^2-dimensional tangent subspace field V on P. At each point $p \in P$, the corresponding tangent subspace is called the **vertical subspace**. The above claim can be seen as follows. Eq. (37.19) implies

$$dx^i = X^i_j \, \theta^j . \tag{37.21}$$

Thus the system (37.20) is equivalent to

$$dx^i = 0 , \quad i = 1, \ldots, n , \tag{37.22}$$

in every coordinate neighborhood $\pi^{-1}(U)$ of P. Clearly, (37.22) is completely integrable, with the maximal integral manifold in P given by

$$x^i = \text{constant} , \quad i = 1, \ldots, n , \tag{37.23}$$

which is precisely the fiber $\pi^{-1}(x)$. Hence the vertical space is the tangent space of the fiber, and is n^2-dimensional.

Now suppose we are given an affine connection D on M, which is a connection in the sense of Def. 35.1 on the vector bundle $T(M)$. (For further discussion on the affine connection see Chapter 39.) For $e_i = X^k_i \dfrac{\partial}{\partial x^k}$, we have, by (35.3) and (35.10),

$$De_i = (dX^k_i + X^j_i \omega^k_j) \otimes \frac{\partial}{\partial x^k} , \tag{37.24}$$

where ω^k_j is the connection matrix of 1-forms of the connection D. Viewing the X^j_i as local coordinates on P, the quantities

$$DX^k_i \equiv dX^k_i + X^j_i \omega^k_j \tag{37.25}$$

are differential 1-forms on $\pi^{-1}(U) \subset P$. It can in fact be shown that $(X^{-1})^j_k \, DX^k_i$ defines a global 1-form. Under the gauge transformation (37.6), the g in (35.17) is given by

$$g^j_i = \frac{\partial x^j}{\partial y^i} . \tag{37.26}$$

Hence, by (35.17), the gauge-transformed affine connection matrix is given by

$$(\omega')^j_i = d\left(\frac{\partial x^k}{\partial y^i}\right) \frac{\partial y^j}{\partial x^k} + \left(\frac{\partial x^k}{\partial y^i}\right) \omega^l_k \left(\frac{\partial y^j}{\partial x^l}\right) . \tag{37.27}$$

Thus

$$DY_i^k = dY_i^k + Y_i^j \, (\omega')_j^k = dX_i^j \, \frac{\partial y^k}{\partial x^j} + X_i^j \, d\left(\frac{\partial y^k}{\partial x^j}\right)$$
$$+ Y_i^j \left(d\left(\frac{\partial x^l}{\partial y^j}\right) \frac{\partial y^k}{\partial x^l} + \left(\frac{\partial x^l}{\partial y^j}\right) \omega_l^m \left(\frac{\partial y^k}{\partial x^m}\right) \right). \quad (37.28)$$

Due to (37.8), the third term on the RHS of the above equation can be written as

$$X_i^h \frac{\partial y^j}{\partial x^h} \left(d\left(\frac{\partial x^l}{\partial y^j}\right) \frac{\partial y^k}{\partial x^l} + \left(\frac{\partial x^l}{\partial y^j}\right) \omega_l^m \left(\frac{\partial y^k}{\partial x^m}\right) \right)$$
$$= -X_i^h \frac{\partial y^j}{\partial x^h} \frac{\partial x^l}{\partial y^j} \, d\left(\frac{\partial y^k}{\partial x^l}\right) + X_i^h \frac{\partial y^j}{\partial x^h} \frac{\partial x^l}{\partial y^j} \omega_l^m \frac{\partial y^k}{\partial x^m} \quad (37.29)$$
$$= -X_i^h \delta_h^l \, d\left(\frac{\partial y^k}{\partial x^l}\right) + X_i^h \delta_h^l \omega_l^m \frac{\partial y^k}{\partial x^m} = -X_i^l d\left(\frac{\partial y^k}{\partial x^l}\right) + X_i^l \omega_l^m \frac{\partial y^k}{\partial x^m}.$$

Eq. (37.28) then yields

$$dY_i^k + Y_i^j \, (\omega')_j^k = (dX_i^j + X_i^l \omega_l^j) \, \frac{\partial y^k}{\partial x^j}, \quad (37.30)$$

or

$$DY_i^k = DX_i^j \, \frac{\partial y^k}{\partial x^j}. \quad (37.31)$$

Let us now write (37.17) and (37.31) in matrix form:

$$X^{-1} = \frac{\partial y}{\partial x} \cdot Y^{-1}, \quad (37.32)$$

$$DY = DX \cdot \frac{\partial y}{\partial x}. \quad (37.33)$$

These imply, in matrix form,

$$DX \cdot X^{-1} = DY \left(\frac{\partial y}{\partial x}\right)^{-1} \left(\frac{\partial y}{\partial x}\right) Y^{-1} = DY \cdot Y^{-1}, \quad (37.34)$$

which is equivalent to

$$(Y^{-1})_k^j \, DY_i^k = (X^{-1})_k^j \, DX_i^k. \quad (37.35)$$

Hence the differential 1-forms

$$\boxed{\theta_i^j \equiv (X^{-1})_k^j \, DX_i^k = (X^{-1})_k^j \, (dX_i^k + X_i^l \omega_l^k)} \quad (37.36)$$

are independent of the choice of local coordinates on P, and are therefore global 1-forms on P. The $n+n^2$ 1-forms θ^i and θ_i^j are linearly independent everywhere on P, and therefore constitute a coframe field on P.

We already know that the Pfaffian system $\theta^i = 0$ determines the n^2-dimensional vertical subspace field $\mathcal{V}$ on P. Similarly, the Pfaffian system

$$\theta^j_i = 0 \,, \quad 1 \leq i, j \leq n \,, \tag{37.37}$$

determines an n-dimensional tangent subspace field $\mathcal{H}$. $\mathcal{H}(p)$, for each $p \in P$, is a subspace of $T_p P$ called the **horizontal subspace**. The vertical and horizontal subspaces have the following properties.

Theorem 37.1. *Suppose (P, M, π) is the frame bundle of an affine space M [M is equipped with a connection D on the tangent bundle $T(M)$]. The global 1-forms θ^i [(37.19)] and θ^j_i [(37.36)] on P then determine the vertical and horizontal subspace fields on P satisfying the following properties.*

1) At each point $p \in P$, $T_p(P)$ can be decomposed as the direct sum of vertical and horizontal subspaces:

$$T_p(P) = \mathcal{V}(p) \oplus \mathcal{H}(p) \,. \tag{37.38}$$

2) The horizontal subspaces of $T_p(P)$ project onto the tangent spaces of the base manifold M, that is, under the projection $\pi(p) = x \in M$, we have the induced projection

$$\pi_*(\mathcal{H}(p)) = T_x(M) \,. \tag{37.39}$$

3) $\mathcal{H}$ is invariant under the left translation L_g on P [$g \in GL(n, \mathbb{R})$]:

$$(L_g)_* \, \mathcal{H}(p) = \mathcal{H}(L_g p) \,. \tag{37.40}$$

Proof. The sum of the dimensions of $\mathcal{V}(p)$ and $\mathcal{H}(p)$ is $n^2 + n$, which is also the dimension of $T_p(P)$. In order to prove property 1), we just need to show that $\mathcal{V}(p) \cap \mathcal{H}(p) = 0$ [the zero vector in $T_p(P)$]. Let $\mathbb{X} \in \mathcal{V}(p) \cap \mathcal{H}(p)$. From the definitions of $\mathcal{V}(p)$ and $\mathcal{H}(p)$, we have

$$\theta^i(\mathbb{X}) = \theta^i_j(\mathbb{X}) = 0 \,, \quad 1 \leq i, j \leq n \,. \tag{37.41}$$

Since $\{\theta^i, \theta^i_j\}$ forms a coframe field, the only vector in $T_p(P)$ satisfying (37.41) is the zero vector.

To prove property 2) we observe that $\pi : P \to M$ is smooth and surjective. Hence $\pi_* : T_p(P) \to T_x(M)$ [$\pi(p) = x$] is a surjective homomorphism. Since by definition of the vertical subspace, $\pi_*(\mathcal{V}(p)) = 0$, $\pi_* : \mathcal{H}(p) \to T_x(M)$ is an isomorphism.

To prove property 3) [or (37.40)], we will show that (1) $(L_g)_* \, \mathcal{H}(p) \subset \mathcal{H}(L_g p)$, and (2) $\mathcal{H}(L_g p) \subset (L_g)_* \, \mathcal{H}(p)$. Suppose $p = (x; e_1, \ldots, e_n)$ and $L_g p = (x; (e')_1, \ldots, (e')_n)$, where

$$(e')_i = g^j_i e_j = g^j_i X^k_j \frac{\partial}{\partial x^k} \equiv (X')^k_i \frac{\partial}{\partial x^k} \,. \tag{37.42}$$

Hence

$$(X')^k_i = g^j_i X^k_j \ ,$$

which implies

$$((X')^{-1})^k_i = (X^{-1})^j_i (g^{-1})^k_j \ . \tag{37.43}$$

In matrix notation,

$$(X')^{-1} = X^{-1} \cdot g^{-1} \ . \tag{37.44}$$

Eq. (37.43) also implies that

$$DX' = g\, DX \ . \tag{37.45}$$

(Remember that here, $g \in GL(n, \mathbb{R})$ is not a function on M.) Thus

$$DX' \cdot (X')^{-1} = g(DX \cdot X^{-1})g^{-1} \ . \tag{37.46}$$

In other words, on recalling the definition of the 1-forms θ^j_i given by (37.36),

$$\boxed{(L_g)^* \, \theta^j_i = g^k_i \theta^l_k (g^{-1})^j_l} \ . \tag{37.47}$$

Suppose $\mathbb{H}' \in (L_g)_* \, \mathcal{H}(p)$. Then

$$\mathbb{H}' = (L_g)_* \, \mathbb{H} \ , \tag{37.48}$$

where $\mathbb{H} \in \mathcal{H}(p)$. Then

$$\langle \, \mathbb{H}', \theta^j_i \, \rangle_{L_g p} = \langle \, (L_g)_* \mathbb{H}, \theta^j_i \, \rangle_{L_g p} = \langle \, \mathbb{H}, (L_g)^* \theta^j_i \, \rangle_p = g^k_i \, \langle \, \mathbb{H}, \theta^l_k \, \rangle_p \, (g^{-1})^j_l = 0 \ , \tag{37.49}$$

since $\mathbb{H} \in \mathcal{H}(p)$. This implies $\mathbb{H}' \in \mathcal{H}(L_g p)$. Thus (1) follows. To prove (2), suppose

$$\mathbb{H}' \in \mathcal{H}(L_g p) \ . \tag{37.50}$$

Then

$$\langle \, \mathbb{H}', \theta^j_i \, \rangle_{L_g p} = 0 \ . \tag{37.51}$$

But $\mathbb{H}'$ must be equal to $(L_g)_* \mathbb{H}$, for some $\mathbb{H} \in T_p(P)$. Thus

$$0 = \langle \, (L_g)_* \mathbb{H}, \theta^j_i \, \rangle_{L_g p} = \langle \, \mathbb{H}, (L_g)^* \theta^j_i \, \rangle_p = g^k_i \, \langle \, \mathbb{H}, \theta^l_k \, \rangle_p \, (g^{-1})^j_l \ . \tag{37.52}$$

It follows that

$$\langle \, \mathbb{H}, \theta^l_k \, \rangle_p = 0 \ , \tag{37.53}$$

which implies $\mathbb{H} \in \mathcal{H}(p)$. Hence

$$\mathbb{H}' \in (L_g)_* \, \mathcal{H}(p) \ . \tag{37.54}$$

This establishes (2). $\qquad\qquad\qquad\qquad\qquad\qquad\qquad\qquad\qquad\qquad\qquad\quad \square$

We can turn the above procedure around and claim the following. If an n-dimensional tangent subspace $\mathcal{H}(p)$ is given smoothly at each point $p \in P$ of the frame bundle such that this field of subspaces satisfies the conditions (1), (2) and (3) of Theorem 37.1, then it determines a connection on the frame bundle P. This connection defined on P then corresponds to an affine connection D on the base manifold M such that $\mathcal{H}$ is the field of horizontal subspaces of the frame bundle P with respect to the connection D.

We are now ready to generalize the frame bundle to the notion of a principal bundle.

Definition 37.2. *A **principal fiber bundle** consists of a manifold P (the **total space**), a Lie group G, a **base manifold** M, and a projection map $\pi : P \to M$ such that the following conditions hold.*

(1) *For each $g \in G$, there is a left action diffeomorphism $L_g : P \to P$ [written $L_g(p) = gp$] such that*

    (a) *$(g_1 g_2)p = g_1(g_2 p)$, for all $g_1, g_2 \in G$ and $p \in P$;*

    (b) *$ep = p$, for all $p \in P$, where e is the identity in G;*

    (c) *if $gp = p$ for some $p \in P$, then $g = e$, that is, there are no fixed points of L_g. Recalling Def. 6.6, we say that G acts freely on P to the left.*

(2) *M is the quotient space of the manifold P with respect to the equivalence relation defined by the group action of G on P; and the projection $\pi : P \to M$ is smooth and surjective (see Fig. 37.1):*

$$\pi^{-1}(\pi(p)) = \{gp \,|\, g \in G\} \, . \tag{37.55}$$

*For $x \in M$, $\pi^{-1}(x)$ is called the **fiber** above x. From condition (1), for each $p \in P$, there is a diffeomorphism $G \to \pi^{-1}(x)$ given by $g \mapsto gp$. Thus all fibers $\pi^{-1}(x)$ are diffeomorphic to G. This map, however, depends on p, and does not lead to a canonical identification of G with $\pi^{-1}(x)$. Thus there is no natural group structure on any fiber $\pi^{-1}(x)$.*

(3) *P is locally trivial: for each $x \in M$ there exists a neighborhood U of x such that $\pi^{-1}(U)$ is diffeomorphic to $U \times G$. In other words, there exists a diffeomorphism $T_U : \pi^{-1}(U) \to U \times G$ given by*

$$T_U(p) = (\pi(p), \varphi_U(p)) \, , \tag{37.56}$$

where the map $\varphi_U : \pi^{-1}(U) \to G$ has the property

$$\varphi_U(gp) = g\varphi_U(p) \, . \tag{37.57}$$

*The map T_U is called a **local trivialization**, or a **choice of gauge** (in physics).*

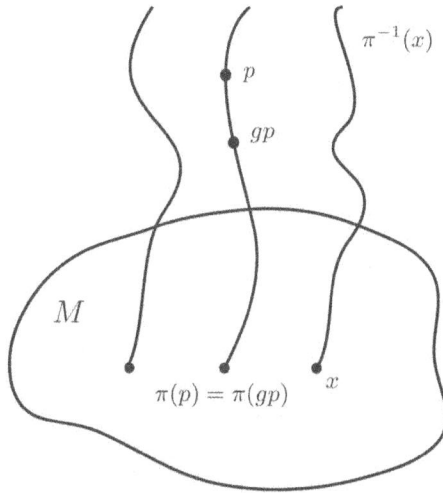

FIGURE 37.1

Definition 37.3. *Suppose T_U and T_V are two local trivializations of a principal fiber bundle $\pi : P \to M$ with group G, and $U \cap V \neq \emptyset$. The **transition function** $G_{UV} : U \cap V \to G$ from T_U to T_V is defined, for $x \in U \cap V$, by*

$$g_{UV}(x) \equiv \varphi_V^{-1}(p) \cdot \varphi_U(p) , \tag{37.58}$$

where p is any element in $\pi^{-1}(x)$.

That g_{UV} is independent of p follows from (37.57). Indeed,

$$\varphi_V^{-1}(gp)\varphi_U(gp) = (g\varphi_V(p))^{-1}g\varphi_U(p)$$
$$= \varphi_V^{-1}(p)\, g^{-1}g\, \varphi_U(p) = \varphi_V^{-1}(p)\varphi_U(p) . \tag{37.59}$$

The transition functions satisfy the following properties:

(1) $g_{UU}(x) = e$, for all $x \in U$; $\qquad\qquad\qquad$ (37.60)

(2) $g_{VU}(x) = (g_{UV}(x))^{-1}$, for all $x \in U \cap V$; $\qquad$ (37.61)

(3) $g_{UV}(x)g_{VW}(x)g_{WU}(x) = e$, for all $x \in U \cap V \cap W$. $\qquad$ (37.62)

$\boxed{\text{Exercise 37.3}}$ Prove the above three properties.

The transition functions are glueing functions which give the instructions on how the different local trivializations $U \times G$, $V \times G$, $W \times G$, ... can be glued together to form the total space P. As in the case of the vector bundles, they determine the structure of a principal fiber bundle.

Definition 37.4. *A principal fiber bundle is said to be **trivial** if there exists a **global trivialization**, that is, a diffeomorphism $T_M : P \to M \times G$. In other words, P is globally a product.*

Definition 37.5. *A **local section** of a principal fiber bundle $\pi : P \to M$ is a map $\sigma : U \to P$, where U is a neighborhood of M, such that $\pi \circ \sigma(x) = x$ for any $x \in U$. A **global section** is a map $\sigma : M \to P$ such that $\pi \circ \sigma(x) = x$ for any $x \in M$.*

Theorem 37.2. *A principal fiber bundle is trivial if and only if it has a global section.*

Proof. It is enough to show that *there is a natural one-to-one correspondence between local (global) sections and local (global) trivializations.* We will show this for the local case; the global case follows exactly the same arguments. Let $\sigma : U \to P$ be a local section. Corresponding to this, we can define a local trivialization $T_U : \pi^{-1}(U) \to U \times G$ as follows. Any $p \in \pi^{-1}(U)$ can be written as $p = g\sigma(\pi(p))$, for some unique $g \in G$. Then set

$$T_U(p) = (\pi(p), g) \, . \tag{37.63}$$

Conversely, given a local trivialization T_U, we can define the corresponding local section to be

$$\sigma(x) = T_U^{-1}(x, e) \, . \tag{37.64}$$

The theorem follows from the definition of triviality given in Def. 37.4. $\square$

Let us now introduce the crucial notion of the connection on a principal fiber bundle. Three standard definitions, which look quite different from each other superficially, will be given, and then shown to be equivalent. We begin with one based on the notions of horizontal and vertical subspaces discussed earlier for the special case of the frame bundle.

Definition 37.6. *A **connection on a principal bundle** $\pi : P \to M$ with group G is a smooth assignment to each $p \in P$ of an n-dimensional subspace $\mathcal{H}(p) \subset T_pP$ of the tangent space to P $(n = \dim M)$ such that*

$$T_pP = \mathcal{H}(p) \oplus \mathcal{V}(p) \, , \tag{37.65}$$

where $\mathcal{V}(p)$ is the subspace

$$\mathcal{V}(p) \equiv \{ \mathbb{X} \in T_pP \,|\, \pi_*(\mathbb{X}) = 0 \} \, . \tag{37.66}$$

The subspace field $\mathcal{H}$ is also required to be invariant under the left action L_g $(g \in G)$:

$$(L_g)_* \, \mathcal{H}(p) = \mathcal{H}(gp) \, , \tag{37.67}$$

*for all $g \in G$. $\mathcal{V}(p)$ and $\mathcal{H}(p)$ are called the **vertical subspace** and **horizontal subspace** of T_pP, respectively.*

Exercise 37.4 Show that it follows from (37.65) and (37.66) that the horizontal subspace $\mathcal{H}(p)$ projects onto the tangent space $T_{\pi(p)}M$ on the base manifold M:

$$\pi_* \, \mathcal{H}(p) = T_{\pi(p)}M \ . \tag{37.68}$$

Recall that in the case of frame bundles, the horizontal subspaces were identified as integral manifolds of (or spaces annihilated by) matrix-valued 1-forms [the θ_i^j of (37.36)] on P. These 1-forms were further shown to satisfy the property (37.47). We are thus led to the following alternative definition of a connection on a principal bundle.

Definition 37.7. *Suppose $\mathcal{G}$ is the Lie algebra of a Lie group G, which is the structure group of a principal fiber bundle $\pi : P \to M$. Let $A \in \mathcal{G}$, and $A^{\#}$, called a **fundamental field**, be the vector field on P defined by*

$$A^{\#}(p) \equiv \frac{d}{dt} \, (\exp(At)\, p)_{t=0} \ . \tag{37.69}$$

*A **connection** on the principal bundle $\pi : P \to M$ is a $\mathcal{G}$-valued one-form ω on P satisfying the following two properties:*

$$(1) \qquad \omega(A^{\#}(p)) = A \tag{37.70}$$

$$(2) \qquad (L_g)^* \, (\omega) = \mathcal{A}d(g)\,(\omega) \ , \tag{37.71}$$

where $\mathcal{A}d(g) : \mathcal{G} \to \mathcal{G}$ is the adjoint representation of G given by (34.81). We say that the connection one-form ω is ad-invariant under a left translation.

Exercise 37.5 Show that condition (2) in the above definition can be restated as

$$\boxed{\omega_{gp}((L_g)_* \, \mathbb{X}_p) = (\mathcal{A}d(g))(\omega_p(\mathbb{X}_p))} \ , \tag{37.72}$$

for all $g \in G$, $p \in P$ and $\mathbb{X}_p \in T_pP$.

Exercise 37.6 Show that for the case of matrix groups G,

$$\mathcal{A}d(g)(A) = gAg^{-1} \ , \tag{37.73}$$

for all $g \in G$ and $A \in \mathcal{G}$ (the Lie algebra of G). The RHS of (37.73) is interpreted in the sense of matrix multiplications. Hint: Write

$$\mathcal{A}d(g)(A) = \frac{d}{dt} \, (ad_g \, \exp(At))_{t=0} \ .$$

We now come to the final definition of the connection on a principal bundle, cast in the form of $\mathcal{G}$-valued one-forms on the base manifold M, instead of one-forms on the total space P. This definition is most suitable in physics applications.

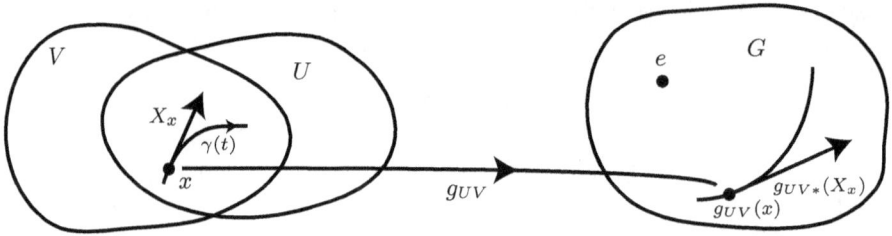

FIGURE 37.2

Definition 37.8. *A* **connection** *on a principal fiber bundle* $\pi : P \to M$ *assigns to each local trivialization* $T_U : \pi^{-1}(U) \to U \times G$ *(a choice of gauge in physics terms) a* $\mathcal{G}$-*valued one-form* ω_U *on* U *which satisfies the following rule for gauge transformations (between different local trivializations):*

$$\boxed{\omega_V(X_x) = (R_{g_{UV}^{-1}(x)})_* \left((g_{UV})_* (X_x)\right) + Ad(g_{UV}(x))(\omega_U(X_x))}, \qquad (37.74)$$

for all $X_x \in T_x M$ *and* $x \in U \cap V$.

The above formula needs to be unpacked somewhat in order for its significance to be understood. Let $\gamma(t)$ be a curve through $x \in U \cap V$ such that $\gamma(0) = x$ and $\gamma'(0) = X_x$ (see Fig. 37.2). Then, in the case of matrix groups G,

$$(R_{g_{UV}^{-1}(x)})_* \left((g_{UV})_*(X_x)\right) = (R_{g_{UV}^{-1}(x)})_* \frac{d}{dt}(g_{UV}(\gamma(t)))_{t=0}$$

$$= \frac{d}{dt}(g_{UV}(\gamma(t)) \, g_{UV}^{-1}(x))_{t=0} = (dg_{UV})(X_x) \, g_{UV}^{-1}(x) \,, \quad (37.75)$$

where dg_{UV} is understood to be a matrix-valued 1-form on $U \cap V$. Also, by (37.73),

$$Ad(g_{UV}(x))(\omega_U(X_x)) = g_{UV}(x)\omega_U(X_x)g_{UV}^{-1}(x) \,. \qquad (37.76)$$

Thus, in the case of matrix groups G, we can write the **gauge transformation rule** (37.74) as

$$\boxed{\omega_V = (dg_{UV}) \, g_{UV}^{-1} + g_{UV} \, \omega_U \, g_{UV}^{-1}}. \qquad (37.77)$$

This is exactly in the same form as (35.17).

Let us now show that the three definitions of the connection on a principal bundle (Defs. 37.6, 37.7, and 37.8) are equivalent.

Suppose we are given a $\mathcal{G}$-valued connection one-form ω, as specified in Def. 37.7. We need to find a field of horizontal subspaces on P that satisfies the condition of Def. 37.6. This we define to be the field of subspaces that are annihilated by ω:

$$\mathcal{H}(p) = \{ \mathbb{X} \in T_p P \,|\, \omega_p(\mathbb{X}) = 0 \} . \tag{37.78}$$

By (37.70), the action of ω on any $\mathbb{Y} \in \mathcal{V}(p)$ (the vertical subspace) is equal to A, for some $A \in \mathcal{G}$, $A \neq 0$. Hence $\mathcal{V}(p) \cap \mathcal{H}(p) = \emptyset$ and $T_p P = \mathcal{V}(p) \oplus \mathcal{H}(p)$. Also, by (37.72) and (37.78), $\mathbb{X} \in \mathcal{H}(p)$ implies $\mathbb{X} \in (L_g)_* \, \mathcal{H}(p)$, and the converse is also true. Thus $(L_g)_* \, \mathcal{H}(p) = \mathcal{H}(gp)$.

Suppose now that a smooth $\mathcal{H}(p)$ is given satisfying the properties specified in Def. 37.6. Let $\mathbb{X} \in \mathcal{H}(p)$ and $A^{\#}(p) \in \mathcal{V}(p)$ as given by (37.69). We define a $\mathcal{G}$-valued one-form ω by

$$\omega(A^{\#} + \mathbb{X}) = A . \tag{37.79}$$

Setting $\mathbb{X} = 0$, we have $\omega(A^{\#}) = A$. This also implies that $\omega(\mathbb{X}) = 0$, for $\mathbb{X} \in \mathcal{H}(p)$. To complete the proof that ω as defined by (37.79) is indeed a connection one-form we need to show that it satisfies (37.72). Any $\mathbb{X} \in T_p P$ can be written $\mathbb{X} = a\mathbb{H} + b\mathbb{V}$, where $\mathbb{H} \in \mathcal{H}(p)$ and $\mathbb{V} \in \mathcal{V}(p)$. So we only need to verify (37.72) separately for arbitrary horizontal and vertical vectors. Suppose $\mathbb{H} \in \mathcal{H}(p)$. Then $\omega(\mathbb{H}) = 0$. Since $\mathcal{H}(p)$ is L_g-invariant [(37.67)], $\omega((L_g)_* \mathbb{H}) = 0$; and (37.72) is verified for $\mathbb{H} \in \mathcal{H}(p)$. Consider $\mathbb{V} = A^{\#}(p) \in \mathcal{V}(p)$, for some $A \in \mathcal{G}$. [Recall the definition of $A^{\#}$ in (37.69).] We have

$$
\begin{aligned}
&\omega_{gp}(\, (L_g)_* A^{\#}(p) \,) \\
&= \omega_{gp} \left((L_g)_* \frac{d}{dt} (\exp(At)p)_{t=0} \right) \quad \text{[by (37.69)]} \\
&= \omega_{gp} \left(\frac{d}{dt} (g \exp(At)p)_{t=0} \right) \\
&= \omega_{gp} \left(\frac{d}{dt} (g \exp(At) \, g^{-1} \cdot gp)_{t=0} \right) \\
&= \omega_{gp} \left(\frac{d}{dt} (ad_g(\exp(At)) \, gp)_{t=0} \right) \quad \text{[by (34.80)]} \\
&= \omega_{gp} \left(\frac{d}{dt} (\exp(Ad(g)At) \, gp)_{t=0} \right) \\
&= \omega_{gp}(\, (Ad(g)A)^{\#} gp \,) \quad \text{[(by (37.69)]} \\
&= Ad(g) \, A = (Ad(g)) \, (\omega(A^{\#}(p))) .
\end{aligned}
\tag{37.80}
$$

Thus (37.72) is proved for a vertical vector also. This completes the demonstration that Defs. 37.6 and 37.7 are equivalent. We turn now to the equivalence of Defs. 37.7 and 37.8.

Let ω be a $\mathcal{G}$-valued connection one-form as specified in Def. 37.7, and $T_U : \pi^{-1}(U) \to U \times G$ be a local trivialization. Associated with T_U there is a local section $\sigma_U : U \to P$ given by $\sigma_U(x) = T_U^{-1}(x, e)$ [c.f. (37.64)]. We define a $\mathcal{G}$-valued one-form ω_U on U by

$$\omega_U = \sigma_U^* \omega , \tag{37.81}$$

and claim that it is the connection one-form in the sense of Def. 37.8. To justify this claim we have to show that the gauge transformation $\omega_U \to \omega_V = \sigma_V^* \omega$ is given by (37.74). Suppose T_U is given by $T_U(p) = (\pi(p), \varphi_U(p))$, then, for $x = \pi(p)$,

$$T_U(\varphi_U(p)\sigma_U(x)) = (x, \ \varphi_U(\varphi_U(p)\sigma_U(x))) = (x, \ \varphi_U(p)\varphi_U(\sigma_U(x)))$$
$$= (x, \ \varphi_U(p)\, e) = (x, \varphi_U(p)) = T_U(p) . \tag{37.82}$$

Thus

$$p = \varphi_U(p)\sigma_U(x) . \tag{37.83}$$

Similarly,

$$p = \varphi_V(p)\sigma_V(x) , \tag{37.84}$$

where $x \in U \cap V$. The above two equations imply

$$\sigma_V(x) = \varphi_V^{-1}(p)\varphi_U(p)\sigma_U(x) = g_{UV}(x)\,\sigma_U(x) , \tag{37.85}$$

where the last equality follows from (37.58). Now let $X_x \in T_x M$, and suppose $\gamma(t)$ is a curve in M with $x = \gamma(0)$ and $X_x = \gamma'(0)$. We can compute the **lift** $(\sigma_V)_*(X_x)$ as follows:

$$(\sigma_V)_* (X_x) = \frac{d}{dt}\left(\sigma_V\left(\gamma(t)\right) \right)_{t=0} = \frac{d}{dt}\left(g_{UV}(\gamma(t))\,\sigma_U(\gamma(t)) \right)_{t=0}$$
$$= \frac{d}{dt}\left(g_{UV}(x)\,\sigma_U(\gamma(t)) \right)_{t=0} + \frac{d}{dt}\left(g_{UV}(\gamma(t))\,\sigma_U(x) \right)_{t=0}$$
$$= (L_{g_{UV}(x)})_* (\sigma_U)_*(X_x) + \frac{d}{dt}\left(g_{UV}(\gamma(t))\,g_{UV}^{-1}(x)\,\sigma_V(x) \right)_{t=0} \tag{37.86}$$
$$= (L_{g_{UV}(x)})_* (\sigma_U)_*(X_x) + \left((R_{g_{UV}^{-1}(x)})_* (g_{UV})_*(X_x) \right)^{\#}_{\sigma_V(x)} .$$

Then the given connection one-form ω on P can be applied to the RHS of the above equation to obtain

$$\omega_V(X_x) = ((\sigma_V)^*\omega)(X_x) = \omega((\sigma_V)_*(X_x))$$
$$= \omega\left(\left((R_{g_{UV}^{-1}(x)})_* (g_{UV})_* (X_x) \right)^{\#}_{\sigma_V(x)} \right) + \omega((L_{g_{UV}(x)})_* (\sigma_U)_*(X_x)) \tag{37.87}$$
$$= (R_{g_{UV}^{-1}(x)})_* ((g_{UV})_*(X_x)) + \mathcal{A}d(g_{UV}(x)) (\omega((\sigma_U)_*(X_x)))$$
$$= (R_{g_{UV}^{-1}(x)})_* ((g_{UV})_*(X_x)) + \mathcal{A}d(g_{UV}(x)) (\omega_U(X_x)) .$$

This is the same as (37.74). We have thus proved that $\omega_U = (\sigma_U)^*\omega$ is the desired local connection one-form on M.

    Conversely, suppose we are given local one-forms ω_U, ω_V, etc. that satisfy the gauge transformation rule of (37.74). Let σ_U be the local section associated with the local trivialization T_U [c.f. (37.64)]. For $x \in U$, $p = \sigma_U(x)$, $X_x \in T_x M$, and $A \in \mathcal{G}$, we define the map $\omega^U : T_p P \to \mathcal{G}$ by

$$\omega^U((\sigma_U)_* X_x + A^\#) = \omega_U(X_x) + A . \tag{37.88a}$$

A local $\mathcal{G}$-valued one-form defined on all of $\pi^{-1}(U)$ can be constructed from this map by extending its domain as follows:

$$\omega^U(\mathbb{X}_{gp}) = (\mathcal{A}d(g))\,\omega^U\left((L_{g^{-1}})_* \mathbb{X}_{gp}\right), \tag{37.88b}$$

for $\mathbb{X}_{gp} \in T_{gp} P$. The resulting one-form on $\pi^{-1}(U)$, also denoted by ω^U, satisfies (37.72) when restricted to the bundle $\pi : \pi^{-1}(U) \to U$, and is thus a connection one-form on this restricted bundle in the sense of Def. 37.7. Suppose T_V is another local trivialization, where $U \cap V \neq \emptyset$. We can similarly define a local connection one-form ω^V on $\pi^{-1}(V)$. Our task is to show that $\omega^U = \omega^V$ on $\pi^{-1}(U \cap V)$, in order for the local one-forms ω^U, ω^V, etc., to collectively define a global connection one-form ω in the sense of Def. 37.7. It is sufficient to show that $\omega^U = \omega^V$ on $\sigma_V(U \cap V)$, for then they would agree on all of $\pi^{-1}(U \cap V)$ by virtue of (37.88b). Since $\omega^U(A^\#) = A = \omega^V(A^\#)$, it is sufficient to check that $\omega^U((\sigma_V)_* X_x) = \omega^V((\sigma_V)_* X_x)$, for $x \in U \cap V$ and $X_x \in T_x M$. By (37.88a), $\omega^V((\sigma_V)_* X_x) = \omega_V(X_x)$, while [recall (37.86)]

$$\omega^U((\sigma_V)_* X_x) = \omega^U\left(((R_{g_{UV}^{-1}(x)})_* (g_{UV})_*(X_x))^\#_{\sigma_V(x)} + (L_{g_{UV}(x)})_* (\sigma_U)_*(X_x)\right)$$

$$= (R_{g_{UV}^{-1}(x)})_* (g_{UV})_*(X_x) + \mathcal{A}d(g_{UV}(x))(\omega_U(X_x)) = \omega_V(X_x), \tag{37.89}$$

where the last equality follows from (37.74). Hence we have constructed a global connection one-form ω on P from the local ones $\omega^U, \omega^V, \ldots$, which are in turn constructed from the local one-forms $\omega_U, \omega_V, \ldots$, on M. Furthermore, it is clear that $\omega_U = (\sigma_U)^* \omega$, $\omega_V = (\sigma_V)^* \omega$, etc. This completes the demonstration of the equivalence of Defs. 37.7 and 37.8.

In the physics literature, the local connection one-forms ω_U on a principal fiber bundle are called **gauge potentials**.

The above development would be rather pointless without the following basic theorem (which will be stated without proof).

Theorem 37.3. *Any principal fiber bundle admits a connection.*

Corresponding to a connection on a principal fiber bundle a curvature two-form on the total space P can be defined. This curvature will be shown to be closely related to the earlier defined curvature 2-form on a vector bundle [c.f. (35.18)]. First we give a preliminary definition.

Definition 37.9. *Let* $\varphi \in \Lambda^i(M, \mathcal{G})$ *and* $\psi \in \Lambda^j(M, \mathcal{G})$ *be* $\mathcal{G}$*-valued exterior differential forms on a manifold* M *of degrees* i *and* j, *respectively. The* **bracket product** *of* φ *and* ψ, *which is an element in* $\Lambda^{i+j}(M, \mathcal{G})$, *and denoted by* $[\varphi, \psi]_{\mathcal{G}}$, *is defined in terms of its action on* $(X_1, \ldots, X_{i+j})$, *where*

$X_1, \ldots, X_{i+j} \in \Gamma(TM)$ are arbitrary vector fields on M, by

$$[\varphi, \psi]_{\mathcal{G}}(X_1, \ldots, X_{i+j})$$
$$\equiv \frac{1}{(i+j)!} \sum_\sigma (sgn\,\sigma) \left[\varphi(X_{\sigma(1)}, \ldots, X_{\sigma(i)}), \psi(X_{\sigma(i+1)}, \ldots, X_{\sigma(i+j)}) \right],$$
$$(37.90)$$

where $\sigma \in \mathcal{S}(i+j)$ is a permutation in the symmetric group $\mathcal{S}(i+j)$, the sum is over all such permutations, and the bracket $[\,,]$ on the RHS of (37.90) is the Lie bracket of the Lie algebra $\mathcal{G}$.

As an example, suppose φ and ψ are both $\mathcal{G}$-valued 1-forms. Then

$$[\varphi, \psi]_{\mathcal{G}}(X_1, X_2) = \frac{1}{2}[\varphi(X_1), \psi(X_2)] - \frac{1}{2}[\varphi(X_2), \psi(X_1)].\qquad(37.91)$$

In particular, if ω is a $\mathcal{G}$-valued 1-form, then

$$[\omega, \omega]_{\mathcal{G}}(X_1, X_2) = \frac{1}{2}([\omega(X_1), \omega(X_2)] - [\omega(X_2), \omega(X_1)]) = [\omega(X_1), \omega(X_2)].$$
$$(37.92)$$

Exercise 37.7   Suppose $\varphi \in \Lambda^i(M, \mathcal{G})$ and $\psi \in \Lambda^j(M, \mathcal{G})$. Let $\{E_1, \ldots, E_f\}$ be a basis of $\mathcal{G}$. Show that there exist unique $\mathbb{R}$-valued i-forms φ^α and j-forms ψ^β, $\alpha, \beta = 1, \ldots, f$, such that

$$\varphi = \varphi^\alpha \otimes E_\alpha,\qquad(37.93)$$
$$\psi = \psi^\beta \otimes E_\beta.\qquad(37.94)$$

Show further that

$$[\varphi, \psi]_{\mathcal{G}} = (\varphi^\alpha \wedge \psi^\beta) \otimes [E_\alpha, E_\beta] = c^\gamma_{\alpha\beta}(\varphi^\alpha \wedge \psi^\beta) \otimes E_\gamma,\qquad(37.95)$$

where the $c^\gamma_{\alpha\beta}$ are the structure constants of $\mathcal{G}$ with respect to the basis $\{E_1, \ldots, E_f\}$.

Exercise 37.8   Use the anti-commutative rule of the exterior product [(28.5)] and the Jacobi identity of a Lie algebra [(4.26)] to show that

(1) $[\psi, \varphi]_{\mathcal{G}} = -(-1)^{ij}[\varphi, \psi]_{\mathcal{G}}$,   (37.96a)

(2) $(-1)^{ik}[[\varphi, \psi], \rho]_{\mathcal{G}} + (-1)^{kj}[[\rho, \varphi], \psi]_{\mathcal{G}} + (-1)^{ji}[[\psi, \rho], \varphi]_{\mathcal{G}} = 0$,
$$(37.96b)$$

where $\varphi \in \Lambda^i(M, \mathcal{G})$, $\psi \in \Lambda^j(M, \mathcal{G})$ and $\rho \in \Lambda^k(M, \mathcal{G})$.

The above two equations show that the algebra of $\mathcal{G}$-valued differential forms on M is a **graded Lie algebra**.

Exercise 37.9   Use the results of Ex. 37.8 to show that, if $\varphi \in \Lambda^i(M, \mathcal{G})$ and $\psi \in \Lambda^j(M, \mathcal{G})$, then

$$d\,[\,\varphi, \psi\,]_{\mathcal{G}} = [\,d\varphi, \psi\,]_{\mathcal{G}} + (-1)^i\,[\,\varphi, d\psi\,]_{\mathcal{G}}\,. \tag{37.97}$$

The curvature on a principal bundle can now be defined, in two steps.

Definition 37.10. *Suppose a connection is given on a principal bundle $\pi : P \to M$, with group G. Then any $\mathbb{X} \in T_pP$ can be written uniquely as the sum of a vertical and a horizontal vector: $\mathbb{X} = \mathbb{V} + \mathbb{H}$, so that if ω is the $\mathcal{G}$-valued one-form corresponding to the connection, $\pi_*\mathbb{V} = 0$ and $\omega(\mathbb{H}) = 0$. Suppose $\varphi \in \Lambda^i(P, \mathcal{G})$, the **exterior covariant derivative** of φ, written $D\varphi$ [$\in \Lambda^{i+1}(P, \mathcal{G})$], is given by*

$$D\varphi \equiv (d\varphi)^H\,, \tag{37.98}$$

where

$$(d\varphi)^H\,(\mathbb{X}_1, \ldots, \mathbb{X}_{i+1}) \equiv d\varphi\,(\mathbb{H}_1, \ldots, \mathbb{H}_{i+1})\,, \tag{37.99}$$

and $\mathbb{H}_1, \ldots, \mathbb{H}_{i+1}$ are the horizontal vectors of $\mathbb{X}_1, \ldots, \mathbb{X}_{i+1} \in T_pP$, respectively.

Definition 37.11. *Given a $\mathcal{G}$-valued connection one-form $\omega \in \Lambda^1(P, \mathcal{G})$ on a principal bundle $\pi : P \to M$, the $\mathcal{G}$-valued **curvature two-form** Ω is given by*

$$\boxed{\Omega \equiv D\omega}\,, \tag{37.100}$$

where D is the exterior covariant derivative corresponding to the given connection.

To make more sense of this rather abstract definition, we will try, in several steps, to project (37.100) down to the base manifold M so that it resembles the more familiar definition of the matrix-valued curvature Ω on a vector bundle [c.f (35.69)]. We will need the following three lemmas (the first one stated without proof).

Lemma 37.1. *Suppose $\pi : P \to M$ is a principal bundle with group G, and ω is a connection 1-form on P. Given a smooth vector field X on M, there exists a unique smooth vector field $\mathbb{X}^H$ on P such that $\omega(\mathbb{X}^H) = 0$ and $\pi_*(\mathbb{X}_p^H) = X_{\pi(p)}$ for all $p \in P$. The field $\mathbb{X}^H$, called the **horizontal lift** of X, is necessarily left-invariant:*

$$(L_g)_*\,\mathbb{X}_p^H = \mathbb{X}_{gp}^H\,, \qquad \text{for all } g \in G\,. \tag{37.101}$$

Lemma 37.2. *If $A, B \in \mathcal{G}$, then we have the following relationship between fundamental fields on P:*

$$[A, B]^\# = [A^\#, B^\#] \equiv A^\# B^\# - B^\# A^\#\,, \tag{37.102}$$

where on the LHS of the first equality, $[\,,\,]$ means the Lie bracket of $\mathcal{G}$. In other words, the map $\# : \mathcal{G} \to \mathcal{V}(p)$ (the vertical subspace at p) preserves the Lie algebra structure of $\mathcal{G}$.

Proof. The map for the fundamental field on P, $\# : \mathcal{G} \to \Gamma(TP)$, given by $\#(A) \equiv A^\#$ [c.f. (37.69)], can also be defined as follows. For every $p \in P$, define the map $\#_p : G \to P$ by $\#_p(g) = gp$, $g \in G$. Then, since $\#_p(e) = p$, for $A \in T_e G$,

$$A^\#(p) = \#(A) = \frac{d}{dt}(\exp(tA)\,p)_{t=0} = (\#_p)_*(A) \,. \tag{37.103}$$

$(\#_p)_* : \mathcal{G} \to T_p P$ is thus a linear map from $\mathcal{G}$ into $T_p P$. Now, by Theorem 34.5 and the definition of the Lie derivative [Def. 34.8 or Eq. (34.60)],

$$[A^\#, B^\#]_p = \lim_{t\to 0} \frac{1}{t} \{B^\# - (L_{g(t)})_* B^\#\}_p \,, \tag{37.104}$$

where $g(t) \equiv \exp(tA)$ and $L_{g(t)}$ (the left action of $g(t) \in G$ on P) is the one-parameter group of diffeomorphisms on P generated by $A^\#$. We have

$$((L_{g(t)})_* B^\#)_p = ((L_{g(t)})_*)\, B^\#_{L_{g^{-1}(t)}\cdot p} = ((L_{g(t)})_*)(\#_{L_{g^{-1}(t)}\cdot p})_*(B) \,, \tag{37.105}$$

where the second equality follows from (37.103). On the other hand, for arbitrary $g' \in G$,

$$\begin{aligned}(L_{g(t)} \circ \#_{L_{g^{-1}(t)}\cdot p})(g') &= g(t)\, g'\, g^{-1}(t) \cdot p \\ &= (ad_{g(t)}\, g') \cdot p = \#_p(ad_{g(t)}\, g') = (\#_p \circ ad_{g(t)})(g') \,.\end{aligned} \tag{37.106}$$

Thus, for $B \in T_e G$,

$$((L_{g(t)})_*)(\#_{L_{g^{-1}(t)}\cdot p})_*(B) = (\#_p)_* \, Ad(g(t))(B) \,, \tag{37.107}$$

where [c.f. (34.81)],

$$Ad(g(t)) \equiv (ad_{g(t)})_* \,.$$

Eq. (37.104) then implies

$$\begin{aligned}[A^\#, B^\#]_p &= \lim_{t\to 0} \frac{1}{t} \{(\#_p)_* B - (\#)_* Ad(g(t))B\} \\ &= (\#_p)_* \left(\lim_{t\to 0} \frac{1}{t} \{B - Ad(g(t))B\} \right) \,.\end{aligned} \tag{37.108}$$

We observe that, since for all $g' \in G$,

$$ad_{g(t)} g' = g(t)\, g'\, g^{-1}(t) = L_{g(t)} R_{g^{-1}(t)}\, g' \,,$$

for $B \in T_e G \sim \mathcal{G}$,

$$Ad(g(t))B = (L_{g(t)})_* \, (R_{g^{-1}(t)})_*\, B \,. \tag{37.109}$$

If we view the Lie algebra $\mathcal{G}$ as the set of right-invariant vector fields on G, so that $(R_{g^{-1}(t)})_* B = B$ [c.f. (34.17)], then

$$Ad(g(t))B = (L_{g(t)})_* B \,. \tag{37.110}$$

This result and (37.108) then finally imply that

$$[A^\#, B^\#]_p = (\#_p)_* \left(\lim_{t \to 0} \frac{1}{t} \{B - (L_{g(t)})_* B\} \right) = (\#_p)_*([A, B]) = [A, B]^\#(p) ,$$

(37.111)

where the second equality again follows from Theorem 34.5, and the third from (37.103). □

Lemma 37.3. *Suppose* X^H *is the horizontal lift of a vector field* X *on the base manifold* M, *and* $A \in \mathcal{G}$. *Then*

$$[A^\#, X^H] = 0 .$$

(37.112)

Proof. Let $g(t) = \exp(tA) \in G$. Then [c.f. (37.101)], X^H is invariant under $(L_g)_*$:

$$(L_{g(t)})_* X^H = X^H .$$

(37.113)

Thus, by Theorem (34.5) and the definition of the Lie derivative [(34.60)],

$$[A^\#, X^H] = \frac{d}{dt} \left((L_{g(t)})_* X^H \right)_{t=0} = \frac{d}{dt} X^H = 0 .$$

(37.114)

The last equality follows from the fact that X^H is independent of t. □

With the above three lemmas, we can now prove the following important theorem, whose result is known as the **structure equation** of the connection one-form on a principal bundle.

Theorem 37.4. *Suppose* ω *is a* $\mathcal{G}$-*valued connection one-form on a principal fiber bundle. Then*

$$\boxed{D\omega = d\omega + \frac{1}{2}[\omega, \omega]_\mathcal{G}} \quad ,$$

(37.115)

or, in view of the definition of the curvature 2-form Ω *[(37.100)],*

$$\boxed{\Omega = d\omega + \frac{1}{2}[\omega, \omega]_\mathcal{G}} \quad .$$

(37.116)

In (37.115), $D\omega$ *is the exterior covariant derivative of the* $\mathcal{G}$-*valued 1-form* ω. *In both of the above equations, the bracket* $[,]_\mathcal{G}$ *on the RHS's is the bracket of* $\mathcal{G}$-*valued forms given by Def. 37.9.*

Proof. By (37.92) and the definition of the exterior covariant derivative [(37.98) and (37.99)], we need to show that

$$d\omega(X_1^H, X_2^H) = d\omega(X_1, X_2) + \frac{1}{2}[\omega(X_1), \omega(X_2)] ,$$

(37.117)

for any two $X_1, X_2 \in \Gamma(TP)$, and X_1^H, X_2^H are the horizontal vector fields of X_1 and X_2, respectively. By linearity of $d\omega$, we need only show (37.117) for the following three cases.

(1) X_1 and X_2 are both horizontal. In this case $X_1 = X_1^H$ and $X_2 = X_2^H$. Also, by the definition of horizontal subspaces, given a connection one-form ω [(37.78)], $\omega(X_1) = \omega(X_1^H) = 0$ and $\omega(X_2) = \omega(X_2^H) = 0$. Eq. (37.117) follows.

(2) X_1 and X_2 are both vertical. We write $X_1 = X_1^V$ and $X_2 = X_2^V$. Then, for each $p \in P$, there exist $A, B \in \mathcal{G}$ such that

$$(X_1)_p = A^\#(p), \quad (X_2)_p = B^\#(p). \tag{37.118}$$

Recall from (28.29) that

$$d\omega(X_1, X_2) = \frac{1}{2}\langle X_1 \wedge X_2, d\omega \rangle. \tag{37.119}$$

We will use (30.39) to evaluate $\langle X_1 \wedge X_2, d\omega \rangle$, which is a consequence of the convention (28.28). We have, then,

$$d\omega(X_1, X_2) = d\omega(X_1^V, X_2^V) = d\omega(A^\#, B^\#) = \frac{1}{2}\langle A^\# \wedge B^\#, d\omega \rangle$$
$$= \frac{1}{2}\{A^\#(\omega(B^\#)) - B^\#(\omega(A^\#)) - \omega([A^\#, B^\#])\}$$
$$= \frac{1}{2}\{A^\#(B) - B^\#(A) - \omega([A, B]^\#)\}$$
$$= -\frac{1}{2}\omega([A, B]^\#) = -\frac{1}{2}[A, B] = -\frac{1}{2}[\omega(A^\#), \omega(B^\#)]$$
$$= -\frac{1}{2}[\omega(X_1^V), \omega(X_2^V)] = -\frac{1}{2}[\omega(X_1), \omega(X_2)],$$
$$\tag{37.120}$$

where in the sixth equality we have used the fact that $A^\#(B) = B^\#(A) = 0$ (since $A, B \in \mathcal{G}$ are constants), and also the result of Lemma 37.2. Thus the RHS of (37.117) vanishes. Clearly, the LHS vanishes also since in this case $X_1^H = X_2^H = 0$.

(3) X_1 is vertical and X_2 is horizontal. By Lemma 37.1, we may assume that X_2, being horizontal, is the horizontal lift of some vector field X on the base manifold M. Suppose also that $X_1 = A^\#$, for some $A \in \mathcal{G}$. Then

$$d\omega(X_1^H, X_2^H) = d\omega(0, X_2) = 0. \tag{37.121}$$

On the other hand,

$$d\omega(X_1, X_2) = d\omega(A^\#, X_2^H) = \frac{1}{2}\langle A^\# \wedge X_2^H, d\omega \rangle$$
$$= \frac{1}{2}\{A^\#(\omega(X_2^H)) - X_2^H(\omega(A^\#)) - \omega([A^\#, X_2^H])\} \tag{37.122}$$
$$= \frac{1}{2}\{A^\#(0) - X_2^H(A) - \omega([A^\#, X_2^H])\} = 0,$$

where in the last equality we have used the fact that $[A^{\#}, X_2^H] = 0$, which follows from Lemma 37.3. Also, it is clear that

$$[\omega(X_1), \omega(X_2)] = [\omega(A^{\#}), \omega(X_2^H)] = [A, 0] = 0. \tag{37.123}$$

Thus, again, both sides of (37.117) vanish.

□

The following are important and useful facts of the curvature two-form Ω on a principal bundle.

Theorem 37.5 (Bianchi identity). *Suppose ω is a $\mathcal{G}$-valued connection one-form on a principal bundle. Then we have [compare with (35.68)]*

$$\boxed{D\Omega = 0} \quad . \tag{37.124}$$

In fact [compare with (35.67) but notice the difference],

$$\boxed{d\Omega = [\Omega, \omega]_{\mathcal{G}}} \quad , \tag{37.125}$$

where the bracket $[\,,\,]_{\mathcal{G}}$ is in the sense of Def. 37.9.

Proof. From (37.99), for arbitrary $X, Y, Z \in T_p P$,

$$(D\Omega)(X, Y, Z) = (d\Omega)(X^H, Y^H, Z^H), \tag{37.126}$$

where X^H, Y^H and Z^H are the horizontal vectors of X, Y and Z, respectively, corresponding to the given connection one-form ω. But since $\omega(X^H) = \omega(Y^H) = 0$, (37.124) follows from (37.125). We will demonstrate the latter as follows. By the structure equation (37.116) for Ω,

$$d\Omega = d\left(d\omega + \frac{1}{2}[\omega, \omega]_{\mathcal{G}}\right) = \frac{1}{2}d\left([\omega, \omega]_{\mathcal{G}}\right) = \frac{1}{2}\left([d\omega, \omega]_{\mathcal{G}} - [\omega, d\omega]_{\mathcal{G}}\right)$$

$$= [d\omega, \omega]_{\mathcal{G}} = [d\omega + \frac{1}{2}[\omega, \omega]_{\mathcal{G}}, \omega] = [\Omega, \omega]_{\mathcal{G}}, \tag{37.127}$$

where the second equality follows from $d^2\omega = 0$, the third from (37.97), the fourth from (37.95), and the fifth from the fact that $[[\omega, \omega], \omega]_{\mathcal{G}} = 0$, which is a consequence of (37.96).

□

Theorem 37.6. *For all $g \in G$,*

$$\boxed{(L_g)^* \Omega = \mathcal{A}d(g)\, \Omega} \quad . \tag{37.128}$$

Proof. The definition of the bracket product [(37.90)] implies that $[\varphi, \psi]_{\mathcal{G}}$ is preserved under a pullback f^*, where f is a map $f : P \to P$, with P being the manifold on which the forms φ and ψ are defined:

$$f^* [\varphi, \psi]_{\mathcal{G}} = [f^*\varphi, f^*\psi]_{\mathcal{G}} .$$

Hence it follows from (37.71) that

$$(L_g)^* \, \Omega = (L_g)^* \, (d\omega + \frac{1}{2} [\omega, \omega]_{\mathcal{G}}) = d((L_g)^*\omega) + \frac{1}{2} [(L_g)^* \, \omega, (L_g)^*\omega]_{\mathcal{G}}$$

$$= d(Ad(g)(\omega)) + \frac{1}{2} [Ad(g)\omega, Ad(g)\omega]_{\mathcal{G}} = Ad(g) \, (d\omega + \frac{1}{2} [\omega, \omega]_{\mathcal{G}}) = Ad(g) \, \Omega .$$

$$\tag{37.129}$$

$\square$

To make contact with our earlier formulas for Ω given in Chapter 35, which were all established for connections on vector bundles (and hence where the structure groups G are matrix groups), we will use Def. 37.8 to identify ω and Ω by their equivalent local $\mathcal{G}$-valued forms defined on the base manifold M. Recall that for a local trivialization $T_U : \pi^{-1}(U) \to U \times G$ and corresponding local section $\sigma_U : U \to P$ given by $\sigma_U(x) = T_U^{-1}(x, e)$, the local connection one-form defined on U is $\omega_U = (\sigma_U)^* \, \omega$. Similarly, we can define the local curvature two-form on U:

$$\Omega_U \equiv (\sigma_U)^* \, \Omega . \tag{37.130}$$

It is easily seen that Ω_U and ω_U satisfy the same relationship that holds between Ω and ω:

$$\Omega_U = d\omega_U + \frac{1}{2} [\omega_U, \omega_U]_{\mathcal{G}} , \tag{37.131}$$

where the $\mathcal{G}$-valued forms Ω_U and ω_U are forms on the base manifold M, and d is the exterior derivative on M. The bracket product is still in the sense of Def. 37.9. Indeed, since d commutes with the pullback $(\sigma_U)^*$ and the bracket product structure is preserved under pullback, we have

$$\Omega_U = (\sigma_U)^* \, \Omega = (\sigma_U)^* \, (d\omega + \frac{1}{2} [\omega, \omega]_{\mathcal{G}})$$

$$= d ((\sigma_U)^* \, \omega) + \frac{1}{2} [(\sigma_U)^* \, \omega, (\sigma_U)^* \, \omega]_{\mathcal{G}} = d\omega_U + \frac{1}{2} [\omega_U, \omega_U]_{\mathcal{G}} . \tag{37.132}$$

We will now consider the important special case of a principal bundle whose structure group G is a matrix group. This also happens to be the case of the most interest in physics applications, where one speaks of the gauge groups in gauge field theory (for example, $U(1)$ in electrodynamics, and $SU(n)$ for other interactions in the standard model).

We recall from (34.44) that if $A, B \in \mathcal{G}$, where $\mathcal{G}$ is the Lie algebra of a matrix group,

$$[A, B] = BA - AB , \tag{37.133}$$

where the LHS is the Lie bracket and the RHS consists of matrix products. Now suppose $\varphi \in \Lambda^i(M, \mathcal{G})$ and $\psi \in \Lambda^j(M, \mathcal{G})$. By Def. 37.9 [(37.90)], we have, for arbitrary $X_1, \ldots, X_{i+j} \in \Gamma(TM)$,

$$[\varphi, \psi]_{\mathcal{G}}(X_1, \ldots, X_{i+j})$$

$$= \frac{1}{(i+j)!} \sum_\sigma (sgn\,\sigma) [\varphi(X_{\sigma(1)}, \ldots, X_{\sigma(i)}), \psi(X_{\sigma(i+1)}, \ldots, X_{\sigma(i+j)})]$$

$$= \frac{1}{(i+j)!} \{ \sum_\sigma (sgn\,\sigma) \psi(X_{\sigma(i+1)}, \ldots, X_{\sigma(i+j)}) \varphi(X_{\sigma(1)}, \ldots, X_{\sigma(i)})$$

$$- \sum_\sigma (sgn\,\sigma) \varphi(X_{\sigma(1)}, \ldots, X_{\sigma(i)}) \psi(X_{\sigma(i+1)}, \ldots, X_{\sigma(i+j)}) \}$$

$$= \frac{1}{(i+j)!} \sum_\sigma (sgn\,\sigma)(-1)^{ij} \psi(X_{\sigma(1)}, \ldots, X_{\sigma(j)}) \varphi(X_{\sigma(j+1)}, \ldots, X_{\sigma(j+i)})$$

$$- (\varphi \wedge \psi)(X_1, \ldots, X_{i+j})$$

$$= ((-1)^{ij} \psi \wedge \varphi - \varphi \wedge \psi)(X_1, \ldots, X_{i+j}),$$

$$(37.134)$$

where in the fourth equality we have used the definition of a general wedge product given by (28.3). Thus, for matrix-valued forms,

$$\boxed{[\varphi, \psi]_{\mathcal{G}} = -(\varphi \wedge \psi - (-1)^{ij} \psi \wedge \varphi)} \quad , \qquad (37.135)$$

where on the RHS, the matrix multiplication is via the wedge. Note that, for matrix-valued forms,

$$[\varphi, \psi]_{\mathcal{G}} = -[\varphi, \psi]_{gr}, \qquad (37.136)$$

where $[\,,\,]_{gr}$, the graded commutator of matrix-valued forms, was defined by (35.65). Since

$$\frac{1}{2}[\omega, \omega]_{\mathcal{G}} = -\frac{1}{2}(\omega \wedge \omega - (-1)\omega \wedge \omega) = -\omega \wedge \omega,$$

the structure equation (37.115) is given by

$$D\omega = d\omega - \omega \wedge \omega, \qquad (37.137)$$

or

$$\Omega = d\omega - \omega \wedge \omega, \qquad (37.138)$$

for matrix-valued connection one-forms and matrix-valued curvature two-forms. Similarly

$$\Omega_U = d\omega_U - \omega_U \wedge \omega_U. \qquad (37.139)$$

The Bianchi identity $d\Omega = [\Omega, \omega]$ [(37.125)] can be written

$$d\Omega = \omega \wedge \Omega - \Omega \wedge \omega. \qquad (37.140)$$

The above three equations have exactly the same form as those given in Chapter 35 for the case of vector bundles, in which the structure groups are matrix groups.

Remark The different kinds of Lie algebra products introduced up to now, especially in this chapter and the last, may be a source of some confusion. We summarize here the notation used in this book for reference.

$$[\,,\,] = \text{Lie bracket in the general sense,}$$
$$[\,,\,]_{\text{gr}} = \text{graded commutator of matrix-valued forms } [(35.65)],$$
$$[\,,\,]_{\mathcal{G}} = \text{Lie bracket of } \mathcal{G}\text{-valued forms } [(37.90)].$$

Thus, for vector fields X, Y,

$$[X, Y] = XY - YX ; \quad X, Y \in \Gamma(TM) .$$

For matrix Lie algebra elements A, B,

$$[A, B] = -(AB - BA) ; \quad A, B \in \mathcal{GL}(n) .$$

For matrix-valued forms $\varphi \in \Lambda^i(M, \mathcal{G}), \psi \in \Lambda^j(M, \mathcal{G})$,

$$[\varphi, \psi]_{\text{gr}} = \varphi \wedge \psi - (-1)^{ij} \psi \wedge \varphi , \quad [\varphi, \psi]_{\mathcal{G}} = -[\varphi, \psi]_{\text{gr}} .$$

From now on we will refer to a principal fiber bundle with Lie group G as a **principal G-bundle**. There is a natural way to associate a principal G-bundle with vector bundles. To illustrate the idea, let us recall the frame bundle (P, M, π) (as a principal G-bundle) introduced at the beginning of this chapter. We have already mentioned that (P, M, π) is associated with the tangent bundle TM, which is a vector bundle. The association is done explicitly as follows. Pick a frame $(x; e_1, \ldots, e_n) = p \in P$, with origin at $x = \pi(p)$. A change of frames at x is effected by a left action L_g, $g \in GL(n, \mathbb{R})$, [given by (37.12) and (37.13)]:

$$L_g\,(x; e_1, \ldots, e_n) = (x; (e')_1, \ldots, (e')_n) \equiv gp = p' , \quad (e')_i = g_i^j\, e_j .$$

Suppose a tangent vector $X \in T_x M$ is given in terms of e_i by $X = X^i e_i$. Define a right action G on $T_x M$, written Xg, by

$$X \to X' = Xg = (X')^i e_i , \tag{37.141}$$

where

$$(X')^i = X^j\, g_j^i .$$

Consider a function $f : \pi^{-1}(x) \to T_x M$ such that, for an arbitrary frame $p \in \pi^{-1}(x)$ and $g \in GL(n, \mathbb{R})$,

$$f(p)g^{-1} = f(gp) . \tag{37.142}$$

In other words, if

$$f(p) = X = X^i e_i , \tag{37.143}$$

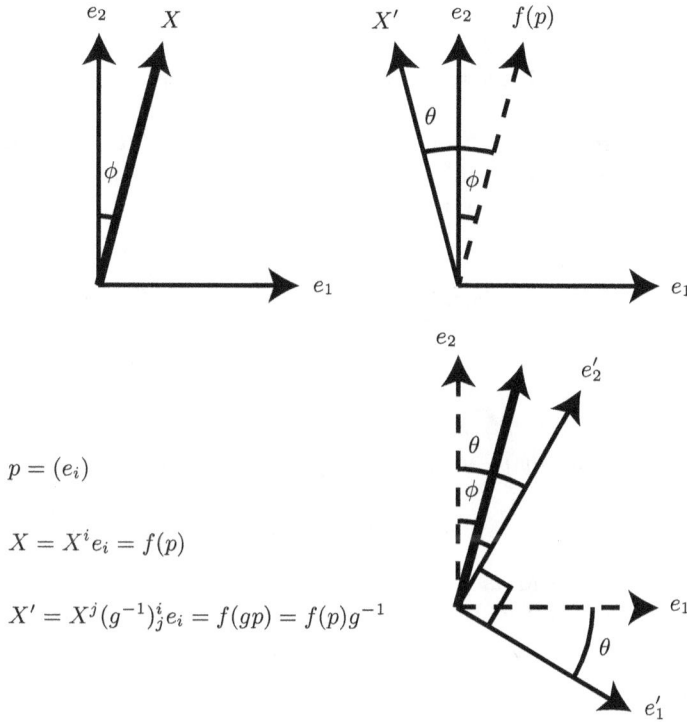

FIGURE 37.3

and

$$(e')_i = g_i^j \, e_j \, , \quad [\text{ or } e_i = (g^{-1})_i^j \, (e')_j \,] \, , \tag{37.144}$$

then

$$f(gp) = X' = (X')^i \, e_i \, , \tag{37.145}$$

where

$$(X')^i = X^j \, (g^{-1})_j^i = (Xg^{-1})^i \, ; \tag{37.146}$$

so that

$$(X')^i \, (e')_i = X^j \, (g^{-1})_j^i \, g_i^k \, e_k = X^j \, e_j \, ,$$

which is the same equation as (1.11), and identifies a unique vector in $T_x M$ (represented by the thick arrows in Fig. 37.3). The function f induces an equivalence relationship $\sim$ in the product set $\mathbb{R}^n \times P$, given by

$$(X, p) \sim (Xg^{-1}, \, gp) \, . \tag{37.147}$$

The quotient set $\mathbb{R}^n \times P / \sim$ [under the equivalence of (37.147), for all $g \in GL(n, \mathbb{R})$] is called the associated bundle of the frame bundle relative to the defining representation of $GL(n, \mathbb{R})$. We can denote this vector bundle by $\pi : \mathbb{R}^n \times_G P \to M$. It is isomorphic to the tangent bundle TM, since the fiber above $x \in M$ in the associated bundle, $\pi^{-1}(x)$, is isomorphic to the tangent space $T_x M$.

We can now replace the frame bundle by a general principal G-bundle, and $T_x M$ by a vector space $\mathbb{V}$ on which G can be represented. The notion of a general associated bundle can then be defined as follows.

Definition 37.12. *Let $r : G \to GL(\mathbb{V})$ be a representation of G, where $GL(\mathbb{V})$ is the general linear group of the vector space $\mathbb{V}$. Let $\pi : P \to M$ be a principal G-bundle. Define a left action of G on $\mathbb{V} \times P$ (and a corresponding right action of G on $\mathbb{V}$) by*

$$g(v, p) \equiv (r(g^{-1})v, gp) \equiv (vg^{-1}, gp) , \qquad (37.148)$$

[compare with (37.142)] and denote the orbit of (v, p) under G by $[v, p]$. Denote the set of all such orbits by $\mathbb{V} \times_G P$. Define a projection $\pi_{\mathbb{V}} : \mathbb{V} \times_G P \to M$ by

$$\pi_{\mathbb{V}}([v, p]) = \pi(p) . \qquad (37.149)$$

*Then $\pi_{\mathbb{V}} : \mathbb{V} \times_G P \to M$ is a vector bundle with typical fiber isomorphic to $\mathbb{V}$; it is called the **associated bundle** of the principal G-bundle P relative to the group representation r of G.*

Exercise 37.10 | Show that the left action given by (37.148) is free. [Recall Def. 6.6].

Exercise 37.11 | Show that the dimension of the associated bundle $\pi_{\mathbb{V}} : \mathbb{V} \times_G P \to M$ is $dim(M) + dim(\mathbb{V})$.

Exercise 37.12 | Show that any two points in the fiber above $\pi(p)$ in the associated bundle have unique representations of the form (v_1, p), (v_2, p), where $v_1, v_2 \in \mathbb{V}$; and one can define the vector addition

$$[v_1, p] + [v_2, p] \equiv [v_1 + v_2, p] . \qquad (37.150)$$

The associated bundle is the proper mathematical set-up for the so-called **wave functions** in physics. Consider a principal G-bundle $\pi : P \to M$ and an associated bundle $\pi_{\mathbb{V}} : \mathbb{V} \times_G P \to M$ with respect to a representation $r : G \to GL(\mathbb{V})$. The wave functions are precisely generalizations of the functions f of (37.142). More specifically, these are functions $\psi : P \to \mathbb{V}$ such that, for any $p \in P$,

$$\psi(gp) = r(g^{-1}) \psi(p) , \qquad \text{for all } g \in G \qquad (37.151)$$

It is customary in physics to project ψ from P to the base manifold M using local trivializations (or equivalently, local sections) of P. In physics, these are called

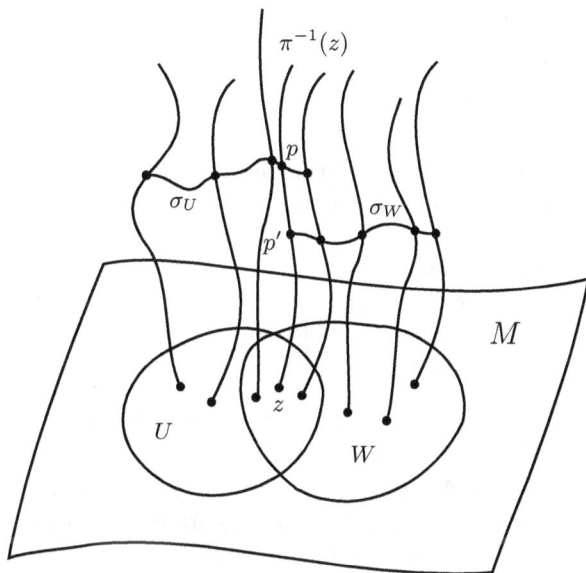

FIGURE 37.4

local choices of gauge. Let $\sigma_U : M \to P$ be a local section on $\pi : P \to M$. Then

$$\psi_U \equiv (\sigma_U)^* \, \psi \, , \tag{37.152}$$

defined in the coordinate neighborhood $(U; x^i)$, is called a **local wave function**. Alternatively, it can also be considered as a local section on the associated bundle:

$$\psi_U(x) = [\, \psi(\sigma_U(x)), \, \sigma_U(x)\,] \, .$$

Suppose $(W; y^i)$ is another coordinate neighborhood with local section σ_W, and $U \cap W \neq \emptyset$. Let $z \in U \cap W$. Then, by (37.63), (37.64) and (37.58) (also see Fig. 37.4)

$$p' = \sigma_W(z) = e^{-1} \cdot \varphi_U(p') \cdot p = (\varphi_W(p'))^{-1} \, \varphi_U(p') \cdot p = g_{UW}(z) \cdot \sigma_U(z) \, , \tag{37.153}$$

where $g_{UW}(z)$ is the transition function defined in (37.58) [recall that $g_{UW}(x)$ is independent of $p \in \pi^{-1}(x)$]. It follows that

$$
\begin{aligned}
\psi_W(z) &= ((\sigma_W)^* \psi)(z) = \psi(\sigma_W(z)) = \psi(\, g_{UW}(z) \cdot \sigma_U(z)\,) \\
&= r((g_{UW}(z))^{-1}) \, \psi(\sigma_U(z)) = r((g_{UW}(z))^{-1}) \, \psi_U(z) \, .
\end{aligned}
\tag{37.154}
$$

The rule for the transformation of a wave function $\psi_U \to \psi_W$ under a local change of gauge is thus

$$\psi_W(z) = r((g_{UW}(z))^{-1}) \, \psi_U(z) \equiv \psi_U(z) \cdot (g_{UW}(z))^{-1} \qquad . \qquad (37.155)$$

Physicists refer to the local connection one-forms ω_U on a principal G-bundle as **gauge potentials**. Electromagnetic potentials again furnish a most basic example. Consider a principal $U(1)$-bundle $\pi : P \to M$ on Minkowski space (M, η), where η is the Lorentz metric. Since $U(1) = \{ e^{i\theta} \,|\, \theta \in \mathbb{R} \}$, its Lie algebra is $\mathcal{U}(1) = \{ i\alpha \,|\, \alpha \in \mathbb{R} \}$. Let a connection ω be given as a collection of local one-forms $\omega_U = (\sigma_U)^*\omega$ (Def. 37.8). Then the **vector potential one-forms** $A^{(U)} \in \Lambda^1(U; \mathcal{U}(1)) = \Lambda^1(U; \mathbb{R})$ are given by

$$\omega_U \equiv -iA^{(U)} = -iA_\mu^{(U)} dx^\mu , \quad \mu = 0, 1, 2, 3, \qquad (37.156)$$

where the real functions $A_\mu^{(U)}(x)$ are the **vector potentials** of electrodynamics. [For the unitary groups $U(n)$, the Lie algebra elements are anti-hermitian matrices. The factor $(-i)$ is inserted to make the A_μ hermitian, to correspond to physical observables.] Under a local change of gauge $\sigma_U \to \sigma_V$, it follows from (37.77) that the gauge transformation rule for $A^{(U)}$ is given by

$$A^{(V)} = i(dg_{UV}) \, g_{UV}^{-1} + g_{UV} \, A^{(U)} \, g_{UV}^{-1} . \qquad (37.157)$$

Equivalently,

$$A_\mu^{(V)} = i\partial_\mu(g_{UV}) \, g_{UV}^{-1} + g_{UV} \, A_\mu^{(U)} \, g_{UV}^{-1} \qquad . \qquad (37.158)$$

Writing $g_{UV}(x) = e^{-i\theta(x)}$ in the present case, the above equation assumes the following familiar form:

$$(A')_\mu(x) = A_\mu(x) + \partial_\mu\theta(x) . \qquad (37.159)$$

In the general case of non-abelian matrix groups G, the matrix-valued **field strength** $F_{\mu\nu}(x)$ is related to the gauge potentials A_μ by (35.43):

$$F_{\mu\nu} = \partial_\mu A_\nu - \partial_\nu A_\mu + i[\, A_\mu, A_\nu \,] . \qquad (37.160)$$

Its components transform tensorially according to [recall (35.25)]

$$F_{\mu\nu}^{(V)} = g_{UV} \, F_{\mu\nu}^{(U)} \, g_{UV}^{-1} \qquad . \qquad (37.161)$$

Since $U(1)$ is abelian, the electromagnetic field strength $F_{\mu\nu}$ is given in terms of the vector potentials A_μ by

$$F_{\mu\nu} = \partial_\mu A_\nu - \partial_\nu A_\mu , \quad \text{(electromagnetic)} , \qquad (37.162)$$

[which is the same as (2.18)], and is invariant under a gauge transformation [by (37.161)].

Let us finally consider the gauge transformation of the directional covariant derivative of a local wave function:

$$D_\mu \psi \equiv D_{\partial_\mu} \psi = \partial_\mu \psi - i\psi A_\mu , \tag{37.163}$$

where we have not specified the coordinate neighborhood for brevity, and $(\psi A_\mu)^i = \psi^j (A_\mu)^i_j$. Under a local gauge transformation [c.f. (37.158) and (37.155)]

$$A_\mu \to (A')_\mu = i(\partial_\mu g)g^{-1} + gA_\mu g^{-1} , \tag{37.164}$$

$$\psi \to \psi' = \psi g^{-1} , \tag{37.165}$$

we have

$$\begin{aligned}
D_\mu \psi &\to (D_\mu \psi)' = \partial_\mu \psi' - i\psi'(A_\mu)' \\
&= \partial_\mu(\psi g^{-1}) - i\psi g^{-1} \left(i(\partial_\mu g)g^{-1} + gA_\mu g^{-1} \right) \\
&= (\partial_\mu \psi)g^{-1} + \psi\partial_\mu(g^{-1}) + \psi g^{-1}(\partial_\mu g)g^{-1} - i\psi A_\mu g^{-1} .
\end{aligned} \tag{37.166}$$

Since $gg^{-1} = 1$, it follows that $(\partial_\mu g)g^{-1} = -g\partial_\mu(g^{-1})$, which implies

$$\psi g^{-1}(\partial_\mu g)g^{-1} = -\psi g^{-1} \cdot g\partial_\mu(g^{-1}) = -\psi\partial_\mu(g^{-1}) .$$

Thus

$$(D_\mu \psi)' = (\partial_\mu \psi - i\psi A_\mu) \cdot g^{-1} = (D_\mu \psi) \cdot g^{-1} , \tag{37.167}$$

confirming the fact that $D_\mu \psi$ transforms as a wave function under a gauge transformation. Hence the directional covariant derivative of a wave function is also a wave function.

In the physics literature, the covariant derivative $D_\mu = \partial_\mu - iA_\mu$, considered as an operator, is sometimes referred to as a **minimal coupling**. For example, the Schrödinger equation describing the coupling between a charged particle (of charge e and mass m) and a classical electromagnetic field is

$$\left(-\frac{\hbar^2}{2m} D_\mu^2 + V(x) \right) \psi(x) = E\psi(x) , \tag{37.168}$$

where

$$D_\mu \equiv \partial_\mu - i\frac{e}{\hbar c} A_\mu(x) , \tag{37.169}$$

and $V(x)$ is the potential energy due to some conservative field. It is clear that (37.168) is invariant under the gauge transformation given by (37.164) and (37.165). This fact is a particular manifestation of the all-important **principle of gauge invariance** in physics.

Chapter 38

Magnetic Monopoles and Molecular Dynamics

In this chapter we will present some examples in non-relativistic quantum mechanics that illustrate the essential role played by the geometry of fiber bundles in physical theories involving gauge invariance. We will begin with the so-called $U(1)$ magnetic monopole.

The magnetic field B due to a **magnetic monopole** of strength (magnetic charge) M is given, according to Coulomb's law, by

$$B = \frac{M\boldsymbol{r}}{r^3} , \qquad (38.1)$$

where $\boldsymbol{r}$ is the radial vector from the position of the monopole to the field point. It is obvious, according to Gauss' law, that the magnetic flux Φ over a sphere S^2 (centered at the monopole) is

$$\Phi = \oint_{S^2} \boldsymbol{B} \cdot d\boldsymbol{a} = 4\pi M . \qquad (38.2)$$

On the other hand, one can use Stokes' theorem to obtain

$$\Phi = \oint_{S^2} (\nabla \times \boldsymbol{A}) \cdot d\boldsymbol{a} = \oint \boldsymbol{A} \cdot d\boldsymbol{l} = \oint (\boldsymbol{A}_N - \boldsymbol{A}_S) \cdot d\boldsymbol{l} = 4\pi M . \qquad (38.3)$$

In the above equation we have divided the sphere into two hemispheres, northern (N) and southern (S), and applied Stokes' theorem separately to each (see Fig. 38.1). $\boldsymbol{A}_N$ and $\boldsymbol{A}_S$ are, respectively, the vector potentials on the northern and southern hemispheres. Eq. (38.3) implies that, along the equator, $\boldsymbol{A}_N \neq \boldsymbol{A}_S$; otherwise, the total flux Φ over the whole sphere would vanish. Thus $\boldsymbol{A}$ cannot be given by the same expression over the entire sphere.

The solution of the problem rests on the basic notion of a differentiable manifold, which, as we have seen, can be regarded as pieces of a Euclidean space smoothly pasted together. In this case, we imagine the 2-sphere S^2 to be

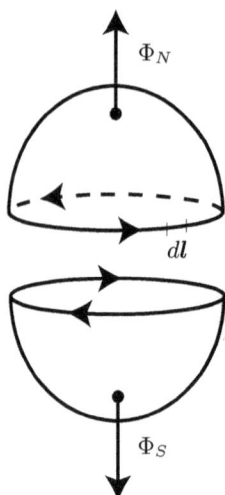
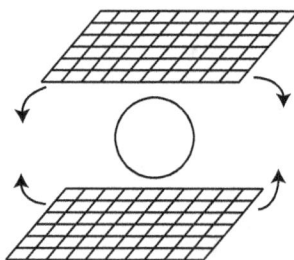

FIGURE 38.1   FIGURE 38.2

covered by two coordinate charts, each homeomorphic to an open region of the Euclidean plane $\mathbb{R}^2$, one covering the northern hemisphere, and the other the southern hemisphere (see Fig. 38.2).

Each point on the sphere is then given a pair of coordinates (x^1, x^2), in such a way that on the overlap, say the equator, the coordinates according to one chart are differentiable functions of the coordinates according to the other. Such a set-up gives a differentiable structure to the 2-sphere, and makes it a differentiable manifold of dimension two. The vector potential $\boldsymbol{A}$ is given by two different expressions, one valid only in the northern hemisphere, and the other valid only in the southern hemisphere. We shall see that, on the overlap (taken to be the equator), these two expressions are related precisely by the gauge transformation rule (37.159) for a connection on a principal G-bundle, where G is abelian.

In fact, we have

$$\boldsymbol{A}_N = \boldsymbol{A}_S + \nabla f , \tag{38.4}$$

where f is a function on the equator. $\boldsymbol{A}_N$ and $\boldsymbol{A}_S$ are given by the so-called **Wu-Yang potential** for the magnetic monopole:

$$\boldsymbol{A}_N(\boldsymbol{r}) = \frac{M(1 - \cos\theta)}{r \sin\theta} \hat{e}_\varphi , \tag{38.5}$$

$$\boldsymbol{A}_S(\boldsymbol{r}) = -\frac{M(1 + \cos\theta)}{r \sin\theta} \hat{e}_\varphi , \tag{38.6}$$

where (r, θ, φ) are the usual spherical coordinates. We note that $\boldsymbol{A}_N$ [as given by (38.5)] is actually valid over the entire sphere except at $\theta = \pi$, whereas $\boldsymbol{A}_S$ [as given by (38.6)] is valid everywhere except at $\theta = 0$. These two regions

$\theta = \pi$ string

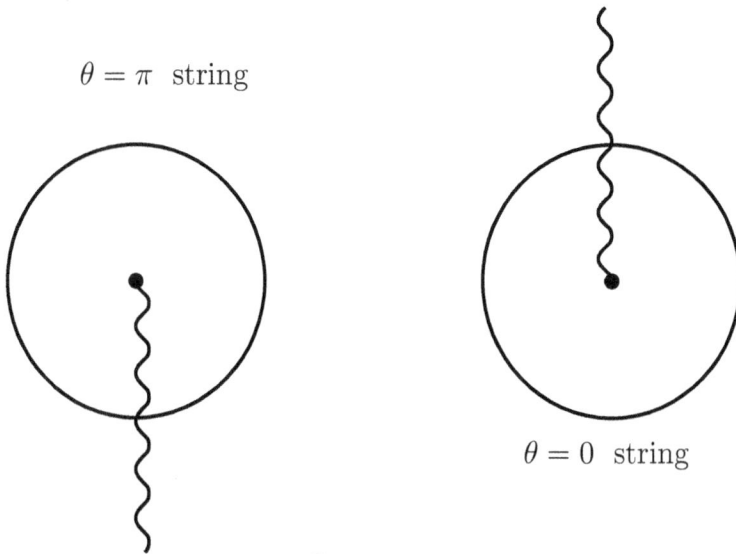

$\theta = 0$ string

FIGURE 38.3

of singularity (Fig. 38.3) are called **Dirac strings** in the physics literature. Thus one can also visualize the two coordinate charts to cover the 2-sphere in the manner shown in Fig. 38.4, where the region of overlap is everywhere on the sphere except at the north and south poles, the two points where the Dirac strings intersect the sphere. It follows easily from (38.5) and (38.6), using ordinary vector calculus, that

$$\boldsymbol{A}_N - \boldsymbol{A}_S = \frac{2M}{r \sin \theta} \hat{e}_\varphi = \nabla \left(2M\varphi \right),\tag{38.7}$$

which is of the form given by (38.4). Thus the line integral around the equator in (38.3) can be evaluated as follows:

$$\oint (\boldsymbol{A}_N - \boldsymbol{A}_S) \cdot d\boldsymbol{l} = \oint \nabla \left(2M\varphi \right) \cdot d\boldsymbol{l}$$
$$= 2M \oint \frac{1}{r \sin \theta} \hat{e}_\varphi \cdot d\boldsymbol{l} = 2M \int_0^{2\pi} \frac{\hat{e}_\varphi \cdot (r \sin \theta d\varphi)\hat{e}_\varphi}{r \sin \theta} = 4\pi M .\tag{38.8}$$

This verifies (38.3).

Exercise 38.1 Show that the Wu-Yang potential can also be expressed in Cartesian components as

$$A_N^x = -\frac{My}{r(r+z)} , \qquad A_N^y = \frac{Mx}{r(r+z)} , \qquad A_N^z = 0 ,\tag{38.9}$$

$$A_S^x = \frac{My}{r(r-z)} , \qquad A_S^y = -\frac{Mx}{r(r-z)} , \qquad A_S^z = 0 .\tag{38.10}$$

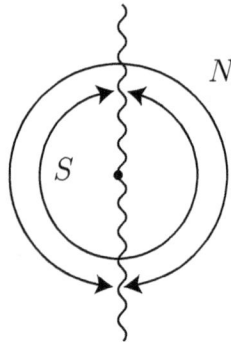

FIGURE 38.4

With these expressions, verify directly that

$$\nabla \times \boldsymbol{A}_N = \frac{M\hat{r}}{r^2} + 4\pi M\delta(x)\delta(y)\theta(-z) , \qquad (38.11)$$

$$\nabla \times \boldsymbol{A}_S = \frac{M\hat{r}}{r^2} + 4\pi M\delta(x)\delta(y)\theta(z) , \qquad (38.12)$$

where $\theta(z)$ is the step function (c.f. Fig. 14.3).

Let us express the above results using the geometric language of fiber bundles. The components of the vector potential $A_i = A^i$, $i = 1, 2\,(\theta, \varphi)$, of a magnetic monopole can be associated with the local connection one-form on a principal $U(1)$-bundle $\pi : P \to S^2$ in the following manner:

$$\omega = -i\left(\frac{e}{\hbar c}\right) A_j\, dx^j , \quad j = 1, 2 . \qquad (38.13)$$

(The physical constant $e/(\hbar c)$ has been inserted to allow for the quantum mechanical description of the interaction between an electron and a magnetic monopole, as will be seen shortly.) From (37.158), (37.159), and (38.7), it is seen that the transition function $g_{SN}(\varphi)$ (relating the north and south coordinate patches) is given by

$$g_{SN}(\varphi) = \exp\left(-2\frac{ie}{\hbar c}M\varphi\right) , \qquad (38.14)$$

where the overlap region is the equator ($\sim S^1$) and thus g_{SN} is a map $g_{SN} : S^1 \to U(1)$. Denoting the northern and southern coordinate patches by U_N and U_S, respectively, our principal bundle has the local structures $U_N \times U(1)$ and $U_S \times U(1)$, with local trivializations $T_N : \pi^{-1}(U_N) \to U_N \times U(1)$ and $T_S : \pi^{-1}(U_S) \to U_S \times U(1)$ given by

$$T_N(p) = (\pi(p),\, \phi_N(p)) , \qquad (38.15)$$
$$T_S(p) = (\pi(p),\, \phi_S(p)) , \qquad (38.16)$$

where $\phi_N : \pi^{-1}(U_N) \to U(1)$ and $\phi_S : \pi^{-1}(U_S) \to U(1)$ are maps such that, for $p \in \pi^{-1}(U_N \cap U_S)$,

$$g_{SN}(\varphi) = (\phi_N)^{-1}(p)\phi_S(p) = \exp\left(-\frac{2ie}{\hbar c}M\varphi\right) . \tag{38.17}$$

Corresponding to the above local trivializations we have the local sections on P:

$$\sigma_S(x) = T_S^{-1}(x, e) , \quad x \in U_S , \tag{38.18}$$

$$\sigma_N(x) = T_N^{-1}(x, e) , \quad x \in U_N . \tag{38.19}$$

In other words, if $p = g\sigma_{N,S}(\pi(p))$, then $\phi_{N,S}(p) = g$. Let $\phi_N(p) = e^{i\gamma_N(p)}$ and $\phi_S(p) = e^{i\gamma_S(p)}$. Then the local coordinates of our principal bundle are given by $(\theta, \varphi; \gamma_N(p))$ for $p \in \pi^{-1}(U_N)$ and $(\theta, \varphi; \gamma_S(p))$ for $p \in \pi^{-1}(U_S)$. The local trivializations, or local choices of gauge, are just different reference points (origins) on the typical fiber [in this case $U(1)$] with respect to which the fiber coordinates $\gamma_N(p)$ and $\gamma_S(p)$ are measured: if $p = e^{i\theta}\sigma_{N,S}(\pi(p))$, then $\gamma_{N,S}(p) = \theta$. Eq. (38.17) implies that

$$\gamma_N(p) - \gamma_S(p) = \frac{2e}{\hbar c}M\varphi , \tag{38.20}$$

where φ is the azimuth coordinate on $U_N \cap U_S$ (that is, the equator). The fact that there is no global gauge (or global trivialization) shows that the $U(1)$-bundle for the magnetic monopole is non-trivial.

Now consider the wave function $\psi(x, t)$ of an electron (of charge e and mass m) interacting with a magnetic monopole. It satisfies the Schrödinger equation

$$\frac{1}{2m}\left(\boldsymbol{p} - \frac{e}{c}\boldsymbol{A}\right)^2 \psi = i\hbar\frac{\partial\psi}{\partial t} , \tag{38.21}$$

where $p_j = \frac{\hbar}{i}\partial_j$. Corresponding to $\boldsymbol{A}_N$ and $\boldsymbol{A}_S$, we have the solutions ψ_N and ψ_S, valid in the coordinate neighborhoods U_N and U_S, respectively. The transformation between these is given by (37.165) and (38.14):

$$\psi_N = \psi_S \cdot g_{SN}^{-1} = \exp\left(\frac{ie}{\hbar c}2M\varphi\right)\psi_S . \tag{38.22}$$

A most interesting and significant physical consequence follows from the fact that the transition function $g_{SN}(\varphi)$ is required to be single-valued on S^1. In order for this condition to be fulfilled, it is required that

$$\frac{2eM}{\hbar c} \cdot 2\pi = 2n\pi , \quad n = 0, \pm1, \pm2, \ldots \in \mathbb{Z} ,$$

from which follows the quantization condition of the magnetic charge:

$$\boxed{\frac{2eM}{\hbar c} \in \mathbb{Z}} . \tag{38.23}$$

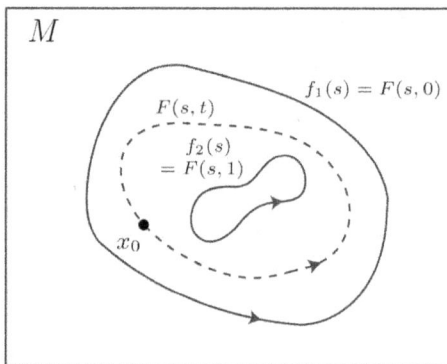

FIGURE 38.5

The wave function satisfying (38.21) is in fact a section of the associated bundle $\pi_{\mathbb{C}^1} : \mathbb{C}^1 \times_{U(1)} P \to S^2$ (recall the notation of Def. 37.12), where $\mathbb{C}^1$ is the standard one-dimensional complex vector space, on which ψ takes values.

On the overlap region $U_N \cap U_S$ (the equator), the fibers have to be "sewn together" according to the transition functions $g_{SN}(\varphi)$, which determine the structure (non-triviality) of the bundle.

The transition functions $g_{SN} : S^1 \to U(1)$ are classified according to **homotopy classes** of mappings from S^1 to S^1 [since $U(1) \sim S^1$]. Let us digress briefly to discuss the topological notion of homotopy. Roughly speaking, the homotopy classes of mappings $f : S^1 \to M$ are classes of homotopic continuous maps from the circle to a topological manifold M, where two maps f_1 and f_2 are said to be **homotopic** (written $f_1 \sim f_2$) if there exists a continuous map $F : I \times I \to M$ (I being the closed interval $[0,1]$) such that

$$F(s,0) = f_1(s) \, , F(s,1) = f_2(s) \, , \qquad\qquad s \in I \, , \qquad\qquad (38.24)$$
$$F(0,t) = F(1,t) = x_0 \in M \, , \qquad\qquad t \in I \, , \qquad\qquad (38.25)$$

where the argument $s \in [0,1]$ in $f(s)$ is a parameter specifying a point on S^1 (see Fig. 38.5). Clearly, two homotopic maps can be thought of as two loops in M that can be continuously deformed into each other. The set of homotopy classes of maps with an element denoted by $[f]$, where f is a representative, can be given a group structure, if we define the following group multiplication:

$$[f_1] \cdot [f_2] \equiv [f_1 \cdot f_2] \, , \qquad\qquad\qquad (38.26)$$

and inverse operation

$$[f]^{-1} \equiv [f^{-1}] \, , \qquad\qquad\qquad (38.27)$$

where on the RHS of (38.26) the product of loops with a common point x_0 is defined as a loop consisting of traversing f_1 from x_0 to x_0 first, and then traversing f_2 from x_0 to x_0 (see Fig. 38.6). If two loops f_1 and f_2 do not have a common point, we can replace one of them on the LHS, say f_2, by a

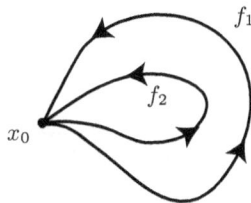

FIGURE 38.6

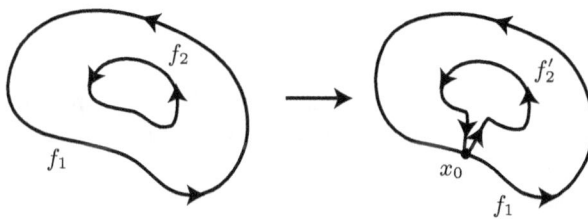

FIGURE 38.7

homotopic map f_2' ($\sim f_2$) which does have a common point with f_1 (Fig. 38.7). In (38.27) f^{-1} simply means a loop traversed in the opposite sense as f.

Definition 38.1. *Let M be a topological space. The set of homotopy classes of loops, or maps $f : S^1 \to M$, such that $f(0) = f(1) = x_0 \in M$ (where the argument parametrizes points on the circle S^1), with the product rule and inverse given by (38.26) and (38.27), is called the* **fundamental group** *(or* **first homotopy group***) of M at x_0, and is denoted by $\pi_1(M, x_0)$.*

$\boxed{\text{Exercise 38.2}}$ Show that the identity of the fundamental group of M at x_0 is the class of loops that can be continuously contracted to the point x_0. Show that $\pi_1(M, x_0)$ is indeed a group, that is, it satisfies the group axioms (of Def. 4.1).

$\boxed{\text{Exercise 38.3}}$ Show that the fundamental groups $\pi_1(M, x_0)$ at different base points $x_0 \in M$ are isomorphic to each other. Thus one can just refer to $\pi_1(M)$.

One can generalize to higher order homotopy groups $\pi_r(M)$ by considering homotopy classes of maps $f : S^r \to M$. The homotopy groups $\pi_r(M)$ are

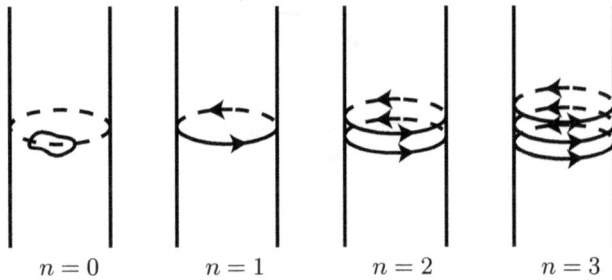

$$n = 0 \qquad n = 1 \qquad n = 2 \qquad n = 3$$

FIGURE 38.8

important because they are **topological invariants** of the topological space M, just as the homology groups $H_r(M)$ [(33.10)] and the de Rham cohomology groups $H^r(M)$ [(33.1)] considered earlier. We will state without proof the following useful theorem relating homotopy and cohomology [see Nash and Sen 1983, Chapter 6].

Theorem 38.1. *If h and g are homotopic maps from M to N and $\beta \in \Lambda^r(N)$ is an r-form on N, then we have the following equality of cohomology classes under pullback:*

$$[h^*\beta] = [g^*\beta] \in H^r(M) .$$

It is seen intuitively that

$$\boxed{\pi_1(S^1) = \mathbb{Z} \quad \text{(the integers)}} \qquad . \qquad (38.28)$$

This result can be understood with the help of Fig. 38.8, which shows a few different ways of looping a band around a pole that cannot be continuously deformed into each other. A principal $U(1)$-bundle on S^2 with transition functions given by

$$g_{UV}(\varphi) = \exp(-in\varphi) , \quad n \in \mathbb{Z} ,$$

is called a **Hopf bundle**.

The quantization condition (38.23) and the result (38.28) show that the allowed quantized values of the magnetic charge are based on the topology of $U(1)$, the structure group of a Hopf bundle, as revealed by $\pi_1(U(1))$. The magnetic charge is an example of what is called a **topological quantum number**.

Further indication of the deep topological origins of the quantization of the magnetic charge can be obtained by considering the local curvature two-form Ω of the local connection one-form ω given by (38.13). Since $U(1)$ is abelian, we have, by virtue of (37.138),

$$
\begin{aligned}
\Omega = d\omega &= -\frac{ie}{\hbar c}\frac{\partial A_j}{\partial x^i}\, dx^i \wedge dx^j \\
&= -\frac{ie}{\hbar c}\left(\frac{\partial A_j}{\partial x^i} - \frac{\partial A_i}{\partial x^j}\right) dx^i \wedge dx^j , \quad i < j .
\end{aligned}
\tag{38.29}
$$

In three dimensions, $\partial_i A_j - \partial_j A_i$ are precisely the components of $\boldsymbol{B} = \nabla \times \boldsymbol{A}$, the magnetic field strength.

The curvature Ω, even though defined locally, can furnish very important global information on the bundle. The fact that it transforms tensorially under a change of gauge: $\Omega' = g\Omega g^{-1}$ [c.f. (35.25)] motivates the consideration of the following invariant (under a gauge transformation):

$$
\det(1 + \frac{i}{2\pi}\Omega) \equiv 1 + C_1(\Omega) + \cdots + C_q(\Omega) ,
\tag{38.30}
$$

where $C_j(\Omega)$ is a $2j$-form on the base manifold. It is obvious that the series terminates (a top-form is obtained when $2j = n =$ dimension of the base manifold). Thus $C_j(\Omega) = 0$ if $2j > n$. It is a most remarkable and significant fact (a result of the so-called Chern-Weil theory, which will be presented in Chapter 41) that $C_j(\Omega)$ is also closed, that is, $dC_j(\Omega) = 0$, and thus determines a real de Rham cohomology class of the base manifold M, *independent of the choice of the connection* on the principal G-bundle under consideration: $C_j(\Omega) \in H^{2j}(M, \mathbb{R})$. The de Rham cohomology classes $C_j(\Omega)$ are the so-called **Chern characteristic classes** (see Chapter 41 below). For the Hopf bundles $\pi : P \to S^2$, $C_j(\Omega) = 0$ if $j > 1$, and $C_1(\Omega) = \frac{i}{2\pi}\Omega$. The integral

$$
c_1 \equiv \int_{S^2} C_1(\Omega) = \frac{i}{2\pi}\int_{S^2}\Omega ,
\tag{38.31}
$$

called the **first Chern number**, is a topological invariant of the Hopf bundles. It can be easily calculated as follows:

$$
\begin{aligned}
c_1 &= \frac{i}{2\pi}\int_{S^2} d\omega = \left(\frac{i}{2\pi}\right)\left(-\frac{ie}{\hbar c}\right)\int_{S^2} dA \quad (A = A_j dx^j) \\
&= \frac{e}{2\pi\hbar c}\oint (\boldsymbol{A}_N - \boldsymbol{A}_S)\cdot dl \quad \text{(by Stokes theorem)} \\
&= \frac{e}{2\pi\hbar c}4\pi M = \frac{2eM}{\hbar c} \in \mathbb{Z} ,
\end{aligned}
\tag{38.32}
$$

as follows from the quantization condition (38.23). Thus the Chern numbers can be associated with topological quantum numbers.

The $U(1)$-magnetic monopole as described above, while very interesting theoretically, has not been observed in nature. We will now discuss real physical systems which exhibit the mathematical structure of a gauge theory, or more specifically, of $U(n)$-magnetic monopoles. Surprisingly, such examples are furnished by molecular dynamics, as well as numerous other types of physical systems in which "fast" and "slow" degrees of freedom can be identified (see A. Shapere and F. Wilczek 1989). In this chapter we will focus on molecular systems.

Denote the collective electronic coordinates by r and the collective nuclear coordinates by R in a molecular system. By a molecular system we mean not only a single molecule, but an arbitrary system involving any number of atoms and/or molecules in interaction. In fact, our formalism will apply not only to bound systems, but also to systems of colliding particles in scattering states. In general, the total Hamiltonian of a molecular system can be written as

$$H(R, r) = -\hbar^2 \sum_I \frac{1}{2M_I} \nabla^2_{R_I} + H_e(R; r) , \qquad (38.33)$$

where the sum is over all the nuclei, and H_e, called the electronic Hamiltonian and parametrized by the nuclear coordinates R, is given by

$$H_e(R; r) = -\frac{\hbar^2}{2m} \sum \nabla^2_r + V_N(R) + V_e(R, r) . \qquad (38.34)$$

The first term on the RHS of the above equation represents the electronic kinetic energy, the second term represents the interaction potential energy between the nuclei, and the last term represents the sum of the electron-electron and electron-nucleus interaction potential energies. The separation of the Hamiltonian according to (38.33) is based on the physical understanding that the electronic degrees of freedom are "fast" ones, while the nuclear degrees of freedom are "slow" ones. The electronic Hamiltonian $H_e(R, r)$ can then be viewed as a "snapshot" Hamiltonian with the nuclear degrees of freedom R frozen.

We assume that the time-independent Schrödinger equation for $H_e(R; r)$ can be solved for fixed R, with a discrete set of eigenstates $|\phi_n(R)\rangle$ and corresponding energy eigenvalues $\epsilon_n(R)$ (both parametrized by R):

$$H_e |\phi_n(R)\rangle = \epsilon_n(R) |\phi_n(R)\rangle . \qquad (38.35)$$

Working with the coordinate representation (r-representation) of $|\phi_n(R)\rangle$, that is, the electronic wave functions given by [recall (15.11)]

$$\phi_n(r, R) = \langle r|\phi_n(R)\rangle . \qquad (38.36)$$

Eq. (38.35) can alternatively be written as

$$H_e \phi_n(r, R) = \epsilon_n(R) \phi_n(r, R) . \qquad (38.37)$$

One can further assume that the set of electronic eigenstates $\{|\phi_n(R)\rangle\}$ are orthonormal and complete [c.f. (12.12)] for each R:

$$\langle \phi_n|\phi_m \rangle = \int dr\, \overline{\phi_n}(r, R)\phi_m(r, R) = \delta_{nm} \quad \text{(orthonormality)} , \qquad (38.38)$$

$$\sum_n |\phi_n(R)\rangle\langle\phi_n(R)| = 1 \quad \text{(completeness)} . \qquad (38.39)$$

Note that in (38.38), the integration for the inner product is over the electronic coordinates only. The set of electronic energy eiganvalues $\epsilon_n(\boldsymbol{R})$ obtained from the solution of (38.35) [or (38.37)], as functions of the nuclear coordinates $\boldsymbol{R}$, are usually referred to as potential energy "surfaces" for nuclear motion.

The time-independent Schrödinger equation for the total Hamiltonian H can be written

$$H\,|\,\Psi\,\rangle = E\,|\,\Psi\,\rangle\,, \tag{38.40}$$

or

$$H\Psi(\boldsymbol{R},\boldsymbol{r}) = E\Psi(\boldsymbol{R},\boldsymbol{r})\,, \tag{38.41}$$

where

$$\Psi(\boldsymbol{R},\boldsymbol{r}) \equiv \langle\,\boldsymbol{R}\,|\,\langle\,\boldsymbol{r}\,|\,\Psi\,\rangle\,. \tag{38.42}$$

Since $\{\phi_n(\boldsymbol{R},\boldsymbol{r}\}$ is a complete set for each $\boldsymbol{R}$, we can expand $\Psi(\boldsymbol{R},\boldsymbol{r})$ as

$$\Psi(\boldsymbol{R},\boldsymbol{r}) = \sum_n \Phi_n(\boldsymbol{R})\phi_n(\boldsymbol{r},\boldsymbol{R})\,. \tag{38.43}$$

It follows from (38.33) that the Schrödinger equation (38.41) can be written as

$$\sum_n \left\{ \sum_I \left(-\frac{\hbar^2}{2M_I}\nabla_{R_I}^2 \right) + H_e \right\} \Phi_n(\boldsymbol{R})\phi_n(\boldsymbol{r},\boldsymbol{R})$$
$$= E\sum_n \Phi_n(\boldsymbol{R})\phi_n(\boldsymbol{r},\boldsymbol{R})\,. \tag{38.44}$$

Using (38.37) and taking the inner product on both sides of the above equation with $\phi_m(\boldsymbol{r},\boldsymbol{R})$ [that is, applying $\int d\boldsymbol{r}\overline{\phi_m}(\boldsymbol{r},\boldsymbol{R})$ to both sides], we obtain

$$\sum_n \int d\boldsymbol{r}\,\overline{\phi_m}(\boldsymbol{r},\boldsymbol{R}) \left(\sum_I -\frac{\hbar^2}{2M_I}\nabla_{R_I}^2 \right) \Phi_n(\boldsymbol{R})\phi_n(\boldsymbol{r},\boldsymbol{R})$$
$$+ \sum_n \Phi_n(\boldsymbol{R})\,\delta_{nm}\,\epsilon_n(\boldsymbol{R}) = E\Phi_m(\boldsymbol{R})\,, \tag{38.45}$$

where the orthonormality of $\{\phi_n\}$ [(38.38)] has also been used. The first term on the LHS can be manipulated as follows [using (30.56) and (30.58)]:

$$\int d\boldsymbol{r}\,\overline{\phi_m}(\boldsymbol{r},\boldsymbol{R})\nabla_{R_I} \cdot \{\nabla_{R_I}(\Phi_n(\boldsymbol{R})\phi_n(\boldsymbol{r},\boldsymbol{R}))\}$$
$$= \int d\boldsymbol{r}\,\overline{\phi_m}\,\nabla_{R_I} \cdot \{\Phi_n\nabla_{R_I}\phi_n + (\nabla_{R_I}\Phi_n)\phi_n\} \tag{38.46}$$
$$= \int d\boldsymbol{r}\,\overline{\phi_m}\{\Phi_n\nabla_{R_I}^2\phi_n + 2(\nabla_{R_I}\Phi_n)\cdot(\nabla_{R_I}\phi_n) + (\nabla_{R_I}^2\Phi_n)\phi_n\}$$
$$= \Phi_n\langle\,\phi_m\,|\,\nabla_{R_I}^2\phi_n\,\rangle + 2\langle\,\phi_m\,|\,\nabla_{R_I}\phi_n\,\rangle\cdot(\nabla_{R_I}\Phi_n) + \delta_{nm}\nabla_{R_I}^2\Phi_n\,,$$

where on the RHS of the last equality, the Dirac bracket of (38.38) has been used. Also note that

$$
\begin{aligned}
\langle \phi_m \,|\, \nabla^2_{R_I} \phi_n \rangle &= \nabla_{R_I} \langle \phi_m \,|\, \nabla_{R_I} \phi_n \rangle - \langle \nabla_{R_I} \phi_m \,|\, \nabla_{R_I} \phi_n \rangle \\
&= \nabla_{R_I} \langle \phi_m \,|\, \nabla_{R_I} \phi_n \rangle - \sum_k \langle \nabla_{R_I} \phi_m \,|\, \phi_k \rangle \langle \phi_k \,|\, \nabla_{R_I} \phi_n \rangle \\
&= \nabla_{R_I} \langle \phi_m \,|\, \nabla_{R_I} \phi_n \rangle + \sum_k \langle \phi_m \,|\, \nabla_{R_I} \phi_k \rangle \langle \phi_k \,|\, \nabla_{R_I} \phi_n \rangle ,
\end{aligned}
\tag{38.47}
$$

where in the second equality we have used the completeness condition for $\{| \phi_k \rangle\}$ [(38.39)] and in the third we have used the fact that

$$
\nabla_{R_I} \langle \phi_m \,|\, \phi_k \rangle = \nabla_{R_I} \delta_{nm} = 0 = \langle \nabla_{R_I} \phi_m \,|\, \phi_k \rangle + \langle \phi_m \,|\, \nabla_{R_I} \phi_k \rangle .
\tag{38.48}
$$

We now define the quantity

$$
\boxed{(\boldsymbol{A}^{(I)})^n_m \equiv i\hbar \langle \phi_m \,|\, \nabla_{R_I} \phi_n \rangle = i\hbar \int dr\, \overline{\phi_m}(\boldsymbol{r}, \boldsymbol{R}) \nabla_{R_I} \phi_n(\boldsymbol{r}, \boldsymbol{R})}
\tag{38.49}
$$

The Schrödinger equation for nuclear motion (38.45) can then be written in the matrix form

$$
H^n_m \, \Phi_n(\boldsymbol{R}) = E\Phi_m(\boldsymbol{R}) ,
\tag{38.50}
$$

where the matrix of operators (H^n_m) are given by

$$
H^n_m(\boldsymbol{R}) = \sum_I \left(-\frac{\hbar^2}{2M_I} \right) (D^{(I)})^k_m \cdot (D^{(I)})^n_k + \delta^n_m \, \epsilon(\boldsymbol{R}) .
\tag{38.51}
$$

In the above equation, the sum over I is over the nuclei and the matrices of vector operators $(D^{(I)})^k_m$ are given by

$$
\boxed{(D^{(I)})^k_m \equiv \delta^k_m \, \nabla_{R_I} - \frac{i}{\hbar} \, (\boldsymbol{A}^{(I)})^k_m(\boldsymbol{R})}
\tag{38.52}
$$

The above two equations reveal that the nuclear motion is described by a Schrödinger equation [(38.50)] that formally resembles that for a charged particle moving in an electromagnetic field, with the operator $(D^{(I)})^k_m$ in (38.52) appearing as a covariant derivative describing a minimal coupling [c.f. (37.169)], and $(\boldsymbol{A}^{(I)})^k_m(\boldsymbol{R})$ as "vector potentials", or gauge potentials. Eq. (38.50) also shows explicitly how the different electronic states are coupled by the nuclear motion.

The gauge potentials $(\boldsymbol{A}^{(I)})^n_m(\boldsymbol{R})$ can be related to the electronic energy difference $\epsilon_n(\boldsymbol{R}) - \epsilon_m(\boldsymbol{R})$, the matrix elements $\langle \phi_m \,|\, (\nabla_{R_I} H_e) \,|\, \phi_n \rangle$, and the gradients of the nuclear potential energies $\nabla_{R_I} \epsilon_n$. We have

$$
\begin{aligned}
\langle \phi_m \,|\, \nabla_{R_I} (H_e \,|\, \phi_n \rangle) &= \langle \phi_m \,|\, \nabla_{R_I} (\epsilon(\boldsymbol{R}) \,|\, \phi_n \rangle) \\
&= \langle \phi_m \,|\, (\nabla_{R_I} H_e) \,|\, \phi_n \rangle + \langle \phi_m \,|\, H_e \,|\, \nabla_{R_I} \phi_n \rangle .
\end{aligned}
\tag{38.53}
$$

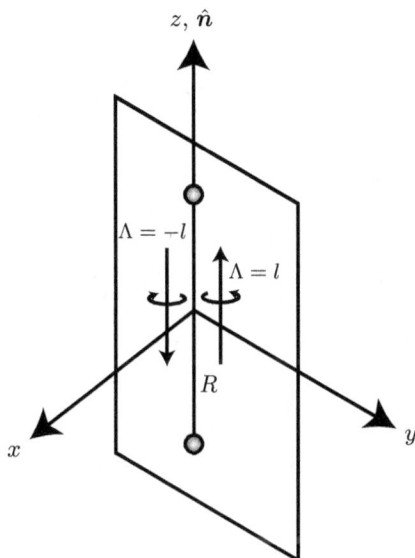

FIGURE 38.9

Hence

$$\epsilon_n(\boldsymbol{R}) \langle \phi_m \,|\, \nabla_{R_I} \phi_n \rangle + \nabla_{R_I} \epsilon_n(\boldsymbol{R}) \delta_{mn}$$
$$= \langle \phi_m \,|\, (\nabla_{R_I} H_e) \,|\, \phi_n \rangle + \epsilon_m(\boldsymbol{R}) \langle \phi_m \,|\, \nabla_{R_I} \phi_n \rangle, \quad (38.54)$$

or

$$(\epsilon_n(\boldsymbol{R}) - \epsilon_m(\boldsymbol{R})) \langle \phi_m \,|\, \nabla_{R_I} \phi_n \rangle$$
$$= \langle \phi_m \,|\, (\nabla_{R_I} H_e) \,|\, \phi_n \rangle - \delta_{mn} \nabla_{R_I} \epsilon_n(\boldsymbol{R}). \quad (38.55)$$

Thus, for non-degenerate states $(\epsilon_m \neq \epsilon_n)$,

$$(\boldsymbol{A}^{(I)})_m^n(\boldsymbol{R}) = i\hbar \frac{\langle \phi_m \,|\, (\nabla_{R_I} H_e) \,|\, \phi_n \rangle}{\epsilon_n(\boldsymbol{R}) - \epsilon_m(\boldsymbol{R})}, \quad (\epsilon_n \neq \epsilon_m). \quad (38.56)$$

Let us now illustrate the above formal development with a specific example in molecular physics: the lambda (Λ) doublets of a homonuclear diatomic molecule (such as O_2), where Λ denotes the projection $\boldsymbol{J} \cdot \hat{n}$ of the total electronic angular momentum $\boldsymbol{J}$ (including spin) of the molecule along the internuclear axis $\hat{n}$. (See Fig. 38.9.) We will show that the gauge potential $\boldsymbol{A}_m^n(\boldsymbol{R})$ for this system has the same mathematical form as that for a magnetic monopole!

The system obviously does not have spherical symmetry, so $J^2 = \boldsymbol{J} \cdot \boldsymbol{J}$ is not conserved (since it does not commute with the electronic Hamiltonian H_e), and J_z is not a good quantum number. However $\boldsymbol{J} \cdot \hat{n}$ is conserved and its eigenvalues

Λ can be used to label the electronic ststes $|\phi_n\rangle$. Moreover, since the system has a mirror symmetry about any plane through the internuclear axis, and $\boldsymbol{J} \cdot \hat{n}$ changes sign under such a parity operation, states with $\boldsymbol{J} \cdot \hat{n} = \Lambda \hbar$ are degenerate with those with $\boldsymbol{J} \cdot \hat{n} = -\Lambda \hbar$. Such degenerate pairs of states are called **lambda doublets**.

For a particular value of Λ, let us label the doublet electronic states $|\phi_\Lambda(\hat{n} = \hat{z})\rangle$ and $|\phi_{-\Lambda}(\hat{n} = \hat{z})\rangle$ with the internuclear axis along the z-direction (Fig. 38.9) by $|\uparrow\rangle$ and $|\downarrow\rangle$, respectively. Supposing the molecule to be modeled by a rigid dumbbell (vibrations frozen) and working in the nuclear center-of-mass (CM) frame, the nuclear coordinates $\boldsymbol{R}$ are just the polar angles (θ, φ) marking points on the surface of a two-sphere S^2. The doublet electronic states with $\hat{n}$ pointed along arbitrary directions [specified by (θ, φ)] are those obtained from $|\uparrow\rangle$ and $|\downarrow\rangle$ by rotations (in nuclear coordinate space). We will label these by $|\uparrow(\theta, \varphi)\rangle$ and $|\downarrow(\theta, \varphi)\rangle$. In fact, recalling Euler's theorem [Eq. (6.25) in Theorem 6.3] and (4.21), we have (on using the symbol $|\uparrow\downarrow\rangle$ to represent either $|\uparrow\rangle$ or $|\downarrow\rangle$),

$$|\uparrow\downarrow(\theta, \varphi)\rangle = e^{-\frac{i}{\hbar}\varphi J_3}\, e^{-\frac{i}{\hbar}\theta J_2}\, e^{\frac{i}{\hbar}\varphi J_3}\,|\uparrow\downarrow\rangle\,, \tag{38.57}$$

where the operator $\exp(\frac{i}{\hbar}\varphi J_3)$ produces no effect on $|\uparrow\downarrow\rangle$ [since a rotation by an angle φ about $\hat{z}$ does not move the north (or south) pole], and has been inserted as a computational device.

The gauge potentials [as defined by (38.49)] are obtained from the quantities

$$i\hbar\langle\uparrow\downarrow(\theta, \varphi)|\nabla_{(\theta, \varphi)}|\uparrow\downarrow(\theta, \varphi)\rangle = \frac{i\hbar}{R}\langle\uparrow\downarrow(\theta, \varphi)|\frac{\partial}{\partial\theta}|\uparrow\downarrow(\theta, \varphi)\rangle\, e_\theta$$
$$+ \frac{i\hbar}{R\sin\theta}\langle\uparrow\downarrow(\theta, \varphi)|\frac{\partial}{\partial\varphi}|\uparrow\downarrow(\theta, \varphi)\rangle\, e_\varphi\,, \tag{38.58}$$

where the constant R is half the internuclear distance.

We proceed to calculate the matrix elements on the RHS of the above equation. We have

$$i\hbar\langle\uparrow\downarrow(\theta, \varphi)|\frac{\partial}{\partial\theta}|\uparrow\downarrow(\theta, \varphi)\rangle$$
$$= i\hbar\langle\uparrow\downarrow|e^{-\frac{i}{\hbar}\varphi J_3}\, e^{\frac{i}{\hbar}\theta J_2}\, e^{\frac{i}{\hbar}\varphi J_3}\,\frac{\partial}{\partial\theta}\,(e^{-\frac{i}{\hbar}\varphi J_3}\, e^{-\frac{i}{\hbar}\theta J_2}\, e^{\frac{i}{\hbar}\varphi J_3})|\uparrow\downarrow\rangle$$
$$= \langle\uparrow\downarrow|e^{-\frac{i}{\hbar}\varphi J_3}\, e^{\frac{i}{\hbar}\theta J_2}\, J_2\, e^{-\frac{i}{\hbar}\theta J_2}\, e^{\frac{i}{\hbar}\varphi J_3}|\uparrow\downarrow\rangle \tag{38.59}$$
$$= \langle\uparrow\downarrow|e^{-\frac{i}{\hbar}\varphi J_3}\, J_2\, e^{\frac{i}{\hbar}\varphi J_3}|\uparrow\downarrow\rangle$$
$$= \langle\uparrow\downarrow|\cos\varphi\, J_2 - \sin\varphi\, J_1|\uparrow\downarrow\rangle\,,$$

where, in the last equality, we have used the operator identity

$$e^{-\frac{i}{\hbar}\alpha J_i}\, J_j\, e^{\frac{i}{\hbar}\alpha J_i} = (\cos\alpha)J_j + (\sin\alpha)\varepsilon_{ij}{}^k J_k\,, \quad \alpha \in \mathbb{R}\,, \tag{38.60}$$

with $\varepsilon_{ij}{}^k$ being the completely antisymmetric tensor.

Exercise 38.4   Show that the above identity results from the general operator identity

$$e^{\xi A} B e^{-\xi A} = B + \xi [A, B] + \frac{\xi^2}{2!} [A, [A, B]]$$

$$+ \frac{\xi^3}{3!} [A, [A, [A, B]]] + \cdots , \quad \xi \in \mathbb{C}, \quad (38.61)$$

and the Lie algebra structure of the angular momentum operators as given by (9.23):

$$[J_i, J_j] = i\hbar \, \varepsilon_{ij}{}^k \, J_k .$$

Similarly, by repeated application of (38.60),

$$i\hbar \, \langle \uparrow\downarrow (\theta, \varphi) | \frac{\partial}{\partial \varphi} | \uparrow\downarrow (\theta, \varphi) \rangle$$

$$= \langle \uparrow\downarrow \, | \, (\cos\theta - 1) J_3 - \sin\theta \cos\varphi \, J_1 - \sin\theta \sin\varphi \, J_2 \, | \uparrow\downarrow \rangle . \quad (38.62)$$

Since the states $| \uparrow\downarrow \rangle$ have internuclear axis along $e_3 = \hat{z}$, we have

$$J_3 | \uparrow \rangle = \Lambda\hbar | \uparrow \rangle , \quad (38.63)$$

$$J_3 | \downarrow \rangle = -\Lambda\hbar | \downarrow \rangle . \quad (38.64)$$

To calculate the matrix elements of J_1 and J_2, we introduce the hermitian conjugate pairs [recall (24.82) and (24.83)]

$$J_\pm \equiv \frac{J_1 \pm i J_2}{\sqrt{2}} , \quad (38.65)$$

which imply

$$J_1 = \frac{1}{\sqrt{2}} (J_+ + J_-) , \quad (38.66)$$

$$J_2 = \frac{1}{i\sqrt{2}} (J_+ - J_-) . \quad (38.67)$$

Temporarily writing the states $| \uparrow \rangle$ and $| \downarrow \rangle$ as $|\Lambda\rangle$ and $|-\Lambda\rangle$, respectively ($\Lambda \geq 0$), we have

$$J_+ | \pm \Lambda \rangle = \alpha_+(\Lambda) | \pm \Lambda + 1 \rangle , \quad (38.68)$$

$$J_- | \pm \Lambda \rangle = \alpha_-(\Lambda) | \pm \Lambda - 1 \rangle , \quad (38.69)$$

where $\alpha_+(\Lambda)$ and $\alpha_-(\Lambda)$ are constants depending on Λ.

Now from our general theory of the representations of $SO(3)$ (Chapter 22), Λ can assume the values $n/2$; $n = 0, 1, 2, 3, \ldots$. Eqs. (38.59) and (38.62) clearly show that the gauge potential A_n^m vanishes for $|\phi_n\rangle$ and $|\phi_m\rangle$ corresponding to different values of $|\Lambda|$. Thus only the Λ-doublet states (for fixed Λ) are possibly coupled by the gauge potential. For $\Lambda \neq 1/2$, the doublet states are actually decoupled (the 2×2 gauge potential matrices are diagonal). In fact, for $\Lambda \neq 1/2$,

$$(A_\theta)^m_n = 0 , \tag{38.70}$$

$$(A_\varphi)^m_n = \begin{pmatrix} -\dfrac{\Lambda\hbar(1 - \cos\theta)}{R\sin\theta} & 0 \\ 0 & \dfrac{\Lambda\hbar(1 - \cos\theta)}{R\sin\theta} \end{pmatrix} , \tag{38.71}$$

where the row and column indices are n and m, respectively, with the values 1 for $\uparrow$ and 2 for $\downarrow$. On comparison with (38.5) for the Wu-Yang potential of a magnetic monopole, we see that each of the nuclear wave functions $\Phi_\uparrow(\theta, \varphi)$ and $\Phi_\downarrow(\theta, \varphi)$ of (38.50) satisfies a Schrödinger equation which formally resembles that describing the motion of a charged particle in a field due to a $U(1)$-magnetic monopole [(38.21)], with $\Lambda\hbar$ playing the role of the magnetic charge for the state $\Phi_\downarrow$, and $-\Lambda\hbar$ playing the same role for $\Phi_\uparrow$.

For $\Lambda = 1/2$, the states $|\uparrow\rangle$ and $|\downarrow\rangle$ are no longer decoupled. In fact, it is straightforward, using (38.63) to (38.69), to show that, for $\Lambda = 1/2$,

$$A_\theta = \hbar \begin{pmatrix} 0 & \lambda e^{-i\varphi}/R \\ \overline{\lambda} e^{i\varphi}/R & 0 \end{pmatrix} , \tag{38.72}$$

$$A_\varphi = \hbar \begin{pmatrix} -(1 - \cos\theta)/(2R\sin\theta) & i\lambda e^{-i\varphi}/R \\ -i\overline{\lambda} e^{i\varphi}/R & (1 - \cos\theta)/(2R\sin\theta) \end{pmatrix} , \tag{38.73}$$

where

$$\hbar\lambda \equiv \langle\uparrow | J_2 | \downarrow\rangle . \tag{38.74}$$

Exercise 38.5   Verify (38.72) and (38.73).

The above vector potentials are those of a $U(2)$-magnetic monopole, mathematically described by a principal $U(2)$-bundle $\pi : P \to S^2$ with base manifold S^2. Note that both A_θ and A_φ are hermitian, so that $-iA_\theta$ and $-iA_\varphi$ are both anti-hermitian, as required to be since $\omega = -iA_\theta d\theta - iA_\varphi d\varphi$ is a $\mathcal{U}(2)$-valued connection one-form, where $\mathcal{U}(2)$ is the Lie algebra of $U(2)$.

We mention finally the physical example of the **integral quantum Hall effect**, which is mathematically described by a $U(1)$-bundle on T^2 (the 2-torus). The so-called **Hall conductance** (ne^2/h, $n \in \mathbb{Z}$) turns out to be just the first Chern number of the bundle, a topological invariant (see M. Kohmoto 1985 and D. J. Thouless 1998).

Chapter 39

Riemannian Geometry

Riemannian geometry is the mathematical basis of Einstein's general theory of relativity, which is the standard classical theory of gravitation. Its object of study is a manifold endowed with a certain quadratic metric.

The notion of a metric has already been introduced informally in Chapter 8. We recall that roughly speaking, it is a prescription for measuring distances and angles on a manifold. In this chapter we will first define the Riemannian metric more precisely, and then explore the relationship between this metric and a unique connection, the Levi-Civita connection, on the tangent bundle that arises from it. The curvature of this connection on four-dimensional spacetime is in fact the gravitational field, according to the general theory of relativity.

Definition 39.1. *Suppose M is an n-dimensional smooth manifold. A **Riemannian metric** G on M is a symmetric, positive-definite, covariant $(0,2)$-type tensor field on M.*

If $(U; x^i)$ is a coordinate neighborhood in M, then G can be expressed locally (on U) as

$$G = g_{ij}\, dx^i \otimes dx^j \,, \tag{39.1}$$

where

$$g_{ij} = g_{ji} \tag{39.2}$$

is a symmetric smooth function on U. At every point $x \in U$, G gives a bilinear function on T_xM : if $X = X^i \dfrac{\partial}{\partial x^i}$ and $Y = Y^i \dfrac{\partial}{\partial x^i}$ are two tangent vectors in T_xM, then [recall (8.6)]

$$G(X,Y) = g_{ij}(x)X^iY^j \,. \tag{39.3}$$

The metric G thus specifies an inner product on T_xM, for all $x \in M$. Positive definiteness means that, for all $X \in T_xM$,

$$G(X,X) \geq 0 \,, \tag{39.4}$$

and the equality holds only when $X = 0$. Note that a positive definite metric is necessarily non-degenerate. [The notion of non-degeneracy of a metric has already been introduced in Chapter 8, in the discussion following (8.6).]

Definition 39.2. *A symmetric, non-degenerate, but not positive-definite, covariant $(0, 2)$-type tensor field on M is called a* **pseudo-Riemannian metric** *on M.*

The Minkowski metric $\eta_{\mu\nu}$ introduced in Chapter 8 is an example of a flat ($\eta_{\mu\nu} =$ constant), pseudo-Riemannian metric on four-dimensional spacetime. The development in this chapter will be applicable to manifolds endowed with non-degenerate metrics (Riemannian or pseudo-Riemannian). We will loosely refer to both of these kinds of manifolds as **Riemannian manifolds**.

If G is positive-definite, then we can define the length of a tangent vector $X = X^i \partial_i \in T_x M$ by

$$|X| \equiv \sqrt{X \cdot X} = \sqrt{G(X, X)}, \tag{39.5}$$

and the angle θ between two tangent vectors $X, Y \in T_x M$ by

$$\cos\theta \equiv \frac{G(X,Y)}{|X|\,|Y|} = \frac{X \cdot Y}{|X|\,|Y|}. \tag{39.6}$$

Also, the **element of arc length** on M is given by

$$ds^2 = g_{ij}\, dx^i dx^j. \tag{39.7}$$

Thus if $C : [t_0, t_1] \to M$ is a continuous and piecewise-smooth curve on M specified by functions $x^i(t)$, $t_0 \leq t \leq t_1$, then the **arc length** of C is defined to be

$$s = \int_{t_0}^{t_1} \left(\frac{ds}{dt}\right) dt = \int_{t_0}^{t_1} \sqrt{g_{ij}\frac{dx^i}{dt}\frac{dx^j}{dt}}\, dt. \tag{39.8}$$

We will state the following basic result without proof.

Theorem 39.1. *There exists a Riemannian metric on any finite-dimensional smooth manifold.*

Riemannian geometry, based on (39.8), may be regarded as a special case of the following situation, where

$$s = \int_{t_0}^{t_1} F\left(x^1, \ldots, x^n; \frac{dx^1}{dt}, \ldots, \frac{dx^n}{dt}\right) dt, \tag{39.9}$$

in which $F(x^1, \ldots, x^n; y^1, \ldots, y^n)$ is required to be a smooth, non-negative function of the $2n$ variables, and has the value 0 only when $y^1 = \cdots = y^n = 0$. It is further required to be either **symmetrically homogeneous** of degree one in the y's:

$$F(x^1, \ldots, x^n; \lambda y^1, \ldots, \lambda y^n) = |\lambda|\, F(x^1, \ldots, x^n; y^1, \ldots, y^n), \quad \lambda \in \mathbb{R}, \tag{39.10}$$

or **positively homogeneous** of degree one in the y's:

$$F(x^1, \ldots, x^n; \lambda y^1, \ldots, \lambda y^n) = \lambda F(x^1, \ldots, x^n; y^1, \ldots, y^n), \quad \lambda > 0. \quad (39.11)$$

$F(x; y)$ is called the **Finsler function** and the corresponding geometry **Finsler geometry**. This case was actually already introduced by Riemann. A more detailed consideration of it is, however, beyond the scope of the present text. (See Chern, Chen and Lam 1999; D. Bao, S. S. Chern and Z. Shen 2000.)

Returning to a Riemannian (or pseudo-Riemannian) metric, we see that, under a change of local coordinates [recall (2.1) and (30.10)],

$$(g')_{ij} = g_{kl} \frac{\partial x^k}{\partial (x')^i} \frac{\partial x^l}{\partial (x')^j}. \quad (39.12)$$

Since G is non-degenerate, the matrix (g_{ij}) has an inverse, which we will denote by (g^{ij}):

$$g^{ik} g_{kj} = g_{jk} g^{ki} = \delta^i_j. \quad (39.13)$$

The transformation rule for g^{ij} under a change of local coordinates can then be shown to be

$$(g')^{ij} = g^{kl} \frac{\partial (x')^i}{\partial x^k} \frac{\partial (x')^j}{\partial x^l}. \quad (39.14)$$

| Exercise 39.1 | Verify the above statement.

Hence g^{ij} is a symmetric contravariant tensor field of rank two [or a $(2,0)$-type tensor field].

As discussed in Chapter 8, the metric G provides an isomorphism between $T_x M$ and its dual $T_x^* M$. In fact, if $X \in T_x M$, then there is a unique element $X^* \in T_x^* M$ such that

$$\langle Y, X^* \rangle = G(Y, X), \quad \text{for all } Y \in T_x M. \quad (39.15)$$

Conversely, since G is non-degenerate, any element in $T_x^* M$ can be expressed in the form X^*, for some $X \in T_x M$. If we write X and X^* in component form with respect to the natural bases:

$$X = X^i \frac{\partial}{\partial x^i}, \quad X^* = X_i \, dx^i, \quad (39.16)$$

then (39.15) implies that

$$X_i = g_{ij} X^j, \quad X^i = g^{ij} X_j. \quad (39.17)$$

Note that X^* (with lower-indexed components) is a covariant vector, while X (with upper-indexed components) is a contravariant vector, according to the terminology introduced in Chapter 2. These vectors can be generated from each other by lowering or raising indices with the help of the metric tensor g_{ij}.

The tensor operations of raising or lowering indices can in fact be extended to arbitrary mixed-type tensors. For example, if $(t^i{}_{jk})$ is a $(1,2)$-type tensor, then

$$t_{ijk} = g_{il}\, t^l{}_{jk}\,, \qquad t^{ij}{}_k = g^{jl}\, t^i{}_{lk} \tag{39.18}$$

are $(0,3)$-type and $(2,1)$-type tensors, respectively.

Let us now introduce the notion of an **affine connection** on a manifold M. It is simply a connection on the tangent bundle TM of M (Def. 35.1). On a coordinate neighborhood $(U; x^i)$ of M, we have the local frame field $\{s_i = \dfrac{\partial}{\partial x^i}\}$, $i = 1, \dots, n$ (n being the dimension of M as a manifold). Then an affine connection D on U is given by [recall (35.8)]

$$Ds_i = \omega_i^j \otimes s_j = \Gamma_{ik}^j\, dx^k \otimes s_j\,, \tag{39.19}$$

where Γ_{ik}^j are smooth functions on U. These are called the **connection coefficients** of D with respect to the local coordinates x^i, or the **Christoffel symbols**. [They play the same role that the connection coefficients $A_{\alpha i}^\beta$ in (35.8) do for a general vector bundle.] Suppose $(W; y^i)$ is another coordinate neighborhood in M, and $(s')_i = \dfrac{\partial}{\partial y^i}$. Let

$$S = {}^t(s_1, \dots, s_n) = {}^t\left(\frac{\partial}{\partial x^1}, \dots, \frac{\partial}{\partial x^n}\right) \tag{39.20}$$

and

$$S' = {}^t((s')_1, \dots, (s')_n) = {}^t\left(\frac{\partial}{\partial y^1}, \dots, \frac{\partial}{\partial y^n}\right). \tag{39.21}$$

Then, on $U \cap W \neq \emptyset$, we have

$$\frac{\partial}{\partial y^i} = \frac{\partial}{\partial x^j}\frac{\partial x^j}{\partial y^i}\,, \tag{39.22}$$

or, in matrix notation,

$$S' = J_{WU}\, S\,, \tag{39.23}$$

where the Jacobian matrix J_{WU} of the change of coordinates is given by

$$J_{WU} = \begin{pmatrix} \dfrac{\partial x^1}{\partial y^1} & \cdots & \dfrac{\partial x^n}{\partial y^1} \\ & \vdots & \\ \dfrac{\partial x^1}{\partial y^n} & \cdots & \dfrac{\partial x^n}{\partial y^n} \end{pmatrix}. \tag{39.24}$$

It follows from (35.17) that, under a local change of coordinates, the transformation rule for the connection matrix of one-forms ω is given by

$$\omega' = dJ_{WU} \cdot J_{WU}^{-1} + J_{WU}\, \omega\, J_{WU}^{-1}\,, \tag{39.25}$$

that is,

$$(\omega')^j_i = d\left(\frac{\partial x^l}{\partial y^i}\right)\frac{\partial y^j}{\partial x^l} + \frac{\partial x^l}{\partial y^i}\omega^m_l\frac{\partial y^j}{\partial x^m}, \tag{39.26}$$

where

$$(\omega')^j_i = (\Gamma')^j_{ik}\,dy^k. \tag{39.27}$$

The transformation rule for the Christoffel symbols under a local change of coordinates is then

$$(\Gamma')^j_{ik} = \frac{\partial x^l}{\partial y^i}\frac{\partial x^p}{\partial y^k}\frac{\partial y^j}{\partial x^m}\Gamma^m_{lp} + \frac{\partial^2 x^l}{\partial y^k\partial y^i}\cdot\frac{\partial y^j}{\partial x^l}. \tag{39.28}$$

Exercise 39.2 Derive the above equation from (39.26) and (39.27).

Only the first term on the RHS of (39.28) conforms with the tensorial transformation rule [(2.1)]. Hence the presence of the second term shows that Γ^j_{ik} is not a tensor field, which implies that the vanishing of the Γ^j_{ik} at a certain point $p \in M$ is not an invariant property under a change of coordinates. In fact, under certain conditions, Γ^j_{ik} can be made to vanish for any point in M under a suitable choice of local coordinates. (See Theorem 39.2 below.)

Let us now see how one can perform the covariant derivative of an arbitrary mixed-type tensor field on M, that is, a section of the tensor bundle $T^r_s(M)$ [c.f. (30.16)]. Start with a vector field $X = X^i\dfrac{\partial}{\partial x^i}$. By (35.2) we have

$$DX = dX^i \otimes \frac{\partial}{\partial x^i} + X^i D\left(\frac{\partial}{\partial x^i}\right) = dX^i \otimes \frac{\partial}{\partial x^i} + X^i\omega^j_i\frac{\partial}{\partial x^j}$$

$$= (dX^i + X^j\omega^i_j) \otimes \frac{\partial}{\partial x^i} = \left(\frac{\partial X^i}{\partial x^j}dx^j + X^k\Gamma^i_{kj}dx^j\right) \otimes \frac{\partial}{\partial x^i} \tag{39.29}$$

$$\equiv X^i_{,j}\,dx^j \otimes \frac{\partial}{\partial x^i},$$

where

$$\boxed{X^i_{,j} \equiv \frac{\partial X^i}{\partial x^j} + X^k\Gamma^i_{kj}} \tag{39.30}$$

DX is a section of the vector bundle $T^*(M) \otimes T(M)$; in other words, it is a $(1,1)$-type tensor-field on M, and is obtained from a $(1,0)$-type tensor field (namely, X). Eq. (39.30) is a basic formula in classical tensor analysis.

As discussed in Chapter 35, given an affine connection on M [a connection on the vector bundle $T(M)$], there is an induced connection on the dual bundle $T^*(M)$ [c.f. (35.52)]. For $(s^*)^i = dx^i$, $1 \le i \le n$, this connection is given by

$$D(s^*)^i = -\omega^i_j(s^*)^j = -\Gamma^i_{jk}dx^k \otimes dx^j. \tag{39.31}$$

For a cotangent vector field X^* with local expression $X^* = X_i dx^i$, we have

$$DX^* = (dX_i - X_j \omega_i^j) \otimes dx^i = X_{i,j} dx^j \otimes dx^i \,,$$

where

$$\boxed{X_{i,j} \equiv \frac{\partial X_i}{\partial x^j} - X_k \Gamma_{ij}^k} \quad .$$

(39.32)

[Compare this with (39.30).] DX^* is a $(0,2)$-type tensor field.

Eqs. (39.30) and (39.32) allow us to perform the covariant derivative of a mixed-type tensor field of arbitrary rank. Suppose T is an (r,s)-type tensor field. Then DT is an $(r, s+1)$-type tensor field. Consider the example of a $(2,1)$-type tensor field, expressed locally as

$$T = T_k^{ij} dx^k \otimes \frac{\partial}{\partial x^i} \otimes \frac{\partial}{\partial x^j} \,.$$

(39.33)

Then

$$DT = T_{k,l}^{ij} \, dx^l \otimes dx^k \otimes \frac{\partial}{\partial x^i} \otimes \frac{\partial}{\partial x^j} \,,$$

(39.34)

where

$$T_{k,l}^{ij} \equiv \frac{\partial T_k^{ij}}{\partial x^l} + \Gamma_{hl}^i T_k^{hj} + \Gamma_{hl}^j T_k^{ih} - \Gamma_{kl}^h T_h^{ij} \,.$$

(39.35)

Exercise 39.3   Prove (39.34) and (39.35) by using the "Leibniz rule" on DT:

$$\begin{aligned} DT &= dT_k^{ij} \otimes dx^k \otimes \frac{\partial}{\partial x^i} \otimes \frac{\partial}{\partial x^j} + T_k^{ij} D(dx^k) \otimes \frac{\partial}{\partial x^i} \otimes \frac{\partial}{\partial x^j} \\ &= T_k^{ij} \otimes dx^k \otimes D\left(\frac{\partial}{\partial x^i}\right) \otimes \frac{\partial}{\partial x^j} + T_k^{ij} \otimes dx^k \otimes \frac{\partial}{\partial x^i} \otimes D\left(\frac{\partial}{\partial x^j}\right) \,, \end{aligned}$$

(39.36)

and by using (39.30) and (39.32) on the individual indices of T_k^{ij}.

Consider a parametrized curve $C : x^i(t)$ in a Riemannian manifold M endowed with an affine connection D, and a tangent vector field $X = X^i \frac{\partial}{\partial x^i}$ on M, that is, a section $s \in \Gamma(T(M))$. Along the curve C, X can be written as

$$X(t) = X^i(t) \left(\frac{\partial}{\partial x^i}\right)_{C(t)} \,.$$

(39.37)

If, along C, the directional covariant derivative (c.f. Def. 35.2) of X vanishes, that is,

$$D_{X_C} X = 0 \,,$$

(39.38)

where

$$X_C = \frac{dx^i(t)}{dt} \frac{\partial}{\partial x^i} \qquad (39.39)$$

is the tangent vector field on the curve C, then $X(t)$ is said to be **parallel** along C (c.f. Def. 35.4), or **parallelly displaced** along C. From (35.3), the condition (39.38) is equivalent to $\langle X_C, DX \rangle = 0$, or, from (39.29),

$$\left\langle \frac{dx^k}{dt} \cdot \frac{\partial}{\partial x^k}, \, X^i_{,j} \, dx^j \otimes \frac{\partial}{\partial x^i} \right\rangle = \left(X^i_{,k} \frac{dx^k}{dt} \right) \frac{\partial}{\partial x^i} = 0 . \qquad (39.40)$$

Since $\{\frac{\partial}{\partial x^i}\}$ are everywhere linearly independent, (39.40) implies $X^i_{,k} \dfrac{dx^k}{dt} = 0$, $i = 1, \ldots, n$. It follows from (39.30) that if X is parallel along C, then

$$\boxed{\frac{dX^i}{dt} + X^j \Gamma^i_{jk} \frac{dx^k}{dt} = 0 , \quad i = 1, \ldots, n} \qquad (39.41)$$

The above condition can also be written as

$$\frac{DX}{dt} = 0 . \qquad (39.42)$$

Eq. (39.41) is a system of first-order ordinary differential equations. Given the initial conditions $X^i(0)$, that is, some $X \in T_{C(0)}M$, $X(t)$ at any point in the curve $C(t)$ is determined. Thus the condition of **parallel displacement** establishes an isomorphism between the tangent spaces $T_{C(t_1)}(M)$ and $T_{C(t_2)}(M)$ at any two points $C(t_1)$ and $C(t_2)$ along the curve $C(t)$.

Of particular interest is the case of **self-parallelism**, that is, the tangent vector field X_C of the curve $C(t)$ is parallel along C. We say that the curve C is **self-parallel**, or a **geodesic**. In this case $X^i = \dfrac{dx^i}{dt}$, and (39.41) gives the following condition for a curve $C : x^i(t)$ being a geodesic:

$$\boxed{\frac{d^2 x^i}{dt^2} + \Gamma^i_{jk} \frac{dx^j}{dt} \frac{dx^k}{dt} = 0 , \quad i = 1, \ldots, n} \qquad (39.43)$$

This is a system of second-order ordinary differential equations. The theory of such equations guarantees that *there exists a unique geodesic through a given point of M which is tangent to a given tangent vector at that point.*

We will now consider the curvature matrix of 2-forms of an affine connection ω on a Riemannian manifold M. Let ω be given locally in terms of the Christoffel symbols by

$$\omega^j_i = \Gamma^j_{ik} \, dx^k . \qquad (39.44)$$

Then, on recalling (35.18),

$$\Omega_i^j = d\omega_i^j - \omega_i^h \wedge \omega_h^j = \frac{\partial \Gamma_{ik}^j}{\partial x^l} dx^l \wedge dx^k - \Gamma_{il}^h \Gamma_{hk}^j dx^l \wedge dx^k$$

$$= \frac{1}{2} \left(\frac{\partial \Gamma_{il}^j}{\partial x^k} - \frac{\partial \Gamma_{ik}^j}{\partial x^l} + \Gamma_{il}^h \Gamma_{hk}^j - \Gamma_{ik}^h \Gamma_{hl}^j \right) dx^k \wedge dx^l . \quad (39.45)$$

The **curvature tensor** R_{ikl}^j of an affine connection is defined by [c.f. (35.45)]

$$\Omega_i^j \equiv \frac{1}{2} R_{ikl}^j dx^k \wedge dx^l , \qquad (39.46)$$

where

$$\boxed{R_{ikl}^j = \frac{\partial \Gamma_{il}^j}{\partial x^k} - \frac{\partial \Gamma_{ik}^j}{\partial x^l} + \Gamma_{il}^h \Gamma_{hk}^j - \Gamma_{ik}^h \Gamma_{hl}^j} \quad . \qquad (39.47)$$

Note that by inspection of the above equation,

$$R_{ikl}^j = -R_{ilk}^j . \qquad (39.48)$$

We need to show that the curvature tensor is globally defined by considering a transformation of local coordinates. Suppose $(W; y^i)$ is another local coordinate system on M. Then in $U \cap W \neq \emptyset$, the local frame field S' [given by (39.21)] is related to the local frame field S [given by (39.20)] by $S' = J_{WU} S$, where J_{WU} is the Jacobian given by (39.24). It follows from (35.13) and (35.25) that, under the coordinate system $(W; y^i)$, the curvature matrix Ω' is given by

$$\Omega' = J_{WU} \cdot \Omega \cdot J_{WU}^{-1} . \qquad (39.49)$$

In terms of components,

$$(\Omega')_i^j = \frac{\partial x^p}{\partial y^i} \Omega_p^q \frac{\partial y^j}{\partial x^q} . \qquad (39.50)$$

Defining the curvature tensor $(R')_{ikl}^j$ by

$$(\Omega')_i^j \equiv \frac{1}{2} (R')_{ikl}^j dy^k \wedge dy^l , \qquad (39.51)$$

we see that

$$(R')_{ikl}^j = R_{prs}^q \frac{\partial y^j}{\partial x^q} \frac{\partial x^p}{\partial y^i} \frac{\partial x^r}{\partial y^k} \frac{\partial x^s}{\partial y^l} . \qquad (39.52)$$

Comparing this equation with (2.1), we see that R_{ikl}^j satisfies the transformation rule for a $(1,3)$-type tensor. Hence the curvature tensor R with components R_{ikl}^j is a global object independent of local coordinates, but which can be expressed locally as

$$R = R_{ikl}^j \frac{\partial}{\partial x^j} \otimes dx^i \otimes dx^k \otimes dx^l . \qquad (39.53)$$

We recall from (35.27a) and (35.27b) that $R(X, Y) : \Gamma(T(M)) \to \Gamma(T(M))$, $X, Y \in \Gamma(T(M))$, maps tangent vector fields to tangent vector fields. Let the tangent vector fields X, Y, Z have local expressions

$$X = X^i \frac{\partial}{\partial x^i}, \quad Y = Y^i \frac{\partial}{\partial x^i}, \quad Z = Z^i \frac{\partial}{\partial x^i}. \tag{39.54}$$

Then (35.27) yields

$$R(X, Y) Z = Z^i \langle X \wedge Y, \Omega_i^j \rangle \frac{\partial}{\partial x^j} = R^j_{ikl} Z^i X^k Y^l \frac{\partial}{\partial x^j}, \tag{39.55}$$

where the second equality follows from (39.46), (28.28), and the symmetry property (39.48).

Exercise 39.4 Show that the second equality of (39.55) is valid.

On replacing X by $\frac{\partial}{\partial x^k}$, Y by $\frac{\partial}{\partial x^l}$, and Z by $\frac{\partial}{\partial x^i}$ in (39.55), we have

$$R^j_{ikl} = \left\langle R\left(\frac{\partial}{\partial x^k}, \frac{\partial}{\partial x^l}\right) \frac{\partial}{\partial x^i}, dx^j \right\rangle. \tag{39.56}$$

For a given affine connection D characterized locally by the Christoffel symbols Γ^j_{ik}, define the quantity

$$T^j_{ik} \equiv \Gamma^j_{ki} - \Gamma^j_{ik}. \tag{39.57}$$

Our discussion earlier [c.f. (39.28)] indicates that Γ^j_{ik} is not a tensor. T^j_{ik}, however, is a $(1, 2)$-type tensor, since (39.28) also implies

$$(T')^j_{ik} = (\Gamma')^j_{ki} - (\Gamma')^j_{ik} = \frac{\partial x^l}{\partial y^k} \frac{\partial x^p}{\partial y^i} \frac{\partial y^j}{\partial x^m} \Gamma^m_{lp} + \frac{\partial^2 x^l}{\partial y^i \partial y^k} \frac{\partial y^j}{\partial x^l}$$

$$- \left(\frac{\partial x^l}{\partial y^i} \frac{\partial x^p}{\partial y^k} \frac{\partial y^j}{\partial x^m} \Gamma^m_{lp} + \frac{\partial^2 x^l}{\partial y^k \partial y^i} \frac{\partial y^j}{\partial x^l}\right) = \frac{\partial y^j}{\partial x^m} \frac{\partial x^p}{\partial y^i} \frac{\partial x^l}{\partial y^k} (\Gamma^m_{lp} - \Gamma^m_{pl})$$

$$= \frac{\partial y^j}{\partial x^m} \frac{\partial x^p}{\partial y^i} \frac{\partial x^l}{\partial y^k} T^m_{pl}, \tag{39.58}$$

which indicates that T^j_{ik} transforms tensorially under a local change of coordinates. The $(1, 2)$-type tensor T^j_{ik} is called the **torsion tensor** of the affine connection D. It is obvious from (39.57) that the torsion tensor has the following symmetry property:

$$T^j_{ik} = -T^j_{ki}. \tag{39.59}$$

The torsion tensor can naturally be viewed as a map from $\Gamma(T(M)) \times \Gamma(T(M))$ to $\Gamma(T(M))$. Indeed, suppose X, Y are any two tangent vector fields on M. Then the torsion tensor

$$T = T^j_{ik} \frac{\partial}{\partial x^j} \otimes dx^i \otimes dx^k \tag{39.60}$$

acts on the pair (X, Y) to yield

$$T(X, Y) = T^j_{ik} X^i Y^k \frac{\partial}{\partial x^j} \ . \tag{39.61}$$

Analogous to the result of Theorem 35.2 [(35.32)], we have

$$\boxed{T(X, Y) = D_X Y - D_Y X - [X, Y]} \ . \tag{39.62}$$

Exercise 39.5   Show that in local coordinates

$$[X, Y] = \left(X^i \frac{\partial Y^j}{\partial x^i} - Y^i \frac{\partial X^j}{\partial x^i} \right) \frac{\partial}{\partial x^j} \ . \tag{39.63}$$

Exercise 39.6   Prove (39.62) using (39.57), (39.61) and the analytical expression for covariant differentiation given by (39.30).

Definition 39.3. *An affine connection D is said to be* **torsion-free** *if the torsion tenor of D vanishes.*

Eq. (39.62) gives a geometrical meaning of the torsion tensor. Consider $X = \epsilon \frac{\partial}{\partial x^i}$ and $Y = \epsilon \frac{\partial}{\partial x^j}$, then $[X, Y] = 0$. If $T = 0$, the result of parallelly displacing the infinitesimal vector $\epsilon \frac{\partial}{\partial x^j}$ along $\epsilon \frac{\partial}{\partial x^i}$ is the same as parallelly displacing $\epsilon \frac{\partial}{\partial x^i}$ along $\epsilon \frac{\partial}{\partial x^j}$, that is,

$$D_{(\epsilon \frac{\partial}{\partial x^i})} (\epsilon \frac{\partial}{\partial x^j}) = D_{(\epsilon \frac{\partial}{\partial x^j})} (\epsilon \frac{\partial}{\partial x^i})$$

(see Fig. 39.1).

Exercise 39.7   Given an affine connection D with Christoffel symbols Γ^j_{ik}, show that one can always construct a torsion-free connection D' with Christoffel symbols

$$\tilde{\Gamma}^j_{ik} = \frac{1}{2} (\Gamma^j_{ik} + \Gamma^j_{ki}) \ . \tag{39.64}$$

Check explicitly that $\tilde{\Gamma}^j_{ik}$ satisfies the transformation rule (39.28).

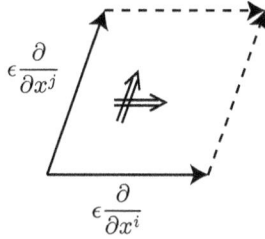

FIGURE 39.1

Suppose D is a torsion-free affine connection with connection coefficients $(\Gamma')^j_{ik}(= (\Gamma')^j_{ki})$ in a local coordinate neighborhood $(W; y^i)$ near a point $p \in M$. Let

$$x^i = y^i + \frac{1}{2}(\Gamma')^i_{mn}(p)(y^m - y^m(p))(y^n - y^n(p)) . \tag{39.65}$$

Then

$$\left.\frac{\partial x^i}{\partial y^j}\right|_p = \delta^i_j , \qquad \left.\frac{\partial^2 x^i}{\partial y^j \partial y^k}\right|_p = (\Gamma')^i_{jk}(p) . \tag{39.66}$$

Thus the matrix $\left(\dfrac{\partial x^i}{\partial y^j}\right)$ is nondegenerate near p, and (39.65) constitutes a local coordinate transformation near p. It then follows from (39.28) that

$$\begin{aligned}
(\Gamma')^j_{ik}(p) &= \delta^l_i \delta^n_k \delta^j_m \, \Gamma^m_{ln}(p) + (\Gamma')^l_{ki}(p)\delta^j_l \\
&= \Gamma^j_{ik}(p) + (\Gamma')^j_{ki}(p) = \Gamma^j_{ik}(p) + (\Gamma')^j_{ik}(p) ,
\end{aligned} \tag{39.67}$$

which implies

$$\Gamma^j_{ik}(p) = 0 . \tag{39.68}$$

We have thus established the following important fact.

Theorem 39.2. *If D is a torsion-free affine connection on M, then at any point $p \in M$, there exists a local coordinate neighborhood $(U; x^i)$ such that the corresponding connection coefficients Γ^j_{ik} vanish at p.*

The following is a useful identity for the curvature tensor of a torsion-free affine connection.

Theorem 39.3. *If D is a torsion-free affine connection on M, then the Bianchi identity assumes the form*

$$\boxed{R^j_{ikl,\,h} + R^j_{ilh,\,k} + R^j_{ihk,\,l} = 0} \qquad . \tag{39.69}$$

Proof. Let ω be the connection matrix of one-forms of a torsion-free affine connection, and $\Omega = d\omega - \omega \wedge \omega$ be the corresponding curvature matrix of 2-forms. The Bianchi identity is given by [c.f. (35.36)]

$$d\Omega = \omega \wedge \Omega - \Omega \wedge \omega . \tag{39.70}$$

In terms of matrix elements,

$$d\Omega_i^j = \omega_i^p \wedge \Omega_p^j - \Omega_i^p \wedge \omega_p^j . \tag{39.71}$$

According to (39.46) and (39.44), the above equation is equivalent to

$$\frac{\partial R_{ikl}^j}{\partial x^h} dx^h \wedge dx^k \wedge dx^l = (\Gamma_{ih}^p R_{p\,kl}^j - \Gamma_{p\,h}^j R_{ikl}^p) \, dx^h \wedge dx^k \wedge dx^l . \tag{39.72}$$

Since [c.f. (39.35)]

$$R_{ikl,\,h}^j = \frac{\partial R_{ikl}^j}{\partial x^h} + \Gamma_{ph}^j R_{ikl}^p - \Gamma_{ih}^p R_{p\,kl}^j - \Gamma_{kh}^p R_{ipl}^j - \Gamma_{lh}^p R_{ikp}^j , \tag{39.73}$$

Eq. (39.72) implies

$$R_{ikl,\,h}^j \, dx^h \wedge dx^k \wedge dx^l = -(\Gamma_{kh}^p R_{ip\,l}^j + \Gamma_{lh}^p R_{ikp}^j) \, dx^h \wedge dx^k \wedge dx^l . \tag{39.74}$$

Making use of the torsion-free property ($\Gamma_{lh}^p = \Gamma_{hl}^p$) and the fact that $R_{ikp}^j = -R_{ipk}^j$ [(39.48)], the second term on the RHS of the above equation can be written

$$\begin{aligned}
\Gamma_{lh}^p R_{ikp}^j \, dx^h \wedge dx^k \wedge dx^l &= -\Gamma_{hl}^p R_{ip\,k}^j \, dx^h \wedge dx^k \wedge dx^l \\
&= -\Gamma_{kh}^p R_{ip\,l}^j \, dx^k \wedge dx^l \wedge dx^h = -\Gamma_{kh}^p R_{ip\,l}^j \, dx^h \wedge dx^k \wedge dx^l .
\end{aligned} \tag{39.75}$$

It follows from (39.74) that

$$R_{ikl,\,h}^j \, dx^h \wedge dx^k \wedge dx^l = 0 . \tag{39.76}$$

By adding two similar terms (each vanishing) obtained by renaming the dummy indices ($k \to l, \, l \to h, \, h \to k$) to the LHS, we obtain

$$(R_{ikl,\,h}^j + R_{ilh,\,k}^j + R_{ihk,\,l}^j) \, dx^h \wedge dx^k \wedge dx^l = 0 . \tag{39.77}$$

Since the quantity within the parentheses is completely antisymmetric with respect to the indices h, k, l, as is the exterior product $dx^h \wedge dx^k \wedge dx^l$, the theorem follows. $\qquad\square$

The curvature tensor R_{ikl}^j and the torsion tensor T_{ik}^j of an affine connection D completely determine the commutation relations for directional covariant derivatives of arbitrary mixed-type tensors. We will illustrate this fact with some examples. First consider a function f (a scalar field) on M. We have the following analog of (39.30),

$$D_{\partial_i} f \equiv f_{,i} = \frac{\partial f}{\partial x^i} \, . \tag{39.78}$$

Write

$$f_{,ij} \equiv (f_{,i})_{,j} \, . \tag{39.79}$$

Then, by (39.32),

$$f_{,ij} = \frac{\partial^2 f}{\partial x^j \partial x^i} - f_{,k} \Gamma^k_{ij} \, . \tag{39.80}$$

Consequently,

$$f_{,ij} - f_{,ji} = f_{,k} \Gamma^k_{ji} - f_{,k} \Gamma^k_{ij} = f_{,k} T^k_{ij} \, . \tag{39.81}$$

Similarly, for a tangent vector field $X = X^i \dfrac{\partial}{\partial x^i}$, we define

$$X^i_{,pq} \equiv (X^i_{,p})_{,q} \, . \tag{39.82}$$

Thus, by (39.30) and (39.32),

$$X^i_{,pq} = \frac{\partial}{\partial x^q}(X^i_{,p}) + X^k_{,p} \Gamma^i_{kq} - X^i_{,l} \Gamma^l_{pq} \, , \tag{39.83}$$

and

$$
\begin{aligned}
& X^i_{,pq} - X^i_{,qp} \\
& = \frac{\partial}{\partial x^q}(X^i_{,p}) - \frac{\partial}{\partial x^p}(X^i_{,q}) + X^k_{,p} \Gamma^i_{kq} - X^i_{,l} \Gamma^l_{pq} - X^k_{,q} \Gamma^i_{kp} + X^i_{,l} \Gamma^l_{qp} \\
& = \frac{\partial}{\partial x^q}\left(\frac{\partial X^i}{\partial x^p} + X^k \Gamma^i_{kp}\right) - \frac{\partial}{\partial x^p}\left(\frac{\partial X^i}{\partial x^q} + X^k \Gamma^i_{kq}\right) \\
& \quad + \left(\frac{\partial X^k}{\partial x^p} + X^l \Gamma^k_{lp}\right) \Gamma^i_{kq} - \left(\frac{\partial X^k}{\partial x^q} + X^l \Gamma^k_{lq}\right) \Gamma^i_{kp} + X^i_{,l} T^l_{pq} \\
& = -X^k \left(\frac{\partial \Gamma^i_{kq}}{\partial x^p} - \frac{\partial \Gamma^i_{kp}}{\partial x^q} + \Gamma^l_{kq} \Gamma^i_{lp} - \Gamma^l_{kp} \Gamma^i_{lq}\right) + X^i_{,l} T^l_{pq} \, .
\end{aligned}
\tag{39.84}
$$

By (39.47), it follows that

$$\boxed{X^i_{,pq} - X^i_{,qp} = -X^k R^i_{kpq} + X^i_{,l} T^l_{pq}} \, . \tag{39.85}$$

Similarly, for a covariant vector field X_i,

$$\boxed{X_{i,pq} - X_{i,qp} = X_l R^l_{ipq} + X_{i,l} T^l_{pq}} \, . \tag{39.86}$$

The commutation relation for an arbitrary mixed-type tensor can be obtained from (39.85) and (39.86). For example, for a $(2,1)$-type tensor T_k^{ij}, we have

$$T_{k,pq}^{ij} - T_{k,qp}^{ij} = -T_k^{lj} R_{lpq}^i - T_k^{il} R_{lpq}^j + T_l^{ij} R_{kpq}^l + T_{k,l}^{ij} T_{pq}^l . \tag{39.87}$$

Exercise 39.8   Verify (39.86) using (39.32); and (39.87) using (39.85) and (39.86).

We now return to a consideration of the Riemannian (or pseudo-Riemannian) metric g_{ij} and its relationship to affine connections.

Definition 39.4. *Suppose M is a Riemannian manifold with a Riemannian (or pseudo-Riemannian) metric $G = (g_{ij})$, and D is an affine connection on M. D is called a **metric-compatible connection** on M if*

$$DG = 0 . \tag{39.88}$$

We will show that metric-compatibility as defined above means that scalar products are preserved under parallel displacements. Writing G as $G = g_{ij}\, dx^i \otimes dx^j$, and expressing the connection matrix of one-forms for D under the local frame field $\left\{ \dfrac{\partial}{\partial x^i} \right\}$ by $\omega = (\omega_i^j)$:

$$D\left(\frac{\partial}{\partial x^i}\right) = \omega_i^j \frac{\partial}{\partial x^j} , \tag{39.89}$$

$$D(dx^i) = -\omega_j^i\, dx^j , \tag{39.90}$$

we have

$$\begin{aligned} DG &= dg_{ij} \otimes dx^i \otimes dx^j - g_{ij}\omega_l^i\, dx^l \otimes dx^j - g_{ij}\omega_l^j\, dx^i \otimes dx^l \\ &= (dg_{ij} - \omega_i^k g_{kj} - \omega_j^k g_{ik}) \otimes dx^i \otimes dx^j . \end{aligned} \tag{39.91}$$

Thus (39.88) is equivalent to

$$dg_{ij} = \omega_i^k g_{kj} + \omega_j^k g_{ik} , \tag{39.92}$$

or, in matrix notation,

$$dG = \omega \cdot G + G \cdot \omega^T , \tag{39.93}$$

where

$$G = \begin{pmatrix} g_{11} & \cdots & g_{1n} \\ & \cdots & \\ g_{n1} & \cdots & g_{nn} \end{pmatrix} , \tag{39.94}$$

$$\omega = \begin{pmatrix} \omega_1^1 & \cdots & \omega_1^n \\ & \cdots & \\ \omega_n^1 & \cdots & \omega_n^n \end{pmatrix} , \tag{39.95}$$

and ω^T is the transpose of ω. In terms of the connection coefficients (Christoffel symbols) Γ^j_{ik}, (39.92) can also be written as

$$dg_{ij} = g_{kj}\Gamma^k_{ih}\,dx^h + g_{ik}\Gamma^k_{jh}\,dx^h \ , \tag{39.96a}$$

or

$$g_{ij,\,h} = 0 \ . \tag{39.96b}$$

The last equation implies that, for a metric-compatible connection, the metric tensor g_{ij} behaves like a constant under covariant differentiation.

Now consider a parallel displacement of vectors $X = X^i\partial_i$ and $Y = Y^i\partial_i$ along a curve $C : x^i(t)$. By (39.41), we have

$$\frac{dX^i}{dt} + X^j\Gamma^i_{jk}\frac{dx^k}{dt} = 0 \ , \qquad \frac{dY^i}{dt} + Y^j\Gamma^i_{jk}\frac{dx^k}{dt} = 0 \ . \tag{39.97}$$

Hence the scalar product of X and Y, $g_{ij}X^iY^j$, varies along the curve C according to

$$\begin{aligned}
\frac{d}{dt}(g_{ij}X^iY^j) &= \frac{dg_{ij}}{dt}X^iY^j + g_{ij}\frac{dX^i}{dt}Y^j + g_{ij}X^i\frac{dY^j}{dt} \\
&= \left(\frac{dg_{ij}}{dt} - g_{kj}\Gamma^k_{ih}\frac{dx^h}{dt} - g_{ik}\Gamma^k_{jh}\frac{dx^h}{dt}\right)X^iY^j \ .
\end{aligned} \tag{39.98}$$

The above vanishes if (39.96) holds. Thus if D is metric-compatible, $g_{ij}X^iY^j =$ constant when X and Y are parallelly displaced.

We now come to the central fact of Riemannian geometry.

Theorem 39.4 (Fundamental Theorem of Riemannian Geometry). *Let M be a Riemannian manifold (with a Riemannian or pseudo-Riemannian metric). Then there exists a unique torsion-free and metric-compatible connection on M, called the **Levi-Civita connection**.*

Proof. In a local coordinate neighborhood $(U; x^i)$, let the connection matrix ω^j_i be given by $\omega^j_i = \Gamma^j_{ik}\,dx^k$. Suppose the connection is both metric-compatible and torsion-free. Then, by (39.92) and (39.57),

$$dg_{ij} = \omega^k_i g_{kj} + \omega^k_j g_{ik} \ , \tag{39.99}$$

$$\Gamma^j_{ik} = \Gamma^j_{ki} \ . \tag{39.100}$$

Define

$$\Gamma_{ijk} \equiv g_{lj}\Gamma^l_{ik} \ , \qquad \omega_{ik} \equiv g_{lk}\omega_i{}^l \ . \tag{39.101}$$

Then it follows from (39.99) and (39.100) that

$$\frac{\partial g_{ij}}{\partial x^k} = \Gamma_{ijk} + \Gamma_{jik} \ , \tag{39.102}$$

$$\Gamma_{ij\,k} = \Gamma_{kj\,i} \ . \tag{39.103}$$

Two more equations analogous to (39.102) corresponding to the even permutations (jki) and (kij) of the indices (ijk) can be obtained:

$$\frac{\partial g_{jk}}{\partial x^i} = \Gamma_{jki} + \Gamma_{kji} , \tag{39.104}$$

$$\frac{\partial g_{ki}}{\partial x^j} = \Gamma_{kij} + \Gamma_{ikj} . \tag{39.105}$$

Calculating (39.104) + (39.105) − (39.102) and using (39.103), we have

$$\boxed{\Gamma_{ikj} = \frac{1}{2} \left(\frac{\partial g_{ik}}{\partial x^j} + \frac{\partial g_{kj}}{\partial x^i} - \frac{\partial g_{ij}}{\partial x^k} \right)} \quad . \tag{39.106}$$

Since, by (39.101),

$$\Gamma_i{}^k{}_j = g^{kl} \Gamma_{ilj} , \tag{39.107}$$

Eq. (39.106) implies

$$\boxed{\Gamma_i{}^k{}_j = \frac{1}{2} g^{kl} \left(\frac{\partial g_{il}}{\partial x^j} + \frac{\partial g_{jl}}{\partial x^i} - \frac{\partial g_{ij}}{\partial x^l} \right)} \quad . \tag{39.108}$$

Thus the Christoffel symbols of a metric-compatible and torsion-free connection are determined by the metric tensor.

Conversely, one can show that, given a metric tensor g_{ij}, the quantities $\Gamma_i{}^k{}_j$ defined by (39.108) are components of a $(1,2)$-type tensor by verifying that they transform according to (39.28) under a local change of coordinates. Thus Γ_{ikj} is a $(0,3)$-type tensor. Furthermore it is quite obvious that Γ_{ikj} as defined by (39.106) satisfies (39.102) and (39.103). Thus (39.108) uniquely specifies the metric-compatible and torsion-free connection on M. $\qquad\square$

The quantities Γ_{ikj} and $\Gamma_i{}^k{}_j$ are called **Christoffel symbols of the first and second kind**, respectively.

Sometimes it is more convenient to use an arbitrary frame field $e_i = X_i^k \dfrac{\partial}{\partial x^k}$ [c.f. (37.3)] instead of the natural one $\left\{ \dfrac{\partial}{\partial x^i} \right\}$, where X_i^k are functions on the local coordinate neighborhood $(U; x^i)$. The one-forms $\theta^i \equiv (X^{-1})_j^i \, dx^j$ [c.f. (37.19)], where X^{-1} is the inverse matrix of X, then constitutes a dual coframe field to $\{e_i\}$, in the sense that

$$\langle e_i, \theta^j \rangle = \delta_i^j . \tag{39.109}$$

With an affine connection given by

$$D\left(\frac{\partial}{\partial x^i} \right) = \omega_i^j \frac{\partial}{\partial x^j} , \tag{39.110}$$

we see that [c.f. (37.36)]

$$\theta_i^j = (X^{-1})_k^j(dX_i^k + X_i^l\omega_l^k) \tag{39.111}$$

are the connection one-forms with respect to the frame field $\{e_i\}$:

$$De_i = \theta_i^j e_j . \tag{39.112}$$

We note that the one-forms θ^i and θ_i^j introduced here are one-forms on M, and are in fact the pullbacks under the map $f : M \to P$, $x \mapsto \{x; e_i\}$, of the similarly labeled one-forms θ^i and θ_i^j defined by (37.19) and (37.36), which are forms on the frame bundle P of M (c.f. Chapter 7). We have

$$dx^i = X_j^i\theta^j , \tag{39.113}$$

$$dX_i^j = -X_i^k\omega_k^j + X_k^j\theta_i^k . \tag{39.114}$$

Exteriorly differentiating (39.113), we have

$$0 = dX_j^i \wedge \theta^j + X_j^i d\theta^j = X_j^i(d\theta^j + \theta_k^j \wedge \theta^k) - X_j^k\omega_k^i \wedge \theta^j$$
$$= X_j^i(d\theta^j - \theta^k \wedge \theta_k^j) - X_j^k\Gamma_{kl}^i\, dx^l \wedge \theta^j = X_j^i(d\theta^j - \theta^k \wedge \theta_k^j) - X_j^k\Gamma_{kl}^i X_m^l\, \theta^m \wedge \theta^j ,$$
$$\tag{39.115}$$

where in the second equality, (39.114) has been used. On multiplying by X^{-1}, we obtain

$$d\theta^j - \theta^k \wedge \theta_k^j = (X^{-1})_r^j X_k^p\Gamma_{pq}^r X_l^q\, \theta^l \wedge \theta^k = \frac{1}{2}(X^{-1})_r^j X_k^p X_l^q T_{pq}^r\, \theta^k \wedge \theta^l , \tag{39.116}$$

where T_{pq}^r is the torsion tensor [(39.57)]. Exterior differentiation of (39.114) yields

$$0 = -dX_i^k \wedge \omega_k^j - X_i^k\, d\omega_k^j + dX_k^j \wedge \theta_i^k + X_k^j\, d\theta_i^k$$
$$= (X_i^k\omega_l^l - X_i^l\theta_i^k) \wedge \omega_l^j - X_i^k\, d\omega_k^j - (X_k^l\omega_l^j - X_j^j\theta_k^l) \wedge \theta_i^k + X_k^j\, d\theta_i^k$$
$$= -X_i^k(d\omega_k^j - \omega_k^l \wedge \omega_l^j) + X_k^j\, d\theta_i^k + X_k^j\, \theta_l^k \wedge \theta_i^l = -X_i^k\Omega_k^j + X_k^j(d\theta_i^k - \theta_l^l \wedge \theta_i^k) ,$$
$$\tag{39.117}$$

where in the second equality we have used (39.114), and in the fourth we have used the definition of the curvature matrix of 2-forms: $\Omega = d\omega - \omega \wedge \omega$ [(35.18)]. Thus

$$d\theta_i^j - \theta_i^l \wedge \theta_l^j = X_i^k\Omega_k^l(X^{-1})_l^j = \frac{1}{2}(X^{-1})_l^j X_i^k R_{kmn}^l\, dx^m \wedge dx^n$$
$$= \frac{1}{2}(X^{-1})_l^j X_i^k X_r^m X_s^n R_{kmn}^l\, \theta^r \wedge \theta^s . \tag{39.118}$$

Writing

$$P_{kl}^j \equiv (X^{-1})_r^j X_k^p X_l^q T_{pq}^r , \tag{39.119}$$

$$S_{ikl}^j \equiv (X^{-1})_q^j X_i^p X_k^r X_l^s R_{prs}^q , \tag{39.120}$$

which are the torsion and curvature tensors, respectively, with respect to $\{e_i\}$ and $\{\theta^i\}$, Eqs. (39.116) and (39.118) become

$$d\theta^j - \theta^k \wedge \theta^j_k = \frac{1}{2} P^j_{kl}\, \theta^k \wedge \theta^l\ , \tag{39.121}$$

$$d\theta^j_i - \theta^l_i \wedge \theta^j_l = \frac{1}{2} S^j_{ikl}\, \theta^k \wedge \theta^l\ . \tag{39.122}$$

These are called the **structure equations** of the connection. [Compare with (31.33) and (31.34).] For a torsion-free connection, the connection one-forms θ^j_i satisfy

$$d\theta^i - \theta^j \wedge \theta^i_j = 0\ . \tag{39.123}$$

Metric-compatibility of ω allows us to define a skew-symmetric $(0,2)$-type curvature tensor of 2-forms. One can exteriorly differentiate (39.93), the metric-compatibility condition, to yield

$$d\omega \cdot G - \omega \wedge dG + dG \wedge \omega^T + G \cdot (d\omega)^T = 0\ , \tag{39.124}$$

which implies, using (39.93) again,

$$d\omega \cdot G - \omega \wedge (\omega \cdot G + G \cdot \omega^T) + (\omega \cdot G + G \cdot \omega^T) \wedge \omega^T + G \cdot (d\omega)^T = 0\ , \tag{39.125}$$

or

$$(d\omega - \omega \wedge \omega) \cdot G + G \cdot ((d\omega)^T + \omega^T \wedge \omega^T) = 0\ . \tag{39.126}$$

Since ω is a matrix of one-forms, $\omega^T \wedge \omega^T = -(\omega \wedge \omega)^T$. This can be seen as follows.

$$(\omega^T \wedge \omega^T)^\beta_\alpha = (\omega^T)^\gamma_\alpha \wedge (\omega^T)^\beta_\gamma = \omega^\alpha_\gamma \wedge \omega^\gamma_\beta$$
$$= -\omega^\gamma_\beta \wedge \omega^\alpha_\gamma = -(\omega \wedge \omega)^\alpha_\beta = -((\omega \wedge \omega)^T)^\beta_\alpha\ .$$

Eq. (39.126) then reads

$$(d\omega - \omega \wedge \omega) \cdot G + G \cdot (d\omega - \omega \wedge \omega)^T = 0\ .$$

By (35.18), this leads to

$$\Omega \cdot G + G \cdot \Omega^T = 0\ .$$

Since $G = (g_{ij})$ is symmetrical, we have

$$\Omega \cdot G + (\Omega \cdot G)^T = 0\ . \tag{39.127}$$

Now define

$$\Omega_{ij} \equiv (\Omega \cdot G)_{ij} = \Omega^k_i g_{kj}\ . \tag{39.128}$$

Eq. (39.127) then becomes

$$\Omega_{ij} + \Omega_{ji} = 0\ . \tag{39.129}$$

Ω_{ij} is the promised skew-symmetric curvature tensor of 2-forms. Eq. (39.128) also implies

$$\Omega_{ij} = (d\omega^k_i - (\omega \wedge \omega)^k_i)\, g_{kj} = (d\omega^k_i)g_{kj} - \omega^l_i \wedge \omega^k_l\, g_{kj}\ . \tag{39.130}$$

On the other hand, if we also define

$$\omega_{ij} \equiv \omega_i^k \, g_{kj} \, , \tag{39.131}$$

then

$$d\omega_{ij} = d(\omega_i^k \, g_{kj}) = (d\omega_i^k) \, g_{kj} - \omega_i^k \wedge dg_{kj} = (d\omega_i^k)g_{kj} - \omega_i^k \wedge (\omega_k^l g_{lj} + \omega_j^l g_{kl})$$
$$= (d\omega_i^k)g_{kj} - \omega_i^l \wedge \omega_l^k \, g_{kj} - \omega_i^k \wedge \omega_{jk} \, , \tag{39.132}$$

where in the third equality, we have employed the metric-compatibility condition (39.92) for dg_{kj}. Thus (39.130) gives

$$\Omega_{ij} = d\omega_{ij} + \omega_i^k \wedge \omega_{jk} \, . \tag{39.133}$$

Let us now define the $(0, 4)$-type curvature tensor

$$R_{ijkl} \equiv R_i^{\,h}{}_{kl} \, g_{hj} \, . \tag{39.134}$$

Then, on recalling (39.46), we have

$$\Omega_{ij} = \frac{1}{2} \, R_{ijkl} \, dx^k \wedge dx^l \, . \tag{39.135}$$

The curvature tensor R_{ijkl} can be calculated in terms of the Christoffel symbols $\Gamma_i{}^j{}_k$ and Γ_{ijk} using (39.47) and (39.101). These equations give

$$R_{ijkl} = R_i^{\,h}{}_{kl} \, g_{hj}$$
$$= g_{hj} \left(\frac{\partial \Gamma_i{}^h{}_l}{\partial x^k} - \frac{\partial \Gamma_i{}^h{}_k}{\partial x^l} + \Gamma_i{}^m{}_l \Gamma_m{}^h{}_k - \Gamma_i{}^m{}_k \Gamma_m{}^h{}_l \right)$$
$$= \frac{\partial \Gamma_{ijl}}{\partial x^k} - \frac{\partial \Gamma_{ijk}}{\partial x^l} - \Gamma_i{}^h{}_l \frac{\partial g_{hj}}{\partial x^k} + \Gamma_i{}^h{}_k \frac{\partial g_{hj}}{\partial x^l} + \Gamma_i{}^m{}_l \Gamma_{mjk} - \Gamma_i{}^m{}_k \Gamma_{mjl} \, . \tag{39.136}$$

Using (39.102) for $\partial g_{hj}/\partial x^k$ and $\partial g_{hj}/\partial x^l$ in the above equation, we obtain

$$\boxed{ R_{ijkl} = \frac{\partial \Gamma_{ijl}}{\partial x^k} - \frac{\partial \Gamma_{ijk}}{\partial x^l} + \Gamma_i{}^h{}_k \Gamma_{jhl} - \Gamma_i{}^h{}_l \Gamma_{jhk} } \, . \tag{39.137}$$

The curvature tensor R_{ijkl} has the following important symmetry properties.

Theorem 39.5. *The curvature tensor R_{ijkl} of a Riemannian manifold satisfies the following:*

$$\begin{array}{lll} 1) & R_{ijkl} = -R_{jikl} = -R_{ijlk} \, ; & (39.138) \\ 2) & R_{ijkl} + R_{iklj} + R_{iljk} = 0 \, ; & (39.139) \\ 3) & R_{ijkl} = R_{klij} \, . & (39.140) \end{array}$$

Proof. 1) This is a direct consequence of (39.129) and (39.137).
2) Since the Levi-Civita connection is torsion-free, we have, from (39.123),

$$dx^i \wedge \omega_i^k = 0 , \tag{39.141}$$

and hence

$$dx^i \wedge \omega_i^k g_{kj} = dx^i \wedge \omega_{ij} = 0 . \tag{39.142}$$

Exteriorly differentiating the above equation and using (39.133) we have

$$dx^i \wedge d\omega_{ij} = dx^i \wedge (\Omega_{ij} - \omega_i^k \wedge \omega_{jk}) = 0 . \tag{39.143}$$

It follows from (39.141) that

$$dx^i \wedge \Omega_{ij} = 0 . \tag{39.144}$$

Using (39.135) for Ω_{ij} in the above equation and the fact that $R_{ijkl} = -R_{jikl}$ [from 1)], we get

$$R_{jikl}\, dx^i \wedge dx^k \wedge dx^l = 0 . \tag{39.145}$$

By adding two similar terms obtained from commuting the indices (i, k, l) in the above equation, we have

$$(R_{jikl} + R_{jkli} + R_{jlik})\, dx^i \wedge dx^k \wedge dx^l = 0 . \tag{39.146}$$

By 1) each R is skew-symmetric in the last two indices. It is then easy to see that the sum of the three R's in the parentheses in the above equation is skew-symmetric in the last three indices: i, k and l. But $dx^i \wedge dx^k \wedge dx^l$ is also skew-symmetric in the same three indices. Thus linear independence of $\{dx^i \wedge dx^k \wedge dx^l \mid i < k < l\}$ implies 2).
3) Write 2) in two different ways:

$$R_{jikl} + R_{jkli} + R_{jlik} = 0 , \tag{39.147}$$
$$R_{ijkl} + R_{iklj} + R_{iljk} = 0 . \tag{39.148}$$

Subtracting (39.147) from (39.148) and using 1), we have

$$2R_{ijkl} + R_{iklj} + R_{ljik} + R_{iljk} + R_{jkil} = 0 . \tag{39.149}$$

Letting $i \leftrightarrow k$, $j \leftrightarrow l$, the above equation appears as

$$2R_{klij} + R_{kijl} + R_{jlki} + R_{kjli} + R_{likj} = 0 . \tag{39.150}$$

But from 1) again, $R_{iklj} = R_{kijl}$, $R_{ljik} = R_{jlki}$, etc. Thus $R_{ijkl} = R_{klij}$.   □

Exercise 39.9   Show that analogous to property 2) of Theorem 39.5, we have

$$R^i_{jkl} + R^i_{klj} + R^i_{ljk} = 0 . \tag{39.151}$$

Exercise 39.10   Show that analogous to (39.69) (in Theorem 39.3), the **Bianchi identity** can also be stated as

$$R_{ijkl,\,h} + R_{ijlh,\,k} + R_{ijhk,\,l} = 0 \qquad . \qquad (39.152)$$

Use (39.96b).

Exercise 39.11   Show that, due to the symmetry properties of Theorem 39.5, the number of independent components of the Riemann curvature tensor R_{ijkl} is $n^2(n^2 - 1)/12$, where n is the dimension of the Riemannian manifold.

The Riemann curvature tensor R_{ijkl}, with its large number of components, is a rather complicated object. (For Lorentzian spacetime, where $n = 4$, it has 20 components.) It turns out that it is possible to characterize R_{ijkl} by a simpler object, called the sectional curvature.

Since R_{ijkl} is a $(0, 4)$-type tensor field, it can be viewed, at each point $p \in M$, as a multilinear function $R : T_p(M) \times T_p(M) \times T_p(M) \times T_p(M) \to \mathbb{R}$, where

$$R = R_{ijkl}\, dx^i \otimes dx^j \otimes dx^k \otimes dx^l \, . \qquad (39.153)$$

Thus, for $X, Y, Z, W \in T_p(M)$ given by $X = X^i \dfrac{\partial}{\partial x^i}$, $Y = Y^i \dfrac{\partial}{\partial x^i}$, $Z = Z^i \dfrac{\partial}{\partial x^i}$ and $W = W^i \dfrac{\partial}{\partial x^i}$,

$$R(X, Y, Z, W) = \langle X \otimes Y \otimes Z \otimes W, R \rangle = R_{ijkl}\, X^i Y^j Z^k W^l \, . \qquad (39.154)$$

In particular

$$R_{ijkl} = R\left(\frac{\partial}{\partial x^i}, \frac{\partial}{\partial x^j}, \frac{\partial}{\partial x^k}, \frac{\partial}{\partial x^l} \right) \, . \qquad (39.155)$$

By (39.55), for $Z, W \in T_p(M)$, $R(Z, W)$ (the curvature operator) is a linear map from $T_p(M)$ to $T_p(M)$:

$$R(Z, W)X = R^j_{ikl}\, X^i Z^k W^l \frac{\partial}{\partial x^j} \, . \qquad (39.156)$$

It follows from (39.134) that

$$R(X, Y, Z, W) = (R(Z, W)X) \cdot Y \, , \qquad (39.157)$$

where the dot means the scalar product with the Riemannian metric g_{ij}.

By Theorem 39.5, the 4-linear function $R(X, Y, Z, W)$ has the following symmetry properties:

1) $\quad R(X, Y, Z, W) = -R(Y, X, Z, W) = -R(X, Y, W, Z) \, ; \qquad (39.158)$
2) $\quad R(X, Y, Z, W) + R(X, Z, W, Y) + R(X, W, Y, Z) = 0 \, ; \qquad (39.159)$
3) $\quad R(X, Y, Z, W) = R(Z, W, X, Y) \, . \qquad (39.160)$

Using the metric tensor G, one can define the 4-linear function

$$G(X, Y, Z, W) \equiv G(X, Z)G(Y, W) - G(X, W)G(Y, Z) . \qquad (39.161)$$

It can be checked directly that $G(X, Y, Z, W)$ also satisfies the symmetry properties 1), 2), 3) above.

Exercise 39.12 If $X, Y \in T_p(M)$, show that

$$G(X, Y, X, Y) = |X|^2 |Y|^2 \sin^2 \theta ,$$

where θ is the angle between X and Y. Thus $G(X, Y, X, Y)$ is the square of the area of the parallelogram bounded by X and Y.

Suppose X, Y span a two-dimensional subspace E of $T_p(M)$. Let X', Y' be another two linearly independent vectors in E. Then X', Y' span E also, and they are related to X, Y by a non-degenerate linear transformation:

$$X' = aX + bY , \quad Y' = cX + dY , \quad (ad - bc \neq 0) .$$

Using properties 1) to 3) above and the fact that R is 4-linear, it is straightforward to show that

$$R(X', Y', X', Y') = (ad - bc)^2 R(X, Y, X, Y) . \qquad (39.162)$$

Exercise 39.13 Verify (39.162).

By the same token,

$$G(X', Y', X', Y') = (ad - bc)^2 G(X, Y, X, Y) . \qquad (39.163)$$

Thus

$$\frac{R(X', Y', X', Y')}{G(X', Y', X', Y')} = \frac{R(X, Y, X, Y)}{G(X, Y, X, Y)} , \qquad (39.164)$$

which means that $R(X, Y, X, Y)/G(X, Y, X, Y)$ only depends on the subspace $E \subset T_p(M)$, and is independent of the choice of the basis $\{X, Y\}$ of E. We call the quantity

$$\boxed{K(X, Y) = K(E) \equiv -\frac{R(X, Y, X, Y)}{G(X, Y, X, Y)}} \qquad (39.165)$$

where $\{X, Y\}$ is any basis of the 2-dimensional subspace $E \subset T_p(M)$, the **sectional curvature** of M at (p, E). A special case of historical importance is when M is a surface in 3-dimensional Euclidean space, where K is the so-called **total curvature**, or **Gaussian curvature** of the surface at a point on the surface. It is given by

$$K = -\frac{R_{1212}}{g} , \qquad (39.166)$$

where

$$g = g_{11}g_{22} - (g_{12})^2 , \tag{39.167}$$

and, by (39.137),

$$R_{1212} = \frac{\partial \Gamma_{122}}{\partial x^1} - \frac{\partial \Gamma_{121}}{\partial x^2} + \Gamma_1{}^h{}_1 \Gamma_{2h2} - \Gamma_1{}^h{}_2 \Gamma_{2h1} . \tag{39.168}$$

Exercise 39.14   For a spherical surface of radius r in 3-dimensional Euclidean space, start with the induced metric (with respect to the natural coordinates θ, ϕ)

$$g_{ij} = \begin{pmatrix} r^2 & 0 \\ 0 & r^2 \sin^2 \theta \end{pmatrix} \tag{39.169}$$

and use (39.106) to (39.108) to calculate the Christoffel symbols Γ_{ijk} and $\Gamma_i{}^j{}_k$, $i, j, k = 1, 2$. Show that

$$\Gamma_1{}^1{}_1 = \Gamma_1{}^2{}_1 = \Gamma_1{}^1{}_2 = \Gamma_{121} = 0 , \tag{39.170}$$

$$\Gamma_1{}^2{}_2 = \cot \theta , \tag{39.171}$$

$$\Gamma_{221} = \Gamma_{122} = r^2 \sin \theta \cos \theta . \tag{39.172}$$

Then use (39.168) to show that

$$R_{1212} = -r^2 \sin^2 \theta , \tag{39.173}$$

and hence that the Gaussian curvature is given by

$$K = \frac{1}{r^2} . \tag{39.174}$$

Exercise 39.15   Recall from (31.36) that, with respect to the orthonormal frame field

$$e_1 = e_\theta = \frac{1}{r} \frac{\partial}{\partial \theta} , \qquad e_2 = e_\phi = \frac{1}{r \sin \theta} ,$$

the connection matrix of the Levi-Civita connection is given by

$$(\theta_i^j) = \begin{pmatrix} 0 & \cos \theta \, d\phi \\ -\cos \theta \, d\phi & 0 \end{pmatrix} . \tag{39.175}$$

Show that the curvature 2-form $\Omega_i^j = d\theta_i^j - \theta_i^k \wedge \theta_k^j$, with respect to (e_1, e_2) is given by

$$(\Omega_i^j) = \begin{pmatrix} 0 & -\sin \theta \, d\theta \wedge d\phi \\ \sin \theta \, d\theta \wedge d\phi & 0 \end{pmatrix} . \tag{39.176}$$

Exercise 39.16   Use (39.169) for g_{ij} and (39.108) for $\Gamma_i{}^k{}_j$ to show that, with respect to the natural frame field $(\partial/\partial\theta, \partial/\partial\phi)$, the connection matrix of the Levi-Civita connection is given by

$$(\omega_i^j) = \begin{pmatrix} 0 & \cot\theta\, d\phi \\ -\sin\theta\cos\theta\, d\phi & \cot\theta\, d\theta \end{pmatrix}. \tag{39.177}$$

Exercise 39.17   The dual frame-field to the orthonormal frame field (e_θ, e_ϕ) is the coframe field (θ^1, θ^2) where [c.f. (31.37)]

$$\theta^1 = r\,d\theta, \qquad \theta^2 = r\sin\theta\,d\phi. \tag{39.178}$$

Write [compare with (39.46)]

$$\Omega_i^j = \frac{1}{2} R_i{}^j{}_{kl}\, \theta^k \wedge \theta^l, \tag{39.179}$$

where Ω_i^j is given by (39.176). Thus

$$\Omega_1^2 = -\sin\theta\, d\theta \wedge d\phi.$$

Show from (39.178) and (39.179) that, with respect to (e_θ, e_ϕ),

$$R_1{}^2{}_{12} = -\frac{1}{r^2}. \tag{39.180}$$

Since (e_θ, e_ϕ) is orthonormal, the Riemannian metric (g_{ij}) with respect to (e_θ, e_ϕ) is simply $g_{ij} = \delta_{ij}$. Thus verify that, with respect to (e_θ, e_ϕ),

$$R_{1212} = R_1{}^2{}_{12} = -\frac{1}{r^2}, \tag{39.181}$$

and hence

$$K = \frac{1}{r^2},$$

in agreement with (39.174).

The importance of the sectional curvature lies in the following theorem, which will be stated without proof.

Theorem 39.6. *The curvature tensor R_{ijkl} of a Riemannian manifold M at a point $p \in M$ is uniquely determined by the sectional curvatures $K(E)$ of all two-dimensional tangent subspaces $E \subset T_p(M)$.*

The Riemannian curvature tensor $R_i{}^j{}_{kl}$ can be contracted to yield other useful tensors.

Definition 39.5. *Let R^j_{ikl} be the Riemannian curvature tensor of a Riemannian manifold M. The* **Ricci curvature tensor** *Ric of M is the contraction of $R_i{}^j{}_{kl}$ given by*

$$R_{ik} \equiv R_i{}^j{}_{kj} . \tag{39.182}$$

The **scalar curvature** *S of M is the contraction of Ric given by*

$$S \equiv g^{ij} R_{ij} . \tag{39.183}$$

If *Ric* is identically zero, M is said to be **Ricci flat**. A flat manifold ($R^j_{ikl} = 0$ identically) is certainly Ricci flat, but the converse is not true.

Lemma 39.1. *The Ricci curvature tensor R_{ij} is symmetric, and is given relative to the natural frame field $\{\partial/\partial x^i\}$ by*

$$\boxed{Ric(X,Y) = \langle\, R(Y,\partial_k)X, \, dx^k \,\rangle} \qquad , \tag{39.184}$$

for any $X, Y \in T_p(M)$.

Proof. By (39.184),

$$R_{ij} = Ric(\partial_i,\, \partial_j) = \langle\, R(\partial_j,\partial_k)\partial_i,\, dx^k \,\rangle = R^l_{ijk} \langle\, \partial_l,\, dx^k \,\rangle = R^l_{ijk}\,\delta^k_l = R_i{}^k{}_{jk} . \tag{39.185}$$

The proof of the symmetry of *Ric* will be given below, immediately after Exercise 39.19. $\qquad\square$

$\boxed{\text{Exercise 39.18}}$ Let $g = \det(g_{ij})$. Show that

$$\frac{\partial}{\partial x^k} \ln |g|^{\frac12} = \frac{1}{2g}\frac{\partial g}{\partial x^k} = \frac12 g^{jm}\frac{\partial g_{mj}}{\partial x^k} . \tag{39.186}$$

Recall (34.106): $\det(e^B) = e^{tr\,B}$. Let $G = (g_{ij}) = e^B$. The metric tensor then satisfies

$$\boxed{\det G = e^{tr\,\ln G}} \qquad . \tag{39.187}$$

$\boxed{\text{Exercise 39.19}}$ Show that

$$\Gamma_k{}^j{}_j = \Gamma_j{}^j{}_k = \frac12 g^{jm}\frac{\partial g_{mj}}{\partial x^k} . \tag{39.188}$$

Use (39.47) to write

$$R_{ik} = R^j_{ikj} = \frac{\partial \Gamma_i{}^j{}_j}{\partial x^k} - \frac{\partial \Gamma_i{}^j{}_k}{\partial x^j} + \Gamma_i{}^h{}_j\Gamma_h{}^j{}_k - \Gamma_i{}^h{}_k\Gamma_h{}^j{}_j , \tag{39.189}$$

$$R_{ki} = R^j_{kij} = \frac{\partial \Gamma_k{}^j{}_j}{\partial x^i} - \frac{\partial \Gamma_k{}^j{}_i}{\partial x^j} + \Gamma_k{}^h{}_j\Gamma_h{}^j{}_i - \Gamma_k{}^h{}_i\Gamma_h{}^j{}_j . \tag{39.190}$$

Thus

$$R_{ik} - R_{ki} = \frac{\partial \Gamma_i{}^j{}_j}{\partial x^k} - \frac{\partial \Gamma_k{}^j{}_j}{\partial x^i} = 0 \ ,$$

where the second equality follows from (39.188).

With respect to an orthonormal frame field $\{e_i\}$ [$g_{ij} = G(e_i, e_j) = \delta_{ij}$], the Ricci curvature tensor can be defined as

$$\boxed{Ric(X, Y) \equiv \sum_k G(R(Y, e_k)X, \ e_k)} \qquad (39.191)$$

for any $X, Y \in T_p(M)$. This is easily seen as follows. Let $X = e_i$, $Y = e_j$. Then

$$R_{ij} = Ric(e_i, e_j) = \sum_k G(R(e_j, e_k)e_i, \ e_k)$$

$$= \sum_k R^l_{ijk} G(e_l, e_k) = \sum_k R_i{}^l{}_{jk} \, \delta_{lk} = R_i{}^l{}_{jl} \ . \qquad (39.192)$$

On the other hand

$$R_{ij} = \sum_k R_i{}^l{}_{jk} \, G(e_l, e_k) = \sum_k R_i{}^l{}_{jk} \, g_{lk} = \sum_k R_{ikjk} = \sum_{k \neq i} R_{ikjk} \ . \qquad (39.193)$$

On the RHS in the last equality, the summation is limited to $k \neq i$ since, by property 1) of Theorem 39.5 [(39.138)], $R_{iijk} = 0$. The above equation allows one to establish a useful relationship between the Ricci curvature and the sectional curvature with respect to an orthonormal frame field $\{e_i\}$. We have, remembering $g_{ij} = \delta_{ij}$,

$$Ric(e_i, e_i) = \sum_{k \neq i} R_{ikik} = \sum_{k \neq i} \frac{R_{ikik}}{g_{ii} g_{kk} - (g_{ik})^2} \ . \qquad (39.194)$$

Thus, from (39.165),

$$\boxed{Ric(e_i, e_i) = - \sum_{k \neq i} K(e_i, e_k)} \quad . \qquad (39.195)$$

The following important fact concerning the "divergence" of the Ricci tensor is the basis of Einstein's equation in general relativity.

Theorem 39.7. *Let R_{ij} be the Ricci curvature tensor [(39.182)] and S the scalar curvature [(39.183)] of a Riemannian manifold M. Then*

$$\boxed{D^i \left(R_{ij} - \frac{1}{2} g_{ij} S\right) = 0} \quad , \qquad (39.196)$$

where $D^i \equiv g^{ij} D_j$ and $D_j \equiv D_{\partial/\partial x^j}$.

Proof. Recall Bianchi's identity (39.69):

$$D_h R^j_{ikl} + D_k R^j_{ilh} + D_l R^j_{ihk} = 0 . \tag{39.197}$$

Contracting the pair of indices (j, l), we have

$$D_h R_{ik} - D_k R_{ih} + D_j R_i{}^j{}_{hk} = 0 . \tag{39.198}$$

Multiplying by g^{ih} (and summing over h) and using the metric-compatibility property $D_k g^{ih} = 0$ [(39.96b)], we have

$$D_h R^h_k - D_k S + D_j (g^{ih} R_i{}^j{}_{hk}) = 0 , \tag{39.199}$$

where

$$R^h_k \equiv g^{ih} R_{ik} . \tag{39.200}$$

The quantity $g^{ih} R_i{}^j{}_{hk}$ contracts to R^j_k. Indeed

$$g^{ih} R_i{}^j{}_{hk} = R^{hj}{}_{hk} = -R^{jh}{}_{hk} = -g^{ij} R_i{}^h{}_{hk} = g^{ij} R_i{}^h{}_{kh} = g^{ij} R_{ik} = R^j_k , \tag{39.201}$$

where in the second equality, we have used an analog of (39.138). Eq. (39.199) then becomes

$$D_j R^j_k - \frac{1}{2} \delta^j_k D_j S = 0 . \tag{39.202}$$

Putting $D_j = g_{ji} D^i$, the above equation is equivalent to

$$D^i R_{ik} - \frac{1}{2} g_{ik} D^i S = 0 .$$

$\square$

The divergence-free tensor $R_{ij} - \frac{1}{2} g_{ij} S$ is called the **Einstein tensor** G_{ij}:

$$\boxed{G_{ij} \equiv R_{ij} - \frac{1}{2} g_{ij} S} . \tag{39.203}$$

In formulating the general theory of relativity, Einstein sought an equation which describes how 4-dimensional spacetime (a Lorentzian manifold) is curved by the presence of matter, or more generally, by anything possessing energy and momentum. The flow of energy and momentum is completely described by a symmetric $(0, 2)$-type tensor $T_{\mu\nu}$, called the **stress-energy tensor**, so that the laws of local conservation of energy and momentum are expressed by

$$D^\mu T_{\mu\nu} = 0 , \tag{39.204}$$

which is the same divergenceless condition that $G_{\mu\nu}$ satisfies. Thus he wrote down the so-called **Einstein's equation**:

$$\boxed{G_{\mu\nu} = 8\pi\kappa T_{\mu\nu}} , \tag{39.205}$$

where $\kappa = 6.67 \times 10^{-11} \, N\text{-}m^2/Kg^2$ is Newton's gravitational constant. This is the simplest equation relating the curvature of spacetime to the presence of matter which automatically implies energy-momentum conservation. This situation is similar to the one in electromagnetism, where the Maxwell's equations $\star d \star \Omega = J$ [(36.23)] automatically implies the equation of continuity [(36.31)] (the conservation of charge).

Exercise 39.20   Consider Minkowski spacetime, where $g_{\mu\nu} = \eta_{\mu\nu}$. Suppose the local coordinates are (x^0, x^i), $i = 1, 2, 3$. The components of the stress-energy tensor are given the following meanings. $T^{00} =$ energy density, $T^{i0} =$ energy flux in the ∂_i direction, $T^{0i} =$ density of the i-th component of momentum, $T^{ij} =$ flux of the j-th component of momentum in the ∂_i direction. Show that (39.204) is equivalent to

$$\frac{\partial T^{00}}{\partial t} + \frac{\partial T^{i0}}{\partial x^i} = 0 \quad \text{(conservation of energy)}, \tag{39.206}$$

and

$$\frac{\partial T^{0j}}{\partial t} + \frac{\partial T^{ij}}{\partial x^i} = 0 \quad \text{(conservation of momentum)}. \tag{39.207}$$

Chapter 40

Complex Manifolds

A **complex manifold** is formally defined in a manner entirely similar to a real manifold [c.f. Chapter 30]. Instead of real coordinate neighborhoods $(U; x^i)$, $x^i \in \mathbb{R}$, we have complex coordinate neighborhoods $(U; z^i)$, $z^i \in \mathbb{C}$, $i = 1, \ldots, n$ (dimension of the manifold). In regions $(U; z^i)$ and $(W; w^i)$ of non-empty overlap, it is required that all the complex functions (of n complex variables)

$$w^i = w^i(z^1, \ldots, z^n), \quad i = 1, \ldots, n, \tag{40.1}$$

are **holomorphic**. The condition of holomorphicity can be expressed in any one of the following three equivalent ways:

1) The **Cauchy-Riemann conditions**. If we write $z^i = x^i + iy^i$, $x^i, y^i \in \mathbb{R}$, and

$$w^i = g^i(x^k; y^l) + ih^i(x^k; y^l), \tag{40.2}$$

where g^i, h^i are real functions, then, for $i = 1, \ldots, n$,

$$\frac{\partial g^i}{\partial x^k} = \frac{\partial h^i}{\partial y^k}; \quad \frac{\partial g^i}{\partial y^k} = -\frac{\partial h^i}{\partial x^k} \quad 1 \le k \le n. \tag{40.3}$$

2) For every $p \in U$ with complex coordinates $(z_0^1, \ldots, z_0^n)$, there exists a neighborhood $V \subset U$, such that for points in V with complex coordinates $z^1, \ldots, z^n$, w^i can be expressed as a convergent series:

$$w^i(z^1, \ldots, z^n) = \sum_{k_1, \ldots, k_n = 0}^{\infty} c_{k_1 \ldots k_n} (z^1 - z_0^1)^{k_1} \ldots (z^n - z_0^n)^{k_n}. \tag{40.4}$$

3) The derivatives $\partial w^i / \partial z^k$, $1 \le i, k \le n$, exist in U.

We further require that the inverse functions $z^i = z^i(w^1, \ldots, w^n)$ exist, and are also holomorphic. Thus

$$\det \frac{\partial(w^1, \ldots, w^n)}{\partial(z^1, \ldots, z^n)} \ne 0, \tag{40.5}$$

491

where $\partial(w^1,\ldots,w^n)/\partial(z^1,\ldots,z^n)$ denotes the Jacobian matrix of the transformation (40.1).

A one-dimensional complex manifold is called a **Riemann surface**. It is the basic object of study in the theory of functions of one complex variable.

On a complex manifold one can impose metrical structures which are generalizations of the Riemannian metrics on real manifolds. To understand how this is done, we need to first develop the concept of an almost complex manifold.

An n-dimensional complex manifold can be viewed as a $2n$-dimensional real manifold. Suppose $(U; z^i), i = 1,\ldots,n$, is a local coordinate neighborhood. Let $z^k = x^k + iy^k$, $x^k, y^k \in \mathbb{R}$. Then $\{x^k, y^k; 1 \le k \le n\}$ is a local coordinate system in a real manifold M; and $\left\{\dfrac{\partial}{\partial x^k}, \dfrac{\partial}{\partial y^k}; 1 \le k \le n\right\}$ gives a local natural frame field on M. Consider a point $x \in M$ whose local coordinates are $(x^1,\ldots,x^n,y^1,\ldots,y^n)$. Define a linear transformation $J_x : T_x(M) \to T_x(M)$ by

$$J_x\left(\frac{\partial}{\partial x^k}\right) = \frac{\partial}{\partial y^k}\; ; \quad J_x\left(\frac{\partial}{\partial y^k}\right) = -\frac{\partial}{\partial x^k}\; ; \quad k = 1,\ldots,n\,. \tag{40.6}$$

J_x obviously satisfies $J_x^2 = -id$, where id is the identity map on $T_x(M)$. This map is independent of the choice of local coordinates z^i. Indeed, suppose $w^k = u^k + iv^k$, $k = 1,\ldots,n$, is another set of local coordinates of $x \in M$. Then the Cauchy-Riemann conditions (40.2) imply that

$$\frac{\partial x^j}{\partial u^k} = \frac{\partial y^j}{\partial v^k}\;, \quad \frac{\partial x^j}{\partial v^k} = -\frac{\partial y^j}{\partial u^k}\,. \tag{40.7}$$

Thus

$$J_x\left(\frac{\partial}{\partial u^k}\right) = J_x\left(\frac{\partial x^j}{\partial u^k}\frac{\partial}{\partial x^j} + \frac{\partial y^j}{\partial u^k}\frac{\partial}{\partial y^j}\right) = \frac{\partial y^j}{\partial v^k}\frac{\partial}{\partial y^j} + \frac{\partial x^j}{\partial v^k}\frac{\partial}{\partial x^j} = \frac{\partial}{\partial v^k}\,, \tag{40.8}$$

$$J_x\left(\frac{\partial}{\partial v^k}\right) = J_x\left(\frac{\partial x^j}{\partial v^k}\frac{\partial}{\partial x^j} + \frac{\partial y^j}{\partial v^k}\frac{\partial}{\partial y^j}\right) = -\frac{\partial y^j}{\partial u^k}\frac{\partial}{\partial y^j} - \frac{\partial x^j}{\partial u^k}\frac{\partial}{\partial x^j} = -\frac{\partial}{\partial u^k}\,. \tag{40.9}$$

There is an induced linear transformation (also denoted J_x) on the dual coframe field $\{dx^k, dy^k; k = 1,\ldots,n\}$ defined by

$$\langle X, J_x Y^* \rangle \equiv \langle J_x X, Y^* \rangle\,, \tag{40.10}$$

for any $X \in T_x(M)$, $Y^* \in T_x^*(M)$. Thus it follows from (40.6) that

$$J_x(dx^k) = -dy^k\; ; \quad J_x(dy^k) = dx^k\; ; \quad k = 1,\ldots,n\,. \tag{40.11}$$

The induced map $J_x : T_x^*(M) \to T_x^*(M)$ also satisfies $J_x^2 = -id$, where id is the identity map on $T_x^*(M)$.

From $z^k = x^k + iy^k$ and $\bar{z}^k = x^k - iy^k$ we have $x^k = \dfrac{1}{2}(z^k + \bar{z}^k)$ and $y^k = \dfrac{1}{2i}(z^k - \bar{z}^k)$. It follows that, for $1 \le k \le n$,

$$\frac{\partial}{\partial z^k} = \frac{1}{2}\left(\frac{\partial}{\partial x^k} - i\frac{\partial}{\partial y^k}\right), \tag{40.12}$$

$$\frac{\partial}{\partial \bar{z}^k} = \frac{1}{2}\left(\frac{\partial}{\partial x^k} + i\frac{\partial}{\partial y^k}\right); \tag{40.13}$$

and

$$dz^k = dx^k + i\,dy^k, \tag{40.14}$$

$$d\bar{z}^k = dx^k - i\,dy^k. \tag{40.15}$$

The set $\left\{\dfrac{\partial}{\partial z^k}, \dfrac{\partial}{\partial \bar{z}^k}\right\}$, $k = 1,\ldots,n$, is seen to be a basis of the complexification of $T_x(M)$, that is, $T_x(M) \otimes \mathbb{C}$. Similarly, $\{dz^k, d\bar{z}^k\}$, $k = 1,\ldots,n$, is a basis of $T_x^*(M) \otimes \mathbb{C}$. In fact, $\left\{\dfrac{\partial}{\partial x^k}, \dfrac{\partial}{\partial y^k}\right\}$ and $\{dx^k, dy^k\}$ are also bases of $T_x(M) \otimes \mathbb{C}$ and $T_x^*(M) \otimes \mathbb{C}$, respectively.

Making use of (40.6) and (40.11), we then have, for $1 \le k \le n$,

$$J_x\left(\frac{\partial}{\partial z^k}\right) = i\frac{\partial}{\partial z^k}, \qquad J_x\left(\frac{\partial}{\partial \bar{z}^k}\right) = -i\frac{\partial}{\partial \bar{z}^k}, \tag{40.16}$$

$$J_x(dz^k) = i\,dz^k, \qquad J_x(d\bar{z}^k) = -i\,d\bar{z}^k. \tag{40.17}$$

Thus the vectors $\partial/\partial z^k$, $k = 1,\ldots,n$, span a subspace $T_x(M)_{\mathbb{C}} \subset T_x(M) \otimes \mathbb{C}$, whose elements are all eigenvectors of J_x with eigenvalue i; and $\partial/\partial \bar{z}^k$, $k = 1,\ldots,n$, span a subspace $\overline{T_x(M)_{\mathbb{C}}} \subset T_x(M) \otimes \mathbb{C}$, whose elements are all eigenvectors of J_x with eigenvalue $-i$. Similarly, $\{dz^k\}$ span $T_x^*(M)_{\mathbb{C}} \subset T_x^*(M) \otimes \mathbb{C}$, the subspace of eigenvectors of J_x with eigenvalue i, and $\{d\bar{z}^k\}$ span $\overline{T_x^*(M)_{\mathbb{C}}} \subset T_x^*(M) \otimes \mathbb{C}$, the subspace of eigenvectors of J_x with eigenvalue $-i$. Furthermore, we have the following vector space direct-sum decomposition:

$$T_x(M) \otimes \mathbb{C} = T_x(M)_{\mathbb{C}} \oplus \overline{T_x(M)_{\mathbb{C}}}, \tag{40.18}$$

$$T_x^*(M) \otimes \mathbb{C} = T_x^*(M)_{\mathbb{C}} \oplus \overline{T_x^*(M)_{\mathbb{C}}}. \tag{40.19}$$

⬛ Exercise 40.1 Verify (40.18) and (40.19).

Vectors (forms) in $T_x(M)_{\mathbb{C}}$ ($T_x^*(M)_{\mathbb{C}}$) are called vectors (forms) of type $(1,0)$; while vectors (forms) in $\overline{T_x(M)_{\mathbb{C}}}$ ($\overline{T_x^*(M)_{\mathbb{C}}}$) are called vectors (forms) of type $(0,1)$.

We can start from an arbitrary even-dimensional real vector space $\mathbb{V}$ and introduce the concept of a complex structure on a real vector space as follows.

Definition 40.1. *Let $\mathbb{V}$ be a $2n$-dimensional real vector space. A **complex structure** on $\mathbb{V}$ is a linear transformation $J : \mathbb{V} \to \mathbb{V}$ such that $J^2 = -\mathrm{id}$.*

It is clear that the eigenvalues of J must be $\pm i$, with i and $-i$ occuring in pairs. The eigenvectors are vectors in the complexified space $V \otimes \mathbb{C}$; those with eigenvalue i constitute a subspace $V_{\mathbb{C}} \subset V \otimes \mathbb{C}$, while those with eigenvalue $-i$ constitute a subspace $\overline{V}_{\mathbb{C}} \subset V \otimes \mathbb{C}$. In general $V \otimes \mathbb{C}$ decomposes into the direct sum

$$V \otimes \mathbb{C} = V_{\mathbb{C}} \oplus \overline{V}_{\mathbb{C}} . \tag{40.20}$$

Choose a basis $\{\xi_i\}$, $i = 1, \ldots, n$, for $V_{\mathbb{C}}$. ξ_j can always be written as

$$\xi_j = \alpha - i\beta , \tag{40.21}$$

where $\alpha, \beta \in V$. The condition that $J\xi_j = i\xi_j$ implies

$$J\alpha = \beta, \qquad J\beta = -\alpha . \tag{40.22}$$

Define

$$\overline{\xi}_j \equiv \alpha + i\beta . \tag{40.23}$$

Then

$$J\overline{\xi}_j = J\alpha + iJ\beta = \beta - i\alpha = -i(\alpha + i\beta) = -i\overline{\xi}_j . \tag{40.24}$$

Thus $\overline{\xi}_j \in \overline{V}_{\mathbb{C}}$; and indeed $\{\overline{\xi}_j\}$, $j = 1, \ldots, n$, span $\overline{V}_{\mathbb{C}}$. Hence, by (40.20), $\{\xi_i, \overline{\xi}_i\}$, $i = 1, \ldots, n$, forms a basis of $V \otimes \mathbb{C}$. Vectors in $V_{\mathbb{C}}$ and $\overline{V}_{\mathbb{C}}$ are called vectors of type-$(1,0)$ and type-$(0,1)$, respectively, with respect to the complex structure J.

Define a set of $2n$ vectors $\{e_i, e_{n+i}\} \in V$, $i = 1, \ldots, n$, by

$$\xi_j = \frac{1}{2}(e_j - ie_{n+j}), \quad j = 1, \ldots, n ; \tag{40.25}$$

$$\overline{\xi}_j = \frac{1}{2}(e_j + ie_{n+j}), \quad j = 1, \ldots, n . \tag{40.26}$$

Then we see that $\{e_i, e_{n+i}\}$ is a basis of V as well as $V \otimes \mathbb{C}$. From (40.22), we also have

$$Je_i = e_{n+i} , \quad Je_{n+i} = -e_i . \tag{40.27}$$

These equations are analogous to (40.6). Let $\{(e^*)^i, (e^*)^{n+i}\}$ be the basis of V^* dual to $\{e_i, e_{n+i}\}$. Define an induced complex structure on V^*, also denoted by J, by

$$\langle v, Ju^* \rangle = \langle Jv, u^* \rangle , \tag{40.28}$$

for any $v \in V$ and $u^* \in V^*$ [c.f. (40.10)]. It is then easy to verify that [compare with (40.11)]

$$J(e^*)^i = -(e^*)^{n+i} , \quad J(e^*)^{n+i} = (e^*)^i , \quad i = 1, \ldots, n . \tag{40.29}$$

$\boxed{\text{Exercise 40.2}}$ Use (40.27) and (40.28) to verify (40.29).

Let

$$\lambda^j = (e^*)^j + i(e^*)^{n+j} , \quad j = 1, \ldots, n , \tag{40.30}$$

$$\overline{\lambda}^j = (e^*)^j - i(e^*)^{n+j} , \quad j = 1, \ldots, n . \tag{40.31}$$

Then, by (40.29),

$$J\lambda^j = i\lambda^j , \quad J\overline{\lambda}^j = -i\overline{\lambda}^j . \tag{40.32}$$

Thus $\lambda^j \in \mathbb{V}_{\mathbb{C}}^*$, $\overline{\lambda}^j \in \overline{\mathbb{V}_{\mathbb{C}}^*}$, where

$$\mathbb{V}^* \otimes \mathbb{C} = \mathbb{V}_{\mathbb{C}}^* \oplus \overline{\mathbb{V}_{\mathbb{C}}^*} . \tag{40.33}$$

Indeed, $\{\lambda^j\}$, $j = 1, \ldots, n$, is a basis of $\mathbb{V}_{\mathbb{C}}^*$, $\{\overline{\lambda}^j\}$, $j = 1, \ldots, n$, is a basis of $\overline{\mathbb{V}_{\mathbb{C}}^*}$, and $\{\lambda^j, \overline{\lambda}^j\}$, $j = 1, \ldots, n$, is a basis of $\mathbb{V}^* \otimes \mathbb{C}$. Note that $\{(e^*)^i, (e^*)^{n+i}\}$ also constitutes a basis of $\mathbb{V}^* \otimes \mathbb{C}$. Furthermore, the basis $\{\lambda^j, \overline{\lambda}^j\}$ and $\{\xi_i, \overline{\xi}_i\}$ are dual to each other:

$$\langle \xi_j, \lambda^k \rangle = \langle \overline{\xi}_j, \overline{\lambda}^k \rangle = \delta_j^k , \tag{40.34}$$

$$\langle \overline{\xi}_j, \lambda^k \rangle = \langle \xi_j, \overline{\lambda}^k \rangle = 0 . \tag{40.35}$$

Exercise 40.3 Verify the above two equations.

We will state the following basic fact without proof.

Theorem 40.1. *Suppose $\mathbb{V}$ is a $2n$-dimensional real vector space with complex structure J. Then any two bases of the form $\{e_i, Je_i\}$, $i = 1, \ldots, n$, give the same orientation to $\mathbb{V}$ (c.f. Def. 29.1).*

We can now define the notion of an almost complex manifold.

Definition 40.2. *Let M be a $2n$-dimensional smooth real manifold. If, for every $x \in M$, there exists a complex structure J_x on the tangent space $T_x(M)$ such that J is a smooth tensor field of $(1,1)$-type on M, then J is called an **almost complex structure** on M. A smooth real manifold endowed with an almost complex structure is called an **almost complex manifold**.*

By Theorem 40.1, we see that *an almost complex manifold must be an orientable manifold of even dimension.* These conditions, however, are not sufficient for a real manifold to possess an almost complex structure. On the other hand, a complex manifold is naturally an almost complex manifold, with the almost complex structure given by (40.6), called the **canonical almost complex structure** of a complex manifold. For an almost complex manifold to be also a complex manifold, the almost complex structure has to satisfy a set of so-called integrability conditions, the details of which will not be presented here. (See Chern 1979; Chern, Chen and Lam 1999.)

On an almost complex manifold M one can impose a **hermitian metric**, which, analogous to the Riemannian metric, is a smooth $(0,2)$-type tensor field H on M such that for each $x \in M$, $H_x : T_x(M) \times T_x(M) \to \mathbb{C}$ is a complex-valued function which satisfies the following conditions:

1) for any $X_1, X_2, Y \in T_x(M)$, $\alpha_1, \alpha_2 \in \mathbb{R}$,

$$H_x(\alpha_1 X_1 + \alpha_2 X_2, Y) = \alpha_1 H_x(X_1, Y) + \alpha_2 H_x(X_2, Y) ; \qquad (40.36)$$

2) for any $X, Y \in T_x(M)$, we have

$$H_x(Y, X) = \overline{H_x(X, Y)} , \qquad (40.37)$$

3) for any $X, Y \in T_x(M)$,

$$H_x(J_x X, Y) = i H_x(X, Y) . \qquad (40.38)$$

[Compare with (8.27) and (8.28).] H_x is called a **hermitian structure** on the real vector space $T_x(M)$, endowed with the complex structure J_x.

Separating into real and imaginary parts, $H_x(X, Y)$ can be given as

$$H_x(X, Y) = F_x(X, Y) + i G_x(X, Y) , \qquad (40.39)$$

where both F_x and G_x are real-valued bilinear functions on $T_x(M)$. Condition 2) then implies

$$F_x(Y, X) + i G_x(Y, X) = F_x(X, Y) - i G_x(X, Y) . \qquad (40.40)$$

Hence

$$F_x(Y, X) = F_x(X, Y) ; \quad G_x(Y, X) = -G_x(X, Y) , \qquad (40.41)$$

that is, F_x is a symmetric bilinear function whereas G_x is an antisymmetric bilinear function. Applying condition 3), we have

$$F_x(J_x X, Y) = -G_x(X, Y) ; \quad G_x(J_x X, Y) = F_x(X, Y) . \qquad (40.42)$$

It follows from (40.41) that

$$F_x(J_x X, J_x Y) = F_x(X, Y) , \qquad (40.43)$$
$$G_x(J_x X, J_x Y) = G_x(X, Y) . \qquad (40.44)$$

Thus a given hermitian structure on $T_x(M)$ is determined by two real-valued bilinear functions on $T_x(M)$ which are invariant under J_x, one symmetric and the other antisymmetric. If F_x corresponding to a hermitian structure H_x is positive definite, then we say that H_x is **positive definite**. A positive definite H_x determines an inner product $(\,,\,)$ on $T_x(M)$ that is invariant under J_x, given by

$$(X, Y) \equiv F_x(X, Y) = \frac{1}{2} \left(H_x(X, Y) + \overline{H_x(X, Y)} \right) , \qquad (40.45)$$

for any $X, Y \in T_x(M)$.

The hermitian structure of a real vector space $\mathbb{V}$ can be expressed in terms of the $(1, 0)$-type and $(0, 1)$-type basis vectors λ^j and $\overline{\lambda}^j$ [c.f. (40.30) and (40.31)] of $V^* \otimes \mathbb{C}$. Suppose $X, Y \in V$ ($\dim V = 2n$). In terms of the basis vectors $\{e_i, e_{n+i}\}$ satisfying (40.27), we can write

$$X = \sum_{j=1}^{n} (X^j e_j + X^{n+j} J e_j) \,, \tag{40.46}$$

$$Y = \sum_{j=1}^{n} (Y^j e_j + Y^{n+j} J e_j) \,, \tag{40.47}$$

where $X^j, Y^j, X^{n+j}, Y^{n+j} \in \mathbb{R}$. Thus, from (40.36) to (40.38),

$$H(X, Y) = \sum_{j,k=1}^{n} (X^j + iX^{n+j})(Y^k - iY^{n+k}) H(e_j, e_k) \,. \tag{40.48}$$

Exercise 40.4 Verify the above equation.

On the other hand, because of the duality between $\{e_j, e_{n+j}\}$ and $\{(e^*)^j, (e^*)^{n+j}\}$, $j = 1, \ldots, n$, we have, from (40.30) and (40.31),

$$\lambda^j(X) = (e^*)^j(X) + i(e^*)^{n+j}(X) = X^j + iX^{n+j} \,, \tag{40.49}$$

$$\overline{\lambda}^k(Y) = Y^k - iY^{n+k} \,. \tag{40.50}$$

It follows from (40.48) that

$$H(X, Y) = h_{j\overline{k}} \lambda^j(X) \overline{\lambda}^k(Y) \,, \tag{40.51}$$

where

$$h_{j\overline{k}} \equiv H(e_j, e_k) \,, \tag{40.52}$$

in which the bar above the index k indicates that it is the index corresponding to the $(0,1)$-type basis vector $\overline{\lambda}^k$. Thus the hermitian structure H can be written as

$$H = h_{j\overline{k}} \lambda^j \otimes \overline{\lambda}^k \,, \tag{40.53}$$

where the components $h_{j\overline{k}}$ satisfy, from (40.37), the hermitian symmetry property

$$\overline{h_{j\overline{k}}} = h_{k\overline{j}} \,. \tag{40.54}$$

In other words, the matrix $h_{k\overline{j}}$ is a hermitian matrix.

Since

$$G(X, Y) = -\frac{i}{2} (H(X, Y) - \overline{H(X, Y)})$$

is a real-valued antisymmetric bilinear function on V, it corresponds to an exterior 2-form $\hat{H}$ defined by

$$\langle X \wedge Y, \hat{H} \rangle \equiv -G(X, Y) \,. \tag{40.55}$$

We thus have, from (40.53),

$$- G(X,Y) = \frac{i}{2}\left(H(X,Y) - \overline{H(X,Y)}\right)$$

$$= \frac{i}{2}\left\{ h_{j\bar{k}}\lambda^j(X)\overline{\lambda}^k(Y) - \overline{h_{k\bar{j}}\lambda^k(X)\overline{\lambda}^j(Y)}\right\} \tag{40.56}$$

$$= \frac{i}{2} h_{j\bar{k}}\left(\lambda^j(X)\overline{\lambda}^k(Y) - \lambda^j(Y)\overline{\lambda}^k(X)\right) = \langle X\wedge Y,\ \frac{i}{2}h_{j\bar{k}}\lambda^j\wedge\overline{\lambda}^k\rangle ,$$

where the last equality follows from (28.28). Eq. (40.55) then gives

$$\boxed{\hat{H} = \frac{i}{2}h_{j\bar{k}}\lambda^j\wedge\overline{\lambda}^k}\quad , \tag{40.57}$$

which is known as the **Kähler form** of the hermitian structure H.

Since an n-dimensional complex manifold M is naturally a $2n$-dimensional almost complex (real) manifold, we can define a hermitian structure on $T_z(M)$ for each $z \in M$. Equivalently, a hermitian structure can be defined on the tangent bundle $T(M)$. In a local coordinate neighborhood $(U; z^i)$, this structure is given by the hermitian matrix

$$H\left(\frac{\partial}{\partial z^i}, \frac{\partial}{\partial z^k}\right) = h_{i\bar{k}} = \overline{h_{k\bar{i}}} . \tag{40.58}$$

If ξ, η are two $(1,0)$-type tangent vector fields on U:

$$\xi = \xi^i\frac{\partial}{\partial z^i} , \quad \eta = \eta^k\frac{\partial}{\partial z^k} , \tag{40.59}$$

where ξ^i, η^k are smooth, complex-valued functions on U, then

$$H(\xi, \eta) = h_{i\bar{k}}\xi^i\overline{\eta}^k . \tag{40.60}$$

Corresponding to a hermitian structure on M, we have a Kähler form on M given by

$$\boxed{\hat{H} = \frac{i}{2}h_{j\bar{k}}dz^j\wedge d\overline{z}^k}\quad . \tag{40.61}$$

By (40.55), this is a real-valued $(1,1)$-type differential form. In general, a (p,q)-type form α is one which can be expressed as

$$\alpha = \alpha_{i_1\dots i_p,\bar{j}_1\dots\bar{j}_q}\,dz^{i_1}\wedge\dots\wedge dz^{i_p}\wedge d\overline{z}^{j_1}\wedge\dots\wedge d\overline{z}^{j_q} . \tag{40.62}$$

The set of all (p,q)-type forms on a complex manifold M, denoted by $A_{p,q}$, is a module over the ring of smooth complex-valued functions of $(z^1,\dots,z^n; \overline{z}^1,\dots,\overline{z}^n)$. Exterior products of forms of different orders and exterior differentiation of (p,q)-type forms can be generalized quite straightforwardly from similar operations on real forms (c.f. Chapters 28 and 30, respectively). We have the following properties:

1) if $\alpha \in A_{p,q}$, then $\bar{\alpha} \in A_{q,p}$;

2) if $\alpha \in A_{p,q}$ and $\beta \in A_{r,s}$, then $\alpha \wedge \beta \in A_{p+r,q+s}$;

3) exterior differentiation satisfies

$$dA_{p,q} \subset A_{p+1,q} \oplus A_{p,q+1} ;$$

4) if either p or $q > n$ (where $n =$ complex dimension of M), then $A_{p,q} = 0$.

Because of property 3) above, we have the important linear maps $\partial : A_{p,q} \to A_{p+1,q}$ and $\bar{\partial} : A_{p,q} \to A_{p,q+1}$, defined by

$$\partial \omega = \prod_{p+1,q} d\omega , \quad \bar{\partial} \omega = \prod_{p,q+1} d\omega , \tag{40.63}$$

where ω is a (p,q)-type form, and $\prod$ is a projection map that picks out the corresponding form of a specific type from $d\omega$. For example, if ω is a $(1,2)$-type form, then $d\omega = \omega^1 \oplus \omega^2$, where $\omega^1 \in A_{2,2}$, $\omega^2 \in A_{1,3}$; and $\partial \omega = \omega^1$, $\bar{\partial} \omega = \omega^2$. It is clear that on complex manifolds

$$d = \partial + \bar{\partial} . \tag{40.64}$$

Suppose $f(z^1, \ldots, z^n; \bar{z}^1, \ldots, \bar{z}^n)$ is a smooth complex-valued function on an n-dimensional complex manifold M. Then (with $z^j = x^j + iy^j$)

$$df = \frac{\partial f}{\partial x^k} dx^k + \frac{\partial f}{\partial y^k} dy^k = \frac{\partial f}{\partial z^k} dz^k + \frac{\partial f}{\partial \bar{z}^k} d\bar{z}^k , \tag{40.65}$$

where $\partial / \partial z^k$ and $\partial / \partial \bar{z}^k$ are given by (40.12) and (40.13), respectively. Thus,

$$\partial f = \frac{\partial f}{\partial z^k} dz^k , \quad \bar{\partial} f = \frac{\partial f}{\partial \bar{z}^k} d\bar{z}^k . \tag{40.66}$$

It follows that for a (p,q)-type form α given by (40.62), we have

$$\partial \alpha = \frac{\partial(\alpha_{i_1 \ldots i_p, \bar{j}_1 \ldots \bar{j}_q})}{\partial z^k} dz^k \wedge dz^{i_1} \wedge \cdots \wedge dx^{i_p} \wedge d\bar{z}^{j_1} \wedge \cdots \wedge d\bar{z}^{j_q} , \tag{40.67}$$

$$\bar{\partial} \alpha = (-1)^p \frac{\partial(\alpha_{i_1 \ldots i_p, \bar{j}_1 \ldots \bar{j}_q})}{\partial \bar{z}^k} dz^{i_1} \wedge \cdots \wedge dz^{i_p} \wedge d\bar{z}^k \wedge d\bar{z}^{j_1} \wedge \cdots \wedge d\bar{z}^{j_q} . \tag{40.68}$$

If we put $f = g + ih$, where g and h are real-valued functions, we have, by (40.13),

$$\frac{\partial f}{\partial \bar{z}^k} = \frac{1}{2} \left(\frac{\partial}{\partial x^k} + i \frac{\partial}{\partial y^k} \right) (g + ih) = \frac{1}{2} \left(\frac{\partial g}{\partial x^k} - \frac{\partial h}{\partial y^k} \right) + \frac{i}{2} \left(\frac{\partial h}{\partial x^k} + \frac{\partial g}{\partial y^k} \right) . \tag{40.69}$$

Thus, from the Cauchy-Riemann conditions [c.f. (40.3)], *a necessary and suffi-cient condition for f to be a holomorphic function is that $\bar{\partial}f = 0$. If $\alpha \in A_{p,0}$ is a $(p,0)$-type form with local expression*

$$\alpha = \alpha_{k_1 \ldots k_p} \, dz^{k_1} \wedge \cdots \wedge dz^{k_p} \,, \tag{40.70}$$

where the $\alpha_{k_1 \ldots k_p}$ are all holomorphic functions, then α is called a **holomorphic p-form**. Since in this case $\bar{\partial}\alpha_{k_1 \ldots k_p} = 0$, we have

$$d\alpha = \partial\alpha = \frac{\alpha_{k_1 \ldots k_p}}{\partial z^k} \, dz^k \wedge dz^{k_1} \wedge \cdots \wedge dz^{k_p} \,. \tag{40.71}$$

The operator ∂ thus maps holomorphic $(p,0)$-forms to holomorphic $(p+1,0)$-forms.

Definition 40.3. *If the Kähler form [c.f. (40.61)] of a hermitian manifold M (a complex manifold M endowed with a hermitian structure) is closed, that is, if*

$$d\hat{H} = 0 \,, \tag{40.72}$$

*then M is called a **Kähler manifold**.*

We conclude this chapter by proving two fundamental facts in the theory of functions of one complex variable (Corollary 40.1 and Theorem 40.3 below). First recall two definitions.

Definition 40.4. *A smooth (C^∞) function $f = g + ih$ on an open set $U \subset \mathbb{C}$, where g and h are real-valued, is said to be holomorphic if it satisfies the Cauchy-Riemann conditions*

$$\frac{\partial g}{\partial x} = \frac{\partial h}{\partial y} \,, \tag{40.73}$$

$$\frac{\partial g}{\partial y} = -\frac{\partial h}{\partial x} \,, \tag{40.74}$$

where $x + iy = z \in U$; or equivalently

$$\frac{\partial f}{\partial \bar{z}} = 0 \,. \tag{40.75}$$

Definition 40.5. *A smooth function f on an open set $U \subset \mathbb{C}$ is said to be **analytic** if, for all $z_0 \in U$, f has a local series expansion in $z - z_0$:*

$$f(z) = \sum_{n=0}^{\infty} a_n (z - z_0)^n \,, \tag{40.76}$$

in some disc $\Delta_\varepsilon(z_0) = \{ z : |z - z_0| < \varepsilon \}$, and the series converges absolutely and uniformly in Δ_ε.

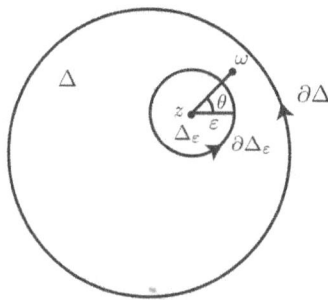

FIGURE 40.1

Theorem 40.2 (Cauchy Integral Formula). *For a disc Δ in $\mathbb{C}$, $z \in \Delta$, and $f \in C^\infty(\overline{\Delta})$, where $\overline{\Delta}$ is the closure of Δ,*

$$f(z) = \frac{1}{2\pi i} \int_{\partial \Delta} \frac{f(w)dw}{w - z} + \frac{1}{2\pi i} \int_{\Delta} \frac{\partial f(w)}{\partial \overline{w}} \frac{dw \wedge d\overline{w}}{w - z} , \tag{40.77}$$

where the integral along the boundary $\partial \Delta$ is in the counterclockwise direction.

Proof. Consider the $(1, 0)$-type form

$$\eta = \frac{1}{2\pi i} \frac{f dw}{w - z} . \tag{40.78}$$

We have, for $z \neq w$,

$$d\eta = \frac{1}{2\pi i} \frac{\partial}{\partial \overline{w}} \left(\frac{f}{w - z} \right) d\overline{w} \wedge dw = -\frac{1}{2\pi i} \frac{\partial f}{\partial \overline{w}} \frac{dw \wedge d\overline{w}}{w - z} . \tag{40.79}$$

Let $\Delta_\varepsilon = \Delta_\varepsilon(z)$ be the disc of radius ε around z (see Fig. 40.1). Then the form η is smooth in $\Delta - \Delta_\varepsilon$. We can thus apply Stokes Theorem (Theorem 32.4) to this region and obtain

$$\int_{\partial(\Delta - \Delta_\varepsilon)} \eta = \int_{\partial \Delta} \eta - \int_{\partial \Delta_\varepsilon} \eta = \int_{\Delta - \Delta_\varepsilon} d\eta , \tag{40.80}$$

or

$$\int_{\partial \Delta_\varepsilon} \eta = \int_{\partial \Delta} \eta - \int_{\Delta - \Delta_\varepsilon} d\eta . \tag{40.81}$$

It follows from (40.78) and (40.79) that

$$\frac{1}{2\pi i} \int_{\partial \Delta_\varepsilon} \frac{f dw}{w - z} = \frac{1}{2\pi i} \int_{\partial \Delta} \frac{f dw}{w - z} + \frac{1}{2\pi i} \int_{\Delta - \Delta_\varepsilon} \frac{\partial f}{\partial \overline{w}} \frac{dw \wedge d\overline{w}}{w - z} . \tag{40.82}$$

Setting $w - z = re^{i\theta}$, we have

$$\frac{1}{2\pi i} \int_{\partial \Delta_\varepsilon} \frac{f(w)dw}{w - z} = \frac{1}{2\pi} \int_0^{2\pi} f(z + \varepsilon e^{i\theta}) d\theta \xrightarrow[\varepsilon \to 0]{} f(z) . \tag{40.83}$$

Moreover, in $\Delta - \Delta_\varepsilon$ (with $w - z = x + iy = re^{i\theta}$, $x = r\cos\theta$, $y = r\sin\theta$),

$$dw \wedge d\overline{w} = -2idx \wedge dy = -2ir\, dr \wedge d\theta . \tag{40.84}$$

Hence

$$\left| \frac{\partial f}{\partial \overline{w}} \frac{dw \wedge d\overline{w}}{w - z} \right| = 2 \left| \frac{\partial f}{\partial \overline{w}} dr \wedge d\theta \right| \le c \left| dr \wedge d\theta \right| , \tag{40.85}$$

where $c \in \mathbb{R}$. Thus $\dfrac{\partial f}{\partial \overline{w}} \dfrac{dw \wedge d\overline{w}}{w - z}$ is absolutely integrable over Δ, and hence integrable over Δ. The above equation also implies

$$\lim_{\varepsilon \to 0} \int_{\Delta_\varepsilon} \frac{\partial f}{\partial \overline{w}} \frac{dw \wedge d\overline{w}}{w - z} = 0 . \tag{40.86}$$

The theorem follows on taking the limit $\varepsilon \to o$ on both sides of (40.82). $\square$

Corollary 40.1. *If* $f \in C^\infty(U)$ *is holomorphic in an open region* U *containing* $\overline{\Delta}$, *where* $\overline{\Delta}$ *is the closure of some disc* Δ, *then, for any* $z \in \Delta$,

$$f(z) = \frac{1}{2\pi i} \int_{\partial \Delta} \frac{f(w)dw}{w - z} . \tag{40.87}$$

This is also known as **Cauchy's integral formula**. It is the foundation of the theory of functions of one complex variable.

Theorem 40.3. *Let* $f \in C^\infty(U)$ *be a smooth function in an open set* $U \subset \mathbb{C}$. *Then* f *is holomorphic in* U *if and only if it is analytic in* U.

Proof. Suppose f is holomorphic, then $\dfrac{\partial f}{\partial \overline{z}} = 0$. By Corollary 40.1, we have, for $z \in \Delta_\varepsilon(z_0)$, $z_0 \in U$, ε sufficiently small,

$$f(z) = \frac{1}{2\pi i} \int_{\partial \Delta_\varepsilon} \frac{f(w)dw}{w - z}$$

$$= \frac{1}{2\pi i} \int_{\partial \Delta_\varepsilon} \frac{f(w)dw}{(w - z_0) - (z - z_0)} = \frac{1}{2\pi i} \int_{\partial \Delta_\varepsilon} \frac{f(w)dw}{(w - z_0)\left(1 - \dfrac{z - z_0}{w - z_0}\right)} \tag{40.88}$$

$$= \sum_{n=0}^\infty \left(\frac{1}{2\pi i} \int_{\partial \Delta_\varepsilon} \frac{f(w)dw}{(w - z_0)^{n+1}} \right) (z - z_0)^n = \sum_{n=0}^\infty a_n(z - z_0)^n ,$$

where

$$a_n = \frac{1}{2\pi i} \int_{\partial \Delta_\varepsilon} \frac{f(w)dw}{(w - z_0)^{n+1}} . \tag{40.89}$$

In the fourth equality, we have used the absolute and uniform convergence of the series

$$\sum_{n=0}^{\infty} \left(\frac{z - z_0}{w - w_0} \right)^n = \left[1 - \left(\frac{z - z_0}{w - w_0} \right) \right]^{-1} , \tag{40.90}$$

with $|z - z_0|/|w - w_0| < 1$. By Def. 40.5, f is analytic in U.

Conversely, suppose f is analytic in U. Then it has a power series expansion

$$f(z) = \sum_{n=0}^{\infty} a_n (z - z_0)^n , \tag{40.91}$$

for $z \in \Delta_\varepsilon(z_0)$, $z_0 \in U$. Since $\frac{\partial}{\partial \bar{z}} (z - z_0)^n = 0$, the partial sums of the expansion satisfy Cauchy's integral formula (40.87); and by the uniform convergence of (40.91), the same is true of $f(z)$, that is,

$$f(z) = \frac{1}{2\pi i} \int_{\partial \Delta} \frac{f(w) dw}{w - z} , \tag{40.92}$$

where Δ is some disc containing z. We can then differentiate under the integral sign to obtain

$$\frac{\partial f(z)}{\partial \bar{z}} = \frac{1}{2\pi i} \int_{\partial \Delta} \frac{\partial}{\partial \bar{z}} \left(\frac{f(w)}{w - z} \right) dw = 0 . \tag{40.93}$$

It follows by Def. 40.4 that $f(z)$ is holomorphic in U. $\qquad\square$

Chapter 41

Characteristic Classes

Characteristic classes are geometric objects which serve to classify the structure of fiber bundles, specifically, the non-triviality of bundles. The general definition is given by Def. 41.3. They have become increasingly important in physics applications in which geometrical and topological considerations play a role, for example, in the so-called topological quantum numbers. One type of such classes, called the Chern classes, is of particular importance in a deep mathematical result known as the Atiyah-Singer index theorem (Chapter 43), which expresses the index of an elliptic operator on a manifold in terms of a geometric invariant. This theorem has also found important applications in modern physics.

We will begin by studying a specific example of characteristic classes, the Chern classes, which arise naturally from complex vector bundles equipped with connections. A **complex vector bundle** $\pi : E \to M$ is simply a vector bundle E with base manifold M, where the typical fiber is a finite-dimensional complex vector space V. The structure group of a general complex bundle is thus $GL(q; \mathbb{C})$, where $q = dim(V)$. The set of all sections on M, denoted $\Gamma(E)$, then has a complex-linear structure, and is in fact a module over the ring of all smooth complex-valued functions on M. The definition of a connection on a complex vector bundle resembles that for a real vector bundle (Def. 35.1), word for word, provided that in places where the real number field $\mathbb{R}$ is implied, the complex number field $\mathbb{C}$ is substituted instead. All the basic results obtained earlier concerning the connection matrix of 1-forms (ω_i^j) and the associated curvature matrix of 2-forms (Ω_i^j) still hold in the present case of complex-valued forms. We recall them below for convenience.

Suppose $\{e_\alpha, \alpha = 1, \ldots, q\}$ is a local frame field on the base manifold M. Then the covariant derivative $D : \Gamma(E) \to \Gamma(E \otimes T^*(M))$ is given by

$$De_\alpha = \omega_\alpha^\beta e_\beta \,, \tag{41.1}$$

or, in matrix notation,

$$De = \omega e \,. \tag{41.2}$$

Under a local change of frame fields

$$g' = ge \, , \tag{41.3}$$

the connection matrix transforms as

$$\omega' = dg \cdot g^{-1} + g\omega g^{-1} \, , \tag{41.4}$$

where $De' = \omega'e'$.

The curvature matrix Ω for the connection ω is

$$\Omega = d\omega - \omega \wedge \omega \, . \tag{41.5}$$

Under the local change of frame fields (41.3), the curvature matrix transforms according to

$$\Omega' = g\Omega g^{-1} \, . \tag{41.6}$$

The covariant derivative D also acts on objects in $\Gamma(E) \otimes A^r(M)$ (section-valued r-forms) and $\Gamma(E \otimes E^*) \otimes A^r(M)$ (matrix-valued r-forms, or equivalently, matrices of r-forms), or more generally, tensorial objects whose components are complex-valued r-forms. We have, for example, recalling (35.59) and (35.66),

$$D\lambda = d\lambda + (-1)^r \lambda \wedge \omega \, , \quad \lambda \in \Gamma(E) \otimes A^r(M) \, , \tag{41.7}$$
$$D\eta = d\eta - [\omega, \eta]_{\mathrm{gr}} \, , \quad \eta \in \Gamma(E \otimes E^*) \otimes A^r(M) \, , \tag{41.8}$$
$$D(\eta_1 \wedge \eta_2) = D\eta_1 \wedge \eta_2 + (-1)^{r_1}\eta_1 \wedge D\eta_2 \, , \tag{41.9}$$

where in the last equation $\eta_1 \in \Gamma(E \otimes E^*) \otimes A^{r_1}(M)$ and $\eta_2 \in \Gamma(E \otimes E^*) \otimes A^{r_2}(M)$, that is, tensorial matrices of r_1- and r_2-forms, respectively.

| Exercise 41.1 | Prove (41.9).

In (41.8), the graded commutator $[\, , \,]_{\mathrm{gr}}$ is defined by (35.65):

$$[\omega, \eta]_{\mathrm{gr}} = \omega \wedge \eta - (-1)^r \eta \wedge \omega \, . \tag{41.10}$$

The Bianchi identity for Ω is

$$d\Omega = \omega \wedge \Omega - \Omega \wedge \omega \, , \tag{41.11}$$

or, by (41.8),

$$D\Omega = 0 \, . \tag{41.12}$$

We also have [c.f. (35.71) and (35.73)],

$$D^2\lambda = \lambda \wedge \Omega \, , \qquad\qquad \lambda \in \Gamma(E) \otimes A^r(M) \, , \tag{41.13}$$
$$D^2\eta = [\eta, \Omega] = [\eta, \Omega]_{\mathrm{gr}} \, , \qquad \eta \in \Gamma(E \otimes E^*) \otimes A^r(M) \, . \tag{41.14}$$

By the transformation property (41.6) for the curvature matrix, Chern was led to consider complex-valued functions of matrices which are invariant under the adjoint transformation exemplified by (41.6). Consider a complex r-linear function $P(A_1, \ldots, A_r)$ of r $q \times q$ matrices $A_1, \ldots, A_r$. Let A_i be expressed in terms of its matrix elements (complex-valued) as

$$A_i = (a_{\alpha\beta}^i), \quad 1 \leq \alpha, \beta \leq q, \; 1 \leq i \leq r. \tag{41.15}$$

Then, in general, we can write P as

$$P(A_1, \ldots, A_r) = \sum_{1 \leq \alpha_i, \beta_i \leq q} \lambda_{\alpha_1 \ldots \alpha_r, \beta_1 \ldots \beta_r} \, a_{\alpha_1 \beta_1}^1 \cdots a_{\alpha_r \beta_r}^r, \tag{41.16}$$

where the $\lambda_{\alpha_1 \ldots \alpha_r, \beta_1 \ldots \beta_r}$ are complex numbers. If for any permutation σ of $\{1, \ldots, r\}$ we have

$$P(A_{\sigma(1)}, \ldots, A_{\sigma(r)}) = P(A_1, \ldots, A_r), \tag{41.17}$$

then P is called a **symmetric polynomial** of $A_1, \ldots, A_r$. If for any $g \in GL(q; \mathbb{C})$, we have

$$P(gA_1 g^{-1}, \ldots, gA_r g^{-1}) = P(A_1, \ldots, A_r), \tag{41.18}$$

then P is called an **ad-invariant polynomial**.

Chern used the following method to construct a sequence of symmetric, ad-invariant polynomials. For a $q \times q$ matrix A write

$$\det\left(1 + \frac{i}{2\pi} A\right) = \sum_{0 \leq j \leq q} P_j(A), \tag{41.19}$$

where $P_j(A)$ is a homogeneous j-th order polynomial of the elements $a_{\alpha\beta}$ of A, and the "1" on the LHS is the $q \times q$ identity matrix. For example, for a 2×2 matrix A, we have

$$\det\left(1 + \frac{i}{2\pi} A\right) = 1 + \frac{i}{2\pi} Tr\, A + \left(\frac{i}{2\pi}\right)^2 \det A,$$

and so

$$P_0(A) = 1, \quad P_1(A) = \left(\frac{i}{2\pi}\right) Tr\, A, \quad P_2(A) = \left(\frac{i}{2\pi}\right)^2 \det A.$$

For a 3×3 matrix A,

$$\det\left(1 + \frac{i}{2\pi} A\right) = 1 + \frac{i}{2\pi} Tr\, A$$

$$+ \left(\frac{i}{2\pi}\right)^2 (a_{11}a_{22} + a_{11}a_{33} + a_{22}a_{33} - a_{12}a_{21} - a_{13}a_{31} - a_{23}a_{32})$$

$$+ \left(\frac{i}{2\pi}\right)^3 \det A,$$

implying

$$P_0(A) = 1 , \quad P_1(A) = \left(\frac{i}{2\pi} \right) Tr\, A ,$$

$$P_2(A) = \left(\frac{i}{2\pi} \right)^2 (a_{11}a_{22} + a_{11}a_{33} + a_{22}a_{33} - a_{12}a_{21} - a_{13}a_{31} - a_{23}a_{32}) ,$$

$$P_3(A) = \left(\frac{i}{2\pi} \right)^3 \det A .$$

For any $g \in GL(q; \mathbb{C})$, we have

$$1 + \frac{i}{2\pi} gAg^{-1} = g \left(1 + \frac{i}{2\pi} A \right) g^{-1} . \tag{41.20}$$

Thus, recalling Theorem 16.6,

$$\det \left(1 + \frac{i}{2\pi} gAg^{-1} \right) = \det \left(1 + \frac{i}{2\pi} A \right) . \tag{41.21}$$

It follows from (41.19) that $P_j(A)$ is ad-invariant:

$$P_j(gAg^{-1}) = P_j(A) . \tag{41.22}$$

Now introduce the so-called **completely polarized polynomial** $P_j(A_1, \ldots, A_j)$ of $P_j(A)$. We require $P_j(A_1, \ldots, A_j)$ to be a symmetric j-linear function of $A_1, \ldots, A_j$ that satisfies

$$P_j(\underbrace{A, \ldots, A}_{j \text{ times}}) = P_j(A) . \tag{41.23}$$

For example,

$$P_2(A_1, A_2) = \frac{1}{2} \{ P_2(A_1 + A_2) - P_2(A_1) - P_2(A_2) \} ,$$

$$P_3(A_1, A_2, A_3) = \frac{1}{6} \{ P_3(A_1 + A_2 + A_3) - P_3(A_1 + A_2) - P_3(A_1 + A_3)$$
$$- P_3(A_2 + A_3) + P_3(A_1) + P_3(A_2) + P_3(A_3) \} .$$

$\boxed{\text{Exercise 41.2}}$ Verify the above two equations using the explicit expressions for the $P_2(A)$ and $P_3(A)$ given earlier.

Since $P_j(A)$ is ad-invariant, the polarization $P_j(A_1, \ldots, A_j)$ is ad-invariant also, and thus by definition, symmetric and ad-invariant.

Suppose $P(A_1, \ldots, A_r)$ is a symmetric, ad-invariant polynomial, and let $g \in GL(q; \mathbb{C})$ be written as

$$g = 1 + g' , \tag{41.24}$$

where 1 is the $q \times q$ identity matrix. Then

$$g^{-1} = 1 - g' + O((g')^2) , \tag{41.25}$$

where $O((g')^2)$ represents the sum of terms involving the second and higher-order terms of elements of g'. Invoking the ad-invariant property (41.18) and using linearity, we have

$$\sum_{1 \leq i \leq r} P(A_1, \ldots, g'A_i - A_ig', \ldots, A_r) = 0 , \tag{41.26}$$

on considering the terms linear in g'.

We now generalize to the situation where the $A_i = (a^i_{\alpha\beta})$ are $q \times q$ matrices of differential forms on M. Following (41.16), $P(A_1, \ldots, A_r)$ is given by

$$P(A_1, \ldots, A_r) = \sum_{1 \leq \alpha_i, \beta_i \leq q} \lambda_{\alpha_1 \ldots \alpha_r, \beta_1 \ldots \beta_r} \, a^1_{\alpha_1 \beta_1} \wedge \cdots \wedge a^r_{\alpha_r \beta_r} . \tag{41.27}$$

The ad-invariant property (41.26) remains valid for the case where the A_i are matrices of differential forms. This fact leads to the following result.

Lemma 41.1. *For $i = 1, \ldots, r$, let A_i be a $q \times q$ matrix of differential d_i-forms. Suppose $P(A_1, \ldots, A_r)$ is a symmetric, ad-invariant polynomial. Then, for any $q \times q$ matrix θ of differential 1-forms,*

$$\sum_{1 \leq i \leq r} (-1)^{d_1 + \ldots + d_{i-1}} P(A_1, \ldots, \theta \wedge A_i, \ldots, A_r)$$

$$+ \sum_{1 \leq i \leq r} (-1)^{d_1 + \ldots + d_i + 1} P(A_1, \ldots, A_i \wedge \theta, \ldots, A_r) = 0 , \tag{41.28}$$

or equivalently,

$$\sum_{i=1}^{r} (-1)^{d_1 + \ldots + d_{i-1}} P(A_1, \ldots, [\theta, A_i]_{gr}, \ldots, A_r) = 0 . \tag{41.29}$$

Proof. We only need to prove the lemma for $\theta = \sum_l (g')_l a_l$, where $(g')_l$ is any $q \times q$ matrix (of complex numbers), and $a_l \in A^1(M)$ (a 1-form on M). In fact, by the multilinear property of P, we only need to show (41.28) for $\theta = g'a$. The LHS of (41.28) then reads

$$\sum_{1 \leq i \leq r} (-1)^{d_1 + \ldots + d_{i-1}} P(A_1, \ldots, A_{i-1}, g'a \wedge A_i, \ldots, A_r)$$

$$+ \sum_{1 \leq i \leq r} (-1)^{d_1 + \cdots + d_i + 1} P(A_1, \ldots, A_i \wedge g'a, \ldots, A_r)$$

$$= a \wedge \{ \sum_{1 \leq i \leq r} P(A_1, \ldots, g'A_i, \ldots, A_r) - \sum_{1 \leq i \leq r} P(A_1, \ldots, A_ig', \ldots, A_r) \} , \tag{41.30}$$

which vanishes, by linearity of P and (41.26). In the above equality, we have used (41.27) and the anti-commutative rule of exterior products (28.5). $\qquad\square$

Because of the ad-invariant property, $P(A_1, \ldots, A_r)$, where A_i is a $q \times q$ matrix of d_i-forms, $1 \leq i \leq r$, is an exterior differential $(d_1 + \cdots + d_r)$-form defined globally on M. A corollary of Lemma 41.1 is the following important fact.

Lemma 41.2. *Let $P(A_1, \ldots, A_r)$ be a symmetric, ad-invariant polynomial. If A_i is a $q \times q$ matrix of d_i-forms $(i = 1, \ldots, r)$ on M, and a connection ω is given on the complex vector bundle $\pi : E \to M$ (where the fiber space is of dimension q), then $P(A_1, \ldots, A_r)$ is a differential $(d_1 + \cdots + d_r)$-form on M, and*

$$dP(A_1, \ldots, A_r) = \sum_{i=1}^{r} (-1)^{d_1 + \cdots + d_{i-1}} P(A_1, \ldots, DA_i, \ldots, A_r), \qquad (41.31)$$

where D is the covariant derivative corresponding to the connection ω.

Proof. We calculate the exterior derivative of $P(A_1, \ldots, A_r)$ directly, and obtain, by using (41.27),

$$dP(A_1, \ldots, A_r) = \sum_{1 \leq \alpha_i, \beta_i \leq q} \lambda_{\alpha_1 \ldots \alpha_r, \beta_1 \ldots \beta_r} \, d(a^1_{\alpha_1 \beta_1} \wedge \cdots \wedge a^r_{\alpha_r \beta_r})$$

$$= \sum_{1 \leq \alpha_i, \beta_i \leq q} \lambda_{\alpha_1 \ldots \alpha_r, \beta_1 \ldots \beta_r} \sum_{i=1}^{r} (-1)^{d_1 + \cdots + d_{i-1}} a^1_{\alpha_1 \beta_1} \wedge \cdots \wedge da^i_{\alpha_i \beta_i} \wedge \cdots \wedge a^r_{\alpha_r \beta_r}$$

$$= \sum_{i=1}^{r} (-1)^{d_1 + \cdots + d_{i-1}} P(A_1, \ldots, dA_i, \ldots, A_r)$$

$$= \sum_{i=1}^{r} (-1)^{d_1 + \cdots + d_{i-1}} P(A_1, \ldots, DA_i + [\omega, A_i]_{\mathrm{gr}}, \ldots, A_r)$$

$$= \sum_{i=1}^{r} (-1)^{d_1 + \cdots + d_{i-1}} P(A_1, \ldots, DA_i, \ldots, A_r)$$

$$+ \sum_{i=1}^{r} (-1)^{d_1 + \cdots + d_{i-1}} P(A_1, \ldots, \omega \wedge A_i + (-1)^{d_i+1} A_i \wedge \omega, \ldots, A_r)$$

$$= \sum_{i=1}^{r} (-1)^{d_1 + \cdots + d_{i-1}} P(A_1, \ldots, DA_i, \ldots, A_r),$$

$$(41.32)$$

where the fourth equality follows from (41.8) and the last from Lemma 41.1. □

We are now ready to state and prove the main theorem of this chapter.

Theorem 41.1 (Chern-Weil Theorem). *Suppose $\pi : E \to M$ is a q-dimensional complex vector bundle on a smooth n-dimensional manifold M, and Ω is the curvature associated with a given connection ω on E. Let $P_j(A_1, \ldots, A_j)$ be a symmetric, ad-invariant polynomial, where A_i, $i = 1, \ldots, j$, are $q \times q$ matrices. Then*

i) the 2j-form $P_j(\Omega)$ is closed:

$$dP_j(\Omega) = 0 , \tag{41.33}$$

where $P_j(A_1, \dots, A_j)$ is the completely polarized polynomial of $P_j(A)$ [c.f. (41.23)].

Suppose $\tilde{\omega}$ is another connection on E with corresponding curvature $\tilde{\Omega}$, then

ii) there exists a $(2j-1)$-differential form Q on M such that

$$P_j(\tilde{\Omega}) - P_j(\Omega) = dQ , \tag{41.34}$$

that is, $P_j(\tilde{\Omega}) - P_j(\Omega)$ is exact.

Proof. Letting $A_1 = \cdots = A_r = \Omega$ in (41.31) we have

$$dP_j(\Omega) = \sum_{i=1}^{j} P_j(\Omega, \dots, \underbrace{D\Omega}_{i\text{-th slot}}, \dots, \Omega) , \tag{41.35}$$

since $d_i = 2$ for $1 \leq i \leq j$. The RHS vanishes by virtue of Bianchi's identity $(D\Omega = 0)$ [c.f. (41.12)] and linearity of P_j. This proves statement i).

To prove statement ii) let

$$\eta \equiv \tilde{\omega} - \omega . \tag{41.36}$$

We show that η is a tensorial matrix of 1-forms as follows. Under a local change of frame fields $e' = ge$,

$$(\tilde{\omega})' = (dg) g^{-1} + g \tilde{\omega} g^{-1} , \tag{41.37}$$

$$\omega' = (dg) g^{-1} + g \omega g^{-1} , \tag{41.38}$$

where $De' = \omega' e'$ and $\tilde{D}e' = (\tilde{\omega})'e'$. Hence

$$\eta' = (\tilde{\omega})' - \tilde{\omega} = g(\tilde{\omega} - \omega) g^{-1} = g \eta g^{-1} . \tag{41.39}$$

Thus the difference between two connections is a tensor. With η we generate a family of connections by

$$\omega_t \equiv \omega + t\eta , \qquad 0 \leq t \leq 1 , \tag{41.40}$$

so that $\omega_0 = \omega$ and $\omega_1 = \tilde{\omega}$. The curvature Ω_t of ω_t is given by

$$\begin{aligned} \Omega_t &= d\omega_t - \omega_t \wedge \omega_t = d\omega + t d\eta - (\omega + t\eta) \wedge (\omega + t\eta) \\ &= (d\omega - \omega \wedge \omega) + t(d\eta - (\eta \wedge \omega + \omega \wedge \eta)) - t^2 \eta \wedge \eta \\ &= \Omega + t D\eta - t^2 \eta \wedge \eta , \end{aligned} \tag{41.41}$$

where in the last equality we have used (41.8). Hence

$$\frac{d\Omega_t}{dt} = D\eta - 2t \, \eta \wedge \eta , \tag{41.42}$$

where D is the covariant derivative for ω.

Corresponding to the given symmetric, ad-invariant polynomial $P_j(A_1, \ldots, A_j)$, we let

$$P_j(A) \equiv P_j(\overbrace{A, \ldots, A}^{j \text{ times}}), \tag{41.43}$$

and

$$Q_j(B, A) \equiv j P_j(B, \overbrace{A, \ldots, A}^{(j-1) \text{ times}}). \tag{41.44}$$

Then

$$\frac{d}{dt} P_j(\Omega_t) = j P_j\left(\frac{d\Omega_t}{dt}, \Omega_t, \ldots, \Omega_t\right) = j P_j(D\eta - 2t\,\eta \wedge \eta, \Omega_t, \ldots, \Omega_t)$$
$$= Q_j(D\eta, \Omega_t) - 2t Q_j(\eta \wedge \eta, \Omega_t). \tag{41.45}$$

Now it follows from (41.41) that

$$D\Omega_t = D\Omega + tD^2\eta - t^2 D(\eta \wedge \eta) = tD^2\eta - t^2 D(\eta \wedge \eta)$$
$$= t[\eta, \Omega] + t^2[\eta, D\eta] = t[\eta, \Omega_t], \tag{41.46}$$

where in the third equality we have used (41.9) and (41.14), and in the fourth we have used (41.41). On the other hand, from Lemma 41.2 [(41.31)] and the symmetry property of P_j, we have

$$dQ_j(\eta, \Omega_t) = j dP_j(\eta, \Omega_t, \ldots, \Omega_t)$$
$$= j P_j(D\eta, \Omega_t, \ldots, \Omega_t) - j(j-1)P_j(\eta, D\Omega_t, \underbrace{\Omega_t, \ldots, \Omega_t}_{(j-2) \text{ times}}). \tag{41.47}$$

Thus, by (41.46),

$$dQ_j(\eta, \Omega_t) = j P_j(D\eta, \Omega_t, \ldots, \Omega_t) - j(j-1)\, t\, P_j(\eta, [\eta, \Omega_t], \Omega_t, \ldots, \Omega_t). \tag{41.48}$$

Meanwhile, letting $\theta = \eta = A_1$ and $A_2 = \cdots = A_j = \Omega_t$ in (41.29), we obtain

$$2P_j(\eta \wedge \eta, \Omega_t, \ldots, \Omega_t) - (j-1)P_j(\eta, [\eta, \Omega_t], \Omega_t, \ldots, \Omega_t) = 0. \tag{41.49}$$

On multiplying the above equation by j and recalling (41.44) we have

$$2Q_j(\eta \wedge \eta, \Omega_t) - j(j-1)P_j(\eta, [\eta, \Omega_t], \Omega_t, \ldots, \Omega_t) = 0. \tag{41.50}$$

Comparing this equation with (41.48) we then obtain

$$dQ_j(\eta, \Omega_t) = Q_j(D\eta, \Omega_t) - 2t\, Q_j(\eta \wedge \eta, \Omega_t) = \frac{dP_j(\Omega_t)}{dt}, \tag{41.51}$$

where the last equality follows from (41.45). Finally, integration of both sides of the above equation over t from 0 to 1 leads to

$$P_j(\tilde{\Omega}) - P_j(\Omega) = d\left(\int_0^1 dt\, Q_j(\eta, \Omega_t)\right). \tag{41.52}$$

The required $(2j-1)$-form Q on M in (41.34) is thus given by

$$Q = \int_0^1 Q_j(\eta, \Omega_t) dt .$$ (41.53)

□

The Chern-Weil Theorem is a remarkable result. Recalling the definition of the de Rham cohomology group [Def. 33.2, Eq. (33.1)], it says that $P_j(\Omega)$ determines a de Rham cohomology class (an element in the de Rham cohomology group $H^{2j}(M)$), *independent of the choice of the connection ω on the complex vector bundle* $\pi : E \to M$. $P_j(\Omega)$ can be calculated by means of local data, namely, the curvature forms, but yet it determines a de Rham cohomology class, which reveals purely global (topological) information on E (and M). This is a marvellous example of the deep relationship between the local and global properties of a manifold.

We are finally in a position to construct the Chern classes, which are real de Rham cohomology classes $[\in H^{2j}(M, \mathbb{R}), \ j = 1, \dots, q]$ constructed from curvature forms on a complex vector bundle $\pi : E \to M$, where q is the dimension of the fiber space.

Recall the notion of a hermitian structure on a complex vector space [c.f. (8.27) and (8.28)]. Suppose a positive definite hermitian structure is assigned smoothly for every $x \in M$ on the fiber space $\pi^{-1}(x)$, that is, for $\lambda_1, \lambda_2 \in \mathbb{C}$, $X, Y, X_1, X_2 \in \pi^{-1}(x)$, and every $x \in M$, there is a complex-valued conjugate-bilinear function $H(X, Y)$ such that

1) $H(\lambda_1 X_1 + \lambda_2 X_2, Y) = \lambda_1 H(X_1, Y) + \lambda_2 H(X_2, Y) ,$ (41.54)

2) $H(X, Y) = \overline{H(Y, X)} ,$ (41.55)

3) $H(X, X) \geq 0 ,$ (41.56)

with the equality holding only for $X = 0$. Note that 1) and 2) imply that [c.f. (8.28b)]

$$H(Y, \lambda_1 X_1 + \lambda_2 X_2) = \overline{\lambda_1} H(Y, X) + \overline{\lambda_2} H(Y, X_2) .$$ (41.57)

Also, if J is a complex structure on a $2n$-dimensional real vector space V, we can define

$$iX = JX , \qquad X \in V ,$$ (41.58)

and turn V into a complex n-dimensional vector space. (See discussion following Def. 40.1.) A complex vector bundle $\pi : E \to M$ equipped with a smooth assignment of positive definite hermitian structures on M is called a **hermitian vector bundle**. In analogy to the Riemannian metric (g_{ij}), we will denote our hermitian structure on E by

$$h_{\alpha \bar{\beta}} = H(s_\alpha, s_\beta) ,$$ (41.59)

with respect to a local frame field $\{s_1, \dots, s_q\}$. Condition 2) above [(41.55)] implies that the matrix $H = (h_{\alpha \bar{\beta}})$ is hermitian, that is

$$^t\overline{H} = H .$$ (41.60)

Furthermore, if $X = X^\alpha s_\alpha$, $Y = Y^\beta s_\beta$, then

$$H(X,Y) = H(X^\alpha s_\alpha, Y^\beta s_\beta) = X^\alpha \overline{Y^\beta}\, h_{\alpha\overline{\beta}} \ . \tag{41.61}$$

Analogous to Theorem 39.1 on the existence of Riemannian metrics, we have the following basic fact (stated without proof).

Theorem 41.2. *There always exists a hermitian structure on a complex vector bundle.*

Analogous to the notion of a metric-compatible connection on a Riemannian manifold, we have the notion of a compatible connection on a hermitian vector bundle.

Definition 41.1. *Let ω be a connection on a hermitian vector bundle $\pi : E \to M$, and D the corresponding covariant derivative. If for any two parallel fields $X, Y \in \Gamma(E)$ (that is, $DX = DY = 0$), $H(X,Y) = $ constant, then ω is called a* **compatible connection** *on the hermitian vector bundle.*

For $X = X^\alpha s_\alpha$, $Y = Y^\beta s_\beta$, the condition of parallelism for X and Y implies that

$$dX^\alpha + X^\gamma \omega^\alpha_\gamma = 0 \ , \tag{41.62}$$

$$dY^\beta + Y^\gamma \omega^\beta_\gamma = 0 \ . \tag{41.63}$$

Thus, from (41.61),

$$\begin{aligned}
dH(X,Y) &= (dh_{\alpha\overline{\beta}}) X^\alpha \overline{Y^\beta} + X^\alpha (d\overline{Y^\beta}) h_{\alpha\overline{\beta}} + (dX^\alpha) \overline{Y^\beta} h_{\alpha\overline{\beta}} \\
&= (dh_{\alpha\overline{\beta}} - h_{\gamma\overline{\beta}} \omega^\gamma_\alpha - h_{\alpha\overline{\gamma}} \overline{\omega^\gamma_\beta}) X^\alpha \overline{Y^\beta} \ .
\end{aligned} \tag{41.64}$$

The compatibility condition for ω is then

$$dh_{\alpha\overline{\beta}} - h_{\gamma\overline{\beta}} \omega^\gamma_\alpha - h_{\alpha\overline{\gamma}} \overline{\omega^\gamma_\beta} = 0 \ . \tag{41.65}$$

[Compare with (39.99), which states the compatibility condition for the Levi-Civita connection.] In matrix notation, (41.65) reads

$$dH = \omega H + H\,{}^t\overline{\omega} \ . \tag{41.66}$$

Exteriorly differentiating this equation we obtain

$$\begin{aligned}
0 &= d\omega \cdot H - \omega \wedge dH + dH \wedge {}^t\overline{\omega} + H \wedge d\,{}^t\overline{\omega} \\
&= (d\omega - \omega \wedge \omega) \cdot H + \omega \wedge \omega H - \omega \wedge (\omega H + H\,{}^t\overline{\omega}) \\
&\quad + (\omega H + H\,{}^t\overline{\omega}) \wedge {}^t\overline{\omega} + H \cdot d\,{}^t\overline{\omega} \\
&= (d\omega - \omega \wedge \omega) H + H(d\,{}^t\overline{\omega} + {}^t\overline{\omega} \wedge {}^t\overline{\omega}) \ .
\end{aligned} \tag{41.67}$$

From the fact that $\Omega = d\omega - \omega \wedge \omega$ [(41.5)], we obtain

$$\,{}^t\overline{\Omega} = d\,{}^t\overline{\omega} + {}^t\overline{\omega} \wedge {}^t\overline{\omega} \ . \tag{41.68}$$

Exercise 41.3   Show that (41.68) follows from (41.5).

Thus, since H is hermitian, $\Omega \cdot H$ is anti-hermitian [compare with (39.127)]:

$$\Omega H + H^t \overline{\Omega} = 0 , \qquad (41.69)$$

or

$$\boxed{{}^t\overline{\Omega} = -H^{-1}\Omega H} \quad . \qquad (41.70)$$

Following (41.19) we write, for a $q \times q$ curvature matrix Ω corresponding to a compatible connection ω on a hermitian vector bundle $\pi : E \to M$ whose fiber dimension is q,

$$\boxed{\det\left(1 + \frac{i}{2\pi}\Omega\right) = 1 + \cdots + C_j(\Omega) + \cdots + C_q(\Omega)} \quad , \qquad (41.71)$$

where $C_j(\Omega)$ is a $2j$-form globally defined on M. According to the Chern-Weil Theorem (Theorem 41.1) $C_j(\Omega)$ determines a $2j$-cohomology class of M. It is in fact real, since

$$\overline{\det\left(1 + \frac{i}{2\pi}\Omega\right)} = \det\left(1 - \frac{i}{2\pi}\overline{\Omega}\right) = \det\left(1 - \frac{i}{2\pi}{}^t\overline{\Omega}\right)$$

$$= \det\left(1 + \frac{i}{2\pi}H^{-1}\Omega H\right) = \det\left[H^{-1}\left(1 + \frac{i}{2\pi}\Omega\right)H\right] = \det\left(1 + \frac{i}{2\pi}\Omega\right) ,$$

$$(41.72)$$

where the second equality follows from the fact that $\det A = \det {}^t A$, the third from (41.70), and the last from Theorem 16.6 [Eq. (16.75)]. Thus

$$\overline{C_j(\Omega)} = C_j(\Omega)$$

determines a real $2j$-de Rham cohomology class on M: $C_j(\Omega) \in H^{2j}(M;\mathbb{R})$, and is independent of the choice of the hermitian structure on $\pi : E \to M$ and the corresponding compatible connection. It is called the j-th **Chern class** (with real coefficients) of the complex vector bundle $\pi : E \to M$. The factor i in the LHS of (41.71) ensures that $C_j(\Omega)$ is real, while the factor 2π is a normalization constant that in fact makes the Chern classes cohomology classes with integral coefficients (see Kobayashi and Nomizu 1969, Vol. II, Chapter XII):

$$\boxed{C_j(\Omega) \in H^{2j}(M;\mathbb{Z})} \quad . \qquad (41.73)$$

By direct expansion of the LHS of (41.71), it is seen that

$$\boxed{C_j(\Omega) = \frac{1}{j!}\left(\frac{i}{2\pi}\right)^j \sum_{1 \le \alpha_r, \beta_r \le q} \delta^{\alpha_1 \, \dots \, \alpha_j}_{\beta_1 \, \dots \, \beta_j} \Omega^{\beta_1}_{\alpha_1} \wedge \cdots \wedge \Omega^{\beta_j}_{\alpha_j}} \quad , \qquad (41.74)$$

where $\delta^{\alpha_1 \dots \alpha_j}_{\beta_1 \dots \beta_j}$ is the generalized Kronecker δ-symbol defined by (28.16).

Exercise 41.4   Make use of (16.77) or (16.80) for the expansion of a determinant to verify (41.74).

The following explicit formulas follow from (41.74):

$$C_0(\Omega) = 1 , \tag{41.75}$$

$$C_1(\Omega) = \left(\frac{i}{2\pi}\right) Tr\,\Omega , \tag{41.76}$$

$$C_2(\Omega) = \frac{1}{2}\left(\frac{i}{2\pi}\right)^2 [Tr\,\Omega \wedge Tr\,\Omega - Tr(\Omega \wedge \Omega)] , \tag{41.77}$$

$$\vdots$$

$$C_q(\Omega) = \left(\frac{i}{2\pi}\right)^q \det \Omega . \tag{41.78}$$

Note that if $dim(M) = n$, $C_j(\Omega)$ with $2j > n$ vanishes.

Exercise 41.5   Derive (41.77) from (41.74).

From the Chern classes we can obtain the so-called Chern characters. The **total Chern character** $Ch(\Omega)$ is defined by

$$Ch(\Omega) \equiv Tr\left(\exp\left(\frac{i}{2\pi}\Omega\right)\right) = \sum_{j=1}^{}\frac{1}{j!}\left(\frac{i}{2\pi}\right)^j Tr\,(\underbrace{\Omega \wedge \cdots \wedge \Omega}_{j \text{ times}}) . \tag{41.79}$$

Since $\Omega \wedge \cdots \wedge \Omega$ vanishes if $2j > n = dim(M)$, $Ch(\Omega)$ is a polynomial of finite order. The j-th **Chern character** $Ch_j(\Omega)$ is

$$Ch_j(\Omega) \equiv \frac{1}{j!}\left(\frac{i}{2\pi}\right)^j Tr\,(\underbrace{\Omega \wedge \cdots \wedge \Omega}_{j \text{ times}}) . \tag{41.80}$$

The Chern characters can be expressed in terms of the Chern classes. For example, by (41.76) and (41.77),

$$Ch_0(\Omega) = q , \tag{41.81}$$

$$Ch_1(\Omega) = C_1(\Omega) , \tag{41.82}$$

$$Ch_2(\Omega) = \frac{1}{2}(C_1(\Omega) \wedge C_1(\Omega) - 2C_2(\Omega)) , \tag{41.83}$$

$$\vdots$$

where q is the fiber dimension.

Chern classes satisfy the following useful properties, some of which will be stated without proof.

Theorem 41.3. *Let $\pi : E \to M$ be a complex vector bundle with a given connection and corresponding curvature Ω, and $f : N \to M$ be a smooth map. Then the **total Chern class***

$$C(\Omega) \equiv \det\left(1 + \frac{i}{2\pi}\Omega\right)$$

and the total Chern character $Ch(\Omega)$ satisfy the following:

$$C(f^*\Omega) = f^*C(\Omega)\,, \qquad (41.84)$$
$$Ch(f^*\Omega) = f^*Ch(\Omega)\,. \qquad (41.85)$$

*These are called the **naturality conditions**.*

Theorem 41.4. *Let $\pi : E \to M$ and $\pi' : F \to M$ be two complex vector bundles with the same base manifold M, and Ω_E, Ω_F be curvatures on E and F, respectively. Then the total Chern class and total Chern character satisfy the following relationships.*

$$\begin{aligned}
i) & & C(\Omega_{E\oplus F}) &= C(\Omega_E) \wedge C(\Omega_F)\,, & (41.86) \\
ii) & & Ch(\Omega_{E\oplus F}) &= Ch(\Omega_E) \oplus Ch(\Omega_F)\,, & (41.87) \\
iii) & & Ch(\Omega_{E\otimes F}) &= Ch(\Omega_E) \wedge Ch(\Omega_F)\,. & (41.88)
\end{aligned}$$

Proof. Suppose the fiber space of E is q_E-dimensional and that of F if q_F-dimensional. Then $\Omega_{E\oplus F}$ can be displayed explicitly in block matrix form as

$$\Omega_{E\oplus F} = \begin{pmatrix} \Omega_E & 0 \\ 0 & \Omega_F \end{pmatrix}\,, \qquad (41.89)$$

where the Ω_E block is $q_E \times q_E$ and the Ω_F block is $q_F \times q_F$. It follows that

$$\begin{aligned}
C(\Omega_{E\oplus F}) &= \det\left(1 + \frac{i}{2\pi}\Omega_{E\oplus F}\right) \\
&= \det\begin{pmatrix} 1 + \frac{i}{2\pi}\Omega_E & 0 \\ 0 & 1 + \frac{i}{2\pi}\Omega_F \end{pmatrix} \qquad (41.90) \\
&= \det\left(1 + \frac{i}{2\pi}\Omega_E\right) \wedge \det\left(1 + \frac{i}{2\pi}\Omega_F\right) = C(\Omega_E) \wedge C(\Omega_F)\,.
\end{aligned}$$

This establishes i).

From the definition of $Ch(\Omega)$ in (41.79) we have

$$Ch(\Omega_{E\oplus F}) = Tr\left(\exp\left(\frac{i}{2\pi}\Omega_{E\oplus F}\right)\right) = \exp\left(Tr\left(\frac{i}{2\pi}\Omega_{E\oplus F}\right)\right)$$

$$= \sum_j \frac{1}{j!}\left(\frac{i}{2\pi}\right)^j Tr\left(\underbrace{\Omega_{E\oplus F}\wedge\cdots\wedge\Omega_{E\oplus F}}_{j \text{ times}}\right). \tag{41.91}$$

From (41.89),

$$Tr\left(\Omega_{E\oplus F}\right) = Tr\left(\Omega_E\right)\oplus Tr\left(\Omega_F\right), \tag{41.92}$$

$$\Omega_{E\oplus F}\wedge\Omega_{E\oplus F} = \begin{pmatrix} \Omega_E & 0 \\ 0 & \Omega_F \end{pmatrix}\wedge\begin{pmatrix} \Omega_E & 0 \\ 0 & \Omega_F \end{pmatrix} = \begin{pmatrix} \Omega_E\wedge\Omega_E & 0 \\ 0 & \Omega_F\wedge\Omega_F \end{pmatrix}, \tag{41.93}$$

$$\underbrace{\Omega_{E\oplus F}\wedge\cdots\wedge\Omega_{E\oplus F}}_{j \text{ times}} = \begin{pmatrix} \overbrace{\Omega_E\wedge\cdots\wedge\Omega_E}^{j \text{ times}} & 0 \\ 0 & \underbrace{\Omega_F\wedge\cdots\wedge\Omega_F}_{j \text{ times}} \end{pmatrix}. \tag{41.94}$$

Thus, (41.91) yields

$$Ch(\Omega_{E\oplus F}) = \sum_j \frac{1}{j!}\left(\frac{i}{2\pi}\right)^j \left(Tr\left(\overbrace{\Omega_E\wedge\cdots\wedge\Omega_E}^{j \text{ times}}\right)\oplus Tr\left(\overbrace{\Omega_F\wedge\cdots\wedge\Omega_F}^{j \text{ times}}\right)\right)$$

$$= Ch(\Omega_E)\oplus Ch(\Omega_F). \tag{41.95}$$

This establishes ii).

To prove iii) we begin by observing that

$$\Omega_{E\otimes F} = (\Omega_E\otimes I_F)\oplus(I_E\otimes\Omega_F), \tag{41.96}$$

where I_F is the $q_F\times q_F$ identity matrix and I_E is the $q_E\times q_E$ identity matrix. Thus

$$Ch(\Omega_{E\otimes F}) = Tr\left(\exp\left(\frac{i}{2\pi}\Omega_{E\otimes F}\right)\right)$$

$$= Tr\left(\exp\left(\frac{i}{2\pi}\{(\Omega_E\otimes I_F)\oplus(I_E\otimes\Omega_F)\}\right)\right)$$

$$= \sum_j \frac{1}{j!}\left(\frac{i}{2\pi}\right)^j Tr\left\{(\Omega_E\otimes I_F)\oplus(I_E\otimes\Omega_F)\right\}^j$$

$$= \sum_j \frac{1}{j!}\left(\frac{i}{2\pi}\right)^j Tr\left[\sum_{m=1}^{j}\binom{j}{m}(\Omega_E\otimes I_F)^m\wedge(I_E\otimes\Omega_F)^{j-m}\right], \tag{41.97}$$

$$\overbrace{}^{j \text{ times}}$$

where $A^j \equiv A \wedge \cdots \wedge A$, for a matrix A of differential forms. Let us write in block form

$$\Omega_E \otimes I_F = \begin{pmatrix} \Omega_E & 0 & 0 & \\ 0 & \Omega_E & 0 & \\ & & \ddots & \\ 0 & & & \Omega_E \end{pmatrix}, \qquad (41.98)$$

where each block is $q_E \times q_E$ and the matrix has $q_F \times q_F$ blocks. Similarly, write

$$I_E \otimes \Omega_F = \begin{pmatrix} (\Omega_F)_1^1 I_E & (\Omega_F)_1^2 I_E & \cdots & (\Omega_F)_1^{q_F} I_E \\ & \vdots & & \\ (\Omega_F)_{q_F}^1 I_E & & & (\Omega_F)_{q_F}^{q_F} I_E \end{pmatrix}. \qquad (41.99)$$

This matrix has the same block structure as that in (41.98): $q_F \times q_F$ blocks and each block is $q_E \times q_E$. For the rest of the proof of this theorem, we will write Tr_q for the trace of $(q \times q)$ matrices. From (41.99), using α and β as block indices, where $\alpha, \beta = 1, \ldots, q_F$, we have

$$X_\alpha^\beta \equiv \overbrace{\left((I_E \otimes \Omega_F)^{j-m} \right)_\alpha^\beta}^{(q_E \times q_E) \text{ matrix}}$$

$$= (\Omega_F)_\alpha^{i_1} \wedge (\Omega_F)_{i_1}^{i_2} \wedge \cdots \wedge (\Omega_F)_{i_{j-m-2}}^{i_{j-m-1}} \wedge (\Omega_F)_{i_{j-m-1}}^\beta \times I_E. \qquad (41.100)$$

Then,

$$Tr_{q_E q_F} \left[\left((\Omega_E) \otimes I_F \right)^m \wedge (I_E \otimes \Omega_F)^{j-m} \right]$$

$$= Tr_{q_E q_F} \begin{pmatrix} (\Omega_E)^m \wedge X_1^1 & \cdots & \\ & \cdots & (\Omega_E)^m \wedge X_2^2 & \\ & & \vdots & \\ & & \cdots & (\Omega_E)^m \wedge X_{q_F}^{q_F} \end{pmatrix}$$

$$= Tr_{q_E} \left[(\Omega_E)^m \wedge (X_1^1 + \cdots + X_{q_F}^{q_F}) \right] = Tr_{q_E} \left[(\Omega_E)^m \wedge (Tr_{q_F} (\Omega_F)^{j-m}) I_E \right]$$

$$= Tr_{q_E} (\Omega_E)^m \wedge Tr_{q_F} (\Omega_F)^{j-m}. \qquad (41.101)$$

It follows from (41.97) that

$$Ch(\Omega_{E \otimes F}) = \sum_j \sum_m \frac{1}{j!} \left(\frac{i}{2\pi}\right)^j \frac{j!}{m!(j-m)!} Tr_{q_E} (\Omega_E)^m \wedge Tr_{q_F} (\Omega_F)^{j-m}$$

$$= \sum_l \sum_m \left(\frac{i}{2\pi}\right)^{m+l} \frac{1}{m!\, l!} Tr_{q_E}(\Omega_E)^m \wedge Tr_{q_F} (\Omega_F)^l = Ch(\Omega_E) \wedge Ch(\Omega_F) \,,$$

$$(41.102)$$

where in the second equality we have let $j - m = l$. This establishes iii). □

Theorem 41.5 (The Splitting Principle). *For the purpose of the calculation of Chern classes, a complex vector bundle (of fiber dimension q) behaves as if it were a sum bundle of q line bundles.*

The proof of this theorem will not be presented here, but its use will be discussed. Suppose $\pi : E \to M$ is indeed a sum of q line bundles L_j, $j = 1, \ldots, q$:

$$E = L_1 \oplus \cdots \oplus L_q \,. \qquad (41.103)$$

(A complex line bundle is a complex vector bundle whose typical fiber is a one-dimensional complex vector space.) Then statement i) of Theorem 41.4 implies that

$$C(\Omega_E) = C(\Omega_{L_1}) \wedge \cdots \wedge C(\Omega_{L_q}) \,, \qquad (41.104)$$

with

$$C(\Omega_{L_j}) = 1 + C_1(\Omega_{L_j}) \equiv 1 + x_j \,. \qquad (41.105)$$

Note that since $q = 1$ for a line bundle, (41.71) implies that

$$\det\left(1 + \frac{i}{2\pi}\Omega_{L_j}\right) = 1 + C_1(\Omega_{L_j}) \,.$$

Then

$$C(\Omega_E) = (1 + x_1) \wedge \cdots \wedge (1 + x_q)$$
$$= 1 + (x_1 + \cdots + x_q) + (x_1 \wedge x_2 + \cdots + x_{q-1} \wedge x_q) + \cdots + (x_1 \wedge \cdots \wedge x_q)$$
$$= 1 + C_1(\Omega_E) + C_2(\Omega_E) + \cdots + C_q(\Omega_E) \,,$$

$$(41.106)$$

with

$$C_0(\Omega_E) = 1 , \tag{41.107}$$

$$C_1(\Omega_E) = x_1 + \cdots + x_q , \tag{41.108}$$

$$C_2(\Omega_E) = x_1 \wedge x_2 + \cdots + x_{q-1} \wedge x_q , \tag{41.109}$$

$$\vdots$$

$$C_j(\Omega_E) = \sum_{i_1 < i_2 < \cdots < i_j} x_{i_1} \wedge x_{i_2} \wedge \cdots \wedge x_{i_j} , \tag{41.110}$$

$$\vdots$$

$$C_q(\Omega_E) = x_1 \wedge \cdots \wedge x_q . \tag{41.111}$$

The same results would have been obtained if the $q \times q$ matrix $X \equiv \dfrac{i}{2\pi} \Omega_E$ were diagonalizable with diagonal elements equal to x_j, $j = 1, \ldots, q$. Since $\det(1+X)$ is ad-invariant, one can take the formal expansion $(1+x_1) \wedge \cdots \wedge (1+x_q)$, compute the j-th order term (which consists of a homogeneous j-th order polynomial involving the x_j's), express it in terms of ad-invariant functions of Ω_E, and then identify it with $C_j(\Omega_E)$. In this way, for example, Eqs. (41.76) to (41.78) result easily from the corresponding equations (41.108) to (41.110). This is perhaps less cumbersome than computing $C_j(\Omega_E)$ directly from (41.74).

The Chern classes are related to two other types of characteristic classes expressible as real cohomology classes. These are the Pontrjagin classes, characterizing real vector bundles (or principal bundles) with structure group $GL(q; \mathbb{R})$ [reducible to $O(q)$ or $SO(q)$]; and the Euler classes, characterizing only those with structure group $SO(q)$. The reduction may be achieved by endowing the bundle E with a fiber metric and introducing orthonormal frames on each fiber. We remind the reader that the Chern classes characterize complex vector bundles with structure group $GL(q; \mathbb{C})$ [reducible to $U(q)$].

The **total Pontrjagin class** of a real $O(q)$ vector bundle $\pi : E \to M$ is defined by

$$p(\Omega) \equiv \det\left(1 + \frac{\Omega}{2\pi}\right) , \tag{41.112}$$

where Ω is the curvature matrix of 2-forms corresponding to a given connection. If Ω_i^j is given with respect to an orthonormal frame field, then $\Omega^T = -\Omega$ [for the structure group $O(q)$, the $\mathcal{O}(q)$-valued (Lie algebra-valued) Ω must be antisymmetric]. It follows that

$$\det\left(1 + \frac{\Omega}{2\pi}\right) = \det\left(1 + \frac{\Omega^T}{2\pi}\right) = \det\left(1 - \frac{\Omega}{2\pi}\right) , \tag{41.113}$$

which implies that $p(\Omega)$ must be an even function of Ω. This remains true when Ω is expressed with respect to any frame field, since $p(\Omega)$ is ad-invariant. Analogous to (41.71) we can then expand

$$p(\Omega) = \det\left(1 + \frac{\Omega}{2\pi}\right) = 1 + p_1(\Omega) + p_2(\Omega) + \cdots , \tag{41.114}$$

where $p(\Omega)$ is a polynomial of order $2j$ in Ω, and is thus a $4j$-form. It is called the j-th **Pontrjagin class**. Note that the series terminates: $p_j(\Omega) = 0$ if either $2j > q$, where q is the fiber dimension, or $4j > n$, where n is the dimension of the base manifold. According to the Chern-Weil Theorem (Theorem 41.1) (which still applies to the present case of real vector bundles),

$$p_j(\Omega) \in H^{4j}(M;\mathbb{R}) \, ,$$

(the $4j$-de Rham cohomology group with real coefficients on the base manifold M). Noting that

$$\det\left(1 - \frac{\Omega}{2\pi}\right) = \det\left(1 + \frac{i}{2\pi}(i\Omega)\right) \, , \qquad (41.115)$$

we can relate the Pontrjagin classes $p_j(\Omega)$ to the Chern classes $C_{2j}(\Omega)$ of the complexified bundle $E \otimes \mathbb{C}$ by

$$p_j(\Omega_E) = i^{2j} C_{2j}(\Omega_{E \otimes \mathbb{C}}) = (-1)^j C_{2j}(\Omega_{E \otimes \mathbb{C}}) \, . \qquad (41.116)$$

Thus we have, from (41.74),

$$p_j(\Omega) = \frac{1}{(2j)! \, (2\pi)^{2j}} \sum_{1 \le \alpha_r, \beta_r \le q} \delta^{\alpha_1 \dots \alpha_{2j}}_{\beta_1 \dots \beta_{2j}} \Omega^{\beta_1}_{\alpha_1} \wedge \dots \wedge \Omega^{\beta_{2j}}_{\alpha_{2j}} \, . \qquad (41.117)$$

Analogous to statement i) of Theorem 41.4, we have the following fact:

$$p(\Omega_{E \oplus F}) = p(\Omega_E) \wedge p(\Omega_F) \, . \qquad (41.118)$$

This can be proved in exactly the same way.

Suppose M is an even-dimensional orientable Riemannian manifold of dimension $q = 2l$, and consider the tangent bundle TM. The structure group can be reduced to $SO(2l)$ by using an orthonormal frame field determined by the Riemannian metric. We can define a $2l$-form e on M, called the **Euler class**, by

$$e(A) \wedge e(A) = p_l(A) \, , \qquad (41.119)$$

where p_l is the l-th Pontrjagin class and A is a $2l \times 2l$ matrix of 2-forms. Note that even though $p_l(\Omega)$ vanishes identically [since $p_l(\Omega) \in H^{4l}(M;\mathbb{R})$ and $dim(M) = 2l$], $e(\Omega)$ is well-defined by (41.119). If $dim(M)$ is odd, we define $e(\Omega) = 0$. We will illustrate this with the simple example $M = S^2$ equipped with the usual metric [(39.169)]. In this case Ω^j_i is given by [(39.176)]

$$(\Omega^j_i) = \begin{pmatrix} 0 & -\sin\theta \, d\theta \wedge d\phi \\ \sin\theta \, d\theta \wedge d\phi & 0 \end{pmatrix} \, .$$

According to (41.116) and (41.77),

$$p_1(\Omega_{T(S^2)}) = -C_2(\Omega) = -\frac{1}{8\pi^2} Tr\,(\Omega \wedge \Omega)$$

$$= \left(\frac{1}{2\pi}\,\sin\theta\,d\theta \wedge d\phi\right) \wedge \left(\frac{1}{2\pi}\,\sin\theta\,d\theta \wedge d\phi\right) . \tag{41.120}$$

This expression vanishes identically; but it allows us, according to (41.119), to identify $e(\Omega)$ as

$$e(\Omega_{T(S^2)}) = \frac{1}{2\pi}\,\sin\theta\,d\theta \wedge d\phi . \tag{41.121}$$

It is interesting to note that

$$\int_{S^2} e(\Omega) = \frac{1}{2\pi} \int_0^{2\pi} d\phi \int_0^{\pi} \sin\theta\,d\theta = 2 , \tag{41.122}$$

which is the Euler characteristic $\chi(S^2)$ of S^2 [c.f. (33.15)]. In fact, (41.122) is a special case of the Gauss-Bonnet Theorem.

Theorem 41.6 (Gauss-Bonnet Theorem). *For a compact orientable Riemannian manifold M,*

$$\boxed{\int_M e(M) = \chi(M)} \quad , \tag{41.123}$$

where $e(M)$ is the Euler class of M and $\chi(M)$ the Euler characteristic of M.

The proof of this historically important theorem is well beyond the scope of this book, and will not be presented here. A beautiful proof due to Chern by an intrinsic method (one that does not assume that M is embedded in a higher dimensional manifold) in fact opened the door to the study of characteristic classes (see S. S. Chern, 1944).

The Euler class $e(\Omega)$ can be written explicitly in terms of the curvature matrix Ω as a result of the following Lemma.

Lemma 41.3. *Suppose A is a skew-symmetric $2l \times 2l$ matrix. Then*

$$\det A = [Pf\,(A)]^2 , \tag{41.124}$$

*where the polynomial $Pf\,(A)$, called the **Pfaffian** of the matrix A, is given by*

$$Pf\,(A) \equiv \frac{(-1)^l}{2^l\,l!} \sum_{\sigma \in S(2l)} (sgn\,\sigma)\,A^{\sigma(2)}_{\sigma(1)} A^{\sigma(4)}_{\sigma(3)} \cdots A^{\sigma(2l)}_{\sigma(2l-1)} . \tag{41.125}$$

The sum in the above equation is over all permutations in the symmetric group $S(2l)$.

Proof. Suppose A is a skew-symmetric $2l \times 2l$ matrix. Then it can be block-diagonalized by some $g \in O(2l)$:

$$g^{-1} A g = g^T A g = \begin{pmatrix} 0 & \lambda_1 & & & 0 \\ -\lambda_1 & 0 & & \cdots & \\ & & \vdots & & \\ & \cdots & & 0 & \lambda_l \\ 0 & & & -\lambda_l & 0 \end{pmatrix} \equiv A' . \tag{41.126}$$

Since the determinant is invariant under similarity transformations (c.f. Theorem 16.6),

$$\det A = \det A' = (\lambda_1)^2 (\lambda_2)^2 \dots (\lambda_l)^2 = \prod_{i=1}^{l} (\lambda_i)^2 . \tag{41.127}$$

According to the definition of a Pfaffian given by (41.125),

$$Pf(A') = (-1)^l (A')_1^2 (A')_3^4 \dots (A')_{2l-1}^{2l} = (-1)^l \prod_{i=1}^{l} \lambda_i . \tag{41.128}$$

It follows from (41.127) that

$$\det A = \det A' = [Pf(A')]^2 . \tag{41.129}$$

| Exercise 41.6 | Use the explicitly given matrix A' in (41.126) to prove (41.128). Note that the only non-vanishing elements of A' are

$$(A')_1^2 = -(A')_2^1 = \lambda_1, \ (A')_3^4 = -(A')_4^3 = \lambda_2, \ \dots, \ (A')_{2l-1}^{2l} = -(A')_{2l}^{2l-1} = \lambda_l .$$

Thus $[Pf(A)]^2$ is an invariant quantity for any skew-symmetric $2l \times 2l$ matrix under $O(2l)$-similarity transformations:

$$[Pf(A)]^2 = [Pf(g^T A g)]^2 , \tag{41.130}$$

and the lemma is established.   □

For the tangent bundle TM of a $2l$-dimensional Riemannian manifold,

$$p_l(\Omega) = \left(\frac{1}{2\pi} \right)^{2l} \det \Omega . \tag{41.131}$$

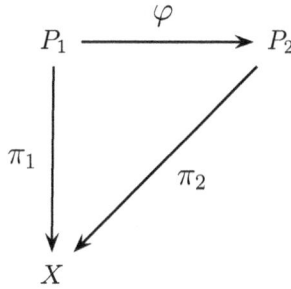

$$\begin{array}{ccc} P_1 & \xrightarrow{\ \varphi\ } & P_2 \\ & & \\ \pi_1 \Big\downarrow & \swarrow \pi_2 & \\ & & \\ X & & \end{array}$$

FIGURE 41.1

[c.f. (41.114) and (41.112)]. Thus, from (41.119), (41.124) and (41.125), we can express the Euler class $e(\Omega)$ as

$$e(\Omega) = \frac{(-1)^l}{l!\,(4\pi)^l} \sum_{\sigma \in S(2l)} (sgn\,\sigma)\,\Omega^{\sigma(2)}_{\sigma(1)} \wedge \Omega^{\sigma(4)}_{\sigma(3)} \wedge \cdots \wedge \Omega^{\sigma(2l)}_{\sigma(2l-1)} \qquad . \qquad (41.132)$$

We will conclude this chapter with a general and somewhat formal outline on how characteristic classes serve to classify principal G-bundles, a member of which is usually written $\pi : P \to X$, with total space P, structure group G, and base manifold X (c.f. Def. 37.2). A basic question in the classification problem is the following: Given a base manifold X and a structure group G, in how many ways, up to equivalence, can these be assembled into principal G-bundles? Two principal G-bundles over X, $\pi_1 : P_1 \to X$ and $\pi_2 : P_2 \to X$ are said to be **equivalent** if there exists a G-**equivariant** homeomorphism $\varphi : P_1 \to P_2$ between them, that is, φ satisfies the property $\pi_1 = \pi_2 \circ \varphi$. Equivalently, the diagram shown in Fig. 41.1 commutes.

It is useful at this point to recall the notion of triviality of a principal bundle and some related facts (c.f. Def. 37.4, Def. 37.5 and Theorem 37.2). The principal G-bundle over X given by $P = X \times G$ with $L_g(x, g') = (x, gg')$ and $\pi(x, g) = x$, $x \in X$, $g, g' \in G$, is called the **product bundle**. Any bundle equivalent to the product bundle is said to be **trivial**. A necessary and sufficient condition for a principal bundle to be trivial is that it admits a global section. We will denote the set of equivalence classes $[P]$ of principal G-bundles over X by $Bndl_G(X)$.

Definition 41.2. *Let* $\pi : P \to X$ *be a principal G-bundle and* $f : Y \to X$ *be a continuous map. The* **pullback bundle** *of P over the base manifold Y, $f^*(P)$, is defined to be the bundle whose total space is*

$$\{\,(p, y) \in P \times Y \mid \pi(p) = f(y)\,\}\,,$$

with the projection $(p, y) \mapsto y$ *and the left* G-*action* $L_g(p, y) = (L_g(p), y)$, $g \in G$.

Theorem 41.7. *Suppose two principal* G-*bundles* $\pi_1 : P_1 \to X$ *and* $\pi_2 : P_2 \to X$ *are equivalent to each other:* $P_1 \sim P_2$, *or* $[P_1] = [P_2] \in Bndl_G(X)$; *and* $f : Y \to X$ *is a continuous map. Then*

$$f^*(P_1) \sim f^*(P_2) , \tag{41.133}$$

that is, pullback bundles of equivalent bundles are equivalent. The bundle pullback map f^* *can then be considered as a map from* $Bndl_G(X)$ *to* $Bndl_G(Y)$, *that is,* $f : Y \to X$ *induces the map*

$$f^* : Bndl_G(X) \to Bndl_G(Y) .$$

Exercise 41.7   Prove the above theorem.

The first crucial fact in the classification theory of principal bundles is the following theorem, which will be stated without proof.

Theorem 41.8 (Covering Homotopy Theorem). *The bundle pullback map* f^* : $Bndl_G(X) \to Bndl_G(Y)$ *depends only on the homotopy class* $[f]$ *of* f.

At this point the general notion of a characteristic class (of which the Chern, Pontrjagin, and Euler classes considered above are special examples) can be introduced.

Definition 41.3. *Suppose* $f : Y \to X$ *is a continuous map between manifolds. A* **characteristic class** C *of principal* G-*bundles over* X *is a map*

$$C : Bndl_G(X) \to H^*(X)$$

that assigns to each principal G-*bundle* P *over* X *an element* $C(P) \in H^*(X)$, *where* $H^*(X)$ *is some cohomology ring of* X *(such as the de Rham cohomology), such that i) the* **naturality condition** *[c.f. (41.84)]*

$$C(f^*(P)) = f^*(C(P)) \tag{41.134}$$

is satisfied; and ii) $C(P) = C(P')$ *if* P *is equivalent to* P'. *The set of all characteristic classes for* G-*bundles is denoted* $Char(G)$.

Lemma 41.4. *A principal* G-*bundle* $\pi : P \to X$ *is trivial if the base manifold* X *is contractible to a point.*

Proof. Suppose X is **contractible** to a point. Then there exists a homotopy class of maps $[f(t)] : X \to X$, $0 \le t \le 1$, such that $f(0)$ is the identity map $[(f(0))(x) = x$, for all $x \in X]$ and $f(1)$ sends every point in X to a fixed point $x_0 \in X$. We note that $f(t)(x_0) = x_0$ for all $t \in [0, 1]$. Consider the pullback bundle $(f(1))^*(P)$. By Def. 41.2, this is just the product bundle $\pi^{-1}(x_0) \times X$, which is trivial. By the Covering Homotopy Theorem (Theorem 41.8), $(f(0))^*(P)$ is equivalent to $(f(1))^*(P)$, since $f(0) \sim f(1)$ (the two maps belong to the same homotopy class). But since $f(0)$ is the identity map in X, $(f(0))^*(P)$ is precisely the bundle $\pi : P \to X$. Thus $\pi : P \to X$ is trivial.   □

For example, since $\mathbb{R}^n$ is contractible to a point, the frame bundle over $\mathbb{R}^n$ is trivial. According to Theorem 37.2, then, there exists a global frame field on $\mathbb{R}^n$.

Lemma 41.5. *If the principal G-bundle $\pi : P \to M$ is trivial, then there exists a continuous map $f : M \to \{p\}$, where $\{p\}$ is a space consisting of a single point p, such that $\pi : P \to M$ is the pullback bundle $f^*(E)$, where $\pi' : E \to \{p\}$ is a G-bundle over the point p.*

Proof. Since $\pi : P \to M$ is trivial, we can, without loss of generality, write $P = M \times G$, with $\pi(x, g) = x$, $x \in M$, $g \in G$. Let $\pi' : E \to \{p\}$ be the G-bundle with the total space $E = \{p\} \times G$, $\pi'(p, g) = p$. Then, by Def. 41.2, the pullback bundle $f^*(E)$ is given by one whose total space is

$$\{ ((p, g), x) \in (\{p\} \times G) \times M \mid \pi'(p, g) = f(x) = p \} \,,$$

which is just $(\{p\} \times G) \times M$, with the projection $((p, g), x) \mapsto x \in M$. This bundle is clearly equivalent to $\pi : M \times G \to M$, or $\pi : P \to M$. $\square$

Under the conditions of the above Lemma, the naturality condition (41.134) implies that, for any $C \in Char\,(G)$,

$$C(f^*(E)) = C(P) = f^*(C(E)) \,. \tag{41.135}$$

But since $\pi' : E \to \{p\}$ is a bundle over a point, $C(E) \in H^*(\{p\})$ is trivial, due to the fact that all the cohomology groups of a space of a single point are trivial (each consisting of a single element, the identity 0). Thus $f^*(C(E))$ is also trivial, and we have the following basic result.

Corollary 41.1. *The characteristic classes of a trivial bundle are all trivial.*

In general, equality of all characteristic classes for two G-bundles is a necessary (but not sufficient) condition for their equivalence; though in many cases, sufficiency also holds.

We will now finally show how the Covering Homotopy Theorem "solves" the classification problem of principal G-bundles by reducing it to the problem of finding all the homotopy classes of continuous maps between manifolds. Let $[X, Z]$ denote the set of all homotopy classes of continuous maps from X to Z, and $f : Y \to X$ be a continuous map. Then, for $[h] \in [X, Z]$,

$$f^*[h] = [h \circ f] \in [Y, Z] \,, \tag{41.136}$$

as seen by the diagram in Fig. 41.2.

Consider any principal G-bundle $\pi : P \to Z$. By the Covering Homotopy Theorem, we have a map $\alpha : [X, Z] \to Bndl_G(X)$ given by

$$\alpha([h]) = [h^*(P)] \,, \tag{41.137}$$

for any $[h] \in [X, Z]$, such that the naturality condition

$$\alpha \circ f^* = f^* \circ \alpha \tag{41.138}$$

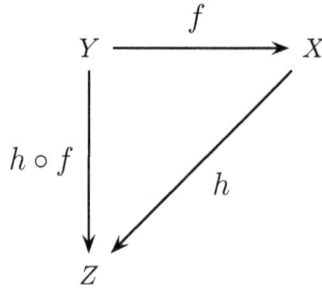

FIGURE 41.2

is satisfied. We now define a so-called **universal G-bundle**, which can be used to classify all principal G-bundles.

Definition 41.4. *A principal G-bundle $\pi : P \to Z$ is said to be* **universal** *if the map $\alpha : [X, Z] \to Bndl_G(X)$ introduced above [(41.137) and (41.138)] is bijective.*

The possibility of classification rests on the following fact, which will be stated without proof.

Theorem 41.9. *Given a group G, universal G-bundles always exist. In fact, if the total space P of a principal G-bundle $\pi : P \to Z$ is contractible, then $\pi : P \to Z$ is universal.*

We will denote a universal G-bundle by $\pi_U : U_G \to B_G$, where the base manifold B_G is called the **classifying space** for the group G. Then, if $\pi : P \to X$ is any principal G-bundle, the universality of $\pi : U_G \to B_G$ implies that there is a unique $[h] \in [X, B_G]$ such that $P \sim h^*(U_G)$ (the given principal G-bundle is equivalent to the pullback of the universal G-bundle by h^*). Any representation h of $[h]$ is then called a **classifying map** for P. It is clear from the diagram in Fig. 41.3 that for any continuous map $f : Y \to X$, $h \circ f$ is a classifying map for the pullback bundle $\pi' : f^*(P) \to Y$. Thus $\pi_G : U_G \to B_G$ is good for the classification of all principal G-bundles, regardless of the base manifold. Note that the classifying map for U_G is just the identity map of B_G onto itself.

The formal strategy to classify principal G-bundles by characteristic classes then consists of the following steps:

1) For a given group G and base manifold X, find a universal G-bundle $\pi_U : U_G \to B_G$.

2) Realize that the set of all characteristic classes for U_G is canonically isomorphic to the cohomology ring of the classifying space B_G:

$$Char\,(G) \sim H^*(B_G)\,. \tag{41.139}$$

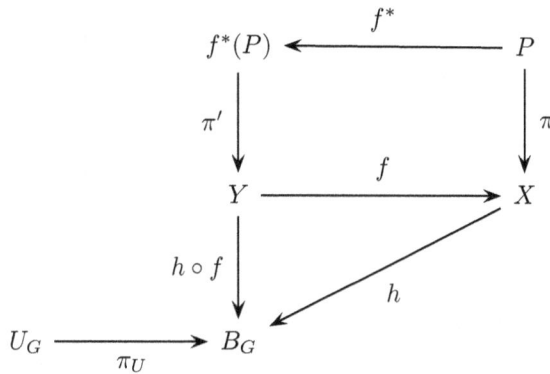

FIGURE 41.3

3) For the principal G-bundle of interest $\pi : P \to X$, find a classifying map h such that $P = h^*(U_G)$.

4) Each characteristic class of the universal bundle $C(U_G) \in H^*(B_G)$ then gives a characteristic class $C(P)$ of the bundle of interest by pullback:

$$C(P) = h^*(C(U_G)) . \tag{41.140}$$

We will state the following interesting facts for universal principal G-bundles without proof. In statements i) to v) below, let $m =$ dimension of the base manifold M in the principal G-bundle of interest $\pi : P \to M$, and the group quotients are all right coset spaces.

i) For $G = O(k)$, the classifying space is given by

$$B_G = Gr\,(N, k, \mathbb{R}) \sim \frac{O(N)}{O(k) \times O(N - k)} , \tag{41.141}$$

where $N \geq m + k + 1$ and $Gr\,(N, k, \mathbb{R})$ is the so-called **Grassmann manifold** of k-dimensional planes through the origin of $\mathbb{R}^N$ (each point in this manifold is a k-dimensional plane through the origin of $\mathbb{R}^N$). The total space is given by

$$U_G = V_{N,\,N-k}(\mathbb{R}) \sim \frac{O(N)}{O(N - k)} , \tag{41.142}$$

called a **Stiefel manifold**. For given N and k, it is in fact the manifold of k-frames in $\mathbb{R}^N$. The projection π in the universal bundle

$$\pi : V_{N,\,N-k}(\mathbb{R}) \to Gr\,(N, k, \mathbb{R})$$

projects each k-frame in $\mathbb{R}^n$ to the k-plane that it spans. The fiber above a particular k-plane is then the set of all k-frames that span it, which, of course, is isomorphic to $O(k)$. Note that the dimension of the Grassmann manifold $Gr\,(N, k, \mathbb{R})$ can be obtained from (41.141) and the facts that

$$dim\,(O(m)) = \frac{m(m-1)}{2}\,, \qquad \text{and}$$

$$dim\,(G/H) = dim\,G - dim\,H\,.$$

Thus

$$dim\,(Gr\,(N, k, \mathbb{R}))$$
$$= \frac{1}{2}N(N-1) - [\frac{1}{2}k(k-1) + \frac{1}{2}(N-k)(N-k-1)] = k(N-k)\,.$$
$$\tag{41.143}$$

Since the total space P of a trivial principal G-bundle $\pi : P \to M$ can be written as $M \times G$, the non-triviality of our universal bundle can be thought of as a measure of the extent to which the following inequality holds:

$$\left(\frac{O(N)}{O(k) \times O(N-k)}\right) \times O(k) \neq \frac{O(N)}{O(N-k)}\,. \tag{41.144}$$

ii) For $G = SO(k)$, the classifying space is

$$B_G = \widetilde{Gr}\,(N, k, \mathbb{R}) \sim \frac{SO(N)}{SO(k) \times SO(N-k)}\,, \tag{41.145}$$

where $N \geq m+k+1$. This is called the **oriented Grassmann manifold**. The total space is

$$U_G = \tilde{V}_{N,\,N-k}(\mathbb{R}) \sim \frac{SO(N)}{SO(N-k)}\,. \tag{41.146}$$

iii) For $G = U(k)$,

$$B_G = Gr\,(N, k, \mathbb{C}) \sim \frac{U(N)}{U(k) \times U(N-k)}\,, \tag{41.147}$$

where $m \geq 2N - 2k$. This is called the **complex Grassmann manifold**, of real dimension $2k(N-k)$. The total space is

$$U_G = \tilde{V}_{N,\,N-k}(\mathbb{C}) \sim \frac{U(N)}{U(N-k)}\,. \tag{41.148}$$

iv) For any compact Lie group G of real dimension k,

$$B_G = \frac{O(N)}{G \times O(N-k)}\,, \tag{41.149}$$

where $N \geq m + k + 1$. The total space is again $U_G = O(N)/(O(N-k))$.

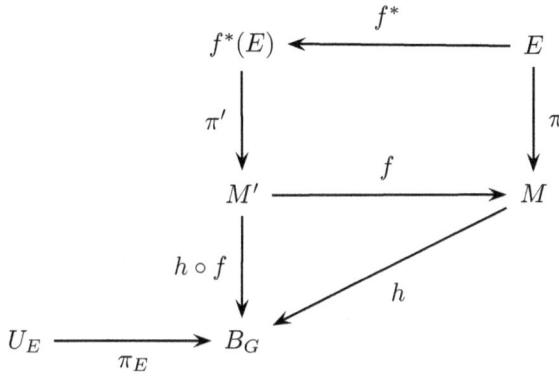

$$\begin{array}{ccc}
f^*(E) & \xleftarrow{\quad f^* \quad} & E \\
\downarrow {\scriptstyle \pi'} & & \downarrow {\scriptstyle \pi} \\
M' & \xrightarrow{\quad f \quad} & M \\
\downarrow {\scriptstyle h \circ f} & {\scriptstyle h} & \\
U_E \xrightarrow{\quad \pi_E \quad} & B_G &
\end{array}$$

FIGURE 41.4

v) For G (of real dimension k) connected but not necessarily compact, let $H \subset G$ be the maximal compact subgroup of G. Then

$$B_G = \frac{O(N)}{H \times O(N-k)}, \tag{41.150}$$

where $N \geq m + k + 1$. The total space is still $U_G = O(N)/(O(N-k))$.

So far we have discussed explicitly only the universal bundles for the classification of principal G-bundles. One can also obtain universal vector bundles $\pi_E : U_E \to B_G$ for the classification of associated vector bundles $\pi : E \to M$ (with structure group G) in a natural way. The classifying spaces B_G are exactly the same as before. The fiber above each point in B_G is simply taken to be the point itself, which is a vector space on which G acts naturally. We have a similar diagram (Fig. 41.4) to that for the universal principal G-bundles.

The characteristic classes of the vector bundles of interest are calculated by pullback [c.f. (41.140)]:

$$C(E) = h^*(C(U_E)). \tag{41.151}$$

In fact, formulas (41.74), (41.117), and (41.132) are explicit algorithms for the calculation of $C(E)$ specifically for the Chern classes, the Pontrjagin classes, and the Euler classes, respectively.

We finally mention one more example of characteristic classes, namely, the **Stiefel-Whitney classes**. These occur in $O(n)$-bundles, but unlike the previous classes considered, they cannot be calculated from curvature 2-forms. For $\pi : E \to M$, they take values in the so-called Čech cohomology $H^*(M; \mathbb{Z}_2)$, with coefficients in $\mathbb{Z}_2 = \{+1, -1\}$. We will not consider the Stiefel-Whitney classes in any detail except to state two important results:

1) A manifold M is orientable if and only if the first Stiefel-Whitney class of the tangent bundle, $w_1(TM)$, is trivial.

2) There exists a spin structure (see Def. 43.10) over a manifold M if and only if the second Stiefel-Whitney class, $w_2(TM)$, is trivial.

Result 2) is obviously important for physics since it imposes a restriction on the kinds of particles that can live on M in terms of the topology of M.

Chapter 42

Chern-Simons Forms

Chern-Simons forms arise from the Chern classes by a procedure called **transgression**, which involves pulling a characteristic form C on the base manifold M of a bundle $\pi : E \to M$ back to the total space E by π^*, and thereby making the pullback form an exact form on E:

$$\pi^* C = d(CS(C)) .$$

The Chern-Simons form $CS(C)$ so obtained has generated numerous important applications in physics, for example, in topological quantum field theory and condensed matter physics (fractional statistics and the quantum Hall effect).

We will begin with an informal discussion on transgression. Following Chern (see Chern 1990) we use the example of the two-dimensional Gauss-Bonnet formula [c.f. (41.123), Theorem 41.6]. Let M be a two-dimensional oriented Riemannian manifold and $\pi : P \to M$ be the orthonormal frame bundle of M. A point in the total space P is then written $(x; e_1 e_2)$, where (e_1, e_2) is an orthonormal frame with origin at $x \in M$, and the projection π sends $(x; e_1 e_2)$ to its origin x. Choose a frame field (e_1^0, e_2^0) with dual coframe field $((\omega^0)^1, (\omega^0)^2)$. Then we have the local expressions

$$e_1 = (\cos \tau) e_1^0 + (\sin \tau) e_2^0 , \quad e_2 = (- \sin \tau) e_1^0 + (\cos \tau) e_2^0 ; \tag{42.1}$$

and

$$\omega^1 = (\cos \tau)(\omega^0)^1 + (\sin \tau)(\omega^0)^2 , \quad \omega^2 = (- \sin \tau)(\omega^0)^1 + (\cos \tau)(\omega^0)^2 , \tag{42.2}$$

where (ω^1, ω^2) is the dual coframe of (e_1, e_2), and τ, the angle between e_1^0 and e_1, is the local fiber coordinate. Eqs. (42.2) show that ω^1 and ω^2 are globally well-defined one-forms on P. Exteriorly differentiating ω^1, we get

$$d\omega^1 = - \sin \tau d\tau \wedge (\omega^0)^1 + \cos \tau d\tau \wedge (\omega^0)^2 + \cos \tau d(\omega^0)^1 + \sin \tau d(\omega^0)^2 . \tag{42.3}$$

We can always write

$$d(\omega^0)^1 = a (\omega^0)^1 \wedge (\omega^0)^2 , \quad d(\omega^0)^2 = b (\omega^0)^1 \wedge (\omega^0)^2 , \tag{42.4}$$

where a and b only depend on base manifold coordinates. By (42.2), we then have

$$d(\omega^0)^1 = a'\,\omega^1 \wedge \omega^2 \,, \qquad d(\omega^0)^2 = b'\,\omega^1 \wedge \omega^2 \,, \tag{42.5}$$

where a' and b' now are quantities depending on both base manifold coordinates and the fiber coordinate τ, that is, functions on P. Substituting (42.5) in (42.3), and performing a similar calculation for $d\omega^2$, we obtain

$$d\omega^1 = d\tau \wedge \omega^2 + A\,\omega^1 \wedge \omega^2 \,, \quad d\omega^2 = \omega^1 \wedge d\tau + B\,\omega^1 \wedge \omega^2 \,, \tag{42.6}$$

where A and B are functions on P. Now define the following one-form on P,

$$\omega_1^2 \equiv d\tau + A\omega^1 + B\omega^2 = -\omega_2^1 \,. \tag{42.7}$$

Then,

$$d\omega^1 = \omega_1^2 \wedge \omega^2 = \omega^2 \wedge \omega_2^1 \,, \tag{42.8}$$

$$d\omega^2 = \omega^1 \wedge \omega_1^2 \,. \tag{42.9}$$

These are precisely the structure equations given by (31.33): they completely determine the 1-form ω_1^2, which, in fact, is the connection 1-form. The structure equations (42.8) and (42.9) express the fact that the connection is torsion-free [c.f. (39.123)]. Taking the exterior derivatives of both sides of (42.8) we find

$$0 = d\omega_1^2 \wedge \omega^2 - \omega_1^2 \wedge d\omega^2 \,. \tag{42.10}$$

But, from (42.6) and (42.7),

$$\omega_1^2 \wedge d\omega^2 = (d\tau + A\omega^1 + B\omega^2) \wedge (\omega^1 \wedge d\tau + B\,\omega^1 \wedge \omega^2)$$
$$= B d\tau \wedge \omega^1 \wedge \omega^2 + B\omega^2 \wedge \omega^1 \wedge d\tau = 0 \,. \tag{42.11}$$

Hence,

$$d\omega_1^2 \wedge \omega^2 = 0 \,. \tag{42.12}$$

Similarly, exterior differentiation of (42.9) yields

$$d\omega_1^2 \wedge \omega^1 = 0 \,. \tag{42.13}$$

The above two equations imply the following basic result of transgression:

$$\boxed{d\omega_1^2 = -K\,\omega^1 \wedge \omega^2} \,. \tag{42.14}$$

It may be suspected that K is in general a function on P, but it is in fact just a function on M, as can be shown easily as follows. Indeed, exterior differentiation of (42.14) gives

$$0 = -dK \wedge \omega^1 \wedge \omega^2 - K d(\omega^1 \wedge \omega^2)$$
$$= dK \wedge \omega^1 \wedge \omega^2 + K\,(d\omega^1 \wedge \omega^2 - \omega^1 \wedge d\omega^2) = dK \wedge \omega^1 \wedge \omega^2 \,, \tag{42.15}$$

where the last equality follows from the expressions of $d\omega^1$ and $d\omega^2$ given by
(42.6). Now (42.7) implies that

$$\omega^1 \wedge \omega^2 \wedge \omega_1^2 = \omega^1 \wedge \omega^2 \wedge d\tau \neq 0 . \tag{42.16}$$

Since ω^1, ω^2 and hence ω_1^2 are all globally defined, the above equation indicates
that P has a coframe field ω^1, ω^2, ω_1^2, and is thus **parallelizable**. If K did
depend on the fiber coordinate τ, dK would have a term proportional to $d\tau$
and (42.16) would then imply that $dK \wedge \omega^1 \wedge \omega^2 \neq 0$, which contradicts (42.15).
Thus K is a function on the base manifold M. We observe that $\omega^1 \wedge \omega^2$ is the
area element on M and K is the Gaussian curvature of M. [c.f. (39.166) and
the results of Exercises 39.14 to 39.17].

The transgression result (42.14) contains the essence of Chern's elegant proof
of the Gauss-Bonnet theorem. Its importance lies in the following observation.
By (42.16) and the fact that K is a function on M, $-K\omega^1 \wedge \omega^2$ is a 2-form
on M, and as such, is not exact. However, when pulled back to P, the LHS of
(42.14) indicates that it becomes the exterior derivative of a 1-form on P (the
connection 1-form ω_1^2) and is thus exact on P. By the Gauss-Bonnet Theorem
the 2-form $\dfrac{1}{2\pi} K\omega^1 \wedge \omega^2$ is in fact the Euler class.

The transgression procedure described above rests on a basic fact: P is
always parallelizable, equivalently, *a global frame field always exists on P, even
though it may not exist on the base manifold M*. This fact is easily seen as
follows. Consider an arbitrary principal G-bundle $\pi : P \to M$. Using the
projection operator π we can construct the pullback bundle $\pi^*(P)$ with base
manifold P, called the **square bundle** of P. This bundle admits the "diagonal"
global section $p \mapsto (p, p)$, and hence, by Theorem 37.2, is always trivial. In this
sense, the total space P is always simpler than the base manifold M.

Let us now apply the transgression procedure to the Chern classes by pulling
them back to the frame bundle $\pi' : P \to M$ associated with a complex vector
bundle $\pi : E \to M$. Suppose in a local coordinate neighborhood of M we have
a local expression ω for a given connection with respect to a choice of a local
frame field $\{e_i\}$ on E. Under a local change of gauge, or a local change of frame
field,

$$(e')_i = g_i^j e_j , \quad (g_i^j) \in G , \tag{42.17}$$

we have [c.f. (35.17)] the following expression for the connection matrix:

$$\varphi = (dg) g^{-1} + g \omega g^{-1} . \tag{42.18}$$

If the elements g_i^j of the matrix g are considered as local fiber coordinates, then
φ becomes the pullback of ω by $(\pi')^*$, that is,

$$\varphi = (\pi')^* (\omega) , \tag{42.19}$$

and is a well-defined connection matrix of one-forms on P. The corresponding
curvature matrix of 2-forms

$$\Phi = d\varphi - \varphi \wedge \varphi = g \Omega g^{-1} , \tag{42.20}$$

where $\Omega = d\omega - \omega \wedge \omega$, is also well-defined on P.

Consider the first Chern class [(41.76)]

$$C_1(\Phi) = \frac{i}{2\pi} Tr \, \Phi \, . \tag{42.21}$$

By (42.20), $Tr \, \Phi$ is given by

$$Tr \, \Phi = Tr \, d\varphi - Tr \, (\varphi \wedge \varphi) = Tr \, (d\varphi) = d(Tr \, \varphi) \, , \tag{42.22}$$

since

$$Tr \, (\varphi \wedge \varphi) = 0 \, , \tag{42.23}$$

by virtue of the fact that

$$Tr \, (\varphi \wedge \varphi) = \varphi_i^j \wedge \varphi_j^i = -\varphi_j^i \wedge \varphi_i^j = -Tr \, (\varphi \wedge \varphi) \, . \tag{42.24}$$

Thus,

$$\boxed{C_1(\Phi) = \frac{i}{2\pi} d(Tr \, \varphi)} \quad . \tag{42.25}$$

In other words, the first Chern class can be written explicitly as an exact form on P.

Now consider the second Chern class [(41.77)]

$$C_2(\Phi) = \frac{1}{2} \left(\frac{i}{2\pi} \right)^2 [Tr \, \Phi \wedge Tr \, \Phi - Tr \, (\Phi \wedge \Phi)] \, . \tag{42.26}$$

For the curvature form Φ on P we have the Bianchi identity [c.f. (35.36)]

$$d\Phi = \varphi \wedge \Phi - \Phi \wedge \varphi \, . \tag{42.27}$$

From (42.22),

$$Tr \, \Phi \wedge Tr \, \Phi = d(Tr \, \varphi) \wedge d(Tr \, \varphi) = d(Tr \, \varphi \wedge d(Tr \, \varphi)) \, . \tag{42.28}$$

To show that the term $Tr \, (\Phi \wedge \Phi)$ in $C_2(\Phi)$ is also exact on P, we perform the following calculations. First we have, using (42.20) for Φ and the Bianchi identity (42.27),

$$\begin{aligned}
d(Tr \, (\varphi \wedge \Phi)) &= Tr \, (d(\varphi \wedge \Phi)) = Tr \, (d\varphi \wedge \Phi - \varphi \wedge d\Phi) \\
&= Tr \, (d\varphi \wedge \Phi) - Tr \, (\varphi \wedge d\Phi) \\
&= Tr \, ((\Phi + \varphi \wedge \varphi) \wedge \Phi) - Tr \, (\varphi \wedge (\varphi \wedge \Phi - \Phi \wedge \varphi)) \\
&= Tr \, (\Phi \wedge \Phi) + Tr \, (\varphi \wedge \Phi \wedge \varphi) = Tr \, (\Phi \wedge \Phi) - Tr \, (\varphi \wedge \varphi \wedge \Phi) \, ,
\end{aligned} \tag{42.29}$$

where in the last equality we have used the fact that

$$Tr \, (\varphi \wedge \Phi \wedge \varphi) = -Tr \, (\varphi \wedge \varphi \wedge \Phi) \, , \tag{42.30}$$

which can be demonstrated as follows:

$$Tr\,(\varphi \wedge \Phi \wedge \varphi) = \varphi_i^j \wedge \Phi_j^k \wedge \varphi_k^i = \varphi_i^j \wedge \varphi_k^i \wedge \Phi_j^k$$
$$= -\varphi_k^i \wedge \varphi_i^j \wedge \Phi_j^k = -Tr\,(\varphi \wedge \varphi \wedge \Phi)\,. \tag{42.31}$$

We also have, again from (42.20) and (42.27),

$$d[Tr\,(\varphi \wedge \varphi \wedge \varphi)] = Tr\,[d(\varphi \wedge \varphi \wedge \varphi)] = Tr\,[d(\varphi \wedge (d\varphi - \Phi))]$$
$$= Tr\,[d(\varphi \wedge d\varphi) - d(\varphi \wedge \Phi)] = Tr\,[d\varphi \wedge d\varphi - (d\varphi \wedge \Phi - \varphi \wedge d\Phi)]$$
$$= Tr\,[(\Phi + \varphi \wedge \varphi) \wedge (\Phi + \varphi \wedge \varphi) - (\Phi + \varphi \wedge \varphi) \wedge \Phi + \varphi \wedge (\varphi \wedge \Phi - \Phi \wedge \varphi)]$$
$$= Tr\,[\Phi \wedge \varphi \wedge \varphi + \varphi \wedge \varphi \wedge \Phi - \varphi \wedge \Phi \wedge \varphi + \varphi \wedge \varphi \wedge \varphi \wedge \varphi]$$
$$= 3\,Tr\,(\varphi \wedge \varphi \wedge \Phi)\,. \tag{42.32}$$

In the last equality, we have used (42.30) and the facts that

$$Tr\,(\Phi \wedge \varphi \wedge \varphi) = Tr\,(\varphi \wedge \varphi \wedge \Phi)\,, \tag{42.33}$$
$$Tr\,(\varphi \wedge \varphi \wedge \varphi \wedge \varphi) = 0\,. \tag{42.34}$$

Exercise 42.1 Prove (42.33) and (42.34). Consult the derivation of (42.30) given by (42.31).

It follows from (42.29) and (42.32) that

$$Tr\,(\Phi \wedge \Phi) = d(Tr\,(\varphi \wedge \Phi)) + Tr\,(\varphi \wedge \varphi \wedge \Phi)$$
$$= d(Tr\,(\varphi \wedge \Phi)) + \frac{1}{3}d(Tr\,(\varphi \wedge \varphi \wedge \varphi)) = d\,[Tr\,(\varphi \wedge \Phi) + \frac{1}{3}Tr\,(\varphi \wedge \varphi \wedge \varphi)]\,; \tag{42.35}$$

and hence

$$\boxed{C_2(\Phi) = d\left[\frac{1}{2}\left(\frac{i}{2\pi}\right)^2 \{Tr\,\varphi \wedge d(Tr\,\varphi) - CS(\varphi)\}\right]}\,, \tag{42.36}$$

where $CS(\varphi)$, known as the **Chern-Simons 3-form**, is given by

$$\boxed{CS(\varphi) \equiv Tr\,(\varphi \wedge \Phi) + \frac{1}{3}Tr\,(\varphi \wedge \varphi \wedge \varphi)}\,. \tag{42.37}$$

Thus, as for $C_1(\Phi)$, the second Chern class $C_2(\Phi)$ is also exact on P. Using (42.20), the Chern-Simons 3-form can also be written as

$$\boxed{CS(\varphi) = Tr\,(\varphi \wedge d\varphi - \frac{2}{3}\varphi \wedge \varphi \wedge \varphi)}\,. \tag{42.38}$$

There is a very simple reason why the pullback forms $C_i(\Phi)$ on P must be exact. First, by naturality of the Chern classes,

$$C_i(\Phi) = C_i((\pi')^* \, \Omega) = (\pi')^* \, (C_i(\Omega)) \, . \tag{42.39}$$

Thus $C_i(\Phi)$ is the pullback of the Chern class $C_i(\Omega)$, where the latter is defined on the frame bundle $\pi' : P \to M$. In other words $C_i(\Phi)$ is a Chern class of the square bundle $(\pi')^* \, (P)$ (with base space P), which, as we have seen, is trivial. Since $C_i(\Phi) \in H^{2i}(P)$, and all characteristic classes of trivial bundles are trivial [by Corollary 41.1], $C_i(\Phi)$ must be exact (as a $2i$-form on P). One can then always write

$$C_i(\Phi) = (\pi')^* \, (C_i(\Omega)) = d(TC_i(\varphi)) \, , \tag{42.40}$$

where $TC_i(\varphi)$ denotes the transgression of the i-th Chern class. $TC_i(\varphi)$ are also referred to as the **Chern-Simons forms**. Note that they are not necessarily characteristic classes, since $TC_i(\varphi)$ are not necessarily closed [$d(TC_i(\varphi)) = C_i(\Phi)$].

The Chern-Simons forms can be given explicit general expressions by the following theorem.

Theorem 42.1. *Let $P_j(\Omega)$ be an ad-invariant polynomial of order j [c.f. (41.43)], where Ω is the curvature form of a $\mathcal{G}$-valued connection ω on a principal G-bundle $\pi' : P \to M$ (c.f. Defs. 37.7 and 37.11). Let $\Phi = (\pi')^* \, (\Omega)$ and set*

$$\Phi_t \equiv t\Phi + \frac{1}{2} \, (t^2 - t) \, [\varphi, \varphi]_\mathcal{G} \, , \tag{42.41}$$

where $[\, , \,]_\mathcal{G}$ is in the sense of (37.90) and $\varphi = (\pi')^* \, (\omega)$. Define the $(2j-1)$-form

$$TP_j(\varphi) \equiv j \int_0^1 P_j(\varphi, \underbrace{\Phi_t, \ldots, \Phi_t}_{(j-1) \ times})dt \, . \tag{42.42}$$

Then

$$d(TP_j(\varphi)) = (\pi')^* \, (P_j(\Omega)) = P_j(\Phi) \, . \tag{42.43}$$

Proof. Let $f(t) \equiv P_j(\Phi_t)$. Then $f(0) = 0$ and $f(1) = P_j(\Phi)$, so that

$$P_j(\Phi) = \int_0^1 \frac{dP_j(\Phi_t)}{dt} dt = \int_0^1 f'(t)dt \, .$$

We need to show that

$$f'(t) = j \, dP_j(\varphi, \overbrace{\Phi_t, \ldots, \Phi_t}^{(j-1) \ times}) \, . \tag{42.44}$$

By the definition of $f(t)$,

$$f'(t) = \frac{dP_j(\Phi_t)}{dt} = j \, P_j(\frac{d\Phi_t}{dt}, \overbrace{\Phi_t, \ldots, \Phi_t}^{(j-1) \ times}) \, . \tag{42.45}$$

From (42.41),

$$\frac{d\Phi_t}{dt} = \Phi + (t - 1/2)\,[\varphi, \varphi]_\mathcal{G} \, . \tag{42.46}$$

Thus

$$f'(t) = j\,P_j(\Phi, \Phi_t, \ldots, \Phi_t) + j\,(t - 1/2)\,P_j([\varphi, \varphi]_\mathcal{G}, \Phi_t, \ldots, \Phi_t) \, . \tag{42.47}$$

On the other hand, recalling the proof of (41.31),

$$j\,dP_j(\varphi, \Phi_t, \ldots, \Phi_t) = j\,P_j(d\varphi, \Phi_t, \ldots, \Phi_t)$$
$$- j(j-1)\,P_j(\varphi, d\Phi_t, \underbrace{\Phi_t, \ldots, \Phi_t}_{(j-2)\text{ times}}) \, . \tag{42.48}$$

Since, by the structural equation (37.116) of the connection φ,

$$\Phi = d\varphi + \frac{1}{2}\,[\varphi, \varphi]_\mathcal{G} \, , \tag{42.49}$$

(42.48) implies that

$$j\,dP_j\,(\varphi, \Phi_t, \ldots, \Phi_t) = j\,P_j(\Phi, \Phi_t, \ldots, \Phi_t)$$
$$- \frac{1}{2}j\,P_j([\varphi, \varphi]_\mathcal{G}, \Phi_t, \ldots, \Phi_t) - j(j-1)\,P_j(\varphi, d\Phi_t, \underbrace{\Phi_t, \ldots, \Phi_t}_{(j-2)\text{ times}}) \, . \tag{42.50}$$

Besides (37.116) we have the Bianchi identity [c.f. (37.125)]:

$$d\Phi = [\Phi, \varphi]_\mathcal{G} \, . \tag{42.51}$$

From these two equations

$$d\Phi_t = d\left(t\Phi + \frac{1}{2}(t^2 - t)\,[\varphi, \varphi]_\mathcal{G}\right) = t\,d\Phi + \frac{1}{2}(t^2 - t)\,d[\varphi, \varphi]_\mathcal{G}$$

$$= t\,[\Phi, \varphi]_\mathcal{G} + \frac{1}{2}(t^2 - t)\,\{[d\varphi, \varphi]_\mathcal{G} - [\varphi, d\varphi]_\mathcal{G}\}$$

$$= t\,[\Phi, \varphi]_\mathcal{G} + \frac{1}{2}(t^2 - t)\left\{[\Phi - \frac{1}{2}[\varphi, \varphi]_\mathcal{G}, \varphi]_\mathcal{G} - [\varphi, \Phi - \frac{1}{2}[\varphi, \varphi]_\mathcal{G}]_\mathcal{G}\right\}$$

$$= t\,[\Phi, \varphi]_\mathcal{G} + \frac{1}{2}(t^2 - t)\,\{[\Phi, \varphi]_\mathcal{G} - [\varphi, \Phi]_\mathcal{G}\}$$

$$= t\,[\Phi, \varphi]_\mathcal{G} + (t^2 - t)\,[\Phi, \varphi]_\mathcal{G} = t^2[\Phi, \varphi]_\mathcal{G} = t\,[\Phi_t, \varphi]_\mathcal{G} \, , \tag{42.52}$$

where in the third equality we have used (37.97) for the calculation of $d[\varphi, \varphi]_\mathcal{G}$, and in the fourth and the last equality we have used the result $[[\varphi, \varphi], \varphi]_\mathcal{G} = 0$, which in turn follows from (37.96). Using (42.52) in (42.50) we have

$$j\,dP_j\,(\varphi, \Phi_t, \ldots, \Phi_t) = j\,P_j(\Phi, \Phi_t, \ldots, \Phi_t) - \frac{1}{2}\,jP_j([\varphi, \varphi]_\mathcal{G}, \Phi_t, \ldots, \Phi_t)$$
$$- j(j-1)\,tP_j(\varphi, [\Phi_t, \varphi]_\mathcal{G}, \underbrace{\Phi_t, \ldots, \Phi_t}_{(j-2)\text{ times}}) \, . \tag{42.53}$$

We will next use the result (41.29), written in a slightly different form as

$$\sum_{i=1}^{j}(-1)^{d_1 + \cdots + d_i} \, P_j(A_1, \ldots, [A_i, \theta]_g, \ldots, A_j) = 0 \, .$$ (42.54)

With $A_1 = \theta = \varphi$, $A_2 = \cdots = A_j = \Phi_t$, it yields

$$-P_j([\varphi, \varphi]_g, \Phi_t, \ldots, \Phi_t) - (j-1)P_j(\varphi, [\Phi_t, \varphi]_g, \Phi_t, \ldots, \Phi_t) = 0 \, .$$ (42.55)

Applying this result to the last term on the RHS of (42.53) we finally have

$$j \, dP_j(\varphi, \Phi_t, \ldots, \Phi_t) = j \, P_j(\Phi, \Phi_t, \ldots, \Phi_t) - \frac{1}{2} jP_j([\varphi, \varphi]_g, \Phi_t, \ldots, \Phi_t)$$
$$+ jtP_j([\varphi, \varphi]_g, \Phi_t, \ldots, \Phi_t) = f'(t) \, ,$$ (42.56)

where the last equality follows from (42.47)   □

As an example let us consider $P_2(\Phi) = Tr \, (\Phi \wedge \Phi)$. Formula (42.42) for the transgression of $P_2(\Phi)$ gives

$$TP_2(\varphi) = 2 \int_0^1 P_2(\varphi, \Phi_t)dt = 2 \int_0^1 dt \, Tr \, (\varphi \wedge \Phi_t) \, .$$ (42.57)

Eq. (42.41) gives

$$\Phi_t = t\Phi - \frac{1}{2} (t^2 - t) \, 2\varphi \wedge \varphi$$
$$= t(d\varphi - \varphi \wedge \varphi) - (t^2 - t) \, \varphi \wedge \varphi = td\varphi - t^2 \, \varphi \wedge \varphi \, ,$$ (42.58)

where we have used (37.135). Thus

$$TP_2(\varphi) = Tr \int_0^1 dt \, (2t \, \varphi \wedge d\varphi - 2t^2 \, \varphi \wedge \varphi \wedge \varphi)$$
$$= Tr \left(\varphi \wedge d\varphi - \frac{2}{3} \varphi \wedge \varphi \wedge \varphi \right) = CS(\varphi) \, ,$$ (42.59)

which is the same result as (42.38).

We note the following properties of transgression forms without proof.

Theorem 42.2. *Let P_l be an ad(G)-invariant symmetric polynomial of degree l and Q_s be the same of degree s. Then*

$$i) \quad P_l Q_s(\overbrace{\Phi, \ldots, \Phi}^{(l+s) \ times}) = P_l(\overbrace{\Phi, \ldots, \Phi}^{l \ times}) \wedge Q_s(\overbrace{\Phi, \ldots, \Phi}^{s \ times}) \, .$$ (42.60)

$$ii) \quad T(P_l Q_s \, (\varphi)) = TP_l(\varphi) \wedge Q_s(\Phi, \ldots, \Phi) + \text{ exact form}$$
$$= TQ_s(\varphi) \wedge P_l(\Phi, \ldots, \Phi) + \text{ exact form} \, .$$ (42.61)

Theorem 42.3. *Let $\omega(t)$ be a smooth one-parameter family of connections on a principal G-bundle $\pi : P \to M$ with $t \in [0,1]$. Let $\varphi(t) = \pi^*(\omega(t))$. Set $\varphi(0) = \varphi$ and*

$$\phi' \equiv \frac{d}{dt}\left. \varphi(t) \right|_{t=0} . \qquad (42.62)$$

If P_j is an $ad(G)$-invariant symmetric polynomial of degree j, then

$$\frac{d}{dt}\left. (TP_j(\varphi(t))) \right|_{t=0} = j\, P_j(\,\varphi', \overbrace{\Phi, \ldots, \Phi}^{(j-1) \ times}\,) + \ exact\ form\,, \qquad (42.63)$$

where $\Phi = d\varphi - \varphi \wedge \varphi$ is the curvature form corresponding to φ.

Suppose $dim(M) = m$ (where M is the base manifold of $\pi' : P \to M$). If $2j > m$, then $P_j(\Omega) = 0$ and so

$$P_j(\Phi) = P_j((\pi')^*\Omega) = (\pi')^*(P_j(\Omega)) = 0 .$$

Thus by (42.43), $TP_j(\varphi)$ is closed, and so defines an element $(TP_j(\varphi)) \in H^{2j-1}(P)$, that is, a cohomology class in P. We have the following basic theorem concerning the transgression forms $TP_j(\varphi)$, which will be stated without proof.

Theorem 42.4. *Let $\pi' : P \to M$ be a principal G-bundle with connection ω and corresponding curvature Ω; $\varphi = (\pi')^*(\omega)$ and $\Phi = (\pi')^*(\Omega)$ be the pullback connection and curvature on P, respectively. Suppose $P_j(\Omega)$ is an ad-invariant polynomial of order j, and $dim(M) = m$. Then*

i) If $2j - 1 > m$, then $TP_j(\varphi)$ is closed and $[TP_j(\varphi)] \in H^{2j-1}(P,\mathbb{R})$ is independent of the choice of the connection φ.

ii) If $2j - 1 = m$, then $TP_j(\varphi)$ is closed and $[TP_j(\varphi)] \in H^m(P,\mathbb{R})$ depends on the connection φ.

For the case $2j - 1 > m$, then, the transgression form $TP_j(\varphi)$ defines a characteristic class called a **secondary characteristic class**. For example, for $dim(M) = m = 3$, the Chern-Simons form $CS(\varphi) = TP_2(\varphi)$ $(j = 2)$ is closed but depends on the choice of the connection φ.

Suppose now that $G = GL(m,\mathbb{R})$ and $\pi : P \to M$ is the (principal) frame bundle of M, with a $\mathcal{GL}(m,\mathbb{R})$-valued connection φ. This connection can be restricted to a subbundle of orthonormal frames, $\pi' : P' \to M$, with structure group $O(m)$. We then have the following theorem (stated without proof).

Theorem 42.5. *Let $\pi : P \to M$ be the frame bundle on M $(dim(M) = m)$ with connection φ. Suppose φ restricts to a connection on an $O(m)$ subbundle of P, and let the curvature form on this subbundle be Φ. Let*

$$Q_j(A_1, \ldots, A_j) = \frac{1}{j!} \sum_{\sigma \in \mathcal{S}_j} Tr\,(\,A_{\sigma(1)} A_{\sigma(2)} \cdots A_{\sigma(j)}\,) \qquad (42.64)$$

*be the **symmetrized trace** of the $m \times m$ matrices $A_1, \ldots, A_j$. Then*

$$\overbrace{(2j+1) \; times}$$

i) $Q_{2j+1}(\overbrace{\Phi, \ldots, \Phi}^{(2j+1) \; times}) = 0 ,$ (42.65)

ii) $TQ_{2j+1}(\varphi)$ *is exact .*   (42.66)

i) and ii) imply that $(TQ_{2j+1}(\varphi))$ is the trivial cohomology class on the subbundle of orthonormal frames on M.

We will now specialize to the case where M is a Riemannian manifold with a Riemannian metric g, and ω the corresponding Levi-Civita connection on TM. Consider the frame bundle $\pi : P \to M$ and let $\varphi = \pi^*(\omega)$ be the pullback of ω on P. φ restricts to a connection (still denoted by φ) on the $O(m)$-subbundle of orthonormal frames $\pi' : P' \to M$. We say that two Riemannian metrics g and $\hat{g}$ are **conformally related** if

$$\hat{g} = e^h g ,$$ (42.67)

where $h \in C^\infty(M)$ (a smooth function on M). The transgression forms of an ad-invariant, symmetric polynomial P_j then exhibit the following fundamental result, stated without proof.

Theorem 42.6. *Let g and $\hat{g}$ be conformally related Riemannian metrics on a Riemannian manifold M, and let φ, Φ, $\hat{\varphi}$, $\hat{\Phi}$ denote the corresponding connection and curvature forms on the principal $O(m)$-bundle of orthonormal frames on M, $\pi' : P' \to M$, where $dim(M) = m$. Then, for any ad-invariant, symmetric polynomial P_j of $m \times m$ matrices,*

i) $TP_j(\hat{\varphi}) - TP_j(\varphi)$ *is exact .*   (42.68)

ii) $P_j(\overbrace{\hat{\Phi}, \ldots, \hat{\Phi}}^{j \; times}) = P_j(\overbrace{\Phi, \ldots, \Phi}^{j \; times}) ,$ (42.69)

iii)   *If $P_j(\Phi, \ldots, \Phi) = 0$, then the cohomology class*

$(TP_j(\varphi)) \in H^{2j-1}(P', \mathbb{R})$ *is a **conformal invariant**.*

Note that iii) follows immediately from i), ii) and Theorem 42.1, while ii) follows immediately from i) and Theorem 42.1.

Let us consider some applications of the Chern-Simons form $CS(\varphi)$ in physics. Instead of φ we will write A, the connection symbol usually used for gauge potentials in physics. It turns out, for instance, that the Chern-Simons form $CS(A)$ is of considerable interest in a quantum field theory in 3 dimensions (Witten 1989). Consider the **Chern-Simons action** of the gauge potential A on a vector bundle $\pi : E \to M$ associated with a principal G-bundle $\pi' : P \to M$:

$$\boxed{S_{CS}(A) = \int_M Tr \left(A \wedge dA - \frac{2}{3} A \wedge A \wedge A \right)} ,$$ (42.70)

where M is a compact, oriented 3-dimensional manifold without boundary. The Chern-Simons form, and hence the corresponding field theory given by (42.70), is defined without reference to a metric in M.

The Chern-Simons action has the following interesting property

Theorem 42.7. *The Chern-Simons action S_{CS} [given by (42.70)] is "almost" gauge-invariant: it is not invariant under a general gauge transformation of A, but is so under a gauge transformation $g \in G$ connected to the identity in G.*

Proof. Let $g_t \in G$, $t \in [0,1]$ be a family of gauge transformations such that $g_0 = 1$ (the identity in G) and $g_1 = g$. In other words, g is connected to the identity. Starting with a connection A, with corresponding curvature 2-form F, the gauge-transformed connection

$$A_t \equiv g_t \, d(g_t)^{-1} + g_t \, A \, (g_t)^{-1} \tag{42.71}$$

is well-defined on the total space P. We need to prove that

$$\frac{d}{dt} S_{CS}(A_t)\Big|_{t=0} = 0 . \tag{42.72}$$

Let

$$T \equiv \frac{d}{dt} g_t \Big|_{t=0} . \tag{42.73}$$

Then, from

$$0 = \frac{d}{dt} g_t(g_t)^{-1} = \left(\frac{dg_t}{dt} \right) (g_t)^{-1} + g_t \left(\frac{d(g_t)^{-1}}{dt} \right) , \tag{42.74}$$

we have

$$\frac{d}{dt} (g_t)^{-1} = -T . \tag{42.75}$$

Thus

$$A' \equiv \frac{d}{dt} A_t \Big|_{t=0} = \frac{d}{dt} \left(g_t \, d(g_t)^{-1} + g_t \, A \, (g_t)^{-1} \right) \Big|_{t=0}$$
$$= (TA(g_t)^{-1} - g_t AT + T \, d(g_t)^{-1} - g_t \, dT)\big|_{t=0} = [T, \, A]_{\mathrm{gr}} - dT , \tag{42.76}$$

where $[T, \, A]_{\mathrm{gr}} \, (= TA - AT)$ was given by (35.65). From Theorem 42.3 we have

$$\frac{d}{dt} S_{CS}(A_t)\Big|_{t=0} = \int_M \frac{d}{dt} CS(A_t)\Big|_{t=0} = 2 \int_M Tr\, (A' \wedge F) + \int_M d\theta , \tag{42.77}$$

where $F = dA - A \wedge A$. By Stokes' Theorem

$$\int_M d\theta = \int_{\partial M} \theta = 0 ,$$

since by assumption, $\partial M = 0$. It follows from (42.76) that

$$\frac{d}{dt} S_{CS}(A_t)\Big|_{t=0}$$

$$= 2 \int_M Tr \left\{ \left([T, A]_{\mathrm{gr}} - dT \right) \wedge (dA - A \wedge A) \right\} \tag{42.78}$$

$$= 2 \int_M Tr \left\{ [T, A]_{\mathrm{gr}} \wedge dA - [T, A]_{\mathrm{gr}} \wedge A \wedge A + dT \wedge A \wedge A \right\} ,$$

where the term $dT \wedge dA$ does not contribute since, again by Stokes' theorem,

$$\int_M Tr \left(dT \wedge dA \right) = \int_M Tr \, d(T \wedge dA) = \int_M d(Tr \left(T \wedge dA \right))$$

$$= \int_{\partial M} Tr \left(T \wedge dA \right) = 0 . \tag{42.79}$$

Next we note that

$$\int_M Tr \left([T, A]_{\mathrm{gr}} \wedge A \wedge A \right) = 0 , \tag{42.80}$$

since by the **graded cyclic property** of the trace [see (42.82) below],

$$Tr \left(TA \wedge A \wedge A \right) = Tr \left(AT \wedge A \wedge A \right) . \tag{42.81}$$

Note that by definition, T is a matrix of 0-forms.

Exercise 42.2 | Prove the graded cyclic property of the trace, which states that

$$Tr \left(\theta \wedge \phi \right) = (-1)^{pq} Tr \left(\phi \wedge \theta \right) , \tag{42.82}$$

where θ and ϕ are matrices (of the same size) of p- and q-forms, respectively.

Eq. (42.78) then implies

$$\frac{d}{dt} S_{CS}(A_t)\Big|_{t=0}$$

$$= 2 \int_M Tr \left\{ [T, A]_{\mathrm{gr}} \wedge dA + dT \wedge A \wedge A \right\}$$

$$= 2 \int_M Tr \left(T \wedge A \wedge dA - A \wedge T \wedge dA + dT \wedge A \wedge A \right) \tag{42.83}$$

$$= 2 \int_M Tr \, d(T \wedge A \wedge A) = 2 \int_M d \left\{ Tr \left(T \wedge A \wedge A \right) \right\}$$

$$= 2 \int_{\partial M} Tr \left(T \wedge A \wedge A \right) = 0 ,$$

where the last two equalities follow from Stokes' Theorem.   □

The following fact is especially relevant for the formulation of a generally **covariant quantum field theory** (one without an a priori choice of a metric on M) based on the Chern-Simons action.

Theorem 42.8. *Let A and A' be connection 1-forms on a principal G-bundle $\pi : P \to M$ related by a gauge transformation $g \in G$:*

$$A' = (dg)g^{-1} + g\,A\,g^{-1} . \tag{42.84}$$

Then

$$\boxed{S_{CS}(A') - S_{CS}(A) = 8\pi^2 n , \quad n \in \mathbb{Z}} , \tag{42.85}$$

where the Chern-Simons action S_{CS} is defined in (42.70).

Proof. Let $A(t)$, $t \in [0,1]$, be a 1-parameter family of connections on $\pi : P \to M$ such that $A(0) = A$ and $A(1) = A'$. Let us consider the 4-dimensional manifold $S^1 \times M$, where M is a compact 3-dimensional manifold without boundary. Defining the projection map $p : S^1 \times M \to M$ by $(t, x) \mapsto x$, $t \in S^1$, $x \in M$, we have the induced bundle $\tilde{P} = p^*(P)$ and the corresponding induced connection $\tilde{A} = p^*(A)$, whose local expression is given by $A(t)$, if S^1 is coordinatized by $t \in [0,1]$ with the two ends of the interval identified. Let $\tilde{F}$ be the corresponding curvature form. Now, according to the fact that the Chern classes $C_j(\tilde{F})$ [as defined by (41.74)] are integral cohomology classes [c.f. (41.73)], it follows from the expression for C_2 [(41.77)] that

$$\frac{1}{8\pi^2} \int_{S^1 \times M} Tr\,(\tilde{F} \wedge \tilde{F}) = n , \quad n \in \mathbb{Z} . \tag{42.86}$$

On the other hand, by (42.35) and the Stokes Theorem

$$\int_{S^1 \times M} Tr\,(\tilde{F} \wedge \tilde{F}) = \int_{[0,1] \times M} d(CS(\tilde{A}))$$

$$= \int_{\partial([0,1] \times M)} CS(\tilde{A}) = S_{CS}(A') - S_{CS}(A) . \tag{42.87}$$

Eqs. (42.86) and (42.87) then imply the theorem. $\qquad\square$

Thus the classical Chern-Simons action is not gauge-invariant. But in a quantum field theory based on the Chern-Simons action formulated in terms of **path integrals**, we can define the **vacuum expectation value** $\langle f \rangle$ of an observable f (considered as a gauge-invariant function of the connection A) as

$$\langle f \rangle = \frac{1}{Z} \int_{A/G} f(A) \exp\left\{ \frac{ik}{4\pi} S_{CS}(A) \right\} DA , \quad k \in \mathbb{Z} , \tag{42.88}$$

$$Z \equiv \int_{A/G} \exp\left\{ \frac{ik}{4\pi} S_{CS}(A) \right\} DA , \tag{42.89}$$

in which $\exp\left\{\dfrac{ik}{4\pi} S_{CS}(A)\right\}$ is clearly gauge-invariant, by the result of Theorem 42.8. In analogy to a similar quantity in statistical mechanics, Z is called the **partition function**. The path measure DA, in the space of gauge potentials (up to gauge transformations) has, in fact, not been rigorously defined. But this has not prevented mathematical physicists from working formally with expressions like (42.88) and (42.89) to obtain extremely interesting and useful results.

Recall that we have studied one physical example of the integrality of Chern classes already, namely, the quantization of the magnetic charge [c.f. (38.32)]. That result is just a consequence of the fact that the first Chern number

$$c_1 = \int_M C_1(\Omega)$$

is an integer. The **integrality condition**

$$C_j(\Omega) \in H^{2j}(M, \mathbb{Z}) \tag{42.90}$$

in particular implies that

$$c_{\frac{m}{2}} = \int_M C_{\frac{m}{2}}(\Omega), \quad m = dim(M), \tag{42.91}$$

is a topological invariant of bundles with even-dimensional base manifolds. When $m = 4$, these are called **instanton numbers**, and are special examples of so-called **Chern numbers**, or **topological quantum numbers** in physics. Thus the magnetic monopole is described by a $U(1)$-bundle over S^2, and the quantized magnetic charge M is given by

$$M = \left(\frac{\hbar c}{2e}\right) c_1, \quad c_1 \in \mathbb{Z}. \tag{42.92}$$

As a further example, we will consider **instantons** described by $SU(2)$-bundles over S^4. For a particular connection A with associated curvature F, the instanton number k is defined in terms of the second Chern character by [c.f. (41.83)]

$$-k \equiv \int_{S^4} Ch_2(F) = -\frac{1}{8\pi^2} \int_{S^4} Tr\,(F \wedge F). \tag{42.93}$$

In fact, for instantons, $\star F = \pm F$ (where $\star$ denotes the Hodge star) [recall the discussion following (36.42)]. Thus the Yang-Mills action functional [(36.36)] can be written as

$$S_{YM} = \mp \int_{S^4} Tr\,(F \wedge F), \tag{42.94}$$

where the signs $(-)$ and $(+)$ give the self-dual and anti self-dual instantons, respectively. S^4 can be considered as the one-point compactification of R^4, and

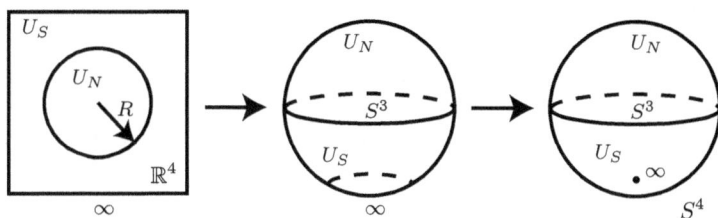

FIGURE 42.1

similar to S^2, can be covered by two coordinate patches, U_N (the "northern hemisphere") and U_S (the "southern hemisphere"), defined by

$$U_N = \{\, x \in \mathbb{R}^4 \,|\, |x| \le R + \epsilon \,\}\,, \quad U_S = \{\, x \in \mathbb{R}^4 \,|\, |x| \ge R - \epsilon \,\}\,, \qquad (42.95)$$

where $R > 0$ (see Fig. 42.1). The region of overlap (in the limit $\epsilon \to 0$) can be contracted to $S^3 \sim SU(2)$ [recall (11.71)]. Thus the connections A can be classified by homotopy classes of transition functions

$$g_{SN} : S^3 \to SU(2) \sim S^3\,,$$

or the homotopy group

$$\pi_3(S^3) = \mathbb{Z}\,. \qquad (42.96)$$

The integer characterizing the homotopy class is called the **degree** of the map g_{SN}. Without loss of generality we can set

$$A^{(S)} = 0\,, \quad x \in U_S\,. \qquad (42.97)$$

Thus

$$A^{(N)} = (dg_{SN})\,(g_{SN})^{-1}\,, \quad x \in U_N\,. \qquad (42.98)$$

We will denote a transition function of degree n by $g_{SN}^{(n)}$. Analogous to (38.14) for the magnetic monopole, we have

$$g_{SN}^{(1)} : S^3 \to \begin{pmatrix} 1 & 0 \\ 0 & 1 \end{pmatrix} \in SU(2) \quad \text{(the constant map)}\,, \qquad (42.99)$$

$$g_{SN}^{(1)}(x) = (\,x^4 + i\,x^j \sigma_j\,)\,, \quad j = 1, 2, 3, \quad \text{(the identity map)}\,, \qquad (42.100)$$

where

$$|x|^2 = (x^1)^2 + (x^2)^2 + (x^3)^2 + (x^4)^2 = 1\,,$$

and σ_i, $i = 1, 2, 3$ are the Pauli matrices given by (11.16), and

$$g_{SN}^{(n)} = (x^4 + ix^j\sigma_j)^n .$$
(42.101)

In the evaluation of the instanton numbers [(42.93)] we can use Stokes' Theorem as follows.

$$\int_{S^4} Tr\,(F \wedge F) = \int_{U_N} Tr\,(F \wedge F) + \int_{U_S} Tr\,(F \wedge F)$$

$$= \int_{U_N} Tr\,(F \wedge F) = \int_{U_N} d\{CS(A)\} = \int_{\partial U_N} CS(A) = \int_{S^3} CS(A) ,$$
(42.102)

where the local Chern-Simons form $CS(A)$ is given by [c.f. (42.37)]

$$\boxed{CS(A) = Tr\,(A \wedge F) + \frac{1}{3} Tr\,(A \wedge A \wedge A)}$$
(42.103)

Note that in the above equation, as distinct from (42.37), the connection A and curvature F are local forms on S^4 (the base manifold), and $Tr\,(F \wedge F)$, regarded as a 4-form on S^4, is only locally exact. Since in U_S, $A^{(S)} = F^{(S)} = 0$, which implies $F^{(N)} = g_{SN} F^{(S)} g_{SN}^{-1} = 0$, Eqs. (42.102) and (42.103) imply that

$$k = \frac{1}{8\pi^2} \int_{S^4} Tr\,(F \wedge F) = \frac{1}{24\pi^2} \int_{S^3} Tr\,(A^{(N)} \wedge A^{(N)} \wedge A^{(N)})$$

$$= \frac{1}{24\pi^2} \int_{S^3} Tr\,(dg\,g^{-1} \wedge dg\,g^{-1} \wedge dg\,g^{-1}) ,$$
(42.104)

where we have written g for g_{SN}. We will show that this integral yields precisely the degree of the map $g : S^3 \to SU(2)$ (where g is understood to be $g_{SN}^{(n)}$). Consider the 3-form

$$\alpha = \frac{1}{24\pi^2} Tr\,(dg\,g^{-1} \wedge dg\,g^{-1} \wedge dg\,g^{-1})$$
(42.105)

on S^3, where both g and dg are expressed in terms of the local coordinates of S^3. It is equal to the pullback of some 3-form β on $SU(2)$:

$$\alpha = g^* \beta .$$
(42.106)

Observe that α is closed on S^3, since $d\alpha$ is a 4-form and S^3 is 3-dimensional. By the integrality of Chern classes, α thus determines a cohomology class $[\alpha] \in H^3(S^3, \mathbb{Z}) = \mathbb{Z}$. Consider a map $h : S^3 \to SU(2)$ homotopic to g ($h \sim g$). Then we have, by Theorem 38.1

$$[g^* \beta] = [h^* \beta] .$$
(42.107)

This implies that

$$g^* \beta = h^* \beta + d\phi ,$$
(42.108)

where ϕ is some 2-form on S^3. By Stokes' Theorem,

$$\int_{S^3} \alpha = \int_{S^3} g^* \beta = \int_{S^3} h^* \beta \,. \tag{42.109}$$

Thus the instanton number only depends on the homotopy class of the map $g : S^3 \to SU(2)$ (which is given by an element of $\pi_3(S^3)$), or the degree of the map g. Now suppose $g = g_{SN}^{(1)}$ (degree 1) and $h = g_{SN}^{(n)}$ (degree n). Let $\alpha^{(1)} = g^* \beta = (g_{SN}^{(1)})^* \beta$ and $\alpha^{(n)} = h^* \beta = (g_{SN}^{(n)})^* \beta$. Since $H^3(S^3, \mathbb{Z}) = \mathbb{Z}$, we see that $[\alpha^{(1)}] = 1$ and $[\alpha^{(n)}] = n$ (as cohomology classes). Thus

$$[\alpha^{(n)} - n\alpha^{(1)}] = [\alpha^{(n)}] - n[\alpha^{(1)}] = 0 \in H^3(S^3, \mathbb{Z}) \,. \tag{42.110}$$

In other words, $\alpha^{(n)} - n\alpha^{(1)}$ is the trivial class, and so must be exact. We then have

$$\alpha^{(n)} - n\alpha^{(1)} = d\eta \,, \tag{42.111}$$

for some 2-form η on S^3. Integration over S^3 gives

$$\int_{S^3} \alpha^{(n)} = n \int_{S^3} \alpha^{(1)} \,. \tag{42.112}$$

We will calculate the integral on the RHS,

$$\int_{S^3} \alpha^{(1)} = \frac{1}{24\pi^2} \int_{S^3} Tr\left(dg_{SN}^{(1)} (g_{SN}^{(1)})^{-1} \wedge dg_{SN}^{(1)} (g_{SN}^{(1)})^{-1} \wedge dg_{SN}^{(1)} (g_{SN}^{(1)})^{-1} \right) , \tag{42.113}$$

explicitly. Using (42.101) we have

$$(g_{SN}^{(1)})^{-1} = x^4 - ix^j \sigma_j \,. \tag{42.114}$$

Exercise 42.3 Write $g_{SN}^{(1)}$ as a 2×2 matrix $[\in SU(2)]$. Then use the explicit expressions for the Pauli matrices σ_j [given by (11.6)] to verify (42.114).

So

$$(dg_{SN}^{(1)})(g_{SN}^{(1)})^{-1} = (dx^4 + i(dx^j)\sigma_j)(x^4 - ix^j \sigma_j) \,. \tag{42.115}$$

The value of the integral (42.113) is not changed if we push the "equator" ($\sim S^3$) in Fig. 42.1 up the "northern hemisphere" towards the north pole (see Fig. 42.2). In each of these retracted boundaries, $dx^4 = 0$.   In the limit of vanishing radius of the S^3 boundary of the northern coordinate patch, this boundary approaches the north pole, at which point $x^4 = 1$, $x^1 = x^2 = x^3 = 0$. We then have, at the north pole,

$$(dg_{SN}^{(1)})(g_{SN}^{(1)})^{-1} = i\sigma_j \, dx^j \,, \tag{42.116}$$

and the integrand of (42.113) becomes

$$Tr\left(A^{(N)} \wedge A^{(N)} \wedge A^{(N)} \right) = i^3 \, Tr\left(\sigma_j \sigma_k \sigma_l \right) dx^j \wedge dx^k \wedge dx^l \,. \tag{42.117}$$

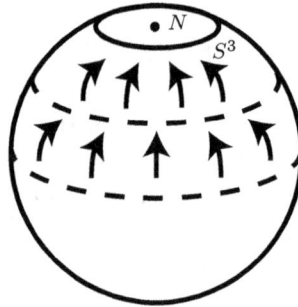

FIGURE 42.2

Using the properties of the σ-matrices (c.f. Exercise 11.5), we have

$$
\begin{aligned}
Tr\left(\sigma_j\sigma_k\sigma_l\right) &= Tr\left(\sigma_j(\delta_{kl} + i\varepsilon_{kl}{}^m\sigma_m)\right) \\
&= i\varepsilon_{kl}{}^m Tr\left(\sigma_j\sigma_m\right) = i\varepsilon_{kl}{}^m Tr\left(\delta_{jm} + i\varepsilon_{jm}{}^n\sigma_n\right) \\
&= i\varepsilon_{kl}{}^m \delta_{jm} Tr\left(e\right) = 2i\varepsilon_{klj} \ .
\end{aligned} \tag{42.118}
$$

Hence

$$
Tr\left(A^{(N)} \wedge A^{(N)} \wedge A^{(N)}\right) = 2\varepsilon_{klj}\, dx^k \wedge dx^l \wedge dx^j = 12\, dx^1 \wedge dx^2 \wedge dx^3 \ , \tag{42.119}
$$

where $dx^1 \wedge dx^2 \wedge dx^3$ is the volume element of S^3. Since the integral in (42.113) cannot depend on the radius of the boundary S^3 [compare the present situation with that of the magnetic monopole given by (38.8)], we can use the volume of the unit S^3, that is,

$$
\int_{S^3} dx^1 \wedge dx^2 \wedge dx^3 = 2\pi^2 \ , \tag{42.120}
$$

and get, finally,

$$
\int_{S^3} \alpha^{(1)} = \frac{1}{24\pi^2} \int_{S^3} Tr\left(A^{(N)} \wedge A^{(N)} \wedge A^{(N)}\right) = \frac{12(2\pi^2)}{24\pi^2} = 1 \ . \tag{42.121}
$$

This result, together with (42.112), show that the instanton number $k^{(n)}$, corresponding to a gauge potential $A^{(n)}$ characterized by $n \in \pi_3(S^3)$ [the degree of the map $g^{(n)} : S^3 \to SU(2)$], is just given by

$$
k^{(n)} = n \ . \tag{42.122}
$$

In conclusion we mention that a quantum field theory with a Chern-Simons action term, in addition to couplings between fermion fields and a gauge potential describing a magnetic flux tube, has been found to be very useful in describing the physics of the **fractional quantum Hall effect**, and the associated phenomenon of **fractional statistics** (see, for example, E. Fradkin

1991). The geometrical setup, analogous to the case of the integral quantum Hall effect, is a $U(1)$-bundle whose base space is a compact $(2+1)$-dimensional space M (2 spatial and 1 time) without boundary. Since the gauge group $U(1)$ is abelian, the Chern-Simons action is

$$S_{CS} = \theta \int_M Tr\,(A \wedge dA)\,, \qquad (42.123)$$

where θ is a "strength" constant. On another front, the study of a topological quantum field theory based on the Chern-Simons action has revealed deep connections between quantum field theory on the one hand, and the topology of three-dimensional spaces on the other, through certain invariants of knot theory (E. Witten 1989, S. Hu, 2001). In short, the study of topological invariants of fiber bundles through the Chern classes and their transgression forms via Chern-Simons theory has brought vast areas of physics and mathematics together in a stunning fashion.

Chapter 43

The Atiyah-Singer Index Theorem

The Atiyah-Singer Index Theorem is a deep and fundamental result relating the local (differential) and global (topological) properties of fiber bundles. It generalizes many important classical results of this nature, such as the Gauss-Bonnet Theorem and the Riemann-Roch-Hirzebruch Theorem. In physics it has found important applications in quantum field theory (especially in the study of anomalies), and in fact, has been shown to be provable by quantum field-theoretic (path integral) methods (for example, see E. Witten 2000). In this chapter, we will present the introductory background to this theorem, state it without proof, and then discuss its application to some special (useful) cases.

We will begin with some aspects of de Rham cohomology theory. Consider the sequence of vector bundles of exterior p-forms on M, $\Lambda^p(M)$, $p = 0, 1, \ldots, m = dim(M)$. Anticipating later development we denote d (the exterior derivative) acting on $\Lambda^p(M)$ by $d^{(p)}$. Since $d^2 = 0$, the sequence of operators $d^{(p)}$ satisfies

$$Im\, d^{(p-1)} \subset ker\, d^{(p)}\,, \quad p = 1, \ldots, m\,. \tag{43.1}$$

The sequence $\{\Lambda^p, d^{(p)}\}$ is then called a **de Rham complex** (see Fig. 43.1). Define a (global) scalar product in $\Lambda^p(M)$ by

$$(\alpha, \beta) \equiv \int_M \alpha \wedge \star \beta\,, \tag{43.2}$$

where $\star\beta$ is the Hodge-star of the p-form β. The adjoint operator of d, denoted δ, is defined by

$$(d\alpha, \beta) = (\alpha, \delta\beta)\,. \tag{43.3}$$

We will show that

$$\boxed{\delta = (-1)^p\, (\star)^{-1}\, d \star}\,, \tag{43.4}$$

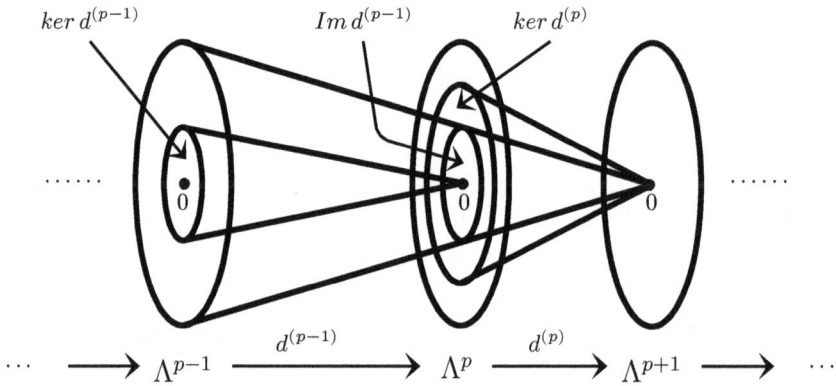

FIGURE 43.1

if M is without boundary, where $(\star)^{-1}$, the inverse operator of $\star$, is given by

$$(\star)^{-1} = (-1)^{p(m-p)} \, sgn(g) \star , \tag{43.5}$$

in which $sgn(g)$ is the sign of $\det(g_{ij})$, with g_{ij} being the metric associated with $\star$. So δ can also be written as

$$\delta = (-1)^p (-1)^{p(m-p)} \, sgn(g) \, \star d \star . \tag{43.6}$$

Indeed, for $\alpha \in \Lambda^{p-1}(M)$, $\beta \in \Lambda^p(M)$,

$$\begin{aligned}
(d\alpha, \beta) &= \int_M d\alpha \wedge \star\beta = \int_M d(\alpha \wedge \star\beta) + (-1)^p \int_M \alpha \wedge d(\star\beta) \\
&= \int_{\partial M} \alpha \wedge \star\beta + (-1)^p \int_M \alpha \wedge (\star)(\star)^{-1} d(\star\beta) = \int_M \alpha \wedge \star\delta\beta ,
\end{aligned} \tag{43.7}$$

where we have used Stokes' Theorem in the third equality (together with the assumption that $\partial M = 0$), and (43.4) in the last. Similar to the property $d^2 = 0$, we have the analogous property for δ:

$$\delta^2 = 0 . \tag{43.8}$$

Indeed, for $\alpha \in \Lambda^p(M)$,

$$\begin{aligned}
\delta^2 \alpha &= \delta(\delta\alpha) = \delta((-1)^p (\star)^{-1} d \star \alpha) \\
&= (-1)^{p-1}(-1)^p (\star)^{-1} d \star (\star)^{-1} d \star \alpha = -(\star)^{-1} d^2 \star \alpha = 0 .
\end{aligned} \tag{43.9}$$

Note that δ sends a p-form to a $(p-1)$-form. Similar to the notation

$$d^{(p)} : \Lambda^p(M) \to \Lambda^{p+1}(M) ,$$

we write

$$\delta^{(p)} : \Lambda^p(M) \to \Lambda^{p-1}(M) \, .$$

We have the so-called **Laplacian operators**

$$\Delta_p : \Lambda^p(M) \to \Lambda^p(M)$$

of the de Rham complex:

$$\boxed{\Delta_p \equiv \delta^{(p+1)} d^{(p)} + d^{(p-1)} \delta^{(p)} = (d^{(p)} + \delta^{(p)})^2} \qquad . \qquad (43.10)$$

A p-form ω such that $\Delta_p \omega = 0$ is called a **harmonic form**. The space of all harmonic p-forms $\in \Lambda^p(M)$ is denoted by $Harm^p(M)$.

A p-form α such that $\delta\alpha = 0$ is said to be **coclosed**; if $\alpha = \delta\beta$ for some $\beta \in \Lambda^{p+1}(M)$, then α is said to be **coexact**. We have the following result.

Theorem 43.1. *A p-form on a compact Riemannian manifold is harmonic if and only if it is closed and coclosed.*

Proof. Let $\omega \in \Lambda^p(M)$. Then

$$
\begin{aligned}
(\omega, \Delta_p \omega) &= (\omega, (\delta d + d\delta)\omega) = (\omega, \delta d\omega) + (\omega, d\delta\omega) \\
&= (d\omega, d\omega) + (\delta\omega, \delta\omega) \geq 0 \, .
\end{aligned}
\qquad (43.11)
$$

Since the inner product is positive definite if M is Riemannian, the theorem follows. □

The following fundamental result can now be stated (without proof).

Theorem 43.2 (Hodge Decomposition Theorem). *Any p-form $\omega \in \Lambda^p(M)$, M being a compact orientable Riemannian manifold without boundary, can always be written as a unique sum*

$$\omega = d\alpha + \delta\beta + \omega_h \, , \qquad (43.12)$$

where $\alpha \in \Lambda^{p-1}(M)$, $\beta \in \Lambda^{p+1}(M)$, $\omega_h \in Harm^p(M)$ is a harmonic form, and the three p-forms on the RHS of (43.12) are mutually orthogonal to each other under the scalar product (,) of (43.2).

Note that the fact that $d\alpha, \delta\beta$ and ω_h are mutually orthogonal to each other can be easily seen. Thus

$$(d\alpha, \delta\beta) = (\alpha, \delta^2\beta) = (d^2\alpha, \beta) = 0 \, , \qquad (43.13)$$

since $d^2 = \delta^2 = 0$. Also, by Theorem 43.1,

$$(d\alpha, \omega_h) = (\alpha, \delta\omega_h) = 0 \, , \qquad (43.14)$$
$$(\delta\beta, \omega_h) = (\beta, d\omega_h) = 0 \, . \qquad (43.15)$$

Now suppose $d\omega = 0$, so that $\omega \in \Lambda^p(M)$ determines a cohomology class $[\omega] \in H^p(M)$. We then have

$$d\omega = 0 = d^2\alpha + d\delta\beta + d\omega_h = d\delta\beta ,\qquad (43.16)$$

where β is the same quantity on the RHS of (43.12). This implies

$$0 = (d\omega, \beta) = (d\delta\beta, \beta) = (\delta\beta, \delta\beta) .\qquad (43.17)$$

Thus $\delta\beta = 0$, and

$$\omega = d\alpha + \omega_h .\qquad (43.18)$$

Since $[d\alpha] = 0 \in H^p(M)$, the above equation immediately implies that

$$[\omega] = [\omega_h] .\qquad (43.19)$$

In other words, if ω is closed ($d\omega = 0$), its harmonic component determines its cohomology class. Let $\tilde{\omega}$ be another closed p-form so that $[\tilde{\omega}] = [\omega]$, or $\tilde{\omega} = \omega + d\eta$, for some $\eta \in \Lambda^{p-1}(M)$. By Theorem 43.2 and the above argument, we can write

$$\tilde{\omega} = d\tilde{\alpha} + \tilde{\omega}_h .\qquad (43.20)$$

Then $\omega + d\eta = d\tilde{\alpha} + \tilde{\omega}_h$, or

$$\omega = d(\tilde{\alpha} - \eta) + \tilde{\omega}_h .\qquad (43.21)$$

Since the decomposition of ω is unique, it follows that $\tilde{\omega}_h = \omega_h$. In other words, for each $[\omega] \in H^p(M)$, there is a unique harmonic representative ω_h. This leads to the following theorem.

Theorem 43.3 (Hodge). *The space of harmonic p-forms, $Harm^p(M)$, on a compact manifold M is isomorphic to the cohomology space $H^p(M)$:*

$$\boxed{H^p(M) \sim Harm^p(M)}\qquad .\qquad (43.22)$$

Recalling the notion of the Betti number [Eq. (33.26) in Corollary 33.1] and the Euler-Poincare Theorem (Theorem 33.1), we have

$$b_p = dim(H^p(M)) = dim(Harm^p(M)) ,\qquad (43.23)$$

$$\chi(M) = \sum_{p=0}^{m} (-1)^p b_p(M) = \sum_{p=0}^{m} (-1)^p dim(Harm^p(M)) ,\qquad (43.24)$$

where b_p is the p-th Betti number of M and $\chi(M)$ is the Euler characteristic of M. By definition,

$$Harm^p(M) = ker\,\Delta_p ,$$

and one can define the so-called **analytic index** of the de Rham complex $\{\, d^{(p)}, \Lambda^p(M)\,\}$ by

$$index\,\{\, d^{(p)}, \Lambda^p(M)\,\} = \sum_{p} (-1)^p\, dim\,ker\,\Delta_p .\qquad (43.25)$$

Eq. (43.24) then reads

$$\chi(M) = index \{ d^{(p)}, \Lambda^p(M) \} . \qquad (43.26)$$

This equation is remarkable in that the LHS concerns only topological information on M, while the RHS is obtained from purely analytical information. It is
a relation of this type that the Atiyah-Singer index theorem seeks to generalize,
from the de Rham complex to an arbitrary so-called elliptic complex.

Definition 43.1. *Let $\pi : E \to M$ and $\pi' : F \to M$ be vector bundles of (fiber)
dimensions l and h, respectively, and let $dim(M) = m$. Suppose U is a local
coordinate neighborhood of M with local coordinates $(x^1, \ldots, x^m)$. Choose a
local frame field $\{e_1, \ldots, e_l\}$ of E in U. An arbitrary section $s \in \Gamma(E)$ can then
be expressed as $s = s^i(x)e_i$. Similarly, with the choice of a local frame field
$\{\epsilon_1, \ldots, \epsilon_h\}$ of F in U, an arbitrary section $s' \in \Gamma(F)$ can be written locally as
$s' = (s')^j(x)\epsilon_j$. A **linear differential operator** $D : \Gamma(E) \to \Gamma(F)$ of order n
is then a linear map which can be expressed locally as*

$$(Ds)^i(x) = \sum_{|k|=0}^{n} a^i_{j\,(k)}(x) \left(\frac{\partial}{\partial x^1} \right)^{k_1} \cdots \left(\frac{\partial}{\partial x^m} \right)^{k_m} s^j(x) , \quad i = 1, \ldots, h .$$

$$(43.27)$$

*In this equation the summation over j is from 1 to l. The symbol $(k) =
(k_1, \ldots, k_m)$ stands for an m-tuple of non-negative integers, and $|k| = k_1 +
\cdots + k_m$. For each (k), the matrix $(a^i_{j\,(k)}(x))$ is $l \times h$, and its elements are
smooth functions of $(x^1, \ldots, x^m)$.*

Definition 43.2. *Let $\xi \in T^*_x(M)$ with components $(\xi_1, \ldots, \xi_m)$ with respect
to the natural basis $\{dx^1, \ldots, dx^m\}$ of $T^*_x(M)$. The **principal symbol** of a
differential operator D at $x \in M$ (corresponding to $\xi \in T^*_x(M)$) is the linear
map*

$$\sigma_x(\xi, D) : E_x \to F_x$$

between fibers at $x \in M$ given by

$$(\sigma_x(\xi, D) s)^i(x) = \sum_{|k|=n} a^i_{j\,(k)}(x) (\xi_1)^{k_1} \cdots (\xi_m)^{k_m} s^j(x) . \qquad (43.28)$$

*The $l \times h$ matrix $\sum_{|k|=n} a^i_{j\,(k)}(\xi_1)^{k_1} \cdots (\xi_m)^{k_m}$ is also referred to as the principal
symbol of D.*

Example 1. Let $\pi : E \to \mathbb{R}^m$ and $\pi' : F \to \mathbb{R}^m$ be real line bundles over $\mathbb{R}^m$.
The 2-nd order differential operator $\Delta : \Gamma(E) \to \Gamma(F)$ defined by

$$\Delta = \frac{\partial^2}{\partial(x^1)^2} + \cdots + \frac{\partial^2}{\partial(x^m)^2} \qquad (43.29)$$

is called the **Laplacian**. In this case $l = h = 1$, and the matrix $a^i_{j\,(k)}$ of (43.27)
has a single element $a^1_{1\,(k)}$ for each (k). The only non-vanishing $a^1_{1\,(k)}$ are for

$$(k) = (2, 0, \ldots, 0), (0, 2, 0, \ldots, 0), \ldots, (0, \ldots, 0, 2) ,$$

each of value 1. Thus the principal symbol of Δ is

$$\sigma_x(\xi, \Delta) = (\xi_1)^2 + \cdots + (\xi_m)^2 .$$

The **d'Alembertian** $\Box : \Gamma(E) \to \Gamma(F)$ is defined by

$$\Box = \frac{\partial^2}{\partial(x^1)^2} + \cdots + \frac{\partial^2}{\partial(x^{m-1})^2} - \frac{\partial^2}{\partial(x^m)^2} . \tag{43.30}$$

Thus

$$\sigma_x(\xi, \Box) = (\xi_1)^2 + \cdots + (\xi_{m-1})^2 - (\xi_m)^2 . \tag{43.31}$$

Example 2. Consider the four-dimensional spin bundle $\pi : E \to M$, where M is four-dimensional Minkowski spacetime. Recall the first-order Dirac operator [c.f. (27.6)] $D : \Gamma(E) \to \Gamma(E)$ given by

$$D = \sum_\mu \gamma_\mu \partial_\mu - im , \tag{43.32}$$

where γ_μ are the 4×4 Dirac matrices given by (27.11). In terms of spinor components $\psi^\alpha(x)$ $(\alpha = 1, 2, 3, 4)$, D is given by

$$(D\psi)^\alpha(x) = \sum_\mu \left[(\gamma_\mu)^\alpha_\beta \partial_\mu - im\delta^\alpha_\beta \right] \psi^\beta(x) . \tag{43.33}$$

The only non-vanishing $a^\alpha_{\beta\,(k)}$ are

$$a^\alpha_{\beta\,(0,0,0,0)} = -im\delta^\alpha_\beta , \qquad a^\alpha_{\beta\,(1,0,0,0)} = (\gamma_0)^\alpha_\beta , \qquad a^\alpha_{\beta\,(0,1,0,0)} = (\gamma_1)^\alpha_\beta ,$$
$$a^\alpha_{\beta\,(0,0,1,0)} = (\gamma_2)^\alpha_\beta , \qquad a^\alpha_{\beta\,(0,0,0,1)} = (\gamma_3)^\alpha_\beta .$$

The principal symbol is the 4×4 matrix

$$\sigma_x(\xi, D) = \xi_1\gamma_0 + \xi_2\gamma_1 + \xi_3\gamma_2 + \xi_4\gamma_3 . \tag{43.34}$$

Definition 43.3. *The differential operator* $D : \Gamma(E) \to \Gamma(F)$ *is said to be* **elliptic** *if for each* $x \in M$ *(the base manifold of both E and F) and* $\xi \neq 0\,(\in T^*_x M)$ *the principal symbol* $\sigma_x(\xi, D) : E_x \to F_x$ *is an isomorphism. An equivalent condition for ellipticity of D is that the principal symbol matrix* $\sum_{|k|=n} a^i_{j\,(k)}(\xi_1)^{k_1} \cdots (\xi_m)^{k_m}$ *for* $\xi \neq 0$ *must be invertible. Note that for D to be elliptic, the fiber dimensions of E and F must be equal.*

Exercise 43.1   Let $M = \mathbb{R}^2$ and consider a 2-nd order differential operator D whose principal symbol is

$$\sigma_x(\xi, D) = a^1_{1\,(2,0)}(\xi_1)^2 + 2a^1_{1\,(1,1)}\xi_1\xi_2 + a^1_{1\,(0,2)}(\xi_2)^2 . \tag{43.35}$$

Show that D is elliptic if and only if $\sigma_x(\xi, D) = 1$ is an ellipse in the ξ_1-ξ_2 plane.

Exercise 43.2 Show that the laplacian Δ of Example 1 above is elliptic but the d'Alembertian $\square$ is not. Show that the Dirac operator of (43.33) is elliptic.

We can now generalize the de Rham complex to the more general notion of the elliptic complex.

Definition 43.4. *Consider a finite sequence of vector bundles $\pi_p : E_p \to M$ over M, and a corresponding sequence of differential operators $D_p : \Gamma(E_p) \to \Gamma(E_{p+1})$. The sequence $\{E_p, D_p\}$ is called a* **complex** *if*

$$Im\, D_{p-1} \subset ker\, D_p \ . \tag{43.36}$$

[c.f. Eq. (43.1) and Fig. 43.1 for the de Rham complex.]

Definition 43.5. *The sequence $\{E_p, D_p\}$ is called an* **elliptic complex** *if $\{E_p, D_p\}$ is a complex and, in addition, for any $x \in M$ and $\xi\,(\neq 0) \in T_x^*M$,*

$$ker\, \sigma_x(\xi, D_{p+1}) = im\, \sigma_x(\xi, D_p) \ . \tag{43.37}$$

If the latter condition is fulfilled, the sequence of linear maps

$$\sigma_x(\xi, D_p) : E_{p,\,x} \to E_{p+1,\,x}$$

is said to be an **exact sequence**.

Suppose the bundles $\pi_p : E_p \to M$ are all equipped with fiber scalar products, denoted in each case by $(\ ,\ )_x\,(x \in M)$ (for example, fiber metrics induced by a Riemannian metric on M). We define a scalar product $(\ ,\ )$ in each $\Gamma(E_p)$ by

$$(u, v) = \int_M (u_x, v_x)\, \sigma \ , \tag{43.38}$$

where σ is the volume element in M and $u, v \in \Gamma(E_p)$. For each $D_p : \Gamma(E_p) \to \Gamma(E_{p+1})$, the adjoint operator $D_p^\dagger : \Gamma(E_{p+1}) \to \Gamma(E_p)$ is defined by

$$\int_M ((D_p u)_x,\, v_x)\sigma = \int_M (u_x,\, (D_p^\dagger v)_x)\sigma \ , \tag{43.39}$$

or

$$(D_p u,\, v) = (u,\, D_p^\dagger v) \ , \tag{43.40}$$

where $u \in \Gamma(E_p)$ and $v \in \Gamma(E_{p+1})$ [c.f. (43.3)].

Exercise 43.3 Show that if $\{E_p, D_p\}$ is an elliptic complex, the sequence $\{E_{p+1}, D_p^\dagger\}$ in the reverse order is also an elliptic complex.

Definition 43.6. *The differential operators* $\Delta_p : E_p \to E_p$ *defined by*

$$\boxed{\Delta_p \equiv D_p^\dagger D_p + D_{p-1} D_{p-1}^\dagger} \qquad (43.41)$$

are called the **Laplacians** *of the complex* $\{E_p, D_p\}$.

We will state without proof the following basic fact (for a proof, see p.398 of Y. Choquet-Bruhat and C. DeWitte-Morette, with M. Dillard-Bleick, 1982).

Theorem 43.4. *The complex* $\{E_p, D_p\}$ *is an elliptic complex if and only if the laplacians* Δ_p *are all elliptic operators.*

We have the following generalization (stated without proof) of the Hodge Decomposition Theorem (Theorem 43.2).

Theorem 43.5. *If* $\{E_p, D_p\}$ *is an elliptic complex over a compact manifold M without boundary, then each* $u_p \in \Gamma(E_p)$ *can be decomposed uniquely as*

$$u_p = D_{p-1} v_{p-1} + D_p^\dagger s_{p+1} + h_p, \qquad (43.42)$$

where $v_{p-1} \in \Gamma(E_{p-1})$, $s_{p+1} \in \Gamma(E_{p+1})$, *and* $\Delta_p h_p = 0$. *The three sections* $[\in \Gamma(E_p)]$ *on the RHS of (43.42) are mutually orthogonal to each other under the scalar product* $(\,,\,)$ *of (43.38).*

Because of (43.36) we can define the cohomology spaces $H^p(E_p, D_p)$ for the complex $\{E_p, D_p\}$ by

$$H^p(E_p, D_p) \equiv \frac{ker\, D_p}{Im\, D_{p-1}}. \qquad (43.43)$$

We also denote the space of harmonic sections $ker\, \Delta_p$ by $Harm^p(E_p, D_p)$. Using similar arguments leading to Hodge's Theorem (Theorem 43.3) we have the following generalization.

Theorem 43.6. *For a complex* $\{E_p, D_p\}$ *we have the following isomorphisms:*

$$\boxed{H^p(E_p, D_p) \sim Harm^p(E_p, D_p)} \quad . \qquad (43.44)$$

Generalization of (43.25) leads to the following definition.

Definition 43.7. *The* **analytical index of an elliptic complex** $\{E_p, D_p\}$ *is given by*

$$\boxed{ind_{an}\{E_p, D_p\} \equiv \sum_p (-1)^p \dim ker\, \Delta_p} \quad . \qquad (43.45)$$

The analytical index of an elliptic complex can be defined in an alternate way. Consider a single elliptic operator $D : \Gamma(E) \to \Gamma(F)$. In addition to the space $ker\, D$ we have the **cokernel** of D defined by

$$coker\, D \equiv \frac{\Gamma(F)}{Im\, D}. \qquad (43.46)$$

Theorem 43.7. *The kernel of $D^\dagger$ is isomorphic to the cokernel of D:*

$$coker\, D \sim ker\, D^\dagger \, . \tag{43.47}$$

Proof. Consider the map $f : ker\, D^\dagger \to coker\, D$ given by $u \mapsto [u]$, that is, f sends $u \in ker\, D^\dagger$ to the element it represents in *coker D*. We will show that this map is a bijection. Given $[s] \in coker\, D$, let

$$s_0 \equiv s - D(D^\dagger D)^{-1} D^\dagger s \, . \tag{43.48}$$

The operator $(D^\dagger D)^{-1} = D^{-1}(D^\dagger)^{-1}$ exists by virtue of the fact that D is elliptic (c.f. Def. 43.3). Clearly $D^\dagger s_0 = 0$ and $[s_0] = [s]$. This establishes surjectiveness. To show that f is injective we will show that if $s_0, (s_0)' \in ker\, D^\dagger$ and $s_0 \neq (s_0)'$, then $[s_0] \neq [(s_0)']$. Assume the contrary, that is, $[s_0] = [(s_0)']$. Then there exists a $u \in \Gamma(E)$ such that $(s_0)' - s = Du$. We then have

$$0 = (u,\, D^\dagger((s_0)' - s)) = (Du,\, (s_0)' - s) = (Du,\, Du) \, , \tag{43.49}$$

which implies $Du = 0$. Consequently $s_0 = (s_0)'$. This establishes the injectiveness of f. $\qquad\square$

Definition 43.8. *A **Fredholm operator** is an elliptic operator D for which both ker D and coker D are finite dimensional.*

Definition 43.9. *The analytical index of an elliptic operator D is given by*

$$\boxed{index\, D \equiv dim\, ker\, D - dim\, coker\, D = dim\, ker\, D - dim\, ker\, D^\dagger} \, . \tag{43.50}$$

The analytical index of a single differential operator D is in fact the same as that of the following complex

$$0 \overset{i}{\longrightarrow} \Gamma(E) \overset{D}{\longrightarrow} \Gamma(F) \overset{\varphi}{\longrightarrow} 0 \, , \tag{43.51}$$

[as given by (43.45)], where i is the inclusion map and φ is the constant map to 0. In fact, according to (43.45), with $D_0 = i$, $D_1 = D$ and $D_2 = \varphi$, we have

$$\begin{aligned}
ind_{an}\{E_p,\, D_p\} = \sum_{p=0}^{2}(-1)^p dim\, H^p(E_p,\, D_p) &= dim\, \frac{ker\, D}{Im\, i} - dim\, \frac{ker\, \varphi}{Im\, D} \\
&= dim\, ker\, D - dim\, Im\, i - (dim\, ker\, \varphi - dim\, Im\, D) \\
&= dim\, ker\, D - (dim\, \Gamma(F) - dim\, Im\, D) \\
&= dim\, ker\, D - dim\, coker\, D = index\, D \, ,
\end{aligned}$$

$$\tag{43.52}$$

where we have used $Im\, i = 0 \in \Gamma(E)$ and $ker\, \varphi = \Gamma(F)$.

Given an elliptic complex $\{E_p, D_p\}$ we can construct the so-called **even bundle** E_+ and **odd bundle** E_-:

$$E_+ \equiv \oplus_q E_{2q}, \quad E_- \equiv \oplus_q E_{2q+1}. \tag{43.53}$$

Define the operator $A : \Gamma(E_+) \to \Gamma(E_-)$ by

$$A \equiv \oplus_q (D_{2q} + D^\dagger_{2q-1}). \tag{43.54}$$

Thus $A^\dagger : \Gamma(E_-) \to \Gamma(E_+)$ is given by

$$A^\dagger = \oplus_q (D_{2q-1} + D^\dagger_{2q}). \tag{43.55}$$

$\boxed{\text{Exercise 43.4}}$ Verify that the domain and range of A (and $A^\dagger$) are as asserted.

Consider the complex

$$0 \xrightarrow{i} E_+ \xrightarrow{A} E_- \xrightarrow{\varphi} 0, \tag{43.56}$$

and set $E_0 = 0$, $E_I = E_+$, $E_{II} = E_-$, with $D_0 = i$, $D_I = A$ and $D_{II} = \varphi$. The Laplacians [as defined by (43.41)] are given by

$$\Delta_0 = 0, \tag{43.57}$$
$$\Delta_I = A^\dagger A + ii^\dagger = A^\dagger A \equiv \Delta_+, \tag{43.58}$$
$$\Delta_{II} = \varphi^\dagger \varphi + AA^\dagger = AA^\dagger \equiv \Delta_-, \tag{43.59}$$

where $\Delta_+ : E_+ \to E_+$ and $\Delta_- : E_- \to E_-$. From (43.54) and (43.55) we have

$$\begin{aligned}
\Delta_+ = A^\dagger A &= \oplus_{p,q} (D_{2q-1} + D^\dagger_{2q})(D_{2p} + D^\dagger_{2p-1}) \\
&= \oplus_q (D_{2q-1} D^\dagger_{2q-1} + D^\dagger_{2q} D_{2q}) = \oplus_q \Delta_{2q},
\end{aligned} \tag{43.60}$$

where Δ_{2q} are the Laplacians of the original complex out of which E_+ and E_- have been constructed. Some remarks on the derivation of (43.60) are in order. When $p \neq q$, it is clear that

$$D_{2q-1} D^\dagger_{2p-1} = D^\dagger_{2q} D_{2p} = 0,$$

since the range of the operator on the right does not intersect the domain of the operator on the left in either case. The cross terms $D_{2q-1} D_{2p}$ and $D^\dagger_{2q} D^\dagger_{2p-1}$ also vanish for any values of p and q. For $p = q$, these vanish for the same reason above. For $p \neq q$, the range and domain will match only for $q = p+1$ for $D_{2q-1} D_{2p}$ and for $q = p-1$ for $D^\dagger_{2q} D^\dagger_{2p-1}$. In the first case, $D_{2q-1} D_{2p} = D_{2p+1} D_{2p}$. But $Im\, D_{2p} \subset ker\, D_{2p+1}$, hence $D_{2p+1} D_{2p} = 0$. Similarly, for $q = p-1$, $D^\dagger_{2q} D^\dagger_{2p-1} = D^\dagger_{2p-2} D^\dagger_{2p} = 0$. Following exactly the same arguments leading to (43.60), we have

$$\Delta_- = AA^\dagger = \oplus_q \Delta_{2q+1}. \tag{43.61}$$

The definition of the index (43.45) applied to the complex (43.56) then finally gives

$$
\begin{aligned}
index\,\{E_+, E_-, A\} &= dim\,ker\,\Delta_+ - dim\,ker\,\Delta_- \\
&= \sum_{q=0} dim\,ker\,\Delta_{2q} - \sum_{q=0} dim\,ker\,\Delta_{2q+1} \\
&= \sum_{p} (-1)^p\,dim\,ker\,\Delta_p = index\,\{E_p, D_p\}\,.
\end{aligned}
\tag{43.62}
$$

Having defined the analytical index of the elliptic complex $\{E_p, D_p\}$, we proceed to give an expression for the so-called **topological index** of $\{E_p, D_p\}$. The original definition is given in terms of K-theory (a cohomology theory of vector bundles), which is beyond the scope of this book. It turns out that the topological index is most practically expressed by a cohomological formulation in terms of the Chern characters of E_p [(41.79)], the so-called **Todd class** of the complexification of TM, $Td(TM \otimes \mathbb{C})$, and the Euler class $e(TM)$ of TM [c.f. (41.119) and (41.132)].

Let us first define the Todd class of a complex vector bundle (using the Splitting Principle of Theorem 41.5). Let q be the fiber dimension, and x_j, $j = 1, \ldots, q$ be defined as in (41.105). Then the Todd class $Td(\Omega)$ is given by

$$
Td(\Omega) \equiv \prod_{j=1}^{q} \frac{x_j}{1 - e^{-x_j}}\,.
\tag{43.63}
$$

This can be expanded formally in powers of x_j as

$$
\begin{aligned}
Td(\Omega) &= 1 + \frac{1}{2}\sum_j x_j + \frac{1}{12}\sum_j (x_j)^2 + \frac{1}{4}\sum_{j<k} x_j x_k + \cdots \\
&= 1 + \frac{1}{2}C_1(\Omega) + \frac{1}{12}[C_1(\Omega)^2 - C_2(\Omega)] + \frac{1}{24}C_1(\Omega)C_2(\Omega) + \cdots\,,
\end{aligned}
\tag{43.64}
$$

where the $C_j(\Omega)$ are the Chern classes.

Exercise 43.5 Verify both equalities of (43.64).

The Todd class satisfies the following basic requirement for any characteristic class:

$$
Td(E \oplus F) = Td(E) \wedge Td(F)\,,
\tag{43.65}
$$

where E and F are any two vector bundles over the same base manifold.

The **topological index** of $\{E_p, D_p\}$ is given by

$$
\boxed{ind_{top}\,\{E_p, D_p\} = (-1)^{\frac{m(m+1)}{2}} \int_M Ch\,(\oplus_p (-1)^p E_p) \wedge \frac{Td(TM \otimes \mathbb{C})}{e(TM)}}\,,
$$

$$
\tag{43.66}
$$

where $m = dim(M)$, $Ch(E)$ is the Chern character of E [(41.79)], and the division on the RHS can be carried out formally using the Splitting Principle (Theorem 41.5). The facts that the integrand is always an m-form and that the RHS always yields an integer may not be immediately obvious, but will be demonstrated by the examples discussed below. We now state (without proof) our central theorem of this chapter.

Theorem 43.8 (Atiyah-Singer Index Theorem). *Let $\{E_p, D_p\}$ be an elliptic complex over an m-dimensional compact manifold M without boundary. Then*

$$\boxed{ind_{an}\{E_p, D_p\} = ind_{top}\{E_p, D_p\}} \quad , \tag{43.67}$$

where the analytical index ind_{an} and the topological index ind_{top} of $\{E_p, D_p\}$ are given by (43.45) and (43.66), respectively. If m is odd, the index vanishes identically.

We will next apply the Atiyah-Singer index theorem to some special cases. Our first example is the de Rham complex introduced at the beginning of this chapter.

Since the topological index is given in terms of the Chern character, which are characteristic classes for complex vector bundles, we will need to use complex-valued forms in $\Lambda^p(M)$. Thus

$$\Lambda^p(M) = \Gamma(M, \Lambda^p T^* M \otimes \mathbb{C}), \tag{43.68}$$

where $T^* M \otimes \mathbb{C}$ is the complexification of the real cotangent bundle $T^* M$. (Note that the base manifold M is not required to be a complex manifold).

| Exercise 43.6 | Use Theorem 43.4 to show that the de Rham complex is an elliptic complex, that is, show that the laplacian $\Delta = \delta d + d\delta$ is an elliptic operator (c.f. Def. 43.3).

The analytical index of the de Rham complex is given by

$$ind_{an}\{\Lambda^p(M), d\} = \sum_p (-1)^p \, dim \, ker \, \Delta_p$$

$$= \sum_p (-1)^p \, dim_{\mathbb{C}} \, H^p(M, \mathbb{C}) \qquad \text{(complex dimension)} \tag{43.69}$$

$$= \sum_p (-1)^p \, dim \, H^p(M, \mathbb{R}) = \sum_p (-1)^p \, b_p(M) = \chi(M),$$

which is the Euler characteristic of M.

Let us now calculate the topological index using (43.66). To compute the Chern character $Ch(\oplus_p (-1)^p \Lambda^p T^* M \otimes \mathbb{C})$ we will use the properties (41.87), (41.88), and the Splitting Principle (Theorem 41.5). By (41.87),

$$Ch \left(\bigoplus_{p=0}^{m} (-1)^p \Lambda^p T^* M \otimes \mathbb{C} \right) = \sum_{p=0}^{m} (-1)^p \, Ch(\Lambda^p T^* M \otimes \mathbb{C}). \tag{43.70}$$

To use the Splitting Principle let us pretend that $TM \otimes \mathbb{C}$ can be split up into a direct sum of m line bundles over M:

$$TM \otimes \mathbb{C} = L_1 \oplus L_2 \oplus \cdots \oplus L_m . \tag{43.71}$$

Then

$$T^*M \otimes \mathbb{C} = L_1^* \oplus L_2^* \oplus \cdots \oplus L_m^* , \tag{43.72}$$

where L_i^* is the dual bundle of L_i, and

$$\Lambda^p T^*M \otimes \mathbb{C} = \bigoplus_{1 \le i_1 < i_2 < \cdots < i_p \le m} (L_{i_1}^* \otimes L_{i_2}^* \otimes \cdots \otimes L_{i_p}^*) \tag{43.73}$$

[c.f. (28.20) and the discussion immediately preceding]. By (41.88), then,

$$Ch(\Lambda^p T^*M \otimes \mathbb{C}) = \sum_{1 \le i_1 < i_2 < \cdots < i_p \le m} Ch(L_{i_1}^*) \wedge \cdots \wedge Ch(L_{i_p}^*) . \tag{43.74}$$

Let

$$C_1(L_j) = \frac{i}{2\pi} Tr(\Omega_{L_j}) = \frac{i}{2\pi} \Omega_{L_j} = x_j . \tag{43.75}$$

Then

$$Ch(L_j) = e^{x_j} , \tag{43.76}$$

and

$$Ch(L_j^*) = e^{-x_j} . \tag{43.77}$$

This is seen by the fact that $L_j \otimes L_j^*$ is a trivial bundle. By Corollary 41.1 all its characteristic classes must be trivial. In particular

$$Ch(L_j \otimes L_j^*) = Ch(L_j) \wedge Ch(L_j^*) = 1 , \tag{43.78}$$

whence (43.77).

Exercise 43.7 Show that $L_j \otimes L_j^*$ is a trivial bundle by showing that it has a global section.

Thus (43.74) and (43.77) yield

$$Ch(\Lambda^p T^*M \otimes \mathbb{C}) = \sum_{1 \le i_1 < i_2 < \cdots < i_p \le m} e^{-x_{i_1}} e^{-x_{i_2}} \ldots e^{-x_{i_p}} . \tag{43.79}$$

It follows from (43.70) that

$$Ch \left(\bigoplus_{p=0}^{m} (-1)^p \Lambda^p T^*M \otimes \mathbb{C} \right)$$

$$= \sum_{p=0}^{m} (-1)^p \sum_{1 \le i_1 < i_2 < \cdots < i_p \le m} e^{-x_{i_1}} e^{-x_{i_2}} \ldots e^{-x_{i_p}}$$

$$= 1 - \sum_{i=1}^{m} e^{-x_i} + \sum_{i<j} e^{-x_i} e^{-x_j} + \cdots + (-1)^m e^{-x_1} e^{-x_2} \ldots e^{-x_m}$$

$$= \prod_{i=1}^{m} (1 - e^{-x_i}) . \tag{43.80}$$

Using the definition for the Todd class given by (43.63), we have

$$Ch(\oplus_p(-1)^p \Lambda^p T^* M \otimes \mathbb{C}) \wedge Td\,(TM \otimes \mathbb{C})$$

$$= \prod_{i=1}^{m}(1 - e^{-x_i}) \prod_{j=1}^{m} \frac{x_j}{1 - e^{-x_j}} = \prod_{j=1}^{m} x_j \,. \quad (43.81)$$

Suppose $m = 2l$. We will use the Splitting Principle again to calculate the Euler class $e(TM)$. We will use the definition of $e(TM)$ given by (41.119) (in terms of the Pontrjagin class p_l) and the relationship between p_l and the Chern classes as given by (41.116) and (41.131). From (41.131) we can write

$$p_l(TM) = (-1)^l \left(\frac{i}{2\pi}\right)^{2l} \det \Omega = (-1)^l C_{2l}(TM \otimes \mathbb{C}) = (-1)^l \prod_{j=1}^{2l} x_j \,, \quad (43.82)$$

where $C_{2l}(TM \otimes \mathbb{C})$ is the top Chern class of the complexified tangent bundle $TM \otimes \mathbb{C}$. By the direct sum decomposition of $TM \otimes \mathbb{C}$ given by (40.18), and assuming (43.71), we can write

$$\frac{i}{2\pi}\Omega_{TM\otimes\mathbb{C}} = \frac{i}{2\pi}\Omega_{TM_\mathbb{C}\oplus\overline{TM_\mathbb{C}}}$$

$$= \begin{pmatrix} x_1 & & & & & \\ & \ddots & & & 0 & \\ & & x_l & & & \\ & & & x_{l+1} & & \\ & 0 & & & \ddots & \\ & & & & & x_{2l} \end{pmatrix} = \begin{pmatrix} x_1 & & & & & \\ & \ddots & & & 0 & \\ & & x_l & & & \\ & & & -x_1 & & \\ & 0 & & & \ddots & \\ & & & & & -x_l \end{pmatrix}.$$

$$(43.83)$$

Hence

$$p_l(TM) = (-1)^l \left(\prod_{j=1}^{l} x_j\right)(-1)^l\left(\prod_{j=1}^{l} x_j\right) = \left(\prod_{j=1}^{l} x_j\right)^2. \quad (43.84)$$

By (41.119),

$$e(TM) = \prod_{j=1}^{l} x_j \,. \quad (43.85)$$

This result, together with (43.81), can be substituted into the RHS of (43.66) to finally give the topological index of the de Rham complex:

$$ind_{top}\{\Lambda^p(M), d\} = (-1)^{2l(2l+1)/2} \int_M \frac{\displaystyle\prod_{j=1}^{2l} x_j}{\displaystyle\prod_{j=1}^{l} x_j}$$

$$= (-1)^{2l^2+l}(-1)^l \int_M \prod_{j=1}^{l} x_j = \int_M e(TM) . \quad (43.86)$$

Together with (43.69), it is seen that *the Atiyah-Singer Index Theorem applied to the de Rham complex yields the celebrated Gauss-Bonnet Theorem* (Theorem 41.6). When m is odd, this theorem still holds since both $e(TM)$ and $\chi(M)$ then vanish [c.f. discussions following (41.119) and (33.31)].

Remark. As in all calculations using the Splitting Principle, the above intermediate results involving x_j are not strictly true unless the vector bundle in question can in fact be split up into a direct sum of line bundles [c.f. (41.103)] (which is by no means always the case). The final results involving only the characteristic classes are, however, always valid. This is, in fact, why the Splitting Principle is so useful.

We will next apply the Atiyah-Singer Index Theorem to the so-called **Dolbeault complex**. This is a generalization of the de Rham complex to the case when M is a complex manifold. It is given by

$$\cdots \xrightarrow{\bar{\partial}} A_{p,q} \xrightarrow{\bar{\partial}} A_{p,q+1} \xrightarrow{\bar{\partial}} \cdots , \quad (43.87)$$

where $A_{p,q}$ is the space of all (p,q)-type forms on M [c.f. (40.62)] and the differential operator $\bar{\partial}$ is defined by (40.63). The corresponding **Dolbeault cohomology groups** [c.f. (43.43)] are given by

$$H_{\bar{\partial}}^{p,q}(M) = \frac{ker\,(\bar{\partial}: A_{p,q} \to A_{p,q+1})}{Im\,(\bar{\partial}: A_{p,q-1} \to A_{p,q})} . \quad (43.88)$$

We note that since $d = \partial + \bar{\partial}$ on a complex manifold

$$d^2 = 0 = \partial^2 + (\partial\bar{\partial} + \bar{\partial}\partial) + \bar{\partial}^2 . \quad (43.89)$$

Let both sides of the above equation act on $\omega \in A_{p,q}$. Then

$$\partial^2\omega \in A_{p+2,q} , \quad (\partial\bar{\partial} + \bar{\partial}\partial)\omega \in A_{p+1,q+1} , \quad \bar{\partial}^2\omega \in A_{p,q+2} .$$

Thus

$$\partial^2 = \partial\bar{\partial} + \bar{\partial}\partial = \bar{\partial}^2 = 0 . \quad (43.90)$$

The last equation makes (43.87) a complex.

Exercise 43.8   Verify that the Dolbeault complex is an elliptic complex for each value of p.

The numbers

$$\boxed{h_{p,q} \equiv dim_{\mathbb{C}} H_{\bar{\partial}}^{p,q}(M)} \tag{43.91}$$

are called **Hodge numbers**. They are refinements of the Betti numbers. We will let $p = 0$ and consider the elliptic complex

$$\cdots \xrightarrow{\bar{\partial}} A_{0,q} \xrightarrow{\bar{\partial}} A_{0,q+1} \xrightarrow{\bar{\partial}} \cdots \quad . \tag{43.92}$$

The analytic index of this complex is then given by

$$ind_{an}\{A_{0,q}, \bar{\partial}\} = \sum_q (-1)^q h_{0,q} . \tag{43.93}$$

This is called the **arithmetic genus** of the complex manifold M.

The calculation of the topological index for the Dolbeault complex is quite similar to that for the de Rham complex. First we need to calculate $Ch(\oplus_q (-1)^q \Lambda^{0,q} T^* M)$, where $\Lambda^{0,q} T^* M$ is the bundle whose sections are $(0, q)$-type forms:

$$\omega = \omega_{i_1 \ldots i_q} d\bar{z}^{i_1} \wedge \cdots \wedge d\bar{z}^{i_q} .$$

In the notation of (40.19), this is precisely $\Lambda^q \overline{T^* M_{\mathbb{C}}}$ (where M is regarded as an even-dimensional real manifold). If a hermitian structure is given on M, then the structure group of $T M_{\mathbb{C}}$ is reduced to $U(n)$, where n is the complex dimension of M: $n = dim_{\mathbb{C}} M$. In this case the dual bundle of a vector bundle over M can be identified with the conjugate bundle, so $\overline{T^* M_{\mathbb{C}}} = T M_{\mathbb{C}}$.

Exercise 43.9   Let $\pi : E \to M$ be a complex vector bundle over a complex manifold M of complex dimension n. If the structure group is $U(n)$, show that $E^* = \overline{E}$, where E^* is the dual bundle and $\overline{E}$ is the conjugate bundle.

We will thus need to evaluate

$$Ch(\oplus_q(-1)^q \Lambda^q T M_{\mathbb{C}}) = \sum_{q=0}^{n} (-1)^q Ch(\Lambda^q T M_{\mathbb{C}}) .$$

By the Splitting Principle, with x_j as given by (43.75),

$$Ch(\Lambda^q T M_{\mathbb{C}}) = \sum_{q=0}^{n} (-1)^q \sum_{1 \le i_1 < i_2 < \ldots < i_q \le n} e^{(x_{i_1} + \cdots + x_{i_q})}$$

$$= \prod_{i=1}^{n} (1 - e^{x_i}) = (-1)^n x_1 \wedge x_2 \wedge \cdots \wedge x_n (-1)^n \prod_{i=1}^{n} \frac{(1 - e^{x_i})}{x_i} . \tag{43.94}$$

By the definition of the Todd class [(43.63)], the result of Exercise 43.9, and realizing that, for $T^*M_{\mathbb{C}}$, we have to replace x_i by $-x_i$, we have

$$(-1)^n \prod_{i=1}^{n} \frac{1 - e^{x_i}}{x_i} = \frac{1}{Td\,(T^*M_{\mathbb{C}})}\,. \tag{43.95}$$

Meanwhile, by (41.78), the top Chern class of $T^*M_{\mathbb{C}}$ is

$$(-1)^n\, x_1 \wedge \cdots \wedge x_n = C_n(T^*M_{\mathbb{C}})\,. \tag{43.96}$$

Hence

$$Ch(\Lambda^q TM_{\mathbb{C}}) = \frac{C_n(T^*M_{\mathbb{C}})}{Td\,(T^*M_{\mathbb{C}})}\,. \tag{43.97}$$

It follows from (43.66) that the topological index of the Dolbeault complex is given by

$$ind_{top}\{A_{0,q}\,,\bar{\partial}\} = (-1)^{n(2n+1)} \int_M \frac{C_n(T^*M_{\mathbb{C}})}{Td\,(T^*M_{\mathbb{C}})} \wedge \frac{Td\,(TM \otimes \mathbb{C})}{e(TM)}\,, \tag{43.98}$$

remembering that in (43.66), M is regarded as an even-dimensional real manifold, m is the real dimension of M and is thus equal to $2n$ in our case, where $dim_{\mathbb{C}} M = n$. But from (40.18) and (43.65)

$$Td\,(TM \otimes \mathbb{C}) = Td\,(TM_{\mathbb{C}} \oplus \overline{TM_{\mathbb{C}}}) = Td\,(TM_{\mathbb{C}}) \wedge Td\,(T^*M_{\mathbb{C}})\,. \tag{43.99}$$

Also, by (43.85) and (43.96),

$$C_n(T^*M_{\mathbb{C}}) = (-1)^n e(TM)\,. \tag{43.100}$$

We finally have

$$ind_{top}\{A_{0,q}\,,\bar{\partial}\} = \int_M Td\,(TM_{\mathbb{C}}) = \int_M Td\,(M)\,, \tag{43.101}$$

where M is regarded as an even-dimensional real manifold. *The Atiyah-Singer Index Theorem applied to the Dolbeault complex* $\{A_{0,q}\,,\bar{\partial}\}$ *then yields*

$$\boxed{\int_M Td\,(M) = \sum_{q=0}^{n} (-1)^q\, h_{0,q}}\,, \tag{43.102}$$

where $dim_{\mathbb{C}} M = n$, and $h_{0,q}$ are the Hodge numbers defined in (43.91). The above statement [(43.102)] is the celebrated **Riemann-Roch-Hirzebruch Theorem**, one of the foundational theorems of algebraic geometry.

As a final example of the application of the Atiyah-Singer Index Theorem, we will look at the **Dirac complex**, which is of great relevance in the physics of fermionic fields. We have alluded to the notion of a spin structure in our brief mention of the Stiefel-Whitney classes at the end of Chapter 41. We need to explain it more carefully here.

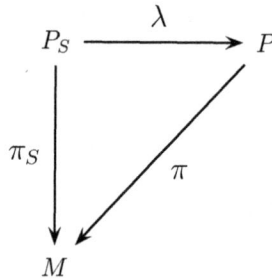

$$
\begin{array}{ccc}
P_S & \xrightarrow{\ \lambda\ } & P \\
{\scriptstyle \pi_S}\big\downarrow & {\scriptstyle \pi} & \\
M & &
\end{array}
$$

FIGURE 43.2

The existence of a spin structure on a manifold M requires M to satisfy two requirements: (1) M is orientable, and (2) the structure group of TM lifts to the universal cover of $SO(n)$, that is, $Spin(n)$, where $n = dim(M)$ (see Def. 43.10 below). Recall, for example, from Chapter 11 that the universal cover of $SO(3)$ is $SU(2)$, that is, $Spin(3) \sim SU(2)$. These are satisfied precisely, and respectively, by the facts that $w_1(M) = 0$ and $w_2(M) = 0$, where $w_1(M)$ and $w_2(M)$ are respectively the first and second Stiefel-Whitney classes. In the following we need to assume that M is endowed with a Euclidean (not Lorentzian) metric [see discussion following (43.109) below] so that we have the principal bundle of orthonormal frames on M, $\pi : P \to M$.

Definition 43.10. *A spin structure for* $\pi : P \to M$ *(the orthonormal frame bundle of M) consists of a principal bundle* $\pi_S : P_S(M) \to M$ *with structure group Spin(n) [the universal cover of $SO(n)$, where $n = dim(M)$] and a map* $\lambda : P_S(M) \to M$ *such that, for all* $p \in P_S(M)$*, (see Fig. 43.2)*

$$
\begin{aligned}
&i) && (\pi \circ \lambda)(p) = \pi_S(p)\,, && and \\
&ii) && \lambda(gp) = \phi(g)\lambda(p)\,, && for\ all\ g \in Spin(n)\,,
\end{aligned}
$$

where $\phi : Spin(n) \to SO(n)$ *is the covering homomorphism of (11.30). Note that gp denotes the left action of Spin(n) on P_S and $\phi(g)\lambda(p)$ denotes the left action of $SO(n)$ on P.*

Each choice of a spin structure determines an element of $H^1(M; \mathbb{Z}_2)$ and vice versa. If M is simply connected, $H^1(M; \mathbb{Z}_2)$ is trivial and the spin structure on M is unique.

We need some basic facts of $Spin(n)$ (which will not be proved here). There is a basic representation of $Spin(n)$, called the **spin representation**, whose representation space S has a dimension given by $dim(S) = 2^{n/2}$ (for n even) and $dim(S) = 2^{(n-1)/2}$ (for n odd). If n is odd, the spin representation is irreducible; but if n is even, it reduces to two irreducible components S^+ and

S^- of equal dimension:

$$S = S^+ \oplus S^- , \quad dim(S^+) = dim(S^-) = 2^{(n/2-1)} = \frac{1}{2} \, dim(S) . \quad (43.103)$$

We define the following **spinor bundles**, which are associated complex vector bundles to P_S (c.f. Def. 37.12):

$$E^+ \equiv S^+ \times_{Spin(n)} P_S , \quad (43.104)$$
$$E^- \equiv S^- \times_{Spin(n)} P_S , \quad (43.105)$$
$$E \equiv E^+ \oplus E^- . \quad (43.106)$$

Elements of $\Gamma(E)$, $\Gamma(E^\pm)$ (that is, sections of E, $E^\pm$) are called **spinors** on M.

Recall the Dirac gamma matrices γ_μ given by (27.11). We change the γ_0 of (27.11a) to $i\gamma_0$, which will be denoted by γ_0 here, so that in the present chapter, the gamma matrices are defined by

$$\gamma_0 = \begin{pmatrix} 0 & i \\ i & 0 \end{pmatrix} , \quad (43.107)$$

$$\gamma_i = \begin{pmatrix} 0 & \sigma_i \\ -\sigma_i & 0 \end{pmatrix} , \quad i = 1, 2, 3 , \quad (43.108)$$

where each entry is a 2×2 block and σ_i are the 2×2 Pauli spin matrices given by (11.16). These gamma matrices satisfy the following relations instead of (27.5):

$$\{ \gamma_a, \gamma_b \} \equiv \gamma_a \gamma_b + \gamma_b \gamma_a = -2\delta_{ab} . \quad (43.109)$$

The change on γ_0 is made to accommodate the fact that the metric on M is Euclidean rather than Lorentzian. (If M is Lorentzian rather than Euclidean, it turns out that the Dirac operator introduced below will not be elliptic.)

The Dirac gamma matrices written down explicitly above [(43.107) and (43.108)] are for $n = dim(M) = 4$. In general, the Dirac γ-matrices are a set of n anti-commuting $2^{n/2} \times 2^{n/2}$ matrices [satisfying (43.109) with $a, b = 1, \ldots, n$] which generate the Clifford algebra of $Spin(n)$. We also need the **chirality operator** γ_{n+1} (only defined when n is even) defined by

$$\gamma_{n+1} \equiv i^{n(n+1)/2} \, \gamma_1 \ldots \gamma_n . \quad (43.110)$$

The chirality operator satisfies

$$\gamma_{n+1}^2 = 1 , \quad (43.111)$$
$$\{ \gamma_{n+1}, \gamma_a \} = 0 , \quad a = 1, \ldots, n . \quad (43.112)$$

γ_a and γ_{n+1} act on $\Gamma(E)$. Eq. (43.111) implies that, with respect to the decomposition $E = E^+ \oplus E^-$, γ_{n+1} is block-diagonalized:

$$\gamma_{n+1} = \begin{pmatrix} 1 & 0 \\ 0 & -1 \end{pmatrix} , \tag{43.113}$$

where each entry is a $2^{n/2-1} \times 2^{n/2-1}$ block. Thus the projections to the $E^{\pm}$ sectors are obtained by

$$P^{\pm} \equiv (1 \pm \gamma_{n+1})/2 . \tag{43.114}$$

Let $\omega = \omega_\mu dx^\mu$ (where ω_μ is matrix-valued, $2^{n/2} \times 2^{n/2}$) be the Riemannian **spin connection** on the spinor bundle $\pi : E \to M$ [where E is given by (43.106)]. This structure will allow us to take covariant derivatives on $\Gamma(E)$ given a Riemannian connection on M. A Riemannian connection on M can be associated with a connection $\tilde{\omega}$ on the principal $SO(n)$-frame bundle P of M. Through the pullback λ^* (c.f. Fig. 43.2) we have a connection ω_S on the principal bundle P_S, which can then be associated with our spin connection ω. Let an orthonormal frame field be given on M by

$$s_a = e^\mu_a(x) \frac{\partial}{\partial x^\mu} , \tag{43.115}$$

where the coefficients $e^\mu_a(x)$ are called **vielbeins**. The orthonormality condition (under a Riemannian metric $g_{\mu\nu}$) implies

$$e^\mu_a e^\nu_b g_{\mu\nu} = \delta_{ab} , \tag{43.116}$$

or, with $e^a_\mu e^\nu_a = \delta^\nu_\mu$, the equivalent condition

$$e^a_\mu e^b_\nu \delta_{ab} = g_{\mu\nu} . \tag{43.117}$$

The **Dirac operator** is then given by the map $\not{D} : \Gamma(M, E) \to \Gamma(M, E)$, where

$$\not{D} = \gamma^a e^\mu_a(x) \left(\frac{\partial}{\partial x^\mu} + \omega_\mu \right) . \tag{43.118}$$

The **chiral Dirac operator** $\not{\partial} : \Gamma(M, E^+) \to \Gamma(M, E^-)$, given by

$$\not{\partial} = \not{D} P^+ , \tag{43.119}$$

then connects $\Gamma(M, E^+)$ to $\Gamma(M, E^-)$. Conversely, we have the adjoint $\not{\partial}^\dagger :$ $\Gamma(M, E^-) \to \Gamma(M, E^+)$ given by

$$\not{\partial}^\dagger = \not{D} P^- . \tag{43.120}$$

| Exercise 43.10 | Verify the above assertions on the domain and range concerning the chiral Dirac operators $\not{\partial}$ and $\not{\partial}^\dagger$ for the case $n = 4$ by making use of the explicit forms for γ_a given by (43.107) and (43.108).

We consider the **Dirac complex**, or **spin complex**, defined as follows [c.f. (43.51)]:

$$0 \xrightarrow{\ i\ } \Gamma(M, E^+) \xrightarrow{\ \partial\!\!\!/\ } \Gamma(M, E^-) \xrightarrow{\ \varphi\ } 0 . \tag{43.121}$$

Exercise 43.11   Verify that $\partial\!\!\!/$ is an elliptic operator.

From (43.50), (43.51) and (43.52), the analytical index of the spin complex is given by

$$ind\,\partial\!\!\!/ = dim\,ker\,\partial\!\!\!/ - dim\,ker\,\partial\!\!\!/^\dagger . \tag{43.122}$$

By (43.43) and Theorem 43.6 we see that, on setting $D_{p-1} = i$ (the inclusion map), the space of harmonic spinors Ψ is given by

$$\Psi \equiv Harm\,(\Gamma(M, E^+), \partial\!\!\!/) = ker\,\partial\!\!\!/ . \tag{43.123}$$

Thus a spinor ψ is a **harmonic spinor** if $\partial\!\!\!/\psi = 0$. Similar to (43.106) Ψ decomposes into positive and negative chirality sectors:

$$\Psi = \Psi^+ \oplus \Psi^- , \tag{43.124}$$

and we have

$$ker\,\partial\!\!\!/ = \Psi^+ , \quad ker\,\partial\!\!\!/^\dagger = \Psi^- . \tag{43.125}$$

Let $dim\,\Psi^\pm = n^\pm$. Then (43.122) implies

$$ind\,\partial\!\!\!/ = n^+ - n^- . \tag{43.126}$$

So by (43.52) the analytical index of the spin complex (43.121) is $n^+ - n^-$.

We proceed to compute the topological index for the spin complex. According to (43.66) this is given by

$$ind_{top}\,(\text{spin complex}) = \int_M (Ch(E^+) - Ch(E^-)) \frac{Td\,(TM \otimes \mathbb{C})}{e(TM)} . \tag{43.127}$$

The following result will be stated without proof:

$$Ch(E^+) - Ch(E^-) = \prod_{i=1}^{n/2} \left(e^{x_i/2} - e^{-x_i/2} \right) , \tag{43.128}$$

where $n = dim(M)$ and the x_i are given by (43.75). [Remember that n has to be even in order for the decomposition (43.103) to be possible]. Using (43.63) for the Todd class, and recalling (43.83) and (43.85) (for the Euler class), we have

$$(Ch(E^+) - Ch(E^-)) \frac{Td\,(TM \otimes \mathbb{C})}{e(TM)}$$

$$= (-1)^{n/2} \prod_{i=1}^{n/2} \left(\frac{e^{x_i/2} - e^{-x_i/2}}{x_i} \right) \frac{x_i}{(1 - e^{-x_i})} \frac{(-x_i)}{(1 - e^{x_i})} = (-1)^{n/2} \prod_{i=1}^{n/2} \frac{(x_i/2)}{\sinh(x_i/2)} . \tag{43.129}$$

Define the so-called **Dirac genus** $\hat{A}$ by

$$\hat{A}(TM) = \prod_{i=1}^{n/2} \frac{(x_i/2)}{\sinh(x_i/2)} = \det\left(\frac{x_i/2}{\sinh(x_i/2)}\right) . \tag{43.130}$$

It has an expression in terms of the Pontrjagin classes p_j [c.f. (41.116)] given by (stated without proof):

$$\hat{A}(TM) = 1 - \frac{1}{24} p_1(TM) + \frac{1}{5760} (7(p_1)^2 - p_2)(TM) + \cdots . \tag{43.131}$$

Since p_j is a $4j$-form on M, the Dirac genus integrated over M vanishes unless $dim(M) = n$ is divisible by 4. In this case $(-1)^{n/2} = 1$, and the Atiyah-Singer Index Theorem applied to the Dirac complex states that

$$index\, \slashed{\partial} = n^+ - n^- = \begin{cases} \int_M \hat{A}(TM) , & \text{if } dim(M) = 0 \bmod 4 , \\ 0 , & \text{otherwise} \end{cases} . \tag{43.132}$$

The above situation describes a free Dirac (fermionic) field (described by spinors) living on a Riemannian manifold. In physics, we frequently consider interactions between free Dirac fields and Yang-Mills gauge potentials (such as the interactions between electrons and photons). In these situations the $\Gamma(M, E^\pm)$ [in (43.121)] need to be replaced by $\Gamma(M, E^\pm \otimes F)$, where F is the vector bundle over M on which the Yang-Mills connection A_μ are defined. The chiral Dirac operator $\slashed{\partial}$ is then replaced by

$$\slashed{\partial}_F = \gamma^a e_a^\mu(x)(\partial_\mu + \omega_\mu + A_\mu)P^+ . \tag{43.133}$$

The complex

$$0 \xrightarrow{i} \Gamma(M, E^+ \otimes F) \xrightarrow{\slashed{\partial}_F} \Gamma(M, E^- \otimes F) \longrightarrow 0 \tag{43.134}$$

is called the **twisted Dirac (spin) complex**. Application of the Atiyah-Singer Index Theorem to the twisted Dirac complex yields the following generalization of (43.132) (stated without proof):

$$index\, \slashed{\partial}_F = (-1)^{n/2} \int_M Ch(F) \wedge \hat{A} . \tag{43.135}$$

There are corresponding twisted versions of the de Rham and Dolbeault complexes, with the following results (again stated without proofs):

$$index\, d_F = m \int_M e(TM) , \quad m = dim(F_x) \quad \text{(twisted de Rham)} , \tag{43.136}$$

$$index\, \bar{\partial}_F = \int_M Ch(F) \wedge Td(M) \quad \text{(twisted Dolbeault)} . \tag{43.137}$$

(For details on twisted complexes, see, for example, C. Nash 1991.)

Finally we mention that the Atiyah-Singer Index Theorem can be generalized to the case where M has a boundary. The result is known as the **Atiyah-Patodi-Singer Index Theorem**. We will not present this here (see, for example, M. Nakahara 1990).

Chapter 44

Symplectic Structures and Hamiltonian Mechanics

The concepts of differential geometry, in particular, differential forms and the tangent and cotangent bundles, provide deep insights into the theoretical structure of classical mechanics and thus facilitate the solutions of many of its problems. In this chapter we will illustrate this fact by providing an introductory discussion of the geometrical basis of Hamiltonian mechanics (for complete details see V. I. Arnold, A. Weinstein and K. Vogtmann 1989 and V. I. Arnold, V. V. Kozlov and A. I. Neishtadt 1997).

Consider a mechanical system described by generalized coordinates q^i, $i = 1, \ldots, n$, which are local coordinates of the **configuration space** (manifold) M of the system. The behavior of the system is completely given by a function $L : TM \times \Delta \to \mathbb{R}$, where TM is the tangent bundle of M and Δ is a closed interval of the time axis. $L(q^i, \dot{q}^i, t)$ is called the **Lagrangian**, with q^i, $\dot{q}^i \equiv dq^i/dt$ being the local coordinates of TM. In fact, the extremals of the variational problem (recall the development at the beginning of Chapter 3)

$$\delta \int_{t_1}^{t_2} L \, dt = 0 \,, \tag{44.1}$$

yield the **Euler-Lagrange equations** of motion:

$$\frac{\partial L}{\partial q^i} - \frac{d}{dt}\left(\frac{\partial L}{\partial \dot{q}^i}\right) = 0 \,, \quad i = 1, \ldots, n \,, \tag{44.2}$$

whose solutions $q^i(t)$ give the classical trajectories. In classical mechanics, the number $n = dim(M)$ (dimension of the configuration space) is called the number of **degrees of freedom** of the mechanical system.

Since $\dot{q}^i \dfrac{\partial}{\partial q^i} \in \Gamma(TM)$ is a local section of TM (or a local tangent vector field on M), the 1-form $p_i dq^i$, where

$$p_i \equiv \frac{\partial L}{\partial \dot{q}^i} \tag{44.3}$$

is a local section of T^*M. The so-called **Legendre transform** of the Lagrangian is given by

$$L(q^i, \dot{q}^i) \longrightarrow H(p_i, q^i) \equiv p_i \dot{q}^i - L , \tag{44.4}$$

where in the last expression, all occurrences of $\dot{q}^i$ are replaced by functions of (p_i, q^i) through (44.3), that is, $\dot{q}^i = \dot{q}^i(p_i, q^i)$. (We consider, for simplicity, that L is not a function of the time t explicitly.) The Legendre transform thus sends a function on TM ($L : TM \to \mathbb{R}$) to a function on T^*M ($H : T^*M \to \mathbb{R}$). The function $H(p_i, q^i)$ is called the **Hamiltonian** of the mechanical system. The crucial point of Hamiltonian mechanics is that there exists a natural symplectic structure on the cotangent bundle T^*M of the configuration space M of a mechanical system. Let us first explain the meaning of the term "symplectic structure", and then explore some of its consequences.

Definition 44.1. *Let N ($dim(N) = 2n$) be an even-dimensional manifold. A* **symplectic structure** *on N is a closed, non-degenerate 2-form ω^2 on N. ω^2 is called the* **symplectic form** *on N. It is thus required to satisfy the following two properties:*

i)

$$d\omega^2 = 0 \qquad (\omega^2 \text{ is closed}) . \tag{44.5}$$

ii) Let $\xi \in \Gamma(TN)$. If $\omega^2(\xi, \eta) = 0$ for all $\eta \in \Gamma(TN)$, then $\xi = 0$. (non-degeneracy)

A manifold endowed with a symplectic structure is called a **symplectic manifold**.

Exercise 44.1   Show that an equivalent condition for non-degeneracy of ω^2 is the following: for all $\xi_1, \xi_2 \in \Gamma(TN)$ such that $\xi_1 \neq \xi_2$, $\omega^2(\xi_1, \eta) \neq \omega^2(\xi_2, \eta)$ for all non-zero $\eta \in \Gamma(TN)$.

We state without proof the following.

Theorem 44.1 (Darboux's Theorem). *In a neighborhood of each point of a symplectic manifold of dimension $2n$, there exist local coordinates (p_i, q^i), $i = 1, \ldots, n$, such that the symplectic structure ω^2 can be expressed in the canonical form*

$$\boxed{\omega^2 = dp_i \wedge dq^i} \qquad . \tag{44.6}$$

The coordinates (p_i, q^i) are called **symplectic** *or* **canonical coordinates**.

A symplectic structure on N induces an isomorphism $\mathcal{I}_x : T_xN \to T_x^*N$, $x \in N$, by the following rule:

$$\mathcal{I}_x(\xi) = \omega_\xi \in T_x^*N , \qquad\qquad \xi \in T_xN , \qquad (44.7)$$

$$\omega_\xi(\eta) = \omega^2(\eta, \xi) , \qquad\qquad \eta \in T_xN . \qquad (44.8)$$

$\boxed{\text{Exercise 44.2}}$ Use the result of Ex. 44.1 to show that the non-degeneracy of ω^2 guarantees that $\mathcal{I}_x$ as defined above is an isomorphism.

The isomorphism $\mathcal{I}_x$, for each $x \in N$, gives rise to the corresponding isomorphism $\mathcal{I} : \Gamma(TN) \to \Gamma(T^*N)$ in an obvious way. Suppose H is a smooth function on N. Then $dH \in \Gamma(T^*N)$, so that $\mathcal{I}^{-1}(dH) \in \Gamma(TN)$, that is, $\mathcal{I}^{-1}(dH)$ is a vector field on N. Such vector fields are called **Hamiltonian vector fields**. The corresponding differential equation, for $x \in N$,

$$\dot{x} = \mathcal{I}^{-1}(dH) \qquad (44.9)$$

is called **Hamilton's equations**. (In this equation, $\dot{x} = dx/dt$.)

Definition 44.2. *Let F and G be smooth functions on a symplectic manifold N with symplectic form ω^2. Then*

$$\boxed{\{ F, G \} \equiv \omega^2(\mathcal{I}^{-1}(dG), \mathcal{I}^{-1}(dF))} \qquad (44.10)$$

is a smooth function on N, called the **Poisson bracket** *of the functions F and G.*

Theorem 44.2. *The Poisson bracket $\{ , \}$ is bilinear and non-degenerate. Non-degeneracy means that if $x \in N$ is not a* **critical point** *of F, that is, if $F(x) \neq 0$, then there exists a smooth non-zero function G such that $\{F,G\}(x) \neq 0$. Furthermore, $\{ , \}$ satisfies the following properties:*

1) $\{F, G\} = -\{G, F\}$ *(skew-symmetry)* , $\qquad (44.11)$
2) $\{F_1 F_2, G\} = F_1\{F_2, G\} + F_2\{F_1, G\}$ *(Leibnitz rule)* , $\qquad (44.12)$
3) $\{\{H, F\}, G\} + \{\{F, G\}, H\} + \{\{G, H\}, F\} = 0$ *(Jacobi identity).* $\qquad (44.13)$

Thus the set of smooth functions on a symplectic manifold constitute a Lie algebra under the Poisson bracket (as the internal multiplication).

$\boxed{\text{Exercise 44.3}}$ Prove the above theorem.

By (44.7) and (44.8), the Poisson bracket as defined by (44.10) can also be written [on replacing G in (44.10) by H]

$$\{F, H\} = dF\left(\mathcal{I}^{-1}(dH)\right) . \qquad (44.14)$$

Using symplectic coordinates (p_i, q^i), we can write the Hamiltonian vector field $\mathcal{I}^{-1}(dH)$ as

$$\mathcal{I}^{-1}(dH) = \sum_{i=1}^{n} \left(\dot{p}_i \frac{\partial}{\partial p_i} + \dot{q}^i \frac{\partial}{\partial q^i} \right) . \tag{44.15}$$

Writing

$$dF = \sum_{i=1}^{n} \left(\frac{\partial F}{\partial p_i} dp_i + \frac{\partial F}{\partial q^i} dq^i \right) , \tag{44.16}$$

we have

$$dF(\mathcal{I}^{-1}(dH)) = \sum_{i=1}^{n} \left(\frac{\partial F}{\partial p_i} \dot{p}_i + \frac{\partial F}{\partial q^i} \dot{q}^i \right) = \dot{F} . \tag{44.17}$$

It follows from (44.14) that Hamilton's equations for the Hamiltonian vector field $\mathcal{I}^{-1}(dH)$ become

$$\dot{F} = \{F, H\} . \tag{44.18}$$

In particular, we have the following system of equations, called **Hamilton's canonical equations**, the solutions of which give the integral curves of the Hamiltonian vector field $\mathcal{I}^{-1}(dH)$:

$$\dot{p}_i = \{p_i, H\} , \quad \dot{q}^i = \{q^i, H\} , \quad 1 \le i \le n . \tag{44.19}$$

Let us now express the Poisson bracket in terms of the symplectic coordinates (p_i, q^i). Suppose $\xi, \eta \in \Gamma(TN)$ are expressed with respect to the natural frame field (in terms of symplectic coordinates) as:

$$\xi = \sum_{i=1}^{n} \left(\xi_{p_i} \frac{\partial}{\partial p_i} + \xi_{q^i} \frac{\partial}{\partial q^i} \right) = (\xi_{p_i}, \xi_{q^i}) , \tag{44.20}$$

$$\eta = \sum_{i=1}^{n} \left(\eta_{p_i} \frac{\partial}{\partial p_i} + \eta_{q^i} \frac{\partial}{\partial q^i} \right) = (\eta_{p_i}, \eta_{q^i}) . \tag{44.21}$$

Then, from (44.8),

$$\omega_\xi(\eta) = \omega^2(\eta, \xi) = \left(\sum_{j=1}^{n} dp_j \wedge dq^j \right) (\eta, \xi)$$

$$= \sum_{j=1}^{n} [(dp_j \wedge dq^j)(\eta, \xi)] = \frac{1}{2} \sum_{j=1}^{n} \begin{vmatrix} dp_j(\eta) & dp_j(\xi) \\ dq^j(\eta) & dq^j(\xi) \end{vmatrix} \tag{44.22}$$

$$= \frac{1}{2} \sum_{j=1}^{n} \begin{vmatrix} \eta_{p_j} & \xi_{p_j} \\ \eta_{q^j} & \xi_{q^j} \end{vmatrix} = \frac{1}{2} \sum_{j=1}^{n} (\eta_{p_j} \xi_{q^j} - \xi_{p_j} \eta_{q^j}) ,$$

where in the fourth equality we have used the evaluation formula for exterior products [(28.14)]. The 1-form ω_ξ can then be written as

$$\omega_\xi = \frac{1}{2} \sum_{j=1}^{n} (\xi_{q^j} dp_j - \xi_{p_j} dq^j) , \qquad (44.23)$$

so that the isomorphism $\xi \mapsto \omega_\xi$ can be written in matrix notation as

$$\frac{1}{2} \begin{pmatrix} \xi_{q^j} \\ -\xi_{p_j} \end{pmatrix} = \frac{1}{2} \begin{pmatrix} 0 & 1 \\ -1 & 0 \end{pmatrix} \begin{pmatrix} \xi_{p_j} \\ \xi_{q^j} \end{pmatrix} . \qquad (44.24)$$

Note that in the above equation the column matrices are $2n \times 1$, and each entry in the square matrix is $n \times n$. The matrix representation of $\mathcal{I}$ (with respect to symplectic coordinates) is thus

$$\mathcal{I} = \frac{1}{2} \begin{pmatrix} 0 & 1 \\ -1 & 0 \end{pmatrix} , \qquad (44.25)$$

from which we have

$$\mathcal{I}^{-1} = 2 \begin{pmatrix} 0 & -1 \\ 1 & 0 \end{pmatrix} . \qquad (44.26)$$

Now write

$$dF = \sum_{i=1}^{n} \left(\frac{\partial F}{\partial p_i} dp_i + \frac{\partial F}{\partial q^i} dq^i \right) , \qquad (44.27)$$

$$dG = \sum_{i=1}^{n} \left(\frac{\partial G}{\partial p_i} dp_i + \frac{\partial G}{\partial q^i} dq^i \right) . \qquad (44.28)$$

The explicit expression for $\mathcal{I}^{-1}$ [(44.26)] then gives

$$\mathcal{I}^{-1}(dF) = 2 \sum_{i=1}^{n} \left(-\frac{\partial F}{\partial q^i} \frac{\partial}{\partial p_i} + \frac{\partial F}{\partial p_i} \frac{\partial}{\partial q^i} \right) , \qquad (44.29)$$

$$\mathcal{I}^{-1}(dG) = 2 \sum_{i=1}^{n} \left(-\frac{\partial G}{\partial q^i} \frac{\partial}{\partial p_i} + \frac{\partial G}{\partial p_i} \frac{\partial}{\partial q^i} \right) . \qquad (44.30)$$

Use of (44.10) together with the evaluation formula (28.14) again finally yields the following standard formula for Poisson brackets:

$$\boxed{ \{F, G\} = \sum_{i=1}^{n} \left(\frac{\partial G}{\partial p_i} \frac{\partial F}{\partial q^i} - \frac{\partial G}{\partial q^i} \frac{\partial F}{\partial p_i} \right) } . \qquad (44.31)$$

With the above formula, Hamilton's canonical equations appear in their conventional form as follows:

$$\dot{p}_i = -\frac{\partial H}{\partial q^i} , \quad \dot{q}^i = \frac{\partial H}{\partial p_i} , \quad i = 1, \ldots, n .\tag{44.32}$$

We return now to the classical mechanics setting, where we have an n-dimensional configuration space M, with local coordinates q^i, $i = 1, \ldots, n$ (the generalized coordinates of Lagrangian mechanics). (p_i, q^i) [with p_i defined by (44.3)] then serve as local coordinates for T^*M (the cotangent bundle of configuration space). Remember that if $dim(M) = n$, then $dim(T^*M) = dim(TM) = 2n$. On T^*M we have the well-defined 1-form

$$\omega = p_i dq^i .\tag{44.33}$$

Define a 2-form on T^*M by

$$\omega^2 \equiv d\omega = dp_i \wedge dq^i .\tag{44.34}$$

This 2-form is obviously closed and also non-degenerate. It is thus a well-defined symplectic form on T^*M, and T^*M is naturally a symplectic manifold. In classical mechanics T^*M is called the **phase space** of the configuration space M.

| Exercise 44.4 | Show that the 2-form ω^2 on T^*M defined by (44.34) is non-degenerate.

Let us consider Hamilton's equations in the form (44.9) a bit more carefully in light of the notion of one-parameter groups of diffeomorphisms on M (c.f. discussion following Def. 34.6). The vector field $\mathcal{I}^{-1}(dH)$ corresponding to a Hamiltonian function H gives a 1-parameter group of diffeomorphisms g_t : $T^*M \to T^*M$ defined by

$$\frac{d}{dt} g_t(x)\Big|_{t=0} = (\mathcal{I}^{-1}(dH))(x) , \quad x \in T^*M .\tag{44.35}$$

The one-parameter group of diffeomorphisms g_t (parametrized by the time t) is called the **Hamilton phase flow** of the Hamiltonian function H. We will state the following important theorem without proof.

Theorem 44.3. *The Hamiltonian phase flow g_t preserves the symplectic structure:*

$$(g_t)^* \omega^2 = \omega^2 ,\tag{44.36}$$

or equivalently, the symplectic structure is invariant under the Hamiltonian phase flow.

The following notions play a crucial role in the study of phase flows.

Definition 44.3. *A differential k-form ω on T^*M is called a **relative integral invariant** of the phase flow g_t if*

$$\int_{g_t C} \omega = \int_C \omega \qquad (44.37)$$

for every closed k-chain (cycle) C in T^*M.

Definition 44.4. *A differential k-form ω on T^*M is called an **absolute integral invariant** of the phase flow g_t if (44.37) is satisfied for every k-chain in T^*M. An absolute integral invariant is also referred to just as an **integral invariant**.*

We have the following useful theorems.

Theorem 44.4. *If ω is a relative integral invariant of g_t, then $d\omega$ is an integral invariant of g_t.*

Proof. Let C be a $(k+1)$-chain in T^*M, and ω be a relative integral invariant of g_t. Then, by Stokes' Theorem,

$$\int_C d\omega = \int_{\partial C} \omega = \int_{g_t(\partial C)} \omega = \int_{\partial(g_t C)} \omega = \int_{g_t C} d\omega . \qquad (44.38)$$

The theorem follows by Def. 44.4. $\qquad\square$

Theorem 44.5. *A k-form ω on T^*M is an integral invariant of g_t if and only if*

$$g_t^* \omega = \omega . \qquad (44.39)$$

Proof. Suppose $g_t^* \omega = \omega$, and C is a k-chain in T^*M. Then

$$\int_{g_t C} \omega = \int_C g_t^* \omega = \int_C \omega , \qquad (44.40)$$

where the first equality follows from (32.29). Thus ω is an integral invariant of g_t. Conversely, suppose ω is an integral invariant of g_t. Then, for any k-chain C in T^*M,

$$\int_C g_t^* \omega = \int_{g_t C} \omega = \int_C \omega , \qquad (44.41)$$

where the first equality follows from (32.29) and the second from the supposition. Since C is an arbitrary k-chain, $g_t^* \omega = \omega$. $\qquad\square$

Corollary 44.1. *The symplectic form ω^2 on T^*M is an integral invariant of the Hamiltonian phase flow g_t.*

Proof. This is a direct consequence of Theorems 44.3 and 44.5. $\qquad\square$

The following will be stated without proof.

Theorem 44.6. *If ω and ω' are both integral invariants of the phase flow g_t, then $\omega \wedge \omega'$ is also an integral invariant of g_t.*

This theorem allows us to build a series of integral invariants of g_t, starting with the symplectic form ω^2:

$$\omega^2, \ \omega^2 \wedge \omega^2, \ (\omega^2)^3, \ \ldots, (\omega^2)^n \ .$$

The form

$$(\omega^2)^n = \underbrace{\omega^2 \wedge \cdots \wedge \omega^2}_{n \text{ times}} \tag{44.42}$$

is a top-form on T^*M and can be used to define a volume element in T^*M. Since it is an integral invariant, we obtain the following theorem.

Theorem 44.7 (Liouville's Theorem). *The Hamiltonian phase flow preserves the volume in phase space.*

In the solution of Hamilton's equations of motion an essential tool is the so-called **canonical transformations** of the canonical coordinates (p_i, q^i) in phase space. It is beyond the scope of this book to enter into the details of this technique, and its use in the powerful Hamilton-Jacobi theory. We will close by simply giving the definition of a canonical map associated with a set of canonical transformations.

Definition 44.5. *A map $\phi : T^*M \to T^*M$ is called **canonical** if the symplectic form ω^2 is invariant under ϕ^*, that is,*

$$\phi^* \omega^2 = \omega^2 \ . \tag{44.43}$$

By Theorem 44.5, $\phi : T^*M \to T^*M$ is canonical if ω^2 is an integral invariant under ϕ:

$$\int_C \omega^2 = \int_{\phi(C)} \omega^2 \tag{44.44}$$

for any 2-chain in T^*M. Another way of stating (44.43) is that a canonical map ϕ preserves the symplectic 2-form:

$$\omega^2 = dp_i \wedge dq^i = dP_i \wedge dQ^i \ , \tag{44.45}$$

if ϕ sends (p_i, q^i) to (P_i, Q^i). Thus

$$d(p_i dq^i) = d(P_i dQ^i) \ , \tag{44.46}$$

which implies

$$\boxed{p_i dq^i - P_i dQ^i = dS} \ , \tag{44.47}$$

for some smooth function $S(p_i, q^i)$. It is easily seen that (44.47) gives an equivalent definition of a canonical transformation. Furthermore, if γ is a closed contour in T^*M that is contractible to a point, then $\partial\gamma = 0$, and by Stokes' Theorem,

$$\oint_\gamma p_i dq^i = \oint_\gamma P_i dQ^i = \oint_{\phi(\gamma)} p_i dq^i \, . \tag{44.48}$$

Thus $p_i dq^i$ is a relative integral invariant under a canonical map (in a simply connected region of T^*M). Finally, from (44.10) it is clear that Poisson brackets are preserved under canonical transformations:

$$\{F, G\}_{P,Q} = \{F, G\}_{p,q} \, . \tag{44.49}$$

As a consequence, Hamilton's equations $\dot{F} = \{F, H\}$ are also invariant under canonical transformations. The main idea of the Hamilton-Jacobi method is to find canonical transformations on the original set of canonical coordinates (p_i, q^i) so that, in the new coordinates (P_i, Q^i), Hamilton's equations becomes easily integrable.

Chapter 45

Quantization via Path Integration

The idea of using functional methods, or *path integrals*, in quantum theory began with Feynman's ingenious adaptation (see, for example, Feynman and Hibbs 1965) of Dirac's prescient proposal to study the quantity

$$\exp\left(\frac{i}{\hbar}\int L(q^i(t), \dot{q}^i(t))\, dt\right). \tag{45.1}$$

In this expression, L is the *Lagrangian*, considered as a function on the *tangent bundle* TM of the *configuration manifold* M (that is, of the canonical coordinates q^i and the velocities $\dot{q}^i$) of a classical mechanical system (c.f. the discussion at the beginning of the previous chapter). The integral in the above equation, known as the *classical action*, is understood to be carried out for a particular trajectory $q^i(t)$, not necessarily the classical one determined by Newton's Law [or the *Euler-Lagrange equations* (c.f. (44.2))]. Our ultimate aim is to understand the importance and the use of the above quantity in *quantum field theory*, which can be considered as the quantum theory (relativistic or not) of a system with infinitely many degrees of freedom, through its use in the quantum mechanics of a system with a finite number of degrees of freedom. We will begin by following the exposition by Faddeev (in Balian and Zinn-Justin 1976), and shall see that this theoretical development, as stressed by Dirac, traces in a very essential manner and directly to the Hamiltonian formulation of classical mechanics that we studied in the previous chapter (see also Lam 2014).

To simplify our discussion, we start with a classical mechanical system with only one degree of freedom, so that its *phase space* (the *cotangent bundle* T^*M) is two-dimensional (with canonical coordinate q and canonical momentum p). Assume that this phase space is just $\mathbb{R}^2$ ($-\infty < q < \infty$ and $-\infty < p < \infty$). The procedure of **canonical quantization** then consists in assigning operators $\hat{q}$ and $\hat{p}$, to the classical functions (on phase space) q and p, respectively, such that when these act on the *coordinate representations* of *Hilbert space* vectors

$|\psi\rangle$, represented by *wave functions* $\psi(q) = \langle q | \psi \rangle$, they are given by

$$\hat{p}\,\psi(q) = \frac{\hbar}{i}\,\frac{\partial\psi(q)}{\partial q}\,, \qquad\qquad \hat{q}\,\psi(q) = q\,\psi(q)\,. \tag{45.2}$$

As discussed in Chapter 15, we can introduce "sharp" position eigenstates $|q\rangle$ of $\hat{q}$, and "sharp" momentum eigenstates $|p\rangle$ of $\hat{p}$, defined by

$$\hat{q}\,|q\rangle = q\,|q\rangle\,, \qquad\qquad \hat{p}\,|p\rangle = p\,|p\rangle\,. \tag{45.3}$$

Through the use of the *Dirac delta function*, we can write the coordinate representations of the above states as [c.f. (15.5) and (15.16)]

$$\langle q' | q \rangle = \delta(q - q')\,, \qquad\qquad \langle q | p \rangle = \frac{1}{\sqrt{2\pi}}\,\exp\left(\frac{i}{\hbar}\,pq\right)\,. \tag{45.4}$$

Suppose the (classical) Hamiltonian function $H(p,q)$ for our system is given by the following polynomial (an analytic function) in p and q:

$$H(p,q) = \sum_{n,m} c_{nm}\,p^n q^m\,, \tag{45.5}$$

where the coefficients c_{nm} are constants, and for the purpose of quantization, we have adopted for convenience an ordering of the factors of p and q such that all the p's stand to the left of all the q's. [The final result of our development, (45.27), is independent of the ordering, but we will not demonstrate this fact explicitly here.] After quantization we obtain the Hamiltonian operator

$$\hat{H}(\hat{p},\hat{q}) = \sum_{n,m} c_{nm}\,\hat{p}^n \hat{q}^m\,. \tag{45.6}$$

Then it follows from (45.3) that

$$\langle p | \hat{H} | q \rangle = \sum_{n,m} c_{nm}\,\langle p | \hat{p}^n \hat{q}^m | q \rangle = \sum_{n,m} c_{nm}\,p^n q^m\,\langle p | q \rangle\,. \tag{45.7}$$

Using (45.4), we arrive at the useful result

$$\langle p | \hat{H} | q \rangle = \frac{1}{\sqrt{2\pi}}\,e^{-\frac{i}{\hbar}pq}\,H(p,q)\,, \tag{45.8}$$

where we note that on the right-hand-side, $H(p,q)$ is a *c-number* (not an operator).

We recall that in quantum mechanics, the time evolution of a physical (measurable) quantity for a system initially specified by a *pure state* (representable by a state vector $|\psi(0)\rangle$) is given by a time-dependent matrix element of an *observable* – a *self-adjoint* operator. For an observable $\hat{O}$, the time-dependence may be carried solely by the state vector (*Schrödinger picture*):

$$|\psi(t)\rangle = \exp(-i\hat{H}t/\hbar)\,|\psi(0)\rangle\,, \tag{45.9}$$

which follows directly from the Schrödinger equation $\hat{H} \, |\psi\rangle = i\hbar \, \partial |\psi\rangle / \partial t$; or solely by the operator (*Heisenberg picture*):

$$\hat{O}(t) = e^{i\hat{H}t/\hbar} \, \hat{O} \, e^{-i\hat{H}t/\hbar} \, , \qquad (45.10)$$

which follows directly from the *Heisenberg equation of motion*

$$\frac{d\hat{O}(t)}{dt} = \frac{[\hat{O}(t), \, \hat{H}]}{i\hbar} + \frac{\partial \hat{O}}{\partial t} \, . \qquad (45.11)$$

Note that in the above equation and in (45.10) the operator $\hat{O}$ [not $\hat{O}(t)$] itself may be explicitly time-dependent. The notation for a so-called *Heisenberg operator* $\hat{O}(t)$ is therefore somewhat ambiguous. We will regard (45.10) as the defining equation for any Heisenberg operator, that is, one that satisfies the equation of motion (45.11). Eqs. (45.9) and (45.10) guarantee that the expectation value of the observable is picture-independent:

$$\langle \psi(0) \, | \, \hat{O}(t) \, | \psi(0) \rangle = \langle \psi(t) \, | \, \hat{O} \, | \, \psi(t) \rangle \, , \qquad (45.12)$$

as it should be.

Let us now define the ket (state vector) $|q, t\rangle$ to be the eigenket of the Heisenberg operator

$$\hat{q}(t) = e^{i\hat{H}t/\hbar} \, \hat{q} \, e^{-i\hat{H}t/\hbar} \qquad (45.13)$$

with eigenvalue q, that is,

$$\hat{q}(t) \, |q, t\rangle = q \, |q, t\rangle \, . \qquad (45.14)$$

As such, it is a state vector in the Heisenberg picture. It then follows from the first equation of (45.3) and (45.10) that the ket $|q, t\rangle$ can be expressed in terms of $|q\rangle$ as

$$|q, t\rangle = e^{i\hat{H}t/\hbar} \, |q\rangle \, . \qquad (45.15)$$

Note the positive sign in the exponent of the above equation. Also, note how the ket $|q, t\rangle$ differs from the one in (45.9) (again by the difference in sign in the exponent).

Problem 45.1 Verify (45.15) by using the defining condition of $|q, t\rangle$ given by (45.14).

The quantity of primary importance in quantum theory (for our system with a single degree of freedom) is the following **propagator function**:

$$\boxed{\langle q', t' \, | \, q, t\rangle = \langle q' \, | \, e^{-i\hat{H}(t'-t)/\hbar} \, | \, q\rangle} \, , \qquad (45.16)$$

where the equality follows directly from (45.15). It has the intuitively obvious physical interpretation as being the *probability amplitude* of finding a particle at position q' at time t' when it is known that the particle is at the position q at the time t. One is then led to the study of the so-called **time-evolution operator**

$$\hat{U}(t', t) \equiv e^{-i\hat{H}(t'-t)/\hbar} . \tag{45.17}$$

It is the attempt to provide a direct algorithm for the calculation of the matrix element in (45.16) (the propagator function) that ushers in the method of functional integration. The idea is to first divide the finite time interval $t' - t$ into a large number (N, say,) of infinitesimal intervals ϵ, so that $t' - t = N\epsilon$. Eventually the limits $N \to \infty$ and $\epsilon \to 0$ will be taken such that $N\epsilon$ remains finite. For small ϵ we have

$$e^{-i\hat{H}\epsilon/\hbar} \approx 1 - i\epsilon\hat{H}/\hbar , \tag{45.18}$$

and we can write

$$\hat{U}(t', t) = e^{-i\hat{H}(t'-t)/\hbar} = \lim_{\substack{N \to \infty \\ \epsilon \to 0}} (1 - i\epsilon\hat{H}/\hbar)^N . \tag{45.19}$$

Rewriting the variables $t \to t', t' \to t'', q \to q', q' \to q''$ in (45.16), we have

$$\langle q'', t'' \,|\, q', t' \rangle = \langle q'' \,|\, \hat{U}(t'', t') \,|\, q' \rangle = \lim_{\substack{N \to \infty \\ \epsilon \to 0}} \langle q'' \,|\, (1 - i\epsilon\hat{H}/\hbar)^N \,|\, q' \rangle . \tag{45.20}$$

In order to manipulate the above matrix element for arbitrary N, we first examine the simple case for $N = 2$ and consider the matrix element $\langle q'' \,|\, (1 - i\epsilon\hat{H}/\hbar)^2 \,|\, q' \rangle$. The trick is to insert the identity operator in the forms of

$$\int_{-\infty}^{\infty} dp \,|p\rangle\langle p| \qquad \text{and} \qquad \int_{-\infty}^{\infty} dq \,|q\rangle\langle q|$$

at appropriate places "inside" the Dirac bracket and use (45.8) and the second equation of (45.4). Thus, on setting $q'' \equiv q_2$ and $q' \equiv q_0$, we have

$$\langle q_2 \,|\, (1 - i\epsilon\hat{H}/\hbar)(1 - i\epsilon\hat{H}/\hbar) \,|\, q_0 \rangle$$
$$= \int dp_2 \int (dp_1 dq_1) \,\langle q_2 \,|\, p_2 \rangle\langle p_2 \,|\, (1 - i\epsilon\hat{H}/\hbar) \,|\, q_1 \rangle\langle q_1 \,|\, p_1 \rangle\langle p_1 \,|\, (1 - i\epsilon\hat{H}/\hbar) \,|\, q_0 \rangle$$
$$= \int dp_2 \int (dp_1 dq_1) \left(\frac{1}{\sqrt{2\pi}} e^{\frac{i}{\hbar} p_2 q_2} \right) \left(\frac{1}{\sqrt{2\pi}} e^{-\frac{i}{\hbar} p_2 q_1} [1 - i\epsilon H(p_2, q_1)/\hbar] \right)$$
$$\times \left(\frac{1}{\sqrt{2\pi}} e^{\frac{i}{\hbar} p_1 q_1} \right) \left(\frac{1}{\sqrt{2\pi}} e^{-\frac{i}{\hbar} p_1 q_0} [1 - i\epsilon H(p_1, q_0)/\hbar] \right) .$$

$$\tag{45.21}$$

Re-expressing the factors involving the Hamiltonian function $H(p, q)$ as exponentials:

$$1 - i\epsilon H(p, q)/\hbar \approx e^{-\frac{i}{\hbar}\epsilon H(p,q)} ,$$

the above result becomes

$$\langle q_2 | (1 - i\epsilon \hat{H}/\hbar)^2 | q_0 \rangle$$

$$= \int \frac{dp_2}{2\pi} \int \frac{dp_1 dq_1}{2\pi} e^{\frac{i}{\hbar} p_2(q_2 - q_1) + \frac{i}{\hbar} p_1(q_1 - q_0)} \exp\left[-\frac{i\epsilon}{\hbar} \{ H(p_2, q_1) + H(p_1, q_0) \} \right].$$

$$(45.22)$$

This immediately allows us to inductively generalize to the case of arbitrary N:

$$\langle q_N | (1 - i\epsilon \hat{H}/\hbar)^N | q_0 \rangle$$

$$= \int \frac{dp_N}{2\pi} \int \frac{dp_{N-1} dq_{N-1}}{2\pi} \cdots \int \frac{dp_1 dq_1}{2\pi} e^{\frac{i}{\hbar} \{ p_N(q_N - q_{N-1}) + \cdots + p_1(q_1 - q_0) \}}$$

$$\times \exp\left[-\frac{i\epsilon}{\hbar} \{ H(p_N, q_{N-1}) + \cdots + H(p_1, q_0) \} \right].$$

$$(45.23)$$

Now make the following time-sequence identifications in the time interval $t' \leq t \leq t''$, $t'' = t' + N\epsilon$:

$$t_0 = t', \qquad t_1 = t_0 + \epsilon, \quad \ldots, \quad t_{N-1} = t_0 + (N-1)\epsilon, \quad t_N = t_0 + N\epsilon = t'' ;$$
$$q' = q_0 = q(t'), \quad q_1 = q(t_1), \quad \ldots, \quad q_{N-1} = q(t_{N-1}), \qquad q_N = q(t'') = q'' ;$$
$$p_1 = p(t'), \quad p_2 = p(t_1), \quad \ldots, \quad p_N = p(t_{N-1}) .$$

The first and last entries in the second line identify the boundary conditions for $q(t)$ for the two fixed times t' and t'':

$$q(t') = q' \text{ (fixed)}, \qquad\qquad q(t'') = q'' \text{ (fixed)} . \qquad (45.24)$$

All the other q-variables (from q_1 to q_{N-1}) are integrated over, as are all the p-variables (from p_1 to p_N), as indicated in (45.23). Our next step is to take the limits $N \to \infty$ and $\epsilon \to 0$ in the exponent of the integrand in (45.23). The sum in this exponent then becomes an integral over the time t:

$$\frac{i}{\hbar} \sum_{n=1}^{N} \{ p_n(q_n - q_{n-1}) - \epsilon H(p_n, q_{n-1}) \}$$

$$= \frac{i}{\hbar} \sum_{n=1}^{N} \{ p(t_{n-1}) (q(t_n) - q(t_{n-1})) - \epsilon H (p(t_{n-1}), q(t_{n-1})) \} \qquad (45.25)$$

$$\xrightarrow[\substack{N \to \infty \\ \epsilon \to 0}]{} \frac{i}{\hbar} \int_{t'}^{t''} dt \, \{ p(t)\dot{q}(t) - H(p(t), q(t)) \} .$$

The integral is precisely the classical mechanical *action functional* appearing in (45.1). The term **functional** refers to the fact that the action is a mapping from *specific* functions $p(t)$ and $q(t)$ to the real numbers ($\mathbb{R}$).

The integrals in (45.23) are completely separate from the time integral in (45.25). They instruct us to integrate over all possible values of the $2N - 1$

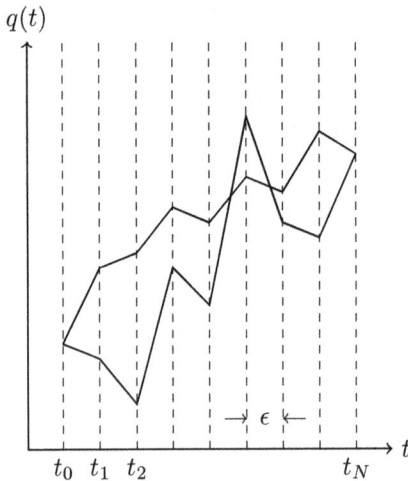

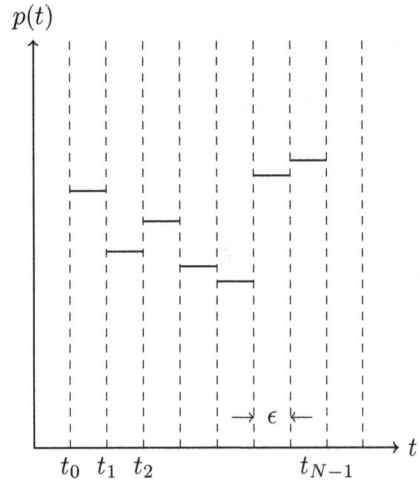

FIGURE 45.1(a)   FIGURE 45.1(b)

variables $q_1 = q(t_1), \ldots, q_{N-1} = q(t_{N-1})$; $p_1 = p(t_0), \ldots, p_N = p(t_{N-1})$. The assignments of the q_i's and p_i's for various discrete times in the interval $t' \leq t \leq t''$ according to the scheme following (45.23) suggest that we use piece-wise continuous linear functions $q(t)$ and piece-wise constant functions $p(t)$ for these integrations. As seen from Figs. 45.1(a) and 45.1(b), this then amounts to, as $N \to \infty$ and $\epsilon \to 0$, choosing all possible functions $q(t)$ $(t' \leq t \leq t'')$ with fixed boundary values $q(t')$ and $q(t'')$, and all possible functions $p(t)$ $(t' \leq t \leq t'')$, with no boundary conditions. This procedure then provides a formal (and heuristic) definition of the so-called **functional integral** over paths $\{p(t), q(t)\}$ in phase-space T^*M [with only boundary conditions on the base (configuration) manifold M], denoted by $[dp\,dq]$:

$$\int \frac{dp_N}{2\pi} \int \frac{dp_{N-1}dq_{N-1}}{2\pi} \cdots \int \frac{dp_1 dq_1}{2\pi} \longrightarrow \lim_{\substack{N \to \infty \\ \epsilon \to 0}} \prod_t \int \frac{dp(t)dq(t)}{2\pi} \equiv \int [dp\,dq] \,.$$
$$(45.26)$$

The quantity involving the limit of a discreet product can be considered as giving a "**measure**" to the functional integration $[dp\,dq]$, in which t is to be thought of as a *label* for the various p's and q's (rather than as a continuous variable). This heuristic device will be even more important when we proceed from quantum mechanics to quantum field theory subsequently (in Chapter 48).

It must be stressed at this point that the foregoing "measure" is by no means rigorously defined [due largely to the presence of the imaginary unit i in (45.25)], but entering into a more mathematically rigorous discussion would be far beyond the scope of this book. We will simply continue to manipulate with it formally, with physical intuition as a guide. Historically this is how physicists have by and large proceeded in their manipulations of path integrals, and managed to achieve spectacular results despite the lack of firm mathematical foundations. In the next chapter, however, we will discuss a procedure, known as *Euclideanization*, by which one can get around this difficulty for a certain class of functional integrals, those of the so-called *Gaussian type*. For more mathematically rigorous expositions in this topic, the interested reader may consult, for example, Glimm and Jaffe (1981) and Vol. II of Reed and Simon (1975).

Eq. (45.20) finally gives rise to

$$\langle q'', t'' \,|\, q', t' \rangle = \int [dp\,dq] \, \exp \left\{ \frac{i}{\hbar} \int_{t'(q')}^{t''(q'')} dt \, [p(t)\dot{q}(t) - H(p(t), q(t))] \right\} \,,$$

$$(45.27)$$

where the notation for the limits of the time-integral serves to specify the boundary conditions (45.24) for the paths $q(t)$ in the functional integration. This is a most remarkable result: whereas the left-hand-side is a Dirac bracket involving vectors in (rigged) Hilbert space, the right-hand-side is devoid of such vectors, and involves only classical mechanical quantities [in particular, the canonically invariant *symplectic form dpdq*, written as $dp \wedge dq$ in (44.45)]! The functional integration introduced above is also referred to picturesquely by Feynman as a "*sum over histories*" of a particle moving in physical space. Whereas in classical mechanics there is only one history, that determined by Newton's Laws, in quantum mechanics there are infinitely many histories, all of which have to be taken into account. According to (45.27), the weight of each history is precisely given by $\exp(\frac{i}{\hbar} \times$ the classical action) [c.f. (45.1)]. As $\hbar \to 0$, the classical path (history), the one that gives rise to a *stationary phase* in the exponential, contributes more and more dominantly in the functional integration, until, in the exact limit, only the classical path contributes. This is one way to understand the transition from quantum mechanics to classical mechanics.

Problem 45.2 From (45.9) and (45.16), show that the solution $\psi(q, t)$ to the time-dependent Schrödinger equation $\hat{H}\psi = i\hbar \dfrac{\partial \psi}{\partial t}$ is given formally by

$$\psi(q, t) = \langle q \,|\, \psi(t) \rangle = \int dq' \langle q \,|\, e^{-i\hat{H}t/\hbar} \,|\, q' \rangle \langle q' \,|\, \psi(0) \rangle \,. \qquad (45.28)$$

Show that the free-particle propagator, for which $\hat{H} = \hat{p}^2/2m$, is given by

$$\langle q, t \,|\, q', 0 \rangle = \langle q \,|\, e^{-i\hat{H}t/\hbar} \,|\, q' \rangle = \left(\frac{m\hbar}{2\pi i t} \right)^{1/2} \exp\left\{ \frac{im|q - q'|^2}{2\hbar t} \right\} \,.$$

The functional integral in (45.27) involving the symplectic form $dp \wedge dq$, however, has to be used with extreme care. The problem is that in Hamiltonian dynamics, the relevant symmetry group is the symplectic group (the group of canonical transformations in phase space). So broadly speaking, a quantization procedure must ideally establish a faithful representation of the symplectic group on phase space (symplectic manifold) by a group of unitary transformations on Hilbert space, a procedure known as *geometric quantization*. Such a representation only exists for certain subgroups of the symplectic group. This introduces delicate mathematical questions into the specification of the measure for $[dpdq]$, on top of those already referred to above. A better approach is to go from functional integration on phase space to functional integration on configuration space. This will lead to Feynman's original proposition, and will be discussed in the next chapter.

We will conclude this chapter with a discussion of the **scattering matrix** or *S*-matrix in the non-relativistic context, and attempt to cast it in terms of functional integrals. We consider a Hamiltonian function with a short-range potential $V(q)$ of the form

$$H(p,q) = H_0(p) + V(q), \quad H_0 = \frac{p^2}{2m}, \quad V(q) \to 0 \text{ as } |q| \to \infty. \quad (45.29)$$

Suppose that as $t \to t'$ and $t \to t''$ the particle is described by *wave packets* whose dynamics are governed by the free-particle Hamiltonian H_0. These wave packets will be designated $\psi_1(q, t')$ and $\psi_2(q, t'')$, respectively. They can be expressed in the momentum representation (Fourier transformed) as follows:

$$\psi_i(q,t) = \langle q, t \,|\, \psi_1 \rangle = \int dp \, \langle q, t \,|\, p \rangle \langle p \,|\, \psi_i \rangle = \int dp \, \langle q, t \,|\, p \rangle \, \tilde{\psi}_i(p)$$
$$= \int dp \, \langle q \,|\, e^{-it\hat{H}_0/\hbar} \,|\, p \rangle \, \tilde{\psi}_i(p) = \int dp \, e^{-i\frac{p^2 t}{2m\hbar}} \langle q \,|\, p \rangle \, \tilde{\psi}_i(p) , \quad (45.30)$$

where in the third equality we have defined

$$\tilde{\psi}_i(p) \equiv \langle p \,|\, \psi_i \rangle , \quad (45.31)$$

and in the fourth we have used (45.15). On using (45.4) for $\langle q \,|\, p \rangle$ we obtain

$$\psi_i(q,t) = \left(\frac{1}{2\pi}\right)^{1/2} \int_{-\infty}^{\infty} dp \, e^{ipq/\hbar} \, e^{-i\frac{p^2 t}{2m\hbar}} \, \tilde{\psi}_i(p)$$
$$\approx \sqrt{\frac{m\hbar}{|t|}} \, e^{i\frac{\pi}{4}\,\text{sign}\,t} \, e^{i\frac{mq^2}{2\hbar t}} \, \tilde{\psi}\left(\frac{mq}{t}\right) , \quad (45.32)$$

where the approximation can be obtained by using the method of *stationary phase* for the evaluation of the dp integral. [For a detailed discussion of this method, see, for example, Copson (1965).]

The *S*-matrix element $\langle \psi_2 \,|\, \hat{S} \,|\, \psi_1 \rangle$ will be defined by

$$\langle \psi_2 \,|\, \hat{S} \,|\, \psi_1 \rangle \equiv \lim_{\substack{t'' \to \infty \\ t' \to -\infty}} \int dq'' \int dq' \, \psi_2^*(q'',t'') \langle q'',t'' \,|\, q',t' \rangle \psi_1(q',t') . \quad (45.33)$$

We can extract an expression for the scattering operator $\hat{S}$ as follows. We have

$$\int dq'' \int dq'\, \psi_2^*(q'',t'')\langle q'',t'' \mid q',t'\rangle \psi_1(q',t')$$

$$= \int dq'' \int dq'\, \langle \psi_2 \mid q'',t''\rangle\langle q'' \mid e^{-i(t''-t')\hat{H}/\hbar} \mid q'\rangle\langle q',t' \mid \psi_1\rangle$$

$$= \int dq'' \int dq'\, \langle \psi_2 \mid e^{it''\hat{H}_0/\hbar} \mid q''\rangle\langle q'' \mid e^{-i(t''-t')\hat{H}/\hbar} \mid q'\rangle\langle q' \mid e^{-it'\hat{H}_0/\hbar} \mid \psi_1\rangle$$

$$= \langle \psi_2 \mid e^{it''\hat{H}_0/\hbar} \left(\int dq'' \mid q''\rangle\langle q'' \mid \right) e^{-i(t''-t')\hat{H}/\hbar} \left(\int dq' \mid q'\rangle\langle q' \mid \right) e^{-it'\hat{H}_0} \mid \psi_i\rangle$$

$$= \langle \psi_2 \mid e^{it''\hat{H}_0/\hbar}\, e^{-i(t''-t')\hat{H}/\hbar}\, e^{-it'\hat{H}_0} \mid \psi_i\rangle \;.$$

$$(45.34)$$

Thus

$$\boxed{\hat{S} = \lim_{\substack{t''\to\infty \\ t'\to-\infty}} e^{it''\hat{H}_0/\hbar}\, e^{-i(t''-t')\hat{H}/\hbar}\, e^{-it'\hat{H}_0/\hbar}} \;. \qquad (45.35)$$

We can also calculate the S-matrix in momentum representation, given by $\langle p'' \mid \hat{S} \mid p'\rangle$. In fact,

$$\langle \psi_2 \mid \hat{S} \mid \psi_1\rangle = \int dp'' \int dp'\, \langle \psi_2 \mid p''\rangle\langle p'' \mid \hat{S} \mid p'\rangle\langle p' \mid \psi_1\rangle$$

$$= \int dp'' \int dp'\, \tilde{\psi}_2^*(p'')\tilde{\psi}_1(p')\,\langle p'' \mid \hat{S} \mid p'\rangle \;. \qquad (45.36)$$

Substituting in (45.33) the approximate expression for $\psi_i(q,t)$ given by (45.32), we have

$$\langle \psi_2 \mid \hat{S} \mid \psi_1\rangle = \lim_{\substack{t''\to\infty \\ t'\to-\infty}} \sqrt{\frac{m\hbar}{|t''|}}\sqrt{\frac{m\hbar}{|t'|}} \left(e^{i\pi/4}\right)^*\left(e^{-i\pi/4}\right)$$

$$\times \int dq'' \int dq'\, \tilde{\psi}_2^*\left(\frac{mq''}{t''}\right)\tilde{\psi}_1\left(\frac{mq'}{t'}\right) e^{-i\frac{mq''^2}{2\hbar t''}}\, e^{i\frac{mq'^2}{2\hbar t'}}\,\langle q'',t'' \mid q',t'\rangle \;. \qquad (45.37)$$

Changing variables from q to p by the substitution $q = pt/m$, we then obtain

$$\langle \psi_2 \mid \hat{S} \mid \psi_1\rangle = \lim_{\substack{t''\to\infty \\ t'\to-\infty}} \left(\frac{\hbar}{m}\right)\sqrt{t''t'} \int dp'' \int dp'\, \tilde{\psi}_2^*(p'')\tilde{\psi}_1(p')$$

$$\times e^{-i\frac{p''^2 t''}{2m\hbar}}\, e^{i\frac{p'^2 t'}{2m\hbar}}\,\left\langle \frac{p''t''}{m},t'' \;\middle|\; \frac{p't'}{m},t' \right\rangle \;. \qquad (45.38)$$

Comparison with (45.36) shows that

$$\langle p'' \mid \hat{S} \mid p'\rangle = \lim_{\substack{t''\to\infty \\ t'\to-\infty}} \left(\frac{\hbar}{m}\right)\sqrt{t''t'}\, e^{-i\frac{p''^2 t''}{2m\hbar}}\, e^{i\frac{p'^2 t'}{2m\hbar}}\,\left\langle \frac{p''t''}{m},t'' \;\middle|\; \frac{p't'}{m},t' \right\rangle \;. \qquad (45.39)$$

The Dirac bracket in this result can be calculated using the functional integral (45.27), where the boundary conditions for $q(t)$ are not fixed points in configuration space, but given by the asymptotic trajectories

$$\lim_{t' \to -\infty} q(t') = \frac{p't'}{m} , \qquad \lim_{t'' \to \infty} q(t'') = \frac{p''t''}{m} . \qquad (45.40)$$

It may not be apparent from the formula (45.39) at this point, but it can be justified (although we will not do so here) that the double limit exists. In the next chapter we will see a related situation concerning infinite limits [see the discussion immediately following (46.34)].

Problem 45.3 Use the method of stationary phase to verify the approximate result for the wave packet $\psi_i(q, t)$ given in (45.32).

Chapter 46

Euclideanization, the Wiener Measure, and the Feynman-Kac Formula

The functional integral in (45.27) over paths $\{p(t), q(t)\}$ in phase space can be reduced to one over only paths $q(t)$ in configuration space, if the Hamiltonian $H(p, q)$ is quadratic in the momentum variable. In particular, for the familiar form

$$H(p, q) = \frac{p^2}{2m} + V(q) , \tag{46.1}$$

where m is the mass of the particle and $V(q)$ is a potential energy, the right-hand-side of (45.27) can be written as

$$\int [dq] \left[\int [dp] \exp\left(\frac{i}{\hbar} \int dt \left\{ p\dot{q} - \frac{p^2}{2m} \right\} \right) \right] \exp\left(-\frac{i}{\hbar} \int dt V(q) \right) . \tag{46.2}$$

The quantity within the square brackets above, the $[dp]$ functional integral (with $q(t)$ being held to be a fixed function), is said to be a *functional integral of the Gaussian type*. Later in this chapter we will explain the reason for this term [see discussion following (46.20)], and in the next one we will study more systematically the method of evaluation for such a type of functional integral, which is the only type that can be evaluated relatively easily. In the meantime we proceed with the reduction of (46.2).

On making the change of functional variables $p(t) \rightarrow p'(t) = p(t) - m\dot{q}(t)$ [this is allowed because the boundary conditions (45.24) only apply to the functions $q(t)$], we have

$$p\dot{q} - \frac{p^2}{2m} = (p' + m\dot{q})\dot{q} - \frac{(p' + m\dot{q})^2}{2m} = \frac{m\dot{q}^2}{2} - \frac{p'^2}{2m} . \tag{46.3}$$

Since $\int[dp] = \int[dp']$ the expression in (46.2) can be written as

$$\left(\int[dp']\exp\left\{-\frac{i}{\hbar}\int dt\,\frac{p'^2}{2m}\right\}\right)\left(\int[dq]\exp\left\{\frac{i}{\hbar}\int dt\left(\frac{1}{2}m\dot{q}^2 - V(q)\right)\right\}\right).$$

$$(46.4)$$

Making use of the functional "measure" defined in (45.26), the result (45.27) assumes the form

$$\langle q'', t''\,|\,q', t'\rangle = \frac{1}{\mathcal{N}}\int\prod_{i=1}^{N-1}dq(t_i)\,\exp\left\{\frac{i}{\hbar}\int_{t'(q')}^{t''(q'')}dt\left(\frac{1}{2}m(\dot{q}(t))^2 - V(q(t))\right)\right\},$$

$$(46.5)$$

where the normalization constant $\mathcal{N}$ is defined by

$$\frac{1}{\mathcal{N}} \equiv \int\prod_{i=1}^{N}\frac{dp(t_i)}{2\pi}\,\exp\left\{-\frac{i}{\hbar}\int_{t'}^{t''}dt\,\frac{(p(t))^2}{2m}\right\}.$$

$$(46.6)$$

Eq. (46.5) is the formula originally given by Feynman, which involves only the functional $\int[dq]$ on paths in configuration space. We will refer to it as **Feynman's formula**. Note that the time-integral in the exponent in this formula is the classical action and the integrand is the Lagrangian function.

The presence of the imaginary unit (the factors i) in the exponents lead to oscillatory phase factors. As we mentioned at the end of the last chapter, the heuristic argument is that most of these coming from the different paths $q(t)$ will tend to cancel out in the functional integration as $\hbar \to 0$ (the *stationary phase approximation*). But because of the imprecision in the definition of the functional measure, this argument cannot be made mathematically rigorous. One will be on firmer mathematical ground, however, if a so-called **Wick rotation** to Euclidean (imaginary) time, or a **Euclideanization** procedure, is made. This consists in *analytically continuing* the propagator function $\langle q''\,|\,e^{-i\hat{H}t/\hbar}\,|\,q'\rangle$ (considered as a function of the complex time t) from the real axis to the imaginary one:

$$e^{-i\hat{H}t/\hbar} \xrightarrow[t\to -it]{} e^{-\hat{H}t/\hbar} \qquad (t\text{ real}).$$

$$(46.7)$$

In other words, on the complex t plane, a value t is rotated from the real axis by $-\pi/2$ to the imaginary one. Formally we can achieve this by letting $dt \to -idt$ in (46.5), with t always real. Thus the exponent in that equation can be transformed as follows:

$$\frac{i}{\hbar}\int_{t'}^{t''}dt\left\{\frac{1}{2}m\left(\frac{dq}{dt}\right)^2 - V(q)\right\} \longrightarrow \frac{i}{\hbar}(-i)\int_{t'}^{t''}dt\left\{\frac{m}{2}\left(\frac{dq}{-idt}\right)^2 - V(q(t))\right\}$$

$$= -\frac{1}{\hbar}\int_{t'}^{t''}dt\left\{\frac{m}{2}\left(\frac{dq}{dt}\right)^2 + V(q(t))\right\}.$$

$$(46.8)$$

The name Euclideanization comes from the fact that if in Minkowski spacetime the substitution $t \to -it$ is made the Minkowski norm $x^2 + y^2 + z^2 - t^2$ of a four-vector turns into the Euclidean norm $x^2 + y^2 + z^2 + t^2$. After the Wick rotation, (46.5) reads

$$\langle q'' | e^{-\hat{H}(t''-t')/\hbar} | q' \rangle$$

$$= \frac{1}{\mathcal{N}_E} \int \prod_{i=1}^{N-1} dq(t_i) \, \exp\left\{ -\frac{1}{\hbar} \int_{t'(q')}^{t''(q'')} dt \, \left(\frac{1}{2} m (\dot{q}(t))^2 + V(q(t)) \right) \right\} , \qquad (46.9)$$

where $\mathcal{N}_E$, the Euclidean normalization constant, has been similarly analytically continued from the $\mathcal{N}$ given by (46.6):

$$\frac{1}{\mathcal{N}_E} \equiv \int \prod_{i=1}^{N} \frac{dp(t_i)}{2\pi} \, \exp\left\{ -\frac{1}{\hbar} \int_{t'}^{t''} dt \, \frac{(p(t))^2}{2m} \right\} . \qquad (46.10)$$

Eq. (46.9) is called the **Feynman-Kac formula**, in which the functional measure, called the conditional *Weiner measure*, can actually be rigorously defined (see development later in this chapter). Again, note that the "measure" in (46.5) (with the factor i) does not have a satisfactory mathematical definition, due largely to the lack of a positivity condition required of all integration measures [refer to the condition (46.21a) below].

At this point, we would like to point out what is probably one of the most remarkable analogies in theoretical physics: the operator $e^{-\hat{H}t/\hbar}$ resembles formally the one appearing in the *partition function Z* in quantum statistical mechanics, when Z is written in the form

$$Z = \text{tr} \, e^{-\beta\hat{H}} , \qquad (46.11)$$

where $\beta \equiv (kT)^{-1}$ (k = Boltzmann's constant, T = absolute temperature), and the trace is over all eigenstates of the Hamiltonian $\hat{H}$ (or over any complete set in the Hilbert space spanned by such eigenstates). Thus *imaginary time in quantum mechanics plays the role of an inverse temperature in quantum statistical mechanics!* Suppose the Hamiltonian $\hat{H}$ is bounded below by a discrete energy eigenvalue E_0 (the ground state energy), so that all excited state energies E_n, with (possible degenerate) eigenstates $|n, \alpha\rangle$, satisfy $E_n > E_0$ for all $n > 0$. Suppose further that the ground state $|0\rangle$ is non-degenerate. We have

$$e^{-\beta\hat{H}} = e^{-\beta\hat{H}} \left(|0\rangle\langle 0| + \sum_{n>0, \, \alpha} |n, \alpha\rangle\langle n, \alpha| \right)$$

$$= e^{-\beta E_0} |0\rangle\langle 0| + \sum_{n>0, \, \alpha} e^{-\beta E_n} |n, \alpha\rangle\langle n, \alpha|$$

$$= e^{-\beta E_0} \left(|0\rangle\langle 0| + \sum_{n>0, \, \alpha} e^{-\beta(E_n - E_0)} |n, \alpha\rangle\langle n, \alpha| \right) \qquad (46.12)$$

$$= e^{-\beta E_0} \left(|0\rangle\langle 0| + O\left(e^{-\beta(E_1 - E_0)} \right) \right) .$$

Keeping in mind that $E_1 - E_0 > 0$, we see that

$$\lim_{\beta \to \infty} e^{-\beta \hat{H}} = e^{-\beta E_0} |0\rangle \langle 0| . \tag{46.13}$$

Thus $e^{-\beta \hat{H}}$ projects onto the ground state as $\beta \to \infty$, and the ground state energy E_0 can be written as

$$E_0 = \lim_{\beta \to \infty} \left(-\frac{1}{\beta} \ln tr\, e^{-\beta \hat{H}} \right) . \tag{46.14}$$

Returning to quantum mechanics and exploiting the analogy described above, we can then use the projection operator

$$\lim_{t \to \infty} e^{-\hat{H}t/\hbar} = \lim_{t \to -i\infty} e^{-i\hat{H}t/\hbar} \tag{46.15}$$

to "pick out" the ground state $|0\rangle$ of the system. This technique will be useful in our subsequent development (see Chapter 49).

To explore the implications of the analogy further we note that, on letting $t \to -it$ in the "free-particle" Schrödinger equation

$$\hat{H}_0 \psi = -\frac{\hbar^2}{2m} \frac{\partial^2 \psi}{\partial q^2} = i\hbar \frac{\partial \psi}{\partial t} , \tag{46.16}$$

one obtains the **heat equation** in classical physics:

$$\frac{\hbar}{2m} \frac{\partial^2 \psi}{\partial q^2} = \frac{\partial \psi}{\partial t} . \tag{46.17}$$

With the initial condition $\psi(q, 0) = f(q)$, the formal solution of this equation is readily seen to be

$$\psi(q, t) = e^{-t\hat{H}_0/\hbar} f(q) ; \tag{46.18}$$

and is given explicitly by

$$\psi(q, t) = \int_{-\infty}^{\infty} dq'\, K_0(q, q'; t) f(q') , \tag{46.19}$$

where $K_0(q, q'; t)$ is the *Gaussian distribution* (the subscript 0 corresponds to the free Hamiltonian $\hat{H}_0$)

$$K_0(q, q'; t) = \sqrt{\frac{m}{2\pi \hbar t}} \exp\left\{ -\frac{m(q - q')^2}{2\hbar t} \right\} . \tag{46.20}$$

This is the reason why functional integrals of the type studied in this chapter are referred to as Gaussian. The quantity K_0 is said to be the **integral kernel** of the differential operator $e^{-t\hat{H}_0/\hbar}$, where $\hat{H}_0 = -(\hbar^2/(2m))\partial^2/\partial q^2$. It has three basic properties:

$$K_0(q, q'; t) > 0 , \tag{46.21a}$$

$$\int_{-\infty}^{\infty} dq' K_0(q, q'; t) = 1 , \tag{46.21b}$$

$$K_0(q, q'; t + s) = \int_{-\infty}^{\infty} dq'' K_0(q, q''; t) K_0(q'', q'; s) . \tag{46.21c}$$

The first two allow one to interpret $K_0(q, q'; t)$ as a *probability density*, while the last ensures that the operators $e^{-t\hat{H}_0}$ form a one-parameter **semigroup**, satisfying the property:

$$e^{-(t+s)\hat{H}_0} = e^{-t\hat{H}_0} e^{-s\hat{H}_0} . \tag{46.22}$$

Problem 46.1 Verify that (46.19) is a solution of the heat equation (46.17) by direct differentiation.

Problem 46.2 Use the explicit form for the kernel $K_0(q, q'; t)$ given by (46.20) to verify the properties (46.21a), (46.21b) and (46.21c).

Problem 46.3 Use (46.18), (46.19) and the property for $K_0(q, q'; t)$ (46.21c) to verify the semigroup property (46.22) for $e^{-t\hat{H}_0}$.

The left-hand-side of (46.9), with $\hat{H}$ replaced by $\hat{H}_0$, is by definition the kernel $K_0(q'', q'; t'' - t')$ [compare the result in Problem 45.2 with (46.19)]. So we can use that equation to write

$$K_0(q'', q'; t'' - t') \equiv \text{kernel} \left(e^{-(t''-t')\hat{H}_0/\hbar} \right)$$

$$= \frac{1}{\mathcal{N}_E} \int \prod_{i=1}^{N-1} dq(t_i) \exp \left\{ -\frac{1}{\hbar} \int_{t'(q')}^{t''(q'')} dt \, \frac{m(\dot{q}(t))^2}{2} \right\} \tag{46.23}$$

$$\equiv \int dW(q'', q'; t'' - t') ,$$

where $dW(q, q'; t)$ denotes the (conditional) *Wiener measure*, and the middle expression (with our usual product sign) is taken to be a symbolic designation of that measure, defined on the space $\mathcal{W}(q, q'; t)$ of continuous paths with the boundary conditions $q(0) = q'$ and $q(t) = q$. (From now on we will refer to the conditional Wiener mesaure simply as the Wiener measure, the difference is that the Wiener measure prescribes only the initial boundary condition whereas the conditional one prescribes boundary conditions at both the initial and final times.) Note that the definition of the Wiener measure involves the kinetic energy term $m\dot{q}^2/2$ in a vital manner, which puts restrictions on the smoothness of allowed paths in the space of paths under consideration. With this notation the Feynman-Kac formula given in (46.9) can also be expressed as

$$K(q'',q';t''-t') \equiv \text{kernel}\left(e^{-(t''-t')\hat{H}/\hbar}\right)$$
$$= \int dW(q'',q';t''-t')\,\exp\left\{-\frac{1}{\hbar}\int_{t'(q')}^{t''(q'')} dt\, V(q(t))\right\}. \tag{46.24}$$

We will now provide a qualitative introduction to the technical definition of the Wiener measure, which will be constructed from the kernels $K_0(p,q;t)$. Our presentation is adapted from Glimm and Jaffe (1981). First consider a subset of $\mathcal{W}(q'',q';t''-t')$ given by

$$C(I_1,t_1) = \{q(s) \in \mathcal{W}(q'',q';t''-t') \,|\, q(t_1) \in I_1 \subset \mathbb{R},\, t' < t_1 < t''\}, \tag{46.25}$$

namely, all paths in the space $\mathcal{W}(q'',q';t''-t')$ of continuous paths which pass through some point in an open interval $I_1 \subset \mathbb{R}$ (known as *Borel subsets* of $\mathbb{R}$ in measure theory) at the time t_1. We define a measure (or "mass") $M(C(I_1,t_1))$ of the subset $C(I_1,t_1)$ by

$$M(C(I_1,t_1)) \equiv \int_{I_1} dq_1\, K_0(q'',q_1;t''-t_1)K_0(q_1,q';t_1-t'). \tag{46.26}$$

If $I_1 = \mathbb{R}$, $C(\mathbb{R},t_1)$ is apparently the entire space $\mathcal{W}(q'',q';t''-t')$, since any path must pass through some point in $\mathbb{R}$ at a given time t_1. Then property (46.21c) of the kernel K_0 implies the following measure-theoretic interpretation of the kernel K_0:

$$M(\mathcal{W}(q'',q';t''-t')) = K_0(q'',q';t''-t'). \tag{46.27}$$

This is also called the "total mass" of the space $\mathcal{W}(q'',q';t''-t')$.

Now consider the time sequence $t' = t_0 < t_1 < t_2 < \cdots < t_{N-1} < t_N = t''$ as in the scheme specified below (45.23). We can generalize from $C(I_1,t_1)$ to the sets $C(I_1,t_1; I_2,t_2;\ldots; I_{N-1},t_{N-1})$, for open intervals (Borel subsets) $I_1, I_2, \ldots, I_{N-1}$ in $\mathbb{R}$, defined by

$$\begin{aligned} &C(I_1,t_1; I_2,t_2;\ldots; I_{N-1},t_{N-1}) \\ &\equiv \{q(s) \in \mathcal{W}(q'',q';t''-t') \,|\, q(t_1) \in I_1, \ldots, q(t_{N-1}) \in I_{N-1}\}. \end{aligned} \tag{46.28}$$

These are so-called **cylinder sets** in $\mathcal{W}(q'',q';t''-t')$. Analogous to (46.26) their measures are given by

$$M(C(I_1,t_1; I_2,t_2;\ldots; I_{N-1},t_{N-1})) \equiv \int_{I_1} dq_1 \int_{I_2} dq_2 \ldots \int_{I_{N-1}} dq_{N-1}$$
$$K_0(q'',q_{N-1};t''-t_{N-1})K_0(q_{N-1},q_{N-2};t_{N-1}-t_{N-2})\ldots K_0(q_1,q';t_1-t'). \tag{46.29}$$

The above equation rigorously defines the **Wiener measure** in the continuous-path space $W(q'', q'; t'' - t')$. Its existence is guaranteed by the following theorem, which we will state without proof.

Theorem 46.1. *The (conditional) Wiener measure given by (46.29) is countably additive on the cylinder subsets of $W(q'', q'; t'' - t')$ and has a unique extension from the cylinder subsets to all Borel subsets of $W(q'', q'; t'' - t')$.*

To conclude this chapter we will verify that the rigorous definition of the Wiener measure (46.29) agrees with the heuristic one introduced in (46.23). On setting $I_1 = I_2 = \cdots = I_{N-1} = \mathbb{R}$, we infer from (46.27) that

$$M(C(\mathbb{R}, t_1; \mathbb{R}, t_2; \ldots; \mathbb{R}, t_{N-1})) = M(W(q'', q'; t'' - t'))$$
$$= K_0(q'', q'; t'' - t') = \text{kernel} \left(e^{-(t''-t')\hat{H}_0/\hbar} \right) . \tag{46.30}$$

Substituting the explicit expression for K_0 given by (46.20) in (46.29), in which we set $I_1 = \ldots I_{N-1} = \mathbb{R}$, we obtain

$$\text{kernel} \exp \left(e^{-(t''-t')\hat{H}_0/\hbar} \right)$$

$$= \int_{-\infty}^{\infty} dq_1 \int_{-\infty}^{\infty} dq_2 \ldots \int_{-\infty}^{\infty} dq_{N-1} \left(\frac{m}{2\pi\hbar} \right)^{N/2} \left(\frac{1}{t'' - t_{N-1}} \right)^{1/2} \cdots \left(\frac{1}{t_1 - t'} \right)^{1/2}$$

$$\times \exp \left\{ -\frac{m(q'' - q_{N-1})^2}{2\hbar(t'' - t_{N-1})} \right\} \exp \left\{ -\frac{m(q_{N-1} - q_{N-2})^2}{2\hbar(t_{N-1} - t_{N-2})} \right\} \ldots \ldots$$

$$\times \exp \left\{ -\frac{m(q_1 - q')^2}{2\hbar(t_1 - t')} \right\}$$

$$= \left(\frac{m}{2\pi\hbar} \right)^{N/2} \int_{-\infty}^{\infty} dq_1 \ldots \int_{-\infty}^{\infty} dq_{N-1} \prod_{j=1}^{N} (t_j - t_{j-1})^{-1/2} \exp \left\{ -\frac{m(q_j - q_{j-1})^2}{2\hbar(t_j - t_{j-1})} \right\}$$

$$= \left(\frac{m}{2\pi\hbar} \right)^{N/2} \int_{-\infty}^{\infty} dq_1 \ldots \int_{-\infty}^{\infty} dq_{N-1} \prod_{j=1}^{N} (\Delta t_j)^{-1/2} \exp \left\{ -\frac{m(\Delta q_j)^2}{2\hbar \Delta t_j} \right\} , \tag{46.31}$$

where

$$\Delta t_j \equiv t_j - t_{j-1} , \qquad \Delta q_j \equiv q_j - q_{j-1} . \tag{46.32}$$

As in the previous chapter, let $\Delta t_j = \epsilon = (t'' - t')/N$. So

$$\prod_{j=1}^{N} (t_j - t_{j-1})^{-1/2} = \underbrace{\left(\frac{1}{\epsilon} \right)^{1/2} \cdots \left(\frac{1}{\epsilon} \right)^{1/2}}_{N \text{ factors}} = \frac{1}{\epsilon^{N/2}} = \left(\frac{N}{t'' - t'} \right)^{N/2} . \tag{46.33}$$

Finally the result (46.31) can be expressed as

$$\text{kernel}\left(e^{-(t''-t')\hat{H}_0/\hbar}\right)$$

$$= \lim_{N\to\infty}\left(\frac{mN}{2\pi\hbar(t''-t')}\right)^{N/2}\left(\prod_{i=1}^{N-1}\int_{-\infty}^{\infty}dq_i\right)\exp\left\{-\frac{m}{2\hbar}\sum_{j=1}^{N}\left(\frac{\Delta q_j}{\Delta t_j}\right)^2\Delta t_j\right\}.$$

$$(46.34)$$

This result formally resembles (46.23). The essence of Theorem 46.1 is that the limit $N \to \infty$ in the above expression actually exists, so that one can manipulate formally (and correctly) with the heuristic result (46.23). The main conclusion to be drawn from this chapter is that when calculating a quantum mechanical propagator function one can first Euclideanize the time-evolution operator $e^{-it\hat{H}/\hbar}$ (by letting $t \to -it$), use the Wiener measure to calculate the Euclideanized propagator function, and finally analytically continue the result $(t \to it)$ to obtain the original quantum mechanical propagator. This procedure is mathematically rigorous when the Euclidean functional integral is of Gaussian type, and is the one most frequently used by physicists.

As a historical note, the Wiener measure was originally formulated by the mathematician Norbert Wiener in 1923 to provide a more mathematically rigorous foundation for a statistical mechanical treatment of the problem of Brownian motion, initiated by Einstein, Smoluchowski, Langevin, and Ornstein and Uhlenbeck, among others, in the early part of the twentieth century. [For an in-depth technical account of this history, the reader can consult Nelson (1967).] Wiener's work was of great importance in the development of the mathematical theory of stochastic processes.

Chapter 47

Functional Integrals of the Gaussian Type

Having developed quantum mechanics using functional integration in the last two chapters, we will use functional methods again to formulate quantum field theory. The fact that this can be done relatively straightforwardly testifies to the power and usefulness of the technique of functional integration. As it turns out, in this more general theory, functional integrals of the Gaussian type also play a prominent role. As preparation, we will first develop in this chapter some basic techniques for working out this type of functional integrals, a typical one of which is the $[dp]$ functional integral inside the square brackets of (46.2) [with the $\dot{q}(t)$ held to be a fixed function of t]. After Euclideanization we rewrite it generically using a different set of symbols as follows:

$$\left(\int \prod_{i=1}^{N} d\phi(t_i) \right) \exp \left\{ -\frac{\alpha}{2} \int_{t'}^{t''} dt \, (\phi(t))^2 + \int_{t'}^{t''} dt \, J(t)\phi(t) \right\} \equiv Z_E^{(0)}[J] \, ,$$

(47.1)

where α is a constant, $\phi(t)$ is the function of t to be functionally integrated, and $J(t)$ is some specified function of t. Essentially, then, a **functional integral of the Gaussian type** is one in which the function to be integrated over functionally appears quadratically in the integrand of an expression of the above type. The symbol Z for the name of this integral is chosen with (46.11) in mind, so we can think of it as a kind of "partition function", a term borrowed from statistical mechanics (recall the analogy between the time evolution operator $e^{-it\hat{H}/\hbar}$ in quantum mechanics and the operator $e^{-\beta\hat{H}}$ in quantum statistical mechanics discussed in the last chapter). The subscript E refers to a Euclideanized functional integral, and the superscript (0) designates the fact that the potential energy term vanishes: $V(\phi) = 0$. In fact, the symbols ϕ and J have been chosen to anticipate their use in quantum field theory in the next chapter, as a *scalar field* and the *source*, respectively.

The key to working out this functional integral is to remember that while performing the integral, in the symbol $\phi(t)$, t is to be formally treated as a label rather than an independent variable. Thus we can write $\phi(t_i) \equiv \phi_i$ ($i = 1, \ldots, N$), and

$$\int \prod_{i=1}^{N} d\phi(t_i) = \int_{-\infty}^{\infty} d\phi_1 \int_{-\infty}^{\infty} d\phi_2 \ldots \int_{-\infty}^{\infty} d\phi_N \ . \tag{47.2}$$

The ranges in the integrals reflect the facts that $\phi(t)$ is a real-valued function of the variable t with unrestricted values in $\mathbb{R}$, and the boundary condition at the end-time $t_N = t''$ has been removed. We can also rewrite the integral in the exponent as a sum (with $t'' - t' = N\epsilon$, and the understanding that $N \to \infty, \epsilon \to 0$):

$$-\frac{\alpha}{2} \int_{t'}^{t''} dt \, (\phi(t))^2 + \int_{t'}^{t''} dt \, J(t)\phi(t) \approx -\frac{\alpha\epsilon}{2} \sum_{i=1}^{N} (\phi(t_i))^2 + \epsilon \sum_{i=1}^{N} J(t_i)\phi(t_i)$$

$$= -\frac{\alpha\epsilon}{2} \sum_{i=1}^{N} (\phi_i)^2 + \epsilon \sum_{i=1}^{N} J_i \phi_i \qquad (J_i \equiv J(t_i)) \quad . \tag{47.3}$$

Then

$$Z_E^{(0)}[J] = \lim_{\substack{N \to \infty \\ \epsilon \to 0}} \int_{-\infty}^{\infty} d\phi_1 \ldots \int_{-\infty}^{\infty} d\phi_N \, \exp\left\{ -\frac{\alpha\epsilon}{2} \sum_{i=1}^{N} (\phi_i)^2 + \epsilon \sum_{i=1}^{N} J_i \phi_i \right\} . \tag{47.4}$$

It will be useful to rewrite the exponent in the form

$$-\frac{1}{2} \sum_{i,j=1}^{N} \phi_i K_{ij} \phi_j + \sum_{i=1}^{N} I_i \phi_i \ , \tag{47.5}$$

where K_{ij} is the symmetric matrix given by $K_{ij} = \alpha\epsilon\delta_{ij}$ and $I_i = \epsilon J_i$. Aside from the limits, that is, considering N and ϵ to be both finite, we are then dealing with an ordinary (not functional) *Gaussian multiple integral* $\mathcal{I}_E[K, I]$:

$$\boxed{\mathcal{I}_E[K, I] \equiv \int_{-\infty}^{\infty} d\phi_1 \ldots \int_{-\infty}^{\infty} d\phi_N \, \exp\left\{ -\frac{1}{2} \sum_{i,j=1}^{N} \phi_i K_{ij} \phi_j + \sum_{i=1}^{N} I_i \phi_i \right\}} \ . \tag{47.6}$$

For future use it will be useful in the above generic expression for a Gaussian integral to let K_{ij} be any $N \times N$ real symmetric matrix with all eigenvalues λ_i non-zero (so that it is diagonalizable and invertible) and satisfying $\lambda_i \geq 0$, and I_i be any real N-vector. (Recall from Chapter 17 that eigenvalues of hermitian and thus real symmetric matrices are always real.)

To evaluate the (ordinary) integral (47.6) we first denote the column matrices $(\phi_1, \ldots, \phi_N)^T$ and $(I_1, \ldots, I_N)^T$ by ϕ and I, respectively, where T means transpose. The exponent in (47.6) can then be written in terms of matrix products as

$$-\frac{1}{2}\phi^T K \phi + I^T \phi , \tag{47.7}$$

where ϕ^T and I^T are row matrices. Introduce the change of variables $\phi_i \to \chi_i$:

$$\phi_i = \sum_j (K^{-1})_{ij} I_j + \chi_i , \tag{47.8}$$

or in matrix notation

$$\phi = K^{-1}I + \chi , \tag{47.9}$$

where χ is the column matrix $(\chi_1, \ldots, \chi_N)^T$. Note that $\sum_j (K^{-1})_{ij} I_j$ are the *stationary values* of the quadratic form (47.7), that is, values of ϕ_i for which

$$\frac{d}{d\phi_k} \left(\sum_{i,j} \frac{1}{2} \phi_i K_{ij} \phi_j - \sum_i I_i \phi_i \right) = 0 . \tag{47.10}$$

Then $\phi^T = I^T (K^{-1})^T + \chi^T = I^T K^{-1} + \chi^T$, since K^{-1} is symmetric, which follows from the fact that K is symmetric by assumption. Indeed, using Einstein's summation convention, we have $K_i^j (K^{-1})_j^k = \delta_i^k$. This implies, since $K_i^j = K_j^i$, $\sum_j K_j^i (K^{-1})_j^k = \delta_i^k = \delta_k^i$. But we also have $\sum_j (K^{-1})_k^j K_j^i = \delta_k^i$. Thus $(K^{-1})_j^k = (K^{-1})_k^j$. We then have

$$
\begin{aligned}
&-\frac{1}{2}\phi^T K \phi + I^T \phi \\
&= -\frac{1}{2}\{(I^T K^{-1} + \chi^T) K (K^{-1}I + \chi)\} + I^T(K^{-1}I + \chi) \\
&= -\frac{1}{2}\{I^T K^{-1}I + \chi^T I + I^T \chi + \chi^T K \chi\} + I^T K^{-1}I + I^T \chi \\
&= \frac{1}{2}(I^T K^{-1}I - \chi^T K \chi) .
\end{aligned} \tag{47.11}
$$

By (47.8),

$$\int_{-\infty}^{\infty} d\phi_1 \ldots \int_{-\infty}^{\infty} d\phi_N = \int_{-\infty}^{\infty} d\chi_1 \ldots \int_{-\infty}^{\infty} d\chi_N , \tag{47.12}$$

so it follows from (47.6) and (47.11) that

$$
\mathcal{I}_E[K, I] = \exp\left\{ \frac{1}{2} \sum_{i,j=1}^{N} I_i (K^{-1})_{ij} I_j \right\}
$$
$$
\times \int_{-\infty}^{\infty} d\chi_1 \ldots \int_{-\infty}^{\infty} d\chi_N \exp\left\{ -\frac{1}{2} \sum_{i,j=1}^{N} \chi_i K_{ij} \chi_j \right\} . \tag{47.13}
$$

To complete the evaluation of the integral we diagonalize K with an orthogonal transformation O (c.f. Corollary 17.2):

$$K = O^T D O , \tag{47.14}$$

where $D = diag\,(\lambda_1, \ldots, \lambda_N)$ is the $N \times N$ diagonal matrix whose diagonal elements are the non-zero eigenvalues λ_i of K. We then have

$$-\frac{1}{2}\chi^T K\chi = -\frac{1}{2}\chi^T O^T DO\chi = -\frac{1}{2}\psi^T D\psi = \frac{1}{2}\sum_{i=1}^{N}\lambda_i(\psi_i)^2 \,, \tag{47.15}$$

where the column vector $\psi \equiv (\psi_1, \ldots, \psi_N)^T = O\chi$. Since O is orthogonal, the change of variables $\chi_i \to \psi_i$ amounts to just a rotation in the N-dimensional Euclidean space $\mathbb{R}^N$, and so

$$\int_{-\infty}^{\infty} d\chi_1 \ldots \int_{-\infty}^{\infty} d\chi_N = \int_{-\infty}^{\infty} d\psi_1 \ldots \int_{-\infty}^{\infty} d\psi_N \,. \tag{47.16}$$

Thus

$$\begin{aligned}
&\int_{-\infty}^{\infty} d\chi_1 \ldots \int_{-\infty}^{\infty} d\chi_N \exp\left\{-\frac{1}{2}\sum_{i,j=1}^{N}\chi_i K_{ij}\chi_j\right\} \\
&= \int_{-\infty}^{\infty} d\psi_1 \ldots \int_{-\infty}^{\infty} d\psi_N \exp\left\{-\frac{1}{2}\sum_{i=1}^{N}\lambda_i(\psi_i)^2\right\} \\
&= \prod_{i=1}^{N}\int_{-\infty}^{\infty} d\psi_i \exp\left\{-\frac{1}{2}\lambda_i(\psi_i)^2\right\} \\
&= \prod_{i=1}^{N}\left(\frac{2\pi}{\lambda_i}\right)^{1/2} = \frac{(2\pi)^{N/2}}{(det\,K)^{1/2}} \,.
\end{aligned} \tag{47.17}$$

Finally,

$$\boxed{\mathcal{I}_E[K, I] = \frac{(2\pi)^{N/2}}{(det\,K)^{1/2}} \exp\left\{\frac{1}{2}\sum_{i,j=1}^{N}I_i(K^{-1})_{ij}I_j\right\}} \,. \tag{47.18}$$

With this result we can go back to (47.4) and rewrite that equation as

$$Z_E^{(0)}[J] = \lim_{\substack{N\to\infty \\ \epsilon\to 0}}\left(\frac{2\pi}{\alpha\epsilon}\right)^{N/2}\exp\left\{\frac{\epsilon^2}{2\alpha}\sum_{i,j=1}^{N}J_i\frac{\delta_{ij}}{\epsilon}J_j\right\} \,. \tag{47.19}$$

It is not at all obvious how the limit is to be taken. But we will resort to the following heuristic steps. First re-express the double sum in the exponent as a double integral over the time variable [recall the reverse step in (47.3)]:

$$\lim_{\substack{N\to\infty \\ \epsilon\to 0}}\frac{\epsilon^2}{2\alpha}\sum_{i,j=1}^{N}J_i\frac{\delta_{ij}}{\epsilon}J_j \approx \frac{1}{2\alpha}\int_{t'}^{t''}d\tau\int_{t'}^{t''}d\tau'\,J(\tau)\delta(\tau - \tau')J(\tau') \,, \tag{47.20}$$

where the Kronecker delta has been replaced by the Dirac delta. Then set $J = 0$ in (47.19) and write

$$Z_E^{(0)}[J = 0] = \lim_{\substack{N \to \infty \\ \epsilon \to 0}} \left(\frac{2\pi}{\alpha\epsilon} \right)^{N/2} , \tag{47.21}$$

which clearly diverges! [Compare with the first factor of the three in (46.34).] But as we will see later, the quantity $Z_E^{(0)}[J = 0]$ is of no physical significance. The important thing is that it factors out from $Z_E^{(0)}[J]$. It can be regarded as a normalization "constant", and will always be cancelled out (or be ignored) in the calculation of physically observable quantities, such as S-matrix elements. We then write (47.4) as

$$Z_E^{(0)}[J] = Z_E^{(0)}[0] \exp\left\{ \frac{1}{2\alpha} \int_{t'}^{t''} d\tau \int_{t'}^{t''} d\tau' \, J(\tau)\delta(\tau - \tau')J(\tau') \right\} . \tag{47.22}$$

Even though one of the time integrals can be immediately done trivially because of the presence of the Dirac delta, we keep the latter in the above expression because of its suggestive form. The Dirac delta here can be interpreted as the inverse of a propagator function from the time τ' to the time τ.

To pursue this point further we re-examine the functional integral in (46.34). Again reconverting the sum in the exponent back to an integral we have

$$-\frac{m}{2\hbar} \sum_{j=1}^{N} \left(\frac{\Delta q_j}{\Delta t_j} \right)^2 \Delta t_j \longrightarrow -\frac{m}{2\hbar} \int_{t'}^{t''} dt \left(\frac{dq(t)}{dt} \right)^2$$

$$= -\frac{m}{2\hbar} \left\{ q(t)\frac{dq}{dt}\Big|_{t'}^{t''} - \int_{t'}^{t''} dt \, q(t)\frac{d^2q}{dt^2} \right\} \tag{47.23}$$

$$= -\frac{m}{2\hbar} \int_{t'}^{t''} d\tau \int_{t'}^{t''} d\tau' \, q(\tau) \left(-\frac{d^2}{d\tau^2}\delta(\tau - \tau') \right) q(\tau') ,$$

where in the second equality it has been assumed that the boundary term vanishes ($q(t') = q(t'') = 0$). We can write this double integral in the form

$$-\frac{1}{2} \int_{t'}^{t''} d\tau \int_{t'}^{t''} d\tau' \, q(\tau)K(\tau - \tau')q(\tau') , \tag{47.24}$$

where now $K(\tau - \tau') = -(m/\hbar)(d^2/d\tau^2)\delta(\tau - \tau')$, which can also be interpreted as a propagator from the time τ' to the time τ. As our development earlier in this chapter demonstrated, the evaluation of the functional integral

$$Z_E^{(0)}[J] = \prod_{i=1}^{N-1} \int_{-\infty}^{\infty} dq_i$$

$$\times \exp\left\{ -\frac{1}{2} \int_{t'}^{t''} d\tau \int_{t'}^{t''} d\tau' \, q(\tau)K(\tau, \tau')q(\tau') + \int_{t'}^{t''} d\tau \, J(\tau)q(\tau) \right\} , \tag{47.25}$$

for the case of a general kernel $K(\tau, \tau')$, involves working out its inverse [c.f. (47.18)]. The result is formally given by [compare with (47.22)]

$$Z_E^{(0)}[J] = Z_E^{(0)}[J = 0] \exp\left\{\frac{1}{2}\int_{t'}^{t''} d\tau \int_{t'}^{t''} d\tau'\, J(\tau) K^{-1}(\tau, \tau') J(\tau')\right\} . \quad (47.26)$$

One can directly apply (47.18) to the calculation of functional integrals of Gaussian type with a complex exponent [for example, the one in (46.5)] without first going through the Euclideanization procedure. Starting with the Euclideanized functional integral in (47.25), one can undo the Wick rotation by letting $d\tau \to id\tau$. Using the example in (47.23), for instance, the exponent in (47.25) can be "un-Wick-rotated" as follows:

$$-\frac{1}{2}\int_{t'}^{t''} d\tau \int_{t'}^{t''} d\tau'\, q(\tau)\, K(\tau, \tau')\, q(\tau') + \int_{t'}^{t''} d\tau\, J(\tau) q(\tau)$$

$$= -\frac{1}{2}\int_{t'}^{t''} d\tau \int_{t'}^{t''} d\tau'\, q(\tau)\left(-\frac{m}{\hbar}\frac{d^2}{d\tau^2}\delta(\tau - \tau')\right) q(\tau') + \int d\tau\, J(\tau) q(\tau)$$

$$\xrightarrow{d\tau \to id\tau} -\frac{1}{2}\int_{t'}^{t''} id\tau \int_{t'}^{t''} d\tau'\, q(\tau)\left(-\frac{m}{\hbar}\frac{d^2}{(id\tau)(id\tau)}\delta(\tau - \tau')\right) q(\tau')$$

$$+ i\int_{t'}^{t''} J(\tau) q(\tau)$$

$$= \frac{i}{2}\int_{t'}^{t''} d\tau \int_{t'}^{t''} d\tau'\, q(\tau)\left(-\frac{m}{\hbar}\frac{d^2}{d\tau^2}\delta(\tau - \tau')\right) q(\tau') + i\int_{t'}^{t''} d\tau\, J(\tau) q(\tau)$$

$$= \frac{i}{2}\int_{t'}^{t''} d\tau \int_{t'}^{t''} d\tau'\, q(\tau) K(\tau, \tau')\, q(\tau') + i\int_{t'}^{t''} d\tau\, J(\tau) q(\tau) .$$

$$(47.27)$$

This transformation can also be achieved directly by making the substitutions

$$K \longrightarrow -iK , \qquad\qquad J \longrightarrow iJ . \qquad\qquad (47.28)$$

Going back to the definition of the ordinary Gaussian integral $\mathcal{I}_E[K, I]$ in (47.6), one can analytically continue from that definition, by letting $K_{lm} \to -iK_{lm}$, $I_l \to iI_l$, to obtain

$$\mathcal{I}[K, I] \equiv \int_{-\infty}^{\infty} d\phi_1 \dots \int_{-\infty}^{\infty} d\phi_N\, \exp\left\{\frac{i}{2}\sum_{l,m=1}^{N} \phi_l K_{lm}\phi_m + i\sum_{l=1}^{N} I_l\phi_l\right\} ,$$

$$(47.29)$$

where K is still a real symmetric matrix and I a real N-vector. Applying (47.28) to (47.18) then yields

$$\mathcal{I}[K, I] = \frac{(2\pi i)^{N/2}}{(det\, K)^{1/2}} \exp\left\{-\frac{i}{2}\sum_{l,m=1}^{N} I_l(K^{-1})_{lm}I_m\right\} . \quad (47.30)$$

The result for the complex Gaussian functional integral corresponding to (47.25) and (47.26) is then

$$
Z^{(0)}[J] = \prod_{i=1}^{N-1} \int_{-\infty}^{\infty} dq_i
$$

$$
\times \exp\left\{ \frac{i}{2} \int_{t'}^{t''} d\tau \int_{t'}^{t''} d\tau'\, q(\tau) K(\tau, \tau') q(\tau') + i \int_{t'}^{t''} d\tau\, J(\tau) q(\tau) \right\}
$$

$$
= Z^{(0)}[J = 0]\, \exp\left\{ -\frac{i}{2} \int_{t'}^{t''} d\tau \int_{t'}^{t''} d\tau'\, J(\tau)\, K^{-1}(\tau, \tau')\, J(\tau') \right\} .
$$

$$(47.31)$$

This result will be generalized in subsequent chapters to applications in quantum field theory.

Chapter 48

Quantum Fields, the Generating Functional, and the Feynman Propagator

It is but a few short steps, if only formally, to transition from quantum mechanics (the quantum theory of finitely many degrees of freedom) to quantum field theory (the quantum theory of infinitely many degrees of freedom) using the techniques of functional integration. Traditionally, in quantum field theory, quantization is usually carried out using a procedure known as **canonical quantization**. which involves treating classical fields (functions of space-time) as operators on Hilbert space, and imposing (equal-time) commutation relationships between the field-theory analogs of canonical momenta and coordinates, similar to those entailed by (45.2). This approach, though rigorous and useful, leads to many results that are not manifestly Lorentz covariant, because space and time are not treated on an equal footing. In contrast, the functional approach, though lacking in rigor, appears to be more straightforward and can be made to be Lorentz covariant from the outset. In this book, we will not dwell on canonical quantization at any length, although it must be mentioned that both it and the functional approach have their respective advantages and drawbacks. Our focus will be entirely on functional methods.

To make the transition we begin with Feynman's formula (46.5), which is the basic functional integral result for a quantum mechanical system with one degree of freedom: $q(t)$. One can readily generalize this result to a system with finitely many degrees of freedom: $q_\alpha(t)$, $\alpha = 1, \ldots, n$. Recall that our heuristic approach to the functional integration measure was to discretize a certain interval (t', t'') of the time axis into a large number of discrete points $t_0, t_1, t_2, \ldots, t_N$ with $t_0 = t'$, $t_N = t''$, $t'' - t' = N\epsilon$ and $t_i = t' + i\epsilon$, and to write

$$\int [dq] = \lim_{\epsilon \to 0} \int \prod_{i=1}^{N-1} dq(t_i) \,. \tag{48.1}$$

611

For the case of finitely many degrees of freedom we would have

$$\int [dq_\alpha] = \lim_{\epsilon \to 0} \int \prod_{\alpha=1}^{n} \prod_{i=1}^{N-1} dq_\alpha(t_i) \,. \tag{48.2}$$

Now imagine the points in physical space to be discretized into a lattice, with lattice size l, and that associated with each lattice point x_α, $\alpha = 1, \ldots, n$ there is an oscillator whose "displacement" is denoted by $q_{x_\alpha}(t) \equiv q_\alpha(t)$. This is called the *mattress paradigm* (imagine jumping up and down on a two-dimensional mattress). As $l \to 0$, the lattice points become more and more dense and $n \to \infty$. Rename q by ϕ, so that

$$q_\alpha(t_i) \longrightarrow \phi(x_\alpha, t_i) \xrightarrow[\substack{l \to 0 \\ \epsilon \to 0}]{} \phi(x, t) \,, \tag{48.3}$$

and voila, we have a field. The functional integration measure becomes

$$\int [d\phi] = \lim_{\substack{l \to 0 \\ \epsilon \to 0}} \int_{-\infty}^{\infty} \prod_{x} d\phi(x) \,, \tag{48.4}$$

where $x = (t, x)$ denotes the collective spacetime coordinates and the product is over all spacetime lattice points in a certain domain. Again, it is useful to think of x as a label for the field and not as a continuous variable, so there are as many "displacements" ϕ as there are spacetime lattice points considered, and each ϕ (one for each x) can take real values between $-\infty$ and ∞. Eqs. (48.3) and (48.4) essentially constitute the transition from quantum mechanics to quantum field theory using functional integration. Usually we consider spacetime to be $(3+1)$-dimensional Minkowskian, although formally it is straightforward to generalize to the $(d+1)$-dimensional case (for any integer d). In quantum mechanics, there is no spatial label even for the case of multiple degrees of freedom, so *one can consider quantum mechanics to be just $(0+1)$-dimensional quantum field theory from the viewpoint of the path integral approach.*

To construct the action functional for field theory we again generalize from Feynman's formula (46.5) and write

$$S[q_\alpha] = \int_{t'}^{t''} dt \left\{ \frac{m}{2} \sum_{\alpha=1}^{n} (\dot{q}_\alpha(t))^2 - V(q_1(t), \ldots, q_n(t)) \right\} \,. \tag{48.5}$$

We assume a potential energy of the form

$$V(q_1, \ldots, q_n) = \frac{k}{2} \sum_{\alpha=1}^{n} (q_{\alpha'} - q_\alpha)^2 + \frac{k'}{2} \sum_{\alpha=1}^{n} q_\alpha^2 + O(q^3) \,, \tag{48.6}$$

where k and k' are constants, and α' denotes the locations in the spatial lattice that are nearest neighbors to α, for a particular α. The first term describes nearest-neighbor interactions along all 3 spatial directions x, y, z, and the second term describes a harmonic external force. Now we make the transition (48.3). In the continuum limit indicated there we have

$$\frac{1}{2}\sum_{\alpha} m \left(\frac{dq_{\alpha}}{dt}\right)^2 \longrightarrow \frac{1}{2}\int d^3x\, \rho \left(\frac{\partial \phi}{\partial t}\right)^2 , \qquad (48.7)$$

$$V(q_1,\dots,q_n) \longrightarrow \frac{1}{2}\int d^3x\, \kappa \left\{\left(\frac{\partial \phi}{\partial x}\right)^2 + \left(\frac{\partial \phi}{\partial y}\right)^2 + \left(\frac{\partial \phi}{\partial z}\right)^2\right\} + \frac{1}{2}\int d^3x\, \kappa' \phi^2$$
$$+ O(\phi^3) , \qquad (48.8)$$

where ρ, κ and κ' are constants. Let $t' \to -\infty$, $t'' \to \infty$, and the volume integral d^3x to extend over all physical space. The action functional can then be written as the following integral over all of spacetime:

$$S[\phi] = \int d^4x \left[\frac{1}{2}\left\{\rho \left(\frac{\partial \phi}{\partial t}\right)^2 - \kappa (\nabla \phi)^2 - \kappa' \phi^2\right\} + O(\phi^3)\right] . \qquad (48.9)$$

It will be convenient to introduce a rescaled field φ defined by $\varphi = \sqrt{\rho}\,\phi$. The action functional above becomes

$$S[\varphi] = \int d^4x \left[\frac{1}{2}\left\{\left(\frac{\partial \varphi}{\partial t}\right)^2 - \frac{\kappa}{\rho}(\nabla \varphi)^2 - \frac{\kappa'}{\rho}\varphi^2\right\} + O(\varphi^3)\right] . \qquad (48.10)$$

Clearly, dimensional analysis shows that $[\kappa/\rho] = (speed)^2$ and $[\kappa'/\rho] = (time)^{-2}$. We then define a characteristic speed c and a characteristic length L of the theory so that

$$\frac{\kappa}{\rho} = c^2 , \qquad \frac{\kappa'}{\rho} = \frac{c^2}{L^2} . \qquad (48.11)$$

The characteristic length L gives rise to a characteristic mass m if we set $L = \hbar/(mc)$ (the *Compton wave length* for a particle of mass m if c is the speed of light). [Note that this mass is not the same as that in (48.5).] Thus

$$S[\varphi] = \int d^4x \left[\frac{1}{2}\left\{\left(\frac{\partial \varphi}{\partial t}\right)^2 - c^2(\nabla \varphi)^2 - \frac{m^2 c^4}{\hbar^2}\varphi^2\right\} + O(\varphi^3)\right] . \qquad (48.12)$$

This form of the action functional is very suggestive, and also gratifying: It is manifestly Lorentz invariant, since both the integrand and the volume element d^4x are. This prompts us to adopt the *natural system of units*, commonly used in high energy physics, in which $\hbar = c = 1$, and the relativistic notation

$$\partial_\mu = \partial/\partial x^\mu = (\partial/\partial t, \nabla) , \qquad \partial^\mu = \eta^{\mu\nu}\partial_\nu = (\partial/\partial t, -\nabla) \qquad (48.13)$$

to finally write

$$S[\varphi] = \int d^4x\, \mathcal{L}(\varphi, \partial_\mu \varphi) , \qquad (48.14)$$

where

$$\mathcal{L}(\varphi, \partial_\mu \varphi) = \mathcal{L}_0(\varphi, \partial_\mu \varphi) - \mathcal{V}'(\varphi) , \qquad (48.15)$$

$$\mathcal{L}_0(\varphi, \partial_\mu \varphi) = \frac{1}{2} \left\{ (\partial_\mu \varphi)(\partial^\mu \varphi) - m^2 \varphi^2 \right\} , \qquad (48.16)$$

and $\mathcal{V}'(\varphi) = O(\varphi^3)$ is in general a polynomial in φ. The quantities $\mathcal{L}$ and $\mathcal{L}_0$ are called the **Lagrangian density** and the *free field* Lagrangian density, respectively. $\mathcal{V}'$ is a potential energy density giving rise to self interactions of the field. Once this last quantity is given, we have a complete *scalar* field theory. The above three equations are usually taken to be the starting point of such a theory. Indeed, the so-called **Landau-Ginsburg** approach to field theory (see Chapter 56) consists in specifying at the beginning a set of desirable symmetries to be satisfied by the Lagrangian density, such as rotational invariance, Lorentz and Poincaré invariance, and certain gauge (internal) symmetries. In particular, as alluded to at the end of Chapter 11, Poincaré invariance imposes very strong restrictions on the types of fields, scalar, tensor, spinor, etc., that are allowable. For simplicity, our development will focus on scalar field theories. In fact, we will even be more restrictive – we will only consider cases where $\mathcal{V}'(\varphi)$ is a monomial in φ, so that only one coupling constant is involved. Denoting this coupling constant by λ, we will write (48.15) in the form

$$\mathcal{L}(\varphi, \partial_\mu \varphi) = \mathcal{L}_0(\varphi, \partial_\mu \varphi) - \lambda \mathcal{V}(\varphi) . \qquad (48.17)$$

A commonly studied and important example is

$$\mathcal{V}(\varphi) = \frac{\varphi^4}{4!} . \qquad (48.18)$$

This is known as the φ^4-theory, and will be our focus in this book. Note that this theory (and others with even powers of φ) possesses the *parity symmetry* $\varphi \to -\varphi$.

The reader is cautioned that the *Minkowski metric* that we have chosen above in (48.13), namely, $\eta_{\mu\nu} = diag(1, -1, -1, -1)$, differs from that given in Chapter 11 [c.f. (11.1)] by a sign, which leads to a difference in sign in the norm $||x||$ of a 4-vector also. Using the present convention, *time-like vectors* will have $||x|| > 0$ and *space-like vectors* will have $||x|| < 0$. As in Chapter 11, *light-like vectors* will still have $||x|| = 0$ (c.f. Fig. 11.5). This is done in conformity with the most common usage in the physics literature in high energy physics and the convention will be adopted for the remainder of this book.

Imposing the boundary condition

$$\varphi(x) \xrightarrow[||x|| \to \infty]{} 0 , \qquad (48.19)$$

one can do an integration by parts to obtain

$$\int d^4 x \, (\partial_\mu \varphi)(\partial^\mu \varphi) = - \int d^4 x \, \varphi \, \partial_\mu \partial^\mu \varphi . \qquad (48.20)$$

The action functional then assumes the form

$$S[\varphi] = -\frac{1}{2} \int d^4x \, \varphi(x)(\partial_x^2 + m^2)\varphi(x) - \lambda \int d^4x V(\varphi(x)) \,, \qquad (48.21)$$

where

$$\partial_x^2 \equiv \partial_\mu \partial^\mu = \frac{\partial}{\partial x^\mu} \frac{\partial}{\partial x_\mu} \,. \qquad (48.22)$$

One can then proceed to write it in a form involving an integral kernel, as suggested by (47.31):

$$S[\varphi] = \frac{1}{2} \int d^4x \int d^4y \, \varphi(x)K(x,y)\varphi(y) - \lambda \int d^4x V(\varphi(x)) \,, \qquad (48.23)$$

where $K(x,y)$ can be read off to be

$$K(x,y) = -\delta^4(x-y) \left(\frac{\partial}{\partial x^\mu} \frac{\partial}{\partial x_\mu} + m^2 \right) \,. \qquad (48.24)$$

With the above setup for the classical action, we can quantize via functional integration. In analogy to (47.31) we define the *partition function* $Z[J]$ for our scalar field theory:

$$Z[J] = \int [d\varphi] \exp\left\{ \frac{i}{2} \int d^4x \int d^4y \, \varphi(x)K(x,y)\varphi(y) + i \int d^4x \, J(x)\varphi(x) \right\}$$

$$\times \exp\left\{ -i\lambda \int d^4x \, V(\varphi(x)) \right\} \,, \qquad (48.25)$$

and the associated quantities

$$Z^{(0)}[J] = \int [d\varphi] \exp\left\{ \frac{i}{2} \int d^4x \int d^4y \, \varphi(x)K(x,y)\varphi(y) + i \int d^4x J(x)\varphi(x) \right\} \,, \qquad (48.26)$$

$$Z[0] = \int [d\varphi] \exp\left\{ \frac{i}{2} \int d^4x \int d^4y \, \varphi(x)K(x,y)\varphi(y) \right\} \exp\left\{ -i\lambda V(\varphi(x)) \right\} \,, \qquad (48.27)$$

$$Z^{(0)}[0] = \int [d\varphi] \exp\left\{ \frac{i}{2} \int d^4x \int d^4y \, \varphi(x)K(x,y)\varphi(y) \right\} \,. \qquad (48.28)$$

Perturbative quantum field theory, which we will begin to study in the next chapter, involves expanding (48.25) in powers of the coupling constant λ. We will often use the notational shorthand

$$\varphi K\varphi \equiv \int d^4x \int d^4y \, \varphi(x)K(x,y)\varphi(y) \,, \qquad (48.29)$$

and write

$$Z[J] = \sum_{n=0}^{\infty} \frac{(-i\lambda)^n}{n!}$$

$$\times \int [d\varphi] \left(\int d^4x\, V(\varphi(x)) \right)^n \exp\left\{ \frac{i}{2}\varphi K \varphi + i \int d^4x\, J(x)\varphi(x) \right\} . \tag{48.30}$$

In the next chapter we will introduce a very clever trick (using so-called *functional derivatives*) to calculate all terms of this expansion in terms of $Z^{(0)}[J]$, and also show how one can calculate all quantities of physical interest for a particular field theory in terms of $Z[J]$ and its three associated quantities given by (48.26) to (48.28). For this reason $Z[J]$ is called the **generating functional** of a field theory. The Gaussian functional integral $Z^{(0)}[J]$ is the continuum analog of the (ordinary) Gaussian integral in (47.29). The result of the ordinary integration given by (47.30) in the continuum limit, just as it did for the last result in (47.31), now points the way to do this functional integral also. It hinges on finding the inverse to the kernel $K(x, y)$. In other words, we need another integral kernel $D(x, y)$ such that

$$\int d^4z\, K(x, z)D(z, y) = \delta^4(x - y) , \tag{48.31}$$

which obviously is the continuum limit of the condition $\sum_l K_{il} D_{lj} = \delta_{ij}$. Substituting the form for $K(x, z)$ given by (48.24) in (48.31) we see that $D(x, y)$ is also the *Green's function* for the differential operator $-(\partial_x^2 + m^2)$:

$$-\left(\frac{\partial}{\partial x^\mu} \frac{\partial}{\partial x_\mu} + m^2 \right) D(x, y) = \delta^4(x - y) = \int \frac{d^4k}{(2\pi)^4} e^{ik(x-y)} . \tag{48.32}$$

With such an inverse in hand, (47.30) leads to

$$Z^{(0)}[J] = Z^{(0)}[0] \exp\left\{ -\frac{i}{2} \int d^4x \int d^4y\, J(x)D(x, y)J(y) \right\} . \tag{48.33}$$

The introduction of the so-called **source function** $J(x)$ for the evaluation of functional integrals arising from the generating functional was due to J. Schwinger.

It can be verified easily that $D(x, y)$, known as the **Feynman propagator**, can be expressed in terms of the following Fourier integral [see Problem 48.1 below]:

$$\boxed{D(x) = \int \frac{d^4k}{(2\pi)^4} \frac{e^{ikx}}{k^2 - m^2 + i\epsilon^+}} , \tag{48.34}$$

where $k^2 = k_\mu k^\mu = (k^0)^2 - \mathbf{k}^2$, $kx = k_\mu x^\mu = k^0 t - \mathbf{k} \cdot \mathbf{x}$ and $i\epsilon^+$ means $\lim_{\epsilon \to 0+} i\epsilon$, ϵ real. The so-called $i\epsilon$-prescription is introduced so the contour can avoid the **mass shell singularity** when performing the k^0 integration in the above Fourier integral, which refers to the poles of the integrand at $k^2 = m^2$. Note immediately from the Fourier integral that $D(x) = D(-x)$ and $D(x, y) = D(x - y)$. Eq. (48.33) suggests that *the Feynman propagator $D(x, y)$ can be interpreted as the amplitude that a disturbance in the quantum field, created by a source at spacetime point y, propagates from y to another spacetime point x, or vice versa.*

Problem 48.1 Show that $D(x - y)$ is indeed given by (48.34) by verifying that it satisfies (48.32). Differentiate under the integral sign in the Fourier integral.

To understand better the nature and implications of the mass shell singularity and the necessity for the $i\epsilon$-prescription we will now do the dk^0 part of the Fourier integral in (48.34). Define

$$\omega_k = +\sqrt{\boldsymbol{k}^2 + m^2} \ . \tag{48.35}$$

The denominator of the integrand can be re-expressed as follows:

$$k^2 - m^2 + i\epsilon^+ = (k^0)^2 - (\boldsymbol{k}^2 + m^2) + i\epsilon^+ = (k^0)^2 - \omega_k^2 + i\epsilon^+$$

$$= (k^0)^2 - (\omega_k^2 - 2i\omega_k|\epsilon|) \approx (k^0)^2 - (\omega_k^2 - 2i\omega_k|\epsilon| - |\epsilon|^2) = (k^0)^2 - (\omega_k - i|\epsilon|)^2$$

$$= (k^0 + k')(k^0 - k') \ ,$$

where $k' \equiv \omega_k - i|\epsilon|$. There are thus two poles in the complex k^0-plane: $k^0 = k' = \omega_k - i|\epsilon|$ (slightly below the real axis in the lower half-plane), and $k^0 = -k' = -\omega_k + i|\epsilon|$ (slightly above the real axis in the upper half-plane). Separating the d^3k integral and the dk^0 integral in $D(x)$ we write

$$D(x) = \int \frac{d^3k}{(2\pi)^3} \int_{-\infty}^{\infty} \frac{dk^0}{2\pi} \frac{e^{i(k^0 t - \boldsymbol{k}\cdot\boldsymbol{x})}}{[k^0 - (\omega_k - i|\epsilon|)][k^0 + (\omega_k - i|\epsilon|)]} \ . \tag{48.36}$$

For $t > 0$ one can close the contour along the real axis from $-\infty$ to ∞ with an infinite semicircle in the upper half-plane to enclose the pole $k^0 = -\omega_k + i|\epsilon|$, and for $t < 0$ the contour can be closed with an infinite semicircle in the lower half-plane to enclose the pole $k^0 = \omega_k - i|\epsilon|$. The *residue theorem* can then be used and one obtains

$$D(x) = -i \int \frac{d^3k}{(2\pi)^3 (2\omega_k)} \left[e^{-i(\omega_k t - \boldsymbol{k}\cdot\boldsymbol{x})}\theta(t) + e^{i(\omega_k t - \boldsymbol{k}\cdot\boldsymbol{x})}\theta(-t) \right] \ , \tag{48.37}$$

where $\theta(t)$ is the *step function*, defined by $\theta(t) = 1$ for $t > 0$, $\theta(t) = 0$ for $t < 0$, and discontinuous at $t = 0$. For $t = 0$, one can define $D(t = 0, \boldsymbol{x})$ by setting $\theta(0) = 1/2$ in both terms of the above equation.

The Feynman propagator $D(x)$ as given by (48.34) is manifestly Lorentz invariant. So it must be a function of $||x||$ and the sign of $x^0 = t$ only, and will behave differently when x is inside the light cones (future and past) and outside. We can demonstrate this by choosing x to be two representative points in these regions. First let $x = (t, \boldsymbol{x} = 0)$, with $t \neq 0$. If $t > 0$, x is in the future light cone; if $t < 0$, x is in the past light cone. In either of these cases, the Feynman propagator is given by

$$D(t, \boldsymbol{x} = 0) = -i \int \frac{d^3k}{(2\pi)^3 (2\omega_k)} \left[e^{-i\omega_k t}\theta(t) + e^{i\omega_k t}\theta(-t) \right] . \tag{48.38}$$

Since this represents the propagation of a disturbance between $x = 0$ and a point inside a light cone, there should be no damping. Indeed, the oscillatory behavior of the propagator as $|t| \to \infty$ confirms this. In contrast, let x be some point $(t = 0, \boldsymbol{x} \neq 0)$. Then, taking $\theta(0) = 1/2$ and $t = 0$ in (48.37), we have

$$D(0, \boldsymbol{x}) = -\frac{i}{2} \int \frac{d^3k}{(2\pi)^3} \frac{e^{i\boldsymbol{k} \cdot \boldsymbol{x}}}{\sqrt{\boldsymbol{k}^2 + m^2}} . \tag{48.39}$$

The somewhat tedious evaluation of this integral will be left as an exercise (see Problem 48.3 below and the many hints given there). The upshot is that the square-root branch structure of the integrand gives rise to an exponential damping factor in $|\boldsymbol{x}|$. The result is

$$D(0, \boldsymbol{x}) = -i\frac{m^2}{4\pi^2} \left(\frac{\pi}{2(m|\boldsymbol{x}|)^3} \right)^{1/2} e^{-m|\boldsymbol{x}|} . \tag{48.40}$$

This is a quantum mechanical result. A particle of mass m can tunnel from inside the light cone to outside with a characteristic decay length of $1/m$. Classically this process is forbidden.

Problem 48.2 Verify the expression for the Feynman propagator $D(x)$ in (48.37) by using contour integration and the residue theorem in the dk^0 integral in (48.36).

Problem 48.3 Do the integral (48.39) and derive the result (48.40). Follow the steps below. In this problem, denote $|\boldsymbol{k}|$ by k and $|\boldsymbol{x}|$ by r.

(a) Use spherical polar coordinates to write $d^3k = 2\pi k^2 \sin\theta dk d\theta$ and $\boldsymbol{k} \cdot \boldsymbol{x} = kr\cos\theta$. Show that

$$D(0, \boldsymbol{x}) = -\frac{1}{8\pi^2 r} \int_{-\infty}^{\infty} dk \frac{k e^{ikr}}{\sqrt{k^2 + m^2}} . \tag{48.41}$$

(b) Show that

$$D(0, \boldsymbol{x}) = \frac{i}{8\pi^2 r} \frac{dI(r)}{dr} , \tag{48.42}$$

where

$$I(r) \equiv \int_{-\infty}^{\infty} dk \frac{e^{ikr}}{\sqrt{k^2 + m^2}} . \tag{48.43}$$

(c) Treat the integral in (48.43) as a contour integral in the complex k-plane. Recognize that the integrand has two square-root *branch points*, at $k = \pm im$. Let one *branch cut* run along the positive imaginary axis from im to ∞ and the other along the negative imaginary axis from $-im$ to ∞. Let C be the contour wrapping around the branch cut in the upper half-plane, as shown in see Fig. 48.1. Justify that the contour along the real axis [in the defining expression for $I(r)$] can be deformed to C, so that

$$I(r) = \int_C dk \, \frac{e^{ikr}}{\sqrt{k^2 + m^2}} \, . \tag{48.44}$$

(d) Realize that, on the two different banks of a cut, the square root differs by a sign. Make the following successive change of variables: $k = i(y + m)$, $u = (y/m) + 1$, $t = \cosh^{-1} u$. Show that

$$I(r) = 2 \int_0^\infty dt \, \exp\{-mr \cosh t\} \, . \tag{48.45}$$

(e) Do the derivative in (48.42). Then make the change of variables $\cosh t = \sqrt{s^2 + 1}$ to show that

$$D(0, \boldsymbol{x}) = -\frac{im}{4\pi^2 r} \int_0^\infty ds \, \exp\left\{-mr\sqrt{1 + s^2}\right\} \, . \tag{48.46}$$

(f) Finally use the *method of steepest descent* [see, for example, Copson (1965)] to do the integral in (48.46) and arrive at the result (48.40).

Another way to gain physical insight on the Feynman propagator is to work with the Fourier transform $\tilde{J}(k)$ of $J(x)$ in (48.33). We have

$$J(x) = \int \frac{d^4 k}{(2\pi)^4} \, e^{ikx} \, \tilde{J}(k) \, , \qquad \tilde{J}(k) = \int d^4 x \, e^{-ikx} \, J(x) \, . \tag{48.47}$$

Note immediately that $\tilde{J}^*(k) = \tilde{J}(-k)$ if $J(x)$ is real, which we always assume to be the case. Eq. (48.33) can then be written

$$
\begin{aligned}
Z^{(0)}[J] = Z^{(0)}[0] \, \exp & \left\{-\frac{i}{2} \int \frac{d^4 k}{(2\pi)^4} \int \frac{d^4 k'}{(2\pi)^4} \, \tilde{J}(k) \tilde{J}(k') \right. \\
& \left. \times \int d^4 x \int d^4 y \, e^{ikx} e^{ik'y} \int \frac{d^4 k''}{(2\pi)^4} \, \frac{e^{ik''(x-y)}}{(k'')^2 - m^2 + i\epsilon^+} \right\} \, .
\end{aligned}
\tag{48.48}
$$

Making use of the integral representation of the 4-dimensional Dirac delta function [c.f. (48.32)] repeatedly, it is not hard to collapse the above five integrals into a single one to obtain

$$Z^{(0)}[J] = Z^{(0)}[0] \, \exp\left\{-\frac{i}{2} \int \frac{d^4 k}{(2\pi)^4} \, \tilde{J}^*(k) \frac{1}{k^2 - m^2 + i\epsilon^+} \tilde{J}(k)\right\} \, . \tag{48.49}$$

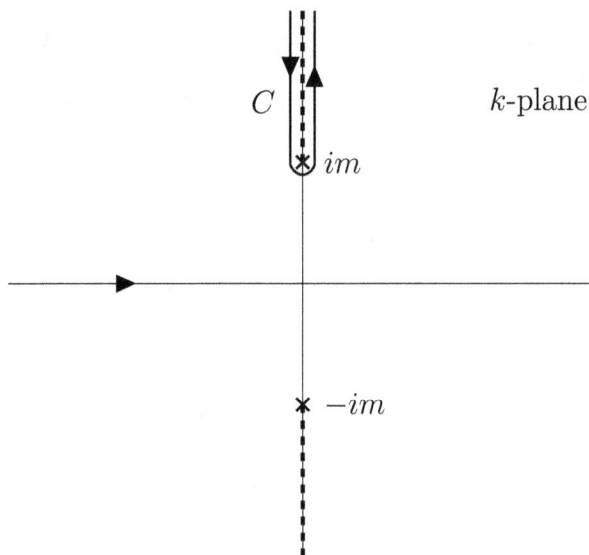

C k-plane

im

$-im$

FIGURE 48.1

This expression clearly shows the presence of a mass shell *resonance* at $k^2 = m^2$, that is, at $(k^0)^2 = \mathbf{k}^2 + m^2$. We can interpret this resonance as the presence of a relativistic particle of mass m moving in spacetime, with energy equal to k^0.

Problem 48.4 Verify (48.49) by working out four of the integrals in (48.48), in the order d^4y, d^4k'', d^4x, d^4k'. Use the integral representation of the 4-dimensional Dirac delta function given in (48.32) in each step.

Chapter 49

Correlation (Green's) Functions and the Functional Derivative

In this chapter we will begin to confront issues related to actual calculations in quantum field theory, within the framework of a scalar field theory that we have developed in the previous chapter. Our first step is to make contact between the operator and the functional integration formalisms of quantum field theory, although, as mentioned before, we will focus on the latter. We begin by introducing a very basic quantity called the n-point *vacuum correlation function*, denoted by $\langle \varphi(x_1) \ldots \varphi(x_n) \rangle$, or n-point **correlation function**, for short. It is defined by

$$\langle \varphi(x_1) \ldots \varphi(x_n) \rangle \equiv \langle 0 \,|\, T[\hat{\varphi}(x_1) \ldots \hat{\varphi}(x_n)] \,|\, 0 \rangle \,, \tag{49.1}$$

where $|0\rangle$ is the unique (non-degenerate) ground state (assumed to exist) of a physical system described by the quantum field theory, $x_1, \ldots, x_n$ are specific spacetime points, $\hat{\varphi}(t, \boldsymbol{x})$ is a (canonically quantized) Heisenberg operator at the time t corresponding to the field $\varphi(t, \boldsymbol{x})$, and T is the **time-ordering operator**, which orders time-dependent operators that it acts on chronologically so that operators corresponding to later times stand to the left of operators corresponding to earlier times. For example,

$$T[\hat{\varphi}(\boldsymbol{x}_1, t_1)\hat{\varphi}(\boldsymbol{x}_2, t_2)] = \hat{\varphi}(\boldsymbol{x}_1, t_1)\hat{\varphi}(\boldsymbol{x}_2, t_2)\,\theta(t_1 - t_2) + \hat{\varphi}(\boldsymbol{x}_2, t_2)\hat{\varphi}(\boldsymbol{x}_1, t_1)\theta(t_2 - t_1) \,.$$

The quantity in (49.1) is sometimes also known as the n-point **Green's function**. Its physical meaning and the reason for its importance is quite apparent from the definition: *The n-point correlation function is the amplitude for the process of quantum excitations to be created at certain of the n specific spacetime points $x_1, \ldots, x_n$ out of the vacuum state, and destroyed at the others to return to the vacuum state.* This quantity thus describes, in a sense, everything that can happen to a system! [For a particularly vivid and colorful description of the processes involved in quantum field theory the reader may refer to A. Zee (2010).] Our development will start with the following theorem.

Theorem 49.1. *For fixed spacetime points $x_1, \ldots, x_n$, the n-point correlation function as defined by (49.1) is given by*

$$\langle \varphi(x_1) \ldots \varphi(x_n) \rangle = \lim_{\beta \to \infty} \frac{tr\, \{e^{-\beta \hat{H}}\, T[\hat{\varphi}(x_1) \ldots \hat{\varphi}(x_n)]\}}{tr\, \{e^{-\beta \hat{H}}\}} . \tag{49.2}$$

In the above equation, $\hat{H}$ is the Hamiltonian operator and tr indicates the trace over a complete set in the Hilbert space spanned by the eigenstates of the Hamiltonian.

Proof. The right-hand-side of (49.2) can be written as

$$\frac{\lim_{\beta \to \infty} tr\, \{e^{-\beta \hat{H}} T[\hat{\varphi}(x_1) \ldots \hat{\varphi}(x_n)]\}}{\lim_{\beta \to \infty} tr\, \{e^{-\beta \hat{H}}\}} .$$

Assuming that the potential function $\mathcal{V}(\varphi)$ [c.f. (47.17) and (47.18)] in our theory leads to a Hamiltonian $\hat{H}$ whose eigenvalues are bounded from below (by a ground state energy E_0), it follows from (46.14) that the denominator is given by

$$\lim_{\beta \to \infty} tr\, \{e^{-\beta \hat{H}}\} = e^{-\beta E_0} . \tag{49.3}$$

To calculate the numerator, we use the identity operator

$$1 = |0\rangle\langle 0| + \sum_{n>0, \alpha} |n, \alpha\rangle\langle n, \alpha| \tag{49.4}$$

expressed in terms of the complete set of orthonormal energy eigenstates of $\hat{H}$ [as in (46.12)] to obtain

$$\lim_{\beta \to \infty} tr\, \{e^{-\beta \hat{H}}\, T[\hat{\varphi}(x_1) \ldots \hat{\varphi}(x_n)]\} = tr\, \{(\lim_{\beta \to \infty} e^{-\beta \hat{H}})\, T[\hat{\varphi}(x_1) \ldots \hat{\varphi}(x_n)]\}$$

$$= tr\, \{e^{-\beta E_0}\, |0\rangle\langle 0|T[\hat{\varphi}(x_1) \ldots \hat{\varphi}(x_n)]\}$$

$$= \langle 0\,|\, e^{-\beta E_0}\,|0\rangle\langle 0\,|\, T[\hat{\varphi}(x_1) \ldots \hat{\varphi}(x_n)]\,|\, 0\rangle$$

$$+ \sum_{n>0, \alpha} \langle n, \alpha\,|\, e^{-\beta E_0}\,|0\rangle\langle 0\,|\, T[\hat{\varphi}(x_1) \ldots \hat{\varphi}(x_n)]\,|\, n, \alpha\rangle$$

$$= e^{-\beta E_0} \langle 0\,|\, T[\hat{\varphi}(x_1) \ldots \hat{\varphi}(x_n)]\,|\, 0\rangle , \tag{49.5}$$

where the second equality follows from (46.13), and the last from the fact that $\langle n, \alpha\,|\, 0\rangle = 0$ for $n > 0$. The theorem then follows. $\qquad\square$

The right-hand-side of (49.2) can be converted to an expression involving functional integrals [the result is given by (49.13) and (49.14) below]. To simplify the development, we will consider the simple case of a 2-point correlation function for a system with one degree of freedom $q(t)$, and then generalize to n-point correlation functions for quantum fields, as we did in the last chapter. We will then start by looking at the following quantity

$$Z \equiv \lim_{\beta \to \infty} \frac{tr\,\{e^{-\beta \hat{H}}\, T[\hat{q}(t_1)\hat{q}(t_2)]\}}{tr\,\{e^{-\beta \hat{H}}\}} . \tag{49.6}$$

Let $0 < t_1 < t_2 < \beta$. Then

$$Z = \lim_{\beta \to \infty} \frac{tr\,\{e^{-\beta \hat{H}}\, \hat{q}(t_2)\hat{q}(t_1)\}}{tr\,\{e^{-\beta \hat{H}}\}} . \tag{49.7}$$

We impose the periodic boundary condition $q(0) = q(\beta)$, and eventually allow β to approach ∞. To calculate the trace we use the complete set of position eigenkets $|q(\beta)\rangle$ for $-\infty < q(\beta) < \infty$. Then, writing $q_\beta = q(\beta)$, $q_2 = q(t_2)$, $q_1 = q(t_1)$ and $q_0 = q(0)$, we have

$$tr\,\{e^{-\beta \hat{H}}\, \hat{q}(t_2)\hat{q}(t_1)\} = \int dq_\beta\, \langle q_\beta\,|\, e^{-\beta \hat{H}} e^{t_2 \hat{H}} \hat{q} e^{-t_2 \hat{H}} e^{t_1 \hat{H}} \hat{q} e^{-t_1 \hat{H}}\,|\, q_0 \rangle$$

$$= \int dq_\beta \int dq_2 \int dq_1\, \langle q_\beta\,|\, e^{-(\beta - t_2)\hat{H}}\,|\, q_2 \rangle \langle q_2\,|\, \hat{q} e^{-(t_2 - t_1)\hat{H}}\,|\, q_1 \rangle \langle q_1\,|\, \hat{q} e^{-t_1 \hat{H}}\,|\, q_0 \rangle$$

$$= \int dq_\beta \int dq_2 \int dq_1\, q_2 q_1\, \langle q_\beta\,|\, e^{-(\beta - t_2)\hat{H}}\,|\, q_2 \rangle \langle q_2\,|\, e^{-(t_2 - t_1)\hat{H}}\,|\, q_1 \rangle$$

$$\times\, \langle q_1\,|\, e^{-(t_1 - 0)\hat{H}}\,|\, q_0 \rangle$$

$$= \int dq_\beta \int dq_2 \int dq_1\, q_2 q_1\, \langle q_\beta, \beta\,|\, q_2\, t_2 \rangle_E \langle q_2, t_2\,|\, q_1, t_1 \rangle_E \langle q_1, t_1\,|\, q_0, 0 \rangle_E ,$$

$$\tag{49.8}$$

where the subscript E in the last line stands for Euclideanized propagator. At this point the Feynman-Kac formula [(46.9)] can be used to write the three Euclidean propagators in terms of path integrals. In each of these path integrals, as they stand, the (conditional) Wiener measure stipulates that the end points q_1 and q_2 are fixed (c.f. Chapter 46). But the integrals dq_1 and dq_2 in front lift these restrictions, and in effect render the entire expression into one path integral on all paths starting from from $t = 0$ to $t = \beta$, without any restrictions even for the end points $q(0)$ and $q(\beta)$ [because of the presence of the integral dq_β and also the boundary condition $q(0) = q(\beta)$]. The entire derivation up to this point can be carried through in the same way if the time-ordering is such that $0 < t_2 < t_1 < \beta$. So we can write

$$tr\,\{e^{-\beta \hat{H}}\, T[\hat{q}(t_1)\hat{q}(t_2)]\} = \int [dq]\, q(t_1)q(t_2)\, \langle q_\beta, \beta\,|\, q_0, 0 \rangle_E , \tag{49.9}$$

where it is understood that all of $\int dq_\beta = \int dq_0$, $\int dq_1 = \int dq(t_1)$ and $\int dq_2 = \int dq(t_2)$, as well as the integrals $\int dq_i$ for all the other discretized time points t_i between $t = 0$ and $t = \beta$, are included in the functional integration measure shorthand $\int [dq]$. We could also have chosen the basic time-interval to be

$(-\beta/2, \beta/2)$ instead of $(0, \beta)$ (giving rise to the same operator $\exp\{-\beta\hat{H}\}$) and written

$$tr\{e^{-\beta\hat{H}}\, T[\hat{q}(t_1)\hat{q}(t_2)]\} = \int [dq]\, q(t_1)q(t_2)\, \langle q_{\beta/2}, \beta/2\,|\,q_{-\beta/2}, -\beta/2\rangle_E\,, \quad (49.10)$$

where $-\beta/2 < t_1, t_2 < \beta/2$. Similarly, we have

$$tr\{e^{-\beta\hat{H}}\} = \int [dq]\, \langle q_{\beta/2}, \beta/2\,|\,q_{-\beta/2}, -\beta/2\rangle_E\,. \quad (49.11)$$

The total propagator $\langle q_{\beta/2}, \beta/2\,|\,q_{-\beta/2}, -\beta/2\rangle_E$ can again be calculated from the Feynman-Kac formula (46.9), but with the restrictions on the end points $q_{\beta/2}$ and $q_{-\beta/2}$ removed from the functional integration measure $\prod_i dq_i$. Analogous to the derivation of (49.8), it is not hard to generalize (49.10) to

$$tr\left\{e^{-\beta\hat{H}}\, T[\hat{q}(t_1)\dots\hat{q}(t_n)]\right\} = \int [dq]\, q(t_1)\dots q(t_n)\, \langle q_{\beta/2}, \beta/2\,|\,q_{-\beta/2}, -\beta/2\rangle_E\,, \quad (49.12)$$

where $-\beta/2 < t_1, \dots, t_n < \beta/2$.

We will not go through the details again on the transition to quantum fields by letting $\int[dq] \to \int[d\varphi]$, which had been given in detail in the last chapter [c.f. (48.3) and (48.4)]. On making the transition, and taking the limit $\beta \to \infty$, we have

$$\langle \varphi(x_1)\dots\varphi(x_n)\rangle = (Z_E[J = 0])^{-1} \int [d\varphi]\, \varphi(x_1)\dots\varphi(x_n)$$

$$\times \exp\left\{-\frac{1}{2}\int d^4x \int d^4y\, \varphi(x)K_E(x,y)\varphi(y) - \lambda\int d^4x V(\varphi(x))\right\}\,, \quad (49.13)$$

where $Z_E[J = 0]$, the quantum-field generalization of (49.11), is given by [c.f. (48.27)]

$$Z_E[0] = \int [d\varphi]\, \exp\left\{-\frac{1}{2}\int d^4x \int d^4y\, \varphi(x)K_E(x,y)\varphi(y) - \lambda\int d^4x V(\varphi(x))\right\}\,. \quad (49.14)$$

In the above two equations $K_E(x,y)$ [given by (49.15) below] is the Euclideanized version of $K(x,y)$ introduced in (48.23). Note that the normalization factor $(\mathcal{N}_E)^{-1}$ in the Feynman-Kac formula [(46.9)] has cancelled in arriving at (49.13) because it is a multiplicative factor in the propagator function $\langle q_{\beta/2}, \beta/2\,|\,q_{-\beta/2}, -\beta/2\rangle$, which is present in both the right-hand-sides of (49.10) and (49.11).

Repeating for the Euclideanized case the development in the last chapter leading up to (48.24), we have the following expression for the Euclidean kernel $K_E(x,y)$:

$$K_E(x,y) = \delta^4(x-y)\left(-\frac{\partial^2}{\partial t^2} - \nabla^2 + m^2\right)\,. \quad (49.15)$$

Analogous to (48.31) we require a Euclidean propagator $D_E(x, y)$ that is inverse to $K_E(x, y)$:

$$\int d^4y\, K_E(x, y) D_E(y, z) = \delta^4(x - z)\,. \qquad (49.16)$$

This requires $D_E(x, y)$ to satisfy

$$\left(-\frac{\partial^2}{\partial t^2} - \nabla^2 + m^2\right) D_E(x, y) = \delta^4(x - y)\,. \qquad (49.17)$$

Analogous to (48.34) the solution to this equation is given by

$$D_E(x - y) = \int \frac{d^4k}{(2\pi)^4} \frac{e^{ik(x-y)}}{k^2 + m^2}\,, \qquad (49.18)$$

where the scalar products are now Euclidean: $k^2 = (k_0)^2 + \boldsymbol{k}^2$ and $kx = k_0 t + \boldsymbol{k} \cdot \boldsymbol{x}$. The kernel $D_E(x)$ is the Euclidean counterpart of the Feynman propagator $D(x)$. Note that here the $i\epsilon$-prescription in the Feynman propagator is not necessary because the integrand does not have any poles on the real axis of the complex k_0-plane. Compare carefully (49.15) with (48.24), and (49.18) with (48.34).

Problem 49.1 Starting with the Feynman-Kac formula (46.9), generalize to quantum fields to verify the expression for the Euclidean kernel $K_E(x, y)$ given by (49.15). Imitate the development for the Minkowski case given at the beginning of Chapter 48, leading up to (48.24).

Problem 49.2 Show that the expression for the Euclidean propagator $D_E(x, y)$ given by (49.18) satisfies (49.17).

We will now use the source function $J(x)$ in the Euclidean generating functional $Z_E[J]$ to compute the quantities in (49.13) and (49.14). Analogous to the generating functional $Z[J]$ defined in (48.25) we have

$$Z_E[J] = \int [d\varphi]\, \exp\left\{-\frac{1}{2}\varphi K_E\varphi - \lambda \int d^4x\, V(\varphi(x)) + \int d^4x\, J(x)\varphi(x)\right\}, \qquad (49.19)$$

where, analogous to (48.29), we define

$$\varphi K_E \varphi \equiv \int d^4x \int d^4y\, \varphi(x) K_E(x, y) \varphi(y)\,. \qquad (49.20)$$

As we will see, what makes $J(x)$ useful is that it can be chosen to be any smooth function of spacetime coordinates x. Our approach will depend on the notion of a *functional derivative*, which we first introduced in (3.9). We recall the definition here. Let $F[f]$ be a functional of f, that is, F maps a certain set of functions $\{f(x)\}$ with specified properties (such as smoothness requirements and asymptotic behavior) to the real numbers, so $F[f] \in \mathbb{R}$. We write $F[f]$ and not $F(f)$ to stress the fact that F acts on a set of functions rather than a set of numbers. Given a small variation function $(\delta f)(x)$ of a function $f(x)$, the **functional derivative** of $F[f]$ with respect to δf, denoted by $\delta F[f]/\delta f(x)$, is defined as follows:

$$F[f + \delta f] - F[f] = \int dx \, (\delta f)(x) \frac{\delta F[f]}{\delta f(x)} + O((\delta f)^2) \, . \tag{49.21a}$$

The first term, involving the functional derivative, is of order δf, and is in fact itself a *linear* functional of δf. For two arbitrary functionals $F_1[f]$ and $F_2[f]$, the functional derivative satisfies the following usual rules of differentiation:

$$\frac{\delta}{\delta f(x)}(F_1[f] + F_2[f]) = \frac{\delta F_1[f]}{\delta f(x)} + \frac{\delta F_2[f]}{\delta f(x)} \, , \tag{49.21b}$$

$$\frac{\delta}{\delta f(x)}(F_1[f]F_2[f]) = F_1[f]\frac{\delta F_2[f]}{\delta f(x)} + F_2[f]\frac{\delta F_1[f]}{\delta f(x)} \, . \tag{49.21c}$$

Let us immediately derive a useful result from this definition. Consider F to be the Dirac delta functional $\delta(x - y)$ which acts on a smooth function $f(x)$ according to

$$F[f] = \int_{-\infty}^{\infty} dx \, f(x)\delta(x - y) = f(y) \in \mathbb{R} \, . \tag{49.22}$$

So

$$F[f + \delta f] - F[f] = (f + \delta f)(y) - f(y) = f(y) + (\delta f)(y) - f(y) = (\delta f)(y) \, . \tag{49.23}$$

According to the definition we then require

$$(\delta f)(y) = \int_{-\infty}^{\infty} dx \, (\delta f)(x)\frac{\delta f(y)}{\delta f(x)} \, . \tag{49.24}$$

This implies

$$\boxed{\frac{\delta f(y)}{\delta f(x)} = \delta(x - y)} \quad . \tag{49.25}$$

This result, which will be the main one on functional differentiation that we will be using below, becomes more apparent if one imagines that the function $f(x)$ is discretized as $f_i = f(x_i)$, so that it merely reduces to the obvious statement $\partial f_i/\partial f_j = \delta_{ij}$. In fact, discretization of continuous functions has been the main formal tool that we used in our treatment of functional integration. In what follows we will be mainly concerned with the situation where a linear functional $F_\varphi[J]$ is defined by

$$F_\varphi[J] = \int_{-\infty}^{\infty} dx \, J(x)\varphi(x) \in \mathbb{R} \, , \tag{49.26}$$

where the smooth function $\varphi(x)$ serves to characterize the functional F_φ. Using (49.25) we immediately have, for a fixed $y \in \mathbb{R}$,

$$\frac{\delta F_\varphi[J]}{\delta J(y)} = \int_{-\infty}^{\infty} dx \, \frac{\delta J(x)}{\delta J(y)} \varphi(x) = \int_{-\infty}^{\infty} dx \, \delta(x-y)\varphi(x) = \varphi(y) \, . \qquad (49.27)$$

Again, this result becomes apparent if one imagines that the variable x is discretized so that

$$\int dx \, J(x)\varphi(x) \longrightarrow \epsilon \sum_i J(x_i)\varphi(x_i) = \epsilon \sum_i J_i \varphi_i \, , \qquad (49.28)$$

where ϵ is a lattice interval along the real axis. Hence, in this example, the functional derivative of the linear functional F_φ with respect to $J(y)$ "picks out" the value $\varphi(y)$.

We are now ready to generate the main quantities of interest in field theory, $\langle \varphi(x_1) \dots \varphi(x_n) \rangle$ [(49.13)] and $Z_E[J=0]$ [(49.14)], from the generating functional $Z_E[J]$ [(49.19)]. Using the result (49.27) it is quite clear that, for n fixed spacetime points $x_1, \dots, x_n$, we have

$$\int [d\varphi] \, \varphi(x_1) \dots \varphi(x_n) \, \exp\left\{ -\frac{1}{2}\varphi K_E \varphi - \lambda \int d^4x \, V(\varphi(x)) \right\}$$

$$= \lim_{J \to 0} \left[\frac{\delta}{\delta J(x_1)} \cdots \frac{\delta}{\delta J(x_n)} \int [d\varphi] \, \exp\left\{ -\frac{1}{2}\varphi K_E \varphi - \lambda \int d^4x \, V(\varphi(x)) \right. \right.$$

$$\left. \left. + \int d^4x \, J(x)\varphi(x) \right\} \right] \qquad (49.29)$$

$$= \left[\frac{\delta}{\delta J(x_1)} \cdots \frac{\delta}{\delta J(x_n)} \, Z_E[J] \right]_{J=0} \, ,$$

where in the second equality we have used (49.19) for $Z_E[J]$. It is important to note that in the last two expressions J is set equal to zero only *after* all the functional derivatives have been taken. By (49.13), therefore,

$$\langle \varphi(x_1) \dots \varphi(x_n) \rangle = (Z_E[0])^{-1} \left[\frac{\delta}{\delta J(x_1)} \cdots \frac{\delta}{\delta J(x_n)} \, Z_E[J] \right]_{J=0} \, . \qquad (49.30)$$

Now, using (49.19) and expanding the exponential $\exp\left\{(-\lambda \int d^4x \, V(\varphi(x)))\right\}$ there, $Z_E[J]$ can be expressed as the following perturbation series:

$$Z_E[J] = \sum_{n=0}^{\infty} \frac{(-\lambda)^n}{n!} \int [d\varphi] \left(\int d^4z \, V(\varphi(z)) \right)^n$$

$$\times \exp\left\{ -\frac{1}{2}\varphi K_E \varphi + \int d^4x \, J(x)\varphi(x) \right\} \, . \qquad (49.31)$$

The crucial step now is to use (49.27) to replace the argument $\varphi(z)$ in $V(\varphi(z))$ by the functional derivative operator $\delta/\delta J(z)$. This clever step removes the necessity for the $\varphi(z)$ in the above expression to be functionally integrated. Thus

$$Z_E[J] = \sum_{n=0}^{\infty} \frac{(-\lambda)^n}{n!} \int [d\varphi] \left(\int d^4z \, \mathcal{V} \left(\frac{\delta}{\delta J(z)} \right) \right)^n$$

$$\exp \left\{ -\frac{1}{2} \varphi K_E \varphi + \int d^4x \, J(x) \varphi(x) \right\}$$

$$= \sum_{n=0}^{\infty} \frac{(-\lambda)^n}{n!} \left(\int d^4z \, \mathcal{V} \left(\frac{\delta}{\delta J(z)} \right) \right)^n \tag{49.32}$$

$$\int [d\varphi] \exp \left\{ -\frac{1}{2} \varphi K_E \varphi + \int d^4x \, J(x) \varphi(x) \right\}$$

$$= \sum_{n=0}^{\infty} \frac{(-\lambda)^n}{n!} \left(\int d^4z \, \mathcal{V} \left(\frac{\delta}{\delta J(z)} \right) \right)^n Z_E^{(0)}[J] \, ,$$

where in the last equality we have used the fact that the non-interacting [with $\mathcal{V}(\varphi) = 0$] Euclidean partition function $Z_E^{(0)}[J]$ is given [in analogy to the Minkowski one in (48.26)], by

$$Z_E^{(0)}[J] = \int [d\varphi] \exp \left\{ -\frac{1}{2} \varphi K_E \varphi + \int d^4x \, J(x) \varphi(x) \right\} \, . \tag{49.33}$$

On taking the continuum limit of (47.18) we can write

$$Z_E^{(0)}[J] = Z_E^{(0)}[0] \exp \left\{ \frac{1}{2} J D_E J \right\} \, , \tag{49.34}$$

where

$$J D_E J \equiv \int d^4x \int d^4y \, J(x) D_E(x, y) J(y) \, , \tag{49.35}$$

with the Euclidean propagator $D_E(x, y)$ given by (49.18). So (49.32) yields

$$Z_E[J] = Z_E^{(0)}[0] \sum_{n=0}^{\infty} \frac{(-\lambda)^n}{n!} \left(\int d^4x \, \mathcal{V} \left(\frac{\delta}{\delta J(x)} \right) \right)^n \exp \left\{ \frac{1}{2} J D_E J \right\} \, . \tag{49.36}$$

As a consequence,

$$Z_E[0] = Z_E^{(0)}[0] \sum_{n=0}^{\infty} \frac{(-\lambda)^n}{n!} \left[\left(\int d^4x \, \mathcal{V} \left(\frac{\delta}{\delta J(x)} \right) \right)^n \exp \left\{ \frac{1}{2} J D_E J \right\} \right]_{J=0} \, . \tag{49.37}$$

Finally, (49.30) yields the central result for the n-point correlation function

$$\langle \varphi(x_1) \dots \varphi(x_n) \rangle$$

$$= \frac{\sum_{m=0}^{\infty} \frac{(-\lambda)^m}{m!} \left[\frac{\delta}{\delta J(x_1)} \cdots \frac{\delta}{\delta J(x_n)} \left(\int d^4x \, \mathcal{V} \left(\frac{\delta}{\delta J(x)} \right) \right)^m \exp \left\{ \frac{1}{2} J D_E J \right\} \right]_{J=0}}{\sum_{m=0}^{\infty} \frac{(-\lambda)^m}{m!} \left[\left(\int d^4x \, \mathcal{V} \left(\frac{\delta}{\delta J(x)} \right) \right)^m \exp \left\{ \frac{1}{2} J D_E J \right\} \right]_{J=0}} \, . \tag{49.38}$$

This formula provides a direct computational algorithm for the correlation functions (see the next chapter for simple examples). In it, the normalization factor $Z_E^{(0)}[0]$, which is of no physical significance, has cancelled out. We see explicitly that the main contents of the theory, aside from the as yet undetermined functional form of $\mathcal{V}(\varphi)$, is entirely determined by the propagator function $D_E(x, y)$. The denominator, which does not involve any external spacetime points x_i, corresponds to the so-called *vacuum* Feynman diagrams; while the numerator contains both diagrams involving external points and vacuum diagrams. The correlation function itself, then, with all the vacuum diagrams divided out by the denominator, only includes contributions from diagrams involving external points. How this happens will be explained in more detail in the next chapter, where we will also explain the genesis and the physical interpretation of the Feynman diagrams, as well as how they are calculated.

Now, using the same Lagrangian density, one could have started off with $Z[J]$, the Minkowski version of the generating functional, which had been defined in the last chapter [c.f. (48.25) and (48.30)]. Analogous to (49.13) we can define a Minkowski n-point correlation function $G(x_1, \ldots, x_n)$ by

$$
G(x_1, \ldots, x_n) \equiv (Z[0])^{-1} \int [d\varphi] \, \varphi(x_1) \ldots \varphi(x_n)
$$
$$
\times \exp\left\{\frac{i}{2}\varphi K \varphi - i\lambda \int d^4 x \, \mathcal{V}(\varphi(x))\right\} , \tag{49.39}
$$

where the quadratic form $\varphi K \varphi$ has been defined in (48.29). Similar to (49.30) we obtain

$$
G(x_1, \ldots, x_n) = \frac{1}{Z[0]} i^{-n} \left[\frac{\delta}{\delta J(x_1)} \cdots \frac{\delta}{\delta J(x_n)} Z[J]\right]_{J=0} . \tag{49.40}
$$

Note the presence of the extra factor i^{-n} due to the presence of the factor i in front of the term $\int d^4 x \, J(x)\varphi(x)$ in (48.30). We claim that, if we let

$$
x = (\boldsymbol{x}, t) , \qquad \bar{x} = (\boldsymbol{x}, -it) , \tag{49.41}
$$

then

$$
G(\bar{x}_1, \ldots, \bar{x}_n) = i^{-n} \langle \varphi(x_1) \ldots \varphi(x_n) \rangle , \tag{49.42}
$$

so one can start with either the Minkowski propagator $K(x, y)$ or the Euclidean propagator $K_E(x, y)$ to do quantum field-theoretic calculations. The respective n-point correlation functions can be converted into each other through analytic continuation in the time variables, via (49.41) and (49.42). This equation can be justified if we use an equation for the Minkowski case corresponding to (49.8) for the Euclidean case, and recall that in order to Euclideanize one has to Wick-rotate (or analytically continue) the operators $e^{-i(\beta-t_2)\hat{H}}$, $e^{-i(t_2-t_1)}$ and $e^{-it_1\hat{H}}$ by letting $\beta \to -i\beta$, $t_2 \to -it_2$ and $t_1 \to -it_1$ [c.f. the discussion surrounding (46.7)].

We will conclude this chapter by pointing out that the n-point correlation functions, in addition to being calculable from the generating functional $Z_E[J]$

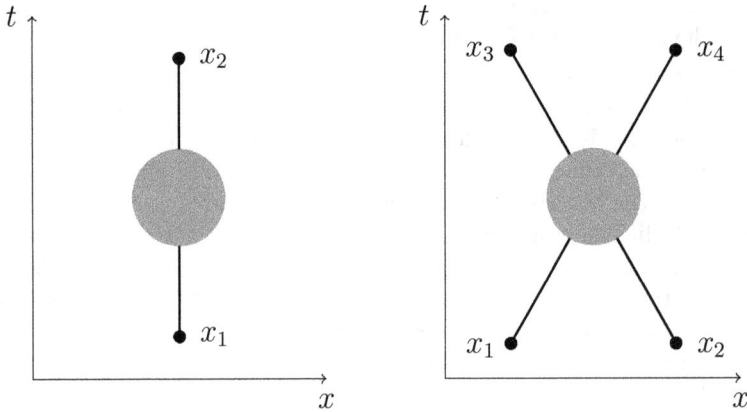

FIGURE 49.1

by (49.30), can also be obtained from $Z_E[J]$ as coefficients in an expansion in powers of the source function $J(x)$. Indeed, on expanding the exponential $\exp\{\int d^4x\, J(x)\varphi(x)\}$ in (49.19), we have

$$Z_E[J] = \int [d\varphi] \sum_{n=0}^{\infty} \frac{1}{n!} \left(\int d^4x\, J(x)\varphi(x) \right)^n \exp\left\{ -\varphi D_E \varphi - \lambda \int d^4x\, V(\varphi(x)) \right\}$$

$$= \sum_{n=0}^{\infty} \frac{1}{n!} \int d^4x_1 \ldots \int d^4x_n\, J(x_1) \ldots J(x_n)$$

$$\times \int [d\varphi]\, \varphi(x_1) \ldots \varphi(x_n) \exp\left\{ -\varphi K_E \varphi - \lambda \int d^4x\, V(\varphi(x)) \right\} .$$

$$(49.43)$$

It follows from (49.29) and (49.30) that

$$Z_E[J] = Z_E[0] \sum_{n=0}^{\infty} \frac{1}{n!} \int d^4x_1 \ldots \int d^4x_n\, J(x_1) \ldots J(x_n) \langle \varphi(x_1) \ldots \varphi(x_n) \rangle .$$

$$(49.44)$$

This equation in fact provides the rationale for the physical interpretation of the correlation functions given at the beginning of this chapter. The various source functions $J(x_1), \ldots, J(x_n)$ can be thought of as physical sources and sinks at the spacetime points $x_1, \ldots, x_n$. Physical sources at certain of these spacetime points create disturbances (excitations of the quantum field) in the vacuum, which then propagate through all regions of spacetime, only to disappear at physical sinks at the other spacetime points (of the set $x_1, \ldots, x_n$). The integrations over these spacetime points allow for the possibilities that these sources and sinks can be anywhere (in space) and can happen at any time. The correlation functions then give precisely the amplitudes that these processes can happen. The actual functional form of $J(x)$ does not matter, because in our calculations $J(x)$ is eventually allowed to vanish. In this way, the source function plays a dual role in the theory: providing a physical interpretation of the correlation functions, and a mathematical "crutch" for their calculation, which can then be discarded after the calculation is done! Figure 49.1 depicts processes described by a 2-point and a 4-point correlation function.

Chapter 50

Perturbative Expansions and Feynman Diagrams

In this chapter we will begin to evaluate an n-point correlation function using (49.38). For concrete results we will consider the so-called φ^4-theory, in which $\mathcal{V}(\varphi) = \varphi^4/4!$ [c.f. (48.17) and (48.18)]. Rather than attempting to directly calculate the n-th order term, we will do the calculations beginning with the lowest orders first. In so doing we will gain, in an inductive manner, a better grasp of the methodology of calculation and the physical meaning of the terms that are calculated. Our aim is to show, as simply and as directly as possible, order by order, how different terms in the result for each order can be represented in a pictorial and intuitive way by so-called Feynman diagrams, and to set up a set of rules for the calculation for each diagram. As long as one adheres to a perturbative approach, as a matter of practical calculations, one often begins with drawing all possible Feynman diagrams, order by order, and then perform calculations for each one of them, to the extent that is practically feasible. We will, of course, not be able to provide anything remotely near an exhaustive account – the subject is simply too vast. The objective is rather to gain some broad and preliminary understanding of how the functional integration approach works in the generation of Feynman diagrams. For more complete and/or specialized treatments the reader is referred to a wealth of excellent and more authoritative accounts in the quantum field theory physics literature. [For sources treating quantum field theory from the functional integration perspective to one extent or another, the interested reader may pursue, for example, Bogoliubov and Shirkov (1959), Faddeev and Slavnov (1980), Itzykson and Zuber (1980), Cheng and Li (1984), Popov (1987), Ramond (1989), Peskin and Schroeder (1995), Weinberg (1995, 1996, 2000), Zinn-Justin (2002), A. Zee (2010), and Huang (2010), among others.]

We will denote by $\langle \varphi(x_1) \ldots \varphi(x_n) \rangle_{(m)}$ the contributions to the n-point correlation function in (49.38) up to the m-th order in λ. To keep things as simple as possible, we start with the 2-point correlation function $\langle \varphi(x_1)\varphi(x_2) \rangle$. We have, for the lowest-order (non-interacting $m = 0$) term,

$$\langle \varphi(x_1)\varphi(x_2) \rangle_{(0)} = \left[\frac{\delta}{\delta J(x_1)} \frac{\delta}{\delta J(x_2)} \exp\left\{ \frac{1}{2} JD_E J \right\} \right]_{J=0} . \tag{50.1}$$

The denominator in (49.38) for $m = 0$ is clearly equal to 1 since there are no functional derivatives to be performed and $\lim_{J\to 0} \exp\{(1/2)JD_E J\} = 1$. Making use of the rules of functional differentiation given by (49.21b), (49.21c) and (49.25), the right-hand-side of the above can be calculated as follows:

$$\left[\frac{\delta}{\delta J(x_1)} \frac{\delta}{\delta J(x_2)} \exp\left\{ \frac{1}{2} JD_E J \right\} \right]_{J=0}$$

$$= \left[\frac{\delta}{\delta J(x_1)} \left(\frac{1}{2} e^{\frac{1}{2} JD_E J} \frac{\delta}{\delta J(x_2)} JD_E J \right) \right]_{J=0}$$

$$= \frac{1}{2} \left[e^{\frac{1}{2} JD_E J} \frac{\delta}{\delta J(x_1)} \frac{\delta}{\delta J(x_2)} (JD_E J) \right.$$

$$\left. + \frac{1}{2} e^{\frac{1}{2} JD_E J} \left(\frac{\delta}{\delta J(x_2)} JD_E J \right) \left(\frac{\delta}{\delta J(x_1)} JD_E J \right) \right]_{J=0}$$

$$= \frac{1}{2} \left[\frac{\delta}{\delta J(x_1)} \frac{\delta}{\delta J(x_2)} JD_E J \right]_{J=0} + \frac{1}{2^2} \left[\frac{\delta}{\delta J(x_2)} JD_E J \right]_{J=0} \left[\frac{\delta}{\delta J(x_1)} JD_E J \right]_{J=0} . \tag{50.2}$$

Using (49.35) for $JD_E J$ and the result (49.27), we will calculate the separate quantities as follows:

$$\frac{\delta}{\delta J(x_1)} JD_E J = \frac{\delta}{\delta J(x_1)} \int d^4x \int d^4y \, J(x) D_E(x-y) J(y)$$

$$= \int d^4x \int d^4y \, \frac{\delta J(x)}{\delta J(x_1)} D_E(x-y) J(y) + \int d^4x \int d^4y \, J(x) D_E(x-y) \frac{\delta J(y)}{\delta J(x_1)}$$

$$= \int d^4x \int d^4y \, \delta^4(x-x_1) D_E(x-y) J(y)$$

$$+ \int d^4x \int d^4y \, J(x) D_E(x-y) \delta^4(y-x_1)$$

$$= \int d^4y \, D_E(x_1-y) J(y) + \int d^4x \, J(x) D_E(x-x_1)$$

$$= 2 \int d^4x \, J(x) D_E(x-x_1) , \tag{50.3}$$

where in the last equality we have used the fact that $D_E(x) = D_E(-x)$ [c.f. (49.18)]. It follows that

$$\left[\frac{\delta}{\delta J(x_1)} JD_E J \right]_{J=0} = \left[\frac{\delta}{\delta J(x_2)} JD_E J \right]_{J=0} = 0 . \tag{50.4}$$

Eq. (50.1) then gives

$$\langle \varphi(x_1)\varphi(x_2)\rangle_{(0)} = \frac{1}{2} \left[\frac{\delta}{\delta J(x_1)} \frac{\delta}{\delta J(x_2)} JD_E J \right]_{J=0} . \tag{50.5}$$

It is quite clear that this second functional derivative will eliminate all the J's. In fact, from (50.3),

$$\frac{\delta}{\delta J(x_2)} \frac{\delta}{\delta J(x_1)} JD_E J = 2 \int d^4x \, \frac{\delta J(x)}{\delta J(x_2)} D_E(x - x_1)$$

$$= 2 \int d^4x \, \delta^4(x - x_2) D_E(x - x_1) = 2D_E(x_2 - x_1) . \tag{50.6}$$

Thus, to 0-th order in λ, the 2-point correlation function is given by

$$\langle \varphi(x_1)\varphi(x_2)\rangle_{(0)} = D_E(x_1 - x_2) = D_E(x_2 - x_1) . \tag{50.7}$$

We represent this quantity by the simplest of Feynman diagrams (Fig. 50.1), which describes the physical process of an excitation of the field propagating freely from spacetime point x_1 to spacetime point x_2, or vice versa, the amplitude of this process being equal to the free Euclidean propagator $D_E(x_1 - x_2)$ [given by (49.18)].

Let us now calculate the 2-point correlation up to first order, denoted by $\langle \varphi(x_1)\varphi(x_2)\rangle_{(1)}$, using (49.38). Begin with the denominator first and denote the contribution up to first order by $\mathcal{D}_{(1)}$ ($\mathcal{D}$ for denominator). It is given by, for $\mathcal{V}(\varphi) = \varphi^4/4!$,

$$\mathcal{D}_{(1)} = 1 - \frac{\lambda}{4!} \left[\int d^4x \left(\frac{\delta}{\delta J(x)} \right)^4 \exp\left\{ \frac{1}{2} JD_E J \right\} \right]_{J=0} , \tag{50.8}$$

where the term 1 is from the contribution for $m = 0$ in the denominator of (49.38). Following the steps used in (50.2) and (50.3), we have

$$\frac{\delta}{\delta J(x)} e^{\frac{1}{2} JD_E J} = e^{\frac{1}{2} JD_E J} (D_E J)(x) , \tag{50.9}$$

where

$$(D_E J)(x) \equiv \int d^4y \, D_E(x - y) J(y) . \tag{50.10}$$

Quite obviously, this quantity satisfies

$$\frac{\delta}{\delta J(y)} (D_E J)(x) = D_E(x, y) . \tag{50.11}$$

Continuing with the higher-order derivatives using (50.9) and (50.11), we obtain straightforwardly the following results:

$$\left(\frac{\delta}{\delta J(x)} \right)^2 e^{\frac{1}{2} JD_E J} = e^{\frac{1}{2} JD_E J} \left\{ ((D_E J)(x))^2 + D_E(x, x) \right\} , \tag{50.12}$$

$$\left(\frac{\delta}{\delta J(x)} \right)^3 e^{\frac{1}{2} JD_E J} = e^{\frac{1}{2} JD_E J} \left\{ ((D_E J)(x))^3 + 3D_E(x, x)(D_E J)(x) \right\} , \tag{50.13}$$

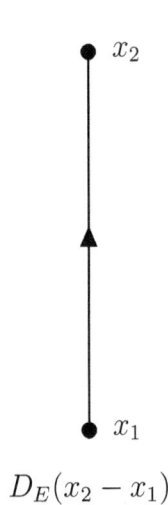

$$D_E(x_2 - x_1)$$

FIGURE 50.1

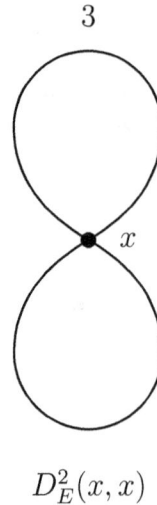

$$D_E^2(x, x)$$

FIGURE 50.2

$$\left(\frac{\delta}{\delta J(x)}\right)^4 e^{\frac{1}{2}JD_E J}$$

$$= e^{\frac{1}{2}JD_E J}\left\{((D_E J)(x))^4 + 6D_E(x, x)((D_E J)(x))^2 + 3(D_E(x, x))^2\right\}.$$ (50.14)

Thus

$$\left[\left(\frac{\delta}{\delta J(x)}\right)^4 e^{\frac{1}{2}JD_E J}\right]_{J=0} = 3(D_E(x, x))^2 .$$ (50.15)

It then follows from (50.8) that

$$\mathcal{D}_{(1)} = 1 - \frac{3\lambda}{4!}\int d^4x\,(D_E(x, x))^2 .$$ (50.16)

The second term above is represented by the Feynman diagram shown in Fig. 50.2. Each loop in the figure "eight" corresponds to one factor of the propagator $D_E(x, x)$, and begins and ends at the same spacetime point x, which is integrated over all spacetime. This kind of integrated point is called a **vertex**, as opposed to **external points**, which are points that are not integrated over. The diagram under consideration, which does not involve any external points (points that are not integrated over), is called a **vacuum diagram**.

On using the previous result (50.7) for the 0-th order contribution, the numerator in (49.38) for the 2-point correlation function up to the first order in

λ, denoted by $\mathcal{N}_{(1)}^{[2]}$ ($\mathcal{N}$ for numerator), is given by

$$\mathcal{N}_{(1)}^{[2]} = D_E(x_2 - x_1) - \frac{\lambda}{4!} \int d^4x \left[\frac{\delta}{\delta J(x_1)} \frac{\delta}{\delta J(x_2)} \left(\frac{\delta}{\delta J(x)} \right)^4 e^{\frac{1}{2} J D_E J} \right]_{J=0}.$$
(50.17)

Let us use (50.14) to calculate the quantity in square brackets. As the calculations will get more involved with more derivatives, we simplify our notation a bit at this point by writing

$$D_J(x) \equiv (D_E J)(x), \qquad \text{and} \qquad D_J^n(x) \equiv ((D_E J)(x))^n.$$
(50.18)

So we have

$$\frac{\delta}{\delta J(x_1)} \frac{\delta}{\delta J(x_2)} \left(\frac{\delta}{\delta J(x)} \right)^4 e^{\frac{1}{2} J D_E J}$$

$$= \frac{\delta}{\delta J(x_1)} \frac{\delta}{\delta J(x_2)} e^{\frac{1}{2} J D_E J} \left\{ D_J^4(x) + 6 D_E(x, x) D_J^2(x) + 3 D_E^2(x, x) \right\}.$$
(50.19)

The derivatives will be done one at a time using the useful results (50.9) and (50.11). First,

$$\frac{\delta}{\delta J(x_2)} \left\{ e^{\frac{1}{2} J D_E J} \left(D_J^4(x) + 6 D_E(x, x) D_J^2(x) + 3 D_E^2(x, x) \right) \right\}$$

$$= \left(e^{\frac{1}{2} J D_E J} D_J(x_2) \right) \left(D_J^4(x) + 6 D_E(x, x) D_J^2(x) + 3 D_E^2(x, x) \right)$$

$$\qquad + e^{\frac{1}{2} J D_E J} \left(4 D_J^3(x) D_E(x, x_2) + (2)(6) D_E(x, x) D_E(x - x_2) D_J(x) \right)$$

$$= e^{\frac{1}{2} J D_E J} \left\{ D_J(x_2) D_J^4(x) + 6 D_E(x, x) D_J(x_2) D_J^2(x) + 3 D_E^2(x, x) D_J(x_2) \right.$$

$$\qquad \left. + 4 D_E(x - x_2) D_J^3(x) + (2)(6) D_E(x, x) D_E(x - x_2) D_J(x) \right\}.$$
(50.20)

Proceeding with the second derivative, then,

$$\frac{\delta}{\delta J(x_1)} \frac{\delta}{\delta J(x_2)} \left(\frac{\delta}{\delta J(x)} \right)^4 e^{\frac{1}{2} J D_E J}$$

$$= \left(e^{\frac{1}{2} J D_E J} D_J(x_1) \right) \left\{ D_J(x_2) D_J^4(x) + 6 D_E(x, x) D_J(x_2) D_J^2(x) \right.$$

$$+ 3 D_E^2(x, x) D_J(x_2) + 4 D_E(x - x_2) D_J^3(x) + (2)(6) D_E(x, x) D_E(x - x_2) D_J(x) \}$$

$$\qquad + e^{\frac{1}{2} J D_E J} \left\{ 4 D_J(x_2) D_J^3(x) D_E(x, x_1) + D_E(x_1, x_2) D_J^4(x) \right.$$

$$\qquad\qquad + 6 D_E(x, x) \left(D_E(x_2, x_1) D_J^2(x) + 2 D_J(x_2) D_J(x) D_E(x, x_1) \right)$$

$$\qquad\qquad + 3 D_E^2(x, x) D_E(x_2 - x_1) + (3)(4) D_E(x, x_2) D_J^2(x) D_E(x, x_1)$$

$$\qquad\qquad \left. + (2)(6) D_E(x, x) D_E(x, x_1) D_E(x, x_2) \right\}.$$
(50.21)

All but two terms in the right-hand-side of the above equation (the first term on the next to last line and the term on the last line) involve D_J raised to some power (and hence J raised to some power), and will vanish as $J \to 0$. The two surviving terms yield

$$\left[\frac{\delta}{\delta J(x_1)} \frac{\delta}{\delta J(x_2)} \left(\frac{\delta}{\delta J(x)} \right)^4 e^{\frac{1}{2} J D_E J} \right]_{J=0} \tag{50.22}$$

$$= 3 D_E^2(x,x) D_E(x_2, x_1) + 12 D_E(x,x) D_E(x, x_1) D_E(x, x_2) \ .$$

This gives, by (50.17),

$$N_{(1)}^{[2]} = D_E(x_1, x_2) + D_E(x_1, x_2) \left[\left(\frac{-3\lambda}{4!} \right) \int d^4x \, D_E^2(x,x) \right] \tag{50.23}$$

$$+ \left(\frac{-12\lambda}{4!} \right) \int d^4x \, D_E(x, x_1) D_E(x, x_2) D_E(x, x) \ .$$

The second term is represented by the Feynman diagram in Fig. 50.3. This is an example of a **disconnected diagram** (the free propagator diagram joining the external points x_1 and x_2 and the figure "eight" vacuum diagram are disconnected from each other). The vacuum diagram by itself, already shown in Fig. 50.2, corresponds to the expression within the square brackets in the term. The third term is represented by the Feynman diagram shown in Fig. 50.4. This is an example of a **connected diagram** (one involving external points and all in one piece).

One can continue with the second order calculations ($m = 2$) in the denominator of (49.38) for the vacuum diagrams with the same functional differentiation techniques shown explicitly thus far to obtain

$$\left[\left(\frac{\delta}{\delta J(y)} \right)^4 \left(\frac{\delta}{\delta J(x)} \right)^4 e^{\frac{1}{2} J D_E J} \right]_{J=0} \tag{50.24}$$

$$= 9 \, D_E^2(x,x) D_E^2(y,y) + 72 \, D_E(x,x) D_E(y,y) D_E^2(x,y) + 24 \, D_E^4(x,y) \ .$$

Incorporating (50.16) for lower order terms and using the above result, we have the following result for the expansion up to second order for the vacuum diagram terms in the denominator of (49.38):

$$\mathcal{D}_{(2)} = 1 - \frac{3\lambda}{4!} \int d^4x \, (D_E(x,x))^2$$

$$+ \frac{1}{2} \left(\frac{\lambda}{4!} \right)^2 \left[\int d^4x \int d^4y \, \{ 9 \, D_E^2(x,x) D_E^2(y,y) \right.$$

$$\left. + 72 \, D_E(x,x) D_E(y,y) D_E^2(x,y) + 24 \, D_E^4(x,y) \} \right] \ . \tag{50.25}$$

The second order terms in the above equation are represented by the three vacuum Feynman diagrams shown in Fig. 50.5, with the corresponding numerical coefficients for each term inside the integral shown.

At this juncture we can already see (or infer) the general features of a Feynman diagram for the φ^4-theory, all of which are consequences of the functional differentiation operations in (49.38). These are listed below.

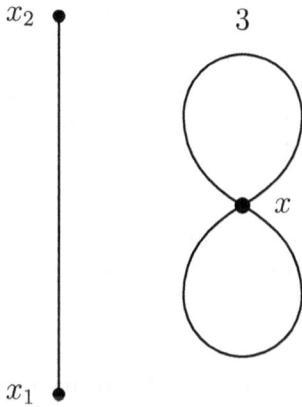

FIGURE 50.3

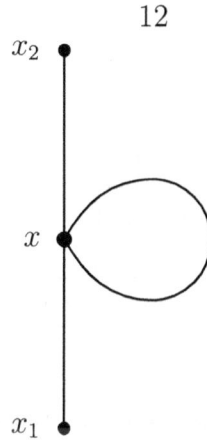

FIGURE 50.4

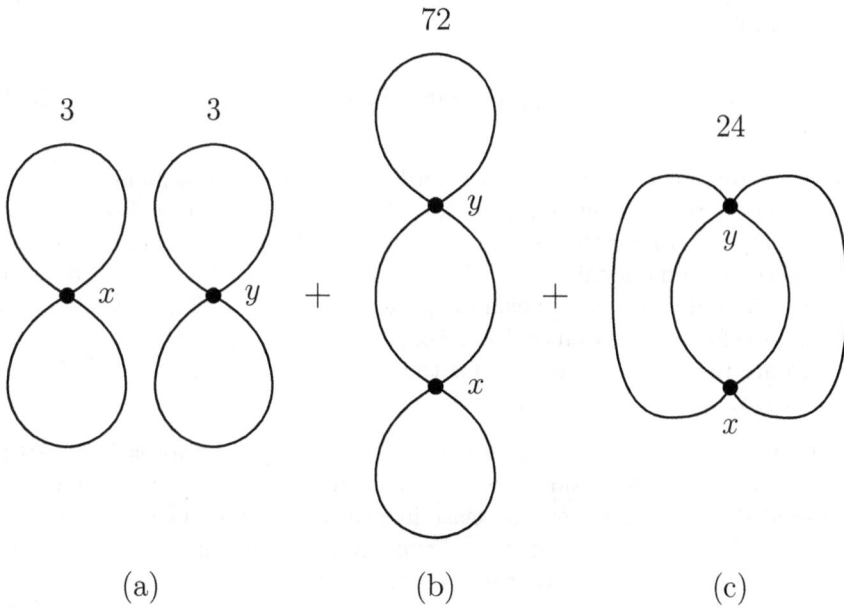

FIGURE 50.5

(a) Each topologically distinct Feynman diagram consists of a number of points, either external points (not integrated over) or vertices (internal points that have to be integrated over), and lines joining them.

(b) Each external point is the abrupt end of a line, but each vertex has 4 line segments emanating from it (only true for the φ^4-theory). These line segments are either connected directly to external points (where they have to end abruptly), or to other line segments emanating from vertices (may be from the same vertex). This rule leads to the possibility of connected and disconnected diagrams, both for diagrams involving external points and vacuum diagrams (see Figs. 50.5 and 50.6).

(c) A diagram for an m-th order term in the numerator of (49.38) for an n-point correlation function has m vertices and n external points. A diagram corresponding to some term in the denominator of the same order, a vacuum diagram, only has m vertices and no external points.

We can also state the rules for the calculation of Feynman diagrams as follows [which are also consequences of the functional differentiation operations in (49.38)]:

(1) Each vertex carries a factor of $-\lambda/4!$.

(2) Each line joining two points x and x' carries a propagator factor $D_E(x, x')$.

(3) Each vertex (internal point) x has to be integrated over all spacetime according to

$$\int d^4x \prod \{D_E(x, x') \text{ propagator factors depending on } x\} . \qquad (50.26)$$

(4) Each topologically distinct m-th order diagram for an n-point correlation function carries a numerical factor $S(n, m)/m!$. $S(n, m)$, called the **symmetry factor** for the diagram, is a combinatorial factor arising from the successive functional differentiations in (49.38). For a vacuum diagram, $n = 0$ by definition. For example, the factors 3 and 12 are the symmetry factors for the diagrams in Figs. 50.3 and 50.4, respectively; and 9, 72, and 24 are the symmetry factors for the (vacuum) diagrams in Figs. 50.5(a), 50.5(b), and 50.5(c), respectively.

The general procedure for the calculation of an n-point correlation function based on (49.38), which sidesteps the laborious explicit calculations for the functional derivatives, is now, at least in principle, clear. One simply draws all the possible topologically distinct Feynman diagrams, up to a certain order in the coupling constant λ according to rules (a), (b) and (c) above, and then proceed to calculate each one by the rules (1), (2) and (3) above. For higher orders, the enumeration of all the possible diagrams is by itself a laborious combinatorial exercise, but at least one can envision doing it. This enumeration is in fact a graphical way of generating all the distinct terms obtainable from doing the functional differentiations explicitly, and was one of Feynman's unique contributions to physics. The only problem at this point is that we do not have an explicit general formula for the symmetry factors $S(n, m)$. We will discuss an alternative way (other than, again, by doing all the functional derivatives explicitly) to obtain these factors later in this chapter. In the meantime, it is instructive to generate the second order diagrams for the 2-point correlation function. With a little trial and error, one can see that all the second order diagrams in the numerator of (49.38) are shown in Fig. 50.6. In these diagrams we have also appended the corresponding symmetry factors for each. Note that the diagrams (a), (b) and (c) in Fig. 50.6 are connected diagrams, while the rest are disconnected. The diagram (g) in Fig. 50.6 is a product of the lower (first) order diagrams shown in Figs. 50.2 and 50.4. In any disconnected diagram, the separate pieces are supposed to be multiplied together. Incidentally, we see that all n-point correlation functions for the φ^4-theory vanish for n odd, as one cannot draw a Feynman diagram for such functions satisfying the above rules. For $n = 1$ and to first order ($m = 1$), this fact is explicitly borne out by examining the result (50.20), since that expression vanishes on letting $J = 0$.

Problem 50.1 Replace x_1 and x_2 in (50.21) by y, then perform the functional differentiation $\delta/\delta J(y)$ twice more, and finally set $J = 0$ in the resulting expression to verify (50.24).

Problem 50.2 Draw some Feynman diagrams according to the rules (a), (b) and (c) above for the φ^4-theory to convince yourself that in order for an n-point correlation function to the m-th order not to vanish, n has to be even for any m.

Problem 50.3 Draw some Feynman diagrams according to the rules (a), (b), and (c) above for the φ^3-theory to convince yourself that in order for an n-point correlation function to the m-th order not to vanish, $n + 3m$ must be even. A modification has to be made here for rule (b): A diagram for this theory must have 3 line segments emanating from each vertex.

Problem 50.4 Draw all third order vacuum diagrams according to the rules (a), (b), and (c) above for the φ^4-theory.

An important feature concerning the relationship between the numerator $\mathcal{N}^{[2]}$ and the denominator $\mathcal{D}$ in (49.38) for the 2-point correlation function begins to emerge at this point: The denominator (sum of vacuum diagrams) cancels out. This can be seen (graphically) as follows. $\mathcal{N}^{[2]}_{(2)}$ is the sum of all the contributions from the diagrams in Figs. 50.1, 50.3, 50.4 and 50.6; while $\mathcal{D}^{[2]}_{(2)}$, explicitly given by (50.25), is the sum of 1 and all the contributions from the (vacuum) diagrams in Figs. 50.2 and 50.5. Up to second order, at least, it is

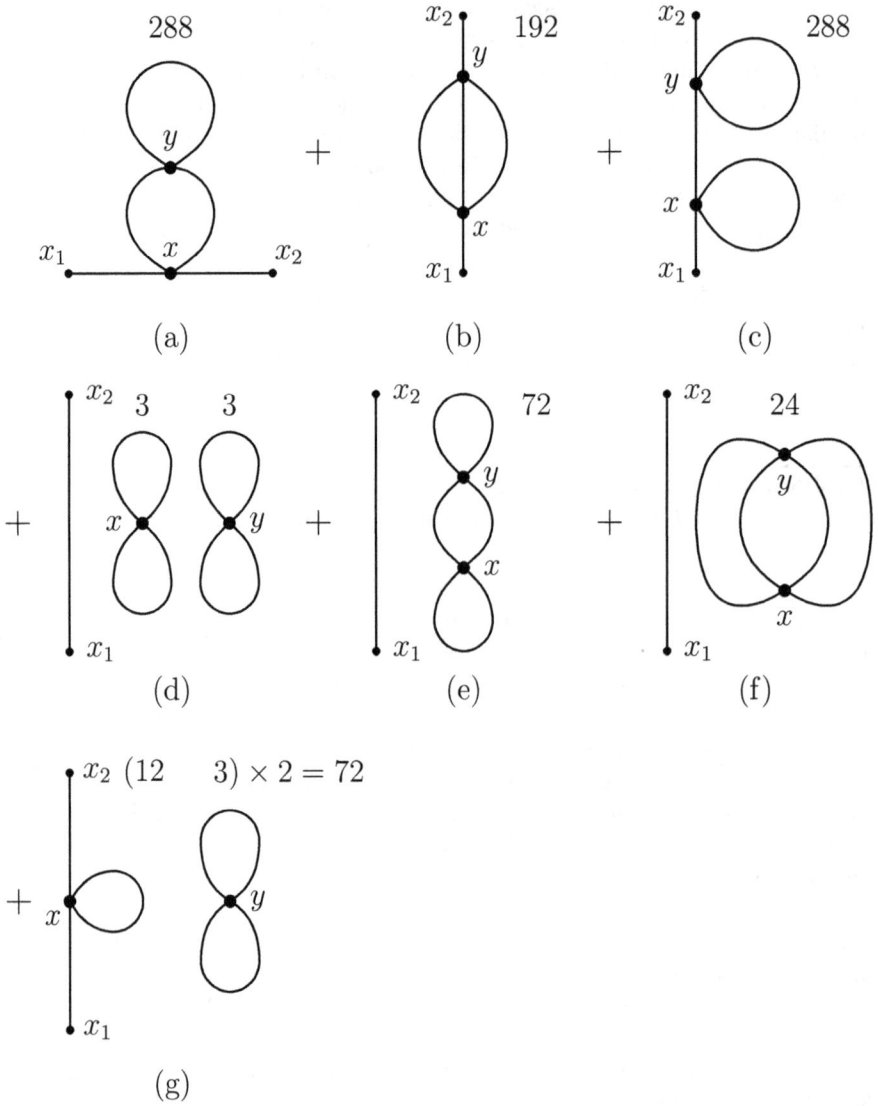

FIGURE 50.6

then easily seen that $\mathcal{N}^{[2]}_{(2)}$ can also be represented as a product of two factors, as shown in Fig. 50.7, when only terms up to second order are retained after the multiplication. The second factor in the figure is clearly $\mathcal{D}_{(2)}$ and cancels out the denominator, while the first factor only includes 2-point non-vacuum diagrams without vacuum-diagram factors. Denoting such a diagram by $diag(nv)^{[2]}_{(m)}$, we can then write, to second order,

$$\langle \varphi(x_1)\varphi(x_2) \rangle_{(2)} = \sum_{m=0}^{2} \left(\sum diag(nv)^{[2]}_{(m)} \right) . \tag{50.27}$$

We claim that *this cancellation of the sum of all vacuum diagrams is a general feature of an arbitrary n-point correlation function to all orders*, so we can write

$$\langle \varphi(x_1) \ldots \varphi(x_n) \rangle = \sum_{m=0}^{\infty} \left(\sum diag(nv)^{[n]}_{(m)} \right) , \tag{50.28}$$

where $diag(nv)^{[n]}_{(m)}$ denotes the contribution from a non-vacuum Feynman diagram for an n-point correlation function of order m (with m vertices and n external points) with no vacuum diagrams as factors, and the second sum is over all such diagrams (both connected and disconnected) of order m. We will not give a proof of (50.28) here. The interested reader is referred to, for example, Amit and Martín-Mayor (2005). As an example of (50.28) for the 4-point correlation function for the φ^4-theory up to second order, we show the diagrammatic representation in Fig. 50.8. Note that here, as opposed to the 2-point function, the diagrams can be both connected and disconnected.

Recalling (49.38) again, we see that since any term in the numerator is obtained from a string of extra functional derivatives $(\delta/\delta J(x_1)) \ldots (\delta/\delta J(x_n))$ acting on the self-interaction terms, which are in turn the ones giving rise to the vacuum diagrams, we expect that the diagrams in (50.28) should be obtainable graphically from the vacuum diagrams in some way. This is indeed the case. We will not prove the general situation but will instead illustrate by some examples. There are in general two kinds of diagrams in the correlation functions: connected and disconnected (see Fig. 50.7 and Fig. 50.8). The connected ones are generated from the connected vacuum diagrams by cutting one or more lines, while the disconnected ones can be obtained from either connected or disconnected vacuum diagrams by cutting one or more lines. A single free propagator may be imagined to be obtained from cutting a single loop without a vertex, which can be treated as a vacuum diagram whose value is 1; and a product of several free propagators can be obtained by cutting several such loops. It is quite obvious that l cuts on a vacuum diagram generate a diagram for a $2l$-point correlation function. Fig. 50.9 illustrates some examples of the generation of correlation function diagrams by cutting vacuum diagrams that are up to second order. Note that some correlation function diagrams may be generated from vacuum diagrams in more than one way, such as the pair (f) and (l), and the pair (j) and (n) in the figure.

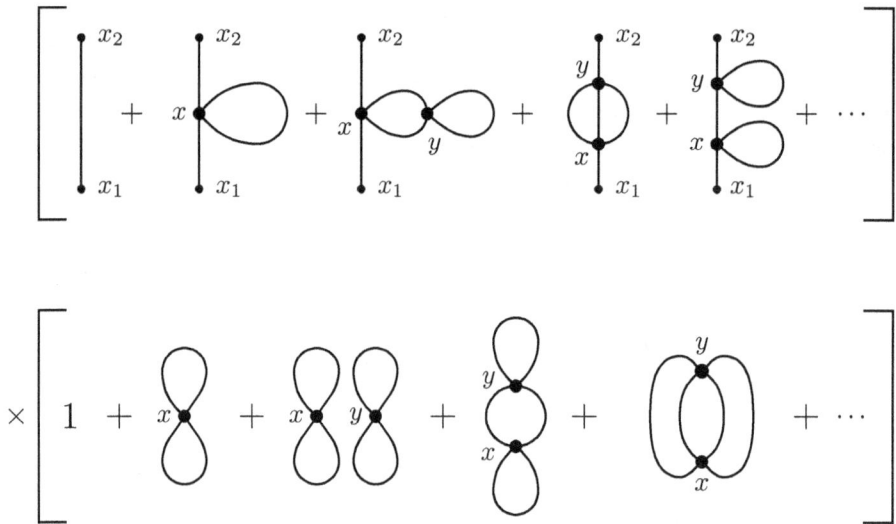

FIGURE 50.7

We will finally address the problem of the calculation of the symmetry factor $S(n, m)$ for a topologically distinct diagram for an n-point correlation function of order m [c.f. Feynman diagram rule (4) above]. A general formula is not available, but we will illustrate the calculation for a specific diagram. The technique introduced should, at least in principle, be applicable to an arbitrary diagram. Essentially we face the combinatorial problem of determining the total number of ways that an even number of points can be paired up, subject to certain constraints depending on the topology of the particular diagram. The pairing requirement can be inferred from results such as (50.23) and (50.25), where the symmetry factors are the coefficients of the various products of propagators $D_E(x, x')$, all of which have the symmetry property $D_E(x, x') = D_E(x', x)$. Let us pick the two-loop connected diagram in Fig. 50.6(a). To begin, it is useful to state a combinatorial fact known as **Wick's Theorem:** *The number of ways that $2n$ objects can be paired up is equal to $(2n-1)!! = (2n-1)(2n-3)\ldots 5\cdot 3\cdot 1$.* To see this, note that the first object can be paired up with the $(2n-1)$ remaining objects. The next unpaired object, if any is remaining, can then be paired up with the $(2n-3)$ still remaining unpaired ones. This process can be repeated until all objects have been paired up, which leads to the theorem. The diagram under consideration has two external points x_1 and x_2, and two vertices (internal points) x and y. Imagine laying them out in a straight line, as shown in Fig. 50.10, with each of x_1 and x_2 occurring once, and x and y each

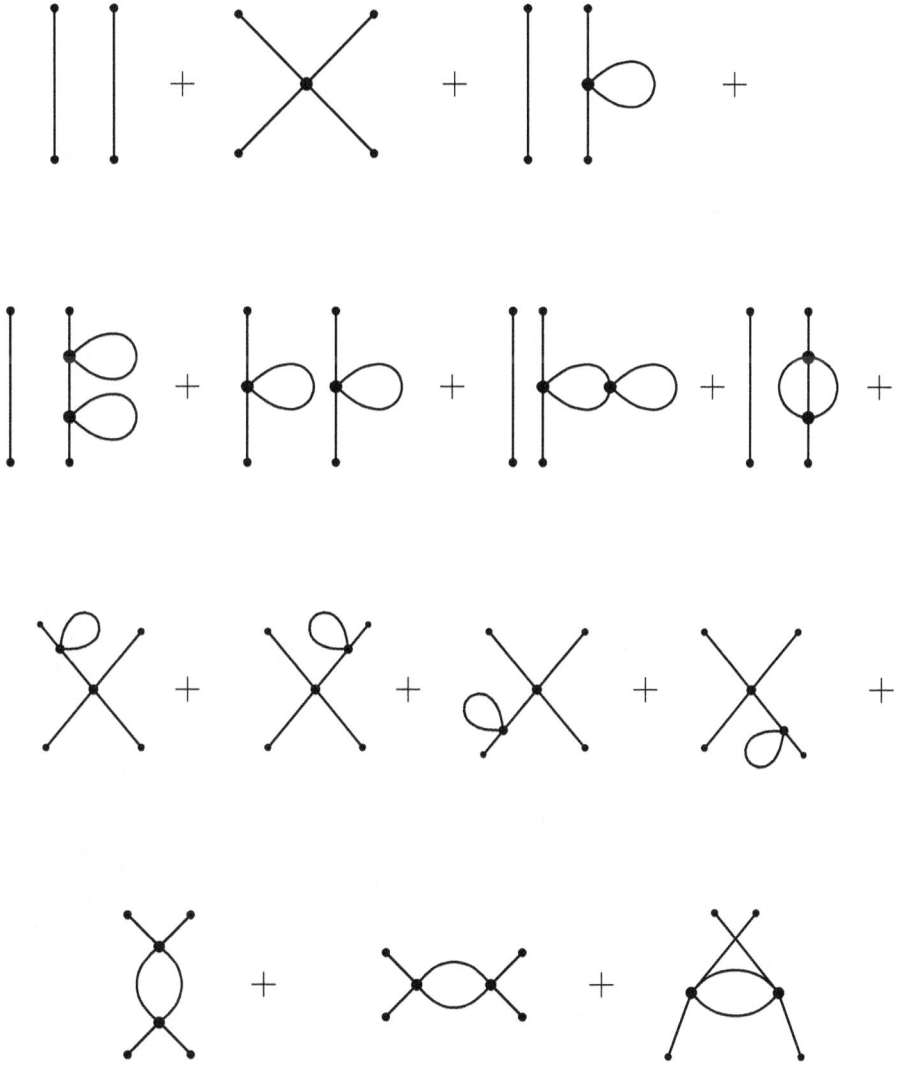

FIGURE 50.8

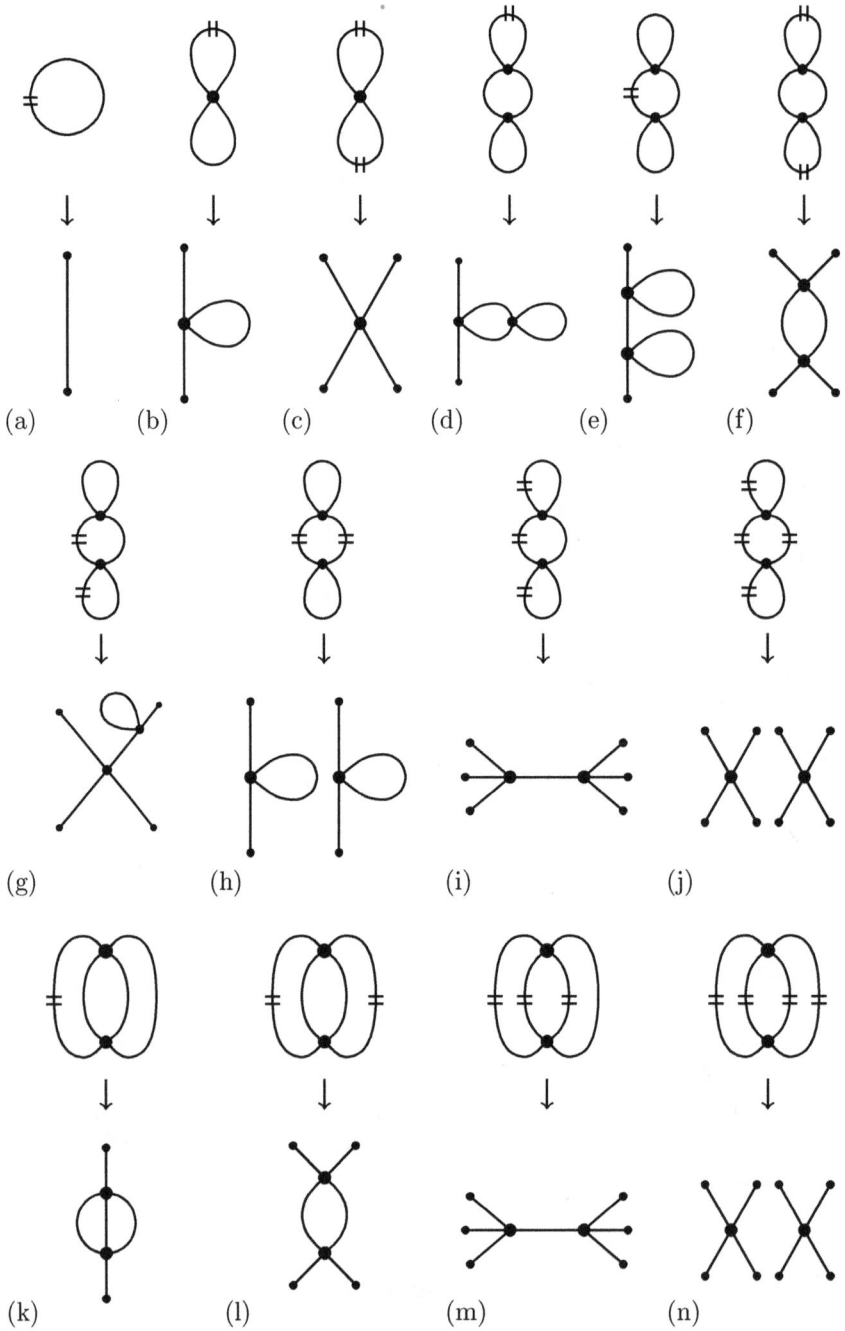

(a)   (b)   (c)   (d)   (e)   (f)

(g)   (h)   (i)   (j)

(k)   (l)   (m)   (n)

FIGURE 50.9

occurring 4 times. This comes from the string of functional derivatives

$$\frac{\delta}{\delta J(x_1)}\frac{\delta}{\delta J(x_2)}\left(\frac{\delta}{\delta J(x)}\right)^4\left(\frac{\delta}{\delta J(y)}\right)^4$$

giving rise to this diagram [c.f. (49.38)]. The diagram stipulates the following patterns of connection between the points:

$$(x_1,x)(x_2,x)(x,y)^2(y,y) \qquad \text{or} \qquad (x_1,y)(x_2,y)(y,x)^2(x,x) \ ,$$

where the order of the pairings is immaterial. The symmetry factor for the diagram is then the total number of ways that these pairings can be carried out. It is quite obvious that the above two patterns each yields the same symmetry factor, so we just need to work with one and multiply the result by 2. Take the first one and start with the point x_1. There are 4 possible ways to pair it with the 4 x's. So the first pairing (x_1,x) in the pattern carries a factor of 4. After this is done the x_2 can only be paired with the remaining 3 x's. So the second pairing in the pattern, (x_2,x), carries a factor of 3. Up to this point, x_1, x_2, and 2 x's have been used up. Let us next work with the pairing (y,y) in the pattern. There are 4 y's left and it is easily seen that they can be paired up in 3 ways (using Wick's Theorem). So the factor (y,y) in the pattern carries a factor of 3. After 2 y's are paired up, only 2 of them remain, together with the 2 x's that are still unpaired up to this point. From 2 x's and 2 y's, there are 4 ways to pair up 1 x and 1 y. So the first factor (x,y) in the pattern carries a factor of 4. Finally there is only 1 x and 1 y left, and these can be paired up in just 1 way, contributing a factor of 1 to the second factor (x,y) in the pattern. The result is that the symmetry factor for the first pattern is $4^2 \times 3^2 = 144$. This number has to be doubled because the second pattern gives exactly the same symmetry factor. So the symmetry factor for the diagram in Fig. 50.6(a) is 288. Fig. 50.10 also shows one way (out of 144) of doing the pairings in the first pattern.

Using the same procedure, one can systematically work out the symmetry factor for any diagram, in particular, all the second order 2-point correlation function diagrams in Fig. 50.6 (where the corresponding symmetry factors are displayed). Note that the symmetry factor for the disconnected free propagator $D_E(x_1,x_2)$ is just 1, quite obviously, because there is only one way to pair the two points x_1, x_2. As a check, we can add up all the symmetry factors for the diagrams in Fig. 50.6 and get

$$288 + 192 + 288 + 9 + 72 + 24 + 72 = 945 \ .$$

This is precisely the number of ways that unrestricted pairings can occur among the 10 objects

$$x_1, x_2, x, x, x, x, y, y, y, y \ ,$$

since, by Wick's Theorem, this is equal to $(2 \times 5 - 1)!! = 9 \times 7 \times 5 \times 3 = 945$.

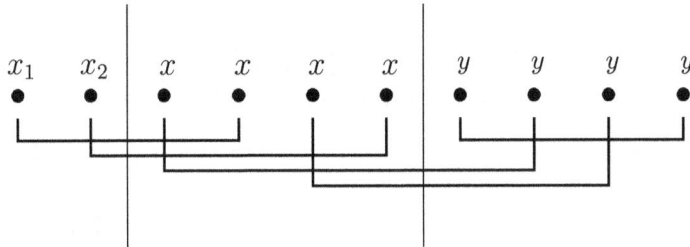

FIGURE 50.10

Problem 50.5 Verify, by combinatorial arguments shown in the text, that the symmetry factors for all the Feynman diagrams in Fig. 50.6 [except for (a)] are as displayed in the figure.

Problem 50.6 Show that the 4-point correlation function for the φ^4-theory due to only connected diagrams (up to second order in λ, these are shown in Fig. 50.8, excluding the disconnected ones), denoted by $\langle\varphi(x_1)\varphi(x_2)\varphi(x_3)\varphi(x_4)\rangle_c$, is related to the full 4-point function and the 2-point functions as follows:

$$
\begin{aligned}
\langle\varphi(x_1)\varphi(x_2)\varphi(x_3)\varphi(x_4)\rangle_c = {} & \langle\varphi(x_1)\varphi(x_2)\varphi(x_3)\varphi(x_4)\rangle \\
& - \langle\varphi(x_1)\varphi(x_2)\rangle\langle\varphi(x_3)\varphi(x_4)\rangle \\
& - \langle\varphi(x_1)\varphi(x_3)\rangle\langle\varphi(x_2)\varphi(x_4)\rangle \\
& - \langle\varphi(x_1)\varphi(x_4)\rangle\langle\varphi(x_2)\varphi(x_3)\rangle \,.
\end{aligned}
\tag{50.29}
$$

Show that this result is in accord with

$$
\langle\varphi(x_1)\varphi(x_2)\varphi(x_3)\varphi(x_4)\rangle_c = \frac{\delta}{\delta J(x_1)}\frac{\delta}{\delta J(x_2)}\frac{\delta}{\delta J(x_3)}\frac{\delta}{\delta J(x_4)} \ln Z_E[J]\Big|_{J=0}, \tag{50.30}
$$

where $Z_E[J]$ is given by (49.36).

The result (50.30) in Problem 50.6 for the 4-point function suggests that it can be generalized to the case of a *connected* n-point correlation function:

$$\langle \varphi(x_1) \ldots \varphi(x_n) \rangle_c = \frac{\delta}{\delta J(x_1)} \cdots \frac{\delta}{\delta J(x_n)} \ln Z[J] \Big|_{J=0} . \qquad (50.31)$$

This is in fact true, but we will not prove it here. The interested reader can consult, for example, Cheng and Li (1984), or Ramond (1989). It should be noted also that the above formula actually applies to both the Euclidean and Minkowski correlation functions (so we have not included the subscript E to the generating functional $Z[J]$). It can be inferred also from (50.31) that the case of the 2-point function is special, since there are no disconnected graphs in $\langle \varphi(x_1)\varphi(x_2) \rangle$ (for graphs up to second order in λ see Fig. 50.7). We have, in fact,

$$\langle \varphi(x_1)\varphi(x_2) \rangle = \langle \varphi(x_1)\varphi(x_2) \rangle_c = \frac{\delta}{\delta J(x_1)} \frac{\delta}{\delta J(x_2)} \ln Z[J] \Big|_{J=0} . \qquad (50.32)$$

Chapter 51

Holomorphic Quantization

In this chapter we return to some fundamental issues on quantization and examine an alternative way to using the phase space variables p_i and q^i in the functional integration $[dp_i dq^i]$ as a route to quantization [as in (45.27)]. This alternative and theoretically very interesting way, based on the complex variables a and a^* (to represent the annihilation and creation operators $\hat{a}$ and $\hat{a}^\dagger$ in quantum mechanics) instead of p and q, is most suitable for harmonic oscillator Hamiltonians of the familiar form

$$\hat{H} = \frac{\hat{p}^2}{2m} + \frac{1}{2}m\omega^2\hat{q}^2 \,, \tag{51.1}$$

where m and ω are the mass and the frequency of the oscillator, respectively. Indeed, the Hamiltonian derived from the free-field Lagrangian density $\mathcal{L}_0$ in (48.16), from which most field theories of interest are built, is a quantum field theory generalization of the Hamiltonian in (51.1). Our presentation will be adapted from Faddeev and Slavnov (1980). For simplicity, we will go back first to the case of a single degree of freedom.

For the moment, let us define the complex variables a and a^* (in terms of the phase space variables p and q) by

$$a = \frac{1}{\sqrt{2\omega}}\left(\sqrt{\frac{m}{\hbar}}\,\omega q + i\frac{1}{\sqrt{m\hbar}}\,p\right) , \qquad a^* = \frac{1}{\sqrt{2\omega}}\left(\sqrt{\frac{m}{\hbar}}\,\omega q - i\frac{1}{\sqrt{m\hbar}}\,p\right) . \tag{51.2}$$

To simplify our expressions we will set $\hbar = m = 1$ and write

$$a = \frac{1}{\sqrt{2\omega}}\,(\omega q + ip) , \qquad a^* = \frac{1}{\sqrt{2\omega}}\,(\omega q - ip) . \tag{51.3}$$

In canonical quantization, the c-numbers p and q are made into operators $\hat{p}$ and $\hat{q}$ on Hilbert space by imposing the familiar commutation relations

$$[\hat{q}, \hat{q}] = [\hat{p}, \hat{p}] = 0 \,, \qquad [\hat{q}, \hat{p}] = i \quad (\hbar = 1) \,. \tag{51.4}$$

The corresponding operators $\hat{a}$ and $\hat{a}^*$, as defined by (51.3), then satisfy

$$[\hat{a}, \hat{a}] = [\hat{a}^*, \hat{a}^*] = 0 , \qquad [\hat{a}, \hat{a}^*] = 1 . \tag{51.5}$$

These are the familiar commutation relations for harmonic oscillator creation and annihilation operators. Note that $\hat{a}^* = \hat{a}^\dagger$, that is, $\hat{a}$ and $\hat{a}^*$ are adjoint to each other. It is significant that the complex factor i is no longer present in the commutation relations.

Just as one sought coordinate representations (in terms of a real coordinate q) for the operators $\hat{q}$ and $\hat{p}$ satisfying the commutation relations (51.4), we now seek coordinate representations (in terms of a complex coordinate z) for the operators $\hat{a}$ and $\hat{a}^*$ satisfying the commutation relations (51.5); and it is not hard to see that the representations

$$\hat{a} = z , \qquad \hat{a}^* = \frac{\partial}{\partial z} , \tag{51.6}$$

as operators on the space of complex *holomorphic functions* $f(z)$ (c.f. Def. 40.4) will do the job. Indeed, for any holomorphic function $f(z)$,

$$[\hat{a}, \hat{a}^*] f(z) = z \frac{\partial f}{\partial z} - \frac{\partial}{\partial z}\{zf(z)\} = f(z) . \tag{51.7}$$

To satisfy the mutual adjointness property, one has to introduce a suitable hermitian inner product $\langle f_1(z) \mid f_2(z)\rangle$ such that

$$\langle f_1 \mid \hat{a} \mid f_2\rangle = \langle \hat{a}^* f_1 \mid f_2\rangle \qquad \text{and} \qquad \langle f_1 \mid \hat{a}^* \mid f_2\rangle = \langle \hat{a} f_1 \mid f_2\rangle . \tag{51.8}$$

One quickly finds that the naive choice for the inner product given by $\int dz\, \overline{f_1(z)} f_2(z)$ (the overbear means complex conjugation) will not work because the integration contour is not specified. The correct choice, as found by Bargmann, Berezin, and Segal, is in fact not straightforward [see Berezin (1966)]. It involves first restricting the space of holomorphic functions of z to the closure of the space of polynomials in z, denoted by $\mathcal{P}$. The required inner product is then given by

$$\langle f_1(z) \mid f_2(z)\rangle = \int \frac{d\bar{z} \wedge dz}{2\pi i} \, \overline{f_1(z)} f_2(z) \, e^{-|z|^2} , \tag{51.9}$$

where, for $z = x + iy = re^{i\theta}$, we understand $d\bar{z} \wedge dz$ by $2i\, dx \wedge dy = 2i\, r dr \wedge d\theta$, and the integral is over the entire z-plane, treated as $\mathbb{R}^2$. This being the case, no questions of the relationships between the orientations of a finite region and its boundary will arise and the integral can be treated as a Riemann integral over the plane [c.f. (32.16)]. From this point on, we will thus dispense with the exterior product and write $d\bar{z} \wedge dz$ as $d\bar{z}dz = dzd\bar{z} = 2idxdy = 2irdrd\theta$, and not worry about orientations.

Considering $\mathbb{C} \sim \mathbb{R}^2 = \mathbb{P}$ (phase space with local coordinates p, q), we note that the differential form $\alpha \equiv -id\bar{z} \wedge dz$ introduced in (51.9) is a $(1, 1)$-type form on $\mathbb{P} \otimes \mathbb{C}$, the complexification of $\mathbb{P}$ [c.f. (40.19)]. The space $\mathcal{P}$ can also be considered as a subspace of the space of *holomorphic sections* on the *complex line bundle* $\pi : \mathbb{L} \to \mathbb{R}^2$ (with typical fiber $\mathbb{C}$), that is [c.f. (30.21) and discussion following], $\mathcal{P} \subset \Gamma(\mathbb{L})$.

Our inner product as defined above clearly satisfies $\langle f_1 \,|\, f_2 \rangle = \overline{\langle f_2 \,|\, f_1 \rangle}$. To see that it is positive definite, it suffices to take the orthonormal basis set $\{z^n\}$ of monomials. Indeed, defining $\phi_n(z) \equiv z^n$, we have

$$\langle \phi_m \,|\, \phi_n \rangle = \frac{1}{\pi} \int_0^\infty r dr\, r^{n+m} e^{-r^2} \int_0^{2\pi} e^{i\theta(n-m)} \,. \tag{51.10}$$

This clearly vanishes for $n \neq m$. For $n = m$, the inner product becomes

$$\langle \phi_n \,|\, \phi_n \rangle = \frac{1}{\pi} \int_0^\infty r dr\, r^{2n} e^{-r^2} \int_0^{2\pi} d\theta = 2 \int_0^\infty r dr\, r^{2n} e^{-r^2}$$
$$= \int_0^\infty dt\, t^n e^{-t} = \Gamma(n+1) = n! \,, \tag{51.11}$$

where in the third equality we have changed the variable of integration from r to $t = r^2$, and $\Gamma(n)$ is the *Gamma function*. This shows that the inner product defined by (51.9) is positive definite, and we can define an orthonormal basis set $\{|\psi_n\rangle\}$ in $\mathcal{P}$ by

$$|\psi_n\rangle \longrightarrow \psi_n(z) \equiv \frac{\phi_n(z)}{\sqrt{n!}} = \frac{z^n}{\sqrt{n!}} \,, \qquad \langle \psi_n \,|\, \psi_m \rangle = \delta_{nm} \,. \tag{51.12}$$

To demonstrate that $\hat{a}$ and $\hat{a}^*$ as defined by (51.6) are mutually adjoint according to the hermitian product given by (51.9), we first test the first condition in (51.8). Let $f_1(z)$ and $f_2(z)$ be two arbitrary functions in the space $\mathcal{P}$. We have

$$\langle f_1 \,|\, \hat{a} \,|\, f_2 \rangle = \int \frac{d\bar{z}dz}{2\pi i} \overline{f_1(z)} z f_2(z) e^{-|z|^2} = \int \frac{d\bar{z}dz}{2\pi i} \overline{f_1(z)} f_2(z) \left(-\frac{\partial}{\partial \bar{z}} e^{-z\bar{z}} \right)$$
$$= \int \int \frac{d\bar{z}dz}{2\pi i} \frac{\partial}{\partial \bar{z}} \left(\overline{f_1(z)} f_2(z) \right) e^{-|z|^2}$$
$$= \int \frac{d\bar{z}dz}{2\pi i} \left\{ \left(\frac{\partial}{\partial \bar{z}} \overline{f_1(z)} \right) f_2(z) + \overline{f_1(z)} \frac{\partial}{\partial \bar{z}} f_2(z) \right\} e^{-|z|^2}$$
$$= \int \frac{d\bar{z}dz}{2\pi i} \frac{\overline{\partial f_1(z)}}{\partial z} f_2(z) e^{-|z|^2} = \langle \hat{a}^* f_1 \,|\, f_2 \rangle \,, \tag{51.13}$$

where in the third equality we have used integration by parts and the fact that the factor $e^{-z\bar{z}}$ causes the boundary term to vanish, and in the next to last equality we have used the *Cauchy-Riemann condition*

$$\frac{\partial f(z)}{\partial \bar{z}} = 0 \tag{51.14}$$

for holomorphic functions [c.f. (40.75)], which the functions in $\mathcal{P}$ certainly are. A similar calculation can be used to check the second condition in (51.8). We have

$$
\begin{aligned}
\langle f_1 \,|\, \hat{a}^* \,|\, f_2 \rangle &= \int \frac{d\bar{z}dz}{2\pi i} \, \overline{f_1(z)} \left(\frac{\partial f_2(z)}{\partial z} \right) e^{-|z|^2} \\
&= \int \frac{d\bar{z}dz}{2\pi i} \, \overline{f_1(z)} \left\{ \frac{\partial}{\partial z} \left(f_2(z) e^{-z\bar{z}} \right) - f_2(z) \frac{\partial}{\partial z} e^{-z\bar{z}} \right\} \\
&= \int \frac{d\bar{z}dz}{2\pi i} \left\{ -\left(\frac{\partial \overline{f_1(z)}}{\partial z} \right) f_2(z) e^{-|z|^2} + \bar{z}\overline{f_1(z)} f_2(z) e^{-|z|^2} \right\} \\
&= \int \frac{d\bar{z}dz}{2\pi i} \, \bar{z}\overline{f_1(z)} f_2(z) e^{-|z|^2} = \langle \hat{a} f_1 \,|\, f_2 \rangle \,,
\end{aligned}
\tag{51.15}
$$

where in the third equality we have used integration by parts and in the last, the Cauchy-Riemann condition in the form $\partial \overline{f(z)}/\partial z = 0$, which follows from (51.13).

The functions $\psi_n(z)$ in (51.12) are the complex-coordinate (holomorphic) representations of the abstract states $|\psi_n\rangle$, which we will now write as $|n\rangle$:

$$
|n\rangle = |\psi_n\rangle \xrightarrow[\substack{\text{holomorphic} \\ \text{representation}}]{} \psi_n(z) = \frac{z^n}{\sqrt{n!}} \,.
\tag{51.16}
$$

Using (51.6), it can be deduced straightforwardly that they satisfy the following relationships:

$$
\begin{array}{lll}
\hat{a}\,\psi_n(z) = \sqrt{n}\,\psi_{n-1}(z) \,, & \hat{a}\,|n\rangle = \sqrt{n}\,|n-1\rangle \,, & (51.17\text{a}) \\[4pt]
\hat{a}^*\,\psi_n(z) = \sqrt{n+1}\,\psi_{n+1}(z) \,, & \hat{a}^*\,|n\rangle = \sqrt{n+1}\,|n+1\rangle \,, & (51.17\text{b}) \\[4pt]
\psi_n(z) = \dfrac{1}{\sqrt{n!}}\,(\hat{a})^n\,\psi_0(z) \,, & |n\rangle = \dfrac{1}{\sqrt{n!}}\,(\hat{a}^*)^n\,|0\rangle \,, & (51.17\text{c}) \\[4pt]
\hat{a}^*\hat{a}\,\psi_n(z) = n\,\psi_n(z) \,, & \hat{a}^*\hat{a}\,|n\rangle = n\,|n\rangle \,, & (51.17\text{d}) \\[4pt]
\hat{a}\,\psi_0(z) = 0 \,, & \hat{a}\,|0\rangle = 0 \,, & (51.17\text{e})
\end{array}
$$

where we have written side by side the action of the operators on the "wave functions" $\psi_n(z)$ and the abstract state vectors $|n\rangle$. These, of course, are the familiar relationships for the quantum harmonic oscillator *number states*. Note that in the present holomorphic (complex-coordinate) representation, the wave functions $\psi_n(z)$ have a much simpler analytical form than the representations of $|n\rangle$ by functions of the real coordinate q, which, as is well known, are given in terms of *Hermite polynomials*.

We would now like to construct the analog of the (non-normalizable) sharp position eigenstates $|q\rangle$, which satisfy $\hat{q}\,|q\rangle = q\,|q\rangle$, in our scheme of **holomorphic quantization** given by (51.6). In other words, for a complex number z, we would like to identify an eigenstate $|z\rangle \in \mathcal{P}$ of $\hat{a}$, which satisfies

$$
\hat{a}\,|z\rangle = z\,|z\rangle \,.
\tag{51.18}
$$

It can be quite readily demonstrated, using the properties (51.17a) and (51.17b), that such a state is given by

$$|z\rangle = e^{z\hat{a}^* - \bar{z}\hat{a}}|0\rangle = e^{-|z|^2/2} \sum_{n=0}^{\infty} \frac{z^n}{\sqrt{n!}} |n\rangle . \tag{51.19}$$

These states, called **coherent states**, were originally developed in the study of *quantum optics*, where $|n\rangle$ is interpreted as the number states for photons of a particular mode. Besides the defining relationship (51.18), it can be readily demonstrated that they satisfy the following properties:

$$|\langle z | w\rangle|^2 = e^{-|z-w|^2} \quad \text{for} \quad z, w \in \mathbb{C} \qquad \text{(normalizability)}, \tag{51.20}$$

$$\int \frac{d\bar{z}dz}{2\pi i} |z\rangle\langle z| = \int \frac{d\bar{z}dz}{2\pi i} |\bar{z}\rangle\langle\bar{z}| = 1 , \qquad \text{(completeness)}, \tag{51.21}$$

$$\langle \bar{z} | n\rangle = \langle n | z\rangle = e^{-|z|^2/2} \frac{z^n}{\sqrt{n!}} = e^{-|z|^2/2} \psi_n(z) \quad \text{(Poisson distribution)} . \tag{51.22}$$

Eq. (51.20) indicates that the coherent states $|z\rangle$, unlike the position states $|q\rangle$, are normalized but not orthonormal. Two coherent states $|z\rangle$ and $|w\rangle$ are, however, nearly orthogonal if z and w are very far from each other on the complex plane. Note also the somewhat unusual result in (51.22) that $\langle z | \psi_n\rangle$ is not equal to $\psi_n(z)$ but to $e^{-|z|^2/2} \overline{\psi_n(z)}$. This is designated the *Poisson distribution* statement because it gives

$$|\langle n | z\rangle|^2 = e^{-|z|^2/2} \frac{(|z|^2)^n}{n!} , \tag{51.23}$$

which, in the context of quantum optics for example, says that, for a photon state described by the coherent state $|z\rangle$, the probability of finding n photons is given by a *Poisson distribution* about the mean $|z|^2$. By using (51.22) and the second completeness condition in (51.21) we can recover the hermitian product introduced in (51.9) as follows:

$$\langle \psi_n | \psi_m\rangle = \langle n | m\rangle = \int \frac{d\bar{z}dz}{2\pi i} \langle n | \bar{z}\rangle\langle \bar{z} | m\rangle$$

$$= \int \frac{d\bar{z}dz}{2\pi i} e^{-|z|^2} \frac{\bar{z}^n}{\sqrt{n!}} \frac{z^m}{\sqrt{m!}} = \int \frac{d\bar{z}dz}{2\pi i} \overline{\psi_n(z)}\psi_m(z) e^{-|z|^2} . \tag{51.24}$$

Problem 51.1 Verify all the properties for $\psi_n(z)$ (alternatively for the states $|n\rangle$) in (51.17).

Problem 51.2 Verify the second equality in (51.19).

Problem 51.3 Use (51.19) to verify the three properties for the coherent states $|z\rangle$ given by (51.20), (51.21) and (51.22).

Problem 51.4 Show that

$$\int \frac{d\bar{z}dz}{2\pi i} e^{\bar{w}z} f(\bar{z}) e^{-|z|^2} = f(\bar{w}) . \tag{51.25}$$

This result shows that $\exp(\bar{w}z)$ plays the role of the Dirac delta function in the holomorphic representation.

Next we wish to determine how a general operator $\hat{A}$ (acting on $\mathcal{P}$) is represented holomorphically. In other words, for $f(z) \in \mathcal{P}$, we wish to determine the function $(\hat{A}f)(z) \in \mathcal{P}$. We begin by observing that, analogus to (51.22), we have

$$\langle \bar{z} \,|\, f \rangle = e^{-|z|^2/2} f(z) , \tag{51.26}$$

while $\hat{A}$ can be written

$$\hat{A} = \sum_{n,m} A_{nm} \,|n\rangle\langle m| , \tag{51.27}$$

where $A_{nm} = \langle n \,|\, \hat{A} \,|\, m \rangle$. Eq. (51.25) implies

$$\langle \bar{z} \,|\, \hat{A}f \rangle = e^{-|z|^2/2} \, (\hat{A}f)(z) ; \tag{51.28}$$

while from (51.26) we have

$$\langle \bar{z} \,|\, \hat{A}f \rangle = \sum_{m,n} A_{nm} \, \langle \bar{z} \,|\, n \rangle\langle m \,|\, f \rangle = \sum_{m,n} A_{nm} \left(e^{-|z|^2/2} \frac{z^n}{\sqrt{n!}} \right) \langle m \,|\, f \rangle$$

$$= e^{-|z|^2/2} \sum_{m,n} A_{nm} \frac{z^n}{\sqrt{n!}} \int \frac{d\bar{\zeta}d\zeta}{2\pi i} \, \langle m \,|\, \bar{\zeta} \rangle\langle \bar{\zeta} \,|\, f \rangle \tag{51.29}$$

$$= e^{-|z|^2/2} \sum_{m,n} A_{nm} \frac{z^n}{\sqrt{n!}} \int \frac{d\bar{\zeta}d\zeta}{2\pi i} \left(e^{-|\zeta|^2/2} \frac{\bar{\zeta}^m}{\sqrt{m!}} \right) \left(e^{-|\zeta|^2/2} f(\zeta) \right) ,$$

where in the third equality we have used the second form of the completeness condition in (51.21). Comparing the last two equations we conclude that the operator $\hat{A}$ can be represented by an integral kernel $A(z, \bar{\zeta})$ as follows:

$$(\hat{A}f)(z) = \int \frac{d\bar{\zeta}d\zeta}{2\pi i} \, A(z, \bar{\zeta}) f(\zeta) \, e^{-|\zeta|^2} , \tag{51.30}$$

where

$$A(z, \bar{\zeta}) = \sum_{m,n} A_{nm} \frac{z^n}{\sqrt{n!}} \frac{\bar{\zeta}^m}{\sqrt{m!}} . \tag{51.31}$$

This being an integral kernel, it follows that the kernel corresponding to the product of two operators, $\hat{A}_1\hat{A}_2$, is given by the *convolution product* of the two kernels A_1 and A_2:

$$(A_1 * A_2)(z, \bar{z}) = \int \frac{d\bar{\zeta}d\zeta}{2\pi i} A_1(z, \bar{\zeta}) A_2(\zeta, \bar{z}) e^{-|\zeta|^2} . \tag{51.32}$$

It is important to realize that, in the above equation, z and $\bar{z}$ (despite the notation) are regarded as independent complex variables, whereas ζ and $\bar{\zeta}$ (occurring together in the differential form $d\bar{\zeta} \wedge d\zeta$), *are* complex conjugates of each other. In general [c.f. (51.30) and (40.4)], an integral kernel $A(z, \bar{z})$ of an arbitrary operator $\hat{A}$ is a holomorphic function of two independent complex variables z and $\bar{z}$.

On expanding $\hat{A}$ in terms of $\hat{a}$ and $\hat{a}^*$ in **normal order** (all $\hat{a}^*$ to the left of all $\hat{a}$):

$$\hat{A} = \sum_{m,n} K_{nm}^{(\hat{A})} (\hat{a}^*)^n \hat{a}^m , \tag{51.33}$$

where $K_{nm}^{(\hat{A})} \in \mathbb{C}$, one can also represent $\hat{A}$ in another way holomorphically besides the representation given by (51.30), namely, by the **normal symbol** $K^{(\hat{A})}(a^*, a)$ for $\hat{A}$, defined by

$$K^{(\hat{A})}(a^*, a) \equiv \sum_{m,n} K_{nm}^{(\hat{A})} (a^*)^n a^m , \tag{51.34}$$

where $a, a^* \in \mathbb{C}$. Thus the normal symbol of an operator is a holomorphic function of two (independent) complex variables, as is the integral kernel $A(z, \bar{\zeta})$. Let us replace z by a^* and $\bar{\zeta}$ by a in (51.30) to write

$$A(a^*, a) = \sum_{m,n} A_{nm} \frac{(a^*)^n}{\sqrt{n!}} \frac{a^m}{\sqrt{m!}} . \tag{51.35}$$

Then we have the following important theorem.

Theorem 51.1. *The integral kernel $A(a^*, a)$ and the normal symbol $K^{(\hat{A})}(a^*, a)$ for an operator $\hat{A}$, as two holomorphic functions of the two complex variables a^* and a, are related by*

$$A(a^*, a) = e^{a^* a} K^{(\hat{A})}(a^*, a) . \tag{51.36}$$

Proof. It suffices to prove the theorem for the monomial $\hat{A} = (\hat{a}^*)^k \hat{a}^l$. In this case the normal symbol is just $K^{(\hat{A})}(a^*, a) = (a^*)^k a^l$. The matrix element $A_{nm} = \langle n | \hat{A} | m \rangle = \langle n | (\hat{a}^*)^k \hat{a}^l | m \rangle$ in (51.34) can be computed in a straightforward manner with the help of (51.17a) and (51.17b). We have, with $\theta(x)$ being the step-function,

$$\hat{a}^l |m\rangle = \theta(m - l)\sqrt{m(m-1)\ldots(m-l+1)} |m - l\rangle , \tag{51.37}$$

while

$$\langle n | (a^*)^k = \theta(n-k)\sqrt{n(n-1)\ldots(n-k+1)}\,\langle n-k | \,. \tag{51.38}$$

Thus (51.34) implies

$$A(a^*,a) = \sum_{m,n} \delta_{n-k,m-l}$$

$$\times \sqrt{n(n-1)\ldots(n-k+1)m(m-1)\ldots(m-l+1)} \tag{51.39}$$

$$\times\, \theta(m-l)\theta(n-k)\,\frac{(a^*)^n}{\sqrt{n!}}\,\frac{a^m}{\sqrt{m!}}\,.$$

Do the sum over m first. The Kronecker delta imposes the restriction that $m = n-k+l$, and so the product of the step functions become $\theta^2(n-k) = \theta(n-k)$. We then have

$$A(a^*,a) = \sum_{n\geq k}\sqrt{n(n-1)\ldots(n-k+1)}\,\sqrt{(n-k+1)\ldots(n-k+l)}$$

$$\frac{(a^*)^n}{\sqrt{n!}}\,\frac{a^{n-k+l}}{\sqrt{(n-k+l)!}}\,. \tag{51.40}$$

Let $s = n - k$ and factor out $(a^*)^k a^l$. The above result can then be re-expressed as

$$A(a^*,a) = (a^*)^k a^l \sum_{s=0}^{\infty}\frac{\sqrt{(s+k)(s+k-1)\ldots(s+1)}\,\sqrt{(s+1)\ldots(s+l)}}{\sqrt{(s+k)!}\,\sqrt{(s+l)!}}\,(a^*)^s a^s$$

$$= (a^*)^k a^l \sum_{s=0}^{\infty}\frac{(a^*)^s}{\sqrt{s!}}\,\frac{a^s}{\sqrt{s!}} = (a^*)^k a^l \sum_{s=0}^{\infty}\frac{(a^*a)^s}{s!} = e^{a^*a}\,(a^*)^k a^l\,.$$

$$\tag{51.41}$$

The theorem is thus proved.   □

There is a formal similarity between the result (51.35) and (45.8) if we let $\hat{A}$ be the Hamiltonian $\hat{H}$ and make the replacement (with $\hbar = 1$)

$$a^*a \longrightarrow -ipq\,.$$

The point is that the roles of the phase space paths $p(t)$ and $q(t)$ are now played by the complex paths $a^*(t)$ and $a(t)$. This is significant because one can now use a similar procedure as the one introduced in Chapter 45 for the derivation of the functional integration formula (45.27) with respect to the phase space variables p and q to arrive at the corresponding result with respect to the variables a^* and a. Consider now a time-evolution operator [c.f. (45.17)] $\hat{U}(a^*,a;t'',t')$ and boundary conditions for the paths $a(t), a^*(t)$ such that

$$a(t') = a\,, \qquad a^*(t'') = a^* \qquad a, a^* \in \mathbb{C}\ (\text{fixed})\,. \tag{51.42}$$

Note that $a(t)$ and $a^*(t)$, as two different complex-valued functions of the (real) time variable, satisfy different boundary conditions, each at a different boundary (time) point. As in Chapter 45, we let $t'' = t' + N\epsilon$ and eventually allow $N \to \infty$ and $\epsilon \to 0$. Let $t_n = t' + n\epsilon$, $(n = 0, \ldots, N)$; $a_n = a(t_n)$, $a_n^* = a^*(t_n)$, $(n = 1, \ldots, N-1)$; $a_0 \equiv a$, $a_N^* \equiv a^*$. Let us see how the calculation for the time-evolution operator proceeds for the simple case $N = 2$ first. The finite interval $t'' - t'$ is partitioned into two equal subintervals with only one intermediate time point t_1. $\hat{U}(a^*, a; t'', t')$ is then composed of $\hat{U}(a^*, a_1; t'', t_1)$ and $\hat{U}(a_1^*, a; t_1, t')$. Theorem 51.1 implies that the integral kernels for these two operators are given by (with $\hbar = 1$)

$$U(a_2^*, a_1) = e^{a_2^* a_1} \, e^{-i\epsilon H(a_2^*, a_1)} \, , \qquad U(a_1^*, a) = e^{a_1^* a} \, e^{-i\epsilon H(a_1^*, a)} \, , \qquad (51.43)$$

where $H(a^*, a)$ is the normal symbol of the Hamiltonian. Writing $z = a^*$, $\bar{z} = a$, $\zeta = a_1^*$ and $\bar{\zeta} = a_1$ in the convolution product result (51.31), we then have

$$U(a^* = a_2, a = a_0; t'', t')$$

$$= \int \frac{da_1 da_1^*}{2\pi i} \left(e^{a_2^* a_1} e^{-i\epsilon H(a_2^*, a_1)} \right) \left(e^{a_1^* a_0} e^{-i\epsilon H(a_1^*, a_0)} \right) e^{-a_1^* a_1}$$

$$= \int \frac{da_1 da_1^*}{2\pi i} \exp\{a_2^* a_1 - a_1^* a_1 + a_1^* a_0\} \exp\{-i\epsilon[H(a_2^*, a_1) + H(a_1^*, a_0)]\} \, .$$

$$(51.44)$$

This expression can be readily generalized to the case for arbitrary $N \, (> 2)$:

$$U(a^* = a_N, a = a_0; t'', t') = \int \prod_{k=1}^{N-1} \frac{da_k da_k^*}{2\pi i}$$

$$\times \exp\left\{a_N^* a_{N-1} - a_{N-1}^* a_{N-1} + a_{N-1}^* a_{N-2} - a_{N-2}^* a_{N-2} + a_{N-2}^* a_{N-3} \right.$$
$$\left. + \cdots \cdots - a_1^* a_1 + a_1^* a_0\right\}$$
$$\times \exp\left\{-i\epsilon \left[H(a_N^*, a_{N-1}) + \cdots + H(a_1^*, a_0)\right]\right\}$$

$$= \int \prod_{k=1}^{N-1} \frac{da_k da_k^*}{2\pi i} \exp\left\{a_N^* a_{N-1} - a_{N-1}^*(a_{N-1} - a_{N-2})\right.$$

$$\left. -a_{N-2}^*(a_{N-2} - a_{N-3}) + \cdots - a_1^*(a_1 - a_0)\right\}$$
$$\times \exp\left\{-i\epsilon \left[H(a_N^*, a_{N-1}) + \cdots + H(a_1^*, a_0)\right]\right\} \, .$$

$$(51.45)$$

On letting $N \to \infty$ and $\epsilon \to 0$ such that $N\epsilon = t'' - t'$, we arrive at the functional integral form for the integral kernel of the time-evolution operator as follows:

$$U(a^*, a; t'', t') = \int \prod_t \frac{da(t) \, da^*(t)}{2\pi i} \exp\{a^*(t'')a(t'')\}$$

$$\times \exp\left\{\int_{t'(a)}^{t''(a^*)} dt \, [-a^*(t)\dot{a}(t) - iH(a^*(t), a(t))]\right\} \, ,$$

$$(51.46)$$

where the limits of the time integral in the second exponent means $a^*(t'') = a^*$, $a(t') = a$ [recall a similar notation in (45.27)]. Again, we remind the reader that in this formula, $a^*(t) = \overline{a(t)}$ for any t in the interval $t' < t < t''$, because in the integral $da(t)da^*(t)/(2\pi i)$ we are only integrating over two real degrees of freedom: $Re(a(t))$ and $Im(a(t))$. But at the end points t' and t'', $a^* = a^*(t'')$ and $a = a(t')$ are two independent complex numbers not necessarily complex conjugates to each other.

The terms in the exponent of the first exponential factor in the right-hand side of (51.44) can be regrouped as follows:

$$(a_N^* - a_{N-1}^*)a_{N-1} + (a_{N-1}^* - a_{N-2}^*)a_{N-2} + \cdots + (a_2^* - a_1^*)a_1 + a_1^* a_0 \, .$$

This implies that the time-evolution operator integral kernel can also be written as

$$U(a^*, a; t'', t') = \int \prod_t \frac{da(t)\, da^*(t)}{2\pi i} \, \exp\{a^*(t')a(t')\}$$

$$\times \exp\left\{ \int_{t'(a)}^{t''(a^*)} dt \, [\dot{a}^*(t)a(t) - iH(a^*(t), a(t))] \right\} \, .$$

$$(51.47)$$

Combining (51.45) and (51.46) we can write a symmetrized form [with respect to the boundary conditions for $a(t)$ and $a^*(t)$] for the time-evolution operator integral kernel as

$$U(a^*, a; t'', t') = \int [da\, da^*] \, \exp\left\{ \frac{1}{2}[a^*(t'')a(t'') + a^*(t')a(t')] \right\}$$

$$\times \exp\left\{ i \int_{t'(a)}^{t''(a^*)} dt \, \left[\frac{1}{2i} (\dot{a}^*(t)a(t) - a^*(t)\dot{a}(t)) - H(a^*(t), a(t)) \right] \right\} \, ,$$

$$(51.48)$$

where the measure of the functional integral $[da\, da^*]$ is given [analogous to (45.26)] by

$$\int \frac{da_{N-1}da_{N-1}^*}{2\pi i} \cdots \frac{da_1 da_1^*}{2\pi i} \xrightarrow[\substack{N\to\infty \\ \epsilon\to 0}]{} \prod_t \int \frac{da(t)da^*(t)}{2\pi i} \equiv \int [da\, da^*] . \quad (51.49)$$

In addition, to allow for the partial freedom for the boundary conditions (51.42), $a(t'')$ and $a^*(t')$ are understood to be integrated over (the complex plane), but not $a^*(t'')$ and $a(t')$, which are kept fixed by the boundary conditions. In this form the time-evolution operator integral kernel most resembles the phase space result (45.27).

The functional integral [expressed as any of (51.46), (51.47), and (51.48)] for the integral kernel of the time-evolution operator is valid for any time-dependent Hamiltonian, not just that for the free oscillator given by (51.1). In view of our emphasis on the generating functional $Z[J]$ with a source term $J(x)$ [c.f. (48.25)], we would like to calculate this integral for the case

$$H(a^*, a, t) = \omega a^* a - J(t)a^* - \overline{J(t)}a , \tag{51.50}$$

where the time-dependent source function $J(t)$ is now complex, and the additional terms to that for the free oscillator involving $J(t)$ represent the "forcing" terms. Because of the quadratic term above (a^*a) the functional integral is of the Gaussian type, and we can use the methods developed in Chapter 47 for its evaluation. Instead, we will use the method of stationary phase (assume $\hbar \to 0$), which posits that the result, up to an unimportant constant and to the lowest order in $\hbar^{1/2}$, is approximately equal to the value of the integrand at the *critical point* – when $a(t)$ and $a^*(t)$ are given by the stationary trajectories. The method is suitable here because these stationary trajectories can be determined without too much difficulty, by setting the variation of the action functional in, for example, (51.46) to be zero. Thus,

$$\delta \left[a^*(t'')a(t'') - \int_{t'}^{t''} dt \, \{a^*(t)\dot{a}(t) + iH\left(a^*(t), a(t), t\right)\} \right] = 0 , \tag{51.51}$$

where $a^*(t'')$ is kept fixed by the boundary conditions (51.42). In this variation, $a(t)$ and $a^*(t)$ are to be varied independently. We thus have

$$a^*(t'')\delta a(t'') - \int_{t'}^{t''} dt \left[a^* \delta \dot{a} + \dot{a}\delta a^* + i\frac{\partial H}{\partial a}\delta a + i\frac{\partial H}{\partial a^*}\delta a^* \right]$$

$$= a^*(t'')\delta a(t'') - \int_{t'}^{t''} a^* \left(\frac{d}{dt}\delta a \right) dt - \int_{t'}^{t''} \dot{a}\delta a^* dt - i \int_{t'}^{t''} dt \left(\frac{\partial H}{\partial a}\delta a + \frac{\partial H}{\partial a^*}\delta a^* \right)$$

$$= a^*(t'')\delta a(t'') - \left(a^*(t'')\delta a(t'') - a^*(t')\delta a(t') - \int_{t'}^{t''} \dot{a}^* \delta a \, dt \right) - \int_{t'}^{t''} \dot{a}\delta a^* dt$$

$$- i \int_{t'}^{t''} dt \left(\frac{\partial H}{\partial a}\delta a + \frac{\partial H}{\partial a^*}\delta a^* \right)$$

$$= \int_{t'}^{t''} dt \left[\delta a \left(\dot{a}^* - i\frac{\partial H}{\partial a} \right) - \delta a^* \left(\dot{a} + i\frac{\partial H}{\partial a^*} \right) \right] = 0 ,$$

$$\tag{51.52}$$

where in the second equality we have used integration by parts for the integral in the second term on the right-hand side of the first equality, and recalled that $\delta a(t') = 0$, because of the boundary condition (51.42). It follows that the *Euler-Lagrange equations* determining the critical trajectories for $a(t)$ and $a^*(t)$ are

$$\dot{a}^* - i\frac{\partial H}{\partial a} = 0 , \qquad \dot{a} + i\frac{\partial H}{\partial a^*} = 0 . \tag{51.53}$$

From (51.50) we have

$$\frac{\partial H}{\partial a} = \omega a^* - \overline{J(t)} , \qquad \frac{\partial H}{\partial a^*} = \omega a - J(t) . \tag{51.54}$$

Hence the equations in (51.33) become

$$\dot{a}^* - i\omega a^* + i\overline{J(t)} = 0 , \qquad \dot{a} + i\omega a - iJ(t) = 0 . \qquad (51.55)$$

These simple equations can be solved immediately to yield the following critical trajectories satisfying the boundary conditions (51.42):

$$a_c^*(t) = a^* e^{i\omega(t-t'')} + i e^{i\omega t} \int_t^{t''} e^{-i\omega s} \overline{J(s)} \, ds , \qquad (51.56)$$

$$a_c(t) = a \, e^{i\omega(t'-t)} + i e^{-i\omega t} \int_{t'}^{t} e^{i\omega s} J(s) \, ds . \qquad (51.57)$$

Note that, significanttly, $a_c(t)$ and $a_c^*(t)$ are not complex conjugates of each other, even though in the functional integration in (51.46) to (51.48), $a^*(t) = \overline{a(t)}$ for $t' < t < t''$. This means that, for the purpose of applying the stationary phase approximation to Gaussian functional integrals, it is legitimate to use critical paths which do not belong to the space of paths that are to be functionally integrated over. We will not dwell further on this mathematical point here.

Problem 51.5 By direct substitution show that the critical trajectories given by (51.56) and (51.57) satisfy the equations (51.55).

On substituting $a_c(t)$ and $a_c^*(t)$ into the integrand in (51.46), the classical action part is equal to

$$\int_{t'}^{t''} dt \left[-a_c^* \dot{a}_c - i \left(\omega a_c^* a_c - J(t) a_c^* - a_c \overline{J(t)} \right) \right]$$

$$= \int_{t'}^{t''} dt \left[-a_c^* \{ \dot{a}_c + i\omega a_c - iJ(t) \} \right] + i \int_{t'}^{t''} ds \, a_c(s) \overline{J(s)} . \qquad (51.58)$$

Since the first term vanishes on account of the second equation in (51.55), the entire exponent in the integrand of (51.46), with $a(t)$ replaced by $a_c(t)$ and $a^*(t)$ by $a_c^*(t)$, then becomes

$$a_c^*(t'') a_c(t'') + i \int_{t'}^{t''} ds \, a_c(s) \overline{J(s)} . \qquad (51.59)$$

On using (51.56) and (51.57), this is equal to

$$
a^* \left(a e^{i\omega(t'-t'')} + i e^{-i\omega t''} \int_{t'}^{t''} e^{i\omega s} J(s) ds \right)
$$

$$
+ i \int_{t'}^{t''} ds \left(a e^{i\omega(t'-s)} + i e^{-i\omega s} \int_{t'}^{s} e^{i\omega u} J(u) du \right) \overline{J(s)}
$$

$$
= a^* \left(a e^{i\omega(t'-t'')} + i e^{-i\omega t''} \int_{t'}^{t''} e^{i\omega s} J(s) ds \right)
$$
(51.60)

$$
+ i a \int_{t'}^{t''} e^{i\omega(t'-s)} \overline{J(s)} ds - \int_{t'}^{t''} ds \int_{t'}^{s} du \, e^{i\omega(u-s)} J(u) \overline{J(s)} .
$$

Finally, the integral kernel for the time-evolution operator is obtained by exponentiating the above. Up to an inessential constant akin to $Z^{(0)}[0]$ [c.f. (48.28)] and to lowest order in $\hbar^{1/2}$, it can be expressed as follows.

$$
U(a^*, a; t'', t') = \exp \left\{ a^* a \, e^{i\omega(t'-t'')} \right.
$$

$$
+ i \left(a^* e^{-i\omega t''} \int_{t'}^{t''} e^{i\omega s} J(s) ds + a \, e^{i\omega t'} \int_{t'}^{t''} e^{-i\omega s} \overline{J(s)} ds \right)
$$

$$
\left. - \frac{1}{2} \int_{t'}^{t''} \int_{t'}^{t''} du ds \, e^{-i\omega|u-s|} \, \overline{J(u)} J(s) \right\} ,
$$
(51.61)

where in the last term we have symmetrized with respect to the u and s integrals. This formula can be immediately generalized to the case of multiple degrees of freedom, in which case, the separate exponential factors, one for each degree of freedom, are simply to be multiplied together.

As was done in Chapter 48, we can even generalize further, to the case of quantum fields. Corresponding to the free-field Lagrangian density $\mathcal{L}_0$ given by (48.16), we can write the classical Lagrangian involving a source $J(x)$ as follows:

$$
L(t) = \int d^3 x \left[\frac{1}{2} \{ (\partial_\mu \varphi(x))(\partial^\mu \varphi(x)) - m^2 \varphi^2) \} + J(x) \varphi(x) \right] ,
$$
(51.62)

where $x = (t, \boldsymbol{x})$, and the integrand is the Lagrangian density $\mathcal{L}$ (with the source). Defining the *conjugate momentum* to φ, denoted π, by

$$
\pi \equiv \frac{\delta \mathcal{L}}{\delta \dot{\varphi}} = \dot{\varphi} ,
$$
(51.63)

the *Hamiltonian density* $\mathcal{H}$ is given by

$$
\mathcal{H}(x) = \pi \dot{\varphi} - \mathcal{L} = \dot{\varphi}^2 - \left[\frac{1}{2} \{ \dot{\varphi}^2 - (\nabla \varphi)^2 - m^2 \varphi^2 \} + J(x) \varphi(x) \right]
$$
(51.64)

$$
= \frac{1}{2} \{ \dot{\varphi}^2 + (\nabla \varphi)^2 + m^2 \varphi^2 \} - J(x) \varphi(x) .
$$

The total Hamiltonian is then

$$H(t) = \int d^3x\, \mathcal{H}(t, \boldsymbol{x})\ . \tag{51.65}$$

To make use of the generalization of (51.61) to the case of infinitely many degrees of freedom we Fourier transform $\varphi(x)$ with respect to the spatial coordinates:

$$\varphi(\boldsymbol{x}, t) = \frac{1}{(2\pi)^{3/2}} \int \frac{d^3k}{\sqrt{2\omega_k}} \left[a(\boldsymbol{k})\, e^{i(\boldsymbol{k}\cdot\boldsymbol{x} - \omega_k t)} + a^*(\boldsymbol{k})\, e^{-i(\boldsymbol{k}\cdot\boldsymbol{x} - \omega_k t)} \right]\ , \tag{51.66}$$

where $\omega_k^2 = \boldsymbol{k}^2 + m^2$. It follows from (51.63) that

$$\pi(\boldsymbol{x}, t) = \frac{i}{(2\pi)^{3/2}} \int d^3k \sqrt{\frac{\omega_k}{2}} \left[a^*(\boldsymbol{k})\, e^{-i(\boldsymbol{k}\cdot\boldsymbol{x} - \omega_k t)} - a(\boldsymbol{k})\, e^{i(\boldsymbol{k}\cdot\boldsymbol{x} - \omega_k t)} \right]\ . \tag{51.67}$$

From (51.64) and (51.65) it is then not hard to show that the Hamiltonian can be written as [c.f. (51.50)]

$$H(t) = \int d^3k \left[\omega_k\, a^*(\boldsymbol{k})a(\boldsymbol{k}) - \tilde{J}(\boldsymbol{k}, t)a^*(\boldsymbol{k}) - \overline{\tilde{J}(\boldsymbol{k}, t)}a(\boldsymbol{k}) \right]\ , \tag{51.68}$$

where

$$\tilde{J}(\boldsymbol{k}, t) \equiv \frac{1}{\sqrt{2\omega_k}\,(2\pi)^{3/2}} \int d^3x\, J(\boldsymbol{x}, t)\, e^{-i\boldsymbol{k}\cdot\boldsymbol{x}} \tag{51.69}$$

is the spatial Fourier transform of the source function $J(\boldsymbol{x}, t)$. Here we can view $\boldsymbol{k}$ as a label for a particular degree of freedom, just as $\boldsymbol{x}$ can be viewed in the same way. Thus the Hamiltonian $H(t)$ given above can be seen as the generalization of $H(a^*, a; t)$ in (51.50) to the case of infinitely many degrees of freedom. The generalization of (51.61) for the time-evolution integral kernel to the quantum field theory case is then

$$U(a^*(\boldsymbol{k}, t''), a(\boldsymbol{k}, t'); t'', t') = \exp\left\{ d^3k \left[a^*(\boldsymbol{k}, t'')a(\boldsymbol{k}, t')\, e^{i\omega_k(t' - t'')} \right. \right.$$

$$+ ia^*(\boldsymbol{k}, t'')\, e^{-i\omega_k t''} \int_{t'}^{t''} ds\, e^{i\omega_k s}\, \tilde{J}(\boldsymbol{k}, s) + ia(\boldsymbol{k}, t')\, e^{i\omega_k t'} \int_{t'}^{t''} ds\, e^{-i\omega_k s}\, \overline{\tilde{J}(\boldsymbol{k}, s)}$$

$$\left. \left. - \frac{1}{2} \int_{t'}^{t''} du \int_{t'}^{t''} ds\, e^{-i\omega_k|u-s|}\, \tilde{J}(\boldsymbol{k}, u)\overline{\tilde{J}(\boldsymbol{k}, s)} \right] \right\}\ . \tag{51.70}$$

We will use this formula to determine the integral kernel of the scattering operator $\hat{S}$ given by (45.35). The calculation will require the convolution product of the three operators $\exp(i\hat{H}_0 t'')$, $\hat{U}(t'', t')$ and $\exp(-i\hat{H}_0 t')$. This will be done with the help of the convolution product formula (51.32), with the integral kernel for the time-evolution operator, $U(t'', t')$, given by (51.70).

Problem 51.6 Use Eqs. (51.64) through (51.67) and (51.69) to verify the expression for the Hamiltonian given by (51.68).

In scattering situations, we can assume $J(x,t) \xrightarrow[t\to\pm\infty]{} 0$. Thus

$$\hat{H}_0 = \hat{H}(J=0) = \int d^3k\, \hat{a}^*(k)\hat{a}(k)\,. \tag{51.71}$$

The integral kernel of $e^{-i\hat{H}_0 t'} = e^{-i\hat{H}_0(t'-0)}$, according to (51.70), is given by

$$U_0(a^*(k,t'), a(k,0); t', 0) = \exp\left\{\int d^3k\, a^*(k,t')a(k,0)\, e^{-i\omega_k t'}\right\}, \tag{51.72}$$

where the subscript 0 in U_0 means that in (51.70) we have set $\tilde{J} = 0$. Similarly, that for $e^{i\hat{H}_0 t''} = e^{-i\hat{H}_0(0-t'')}$, is given by

$$U_0(a^*(k,0), a(k,t''); 0, t'') = \exp\left\{\int d^3k\, a^*(k,0)a(k,t'')\, e^{i\omega_k t''}\right\}. \tag{51.73}$$

Denote these kernels, respectively, by

$$U_0(a_k^*(t'), a_k; t') \equiv \exp\left\{\int d^3k\, a_k^*(t')\, a_k\, e^{-i\omega_k t'}\right\} = \exp\left\{\int d^3k\, V_k(t')\right\}, \tag{51.74}$$

$$U_0^{-1}(a_k^*, a_k(t''); t'') \equiv \exp\left\{\int d^3k\, a_k^*\, a_k(t'')\, e^{i\omega_k t''}\right\} = \exp\left\{\int d^3k\, W_k(t'')\right\}, \tag{51.75}$$

where we have written

$$a_k \equiv a(k,0)\,, \quad a_k^* \equiv a^*(k,0)\,; \quad a_k^*(t') \equiv a^*(k,t')\,, \quad a_k(t'') \equiv a(k,t'')\,, \tag{51.76}$$

and made the following definitions:

$$V_k(t') \equiv a_k^*(t')a_k\, e^{-i\omega_k t'}\,, \quad W_k(t'') \equiv a_k^*a_k(t'')\, e^{i\omega_k t''}\,. \tag{51.77}$$

Write the quantity within the square brackets in (51.70) as $(A_J)_k(t'', t')$, and discretize the label k for convenience for the time being. Then the integral kernel of the scattering operator $\hat{S}(t'', t')$ can be written as the convolution product

$$S_J(t'', t') \longrightarrow \exp\left\{\sum_k W_k(t'')\right\} * \exp\left\{\sum_k (A_J)_k(t'', t')\right\} * \exp\left\{\sum_k V_k(t')\right\}$$

$$= \left(\prod_k e^{W_k(t'')}\right) * \left(\prod_k e^{(A_J)_k(t'', t')}\right) * \left(\prod_k e^{V_k(t')}\right)$$

$$= \prod_k \left(e^{W_k(t'')} * e^{(A_J)_k(t'', t')} * e^{V_k(t')}\right). \tag{51.78}$$

We can thus convolve mode by mode and multiply the results. To make use of the convolution formula (51.32) efficiently, we let

$$z_k \equiv a_k^*, \qquad \overline{z_k} \equiv a_k, \qquad \zeta_k(t') \equiv a_k^*(t'), \qquad \overline{\xi_k}(t'') = a_k(t''). \qquad (51.79)$$

It will prove convenient to first calculate the following convolution:

$$e^{W_k(t'')} * A(\xi_k(t''), \overline{\zeta_k}(t')) * e^{V_k(t')}, \qquad (51.80)$$

where $A(u, \overline{v})$ is a general function of its arguments. Using (51.77) and (51.32) twice, we see that the above quantity is equal to

$$\exp\left\{ \left(z_k e^{i\omega_k t''} \right) \overline{\xi_k}(t'') \right\} * A(\xi_k(t''), \overline{\zeta_k}(t')) * \exp\left\{ \zeta_k(t') \left(\overline{z_k}\, e^{-i\omega_k t'} \right) \right\}$$

$$= \int \frac{d\xi_k(t'') d\overline{\xi_k}(t'')}{2\pi i} \exp\left\{ \left(z_k e^{i\omega_k t''} \right) \overline{\xi_k}(t'') \right\} e^{-\xi_k(t'')\overline{\xi_k}(t'')}$$

$$\times \int \frac{d\zeta_k(t') d\overline{\zeta_k}(t')}{2\pi i} \exp\left\{ \zeta_k(t') \left(\overline{z_k}\, e^{-i\omega_k t'} \right) \right\} A(\xi_k(t''), \overline{\zeta_k}(t')) e^{-\zeta_k(t')\overline{\zeta_k}(t')}$$

$$= \int \frac{d\xi_k(t'') d\overline{\xi_k}(t'')}{2\pi i} \exp\left\{ \left(z_k e^{i\omega_k t''} \right) \overline{\xi_k}(t'') \right\} e^{-\xi_k(t'')\overline{\xi_k}(t'')} A\left(\xi_k(t''), \overline{z_k}\, e^{-i\omega_k t'} \right)$$

$$= A\left(z_k e^{i\omega_k t''}, \overline{z_k}\, e^{-i\omega_k t'} \right) = A\left(a_k^* e^{i\omega_k t''}, a_k e^{-i\omega_k t'} \right), \tag{51.81}$$

where in the second and third equalities we have used (51.25). This remarkable result implies that on doing the convolution for each k in (51.78), the result is obtained by replacing the quantities $a_k^*(t'')$ and $a_k(t')$ in (51.70) by $a_k^* e^{i\omega_k t''}$ and $a_k e^{-i\omega_k t'}$, respectively. On making k continuous again in (51.78), we obtain the integral kernel for the scattering operator (with source term in the Hamiltonian but no self-interacting potential) as follows:

$$S_J^{(0)}(a_k^*, a_k) = \lim_{\substack{t'' \to \infty \\ t' \to -\infty}} S_J^{(0)}(a_k^*, a_k; t'', t') = \exp\left\{ \int d^3 k \left[a_k^* a_k \right. \right.$$

$$+ i a_k^* \int_{-\infty}^{\infty} ds\, e^{i\omega_k s}\, \tilde{J}(k, s) + i a_k \int_{-\infty}^{\infty} ds\, e^{-i\omega_k s}\, \overline{\tilde{J}(k, s)} \tag{51.82}$$

$$\left. \left. - \frac{1}{2} \int_{-\infty}^{\infty} du \int_{-\infty}^{\infty} ds\, e^{-i\omega_k |u-s|}\, \tilde{J}(k, u)\overline{\tilde{J}(k, s)} \right] \right\}.$$

It is apparent, then, by Theorem 51.1, that the normal symbol for $\hat{S}_J^{(0)}$, denoted $K^{(\hat{S}_J^{(0)})}$, is obtained from the above expression by just dropping the term $\int d^3 k\, a_k^* a_k$ in the exponent. Re-expressing $\tilde{J}(k, t)$ in terms of $J(x, t)$ using

(51.69), we have

$$
K^{(\hat{S}_J^{(0)})}(a_{\bm{k}}^*, a_{\bm{k}}) = \exp\Big\{ i \int_{-\infty}^{\infty} ds \int d^3x\, J(\bm{x}, s)
$$
$$
\times \Big[\frac{1}{(2\pi)^{3/2}} \int \frac{d^3k}{\sqrt{2\omega_k}} \left(a_{\bm{k}}^*\, e^{i(\omega_k s - \bm{k}\cdot\bm{x})} + a_{\bm{k}}\, e^{-i(\omega_k s - \bm{k}\cdot\bm{x})} \right) \Big]
$$
$$
- \frac{i}{2} \int_{-\infty}^{\infty} du \int d^3x \int_{-\infty}^{\infty} ds \int d^3y\, J(\bm{x}, u) J(\bm{y}, s)
$$
$$
\times \left(\frac{-i}{(2\pi)^3} \int \frac{d^3k}{(2\omega_k)}\, e^{-i\omega_k(|u-s| - \bm{k}\cdot(\bm{x}-\bm{y}))} \right) \Big\} .
$$

$$(51.83)$$

The quantity within the square brackets above is recognized as the free field $\varphi^{(0)}(x)$ corresponding to the free-field Lagrangian density $\mathcal{L}_0$ (no source term and no self-interacting potential) [compare with $\varphi(x)$ in (51.66), which is the field corresponding to the Lagrangian with the source term in (51.62)]:

$$
\varphi^{(0)}(\bm{x}, t) = \frac{1}{(2\pi)^{3/2}} \int \frac{d^3k}{\sqrt{2\omega_k}} \left(a_{\bm{k}}^*\, e^{i(\omega_k t - \bm{k}\cdot\bm{x})} + a_{\bm{k}}\, e^{-i(\omega_k t - \bm{k}\cdot\bm{x})} \right) . \qquad (51.84)
$$

This is so because $a_{\bm{k}}^*$ and $a_{\bm{k}}$, as defined in (51.76), are the normal symbols for the creation and annihilation operators $\hat{a}_{\bm{k}}^*$ and $\hat{a}_{\bm{k}}$, respectively, corresponding to the free Hamiltonian $\hat{H}_0$ [c.f. (51.71) to (51.73)]. Meanwhile, the quantity within the parentheses in the last line of (51.83) is precisely the *Feynman propagator* $D(x - y)$, as given in (48.37), where $x = (\bm{x}, s)$ and $y = (\bm{y}, u)$. So (51.83) can be expressed in the compact form

$$
K^{(\hat{S}_J^{(0)})}(a_{\bm{k}}^*, a_{\bm{k}}) = \exp\Big\{ i \int d^4x\, J(x)\varphi^{(0)}(x) - \frac{i}{2} \int d^4x \int d^4y\, J(x) D(x - y) J(y) \Big\} .
$$

$$(51.85)$$

The reader should compare this expression with (49.34) and (49.35) for the generating functional $Z_E^{(0)}[J]$.

Analogous to (49.36) for $Z_E[J]$, the corresponding normal symbol for $\hat{S}_J$ [with a self-interacting potential $\lambda V(\varphi(x))$] is given by

$$
K^{(\hat{S})}(a_{\bm{k}}^*, a_{\bm{k}}) = \lim_{J \to 0} \exp\left\{ -i\lambda \int d^4x\, V\left(-i\frac{\delta}{\delta J(x)} \right) \right\} K^{(\hat{S}_J^{(0)})}(a_{\bm{k}}^*, a_{\bm{k}}) . \quad (51.86)
$$

The perturbative procedure in powers of the coupling constant λ then yields the Feynman diagrams and their corresponding rules of calculation as discussed in the last chapter.

Chapter 52

Perturbative Renormalization

No sooner had we set up the rules for the computation of the Feynman diagrams (in Chapter 50) than we discover that some diagrams involving loops lead to various degrees of infinities. This situation renders the theory useless unless the infinities can be controlled in a consistent way, order by order, to all orders in our perturbative approach. Furthermore, for the theory to be predictive, the outcomes of the calculations must depend on the input parameters of the theory, in our case the mass m and the coupling constant λ. The program of **perturbative renormalization** then consists of **regularization schemes** to render all divergent quantities finite so that we can formally manipulate with them, and then "absorbing" these regularized (but really infinite) quantities into the **bare parameters** (those which enter the Lagrangian initially) to construct observable (measurable) parameters which are finite, in terms of which the results of the theory are expressed. In this chapter we will give a brief introduction to see how such a "miracle" can be achieved. Our presentation has been adapted from the accounts given in Callan (1976), Cheng and Li (1984), Coleman (1985), and Amit and Martin-Mayor (2005).

We will first illustrate by some simple examples from the φ^4-theory to see how the aforementioned infinities can arise. Instead of the n-point Green's functions in *real space* $G(x_1, x_2, \ldots, x_n)$ introduced earlier [c.f. (49.1) and (49.42)] it will be convenient to work with the associated Green's functions $\tilde{G}(k_1, k_2, \ldots, k_n)$ in *momentum space*, defined as the Fourier transforms of the former, in the following way:

$$
\begin{aligned}
(2\pi)^4 \, \delta^4(k_1 + &\cdots + k_n) \, \tilde{G}(k_1, \ldots, k_n) \\
&= \int d^4x_1 \ldots \int d^4x_n \, e^{-i(k_1 x_1 + \cdots + k_n x_n)} \, G(x_1, \ldots, x_n) \, .
\end{aligned}
\tag{52.1}
$$

On using the integral representations of the Feynman propagator $D(x - y)$ and the 4-dimensional Dirac delta function given in (48.34) and (48.32), respectively, it is not difficult to see that the overall Dirac delta function appearing on the left-hand side of the above equation always results, which simply expresses the law of conservation of 4-momentum, that is, conservation of energy and momentum due to spacetime translation invariance of the Lagrangian density $\mathcal{L}$. To conform with the physics literature in particle physics, where perturbative renormalization is usually applied, we will work with Minkowski propagators [which the Feynman propagator $D(x)$ is one] rather than the Euclidean propagators D_E in this chapter. Comparing Eqs. (48.33) and (49.34) for the generating functionals, one sees that to transition from the Euclidean to the Minkowski case, one simply lets $J \to iJ$ and $D_E \to iD$ in all Euclidean formulas. In computing a Feynman diagram for $\tilde{G}(k_1, \ldots, k_n)$, then, the same basic rules similar to rules (1) to (4) (for the Euclidean case) given in Chapter 50 still apply, but have to be revised as follows.

(1') Each vertex carries a factor of $-i\lambda/4!$.

(2') Each line is labeled by a 4-momentum k, and carries a propagator factor $i\tilde{D}(k^2)$, where $\tilde{D}(k^2)$ is given by [c.f. (48.34)]

$$\tilde{D}(k^2) = \frac{1}{k^2 - m^2 + i\epsilon^+} . \tag{52.2}$$

(3') At each vertex, 4-momentum is conserved: $l_1 + l_2 + l_3 + l_4 = 0$ (there are always 4 lines emanating from a vertex in the φ^4-theory), so that not all internal lines are labeled by independent momenta.

(4') Each independent internal 4-momentum l is to be integrated over, with the measure $\int d^4l/(2\pi)^4$. Any line that terminates at a vertex is assigned an arrow of arbitrary direction (towards or away from the vertex) for the purpose of accounting for momentum conservation: 4-momenta corresponding to oppositely directed arrows (one towards and one away from the vertex) are assigned opposite signs.

Rule (4) for the symmetry factors remain unchanged.

For example, we can redraw the one-loop diagram for the two-point function $\tilde{G}(k_1, k_2)$ as in Fig. 52.1(a) (c.f. Fig. 50.4), also known as the *tadpole diagram*, with the 4-momentum labels shown. As in Fig. 50.4, this diagram also carries a symmetry factor of 12. According to the above rules, the one-loop tadpole diagram has the value

$$12 \left(-i\frac{\lambda}{4!}\right) \frac{i^2}{(k^2 - m^2 + i\epsilon^+)^2} \int \frac{d^4l}{(2\pi)^4} \frac{i}{l^2 - m^2 + i\epsilon^+} . \tag{52.3}$$

The factor $i\tilde{D}(k)$ occurs twice because of the overall momentum conservation: $k_1 = k_2 = k$. By straightforward power-counting (of powers of l) we see that the integral is *quadratically divergent* – the numerator of the integrand has four powers of the integrated momentum and the denominator has two. [Recall that on setting $\hbar = c = 1$, all dimensions can be reduced to that of length, or

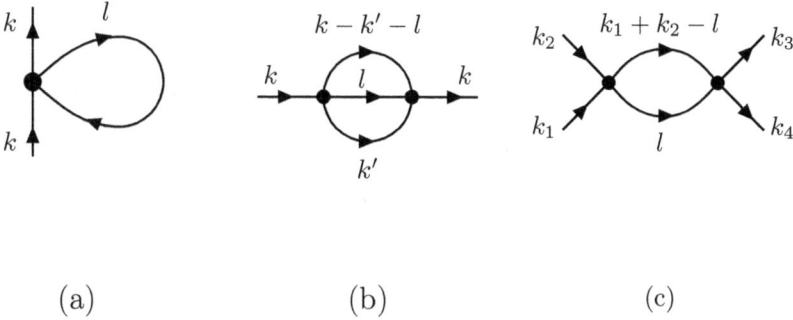

(a)　　　　　　　　　(b)　　　　　　　　　(c)

FIGURE 52.1

momentum ($= 1/[\text{length}])$. In this case the so-called **superficial degree of divergence** is $D = 2$ (more on this later). As another example, consider the two-loop diagram of the 2-point function shown in Fig. 52.1(b), also known as the *setting-sun diagram*. This diagram has two vertices and two independent internal momenta, k' and l, to be integrated over, on observing the rule of momentum conservation at each vertex [rule (3')]. Recalling that the symmetry factor for this diagram is 192 [c.f. Fig. 50.6(b)] it has the value

$$\frac{192}{2!} \left(\frac{-i\lambda}{4!} \right)^2 \frac{i^2}{(k^2 - m^2 + i\epsilon^+)^2} \int \frac{d^4 l}{(2\pi)^4} \int \frac{d^4 k'}{(2\pi)^4}$$

$$\times \frac{i}{l^2 - m^2 + i\epsilon^+} \frac{i}{k'^2 - m^2 + i\epsilon^+} \frac{i}{(k - k' - l)^2 - m^2 + i\epsilon^+} \,. \tag{52.4}$$

Simple power counting indicates that for this diagram also, $D = 2$. Next we consider the one-loop diagram for the 4-point function, also known as the *fish diagram*, shown in Fig. 52.1(c) with the independent momentum labels. This diagram has two vertices and one independent internal momentum l to be integrated over. Up to a symmetry factor and dropping the external propagators, it has the value

$$\left(\frac{-i\lambda}{4!} \right)^2 \int \frac{d^4 l}{(2\pi)^4} \frac{i}{l^2 - m^2 + i\epsilon^+} \frac{i}{(k_1 + k_2 - l)^2 - m^2 + i\epsilon^+} \,. \tag{52.5}$$

This diagram has $D = 0$. It is thus *logarithmically divergent*.

As already apparent from the above three examples, the ultimate reason why infinities emerge in the calculation of any Feynman diagram is that the magnitude of the so-called momentum **cutoff** is taken to be infinite in the integrals. That means the high energies at which known physics will most likely fail are sampled, thus leading to unreasonable results. This immediately suggests that one of the regularization schemes is to impose a finite momentum cutoff Λ. This procedure will be discussed in more detail later in the next chapter. The introduction of a momentum cutoff will also play a central role in our discussion of the *renormalization group* later on in this book.

Problem 52.1 Do the Fourier transform specified by (52.1) for the one-loop diagram in a 2-point Green's function to see that the overall delta function $\delta^4(k_1 + k_2)$ emerges, and to verify the expression (52.3).

Let us now introduce the notion of n-**particle irreducible diagrams** (nPI), which are defined as Feynman diagrams that cannot be rendered disconnected by cutting up to n internal lines. For example, for a 2-point Green's function, the two-loop diagram in Fig. 50.6(b) is 1PI (not made disconnected by cutting any internal line), while that in Fig. 50.6(c) is 1-particle reducible (made disconnected by cutting an internal line). A contribution due to a sum of 1PI diagrams is called a **vertex function**. For the 2-point function we will denote the vertex function that is the sum of all 1PI diagrams without any external lines by $-i\Sigma(k^2)$, where k is the external 4 momentum, and the general dependence on k^2 can be deduced from the requirement of Lorentz invariance. We can represent this quantity, called the **self-energy graph** or the **mass operator**, diagrammatically by a shaded blob in Fig. 52.2, in which the (removed) external lines are shown as dotted lines. The complete 2-point function, which will include all the one-particle reducible diagrams as well, can then be represented in Fig. 52.3. Since each shaded blob can only have two external lines attached to it, one can readily see that, topologically, the shaded blobs in sequence is the only way that they can enter: there are no nPI diagrams for a 2-point function for $n > 1$. The leading term in $-i\Sigma(k^2)$, as seen from Fig. 52.2, is precisely the integral in (52.3), which, as we have seen, is quadratically divergent and actually independent of k^2. We assume then that a regularization procedure has been applied, so we can formally manipulate it as a finite quantity. Using rule (2') above we have

$$\tilde{G}(k^2) = i\tilde{D}(k^2) + \left(i\tilde{D}(k^2)\right)\left(-i\Sigma(k^2)\right)\left(i\tilde{D}(k^2)\right)$$

$$+ \left(i\tilde{D}(k^2)\right)\left(-i\Sigma(k^2)\right)\left(i\tilde{D}(k^2)\right)\left(-i\Sigma(k^2)\right)\left(i\tilde{D}(k^2)\right) + \cdots\cdots$$

$$= i\tilde{D}(k^2)\left[1 + \left(-i\Sigma(k^2)\right)\left(i\tilde{D}(k^2)\right) + \left(-i\Sigma(k^2)\right)^2\left(i\tilde{D}(k^2)\right)^2 + \cdots\cdots\right].$$

$$(52.6)$$

The terms inside the square brackets can be summed formally as a geometric series to yield

$$\tilde{G}(k^2) = i\tilde{D}(k^2) \left[\frac{1}{1 - (-i\Sigma(k^2))\left(i\tilde{D}(k^2)\right)} \right] . \tag{52.7}$$

Using (52.2) for $\tilde{D}(k^2)$ we finally have

$$\tilde{G}(k^2) = \left(\frac{i}{k^2 - m^2 + i\epsilon^+} \right) \left[1 - \frac{\Sigma(k^2)}{k^2 - m^2 + i\epsilon^+} \right]^{-1} , \tag{52.8}$$

or

$$\tilde{G}(k^2) = \frac{i}{k^2 - m^2 - \Sigma(k^2) + i\epsilon^+} . \tag{52.9}$$

The self-energy term $\Sigma(k^2)$ is where the infinities reside. We will control it by isolating its finite parts through Taylor expanding this quantity about an arbitrary point $k^2 = \mu^2$. So we write

$$\Sigma(k^2) = \Sigma(\mu^2) + \left. \frac{\partial \Sigma(k^2)}{\partial k^2} \right|_{\mu^2} (k^2 - \mu^2) + \frac{1}{2} \left. \frac{\partial^2 \Sigma(k^2)}{\partial (k^2)^2} \right|_{\mu^2} (k^2 - \mu^2)^2 + \cdots .$$
$$\tag{52.10}$$

To calculate the derivative terms we observe that, for any function $\Gamma(k^2)$,

$$k^\mu \frac{\partial \Gamma(k^2)}{\partial k^\mu} = k^\mu \frac{\partial \Gamma(k^2)}{\partial k^2} \frac{\partial k_\nu k^\nu}{\partial k^\mu} = 2k_\mu k^\mu \frac{\partial \Gamma(k^2)}{\partial k^2} . \tag{52.11}$$

Hence

$$\frac{\partial}{\partial k^2} \Gamma(k^2) = \frac{1}{2k^2} k^\mu \frac{\partial}{\partial k^\mu} \Gamma(k^2) . \tag{52.12}$$

This equation, however, can be used easily only for 1-loop diagrams, for which there is only a single integration, such as the tadpole diagram and the fish diagram. As we noted before, the tadpole diagram is independent of k^2, so the Taylor expansion for it is simply $\Sigma(k^2) = \Sigma(\mu^2)$. The setting-sun diagram contribution to $\Sigma(k^2)$ (two-loop), and the fish diagram contribution to the loop integration for a 4-point function, however, have non-trivial k^2 dependences. Consider first the logarithmically divergent fish diagram integral in (52.5). Denoting it by I, (52.12) gives

$$\frac{\partial I}{\partial k^2} = \left(-\frac{\lambda}{4!} \right)^2 \left(\frac{i^2}{k^2} \right) \int \frac{d^4 l}{(2\pi)^4} \frac{(k_1 + k_2 - l) \cdot k}{(l^2 - m^2 + i\epsilon^+)[(k_1 + k_2 - l)^2 - m^2 + i\epsilon^+]^2} ,$$
$$\tag{52.13}$$

which has $D = -1$ and is thus finite. In general, the superficial degree of divergence D is decreased by one unit on each differentiation $\partial/\partial k^\mu$. Starting with an integral with $D \geq 0$, at some point, after an enough number of derivatives have been taken with respect to the external momenta, the result will become convergent. In other words, after a finite number of terms in the Taylor series,

$$-i\Sigma(k^2) =$$

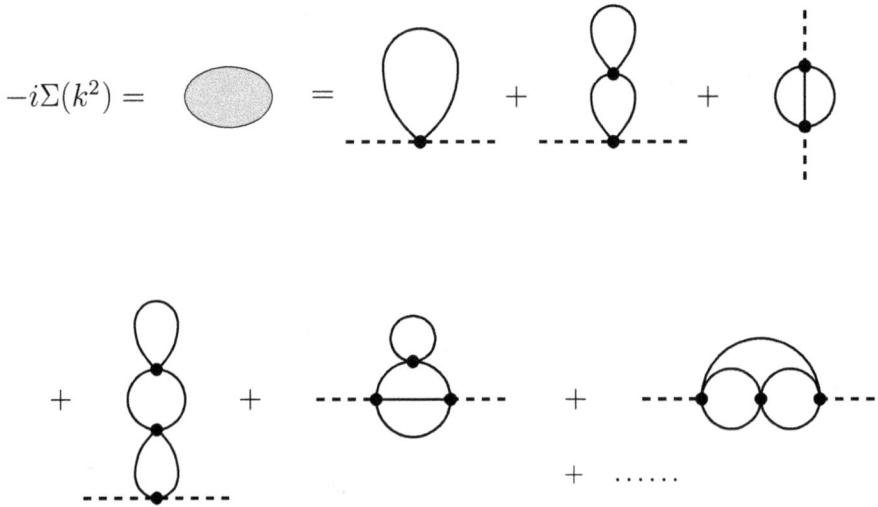

FIGURE 52.2

$$\tilde{G}(k^2) =$$

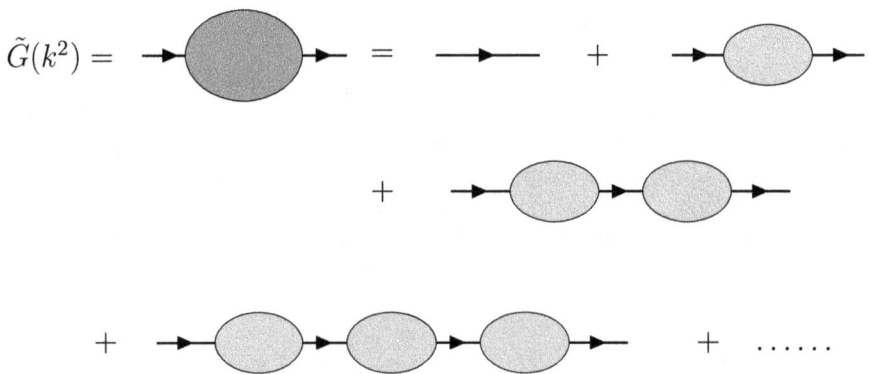

FIGURE 52.3

the remaining terms will all consist of finite integrals. As we have just seen, the Taylor expansion for the fish diagram leads to the sum of a logarithmicaly divergent part and a finite part. We will make use of this important fact below [c.f. (52.42)].

Let us now return to the vertex function $\Sigma(k^2)$ and consider 2-loop contributions, which are due to the second and third diagrams in the top row on the right-hand side of the equation in Fig. 52.2. Instead of using (52.12) we will Taylor expand directly in powers of $k^\mu - k_0^\mu$, with $k_0^2 = \mu^2$:

$$
\Sigma(k^2) = \Sigma(\mu^2) + (k^\mu - k_0^\mu) \left. \frac{\partial \Sigma(k^2)}{\partial k^\mu} \right|_{k_0^\mu}
$$

$$
+ \frac{1}{2}(k^\mu - k_0^\mu)(k_\nu - k_{0\nu}) \left. \frac{\partial}{\partial k^\mu} \frac{\partial}{\partial k_\nu} \Sigma(k^2) \right|_{k_0^\mu} + \cdots .
$$

We can thus write, up to $O(\lambda^2)$, that is, including one- and two-loop contributions,

$$
\Sigma(k^2) = \Sigma(\mu^2) + (k^2 - \mu^2)\Sigma'(\mu^2) + \tilde{\Sigma}(k^2) , \tag{52.14}
$$

where the first-order term [proportional to $(k^\mu - k_0^\mu)$] has been eliminated due to the requirement that $\Sigma(\mu^2)$ is a Lorentz scalar, and the second-order term [proportional to $(k^2 - \mu^2)$] has been written in conformity with (52.10), with $\Sigma'(\mu^2) \equiv (1/2)\partial_\mu\partial^\mu\Sigma(k^2)|_{k^2=\mu^2}$. In (52.14), the first term $\Sigma(\mu^2)$ includes the quadratically divergent one-loop contribution as well as the quadratically divergent part of the two-loop contribution. Since Σ' involves two derivatives of the quadratically divergent part of the two-loop contribution, it is logarithmically divergent. The last term $\tilde{\Sigma}(k^2)$ is then finite. On setting $k^2 = \mu^2$ in (52.14), and on differentiating that equation once with respect to k^2 and setting $k^2 = \mu^2$ in the result, we see that the finite part $\tilde{\Sigma}(k^2)$ satisfies the conditions

$$
\tilde{\Sigma}(\mu^2) = 0 , \qquad \tilde{\Sigma}'(\mu^2) \equiv \left. \frac{\partial \tilde{\Sigma}(k^2)}{\partial k^2} \right|_{k^2=\mu^2} = 0 . \tag{52.15}
$$

Now let us substitute (52.14) into (52.9). We find that

$$
\tilde{G}(k^2) = \frac{i}{k^2 - m^2 - \Sigma(\mu^2) - (k^2 - \mu^2)\Sigma'(\mu^2) - \tilde{\Sigma}(k^2) + i\epsilon^+} . \tag{52.16}
$$

At this point we can define the *physical mass* as the pole of the above propagator, and have the freedom to determine this quantity in terms of μ^2, which up to now is an arbitrary parameter. We fix μ^2 by the **mass renormalization** condition as follows:

$$
m^2 + \Sigma(\mu^2) = \mu^2 . \tag{52.17}
$$

Then

$$
\tilde{G}(k^2) = \frac{i}{(k^2 - \mu^2)\{1 - \Sigma'(\mu^2)\} - \tilde{\Sigma}(k^2) + i\epsilon^+} , \tag{52.18}
$$

which has a pole at $k^2 = \mu^2$ by virtue of (52.15). Thus μ becomes the physical mass, which indeed is finite. But since $\Sigma(\mu^2)$ is infinite, condition (52.17) requires that the **bare mass** m must be taken to be infinite also. Every quantity in the propagator $\tilde{G}(k^2)$ is now finite except for $\Sigma'(\mu^2)$, which can be turned into a multiplicative factor as follows. Observe that in (52.14) both $\Sigma'(\mu^2)$ and $\tilde{\Sigma}(k^2)$ are of order λ^2, since both terms are due to two-loop contributions. So we can write, when $\Sigma'(\mu^2)$ is considered as a regularized, and hence finite, quantity,

$$\tilde{\Sigma}(k^2) \approx \tilde{\Sigma}(k^2)\{1 - \Sigma'(\mu^2)\}\,. \tag{52.19}$$

Putting the right-hand side of this equation in place of $\tilde{\Sigma}(k^2)$ in the denominator of the right-hand side of (52.18) we obtain

$$\tilde{G}(k^2) = \frac{i}{\{1 - \Sigma'(\mu^2)\}\{k^2 - \mu^2 - \tilde{\Sigma}(k^2)\} + i\epsilon^+}\,, \tag{52.20}$$

or

$$\tilde{G}(k^2) = \frac{iZ_\varphi}{k^2 - \mu^2 - \tilde{\Sigma}(k^2) + i\epsilon^+}\,, \tag{52.21}$$

where

$$Z_\varphi \equiv \frac{1}{1 - \Sigma'(\mu^2)} = 1 + \Sigma'(\mu^2) + O(\lambda^4)\,. \tag{52.22}$$

Thus all the infinities in $\tilde{G}(k^2)$, up to order λ^2, have been "absorbed" into the multiplicative constant Z_φ.

Now, from (52.1) and (49.1), we can write, remembering that for the 2-point function $k = k_1 = -k_2$,

$$(2\pi)^4\delta^4(k_1 + k_2)\tilde{G}(k^2) = \int d^4x_1 \int d^4x_2\, e^{-ik_1x_1 - ik_2x_2} \langle 0 \,|\, T[\hat{\varphi}(x_1)\hat{\varphi}(x_2)] \,|\, 0\rangle\,. \tag{52.23}$$

By (52.21) this implies

$$(2\pi)^4\delta^4(k_1 + k_2)\left(\frac{i}{k^2 - \mu^2 + \tilde{\Sigma}(k^2) + i\epsilon^+}\right) = \int d^4x_1 \int d^4x_2\, e^{-ik_1x_1 - ik_2x_2}$$
$$\times \left\langle 0 \,\left|\, T\left[\frac{\hat{\varphi}(x_1)}{\sqrt{Z_\varphi}}\frac{\hat{\varphi}(x_2)}{\sqrt{Z_\varphi}}\right]\,\right|\, 0\right\rangle\,. \tag{52.24}$$

So if we define a *renormalized* field (or wave function) $\varphi_R(x)$ and a *renormalized momentum space* 2-point Green's function $\tilde{G}_R(k^2)$ by

$$\varphi_R(x) \equiv Z_\varphi^{-1/2}\,\varphi(x)\,, \tag{52.25}$$

$$\tilde{G}_R(k^2) \equiv \frac{i}{k^2 - \mu^2 - \tilde{\Sigma}(k^2) + i\epsilon^+} = \frac{1}{\sqrt{Z_\varphi}\sqrt{Z_\varphi}}\tilde{G}(k^2)\,, \tag{52.26}$$

we have

$$(2\pi)^4\delta^4(k_1 + k_2)\,\tilde{G}_R(k^2) = \int d^4x_1 \int d^4x_2\, e^{-ik_1x_1 - ik_2x_2}\,G_R(x_1, x_2)\,, \tag{52.27}$$

where the *renormalized real space* 2-point Green's function $G_R(x_1, x_2)$ is given by

$$G_R(x_1, x_2) \equiv \langle 0 \, | \, T[\hat{\varphi}_R(x_1) \hat{\varphi}_R(x_2)] \, | \, 0 \rangle = \frac{1}{\sqrt{Z_\varphi}\sqrt{Z_\varphi}} G(x_1, x_2) \, . \qquad (52.28)$$

Eq. (52.25) is the reason why Z_φ is called the **field** (or **wave function**) **renormalization constant**. The generalization of (52.28) to the n-point case is straightforward:

$$G_R(x_1, \ldots, x_n) = Z_\varphi^{-n/2} G(x_1, \ldots, x_n) \, . \qquad (52.29)$$

We also have the generalization of (52.26) to the n-point case for the Fourier transform:

$$\tilde{G}_R(k_1, \ldots, k_n) = Z_\varphi^{-n/2} \tilde{G}(k_1, \ldots, k_n) \, . \qquad (52.30)$$

Looking at (52.21) we see that the (unrenormalized) 2-point function $\tilde{G}(k^2)$ has the form of a free propagator (in the sense that there are no integrals over loop momenta) after taking into account all the 1PI 2-point diagrams with external legs amputated (c.f. Fig. 52.3). This quantity, if left unregulated, is infinite (solely due to the presence of the factor Z_φ). It can then be understood as the propagator for an external line for the part of $\tilde{G}_R(k_1, \ldots, k_n)$ coming from the contributions of only 1PI diagrams. We will designate such quantities by $\tilde{G}_{(1PI)}(k_1, \ldots, k_n)$. When contributions from the external legs are divided out from them, the corresponding quantities are known as 1PI **amputated Green's functions**, denoted by $\Gamma(k_1, \ldots, k_n)$. Thus we have

$$\Gamma(k_1, \ldots, k_n) = \frac{\tilde{G}_{(1PI)}(k_1, \ldots, k_n)}{\tilde{G}(k_1^2) \ldots \tilde{G}(k_n^2)} \, . \qquad (52.31)$$

So we can define the renormalized counterpart $\Gamma_R(k_1, \ldots, k_n)$ by

$$\Gamma_R(k_1, \ldots, k_n) \equiv \frac{\tilde{G}_{R(1PI)}(k_1, \ldots, k_n)}{\tilde{G}_R(k_1^2) \ldots \tilde{G}_R(k_n^2)} \, , \qquad (52.32)$$

where $\tilde{G}_R(k_i^2)$ is given by (52.26). By (52.30) and (52.26), the right-hand side of the above equation becomes $Z_\varphi^{-n/2} \tilde{G}_{(1PI)}(k_1, \ldots, k_n)/[Z_\varphi^{-n} \tilde{G}(k_1^2) \ldots \tilde{G}(k_n^2)]$. From (52.31) it then follows that the renormalized 1PI amputated Green's functions are related to the unrenormalized ones by

$$\Gamma_R(k_1, \ldots, k_n) = Z_\varphi^{n/2} \Gamma(k_1, \ldots, k_n) \, . \qquad (52.33)$$

The 2-point amputated function $\Gamma(k_1, k_2)$ is special because it turns out to be the inverse of the full 2-point propagator $\tilde{G}(k^2)$. This can be seen from (52.31) as follows. A glance at Fig. 52.3 shows that $\tilde{G}_{(1PI)}(k_1, k_2)$ consists of contributions from only the first two diagrams on the right-hand side of the equation shown there. Thus [c.f. (52.6)]

$$\begin{aligned}
\tilde{G}_{(1PI)}(k_1, k_2) = \tilde{G}_{(1PI)}(k^2) &= \left(i\tilde{D}(k^2)\right) + \left(i\tilde{D}(k^2)\right)\left(-i\Sigma(k^2)\right)\left(i\tilde{D}(k^2)\right) \\
&= \left(i\tilde{D}(k^2)\right)\left[1 + \left(-i\Sigma(k^2)\right)\left(i\tilde{D}(k^2)\right)\right] \\
&\approx \left(i\tilde{D}(k^2)\right)\left[1 - \left(\Sigma(k^2)\right)\left(\tilde{D}(k^2)\right)\right]^{-1} = \tilde{G}(k^2) ,
\end{aligned}$$

$$(52.34)$$

where the last equality follows from (52.8). Thus (52.31) implies that

$$\Gamma(k_1, k_2) = \Gamma(k^2) = \frac{1}{\tilde{G}(k^2)} , \qquad (52.35)$$

as claimed.

We will now proceed to consider the infinities in 4-point functions up to $O(\lambda^2)$. Fig. 50.8 shows the Feynman diagrams in this approximation. Disregarding the disconnected diagrams, which consist of 2-point functions we have dealt with already, one is left with the diagrams shown in Fig. 52.4, all of which, except for the first one, are $O(\lambda^2)$. Of the second order diagrams, the ones in the top row (the fish diagrams) are $1PI$, while the ones in the bottom row are 1P reducible. The first diagram is known as a **tree diagram**, so named because of its shape. Momentum labels for internal momenta have been indicated in Fig. 52.4 to reflect the rule of 4-momentum conservation at each vertex. We will first compute the contributions to $\Gamma_{one\ loop}(k_1, k_2, k_3, k_4)$ (up to the one-loop approximation) due to the fish diagrams, using the so-called **Mandelstam variables** s, t and u, each suitable for one of these diagrams. They are defined by

$$s \equiv (k_1 + k_2)^2 , \qquad t \equiv (k_1 - k_3)^2 , \qquad u \equiv (k_1 - k_4)^2 . \qquad (52.36)$$

It is not hard to show, using the method discussed immediately above Problem 50.5, that the symmetry number for each of the three fish diagrams is 576. So each of these diagrams carries a numerical factor of $576/[2!(4!)^2] = 1/2$. Because of overall momentum conservation: $k_1 + k_2 = k_3 + k_4$ in terms of the momentum labels in Fig. 52.4, the amputated functions Γ for the fish diagrams from left to right in the figure (with the correct symmetry numbers), are then $\Gamma(s), \Gamma(t)$ and $\Gamma(u)$, respectively, where

$$\Gamma(k^2) = \frac{(-i\lambda)^2}{2} \int \frac{d^4l}{(2\pi)^4} \frac{i}{l^2 - m^2 + i\epsilon^+} \frac{i}{(l - k)^2 - m^2 + i\epsilon^+} . \qquad (52.37)$$

By counting powers of l, we see that $\Gamma(s), \Gamma(t)$ and $\Gamma(u)$ all have superficial degree of divergence given by $D = 0$ (logarithmically divergent), as we have seen before.

Again we assume that $\Gamma(k^2)$ is regularized (made finite) in some way, so we can manipulate with it formally as a finite quantity. Following the strategy used for $\Sigma(k^2)$ [c.f. (52.10)] we will Taylor expand $\Gamma(k^2)$ around a suitable point to

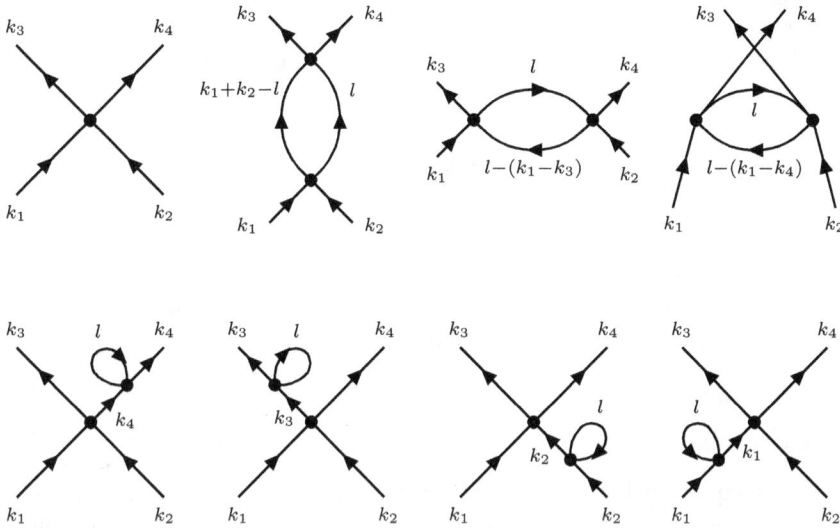

FIGURE 52.4

separate the infinite and finite parts. Having determined the physical mass μ [c.f. (52.17)], we see that for on-shell external momenta, that is $k_i^2 = \mu^2$ ($i = 1, 2, 3, 4$), the Mandelstam variables satisfy the condition

$$s + t + u = 4\mu^2 .$$ (52.38)

Indeed, by (52.36) and recalling that $k_1 + k_2 = k_3 + k_4$, we have

$$
\begin{aligned}
s + t + u &= (k_1 + k_2)^2 + (k_1 - k_3)^2 + (k_1 - k_4)^2 \\
&= (k_1^2 + k_2^2 + 2k_1 \cdot k_2) + (k_1^2 + k_3^2 - 2k_1 \cdot k_3) + (k_1^2 + k_4^2 - 2k_1 \cdot k_4) \\
&= 3k_1^2 + k_2^2 + k_3^2 + k_4^2 + 2k_1 \cdot (k_2 - k_3 - k_4) \\
&= 3k_1^2 + k_2^2 + k_3^2 + k_4^2 - 2k_1^2 = 4\mu^2 .
\end{aligned}
$$ (52.39)

It is thus convenient to expand $\Gamma(k^2)$ around the so-called *symmetric point*, given by

$$s_0 = t_0 = u_0 \equiv \frac{4\mu^2}{3} .$$ (52.40)

So we write

$$\Gamma(k^2) = \Gamma(s_0) + (k^2 - s_0) \left. \frac{\partial \Gamma(k^2)}{\partial k^2} \right|_{s_0} + \frac{1}{2}(k^2 - s_0)^2 \left. \frac{\partial^2 \Gamma(k^2)}{\partial(k^2)^2} \right|_{s_0} + \cdots ,$$ (52.41)

with similar expressions on replacing s_0 by t_0 or u_0. Using (52.12) we again see that each derivative decreases D by one unit. Since $\Gamma(s_0)$ is already log-arithmically divergent, the terms with derivatives must all be finite. We thus write

$$\Gamma(k^2) = \Gamma(s_0) + \tilde{\Gamma}(k^2) , \tag{52.42}$$

where $\tilde{\Gamma}(k^2)$ is the finite part, which satisfies

$$\tilde{\Gamma}(s_0) = 0 . \tag{52.43}$$

We will denote the contributions to $\Gamma(k_1, k_2, k_3, k_4)$ from the tree diagram and the three fish diagrams by Γ_{tree} and Γ_{fish}, respectively. Thus, with the symmetry factor for the tree diagram equal to 4!,

$$\Gamma_{tree} = -i\lambda , \tag{52.44}$$

$$\Gamma_{fish}(s, t, u) = \Gamma(s) + \Gamma(t) + \Gamma(u) = 3\Gamma(s_0) + \tilde{\Gamma}(s) + \tilde{\Gamma}(t) + \tilde{\Gamma}(u) . \tag{52.45}$$

In the last equation $\Gamma(s_0)$ is logarithmically divergent (before regularization) and the rest of the terms are finite. From (52.33) we see that the renormalized 1PI amputated 4-point Green's function, up to $O(\lambda^2)$ (or up to the one-loop approximation) and denoted by $\Gamma_{R(one\,loop)}$, is given by

$$\Gamma_{R(one\,loop)}(s, t, u) = Z_\varphi^2 \{\Gamma_{tree} + \Gamma_{fish}(s, t, u)\} = Z_\varphi^2 \Gamma_{one\,loop}(s, t, u) . \tag{52.46}$$

We then define the (*physical*) **renormalized coupling constant** λ_R, which is necessarily finite, by

$$-i\lambda_R \equiv \Gamma_{R(one\,loop)}(s_0, t_0, u_0) . \tag{52.47}$$

It follows from (52.33), (52.43), (52.45) and (52.46) that

$$-i\lambda_R = Z_\varphi^2 \left[-i\lambda + 3\Gamma(s_0)\right] . \tag{52.48}$$

In this equation both Z_φ [c.f. (52.22)] and $\Gamma(s_0)$ [c.f. (52.37)] are divergent. So we further introduce the **vertex renormalization constant** Z_λ:

$$-iZ_\lambda^{-1}\lambda = -i\lambda + 3\Gamma(s_0) , \tag{52.49}$$

to absorb the infinities into a multiplicative constant and write (52.48) as

$$\lambda_R = Z_\varphi^2 Z_\lambda^{-1} \lambda . \tag{52.50}$$

Meanwhile, by (52.46) and (52.49),

$$\Gamma_{R(one\,loop)}(s, t, u) = Z_\varphi^2 \left\{-i\lambda + 3\Gamma(s_0) + \tilde{\Gamma}(s) + \tilde{\Gamma}(t) + \tilde{\Gamma}(u)\right\}$$

$$= Z_\varphi^2 \left(-iZ_\lambda^{-1}\lambda\right) + Z_\varphi^2 \left(\tilde{\Gamma}(s) + \tilde{\Gamma}(t) + \tilde{\Gamma}(u)\right) . \tag{52.51}$$

Thus, by (52.50),

$$\Gamma_{R(one\,loop)}(s, t, u) = -i\lambda_R + Z_\varphi^2 \left\{\tilde{\Gamma}(s) + \tilde{\Gamma}(t) + \tilde{\Gamma}(u)\right\} . \tag{52.52}$$

Now recall that $\tilde{\Gamma}(s), \tilde{\Gamma}(t)$ and $\tilde{\Gamma}(u)$ are all $O(\lambda^2)$ [c.f. (52.37)] and $Z_\varphi \sim 1 + O(\lambda^2)$ [c.f. (52.22)]. So we can rewrite the above equation as

$$\Gamma_{R(one\,loop)}(s,t,u) = -i\lambda_R + \left\{\tilde{\Gamma}(s) + \tilde{\Gamma}(t) + \tilde{\Gamma}(u)\right\} + O(\lambda^3)\,, \qquad (52.53)$$

where the explicitly given terms are up to order λ^2, and are completely finite.

As mentioned above for the bare mass m, Eq. (52.50) indicates that, in order for the renormalized coupling constant λ_R to be finite, the bare coupling constant λ has to be infinite also, since both renormalization constants Z_φ and Z_λ are infinite. This implies that, to calculate even the finite parts $\tilde{\Sigma}(k^2)$ and $\tilde{\Gamma}(k^2)$ of vertex functions, one cannot start off with the (infinite) bare quantities m and λ. This conundrum will be resolved towards the end of this chapter, when we discuss the *multiplicative renormalization* and the so-called BPH renormalization procedures. We will see how to "work in reverse" and start the renormalization process with the calculation of the vertex functions (both finite and infinite parts) using the physical or some other finite renormalized masses and coupling constants.

Having renormalized the 1PI amputated 4-point Green's functions we will now try to renormalize the full 4-point Green's function $\tilde{G}(k_1, k_2, k_3, k_4)$, again up to the one-loop approximation. In addition to the 1-PI diagrams in Fig. 52.4 (the ones in the top row), we have to include also the one-loop 1P reducible diagrams in that figure (the ones in the bottom row). In Fig. 52.5 we have shown the amputated contributions from all the diagrams, with the amputated external lines drawn as dashed lines. To compute the full (unrenormalized) Green's function, all we have to do is to append the product of the four external propagators to the contribution from each term in the figure (including each vertex contribution, shown as a solid dot).

Aside from the external legs, the contribution from the leftmost 1P reducible diagram in Fig. 52.5 is thus

$$-i\lambda \left(\frac{i}{k_4^2 - m^2 + i\epsilon^+}\right) \text{ (one loop contribution to } -i\Sigma(k_4^2))\,. \qquad (52.54)$$

Similar expressions apply to the other three, when k_4 is replaced by k_1, k_2 or k_3. We note that, from Fig. 52.2 and (52.14), the one-loop contribution to $\Sigma(k_i^2)$ is just $\Sigma(\mu^2)$, since the external momenta k_i are supposed to be on mass shell, that is, $k_i^2 = \mu^2$. So we can simply write the last factor in the above equation as $-i\Sigma(k_4^2)$. Recalling (52.44) and (52.45), then, we have

$$\tilde{G}_{one\,loop}(k_1, k_2, k_3, k_4) = \left\{\prod_{i=1}^{4} \frac{i}{k_i^2 - m^2 + i\epsilon^+}\right\}$$

$$\times \left[-i\lambda + \left(3\Gamma(s_0) + \tilde{\Gamma}(s) + \tilde{\Gamma}(t) + \tilde{\Gamma}(u)\right) + (-i\lambda)\sum_{j=1}^{4}\left(\frac{i}{k_j^2 - m^2 + i\epsilon^+}\right)(-i\Sigma(k_j^2))\right]\,.$$

$$(52.55)$$

We will consider the terms involving $\Gamma(s_0)$ and the $\tilde{\Gamma}$'s first, and write their contributions to $\tilde{G}(k_1, k_2, k_3, k_4)$ as

$$\left(\left\{ \prod_{i=1}^{4} \left(\frac{i}{k_i^2 - m^2 - \Sigma(k_i^2) + i\epsilon^+} \right) \right\} \left[3\Gamma(s_0) + \tilde{\Gamma}(s) + \tilde{\Gamma}(t) + \tilde{\Gamma}(u) \right] \right)_{(\lambda^2)} ,$$

(52.56)

where the subscript label (λ^2) means that only terms *up to* order λ^2 are retained. This is permitted since these contributions in the original expression in (52.55) are manifestly $O(\lambda^2)$, due to the fact that $\Gamma(s_0), \tilde{\Gamma}(s), \tilde{\Gamma}(t), \tilde{\Gamma}(u)$ are all $O(\lambda^2)$ (being one-loop contributions with two vertices), and $\Sigma(k_i^2) \sim O(\lambda)$ (being a one-loop contribution with one vertex). The rest of the terms in (52.55), which are also manifestly up to order λ^2, can be combined in the following way:

$$- i\lambda \left\{ \prod_{i=1}^{4} \left(\frac{i}{k_i^2 - m^2 + i\epsilon^+} \right) \right\} \left[1 + \sum_{j=1}^{4} \frac{\Sigma(k_j^2)}{k_j^2 - m^2 + i\epsilon^+} \right]$$

$$= -i\lambda \left\{ \prod_{i=1}^{4} \left(\frac{i}{k_i^2 - m^2 + i\epsilon^+} \right) \right\} \left(\left[1 - \sum_{j=1}^{4} \frac{\Sigma(k_j^2)}{k_j^2 - m^2 + i\epsilon^+} \right]^{-1} \right)_{(\lambda)}$$

(52.57)

$$= -i\lambda \left\{ \prod_{i=1}^{4} \left(\frac{i}{k_i^2 - m^2 + i\epsilon^+} \right) \right\} \left(\prod_{j=1}^{4} \left\{ 1 - \frac{\Sigma(k_j^2)}{k_j^2 - m^2 + i\epsilon^+} \right\}^{-1} \right)_{(\lambda)}$$

$$= -i\lambda \left(\prod_{i=1}^{4} \frac{i}{k_i^2 - m^2 - \Sigma(k_i^2) + i\epsilon^+} \right)_{(\lambda)} ,$$

where (λ) in the subscript position means retaining terms up to first order in λ. Combining this result and that from (52.56), we have

$$\tilde{G}_{one\,loop}(k_1, k_2, k_3, k_4)$$

$$= \left(\left\{ \prod_{j=1}^{4} \frac{i}{k_j^2 - m^2 - \Sigma(k_j^2) + i\epsilon^+} \right\} \left[-i\lambda + 3\Gamma(s_0) + \tilde{\Gamma}(s) + \tilde{\Gamma}(t) + \tilde{\Gamma}(u) \right] \right)_{(\lambda^2)}$$

$$= \left(\left\{ \prod_{j=1}^{4} \tilde{G}(k_j^2) \right\} \left[\Gamma_{tree} + \Gamma_{fish}(s, t, u) \right] \right)_{(\lambda^2)} ,$$

(52.58)

where the second equality follows from (52.9). We then have, by (52.9), the following expression for the renormalized 4-point Green's function in the one-

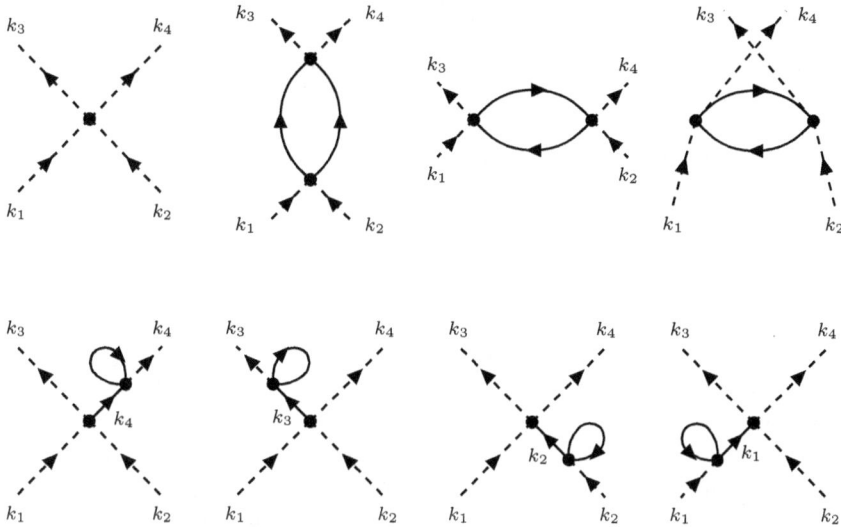

FIGURE 52.5

loop approximation:

$$\tilde{G}_{R(one\ loop)}(k_1, k_2, k_3, k_4)$$

$$= \left(Z_\varphi^{-2} \left\{ Z_\varphi^4 \prod_{j=1}^{4} \tilde{G}_R(k_j^2) \right\} Z_\varphi^{-2} \Gamma_{R(one\ loop)}(s, t, u) \right)_{(\lambda^2)} \qquad (52.59)$$

$$= \left(\left\{ \prod_{j=1}^{4} \tilde{G}_R(k_j^2) \right\} \left\{ \Gamma_{R(one\ loop)}(s, t, u) \right\} \right)_{(\lambda^2)} ,$$

where we have used (52.26) to relate $\tilde{G}_R(k^2)$ to $\tilde{G}(k^2)$, (52.30) to relate $\tilde{G}_R(k_1, k_2, k_3, k_4)$ to $\tilde{G}(k_1, k_2, k_3, k_4)$, and (52.46) to relate $\Gamma_{R(one\ loop)}$ to $\Gamma_{one\ loop}$. Eqs. (52.26) and (52.53) indicate that the renormalized quantities $\tilde{G}_R(k_j^2)$ and $\Gamma_{R(one\ loop)}$ are completely finite. Hence the renormalized 4-point Green's function $\tilde{G}_{R(one\ loop)}(k_1, k_2, k_3, k_4)$ in the one-loop approximation is finite also.

The 2-point and 4-point Green's functions that we have been studying in some detail up to this point actually play a special role in the renormalizability of the φ^4-theory. This can be seen from the notion of the superficial degree of divergence D introduced earlier, which can be deduced for any Feynman diagram quite simply. Suppose a particular diagram has n external lines, I internal lines, V vertices, and L loops. Then the number L of independent internal momenta

k_i that must be integrated over is given by $L = I - V + 1$. This is so because each vertex carries one condition for momentum conservation, so V vertices will carry V conditions. But these are not all independent because they imply the overall external momentum conservation condition, which makes the total number of independent momentum conservation conditions $V - 1$. The general form of the integral due to the contribution of the diagram to the n-point $\Gamma^{(n)}$, apart from powers of i, λ and the symmetry number, is then

$$\int \prod_{i=1}^{L} \frac{d^4 k_i}{(2\pi)^4} \prod_{j=1}^{I} \frac{1}{l_j^2 - m^2 + i\epsilon^+} , \qquad (52.60)$$

where l_j is in general a linear combination of the external momenta and the loop momenta as a result of momentum conservation at each vertex. Examples of this expression occur in (52.3), (52.4) and (52.5). The superficial degree of divergence for this diagram is then $D = 4L - 2I$. So we have $D = 4(I - V + 1) - 2I = 2I - 4V + 4$. But since each vertex is connected to 4 lines (only in the φ^4-theory), either internal or external, and each internal line connects 2 vertices, we have the topological condition $4V = 2I + n$. This then gives

$$D = 4 - n . \qquad (52.61)$$

Since n must be even in the φ^4-theory, we conclude that in this theory, the only n-point functions with $D \geq 0$ (divergent) are the 2-point functions ($D = 2$) and the 4-point functions ($D = 0$), hence our focus on these functions up to now. It is highly significant that for the φ^4-theory, D is independent of the order of the perturbation λ. This in fact is a necessary condition for the renormalizability of the theory.

The fact that $D < 0$ for all n-point functions with $n > 4$, however, does not guarantee their convergence. This can be seen in the example of the 6-point 1PI but 2P reducible diagram (made disconnected by cutting two internal lines) shown in Fig. 52.6, with a boxed-off 4-point subgraph that is clearly divergent (logarithmically), as we have shown earlier. In the 1PI but also 2P reducible 6-point diagram shown in Fig. 52.7, we have a 4-point (logarithmically divergent) one-loop subgraph as well as a 2-point (quadratically divergent) two-loop subgraph, also boxed off. The appearance of divergent subgraphs for an overall divergent diagram with $D < 0$ is the reason for the adjective "superficial" in the appellation superficial divergence. To hunt for divergent diagrams with divergent subgraphs for $n > 4$ in general, one classifies them according to whether they are 1P reducible, 2P reducible, or 3P reducible. Any diagram is always 4P reducible because each vertex is necessarily connected to 4 lines (internal or external). 1PI diagrams for $n > 4$ are always convergent since $D < 0$ for such diagrams and they contain no subgraphs by cutting internal lines. One keeps reducing the reducible diagrams until divergent subgraphs show up. In all cases they will turn out to be either 2-point or 4-point divergences. The important point is that they can all be renormalized to any order in λ, which we expect since D is independent of λ. We have shown how to do this explicitly in the

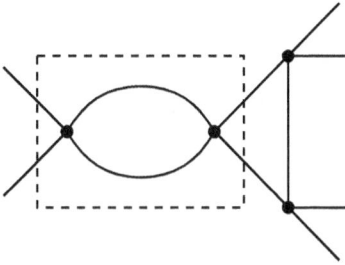 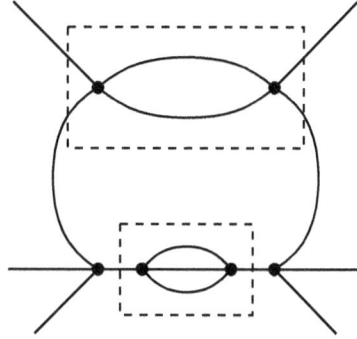

FIGURE 52.6   FIGURE 52.7

one-loop approximation for 4-point functions and the two-loop approximation for 2-point functions. That this can be done to all orders in λ is what makes for the **renormalizability** of the theory.

Superficially convergent graphs ($n > 4$) with divergent subgraphs (such as that in Fig. 52.7) can always be constructed from so-called skeleton graphs. A **skeleton graph** is defined as a graph with no divergent subgraphs. An example of this is the 6-point graph shown in Fig. 52.8 (without the dotted boxes). It is quite clear that if we take the box enclosing a vertex and replace it by a divergent 4-point graph, and/or the box surrounding part of an internal propagator and replace it by a divergent 2-point graph, we will obtain a superficially convergent graph with divergent subgraphs. Such a procedure is called a **skeleton expansion**. Besides the example in Fig. 52.7, we show another more complicated 6-point graph with divergent subgraphs in Fig. 52.9, constructed from the skeleton graph in Fig. 52.8.

Based on the mathematical details on the renormalizability of the φ^4-theory up to low orders in λ (one-loop and two-loop approximations) presented so far, we can now describe briefly (without giving detailed proofs) two equivalent procedures to achieve renormalizability to all orders. The first is known as **multiplicative renormalization**, while the second as **BPH renormalization**, invented by Bogoliubov, Parasyuk, and Hepp [see Bogoliukov and Shirkov (1959)]. These methods work for all renormalizable theories (except gauge field theories with massless particles) but we will illustrate the main points by the

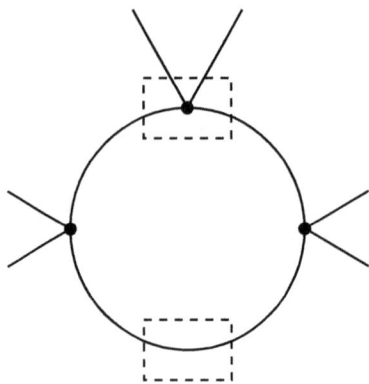

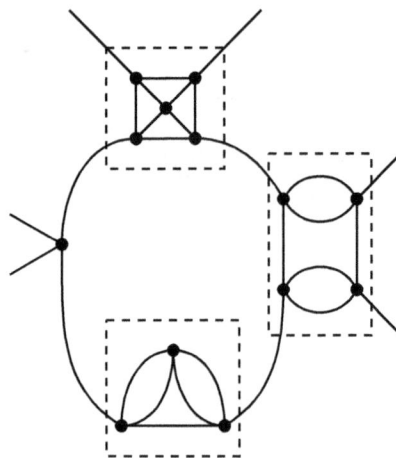

FIGURE 52.8   FIGURE 52.9

φ^4-theory. In the first method, one first regularizes all propagators so that they vanish quickly as the integrated-momentum cutoff $\Lambda \to \infty$, thus causing the integrals to become convergent, and such that the regularized integral approaches the unregularized one in the same limit. (The main regularization schemes will be discussed in the next chapter.) For example, one may replace a free propagator as follows:

$$\frac{1}{k^2 - \mu^2} \longrightarrow \frac{1}{k^2 - \mu^2} - \frac{1}{k^2 - \mu^2 - \Lambda^2} = -\frac{\Lambda^2}{(k^2 - \mu^2)(k^2 - \mu^2 - \Lambda^2)}, \quad (52.62)$$

where it (as part of an integrand) has gone from $O(k^{-2})$ to $O(k^{-4})$. It is then argued that since all divergencies originate from just a finite number of Green's functions, specifically, 2-point and 4-point functions, they can be removed by adding a *finite* number of so-called **counterterms** to the original Lagrangian, which we will write as

$$\mathcal{L}' = \frac{1}{2} \left[(\partial_\mu \varphi')^2 - \mu'^2 \varphi'^2 \right] - \frac{\lambda'}{4!} \varphi'^4 . \quad (52.63)$$

Some explanation of the notation is warranted here. Up to now we have distinguished between the unrenormalized and renormalized fields $\varphi(x)$ and $\varphi_R(x)$, respectively; the unrenormalized and renormalized (physical) masses m and μ, respectively; and the unrenormalized and renormalized coupling constants, λ

and λ_R, respectively. But a crucial point is that renormalized quantities depend on the choice of an arbitrary point – the subtraction point – in the k^2 complex plane about which $\Sigma(k^2)$ and $\Gamma(k^2)$, as *analytic functions* of k^2, are Taylor expanded [c.f. (52.14) and (52.42)]. (So far, we have chosen that point to be $k^2 = \mu^2$, where μ is the physical mass.) They also depend on the powers of λ to which the perturbative renormalization procedure is carried out [we have done it explicitly up to $O(\lambda^2)$ so far]; and as regularized quantities, also on the momentum cutoff Λ. Keeping these facts in mind, the quantities φ', μ' and λ' in (52.63) are then referred to as the *intermediate field*, the *intermediate mass*, and the *intermediate coupling constant*, respectively. They are intermediate between the bare quantities (φ, m, λ) and the physical ones $(\varphi_R, \mu, \lambda_R)$, and are obtained by carrying out the aforementioned Taylor expansions about some *other* point in the k^2 complex plane than $k^2 = \mu^2$, such as $k^2 = 0$, for example.

On calculating the Feynman diagrams for the Lagrangian density in (52.63), infinities will arise as we have shown. It is then conjectured that counterterms can be added to this Lagrangian so that the infinities can be cancelled to all orders in λ'. The infinities arise from 2-point and 4-point functions, proportional to φ'^2 and φ'^4, respectively. Since $\mathcal{L}'$ already contains terms to these powers of φ', we hope that the counterterms will only contain terms to these powers of φ' also. Denoting the sum of the counterterms by $\delta\mathcal{L}'$, we write

$$\delta\mathcal{L}' = \frac{1}{2}\left[B(\partial_\mu \varphi')^2 - C\varphi'^2\right] - \frac{A}{4!}\varphi'^4 , \qquad (52.64)$$

where A, B and C are renormalization "constants" that can be expressed as power series in λ' whose coefficients are Λ-dependent, and are to be chosen so that

$$\mathcal{L}' + \delta\mathcal{L}' = \mathcal{L} = \frac{1}{2}\left[(\partial_\mu \varphi)^2 - m^2\varphi^2\right] - \frac{\lambda}{4!}\varphi^4 , \qquad (52.65)$$

where φ is the unrenormalized field, and m and λ are the bare mass and bare coupling constant, respectively [c.f. (48.15) to (48.18)]. It is straightforward to verify that, in order to satisfy (52.65), the quantities A, B and C in (52.64) have to satisfy

$$\varphi = (1 + B)^{1/2}\, \varphi' , \qquad (52.66a)$$

$$m^2 = \frac{\mu'^2 + C}{1 + B} , \qquad (52.66b)$$

$$\lambda = \frac{\lambda' + A}{(1 + B)^2} . \qquad (52.66c)$$

Eqs. (52.66b) and (52.66c) suggest that we rewrite

$$C \equiv \delta\mu'^2 = Zm^2 - \mu'^2 , \qquad A \equiv \delta\lambda' = Z^2\lambda - \lambda' , \qquad (52.67)$$

where the renormalization constant Z [similar to the Z_φ in (52.25)] is defined by

$$Z \equiv (1 + B) . \qquad (52.68)$$

So (52.66) can be rewritten as

$$\varphi = Z^{1/2}\,\varphi'\,, \qquad m^2 = (\mu'^2 + \delta\mu'^2)\,Z^{-1}\,, \qquad \lambda = (\lambda' + \delta\lambda')\,Z^{-2}\,, \quad (52.69)$$

and the counterterms in $\delta\mathcal{L}'$ appear as

$$\delta\mathcal{L}' = \frac{1}{2}\left[(Z-1)(\partial_\mu\varphi')^2 - \delta\mu'^2\varphi'^2\right] - \frac{\delta\lambda'}{4!}\varphi'^4\,, \qquad (52.70)$$

with $\delta\mu'^2, \delta\lambda'$ and Z given by (52.67) and (52.68). Renormalizability of the theory means that, to all orders in λ, an equation analogous to (52.33) applies, that is,

$$\overline{\Gamma}^{(n)} = Z^{n/2}\,\Gamma^{(n)}\,, \qquad (52.71)$$

where $\overline{\Gamma}^{(n)}$ ($\Gamma^{(n)}$) is the renormalized (unrenormalized) n-point 1P1 amputated Green's function, with the understanding that $\overline{\Gamma}^{(n)}$ is calculated from $\mathcal{L}' + \delta\mathcal{L}'$ and $\Gamma^{(n)}$ from $\mathcal{L}'$. Eq. (52.71) is the reason for the label "multiplicative renormalization". The three renormalization constants $Z, \delta\mu'^2$ and $\delta\lambda'$ can be fixed by setting two conditions on the quantity $\overline{\Gamma}^{(2)}(k^2)$ and one condition on the quantity $\overline{\Gamma}^{(4)}(k^2)$ [where $k^2 = s, t$ or u, the Mandelstam variables, c.f.(52.36)]. One choice is to set these as follows, with the subtraction point $k^2 = 0$ (but it is equally valid to use other subtraction points):

$$\overline{\Gamma}^{(4)}(0) = -i\lambda'\,, \qquad \overline{\Gamma}^{(2)}(0) = i\mu'^2\,, \qquad \frac{d}{dk^2}\overline{\Gamma}^{(2)}(k^2)\bigg|_{k^2=0} = -i\,. \qquad (52.72)$$

In Chapter 54 we will give a qualitative overview of the proof of multiplicative renormalizability based on the so-called *Callan-Symanzik equation*, which is a special *renormalization group equation*.

Problem 52.2 Verify that the renormalization constants A, B and C as determined by the expressions in (52.66) satisfy (52.65).

We will end this chapter with a brief discussion of the BPH renormalization procedure. It is an order by order (in λ') perturbative procedure whose starting point is the scheme based on (52.64) and (52.65). We will pursue this up to order λ'^2, following closely the development for the absorption of infinities presented earlier in this chapter. Up to this order, the counterterm Lagrangian $\delta\mathcal{L}'_{(2)}$ can be written as [c.f. (52.70)]

$$\delta\mathcal{L}'_{(2)} = \frac{1}{2}\left[(Z_{(2)}-1)\,(\partial_\mu\varphi')^2 - \delta(\mu'^2)_{(2)}\,\varphi'^2\right] - \frac{1}{4!}\,(\delta\lambda')_{(2)}\,\varphi'^4\,, \qquad (52.73)$$

where the subscript (2) means that the correction terms are given to order λ'^2; and the corrected Lagrangian as [c.f. (52.65)]

$$\mathcal{L}'_{(2)} = \mathcal{L}' + \delta\mathcal{L}'_{(2)} = \frac{1}{2}\left[(\partial_\mu\varphi_{(2)})^2 - m^2_{(2)}\varphi^2_{(2)}\right] - \frac{1}{4!}\lambda_{(2)}\varphi^4_{(2)} . \tag{52.74}$$

We will also expand $\Sigma(k^2)$ around $k^2 = \mu'^2$ and write

$$\Sigma(k^2) = \Sigma(\mu'^2) + (k^2 - \mu'^2)\Sigma'(\mu'^2) + \tilde{\Sigma}(k^2) . \tag{52.75}$$

In (52.74), the unrenormalized mass $m_{(2)}$ and the unrenormalized coupling constant $\lambda_{(2)}$ are given by [c.f. (52.69)]

$$m^2_{(2)} = (\mu'^2 + \delta(\mu'^2)_{(2)})\, Z^{-1}_{(2)} , \qquad \lambda_{(2)} = (\lambda' + (\delta\lambda')_{(2)})\, Z^{-2}_{(2)} , \tag{52.76}$$

and the field renormalization constant $Z_{(2)}$ by [c.f. (52.22)]

$$Z_{(2)} = \frac{1}{1 - \Sigma'(\mu'^2)} . \tag{52.77}$$

In order to see explicitly how the counterterms can be chosen to cancel the divergent quantities it is more useful to write the corrected Lagrangian $\mathcal{L}'_{(2)}$ as

$$\mathcal{L}'_{(2)} = \frac{1}{2}\left[Z_{(2)}(\partial_\mu\varphi')^2 - (\mu'^2 + \delta(\mu'^2)_{(2)})\,\varphi'^2\right] - \frac{1}{4!}\left(\lambda' + (\delta\lambda')_{(2)}\right)\varphi'^4 . \tag{52.78}$$

This will be used to calculate the 2-point function $\Gamma^{(2)}(k^2)$ and the 4-point function $\Gamma^{(4)}(s, t, u)$. We first note that because of the $Z_{(2)}$ factor preceding $(\partial_\mu\varphi')^2$ in the kinetic term of the Lagrangian, the Feynman propagator for $\mathcal{L}'_{(2)}$ is

$$\tilde{D}(k^2) = \frac{1}{Z_{(2)}k^2 - (\mu'^2 + \delta(\mu'^2)_{(2)}) + i\epsilon^+} . \tag{52.79}$$

It follows from (52.16), (52.35) and (52.75) that

$$\Gamma^{(2)} = \frac{1}{\tilde{G}(k^2)}$$
$$= -i\left[Z_{(2)}k^2 - (\mu'^2 + \delta(\mu'^2)_{(2)}) - \Sigma(\mu'^2) - (k^2 - \mu'^2)\Sigma'(\mu'^2) - \tilde{\Sigma}(k^2)\right] . \tag{52.80}$$

We discover that in order for the infinite quantities $\Sigma(\mu'^2), \Sigma'(\mu'^2)$ and $Z_{(2)}$ to cancel out in the above expression, we need to set

$$-\delta(\mu'^2)_{(2)} = \mu'^2\left(1 - Z_{(2)}\right) + \Sigma(\mu'^2) . \tag{52.81}$$

Indeed, with this choice for $\delta(\mu'^2)_{(2)}$, we have

$$\Gamma^{(2)}(k^2) = -i\left[Z_{(2)}k^2 - Z_{(2)}\mu'^2 + \Sigma(\mu'^2) - \Sigma(\mu'^2) - (k^2 - \mu'^2)\Sigma'(\mu'^2) - \tilde{\Sigma}(k^2)\right]$$

$$= -iZ_{(2)}\left[(k^2 - \mu'^2) - (k^2 - \mu'^2)Z_{(2)}^{-1}\Sigma'(\mu'^2) - Z_{(2)}^{-1}\tilde{\Sigma}(k^2)\right]$$

$$= (-i)\left(1 - \Sigma'(\mu'^2)\right)^{-1}$$
$$\times \left[(k^2 - \mu'^2)\left\{1 - \left(1 - \Sigma'(\mu'^2)\right)\Sigma'(\mu'^2)\right\} - \left(1 - \Sigma'(\mu'^2)\right)\tilde{\Sigma}(k^2)\right]$$

$$\approx (-i)\left(1 - \Sigma'(\mu'^2)\right)^{-1}\left[(k^2 - \mu'^2)\left(1 - \Sigma'(\mu'^2)\right) - \left(1 - \Sigma'(\mu'^2)\right)\tilde{\Sigma}(k^2)\right]$$

$$= -i\left(k^2 - \mu'^2 - \tilde{\Sigma}(k^2)\right),$$

$$\tag{52.82}$$

which is completely finite.

The 4-point function $\Gamma^{(4)}(s,t,u)$, as calculated from the original (uncorrected) Lagrangian $\mathcal{L}'$, is given by (52.44) and (52.45) to be

$$\Gamma^{(4)}(s,t,u) = -i\lambda' + 3\Gamma(0) + \tilde{\Gamma}(s) + \tilde{\Gamma}(t) + \tilde{\Gamma}(u), \tag{52.83}$$

if we expand $\Gamma_{fish}(s,t,u)$ about the subtraction point $s = t = u = 0$:

$$\Gamma_{fish}(s,t,u) = 3\Gamma(0) + \tilde{\Gamma}(s) + \tilde{\Gamma}(t) + \tilde{\Gamma}(u), \tag{52.84}$$

where $\Gamma(k^2)$ is the amputated function for the fish diagram and $\Gamma(0)$ is its logarithmically divergent part. It readily follows that in order for $\Gamma(0)$ to be cancelled out, it is sufficient to choose in (52.78)

$$\delta\lambda' = -3i\Gamma(0). \tag{52.85}$$

As calculated from the corrected Lagrangian $\mathcal{L}'_{(2)}$, then, the 4-point function is given by

$$\Gamma^{(4)}(s,t,u) = -i\lambda' + \tilde{\Gamma}(s) + \tilde{\Gamma}(t) + \tilde{\Gamma}(u), \tag{52.86}$$

which is, again, completely finite.

To summarize, the three terms in the counterterm Lagrangian (52.73) serve the following cancellation purposes. The term proportional to $(\partial_\mu\varphi')^2$ cancels the *logarithmically divergent* part involving $\Sigma'(\mu'^2)$ in the Taylor expansion of the vertex function $\Sigma(k^2)$, while the term proportional to φ'^2 cancels the *quadratically divergent* part involving $\Sigma(\mu'^2)$ in the same Taylor expansion. These arise from the first three terms (diagrams) in the right-hand side of the equation depicted in Fig. 52.2. Lastly, the term proportional to φ'^4 cancels the logarithmically divergent contribution $3\Gamma(0)$ of the 4-point (one-loop) fish diagram to the Taylor expansion of the vertex function $\Gamma(k^2)$.

The general idea of the BPH procedure is to continue this cancellation scheme to higher loop orders (or higher orders in λ'). Thus the next step beyond the order λ'^2 would be to use the new Lagrangian $\mathcal{L}' + \delta\mathcal{L}'_{(2)}$ to calculate vertex functions to order λ'^3. The infinities encountered will then need to be cancelled by the next higher-order counterterm Lagrangian, and so on. The details, depending on the peculiarities of the higher-order Feynman diagrams, such as overlapping or nested divergent subgraphs, are rather complicated; and will not be presented here. The upshot is that divergences can be eliminated order by order, in such a way that in the limit of infinitely high order (in λ' or number of loops), the resulting vertex functions or Green's functions for the theory are independent of the momentum cutoff Λ, as $\Lambda \to \infty$. The mathematical statement guaranteeing the validity of the BPH procedure is known as **Hepp's Theorem**. A proof of this theorem is quite beyond the scope of this text. The interested reader can consult, for example, the relevant articles in Coleman (1985), and the references to original sources given there.

Problem 52.3 Show that if we had expanded $\Sigma(k^2)$ around $k^2 = 0$ instead of $k^2 = \mu'^2$ as in (52.75), the term $\delta(\mu'^2)_{(2)}$ in the counterterm Lagrangian $\delta\mathcal{L}'_{(2)}$ would have to be given by

$$-\delta(\mu'^2)_{(2)} = \mu'^2 \left[1 - Z_{(2)} + \Sigma'(0)\right] + \Sigma(0), \tag{52.87}$$

in order for $\Gamma^{(2)}(k^2)$ to be the finite quantity given by the last expression in (52.82).

Problem 52.4 Consider the φ^5-theory with the Lagrangian

$$\mathcal{L} = \frac{1}{2}\left[(\partial_\mu\varphi)^2 - m^2\varphi^2\right] - \lambda\frac{\varphi^5}{5!}. \tag{52.88}$$

For a Feynman diagram with n external lines, I internal lines, V vertices and L loops, show that the superficial degree of divergence D is given by

$$D = V + 4 - n. \tag{52.89}$$

This result indicates that D increases with the order of the perturbation and implies that the theory is non-renormalizable. (*Hint:* The *rule of conservation of boson ends*, which in the case of the φ^4-theory is expressed by $4V = 2I + n$, is now expressed by $5V = 2I + n$.)

Chapter 53

Regularization Schemes for Feynman Integrals

In this chapter we will present two schemes for regularizing infinite integrals that appear in the calculations for one-particle irreducible (1PI) Feynman diagrams (those with superficial degree of divergence $D \geq 0$). There are two commonly used schemes. The first one is known as **Pauli-Villars** or **covariant regularization** and the second one as **dimensional regularization**. Both schemes must be carried out in such a way to preserve the Lorentz invariance and other symmetries inherent in the Lagrangian of the theory. Our presentation will be based on that of Cheng and Li (1984).

In the Pauli-Villars regularization scheme one begins by replacing the free-particle Feynman propagator by one that depends explicitly on a momentum cutoff $\Lambda \gg \mu$ (μ being the physical mass) and at the same time increases the power of the momentum in the denominator. The new propagator must also approach the original one as $\Lambda \to \infty$. See, for example, the replacement shown in (52.62). More generally, one can modify the free propagator as follows:

$$\frac{1}{l^2 - \mu^2 + i\epsilon^+} \longrightarrow \frac{1}{l^2 - \mu^2 + i\epsilon^+} + \sum_i \frac{a_i}{l^2 - \Lambda_i^2 + i\epsilon^+}, \tag{53.1}$$

where $\Lambda_i^2 \gg \mu^2$ and the constants a_i are chosen so that, with the new propagator, a Feynman integral will have sufficient powers of the integrated momentum l in the denominator to make the integral convergent. We will illustrate this process with the vertex functions $\Gamma(k^2)$ [c.f. (52.37) for the fish diagram] and $\Sigma(k^2)$ [c.f. (52.3)] for the tadpole diagram for the φ^4-theory, introduced in the last chapter.

Note that in this chapter, in the spirit of the BPH renormalization procedure, all calculations are begun with the finite physical mass μ and the finite physical coupling constant, which we now denote by λ. This may cause some confusion with the notation used in the last chapter, where λ denoted the bare coupling constant instead, and m the bare mass. Hopefully the reader will discern the advantage of this seemingly confusing choice as we proceed. In the meantime, for ease of reference, we list below the correspondences between the notation used in the last chapter and that used in this one, where the left column (right column) contains the symbols used in Chapter 52 (Chapter 53):

λ (bare coupling constant)   does not appear

m (bare mass)   does not appear

λ_R (renormalized coupling constant) λ (physical coupling constant)

μ (physical mass) μ (physical mass)

λ' (intermediate coupling constant)   chosen to be λ

μ' (intermediate mass)   chosen to be μ

Consider $\Gamma(k^2)$ first, whose equation, given by (52.37), will be reproduced here for convenience:

$$\Gamma(k^2) = \frac{(-i\lambda)^2}{2} \int \frac{d^4l}{(2\pi)^4} \frac{i}{l^2 - \mu^2 + i\epsilon^+} \frac{i}{(l-k)^2 - \mu^2 + i\epsilon^+}. \tag{53.2}$$

This integral is clearly logarithmically divergent by simple power counting. So any increase in the power of l in the denominator of the integrand would render the integral convergent. We use the following implementation of (53.1):

$$\frac{1}{l^2 - \mu^2 + i\epsilon^+} \longrightarrow$$
$$\frac{1}{l^2 - \mu^2 + i\epsilon^+} - \frac{1}{l^2 - \Lambda^2 + i\epsilon^+} = \frac{\mu^2 - \Lambda^2}{(l^2 - \mu^2 + i\epsilon^+)(l^2 - \Lambda^2 + i\epsilon^+)}. \tag{53.3}$$

Clearly, the power of the integrated momentum l has been increased by two in the denominator of the new propagator, which also approaches the original one as $\Lambda \to \infty$. Since $\Lambda^2 \gg \mu^2$ by assumption, we can drop μ^2 in the numerator of the new propagator and write

$$\Gamma(k^2) = -\frac{\lambda^2 \Lambda^2}{2} \int \frac{d^4l}{(2\pi)^4} \frac{1}{\{(l-k)^2 - \mu^2 + i\epsilon^+\}(l^2 - \mu^2 + i\epsilon^+)(l^2 - \Lambda^2 + i\epsilon^+)}. \tag{53.4}$$

If we expand the above in a Taylor series about $k^2 = 0$ (or make a subtraction at $k^2 = 0$) we have

$$\Gamma(k^2) = \Gamma(0) + \tilde{\Gamma}(k^2), \tag{53.5}$$

with

$$\Gamma(0) = -\frac{\lambda^2 \Lambda^2}{2} \int \frac{d^4l}{(2\pi)^4} \frac{1}{(l^2 - \mu^2 + i\epsilon^+)^2(l^2 - \Lambda^2 + i\epsilon^+)}. \tag{53.6}$$

It follows from (53.2) that

$$\tilde{\Gamma}(k^2) = \Gamma(k^2) - \Gamma(0)$$

$$= -\frac{\lambda^2 \Lambda^2}{2} \int \frac{d^4 l}{(2\pi)^4} \frac{1}{(l^2 - \mu^2 + i\epsilon^+)(l^2 - \Lambda^2 + i\epsilon^+)}$$

$$\times \left\{ \frac{1}{(l-k)^2 - \mu^2 + i\epsilon^+} - \frac{1}{l^2 - \mu^2 + i\epsilon^+} \right\}$$

$$= -\frac{\lambda^2 \Lambda^2}{2} \int \frac{d^4 l}{(2\pi)^4} \frac{1}{(l^2 - \mu^2 + i\epsilon^+)(l^2 - \Lambda^2 + i\epsilon^+)}$$

$$\left[\frac{2l \cdot k - k^2}{\{(l-k)^2 - \mu^2 + i\epsilon^+\}(l^2 - \mu^2 + i\epsilon^+)} \right]$$

$$= -\frac{\lambda^2}{2} \int \frac{d^4 l}{(2\pi)^4} \frac{2l \cdot k - k^2}{(l^2 - \mu^2 + i\epsilon^+)^2 \{(l-k)^2 - \mu^2 + i\epsilon^+\}} \left(\frac{\Lambda^2}{l^2 - \Lambda^2 + i\epsilon^+} \right)$$

$$= \frac{\lambda^2}{2} \int \frac{d^4 l}{(2\pi)^4} \frac{2l \cdot k - k^2}{(l^2 - \mu^2 + i\epsilon^+)^2 \{(l-k)^2 - \mu^2 + i\epsilon^+\}} ,$$

$$(53.7)$$

where in the left-hand side of the last equality we have taken the limit $\Lambda \to \infty$ of the quantity within the brackets. Note that $D = -1$ for this integral, which makes it convergent.

A useful identity to evaluate convergent Feynman integrals is the following identity:

$$\frac{1}{a_1 a_2 \dots a_n} = (n-1)! \int_0^1 dx_1 \dots \int_0^1 dx_n \frac{\delta \left(1 - \sum_{i=1}^n x_i \right)}{(a_1 x_1 + a_2 x_2 + \dots + a_n x_n)^n} \quad (n > 1).$$

$$(53.8)$$

This can be differentiated with respect to a_1, for example, to yield

$$\frac{1}{a_1^2 a_2 \dots a_n} = n! \int_0^1 dx_1 \dots \int_0^1 dx_n \frac{x_1 \delta \left(1 - \sum_{i=1}^n x_i \right)}{(a_1 x_1 + a_2 x_2 + \dots + a_n x_n)^{n+1}} \quad (n > 1),$$

$$(53.9)$$

with a similar formula for other a_i's ($i \neq 1$) that are squared on the left-hand side. In the above formulas, the quantities a_i are understood to be $a_i + i\epsilon^+$ (a_i real), and the quantities $x_1, \dots, x_n$ are called the **Feynman parameters**. For the proof of (53.8) the reader is referred to Problem 53.1 below.

Problem 53.1 Prove (53.8) by following the steps below.

(a) Verify that

$$\frac{1}{a + i\epsilon^+} = -i \int_0^\infty dt \, e^{it(a + i\epsilon^+)}.$$

$$(53.10)$$

So it follows that

$$\frac{1}{(a_1 + i\epsilon^+) \dots (a_n + i\epsilon^+)} = (-i)^n \int_0^\infty dt_1 \dots \int_0^\infty dt_n \exp \left\{ i \sum_{i=1}^n t_i (a_i + i\epsilon^+) \right\}.$$

$$(53.11)$$

(b) Use the fact

$$\delta(f(x)) = \sum_n \frac{\delta(x - x_n)}{|f'(x_n)|} , \qquad (53.12)$$

where x_n are the roots of $f(x) = 0$, to show that

$$\delta\left[1 - \frac{1}{t}(t_1 + \cdots + t_n)\right] = t\,\delta[t - (t_1 + \cdots + t_n)] , \qquad (53.13)$$

where t is the variable in the arguments of the delta functions. Hence show that

$$1 = \int_0^\infty \frac{dt}{t}\,\delta\left[1 - \frac{1}{t}(t_1 + \cdots + t_n)\right] . \qquad (53.14)$$

(c) Let $t_i = tx_i$ in (53.11) and insert the factor 1 [as given by (53.14)] in the right-hand side. Show that the equation can be written as

$$\frac{1}{(a_1 + i\epsilon^+)\ldots(a_n + i\epsilon^+)} = (-i)^n \int_0^\infty dx_1 \ldots \int_0^\infty dx_n\,\delta\left(1 - \sum_{i=1}^n x_i\right)$$

$$\times \int_0^\infty dt\,t^{n-1}\,\exp\left\{it\left(\sum_{i=1}^n x_i(a_i + i\epsilon^+)\right)\right\} .$$

$$(53.15)$$

(d) Use the fact for the Gamma function

$$\Gamma(n) = (n - 1)! = \int_0^\infty dt\,t^{n-1}e^{-t} \qquad (53.16)$$

to show that

$$\frac{(n - 1)!}{z^n} = \int_0^\infty dt\,t^{n-1}e^{-tz} , \qquad (53.17)$$

provided $Re(z) > 0$. Finally let $z = -i\sum_{i=1}^n x_i(a_i + i\epsilon^+)$ in (53.15) and use the delta function there to limit the range of integration for each x_i between 0 and 1 to complete the proof.

Using (53.9) the denominator factors in the integrand of (53.7) can be expressed as

$$\frac{1}{(l^2 - \mu^2 + i\epsilon^+)^2}\,\frac{1}{(l - k)^2 - \mu^2 + i\epsilon^+}$$

$$= 2\int_0^1 dx_1 \int_0^1 dx_2 \frac{x_1\delta(1 - (x_1 + x_2))}{[(l^2 - \mu^2 + i\epsilon^+)x_1 + \{(l - k)^2 - \mu^2 + i\epsilon^+\}x_2]^3}$$

$$= 2\int_0^1 dx_2 \frac{1 - x_2}{[(l^2 - \mu^2 + i\epsilon^+)(1 - x_2) + \{(l - k)^2 - \mu^2 + i\epsilon^+\}x_2]^3} \qquad (53.18)$$

$$= 2\int_0^1 dx \frac{1 - x}{[(l^2 - \mu^2 + i\epsilon^+)(1 - x) + \{(l - k)^2 - \mu^2 + i\epsilon^+\}x]^3}$$

$$= 2\int_0^1 dx \frac{1 - x}{[(l - xk)^2 - a^2(x, k^2) + i\epsilon^+]^3} ,$$

where

$$a^2(x, k^2) \equiv \mu^2 - x(1-x)k^2 . \tag{53.19}$$

Then, by (53.7), we have

$$\tilde{\Gamma}(k^2) = \lambda^2 \int \frac{d^4 l}{(2\pi)^4} \int_0^1 dx \, \frac{(1-x)(2l \cdot k - k^2)}{[(l-xk)^2 - a^2(x, k^2) + i\epsilon^+]^3}$$

$$= \lambda^2 \int_0^1 dx \, (1-x) \int \frac{d^4 l'}{(2\pi)^4} \frac{2(l'+xk) \cdot k - k^2}{[l'^2 - a^2(x, k^2) + i\epsilon^+]^3} \tag{53.20}$$

$$= \lambda^2 \int_0^1 dx \, (1-x) \int \frac{d^4 l}{(2\pi)^4} \frac{2l \cdot k + 2xk^2 - k^2}{[l^2 - a^2(x, k^2) + i\epsilon^+]^3} ,$$

where in the second equality we have changed variables of integration and let
$l' = l - xk$. In the integral above the term $2l \cdot k$ does not contribute upon
integration over the entire 4-dimensional l-space (because of its odd parity).
We then have

$$\tilde{\Gamma}(k^2) = \lambda^2 k^2 \int_0^1 dx \, (1-x)(2x-1) \int \frac{d^4 l}{(2\pi)^4} \frac{1}{[l^2 - a^2(x, k^2) + i\epsilon^+]^3} . \tag{53.21}$$

The l-integral here is much simpler than in (53.7); and can be evaluated readily
by the *Wick rotation* process that was introduced in Chapter 46. In the present
situation, this consists in treating the time-component of l, l^0, as a complex
variable and rotating the contour of the l^0 integration along the real axis ($-\infty <
l^0 < \infty$) by $\pi/2$ in the counter clockwise sense (see Fig. 53.1). (Recall that doing
the time-component integration first by complexification of that component in
an integration over the entirety of Minskowski spacetime was the same technique
used in the evaluation of the Feynman propagator in Chapter 48.) In the contour
$\mathcal{C}$ shown in Fig. 53.1, the portions along the two (infinite radius) quarter-circular
arcs do not contribute, as can be readily seen by power-counting of l^0 [the
numerator $\sim dl^0$ while the denominator $\sim (l^0)^6$]. On the other hand, the factor
$l^2 - a^2 + i\epsilon^+$ in the integrand of the $d^4 l$ integral in (53.21) can be written as

$$l^2 - a^2 + i\epsilon^+ = (l^0)^2 - (l_1^2 + l_2^2 + l_3^2) - a^2 + i\epsilon^+ = (l^0)^2 - (\mathbf{l}^2 + a^2 - i\epsilon^+)$$

$$= (l^0)^2 - [(\mathbf{l}^2 + a^2)^{1/2} - i\epsilon^+]^2 , \tag{53.22}$$

where $\mathbf{l}^2 = l_1^2 + l_2^2 + l_3^2$. Thus the poles in the dl^0 integration are

$$l^0 = \pm[(\mathbf{l}^2 + a^2)^{1/2} - i\epsilon^+] , \tag{53.23}$$

as indicated in Fig. 53.1. We thus see that the contour $\mathcal{C}$ does not enclose any
poles, and so, by the *residue theorem*,

$$\int_{\mathcal{C}} dl^0 = \int_{-\infty}^{\infty} dl^0 + \int_{i\infty}^{-i\infty} dl^0 = 0 . \tag{53.24}$$

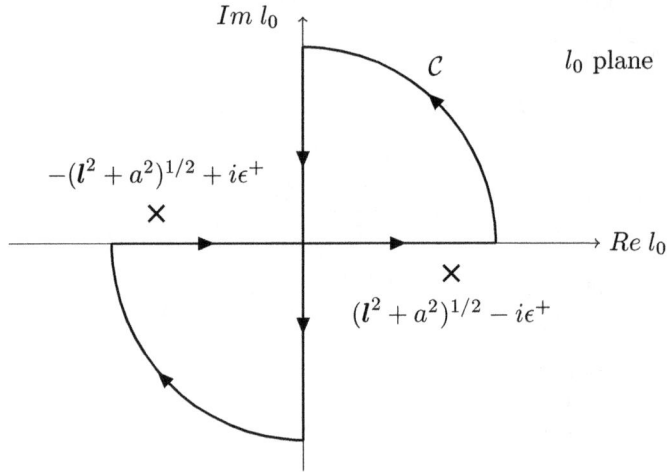

FIGURE 53.1

This implies

$$\int_{-\infty}^{\infty} dl^0 = -\int_{i\infty}^{-i\infty} dl^0 = \int_{-i\infty}^{i\infty} dl^0 \,. \tag{53.25}$$

We now introduce a real variable l_4 and let $l^0 = il_4$ for values of l^0 on the imaginary l^0-axis. Thus

$$\int_{-\infty}^{\infty} dl^0 = i \int_{-\infty}^{\infty} dl_4 \,, \tag{53.26}$$

and

$$l^2 - a^2 + i\epsilon^+ = -(l^2 + l_4^2 + a^2 - i\epsilon^+) = -(l_1^2 + l_2^2 + l_3^2 + l_4^2 + a^2 - i\epsilon^+) \,. \tag{53.27}$$

The d^4l integral in (53.21) can be written

$$\int \frac{d^4l}{(2\pi)^4} \frac{1}{(l^2 - a^2 + i\epsilon^+)^3} = -i \int_{-\infty}^{\infty} \frac{dl_1 dl_2 dl_3 dl_4}{(2\pi)^4} \frac{1}{[(l_1^2 + l_2^2 + l_3^2 + l_4^2) + a^2 - i\epsilon^+]^3} \,, \tag{53.28}$$

which is an integral over all of 4-dimensional Euclidean space. Since the integrand only depends on the length squared L^2 of the 4-Euclidean vector $\boldsymbol{L} = (l_1, l_2, l_3, l_4)$, it is convenient to use polar coordinates and the general formula for the case of the n-dimensional Euclidean space

$$\int_{-\infty}^{\infty} dl_1 \ldots dl_n \, f(L) = nC_n \int_0^{\infty} dL \, L^{n-1} f(L) \,, \tag{53.29}$$

where

$$C_n = \frac{\pi^{n/2}}{\Gamma(n/2+1)} , \tag{53.30}$$

with $\Gamma(z)$ being the **gamma function**, a function of the complex variable z, defined by [c.f. (53.16)]

$$\Gamma(z) = \int_0^\infty dt\, t^{z-1} e^{-t} , \tag{53.31}$$

and $L = |\mathbf{L}| = (l_1^2 + \cdots + l_n^2)^{1/2}$ (see Problem 53.2). With $C_4 = \pi^2/\Gamma(3) = \pi^2/2$ we have

$$\int \frac{d^4 l}{(2\pi)^4} \frac{1}{(l^2 - a^2 + i\epsilon^+)^3} = \frac{(-i)(2\pi^2)}{(2\pi)^4} \int_0^\infty dL \frac{L^3}{(L^2 + a^2 - i\epsilon^+)^3}$$

$$= \frac{-i}{16\pi^2} \int_0^\infty d(L^2) \frac{L^2}{(L^2 + a^2 - i\epsilon^+)^3} = \frac{-i}{16\pi^2} \int_0^\infty dt \frac{t}{(t + a^2 - i\epsilon^+)^3} . \tag{53.32}$$

While this integral can be evaluated by elementary means (for $a^2 > 0$), we will do it by a more elegant method through the use of the **beta function** $B(r,s)$ (a function of two complex variables r and s) defined by

$$B(r,s) \equiv \frac{\Gamma(r)\Gamma(s)}{\Gamma(r+s)} . \tag{53.33}$$

For $Re(r) > 0$ and $Re(s) > 0$, it can be shown, using (53.31), that (see Problem 53.3)

$$B(r,s) = \int_0^1 dt\, t^{r-1}(1-t)^{s-1} . \tag{53.34}$$

The property of this function that we need for the evaluation of the dt integral in (53.32) is (see Problem 53.4)

$$\int_0^\infty dt \frac{t^{m-1}}{(t+a^2)^n} = \frac{1}{(a^2)^{n-m}} B(m, n-m) \qquad (a^2 > 0) . \tag{53.35}$$

By setting $m = 2$ and $n = 3$, and using the factorial property of the gamma function that $\Gamma(n) = (n-1)!$ [recall (53.16)], we have

$$\int \frac{d^4 l}{(2\pi)^4} \frac{1}{(l^2 - a^2 + i\epsilon^+)^3} = \frac{-i}{16\pi^2(a^2 - i\epsilon^+)} \frac{\Gamma(2)\Gamma(1)}{\Gamma(3)} = \frac{-i}{32\pi^2(a^2 - i\epsilon^+)} . \tag{53.36}$$

Thus, on recalling the definition of a^2 given by (53.19), it follows from (53.21) that

$$\tilde{\Gamma}(k^2) = \frac{(-i)\lambda^2 k^2}{32\pi^2} \int_0^1 dx \frac{(1-x)(2x-1)}{\mu^2 - x(1-x)k^2} \qquad (k^2 < 4\mu^2) . \tag{53.37}$$

In the above expression the $-i\epsilon^+$ has been dropped from the denominator of the integrand (provided $k^2 < 4\mu^2$) since the latter does not have any poles for

$0 < x < 1$. Indeed, the maximum value for $x(1-x)$ in this range is $1/4$; and so, as long as $\mu^2 > k^2/4$, the denominator does not vanish. This condition is true for any of the three Mandelstam variables $k^2 = s, t$ or u, with the external momenta being on mass-shell [c.f. (52.36) and (52.38)]. We will not give the analytical expression for the integral in (53.37), which can be evaluated by elementary means. Suffice it to say that it is clearly finite, which corroborates the fact that $\tilde{\Gamma}(k^2)$ is finite. [*Note:* The reader should beware of confusing the gamma function $\Gamma(z)$ with the vertex function $\Gamma(k^2)$. The context of the presentation should make it clear which one is meant.]

Problem 53.2 Prove (53.29) and (53.30) by following the steps below.

(a) Consider the volume $\Omega_n(L)$ of an n-sphere of radius L given by:

$$\Omega_n(L) = \int_{l_1^2 + \cdots + l_n^2 < L^2} dl_1 \ldots dl_n . \tag{53.38}$$

Justify that

$$\Omega_n(L) = C_n L^n , \tag{53.39}$$

where C_n is a constant. Also justify that the surface area $S_n(L)$ of the n-sphere is given by

$$S_n(L) = \frac{d\Omega_n(L)}{dL} . \tag{53.40}$$

(b) Consider the following Gaussian-integral identity

$$\int_{-\infty}^{\infty} dl_1 \ldots \int_{-\infty}^{\infty} dl_n \, e^{-(l_1^2 + \cdots + l_n^2)} = \left(\int_{-\infty}^{\infty} dx \, e^{-x^2} \right)^n = \pi^{n/2} . \tag{53.41}$$

Justify that the left-hand side can be written

$$\int_{-\infty}^{\infty} dl_1 \ldots \int_{-\infty}^{\infty} dl_n \, e^{-(l_1^2 + \cdots + l_n^2)} = \int_0^{\infty} dL \, S_n(L) e^{-L^2} . \tag{53.42}$$

(c) Let $L^2 = t$ on the right-hand side of (53.16), and use (53.39) and (53.40) to show that

$$\int_{-\infty}^{\infty} dl_1 \ldots \int_{-\infty}^{\infty} dl_n \, e^{-(l_1^2 + \cdots + l_n^2)} = nC_n \int_0^{\infty} dL \, L^{n-1} e^{-L^2} = \frac{1}{2} nC_n \Gamma(n/2) . \tag{53.43}$$

(d) Use (53.41) to calculate C_n and conclude the proof. *Hint:* Use the following recursion relation for the gamma function,

$$\Gamma(z+1) = z\Gamma(z) . \tag{53.44}$$

Problem 53.3 Use the definition of the gamma function given by (53.31) to verify the integral representation of the beta function given by (53.34). *Hints:* First write

$$\Gamma(r)\Gamma(s) = \int_0^{\infty} dx \, x^{r-1} e^{-x} \int_0^{\infty} dy \, y^{s-1} e^{-y} . \tag{53.45}$$

Then let $x + y = u$ to obtain

$$\Gamma(r)\Gamma(s) = \int_0^\infty du \int_0^u dx\, x^{r-1}(u-x)^{s-1}e^{-u} . \qquad (53.46)$$

Finally let $x = ut$ to write

$$\Gamma(r)\Gamma(s) = \int_0^\infty du\, e^{-u} u^{r+s-1} \int_0^1 dt\, t^{r-1}(1-t)^{s-1} . \qquad (53.47)$$

$\boxed{\textbf{Problem 53.4}}$ Use the integral representation of the beta function given by (53.34) to verify the property (53.35). *Hints:* First show that

$$(a^2)^{n-m} \int_0^\infty dt\, \frac{t^{m-1}}{(t+a^2)^n} = \int_0^\infty dx\, \frac{x^{m-1}}{(1+x)^n} . \qquad (53.48)$$

Then use (53.34) for $B(m, n-m)$ to write

$$B(m, n-m) = \int_0^1 dt\, t^{m-1}(1-t)^{n-m-1} , \qquad (53.49)$$

and change the integration variable by letting $t = x/(1+x)$.

Exactly the same integration procedures that have been used above for $\tilde{\Gamma}(k^2)$ [the finite part of $\Gamma(k^2)$] can also be used for $\Gamma(0)$ (the infinite, cutoff-dependent part) [given by (53.6)]. These involve, again, the use of (53.9), the Wick rotation technique, and the use of the beta-function property (53.35). The details will not be shown here (but will be left as an exercise in Problem 53.5). In analogy to (53.37) for $\tilde{\Gamma}(k^2)$ we wind up with

$$\Gamma(0) = \frac{i\lambda^2\Lambda^2}{32\pi^2} \int_0^1 dx\, \frac{1-x}{\mu^2 - x(\mu^2 - \Lambda^2)} . \qquad (53.50)$$

On changing the variable of integration to $x' = 1 - x$, and rewriting x' as x again, we find

$$\Gamma(0) = \frac{i\lambda^2\Lambda^2}{32\pi^2} \int_0^1 dx\, \frac{x}{x(\mu^2 - \Lambda^2) + \Lambda^2} . \qquad (53.51)$$

The integral is readily evaluated to be

$$\left(\frac{1}{\mu^2 - \Lambda^2}\right)\left(1 + \frac{\Lambda^2}{\mu^2 - \Lambda^2}\ln\frac{\Lambda^2}{\mu^2}\right) \xrightarrow[\Lambda\to\infty]{} \frac{1}{\Lambda^2}\ln\frac{\Lambda^2}{\mu^2} . \qquad (53.52)$$

Thus

$$\Gamma(0) \xrightarrow[\Lambda\to\infty]{} \frac{i\lambda^2}{32\pi^2}\ln\frac{\Lambda^2}{\mu^2} , \qquad (53.53)$$

which, indeed, diverges logarithmically. In analogy to (52.45) we find that

$$\Gamma_{fish\,(1\,loop)}(s, t, u) = 3\Gamma(0) + \tilde{\Gamma}(s) + \tilde{\Gamma}(t) + \tilde{\Gamma}(u) , \qquad (53.54)$$

where s, t, u are the Mandelstam variables given by (52.36), and $\Gamma(0)$ and $\tilde{\Gamma}(k^2 = s, t, u)$ are given by (53.53) and (53.37), respectively. The difference between this formula and (52.45) is that here the Taylor series for $\Gamma_{fish}(k^2)$ is expanded around $s, t, u = 0$ rather than $s, t, u = s_0 = t_0 = u_0 = 4\mu^2/3$ [c.f. (52.40)].

Analogous to (52.44) and (52.45) we have, up to one-loop contributions,

$$\Gamma_{(one\,loop)}(s, t, u) = -i\lambda + 3\Gamma(0) + \tilde{\Gamma}(s) + \tilde{\Gamma}(t) + \tilde{\Gamma}(u) . \qquad (53.55)$$

The inclusion of the term proportional to φ'^4 in the counterterm Lagrangian (52.80) has the effect of replacing λ in the above equation by $\lambda - 3i\Gamma(0)$. We thus obtain the renormalized vertex function [to $O(\lambda^2)$]

$$\overline{\Gamma}_{(one\,loop)}(s, t, u) = -i\lambda + \tilde{\Gamma}(s) + \tilde{\Gamma}(t) + \tilde{\Gamma}(u) , \qquad (53.56)$$

which is completely finite [the infinite cutoff-dependent part $3\Gamma(0)$ in (53.55) has been cancelled]. Analogous to (52.49) the terms $-i\lambda + 3\Gamma(0)$ in the right-hand side of (53.55) can be lumped into the renormalization constant Z_λ as follows:

$$-iZ_\lambda^{-1}\lambda = -i\lambda + 3\Gamma(0) . \qquad (53.57)$$

It follows from (53.53) that

$$Z_\lambda^{-1} = 1 + \frac{3i\Gamma(0)}{\lambda} = 1 - \frac{3\lambda}{32\pi^2} \ln \frac{\Lambda^2}{\mu^2} . \qquad (53.58)$$

We will next look at the vertex function $\Sigma(k^2)$ due only to the (one-loop) tadpole diagram (first diagram on the right-hand side of the equation shown in Fig. 52.2), whose contribution is given by [c.f. (52.3)]

$$i\Sigma_{(one\,loop)}(k^2) = -\frac{i\lambda}{2} \int \frac{d^4l}{(2\pi)^4} \frac{i}{l^2 - \mu^2 + i\epsilon^+} . \qquad (53.59)$$

It is clear that this quantity is independent of k, and that the integral in it is quadratically divergent. According to the Pauli-Villars prescription in (53.1), we will make two additions to the propagator in the integrand in an attempt to decrease the superficial degree of divergence D of the integral:

$$\frac{1}{l^2 - \mu^2 + i\epsilon^+} \longrightarrow \frac{1}{l^2 - \mu^2 + i\epsilon^+} + \frac{a_1}{l^2 - \Lambda_1^2 + i\epsilon^+} + \frac{a_2}{l^2 - \Lambda_2^2 + i\epsilon^+} . \qquad (53.60)$$

On choosing

$$a_1 = \frac{\mu^2 - \Lambda_2^2}{\Lambda_2^2 - \Lambda_1^2} , \qquad a_2 = -\frac{\mu^2 - \Lambda_1^2}{\Lambda_2^2 - \Lambda_1^2} , \qquad (53.61)$$

it is readily seen that the replacement in (53.60) becomes

$$\frac{1}{l^2 - \mu^2 + i\epsilon^+} \longrightarrow \frac{(\Lambda_1^2 - \mu^2)(\Lambda_2^2 - \mu^2)}{(l^2 - \mu^2 + i\epsilon^+)(l^2 - \Lambda_1^2 + i\epsilon^+)(l^2 - \Lambda_2^2 + i\epsilon^+)}$$

$$\longrightarrow \frac{\Lambda^4}{(l^2 - \mu^2 + i\epsilon^+)(l^2 - \Lambda^2 + i\epsilon^+)^2} , \qquad (53.62)$$

where in the last expression both Λ_1 and Λ_2 have been allowed to approach the large quantity $\Lambda \gg \mu$. This is $O(l^{-6})$, and so makes $D = -2$ for the integral in (53.59). Thus

$$i\Sigma_{(one\,loop)}(k^2) = \frac{\lambda\Lambda^4}{2} \int \frac{d^4l}{(2\pi)^4} \frac{1}{(l^2 - \Lambda^2 + i\epsilon^+)^2 (l^2 - \mu^2 + i\epsilon^+)} \cdot \qquad (53.63)$$

As for $\tilde{\Gamma}(k^2)$ and $\Gamma(0)$, exactly the same integration techniques can be used here; and we find

$$i\Sigma_{(one\,loop)}(k^2) = -\frac{i\lambda\Lambda^4}{32\pi^2} \int_0^1 dx \frac{x}{\mu^2 + (\Lambda^2 - \mu^2)x} \cdot \qquad (53.64)$$

Working out the integral we have

$$i\Sigma_{(one\,loop)}(k^2) = -\frac{i\lambda\Lambda^4}{32\pi^2} \left\{ \frac{1}{\Lambda^2 - \mu^2} - \frac{\mu^2}{(\Lambda^2 - \mu^2)^2} \ln \frac{\Lambda^2}{\mu^2} \right\}$$
$$\xrightarrow[\Lambda\to\infty]{} -\frac{i\lambda}{32\pi^2} \left\{ \Lambda^2 - \mu^2 \ln \frac{\Lambda^2}{\mu^2} \right\} \cdot \qquad (53.65)$$

Retaining only the dominant term in Λ we finally obtain

$$\Sigma_{(one\,loop)}(k^2) = \Sigma_{(one\,loop)}(0) = -\frac{\lambda}{32\pi^2}\Lambda^2 , \qquad (53.66)$$

which is indeed quadratically divergent. Since $\Sigma_{(one\,loop)}(k^2)$ is k^2-independent, we also have

$$\Sigma'_{(one\,loop)}(0) = 0 . \qquad (53.67)$$

It follows from (52.76) that, up to the one-loop approximation, the wavefunction renormalization constant is given by

$$Z_\varphi = 1 \qquad \text{(one-loop approximation)} . \qquad (53.68)$$

To summarize the results of our regularization procedure so far, the renormalization constants to be used in the counterterm Lagrangian, up to the one-loop approximation, are given by [c.f. (52.70) and (52.80)]

$$Z_\varphi^{-1} - 1 = 1 - \Sigma'(0) - 1 = 0 , \qquad (53.69)$$

$$\delta\mu^2 = -\Sigma(0) = \frac{\lambda}{32\pi^2}\Lambda^2 , \qquad (53.70)$$

$$\delta\lambda = -3i\Gamma(0) = \frac{3\lambda^2}{32\pi^2} \ln \frac{\Lambda^2}{\mu^2} . \qquad (53.71)$$

By (52.26) the renormalized two-point amputated Green's function is given by

$$\bar{\Gamma}^{(2)}(k^2) = \frac{1}{\tilde{G}_R(k^2)} = -i(k^2 - \mu^2 + \tilde{\Sigma}(k^2)) . \qquad (53.72)$$

From the Taylor expansion (52.75) of $\Sigma(k^2)$ around $k^2 = 0$, we have, in analogy to (52.15),

$$\tilde{\Sigma}(0) = 0 \,, \qquad \tilde{\Sigma}'(0) = \left.\frac{d\tilde{\Sigma}(k^2)}{dk^2}\right|_{k^2=0} = 0 \,. \qquad (53.73)$$

Then (53.72) implies

$$\overline{\Gamma}^{(2)}(0) = i\mu^2 \,, \qquad \left.\frac{d}{dk^2}\overline{\Gamma}^{(2)}(k^2)\right|_{k^2=0} = -i - i\tilde{\Sigma}'(0) = -i \,. \qquad (53.74)$$

On the other hand, the renormalized four-point function result (53.56) implies

$$\overline{\Gamma}^{(4)}(s = 0,\, t = 0,\, u = 0) = -i\lambda \,. \qquad (53.75)$$

The conditions stated in the above two equations are in conformity with those in (52.72). It is important to note again that similar sets of conditions corresponding to different choices of expansion points for k^2 in the Taylor expansions for $\Sigma(k^2)$ and $\Gamma(k^2)$ will give different values for the renormalization constants. This is a crucial point for an appreciation of the importance of the *renormalization group*, to be introduced in the next chapter.

Problem 53.5 Verify the results (53.50) and (53.64) for $\Gamma(0)$ and $\Sigma_{(one\ loop)}(k^2)$, respectively, using the same integration procedure described in the text to evaluate $\tilde{\Gamma}(k^2)$.

We will now proceed to discuss a second method of regularization, known as **dimensional regularization**. In this method, the dimension n of the Minkowski space over which an independent internal momentum is integrated is allowed to deviate from four and to assume, at first, an arbitrary positive-integer value. This is done in order to take advantage of the fact that the superficial degree of divergence D of a Feynman integral clearly depends on n: the smaller n becomes, the more convergent the integral will be. No momentum cutoffs Λ need to be introduced here, whose assumption of large values will lead to infinities in vertex functions. Instead, it will be shown that these infinities will eventually appear as poles of the vertex functions, when they are considered as functions of the complex variable n. We will illustrate this procedure first with the vertex function $\Gamma(k^2)$, given by (53.2). Thus we write

$$\Gamma(k^2; n) = \frac{\lambda^2}{2(2\pi)^4}\int d^n l \, \frac{1}{l^2 - \mu^2 + i\epsilon^+}\frac{1}{(l-k)^2 - \mu^2 + i\epsilon^+} \,. \qquad (53.76)$$

In the integrand, the integrated internal momentum l will have n non-vanishing components in general: $l^\mu = (l^0, l^1, \ldots, l^{n-1})$, while the external momentum k

will have vanishing components beyond the first four: $k^\mu = (k^0, k^1, k^2, k^3, 0, \ldots, 0)$. So $d^n l = dl^0 dl^1 \ldots dl^{n-1}$. Our objective is to first evaluate the above integral and then *analytically continue* the vertex function $\Gamma(k^2; n)$ to other (complex) values z beyond the positive integers n.

The integral is first recast using (53.8). We obtain

$$\Gamma(k^2) = \frac{\lambda^2}{2(2\pi)^4} \int d^n l$$

$$\times \int_0^1 dx_1 \int_0^1 dx_2 \frac{\delta(1 - (x_1 + x_2))}{[(l^2 - \mu^2 + i\epsilon^+)x_1 + \{(l - k)^2 - \mu^2 + i\epsilon^+\}x_2]^2}$$

$$= \frac{\lambda^2}{2(2\pi)^4} \int d^n l \int_0^1 dx \frac{1}{[(l^2 - \mu^2 + i\epsilon^+)(1 - x) + \{(l - k)^2 - \mu^2 + i\epsilon^+\}x]^2} .$$

$$(53.77)$$

The quantity within the square brackets in the denominator of the integrand can be expressed as

$$l^2 - 2(l \cdot k)x + k^2 x - \mu^2 + i\epsilon^+ = (l - kx)^2 - \{\mu^2 - x(1 - x)k^2 - i\epsilon^+\} .$$

So we have

$$\Gamma(k^2) = \frac{\lambda^2}{2(2\pi)^4} \int_0^1 dx \int d^n l \frac{1}{[(l - kx)^2 - \alpha^2 + i\epsilon^+]^2}$$

$$= \frac{\lambda^2}{2(2\pi)^4} \int_0^1 dx \int d^n l \frac{1}{[l^2 - \alpha^2 + i\epsilon^+]^2} ,$$

$$(53.78)$$

where

$$\alpha^2 \equiv \mu^2 - x(1 - x)k^2 . \tag{53.79}$$

Next we perform a Wick rotation on the dl^0-integration (set $l^0 = il^n$ and rotate the contour of the l^0 integration counter-clockwise by $\pi/2$) to obtain

$$\Gamma(k^2) = \frac{i\lambda^2}{2(2\pi)^4} \int_0^1 dx \int_{-\infty}^\infty dl_1 \ldots dl_n \frac{1}{[(l_1^2 + \cdots + l_n^2)^2 + \alpha^2 - i\epsilon^+]^2}$$

$$= \frac{i\lambda^2}{2(2\pi)^4} \frac{n\pi^{n/2}}{\Gamma(n/2 + 1)} \int_0^1 dx \int_0^\infty dL \frac{L^{n-1}}{[L^2 + \alpha^2 - i\epsilon^+]^2} ,$$

$$(53.80)$$

where in the second equality we have used (53.29) and (53.30). Since $\Gamma(n/2 + 1) = (n/2)\Gamma(n/2)$ [c.f. (53.44)] we have

$$\Gamma(k^2) = \frac{i\lambda^2}{2(2\pi)^4} \frac{2\pi^{n/2}}{\Gamma(n/2)} \int_0^1 dx \int_0^\infty dL \frac{L^{n-1}}{(L^2 + \alpha^2 - i\epsilon^+)^2} . \tag{53.81}$$

As it stands, the dL integral is convergent for $0 < Re(n) < 4$. The lower limit $0 < Re(n)$ is due to *infrared (IR) divergence* as $L \to 0$, while the upper limit $Re(n) < 4$ is due to *ultraviolet (UV) divergence* as $L \to \infty$. However, we can analytically continue it, as a function of the complex variable n, to the left of the imaginary n-axis. If we do the integral by parts once, we obtain

$$
\frac{1}{\Gamma(n/2)} \int_0^\infty dL \, \frac{L^{n-1}}{(L^2 + a^2 - i\epsilon^+)^2}
$$
$$
= \frac{1}{\Gamma(n/2)} \frac{L^n}{n(L^2 + a^2 - i\epsilon^+)^2} \Big|_0^\infty - \frac{1}{n\Gamma(n/2)} \int_0^\infty dL \, L^n \frac{d}{dL} \left(\frac{1}{(L^2 + a^2 - i\epsilon^+)^2} \right).
$$
$$(53.82)$$

The boundary term vanishes for $0 < Re(n) < 4$. So we have

$$
\frac{1}{\Gamma(n/2)} \int_0^\infty dL \, \frac{L^{n-1}}{(L^2 + a^2 - i\epsilon^+)^2} = \frac{2}{\Gamma(n/2 + 1)} \int_0^\infty dL \, \frac{L^{n+1}}{(L^2 + a^2 - i\epsilon^+)^3} ,
$$
$$(53.83)$$

where we have used the property (53.44) for the gamma function again. The integral on the right-hand side is now convergent for $-2 < Re(n) < 4$, the lower limit due to the infrared divergence of the integrand as well as to the fact that $\Gamma(z)$ has first-order poles at $z = 0, -1, -2, \ldots$. By doing the integration by parts m times the dL integral in (53.81) can be analytically continued to the region $-2m < Re(n) < 4$. As $m \to \infty$, then, the integral can be analytically to the domain $Re(n) < 4$. With this understanding we will rewrite the integral, by making the change of variables $L^2 = t$, as

$$
\int_0^\infty dL \, \frac{L^{n-1}}{(L^2 + a^2 - i\epsilon^+)^2} = \frac{1}{2} \int_0^\infty dt \, \frac{t^{n/2-1}}{(t + a^2 - i\epsilon^+)^2}
$$
$$
= \frac{1}{2(a^2 - i\epsilon^+)^{2-n/2}} \frac{\Gamma(n/2)\Gamma(2 - n/2)}{\Gamma(2)} ,
$$
$$(53.84)$$

where we have used (53.35) and (53.33). It follows from (53.81) that

$$
\Gamma(k^2; n) = \frac{i\lambda^2}{2(2\pi)^4} \pi^{n/2} \Gamma(2 - n/2) \int_0^1 dx \, \frac{1}{(a^2 - i\epsilon^+)^{2-n/2}} .
$$
$$(53.85)$$

Since $\Gamma(z)$ has a simple pole at $z = 0$, $\Gamma(2 - n/2)$ has a simple pole at $n = 4$. Thus $\Gamma(k^2; n)$ has a simple pole at $n = 4$. We can see this explicitly by using (53.44) to write

$$
(2 - n/2)\Gamma(2 - n/2) = \Gamma(3 - n/2) ,
$$
$$(53.86)$$

which implies

$$
\Gamma(2 - n/2) = \frac{\Gamma(3 - n/2)}{2 - n/2} = \frac{2\Gamma(3 - n/2)}{4 - n} \xrightarrow[n \to 4]{} \frac{2}{4 - n} .
$$
$$(53.87)$$

To get the regular part of $\Gamma(k^2; n)$ (as a *meromorphic function* of n), we make the following expansions about $(4 - n)$:

$$
\Gamma(2 - n/2) = \frac{2}{4 - n} + A + B(4 - n) + C(4 - n)^2 + \cdots ,
$$
$$(53.88)$$

$$
\frac{1}{a^{4-n}} = a^{n-4} = e^{(n-4) \ln a} = 1 + \frac{n - 4}{2} \ln a^2 + \cdots ,
$$
$$(53.89)$$

where A, B, C are constants independent of n. Thus, recalling the definition of α^2 given by (53.79), we have

$$\Gamma(2-n/2) \int_0^1 \frac{dx}{(\alpha^2 - i\epsilon^+)^{2-n/2}} \approx \left(\frac{2}{4-n} + A + B(4-n) + C(4-n) + \cdots \right)$$
$$\times \left(1 - \frac{4-n}{2} \int_0^1 dx \, \ln\left[\mu^2 - x(1-x)k^2 \right] + \cdots \right)$$
$$= \frac{2}{4-n} - \int_0^1 dx \, \ln\left[\mu^2 - x(1-x)k^2 \right] + A + O(4-n) .$$
(53.90)

Eq. (53.85) then yields

$$\Gamma(k^2; n) = \frac{\lambda^2}{32\pi^2} \left\{ \frac{2i}{4-n} - i \int_0^1 dx \, \ln\left[\mu^2 - x(1-x)k^2 \right] + iA + O(4-n) \right\} .$$
(53.91)

We see that in the dimensional regularization scheme the divergence of the vertex function $\Gamma(k^2; n)$ shows up as a simple pole at $n = 4$. The finite part can then be determined by

$$\tilde{\Gamma}(k^2) = \Gamma(k^2; n) - \Gamma(0; n) .$$
(53.92)

From (53.91),

$$\Gamma(0; n) = \frac{\lambda^2}{32\pi^2} \left\{ \frac{2i}{4-n} - i \ln(\mu^2) + iA \right\} \approx \frac{i\lambda^2}{16\pi^2(4-n)} .$$
(53.93)

It then follows from (53.91) and (53.92) that

$$\tilde{\Gamma}(k^2) \approx \frac{-i\lambda^2}{32\pi^2} \int_0^1 dx \, \ln\left[\frac{\mu^2 - x(1-x)k^2}{\mu^2} \right] .$$
(53.94)

It is readily seen that a by-parts integration yields

$$\int_0^1 \ln\left[1 - x(1-x)\frac{k^2}{\mu^2} \right] = k^2 \int_0^1 dx \, \frac{x(1-2x)}{\mu^2 - x(1-x)k^2}$$
$$= k^2 \int_0^1 dx \, \frac{(1-x)(2x-1)}{\mu^2 - x(1-x)k^2} ,$$
(53.95)

where the first equality follows from the fact that the boundary term from the by-parts integration vanishes, and the second equality results from a change of variables $x = 1 - x'$. This result, when substituted in (53.94), yields exactly the same result for the finite part $\tilde{\Gamma}(k^2)$ as in (53.37), as it should. We thus see that *the finite part of the vertex function* $\Gamma(k^2)$ *does not depend on the regularization scheme, but only on the Taylor series subtraction point for* k^2 (in our case it is $k^2 = 0$).

Problem 53.6 Use the dimensional regularization scheme to evaluate the quadratically divergent vertex function $\Sigma(k^2)$ due to the tadpole diagram (c.f. the first diagram
on the right-hand side of the equation in Fig. 52.2), given by [c.f. (52.3)]

$$\Sigma_{(one\ loop)}(k^2;n) = \frac{-i\lambda^2}{2} \int \frac{d^n l}{(2\pi)^4} \frac{1}{l^2 - \mu^2 + i\epsilon^+} . \tag{53.96}$$

Use the Wick rotation procedure to show that

$$\Sigma_{(one\ loop)}(k^2;n) = -\frac{\lambda}{32\pi^4} \frac{\pi^{n/2}\,\Gamma(1-n/2)}{(\mu^2)^{1-n/2}} . \tag{53.97}$$

Use the property of the gamma function (53.44) to show that $\Sigma_{(one\ loop)}(k^2;n)$ has two
simple poles, at $n = 2$ and at $n = 4$. Then show that

$$\Sigma_{(one\ loop)}(k^2) \xrightarrow[n \to 4]{} -\frac{\lambda}{32\pi^2} \frac{(2\mu^2)}{(4-n)} . \tag{53.98}$$

Compare this result with (53.66) for the same quantity obtained by the covariant
(Pauli-Villars) regularization scheme.

Chapter 54

The Callan-Symanzik Equation and the Renormalization Group

Different sets of renormalization conditions of the form (52.72) for the φ^4-theory, each of which depends on different *subtraction points* in the Taylor expansions of the vertex functions $\Gamma^{(2)}(k^2)$ and $\Gamma^{(4)}(s,t,u)$, give rise to different values of *renormalized* (physical) parameters: the renormalized mass μ and the renormalized coupling constant λ. In general, the subtraction points for each of the variables in the vertex functions may even be chosen differently, so we may write the following renormalization conditions analogous to (52.72):

$$\overline{\Gamma}^{(4)}(s',t',u') = -i\lambda \,,$$

$$\overline{\Gamma}^{(2)}(k'^2) = i\mu^2 \,, \qquad \frac{d}{dk^2}\overline{\Gamma}^{(2)}(k^2)\bigg|_{k^2=k''^2} = -i \,, \tag{54.1}$$

where the overbear denotes renormalized quantities and the set of subtraction points s',t',u',k'^2,k''^2 are (in general different) fixed points on the complex k^2-plane. The special choice in (52.72) is thus $s'=t'=u'=k'^2=k''^2=0$. In this chapter, we will designate a particular choice of subtraction points by R, and denote, as we did in Chapter 52, the renormalized mass and renormalized coupling constant corresponding to R by μ_R and λ_R, respectively, and the renormalized field by φ_R. The unrenormalized (bare) quantities will carry a subscript 0. Thus we have μ_0, λ_0 and φ_0 for the bare mass, the bare coupling constant, and the unrenormalized field, respectively. The unsubscripted symbols, μ, λ and φ will be reserved for variable quantities corresponding to variable sets of subtraction points.

Using the present notation we can rewrite (52.69) as follows

$$\varphi_R(x) = Z_\varphi^{-1/2}(R)\varphi_0(x) , \qquad Z_\varphi(R) = \frac{1}{1 - \Sigma'(R)} , \qquad (54.2)$$

$$\mu_R^2 = Z_\varphi^{-1}(R)\mu_0^2 - \delta\mu^2(R) , \qquad \delta\mu^2(R) = -\Sigma(R) , \qquad (54.3)$$

$$\lambda_R = Z_\varphi^{-2}(R)\lambda_0 - \delta\lambda(R) , \qquad \delta\lambda(R) = -3i\Gamma(R) , \qquad (54.4)$$

where the results on the right column follow from (52.76) and (52.80). Suppose R' is another choice of subtraction points. Then

$$\varphi_{R'}(x) = Z_\varphi^{-1/2}(R')\varphi_0(x) , \qquad (54.5)$$

which, together with the first equation of (54.2), imply

$$\varphi_{R'}(x) = Z_\varphi^{-1/2}(R')Z_\varphi^{1/2}(R)\,\varphi_R(x) = Z_\varphi^{-1/2}(R',R)\,\varphi_R(x) , \qquad (54.6)$$

where

$$Z_\varphi(R',R) \equiv Z_\varphi(R')Z_\varphi^{-1}(R) . \qquad (54.7)$$

Note from (54.6) that, since $\varphi'_R(x)$ and $\varphi_R(x)$ are both finite (being renormalized fields), the *renormalization transformation function* $Z_\varphi(R',R)$ between R and R' must be finite also, *if* the quantity defined on the right-hand side of (54.7) exists. By (54.7) this is possible (but not guaranteed) even though both $Z_\varphi(R)$ and $Z_\varphi(R')$ are infinite (before regularization), because it is a ratio of two infinite quantities. Assuming that $Z_\varphi(R',R)$ is well-defined, it would appear that we can write, conversely,

$$\varphi_R(x) = Z_\varphi^{-1/2}(R)Z_\varphi^{1/2}(R')\,\varphi_{R'}(x) = Z_\varphi^{-1/2}(R,R')\,\varphi_{R'}(x) , \qquad (54.8)$$

where

$$Z_\varphi(R,R') = Z_\varphi(R)Z_\varphi^{-1}(R') = Z_\varphi^{-1}(R',R) . \qquad (54.9)$$

But the middle quantity in the above equation may not be well-defined, even if the right-hand side of (54.7) is, precisely because of the fact that both $Z_\varphi(R)$ and $Z_\varphi(R')$ are infinite. So *the inverse of a well-defined transformation $Z_\varphi(R',R)$ may not exist*. On the other hand, if two transformations $Z_\varphi(R',R)$ and $Z_\varphi(R'',R')$ are both well-defined (for three choices of subtraction points R, R' and R''), then the quantity

$$Z_\varphi(R'')Z_\varphi^{-1}(R) = Z_\varphi(R'')Z_\varphi^{-1}(R')Z_\varphi(R')Z_\varphi^{-1}(R) = Z_\varphi(R'',R')Z_\varphi(R',R) \tag{54.10}$$

is well-defined, because the right-hand side of the last equality is, by assumption. Hence $Z_\varphi(R'',R) = Z_\varphi(R'')Z_\varphi^{-1}(R)$ is well-defined, and constitutes the transformation function from $\varphi_R(x)$ to $\varphi_{R''}$. Thus the composition of two well-defined renormalization transformation functions is another well-defined renormalization transformation function. In addition, it is obvious from (54.7) that $Z_\varphi(R,R)$ constitutes the identity transformation. We then conclude that *the set of all well-defined renormalization transformation functions $Z_\varphi(R',R)$, lacking* *inverses in general, possesses a **semigroup structure**.* Such a set is given the name **renormalization group** (RG) in the physics literature. It plays a central role in understanding the importance of **scale changes** underlying a variety of physical phenomena.

Problem 54.1 From (54.3) show that renormalized masses are related by

$$\mu_{R'}^2 - \mu_R^2 = \{Z_\varphi(R') - Z_\varphi(R)\}\mu_0^2 + \{\delta\mu^2(R') - \delta\mu^2(R)\}$$
$$\approx \{\Sigma'(R') - \Sigma'(R)\} + + \{\delta\mu^2(R') - \delta\mu^2(R)\} \tag{54.11}$$
$$\approx \{\delta\mu^2(R') - \delta\mu^2(R)\} = \Sigma(R) - \Sigma(R'),$$

and justify the approximation signs to lowest order in λ_0. This result is usually presented as

$$\mu_{R'}^2 = \mu_R^2 + \delta\mu^2(R', R), \tag{54.12}$$

where

$$\delta^2\mu(R', R) \equiv \delta\mu^2(R') - \delta\mu^2(R). \tag{54.13}$$

Problem 54.2 From (52.50) relating λ_R and λ_0, show that renormalized coupling constants for different R are related by

$$\lambda_{R'} = Z_\varphi^2(R', R)Z_\lambda^{-1}(R', R)\lambda_R, \tag{54.14}$$

where

$$Z_\lambda(R', R) \equiv Z_\lambda(R')Z_\lambda^{-1}(R). \tag{54.15}$$

In (54.14) it is assumed that $Z_\varphi(R', R)$ as given by (54.7) and $Z_\lambda(R', R)$ as given by (54.15) are both well defined.

In any quantum field theory, in particular the φ^4-theory that we have been focusing on, the basic quantities of interest are the renormalized amputated 1PI n-point Green's functions $\overline{\Gamma}(k_1, \ldots, k_n; \mu, \lambda)$, where $k_1, \ldots, k_n$ are the external momenta, and μ and λ are the renormalized mass and renormalized coupling constant, respectively. The existence of a renormalization group implies that these Green's functions with different μ and λ are related to each other by renormalization transformations. Since μ and λ are continuous parameters, we expect the relationship to appear in the form of a differential equation. Such an equation is known as the *Callan-Symanzik equation*. In order to derive it we first need to make a detour to develop the procedure of the *insertion of composite operators* in an n-point Green's function.

In general, a **composite operator** is defined to be a monomial of products of fields and their derivatives, for example, $\varphi^2(x), \varphi^3(x), \partial_\mu\partial^\mu\varphi(x)$, etc. In a general theory with more than one kind of quantum field, the monomial may involve products of the different kinds of fields and their derivatives. But again we will focus our attention on the φ^4-theory, and consider an insertion of the form

$$\theta(y_1)\ldots\theta(y_l) \equiv 2^{-l}\varphi^2(y_1)\ldots\varphi^2(y_l)\,, \qquad \text{with } \theta(x) \equiv \varphi^2(x)/2 \quad (54.16a)$$

into an ordinary n-point Green's function. This gives rise to the following Green's function with insertions:

$$G^{(n)}_{\theta^l}(y_1,\ldots,y_l;x_1,\ldots,x_n) \equiv \langle 0\,|\,T[2^{-l}\hat{\varphi}^2(y_1)\ldots\hat{\varphi}^2(y_l)\hat{\varphi}(x_1)\ldots\hat{\varphi}(x_n)]\,|\,0\rangle\,,$$
$$(54.16b)$$

where T is the time-ordering operator [c.f. (49.1)], with the corresponding momentum-space function $\tilde{G}^{(n)}_{\theta^l}(p_1,\ldots,p_l;k_1,\ldots,k_n)$ defined by the Fourier transform [c.f. (52.1)]:

$$(2\pi)^4\delta^4(p_1+\cdots+p_l+k_1+\cdots+k_n)\,\tilde{G}^{(n)}_{\theta^l}(p_1,\ldots,p_l;k_1,\ldots,k_n)$$
$$= \int d^4y_1 \ldots \int d^4y_l\, e^{-i(p_1\cdot y_1+\cdots+p_l\cdot y_l)}$$
$$\times \int d^4x_1 \ldots \int d^4x_n\, e^{-i(k_1\cdot x_1+\cdots+k_n\cdot x_n)}\,G^{(n)}_{\theta^l}(y_1,\ldots,y_l;x_1,\ldots,x_n)\,.$$
$$(54.17)$$

For the purpose at hand, we will be particularly interested in the functions

$$G^{(n)}_{\theta}(x;x_1,\ldots,x_n) = \langle 0\,|\,T\,[2^{-1}\hat{\varphi}^2(x)\hat{\varphi}(x_1)\ldots\hat{\varphi}(x_n)]\,|\,0\rangle\,, \qquad (54.18a)$$
$$G^{(n)}_{\theta^2}(x,y;x_1,\ldots,x_2) = \langle 0\,|\,T[2^{-2}\hat{\varphi}^2(x)\hat{\varphi}^2(y)\hat{\varphi}(x_1)\ldots\hat{\varphi}(x_n)]\,|\,0\rangle\,; \quad (54.18b)$$

and their Fourier transforms $\tilde{G}^{(n)}_{\theta}(k;k_1,\ldots,k_n)$ and $\tilde{G}^{(n)}_{\theta^2}(p,q;k_1,\ldots,k_n)$ defined by

$$(2\pi)^4\delta^4(k+k_1+\cdots+k_n)\,\tilde{G}^{(n)}_{\theta}(k;k_1,\ldots,k_n)$$
$$= \int d^4x \int d^4x_1 \ldots \int d^4x_n\, e^{-i(k\cdot x+k_1\cdot x_1+\cdots+k_n\cdot x_n)}\,G^{(n)}_{\theta}(x;x_1,\ldots,x_n)\,,$$
$$(54.19a)$$

$$(2\pi)^4\delta^4(p+q+k_1+\cdots+k_n)\,\tilde{G}^{(n)}_{\theta^2}(p,q;k_1,\ldots,k_n)$$
$$= \int d^4x \int d^4y\, e^{-i(p\cdot x+q\cdot y)}$$
$$\times \int d^4x_1 \ldots \int d^4x_n\, e^{-i(k_1\cdot x_1+\cdots+k_n\cdot x_n)}G^{(n)}_{\theta^2}(x,y;x_1,\ldots,x_n)\,.$$
$$(54.19b)$$

Recalling the functional derivative formalism as established in Chapter 49 and using the Minkowski version of (49.38) (replacing D_E by the Feynman propagator D and J by $-iJ$), we have, for example, the result for the 2-point Green's function with the insertion of the composite operator $\varphi^2/2$ to zeroth-order in λ_0 given by

$$\langle 2^{-1}\varphi^2(x)\varphi(x_1)\varphi(x_2)\rangle_{(0)}$$
$$= \frac{(-i)^4}{2}\left[\left(\frac{\delta}{\delta J(x)}\right)^2 \frac{\delta}{\delta J(x_1)}\frac{\delta}{\delta J(x_2)}\exp\left\{-\frac{i}{2}JDJ\right\}\right]_{J=0}\,, \qquad (54.20)$$

where $D(x, y) = D(x-y)$ and $D(x)$ is the Feynman propagator given by (48.34), and

$$JDJ \equiv \int d^4x \int d^4y\, J(x) D(x-y) J(y) . \tag{54.21}$$

Working out the functional derivatives following the examples shown in Chapter 50, especially with the help of (50.9), (50.10) and (50.11), we obtain

$$\langle 2^{-1}\varphi^2(x)\varphi(x_1)\varphi(x_2)\rangle_{(0)} = (iD(x,x))(iD(x_1,x_2)) + 2(iD(x,x_1))(iD(x,x_2)) . \tag{54.22}$$

The first term on the right-hand side corresponds to a disconnected diagram, and its Fourier transform, as defined by (54.19), vanishes due to the momentum-conservation delta function in that equation. Working out the Fourier transform of the second term we have

$$\left\{ \tilde{G}_\theta^{(2)}(k; k_1, k_2) \right\}_{(0)} = \left\{ \tilde{G}_\theta^{(2)}(k; k_1, -k - k_1) \right\}_{(0)}$$

$$= \left(\frac{i}{k_1^2 - \mu_0^2 + i\epsilon^+} \right) \left(\frac{i}{(k + k_1)^2 - \mu_0^2 + i\epsilon^+} \right) . \tag{54.23}$$

The two terms on the right-hand side are free propagators, but the relationship between the free momenta (k_1 and $k_1 + k$) indicates a corresponding diagrammatic structure shown in Fig. 54.1. This diagram can be interpreted to mean that the composite operator $\theta(x)$ generates a special 3-point vertex (with three momentum-bearing lines feeding into it) and carries a free momentum k (shown by a dotted line with direction). Note that 4-momentum is conserved at this special vertex. Eq. (54.23) immediately implies that the amputated 2-point function (to zeroth-order in λ_0) with insertion of $\theta(x)$ is given by

$$\left\{ \Gamma_\theta^{(2)}(k; k_1, -k - k_1) \right\}_{(0)} = 1 . \tag{54.24}$$

Problem 54.3 Verify (54.22) for the spatial 2-point Green's function with insertion $\varphi^2/2$ to zeroth-order in λ_0, and justify that the Fourier transform, according to (54.19), of the first term in the right-hand side of (54.22) vanishes.

Problem 54.4 Verify that the Fourier transform according to (54.19) of the second term in the right-hand side of (54.22) yields the expression (54.23) for the zeroth-order (in λ_0) 2-point Green's function in momentum space with insertion $\varphi^2/2$.

Problem 54.5 With the help of Fig.54.1 justify that the Feynman diagram for the first-order (in λ_0) 2-point Green's function $\left\{ \tilde{G}_\theta^{(2)}(k; k_1, -k - k_1) \right\}_{(1)}$ in momentum space with insertion $\theta = \varphi^2/2$ is as shown in Fig. 54.2. Then show that the corresponding amputated Green's function is given by the following expression:

$$\left\{ \Gamma_\theta^{(2)}(k; k_1, -k - k_1) \right\}_{(1)} = \frac{-i\lambda_0}{2} \int \frac{d^4l}{(2\pi)^4} \frac{i}{l^2 - \mu_0^2 + i\epsilon^+} \frac{i}{(k - l)^2 - \mu_0^2 + i\epsilon^+} . \tag{54.25}$$

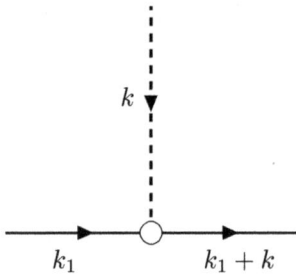

FIGURE 54.1

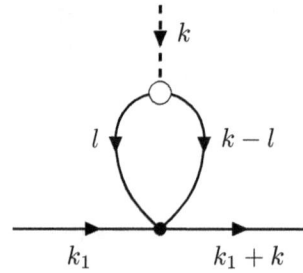

FIGURE 54.2

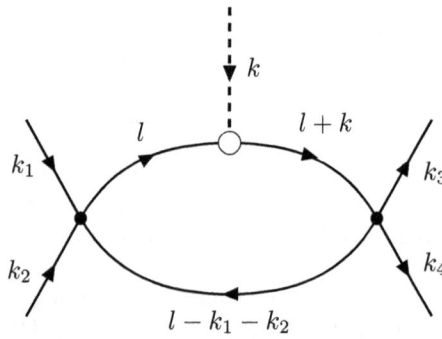

FIGURE 54.3

Problem 54.6 Justify that a Feynman diagram for the second-order (in λ_0) 4-point Green's function in momentum space with insertion $\theta = \varphi^2/2$ is as shown in Fig. 54.3. Then show that the corresponding amputated Green's function for this diagram is given by

$$\left\{\Gamma_\theta^{(4)}(k; k_1, k_2, k_3, k_4)\right\}_{(2)}$$
$$= \frac{(-i\lambda_0)^2}{2} \int \frac{d^4l}{(2\pi)^4} \frac{i}{l^2 - \mu_0^2 + i\epsilon^+} \frac{i}{(l+k)^2 - \mu_0^2 + i\epsilon^+} \frac{i}{(l - k_1 - k_2)^2 - \mu_0^2 + i\epsilon^+} .$$
$$(54.26)$$

It is apparent from Figs. 54.2 and 54.3 that the Feynman diagrams for $\Gamma_\theta^{(n)}$ differ from those for $\Gamma^{(n)}$ by the addition of one internal propagator. In fact, since

$$\frac{\partial}{\partial \mu_0^2}\left(\frac{i}{l^2 - \mu_0^2 + i\epsilon^+}\right) = \frac{i}{l^2 - \mu_0^2 + i\epsilon^+}(-i)\frac{i}{l^2 - \mu_0^2 + i\epsilon^+} , \qquad (54.27)$$

we see that the differentiation of $\Gamma^{(n)}$ with respect to μ_0^2, holding λ_0 and the momentum cutoff Λ constant, is equivalent to making an insertion of the composite operator $\theta = \varphi^2/2$ carrying zero 4-momentum in $\Gamma^{(n)}$, and multiplying by the factor $-i$. This is clear for the case of the 2-point function ($n = 2$) from (54.23), on setting $k = 0$ in that equation. Thus we have the following important relationship:

$$\frac{\partial}{\partial \mu_0^2}\Gamma^{(n)}(k_1, \ldots, k_n) = -i\Gamma_\theta^{(n)}(0; k_1, \ldots, k_n) . \qquad (54.28)$$

This relationship will constitute the basis for the derivation of the Callan-Symanzik equation. The above two equations also indicate that the superficial degree of divergence D of $\Gamma_\theta^{(n)}$ is reduced from that of $\Gamma^{(n)}$ by 2. Since the latter has $D = 4 - n$, the superficial degree of divergence of $\Gamma_\theta^{(n)}$ is $2 - n$. It follows then that we only have to consider $\Gamma_\theta^{(2)}$ for renormalization procedures. Eq. (54.25) shows that, to lowest non-vanishing order in λ_0, $\Gamma_\theta^{(2)}$ is logarithmically divergent. So one subtraction will suffice for the isolation of the finite part $\tilde{\Gamma}_\theta^{(2)}$, and we can write, if we choose to subtract at zero momentum values,

$$\left\{\Gamma_\theta^{(2)}(k; k_1, -k - k_1)\right\}_{(1)} = \left\{\Gamma_\theta^{(2)}(0; 0, 0)\right\}_{(1)} + \left\{\tilde{\Gamma}_\theta^{(2)}(k; k_1, -k - k_1)\right\}_{(1)} .$$
$$(54.29)$$

Setting all 4-momentum values to be zero, we obtain

$$\left\{\tilde{\Gamma}_\theta^{(2)}(0; 0, 0)\right\}_{(1)} = 0 . \qquad (54.30)$$

But (54.24) shows that there is a zeroth-order (in λ_0) contribution to $\Gamma_\theta^{(2)}$ as well, which is independent of momentum values. We will then set the renormalized 2-point Green's function with insertion, $\overline{\Gamma}_\theta^{(2)}$, to be, up to first order in λ_0,

$$\overline{\Gamma}_\theta^{(2)}(k; k_1, -k - k_1) = 1 + \left\{ \tilde{\Gamma}_\theta^{(2)}(k; k_1, -k - k_1) \right\}_{(1)} , \qquad (54.31)$$

which, according to (54.30), satisfies the renormalization condition

$$\overline{\Gamma}_\theta^{(2)}(0; 0, 0) = 1 . \qquad (54.32)$$

Since we are at liberty to choose renormalization conditions [recall (54.1)], we will in fact choose the above equation to the renormalization condition for $\overline{\Gamma}_\theta^{(2)}$ up to all orders in λ_0. Eq. (52.33) (relating the unrenormalized and renormalized amputated Green's functions without insertion) then suggests that, to accommodate the insertion factor $\theta = \varphi^2$, the unrenormalized and renormalized n-point amputated Green's functions with insertion are related as follows:

$$\overline{\Gamma}_\theta^{(n)}(k; k_1, \ldots, k_n; \lambda, \mu)$$
$$= (Z_\theta(\lambda_0, \mu_0, \Lambda))^{-1} (Z_\varphi(\lambda_0, \mu_0, \Lambda))^{n/2} \, \Gamma_\theta^{(n)}(k; k_1, \ldots, k_n; \lambda_0, \mu_0) , \qquad (54.33)$$

where Z_θ is a new cutoff-dependent renormalization constant necessitated by the insertion $\theta = \varphi^2/2$, Z_φ is the cutoff-dependent field renormalization constant for Green's functions without insertion [c.f. (54.2)], and μ and λ are the renormalized mass and renormalized coupling constant, respectively. In the above equation the dependences of the Green's functions and the renormalization constants on the (renormalized and bare) masses and coupling constants, as well as the momentum cutoff Λ, have been shown explicitly for clarity. For ease of reference we will also present here again (52.33), with the explicit dependences on the various arguments,

$$\boxed{\overline{\Gamma}^{(n)}(k_1, \ldots, k_n; \lambda, \mu) = (Z_\varphi(\lambda_0, \mu_0, \Lambda))^{n/2} \, \Gamma^{(n)}(k_1, \ldots, k_n; \lambda_0, \mu_0)} .$$
$$(54.34)$$

With the above background on the insertion of the composite operator $\varphi^2/2$ in Green's functions, we can now proceed directly to the derivation of the Callan-Symanzik equation. Simply by the chain rule of differentiation, we first write

$$\frac{\partial}{\partial \mu_0^2} \overline{\Gamma}^{(n)}(k_1, \ldots, k_n; \lambda, \mu) = \left(\frac{\partial \mu^2}{\partial \mu_0^2} \frac{\partial}{\partial \mu^2} + \frac{\partial \lambda}{\partial \mu_0^2} \frac{\partial}{\partial \lambda} \right) \overline{\Gamma}^{(n)}(k_1, \ldots, k_n; \lambda, \mu) .$$
$$(54.35)$$

On using (54.34) the left-hand side can be evaluated as follows:

$$\frac{\partial}{\partial \mu_0^2} \overline{\Gamma}^{(n)}(k_1, \ldots, k_n; \lambda, \mu) = \frac{\partial}{\partial \mu_0^2} \left(Z_\varphi^{n/2} \Gamma^{(n)}(k_1, \ldots, k_n; \lambda_0, \mu_0) \right)$$
$$= \left(\frac{n}{2} Z_\varphi^{n/2-1} \frac{\partial Z_\varphi}{\partial \mu_0^2} \right) \Gamma^{(n)} + Z_\varphi^{n/2} \frac{\partial \Gamma^{(n)}}{\partial \mu_0^2} \qquad (54.36)$$
$$= \frac{n}{2} \frac{\partial \ln Z_\varphi}{\partial \mu_0^2} \overline{\Gamma}^{(n)}(k_1, \ldots, k_n; \lambda, \mu) - i Z_\theta \overline{\Gamma}_\theta^{(n)}(0; k_1, \ldots, k_n; \lambda_0, \mu_0) ,$$

where in the last equality we have rewritten $\Gamma^{(n)}$ in terms of $\overline{\Gamma}^{(n)}$ using (54.34) in the first term, and made use of (54.28) and (54.33) for the second term. Substituting this result in (54.35) and dividing by $\partial\mu^2/\partial\mu_0^2$, rewriting $\partial/\partial\mu^2$ as $(1/2\mu)\partial/\partial\mu$, and finally multiplying through by $2\mu^2$, we arrive at the so-called **Callan-Symanzik equation** (CS equation) for the renormalized n-point amputated Green's functions:

$$
\left(\mu\frac{\partial}{\partial\mu} + \beta(\lambda)\frac{\partial}{\partial\lambda} - n\gamma(\lambda)\right)\overline{\Gamma}^{(n)}(k_1,\ldots,k_n;\lambda,\mu)
$$
$$
= -i\mu^2\alpha(\lambda)\,\overline{\Gamma}_\theta^{(n)}(0;k_1,\ldots,k_n;\lambda,\mu)\,, \quad (54.37)
$$

where

$$
\beta(\lambda) \equiv 2\mu^2\frac{\partial\lambda/\partial\mu_0^2}{\partial\mu^2/\partial\mu_0^2}\,, \qquad \gamma(\lambda) \equiv \mu^2\frac{\partial\ln Z_\varphi/\partial\mu_0^2}{\partial\mu^2/\partial\mu_0^2}\,, \qquad \alpha(\lambda) \equiv \frac{2Z_{\varphi^2}}{\partial\mu^2/\partial\mu_0^2}\,.
$$
$$(54.38)$$

The CS equation is an example of a **renormalization-group equation**, which gives a *scaling law* for renormalized Green's functions, in the sense that the equation describes how the Green's functions vary as the scaling parameter μ of the theory (the one carrying dimensions) changes. It also describes how the Green's functions change as functions of the coupling constant λ (a non-scaling parameter in the case of the φ^4-theory). Thus it provides direct information on the physical predictions of the theory as one varies the length (or energy) scale at which the theory is intended to be relevant, and at the same time, not surprisingly, contains the key to a proof of the renormalizability of a theory. It is this basic property of the equation which confers to it a fundamental status and unique generality.

The sole dependences on λ of the functions in (54.38) can be justified as follows. We first demonstrate that α and γ are not independent. In fact, on letting $n = 2$ in (54.37), we have

$$
\left(\mu\frac{\partial}{\partial\mu} + \beta\frac{\partial}{\partial\lambda} - 2\gamma\right)\overline{\Gamma}^{(2)}(k^2;\lambda,\mu) = -i\mu^2\alpha\,\overline{\Gamma}_\theta^{(2)}(0;k^2;\lambda,\mu)\,. \quad (54.39)
$$

If we set $k^2 = 0$ and choose the renormalization conditions [c.f. the middle equation of (52.72) and (54.32)]

$$
\overline{\Gamma}^{(2)}(0;\lambda,\mu) = i\mu^2\,, \qquad \overline{\Gamma}_\theta^{(2)}(0;0,0;\lambda,\mu) = 1\,, \quad (54.40)
$$

Eq. (54.39) immediately yields

$$
\alpha = 2(\gamma - 1)\,. \quad (54.41)
$$

Now, from the left-hand side of (54.37), β must be of dimension $[\lambda]$ and γ must be dimensionless. It follows from (54.41) that α must also be dimensionless.

But for the φ^4-theory, the coupling constant λ is dimensionless, since the action $\mathcal{L}\,d^4x$ must be dimensionless and so $[\mu^2\varphi^2] \sim [L]^{-4}$ implies $[\varphi^4] \sim [L]^{-4}$. So α, β and γ are all dimensionless quantities, which implies that they can only depend on dimensionless arguments, of which there are only two in the present situation: λ and μ^2/Λ^2. We argue that they cannot depend on the cutoff Λ as follows. On differentiating (54.39) with respect to k^2 and then evaluating the result at $k^2 = 0$ we have

$$\left(\mu\frac{\partial}{\partial\mu} + \beta\frac{\partial}{\partial\lambda} - 2\gamma\right)\frac{\partial\overline{\Gamma}^{(2)}(k^2;\lambda,\mu)}{\partial k^2}\bigg|_{k^2=0} = -i\mu^2\alpha\,\frac{\partial\overline{\Gamma}_\theta^{(2)}(0;k^2;\lambda,\mu)}{\partial k^2}\bigg|_{k^2=0}.$$
$$(54.42)$$

Choosing the renormalization condition [c.f. (52.72)]

$$\frac{\partial\overline{\Gamma}^{(2)}(k^2;\lambda,\mu)}{\partial k^2}\bigg|_{k^2=0} = -i\,,\qquad(54.43)$$

and recalling (54.41), the above equation reduces to

$$\gamma = \mu^2(1-\gamma)\,\frac{\partial\overline{\Gamma}_\theta^{(2)}(0;k^2;\lambda,\mu)}{\partial k^2}\bigg|_{k^2=0}.\qquad(54.44)$$

Since $\overline{\Gamma}_\theta^{(2)}$ (being a renormalized Green's function) is independent of the cutoff Λ, the above equation implies that γ must be independent of Λ. It follows from (54.41) that α must be independent of Λ also, as must $\overline{\Gamma}^{(2)}(k^2;\lambda,\mu)$ (again being a renormalized Green's function). Thus we see that in (54.39) every quantity except β has been shown to be Λ-independent. It follows that β must of necessity share this property also. We have thus succeeded in showing that α, β and γ can only depend on the dimensionless quantities λ and μ^2/Λ^2, but yet cannot depend on Λ. The only possibility then is that α, β and γ can each only depend on λ, as expressed on the left-hand sides of (54.38), and in the CS equation (54.37).

Besides the CS equation (54.37) relating $\overline{\Gamma}^{(n)}$ to $\overline{\Gamma}_\theta^{(n)}$, we also have a similar renormalization-group equation relating $\overline{\Gamma}_\theta^{(n)}$ to $\overline{\Gamma}_{\theta^2}^{(n)}$ that will turn out to be useful in the proof of renormalizability. The basic premise of this equation is one analogous to (54.28), and can similarly be justified:

$$\frac{\partial}{\partial\mu_0^2}\Gamma_\theta^{(n)}(k;k_1,\dots,k_n) = -i\Gamma_{\theta^2}^{(n)}(0,k;k_1,\dots,k_n)\,.\qquad(54.45)$$

There is also an equation analogous to (54.33) relating the renormalized and unrenormalized Green's functions with 2-point insertions:

$$\overline{\Gamma}_{\theta^2}^{(n)}(p,q;k_1,\dots,k_n;\lambda,\mu)$$
$$= (Z_\theta(\lambda_0,\mu_0,\Lambda))^{-2}(Z_\varphi(\lambda_0,\mu_0,\Lambda))^{n/2}\,\Gamma_{\theta^2}^{(n)}(p,q;k_1,\dots,k_n;\lambda_0,\mu_0)\,.$$
$$(54.46)$$

Essentially the same derivational steps leading to the CS equation (54.37) will yield the aforementioned renormalization-group equation for $\overline{\Gamma}_\theta^{(n)}$ as follows:

$$\left(\mu\frac{\partial}{\partial\mu} + \beta(\lambda)\frac{\partial}{\partial\lambda} - n\gamma(\lambda) + \gamma_\theta(\lambda)\right)\overline{\Gamma}_\theta^{(n)}(k;k_1,\dots,k_n;\lambda,\mu)$$

$$= -i\mu^2\alpha(\lambda)\overline{\Gamma}_{\theta^2}^{(n)}(0,k;k_1,\dots,k_n;\lambda,\mu)\,, \quad (54.47)$$

where

$$\gamma_\theta(\lambda) \equiv 2\mu^2\frac{\partial\ln Z_\theta/\partial\mu_0^2}{\partial\mu^2/\partial\mu_0^2}\,. \quad (54.48)$$

The sole dependence of γ_θ on λ can be justified along similar lines as for $\gamma(\lambda)$. We will leave this as an exercise (see Problem 54.7).

The CS equation has many important applications, chief among which is its use in an inductive proof of multiplicative renormalizability to any order in λ of a renormalizable theory. Another is its role in the calculation of critical exponents in the theory of second-order phase transitions (to be discussed in Chapter 58). We will present the details of the proof of renormalizabilty (due to Callan) in due course [following the elegant presentation by Callan (1976)]. Before proceeding to it directly, however, several preparatory facts need to be established.

First we need to recall more concretely some basic notions and facts on Feynman diagrams and Green's functions first introduced in Chapter 52, namely, those involving skeleton diagrams and skeleton expansions. For our present purpose, we will generalize them to include Green's functions with point insertion $\theta(x) = \varphi^2(x)/2$. The generalizations can be summarized in the following three definitions and one theorem (which will be presented without proof).

Definition 54.1. *A **skeleton diagram** for an amputated $1PI$ n-point Green's function $\Gamma^{(n)}$ is a Feynman diagram with no subdiagrams identifiable as contributions to $\Gamma^{(2)}$ or $\Gamma^{(4)}$.*

Definition 54.2. *A **skeleton diagram** for an amputated $1PI$ n-point Green's function $\Gamma_\theta^{(n)}$ with point insertion $\theta(x) = \varphi^2/2$ is a Feynman diagram with no subdiagrams identifiable as contributions to $\Gamma^{(2)}, \Gamma^{(4)}$ or $\Gamma_\theta^{(2)}$.*

Definition 54.3. *A **skeleton expansion** on a skeleton diagram S is obtained from S by replacing vertices in S by graphical contributions to $\Gamma^{(4)}$, point insertions of $\theta(x) = \varphi^2(x)/2$ in S by graphical contributions to $\Gamma_\theta^{(2)}$, and internal lines in S by graphical contributions to $\Gamma^{(2)}$.*

Theorem 54.1. *All $1PI$ n-point Feynman diagrams contributing to $\Gamma^{(n)}$ for $n > 4$ and all $1PI$ n-point diagrams with point insertion $\theta(x) = \varphi^2/2$ contributing to $\Gamma_\theta^{(n)}$ for $n > 2$ can be obtained from skeleton expansions on skeleton diagrams for these Green's functions.*

Problem 54.7 Justify that γ_θ depends on λ only. Follow the steps below.

(a) Justify that γ_θ is dimensionless, and so must be a function of only the dimensionless parameters λ and μ_0/Λ.

(b) Set $n = 2$ in the renormalization-group equation (54.47), differentiate it with respect to k_1^2 ($= k_2^2$ since we are considering a 2-point function) and evaluate the derivatives at $k_1^2 = 0$.

(c) Recall that α, β and γ are all functions of λ only. Then use the result in (b) and (54.44) to justify that γ_θ cannot depend on the cutoff Λ.

Problem 54.8 Verify by means of some Feynman diagrams that the dimension (in powers of mass) of $\Gamma^{(n)}$ is $4 - n$ in the φ^4-theory.

Continuing with the preparation for the inductive proof of renormalizability, we will present the following lemmas and most of their proofs concerning the leading orders of dependence on λ for the power series expansions of the various functions of interest, $\alpha(\lambda), \beta(\lambda), \gamma(\lambda)$ and $\gamma_\theta(\lambda)$, in the use of the CS equation (53.37) and the renormalization-group equation (53.47). In the following the symbol $O(\lambda^n)$ will mean a power series in λ beginning with a term proportional to λ^n.

Lemma 54.1. *With the renormalization conditions [c.f. (52.72)]*

$$\overline{\Gamma}^{(2)}(0) = i\mu^2 , \qquad \frac{d}{dk^2}\overline{\Gamma}^{(2)}(k^2)\bigg|_{k^2=0} = -i , \qquad \overline{\Gamma}^{(4)}(0) = -i\lambda , \qquad (54.49)$$

for $\overline{\Gamma}^{(2)}(k^2)$ and $\overline{\Gamma}^{(4)}(k^2)$, the following statements hold:

(i) *The field renormalization constant Z_φ is given by*

$$Z_\varphi^{-1} = i(\Gamma^{(2)})'(0) . \qquad (54.50)$$

(ii) *The renormalized mass μ is given by*

$$\mu^2 = -\frac{\Gamma^{(2)}(0)}{(\Gamma^{(2)})'(0)} . \qquad (54.51)$$

(iii) *The renormalized coupling constant λ is given by*

$$\lambda = -\frac{i\Gamma^{(4)}(0)}{[(\Gamma^{(2)})'(0)]^2} . \qquad (54.52)$$

Proof. By setting $n = 2$ and $k^2 = 0$ in (54.34) we have

$$\Gamma^{(2)}(k^2) = Z_\varphi^{-1} \overline{\Gamma}^{(2)}(k^2) . \tag{54.53}$$

Differentiating both sides of this equation with respect to k^2 and then setting $k^2 = 0$ in the derivatives we obtain

$$(\Gamma^{(2)})'(0) = Z_\varphi^{-1} (\overline{\Gamma}^{(2)})'(0) = -iZ_\varphi^{-1} , \tag{54.54}$$

where the last equality follows by applying the second renormalization condition in (54.49). Statement (i) [(54.50)] follows.

Setting $k^2 = 0$ in (54.53) and applying the renormalization condition given by the first equation in (54.49), we have

$$\Gamma^{(2)}(0) = Z_\varphi^{-1} \overline{\Gamma}^{(2)}(0) = i\mu^2 Z_\varphi^{-1} . \tag{54.55}$$

Statement (ii) [(54.51)] then follows by using (54.50) for Z_φ^{-1} in this result.

From the normalization condition given by the third equation in (54.49) and the fact that

$$\overline{\Gamma}^{(4)}(k^2) = Z_\varphi^2 \Gamma^{(4)}(k^2) , \tag{54.56}$$

which results from setting $n = 4$ in (54.34), we have

$$\lambda = iZ_\varphi^2 \Gamma^{(4)}(0) . \tag{54.57}$$

Statement (iii) [(54.52)] then follows from using (54.50) for Z_φ in this result. $\square$

Lemma 54.2. *With the renormalization conditions given in Lemma 54.1 [c.f. (54.9)], the following leading-term behavior for the power series expansions of $\Gamma^{(2)}(0)$, $(\Gamma^{(2)})'(0)$ and $\Gamma^{(4)}(0)$ in powers of λ hold for the 2-point and 4-point functions:*

$$\Gamma^{(2)}(0) = i\mu_0^2(1 + O(\lambda_0)) , \tag{54.58}$$
$$(\Gamma^{(2)})'(0) = -i(1 + O(\lambda_0^2)) , \tag{54.59}$$
$$\Gamma^{(4)}(0) = -i\lambda_0(1 + O(\lambda_0)) , \tag{54.60}$$

where the terms $O(\lambda_0)$ may depend on μ_o and the cutoff Λ, typically through powers of $\ln(\Lambda/\mu_0)$ and Λ/μ_0.

Proof. Using the expansion

$$\Sigma(k^2) = \Sigma(0) + k^2\Sigma'(0) + \tilde{\Sigma}(k^2) \tag{54.61}$$

in (52.9) we have

$$\Gamma^{(2)}(k^2) = \frac{1}{\tilde{G}(k^2)} = -i\left\{k^2 - \mu_0^2 - \Sigma(0) - k^2\Sigma'(0) - \tilde{\Sigma}(k^2)\right\} . \tag{54.62}$$

Setting $k^2 = 0$ in the above we obtain

$$\Gamma^{(2)}(0) = -i\left(-\mu_0^2 - \Sigma(0) - \Sigma'(0)\right) = i\mu_0^2\left(1 + \frac{\Sigma(0)}{\mu_0^2} + \frac{\Sigma'(0)}{\mu_0^2}\right), \qquad (54.63)$$

since $\tilde{\Sigma}(0) = 0$. The lowest order contribution to $\Sigma(0)$ is from the one-loop tadpole diagram (which is independent of k^2), while the lowest order contributions to $\Sigma'(0)$ are from two-loop diagrams. So we have

$$\Sigma(0) \sim O(\lambda_0), \qquad \Sigma'(0) \sim O(\lambda_0^2). \qquad (54.64)$$

Eq. (54.58) follows. Note that the first equation above is corroborated by (53.59) and (53.66).

Differentiating (54.61) with respect to k^2 and then setting $k^2 = 0$ we have

$$\tilde{\Sigma}'(0) = 0. \qquad (54.65)$$

Differentiating (54.62) with respect to k^2 and then setting $k^2 = 0$ and noting the above result we obtain

$$(\Gamma^{(2)})'(0) = \left.\frac{d\Gamma^{(2)}}{dk^2}\right|_{k^2=0} = -i(1 - \Sigma'(0)). \qquad (54.66)$$

From the second equation of (54.64), (54.59) then follows.

Finally, from (52.83) and (52.84), the leading terms in $\Gamma^{(4)}(0)$, up to the contributions due to the tree diagram and a one-loop (fish) diagram, are given by

$$\Gamma^{(4)}(0) = -i\lambda_0 + \Gamma(0) + \tilde{\Gamma}(0) = -i\lambda_0 + \Gamma(0), \qquad (54.67)$$

since $\tilde{\Gamma}(0) = 0$ when the contribution $\Gamma(k^2)$ from a fish diagram is expanded around $k^2 = 0$ as

$$\Gamma(k^2) = \Gamma(0) + \tilde{\Gamma}(k^2). \qquad (54.68)$$

The logarithmically divergent part $\Gamma(0)$, being due to a one-loop contribution, obviously satisfies

$$\Gamma(0) \sim O(\lambda_0^2). \qquad (54.69)$$

Eq. (54.60) then follows from (54.67).   □

Lemma 54.3. *The renormalized and bare coupling constants are related by*

$$\lambda \sim O(\lambda_0), \qquad \lambda_0 \sim O(\lambda). \qquad (54.70)$$

Proof. It follows from (54.52), (54.59) and (54.60) that

$$\lambda \sim \frac{\Gamma^{(4)}(0)}{[(\Gamma^{(2)})'(0)]^2} \sim \frac{\lambda_0(1 + O(\lambda_0))}{[1 + O(\lambda_0^2)]^2} \sim \lambda_0(1 + O(\lambda_0))(1 + O(\lambda_0^2)) \qquad (54.71)$$
$$\sim \lambda_0(1 + O(\lambda_0)) \sim O(\lambda_0).$$

The second equation in (54.70) follows directly from the first.   □

Lemma 54.4. *The functions $\alpha(\lambda), \beta(\lambda), \gamma(\lambda)$ and $\gamma_\theta(\lambda)$ appearing in the Callan-Symanzik equation (54.37) and the renormalization group equation (54.47), and as defined by (54.38) and (54.48), have the following leading-term dependences on λ:*

$$\alpha(\lambda) \sim -2 + \mu^2 O(\lambda^2), \tag{54.72}$$

$$\beta(\lambda) \sim \mu^2 O(\lambda^2), \tag{54.73}$$

$$\gamma(\lambda) \sim \mu^2 O(\lambda^2), \tag{54.74}$$

$$\gamma_\theta(\lambda) \sim \mu^2 O(\lambda). \tag{54.75}$$

Proof. We will start with $\beta(\lambda)$ first. From its definition given by (54.38), the expressions for λ and μ^2 given in Lemma 54.1 in terms of $\Gamma^{(2)}(0), (\Gamma^{(2)})'(0)$ and $\Gamma^{(4)}(0)$, and the leading-power (in λ_0) expansions for these three quantities given in Lemma 54.2, we have

$$\frac{\beta(\lambda)}{\mu^2} = \frac{2i\dfrac{\partial}{\partial\mu_0^2}\left(\dfrac{\Gamma^{(4)}(0)}{[(\Gamma^{(2)})'(0)]^2}\right)}{\dfrac{\partial}{\partial\mu_0^2}\left(\dfrac{\Gamma^{(2)}(0)}{(\Gamma^{(2)})'(0)}\right)} \sim \frac{\dfrac{\partial}{\partial\mu_0^2}\left(\dfrac{\lambda_0(1 + O_1(\lambda_0))}{(1 + O_2(\lambda_0^2))^2}\right)}{\dfrac{\partial}{\partial\mu_0^2}\left(\dfrac{\mu_0^2(1 + O_3(\lambda_0))}{1 + O_2(\lambda_0^2)}\right)}, \tag{54.76}$$

where

$$O_i(\lambda_0^n) = \lambda_0^n\, f_i\left(\lambda_0, \frac{\Lambda}{\mu_0}\right) \tag{54.77}$$

refer to different power series expansions in λ_0. We thus have

$$\frac{\beta(\lambda)}{\mu^2} \sim \frac{\dfrac{\partial}{\partial\mu_0^2}(\lambda_0[1 + O_4(\lambda_0)])}{\dfrac{\partial}{\partial\mu_0^2}([\mu_0^2(1 + O_5(\lambda_0))])} \sim \frac{\lambda_0\dfrac{\partial O_4(\lambda_0)}{\partial\mu_0^2}}{1 + O_5(\lambda_0) + \mu_0^2\dfrac{\partial O_5(\lambda_0)}{\partial\mu_0^2}}. \tag{54.78}$$

From (54.77),

$$\frac{\partial O_i(\lambda_0)}{\partial\mu_0^2} = \lambda_0\frac{\partial}{\partial\mu_0^2}f_i\left(\lambda_0, \frac{\Lambda}{\mu_0}\right) = \lambda_0 f_i''\left(\lambda_0, \frac{\Lambda}{\mu_0}\right), \tag{54.79}$$

where $f_i''(\lambda_0, \Lambda/\mu_0) \equiv (\partial/\partial\mu_0^2)f_i(\lambda_0, \Lambda/\mu_0)$. It follows that

$$\frac{\beta(\lambda)}{\mu^2} \sim \frac{\lambda_0^2 f_4''\left(\lambda_0, \dfrac{\Lambda}{\mu_0}\right)}{1 + \lambda_0 f_5\left(\lambda_0, \dfrac{\Lambda}{\mu_0}\right) + \lambda_0\mu_0^2 f_5''\left(\lambda_0, \dfrac{\Lambda}{\mu_0}\right)} \sim O(\lambda_0^2) \sim O(\lambda^2), \tag{54.80}$$

where the last equivalence follows from Lemma 54.3 [the second equation in (54.70)]. We have thus established (54.73).

Next we consider $\gamma(\lambda) = (\partial(\ln Z_\varphi)/\partial\mu_0^2)/(\partial\mu^2/\partial\mu_0^2)$, as defined by (54.38). By Lemma 54.1 [(54.50)] and (54.59) in Lemma 54.2,

$$Z_\varphi = -i[(\Gamma^{(2)})'(0)]^{-1} = (1 + O(\lambda_0^2))^{-1} \sim 1 + O(\lambda_0^2). \tag{54.81}$$

Hence

$$\ln Z_\varphi = \ln(1 + O(\lambda_0^2)) \sim O(\lambda_0^2) = \lambda_0^2 \, f\left(\lambda_0, \frac{\Lambda}{\mu_0}\right) , \qquad (54.82)$$

and

$$\frac{\partial \ln Z_\varphi}{\partial \mu_0^2} = \lambda_0^2 \, f'\left(\lambda_0, \frac{\Lambda}{\mu_0}\right) \sim O(\lambda_0^2) . \qquad (54.83)$$

Meanwhile, as we have determined earlier [see the denominator of the expression on the right-hand side of (54.78)],

$$\partial \mu^2 / \partial \mu_0^2 \sim 1 + O(\lambda_0) . \qquad (54.84)$$

The result (54.74) for $\gamma(\lambda)$ then follows. The result (54.72) for $\alpha(\lambda)$ immediately follows from (54.41).

Lastly, we consider $\gamma_\theta(\lambda) = 2\mu^2(\partial \ln Z_\theta/\partial \mu_0^2)/(\partial \mu^2/\partial \mu_0^2)$, as defined by (54.48). Using (54.29) to (54.33) we have

$$1 + \tilde{\Gamma}_\theta^{(2)} = \overline{\Gamma}_\theta^{(2)} = Z_\theta^{-1} Z_\varphi \, \Gamma_\theta^{(2)} = Z_\theta^{-1} Z_\varphi \left[1 + \Gamma_\theta^{(2)}(0) + \tilde{\Gamma}_\theta^{(2)}\right] , \qquad (54.85)$$

where we have dropped the subscript (1) in (54.29) to (54.31), and note that the quantities

$$\tilde{\Gamma}_\theta^{(2)} \equiv \tilde{\Gamma}_\theta^{(2)}(k; k_1, -k - k_1) , \qquad \Gamma_\theta^{(2)}(0) \equiv \Gamma_\theta^{(2)}(0; 0, 0) \qquad (54.86)$$

are both $O(\lambda_0)$, with lowest order contributions coming from the diagram with insertion shown in Fig. 54.2. Since $Z_\varphi \sim 1 + O(\lambda_0^2)$ we can ignore it in (54.85) and rewrite the equation as

$$1 + \tilde{\Gamma}_\theta^{(2)} \approx Z_\theta^{-1} \left[1 + \Gamma_\theta^{(2)}(0) + \tilde{\Gamma}_\theta^{(2)}\right] . \qquad (54.87)$$

To $O(\lambda_0)$ this is satisfied by $Z_\theta^{-1} = 1 - \Gamma_\theta^{(2)}(0)$, which implies

$$Z_\theta \sim 1 + O(\lambda_0) . \qquad (54.88)$$

Thus $\ln Z_\theta \sim O(\lambda_0)$ and, as shown above [c.f. (54.82) and (54.83)], $\partial \ln Z_\theta/\partial \mu_0^2 \sim O(\lambda_0)$. This fact and (54.84) imply (54.75). □

We will state one more lemma (without proof) before we delve into the proof of renormalizability.

Lemma 54.5. *Assume that the Green's functions $\Gamma^{(2)}, \Gamma^{(4)}$ and $\Gamma_\theta^{(2)}$ are renormalizable to all orders in λ. If their renormalized counterparts $\overline{\Gamma}^{(2)}, \overline{\Gamma}^{(4)}$ and $\overline{\Gamma}_\theta^{(2)}$ are inserted into skeleton diagrams to compute superficially convergent Green's functions $(\overline{\Gamma}^{(n)}$ for $n > 4$ and $\overline{\Gamma}_\theta^{(n)}$ for $n > 2$), then all Feynman integrals encountered will continue to converge.*

We are now ready to state and prove the following central theorem in quantum field theory.

Theorem 54.2. *The φ^4-theory with the Lagrangian density*

$$\mathcal{L}(\varphi_0) = \frac{1}{2}\left\{(\partial_\mu \varphi_0)^2 - \mu_0^2 \varphi_0^2\right\} - \frac{\lambda_0^4}{4!}\varphi_0^4 \,,$$

where $\varphi_0(x)$ is the unrenormalized field, and μ_0, λ_0 are the bare mass and the bare coupling constant, respectively, is renormalizable to all orders in λ_0.

Proof. The proof will be by induction. By Lemma 54.5 it is sufficient to prove that $\Gamma^{(2)}, \Gamma^{(4)}$ and $\Gamma_\theta^{(2)}$ are renormalizable to all orders in λ. It has been shown earlier (in the last chapter and this one) that these Green's functions are renormalizable to $O(\lambda), O(\lambda^2)$ and $O(\lambda)$, respectively. So, as the induction premise, we assume that $\Gamma^{(2)}, \Gamma^{(4)}$ and $\Gamma_\theta^{(2)}$ are renormalizable to $O(\lambda^r), O(\lambda^{r+1})$ and $O(\lambda^r)$, respectively, for a positive integer $r > 1$. This means that the regularizations of these functions calculated to the indicated orders in λ, as functions of the momentum cutoff Λ, will approach finite limits as $\Lambda \to \infty$. These limits will accordingly be denoted by $[\overline{\Gamma}^{(2)}]_{(r)}, [\overline{\Gamma}^{(4)}]_{(r+1)}$ and $[\overline{\Gamma}_\theta^{(2)}]_{(r)}$. In fact, $[f]_{(r)}$ will denote any quantity f computed to order λ^r. In what follows, by saying that $[f]_{(r)}$ exists, we mean that it has a finite limit as $\Lambda \to \infty$.

By Lemmas 54.2 and 54.3, we have

$$\overline{\Gamma}^{(2)} = Z_\varphi \Gamma^{(2)} \sim i\mu_0^2(1 + O(\lambda^2))(1 + O(\lambda)) \sim O(\lambda^0) \,, \tag{54.89}$$

$$\overline{\Gamma}^{(4)} = Z_\varphi^2 \Gamma^{(4)} \sim -i\lambda_0(1 + O(\lambda^2))(1 + O(\lambda)) \sim O(\lambda^1) \,. \tag{54.90}$$

Eq. (54.89) implies that

$$\overline{\Gamma}_\theta^{(2)} \sim O(\lambda^0) \,. \tag{54.91}$$

Now consider the CS equation (54.37) and apply it to the case $n = 4$. We have

$$\left(\mu\frac{\partial}{\partial\mu} + \beta(\lambda)\frac{\partial}{\partial\lambda} - 4\gamma(\lambda)\right)\overline{\Gamma}^{(4)} = -i\mu^2\alpha(\lambda)\overline{\Gamma}_\theta^{(4)} \,. \tag{54.92}$$

Since $[\mu(\partial\overline{\Gamma}^{(4)}/\partial\mu)]_{(r+1)}$ exists (because $[\overline{\Gamma}^{(4)}]_{(r+1)}$ exists by the induction premise), we expect to be able to compute both sides of the above equation to $O(\lambda^{r+1})$. Consider the right-hand side first. Since $\alpha(\lambda) \sim O(\lambda^0)$, it will involve $[\overline{\Gamma}_\theta^{(4)}]_{(r+1)}$. We argue that this quantity exists because $\overline{\Gamma}_\theta^{(4)}$ has a skeleton expansion. The reason is that one can replace, in a skeleton expansion of $\overline{\Gamma}_\theta^{(4)}$, a skeleton vertex by $[\overline{\Gamma}^{(4)}]_{(r)}$ (which exists since $[\overline{\Gamma}^{(4)}]_{(r+1)}$ is assumed to exist), or replace a skeleton propagator by $[\overline{\Gamma}^{(2)}]_{(r-1)}$ or $[\overline{\Gamma}_\theta^{(2)}]_{r-1}$ (both exist by the assumptions that $[\overline{\Gamma}^{(2)}]_{(r)}$ and $[\overline{\Gamma}_\theta^{(2)}]_{(r)}$ exist), to obtain $[\overline{\Gamma}_\theta^{(4)}]_{(r+1)}$. The CS equation (54.92) then implies that the quantity

$$\left[\left(\beta(\lambda)\frac{\partial}{\partial\lambda} - 4\gamma(\lambda)\right)\overline{\Gamma}^{(4)}\right]_{(r+1)} \tag{54.93}$$

exists. But by Lemma 54.2 [c.f. (54.60)], $\overline{\Gamma}^{(4)} \sim O(\lambda)$. Hence the above quantity will necessarily involve $[\beta]_{(r+1)}$ and $[\gamma]_{(r)}$, which must then exist also.

To achieve the proof by induction, we need to establish that all relevant quantities considered above, which are either assumed or established to exist to a certain order in λ, must exist in the next higher order in λ. In other words, we need to show that the following quantities all exist:

$$[\overline{\Gamma}^{(2)}]_{(r+1)} , \quad [\overline{\Gamma}^{(4)}]_{(r+2)} , \quad [\overline{\Gamma}^{(2)}_\theta]_{(r+1)} , \quad [\beta]_{(r+2)} , \quad [\gamma]_{(r+1)} .$$

We will work with $[\overline{\Gamma}^{(4)}]_{(r+2)}$ first. Again use the CS equation (54.37) for $n = 4$, but now calculated to $O(\lambda^{r+2})$. We have

$$\mu \frac{\partial}{\partial \mu} [\overline{\Gamma}^{(4)}]_{(r+2)} = -i\mu^2 [\overline{\Gamma}^{(4)}_\theta]_{(r+2)} - \left[\left(\beta(\lambda) \frac{\partial}{\partial \lambda} - 4\gamma(\lambda) \right) \overline{\Gamma}^{(4)} \right]_{(r+2)} , \qquad (54.94)$$

where, as before, the factor $\alpha(\lambda)$ in the term proportional to μ^2 has been dropped because it is $O(\lambda^0)$. We argue as before that $[\overline{\Gamma}^{(4)}_\theta]_{(r+2)}$ exists because $\overline{\Gamma}^{(4)}_\theta$ has a skeleton expansion: replace a skeleton vertex by $[\overline{\Gamma}^{(4)}]_{(r+1)}$, or a skeleton propagator by $[\overline{\Gamma}^{(2)}]_{(r)}$ or $[\overline{\Gamma}^{(2)}_\theta]_{(r)}$. Next we need to establish the existence of the second term on the right-hand side of (54.94). Since $\beta \sim O(\lambda^2)$ and $\gamma \sim O(\lambda^2)$, this term will involve at most $[\overline{\Gamma}^{(4)}]_{(r+1)}$. But since $[\overline{\Gamma}^{(4)}]_{(r+1)} \sim O(\lambda)$ (the leading term is of order λ), $[\beta(\partial \overline{\Gamma}^{(4)}/\partial \lambda)]_{(r+2)}$ will involve $[\beta]_{(r+2)}$ and $[\gamma \overline{\Gamma}^{(4)}]_{(r+2)}$ will involve $[\gamma]_{(r+1)}$, neither of which has been shown to exist yet. These quantities, however, are not required *separately* at this point, only a certain linear combination of them. This can be seen as follows. Imposing the renormalization condition $\overline{\Gamma}^{(4)}(0) = -i\lambda$ [c.f. the last equation in (54.49)] and evaluating (54.94) at zero momentum we get

$$0 = -i\mu^2 [\overline{\Gamma}^{(4)}_\theta(0)]_{(r+2)} - \left[\left(\beta \frac{\partial}{\partial \lambda} - 4\gamma \right) (-i\lambda) \right]_{(r+2)} , \qquad (54.95)$$

or

$$[\beta]_{(r+2)} - 4\lambda[\gamma]_{(r+1)} = \mu^2 [\overline{\Gamma}^{(4)}_\theta(0)]_{(r+2)} . \qquad (54.96)$$

Since the right-hand side exists, as we have just seen [see argument immediately below (54.94)], the linear combination in the left-hand side must exist as well. But this combination is precisely the only quantity involving $[\beta]_{(r+2)}$ and $[\gamma]_{(r+1)}$ required to compute the second term on the right-hand side of (54.94). Indeed, to involve $[\beta]_{(r+2)}$ and $[\gamma]_{(r+1)}$ at all in that computation, only the leading term of $[\overline{\Gamma}^{(4)}]_{(r+1)}$, which is proportional to λ, will be needed. [All the other terms will involve only lower order contributions of $\beta(\lambda)$ and $\gamma(\lambda)$.] The contribution from this leading term is exactly the linear combination in the left-hand side of (54.96), which exists by virtue of (54.96), as we have just seen. The above chain of reasoning thus establishes the existence of the right-hand side of (54.94): it approaches a finite limit as $\Lambda \to \infty$. To find $[\overline{\Gamma}^{(4)}]_{(r+2)}$, it remains to integrate that equation.

To do the integration we first rewrite the equation to be integrated as follows:

$$\mu \frac{\partial}{\partial \mu} \left[\overline{\Gamma}^{(4)} \left(\frac{k_i}{\mu}; \frac{\Lambda}{\mu}, \lambda \right) \right]_{(r+2)} = \left[\Phi \left(\frac{k_i}{\mu}; \frac{\Lambda}{\mu}, \lambda \right) \right]_{(r+2)} , \tag{54.97}$$

where the symbol Φ on the right-hand side simply refers to everything on the right-hand side of (54.94), k_i ($i = 1, \ldots, 4$) are the external momenta of the 4-point function $\overline{\Gamma}^{(4)}$, and the dependences on all the relevant variables are indicated explicitly. These all appear as dimensionless quantities because each term in (54.94) is dimensionless, and hence can only depend on dimensionless quantities. The reason behind the above assertion is that λ is dimensionless and $\overline{\Gamma}^{(4)}$ is readily seen to be dimensionless (the physical dimensions of $\overline{\Gamma}^{(n)}$ is μ raised to a power equal to its superficial degree of divergence, that is, $[\mu]^{4-n}$), while $\overline{\Gamma}_\theta^{(4)}$, with an additional factor of dimensions $[\mu]^{-2}$, has dimensions of $[\mu]^{4-4-2} = [\mu]^{-2}$ [c.f. (54.26)]. Renaming the integration variable in (54.97) by μ' and directly integrating that equation from $\mu' = \mu$ to $\mu' = \infty$ (where μ is fixed), we obtain

$$\left[\overline{\Gamma}^{(4)} \left(\frac{k_i}{\mu'}; \frac{\Lambda}{\mu'}, \lambda \right) \right]_{(r+2)} \Bigg|_{\mu'=\mu}^{\mu'=\infty} = \int_\mu^\infty \frac{d\mu'}{\mu'} \left[\Phi \left(\frac{k_i}{\mu'}; \frac{\Lambda}{\mu'}, \lambda \right) \right]_{(r+2)} \tag{54.98}$$
$$= \int_0^1 \frac{d\alpha}{\alpha} \left[\Phi \left(\alpha \frac{k_i}{\mu}; \alpha \frac{\Lambda}{\mu}, \lambda \right) \right]_{(r+2)} ,$$

where in the last equality we have made the change of variables $\alpha = \mu/\mu'$. The above equation can be rewritten

$$\left[\overline{\Gamma}^{(4)} \left(\alpha \frac{k_i}{\mu}; \alpha \frac{\Lambda}{\mu}, \lambda \right) \right]_{(r+2)} \Bigg|_{\alpha=1}^{\alpha=0} = \int_0^1 \frac{d\alpha}{\alpha} \left[\Phi \left(\alpha \frac{k_i}{\mu}; \alpha \frac{\Lambda}{\mu}, \lambda \right) \right]_{(r+2)} . \tag{54.99}$$

Imposing the renormalization condition $\overline{\Gamma}^{(4)}(0) = -i\lambda$, which yields the integration constant, we have

$$\left[\overline{\Gamma}^{(4)} \left(\frac{k_i}{\mu'}; \frac{\Lambda}{\mu}, \lambda \right) \right]_{(r+2)} = -i\lambda - \int_0^1 \frac{d\alpha}{\alpha} \left[\Phi \left(\alpha \frac{k_i}{\mu}; \alpha \frac{\Lambda}{\mu}, \lambda \right) \right]_{(r+2)} . \tag{54.100}$$

It may appear that there is some question regarding the convergence of the above integral at $\alpha = 0$. But this is not a concern since $[\Phi(\alpha k_i/\mu)]_{(r+2)}$ is assumed to be analytic at $\alpha = 0$ and guaranteed to vanish at that point, that is, for zero momentum. This is seen by evaluating both sides of (54.97) for $k_i = 0$ and using the renormalization condition $\overline{\Gamma}^{(4)}(0) = -i\lambda$. Recalling (54.94) and the thrust of our arguments up to this pout, $[\Phi]_{(r+2)}$ involves only quantities which have a finite limit as $\Lambda \to \infty$. So $[\overline{\Gamma}^{(4)}]_{(r+2)}$ must have a finite limit as

$\Lambda \to \infty$ and is in fact given explicitly by the following formula:

$$\lim_{\Lambda \to \infty} \left[\overline{\Gamma}^{(4)} \left(\frac{k_i}{\mu}; \frac{\Lambda}{\mu}, \lambda \right) \right]_{(r+2)} = -i\lambda - \int_0^1 \frac{d\alpha}{\alpha} \left\{ \lim_{\Lambda \to \infty} \left[\Phi \left(\alpha \frac{k_1}{\mu}; \alpha \frac{\Lambda}{\mu}, \lambda \right) \right]_{(r+2)} \right\}.$$
(54.101)

We still need to prove that $[\overline{\Gamma}^{(2)}]_{(r+1)}$ and $[\overline{\Gamma}_\theta^{(2)}]_{(r+1)}$ exist. To calculate the first quantity we might at first be led to the CS equation (54.37) for $n = 2$ and calculated to order $O(\lambda^{r+1})$:

$$\mu \frac{\partial}{\partial \mu} \left[\overline{\Gamma}^{(2)} \right]_{(r+1)} = -i\mu^2 \left[\alpha \overline{\Gamma}_\theta^{(2)} \right]_{(r+1)} - \left[\left(\beta \frac{\partial}{\partial \lambda} - 2\gamma \right) \overline{\Gamma}^{(2)} \right]_{(r+1)}. \quad (54.102)$$

Since $\alpha(\lambda) \sim O(\lambda^0)$, $[\overline{\Gamma}_\theta^{(2)}]_{(r+1)}$ is required on the right-hand side. But this is one of the quantities whose existence we want to prove; so (54.102) cannot be used at this point to determine $[\overline{\Gamma}^{(2)}]_{(r+1)}$. Instead, we will need the renormalization equation (54.47) for $n = 2$ and calculated to order $O(\lambda^{(r+1)})$ to determine $[\overline{\Gamma}_\theta^{(2)}]_{(r+1)}$ first:

$$\mu \frac{\partial}{\partial \mu} \left[\overline{\Gamma}_\theta^{(2)} \right]_{(r+1)} = -i\mu^2 \left[\alpha \overline{\Gamma}_{\theta^2}^{(2)} \right]_{(r+1)} - \left[\left(\beta \frac{\partial}{\partial \lambda} - 2\gamma + \gamma_\theta \right) \overline{\Gamma}_\theta^{(2)} \right]_{(r+1)}.$$
(54.103)

Before proceeding, we will use (54.47) for $n = 2$ and calculated to $O(\lambda^r)$ to argue as before that $[\gamma_\theta]_{(r)}$ exists, based on the assumed existence of $[\overline{\Gamma}_\theta^{(2)}]_{(r)}$, and hence of $[\overline{\Gamma}_\theta^{(2)}]_{(r)}$. Essentially the same strategy will now be used as that for the determination of $[\overline{\Gamma}^{(4)}]_{(r+2)}$ earlier. The 2-point function $\overline{\Gamma}_{\theta^2}^{(2)}$ with a 2-point insertion, just as $\overline{\Gamma}_\theta^{(2)}$, has a skeleton expansion, so $[\overline{\Gamma}_{\theta^2}^{(2)}]_{(r+1)}$ can be obtained by replacing a skeleton vertex with $[\overline{\Gamma}^{(4)}]_{(r+1)}$, or a skeleton propagator with $[\overline{\Gamma}^{(2)}]_{(r)}, [\overline{\Gamma}_\theta^{(2)}]_{(r)}$ or $[\overline{\Gamma}_{\theta^2}^{(2)}]_{(r)}$, all of which exist by virtue of the induction premise. Thus $[\overline{\Gamma}_{\theta^2}^{(2)}]_{(r+1)}$ exists. Using Lemma 54.4, we see that only $[\overline{\Gamma}^{(2)}]_{(r)}$ is required in the computation of the right-hand side of (54.103). Since $\overline{\Gamma}_\theta^{(2)} \sim O(\lambda^0)$ (the leading term is independent of λ), we only need at most $[\beta]_{(r+1)}$, which has been shown to exist. The apparent difficulty is that $[\gamma]_{(r+1)}$ and $[\gamma_\theta]_{(r+1)}$ are also required, neither of which has been shown to exist yet. However, the renormalization condition $\overline{\Gamma}_\theta^{(2)}(0) = 1$ [c.f. (54.32)] comes to the rescue. For, evaluating (54.103) at zero momentum with the aid of this condition, we obtain

$$[2\gamma - \gamma_\theta]_{(r+1)} = i\mu^2 \left[\alpha \overline{\Gamma}_{\theta^2}^{(2)}(0) \right]_{(r+1)}. \quad (54.104)$$

We have just shown that the right-hand side of this equation exists, so the left-hand side does as well. Yet it is precisely this linear combination of $[\gamma]_{(r+1)}$ and $[\gamma_\theta]_{(r+1)}$ that is required in calculating the right-hand side of (54.103), since

$\overline{\Gamma}_\theta^{(2)} \sim O(\lambda^0)$. Pulling the above considerations together, we conclude that the entire right-hand side of (54.103) exists. We can then integrate that equation to obtain an expression for $[\overline{\Gamma}_\theta^{(2)}]_{(r+1)}$ which has a finite limit as $\Lambda \to \infty$, as we did earlier for $[\overline{\Gamma}^{(4)}]_{(r+2)}$.

Let us now go back to (54.102). Having determined that $[\overline{\Gamma}_\theta^{(2)}]_{(r+1)}$ exists, we focus our attention on the terms involving β and γ on the right-hand side. Since $\beta \sim O(\lambda^2)$ and $\gamma \sim O(\lambda^2)$, at most $[\overline{\Gamma}^{(2)}]_{(r)}$ is required, which exists by assumption. But since $\overline{\Gamma}^{(2)} \sim O(\lambda^0)$, $[\beta]_{(r+1)}$ and $[\gamma]_{(r+1)}$ are required also. We know already that $[\beta]_{(r+1)}$ exists. That $[\gamma]_{(r+1)}$ exists also can be justified by recalling (54.44). Evaluating both sides of that equation to $O(\lambda^{r+1})$ we obtain

$$[\gamma]_{(r+1)} = \mu^2 \left[(1 - \gamma) \left. \frac{\partial \overline{\Gamma}_\theta^{(2)}(0; k^2)}{\partial k^2} \right|_{k^2=0} \right]_{(r+1)} . \tag{54.105}$$

Since the leading term of $\overline{\Gamma}_\theta^{(2)}$ [of order $O(\lambda^0)$] is momentum independent [c.f. (54.28)], the leading term of the derivative in the above equation is of $O(\lambda)$. Hence in the right-hand side, only $[\gamma]_{(r)}$ is required, which we know exists. So the entire right-hand side is finite as $\Lambda \to \infty$, which implies that the quantity on the left-hand side, $[\gamma]_{(r+1)}$, has the same property. This allows us to conclude finally that the entire right-hand side of (54.102) exists, and that equation can be integrated to yield $[\overline{\Gamma}^{(2)}]_{(r+1)}$. Incidentally, the existence of $[\gamma]_{(r+1)}$ implies the existence of $[\beta]_{(r+2)}$, as follows from (54.96).

To summarize, on the assumption that $[\overline{\Gamma}^{(2)}]_{(r)}$, $[\overline{\Gamma}_\theta^{(2)}]_{(r)}$ and $[\overline{\Gamma}^{(4)}]_{(r+1)}$ have finite limits as $\Lambda \to \infty$, we have proved that $[\overline{\Gamma}^{(2)}]_{(r+1)}$, $[\overline{\Gamma}_\theta^{(2)}]_{(r+1)}$ and $[\overline{\Gamma}^{(4)}]_{(r+2)}$ also have the same property. Thus the proof of renormalizability by induction is achieved. $\qquad \square$

We will close this rather long chapter by presenting the solution of the *homogeneous* CS equation, that is, (54.37) with the right-hand side set equal to zero, which describes the asymptotic (high energy and high momenta) behavior of Green's functions. The following theorem (stated without proof), known as *Weinberg's Theorem*, gives the conditions under which this approximation can be made.

Theorem 54.3. (*Weinberg's Theorem*) *Assume that the external momenta of an n-point function* $\overline{\Gamma}^{(n)}(k_1, \ldots, k_n; \lambda, \mu)$ *do not satisfy the condition that* $k_{i_1} + k_{i_2} + \cdots + k_{i_l} = 0$, *where* $\{i_1, i_2, \ldots, i_l\}$ *is a proper subset of* $\{1, 2, \ldots, n\}$. *If the time components of all the* k_i's *are analytically continued so that they are imaginary:* $k_i^0 = i(k_i)_4$, $(k_i)_4$ *real,* $i = 1, \ldots, n$ *(the* k_i's *are then said to be in the Euclidean region), then, under simultaneous scaling of all the* k_i's *by a dimensionless real parameter* σ,

$$\overline{\Gamma}^{(n)}(\sigma k_i; \lambda, \mu) \xrightarrow[\sigma \to \infty]{} \sigma^{4-n} \left[a_0 (\ln \sigma)^{b_0} + a_1 (\ln \sigma)^{b_1} \lambda + \cdots \right] , \tag{54.106}$$

to any finite order in λ, where the a_i's and b_i's are quantities independent of σ and λ. The n-point function with insertion $\overline{\Gamma}_{\theta}^{(n)}$, in the same limit, has the same logarithmic dependence on σ, but its power dependence on σ is σ^{2-n} instead.

Thus Weinberg's theorem is concerned with values of the external momenta in the **deep Euclidean region** ("deep" because, as the scaling parameter becomes large, the energy and momentum values all become large) for which no partial sum vanishes. The reason for the designation *Euclidean* is that for a 4-vector $k^{\mu} = (ik_4, k^1, k^2, k^3)$ in Minkowski space, $k^2 = -[(k_4)^2 + (k^1)^2 + (k^2)^2 + (k^3)^2]$, which equals the negative of the squared-length of a Euclidean 4-vector $\boldsymbol{k} = (k^1, k^2, k^3, k_4)$. In the deep Euclidean region the external 4-momenta are all far off mass shell and are thus maximally unphysical. If the expression within the square brackets in (54.106) (involving the logarithmic terms) sums to a quantity proportional to some power of σ, such as σ^{δ}, for example, then the dimension of $\overline{\Gamma}^{(n)}$ will be altered from $4 - n$ to $4 - n + \delta$. The quantity δ is thus called the **anomalous dimension** of $\overline{\Gamma}^{(n)}$. We note (without proof) that Weinberg's Theorem, provided that it is true to all orders in λ, is a sufficient condition for Lemma 54.5 to hold.

We will indeed assume that Weinberg's Theorem is true to all orders in λ. It is then seen that in the deep Euclidean region (for the external momenta k_i), $\overline{\Gamma}^{(n)} \gg \overline{\Gamma}_{\theta}^{(n)}$ as $\sigma \to \infty$, to all orders in λ. This is apparent since

$$\frac{\overline{\Gamma}_{\theta}^{(n)}}{\overline{\Gamma}^{(n)}} \xrightarrow[\sigma \to \infty]{} O(\sigma^{-2}) . \tag{54.107}$$

The right-hand side in the CS equation (54.37) can then be neglected and the following *homogeneous Callan-Symanzik equation* is obtained:

$$\left(\mu \frac{\partial}{\partial \mu} + \beta(\lambda) \frac{\partial}{\partial \lambda} - n\gamma(\lambda) \right) \overline{\Gamma}_{as}^{(n)}(k_i; \mu, \lambda) = 0 , \tag{54.108}$$

where $\overline{\Gamma}_{as}^{(n)}$ denotes the asymptotic ($\sigma \to \infty$) limit of $\overline{\Gamma}^{(n)}$. We will see that the solution of this homogeneous equation will provide information on the anomalous dimension of the Green's function $\overline{\Gamma}^{(n)}$.

The method of solution of (54.108) begins with the definition of a dimensionless Green's function $\overline{\mathcal{G}}^{(n)}(k_i/\mu, \lambda)$ depending on the indicated dimensionless quantities:

$$\overline{\Gamma}_{as}^{(n)}(k_i; \mu, \lambda) = \mu^{4-n} \overline{\mathcal{G}}^{(n)}(k_i/\mu, \lambda) . \tag{54.109}$$

Next we will scale all the momenta k_i with a real parameter σ and write $k_i = \sigma\kappa_i$, where $\{\kappa_i\}$ is a fixed set. Thus we write

$$\overline{\mathcal{G}}^{(n)}(k_i/\mu, \lambda) = \overline{\mathcal{G}}^{(n)}(\kappa_i(\sigma/\mu), \lambda) . \tag{54.110}$$

Setting $z = \sigma/\mu$, we have

$$\mu \frac{\partial \overline{\mathcal{G}}^{(n)}}{\partial \mu} = \mu \frac{\partial \overline{\mathcal{G}}^{(n)}}{\partial z} \frac{\partial z}{\partial \mu} = -\frac{\sigma}{\mu} \frac{\partial \overline{\mathcal{G}}^{(n)}}{\partial z} = -\sigma \frac{\partial z}{\partial \sigma} \frac{\partial \overline{\mathcal{G}}^{(n)}}{\partial z} = -\sigma \frac{\partial \overline{\mathcal{G}}^{(n)}}{\partial \sigma} , \tag{54.111}$$

or, on using (54.109),

$$\left(\mu \frac{\partial}{\partial \mu} + \sigma \frac{\partial}{\partial \sigma}\right)\left(\mu^{n-4}\overline{\Gamma}^{(n)}_{as}\right) = 0 . \qquad (54.112)$$

Carrying out the differentiation with respect to μ, we obtain

$$\left(\mu \frac{\partial}{\partial \mu} + \sigma \frac{\partial}{\partial \sigma} + (n-4)\right)\overline{\Gamma}^{(n)}_{as}(\sigma(\kappa_i/\mu), \lambda) = 0 . \qquad (54.113)$$

This equation shows that, for the asymptotic Green's function, a small change in the mass μ can be compensated for by a small change in the momentum scaling parameter σ. Combined with the homogeneous CS equation (54.108), it implies

$$\left(\sigma \frac{\partial}{\partial \sigma} - \beta(\lambda)\frac{\partial}{\partial \lambda} + n\gamma(\lambda) + (n-4)\right)\overline{\Gamma}^{(n)}_{as}(\sigma k_i; \lambda, \mu) = 0 , \qquad (54.114)$$

where we have re-expressed κ_i by k_i in the arguments of the asymptotic Green's function. The next step in the solution of (54.108) consists in the introduction of a new function $\overline{F}^{(n)}(\sigma k_i; \lambda, \mu)$ defined by

$$\overline{\Gamma}^{(n)}_{as}(\sigma k_i; \lambda, \mu) = \sigma^{4-n} \exp\left[n \int_0^\lambda d\lambda' \frac{\gamma(\lambda')}{\beta(\lambda')}\right] \overline{F}^{(n)}(\sigma k_i; \lambda, \mu) . \qquad (54.115)$$

Then

$$\left(\sigma \frac{\partial}{\partial \sigma} - \beta \frac{\partial}{\partial \lambda}\right)\overline{\Gamma}^{(n)}_{as} = \sigma^{4-n} \exp\left[n \int_0^\lambda d\lambda' \frac{\gamma(\lambda')}{\beta(\lambda')}\right]\left(\sigma \frac{\partial}{\partial \sigma} - \beta \frac{\partial}{\partial \lambda}\right)\overline{F}^{(n)}$$
$$+ \left\{\left(\sigma \frac{\partial}{\partial \sigma} - \beta \frac{\partial}{\partial \lambda}\right)\left(\sigma^{4-n} \exp\left[n \int_0^\lambda d\lambda' \frac{\gamma(\lambda')}{\beta(\lambda')}\right]\right)\right\}\overline{F}^{(n)} . \qquad (54.116)$$

Straightforward calculations show that the second term on the right-hand side is equal to $-[(n-4)+n\gamma(\lambda)]\overline{\Gamma}^{(n)}_{as}$. Eq. (54.116) and the homogeneous CS equation (54.108) then imply

$$\sigma^{4-n} \exp\left[n \int_0^\lambda d\lambda' \frac{\gamma(\lambda')}{\beta(\lambda')}\right]\left(\sigma \frac{\partial}{\partial \sigma} - \beta \frac{\partial}{\partial \lambda}\right)\overline{F}^{(n)} = 0 , \qquad (54.117)$$

which in turn yields

$$\left(\sigma \frac{\partial}{\partial \sigma} - \beta(\lambda)\frac{\partial}{\partial \lambda}\right)\overline{F}^{(n)}(\sigma k_i; \lambda, \mu) = 0 . \qquad (54.118)$$

Define $t \equiv \ln \sigma$, so that $\sigma \partial/\partial \sigma = \partial/\partial t$. So we rewrite the above equation as

$$\left(\frac{\partial}{\partial t} - \beta(\lambda)\frac{\partial}{\partial \lambda}\right)\overline{F}^{(n)}(e^t k_i; \lambda, \mu) = 0 . \qquad (54.119)$$

At this point we introduce the so-called **effective coupling constant** $\lambda_e(t, \lambda)$ defined by

$$\frac{d}{dt}\lambda_e(t, \lambda) = \beta(\lambda_e) , \qquad (54.120)$$

with initial condition $\lambda_e(0, \lambda) = \lambda$. Integrating the above equation we obtain

$$\int_\lambda^{\lambda_e(t,\lambda)} \frac{d\lambda_e}{\beta(\lambda_e)} = \int_0^t dt = t . \qquad (54.121)$$

Differentiating both sides of the above equation with respect to λ we get

$$0 = \frac{\partial}{\partial\lambda}\left(\int_\lambda^{\lambda_e(t,\lambda)} \frac{d\lambda_e}{\beta(\lambda_e)}\right) = \frac{1}{\beta(\lambda_e)}\frac{\partial\lambda_e}{\partial\lambda} - \frac{1}{\beta(\lambda)} . \qquad (54.122)$$

Multiplying by $\beta(\lambda_e)\beta(\lambda)$ and using (54.120), the above equation can be rewritten as

$$\left(\frac{\partial}{\partial t} - \beta(\lambda)\frac{\partial}{\partial\lambda}\right)\lambda_e(t, \lambda) = 0 , \qquad (54.123)$$

which is of the same form as (54.119). Thus, if the dependence of $\overline{F}^{(n)}(\sigma(t)k_i; \lambda, \mu)$ on λ and t is through the function $\lambda_e(t, \lambda)$, that is, if

$$\overline{F}^{(n)}(\sigma(t)k_i; \lambda, \mu) = \overline{F}^{(n)}(k_i; \lambda_e(t, \lambda), \mu) , \qquad (54.124)$$

then $\overline{F}^{(n)}(k_i; \lambda_e(t, \lambda), \mu)$ will be a solution of (54.118). This is so since it will then be expressible as a power series in λ_e:

$$\overline{F}^{(n)}(k_i; \lambda_e(t, \lambda), \mu) = \sum_m a_m \lambda_e^m(t, \lambda) , \qquad (54.125)$$

where the coefficients a_m are independent of t and λ, and

$$\left(\frac{\partial}{\partial t} - \beta\frac{\partial}{\partial\lambda}\right)\overline{F}^{(n)}(k_i; \lambda_e(t, \lambda), \mu) = \sum_m a_m \left(\frac{\partial}{\partial t} - \beta\frac{\partial}{\partial\lambda}\right)\lambda_e^m(t, \lambda) = 0 ,$$
$$(54.126)$$

by (54.123). We can then, according to (54.115), write the asymptotic solution $\overline{\Gamma}_{as}^{(n)}$ as

$$\overline{\Gamma}_{as}^{(n)}(\sigma k_i; \lambda, \mu) = \sigma^{4-n}\exp\left[n\int_0^\lambda d\lambda' \frac{\gamma(\lambda')}{\beta(\lambda')}\right]\overline{F}^{(n)}(k_i; \lambda_e(t, \lambda), \mu) . \qquad (54.127)$$

On the right-hand side of the above equation, $\overline{F}^{(n)}(k_i; \lambda_e(t, \lambda), \mu))$ can in fact be regarded as an arbitrary function of the indicated variables, as long as its dependence on λ and $t = \ln\sigma$ are through the effective coupling function $\lambda_e(t, \lambda)$ defined by (54.120). To see the effect of the scaling of the external momenta on the asymptotic solution $\overline{\Gamma}_{as}^{(n)}$ more explicitly we write the exponential factor as follows:

$$
\exp\left[n\int_0^\lambda d\lambda'\,\frac{\gamma(\lambda')}{\beta(\lambda')}\right] = \exp\left[n\int_0^{\lambda_e} d\lambda'\,\frac{\gamma(\lambda')}{\beta(\lambda')} + n\int_{\lambda_e}^\lambda d\lambda'\,\frac{\gamma(\lambda')}{\beta(\lambda')}\right]
$$

$$
= \exp\left[n\int_0^{\lambda_e} d\lambda'\,\frac{\gamma(\lambda')}{\beta(\lambda')}\right]\exp\left[n\int_{\lambda_e}^\lambda d\lambda'\,\frac{\gamma(\lambda')}{\beta(\lambda')}\right].
$$
(54.128)

We define

$$
H(\lambda_e(t,\lambda)) \equiv \exp\left[n\int_0^{\lambda_e(t,\lambda)} d\lambda'\,\frac{\gamma(\lambda')}{\beta(\lambda')}\right],
$$
(54.129)

and rewrite the second exponential as

$$
\exp\left[n\int_t^0 dt'\,\frac{d\lambda'}{dt'}\frac{\gamma(\lambda')}{\beta(\lambda')}\right] = \exp\left[-n\int_0^t dt'\,\gamma(\lambda_e(t',\lambda))\right],
$$
(54.130)

where we have made use of (54.120). Eq. (54.127) can then be rewritten as

$$
\overline{\Gamma}_{as}^{(n)}(\sigma k_i;\lambda,\mu)
$$
$$
= \sigma^{4-n}\,H(\lambda_e(t,\lambda))\exp\left[-n\int_0^t dt'\,\gamma(\lambda_e(t',\lambda))\right]\overline{F}^{(n)}(k_i;\lambda_e(t,\lambda),\mu).
$$
(54.131)

If we set $t=0$ in this equation we obtain

$$
\overline{\Gamma}_{as}^{(n)}(k_i;\lambda,\mu) = H(\lambda)\overline{F}^{(n)}(k_i;\lambda,\mu),
$$
(54.132)

on recalling $\sigma = e^t$ and $\lambda_e(0,\lambda) = \lambda$. Thus we have

$$
\overline{\Gamma}_{as}^{(n)}(\sigma k_i;\lambda,\mu) = \sigma^{4-n}\exp\left[-n\int_0^{\ln\sigma} dt'\,\gamma(\lambda_e(t',\lambda))\right]\overline{\Gamma}_{as}^{(n)}(k_i;\lambda_e(\ln\sigma,\lambda),\mu).
$$
(54.133)

This equation shows clearly that the effect of scaling the external momenta k_i in the asymptotic solution $\overline{\Gamma}_{as}^{(n)}(k_i;\lambda,\mu)$ by a scale factor σ, apart from some multiplicative factors, is to replace the coupling constant λ by a σ-dependent effective coupling constant λ_e. Furthermore, the exponential factor in the above equation gives the *anomalous dimension* of $\overline{\Gamma}^{(n)}$, which is controlled by the function $\gamma(\lambda)$.

Problem 54.9 Carry out the differentiations in the second term on the right-hand side of (54.116) and verify the statement immediately after that equation.

Chapter 55

The Wilsonian Approach to the Renormalization Group

The techniques of perturbative renormalization that we have developed so far were aimed at the elimination of infinite quantities that appeared in the calculation of Feynman integrals. These were at first rendered finite by regularization procedures, either through the introduction of a large momentum cutoff Λ, a variable spacetime dimension parameter n (which for Minkowski spacetime equals 4), or some other means. The regularized quantities are then manipulated so that the originally infinite quantities, to all orders in the coupling constant λ, can either be lumped together in a multiplicative factor Z, as in the method of multiplicative renormalization, or be canceled by the order-by order introduction of counterterms in the Lagrangian, as in the method of BPH renormalization. The important point is that all these different mathematical techniques ultimately led to the concept of the renormalization group (RG) and the existence of renormalization group equations (of which the Callan-Symanzik equation is a prototype) that govern the behavior of physical quantities under scale changes of the parameters entering into the theory, and yield the same physical results that are meant to be subjected to experimental tests. But at the end, a quantity such as the momentum cutoff Λ, which in high energy physics is supposed to approach infinity eventually, still has the flavor of a necessary but undesirable mathematical artifact or device, and a quantity that is not germane to the conceptual foundation of the theory. This odd situation has in fact historically generated substantial controversy among leading theoretical physicists. Luminaries such as Dirac and Landau, for example, have cast serious doubt on the whole renormalization procedure.

Towards the latter half of the twentieth century, when the achievements of quantum field theory in high energy physics became incontrovertible (especially in quantum electrodynamics and other gauge field theories), quantum field-theoretic methods were also applied to condensed matter physics with astounding success, especially in the difficult problem of *critical phenomena in second-order phase transitions.* This latter development, begun by Ginsburg and Landau, culminated in the early nineteen seventies in the work of K. Wilson, E. Brézin, Le Guillou, J. Zinn-Justin, Fisher and Kadanoff, among others (see, for example, Wilson and Kogut 1974, Brézin 1976, Kadanoff 1977, Huang 1987, Huang 2010, Itzykson and Drouffe 1989, Zinn-Justin 2002, and Amit and Martín-Mayor 2005). The contributions by Wilson, in particular, had a profound impact on the understanding of the theoretical foundations and interpretations of the renormalization group. We will try to provide an introduction to the so-called Wilsonian approach to the RG in this and subsequent chapters.

The Wilsonian approach, although leading to the same results as perturbative renormalization in areas of overlapping applications, differs from the latter in both mathematical technique and conceptual understanding. The most important difference is that, *whereas in perturbative renormalization, the momentum cutoff Λ is a mathematical device, in the Wilsonian approach it is a physically relevant quantity, being the physical energy scale beyond which the theory under consideration is supposed to be irrelevant.* The renormalization group, then, is a semigroup of transformations of physical parameters entering into a theory; and the RG equations give rise to so-called *RG flows* along certain trajectories in parameter-space [or even more generally, in the (mathematicaly ill-defined) space of Lagrangians, or theories], in such a way that theories can be mapped into each other under scale changes of the parameters. Of particular importance in the theoretical framework of the Wilsonian approach are the *fixed points* of the RG flows. These demarcate, in their neighborhoods, regions where theories are "stable" under scale-change perturbations, and consequently provide information on, very loosely speaking, what "correct" theories are possible. For example, in the context of high energy physics, the so-called *Gaussian fixed points* imply the existence of *asymptotically free* theories; while in the context of statistical mechanics, the so-called *infrared fixed points* mark out *critical regions* near second-order phase transitions. We will elaborate on these applications, especially with regard to critical phenomena, in due course.

We find that the φ^4-theory will provide a relatively simple yet sufficiently rich model for the explication of the aforementioned ideas, and will continue to exploit it for our purposes. We will use the *Euclidean weight* in functional integrations (which is more relevant for statistical mechanical applications), and allow the dimension of the underlying Euclidean space of the field theory to assume an arbitrary value d. (Recall our discussion in Chapter 46 concerning the remarkable analogy between the quantum mechanical time evolution operator and the statistical mechanical operator appearing in the partition function, and the process of Euclideanization.) The formal development in this chapter is adapted from that in Huang 2010. We begin by writing the classical action $S[\varphi]$ in the φ^4-theory in the general form

$$S[\varphi] = \int d^d x \left[\frac{1}{2} (\partial \varphi)^2 + g_2 \varphi^2 + g_4 \varphi^4 \right] , \qquad (55.1)$$

where, under the Euclidean metric,

$$(\partial\varphi)^2 = (\partial_1\varphi)^2 + \cdots + (\partial_d\varphi)^2 \qquad (\partial_i = \partial/\partial x_i) \, ; \tag{55.2}$$

and g_2, g_4 are the physical parameters entering into the theory. The factor $1/2$ in the first term of the Lagrangian density sets the scale of the theory. [The action $S[\varphi]$ must be dimensionless and we use natural units where $\hbar = c = 1$; so the only dimension is *length* or *momentum* $= (length)^{-1}$.] The sum of the terms to order φ^2 in the Lagrangian density is called the **kinetic term**. The *generating function* (or *partition function*) Z, with positive (or *Euclidean measure*), is given in terms of a functional integral by [c.f. (48.4) and (48.14)]

$$Z = \int [d\varphi]\, e^{-S[\varphi]} \, . \tag{55.3}$$

The physical system described by the theory is imagined to be placed in a d-dimensional *hypercube* of edge length L and volume $V = L^d$, with periodic boundary conditions, so that momentum values k_i ($i = 1, \ldots, d$) will be discretized. Eventually the limit $L \to \infty$ will be taken. For analytical purposes, it will be more convenient to write everything in momentum space, so as to display at first the momentum cutoff Λ more explicitly. Functions with continuous variables, such as the field in terms of the (continuous) spatial variables $x = (x_1, \ldots, x_d)$, will be written $\varphi(x)$; while those with discrete variables, such as the Fourier transform of $\varphi(x)$, will be written φ_k. We have the following Fourier-transform pair:

$$\varphi(x) = \frac{1}{V} \sum_k e^{-ik\cdot x} \varphi_k \, , \qquad \varphi_k = \int_V d^d x \, e^{ik\cdot x} \varphi(x) \, , \tag{55.4}$$

where $k \cdot x = k_1 x_1 + \cdots + k_d x_d$, $\int_V$ means integration over the hypercube of volume V, and $\sum_k$ means summing over all allowed discretized values of the d-dimensional vector k: $k_i = 2n_i\pi/L$ ($n_i = 0, \pm 1, \pm 2, \ldots$). In the continuum limit $L \to \infty$ we have

$$\sum_k \xrightarrow{\;L\to\infty\;} \frac{V}{(2\pi)^d} \int d^d k \, . \tag{55.5}$$

Note that while $\varphi(x)$ is real-valued, φ_k is complex; and the fact that $\varphi(x)$ is real implies that $\varphi_k^* = \varphi_{-k}$. In terms of the Fourier field components φ_k, and with the introduction of a large momentum cutoff Λ, we can write, for $k^2 = |k|^2 = k_1^2 + \cdots + k_d^2$,

$$S[\varphi] = \frac{1}{2} \sum_{|k|<\Lambda} (k^2 + 2g_2)\, \varphi_k \varphi_{-k} + S_I[\varphi] \, . \tag{55.6}$$

In the above expression, $S_I[\varphi]$, the interaction term, is given by

$$S_I[\varphi] = \frac{g_4}{V} \sum_{|k_i|<\Lambda} (\delta_{k_1+k_2+k_3+k_4})\, \varphi_{k_1} \varphi_{k_2} \varphi_{k_3} \varphi_{k_4} \, , \tag{55.7}$$

where $\delta_{\sum_i k_i}$ is the (dimensionless) Kronecker delta symbol: $\delta_p = 1$ if $p = 0$, and equal to zero otherwise. It has the integral representation

$$\delta_p = \frac{1}{V} \int_V d^d x \, e^{i p \cdot x} . \tag{55.8}$$

In terms of the Fourier field φ_k, the partition function Z is given by the same expression as in (55.3), but with the functional integration measure given by

$$\int [d\varphi] = \mathcal{N} \prod_{|k|<\Lambda} \int d\varphi_k \int d\varphi_k^* , \tag{55.9}$$

where $\mathcal{N}$ is a normalization constant (of no physical importance), and each integration under the product is over the complex φ_k-plane. [Equivalently one can integrate over $Re(\varphi_k)$ and $Im(\varphi_k)$.]

Problem 55.1 Do the necessary Fourier transforms specified by (55.4) to verify the results (55.6) and (55.7). *Hint*: Use (55.8).

At this point, it will be important to do some dimensional analysis. We choose $[momentum] = [length]^{-1}$ as the basic dimension, and write $[X] \sim \Lambda^{Dx}$ to mean that a certain quantity X has dimensions $(momentum)^{Dx}$. Requiring that the action is dimensionless: $S[\varphi] \sim \Lambda^0$, it is easy to establish the following dimensional relationships from (55.1) and (55.6):

$$[\varphi(x)] \sim \Lambda^{d/2-1} , \qquad [\varphi_k] \sim \Lambda^{-1} , \qquad [g_2] \sim \Lambda^2 , \qquad [g_4] \sim \Lambda^{4-d} . \tag{55.10}$$

The above dimensions of various quantities, established by simple power counting, are called **canonical dimensions**. It will be convenient to introduce the following dimensionless quantities:

$$\begin{align}
\text{momentum}: \qquad & \tilde{k} = k/\Lambda , & \text{(55.11a)} \\
\text{Fourier field component}: \qquad & \tilde{\varphi}_{\tilde{k}} = \Lambda \varphi_k , & \text{(55.11b)} \\
\text{coupling coefficients}: \qquad & \tilde{g}_2 = \Lambda^{-2} g_2 , \quad \tilde{g}_4 = \Lambda^{d-4} g_4 . & \text{(55.11c)}
\end{align}$$

The partition function Z can then be written in terms of the dimensionless field quantities as

$$Z = \int [d\tilde{\varphi}] \, e^{-S[\tilde{\varphi}]} , \tag{55.12}$$

where

$$S[\tilde{\varphi}] = \frac{1}{2} \sum_{|\tilde{k}|<1} \left(\tilde{k}^2 + 2\tilde{g}_2(\Lambda) \right) \tilde{\varphi}_{\tilde{k}} \tilde{\varphi}_{-\tilde{k}} + S_I[\tilde{\varphi}] , \tag{55.13}$$

with

$$S_I[\tilde{\varphi}] = \tilde{g}_4(\Lambda) \sum_{|\tilde{k}_i|<1} \left(\delta_{\tilde{k}_1+\tilde{k}_2+\tilde{k}_3+\tilde{k}_4}\right) \tilde{\varphi}_{\tilde{k}_1} \tilde{\varphi}_{\tilde{k}_2} \tilde{\varphi}_{\tilde{k}_3} \tilde{\varphi}_{\tilde{k}_4} \,. \tag{55.14}$$

Written in this form, the cutoff Λ has disappeared from consideration for the (dimensionless) momentum variables $\tilde{k}_i$. Its change is solely relegated to the changes in the (dimensionless) coupling coefficients. This will be an important point in our development to follow. A certain φ^4-field theory is then specified by a specific energy scale Λ, whose action is given by (55.13) and (55.14), with Λ-dependent (dimensionless) coupling coefficients $\tilde{g}_2(\Lambda)$ and $\tilde{g}_4(\Lambda)$. Wilson provided the following definition for a *renormalization transformation:*

Definition 55.1. *A **renormalization transformation** for a φ^4-theory specified by an energy scale Λ and coupling coefficients $\tilde{g}_2(\Lambda)$ and $\tilde{g}_4(\Lambda)$ consists in the following transformations of the coupling coefficients in the theory:*

$$\tilde{g}_\alpha(\Lambda) \longrightarrow \tilde{g}'_\alpha(\Lambda/b) \qquad (\alpha = 2, 4) \,, \tag{55.15}$$

with $b > 1$.

Thus, in the execution of one renormalization transformation, the momentum cutoff, or the energy scale of the theory, has been made smaller. Equivalently, the length scale of the theory has been made larger. In colloquial terms, we put on "blurry glasses" and decrease the resolution of the physical picture that we see. Such an interpretation provides a much more intuitive and physical understanding of the meaning of renormalization than that entailed by perturbative renormalization.

The implementation of a renormalization teansformation in momentum space, as proposed by Wilson, consists of three mathematical steps. Roughly speaking, Step (1) involves integrating out large (fast) momentum values in a certain interval immediately below the cutoff, Step (2), the so-called *coarse-graining* step, re-expresses the action in terms of the slower degrees of freedom, and finally, Step (3) rescales the theory to its former scale. The mathematical details for the three steps are presented below, with reference to Fig. 55.1.

(1) *Separation into fast and slow modes.* We decompose an arbitrary Fourier field $\tilde{\varphi}_{\tilde{k}}$ into a "slow" part $\tilde{\varphi}_{\tilde{k}}^{(s)}$ and a "fast" part $\tilde{\varphi}_{\tilde{k}}^{(f)}$ as follows:

$$\tilde{\varphi}_{\tilde{k}} = \tilde{\varphi}_{\tilde{k}}^{(s)} + \tilde{\varphi}_{\tilde{k}}^{(f)} \,, \tag{55.16}$$

where

$$\tilde{\varphi}_{\tilde{k}}^{(s)} \equiv \tilde{\varphi}_{\tilde{k}}\, \theta(1/b - |\tilde{k}|) \,, \qquad \tilde{\varphi}_{\tilde{k}}^{(f)} \equiv \tilde{\varphi}_{\tilde{k}}\, \theta(|\tilde{k}| > 1/b) \qquad (b > 1) \,, \tag{55.17}$$

with $\theta(x)$ being the step function ($= 1$ for $x > 0$, $= 0$ for $x < 0$). Since $\tilde{\varphi}_{\tilde{k}}^{(s)}$ and $\tilde{\varphi}_{\tilde{k}}^{(f)}$ always have disjoint supports (in $\tilde{k}$ space), the functional integration for the partition function Z given by (55.12) can be written

$$Z = \int [d\tilde{\varphi}^{(s)}] \int [d\tilde{\varphi}^{(f)}] \, e^{-S[\tilde{\varphi}^{(s)} + \tilde{\varphi}^{(f)}]} \,. \tag{55.18}$$

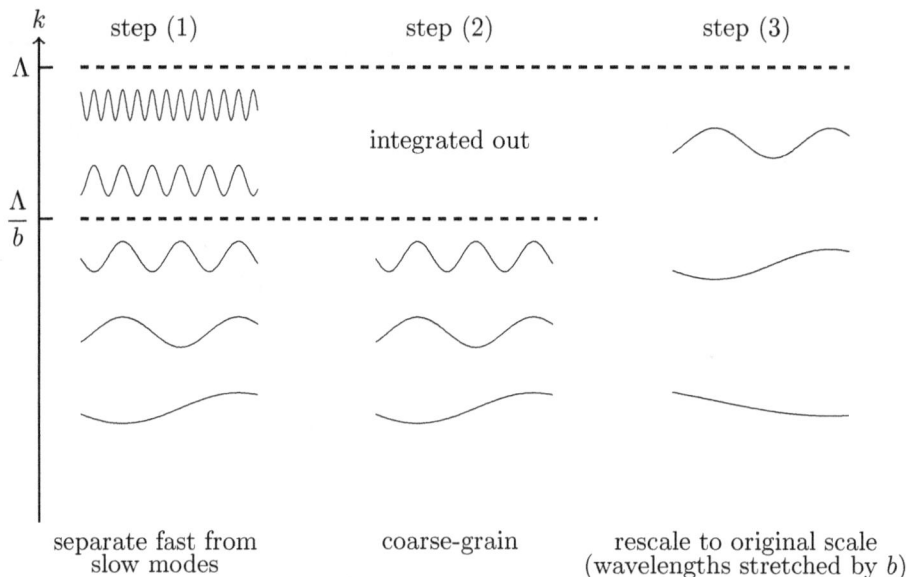

$$\text{FIGURE } 55.1$$

(2) *Coarse graining.* Imagine only doing the functional integration $\int [d\tilde{\varphi}^{(f)}]$ first in the above equation, that is, integrating out the fast field modes. One obtains a quantity depending only on $\tilde{\varphi}_{\tilde{k}}^{(s)}$. Express this quantity in the following way:

$$\int [d\tilde{\varphi}^{(f)}] \, e^{-S[\tilde{\varphi}^{(s)} + \tilde{\varphi}^{(f)}]} = e^{-(S_0 + S'[\tilde{\varphi}^{(s)}])} , \qquad (55.19)$$

where S_0 is independent of any fields, and the action functional $S'[\tilde{\varphi}^{(s)}]$ is written in the form

$$S'[\tilde{\varphi}^{(s)}] = \frac{1}{2} \sum_{|\tilde{k}| < 1/b} \left(z\tilde{k}^2 + r_1 \right) \tilde{\varphi}_{\tilde{k}}^{(s)} \tilde{\varphi}_{-\tilde{k}}^{(s)} + S'_I[\tilde{\varphi}^{(s)}] . \qquad (55.20)$$

This equation effectively defines the constants z, r_1 and the new coupling constants in $S'_I[\tilde{\varphi}^{(s)}]$. The constant z has to be inserted in front of the $\tilde{k}^2$ term because, in this coarse graining step, the scale of the field $\tilde{\varphi}^{(s)}$ has been changed from 1 to $1/b$. The constant z, however, approaches 1 as $b \to 1$. We will assume that

$$z = b^{-\eta} . \qquad (55.21)$$

The quantity η is known as the **anomalous dimension**. The partition function Z in (55.18) can now be written as

$$Z = e^{-S_0} \int [d\tilde{\varphi}^{(s)}] \, e^{-S'[\tilde{\varphi}^{(s)}]} \, . \tag{55.22}$$

(3) *Restoring original scale.* In this last step, the dimensionless momentum cutoff is restored to 1, and the field quantities are renormalized such that the coefficient of the quadratic term is again $1/2$, so as to bring the action functional back to the same mathematical form as (55.13). This is achieved by introducing new dimensionless momenta $\tilde{k}'$ and a new dimensionless field $\tilde{\varphi}'_{\tilde{k}'}$ as follows:

$$\tilde{k}' \equiv b\tilde{k} \, , \qquad \tilde{\varphi}'_{\tilde{k}'} \equiv b^{-1-\eta/2} \, \tilde{\varphi}^{(s)}_{\tilde{k}'/b} \, . \tag{55.23}$$

The renormalized action functional $S'[\tilde{\varphi}']$ is then given by (55.20) to be

$$\begin{aligned} S'[\tilde{\varphi}'] &\equiv S'[\tilde{\varphi}^{(s)}(\tilde{\varphi}')] \\ &= \frac{1}{2} \sum_{|\tilde{k}'|<1} \left(b^{-\eta} b^{-2} \tilde{k}'^2 + r_1 \right) b^{\eta+2} \tilde{\varphi}'_{\tilde{k}'} \tilde{\varphi}'_{-\tilde{k}'} + S'_I[\tilde{\varphi}^{(s)}(\tilde{\varphi}')] \\ &= \frac{1}{2} \sum_{|\tilde{k}'|<1} \left(\tilde{k}'^2 + b^{\eta+2} r_1 \right) \tilde{\varphi}'_{\tilde{k}'} \tilde{\varphi}'_{-\tilde{k}'} + S'_I[\tilde{\varphi}'] \, . \end{aligned} \tag{55.24}$$

Rewriting the dummy summation variable $\tilde{k}'$ as $\tilde{k}$ we have

$$S'[\tilde{\varphi}'] = \frac{1}{2} \sum_{|\tilde{k}|<1} \left(\tilde{k}^2 + r' \right) \tilde{\varphi}'_{\tilde{k}} \tilde{\varphi}'_{-\tilde{k}} + S'_I[\tilde{\varphi}'] \, , \tag{55.25}$$

where

$$r' \equiv b^{\eta+2} r_1 \equiv 2\tilde{g}'_2 \, . \tag{55.26}$$

Analogous to (55.14) we write

$$S'_I[\tilde{\varphi}'] = \tilde{g}'_4 \sum_{|\tilde{k}_i|<1} (\delta_{\tilde{k}_1+\tilde{k}_2+\tilde{k}_3+\tilde{k}_4}) \, \tilde{\varphi}'_{\tilde{k}_1} \tilde{\varphi}'_{\tilde{k}_2} \tilde{\varphi}'_{\tilde{k}_3} \tilde{\varphi}'_{\tilde{k}_4} \, . \tag{55.27}$$

This equation determines the new (renormalized) coupling constant $\tilde{g}'_4$ in terms of the old ones $\tilde{g}_2$ and $\tilde{g}_4$. An RG transformation in the φ^4-theory thus amounts to a transformation of the coupling constants

$$\{\tilde{g}_2, \tilde{g}_4\} \longrightarrow \{\tilde{g}'_2(b; \tilde{g}_2, \tilde{g}_4), \, \tilde{g}'_4(b; \tilde{g}_2, \tilde{g}_4)\} \, . \tag{55.28}$$

The continuous RG group parameter is identified as the (dimensionless) scale variable b. In practice it is more advantageous to express it in terms of an *additive* parameter t defined by

$$b = e^{t-t_0} \, , \tag{55.29}$$

where t_0 is some fixed real number (not necessarily positive). It has the meaning that, at $t = t_0$, the cutoff momentum for the theory is Λ and the coupling constants assume their bare values. The RG equations can then be expressed as a set of coupled first-order differential equations

$$\frac{d\tilde{g}_i}{dt} = \beta_i(t; \tilde{g}_2, \tilde{g}_4) \qquad (i = 2, 4) . \tag{55.30}$$

The functions $\beta_i(t; \tilde{g}_2, \tilde{g}_4)$ above (calculable by the three RG steps identified earlier), are known as the **beta functions** in the physics RG literature. Mathematically, they give the *tangent vector field* to the RG flow in the coupling-constant manifold. The key to studying the qualitative features of the flow is then the identification of the *fixed points* of the flow, determined by the zeros of the beta functions. We will examine the mathematical details of this procedure by specific examples subsequently.

It is seen that, in the flow direction with $t - t_0 > 0$, $b > 1$, and hence the momentum cutoff is scaled downwards. Equivalently, in this direction, the flow is towards theories characterized by lower energy scales, or larger physical length dimensions. The fixed points, in the vicinity of which such a flow may wind up as t increases, are then known as **infrared fixed points**. Such fixed points are the relevant ones in the study of *critical phenomena* in statistical physics, since, in the critical region (near *phase transition* boundaries) of the thermodynamic phase space, the so-called *correlation length* of the system approaches infinity. On the other hand, fixed points from which RG flow lines emanate in all directions as $t - t_0$ increases are known as **ultraviolet fixed points**. Such fixed points are relevant, quite obviously, in high energy physics. We will consider critical phenomena in more detail in the next chapter.

In general the mathematical machinery of *dynamical systems* will be suitable for the study of a system of coupled equations such as (55.30). Indeed the techniques therein will be useful for higher-dimensional $(n > 2)$ coupling constant spaces also. Focusing on our 2-dimensional case, around a fixed point $\tilde{\boldsymbol{g}}_* = (\tilde{g}_{2*}, \tilde{g}_{4*})$ we can linearize the equations (55.30) by Taylor expanding the beta functions β_i around $\tilde{\boldsymbol{g}}_*$:

$$\beta_i \approx \sum_j \left.\frac{\partial \beta_i}{\partial \tilde{g}_j}\right|_{\tilde{\boldsymbol{g}}_*} (\tilde{g}_j - \tilde{g}_{j*}) = \sum_j R_{ij} (\tilde{g}_j - \tilde{g}_{j*}) \qquad (i, j = 2, 4) , \tag{55.31}$$

where the constant matrix R_{ij} is called the **renormalization matrix** (RG matrix). So near a fixed point in coupling constant space we have

$$\frac{d}{dt}(\tilde{g}_i - \tilde{g}_{i*}) = \sum_j R_{ij} (\tilde{g}_j - \tilde{g}_{j*}) . \tag{55.32}$$

The problem then reduces to the linear algebra one of the determination of eigenvectors and eigenvalues of linear operators. The reader may want to review Chapter 17 of this text, or see, for example, Lam 2014 for a detailed discussion of the 2-dimensional case. Suppose, with respect to a fixed point $\tilde{\boldsymbol{g}}_*$ in coupling

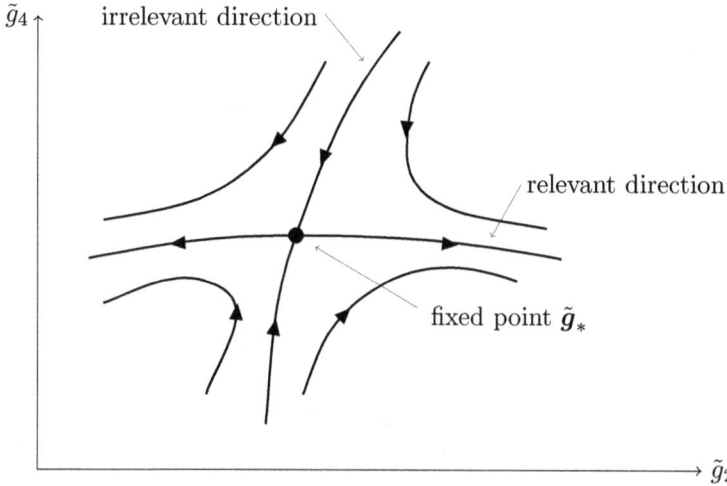

$\tilde{g}_4$ — irrelevant direction — relevant direction — fixed point $\tilde{g}_*$ — $\tilde{g}_2$

FIGURE 55.2

constant space, the vector $\boldsymbol{u}$ is an eigenvector of R with real eigenvalue λ, that is,

$$Ru = \lambda u \ . \tag{55.33}$$

Then, along an eigen-direction of $\boldsymbol{u}$,

$$\tilde{g}(t) - \tilde{g}_* = c\, e^{\lambda t} \ , \tag{55.34}$$

where c is some constant vector. The following terminology is commonly used int the physics RG literature. If $\lambda > 0$, the corresponding eigen-direction is called a **relevant direction**. If $\lambda < 0$, the corresponding eigen-direction is called an **irrelevant direction**. The situation $\lambda = 0$ corresponds to a so-called **marginal direction**, for which the linear approximation (55.32) no longer suffices. An RG trajectory then flows away from a fixed point along a relevant direction, and towards it along an irrelevant direction (see Fig. 55.2).

The Wilsonian approach to the RG can be shown to be equivalent, in terms of mathematical outcomes, to the procedure of perturbative renormalization discussed in earlier chapters. Let us consider the following (unrenormalized) n-point Green's function:

$$G^{(n)}(k_1, \ldots, k_n; g_2, g_4, \Lambda) = \frac{\int [d\varphi]\, e^{-S[\varphi]}\, (\varphi_{k_1}^{(s)} \cdots \varphi_{k_n}^{(s)})}{\int [d\varphi]\, e^{-S[\varphi]}} \ , \tag{55.35}$$

where only correlations among slow momentum modes $0 < |\mathbf{k}_i| < \Lambda/b$ ($b > 0$) are of interest, $S[\varphi]$ is given by (55.6), and the coupling constants g_2 and g_4 appearing in the left-hand side are the bare coupling constants in (55.1). The denominator on the right-hand side is the partition function Z of (55.3). Expressing the right-hand side in terms of dimensionless quantities via (55.11) we have

$$G^{(n)}(\mathbf{k}_i; g_2, g_4, \Lambda) = \Lambda^{-n}\mathcal{G}^{(n)}(\tilde{\mathbf{k}}_i; \tilde{g}_2, \tilde{g}_4) , \tag{55.36}$$

where the dimensionless quantities $\tilde{\mathbf{k}}_i, \tilde{g}_2$ and $\tilde{g}_4$ are defined by (55.11); and

$$\mathcal{G}^{(n)}(\tilde{\mathbf{k}}_i; \tilde{g}_2, \tilde{g}_4) = \frac{\int [d\tilde{\varphi}]\, e^{-S[\tilde{\varphi}]}\, (\tilde{\varphi}^{(s)}_{\tilde{\mathbf{k}}_1} \cdots \tilde{\varphi}^{(s)}_{\tilde{\mathbf{k}}_n})}{\int [d\tilde{\varphi}]\, e^{-S[\tilde{\varphi}]}} . \tag{55.37}$$

In this equation $S[\tilde{\varphi}]$ is given by (55.13) and (55.14). We now implement Step (2) above (the coarse-graining step, integrating out the fast momentum modes with magnitudes between Λ/b and Λ), and obtain

$$\mathcal{G}^{(n)}(\tilde{\mathbf{k}}_i; \tilde{g}_2, \tilde{g}_4) = \frac{\int [d\tilde{\varphi}^{(s)}](\tilde{\varphi}^{(s)}_{\tilde{\mathbf{k}}_1} \cdots \tilde{\varphi}^{(s)}_{\tilde{\mathbf{k}}_n}) \int [d\tilde{\varphi}^{(f)}]\, e^{-S[\tilde{\varphi}^{(s)} + \tilde{\varphi}^{(f)}]}}{\int [d\tilde{\varphi}^{(s)}] \int [d\tilde{\varphi}^{(f)}]\, e^{-S[\tilde{\varphi}^{(s)} + \tilde{\varphi}^{(f)}]}}$$
$$= \frac{\int [d\tilde{\varphi}^{(s)}]\, (\tilde{\varphi}^{(s)}_{\tilde{\mathbf{k}}_1} \cdots \tilde{\varphi}^{(s)}_{\tilde{\mathbf{k}}_n})\, e^{-S'[\tilde{\varphi}^{(s)}]}}{\int [d\tilde{\varphi}^{(s)}]\, e^{-S'[\tilde{\varphi}^{(s)}]}} , \tag{55.38}$$

where the e^{-S_0} factor in (55.19) appears in both the numerator and the denominator and thus cancels out, and $S'[\tilde{\varphi}^{(s)}]$ is given by (55.20). Next we implement Step (3) above and use (55.23) to write

$$\tilde{\varphi}^{(s)}_{\tilde{\mathbf{k}}_i} = \tilde{\varphi}^{(s)}_{\tilde{\mathbf{k}}'_i/b} = b^{\eta/2+1}\, \tilde{\varphi}'_{b\tilde{\mathbf{k}}_i} . \tag{55.39}$$

Hence

$$\tilde{\varphi}^{(s)}_{\tilde{\mathbf{k}}_1} \cdots \tilde{\varphi}^{(s)}_{\tilde{\mathbf{k}}_n} = b^n b^{\eta n/2}\tilde{\varphi}'_{b\mathbf{k}_1} \cdots \tilde{\varphi}'_{b\mathbf{k}_n} = (z^{-1})^{n/2}b^n\, \tilde{\varphi}'_{b\mathbf{k}_1} \cdots \tilde{\varphi}'_{b\mathbf{k}_n} , \tag{55.40}$$

where we have used (55.21) for z. So we can write

$$\mathcal{G}^{(n)}(\tilde{\mathbf{k}}_i; \tilde{g}_2, \tilde{g}_4) = z^{-n/2}b^n \frac{\int [d\tilde{\varphi}']\, (\tilde{\varphi}'_{b\mathbf{k}_1} \cdots \tilde{\varphi}'_{b\mathbf{k}_n})\, e^{-S'[\tilde{\varphi}']}}{\int [d\tilde{\varphi}']\, e^{-S'[\tilde{\varphi}']}} , \tag{55.41}$$

where the action functional $S'[\tilde{\varphi}']$ is given by (55.25) to (55.27). This gives the following RG transformation rule for $\mathcal{G}^{(n)}(\tilde{\mathbf{k}}_i; \tilde{g}_2, \tilde{g}_4)$:

$$\mathcal{G}^{(n)}(\tilde{\mathbf{k}}_i; \tilde{g}_2, \tilde{g}_4) = z^{-n/2}b^n\, \mathcal{G}^{(n)}(b\tilde{\mathbf{k}}_i; \tilde{g}'_2, \tilde{g}'_4) = b^{n+\eta n/2}\, \mathcal{G}^{(n)}(b\tilde{\mathbf{k}}_i; \tilde{g}'_2, \tilde{g}'_4) , \tag{55.42}$$

where $\tilde{g}'_2$ and $\tilde{g}'_4$ are the renormalized coupling constants. Now choose

$$b = \frac{\Lambda}{\mu} , \tag{55.43}$$

where μ is the physical mass, for example. But it can be any finite quantity fixing the renormalization scale. To bring the Wilsonian approach into contact with perturbative renormalization, we let $\Lambda \to \infty$ while holding μ fixed. This makes $b \to \infty$, and so the RG flow starts from an ultraviolet fixed point. Multiplying both sides of (55.38) by $\Lambda^{-n} = b^{-n}\mu^{-n}$, we have

$$\Lambda^{-n}\, \mathcal{G}^{(n)}(\tilde{k}_i; \tilde{g}_2, \tilde{g}_4) = z^{-n/2}\mu^{-n}\, \mathcal{G}^{(n)}(b\tilde{k}_i; \tilde{g}_2', \tilde{g}_4')\,, \tag{55.44}$$

and so it follows from (55.32) that

$$G^{(n)}(\boldsymbol{k}_i; g_2, g_4, \Lambda) = z^{-n/2}\mu^{-n}\, \mathcal{G}^{(n)}(b\tilde{k}_i; \tilde{g}_2', \tilde{g}_4')\,. \tag{55.45}$$

Noting that

$$b\tilde{\boldsymbol{k}}_i = \frac{\Lambda}{\mu}\tilde{\boldsymbol{k}}_i = \frac{\Lambda}{\mu}\frac{\boldsymbol{k}_i}{\Lambda} = \frac{\boldsymbol{k}_i}{\mu}\,, \tag{55.46}$$

we can write (55.41) as

$$G^{(n)}(\boldsymbol{k}_i; g_2, g_4, \Lambda) = \left(b^{-\eta}\right)^{-n/2}\mu^{-n}\mathcal{G}^{(n)}(\boldsymbol{k}_i/\mu; \tilde{g}_2', \tilde{g}_4')\,. \tag{55.47}$$

Defining the renormalized n-point Green's function $\overline{G}^{(n)}(\boldsymbol{k}_i; \tilde{g}_2', \tilde{g}_4', \mu)$ by

$$\overline{G}^{(n)}(\boldsymbol{k}_i; \tilde{g}_2', \tilde{g}_4', \mu) = \mu^{-n}\mathcal{G}^{(n)}(\boldsymbol{k}_i/\mu; \tilde{g}_2', \tilde{g}_4', \mu)\,, \tag{55.48}$$

Eq. (55.43) can be rewritten as

$$G^{(n)}(\boldsymbol{k}_i; g_2, g_4, \Lambda) = \left[\left(\frac{\Lambda}{\mu}\right)^{\eta}\right]^{n/2}\overline{G}^{(n)}(\boldsymbol{k}_i; \tilde{g}_2', \tilde{g}_4', \mu)\,. \tag{55.49}$$

On the right-hand side the cutoff Λ appears explicitly only in the renormalization factor in front, while the renormalized Green's function $\overline{G}^{(n)}$ only depends on the renormalized coupling constants and the physical mass μ. This equation is precisely of the same form as (52.30), which is a statement of multiplicative renormalizability. We have thus established the equivalence between the Wilsonian approach to renormalization and perturbative renormalization.

Corresponding to the momentum space RG three-step procedure introduced by Wilson as described above there is the so-called **real space RG** three-step procedure. This physically more intuitive procedure originated from Kadanoff's **block-spin transformation** in the study of spin systems on a discrete lattice (Kadanoff 1977). In the present context we will call the original lattice points microscopic sites. At each microscopic site $\boldsymbol{x}$ (in the d-dimensional lattice) is associated a (discretized) field $\varphi_{\boldsymbol{x}}$ (a real-valued field defined only at the discretized lattice points). Let $a = 1/\Lambda$ be the microscopic lattice spacing. Going from the continuum space ($a \to 0, \Lambda \to \infty$) operations to discrete ones we have

$$\int d^d x \longrightarrow a^d \sum_{\boldsymbol{x}}\,, \qquad \partial\varphi(\boldsymbol{x}) \longrightarrow \frac{1}{a}(\varphi_{\boldsymbol{x}} - \varphi_{\boldsymbol{y}})\,, \tag{55.50}$$

where x and y above are nearest neighbor microscopic sites. Then, corresponding to (55.1) we have the following discretized action functional for the φ^4-theory:

$$S[\varphi] = \frac{a^{d-2}}{2} \sum_{\langle x,y \rangle} (\varphi_x - \varphi_y)^2 + a^d \sum_x \left(g_2 \varphi_x^2 + g_4 \varphi_x^4 \right) , \tag{55.51}$$

where the notation $\langle x, y \rangle$ in the sum for the kinetic term means summing over all nearest-neighbor pairs. Requiring $S[\varphi]$ to be dimensionless, we have $[\varphi_x] \sim a^{1-d/2}$, $[g_2] \sim a^{-2}$ and $g_4 \sim a^{d-4}$ [c.f. (55.10)]. Thus we introduce the following dimensionless quantities [c.f. (55.11)]:

$$\tilde{\varphi}_x = a^{d/2-1} \varphi_x , \qquad \tilde{g}_2 = a^2 g_2 , \qquad \tilde{g}_4 = a^{4-d} g_4 , \tag{55.52}$$

and write (55.51) as

$$S[\tilde{\varphi}] = \frac{1}{2} \sum_{\langle x,y \rangle} (\tilde{\varphi}_x - \tilde{\varphi}_y)^2 + \sum_x \left(\tilde{g}_2 \tilde{\varphi}_x^2 + \tilde{g}_4 \tilde{\varphi}_x^4 \right) . \tag{55.53}$$

The partition function Z is then given by the same formal equation as (55.12), with the understanding that now the field is defined on (discretized) real space rather than momentum space. We will repeat the equation below:

$$Z = \int [d\tilde{\varphi}] \, e^{-S[\tilde{\varphi}]} , \tag{55.54}$$

where the functional integration measure is given by

$$\int [d\tilde{\varphi}] = \prod_x \int d\tilde{\varphi}_x . \tag{55.55}$$

We will now describe the real-space RG three-step procedure now, with reference to Fig. 55.3.

(1) Grouping microscopic sites into blocks. Make blocks in the microscopic lattice, with l consecutive sites on a side of each block. Define the average field $\langle \tilde{\varphi} \rangle_X$ inside a block (specified by some mean position X inside it), by

$$\langle \tilde{\varphi} \rangle_X \equiv \frac{1}{l^d} \sum_{x \in block\ X} \tilde{\varphi}_x . \tag{55.56}$$

Note that the above definition of the average field is not unique. For a particular *microscopic-field configuration* $\{\tilde{\varphi}_x\}$ (meaning a specified set of values $\tilde{\varphi}_x$ for all x), the above definition yields a particular *block-field configuration* $\{\tilde{\varphi}'_X\}$ given by

$$\tilde{\varphi}'_X \equiv \langle \tilde{\varphi} \rangle_X . \tag{55.57}$$

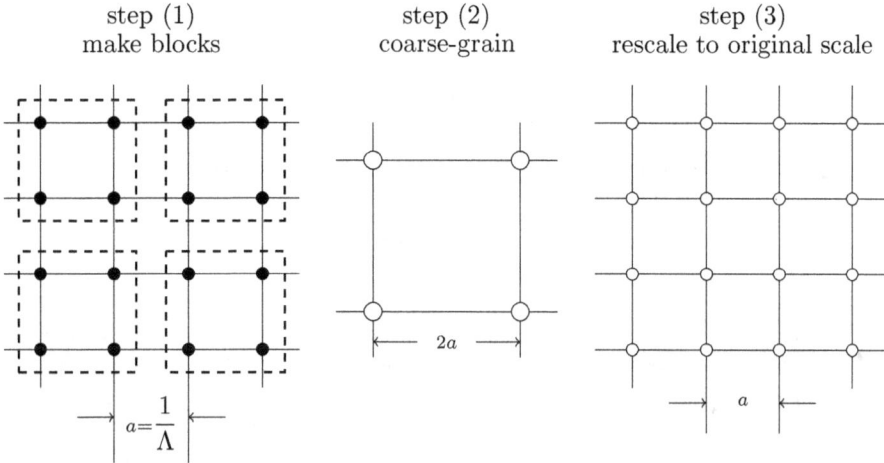

step (1)
make blocks

step (2)
coarse-grain

step (3)
rescale to original scale

FIGURE 55.3

For a general block field configuration $\tilde{\varphi}'_{\mathbf{X}}$, however, there may in general be several *different* microscopic field configurations giving rise to it. We then introduce a *weight function*, $P(\{\tilde{\varphi}'_{\mathbf{X}}\}, \{\tilde{\varphi}_{\mathbf{x}}\})$, mapping a pair of block-field and microscopic-field configurations to $\{0, 1\}$, by

$$P(\{\tilde{\varphi}'_{\mathbf{X}}\}, \{\tilde{\varphi}_{\mathbf{x}}\}) \equiv \prod_{\mathbf{X}} \delta_{\tilde{\varphi}'_{\mathbf{X}} - \langle \tilde{\varphi} \rangle_{\mathbf{X}}} , \tag{55.58}$$

where δ denotes the Kronecker delta. It ensures that the product (over all block-field sites $\mathbf{X}$) on the right-hand side gives 1 when the microscopic-field configuration $\{\tilde{\varphi}_{\mathbf{x}}\}$ gives rise to the block-field configuration $\{\tilde{\varphi}'_{\mathbf{X}}\}$, and gives 0 otherwise. The weight function is so called because it satisfies the following conditions, for a particular microscopic-field configuration $\{\tilde{\varphi}_{\mathbf{x}}\}$,

$$P(\{\tilde{\varphi}'_{\mathbf{X}}\}, \{\tilde{\varphi}_{\mathbf{x}}\}) \geq 0 , \tag{55.59a}$$

$$\sum_{\{\tilde{\varphi}'_{\mathbf{X}}\}} P(\{\tilde{\varphi}'_{\mathbf{X}}\}, \{\tilde{\varphi}_{\mathbf{x}}\}) = 1 , \tag{55.59b}$$

where in the second condition the sum is over all block-field configurations. Expressing this sum as the functional integral

$$\sum_{\{\tilde{\varphi}'_{\mathbf{X}}\}} = \int [d\tilde{\varphi}'] = \prod_{\mathbf{X}} d\tilde{\varphi}'_{\mathbf{X}} , \tag{55.60}$$

Eq. (55.59b), by virtue of (55.58), can be written as

$$\int [d\tilde{\varphi}'] \prod_X \delta_{\tilde{\varphi}'_X - \langle \tilde{\varphi} \rangle_X} = 1 . \tag{55.61}$$

The partition function Z [c.f. (55.54)] can then be written

$$Z = \int [d\tilde{\varphi}'] \int [d\tilde{\varphi}] \prod_X \delta_{\tilde{\varphi}'_X - \langle \tilde{\varphi} \rangle_X} \, e^{-S[\tilde{\varphi}]} . \tag{55.62}$$

(2) *Coarse-graining the microscopic field.* Holding the block-field $\tilde{\varphi}'_X$ fixed in (55.62), we first integrate over the microscopic field $\tilde{\varphi}_x$. Analogous to (55.19) we define a new action functional $S'[\tilde{\varphi}']$ by

$$\int [d\tilde{\varphi}] \prod_X \delta_{\tilde{\varphi}'_X - \langle \tilde{\varphi} \rangle_X} \, e^{-S[\tilde{\varphi}]} = e^{-S_0} \, e^{-S'[\tilde{\varphi}']} , \tag{55.63}$$

where S_0 is a constant [different from the S_0 in (55.19) and independent of field quantities]. Then, analogous to (55.22), we obtain

$$Z = e^{-S_0} \int [d\tilde{\varphi}'] \, e^{-S'[\tilde{\varphi}']} . \tag{55.64}$$

Analogous to (55.20), we express the new (coarse-grained) action functional $S'[\tilde{\varphi}']$ as

$$S'[\tilde{\varphi}'] = \frac{z}{2} \sum_{\langle X, Y \rangle} (\tilde{\varphi}'_X - \tilde{\varphi}'_Y)^2 + \sum_X \left\{ \tilde{g}''_2 (\tilde{\varphi}'_X)^2 + \tilde{g}''_4 (\tilde{\varphi}'_X)^4 \right\} , \tag{55.65}$$

where $\langle X, Y \rangle$ denotes summation over all nearest-neighbor block-field sites, and, analogous to (55.21), we can set

$$z \equiv l^{-\eta} , \tag{55.66}$$

with η being the anomalous dimension.

(3) *Restoring original microscopic scale.* After coarse-graining the (block-field) lattice size is increased from that of the microscopic lattice size by a factor of $l \, (> 1)$. We will now restore the original microscopic scale by setting

$$x' \equiv X/l , \qquad \tilde{\varphi}'_{x'} \equiv l^{-\eta/2} \tilde{\varphi}'_{lx'} , \tag{55.67}$$

so as to bring the new action functional to the same mathematical form as the original one:

$$S'[\tilde{\varphi}'] = \frac{1}{2} \sum_{\langle x, y \rangle} (\tilde{\varphi}'_x - \tilde{\varphi}'_y)^2 + \sum_x (\tilde{g}'_2 (\tilde{\varphi}'_x)^2 + \tilde{g}'_4 (\tilde{\varphi}'_x)^4) , \tag{55.68}$$

where (x', y') have been written as (x, y) because these are dummy summation variables, $\langle x, y \rangle$ denotes summation over nearest-neighbor microscopic sites, and $(\tilde{g}'_2, \tilde{g}'_4)$ are the renormalized coupling constants. The
partition function is still given by (55.64), but now the functional integration measure is given by

$$\int [d\tilde{\varphi}'] = \prod_x \int d\tilde{\varphi}'_x \,, \qquad (55.69)$$

where the product is over the microscopic lattice points again. Comparing
(55.53) and (55.68), we see that, as in the momentum-space three-step RG
transformation, the real-space three-step RG transformation also leads to
a transformation of the coupling constants [compare with (55.28)]:

$$\{\tilde{g}_2 . \tilde{g}_4\} \longrightarrow \{\tilde{g}'_2(l; \tilde{g}_2, \tilde{g}_4), \tilde{g}'_4(l; \tilde{g}_2, \tilde{g}_4)\} \,. \qquad (55.70)$$

From this point on, the analysis for an RG transformation proceeds as that
given after (55.28). But we will take advantage of the more physically intuitive
picture offered by the real-space RG procedure to highlight some additional
important concepts, in the context of a more general theory (than the φ^4-theory)
with N coupling constants for $N \geq 2$.

In the theory of phase transitions in statistical mechanics, the concept of
the **correlation length** ξ plays a key role. It is defined in terms of a 2-point
(real-space) Green's function by

$$\langle \varphi(x)\varphi(y) \rangle \xrightarrow[|x-y| \to \infty]{} Ce^{-|x-y|/\xi} \,, \qquad (55.71)$$

where the angular brackets usually denote an ensemble average [c.f. (49.2)] and
C is a quantity depending on some power of $|x - y|$ at most. A corresponding
dimensionless correlation length $\tilde{\xi}$ can be defined as

$$\xi = \tilde{\xi} \Lambda \,. \qquad (55.72)$$

In real-space RG, a renormalization step is defined by a change of length scale
$|x| \to |x|/l$ $(l > 1)$. At a fixed point, then, in the coupling-constant space, a
physical system becomes invariant under a change of length scale. This implies
that, at a fixed point, the correlation length ξ is either 0 or ∞. In statistical
mechanics, a correlation length of $\xi = \infty$, defines a so-called *critical point* in
thermodynamic phase space in the theory of second-order phase transitions.
We will see later that the case $\xi = 0$ corresponds to the situation where the
temperature $T \to \infty$, and is thus not physically reelvant.

To simplify the notation, we can choose the fixed point $\tilde{g}_*$ to be at the origin
of the N-dimensional coupling-constant space. So (55.32) implies, near the fixed
point,

$$\frac{d}{dt}\tilde{g}_i = \sum_j R_{ij}\tilde{g}_j \equiv \tilde{h}_i \qquad (i, j = 1, \ldots, N) \,, \qquad (55.73)$$

where R_{ij} is the RG transformation matrix given by (55.31). Let us use the Dirac notation to write

$$| \tilde{\boldsymbol{h}} \rangle \equiv (\tilde{h}_1, \ldots, \tilde{h}_N)^T , \qquad | \tilde{\boldsymbol{g}} \rangle = (\tilde{g}_1, \ldots, \tilde{g}_N)^T , \qquad (55.74)$$

where T means transpose. So we have

$$| \tilde{\boldsymbol{h}} \rangle = R | \tilde{\boldsymbol{g}} \rangle , \qquad (55.75)$$

where R denotes the linear operator in coupling-constant space represented by the matrix R_{ij}. Introduce the left-eigenvectors $\langle \phi_\beta | = (\phi_{\beta 1}, \ldots, \phi_{\beta N})$ of R defined by

$$\langle \phi_\beta | R = \lambda'_\beta \langle \phi_\beta | \qquad (\beta = 1, \ldots, N) ; \qquad (55.76)$$

or, in matrix notation,

$$\sum_{\gamma=1}^{N} \phi_{\beta\gamma} R_{\gamma\alpha} = \lambda'_\beta \, \phi_{\beta\alpha} \qquad (\alpha, \gamma = 1, \ldots, N) . \qquad (55.77)$$

Then

$$\langle \phi_\beta | \tilde{\boldsymbol{h}} \rangle = \langle \phi_\beta | R | \tilde{\boldsymbol{g}} \rangle = \lambda'_\beta \langle \phi_\beta | \tilde{\boldsymbol{g}} \rangle . \qquad (55.78)$$

We will define the so-called **scaling fields** (defined on the coupling-constant space, which we will now denote by M_{cc}) by

$$v'_\beta(\tilde{\boldsymbol{h}}) \equiv \langle \phi_\beta | \boldsymbol{h} \rangle , \qquad v_\beta(\tilde{\boldsymbol{g}}) \equiv \langle \phi_\beta | \tilde{\boldsymbol{g}} \rangle . \qquad (55.79)$$

Thus an RG transformation can be specified by the following transformation of the scaling fields:

$$v_\beta \longrightarrow v'_\beta = \lambda'_\beta v_\beta \qquad (\beta = 1, \ldots, N) . \qquad (55.80)$$

In the right-hand side of the above there is no sum over β. Under an RG transformation a scaling field scales with a factor λ', hence the name. Note that the eigenvalues λ'_β are the same as the λ in (55.33), assuming both are real. In general, we can write

$$\lambda'_\beta = l^{y_\beta(\Lambda)} , \qquad (55.81)$$

where y_β plays the role of an anomalous dimension. In analogy to the earlier definitions for relevant, irrelevant, and marginal directions, we have the following equivalent definitions for scaling fields, which can be thought of as *cotangent vectors* on M_{cc}, dual to the tangent vectors $\boldsymbol{u}$ of (55.33). Note that we assume that the coupling-constant space M_{cc} is a linear vector space, for simplicity. If $\lambda'_\beta > 1$, v_β is called a **relevant scaling field**. If $\lambda'_\beta < 1$, v_β is called an **irrelevant scaling field**. If $\lambda'_\beta = 0$, the v_β is called a **marginal scaling field**.

The notion of scaling fields gives rise to that of *critical surfaces* in the coupling-constant manifold.

Definition 55.2. *The **critical surface** S_c for a particular fixed point $\tilde{g}_*$ in a coupling-constant space M_{cc} is the surface containing $\tilde{g}_*$ on which all relevant scaling fields v_β (with $\lambda'_\beta > 0$) vanish.*

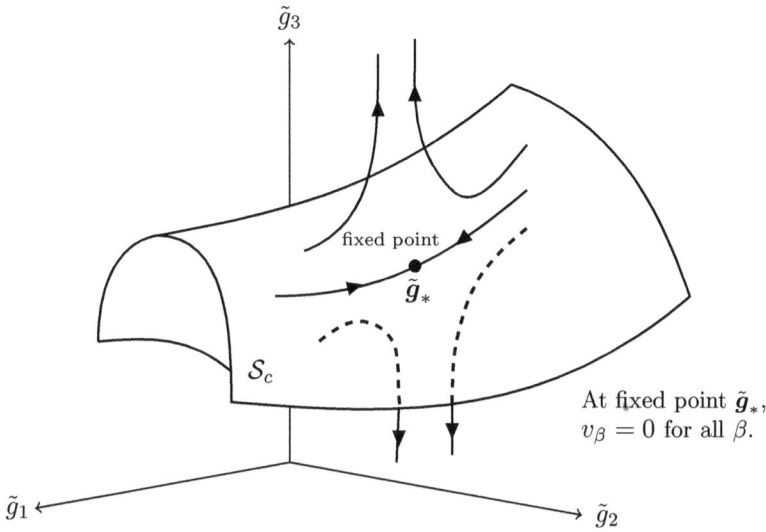

FIGURE 55.4

We show in Fig.55.4 a schematic picture of RG flows on and near the critical surface of an infrared fixed point in a 3-dimensional coupling-constant space. Note that if $dim(M_{cc}) = 2$, the critical surface becomes a critical line.

We will see in the next chapter that in the theory of second-order phase transitions in statistical mechanics, points on a critical surface correspond to physical systems in the same *universality class*. Such systems share the same *critical exponents* in their second-order phase transitions.

Problem 55.2 Justify (55.81) by assuming that a scaling field v has dimension $[v] \sim [length]^{-y}$ and then use (55.80). Also recall that under a real-space RG transformation step,

$$\text{(old length)} \longrightarrow \text{(new length)} = \text{(old length)}/l \qquad (l > 1) . \tag{55.82}$$

Chapter 56

Critical Phenomena, the Landau-Ginsburg Theory, Mean-Field Theory

Starting with this chapter we will use the language of statistical mechanics, in particular, with respect to the problem of *critical phenomena* in *second-order phase transitions*, to continue to explore the applications of the Wilsonian approach to the renormalization group in quantum field theory. Historically, this was actually how Wilson's approach to RG was first applied (Wilson and Kogut 1974). A brief review of the theoretical underpinnings of critical phenomena is then in order before we proceed to the mathematical details. To do this in a focused manner and avoid long digressions into a discussion of the rich array of physical systems that exhibit this kind of phenomena, we will choose a single and commonly used example, that of *ferromagnetism*, which, unlike some others, such as *superfluidity* and *superconductivity*, has the advantage of being intuitively appealing. (For more detailed introductions on the physics of this problem see, for example, Ma 1976.) The presentation here is based on Huang 1987.

In the thermodynamic description of ferromagnetism, the thermodynamic variables are the triplet (M, H, T), where M is the magnetization, H is the externally applied magnetic field, and T is the absolute temperature. There is an *equation of state* of the form

$$f(M, H, T) = 0 \,, \tag{56.1}$$

which gives a two-dimensional surface in the three-dimensional (M, H, T) thermodynamic phase space (see Fig. 56.1). A point on this surface represents a *thermodynamic state* of the system. The central thermodynamic function of importance is the **Gibbs free energy** $G(H, T)$, defined in terms of the *partition function* $Q(H, T)$ by

$$Q(H, T) = e^{-G(H,T)/(kT)} = Tr\left\{-\mathcal{H}/(kT)\right\} \,, \tag{56.2}$$

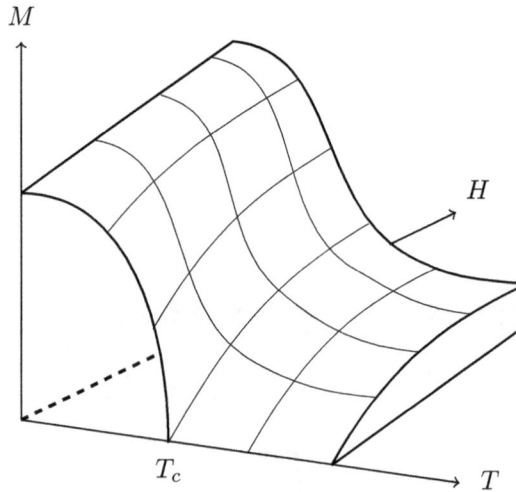

FIGURE 56.1

where $\mathcal{H}$ is the Hamiltonian of the system, the trace is over all states in the *canonical ensemble*, and k is *Boltzmann's constant*. From the Gibbs free energy one can calculate measurable thermodynamic *response functions* of interest such as the *magnetization* M, the *susceptibility* χ, and the *specific heat* C:

$$M = -\frac{\partial G}{\partial H}, \qquad \chi = \frac{1}{V}\frac{\partial M}{\partial H} = -\frac{1}{V}\frac{\partial^2 G}{\partial H^2}, \qquad C = -T\frac{\partial^2 G}{\partial T^2}. \qquad (56.3)$$

In the second equation above, V is the real-space volume of the system. We note that in a ferromagnetic system the ordered triplet of variables (H, M, T) plays the same role as the ordered triplet (P, V, T) in the more familiar liguid-gas systems. In thermodynamics, the variables H and P, which do not depend on the size of the system, are known *intensive variables*, while the variables M and V, which depend on the size of the system, are known as *extensive variables*.

In general critical phenomena are those exhibited by a system near the so-called **critical temperature** T_c, at which **phase transitions** take place. Referring to Fig. 56.1, for example, with zero magnetic field M, the system shows an abrupt change in the magnetization M (from zero to non-zero values) as it passes the critical temperature T_c. This is an example of a so-called *second-order phase transition*. Traditionally, *phase transitions are classified according to the degree of singularity of the derivatives of the Gibbs free energy at the critical temperature with respect to its natural variables.* Systems for which the first derivatives are discontinuous are said to exhibit **first-order phase transi-**

tions, while systems for which the first derivatives at T_c are continuous, but the second derivatives at T_c are discontinuous, are said to exhibit **second-order phase transitions**. Strictly speaking, according to this scheme, there may be third-order, fourth-order phase transitions, and so on. But there is an important feature that separates first-order from higher-order phase transitions: A preponderance of experimental evidence based on diverse systems suggest that *for orders of phase transitions higher than the first, near the critical point, the correlation length ξ is the only characteristic length of the system, and $\xi \to \infty$ as $T \to T_c$ for these systems.* In fact, this statement is a central hypothesis known as the **scaling hypothesis** in the study of second-order phase transitions. Physically, this means that field disturbances at very distant locations in the physical system have very large effects on each other, and specific local features of individual systems wash out near the critical point. Equivalently, fluctuations of very disparate wavelengths are equally important. This is the basic reason behind the so-called phenomenon of *universality* in systems exhibiting second-order phase transitions. In the physics literature, the term "second-order phase transitions" is often used in the sense of "not first-order". We will adopt this usage in our presentation.

Critical phenomena are usually described mathematically by the behavior of thermodynamic response functions such as M, χ and C near the critical temperature T_c. For this description it will be convenient to define the "dimensionless" temperature t:

$$t \equiv \frac{T - T_c}{T_c} .\tag{56.4}$$

We put the term "dimensionless" in quotes here because, as will be shown later, it will be useful to associate a so-called *canonical dimension D_t* with the quantity t [see (56.48)]. In the limit $t \to 0$, experimental evidence and exact calculations based on specific models (such as the 2-dimensional *Ising spin system*) suggest the following *universal* singular behavior for various response functions:

$$\text{heat capacity}: \qquad C \sim |t|^{-\alpha} \qquad\qquad (H = 0), \tag{56.5}$$

$$\text{magnetization}: \qquad M \sim |t|^{\beta} \qquad\qquad (H = 0, t < 0), \tag{56.6}$$

$$\text{susceptibility}: \qquad \chi \sim |t|^{-\gamma} \qquad\qquad (H = 0), \tag{56.7}$$

$$\text{equation of state}: \qquad M \sim H^{1/\delta} \qquad\qquad (H \neq 0, t = 0), \tag{56.8}$$

where the symbol $\sim$ means the singular parts of the various thermodynamic quantities in the above equations as $|t| \to 0$, if they do not vanish (these quantities may have regular parts also as $|t| \to 0$). The exponents α, β, γ and δ are known as **critical exponents**. In addition to these there are two more, ν and η, defined in the following way. We first define the local mangetization density $m(\boldsymbol{x})$ corresponding to the thermodynamic variable M by

$$M = \left\langle \int d^d x\, m(\boldsymbol{x}) \right\rangle , \tag{56.9}$$

where the angular brackets denote the ensemble average. The quantity $m(\boldsymbol{x})$, analogous to the field $\varphi(\boldsymbol{x})$ in the last chapter, is known as an **order parameter** in the context of statistical physics. Recall the 2-point correlation function introduced in (55.71) in the last chapter. Here, assuming translational invariance for the physical system, we introduce the related correlation function [c.f. (50.29)]

$$\Gamma(\boldsymbol{x}) \equiv \langle m(\boldsymbol{x})m(0)\rangle - \langle m(\boldsymbol{x})\rangle\langle m(0)\rangle , \qquad (56.10)$$

where the angular brackets again denote ensemble averages. It is usually assumed that $\Gamma(\boldsymbol{x})$ has the so-called **Ornstein-Zernike form** in the critical limit:

$$\Gamma(\boldsymbol{x}) \xrightarrow[t \to 0]{} |\boldsymbol{x}|^{-p} e^{-|\boldsymbol{x}|/\xi} . \qquad (56.11)$$

Then, the last two critical exponents ν and η, are defined as follows:

$$\text{correlation length}: \qquad \xi \sim |t|^{-\nu} , \qquad (56.12)$$
$$\text{power-law decay}: \qquad p = d - 2 + \eta . \qquad (56.13)$$

In the last equation, d is the dimension of the Euclidean space in which the physical system resides. As we will see later, critical exponents depend strongly on the value that d assumes. Values for the six critical exponents $(\alpha, \beta, \gamma, \delta, \nu, \eta)$ serve to classify a large number of diverse physical systems. Different systems which share the same set of values for these critical exponents are said to belong to a particular **universality class**.

The above-identified six critical exponents are, however, not independent. In fact, there exist four relationships among them, known as **scaling laws**. These imply that there are only two independent critical exponents among the six. We will first state these scaling laws (identified by their discoverers), and then derive them on the basis of the scaling hypothesis stated above, with the help of dimensional analysis. The scaling laws are the following statements:

$$\gamma = \nu(2 - \eta) \qquad \text{(Fisher)}, \qquad (56.14)$$
$$\alpha + 2\beta + \gamma = 2 \qquad \text{(Rushbrooke)}, \qquad (56.15)$$
$$\gamma = \beta(\delta - 1) \qquad \text{(Widom)}, \qquad (56.16)$$
$$\nu d = 2 - \alpha \qquad \text{(Josephson)}. \qquad (56.17)$$

The scaling hypothesis will be implemented as follows: Near the critical point, a quantity X with dimensions $[X] \sim L^{-D_X}$ (where L is length) will satisfy the relationship

$$X \approx \tilde{X} \xi^{-D_X} , \qquad (56.18)$$

where ξ is the correlation length and $\tilde{X}$ is a dimensionless proportionality constant. The quantity D_X in the above equation is known as the "dimension" of the quantity X. To prove the above scaling laws we will first calculate the dimensions of the quantities $g \equiv G/(kTV)$ (Gibbs free energy per unit volume in units of kT), $\Gamma(\boldsymbol{x})$ (correlation function), M/V (magnetization per unit volume), $kT\chi$ (χ being the magnetic susceptibility) and $H/(kT)$ (H being the external magnetic field).

We first write the Hamiltonian (in the presence of an external magnetic field H) as

$$\mathcal{H} = \mathcal{H}_0 - H \int d^d x \, m(\boldsymbol{x}) \,, \tag{56.19}$$

where $\mathcal{H}_0$ is the Hamiltonian of the system in the absence of H. Then by (56.9),

$$M = \frac{Tr\left\{ \left(\int d^d x \, m(\boldsymbol{x}) \right) e^{-\mathcal{H}/kT} \right\}}{Tr\left\{ e^{-\mathcal{H}/kT} \right\}} = \int d^d x \, \frac{Tr\left\{ m(\boldsymbol{x}) e^{-\mathcal{H}/kT} \right\}}{Tr\left\{ e^{-\mathcal{H}/kT} \right\}} = \int d^d x \, \langle m(\boldsymbol{x}) \rangle \,. \tag{56.20}$$

Assuming translational invariance of the system we have $\langle m(\boldsymbol{x}) \rangle = \langle m(0) \rangle$, which implies $M = V \langle m(0) \rangle$. By (56.3), the magnetic susceptibility χ is then given by

$$\chi = \frac{\partial}{\partial H} \left[\frac{Tr\left\{ m(0) e^{-\mathcal{H}/kT} \right\}}{Tr\left\{ e^{-\mathcal{H}/kT} \right\}} \right] \,. \tag{56.21}$$

Using the form of $\mathcal{H}$ given by (56.19), it is not difficult to show that

$$\chi = \frac{1}{kT} \int d^d x \, \left(\langle m(\boldsymbol{x}) m(0) \rangle - \langle m(0) \rangle^2 \right) \,. \tag{56.22}$$

By the definition of the correlation function given by (56.10) it follows that

$$kT\chi = \int d^d x \, \Gamma(\boldsymbol{x}) \,, \tag{56.23}$$

which is a special case of a central theorem in statistical mechanics known as the **fluctuation-dissipation theorem**. By (56.11) and (56.13),

$$[\Gamma(\boldsymbol{x})] \sim L^{2-d-\eta} \,. \tag{56.24}$$

It follows from (56.23) that $[kT\chi] \sim D^d L^{2-d-\eta}$, or,

$$[kT\chi] \sim L^{2-\eta} \,. \tag{56.25}$$

Since $G/(kT)$ is dimensionless, $[G/(kT)] \sim L^0$. Hence, with

$$g \equiv \frac{G}{kTV} \,, \tag{56.26}$$

we have

$$[g] \sim L^{-d} \,. \tag{56.27}$$

By (56.9), $[M/V] = [m(\boldsymbol{x})]$. But by (56.10), $[m(\boldsymbol{x})] = [\Gamma(\boldsymbol{x})]^{1/2}$. Hence, by (56.24),

$$[M/V] \sim L^{(2-d-\eta)/2} \tag{56.28}$$

By the first equation of (56.3),

$$[M] = \frac{[G]}{[H]} = \frac{[G/(kT)]}{[H/(kT)]} = \frac{[gV]}{[H/(kT)]} \, . \tag{56.29}$$

Hence $[H/(kT)] = [g]/[M/V]$. By (56.27) and (56.28), we have

$$[H/(kT)] = L^{-(2+d-\eta)/2} \, . \tag{56.30}$$

Having obtained the dimensional relationships above, we can now invoke the scaling hypothesis [in the form (56.18)] to derive the scaling laws (56.14) to (56.17). First work with the heat capacity per unit volume $C/(kV)$. The definition of the critical exponent α given by (56.5) implies that

$$\frac{C}{kV} \sim |t|^{-\alpha} \, , \tag{56.31}$$

where t is given by (56.4). On the other hand, the third equation of (56.3) (for the definition of the heat capacity) gives, near the critical temperature T_c,

$$\frac{C}{kV} \approx -T_c \frac{\partial^2}{\partial T^2}\left(\frac{G}{kV}\right) \approx -T_c^2 \frac{\partial^2}{\partial T^2}\left(\frac{G}{kVT}\right) = -T_c^2 \frac{\partial^2 g}{\partial T^2} = -\frac{\partial^2 g}{\partial t^2} \, . \tag{56.32}$$

By the scaling hypothesis and (56.27), we can write

$$g \approx \tilde{g}\, \xi^{-d} \, . \tag{56.33}$$

By the definition of the critical exponent ν [c.f. (56.12)] we have

$$\xi = \xi_0 \, |t|^{-\nu} \, . \tag{56.34}$$

Hence

$$g \approx (constant) \, |t|^{\nu d} \, ; \tag{56.35}$$

and it follows from (56.32) that

$$\frac{C}{kV} \approx \frac{\partial^2 g}{\partial t^2} \sim |t|^{\nu d - 2} \, . \tag{56.36}$$

Comparing with (56.31) we obtain $\alpha = 2 - \nu d$, which is precisely Josephson's scaling law (56.17). Next examine the quantity M/V. By (56.28) and the scaling hypothesis, we can write

$$\frac{M}{V} \approx \tilde{A}\, \xi^{(2-d-\eta)/2} \, , \tag{56.37}$$

where $\tilde{A}$ is dimensionless constant. Eq. (56.34) then implies

$$\frac{M}{V} \approx (constant) \, |t|^{-\nu(2-d-\eta)/2} \, . \tag{56.38}$$

Comparison with the definition of the critical exponent β given by (56.6) then yields

$$\beta = -\frac{\nu}{2}(2 - d - \eta) .$$ (56.39)

Moving on to the quantity $kT\chi$, the scaling hypothesis and (56.25) allow us to write, near the critical temperature T_c,

$$kT_c\chi \approx \tilde{B}\,\xi^{2-\eta} \approx (constant)\,|t|^{-\nu(2-\eta)} ,$$ (56.40)

where $\tilde{B}$ is a dimensionless constant. Comparing this with the definition of the critical exponent γ given by (56.7), we get $\gamma = \nu(2 - \eta)$, which is precisely Fisher's scaling law (56.14). Finally, applying the same reasoning to the quantity $H/(kT)$, we can write, with the help of (56.30),

$$\frac{H}{kT_c} \approx \tilde{F}\,\xi^{-(2+d-\eta)/2} \approx (constant)\,|t|^{\nu(2+d-\eta)/2} ,$$ (56.41)

where $\tilde{F}$ is a dimensionless constant. By virtue of the definition of the critical exponent δ given by (56.8), and using (56.6) again, we have

$$H \sim M^\delta \sim |t|^{\beta\delta} .$$ (56.42)

Comparison of the above two equations gives

$$\beta\delta = \frac{\nu}{2}(2 + d - \eta) .$$ (56.43)

Subtracting (56.39) from (56.43), we have

$$\beta(\delta - 1) = \nu(2 - \eta) .$$ (56.44)

The right-hand side is equal to γ, by Fisher's scaling law (56.14). Hence we obtain precisely Widom's scaling law (56.16): $\beta(\delta - 1) = \gamma$. Adding (56.39) and (56.43) we get

$$\beta(\delta + 1) = \nu d .$$ (56.45)

Widom's scaling law implies $\delta + 1 = 2 + \gamma/\beta$. Substituting this into the left-hand side of (56.45) and using Josephson's scaling law (56.17) for $\nu d\,(= 2 - \alpha)$ on the right-hand side, we obtain precisely Rushbrooke's scaling law (56.15): $\alpha + 2\beta + \gamma = 2$. Thus the four scaling laws (56.14) to (56.17) are all proved.

The six critical exponents appearing in the scaling laws are very often conveniently expressed in term of the two quantities D_t and D_h, the dimensions of t and h, respectively [c.f. (56.18) for the definition of the dimension of a quantity], where t is defined by (56.4) and $h \equiv H/(kT)$. Eq. (56.30) immediately gives

$$D_h = \frac{2 + d - \eta}{2} .$$ (56.46)

Even though by (56.4) t is strictly speaking dimensionless, we can use (56.34) to write

$$|t| = \xi_0^{-1/\nu}\,\xi^{-1/\nu} .$$ (56.47)

With the understanding that wherever t occurs it is always multiplied by the factor $\xi_0^{1/\nu}$ that converts it to $\xi^{-1/\nu}$, we designate

$$D_t = \frac{1}{\nu}. \tag{56.48}$$

It can then be easily verified that the scaling laws (56.14) to (56.17) yield the following expressions for the critical exponents:

$$\alpha = 2 - \frac{d}{D_t}, \qquad \beta = \frac{d - D_h}{D_t}, \qquad \gamma = \frac{2D_h - d}{D_t}, \qquad \delta = \frac{D_h}{d - D_h},$$

$$\nu = \frac{1}{D_t}, \qquad \eta = 2 + d - 2D_h. \tag{56.49}$$

Problem 56.1 Show that the following scaling law can be used as an alternative to Rushbrooke's scaling law, and comprises, together with the other three, a complete set of four independent scaling laws:

$$\delta = \frac{d + 2 - \eta}{d - 2 + \eta}. \tag{56.50}$$

Hint: Use (56.39) and (56.43).

Problem 56.2 Deduce the expressions for the critical exponents given in (56.49) by using the scaling laws (56.14) to (56.17).

The **Landau-Ginsburg (LG) theory** of critical phenomena is a phenomenological theory which takes as its starting point the existence of an energy functional $E[m(\boldsymbol{x}), H(\boldsymbol{x})]$ (here written in the context of ferromagnetic systems), such that, near the critical temperature T_c, the partition function $Q(H, T)$ of (56.2) is expressed as the following functional integral:

$$Q(H, T) = \int [dm]\, e^{-E[m(\boldsymbol{x}), H(\boldsymbol{x})]}. \tag{56.51}$$

The energy functional $E[m(\boldsymbol{x}), H(\boldsymbol{x})]$ is expressed as the spatial integral

$$E[m(\boldsymbol{x}), H(\boldsymbol{x})] = E_0 + \int d^d x\, \mathcal{E}(m(\boldsymbol{x}), H(\boldsymbol{x})), \tag{56.52}$$

where E_0 is an additive constant independent of any field quantities, and the function $\mathcal{E}(m(\boldsymbol{x}), H(\boldsymbol{x}))$ is known as the **Landau free energy**. In light of our earlier developments in the past several chapters, Eq. (56.51) immediately puts the LG theory in the category of quantum field theories, because it is identical in form to the expression for the generating functional $Z[J]$ given by (49.19). There is considerable freedom in the choice of the function $\mathcal{E}(m(\boldsymbol{x}), H(\boldsymbol{x}))$. But the thrust of our subsequent developments for the rest of this book will be to demonstrate that the m^4-theory (to be specified below and obviously analogous to the φ^4-theory in 4 spacetime dimensions) is a particularly fruitful theory in the theoretical investigation of critical phenomena.

Eq. (56.51) can be justified on the basis of statistical mechanics and thermodynamics as follows. We assume that, starting with a microscopic configuration in the canonical ensemble, one can, after an averaging (such as the block-spin procedure discussed in the last chapter) arrive at an order-parameter configuration specified by some function $m(\boldsymbol{x})$. This implies that there may be multiple microscopic configurations corresponding to a single order-parameter configuration. We will denote by $W[m]$ the number of microscopic configurations corresponding to an order-parameter configuration $m(\boldsymbol{x})$. Now the trace operation in (56.2) is with respect to the microscopic configurations. So we can write

$$Q = Tr\left\{e^{-\beta\mathcal{H}}\right\} = \int [dm] \left(Tr\, e^{-\beta\mathcal{H}}\right)_m , \qquad (56.53)$$

where β is the inverse temperature $1/(kT)$, and the trace in the integrand means summing over all microscopic configurations corresponding to a fixed order-parameter configuration $m(\boldsymbol{x})$. Assume that the total energy corresponding to the fixed order-parameter configuration $m(\boldsymbol{x})$ is $\epsilon[m]$. Then

$$Q = \int [dm]\, W[m]\, e^{-\beta\epsilon[m]} = \int [dm]\, e^{\ln W[m] - \beta\epsilon[m]} . \qquad (56.54)$$

We will define an *entropy*-like quantity $S[m]$ by

$$S[m] \equiv \ln W[m] , \qquad (56.55)$$

and write

$$E[m(\boldsymbol{x}), H(\boldsymbol{x})] \equiv \beta\epsilon[m] - S[m] . \qquad (56.56)$$

Eq. (56.54) then appears as (56.51). The quantity E_0 in (56.52) then plays the role of the negative of an additive constant (independent of field quantities) in the "entropy" term $S[m]$. Note that (56.56) is reminiscent of the familiar *Legendre transform* relationship $A = U - TS$ in thermodynamics, where A is the *Helmholtz free energy*, U is the internal energy, and S is the real entropy.

Let us illustrate how a Landau-Ginsburg partition function of the form (56.51) can be derived from a specific discrete model, such as the Ising spin model. Consider a d-dimensional lattice of spin variables s_i, $(i = 1, \ldots, N$ is a site index), where the total Hamiltonian $\mathcal{H}$ is given by

$$\frac{\mathcal{H}(\{s\})}{kT} = -\frac{1}{2}\sum_{i,j} s_i V_{ij} s_j - \sum_i H_i s_i . \qquad (56.57)$$

In the above equation, V_{ij} is an interaction potential matrix (between spins at different sites), and H_i can be interpreted as the external magnetic field at site *i.* Writing $s = (s_1, \ldots, s_N)^T$, $H = (H_1, \ldots, H_N)^T$ (as column vectors), and denoting the linear operator represented by the matrix V_{ij} by V, the partition function Q can be written as follows:

$$Q = \sum_{\{s\}} \exp \left\{ \frac{1}{2} s \cdot V s + H \cdot s \right\} , \tag{56.58}$$

where the sum is over all spin configurations. Recalling our earlier results on Gaussian functional integrals [specifically (47.6) and (47.18)], we see that the quantity $\exp\{(1/2)s \cdot V s\}$ above can be expressed as a functional integral and we have

$$Q = \sum_{\{s\}} \frac{\det \left(V^{-1} \right)^{1/2}}{(2\pi)^{N/2}} \int [d\varphi] \exp \left\{ -\frac{1}{2} \varphi V^{-1} \varphi + s \cdot \varphi \right\} \exp \left(H \cdot s \right)$$

$$= \frac{1}{(2\pi)^{N/2} (\det V)^{1/2}} \int [d\varphi] \exp \left\{ -\frac{1}{2} \varphi \cdot V^{-1} \varphi \right\} \sum_{\{s\}} \exp \left\{ s \cdot (\varphi + H) \right\}$$

$$= \frac{1}{(2\pi)^{N/2} (\det V)^{1/2}} \int [d\varphi] \exp \left\{ -\frac{1}{2} (\varphi - H) \cdot V^{-1} (\varphi - H) \right\}$$

$$\times \sum_{\{s\}} \exp \left\{ s \cdot \varphi \right\} , \tag{56.59}$$

where φ is viewed as the column vector $(\varphi_1, \ldots, \varphi_N)^T$ and the functional integration measure is given by

$$\int [d\varphi] = \int_{-\infty}^{\infty} d\varphi_1 \ldots \int_{-\infty}^{\infty} d\varphi_N . \tag{56.60}$$

Defining the constant E_0 and the quantity $S[\varphi]$ by

$$e^{-E_0} \equiv \frac{1}{(2\pi)^{N/2} (\det V)^{1/2}} , \tag{56.61}$$

$$e^{S[\varphi]} \equiv \sum_{\{s\}} e^{s \cdot \varphi} , \qquad S[\varphi] = \ln \sum_{\{s\}} e^{s \cdot \varphi} , \tag{56.62}$$

we have the Landau-Ginsburg form (56.51) for the partition function Q:

$$Q = \int [d\varphi] \, e^{-E[\varphi, H]} , \tag{56.63}$$

where

$$E[\varphi, H] = E_0 + \frac{1}{2} \left\{ (\varphi - H) \cdot V^{-1} (\varphi - H) \right\} - S[\varphi] . \tag{56.64}$$

More specifically, for the **Ising model** in any dimension d,

$$\boldsymbol{s} = (s_1, \ldots, s_N)^T \qquad (s_i = \pm 1) , \tag{56.65}$$

$$V_{ij} = \begin{cases} V & \text{[if } i \text{ and } j \text{ are nearest-neighbor sites]} \\ 0 & \text{(otherwise)} \end{cases} . \tag{56.66}$$

Then

$$\sum_{\{s\}} e^{\boldsymbol{s} \cdot \boldsymbol{\varphi}} = \sum_{\{s\}} e^{s_1 \varphi_1 + \cdots + s_N \varphi_N} = \sum_{s_1} e^{s_1 \varphi_1} \cdots \sum_{s_N} e^{s_N \varphi_N}$$

$$= \left(e^{\varphi_1} + e^{-\varphi_1} \right) \ldots\ldots \left(e^{\varphi_N} + e^{-\varphi_N} \right) = 2^N \prod_{i=1}^{N} \cosh \varphi_i , \tag{56.67}$$

and thus

$$S[\boldsymbol{\varphi}] = \ln \left(2^N \prod_{i=1}^{N} \cosh \varphi_i \right) = N \ln 2 + \sum_{i=1}^{N} \ln \left(\cosh \varphi_i \right) . \tag{56.68}$$

The constant $N \ln 2$ can conveniently be absorbed into the constant E_0 in (56.64). We note that the Ising model can be solved exactly in the $d \le 2$ cases, and has furnished very important theoretical checks against the Landau-Ginsburg approach (see, for example, Baxter 2007).

We will now introduce the m^4-Landau free energy function $\mathcal{E}(m(\boldsymbol{x}), H(\boldsymbol{x}))$ given by [c.f. (48.16) to (48.18)]

$$\mathcal{E}(m(\boldsymbol{x}), H(\boldsymbol{x})) = \frac{1}{2} \left[(\nabla m(\boldsymbol{x}))^2 + a(t) m^2(\boldsymbol{x}) \right] + \frac{b(t)}{4!} m^4(\boldsymbol{x}) - \frac{1}{kT} m(\boldsymbol{x}) H(\boldsymbol{x}) , \tag{56.69}$$

where the temperature-dependent coefficients $a(t)$ and $b(t)$ [with t defined by (56.4)] are in general expanded in power series as

$$a(t) = a_0 t + a_1 t^2 + \cdots , \qquad b(t) = b_0 + b_1 t + \cdots \qquad (a_0, b_0 > 0) . \tag{56.70}$$

[The reason why the constant term in $a(t)$ is set to zero will become clear shortly.] Note that $a(t)$ plays the role of mass-squared, and $b(t)$ that of the coupling constant, in quantum field theory. Our first application of the m^4-theory will be the so-called **mean-field theory**, the premise of which is the somewhat unrealistic assumption that local variations of the order parameter $m(\boldsymbol{x})$ can be ignored, so that a (constant) mean field $\overline{m}$ can be used in the functional integral in (56.51). We will see that, although unrealistic and quantitatively deficient, the mean-field theory has the appeal of having the capacity to yield a set of universal critical exponents in a simple and elegant manner. Mathematically, this theory amounts to using the so-called **saddle-point method** in the evaluation of the functional integral, which is equivalent to replacing the integral by the maximum value of the integrand (up to physically unimportant constants).

Since $m(x)$ is taken to be constant in the mean-field approximation, the gradient term on the right-hand side of (56.69) vanishes, and the Landau free energy $\mathcal{E}$ can be written as

$$\mathcal{E}(m(x), H(x)) = \frac{a(t)\, m^2(x)}{2} + \frac{b(t)\, m^4(x)}{4!} - \frac{1}{kT} m(x) H(x) \,. \tag{56.71}$$

The maximum of the integrand in (56.51) is then determined by the functional derivative condition [c.f. (49.21a)]

$$\frac{\delta E[m(x), H(x)]}{\delta m(x)} = 0 \,, \tag{56.72}$$

or the condition

$$\frac{\partial \mathcal{E}}{\partial m}\bigg|_{\overline{m}} = 0 \,, \tag{56.73}$$

which determines the mean field $\overline{m}$ to be used. Differentiating (56.71), we have, for constant magnetic field H, the following equation for the mean field $\overline{m}$:

$$a(t)\overline{m} + \frac{b(t)}{6}\overline{m}^3 - \frac{H}{kT} = 0 \,, \tag{56.74}$$

which also serves as the *equation of state*. Eqs. (56.2), (56.51) and (56.52) then imply

$$e^{-G/(kT)} = e^{-E_0} \exp\left\{ -\int d^d x\, \mathcal{E}(\overline{m}, H) \right\}$$

$$= (constant)\exp\left\{ -V\left(\frac{a(t)}{2}\overline{m}^2 + \frac{b(t)}{4!}\overline{m}^4 - \frac{\overline{m}H}{kT} \right) \right\} \,, \tag{56.75}$$

which gives, up to an additive constant,

$$g \equiv \frac{G(H, t)}{kTV} = \frac{a(t)}{2}\overline{m}^2 + \frac{b(t)}{4!}\overline{m}^4 - \frac{\overline{m}H}{kT} \,. \tag{56.76}$$

The first equation in (56.3) then gives, for the total magnetization,

$$M = \overline{m}V \,, \tag{56.77}$$

as expected. For zero magnetic field $(H = 0)$ we have

$$a(t)\overline{m} + \frac{b(t)}{6}\overline{m}^3 = 0 \,. \tag{56.78}$$

For $0 < t \ll 1$, $a(t) \approx a_0 t > 0$ and $B(t) \approx b_0 > 0$ [c.f. (56.70)], and the only solution is $\overline{m} = 0$. For $t < 0$ and $|t| \ll 1$, however, $a(t) \approx -a_0|t|$ and a non-zero mean field $\overline{m}$ (and magnetizationn M) results. We thus have

$$M = \begin{cases} 0 & (t > 0,\ H = 0) \\ \pm V|\overline{m}| = \pm V\sqrt{\dfrac{6a_0}{b_0}}\, |t|^{1/2} & (t \leq 0,\ H = 0) \end{cases} \,. \tag{56.79}$$

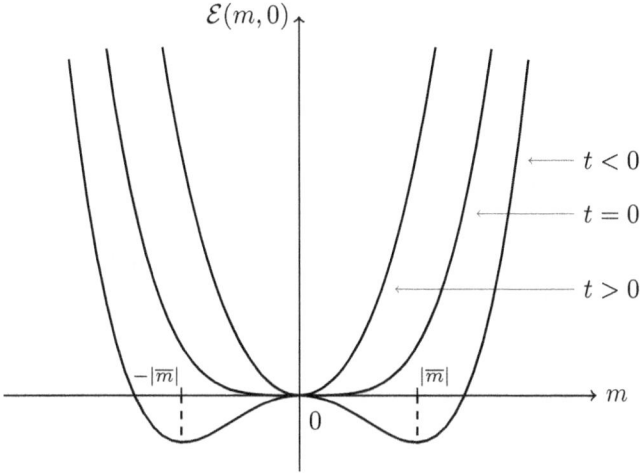

FIGURE 56.2

Comparison with (56.6) immediately yields

$$\beta = \frac{1}{2} \qquad \text{(mean-field result)} . \qquad (56.80)$$

The desired result (56.79) for spontaneous magnetization for $T < T_c$ (see Fig. 56.1) is the reason why the form (56.70) has been picked for $a(t)$. Schematic graphs of $\mathcal{E}(m, H = 0)$ for various values of t are shown in Fig. 56.2. The figure shows clearly that, for $H = 0$, the Landau free energy has a single minimum at $m = 0$ for $t > 0$, and two distinct minima at $m = \pm|\overline{m}|$ for $t < 0$. Note that terms in higher powers of t than the leading ones in the expressions for $a(t)$ and $b(t)$ in (56.70) do not lead to a different result for the critical exponent β.

Next we calculate the magnetic susceptibility χ in the critical region using the formula

$$\chi = \frac{1}{V} \frac{\partial M}{\partial H}\bigg|_{H=0} = \frac{\partial \overline{m}}{\partial H}\bigg|_{H=0} . \qquad (56.81)$$

The last derivative is obtained by differentiating the equation of state (56.74) with respect to H to yield

$$a(t)\frac{\partial \overline{m}}{\partial H} + \frac{b(t)}{2}\overline{m}^2\frac{\partial \overline{m}}{\partial H} - \frac{1}{kT} = 0 , \qquad (56.82)$$

and then evaluating this expression at $H = 0$. From (56.79) we see that, for

$t > 0, H = 0, \overline{m} = 0$. Hence, near $t = 0$,

$$a(t) \left.\frac{\partial \overline{m}}{\partial H}\right|_{H=0} \approx a_0 t \left.\frac{\partial \overline{m}}{\partial H}\right|_{H=0} = \frac{1}{kT} \qquad (t > 0) . \qquad (56.83)$$

For $t < 0$, (56.82) gives

$$\left.\frac{\partial \overline{m}}{\partial H}\right|_{H=0} \left(a(t) + \frac{b(t)}{2} \overline{m}^2(t, H = 0) \right) = \frac{1}{kT} \qquad (t < 0) . \qquad (56.84)$$

Using the result (56.79) for $\overline{m}$ for $t < 0$, the quantity within the parentheses in the above equation is approximately equal to

$$a_0 t + \frac{b_0}{2} \frac{6 a_0}{b_0} (-t) = -2 a_0 t .$$

Thus

$$\left.\frac{\partial \overline{m}}{\partial H}\right|_{H=0} (-2 a_0 t) = \frac{1}{kT} \qquad (t < 0) . \qquad (56.85)$$

Combining the results (56.83) and (56.85), and using (56.81) for χ, we have

$$\chi = \begin{cases} (a_0 kT)^{-1} t^{-1} & (t > 0) \\ \frac{1}{2} (a_0 kT)^{-1} (-t)^{-1} & (t < 0) \end{cases} . \qquad (56.86)$$

Eq. (56.7) then implies that

$$\gamma = 1 \qquad \text{(mean-field result)} . \qquad (56.87)$$

From (56.74) we have, at $T = T_c$,

$$\frac{b_0}{6} \overline{m}^3 = \frac{H}{kT} \qquad (t = 0) . \qquad (56.88)$$

This implies

$$M \sim H^{1/3} \qquad (t = 0) . \qquad (56.89)$$

Comparison with (56.8) then gives

$$\delta = 3 \qquad \text{(mean-field result)} . \qquad (56.90)$$

To calculate the specific heat C, we use (56.32), which gives

$$C \approx -kV \frac{\partial^2 g}{\partial t^2} , \qquad (56.91)$$

where g is given by (56.76). Using (52.79) for $\overline{m}$ we have, near $T = T_c$,

$$g = \begin{cases} 0 & (t > 0) \\ \dfrac{a_0 t}{2} \dfrac{6 a_0}{b_0} |t| + \dfrac{b_0}{4!} \left(\dfrac{6 a_0}{b_0} \right)^2 |t|^2 - \dfrac{1}{kT_c} \sqrt{\dfrac{6 a_0}{b_0}} |t|^{1/2} & (t < 0) \end{cases}$$

$$\approx \begin{cases} 0 & (t > 0) \\ -\dfrac{3 a_0^2}{2 b_0} t^2 & (t < 0) \end{cases} , \qquad (56.92)$$

where, for $t < 0$, we have neglected the term proportional to $|t|^{1/2}$ on the right-hand side of the first equality. Eq. (56.91) then gives

$$C = \begin{cases} 0 & (t > 0) \\ 3kV\dfrac{a_0^2}{b_0} & (t < 0) \end{cases} . \tag{56.93}$$

This result indicates that the specific heat C has a discontinuity at $T = T_c$. At the same time, comparison with (56.5) shows that

$$\alpha = 0 \qquad \text{(mean-field result)} . \tag{56.94}$$

We note that the mean-field results for the four critical exponents α, β, γ and δ obtained above satisfy Rushbrooke's scaling law (56.15) and Widom's scaling law (56.16). Since these results are independent of d (the dimension of physical space), Josephson's scaling law (56.17), which involves d, cannot be satisfied in general; but it is interesting to note that Josephson's law with $d = 4$ yields $\nu = 1/2$. We will now proceed to justify this result for ν within mean-field theory.

Since the critical exponent ν specifies the behavior of the correlation length ξ near the critical point $t = 0$, which in turn specifies the asymptotic behavior of the 2-point correlation function $\langle m(\boldsymbol{x})m(0)\rangle$, we have to investigate the behavior of the latter as $|\boldsymbol{x}| \to \infty$ in order to determine ν. We cannot take $m(\boldsymbol{x})$ to be completely constant (independent of $\boldsymbol{x}$) as we did earlier, for a constant m would not yield any information on ξ. So we put in a local source, a localized magnetic field $H(\boldsymbol{x}) = \mathfrak{h}\delta(\boldsymbol{x})$, and attempt to calculate how $\overline{m}(\boldsymbol{x}) = \overline{m} + \delta m(\boldsymbol{x})$ (the new position-dependent mean field) responds as a result, where $\overline{m}$ is the constant mean field determined earlier. Within the spirit of the mean-field approximation, this new mean field is still determined by the saddle-point method in the evaluation of the functional integral in (56.51), except that now we have to include the gradient term in the Landau free energy in (56.69). We recall that, in the Landau-Ginsburg theory, the ensemble average of a quantity $f(m(\boldsymbol{x}))$ is given by

$$\langle f(m(\boldsymbol{x}))\rangle = \frac{\int[dm]\, e^{-E[m(\boldsymbol{x})]} f(m(\boldsymbol{x}))}{\int[dm]\, e^{-E[m(\boldsymbol{x})]}} , \tag{56.95}$$

where the functional $E[m(\boldsymbol{x})]$ is given by (56.52). With the saddle point approximation

$$\int[dm]\, e^{-E[m(\boldsymbol{x})]} = (constant)\, e^{-E[\overline{m}]} , \tag{56.96a}$$

$$\int[dm]\, e^{-E[m(\boldsymbol{x})]} f(m(\boldsymbol{x})) = (constant)\, e^{E[\overline{m}(\boldsymbol{x})]} f(\overline{m}(\boldsymbol{x})) , \tag{56.96b}$$

where the function $\overline{m}(\boldsymbol{x})$ satisfies (56.72). We thus have, in the saddle point approximation,

$$\langle m(\boldsymbol{x})m(0)\rangle = \overline{m}(\boldsymbol{x})\overline{m}(0) = \langle m(\boldsymbol{x})\rangle\langle m(0)\rangle . \tag{56.97}$$

The appropriate 2-point correlation function to use in this approximation for the investigation of the critical behavior of the correlation length ξ is then

$$G(\boldsymbol{x}) \equiv \langle m(\boldsymbol{x})m(0)\rangle = \overline{m}(0)\overline{m}(\boldsymbol{x}) . \tag{56.98}$$

Performing the functional derivative in (56.72), with $E[m(\boldsymbol{x})]$ given by [ignoring the constant term E_0 in (56.72)]

$$E[m(\boldsymbol{x}), H(\boldsymbol{x})] = \int d^d x \, \mathcal{E}(m(\boldsymbol{x}), H(\boldsymbol{x})) , \tag{56.99}$$

and $\mathcal{E}(m(\boldsymbol{x}), H(\boldsymbol{x}))$ by (56.69), we find that the new mean field $\overline{m}(\boldsymbol{x})$ satisfies the differential equation

$$\left[\nabla^2 - a(t)\right] \overline{m}(\boldsymbol{x}) - \frac{b(t)}{6} \overline{m}^3(\boldsymbol{x}) = \frac{\mathfrak{h}}{kT}\delta(\boldsymbol{x}) , \tag{56.100}$$

where on the right-hand side $\mathfrak{h}\delta(\boldsymbol{x})$ is the magnetic field $H(\boldsymbol{x})$. We assume $\mathfrak{h}$ is a small parameter. For $\mathfrak{h} = 0$ the solution is $\overline{m}$ [the old mean field given by (56.79)]. So we write

$$\overline{m}(\boldsymbol{x}) = \overline{m} + \delta m(\boldsymbol{x}) , \qquad \delta m(\boldsymbol{x}) \sim O(\mathfrak{h}) . \tag{56.101}$$

Expanding up to only first order in $\delta m(\boldsymbol{x})$, (56.100) becomes

$$\left[\nabla^2 - \left(a(t) + \frac{b(t)}{2} \overline{m}^2\right)\right] \delta m(\boldsymbol{x}) = \frac{\mathfrak{h}}{kT}\delta(\boldsymbol{x}) + a(t)\overline{m} + \frac{b(t)}{6}\overline{m}^3 . \tag{56.102}$$

For $t > 0$, $\overline{m} = 0$, so the above equation becomes

$$\left[\nabla^2 - a(t)\right] \delta m(\boldsymbol{x}) = \frac{\mathfrak{h}}{kT}\delta(\boldsymbol{x}) \qquad (t > 0) . \tag{56.103}$$

Fourier transforming we have

$$- \left[k^2 + a(t)\right] \delta\tilde{m}(\boldsymbol{k}) = \frac{\mathfrak{h}}{(2\pi)^d kT} , \tag{56.104}$$

where

$$\delta m(\boldsymbol{x}) = \int d^d k \, e^{i\boldsymbol{k}\cdot\boldsymbol{x}} \, \delta\tilde{m}(\boldsymbol{k}) , \qquad \delta(\boldsymbol{x}) = \int \frac{d^d k}{(2\pi)^d} \, e^{i\boldsymbol{k}\cdot\boldsymbol{x}} . \tag{56.105}$$

We thus have

$$\delta m(\boldsymbol{x}) = -\frac{\mathfrak{h}}{(2\pi)^d kT} \int d^d k \, \frac{e^{i\boldsymbol{k}\cdot\boldsymbol{x}}}{k^2 + a(t)} \qquad (t > 0) . \tag{56.106}$$

From dimensional arguments it is clear that the dimension of the integral is k^{d-2}, or L^{2-d}. So the integral must be proportional to $|\boldsymbol{x}|^{2-d}$. The integrand, viewed as a function of the complex variable $k = |\boldsymbol{k}|$, has first-order poles at $k = \pm i\sqrt{a(t)}$. By contour integration, we see that

$$\delta m(\boldsymbol{x}) \sim |\boldsymbol{x}|^{2-d} e^{-\sqrt{a(t)}\,|\boldsymbol{x}|} \approx |\boldsymbol{x}|^{2-d} e^{-\sqrt{a_0 t}\,|\boldsymbol{x}|} \qquad (t>0) . \qquad (56.107)$$

This is precisely the Ornstein-Zernike form [c.f. (56.11)], from which we immediately deduce that the correlation length ξ is given by

$$\xi = (a_0 t)^{-1/2} \qquad (t>0) . \qquad (56.108)$$

Furthermore, comparison with (56.12) and (56.13) yields

$$\nu = \frac{1}{2}, \qquad \eta = 0 \qquad \text{(mean-field result)} . \qquad (56.109)$$

For $t < 0$, $\overline{m} = \sqrt{6a_0/b_0}\,|t|^{1/2}$ [from (56.79)]. Thus, near the critical point $t = 0$,

$$a(t) + \frac{b(t)}{2}\overline{m}^2 \approx a_0 t + \frac{b_0}{2}\left(\frac{6a_0}{b_0}\right)|t| = -2a_0 t \qquad (t<0) , \qquad (56.110)$$

$$a(t)\overline{m} + \frac{b(t)}{6}\overline{m}^3 \approx (a_0 t)\sqrt{\frac{6a_0}{b_0}}\,|t|^{1/2} + \frac{b_0}{6}\left(\frac{6a_0}{b_0}\right)^{3/2}|t|^{3/2} = 0 \qquad (t<0) . \qquad (56.111)$$

Eq. (56.102) then becomes

$$\left(\nabla^2 - 2a_0|t|\right)\delta m(\boldsymbol{x}) = \frac{\hbar}{kT}\delta(\boldsymbol{x}) \qquad (t<0) . \qquad (56.112)$$

This has exactly the same form as (56.103), except that the term $a_0|t|$ in the latter has become $2a_0|t|$. The solution $\overline{m}(\boldsymbol{x})$ now has the form

$$\overline{m}(\boldsymbol{x}) = \overline{m} + (constant)\,|\boldsymbol{x}|^{2-d} e^{-\sqrt{2a_0|t|}\,|\boldsymbol{x}|} \qquad (t<0) , \qquad (56.113)$$

where the spatial dependence appears entirely in the second term. Instead of (56.108) we have

$$\xi = (2a_0|t|)^{-1/2} \qquad (t<0) . \qquad (56.114)$$

The critical exponents ν and η are thus still the same as given by (56.109). We note that the mean-field results for the critical exponents derived above, significantly, do not depend on the values of the coupling constants $a(t)$ and $b(t)$ in the Landau free energy, nor do they depend on the dimension d of the physical space. This gives rise to the phenomenon of *universality* in second order phase transitions.

In Table 56.1, we compare the Landau-Ginsburg mean-field results for the critical exponents with those obtained by an exact solution of the well known and exhaustively studied two-dimensional Ising model (see, for example, Huang 1987). It is clear that the mean-field values, while providing good approximations in some instances, require a lot of improvement. We stress again, however, that the mean field theory is still theoretically important because it provides a theoretical pathway to the explanation of universality in a relatively simple manner, and a starting point for more sophisticated treatments. In the next chapter we will introduce some of these based on renormalization group methods.

	α	β	γ	δ	ν	η
mean field	0	$\frac{1}{2}$	1	3	$\frac{1}{2}$	0
2-d Ising	0	$\frac{1}{8}$	$\frac{7}{4}$	15	1	$\frac{1}{4}$

Table 56.1

Problem 56.3 Use (56.74) to show that, within the mean-field theory, the equation of state of a ferromagnetic system in the critical region (near $T = T_c$) can be written in the form

$$\frac{H}{M^\delta} = \left(\frac{kT_c}{V}\right) f\left(\frac{t}{M^{1/\beta}}\right) , \tag{56.115}$$

where f is a linear function of the indicated argument, and the critical exponents β and δ are given by the mean-field values of (56.80) and (56.90), respectively.

Problem 56.4 Define the 2-point correlation function in the critical region (near $T = T_c$) by

$$G(\mathbf{k}, t) = \int d^d x \, \langle m(\mathbf{x}) m(0) \rangle \, e^{i\mathbf{k} \cdot \mathbf{x}} . \tag{56.116}$$

In the limit of vanishing coupling constant $[b(t) \to 0]$ justify that

$$G(\mathbf{k}, t; b(t) = 0) \sim \frac{1}{k^2 + a(t)} , \tag{56.117}$$

where $a(t)$ and $b(t)$ are given by (56.69) and (56.70). Then use the fluctuation-dissipation theorem [c.f. (56.23)] and the definition of the critical exponent ν given by (56.12) to show that $\gamma = 1$ and $\nu = 1/2$.

Problem 56.5 Recall that [immediately above (56.18)] the dimension of a quantity X is defined to be D_X if $[X] \sim L^{-D_X}$, L being a length. Define a **homogeneous function** of X, $f(X)$, to be one that satisfies

$$f(b^{D_X} X) = b^{D_f} f(X) , \tag{56.118}$$

where b is a positive constant. Show that the free energy density $g(h, t) = G/(kTV)$, where $h \equiv H/(kT)$, is a homogeneous function of h and t and satisfies

$$g\left(b^{D_h} h, b^{D_t} t\right) = b^d \, g(h, t) , \tag{56.119}$$

where d is the dimension of physical space.

Chapter 57

The Wilsonian Approach and Critical Exponents

Near the critical point the correlation length ξ of a system approaches infinity. This implies that in real space one has to consider, in principle, infinitely many microscopic degrees of freedom acting in concert, a seemingly impossible theoretical task. The conventional reductionistic thermodynamic approach of calculating average properties of smaller subsystems of finite volume V first (and hence taking into account initially only a finite number of degrees of freedom), and then taking the limit $V \to \infty$, only works far from the critical region, and breaks down as $\xi \to L$, where L is the spatial size of the system. In statistical physics, a physical system typically has a (small) fundamental length scale a, such as the length separation between spin or atomic sites, for example. In dealing with critical phenomena, one then has to consider fluctuations involving momentum scales all the way from 0 up to $\Lambda \approx 1/a$. In the case of statistical physics, the high momentum cutoff Λ is physically imposed (by the existence of a fundamental small length scale a), and critical phenomena entails non-analytical behavior of thermodynamic quantities (as a function of $|T - T_c|$) (with T_c being the critical temperature) brought about by the effects of long-range (large wavelength or low momenta, or *infrared*) fluctuations. On the other hand, in high energy physics, there seems to be no upper limit to the energy scale (or equivalently, no lower limit to the length scale) one needs to consider. Infinities show up in the Feynman integrals, as we saw earlier, precisely because of this fact. But the imposition of a high momentum cutoff Λ as a mathematical device in a particular regularization scheme necessarily implies, mathematically, the existence of a discrete spatial lattice structure of lattice separation $1/\Lambda$. Thus high energy physics and statistical physics seems to have at root the same theoretical difficulty: the necessity to deal with very disparate scales in length (or momentum). The renormalization group approach (in either real space or momentum space) to treat this problem, as formulated by Wilson (and introduced in Chapter 55), can be understood now as a step-by-step strategy to handle the multi-scale problem one spatial scale range (or, equivalently, one momentum scale range) at a time, starting with either small spatial scales, or large momentum scales, until a stable effective theory (critical theory) emerges (if it does exist). As we mentioned earlier, the real-space formulation is more intuitively clear and more susceptible to numerical simulations (such as Monte-Carlo methods), while the momentum space formulation is more analytically tractable. We will focus on the latter approach, in keeping with our earlier developments, but stress that numerical simulations in this field are no less important.

In this chapter we will apply the Wilsonian approach to the renormalization group for the Landau-Ginsburg m^4 theory in more detail than was done in Chapter 55, for the purpose of extracting the critical exponents. This will be followed in the next one where it will be shown how the same results can be obtained using the Callan-Symanzik equation introduced in Chapter 54, which, as we have seen, is based on mutiplicative renormalization, and represents an alternative approach to the study of the renormalization group. As it turns out, the crucial small parameter in terms of which the results can be expanded in power series is the dimensional deficit $\epsilon = 4 - d$. It may be hard to appreciate from the outset, but in the course of our presentation, it will become more transparent why the renormalization group is important and so useful in the study of critical phenomena. Our account in this chapter will be mainly adapted from Wilson 1983 and Huang 1987. For other in-depth treatments see Wilson and Kogut 1974, Parisi 1988, Wilson 1975, Zinn-Justin 2002, and Amit and Martín-Mayor 2005, among other sources.

As a warm-up exercise and to lay the groundwork for subsequent calculations, we will first apply Wilson's method to the Gaussian model, which is usually referred to as the "trivial" case, because it can be solved exactly, as will be demonstrated. Recall that this model is defined by the following Landau free energy [by setting $b(t) = 0$ in (56.69), the coupling constant that is the small expansion parameter in perturbative renormalization]:

$$\mathcal{E}(m(\boldsymbol{x}), H(\boldsymbol{x})) = \frac{1}{2}\left[(\nabla m(\boldsymbol{x}))^2 + a(t)m^2(\boldsymbol{x})\right] - h(\boldsymbol{x})m(\boldsymbol{x}) , \qquad (57.1)$$

where $h(\boldsymbol{x}) \equiv H(\boldsymbol{x})/(kT)$, with $H(\boldsymbol{x})$ being the magnetic field. Thus

$$E[m(\boldsymbol{x}), H(\boldsymbol{x})] = E_0 + \int d^d x \left[\frac{1}{2}m(\boldsymbol{x})\left(-\nabla^2 + a(t)\right)m(\boldsymbol{x}) - m(\boldsymbol{x})h(\boldsymbol{x})\right] , \qquad (57.2)$$

where we have replaced $(\nabla m(\boldsymbol{x}))^2$ by $-m(\boldsymbol{x})\nabla^2 m(\boldsymbol{x})$ in the integrand on ignoring the surface term after doing an integration by parts. It follows from (56.51) that the partition function Q is given by

$$Q = \mathfrak{N} \int [dm] \exp\left[-\frac{1}{2}\int d^d x\, m(\boldsymbol{x})\left(-\nabla^2 + a(t)\right)m(\boldsymbol{x}) + \int d^d x\, h(\boldsymbol{x})m(\boldsymbol{x})\right] , \qquad (57.3)$$

where $\mathfrak{N} \equiv e^{-E_0}$ is a physically irrelevant constant. This functional integral is of Gaussian type and can be evaluated exactly. Recalling (47.6) and (47.18) we obtain

$$Q = \mathcal{N} \left(\det K \right)^{-1/2} \exp \left(\frac{1}{2} h, K^{-1} h \right) , \tag{57.4}$$

where $\mathcal{N} = \lim_{N \to \infty} (2\pi)^{N/2} \mathfrak{N}$ is an infinite but physicaly irrelevant constant, K is the differential operator $K \equiv -\nabla^2 + a(t)$, and

$$\left(h, K^{-1} h \right) \equiv \int d^d x \, h(\boldsymbol{x}) \left(-\nabla^2 + a(t) \right)^{-1} h(\boldsymbol{x}) . \tag{57.5}$$

It is quite clear that $e^{i \boldsymbol{k} \cdot \boldsymbol{x}}$ is an eigenfunction of K with eigenvalue $k^2 + a(t)$. Hence

$$\det K = \prod_k \left(k^2 + a(t) \right) . \tag{57.6}$$

On the other hand, it is not difficult to show that (see Problem 57.1 below)

$$\left(h, K^{-1} h \right) = \frac{1}{V} \sum_k \frac{\tilde{h}_{-k} \tilde{h}_k}{k^2 + a(t)} , \tag{57.7}$$

where $\tilde{h}_k$ is the Fourier transform of $h(\boldsymbol{x})$. It follows from (57.4) that, up to a physically irrelevant additive constant,

$$\ln Q = -\frac{1}{2} \sum_k \ln \left(k^2 + a(t) \right) + \frac{1}{2V} \sum_k \frac{\tilde{h}_{-k} \tilde{h}_k}{k^2 + a(t)} . \tag{57.8}$$

Problem 57.1 By Fourier transforming $h(\boldsymbol{x})$ as in (55.4) and making use of the integral representation of the Kronecker delta function given by (55.8), prove the result (57.7). Note that $\tilde{h}_{-k} = \tilde{h}_k^*$ since $h(\boldsymbol{x})$ is real.

We set $h(\boldsymbol{x}) = 0$ and assume $t \to 0$ (which are the appropriate limits in the calculation of critical exponents), so that the second term in (57.8) vanishes and $a(t) \approx a_0 t$. The summation over $\boldsymbol{k}$ in the first term in (57.8) presents us with the situation of **ultraviolet divergence** as $|\boldsymbol{k}| \to \infty$. So we will impose a high momentum cutoff Λ and write

$$Q(a_0 t, \Lambda) = \mathcal{N} \prod_{|\boldsymbol{k}| < \Lambda} \left(k^2 + a_0 t \right)^{-1/2} , \tag{57.9}$$

where $\Lambda \sim 1/(lattice \ spacing)$. At this point we will carry out Wilson's three-step renormalization procedure. To implement the first step, we separate the slow modes ($|\boldsymbol{k}| < \Lambda/b$) from the fast modes ($\Lambda/b < |\boldsymbol{k}| < \Lambda$, $b > 0$) and write

$$Q(a_0 t, \Lambda) = \mathcal{N} \left(\prod_{|\boldsymbol{k}| < \Lambda/b} \left(k^2 + a_0 t \right)^{-1/2} \right) \left(\prod_{\Lambda/b < |\boldsymbol{k}| < \Lambda} \left(k^2 + a_0 t \right)^{-1/2} \right) (b > 1) .$$

$$\tag{57.10}$$

The second, or coarse-graining step, is trivial, since the second product above is regular (non-singular) as $t \to 0$, and just amounts to some constant. It can then be absorbed into the constant $\mathcal{N}$ to yield a new constant $\mathcal{N}'$. After this step we have

$$Q(a_0 t, \Lambda) = \mathcal{N}' \prod_{|\boldsymbol{k}| < \Lambda/b} \left(k^2 + a_0 t\right)^{-1/2} . \qquad (57.11)$$

Note that now the product is singular as $t \to 0$, because of the lower limit in $|\boldsymbol{k}|$. This is referred to as **infrared divergence**. The third step involves the rescaling $k = k'/b$, which restores the cutoff to Λ. Thus

$$Q(a_0 t, \Lambda) = \mathcal{N}' \prod_{|\boldsymbol{k}'| < \Lambda} \left(\frac{k'^2}{b^2} + a_0 t\right)^{-1/2} = \mathcal{N}' \prod_{|\boldsymbol{k}'| < \Lambda} b \left(k'^2 + a_0 b^2 t\right)^{-1/2}$$

$$= \mathcal{N}' b^{N(\Lambda)} \prod_{|\boldsymbol{k}'| < \Lambda} \left(k'^2 + a_0 b^2 t\right)^{-1/2} ,$$

$$(57.12)$$

where $N(\Lambda)$ is the number of discrete $\boldsymbol{k}'$ points in a d-sphere of radius Λ. Rewriting the dummy variable $\boldsymbol{k}'$ as $\boldsymbol{k}$, we have

$$Q(a_0 t, \Lambda) = b^{N(\Lambda)} \left(\mathcal{N}' \prod_{|\boldsymbol{k}| < \Lambda} \left(k^2 + a_0 b^2 t\right)^{-1/2} \right) . \qquad (57.13)$$

Comparing with (57.4) we have

$$\prod_{|\boldsymbol{k}| < \Lambda/b} \left(k^2 + a_0 t\right)^{-1/2} = b^{N(\Lambda)} \prod_{|\boldsymbol{k}| < \Lambda} \left(k^2 + a_0 b^2 t\right)^{-1/2} . \qquad (57.14)$$

Multiplying by $\mathcal{N}$ on both sides [the original constant in (57.2)], we have

$$Q(a_0 t, \Lambda/b) = b^{N(\Lambda)} Q(a_0 b^2 t, \Lambda) . \qquad (57.15)$$

This implies

$$Q(a_0 t, \Lambda) = b^{-N(\Lambda)} Q\left(\frac{a_0}{b^2} t, \frac{\Lambda}{b}\right) . \qquad (57.16)$$

Now choose $b = \Lambda/\lambda$, where $\lambda < \Lambda$ is an arbitrary finite constant having the dimension of momentum. We then arrive at

$$Q(a_0 t, \Lambda) = \left(\frac{\Lambda}{\lambda}\right)^{-N(\Lambda)} Q\left(\frac{a_0 \lambda^2}{\Lambda^2} t, \lambda\right) = Z(\Lambda, \lambda) Q(a_0' t, \lambda) , \qquad (57.17)$$

where

$$Z(\Lambda, \lambda) \equiv \left(\frac{\Lambda}{\lambda}\right)^{-N(\Lambda)} , \qquad a_0' \equiv \frac{a_0 \lambda^2}{\Lambda^2} . \qquad (57.18)$$

We will justify subsequently that the renormalization constant $Z(\Lambda, \lambda)$ is irrelevant in the calculation of thermodynamic functions from Q. The quantity

a_0' can be regarded as a renormalized coupling constant, and set equal to a (physically observable) finite quantity. This is the equivalent of a renormalized mass discussed in Chapter 52. $Q(a_0't, \lambda)$ is then the renormalized partition function, from which the high momentum cutoff Λ has disappeared. We have thus achieved multiplicative renormalization of the partition function by Wilson's procedure.

The free energy per unit volume $g = G/(kTV)$ is given by

$$g = -\frac{1}{V} \ln Q(a_0 t, \Lambda) = -\frac{1}{V} \{\ln Z(\Lambda, \lambda) + \ln Q(a_0't, \lambda)\} . \quad (57.19)$$

The first term is an (irrelevant) additive constant, and the second term is independent of the cutoff Λ. It can be written as

$$-\frac{1}{V} \ln Q(a_0't, \lambda) = -\frac{1}{V} \ln \left\{ \mathcal{N} \prod_{|\mathbf{k}|<\lambda} \left(k^2 + a_0't\right)^{-1/2} \right\}$$

$$= -\frac{1}{V} \ln \mathcal{N} - \frac{1}{V} \ln \left\{ \prod_{|\mathbf{k}|<\lambda} \left(k^2 + a_0't\right)^{-1/2} \right\} . \quad (57.20)$$

The first term above is another additive constant which can be dropped. So we have, up to an overall additive constant,

$$g = \frac{1}{2V} \sum_{|\mathbf{k}|<\lambda} \ln \left(k^2 + a_0't\right) \xrightarrow[V \to \infty]{} \frac{1}{2} \int_{|\mathbf{k}|<\lambda} \frac{d^d k}{(2\pi)^d} \ln(k^2 + a_0't)$$

$$= \frac{1}{2(2\pi)^d} \int_0^\lambda dk\, S_d(k)\, \ln(k^2 + a_0't) , \quad (57.21)$$

where

$$S_d = \frac{d\pi^{d/2} k^{d-1}}{\Gamma(d/2 + 1)} \quad (57.22)$$

is the surface area of a d-sphere of radius k [c.f. (53.30) and (53.40)]. The arbitrary constant λ is not relevant near the critical region $t \to 0$, since in this limit the integral diverges at the lower limit $|\mathbf{k}| \to 0$, giving rise to *infrared divergence*; and it is this divergence that determines the critical behavior. We will explain and justify this claim as follows. Indeed, using (56.32) to calculate the specific heat C from g we obtain

$$\frac{C}{kV} = -\frac{\partial^2 g}{\partial t^2} = A \int_0^\lambda dk\, \frac{k^{d-1}}{(k^2 + a_0't)^2} , \quad (57.23)$$

where

$$A \equiv \frac{(a_0')^2 d}{2^{d+1}\pi^{d/2}\Gamma(d/2 + 1)} . \quad (57.24)$$

Changing the integration variable in the above by setting $s^2 = k^2/(a'_0 t)$ it is straightforward to show that

$$\frac{C}{kV} = A(a'_0)^{d/2-2} t^{(d-4)/2} \int_0^{\frac{\lambda}{\sqrt{a'_0 t}}} ds \frac{s^{d-1}}{(s^2+1)^2} . \qquad (57.25)$$

This expression allows us to determine the critical exponent α associated with the specific heat by analysis of its behavior as $t \to 0$. We claim that in this case (Gaussian model), the singular behavior $C \sim |t|^{-\alpha}$ [c.f. (56.5)] is already totally isolated by the factor $t^{(d-4)/2}$ in the above expression, and that the appearance of t in the upper limit of the integral plays no role in this singular behavior. Clearly, the dimension d plays an important role. We distinguish between three cases: $d < 4$, $d = 4$ and $d > 4$, and analyze the behavior of the above integral in each case as $t \to 0$ while λ remains fixed and finite, that is, as the upper limit $\lambda/\sqrt{a'_0 t} \to \infty$. By simple power counting of powers of s we see that the degree of divergence D of the integral is $D = d-4$. So for $d < 4$ the integral converges and indeed the factor $t^{(d-4)/2}$ gives the full singular behavior. For $d = 4$ the integral is logarithmically divergent and behaves as $-(1/2) \ln t$ as $t \to 0$, while the factor $t^{(d-4)/2}$ approaches 1. So for $d = 4$, $C \sim (constant) \ln t$ as $t \to 0$. This situation, however, is still consistent with the power-law singularity description for C given by (56.5). Finally, for $d > 4$, the integral diverges like $(\lambda/\sqrt{a'_0 t})^{d-4} \sim t^{-(d-4)/2}$. But this precisely cancels the factor $t^{(d-4)/2}$, and so $C \sim constant$ as $t \to 0$ (the singular part vanishes). We thus see that for all values of d, $\alpha = (4-d)/2$. By the first equation of (56.49) this immediately yields $D_t = 2$. So we can state that

$$\alpha = \frac{4-d}{2} , \qquad \nu = \frac{1}{D_t} = \frac{1}{2} \qquad \text{(Gaussian model)} . \qquad (57.26)$$

Note that these two critical exponents are not independent as they are related by Josephson's scaling law (56.17).

Next we will implement Wilson's three-step momentum space renormalization procedure in more detail (than was done in Chapter 55) for the Gaussian model in Landau-Ginsburg theory, and demonstrate how to determine all the critical exponents for this model. Following (55.4) we write the Fourier transform pair for the order parameter $m(x)$ as:

$$\tilde{m}(\boldsymbol{k}) = \int_V d^d x \, e^{i\boldsymbol{k}\cdot\boldsymbol{x}} \, m(\boldsymbol{x}) , \qquad k_i = \frac{2\pi n_i}{L} \quad (n_i = 0, \pm 1, \pm 2, \ldots) , \qquad (57.27\text{a})$$

$$m(\boldsymbol{x}) = \frac{1}{V} \sum_k e^{-i\boldsymbol{k}\cdot\boldsymbol{x}} \, \tilde{m}_{\boldsymbol{k}} \xrightarrow[V \to \infty]{} \int \frac{d^d k}{(2\pi)^d} e^{-i\boldsymbol{k}\cdot\boldsymbol{x}} \, \tilde{m}(\boldsymbol{k}) , \qquad (57.27\text{b})$$

where V is the volume of the hypercube in which the system is placed and L is the length of each side of this cube. Assuming $h(\boldsymbol{x}) = h$ is constant and expressing the right-hand side of (57.2) in terms of the Fourier fields $\tilde{m}(\boldsymbol{k})$ we have

$$E[\tilde{m}] = \frac{1}{2} \int d^d k \, (k^2 + a(t)) \, |\tilde{m}(\boldsymbol{k})|^2 - h\tilde{m}(0) , \qquad (57.28)$$

where the term E_0 in (57.2) has been dropped since it is physically irrelevant, and $|\tilde{m}(\boldsymbol{k})|^2 = \tilde{m}^*(\boldsymbol{k})\tilde{m}(\boldsymbol{k})$ appears instead of $\tilde{m}(-\boldsymbol{k})\tilde{m}(\boldsymbol{k})$ since, for real $m(\boldsymbol{x})$, $\tilde{m}^*(\boldsymbol{k}) = \tilde{m}(-\boldsymbol{k})$. The term $h\tilde{m}(0)$ appears since

$$h \int d^d x\, m(\boldsymbol{x}) = h \int d^d x \int \frac{d^d k}{(2\pi)^d} e^{-i\boldsymbol{k}\cdot\boldsymbol{x}}\, \tilde{m}(\boldsymbol{k})$$

$$= h \int d^d k \left(\int \frac{d^d x}{(2\pi)^d} e^{-i\boldsymbol{k}\cdot\boldsymbol{x}} \right) \tilde{m}(\boldsymbol{k}) = h \int d^d k\, \delta^d(\boldsymbol{k})\tilde{m}(\boldsymbol{k}) = h\tilde{m}(0)\,. \tag{57.29}$$

The partition function Q (which we will not calculate explicitly) is given by

$$Q = e^{-G/(kT)} = \int [d\tilde{m}]\, e^{-E[\tilde{m}]} = \prod_{k<\Lambda}{}' \int d\tilde{m}^*(\boldsymbol{k})d\tilde{m}(\boldsymbol{k})\, e^{-E[\tilde{m}]}\,, \tag{57.30}$$

where the prime next to the product sign means that the product is only taken over half of the lattice points in $\boldsymbol{k}$-space. This is done because for real $m(\boldsymbol{x})$, $\tilde{m}(-\boldsymbol{k}) = \tilde{m}^*(\boldsymbol{k})$. (One can also dispense with the prime but take the square root of the unrestricted product.) We have to integrate over both $\tilde{m}(\boldsymbol{k})$ and $\tilde{m}^*(\boldsymbol{k})$ for each $\boldsymbol{k}$ because $\tilde{m}(\boldsymbol{k})$ is complex.

Now we begin the momentum space three-step renormalization process.

(1) *Separation into fast and slow modes.* Write the partition partition Q as [c.f. (55.18)]

$$Q = \int [d\tilde{m}^{(s)}(\boldsymbol{k})] \int [d\tilde{m}^{(f)}(\boldsymbol{k})]\, e^{-E[\tilde{m}^{(s)}+\tilde{m}^{(f)}]}\,, \tag{57.31}$$

where $\tilde{m}^{(s)}(\boldsymbol{k})$ and $\tilde{m}^{(f)}(\boldsymbol{k})$ are defined by (55.16) and (55.17).

(2) *Coarse graining.* Define a new Hamiltonian functional $E'[\tilde{m}(\boldsymbol{k})]$ consisting of only the slow modes $|\boldsymbol{k}| < \Lambda/b$ $(b > 1)$ by (functionally) integrating out the fast modes $\Lambda/b < |\boldsymbol{k}| < \Lambda$ [c.f. (55.19)] as follows:

$$\int [d\tilde{m}^{(f)}(\boldsymbol{k})]\, e^{-E[\tilde{m}(\boldsymbol{k})]} = \prod_{\Lambda/b<|\boldsymbol{k}|<\Lambda}{}' d\tilde{m}(\boldsymbol{k})d\tilde{m}^*(\boldsymbol{k})\, e^{-E[\tilde{m}(\boldsymbol{k})]}$$

$$= e^{-\mathfrak{E}} e^{-E'[\tilde{m}(\boldsymbol{k})]}\,, \tag{57.32}$$

where $\mathfrak{E} = \mathfrak{E}(\Lambda, b, a(t))$ is a function of the cutoff Λ and the rescaling parameter $b(> 1)$, as well as the coupling constant $a(t)$ [see (57.28)]. We can determine $E'[\tilde{m}]$ easily for the Gaussian model. Writing $\tilde{m} = \tilde{m}^{(s)}+\tilde{m}^{(f)}$, where $\tilde{m}^{(s)} = 0$ for $\Lambda/b < |\boldsymbol{k}| < \Lambda$ and $\tilde{m}^{(f)} = 0$ for $|\boldsymbol{k}| < \Lambda/b$, we have, by (57.28),

$$E[\tilde{m}(\boldsymbol{k})] = \frac{1}{2} \int_{|\boldsymbol{k}|<\Lambda/b} d^d k\, (k^2 + a(t))\, |\tilde{m}^{(s)}(\boldsymbol{k})|^2$$

$$+ \int_{\Lambda/b<|\boldsymbol{k}|<\Lambda} d^d k\, (k^2 + a(t))\, |\tilde{m}^{(f)}(\boldsymbol{k})|^2 - h\tilde{m}(0)\,. \tag{57.33}$$

Hence

$$\int [d\tilde{m}^{(f)}(\boldsymbol{k})]\, e^{-E[\tilde{m}(\boldsymbol{k})]}$$

$$= \exp\left[-\left\{\frac{1}{2}\int_{|\boldsymbol{k}|<\Lambda/b} d^d k \,\left(k^2 + a(t)\right) |\tilde{m}^{(s)}(\boldsymbol{k})|^2 - h\tilde{m}(0)\right\}\right]$$

$$\times \int [d\tilde{m}^{(f)}]\, \exp\left[-\frac{1}{2}\int_{\Lambda/b<|\boldsymbol{k}|<\Lambda} d^d k \,\left(k^2 + a(t)\right) |\tilde{m}^{(f)}(\boldsymbol{k})|^2\right].$$

$$(57.34)$$

The functional integral in the last line is of Gaussian type and can be done exactly. We will denote it by $e^{-\mathcal{E}(\Lambda,b,a(t))}$. Combining with (57.32) we obtain the following expression for the new Hamiltonian functional $E'[\tilde{m}(\boldsymbol{k})]$:

$$E'[\tilde{m}(\boldsymbol{k})] = \frac{1}{2}\int_{|\boldsymbol{k}|<\Lambda/b} d^d k \,\left(k^2 + a(t)\right) |\tilde{m}(\boldsymbol{k})|^2 - h\tilde{m}(0) . \qquad (57.35)$$

(3) *Restoring original scale.* Let $\boldsymbol{k}' = b\boldsymbol{k}$ $(b > 1)$. Then (57.35) gives

$$E'[\tilde{m}] = \frac{b^{-d}}{2}\int_{|\boldsymbol{k}'|<\Lambda} d^d k' \,\left(\frac{k'^2}{b^2} + a(t)\right) \left|\tilde{m}\left(\frac{\boldsymbol{k}'}{b}\right)\right|^2 - h\tilde{m}(0) . \quad (57.36)$$

Rewriting the dummy integration variable $\boldsymbol{k}'$ as $\boldsymbol{k}$, the above equation can be rewritten as

$$E'[\tilde{m}] = \frac{b^{-d}}{2}\int_{|\boldsymbol{k}|<\Lambda} d^d k \,\left(\frac{k^2}{b^2} + a(t)\right) \left|\tilde{m}\left(\frac{\boldsymbol{k}}{b}\right)\right|^2 - h\tilde{m}(0)$$

$$= \frac{1}{2}\int_{|\boldsymbol{k}|<\Lambda} d^d k \,\left(k^2 + b^2 a(t)\right) \left|\sqrt{\frac{1}{b^{d+2}}}\,\tilde{m}\left(\frac{\boldsymbol{k}}{b}\right)\right|^2 \qquad (57.37)$$

$$- \left(\sqrt{b^{d+2}}\,h\right)\left(\sqrt{\frac{1}{b^{d+2}}}\,\tilde{m}(0)\right) .$$

We can then define the renormalized field (order parameter) $\tilde{m}'(\boldsymbol{k})$ by

$$\tilde{m}'(\boldsymbol{k}) \equiv \sqrt{\frac{1}{b^{d+2}}}\,\tilde{m}\left(\frac{\boldsymbol{k}}{b}\right) ; \qquad (57.38)$$

and write

$$E'[\tilde{m}'(\boldsymbol{k})] = \frac{1}{2}\int_{|\boldsymbol{k}|<\Lambda} d^d k \,\left(k^2 + a'(t)\right) |\tilde{m}'(\boldsymbol{k})|^2 + h'\tilde{m}'(0) , \qquad (57.39)$$

where

$$a'(t) \equiv b^2\, a(t) \xrightarrow[t\to 0]{} b^2\, a_0 t \qquad \text{and} \qquad h' \equiv b^{(d+2)/2}\, h \qquad (57.40)$$

are the renormalized coupling constant (equivalent to the mass-squared) and the renormalized magnetic field, respectively. Eq. (57.39) is to be compared with (55.25). Note that $E'[\tilde{m}'(\mathbf{k})]$ has exactly the same form as $E[\tilde{m}(\mathbf{k})]$, except that $a(t)$ and h acquire new values as above. The three steps of momentum space renormalization are now completed.

Eq. (57.40) and the homogeneity condition immediately imply

$$D_t = 2 , \qquad D_h = \frac{d+2}{2} \qquad \text{(Gaussian model)} . \qquad (57.41)$$

These values then allow all the critical exponents to be determined via (56.49). It is interesting to note that they agree with the mean-field values for $d = 4$.

We are now ready to carry out the Wilson three-step procedure for the full m^4 model (first described in Wilson 1983). To conform with the notation used in the original literature (see Wilson and Kogut 1974), we will change the symbols for the coupling constants in (56.69), using $r(t)$ instead of $a(t)$ and $u(t)$ instead of $b(t)/4!$, and rewrite the Hamiltonian functional $E[m]$ (for $H = 0$) as

$$E[m(\mathbf{x})] = \int d^d x \left\{ \frac{1}{2} \left(|\nabla m(\mathbf{x})|^2 + r(t)m^2(\mathbf{x}) \right) + u(t)m^4(\mathbf{x}) \right\} . \qquad (57.42)$$

(1) *Separation into fast and slow modes.* First separate out the fast k modes in the shell $\Lambda/b < |\mathbf{x}| < \Lambda$ ($b > 1$). We will eventually let b be a continuous variable and write $b \approx 1 + x \approx 1$ ($0 < x \ll 1$). Then

$$\Delta k \equiv \Lambda - \frac{\Lambda}{b} = \Lambda \left(\frac{b-1}{b} \right) = \frac{\Lambda x}{b} \approx \Lambda x , \qquad (57.43)$$

and

$$\frac{\Delta k}{\Lambda} \approx x \approx \ln(1+x) = \ln b = \Delta(\ln \Lambda) . \qquad (57.44)$$

The number of discrete $\mathbf{k}$ modes in the k-shell $\Lambda/b < |\mathbf{k}| < \Lambda$, N, is given by

$$N = \frac{V}{2(2\pi)^d} \times \text{(volume of } k\text{-shell)} \approx \frac{V}{2(2\pi)^d} S_d(\Lambda)\Delta k , \qquad (57.45)$$

where V is the physical volume of the system, $S_d(\Lambda)$ is the surface area of a d-sphere of radius Λ, and the reason for the extra factor of 2 in the denominator has been given immediately after (57.30). Decompose the order-parameter field $m(\mathbf{x})$ into slow and fast components according to

$$m(\mathbf{x}) = m^{(s)}(\mathbf{x}) + m^{(f)}(\mathbf{x}) , \qquad (57.46)$$

where $m^{(f)}(\mathbf{x})$ contains only the Fourier components to be integrated out, that is, in the k-shell specified above. Because of the presence of the term proportional to m^4 in the Hamiltonian functional (57.42), the fast component $m^{(f)}(\mathbf{x})$ has to be treated carefully. Wilson expands it as a linear superposition of N [given by (57.45)] roughly non-overlapping real wave packets $\phi_1(\boldsymbol{x}), \ldots, \phi_N(\boldsymbol{x})$ with centers $\boldsymbol{x}_i$ such that, together, they occupy the physical volume of the system, and each packet is an extended object [with size in each spatial dimension $\Delta x \approx 1/\Delta k \approx 1/(\Lambda \ln b)$] whose Fourier components have short wavelengths $\sim 1/\Lambda$. We write

$$m^{(f)}(\boldsymbol{x}) = \sum_{i=1}^{N} c_i \phi_i(\boldsymbol{x}) \,. \tag{57.47}$$

The wave packets $\phi_i(\boldsymbol{x})$ are also required to satisfy the following properties:

$$\int d^d x \, \phi_i(\boldsymbol{x}) \phi_j(\boldsymbol{x}) \approx \delta_{ij} \,, \tag{57.48}$$

$$\int d^d x \, [\phi_i(\boldsymbol{x})]^n \approx 0 \qquad (n \text{ odd}) \,, \tag{57.49}$$

$$m^{(s)}(\boldsymbol{x}) \approx m^{(s)}(\boldsymbol{x}_i) \approx constant \qquad (\text{over } \Delta V_i) \,, \tag{57.50}$$

where in the last condition, $\Delta_i \approx (\Delta x)^d$ is the spatial extent of the i-th packet. Each of the three terms in (57.42) will now be analyzed individually. We have, for the first term,

$$\int d^d x \, \left| \nabla \left\{ m^{(s)}(\boldsymbol{x}) + m^{(f)}(\boldsymbol{x}) \right\} \right|^2 = \int d^d x \, \left| \nabla m^{(s)}(\boldsymbol{x}) + \nabla m^{(f)}(\boldsymbol{x}) \right|^2$$

$$= \int d^d x \, \left| m^{(s)}(\boldsymbol{x}) \right|^2 + 2 \int d^d x \, \nabla m^{(s)}(\boldsymbol{x}) \cdot \nabla m^{(f)}(\boldsymbol{x})$$

$$+ \int d^d x \, \left| \nabla m^{(f)}(\boldsymbol{x}) \right|^2 \,. \tag{57.51}$$

By virtue of (57.51), $\nabla m^{(s)} \approx 0$ everywhere, so the second term on the right-hand side of the above equation vanishes. Hence

$$\int d^d x \, \left| \nabla \left\{ m^{(s)}(\boldsymbol{x}) + m^{(f)}(\boldsymbol{x}) \right\} \right|^2 \approx \int d^d x \, \left| \nabla m^{(s)}(\boldsymbol{x}) \right|^2 + \int d^d x \, \left| m^{(f)}(\boldsymbol{x}) \right|^2 \,. \tag{57.52}$$

By (57.47) the second term in the right-hand side above can be evaluated as follows:

$$\int d^d x \, \left| \nabla m^{(f)}(\boldsymbol{x}) \right|^2 = \int d^d x \, \left| \sum_{i=1}^{N} c_i \nabla \phi_i(\boldsymbol{x}) \right|^2 \tag{57.53}$$

$$= \sum_{i=1}^{N} c_i^2 \int d^d x \, |\nabla \phi_i(\boldsymbol{x})|^2 + \sum_{i \neq j}^{N} c_i c_j \int d^d x \, \nabla \phi_i(\boldsymbol{x}) \cdot \nabla \phi_j(\boldsymbol{x}) \,.$$

We assert that (57.48) implies that (see Problem 57.2 below)

$$\int d^d x \, \nabla \phi_i(\boldsymbol{x}) \cdot \nabla \phi_j(\boldsymbol{x}) \approx \Lambda^2 \delta_{ij} \,. \tag{57.54}$$

Then it follows from (57.52) that

$$\int d^d x \, |\nabla m(\boldsymbol{x})|^2 \approx \int d^d x \, \left|\nabla m^{(s)}(\boldsymbol{x})\right|^2 + \Lambda^2 \sum_{i=1}^{N} c_i^s \, . \tag{57.55}$$

Next we consider the second term in (57.42). This is proportional to

$$\int d^d x \, \left[m^{(s)}(\boldsymbol{x}) + m^{(f)}(\boldsymbol{x})\right]^2$$
$$= \int d^d x \, [m^{(s)}(\boldsymbol{x})]^2 + 2 \int d^d x \, m^{(s)}(\boldsymbol{x}) m^{(f)}(\boldsymbol{x}) + \int d^d x \, [m^{(f)}(\boldsymbol{x})]^2 \, . \tag{57.56}$$

The second term on the right-hand side vanishes since it can be written, apart from a factor of 2, as

$$\sum_{i=1}^{N} c_i \int d^d x \, m^{(s)}(\boldsymbol{x}) \phi_i(\boldsymbol{x}) \approx \sum_{i=1}^{N} c_i m^{(s)}(\boldsymbol{x}_i) \int d^d x \, \phi_i(\boldsymbol{x}) \approx 0 \, ,$$

where the first approximation follows from (57.50) and the integral on the right-hand side of this approximation vanishes by virtue of (57.49). The third term on the right-hand side of (57.56) can be manipulated as follows:

$$\int d^d x \, [m^{(f)}(\boldsymbol{x})]^2 = \sum_{i,j=1}^{N} c_i c_j \int d^d x \, \phi_i(\boldsymbol{x}) \phi_j(\boldsymbol{x}) \approx \sum_{i,j=1}^{N} c_i c_j \delta_{ij} = \sum_{i=1}^{N} c_i^2 \, , \tag{57.57}$$

where the approximation follows from (57.48). The second term in (57.42) then appears as

$$\int d^d x \, m^2(\boldsymbol{x}) \approx \int d^d x \, [m^{(s)}(\boldsymbol{x})]^2 + \sum_{i=1}^{N} c_i^2 \, . \tag{57.58}$$

Finally, the third term in (57.42), the non-Gaussian one involving $m^4(\boldsymbol{x})$, is proportional to

$$\int d^d x \, m^4(\boldsymbol{x}) = \int d^d x \, [m^{(s)}(\boldsymbol{x}) + m^{(f)}(\boldsymbol{x})]^4$$
$$\approx \int d^d x \, \left[(m^{(s)})^4 + 4(m^{(s)})^3(m^{(f)}) + 6(m^{(s)})^2(m^{(f)})^2\right] \, , \tag{57.59}$$

where on the right-hand side we have ignored the terms proportional to $(m^{(f)})^3$ and $(m^{(f)})^4$. [The dropping of the $(m^{(f)})^3$ term can be justified along the same lines of reasoning that we will be using for the handling of the $m^{(f)}$ and $(m^{(f)})^2$ terms (see immediately below and Problem 57.3).

The discarding of the $\left(m^{(f)}\right)^4$ term is partly based on our desire to keep the perturbative contributions to as low an order in $u(t)$ as possible and still obtain results that are beyond those of the (trivial) Gaussian model. In fact, as we will see, this step will keep the functional integration $\int [dm^{(f)}]$ as a Gaussian one {see (57.66) and (57.67) below}. If this approximation were not made, we would have to use a Feynman graph perturbative approach to do the $\int [dm^{(f)}]$ functional integration in a renormalization procedure.] The second integral on the right-hand side of the above approximation [in (57.79)] vanishes since it equals

$$\sum_{i=1}^{N} c_i \int d^d x \left(m^{(s)}(\boldsymbol{x})\right)^3 \phi_i(\boldsymbol{x}) \approx \sum_{i=1}^{N} c_i \left(m^{(s)}(\boldsymbol{x}_i)\right)^3 \int d^d x \, \phi_i(\boldsymbol{x}) \approx 0 \,,$$

(57.60)

where the last approximation follows from (57.49) and (57.50). The third integral on the right-hand side of the approximation in (57.59) can be similarly manipulated:

$$\sum_{i,j}^{N} c_i c_j \int d^d x \left(m^{(s)}(\boldsymbol{x})\right)^2 \phi_i(\boldsymbol{x}) \phi_j(\boldsymbol{x})$$

$$= \sum_{i=1}^{N} c_i^2 \int d^d x \left(m^{(s)}(\boldsymbol{x})\right)^2 (\phi_i(\boldsymbol{x}))^2$$

$$+ \sum_{i \neq j}^{N} c_i c_j \int d^d x \left(m^{(s)}(\boldsymbol{x})\right)^2 \phi_i(\boldsymbol{x}) \phi_j(\boldsymbol{x})$$

(57.61)

$$\approx \sum_{i=1}^{N} c_i^2 \left(m^{(s)}(\boldsymbol{x}_i)\right)^2 \,,$$

where the second integral on the right-hand side of the first equality is approximated to be zero due to the fact that (57.48) implies that different wave packets ϕ_i and ϕ_j are non-overlapping in space, and the approximation follows from (57.48) and (57.50). Eqs. (57.59) to (57.61) then yield

$$\int d^d x \, m^4(\boldsymbol{x}) \approx \int d^d x \left(m^{(s)}(\boldsymbol{x})\right)^4 + 6 \sum_{i=1}^{N} c_i^2 \left(m^{(s)}(\boldsymbol{x}_i)\right)^2 . \quad (57.62)$$

Combining (57.42), (57.55), (57.58) and (57.62), we have the following expression for the Hamiltonian functional $E[m(\boldsymbol{x})]$, after having separated $m^{(s)}(\boldsymbol{x})$ from $m^{(f)}(\boldsymbol{x})$:

$$E[m(\boldsymbol{x})] = E[m^{(s)}(\boldsymbol{x})] + \sum_{i=1}^{N} \left[\frac{1}{2}\left(\Lambda^2 + r(t)\right) + 6u(t)\left(m^{(s)}(\boldsymbol{x}_i)\right)^2\right] c_i^2$$

(57.63)

where

$$E[m^{(s)}(\boldsymbol{x})]$$
$$= \int d^d x \left[\frac{1}{2} \left\{ \left(\nabla m^{(s)}(\boldsymbol{x}) \right)^2 + r(t) \left(m^{(s)}(\boldsymbol{x}) \right)^2 \right\} + u(t) \left(m^{(s)}(\boldsymbol{x}) \right)^4 \right].$$
(57.64)

(2) *Coarse graining.* We define the renormalized Hamiltonian functional $E'[m^{(s)}(\boldsymbol{x})]$ by functionally integrating out $m^{(f)}(\boldsymbol{x})$, that is,

$$e^{-E'[m^{(s)}(\boldsymbol{x})]} = \int [dm^{(f)}(\boldsymbol{x})] e^{-E[m(\boldsymbol{x})]}, \qquad (57.65)$$

where, due to (57.47),

$$\int [dm^{(f)}(\boldsymbol{x})] = \prod_{i=1}^{N} \int_{-\infty}^{\infty} dc_i . \qquad (57.66)$$

We then have

$$e^{-E'[m^{(s)}(\boldsymbol{x})]} = e^{-E[m^{(s)}(\boldsymbol{x})]}$$

$$\times \prod_{i=1}^{N} \int_{-\infty}^{\infty} dc_i \left(\exp \left\{ - \sum_{i=1}^{N} c_i^2 \left[\frac{1}{2} (\Lambda^2 + r(t)) + 6u(t) \left(m^{(s)}(\boldsymbol{x}_i) \right)^2 \right] \right\} \right)$$

$$= e^{-E[m^{(s)}(\boldsymbol{x})]}$$

$$\times \prod_{i=1}^{N} \left(\int_{-\infty}^{\infty} dc_i \exp \left\{ -c_i^2 \left[\frac{1}{2} (\Lambda^2 + r(t)) + 6u(t) \left(m^{(s)}(\boldsymbol{x}_i) \right)^2 \right] \right\} \right).$$
(57.67)

Using the basic Gaussian integral formula

$$\int_{-\infty}^{\infty} dx \, e^{-ax^2/2} = \sqrt{\frac{2\pi}{a}}, \qquad (57.68)$$

the c_i integral on the right-hand side of the second equality in (57.67) is equal to

$$\sqrt{\frac{2\pi}{\Lambda^2 + r(t) + 12u(t) \left(m^{(s)}(\boldsymbol{x}_i) \right)^2}} .$$

It follows from (57.67) that

$$e^{-E'[m^{(s)}(\boldsymbol{x})]} = e^{-E[m^{(s)}(\boldsymbol{x})]}$$

$$\times \exp \left[\ln \left\{ \prod_{i=1}^{N} \left(\frac{2\pi}{\Lambda^2 + r(t) + 12u(t) \left(m^{(s)}(\boldsymbol{x}_i) \right)^2} \right)^{1/2} \right\} \right]. \qquad (57.69)$$

The exponential in the last line can be further manipulated to yield

$$\exp\left[-\frac{1}{2}\sum_{i=1}^{N}\ln\left(\frac{\Lambda^2+r(t)+12u(t)\left(m^{(s)}(\boldsymbol{x}_i)\right)^2}{2\pi}\right)\right]$$

$$=\exp\left[-\frac{1}{2}\sum_{i=1}^{N}\ln\left\{\left(\frac{\Lambda^2}{2\pi}\right)\left(1+\frac{r(t)}{\Lambda^2}+\frac{12u(t)}{\Lambda^2}\left(m^{(s)}(\boldsymbol{x}_i)\right)^2\right)\right\}\right].$$

$$(57.70)$$

Eq. (57.69) then implies that

$$E'[m^{(s)}(\boldsymbol{x})]=E[m^{(s)}(\boldsymbol{x})]$$

$$+\frac{1}{2}\sum_{i=1}^{N}\ln\left\{\left(\frac{\Lambda^2}{2\pi}\right)\left(1+\frac{r(t)}{\Lambda^2}+\frac{12u(t)}{\Lambda^2}\left(m^{(s)}(\boldsymbol{x}_i)\right)^2\right)\right\}.$$

$$(57.71)$$

Note that the coefficient of $|\nabla m^{(s)}(\boldsymbol{x})|^2$ $(=1/2)$ has not been changed in this coarse graining step. [Compare with the general situation in (55.20), where a factor z involving the anomalous dimension η has to be inserted.] The next step is to replace the sum in the above expression by a spatial integral, in the limit $V\to\infty$, using (57.45):

$$\sum_{i=1}^{N}\xrightarrow[V\to\infty]{}\frac{V}{2(2\pi)^d}S_d(\Lambda)\Delta k\approx\frac{V}{2(2\pi)^d}S_d(\Lambda)\Lambda\ln b,\qquad(57.72)$$

where we have used (57.44) for Δk. Writing the spatial volume V as $\int d^d x$ and using (57.22) for $S_d(\Lambda)$, we have

$$\sum_{i=1}^{N}\xrightarrow[V\to\infty]{}C_d\Lambda^d\ln b\int d^d x,\qquad(57.73)$$

where

$$C_d\equiv\frac{d}{2^{d+1}\pi^{d/2}\Gamma(d/2+1)}.\qquad(57.74)$$

Invoking the elementary expansion

$$\ln(1+x)=x-\frac{x^2}{2}+\frac{x^3}{3}-\cdots\qquad(|x|<1),\qquad(57.75)$$

the logarithm term for each i in the sum in (57.71) can be expanded as

$$\frac{1}{2}\ln\left(\frac{\Lambda^2}{2\pi}\right)+\frac{1}{2}\left(1+\frac{r(t)}{\Lambda^2}+\frac{12u(t)}{\Lambda^2}\left(m^{(s)}(\boldsymbol{x}_i)\right)^2\right)$$

$$\approx\frac{1}{2}\left[\ln\left(\frac{\Lambda^2}{2\pi}\right)+\frac{r(t)}{\Lambda^2}-\frac{r^2(t)}{2\Lambda^4}\right]\qquad(57.76)$$

$$+6\left(\frac{u(t)}{\Lambda^2}-\frac{r(t)u(t)}{\Lambda^4}\right)(m(s)(\boldsymbol{x}_i))^2-\frac{36u^2(t)}{\Lambda^4}\left(m^{(s)}(\boldsymbol{x}_i)\right)^4.$$

The first term on the right-hand side above (with the square brackets) does not involve the field quantity $m^{(s)}(x_i)$ [similar to the term E_0 in (57.2)], and can be dropped as a physically irrelevant quantity on renormalization. Then (57.71) can be written as

$$E'[m^{(s)}(x)] = E[m^{(s)}(x)] + C_d \Lambda^d (\ln b)$$
$$\times \int d^d x \left[6 \left(\frac{u(t)}{\Lambda^2} - \frac{r(t)u(t)}{\Lambda^4} \right) \left(m^{(s)}(x) \right)^2 - \frac{36u^2(t)}{\Lambda^4} \left(m^{(s)}(x) \right)^4 \right],$$
$$(57.77)$$

where on the right-hand side we have replaced x_i by x, since on converting from the sum over i in (57.71) to a spatial integral we have moved to the continuum limit. On using (57.64) for $E[m^{(s)}(x)]$ the above equation appears as, up to a physically irrelevant additive constant,

$$E'[m^{(s)}(x)]$$
$$= \int d^d x \left[\frac{1}{2} \left\{ |\nabla m^{(s)}(x)|^2 + \tilde{r}(t) \left(m^{(s)}(x) \right)^2 \right\} + \tilde{u}(t) \left(m^{(s)}(x) \right)^4 \right],$$
$$(57.78)$$

where

$$\tilde{r}(t) \equiv r(t) + 12 C_d (\ln b) \left(\Lambda^{d-2} u(t) - \Lambda^{d-4} r(t) u(t) \right),\qquad (57.79)$$
$$\tilde{u}(t) \equiv u(t) - 36 C_d (\ln b) \Lambda^{d-4} u^2(t).\qquad (57.80)$$

The coarse-graining step is now completed, since only the slow field $m^{(s)}(x)$, with only Fourier modes $|k| < \Lambda/b$, appears in the Hamiltonian functional.

(3) *Restoring original scale.* Let $x' = x/b$. Then

$$\nabla_i = \frac{\partial}{\partial x^i} = \frac{\partial}{\partial x'^j} \frac{\partial x'^j}{\partial x^i} = \frac{1}{b} \frac{\partial}{\partial x'^i} = \frac{1}{b} \nabla'_i,$$

and (57.78) gives

$$E'[m^{(s)}(x')]$$
$$= b^d \int d^d x' \left[\frac{1}{2b^2} \left| \nabla' m^{(s)}(bx') \right|^2 \right. \qquad (57.81)$$
$$\left. + \frac{\tilde{r}(t)}{2} \left(m^{(s)}(bx') \right)^2 + \tilde{u}(t) \left(m^{(s)}(bx') \right)^4 \right].$$

Define the renormalized field $m'(x')$ by

$$m'(x') \equiv \sqrt{b^{d-2}}\, m^{(s)}(bx').\qquad (57.82)$$

Then we have

$$E'[m'(x')]$$
$$= \int d^d x' \left[\frac{1}{2} |\nabla' m'(x')|^2 + \frac{b^2 \tilde{r}(t)}{2} (m'(x'))^2 + b^{4-d} \tilde{u}(t) (m'(x'))^4 \right].$$
$$(57.83)$$

Defining the renormalized coupling constants $r'(t)$ and $u'(t)$ by

$$r'(t) \equiv b^2 \, \tilde{r}(t) \, , \qquad u'(t) \equiv b^{4-d} \, \tilde{u}(t) \, , \qquad (57.84)$$

and writing the integration variable x' in (57.83) as x again, we finally have the following expression for the renormalized Hamiltonian:

$$E'[m'(x)] = \int d^d x \left[\frac{1}{2} \left\{ |\nabla m'(x)|^2 + r'(t)(m'(x))^2 \right\} + u'(t) \, (m'(x))^4 \right] \, . \qquad (57.85)$$

This result is to be compared with (57.42), the expression for the Hamiltonian functional before renormalization. We see that the renormalized Hamiltonian has the same form as the bare one, and only the coupling constants $r(t)$ and $u(t)$ have been changed. This concludes the three-step renormalization procedure.

Problem 57.2 Justify (57.54) by working with the Fourier transform $\tilde{\phi}_i(k)$ of $\phi_i(x)$, where the Fourier transform pair is related by

$$\phi_i(x) \approx \frac{1}{V} \sum_{\Lambda/b < |k| < \Lambda} e^{-ik \cdot x} \, \tilde{\phi}_{ik} \xrightarrow[V \to \infty]{} \frac{1}{(2\pi)^d} \int_{\Lambda/b < |k| < \Lambda} e^{-ik \cdot x} \, \tilde{\phi}_i(k) \, . \qquad (57.86)$$

Hints: Recall that the d-dimensional Dirac delta function $\delta^d(k)$ has the integral representation

$$\delta^d(k) = \frac{1}{(2\pi)^d} \int d^d x \, e^{-ik \cdot x} \, , \qquad (57.87)$$

and assume that, inside the shell $\Lambda/b < |k| < \Lambda$, $k^2 \approx \Lambda^2$.

Problem 57.3 Justify that in (57.59), the term proportional to $\left(m^{(f)} \right)^3$ can be ignored on the basis of the properties of the wave packets $\phi_i(x)$ specified by (57.48) to (57.50). *Hint*: Review the derivation of (57.62).

We will now examine the relationship between the renormalized coupling constants $\{r'(t), u'(t)\}$ and the bare ones $\{r(t), u(t)\}$ in more detail. Recall from (57.44) that on writing $b = 1 + x$ ($|x| \ll 1$), we have $\ln b = \ln(1 + x) \approx x$. Thus $b \approx 1 + \ln b$, for $b \approx 1$. From (57.79) and (57.84), we then have

$$r'(t) = b^2 \, \tilde{r}(t) \approx (1 + 2 \ln b) \left[r(t) + 12 C_d (\ln b) \left(\Lambda^{d-2} u(t) - \Lambda^{d-4} r(t) u(t) \right) \right] \, , \qquad (57.88)$$

from which we obtain, up to first order in $\ln b$,

$$r'(t) - r(t) \approx \left[2r(t) + 12 C_d \left(\Lambda^{d-2} u(t) - \Lambda^{d-4} r(t) u(t) \right) \right] (\ln b) \, . \qquad (57.89)$$

Similarly, by (57.80) and (57.84), we have, up to first order in $\ln b$,

$$u'(t) - u(t) \approx \left[(4-d)u(t) - 36 C_d \Lambda^{d-4} u^2(t) \right] (\ln b) . \tag{57.90}$$

We will try to absorb the constant factor C_d [c.f. (57.74)] into the cutoff Λ as follows. First recall from (56.70) that near the critical temperature ($t \approx 0$), the coupling constant $r(t)$ [denoted as $a(t)$ in (56.70)] is only required to be proportional to t, so it is defined only up to a finite multiplicative constant, α, say. Let

$$\mathfrak{r}(t) \equiv \alpha r(t) , \qquad \mathfrak{r}'(t) \equiv \alpha r'(t) . \tag{57.91}$$

Then, on multiplying (57.89) by α, we have

$$\mathfrak{r}'(t) - \mathfrak{r}(t) = \left[2\alpha r(t) + 12\alpha C_d \left(\Lambda^{d-2} u(t) - \Lambda^{d-4} r(t) u(t) \right) \right] (\ln b) . \tag{57.92}$$

Introduce a new cutoff Λ' which satisfies

$$\Lambda'^{d-2} = \alpha C_d \Lambda^{d-2} , \qquad \Lambda'^{d-4} = C_d \Lambda^{d-4} . \tag{57.93}$$

Dividing the first of the above equations by the second we obtain

$$\Lambda' = \sqrt{\alpha} \, \Lambda . \tag{57.94}$$

Eq. (57.93) are then satisfied if we set

$$\alpha^{d/2-1} = C_d . \tag{57.95}$$

Using (57.93) in (57.92) and (57.90) we have

$$\mathfrak{r}'(t) - \mathfrak{r}(t) = \left[2\mathfrak{r}(t) + 12\Lambda'^{d-2} u(t) - 12\Lambda'^{d-4} \mathfrak{r}(t) u(t) \right] (\ln b) , \tag{57.96}$$
$$u'(t) - u(t) = \left[(4-d) u(t) - 36\Lambda'^{d-4} u^2(t) \right] (\ln b) . \tag{57.97}$$

To tun these difference equations into differential equations we will write $b = 1 + \Delta\tau$, so that

$$\ln b = \ln (1 + \Delta\tau) \approx \Delta\tau , \qquad b \approx e^{\Delta\tau} . \tag{57.98}$$

Then

$$\frac{\mathfrak{r}'(t) - \mathfrak{r}(t)}{\ln b} = \frac{\Delta\mathfrak{r}}{\Delta\tau} \xrightarrow{\Delta\tau \to 0} \frac{d\mathfrak{r}}{d\tau} , \qquad \frac{u'(t) - u(t)}{\ln b} \xrightarrow{\Delta\tau \to 0} \frac{du}{d\tau} . \tag{57.99}$$

If we also relabel Λ' by Λ and $\mathfrak{r}$ by r again in (57.96) and (57.97), these equations become

$$\frac{dr}{d\tau} = 2r + 12\Lambda^{d-2} u - 12\Lambda^{d-4} ru , \tag{57.100a}$$

$$\frac{du}{d\tau} = (4-d) u - 36\Lambda^{d-4} u^2 . \tag{57.100b}$$

These are the renormalization group equations for the Landau-Ginsburg m^4-model.

Since $E[m]$ is dimensionless, that is, $[E] = L^0$, dimensional analysis using (57.85) leads to

$$[m] = L^{-(d-2)/2} , \qquad [r(t)] = L^{-2} , \qquad [u(t)] = L^{-(4-d)} . \tag{57.101}$$

On defining

$$x \equiv \frac{r}{\Lambda^2} , \qquad y \equiv \frac{u}{\Lambda^{4-d}} . \tag{57.102}$$

The RG equations (57.100) can finally be written in dimensionless form as [c.f. (55.30)]

$$\frac{dx}{d\tau} = 2x - 12xy + 12y , \tag{57.103a}$$

$$\frac{dy}{d\tau} = \epsilon y - 36y^2 , \tag{57.103b}$$

where

$$\epsilon \equiv 4 - d , \tag{57.104}$$

the dimension deficit, is to be considered as a small parameter. Note that the RG equations (57.103) are nonlinear.

As discussed in Chapter 55, we would like to study the solutions of the RG equations near the *fixed points* (x^*, y^*), defined by setting the right-hand sides of (57.103) equal to zero. These are then found by the solutions of the following simultaneous equations:

$$2x^* - 12x^*y^* + 12y^* = 0 , \qquad \epsilon y^* - 36y^{*2} = 0 . \tag{57.105}$$

There are two solutions:

$$x^* = y^* = 0 \qquad \qquad \text{(Gaussian fixed point)} , \qquad (57.106a)$$

$$x^* = -\frac{\epsilon/6}{1 - \epsilon/6} \approx -\frac{\epsilon}{6} , \qquad y^* = \frac{\epsilon}{36} \qquad \text{(nontrivial fixed point)} . \qquad (57.106b)$$

Note that the nontrivial fixed point approaches the Gaussian fixed point as $\epsilon \to 0$, that is, as $d \to 4$. The RG equations (57.103) can be linearized around the fixed points. Define the column vectors

$$\boldsymbol{X} = \begin{pmatrix} x(\tau) \\ y(\tau) \end{pmatrix} = \boldsymbol{X}^* + \delta\boldsymbol{X} , \qquad \boldsymbol{X}^* = \begin{pmatrix} x^* \\ y^* \end{pmatrix} , \qquad \delta\boldsymbol{X} = \begin{pmatrix} \delta x(\tau) \\ \delta y(\tau) \end{pmatrix} . \tag{57.107}$$

Then (57.103) can be written as

$$\frac{d\boldsymbol{X}}{d\tau} = \boldsymbol{f}(\boldsymbol{X}) , \tag{57.108}$$

where

$$f(X) = \begin{pmatrix} f_1(x,y) \\ f_2(x,y) \end{pmatrix} = \begin{pmatrix} 2x - 12xy + 12y \\ \epsilon y - 36y^2 \end{pmatrix}. \tag{57.109}$$

Taylor expand $f(X)$ around X^* to obtain

$$f(X) \approx f(X^*) + Df(X)|_{X=X^*}(X - X^*) = R(X - X^*), \tag{57.110}$$

where the equality follows from the fact that $f(X^*) = 0$ and the linear operator R is given by

$$R \equiv Df(X)|_{X=X^*} = \begin{pmatrix} \dfrac{\partial f_1}{\partial x} & \dfrac{\partial f_1}{\partial y} \\ \dfrac{\partial f_2}{\partial x} & \dfrac{\partial f_2}{\partial y} \end{pmatrix}_{X=X^*} = \begin{pmatrix} 2(1-6y^*) & 12(1-x^*) \\ 0 & \epsilon - 72y^* \end{pmatrix}, \tag{57.111}$$

where the last equality follows from differentiation of (57.109). The linearized RG equations then assume the following matrix form:

$$\frac{d}{d\tau}\begin{pmatrix} \delta x(\tau) \\ \delta y(\tau) \end{pmatrix} = \begin{pmatrix} 2(1-6y^*) & 12(1-x^*) \\ 0 & \epsilon - 72y^* \end{pmatrix}\begin{pmatrix} \delta x(\tau) \\ \delta y(\tau) \end{pmatrix}. \tag{57.112}$$

The left eigenvectors (ϕ_1 ϕ_2) (row vectors) and eigenvalues λ_1, λ_2 of R [as defined by (55.77)] are then determined by the following matrix equation:

$$(\phi_1 \ \phi_2)\begin{pmatrix} 2(1-6y^*) & 12(1-x^*) \\ 0 & \epsilon - 72y^* \end{pmatrix} = (\lambda\phi_1 \ \lambda\phi_2), \tag{57.113}$$

which are equivalent to

$$2(1-6y^*)\phi_1 = \lambda\phi_1, \qquad 12(1-x^*)\phi_1 + (\epsilon - 72y^*)\phi_2 = \lambda\phi_2. \tag{57.114}$$

The eigenvalues are determined by

$$\begin{vmatrix} 2(1-6y^*) - \lambda & 12(1-x^*) \\ 0 & (\epsilon - 72y^*) - \lambda \end{vmatrix} = 0, \tag{57.115}$$

which yields

$$\lambda_1 = 2(1-6y^*), \qquad \lambda_2 = \epsilon - 72y^*. \tag{57.116}$$

When $\lambda = \lambda_1$, the first equation of (57.114) implies that $\phi_1 = \alpha$, where α is any real number. We will choose $\phi_1 = 1$. The second equation of (57.114) then gives

$$\phi_2 = \frac{12(1-6y^*)}{2 + 60y^* - \epsilon}. \tag{57.117}$$

When $\lambda = \lambda_2$, the second equation of (57.114) implies that $\phi_1 = 0$. Then we can choose $\phi_2 = 1$. Using the notation of (55.76) and (55.77) for the left eigenvectors of R, we can summarize our results as follows.

$$\lambda = \lambda_1 = 2(1 - 6y^*) ; \quad \langle \phi_1 | = (\phi_{11}, \phi_{12}) = \left(1, \frac{12(1 - 6y^*)}{2 + 60y^* - \epsilon} \right) , \qquad (57.118a)$$

$$\lambda = \lambda_2 = \epsilon - 72y^* ; \quad \langle \phi_2 | = (\phi_{21}, \phi_{22}) = (0, 1) . \qquad (57.118b)$$

Using the Dirac notation in an obvious way, the linearized RG equations (57.112) can be written as

$$\frac{d}{d\tau} | \delta \boldsymbol{X} \rangle = R | \delta \boldsymbol{X} \rangle . \qquad (57.119)$$

Multiplying by the left eigenvectors $\langle \phi_i |$ of R on the left, we have

$$\frac{d}{d\tau} \langle \phi_i | \delta \boldsymbol{X} \rangle = \langle \phi_i | R | \delta \boldsymbol{X} \rangle = \lambda_i \langle \phi_i | \delta \boldsymbol{X} \rangle , \qquad (57.120)$$

where the matrix elements

$$\langle \phi_i | \delta \boldsymbol{X} \rangle = \phi_{i1} \, \delta x + \phi_{i2} \, \delta y \equiv v_i(\delta x, \delta y) \qquad (57.121)$$

are precisely the *scaling fields* defined in (55.79). So the linearized RG equations boil down to

$$\frac{dv_i}{d\tau} = \lambda_i v_i \qquad (i = 1, 2) , \qquad (57.122)$$

with the obvious solutions

$$v_i(\tau) = v_i(\tau_0) \, e^{\lambda_i (\tau - \tau_0)} \approx b^{\lambda_i} v_i(0) \qquad (i = 1, 2) , \qquad (57.123)$$

where we have used (57.98) and written $\tau - \tau_0$ for $\Delta\tau$ in that equation.

Using the values for the critical-point coordinates (x^*, y^*) given by (57.106), we can easily calculate from (57.116) that the eigenvalues λ_1 and λ_2 are given, to first order in ϵ, by

$$\lambda_1 = 2 - \frac{\epsilon}{3} , \qquad \lambda_2 = -\epsilon . \qquad (57.124)$$

For $d < 4$, that is, $\epsilon > 0$, the scaling field v_1 is then a *relevant scaling field*, and v_2 is an irrelevant scaling field. Note that the above formulas apply to the nontrivial fixed point for $\epsilon \neq 0$, and to the Gaussian fixed point for $\epsilon = 0$. Using (57.106) in (57.118), the scaling fields can also be easily calculated, to first order in ϵ, to be

$$v_1 = \delta x + (6 - \epsilon) \, \delta y , \qquad v_2 = \delta y . \qquad (57.125)$$

Again, the Gaussian scaling fields are simply obtained from the above by setting $\epsilon = 0$. Fig. 57.1 shows the flow lines in the present two-dimensional coupling space [(x, y)-space] for the m^4-model near the Gaussian and nontrivial fixed points, as inferred by the solutions (57.123) to the linearized RG equations. In the figure, the v_1 axis is determined by $v_2 = \delta y = 0$, so it coincides with the

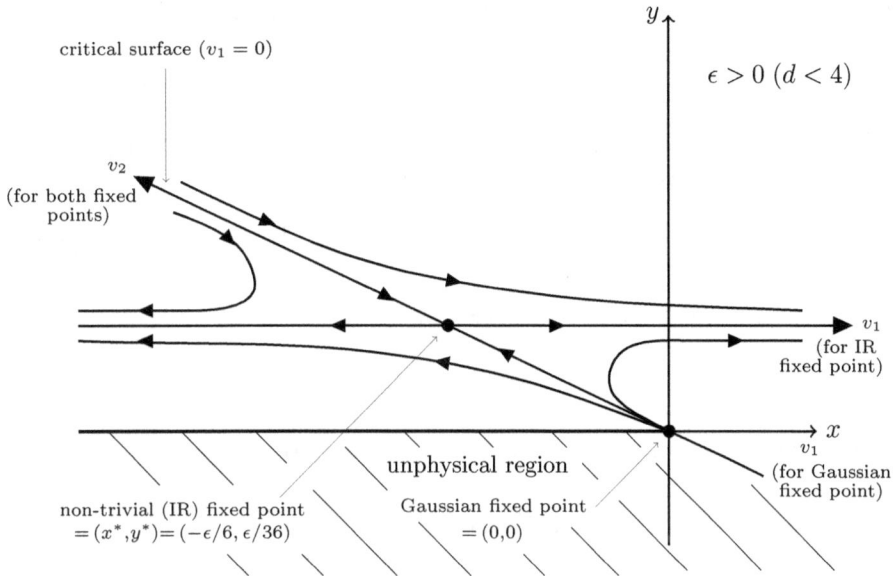

FIGURE 57.1

x-axis, while the v_2 axis is determined by $v_1 = \delta x + (6 - \epsilon)\,\delta y = 0$, so it is obliquely oriented, as shown.

Problem 57.4 Verify the result (57.124) for the eigenvalues λ_1 and λ_2 and the result (57.125) for the scaling fields v_1 and v_2 by using (57.106), (57.118) and (57.121).

Now we come to the final objective of our somewhat lengthy development in this chapter for the m^4 non-Gaussian model – the determination of the critical exponents. And this will be swift, with all the groundwork that has been laid. We merely observe that, along the v_1 direction, that is, for $v_2 = \delta y = 0$, $v_1 = \delta x$ [by (57.125)], and so by (57.123),

$$\delta x(\tau) = v_1(\tau) = b^{\lambda_1}\, v_1(\tau_0) = b^{\lambda_1}\, \delta x(\tau_0) \,. \tag{57.126}$$

Now recall (56.70) and write $\delta x \approx a_0 t$ for $t \approx 0$. Also recall (56.34), (56.47) and (56.48) and write $t \sim \xi^{-D_t}$, where ξ is the correlation length. Under an RG transformation, $\xi \to \xi/b$; and so

$$t(\tau_0) \xrightarrow[\text{RG transf.}]{} t(\tau) \sim \left(\frac{\xi}{b}\right)^{-D_t} = b^{D_t}\, t(\tau_0) \,. \tag{57.127}$$

Comparing with (57.126) and recalling (57.124) we immediately see that, to first order in ϵ,

$$D_t = 2 - \frac{\epsilon}{3} \qquad (m^4\text{-model}) . \qquad (57.128)$$

We will now argue that, up to first order in ϵ, the value of D_h for the m^4-model is unchanged from that for the Gaussian model [given by (57.41)], that is,

$$D_h = \frac{d+2}{2} \qquad (m^4\text{-model}) . \qquad (57.129)$$

It will be sufficient to consider the case of a constant magnetic field so that we add the term

$$-h \int d^d x \, m(\boldsymbol{x}) \qquad (h = constant)$$

to the Hamiltonian functional $E[m(\boldsymbol{x})]$ in (57.42). We see that

$$-h \int d^d x \, m(\boldsymbol{x}) = -h \int d^d x \left(m^{(s)}(\boldsymbol{x}) + m^{(f)}(\boldsymbol{x}) \right)$$

$$= -h \int d^d x \, m^{(s)}(\boldsymbol{x}) - h \sum_{i=1}^{N} c_i \int d^d x \, \phi(\boldsymbol{x}) \approx -h \int d^d x \, m^{(s)}(\boldsymbol{x}) , \qquad (57.130)$$

where we have used the expansion (57.47) for $m^{(f)}(\boldsymbol{x})$, and the term involving the sum has been dropped in accordance with (57.49). It follows that after Step (1), (57.63) remains valid if we add the term

$$-h \int d^d x \, m^{(s)}(\boldsymbol{x})$$

to the right-hand side of (57.64); and after Step (2), (57.78) remains valid if we add the same term to its right-hand side. The addition of this term implies that, after Step (3), we have to add the term

$$-h b^d \int d^d x' \, m^{(s)}(b\boldsymbol{x}')$$

to the right-hand side of (57.81). Applying (57.82), the above quantity can be written as

$$-h b^d \int d^d x \, \frac{1}{\sqrt{b^{d-2}}} m'(\boldsymbol{x}) = -b^{(d+2)/2} h \int d^d x \, m(\boldsymbol{x}) .$$

Thus, under an RG transformation

$$h \xrightarrow[\text{RG transf.}]{} b^{(d+2)/2} \, h . \qquad (57.131)$$

This directly confirms (57.129). With D_t and D_h in hand, (56.49) can be used to obtain all the critical exponents. These are given, up to first order in ϵ for the m^4 Landau-Ginsburg theory, as follows:

$$\alpha = 2 - d\left(\frac{1}{2} + \frac{\epsilon}{12}\right) , \qquad \delta = \frac{d+2}{d-2} , \qquad (57.132a)$$

$$\beta = \left(\frac{d-2}{4}\right)\left(1 + \frac{\epsilon}{6}\right) , \qquad \nu = \frac{1}{2} + \frac{\epsilon}{12} , \qquad (57.132b)$$

$$\gamma = 1 + \frac{\epsilon}{6} , \qquad \eta = 0 . \qquad (57.132c)$$

Chapter 58

The Callan-Symanzik Equation and Critical Exponents

In the previous chapter Wilson's approach to the renormalization group was used to calculate the critical exponents in the m^4 Landau-Ginsburg theory to first order in the dimension deficit $\epsilon = 4 - d$. This was achieved, somewhat artificially, by imposing the condition that the renormalized Hamiltonian $E'[m']$ has the same form as the bare one, $E[m]$, under the assumption that the coupling constant $u(t)$ is much smaller than $r(t)$ [c.f. (57.42)]. This obviated the necessity to enter into the evaluation of various higher-order Feynman-diagram contributions in the coarse-graining renormalization step (the functional integration over the fast modes $m^{(f)}$), which would have produced a much more complicated renormalized Hamiltonian functional than the one in (57.85). Furthermore, the RG equations (57.103) were linearized in order to obtain closed-form solutions in a simple manner near the fixed points. Although the Wilsonian approach can be applied with full force to incorporate the diagrammatic analysis systematically in a perturbative manner to include higher-order contributions and to handle the nonlinear nature of the renormalization group equations (for a detailed and exhaustive reference see Wilson and Kogut 1974), we will not pursue this path in this chapter. Instead, in order to re-establish contact with more conventional approaches to renormalization studied earlier in this book, we will attempt to illustrate, in a skeleton fashion and without overly detailed calculations, how the critical exponents can also be calculated as a perturbation expansion in ϵ by going back to the Callan-Symanzik equation introduced in Chapter 54. This method may hide the clear physical interpretation of the Wilsonian approach, but it has the merits of formal coherence and elegance. Our introductory account is based on the in-depth treatment given in Itzykson and Drouffe 1989.

We will begin with some general remarks. It seems that in the study of critical phenomena, there is no a priori reason to expect the m^4-term coupling constant $u(t)$ to be small in comparison to $r(t)$, especially when $d < 4$. This is because when $d < 4$, $u(t)$ becomes a dimensional quantity: $[u(t)] = L^{-(4-d)}$. If one assumes that $u(t) \sim 1/a^{4-d}$, where a is the microscopic lattice separation, then $u(t)$ can no longer be guaranteed to be small; and naively, a perturbative expansion in $u(t)$ may not be suitable. Also for $d < 4$, the infrared divergence as $t \to 0$ (and thus $r(t) \to 0$) becomes more severe as the perturbative order increases. This is so even for diagrams that are ultraviolet convergent. For instance, for the 4-point (one loop) fish diagram, we have the Feynman integral

$$\int \frac{d^d k}{(2\pi)^d} \frac{1}{(k^2 - a_0 t)((k_1 + k_2 - k)^2 - a_0 t)} \xrightarrow{t \to 0} \int \frac{d^d k}{(2\pi)^d} \frac{1}{k^2 (k_1 + k_2 - k)^2} \,,$$

which diverges for $d < 4$ like $|k|^{d-4}$ as $k \to 0$. Thus, to control infrared divergences, we have to keep a small but non-zero mass term, that is, a non-zero $r(t)$ for $d < 4$. To make our notation more dimensionally transparent, we will write $m_0 \equiv \sqrt{|r(t)|}$, where m_0 is considered as a bare mass in the theory. Critical phenomena are thus concerned with the momentum regions

$$\frac{k}{\Lambda} \ll 1 \qquad \text{and} \qquad \frac{k}{m} \gg 1 \,, \tag{58.1}$$

where m is a renormalized mass. We have already seen how the high momentum-cutoff Λ can be absorbed in the constant a_0 in $r(t) = a_0 t$ [c.f. (57.18)]. The physical length scale provided by the correlation length ξ can now be used to set the scale for the small but non-vanishing renormalized mass m. So we define

$$m = \frac{1}{\xi} \sim |t|^\nu \,, \tag{58.2}$$

which becomes a fundamental parameter in the theory. The second condition in (58.1), which implies $k^{-1} \ll m^{-1} = \xi$, then says that in studying critical phenomena, one needs to analyze the short-distance (or large momentum) behavior of the renormalized theory under the length scale provided by ξ. On examining the relationships

$$m_0^2 = a_0 t \,, \qquad\qquad m^2 = \xi^{-2} \sim |t|^{2\nu} \,, \tag{58.3}$$

and further making the notation change $u(t) \to g_0$, where g_0 is taken to be the bare coupling constant, one is led back to the procedure of multiplicative renormalization studied in earlier chapters, where we had the analytical functional relationships

$$m_0 = m_0(m, g) \,, \qquad\qquad g_0 = g_0(m, g) \,, \tag{58.4}$$

in which g is a finite renormalized coupling constant. [In this chapter the symbols m (mass) and g (coupling constant) replace μ and λ, respectively, used in earlier chapters.] Perturbative expansions can then be carried out in powers of the finite coupling constant g, no matter what value g_0 takes. Thus, for $d < 4$, one keeps a finite mass m in the **symmetric phase** ($t > 0$) and investigate the ultraviolet behavior of Feynman integrals for $|k| \gg m$. [The reader should beware of the apparent contradiction here, since we have just been saying that critical behavior has to do with infrared divergences! One should always keep in mind the momentum region of interest stated in (58.1) in order to avoid this confusion.] On the other hand, the renormalized theory approaches the Gaussian model as $d \to 4$ (equivalently $\epsilon \to 0$), as we saw at the end of the previous chapter. The upshot is that we are justified to proceed with double expansions in g and $\epsilon = 4 - d$ in the calculations of critical exponents.

To start the calculation of the critical exponents, we recall the *fluctuation dissipation theorem* relating the magnetic susceptibility χ to the 2-point correlation function given by (56.22):

$$kT\chi = \int d^d x \, G^{(2)}(\boldsymbol{x}) = \tilde{G}^{(2)}(\boldsymbol{k} = 0) = \frac{1}{\Gamma^{(2)}(\boldsymbol{k} = 0)} \,, \qquad (58.5)$$

where (i) the regular part of χ, given by the second term on the right-hand side in (56.22) [the one proportional to $V\langle m(0)\rangle^2]$, has been ignored, (ii) $\tilde{G}(\boldsymbol{k})$ is the Fourier transform of the 2-point correlation $G^{(2)}(\boldsymbol{x}) = \langle m(\boldsymbol{x})m(0)\rangle$, and (iii) the last equality follows from (52.35), with $\Gamma^{(2)}(k^2)$ being the amputated 2-point correlation function (in momentum space). By (54.34) we can write

$$\frac{1}{\Gamma^{(2)}(0)} = \frac{1}{Z_\varphi^{-1}\overline{\Gamma}^{(2)}(0)} = \frac{Z_\varphi}{\overline{\Gamma}^{(2)}(0)} \,, \qquad (58.6)$$

where Z_φ is the field renormalization constant originally given by (52.22). Using the second of the renormalization conditions in (52.72) [for $\overline{\Gamma}^{(2)}(0)$, with the renormalized mass μ' rewritten as m] and the definition of the critical exponent γ given by (56.7), we obtain

$$\frac{Z_\varphi}{m^2} \sim |t|^{-\gamma} \,. \qquad (58.7)$$

Now it is not difficult to see that the Callan-Symanzik equation (54.37), together with the renormalization conditions (54.1) [for $\overline{\Gamma}^{(2)}(\boldsymbol{k} = 0; m, g)$] and (54.32) [for $\overline{\Gamma}_\theta^{(2)}(0; \boldsymbol{k} = 0; m, g)$] imply the following (see Problem 58.1 below):

$$Z_\theta = \frac{\partial m^2}{\partial m_0^2}\bigg|_{g_0, \Lambda} \left(1 - \frac{\gamma_1(g)}{2}\right) \,, \qquad (58.8)$$

where Z_θ is the renormalization constant appearing in (54.33), $\gamma_1(g) = 2\gamma(g)$, with $\gamma(g)$ being given by (54.38), and the subscripts next to the partial derivative indicate the variables to be held fixed. On the other hand, from (58.3),

$$\frac{\partial m^2}{\partial m_0^2} \sim \frac{\partial m^2}{\partial t} \sim \frac{\partial t^{2\nu}}{\partial t} \sim t^{2\nu - 1} \,. \qquad (58.9)$$

Hence (58.8) implies

$$Z_\theta \sim t^{2\nu - 1} \,. \qquad (58.10)$$

This equation and (58.7) are the keys to the determination of the critical exponents. To proceed, we need to investigate how the renormalization constants

Z_ϕ and Z_θ behave as functions of the renormalized mass m as $m \to 0$, that is, as $t \to 0$ [c.f. (58.3)]. This can best be done by considering the functions $\beta(g), \gamma(g)$ and $\gamma_\theta(g)$ appearing in the CS equation (54.37) and its associated equation (54.47). These have already been defined in (54.38) and (54.48). But we will give them here in slightly different forms and notation. As mentioned above, we will set $\gamma(g) = \gamma_1(g)/2$, and rename $\gamma_\theta(g)$ by $\gamma_2(g)$. We have

$$\beta(g) = m \left. \frac{\partial g}{\partial m} \right|_{g_0,\Lambda} , \qquad \gamma_1(g) = \frac{m}{Z_\phi} \left. \frac{\partial Z_\phi}{\partial m} \right|_{g_0,\Lambda} , \qquad \gamma_2(g) = \frac{m}{Z_\theta} \left. \frac{\partial Z_\theta}{\partial m} \right|_{g_0,\Lambda} .$$

$$(58.11)$$

The reader should verify that the above equations are equivalent to (54.38) and (54.48). As will be seen below, in the limit $\Lambda \to \infty$, g_0/m^{4-d} is a function of g only for $d < 4$. We thus have

$$\frac{d}{dg} \left(\frac{g_0}{m^{4-d}} \right) = \left\{ \frac{d}{dm} \left(\frac{g_0}{m^{4-d}} \right) \right\} \left(\frac{\partial m}{\partial g} \right) = g_0(d-4)m^{d-4} \frac{1}{m} \frac{\partial m}{\partial g}$$
$$= -\frac{(4-d)}{\beta} \frac{g_0}{m^{4-d}} ,$$

$$(58.12)$$

where in the last equality we have made use of the definition of β given by the first equation of (58.11). This implies

$$\left\{ (4-d) + \beta(g) \frac{d}{dg} \right\} \left(\frac{g_0}{m^{4-d}} \right) = 0 .$$

$$(58.13)$$

This equation and the last two of (58.11) can be integrated to obtain g_0/m^{4-d}, Z_ϕ and Z_θ, in terms of the Callan-Symanzik functions $\gamma_1(g), \gamma_2(g)$ and $\beta(g)$.

Dividing the second equation of (58.11) by the first, we have

$$\frac{\gamma_1(g)}{\beta(g)} = \frac{1}{Z_\phi} \frac{\partial Z_\phi}{\partial m} \frac{\partial m}{\partial g} = \frac{\partial \ln Z_\phi}{\partial g} .$$

$$(58.14)$$

It follows that

$$Z_\phi = \exp \left(\int_0^g dg' \frac{\gamma_1(g')}{\beta(g')} \right) ,$$

$$(58.15)$$

using the boundary condition $Z_\phi(g = 0) = 1$. Similarly, dividing the third equation of (58.11) by the first and integrating, we obtain

$$Z_\theta = \exp \left(\int_0^g dg' \frac{\gamma_2(g')}{\beta(g')} \right) .$$

$$(58.16)$$

Eq. (58.13) can be rewritten as

$$\frac{1}{g_0/m^{4-d}} \frac{d}{dg} \left(\frac{g_0}{m^{4-d}} \right) = - \left(\frac{4-d}{\beta(g)} \right) .$$

$$(58.17)$$

This can be integrated to obtain

$$\left(\frac{g_0(g)}{m^{4-d}} \right) = \exp\left(-\int_0^g dg' \, \frac{4-d}{\beta(g')} \right) \exp\left(\ln \frac{g_0(0)}{m^{4-d}} \right). \tag{58.18}$$

It will be seen below [(58.73)] that $g_0(0)/m^{4-d} = 0$. The last exponential can then be written as

$$e^{\ln(0)} = \frac{g}{e^{\ln g}} \, e^{\ln(0)} = g \left\{ \exp\left(-\ln g + \ln(0) \right) \right\} = g \exp\left(-\int_0^g \frac{dg'}{g'} \right). \tag{58.19}$$

The result (58.18) can thus be written as

$$\frac{g_0}{m^{4-d}} = g \exp\left\{ -\int_0^g dg' \left(\frac{4-d}{\beta(g')} + \frac{1}{g'} \right) \right\}. \tag{58.20}$$

For $d < 4$ this equation reveals the following interesting fact. As g varies from 0 to the first zero of $\beta(g)$, say g_c, on the right-hand side, $m \to 0^+$ on the left-hand side, assuming that g_0 is held fixed. Eq. (58.3) then implies that near the critical point, $g \to g_c$. We thus want to expand the above result around $g = g_c$. The result is (see Problem 58.2 below)

$$\frac{g_0}{m^{4-d}} \approx A \left(1 - \frac{g}{g_c} \right)^{-(4-d)/\omega} \qquad (d < 4), \tag{58.21}$$

where g_c is the root of $\beta(g) = 0$, $\omega \equiv \left. \dfrac{d\beta(g)}{dg} \right|_{g=g_c}$, and A is a constant given by

$$A \equiv g_c \exp\left\{ -\int_0^{g_c} dg' \left(\frac{4-d}{\beta(g')} + \frac{1}{g'} - \frac{4-d}{\omega(g' - g_c)} \right) \right\}. \tag{58.22}$$

It is easily seen that (58.21) is equivalent to

$$g \approx g_c \left[1 - \left(\frac{Am^{4-d}}{g_0} \right)^{\omega/(4-d)} \right], \tag{58.23}$$

from which we obtain

$$g \approx g_c + A' \left(m g_0^{1/(d-4)} \right)^{\omega}, \tag{58.24}$$

where A' is a constant. Now we will evaluate Z_φ in the neighborhood of g_c. Use (58.15) to write

$$Z_\varphi(g \approx g_c) = \exp\left[\int_0^{g_c} dg' \, \frac{\gamma_1(g')}{\beta(g')} + \int_{g_c}^g dg' \, \frac{\gamma_1(g')}{\beta(g')} \right] = C_1' \exp\left(\int_{g_c}^g dg' \, \frac{\gamma_1(g')}{\beta(g')} \right), \tag{58.25}$$

where the constant C_1' is given by

$$C_1' \equiv \exp\left(\int_0^{g_c} dg' \, \frac{\gamma_1(g')}{\beta(g')} \right). \tag{58.26}$$

Next we use the definition of $\beta(g)$ given by (58.11) to extract the m-dependence of the exponential on the right-hand side of (58.25) as follows:

$$\exp\left(\int_{g_c}^{g} dg' \frac{\gamma_1(g')}{\beta(g')}\right) \approx \exp\left\{\gamma_1(g_c) \int_{g_c}^{g} dg' \left(m\frac{dg'}{dm}\right)^{-1}\right\}$$

$$= \exp\left\{\gamma_1(g_c) \int_{m(g_c)}^{m} \frac{dm}{m}\right\} = \exp\left\{\gamma_1(g_c) \ln m - \gamma_1(g_c) \ln(m(g_c))\right\},$$

(58.27)

where we have assumed that $\gamma_1(g)$ has no singular behavior near $g = g_c$. So we can write

$$\exp\left(\int_{g_c}^{g} dg' \frac{\gamma_1(g')}{\beta(g')}\right) \approx B \exp\left\{\gamma_1(g_c) \ln m\right\} = B\, m^{\gamma_1(g_c)},$$

(58.28)

where $B \equiv \exp\left\{-\gamma_1(g_c) \ln(m(g_c))\right\}$ is a constant. Note that actually $B \to \infty$, since $m(g_c) \to 0$; but this does not affect the m-dependence of the last equation, which is what we are interested in. Eq. (58.25) then implies

$$Z_\varphi \approx C_1\, m^{\gamma_1(g_c)},$$

(58.29)

where $C_1 \equiv BC_1'$ is another constant (independent of m). Assuming also that $\gamma_2(g)$ does not have any singular behavior near $g = g_c$, exactly the same reasoning that led to (58.29) and the use of (58.16) allow us to conclude that

$$Z_\theta \approx C_2\, m^{\gamma_2(g_c)},$$

(58.30)

where C_2 is a constant independent of m.

We are in a position now to obtain some concrete, if only intermediate, results for the critical exponents. Observe that

$$t^{-\gamma} \sim \frac{Z_\varphi}{m^2} \sim m^{\gamma_1(g_c)-2} \sim t^{\nu\{\gamma_1(g_c)-2\}},$$

(58.31)

where the first relational sign follows from (58.7), the second from (58.29), and the third from (58.3). Comparing exponents between the first quantity and the last one, we arrive at

$$\frac{\gamma}{\nu} = 2 - \gamma_1(g_c).$$

(58.32)

Analogous to (58.31) we have

$$t^{2\nu-1} \sim Z_\theta \sim m^{\gamma_2(g_c)} \sim t^{\nu\gamma_2(g_c)},$$

(58.33)

where the first relational sign follows from (58.10), the second from (58.30), and the last one from (58.3) again. Comparison of exponents leads to

$$\nu = \frac{1}{2 - \gamma_2(g_c)}.$$

(58.34)

This equation and (58.32) allow us to calculate the critical exponents γ and ν in terms of $\gamma_1(g_c)$ and $\gamma_2(g_c)$. By the scaling laws [c.f. (56.14) to (56.17)], all six critical exponents are then determined by $\gamma_1(g_c)$ and $\gamma_2(g_c)$. In fact it follows immediately from (58.32) and Fisher's scaling law (56.14) that

$$\eta = \gamma_1(g_c) . \tag{58.35}$$

Problem 58.1 Show that the Callan-Symanzik equation (54.37) can be written in the following form (using the notation of this chapter):

$$\left(m\frac{\partial}{\partial m} + \beta(g)\frac{\partial}{\partial g} - \frac{n}{2}\gamma_1(g) \right) \overline{\Gamma}^{(n)}(k;m,g) = Z_\theta \left(m \left. \frac{\partial m_0^2}{\partial m} \right|_{g_0,\Lambda} \right) \overline{\Gamma}_\theta^{(n)}(0;k;m,g) . \tag{58.36}$$

Then set $n = 2$ and use the renormalization conditions (54.1) and (54.32) in the forms

$$\overline{\Gamma}^{(2)}(k=0;m,g) = m^2 , \qquad \overline{\Gamma}_\theta^{(2)}(0;k=0;m,g) = 1 \tag{58.37}$$

to verify (58.8).

Problem 58.2 Verify (58.21) by following the steps below:

(a) Taylor expand (58.20) around $g = g_c$ to obtain formally, to first order in g,

$$\frac{g_0}{m^{4-d}} \approx \left. \frac{g_0}{m^{4-d}} \right|_{g_c} + \left. \frac{d}{dg}\left(\frac{g_0}{m^{4-d}} \right) \right|_{g_c} (g - g_c)$$

$$= g_c \exp\left\{ -\int_0^{g_c} dg' \left(\frac{4-d}{\beta(g')} + \frac{1}{g'} \right) \right\} \left[\frac{g}{g_c} - (g - g_c)\left(\frac{4-d}{\beta(g_c)} + \frac{1}{g_c} \right) \right]$$

$$= A \exp\left(-\int_0^{g_c} dg' \frac{4-d}{\omega(g' - g_c)} \right) \left[\frac{g}{g_c} - (g - g_c)\left(\frac{4-d}{\beta(g_c)} + \frac{1}{g_c} \right) \right] , \tag{58.38}$$

where A is given by (58.22).

(b) Show that the exponential on the right-hand side of the last equality in (58.25) can be formally written as

$$\exp\left(-\int_0^{g_c} dg' \frac{4-d}{\omega(g' - g_c)} \right) = \left(-\frac{\beta(g_c)}{\omega g_c} \right)^{-(4-d)/\omega} . \tag{58.39}$$

(c) Show that the quantity within the square brackets on the right-hand side of the last equality in (58.25) can be formally approximated by

$$\left[1 + \frac{\omega}{\beta(g_c)}(g - g_c) \right]^{-(4-d)/\omega} , \tag{58.40}$$

and then complete the problem.

It remains to calculate the anomalous-dimension Callan-Symanzik functions $\gamma_1(g)$ and $\gamma_2(g)$. This will be done by means of their definitions given by the second and third equations in (58.11). But before we are ready to do that, we will need to revisit the basic perturbative renormalization procedures introduced in Chapter 52, in particular, the BPH method discussed at the end of that chapter. For ease of reference we will restate here using the notation of the present chapter the renormalization conditions analogous to (52.72), adapted to Euclidean generating Green's functions:

$$\overline{\Gamma}^{(2)}(\boldsymbol{k} = \boldsymbol{0}) = m^2 \,, \qquad \frac{d}{dk^2}\overline{\Gamma}^{(2)}(\boldsymbol{k} = \boldsymbol{0}) = 1 \,, \tag{58.41a}$$

$$\overline{\Gamma}^{(4)}(\boldsymbol{k}_1 = \boldsymbol{k}_2 = \boldsymbol{k}_3 = \boldsymbol{k}_4 = \boldsymbol{0}) = m^{4-d}g \,. \tag{58.41b}$$

Note that, unlike (52.72), $\overline{\Gamma}^{(4)}$ has a dimensional scaling factor m^{4-d} for $d < 4$, and factors of i have been dispensed with. Eqs. (52.30) and (52.33) are also reproduced here as

$$\tilde{G}^{(2)}(\boldsymbol{k}; m_0, g_0, \Lambda) = Z_\varphi \tilde{G}^{(2)}_R(\boldsymbol{k}; m, g) \,, \tag{58.42}$$

$$\Gamma^{(2)}(\boldsymbol{k}; m_0, g_0, \Lambda) = Z_\varphi^{-1}\overline{\Gamma}^{(2)}(\boldsymbol{k}; m, g) \,, \tag{58.43}$$

where renormalized quantities appear on the right-hand side. Eqs. (58.41) and (58.43) enable us to calculate the bare quantities m_0, g_0 and the field renormalization constant Z_φ as functions of the renormalized quantities m, g and the momentum cutoff Λ, as perturbative series in the number of loops k. Formally we can write

$$m_0^2 = m_0^{2[0]} + m_0^{2[1]} + m_0^{2[2]} + \cdots \,, \tag{58.44}$$

$$g_0 = g_0^{[0]} + g_0^{[1]} + g_0^{[2]} + \cdots \,, \tag{58.45}$$

$$Z_\varphi = Z_\varphi^{[0]} + Z_\varphi^{[1]} + Z_\varphi^{[2]} + \cdots \,. \tag{58.46}$$

where $[k]$ in a particular term means that that term is calculated with a k-loop contribution, and it is understood that all quantities on the right-hand sides are functions of m, g and Λ. The renormalization conditions (58.41) ensure that. to order $[0]$, the bare and renormalized quantities are identical, except for the scaling factor m^{4-d}:

$$m_0^{2[0]} = m^2 \,, \qquad g_0^{[0]} = m^{4-d}g \,, \qquad Z_\varphi^{[0]} = 1 \,. \tag{58.47}$$

Thus, to order $[0]$ (zero-loop),

$$\overline{\Gamma}^{(2)[0]}(\boldsymbol{k}) = k^2 + m^2 \,, \qquad \overline{\Gamma}^{(4)[0]}(\boldsymbol{k}_i) = m^{4-d}g \,. \tag{58.48}$$

Higher $[k]$ order Green's functions ($k > 0$) can be calculated systematically, order by order, with the following BPH Lagrangian with counterterms to all orders [compare with (52.63) and (52.64)]:

$$
\begin{aligned}
\mathcal{L} &= \frac{Z_\varphi}{2}\left[(\partial\varphi_R)^2 + m_0^2\varphi_R^2\right] + \frac{g_0}{4!}Z_\varphi^2\varphi_R^4 \\
&= \frac{1}{2}\left[(\partial\varphi_R)^2 + m^2\varphi_R^2\right] + \frac{1}{4!}m^{4-d}g\,\varphi_R^4 \\
&\quad + \sum_{k=1}^{\infty}\left[\frac{Z_\varphi^{[k]}}{2}(\partial\varphi_R)^2 + \frac{1}{2}\left(Z_\varphi m_0^2\right)^{[k]}\varphi_R^2 + \frac{1}{4!}\left(g_0Z_\varphi^2\right)^{[k]}\varphi_R^4\right],
\end{aligned}
\tag{58.49}
$$

where the last line consists of the counterterms. For example, the order [1] two-point and 4-point renormalized Green's functions are given by

$$
\overline{\Gamma}^{(2)[1]}(k) = \left(Z_\varphi m_0^2\right)^{[1]} + Z_\varphi^{[1]}k^2 + \frac{1}{2}m^{4-d}g\int_\Lambda \frac{d^dk'}{(2\pi)^d}\frac{1}{k'^2+m^2},
\tag{58.50}
$$

and

$$
\overline{\Gamma}^{(4)[1]}(k_1,k_2,k_3,k_4) = \left(g_0Z_\varphi^2\right)^{[1]}
$$
$$
- \frac{1}{2}m^{8-2d}g^2\sum_{i=2}^{4}\int_\Lambda \frac{d^dk'}{(2\pi)^d}\frac{1}{(k'^2+m^2)\left\{(k'-k_1-k_i)^2+m^2\right\}}.
\tag{58.51}
$$

The first terms (not involving integrals) in the above equations are contributions from the order [1] counterterms in (58.49). The term involving the integral in (58.50) is the one-loop contribution for the 2-point function from the part of $\mathcal{L}$ without the counterterms, and the term involving the sum of integrals in (58.51) is the one-loop contribution due to the three fish diagrams, from the same Lagrangian without the counterterms.

Using (58.48), (58.50) and (58.51), it is easily seen that, up to order [1],

$$
\overline{\Gamma}^{(2)}(k=0) = m^2 + \left(Z_\varphi m_0^2\right)^{[1]} + \frac{1}{2}m^{4-d}g\int_\Lambda \frac{d^dk'}{(2\pi)^d}\frac{1}{k'^2+m^2} = m^2,
\tag{58.52}
$$

$$
\frac{d}{dk^2}\overline{\Gamma}^{(2)}(k=0) = 1 + Z_\varphi^{[1]} = 1,
\tag{58.53}
$$

$$
\overline{\Gamma}^{(4)}(k_i=0) = m^{4-d}g + \left(g_0Z_\varphi^2\right)^{[1]} - \frac{3}{2}m^{8-2d}g^2\int_\Lambda \frac{d^dk'}{(2\pi)^d}\frac{1}{(k'^2+m^2)^2}
\tag{58.54}
$$
$$
= m^{4-d}g,
$$

where the last equalities in all three equations are required by the renormalization conditions (58.41). It follows that

$$
Z_\varphi^{[1]} = 0,
\tag{58.55}
$$

$$
\left(Z_\varphi m_0^2\right)^{[1]} = m_0^{2[1]} = -\frac{1}{2}m^{4-d}g\int_\Lambda \frac{d^dk'}{(2\pi)^d}\frac{1}{k'^2+m^2},
\tag{58.56}
$$

$$
\left(g_0Z_\varphi^2\right)^{[1]} = g_0^{[1]} = \frac{3}{2}m^{8-2d}g^2\int_\Lambda \frac{d^dk'}{(2\pi)^d}\frac{1}{(k'^2+m^2)^2},
\tag{58.57}
$$

where the three equations, in order, follow from (58.53), (58.52), and (58.54), respectively. These results then imply

$$\overline{\Gamma}^{(2)[1]}(k) = 0 \, , \tag{58.58}$$

$$\overline{\Gamma}^{(4)[1]}(k_1, k_2, k_3, k_4) = -\frac{1}{2}m^{8-2d}g^2 \sum_{i=2}^{4} \int_\Lambda \frac{d^d k'}{(2\pi)^d} \frac{1}{k'^2 + m^2}$$
$$\times \left\{ \frac{1}{(k' - k_1 - k_i)^2 + m^2} - \frac{1}{k'^2 + m^2} \right\} \, . \tag{58.59}$$

We will not show calculations for higher $[k]$ order $(k > 1)$ contributions, which are much more involved.

The perturbation expansion for Z_θ [defined in (54.33)] can be similarly obtained as that for Z_φ. We first reproduce the definition in terms of the 2-point Green's function:

$$\Gamma_\theta^{(2)}(k; k_1, k_2; g_0, m_0, \Lambda) = Z_\varphi^{-1} Z_\theta \, \overline{\Gamma}_\theta^{(2)}(k; k_1, k_2; g, m) \qquad (k + k_1 + k_2 = 0) \, , \tag{58.60}$$

and also the renormalization condition (54.32):

$$\overline{\Gamma}_\theta^{(2)}(0; 0, 0) = 1 \, . \tag{58.61}$$

An additional important fact to recall is that [c.f. (54.24)]

$$\Gamma_\theta^{(2)[0]}(k; k_1, -k - k_1) = \overline{\Gamma}_\theta^{(2)[0]}(k; k_1, -k - k_1) = 1 \, . \tag{58.62}$$

Analogous to (58.46) we write

$$Z_\theta = Z_\theta^{[0]} + Z_\theta^{[1]} + Z_\theta^{[2]} + \cdots \, . \tag{58.63}$$

On keeping only order $[0]$ terms in (58.60) and recalling that $Z_\varphi^{[0]} = 1$ [(58.47)], the (58.62) implies that

$$Z_\theta^{[0]} = 1 \, . \tag{58.64}$$

Making use of our earlier result that $Z_\varphi^{[1]} = 0$ [c.f. (58.55)], we can write (58.60) to order $[1]$ as follows:

$$\Gamma_\theta^{(2)[0]} + \Gamma_\theta^{(2)[1]} = 1 + \Gamma_\theta^{(2)[1]} = \left(1 + Z_\theta^{[1]} \right) \left(\overline{\Gamma}_\theta^{(2)[0]} + \overline{\Gamma}_\theta^{(2)[1]} \right)$$
$$= \left(1 + Z_\theta^{[1]} \right) \left(1 + \overline{\Gamma}_\theta^{(2)[1]} \right) \approx 1 + Z_\theta^{[1]} + \overline{\Gamma}_\theta^{[1]} \, . \tag{58.65}$$

Analogous to (54.25), the unrenormalized Green's function with insertion $\Gamma_\theta^{(2)[1]}$ can be written as

$$\Gamma_\theta^{(2)[1]}(k; k_1, -k - k_1; g_0, m_0)$$
$$= -\frac{1}{2}g_0^{[0]} \int \frac{d^d k'}{(2\pi)^d} \frac{1}{\left(k'^2 + m_0^{2[0]} \right) \left\{ (k + k')^2 + m_0^{2[0]} \right\}} \tag{58.66}$$
$$= -\frac{1}{2}m^{4-d}g \int \frac{d^d k'}{(2\pi)^d} \frac{1}{(k'^2 + m^2)\left\{ (k + k')^2 + m^2 \right\}} \, ,$$

where we have used (58.47) for $m_0^{2[0]}$ and $g_0^{[0]}$. Eq. (58.65) can then be written as

$$1 - \frac{1}{2}m^{4-d}g \int \frac{d^d k'}{(2\pi)^d} \frac{1}{(k'^2 + m^2)\left\{(\boldsymbol{k} + \boldsymbol{k}')^2 + m^2\right\}} \qquad (58.67)$$

$$= 1 + Z_\theta^{[1]} + \overline{\Gamma}_\theta^{(2)[1]}(\boldsymbol{k}; \boldsymbol{k}_1, \boldsymbol{k}_2; g, m) \, .$$

Setting $\boldsymbol{k} = \boldsymbol{k}_1 = \boldsymbol{k}_2 = \boldsymbol{0}$ and recalling (58.61) and (58.61) we have

$$Z_\theta^{[1]} = -\frac{1}{2}m^{4-d}g \int \frac{d^d k'}{(2\pi)^d} \frac{1}{(k'^2 + m^2)^2} \, . \qquad (58.68)$$

Then (58.67) implies

$$\overline{\Gamma}_\theta^{(2)[1]}(\boldsymbol{k}; \boldsymbol{k}_1, \boldsymbol{k}_2; g, m) = -\frac{1}{2}m^{4-d}g \int \frac{d^d k'}{(2\pi)^d} \frac{1}{k'^2 + m^2}$$

$$\times \left\{ \frac{1}{(\boldsymbol{k} + \boldsymbol{k}')^2 + m^2} - \frac{1}{k'^2 + m^2} \right\} \, . \qquad (58.69)$$

We are now in a position to write down explicit expansions for m_0^2, g_0, Z_φ and Z_θ to order [1] (1 loop) [c.f. (58.44) to (58.46), and (58.63)]. The necessary information is given by (58.56) for $m_0^{2[1]}$, (58.57) for $g_0^{[1]}$, (58.55) for $Z_\varphi^{[1]}$, and (58.68) for $Z_\theta^{[1]}$. The following two Feynman integrals are required:

$$I_1(d, m) \equiv \int \frac{d^d k}{(2\pi)^d} \frac{1}{k^2 + m^2} \, , \qquad I_2(d, m) \equiv \int \frac{d^d k}{(2\pi)^d} \frac{1}{(k^2 + m^2)^2} \, . \qquad (58.70)$$

Note that when $d = 4$, I_1 is quadratically divergent and I_2 is logarithmically divergent. We have studied these types of ultraviolet divergences already in Chapter 53, when we discuss regularization procedures. In the present context, $d < 4$ and formally may not even be integral. The dimensional regularization procedure introduced in Chapter 53 is then the most appropriate method of evaluation, which is based on analytic continuation of I_1 and I_2 considered as meromorphic functions of the complex variable d. We will not show the details here but merely give the results:

$$I_1(d, m) = \frac{m^{d-2}}{(4\pi)^{d/2}} \Gamma(1 - d/2) \, , \qquad I_2(d, m) = \frac{m^{d-4}}{(4\pi)^{d/2}} \Gamma(2 - d/2) \, , \qquad (58.71)$$

where $\Gamma(z)$ is the gamma function considered as a meromorphic function of the complex variable z. Because $\Gamma(z)$ has simple poles at $z = 0, -1, -2, -3, \ldots$, the above results are not valid for d even, in which case a finite cutoff Λ will be required explicitly. With these results, and also recalling (58.47) and (58.64) for the order [0] contributions, we have, for $d < 4$,

$$m_0^2 = m^2 \left\{ 1 - \frac{\Gamma(1 - d/2)}{2(4\pi)^{d/2}} g + \cdots \right\}, \tag{58.72}$$

$$g_0 = m^{4-d} \left\{ g + \frac{3}{2} \frac{\Gamma(2 - d/2)}{(4\pi)^{d/2}} g^2 + \cdots \right\}, \tag{58.73}$$

$$Z_\varphi = 1 + O(g^2), \tag{58.74}$$

$$Z_\theta = 1 - \frac{\Gamma(2 - d/2)}{2(4\pi)^{d/2}} g + \cdots. \tag{58.75}$$

It is not difficult to invert (58.73) to get

$$g = \frac{g_0}{m^{4-d}} - \frac{3}{2} \frac{\Gamma(2 - d/2)}{(4\pi)^{d/2}} \left(\frac{g_0}{m^{4-d}} \right)^2 + \cdots. \tag{58.76}$$

By direct differentiation of the above we have

$$m \frac{\partial g}{\partial m} \bigg|_{g_0, \Lambda} = -(4 - d) \frac{g_0}{m^{4-d}} + \frac{3}{2}(4 - d) \frac{\Gamma(2 - d/2)}{(4\pi)^{d/2}} \left(\frac{g_0}{m^{4-d}} \right)^2 + \cdots. \tag{58.77}$$

Using the expansion for g_0/m^{4-d} given by (58.73) in the above, we get

$$m \frac{\partial g}{\partial m} \bigg|_{g_0, \Lambda} = -(4 - d)g + \frac{3}{2}(4 - d) \frac{\Gamma(2 - d/2)}{(4\pi)^{d/2}} g^2 + \cdots$$

$$= -(4 - d)g + 3(2 - d/2) \frac{\Gamma(2 - d/2)}{(4\pi)^{d/2}} g^2 + \cdots. \tag{58.78}$$

By the first equation of (58.11), the above is precisely the function $\beta(g)$. Recalling the property of the gamma function $\Gamma(z + 1) = z\Gamma(z)$, we have

$$\beta(g) = -(4 - d)g + \frac{3\Gamma(3 - d/2)}{(4\pi)^{d/2}} g^2 + \cdots. \tag{58.79}$$

Setting the right-hand side to be zero and solving, we obtain

$$g_c \approx \frac{(4\pi)^{d/2}(4 - d)}{3\Gamma(3 - d/2)}. \tag{58.80}$$

By the second equation of (58.11) and (58.74),

$$\gamma_1(g) = \frac{m}{Z_\varphi} \frac{\partial Z_\varphi}{\partial m} \bigg|_{g_0, \Lambda} \sim O(g^2). \tag{58.81}$$

By differentiation of (58.75),

$$m \frac{\partial Z_\theta}{\partial m} = -\frac{\Gamma(2 - d/2)}{2(4\pi)^{d/2}} \left(m \frac{\partial g}{\partial m} \right) = -\frac{\Gamma(2 - d/2)}{2(4\pi)^{d/2}} \beta(g)$$

$$= -\frac{\Gamma(2 - d/2)}{2(4\pi)^{d/2}} \left[-(4 - d)g + \frac{3\Gamma(3 - d/2)}{(4\pi)^{d/2}} g^2 + \cdots \right], \tag{58.82}$$

where in the last line, we have used the expansion for $\beta(g)$ given by (58.79). By the third equation of (58.11), we than have

$$
\begin{aligned}
\gamma_2(g) = \frac{m}{Z_\theta} \frac{\partial Z_\theta}{\partial m}\bigg|_{g_0,\Lambda} &= \left\{1 - \frac{\Gamma(2-d/2)}{2(2\pi)^{d/2}} g + \cdots \right\}^{-1} \\
&\times \frac{\Gamma(2-d/2)}{2(4\pi)^{d/2}} \left\{(4-d)g - \frac{3\Gamma(3-d/2)}{(4\pi)^{d/2}} g^2 + \cdots \right\} \\
&= \frac{(2-d/2)\Gamma(2-d/2)}{(4\pi)^{d/2}} g + \cdots ,
\end{aligned}
\tag{58.83}
$$

or

$$
\gamma_2(g) = \frac{\Gamma(3-d/2)}{(4\pi)^{d/2}} g + \cdots .
\tag{58.84}
$$

Using the value of g_c given by (58.80), we obtain

$$
\gamma_2(g_c) = \frac{4-d}{3} + \cdots .
\tag{58.85}
$$

Recalling the definition of the dimension deficit $\epsilon = 4 - d$ and the result (58.81) for $\gamma_1(g)$, our efforts in this chapter culminate in the following statements:

$$
\gamma_1(g_c) = O(\epsilon^2) , \qquad \gamma_2(g_c) = \frac{\epsilon}{3} + \cdots \qquad (\epsilon > 0) .
\tag{58.86}
$$

The critical exponents finally follow from (58.32), (58.34) and (58.35). We have the following three results, given explicitly to first order in ϵ,

$$
\nu = \frac{1}{2 - \gamma_2(g_c)} = \frac{1}{2 - \frac{\epsilon}{3} + \cdots} = \frac{1}{2}\left(1 - \frac{\epsilon}{6} + \cdots\right)^{-1} = \frac{1}{2} + \frac{\epsilon}{12} + O(\epsilon^2) ,
\tag{58.87}
$$

$$
\gamma = \nu \left(2 - \gamma_1(g_c)\right) = \left\{\left(\frac{1}{2} + \frac{\epsilon}{12}\right) + O(\epsilon^2)\right\} \{2 - O(\epsilon^2)\} = 1 + \frac{\epsilon}{6} + O(\epsilon^2) ,
\tag{58.88}
$$

$$
\eta = O(\epsilon^2) .
\tag{58.89}
$$

The others can be calculated from the scaling laws (56.14) to (56.17). Amazingly, the above results agree perfectly from those derived in the last chapter [c.f. (57.132)] using the methodologically quite different Wilsonian approach.

For reference we list below some results obtained by very elaborate higher order calculations using Feynman-graph techniques, first reported in Wilson 1972, and Brezin et. al. 1973. They can be found, for example, in Wilson and Kogut 1974, and Itzykson and Drouffe 1989. These are for a more general $O(n)$ symmetric n-component vector-field theory, with the vector field $\varphi(\boldsymbol{x}) = (\varphi_1(\boldsymbol{x}), \ldots, \varphi_n(\boldsymbol{x}))$. The Landau free energy $\mathcal{E}(\varphi(\boldsymbol{x}))$ for this φ^4-theory is given by [c.f. (56.69)]

$$
\mathcal{E}(\varphi(\boldsymbol{x})) = \frac{1}{2} \left[(\nabla \cdot \varphi)^2 + a(t)|\varphi|^2\right] + \frac{b(t)}{4!} \left(|\varphi|^2\right)^2 .
\tag{58.90}
$$

We will only state the results for the Callan-Symanzik functions $\gamma_1(g_c), \gamma_2(g_c)$ and the critical exponent γ:

$$\gamma_1(g_c) = \eta = \frac{n+2}{2(n+8)^2}\epsilon^2 + \frac{n+2}{8(n+8)^4}\left[24(3n+14) - (n+8)^2\right]\epsilon^3$$
$$+ \frac{n+2}{2(n+8)^6}\left[-\frac{5}{16}n^4 - \frac{115}{8}n^3 + \frac{281}{4}n^2 + 1120n + 2884\right.$$
$$\left. - 24(n+8)(5n+22)\zeta(3)\right]\epsilon^4 + O(\epsilon^5) \,, \tag{58.91}$$

$$\gamma_2(g_c) = \frac{n+2}{n+8}\epsilon + \frac{n+2}{2(n+8)^3}(13n+44)\epsilon^2$$
$$+ \frac{n+2}{(n+8)^4}\left[\frac{36(3n+14)(n+3)}{n+8} - \frac{3n^2 + 388n + 848}{8}\right.$$
$$\left. - 12(5n+22)\zeta(3)\right]\epsilon^3 + O(\epsilon^4) \tag{58.92}$$
$$= 2 - \frac{1}{\nu} \,,$$

$$\gamma = 1 + \frac{n+2}{2(n+8)}\epsilon + \frac{(n+2)(n^2 + 22n + 52)}{4(n+8)^3}\epsilon^2 + \frac{n+2}{(n+8)^5}$$
$$\times \left[\frac{n^4}{8} + \frac{11n^3}{3} + 83n^2 + 312n + 388 - 6(n+8)(5n+22)\zeta(3)\right]\epsilon^3 + O(\epsilon^4)$$
$$= \nu(2 - \eta) \,. \tag{58.93}$$

In the above equations $\zeta(z)$ is the Riemann-zeta function, with

$$\zeta(3) = \sum_{j=1}^{\infty}\frac{1}{j^3} = 1.20206.... \qquad . \tag{58.94}$$

Bibliography

D. J. Amit and V. Martín-Mayor, *Field Theory, the Renormalization Group, and Critical Phenomena*, Third Edition, World Scientific (2005).

V. I. Arnold, V. V. Kozlov and A. I. Neishtadt, *Mathematical Aspects of Classical and Celestial Mechanics* (Springer, Berlin, 1997).

V. I. Arnold, A. Weinstein and K. Vogtmann, *Mathematical Methods of Classical Mechanics*, 2nd edition, Graduate Texts in Mathematics, No. 60 (Springer-Verlag, New York, 1989).

R. Balian and J. Zinn-Justin, ed., *Methods in Field Theory, Les Houches Session XXVIII 1975*, North Holland and World Scientific (1976).

D. Bao, S. S. Chern and Z. Shen, *An Introduction to Riemann-Finsler Geometry*, Graduate Texts in Mathematics, No. 200 (Springer-Verlag, New York, 2000).

R. J. Baxter, *Exactly Solved Models in Statistical Mechanics*, Dover (2007).

F. A. Berezin, *Method of the Second Quantization*, Academic Press (1966).

N. N. Bogoliubov and D. V. Shirkov, *Introduction to the Theory of Quantized Fields*, Interscience Publishers (1959).

E. Brézin, *"Applications of the Renormalization Group to Critical Phenomena"*, in *Methods in Field Theory, Les Houches Session XXVIII 1975*, ed. by R. Balian and J. Zinn-Justin, North Holland and World Scientific (1976).

E. Brézin, J. C. Le Guillou, J. Zinn-Justin and B. G. Nickel, Phys. Lett. Vol. 44A, No.3, 227 (1973).

C. G. Callan, Jr., *"Introduction to Renormalization Theory"*, in *Methods in Field Theory, Les Houches Session XXVIII 1975*, ed. by R. Balian and J. Zinn-Justin, North Holland and World Scientific (1976).

T.-P. Cheng and L.-F. Li, *Gauge Theory of Elementary Particle Physics*, Oxford University Press (1984).

S. S. Chern, *A Simple Intrinsic Proof of the Gauss-Bonnet Formula for Closed Riemannian Manifolds*, Annals of Math., Vol. **45** (1944), 747–752. Reprinted in *Shiing-Shen Chern Selected Papers* (Springer-Verlag, New York, 1978), 83–88.

S. S. Chern, *Complex Manifolds Without Potential Theory*, 2nd edition, revised (Springer-Verlag, New York, 1979).

S. S. Chern, *Historical Remarks on Gauss-Bonnet*, in *Analysis, et cetera*, Volume in Honor of Jurgen Moser (Academic Press, London, 1990), 209–217. Reprinted in *A Mathematician and His Mathematical Work: Selected Papers of S. S. Chern*, edited by S. Y. Cheng, P. Li and G. Tian (World Scientific, Singapore, 1996), 539–547.

S. S. Chern, W. H. Chen and K. S. Lam, *Lectures on Differential Geometry* (World Scientific, Singapore, 1999).

Y. Choquet-Bruhat and C. DeWitt-Morette, with M. Dillard-Bleick, *Analysis, Manifolds and Physics*, revised edition (North-Holland, Amsterdam, 1982).

Y. Choquet-Bruhat and C. DeWitte-Morette, *Analysis, Manifolds and Physics, Part II: 92 Applications* (North-Holland, Amstedam, 1989).

S. Coleman, *Aspects of Symmetry, Selected Erice Lectures of Sidney Coleman*, Cambridge University Press (1985).

E. T. Copson, *Asymptotic Expansions*, Cambridge University Press (1965).

P. Deligne, P. Etingof, D. S. Freed, L. C. Jeffrey, D. Kazhdan, J. Morgan, D. R. Morrison, and E. Witten, (Editors), *Quantum Fields and Strings: A Course for Mathematicians, Vol. 1*, American Mathematical Society, Institute for Advanced Study (1999).

S. K. Donaldson and P. B. Kronheimer, *The Geometry of Four-Manifolds* (Clarendon Press, Oxford, New York, 1991).

L. D. Faddeev and A. A. Slavnov, *Gauge Fields: Introduction to Quantum Theory*, Benjamin/Cummings (1980).

R. P. Feynman and A. R. Hibbs, *Quantum Mechanics and Path Integrals*, McGraw Hill (1965).

W. Foulton and J. Harris, *Representation Theory: A First Course*, Graduate Texts in Mathemetics, No. 129 (Springer-Verlag, New York 1991).

E. Fradkin, *Field Theories of Condensed Matter Systems*, (Addison-Wesley, Redwood City, 1991).

J. Glimm and A. Jaffe, *Quantum Physics: A Functional Integral Point of View*, Springer-Verlag (1981).

M. W. Hirsch and S. Smale, *Differential Equations, Dynamical Systems, and Linear Algebra* (Academic Press, New York, 1974).

S. Hu, *Lecture Notes on Chern-Simons-Witten Theory* (World Scientific, Singapore, 2001).

K. Huang, *Statistical Mechanics*, Second Edition, John Wiley and Sons (1987).

K. Huang, *Quantum Field Theory: From Operators to Path Integrals*, Wiley-VCH (2010).

C. Itzykson and J.-B. Zuber, *Quantum Field Theory*, McGraw-Hill (1980).

C. Itzykson and J.-M. Drouffe, *Statistical Field Theory, Vol. 1: From Brownian Motion to Renormalization and Lattice Gauge Theory*, Cambridge University Press (1989).

N. Jacobson, *Lie Algebras* (Dover, New York, 1962).

L. P. Kadanoff, Rev. Mod. Phys. Vol. 49, 267 (1977).

S. Kobayashi and K. Nomizu, *Foundations of Differential Geometry* (Interscience Publishers, New York, Vol. I 1963, Vol. II 1969).

M. Kohmoto, *Topological Invariant and the Quantization of the Hall Conductance*, Annals of Physics, Vol. **160** (1985) 343–354. Reprinted in *Topological Quantum Numbers in Nonrelativistic Physics*, by D. J. Thouless (World Scientific, Singapore, 1998).

K. S. Lam, *Fundamental Principles of Classical Mechanics: A Geometrical Perspective*, World Scientific (2014).

S.-K. Ma, *Modern Theory of Critical Phenomena*, Benjamin/Cummings (1976).

M. Nakahara, *Geometry, Topology and Physics* (Adam Hilger, Bristol, 1990).

C. Nash, *Differential Topology and Quantum Field Theory* (Academic Press, London, 1991).

C. Nash and S. Sen, *Topology and Geometry for Physicists* (Academic Press, London, 1983).

E. Nelson, *Dynamical Theories of Brownian Motion*, Princeton University Press (1967).

S. Okubo, *Introduction to Group Theory*, unpublished notes, Department of Physics and Astronomy, University of Rochester (1980).

G. Parisi, *Statistical Field Theory*, Addison-Wesley (1988).

W. Pauli, Z. Physik, *Über das Wasserstoffspektrum vom Standpunkt der neuen Quantenmechanik (On the Hydrogen Spectrum from the Standpoint of the New Quantum Mechanics)*, Vol. **36**, 336–363 (1926). English translation in *Sources of Quantum Mechanics*, edited by B. L. Van der Waerden (Dover, New York, 1967).

M. E. Peskin and D. V. Schroeder, *An Introduction to Quantum Field Theory*, Perseus Books (1995).

V. N. Popov, *Functional Integrals and Collective Excitations*, Cambridge University Press (1987).

P. Ramond, *Field Theory: A Modern Primer*, Second Edition, Addison-Wesley (1989).

M. Reed and B. Simon, *Methods of Modern Mathematical Physics, Vol. II: Fourier Analysis, Self-Adjointness*, Academic Press (1975).

D. H. Sattinger and O. L. Weaver, *Lie Groups and Algebras with Applications to Physics, Geometry, and Mechanics*, Applied Mathematical Sciences, No. 61 (Springer-Verlag, New York, 1986).

A. Shapere and F. Wilczek, editors, *Geometric Phases in Physics* (Word Scientific, Singapore, 1989).

I. M. Singer and J. A. Thorpe, *Lecture Notes in Elementary Topology and Geometry*, Undergraduate Texts in Mathematics (Springer-Verlag, New York, 1976).

S. Sternberg, *Group Theory and Physics*, (Cambridge University Press, Cambridge, 1995).

D. J. Thouless, *Topological Quantum Numbers in Nonrelativistic Physics* (World Scientific, Singapore, 1998).

W.-K. Tung, *Group Theory in Physics* (World Scientific, Singapore, 1985).

J. von Neumann, *Mathematical Foundations of Quantum Mechanics* (Princeton University Press, Princeton, 1955).

S. Weinberg, *The Quantum Theory of Fields*, Cambridge University Press (Vol. I: 1995, Vol. II: 1996, Vol. III: 2000).

K. Wilson, Phys. Rev. Lett., Vol. 28, No.9, 548 (1972).

K. G. Wilson and J. Kogut, Physics Reports, Vol. 12, No. 2, 75 - 200 (1974).

K. G. Wilson, Rev. Mod. Phys. , Vol. 47, No. 4, 773 (1975).

K. G. Wilson, Rev. Mod. Phys. , Vol. 55, No. 3, 583 (1983).

E. Witten, *Quantum Field Theory and the Jones Polynomial*, Commun. Math. Phys., Vol. **121** (1989) 351–399. Reprinted in *Braid Group, Knot Theory and Statistical Mechanics*, edited by C. N. Yang and M. L. Ge (World Scientific, Singapore, 1989) 239–329.

E. Witten, *Index of Dirac Operators*, in *Quantum Fields and Strings: A Course for Mathematicians*, Vol. 1, edited by P. Deligne, P. Etingof, D. S. Freed, L. C. Jeffrey, D. Kazhdan, J. W. Morgan, D. R. Morrison and E. Witten (American Mathematical Society/Institute for Advanced Study, 2000) 475–511.

A. Zee, *Quantum Field Theory in a Nutshell*, Second Edition, Princeton University Press (2010).

J. Zinn-Justin, *Quantum Field Theory and Critical Phenomena*, Fourth Edition, Oxford University Press (2002).

Index

www.ingramcontent.com/pod-product-compliance
Lightning Source LLC
Chambersburg PA
CBHW081207220326
41598CB00037B/6698